HANDBOOK OF
NANOPHYSICS

Handbook of Nanophysics

Handbook of Nanophysics: Principles and Methods

Handbook of Nanophysics: Clusters and Fullerenes

Handbook of Nanophysics: Nanoparticles and Quantum Dots

Handbook of Nanophysics: Nanotubes and Nanowires

Handbook of Nanophysics: Functional Nanomaterials

Handbook of Nanophysics: Nanoelectronics and Nanophotonics

Handbook of Nanophysics: Nanomedicine and Nanorobotics

HANDBOOK OF
NANOPHYSICS

Functional Nanomaterials

Edited by

Klaus D. Sattler

CRC Press
Taylor & Francis Group
Boca Raton   London   New York

CRC Press is an imprint of the
Taylor & Francis Group, an **informa** business

CRC Press
Taylor & Francis Group
6000 Broken Sound Parkway NW, Suite 300
Boca Raton, FL 33487-2742

First issued in paperback 2020

© 2011 by Taylor and Francis Group, LLC
CRC Press is an imprint of Taylor & Francis Group, an Informa business

No claim to original U.S. Government works

ISBN 13: 978-1-138-11193-6 (pbk)
ISBN 13: 978-1-4200-7552-6 (hbk)

Library of Congress Cataloging-in-Publication Data

Handbook of nanophysics. Functional nanomaterials / editor, Klaus D. Sattler.
    p. cm.
  "A CRC title."
  Includes bibliographical references and index.
  ISBN 978-1-4200-7552-6 (alk. paper)
  1. Nanostructured materials--Handbooks, manuals, etc. 2. Nanoelectromechanical systems--Handbooks, manuals, etc. 3. Nanocomposites (Materials)--Handbooks, manuals, etc. I. Sattler, Klaus D. II. Title.

TA418.9.N35H3645 2010
620.1'1--dc22
        2010001109

Visit the Taylor & Francis Web site at
http://www.taylorandfrancis.com

and the CRC Press Web site at
http://www.crcpress.com

Contents

PART I Nanocomposites

PART II Nanoporous and Nanocage Materials

PART III Nanolayers

PART IV Indentation and Patterning

Preface

The *Handbook of Nanophysics* is the first comprehensive reference to consider both fundamental and applied aspects of nanophysics. As a unique feature of this work, we requested contributions to be submitted in a tutorial style, which means that state-of-the-art scientific content is enriched with fundamental equations and illustrations in order to facilitate wider access to the material. In this way, the handbook should be of value to a broad readership, from scientifically interested general readers to students and professionals in materials science, solid-state physics, electrical engineering, mechanical engineering, computer science, chemistry, pharmaceutical science, biotechnology, molecular biology, biomedicine, metallurgy, and environmental engineering.

What Is Nanophysics?

Modern physical methods whose fundamentals are developed in physics laboratories have become critically important in nanoscience. Nanophysics brings together multiple disciplines, using theoretical and experimental methods to determine the physical properties of materials in the nanoscale size range (measured by millionths of a millimeter). Interesting properties include the structural, electronic, optical, and thermal behavior of nanomaterials; electrical and thermal conductivity; the forces between nanoscale objects; and the transition between classical and quantum behavior. Nanophysics has now become an independent branch of physics, simultaneously expanding into many new areas and playing a vital role in fields that were once the domain of engineering, chemical, or life sciences.

This handbook was initiated based on the idea that breakthroughs in nanotechnology require a firm grounding in the principles of nanophysics. It is intended to fulfill a dual purpose. On the one hand, it is designed to give an introduction to established fundamentals in the field of nanophysics. On the other hand, it leads the reader to the most significant recent developments in research. It provides a broad and in-depth coverage of the physics of nanoscale materials and applications. In each chapter, the aim is to offer a didactic treatment of the physics underlying the applications alongside detailed experimental results, rather than focusing on particular applications themselves.

The handbook also encourages communication across borders, aiming to connect scientists with disparate interests to begin interdisciplinary projects and incorporate the theory and methodology of other fields into their work. It is intended for readers from diverse backgrounds, from math and physics to chemistry, biology, and engineering.

The introduction to each chapter should be comprehensible to general readers. However, further reading may require familiarity with basic classical, atomic, and quantum physics. For students, there is no getting around the mathematical background necessary to learn nanophysics. You should know calculus, how to solve ordinary and partial differential equations, and have some exposure to matrices/linear algebra, complex variables, and vectors.

External Review

All chapters were extensively peer reviewed by senior scientists working in nanophysics and related areas of nanoscience. Specialists reviewed the scientific content and nonspecialists ensured that the contributions were at an appropriate technical level. For example, a physicist may have been asked to review a chapter on a biological application and a biochemist to review one on nanoelectronics.

Organization

The *Handbook of Nanophysics* consists of seven books. Chapters in the first four books (*Principles and Methods*, *Clusters and Fullerenes*, *Nanoparticles and Quantum Dots*, and *Nanotubes and Nanowires*) describe theory and methods as well as the fundamental physics of nanoscale materials and structures. Although some topics may appear somewhat specialized, they have been included given their potential to lead to better technologies. The last three books (*Functional Nanomaterials*, *Nanoelectronics and Nanophotonics*, and *Nanomedicine and Nanorobotics*) deal with the technological applications of nanophysics. The chapters are written by authors from various fields of nanoscience in order to encourage new ideas for future fundamental research.

After the first book, which covers the general principles of theory and measurements of nanoscale systems, the organization roughly follows the historical development of nanoscience. *Cluster scientists* pioneered the field in the 1980s, followed by extensive work on *fullerenes*, *nanoparticles*, and *quantum dots* in the 1990s. Research on *nanotubes* and *nanowires* intensified in subsequent years. After much basic research, the interest in applications such as the *functions of nanomaterials* has grown. Many bottom-up and top-down techniques for nanomaterial and nanostructure generation were developed and made possible the development of *nanoelectronics* and *nanophotonics*. In recent years, real applications for *nanomedicine* and *nanorobotics* have been discovered.

Acknowledgments

Many people have contributed to this book. I would like to thank the authors whose research results and ideas are presented here. I am indebted to them for many fruitful and stimulating discussions. I would also like to thank individuals and publishers who have allowed the reproduction of their figures. For their critical reading, suggestions, and constructive criticism, I thank the referees. Many people have shared their expertise and have commented on the manuscript at various stages. I consider myself very fortunate to have been supported by Luna Han, senior editor of the Taylor & Francis Group, in the setup and progress of this work. I am also grateful to Jessica Vakili, Jill Jurgensen, Joette Lynch, and Glenon Butler for their patience and skill with handling technical issues related to publication. Finally, I would like to thank the many unnamed editorial and production staff members of Taylor & Francis for their expert work.

Klaus D. Sattler
Honolulu, Hawaii

Editor

Klaus D. Sattler pursued his undergraduate and master's courses at the University of Karlsruhe in Germany. He received his PhD under the guidance of Professors G. Busch and H.C. Siegmann at the Swiss Federal Institute of Technology (ETH) in Zurich, where he was among the first to study spin-polarized photoelectron emission. In 1976, he began a group for atomic cluster research at the University of Konstanz in Germany, where he built the first source for atomic clusters and led his team to pioneering discoveries such as "magic numbers" and "Coulomb explosion." He was at the University of California, Berkeley, for three years as a Heisenberg Fellow, where he initiated the first studies of atomic clusters on surfaces with a scanning tunneling microscope.

Dr. Sattler accepted a position as professor of physics at the University of Hawaii, Honolulu, in 1988. There, he initiated a research group for nanophysics, which, using scanning probe microscopy, obtained the first atomic-scale images of carbon nanotubes directly confirming the graphene network. In 1994, his group produced the first carbon nanocones. He has also studied the formation of polycyclic aromatic hydrocarbons (PAHs) and nanoparticles in hydrocarbon flames in collaboration with ETH Zurich. Other research has involved the nanopatterning of nanoparticle films, charge density waves on rotated graphene sheets, band gap studies of quantum dots, and graphene foldings. His current work focuses on novel nanomaterials and solar photocatalysis with nanoparticles for the purification of water.

Among his many accomplishments, Dr. Sattler was awarded the prestigious Walter Schottky Prize from the German Physical Society in 1983. At the University of Hawaii, he teaches courses in general physics, solid-state physics, and quantum mechanics.

In his private time, he has worked as a musical director at an avant-garde theater in Zurich, composed music for theatrical plays, and conducted several critically acclaimed musicals. He has also studied the philosophy of Vedanta. He loves to play the piano (classical, rock, and jazz) and enjoys spending time at the ocean, and with his family.

Contributors

Lionel Aigouy
Laboratoire de Physique et d'Etude des
    Matériaux
Ecole Supérieure de Physique et de
    Chimie Industrielles
Paris, France

Julio A. Alonso
Departamento de Física Teórica,
    Atómica y Óptica
Universidad de Valladolid
Valladolid, Spain
and
Departamento de Física de
    Materiales
Facultad de Química
Universidad del País Vasco
and
Donostia International Physics
    Center
San Sebastián, Spain

Bibin T. Anto
Department of Physics
National University of Singapore
Singapore, Singapore

Vadym Apalkov
Department of Physics and Astronomy
Georgia State University
Atlanta, Georgia

Hidetaka Asoh
Department of Applied Chemistry
Faculty of Engineering
Kogakuin University
Tokyo, Japan

Duangkamon Baowan
Department of Mathematics
Faculty of Science
Mahidol University
Bangkok, Thailand

Asa H. Barber
Centre for Materials Research
and
School of Engineering and Materials
    Science
Queen Mary University of London
London, United Kingdom

Eduardo B. Barros
Departamento de Física
Universidade Federal do Ceará
Fortaleza, Brazil

Çağlar Çelik Bayar
Department of Chemistry
Middle East Technical University
Ankara, Turkey

Pierre Bénard
Hydrogen Research Institute
Université du Québec à Trois-Rivières
Trois-Rivières, Quebec, Canada

Claire Berger
School of Physics
Georgia Institute of Technology and
    National Center for Scientific Research
Atlanta, Georgia

Francisco Javier Bermejo
Instituto de Estructura de la Materia
Consejo Nacional de Investigaciones
    Científicas
Madrid, Spain
and
Department of Electricity
    and Electronics
University of the Basque Country
Bilbao, Spain

Martin S. Bojinov
Department of Physical Chemistry
University of Chemical Technology
    and Metallurgy
Sofia, Bulgaria

Luis Brey
Instituto de Ciencia de Materiales de
    Madrid
Consejo Superior de Investigaciones
    Cientificas
Cantoblanco, Spain

Iván Cabria
Departamento de Física Teórica,
    Atómica y Óptica
Universidad de Valladolid
Valladolid, Spain

Thomas Cecil
National Institute of Standards
    and Technology
Boulder, Colorado

Richard Chahine
Hydrogen Research Institute
Université du Québec à Trois-Rivières
Trois-Rivières, Quebec, Canada

Bin Chu
College of Chemistry and Molecular
    Sciences
Wuhan University
Wuhan, People's Republic of China

Lay-Lay Chua
Department of Chemistry
National University of Singapore
Singapore, Singapore

Andrew N. Cleland
Department of Physics
University of California
Santa Barbara, California

Barry J. Cox
School of Mathematics and Applied
    Statistics
University of Wollongong
Wollongong, New South Wales, Australia

Walt A. de Heer
School of Physics
Georgia Institute of Technology
Atlanta, Georgia

Hua Deng
State Key Laboratory of Polymer
    Materials Engineering
College of Polymer Science
    and Engineering
Sichuan University
Sichuan, China
and
Centre for Materials Research
and
School of Engineering and Materials
    Science
Queen Mary University of London
London, United Kingdom

Nada M. Dimitrijevic
Center for Nanoscale Materials
Argonne National Laboratory
Argonne, Illinois

Tyler Dunn
Department of Physics
Boston University
Boston, Massachusetts

Adam Elhofy
Feinberg School of Medicine
Northwestern University
Chicago, Illinois

Eugene A. Eliseev
Institute for Problems of Material
    Science
National Academy of Science of Ukraine
Kiev, Ukraine

Klaus Ensslin
Solid State Physics Laboratory
ETH Zürich
Zürich, Switzerland

Jens Falta
Institute of Solid State Physics
University of Bremen
Bremen, Germany

Te-Hua Fang
Institute of Mechanical &
    Electromechanical Engineering
National Formosa University
Yunlin, Taiwan

Zhi-Qiang Feng
Laboratoire de Mécanique et
    d'Energétique d'Evry
Université d'Évry-Val d'Essonne
Evry, France

Maurizio Fermeglia
Molecular Simulation Engineering
    Laboratory
Department of Chemical,
    Environmental, and Raw Materials
    Engineering
University of Trieste
Trieste, Italy

Felix Fernandez-Alonso
ISIS Facility
Rutherford Appleton Laboratory
Science and Technology Facilities
    Council
Oxfordshire, United Kingdom

and

Department of Physics and Astronomy
University College London
London, United Kingdom

Joaquin Fernández-Rossier
Departamento de Física Aplicada
Universidad de Alicante
Alicante, Spain

Herb A. Fertig
Department of Physics
Indiana University
Bloomington, Indiana

Mehrdad N. Ghasemi-Nejhad
Hawaii Nanotechnology and Intelligent
    and Composite Materials Laboratories
Department of Mechanical Engineering
University of Hawaii at Manoa
Honolulu, Hawaii

Thomas Greber
Physik Institut
Universität Zürich
Zürich, Switzerland

Diego Guerra
Department of Physics
Boston University
Boston, Massachusetts

Johannes Güttinger
Solid State Physics Laboratory
ETH Zürich
Zürich, Switzerland

Frank Hagelberg
Department of Physics and Astronomy
East Tennessee State University
Johnson City, Tennessee

Tsuyoshi Hamaguchi
Toyota Central R&D Labs, Inc.
Aichi-Gun, Japan

Qi-Chang He
Laboratoire de Modélisation et
    Simulation Multi Echelle
Université Paris-Est
Marne-la-Vallée, France

Henning Heiberg-Andersen
Physics Department
Institute for Energy Technology
Kjeller, Norway

Ranko Heindl
National Institute of Standards
    and Technology
Boulder, Colorado

Sarah Hellmüller
Solid State Physics Laboratory
ETH Zürich
Zürich, Switzerland

Prabath Hewageegana
Department of Physics
University of Kelaniya
Kelaniya, Sri Lanka

James M. Hill
School of Mathematics and Applied
    Statistics
University of Wollongong
Wollongong, New South Wales, Australia

Peter K. H. Ho
Department of Physics
National University of Singapore
Singapore, Singapore

Thomas Ihn
Solid State Physics Laboratory
ETH Zürich
Zürich, Switzerland

Matthias Imboden
Department of Physics
Boston University
Boston, Massachusetts

Arnhild Jacobsen
Solid State Physics Laboratory
ETH Zürich
Zürich, Switzerland

Pierre Joli
Laboratoire de Mécanique et
    d'Energétique d'Evry
Université d'Évry-Val d'Essonne
Evry, France

Mikhail I. Katsnelson
Institute for Molecules and Materials
Radboud University of Nijmegen
Nijmegen, the Netherlands

Raoul Kopelman
Department of Chemistry
University of Michigan
Ann Arbor, Michigan

Toyoki Kunitake
The Institute of Physical and Chemical
    Research
and
Nanomembrane Technologies
Saitama, Japan

Yong-Eun Koo Lee
Department of Chemistry
University of Michigan
Ann Arbor, Michigan

Xin Jian Li
Laboratory of Material Physics
Department of Physics
Zhengzhou University
Zhengzhou, People's Republic of China

Alexander I. Lichtenstein
I. Institute for Theoretical Physics
Hamburg University
Hamburg, Germany

Yu-Cheng Lin
Department of Engineering Science
National Cheng Kung University
Tainan, Taiwan

Yuan Liu
College of Science
University of Shanghai for Science and
    Technology
Shanghai, People's Republic of China

María J. López
Departamento de Física Teórica, Atómica
    y Óptica
Universidad de Valladolid
Valladolid, Spain

Xuejun Lu
Department of Electrical and Computer
    Engineering
University of Massachusetts Lowell
Lowell, Massachusetts

Kougen Ma
Hawaii Nanotechnology and Intelligent
    and Composite Materials Laboratories
Department of Mechanical Engineering
University of Hawaii at Manoa
Honolulu, Hawaii

Juan Matos
Engineering of Materials
    and Nanotechnology Centre
Venezuelan Institute for Scientific
    Research
Caracas, Venezuela

Patrice Mélinon
Laboratoire de Physique de la Matière
    Condensée et Nanostructures
Université de Lyon 1 et Centre National
    de la Recherche Scientifique
Université de Lyon
Villeurbanne, France

Josue Mendes Filho
Departamento de Física
Universidade Federal do Ceará
Fortaleza, Brazil

Pritiraj Mohanty
Department of Physics
Boston University
Boston, Massachusetts

Françoise Molitor
Solid State Physics Laboratory
ETH Zürich
Zürich, Switzerland

Anna N. Morozovska
Institute of Semiconductor Physics
National Academy of Science of Ukraine
Kiev, Ukraine

Michel Mortier
Laboratoire de Chimie de la Matière
    Condensée de Paris
Centre national de la recherche
    scientifique
Ecole Nationale Supérieure de Chimie
    de Paris
Paris, France

Stine Nalum Naess
Department of Physics
Norwegian University of Science
    and Technology
Trondheim, Norway

Sachiko Ono
Department of Applied Chemistry
Faculty of Engineering
Kogakuin University
Tokyo, Japan

Juan Jose Palacios
Departamento de Física de la Materia
    Condensada
Facultad de Ciencias
Universidad Autonoma
Madrid, Spain

Ton Peijs
Centre for Materials Research
and
School of Engineering and Materials
    Science
Queen Mary University of London
London, United Kingdom
and
Eindhoven Polymer Laboratories
Eindhoven University of Technology
Eindhoven, the Netherlands

Vladimir M. Petrov
Physics Department
Oakland University
Rochester, Michigan

and

Institute for Electronic Information
    Systems
Novgorod State University
Veliky Novgorod, Russia

Rui-Qi Png
Department of Physics
National University of Singapore
Singapore, Singapore

Paola Posocco
Molecular Simulation Engineering
    Laboratory
Department of Chemical,
    Environmental, and Raw Materials
    Engineering
University of Trieste
Trieste, Italy

Sabrina Pricl
Molecular Simulation Engineering
    Laboratory
Department of Chemical,
    Environmental, and Raw Materials
    Engineering
University of Trieste
Trieste, Italy

Tijana Rajh
Center for Nanoscale Materials
Argonne National Laboratory
Argonne, Illinois

Marc-André Richard
Hydrogen Research Institute
Université du Québec à Trois-Rivières
Trois-Rivières, Quebec, Canada

William H. Rippard
National Institute of Standards
    and Technology
Boulder, Colorado

Debdulal Roy
National Physical Laboratory
Teddington, United Kingdom

Elena Rozhkova
Center for Nanoscale Materials
Argonne National Laboratory
Argonne, Illinois

Stephen E. Russek
National Institute of Standards
    and Technology
Boulder, Colorado

Marie-Louise Saboungi
Centre de Recherche sur la Matière
    Divisée
Centre National de la Recherche
    Scientifique—Université d'Orléans
Orléans, France

Alfonso San Miguel
Laboratoire de Physique de la Matière
    Condensée et Nanostructures
Université de Lyon 1 et Centre National
    de la Recherche Scientifique
Université de Lyon
Villeurbanne, France

Thomas Schmidt
Institute of Solid State Physics
University of Bremen
Bremen, Germany

Stephan Schnez
Solid State Physics Laboratory
ETH Zürich
Zürich, Switzerland

Eric Schurtenberger
Solid State Physics Laboratory
ETH Zürich
Zürich, Switzerland

Giulio Scocchi
Molecular Simulation Engineering
    Laboratory
Department of Chemical,
    Environmental, and Raw Materials
    Engineering
University of Trieste
Trieste, Italy

Jun Shen
Pohl Institute of Solid State Physics
Tongji University
Shanghai, People's Republic of China

Sankaran Sivaramakrishnan
Department of Physics
National University of Singapore
Singapore, Singapore

Ame Torbjørn Skjeltorp
Physics Department
Institute for Energy Technology
Kjeller, Norway

and

Department of Physics
University of Oslo
Oslo, Norway

Antonio G. Souza Filho
Departamento de Física
Universidade Federal do Ceará
Fortaleza, Brazil

Gopalan Srinivasan
Physics Department
Oakland University
Rochester, Michigan

Christoph Stampfer
Solid State Physics Laboratory
ETH Zürich
Zürich, Switzerland

Jiyu Sun
The Key Laboratory of Bionic
    Engineering and College of Biological
    and Agricultural Engineering
Jilin University
Changchun, People's Republic of China

Ngamta Thamwattana
School of Mathematics and Applied
  Statistics
University of Wollongong
Wollongong, New South Wales, Australia

Jin Tong
The Key Laboratory of Bionic
  Engineering and College of Biological
  and Agricultural Engineering
Jilin University
Changchun, People's Republic of China

Lemi Türker
Department of Chemistry
Middle East Technical University
Ankara, Turkey

Masayoshi Uno
Research Institute of Nuclear
  Engineering
University of Fukui
Fukui, Japan

Gavin Stuart Walker
Department of Mechanical, Materials &
  Manufacturing Engineering
University of Nottingham
Nottingham, United Kingdom

Tong Hong Wang
Department of Engineering Science
National Cheng Kung University
Tainan, Taiwan

and

Stress-Thermal Laboratory
Advanced Semiconductor
  Engineering, Inc.
Kaohsiung, Taiwan

Zheng-Ming Wang
Energy Technology Research Institute
National Institute of Advanced Industrial
  Science and Technology
Tsukuba, Japan

Hirohmi Watanabe
The Institute of Physical and Chemical
  Research
Saitama, Japan

Tim O. Wehling
Institute for Theoretical Physics
Hamburg University
Hamburg, Germany

and

Institute for Molecules and Materials
Radboud University of Nijmegen
Nijmegen, the Netherlands

Josef-Stefan Wenzler
Department of Physics
Boston University
Boston, Massachusetts

Loke-Yuen Wong
Department of Physics
National University of Singapore
Singapore, Singapore

Xiaosong Wu
School of Physics
Georgia Institute of Technology
Atlanta, Georgia

Shinsuke Yamanaka
Division of Sustainable Energy and
  Environmental Engineering
Graduate School of Engineering
Osaka University
Suita, Japan

Oleg V. Yazyev
Institute of Theoretical Physics
Ecole Polytechnique Fédérale de
  Lausanne (EPFL)
and
Institut Romand de Recherche
  Numérique en Physique des
  Matériaux (IRRMA)
Lausanne, Switzerland

Qingfeng Zeng
National Key Laboratory of
  Thermostructural Composite
  Materials
Northwestern Polytechnical
  University
Xi'an, People's Republic of China

Qing Zhang
School of Chemistry and Chemical
  Engineering
Shanghai Jiao-Tong University
Shanghai, People's Republic of China

Jiang Zhou
The Key Laboratory of Bionic
  Engineering and College of Biological
  and Agricultural Engineering
Jilin University
Changchun, People's Republic of China

I

Nanocomposites

1

Carbon Nanotube/Polymer Composites

Hua Deng
Sichuan University

and

Queen Mary University of London

Asa H. Barber
Queen Mary University of London

Ton Peijs
Queen Mary University of London

and

Eindhoven University of Technology

1.1 Introduction

Carbon nanotubes (CNTs) have attracted great enthusiasm among researchers due to their extraordinary mechanical, electrical, and thermal properties. The use of polymers filled with CNTs is especially interesting as an extension of research on composites. The number of publications on CNTs and CNT/polymer composites has increased dramatically in recent years (see Figure 1.1). Size in nanotubes has an immediate advantage; for example, the increase in surface area of CNTs relative to traditional fibers leads to greater interaction between the fiber and the surrounding polymer matrix when they are incorporated within polymer composites. Considerable mechanical reinforcement of polymers with CNTs is of particular advantage and is a result of composites because of the inherent mechanical properties of the nanotubes themselves. The high aspect ratio (near 1000) and outstanding conductivity of CNTs can result in ultralow electrical percolation threshold in polymer composites. Finally, due to the one-dimensional structure of CNTs, oriented CNT/polymer composite fibers or tapes have also generated interest, as such oriented systems can result in high reinforcing efficiency as well as excellent uniaxial conductivity [1]. A number of review papers and books have been published on CNT-reinforced polymer composites for further reference [2–8].

The structure of this chapter is set as follows. Firstly, the wetting and adhesion between polymer and CNTs is reviewed. Secondly, the dispersion of CNTs in polymers is discussed.

Thirdly, the mechanical reinforcement of CNTs in polymer composites is reviewed. Fourthly, the electrical properties of CNT/polymer composites in both isotropic and oriented systems are discussed and, finally, the sensing properties and thermal stability of CNT-based polymer composites are addressed.

1.2 Adhesion between Polymers and CNTs

Key issues for the performance of CNT-based polymer composites include the extent to which the CNTs can be wetted by a given polymer and the resultant adhesion between the nanotube and the surrounding polymer matrix material [9]. Therefore, the interface between the polymer matrix and the reinforcement plays a crucial role in the physical properties of the composites. Wetting studies have often been used to evaluate if strong adhesion exists between a fiber and a polymer. Good wetting is a requirement for effective adhesion and usually attempts to measure if the polymer in a liquid phase, which often occurs at some point in the processing of the composite, has a tendency to cover the surface of the fiber. For CNTs, understanding the wetting process is challenging due to the relatively small size of the nanotube. The observation of liquid contact angles at nanotube surfaces has been possible by using electron microscopy, as shown in Figure 1.2. The small contact angle made between the liquid and the nanotube indicates a good wetting state, but

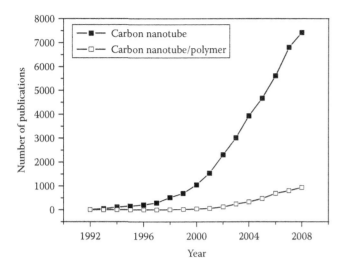

FIGURE 1.1 The number of publications for CNTs and CNT/polymer composites in recent years.

these initial measurements [9] were limited when measuring the contact angle accurately, as is the case in larger systems, and the influence of the electron beam and vacuum condition of the microscope during the observation. Improvements were made by attaching individual CNTs to the end of an atomic force microscope (AFM) and dipping the nanotube into a number of organic liquids in a setup analogous to the Wilhelmy balance

technique [10,11]. The resultant wetting force was used to measure contact angles accurately and indicate a surprising preference for nanotubes to be wetted by a number of different liquids. This included polar liquids despite the surface of the nanotube being predicted to be somewhat hydrophobic. The wetting studies indicated how strong adhesion could be possible between CNTs and polymers, but further work to evaluate the mechanical performance of this adhesion and the resultant composite properties was needed.

As in composite materials, reinforcement is only effective when the fiber is able to deform ideally in unison with its surrounding matrix material. Since the reinforcement is usually stiffer than the matrix, more force is required to deform the fiber than the matrix. This causes the stiffer reinforcing phase to carry most of the force. For a noncontinuous reinforcing phase, the force or load is transferred from the deforming matrix to the reinforcement across the interface. The effective transfer of load is critically dependent on the adhesion between the fiber and the matrix material. Thus, poor interfacial adhesion causes inefficient load transfer and the failure of the interface itself at low deformation, which significantly hinders this load transfer. As a result, the fiber will bear less of the load and become less effective as a mechanical reinforcing phase. It is because of this that researchers seek to understand the properties of the interface and improve adhesion between CNTs and the supporting polymer matrix [12].

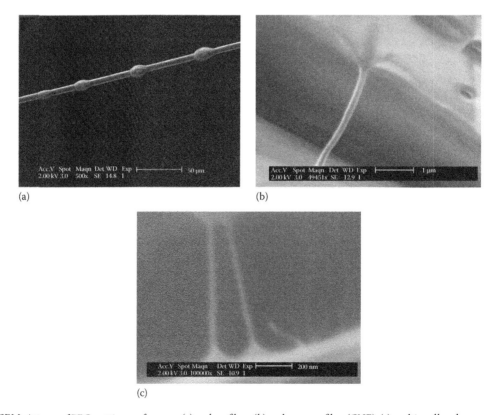

FIGURE 1.2 ESEM pictures of PEG wetting surfaces on: (a) carbon fiber; (b) carbon nanofiber (CNF); (c) multi-wall carbon nanotubes (MWNTs), from which the contact angle could be measured. (From Nuriel, S. et al., *Chem. Phys. Lett.*, 404, 263, 2005. With permission.)

Initial studies on observing the failure in CNT/polymer composites indicated that CNT pullout from the polymer was prevalent [13]. This work indicated qualitatively that the strength of the CNTs was significantly higher than the CNT/polymer adhesion strength and that failure at the interface was preferred. A number of studies have demonstrated similar adhesion or interphase between CNTs and polymer matrices [9,14,15] (see Figure 1.2). However, more quantitative studies on CNT composites was achieved by Barber et al. [12,16] using individual CNT composites. In these works, an AFM is used to pull individual CNTs out of polymer in order to measure the interfacial adhesion between the nanotube and the polymer matrix directly, as shown in Figure 1.3. The pullout force increases with pullout area [16]. Increasing the nanotube embedded length causes the nanotube to break instead of pullout from the polymer, and a tensile strength of 133 ± 73 GPa is calculated for individual CNTs. Furthermore, the chemical modification of CNT surfaces has been shown to increase the interfacial adhesion with a polymer matrix compared to pristine (unmodified) CNT surfaces using the same technique [12]. The conclusions from this work were promising and indicated that the strong interfacial adhesion between CNTs and polymers, especially using the chemical

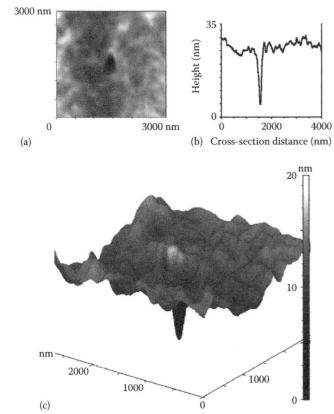

(a)

(b) Cross-section distance (nm)

(c)

FIGURE 1.3 AFM topography imaging of the pullout area reveals an exit hole (a) in the polymer surface, the depth of which can be measured (b) and used to calculate the embedded length. The geometry of the pullout area, clearly showing the hole previously containing the nanotube, is displayed from the AFM height data (c). (From Barber, A.H. et al., *Appl. Phys. Lett.*, 82, 4140, 2003. With permission.)

modification of the nanotube surface, could result in the fabrication of high-performance nanotube-reinforced polymer composites.

1.3 Dispersion Methods for CNTs in Polymers

The incorporation of CNTs as reinforcement into polymer matrices requires composite manufacturing where the CNTs are dispersed throughout the composite material. The clustering of CNTs is particularly troublesome as significant intertube sliding will occur under an external force, which will reduce the effective load transfer and result in a poor overall mechanical performance of the composite. Dispersion methods are mainly physical processing methods (such as melt compounding or ultrasonication), chemical-based processing methods (including covalent method, noncovalent method with surfactant or polymer, and so on), and in situ polymerization methods. However, other techniques such as latex technology [18], solid-state shear pulverization [19], and layer-by-layer assembly techniques [20] have proved to be efficient at dispersing CNTs into polymer matrices as well. In the following section, processing methods, including physical and chemical methods, will be reviewed.

1.3.1 Physical Processing

CNTs are tightly held by van der Waals forces, and it requires a mechanical force to separate them in the case of physical methods of dispersion. There are three main mechanical processes that have been successful in dispersing tubes in various media: melt compounding, ultrasonication, and milling and grinding.

1.3.1.1 Melt Compounding

High shear mixing is an effective way of dispersing CNTs in a viscous polymer melt. The advantages of the technique are: direct dispersion into the host matrix, no chemical modification is required, and nanotubes are prevented from re-aggregation by viscous forces [2]. Moreover, it fits well with current industrial practices.

Many studies have demonstrated the successful application of melt shear mixing in dispersing CNTs in polymer matrices. As one of the first studies in this field, Andrews et al. processed multi-walled CNTs/polystyrene (MWNT/PS) nanocomposites in a twin-blade mixer [21]. They studied the effect of shear mixing energy on the tube length by a transmission electron microscope (TEM). It was found that the tube length diminishes with increasing mixing energy, and the rate of tube breakage was reduced as the tubes were better dispersed. Soon after, Potschke et al. demonstrated in a series of studies on melt processing nanocomposites based on CNTs [22–28] of an interesting relation between nanotube connectivity and the onset of non-Newtonian nanotube–polymer behavior.

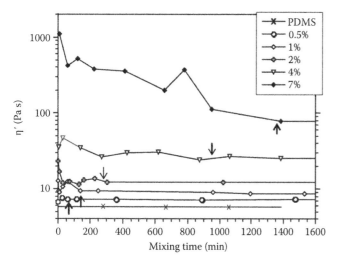

FIGURE 1.4 Plot of the real viscosity, η', at 50 Hz, against the time of mixing, for a range of different weight fractions of nanotubes in the composite. The arrows mark the critical time t^* for each concentration. (From Huang, Y.Y. et al., *Phys. Rev. B*, 73, 125421, 2006. With permission.)

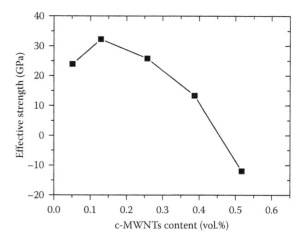

FIGURE 1.5 Effective tensile stress as back-calculated from the rule of mixtures for oriented ($\lambda = 10$) nanocomposite tapes based on PP composites at different HDPE-coated MWNTs contents. (From Deng, H. et al., *J. Appl. Polym. Sci.*, accepted 2010. With permission.)

Recently, Huang et al. [29] reported a study involving melt compounding PDMS (polydimethylsiloxane) with MWNTs. The real part of composite viscosity (η) is recorded during mixing. Measured viscosity changes as a function of nanotube–polymer mixing time gave some quantitative understanding of CNT dispersion in the matrix polymer. Calculations showed that the shear stresses in the mixer are far below the ultimate tensile strength of MWNTs, indicating that the direct scission of MWNTs is unlikely to occur during mixing.

Obviously, every batch of the same concentration tends to exhibit similar levels of dispersion when mixed long enough; $t > t^*$. The indication of good dispersion is a plateau in the viscosity response, as shown in Figure 1.4. The higher the concentration, the longer the t^* needed to achieve a good dispersion.

As a result of that, a worse dispersion of CNTs is expected at higher loadings when the same melt processing condition is applied to a thermoplastic polymer matrix with different CNT contents. Therefore, the effective reinforcement efficiency of CNT is decreased with increasing CNT content (see Figure 1.5).

Overall, shear mixing is an effective and efficient way for dispersing CNTs in polymer matrices. However, attention needs to be paid to critical mixing time and the shear stress inside the mixer. Too high shear stresses are not recommended due to tube breakage. For high shear mixing processes, an optimum mixing time needs to be found to optimize the properties of nanocomposites. More work is needed to find a shear mixing method that will avoid tube breakage as well as degradation of the matrix.

1.3.1.2 Ultrasonication

In an ultrasonication process, the polymer and CNTs need to be dissolved in a solvent, and then placed into an ultrasonic regime. There are two main instruments for doing this: ultrasonic bath and ultrasonic tip. Ultrasonic processes deliver high levels

of vibration energy to the system. The process starts from the outside and works its way in with each layer fraction independently from preceding and succeeding layers. So it has the possibility to make the tube bundles thinner as well as shorter, which may not always lead to the high aspect ratios needed for optimal mechanical and electrical properties of polymer composites.

There are three mechanisms in ultrasonic processing: bubble nucleation and subsequent implosion, localized heating, and the formation of free radicals [2]. Cavitation mechanism caused much of the dispersion as well as tube damage. Lower frequencies can produce larger bubbles that lead to a larger energy distribution as they collapse. Ultrasonic horns normally work at lower frequencies compared with baths. Thus, high-energy cavity formation is likely to happen. In the case of ultrasonic tips or horn, dispersive energy is high and localized around the tip. The ultrasonic bath has higher frequencies and lower energy than the tip, and the cavitation zone is not well defined.

1.3.1.3 Milling and Grinding

This method uses one or several rotating cylinder(s) filled with iron balls, which work as grinding media to break down the aggregated tubes. Ball milling can be used to break up MWNT aggregates and reduce their length and diameter distribution [31]. Ball milling is also the most destructive method for processing CNTs as large amounts of amorphous carbon can be produced [32]. However, it might be considered as a "cheap and fast" method for industrial processes.

1.3.2 Chemical Method

Physical techniques such as melt compounding can lead to scalable dispersion, but local homogeneous dispersion states are difficult to achieve without breaking down the CNTs. Melt compounding is also more applicable to composites based on MWNTs due to their more rigid-rod behavior [2]. Thus, different

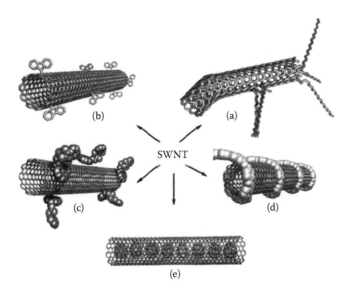

FIGURE 1.6 Functionalization possibilities for SWNTs: (a) defect-group functionalization, (b) covalent sidewall functionalization, (c) noncovalent exohedral functionalization with surfactants, (d) noncovalent exohedral functionalization with polymers, and (e) endohedral functionalization with, for example, C_{60}. (From Hirsch A., *Angew. Chem. Int. Ed.* 41, 1853, 2002. With permission.)

methods need to be considered for different types of CNTs, such as solvent processing. Nevertheless, the sp^2-hybridized structure of nanotubes makes them insoluble in common organic solvent. The introduction of functionalization on CNT surfaces might give solubility to CNTs and strengthen the interaction between CNTs and the matrix in the composites. Nevertheless, choosing the right level and type of functionalization can have a very important effect on composite properties (as shown in Figure 1.6).

Due to above-mentioned reasons, there has been much interest in using the chemical functionalization of CNT surfaces to make the tubes more soluble and separable in a given solvent or polymer matrix. In the case of organic solvent mixing method, homogeneous dispersion throughout the solvent, and thus the host matrix, can be achieved. It also gives much technical flexibility for CNT/polymer composite manufacturers. Two categories of chemical methods for dispersion are considered: covalent and noncovalent.

1.3.2.1 Covalent Method

Covalent functionalization and the surface chemistry of CNTs have been envisaged as very important factors for nanotube processing and applications [33]. Covalent methods refer to functionalization treatments involving covalent bond breakage across the nanotube surface, which disrupts the delocalized π-electrons and σ-bonds and, hence, incorporates other species across the exterior of the shell. The systematic development of fullerene chemistry has shown that their reactivity in addition reactions depends on the curvature of the fullerene. An increase of the curvature of the carbon network results in a more pronounced pyramidalization of the sp^2-hybridized C atoms and,

therefore, an increased tendency to undergo additional reactions. Thus, the mechanisms may preferentially occur at defect sites or where tube curvature is the highest [34].

The functionalized tubes are more soluble in an organic solvent; it is the sidewalls of the CNTs where the chemical attack during oxidation happens [35,36]. The bundles of CNTs are broken up into smaller groups or individual tubes during functionalization. There is more space available within most of the tubes to allow the uptake of solvent molecules. The increased solubility of CNTs in organic solvents such as tetrahydrofuran (THF) [35,37–39], chloroform, methylene chloride [40], and dimethylformamide (DMF) [37] has been shown after covalently attaching alkene groups. Shaffer et al. [41] have shown that acid/oxidative treatments enable the stable aqueous solutions of catalytically produced MWNT to be produced by way of introducing oxygen-containing surface groups, leading to the formation of viscoelastic gels at high concentration.

1.3.2.2 Noncovalent Method with Surfactant or Polymer

Noncovalent method generally involves the use of surfactant to bind ionically to the nanotube surface and prevent the tubes aggregating. This is a much less destructive method for good dispersion than covalent method. Surface active molecules such as sodium dodecylsulfate (SDS) or benzylalkonium chloride have proven successful in dispersing nanotubes into an aqueous phase [42–44]. The mechanism is thought to involve the CNTs residing in the hydrophobic interiors of the micelles [34]. If the hydrophobic region of the amphiphile contains an aromatic group, π–π stacking interactions occur with the nanotube sidewalls, as demonstrated by Chen et al. [45].

Another noncovalent method is associated with polymer wrapping. The polymer wraps around the nanotubes in order to separate them from each other. A number of polymers have been used to wrap CNT by O'Connell et al. [46]. SWNTs were reversibly solubilized in water by wrapping them with a variety of linear polymers. The process works most successfully with polyvinyl pyrrolidone (PVP) and polystyrene sulfonate (PSS), and provides a route to more precise manipulation purification, fractionation, and functionalization. Lately, Wang et al. coated solvent-resistant polymers on MWNT in supercritical carbon dioxide (scCO₂), which allows the selective deposition of high molecular weight fluorinated graft poly(methyl vinyl ether-alt-maleic anhydride) polymer onto MWNT in scCO₂ [47]. This allows the coating polymer to stay intact during further process where normally organic solvents are used. The uniform thickness of the insulating polymer coating on the nanotubes allows the coated nanotubes to have a range of important applications, such as nanoelectronics.

Finally, high-density polyethylene (HDPE) was wrapped on MWNTs by Bonduel et al. [48]. HDPE was homogeneously coated on the MWNT surface by in situ polymerization. The polymerization of ethylene was catalyzed by a highly active metallocene complex physicochemically anchored onto the nanotube surface (see Figure 1.7). The amount of coating layer can be controlled by adjusting the polymerization time.

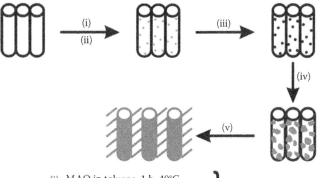

(i)   MAO in toluene, 1 h, 40°C   } MAO fixation

(ii)   Solvent evaporation, 2 h, 150°C

(iii)   Catalyst fixation: *n*-heptane, $Cp_2^*ZrCl_2$, 0.5 h, 50°C

(iv)   Polymerization: C_2H_4 (2.7 bars), 50°C

(v) C_2H_4 (2.7 bars), further polymerization at 50°C

FIGURE 1.7 Schematic of the polymerization-filling technique (PFT) applied to MWNTs from Bonduel et al. (From Bonduel, D. et al., *Chem. Commun.*, 781, 2004. With permission.)

Soon after, the same group compounded ethylene vinyl acetate copolymer (EVA) with HDPE-coated MWNT (c-MWNT) by simple melt blending [6]. Much better dispersion was observed for c-MWNTs in EVA compared to neat MWNTs.

In their studies [6,48], successful HDPE coating was applied on MWNT surfaces by Dubois and coworkers. They have shown it helps with the dispersion not only in miscible systems but also in immiscible polymer systems. Subsequent studies have been carried out to reveal the real potential of HDPE-coated MWNTs (commercialized by Nanocyl S.A. under the trade name of Nanocyl 9000®) in polypropylene (PP)-based composites [30].

1.3.3 In Situ Polymerization

In situ polymerization is another technique to introduce CNTs in polymer matrices. The advantage of this process is that the polymer chains and CNTs can be dispersed and grafted at a molecular scale. This gives excellent CNT dispersion and potentially good interfacial adhesion between CNTs and the polymer matrix. Successful investigations have been reported in literature from different groups [49–51]. The uniform dispersion of CNTs was obtained and improvement in both mechanical and electrical properties was observed.

1.4 Mechanical Properties of CNT/Polymer Composites

Many investigations have been published with a range of improvements of different properties. The first proof of the concept of a nanotube–polymer composite was published by Ajayan et al. [52]. They were originally looking for a way to align CNTs. Since then, a large number of papers have been published in this area, including a number of review articles focused on various aspects of CNT/polymer composites [47–50].

1.4.1 Strength and Toughness Potential

CNTs are widely considered as the next generation of carbon fiber. The clear advantage of CNTs over any other fiber is their ultrahigh strength [53], as the modulus of CNT is rather close to the modulus of carbon fiber (modulus of CNT is in the range of 1 TPa [54], while the modulus of carbon fiber can be as high as 600–800 GPa [55]). Tensile strengths in the range of 20–150 GPa are obtained for individual CNTs, as shown in Figure 1.8 [56], and are much stronger than the strongest carbon fiber (Toray's T1000 fiber has

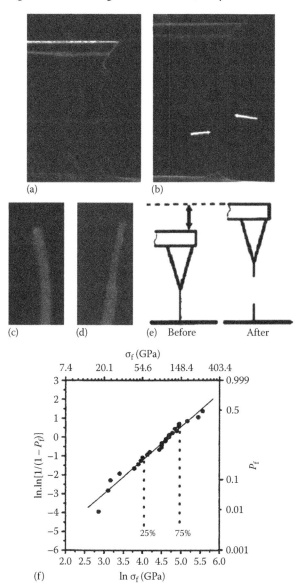

FIGURE 1.8 (a) MWNTs partially embedded in epoxy glue were tested in tension until (b) fracture of the nanotube occurred (fracture ends indicated by arrows). The ends of the nanotube exhibited either (c) a clear break or (d) a thinning of the nanotube diameter over a small length (~400 nm). The total cantilever deflection used to calculate fracture force was measured (e) from the double black arrow, (f) Weibull plot for the tensile strength of CVD-grown MWNTs. Dotted lines show the stress levels where 25% and 75% of nanotubes fail. (From Barber, A.H. et al., *Appl. Phys. Lett.*, 87, 203106, 2005. With permission.)

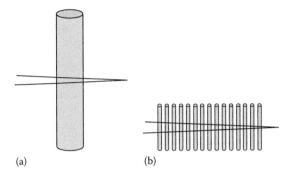

FIGURE 1.9 Schematic view of a crack propagating in a composite: (a) through a microfiber (carbon fiber) of given volume V_{CF} and critical length l_{cCF}; (b) through a series of nanotubes (CNTs) of total volume equivalent to the volume of the microfiber in (a), and of critical length l_{cCNT}. (From Wichmann, M.H.G. et al., *Comp. Sci. Technol.*, 68, 329, 2008. With permission.)

a strength near 7 GPa [57]). However, often the effective strength of CNTs in polymer composites is relatively low compared to the values expected, with few exceptions obtaining effective strength of 88 GPa for SWNT in polyvinyl alcohol (PVA) fiber [1] and 56 GPa for SWNT in PP fiber [58]. However, often the amount of CNTs used in these highly efficient polymer composites is relatively small (typically <2 wt.%). The larger amounts of CNTs in the polymer systems result in worsening of the dispersion and significantly reduced nanotube efficiencies [30,58]. Polymer fibers containing high CNT content or 100% CNTs are demonstrated to have much lower effective strength [59,60] than the values expected, indicating that significant efforts are still needed to obtain the true next high performance fiber based on CNTs.

Another clear advantage of CNTs over any other micronsized fibers is their much larger surface area, which results in much larger interaction with the polymer matrix, and potentially a much better toughness of composites compared with traditional fibers. As discussed by Wichmann et al. [61], the same volume of CNTs and carbon fibers are considered in a crack propagating scenario by modeling (see Figure 1.9), where the composites with CNTs achieved 4–100 times higher toughness than the composites containing carbon fiber. It indicates the great potential of CNTs in polymer composites with outstanding toughness.

1.4.2 Effective Mechanical Properties of CNTs in Polymer Composites

A large number of theoretical studies have been carried out since the 1950s with the aim of modeling the mechanical properties of fiber-reinforced composites. Some of these models are very sophisticated; two of the most common models are the rule of mixtures and the Halpin–Tsai equations.

In its simplest case, a composite can be considered as an isotropic elastic matrix filled with aligned elastic fibers that span the full length of the specimen [4]. The matrix and fiber are assumed to be well bonded. Thus, the matrix and fiber will be equally strained when they are under stress in the fiber direction. Under these circumstances, the composite tensile modulus in the alignment direction, E_c, is given by

$$E_c = V_f E_f + (1 - V_f) E_m \tag{1.1}$$

where
   E_f is the fiber modulus
   E_m is the matrix modulus
   V_f is the fiber volume fraction

This is the well-known rule of mixtures [4].

The stress transferred to the fiber builds up to its maximum value (σ_f^*, that will cause fiber breakage) over a distance of $l_c/2$ from the end of fiber. This indicates that short fibers carry load less effectively than long fibers, which means that they can be considered as less effective reinforcement (see Figure 1.10). This was first considered by Cox [62], and it was shown that for aligned fibers, the composite modulus is given by

$$E_C = \eta_L V_f E_f + (1 - V_f) E_m \tag{1.2}$$

where η_L is the length efficiency factor, which approaches 1 for $l/D > 10$, indicating the fact that high aspect ratio fillers are preferred.

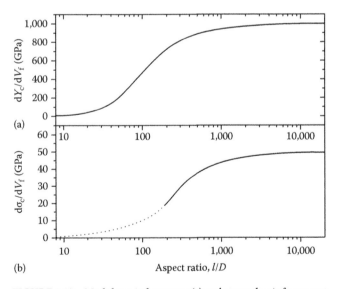

FIGURE 1.10 Modulus reinforcement (a) and strength reinforcement (b) as a function of filler aspect ratio for aligned composites, as calculated by the rule of mixtures. Modulus reinforcement is calculated with $Y_f = 1$ TPa, $Y_m = 1$ GPa for a volume fraction of 1%. In part (b), the strength is calculated for two regimes; $l < l_c$ (dotted line) and $l > l_c$ (solid line), respectively. The parameters used were $\sigma_f = 50$ GPa, $\sigma_m = 10$ MPa, and $\tau = 100$ MPa. Using these parameters, the critical aspect ratio is $(l/D)_c = 250$. In both parts (a) and (b), these parameters are appropriate for composites of arc discharge MWNT in a polymer matrix with good interfacial shear strength. (From Coleman, J.N. et al., *Carbon*, 44, 1624, 2006. With permission.)

When the fibers are not aligned, the short fiber composite modulus is given by

$$E_C = \eta_L \eta_0 V_f E_f + (1 - V_f) E_m \tag{1.3}$$

where η_0 is the orientation efficiency factor. This has the values of $\eta_0 = 1$ for aligned fibers, $\eta_0 = 3/8$ for fibers aligned in one plane, and $\eta_0 = 1/5$ for randomly oriented fibers in three-dimensional (3D) space [4,63].

Similar calculations can be used to derive an equation for the composite strength. It is described by

$$\sigma_C = \eta_L \eta_0 V_f \sigma_f + (1 - V_f) \sigma_m \tag{1.4}$$

where σ_C, σ_f, and σ_m are the composite, fiber, and matrix strengths, respectively.

Then, taking $\eta_0 = 1$ and $\eta_L = 1$ for aligned CNT/polymer fiber, the effective modulus and strength of CNTs can be back-calculated using Equations 1.3 and 1.4, respectively. Table 1.1 lists the calculated effective modulus and strength (or stress) from data reported in literature. There are two categories of highly oriented CNT fibers: one are fibers that mainly or only consist of CNTs (such as work from Dalton et al. [59], Zhang et al. [64], and Koziol et al. [60]); another type are polymer fibers containing small amounts of CNTs (typically <10 wt.%).

The highest mechanical properties of neat CNT spun fiber ever been reported is from Koziol et al. [60], where a modulus of 350 GPa and strength of 9 GPa are obtained. Its modulus is similar to intermediate modulus carbon fiber, while its best tensile strength is indeed higher than any other man-made fiber. However, it needs to be noted that there is a large amount of scatter in their experimental data. Nevertheless, even these interesting properties are still far removed from the theoretical values of CNTs. The second category of polymer fibers contains relatively small amounts of CNTs. The first report of such a fiber is by Andrews et al. [65], where the modulus increased 150% and strength increased 90% after adding 5 wt.% SWNT in a petroleum pitch matrix. It results in a back-calculated modulus for the CNTs of 1296 GPa and strength of 13 GPa. The mechanical properties of CNT/polymer fibers based on polyphenyl-enebenzobisoxazole (PBO) [49], PVA [1,13,77,81], PP [4,66,67], polyacrylonitrile (PAN) [68,69], and so on are investigated in the literature and are listed in Table 1.1. The great advantage of CNTs as mechanical reinforcing fiber is its extremely high tensile strength of upto 150 GPa compared with 7 GPa for conventional carbon fiber. As listed in Table 1.1, only a few studies have achieved effective tensile strengths above 10 GPa [1,28,69–72]. The highest effective reinforcement in strength was achieved by Wang et al., where a threefold increase in strength was obtained by adding 1 wt.% SWNT in an oriented PVA matrix (see Figure 1.11). It was concluded that a high level of dispersion,

TABLE 1.1 Calculated Effective Mechanical Contribution (Using Equations 1.3 and 1.4) of CNTs in Oriented Polymer Fibers or Tapes Reported in Literature

Type of CNT	Matrix	E_f [GPa]	σ_f [GPa]	Reference
SWNT	Pitch	1296	13	Andrews et al. [65]
SWNT	PVA	1120	45	Ciselli [71]
MWNT	UHMW-PE	868	4	Ruan et al. [73]
MWNT	PVA	850	—	Wang et al. [74]
SWNT	PP	610	56	Kearns and Shambaugh [58]
SWNT	PVA	600	88	Wang et al. [1]
MWNT	PP	497	5.5	Dondero and Gorga [75]
SWNT	PBO	449	19	Kumar et al. [49]
SWNT	PVA	406	8	Zhang et al. [76]
MWNT	—	350	9	Koziol et al. [60]
MWNT	PP	305	15.3	Jose et al. [77]
MWNT	—	195	1.9	Zhang et al. [64]
SWNT	PA	153	36	Gao et al. [78]
MWNT	—	150	6	Koziol et al. [60]
SWNT	PAN	149	2	Sreekumar et al. [69]
SWNT	PAN	149	2	Chae et al. [68]
SWNT	PVA	147	3	Dalton et al. [59]
MWNT	PAN	110	6	Chae et al. [68]
SWNT	PP	100	1.5	Chang et al. [79]
MWNT	—	65	1	Koziol et al. [60]
DWNT	PAN	61	2	Chae et al. [68]
SWNT	PMMA	55	—	Haggenmueller et al. [80]
MWNT	PC	48	−11	Potschke et al. [28]
MWNT	PC	48	1	Fornes et al. [81]

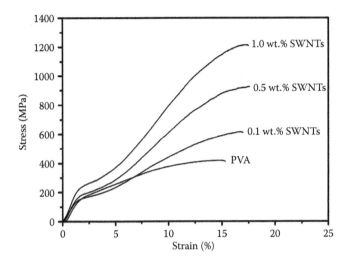

FIGURE 1.11 Stress–strain curves for oriented PVA/SWNT tapes with a draw ratio of 5, showing a strong increase in tensile strength with the addition of small amounts of SWNTs to the polymer. (From Wang, Z. et al., *Nanotechnology*, 18, 455709, 2007. With permission.)

interfacial interaction, and alignment of the nanofillers was essential for achieving true mechanical reinforcement by CNTs in composites [1].

1.5 Electrical Properties of CNT/Polymer Composites

1.5.1 Isotropic Polymer Composites

The demand for polymer composites is increasing every year due to their outstanding performance. There are some applications such as electrostatic discharge (ESD), electrostatic painting, and electromagnetic-radio frequency interference (EMI) protection that require the composites to have certain levels of conductivity ($>10^{-6}$ S/m) [82]. Conductive polymer composites (CPCs) are

conventionally made by adding carbon black, metal powder, or carbon fiber into a polymer matrix. Even with the polymer matrix being an insulator, the conductivity of the composites can demonstrate a sudden jump when a critical filler content is reached. This phenomenon is often described as percolation [83]. The percolation threshold of composites has been shown both experimentally and theoretically to decrease with filler aspect ratio (see Figure 1.12) [81–83]. Typically, 5–20 wt.% of conventional conductive filler (such as carbon black) is added into the polymer matrix to achieve a percolating network [84]. Such high filler contents quite often negatively affect the mechanical properties and processability of the resulting composites.

CNTs are one of the most interesting fillers for CPCs due to their high aspect ratio and excellent conductivity. In one of the first studies on CPCs containing CNTs, Sandler et al. [82] dispersed MWNTs in epoxy matrix, and the percolation threshold they obtained was as low as 0.04 vol.%. Soon after, they decreased the percolation threshold even further to 0.0025 wt.% by using aligned CVD-MWNT in the epoxy matrix [86]. Similarly, percolation thresholds as low as 0.052 vol.% and between 0.05 and 0.023 wt.% were later reported for single-walled nanotubes (SWNTs)/epoxy composites by Bryning et al. [70] and Moisala et al. [87], respectively. They reported that the lowest percolation threshold was obtained when the SWNTs were allowed to re-aggregate. Recently, Kovacs et al. [88] obtained different percolation thresholds (0.1, 0.01, and 0.02 wt.%) in the same MWNT/epoxy system. Processing conditions were found to be responsible for these differences in percolation threshold. Similar results on the process-induced aggregation of CNTs have also been reported in other solvent-based systems in literature [89].

For thermoplastic polymers, Potschke et al. [24] reported a percolation threshold of 1.4 wt.% for melt-compounded MWNT/PC composites. Percolation threshold values between 1 and 2 wt.% are commonly seen in melt-compounded MWNT/PP systems [90,91]. Solvent-processed CPCs exhibit lower percolation thresholds than melt-compounded systems, and a value of 0.05 vol.%

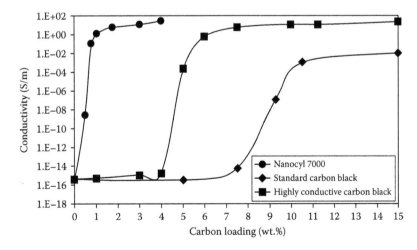

FIGURE 1.12 A comparison of percolation thresholds of polymer composites containing CNTs and carbon black. (Courtesy of Nanocyl, SA Sambreville, Belgium, www.nanocyl.com. With permission.)

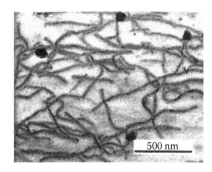
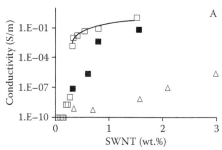

FIGURE 1.13 (Left) SEM of a PS nanocomposite containing 0.3 wt.% SDS-dispersed SWNTs, and (right) the percolation threshold of composite films containing SDS-dispersed SWNT in PS (open square) and in PMMA (in filled square) matrices. The conductivity of the composites from gum Arabic-dispersed SWNT in PS is shown in open triangle. (From Regev, O. et al., *Adv. Mater.*, 16, 248, 2004. With permission.)

was obtained for SWNT/polyimide composites by Ounaies et al. [92]. A percolation threshold of nearly 0.3 wt.% is obtained by a latex-based method [18,93] (see Figure 1.13). Similar percolation thresholds (near 0.1 wt.%) in solvent-processed CPCs containing either SWNT [94] or MWNT [95] are quite often obtained.

Such low percolation thresholds (<1.5 wt.%) of CNTs in various polymer systems through various processing methods, as reviewed above, can result in very good processability and conductivity for the CPCs. Currently, CPCs based on MWNTs are already used in various antistatic applications [85].

1.5.2 Oriented Polymer Composites

Due to the one-dimensional structure of CNTs, oriented CNT/ polymer composite fibers or tapes generate intense interest as such oriented systems could result in high mechanical reinforcing efficiency. Ajayan et al. were the first to consider drawing as a method to align CNTs in a polymer matrix [52]. Since then, oriented polymer fibers or tapes containing CNTs are investigated extensively.

The conductivity of such oriented CNT/polymer fibers or tapes is reported to decrease upon drawing for various polymer matrices [49,96,97]. The aspect ratio of the conductive filler is found

to play an important role in the conductivity of the oriented composites [96,97]. The drawing process applied to the composites is shown to align the CNT network (see Figure 1.14). Recently, the effect of CNT orientation on the conductivity of polymer fibers was extensively studied by Du et al. [98]. They made a series of SWNT/PMMA fibers with different degrees of nanotube alignment by controlling the melt spinning conditions. The degree of alignment was quantified using the full-width at half-maximum (FWHM) of the SWNT obtained from x-ray study, where the higher FWHM corresponds to less alignment. As shown in Figure 1.15, the conductivity decreases with increasing alignment, and they form orientation percolation between 20° and 40°. It also shows that intermediate levels of orientation give higher conductivity than isotropic samples.

The percolation threshold of highly oriented CPCs based on CNTs is relatively high (~5 wt.% [96]) compared to the values obtained for isotropic systems, as discussed previously. However, the conductivity of oriented fibers or tapes is also found to increase during thermal treatments or annealing [99–102]. This increase is explained by an improvement of local contacts between conductive regions caused by thermal energy.

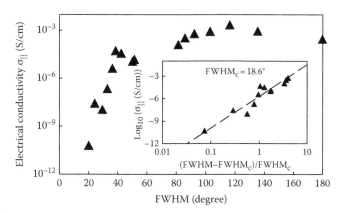

FIGURE 1.15 Electrical conductivity of a 2 wt.% SWNT/PMMA composite along the alignment direction with increasing nanotube isotropy. Nanotube alignment is assessed using x-ray scattering, where FWHM = 0 is perfectly aligned and FWHM = 180 is isotropic. Inset: a log–log plot of electrical conductivity vs. reduced FWHM determines the critical alignment, $FWHM_c$.

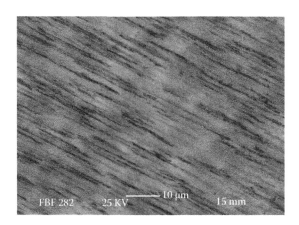

FIGURE 1.14 SEM of highly oriented MWNT networks on the surface of MWNT/PP composite tape (5 wt.% MWNTs, $\lambda \approx 8$).

Therefore, it is possible to reduce the percolation threshold in oriented systems by thermal treatment.

1.5.3 Modeling of Percolation Threshold

The increasing conductivity of composite materials with conductive filler content can be described by a scaling law according to classical percolation theory, as follows:

$$\sigma = \sigma_0 (P - P_c)^t \tag{1.5}$$

where
  σ_0 is a scaling factor
  P_c is the percolation threshold
  σ is the conductivity of the CPC
  P is the content of the filler in the CPC [83]

The exponent t is an exponent that depends on the dimensionality of the conductive network. It is expected to vary with different materials with calculated values of $t \approx 1.3$ and $t \approx 2.0$ in two and three dimensions, respectively. Therefore, the percolation threshold of the CPCs can be accurately determined, and information on the dimensionality can be obtained for the conducting network before and after drawing by fitting classical percolation theory to conductivity data obtained experimentally.

To model the effect of filler aspect ratio and filler orientation on the percolation threshold of CPCs, analytical models such as excluded volume theory [103] can be used to predict the percolation threshold of high aspect ratio rod-shaped filler-based polymer composites. The excluded volume of an object is defined as the region of space into which the center of another similar object is not allowed to enter if overlapping of these two is to be avoided [104]. It is capable of predicting the percolation threshold of high aspect ratio cylindrical particles embedded in a noninteracting matrix.

It has been extensively used to describe both micro-size and nano-size composite systems [103–105]. Here, the nanotube is modeled as the capped cylinders of diameter W and length L. The excluded volume is expressed as [103]

$$\langle V_e \rangle = \frac{4\pi}{3} W^3 + 2\pi W^2 L + 2WL^2 \langle \sin\gamma \rangle \tag{1.6}$$

where
  γ is the angle between the two cylinders in contact with each other
  $\langle \sin\gamma \rangle$ is calculated to be $\pi/4$ for randomly oriented cylinders [104]

The value of $\langle \sin\gamma \rangle$ for oriented systems with a maximum disorientation angle of 0°, 30°, 45°, 90°; $\langle \sin\gamma \rangle$ is given as: 0, 0.44, 0.60, 0.78 = $\pi/4$, respectively [103]. Thus, the exclude volume of an isotropic system can be written as

$$\langle V_e \rangle = \frac{4\pi}{3} W^3 + 2\pi W^2 L + \frac{\pi}{2} WL^2 \tag{1.7}$$

It was shown by Celzard et al. [103] that the percolation threshold of an infinitely thin cylinder system is expressed as

$$1 - \exp\left(-\frac{1.4V}{\langle V_e \rangle}\right) \leq \phi_c \leq 1 - \exp\left(-\frac{2.8V}{\langle V_e \rangle}\right) \tag{1.8}$$

where the volume of the MWNT is written as

$$\langle V \rangle = \frac{\pi}{6} W^3 + \frac{\pi}{4} W^2 L \tag{1.9}$$

Hence, the percolation threshold of such a high aspect ratio filler system can be calculated according to Equations 1.6 through 1.9 (Figure 1.16).

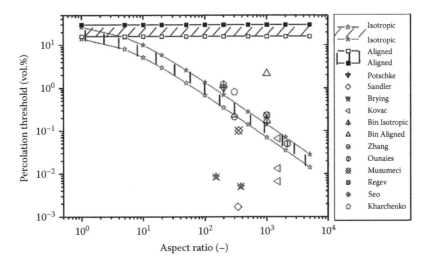

FIGURE 1.16 Theoretical percolation threshold for aligned and isotropic composite systems (the shadowed area) according to Celzard et al. [103], together with experimental data reported from Potschke et al. [24], Sandler et al. [86], Brying et al. [70], Kovacs et al. [88], Bin et al. [96], Zhang et al. [106], Ounaies et al. [92], Musumeci et al. [95], Regev et al. [18], Seo et al. [91], and Kharchenko et al. [90], respectively.

As shown in Figure 1.16, some of the results reported in literature [66,86,88] lie well below the percolation threshold predicted by excluded volume theory. In such systems, particle movement and re-aggregation is allowed to induce the formation of a conductive network [107]. These systems can be defined as systems with a kinetic percolation threshold [88]. Other data fit the theory quite well and are defined as systems with a statistical percolation threshold [88], in which the conductive phase is randomly dispersed. It is worthy to mention that solution processed composites (such as the data reported in [18,70,86,88]) have lower percolation thresholds than melt-compounded composites [24,90,91]. There are two possible reasons: one is that the solvent evaporation process takes quite a long time (typically 24 h), which may allow for the re-aggregation of the conductive filler in the low viscosity system to reduce contact resistance. Another reason might be the reduction of tube length during high shear melt compounding. Finally, it is noted that the percolation threshold of oriented polymer systems reported in literature [96] lies well between the values for perfect aligned systems and isotropic system predicted by Celzard et al. [103].

1.6 Sensing Properties

CNT/polymer composites have been studied to sense external stimuli such as biomolecules, chemicals [108–112], gases [108–118], vapor [107,118–120], mechanical stress or strain [106,110,113,114], and temperature [105,107–109]. The exposure of the CPCs to external stimuli can result in changes of the electrical properties, which can be considered as a signal. As shown in Figure 1.17 [106], a mechanical strain applied to a CNT/elastomer composites can result in a clear electric signal that can be used for sensing. Applications for sensors based on CNT/polymer composites are expected in a wide range of fields, such as building application, medical application, and protective clothing [115].

Damage sensing in structural composites is another important possible application for CNT/polymer composites [116]. It is demonstrated that conducting carbon nanotube networks formed in a thermoset polymer matrix can be used as highly sensitive sensors for detecting the onset, nature, and evolution of damage in

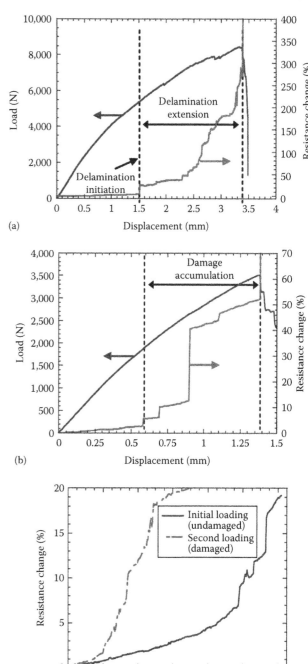

FIGURE 1.18 (a) Load-displacement and resistance curves for the 0° specimen with center ply cut to initiate delamination, (b) load-displacement and resistance curves for the 0/90 specimen, (c) resistance curves for initial loading (undamaged) and reloading (damaged) laminates. (From Thostenson, E.T. and Chou, T.W., *Adv. Mater.*, 18, 2837, 2006. With permission.)

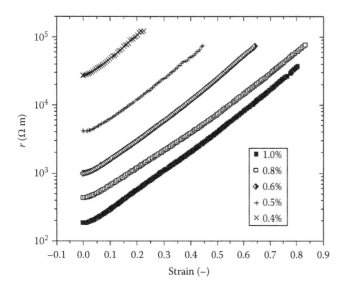

FIGURE 1.17 Resistivity vs. strain for various CNT concentrations. (From Zhang, R. et al., *Phys. Rev. B*, 76, 195433, 2007. With permission.)

advanced composites. The internal damage accumulation can be monitored in situ using electrical measurements (see Figure 1.18) [117]. After the onset of damage and subsequent reloading of the damaged structure, there is a remarkable shift in the sensing curve, indicating irreversible damage. These results demonstrate promises for the evaluation of automatic self-healing approaches for polymer composites and the development of enhanced life prediction methodologies [118–121].

1.7 Thermal Stability

Improved thermal stability observed by the thermogravimetric analysis (TGA) of nanotube/polymer composites has been reported for different polymer matrix systems [49,75,122–127]. The onset decomposition temperature, T_{onset}, and the maximum weight loss rate, T_{peak}, are found to be higher for CNT/polymer composites compared with neat polymer matrix. A number of mechanisms for increased thermal stability have been suggested. Well-dispersed nanotubes might hinder the flux of degradation product and, therefore, delay the onset of degradation. Polymers near the nanotubes may degrade more slowly, which would shift T_{peak} to higher temperatures [7]. Another possible mechanism is improved thermal conductivity, which improves the heat dissipation within the composites and, therefore, the thermal stability [128].

In a study on PP/MWNT carried out by Kashiwagi et al. [129], T_{peak} increased from 430° to 442° and 443° for 1 and 2 wt.%, respectively, when the experiment was carried out in nitrogen. A more impressive increase was observed when the experiment was done in air: from 298° to 355° and 376° for 1 and 2 wt.%, respectively.

The observed improvement of thermal stability potentially means that these materials can be used as fire-retardant additives in polymer matrix. The flammability can be evaluated by cone calorimeter or micro-calorimeter [127,130]. The retardancy is also found to improve with better dispersion, higher interfacial area, and higher loading of CNTs [125].

1.8 Conclusion and Outlook

A general review on CNT/polymer composites has been carried out in this chapter, and main efforts have been focused on the mechanical and electrical properties of CNT/polymer composites. It is demonstrated that CNTs have potential for a wide range of applications in the composite field. Their large aspect ratio, excellent conductivity, and ultrahigh mechanical and thermal properties make them an outstanding candidate for a multifunctional nanofiller.

Multifunctional polymer composites are becoming an interesting topic in the field of CNT/polymer composites. Thanks to their intrinsic multifunctionality, CNTs have demonstrated to be able to provide polymer composites with electrical properties, mechanical properties, sensing ability, self-healing properties, thermal conductivity and improved thermal stability, etc. A combination of these properties could be obtained in specific composites, whereas the properties can be tailored by

engineering these materials from the nano to macro level in order to fulfill desired applications. However, due to their nano-size and large aspect ratio, there are still difficulties in dispersing them into polymer matrix, especially at higher loadings.

References

1. Wang Z, Ciselli P, and Peijs T. *Nanotechnology* 2007; 18:455709.
2. Ahir SV and Terentjev EM. Polymer containing carbon nanotubes: Active composite materials. In: H.S. Nalwa (ed.), *Polymeric Nanostructures and Their Applications*, American Scientific Publishers, Los Angeles, CA, 2005.
3. Ciselli P, Wang Z, and Peijs T. *Materials Technology* 2007; 22:10–21.
4. Coleman JN, Khan U, Blau WJ, and Gun'ko YK. *Carbon* 2006; 44:1624–1652.
5. Coleman JN, Khan U, and Gun'ko YK. *Advanced Materials* 2006; 18:689–706.
6. Dubois P and Alexandre M. *Advanced Engineering Materials* 2006; 8:147–154.
7. Moniruzzaman M and Winey K. *Macromolecules* 2006; 39:5194–5205.
8. Thostenson ET, Ren Z, and Chou TW. *Composites Science and Technology* 2001; 61:1899–1912.
9. Nuriel S, Liu L, Barber AH, and Wagner HD. *Chemical Physics Letters* 2005; 404:263–266.
10. Barber AH, Cohen SR, and Wagner HD. *Physical Review Letters* 2004; 92(18):4.
11. Barber AH, Cohen SR, and Wagner HD. *Physical Review B* 2005; 71(11):5.
12. Barber AH, Cohen SR, Eitan A, Schadler LS, and Wagner HD. *Advanced Materials* 2006; 18:83–87.
13. Qian D, Dickey EC, Andrews R, and Rantell T. *Applied Physics Letters* 2000; 76:2868–2870.
14. Cooper CA, Cohen SR, Barber AH, and Wagner HD. *Applied Physics Letters* 2002; 81:3873–3875.
15. Hwang GL, Shieh YT, and Hwang KC. *Advanced Functional Materials* 2004; 14:487–491.
16. Assouline E, Lustiger A, Barber AH, Cooper CA, Klein E, Wachtel E, and Wagner HD. *Journal of Polymer Science: Part B: Polymer Physics* 2003; 41:520–527.
17. Barber AH, Cohen RE, and Wagner HD. *Applied Physics Letters* 2003; 82:4140–4142.
18. Regev O, Elkati NB, Loos J, and Koning CE. *Advanced Materials* 2004; 16:248–251.
19. Xia HS, Wang Q, Li KS, and Hu GH. *Journal of Applied Polymer Science* 2004; 93:378–386.
20. Mamedov AA, Kotov NA, Prato M, Guldi DM, Wicksted JP, and Hirsch A. *Nature Materials* 2002; 1:190–194.
21. Andrews R, Jacques D, Qian DL, and Rantell T. *Accounts of Chemical Research* 2002; 35:1008–1017.
22. Potschke P, Fornes TD, and Paul DR. *Polymer* 2002; 43:3247–3255.

23. Potschke P, Bhattacharyya AR, Janke A, and Goering H. *Composite Interfaces* 2003; 10:389–404.

24. Potschke P, Dudkin SM, and Alig I. *Polymer* 2003; 44:5023–5030.

25. Potschke P, Abdel-Goad M, Alig I, Dudkin S, and Lellinger D. *Polymer* 2004; 45:8863–8870.

26. Potschke P, Bhattacharyya AR, and Janke A. *European Polymer Journal* 2004; 40:137–148.

27. Potschke P, Bhattacharyya AR, Janke A, and Pegel S. *Fullerenes, Nanotubes, and Carbon Nanostructures* 2005; 13:211–224.

28. Potschke P, Brunig H, Janke A, Fisher D, and Jehnichen D. *Polymer* 2005; 46:10335–10363.

29. Huang YY, Ahir SV, and Terentjev EM. *Physical Review B* 2006; 73:125421–125429.

30. Deng H, Bilotti E, Zhang R, and Peijs T. *Journal of Applied of Polymer Science* 2010; Accepted.

31. Kim YA, Hayashi T, Fukai Y, Endo M, Yanagisawa T, and Dresselhaus MS. *Chemical Physics Letters* 2002; 355:3–4.

32. Pierard N, Fonseca A, Konya Z, Willems I, Van Tendeloo G, and Nagy JB. *Chemical Physics Letters* 2001; 335:1–2.

33. Banerjee S, Hemraj-Benny T, and Wong SS. *Advanced Materials* 2005; 17:17–29.

34. Hirsch A. *Angewandte Chemie-International Edition* 2002; 41:1853–1859.

35. Chen J, Hamon MA, Hu H, Chen YS, Rao AM, Eklund PC, and Haddon RC. *Science* 1998; 282:95–98.

36. Hamon MA, Hu H, Bhowmik P, Niyogi S, Zhao B, Itkis ME, and Haddon RC. *Chemical Physics Letters* 2001; 347:8–12.

37. Georgakilas V, Kordatos K, Prato M, Guldi DM, Holzinger M, and Hirsch A. *Journal of the American Chemical Society* 2002; 124:760–761.

38. Georgakilas V, Voulgaris D, Vazquez E, Prato M, Guldi DM, Kukovecz A, and Kuzmany H. *Journal of the American Chemical Society* 2002; 124:14318–14319.

39. Holzinger M, Vostrowsky O, Hirsch A, Hennrich F, Kappes M, Weiss R, and Jellen F. *Angewandte Chemie-International Edition* 2001; 40:4002–4005.

40. Boul PJ, Liu J, Mickelson ET, Huffman CB, Ericson LM, Chiang IW, Smith KA, Colbert DT, Hauge RH, Margrave JL, and Smalley RE. *Chemical Physics Letters* 1999; 310:367–372.

41. Shaffer MSP, Fan X, and Windle AH. *Carbon* 1998; 36:1603–1612.

42. Bandow S, Rao AM, Williams KA, Thess A, Smalley RE, and Eklund PC. *Journal of Physical Chemistry B* 1997; 101:8839–8842.

43. Duesberg GS, Burghard M, Muster J, Philipp G, and Roth S. *Chemical Communications* 1998; 3:435–436.

44. Krstic V, Duesberg GS, Muster J, Burghard M, and Roth S. *Chemistry of Materials* 1998; 10:2338.

45. Chen RJ, Zhan YG, Wang DW, and Dai HJ. *Journal of the American Chemical Society* 2001; 123:3838–3839.

46. O'Connell MJL, Boul P, Ericson LM, Huffman C, Wang YH, Haroz E, Kuper C, Tour J, Ausman KD, and Smalley RE. *Chemical Physics Letters* 2001; 342:265–271.

47. Wang JW, Khlobystov AN, Wang WX, Howdle SM, and Poliakoff M. *Chemical Communications* 2006; 1670–1672.

48. Bonduel D, Mainil M, Alexandre M, Monteverde F, and Dubois P. *Chemical Communications* 2004; 781–783.

49. Kumar S, Dang TD, Arnold FE, Bhattacharyya AR, Min BG, Zhang X, Vaia RA et al. *Macromolecules* 2002; 35:9039–9043.

50. Park C, Ounaies Z, Watson KA, Crooks RE, Smith J, Lowther SE, Connell JW, Siochi EJ, Harrison JS, and Clair TLS. *Chemical Physics Letters* 2002; 364:303–308.

51. Saeed K and Park SY. *Journal of Applied Polymer Science* 2007; 106:3729–3735.

52. Ajayan PM, Stephan O, Colliex C, and Trauth D. *Science* 1994; 265:1212–1214.

53. Chae HG and Kumar S. *Science* 2008; 319:908–909.

54. Salvetat JP, Briggs GAD, Bonard JM, Bacsa RR, Kulik AJ, Stockli T, Burnham NA, and Forro L. *Physical Review Letters* 1999; 82:944–947.

55. http://www.netcomposites.com/

56. Barber AH, Andrews R, Schadler LS, and Wagner HD. *Applied Physics Letters* 2005; 87:203106.

57. http://www.toray.com

58. Kearns JC and Shambaugh RL. *Journal of Applied Polymer Science* 2002; 86:2079–2084.

59. Dalton AB, Collins S, Munoz E, Razal JM, Ebron VH, Ferraris JP, Coleman JN, Kim BG, and Baughman RH. *Nature* 2003; 423:703.

60. Koziol K, Vilatela J, Moisala A, Motta M, Cunniff P, Sennett M, and Windle AH. *Science* 2007; 21:1892–1895.

61. Wichmann MHG, Schulte K, and Wagner HD. *Composites Science and Technology* 2008; 68:329–331.

62. Cox HL. *Brazilian Journal of Applied Physics* 1952; 3:72–79.

63. Krenchel H. *Fibre Reinforcement*, Copenhagen, Denmark, Akademisk Forlag, 1964.

64. Zhang XF, Li QW, Holesinger TG, Arendt PN, Huang JY, Kirven PD, Clapp TG et al. *Advanced Materials* 2007; 19:4198–4201.

65. Andrews R, Jacques D, Rao AM, Rantell T, Derbyshire F, Chen Y, Chen J, and Haddon RC. *Applied Physics Letters* 1999; 75:1329–1331.

66. Ganb M, Satapathy BK, Thunga M, Weidisch R, Potschke P, and Jehnichen D. *Acta Materialia* 2008; 56(10):2247–2261.

67. Manchado MAL, Valentini L, Biagiotti J, and Kenny JM. *Carbon* 2005;43:1499–1505.

68. Chae HG, Minus ML, and Kumar S. *Polymer* 2006; 47:3494–3504.

69. Sreekumar TV, Liu T, Min BG, Guo HN, Kumar S, Hauge RH, and Smalley RE. *Advanced Materials* 2004; 16:58–61.

70. Bryning MB, Islam MF, Kikkawa JM, and Yodh AG. *Advanced Materials* 2005; 17:1186–1191.

71. Ciselli P. The potential of carbon nanotubes in polymer composites, PhD thesis, Eindhoven University of Technology, Eindhoven, the Netherlands, 2007.

72. Valentini L, Biagiotti J, Kenny JM, and Santucci S. *Composites Science and Technology* 2003; 63:1149–1153.

73. Ruan S, Gao P, and Yu TX. *Polymer* 2006; 47:1604–1611.

74. Wang W, Ciselli P, Kuznetsov E, Peijs T, and Barber AH. *Philosophical Transactions of the Royal Society of London A*, 2008; 18192168.

75. Dondero WE and Gorga RE. *Journal of Polymer Science Part B: Polymer Physics* 2005; 44:864–878.

76. Zhang XF, Liu T, Sreekumar TV, Kumar S, Hu XD, and Smith K. *Polymer* 2004; 45:8801–8807.

77. Jose MV, Dean D, Tyner J, Price G, and Nyairo E. *Journal of Applied Polymer Science* 2007; 103:3844–3850.

78. Gao J, Itkis ME, Yu A, Bekyarova E, Zhao B, and Haddon RC. *Journal of the American Chemical Society* 2005; 127:3847–3854.

79. Chang TE, Jensen LR, Kisliuk A, Pipes RB, Pyrz R, and Sokolov AP. *Polymer* 2005; 46:439–444.

80. Haggenmueller R, Gommans HH, Rinzler AG, Fischer JE, and Winey KI. *Chemical Physics Letters* 2000; 330:219–225.

81. Fornes TD, Baur JW, Sabba Y, and Thomas EL. *Polymer* 2006; 47:1704–1714.

82. Sandler J, Shaffer MSP, Prasse T, Bauhofer W, Schulte K, and Windle AH. *Polymer* 1999; 40:5967–5971.

83. Stauffer D and Aharony A. *Introduction to Percolation Theory*, Taylor & Francis, Boca Raton, FL, 1985.

84. Probst N. Conducting carbon black. In: Donnet JB, Bansal RC, and Wang M (eds.). *Carbon Black Science and Technology*, 2nd edn., Marcel Dekker, New York, 1993, pp. 271–288.

85. www.nanocyl.com

86. Sandler JKW, Kirk JE, Kinloch IA, Shaffer MSP, and Windle AH. *Polymer* 2003; 44:5893–5899.

87. Moisala A, Li Q, Kinloch IA, and Windle AH. *Composites Science and Technology* 2006; 66:1285–1288.

88. Kovacs JZ, Velagalaa BS, Schulte K, and Bauhofer W. *Composites Science and Technology* 2007; 67:922–928.

89. Schmidt RH, Kinloch IA, Burgess AN, and Windle AH. *Langmuir* 2007; 23(10):5707–5712.

90. Kharchenko SB, Douglas JF, Obrzut J, Grulke E, and Migler KB. *Nature Materials* 2004; 3:564–568.

91. Seo M and Park S. *Chemical Physics Letters* 2004; 395:44–48.

92. Ounaies Z, Park C, Wise KE, Siochi EJ, and Harrison JS. *Composites Science and Technology* 2003; 63:1637–1646.

93. Grossiord N, Miltner HE, Loos J, Meuldijk J, Mele BV, and Koning CE. *Chemistry of Materials* 2007; 19:3787–3792.

94. Ramasubramaniam R, Chen J, and Liu HY. *Applied Physics Letters* 2003; 83:2928–2930.

95. Musumeci AW, Silva GG, Liu JW, Martens WN, and Waclawik ER. *Polymer* 2007; 48:1667–1678.

96. Bin Y, Mine M, Ai K, Jiang X, and Masaru M. *Polymer* 2006; 47:1308–1317.

97. Deng H, Zhang R, Bilotti E, Peijs T, and Loos J. *Journal of Applied Polymer Science* 2009; 113:742–751.

98. Du FM, Fischer JE, and Winey KI. *Physical Review B* 2005; 72:121404/121401–121404/121404.

99. Bin Y, Chen QY, Tashiro K, and Matsuo M. *Physical Review B* 2008; 77:035419.

100. Deng H, Zhang R, Reynolds CT, Bilotti E, Peijs T. *Macromolecular Materials and Engineering* 2009; 294:749–755.

101. Miaudet P, Bartholome C, Derre A, Maugey M, Sigaud G, Zakri C, and Poulin P. *Polymer* 2007; 48:4068–4074.

102. Zhang R, Dowden A, Deng H, Baxendale M, and Peijs T. *Composites Science and Technology* 2009; 69:1499–1504.

103. Celzard A, McRae E, Deleuze C, Dufort M, Furdin G, and Mareche JF. *Physical Review B* 1996; 53:6209–6214.

104. Balberg I, Andrerson CH, Alexander S, and Wagner N. *Physical Review B* 1984; 30:3933–3943.

105. Dalmas F, Dendievel R, Chazeau L, Cavaille JY, and Gauthier C. *Acta Materialia* 2006; 54:2923–2931.

106. Zhang R, Baxendale M, and Peijs T. *Physical Review B* 2007; 76:195433.

107. Martin CA, Sandler JKW, Shaffer MSP, Schwarz MK, Bauhofer W, Schulte K, and Windle AH. *Composites Science and Technology* 2004; 64:2309–2316.

108. Agui L, Yanez-Sedeno P, and Pingarron JM. *Analytica Chimica Acta* 2008; 622(1–2):11–47.

109. Merkoci A, Pumera M, Llopis X, Perez B, del Valle M, and Alegret S. *TRAC—Trends in Analytical Chemistry* 2005; 24(9):826–838.

110. Kang IP, Heung YY, Kim JH, Lee JW, Gollapudi R, Subramaniam S, Narasimhadevara S et al. *Composites Part B—Engineering* 2006; 37(6):382–394.

111. Narkis M, Srivastava S, Tchoudakov R, and Breuer O. *Synthetic Metals* 2000; 113(1–2):29–34.

112. Kobashi K, Villmow T, Andres T, Haussler L, and Poetschke P. *Smart Materials & Structures* 2009; 18(3):15.

113. Wood JR, Zhao Q, Frogley MD, Meurs ER, Prins AD, Peijs T, Dunstan DJ, and Wagner HD. *Physical Review B* 2000; 62(11):7571–7575.

114. Li C, Thostenson ET, and Chou TW. *Composites Science and Technology* 2008; 68(6):1227–1249.

115. http://www.inteltex.eu/

116. Fiedler B, Gojny FH, Wichmann MHG, Bauhofer W, and Schulte K. *Annales De Chimie-Science Des Materiaux* 2004; 29(6):81–94.

117. Thostenson ET and Chou TW. *Advanced Materials* 2006; 18:2837–2841.

118. Gao LM, Thostenson ET, Zhang Z, and Chou TW. *Advanced Functional Materials* 2009; 19(1):123–130.

119. Thostenson ET and Chou TW. Multifunctional composites with self-sensing capabilities: Carbon nanotube-based networks—art. no. 65261X. In: Dapino MJ (ed.). *Conference on Behavior and Mechanics of Multifunctional and Composite Materials.* San Diego, CA, 2007, pp. X5261–X5261.

120. Thostenson ET and Chou TW. *Composites Science and Technology* 2008; 68(12):2557–2561.

121. Thostenson ET and Chou TW. *Nanotechnology* 2008; 19(21):215713.

122. Bhattacharyya AR, Sreekumar TV, Liu T, Kumar S, Ericson LM, Hauge H, and Smalley RE. *Polymer* 2003; 44:2373–2377.

123. Chen GX, Kim HS, Park BH, and Yoon JS. *Polymer* 2006; 47:4760–4767.

124. Ge JJ, Hou HQ, Li Q, Graham MJ, Greiner A, Reneker DH, Harris FW, and Cheng SZD. *Journal of the American Chemical Society* 2004; 126:15754–15761.

125. Kashiwagi T, Du FM, Winey KI, Groth KA, Shields JR, Bellayer SP, Kim H, and Douglas JF. *Polymer* 2005; 46:471–481.

126. Li J, Tong LF, Fang ZP, Gu AJ, and Xu ZB. *Polymer Degradation and Stability* 2006; 91:2046–2052.

127. Mahfuz H, Adnan A, Rangari VK, Hasan MM, Jeelani S, Wright WJ, and DeTeresa SJ. *Applied Physics Letters* 2006; 88:083119.

128. Huxtable ST, Cahill DG, Shenogin S, Xue LP, Ozisik R, Barone P, Usrey M, Strano MS, Siddons G, Shim M, and Keblinski P. *Nature Materials* 2003; 2:731–734.

129. Kashiwagi T, Grulke E, Hilding J, Harris R, Awad W, and Douglas J. *Macromolecular Rapid Communications* 2002; 23:761–765.

130. Hapuarachchi TD and Peijs T. *Composites Part A: Applied Science and Manufacturing* 2010, accepted; DOI: 10.1016/j.compositesa.2010.03.004.

2

Printable Metal Nanoparticle Inks

Bibin T. Anto
National University of Singapore

Loke-Yuen Wong
National University of Singapore

Rui-Qi Png
National University of Singapore

Sankaran
Sivaramakrishnan
National University of Singapore

Lay-Lay Chua
National University of Singapore

Peter K. H. Ho
National University of Singapore

2.1 Overview

2.1.1 Printable Electronics

The field of printed electronics aims to ultimately fully print most if not all the dielectric, semiconductor, and metal layers to give electrically functional and integrated devices on a large scale on flexible substrates. This is often associated with plastic electronics, which employs organic (often polymer) semiconductors in the devices. Printed electronics is widely seen to be a disruptive technology that revolutionizes the application of electronics through the development of super-large displays, flexible or wearable displays, smart identification tags, electronic books, etc.

Printing holds a particular attraction as a potential manufacturing method because it is scalable, has good throughput, and short design-to-production turnaround time. It is also an additive processing method in which the material is placed exactly where it is required. This provides a fundamentally more efficient use of materials, which simplifies manufacturing steps and is potentially more environmentally friendly. More crucially, printed electronics permits the use of new "soft" electronic materials such as molecules, polymers, and nano-materials on large-area and flexible substrates for the creation of electronics and products that transcend traditional boundaries.

Of the established printing technologies[1] potentially suitable for printable electronics, inkjet printing (IJP) is one of the most widely investigated for numerous applications compared with gravure, offset, flexographic, or other forms of printing. In IJP, 1–100 pL droplets of "ink" containing functional materials, either molten or more often in a suitable solvent dispersion, are ejected across an air gap by thermal, piezoelectric, or Rayleigh breakup methods

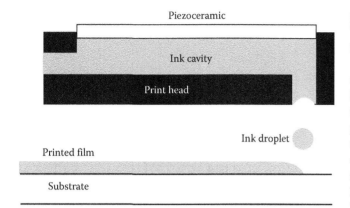

FIGURE 2.1 Schematic of inkjet printing.

onto a substrate where they solidify or dry. IJP has several unique advantages: versatility, droplet-on-demand, real-time control, noncontact operation (hence no dust pickup or defect generation), surface-relief tolerance, and ready availability of mature low-cost research units. The IJP technology has already been widely used for the printing of edible images on cakes, conductive tracks for circuits, and color filters in LCD and plasma displays. Figure 2.1 shows the schematic representation of inkjet printing.

Gravure, offset, flexographic, screen, and xerographic printing are all contact printing methods. The first three require the production first of a print master on a cylinder, which is thus compatible with reel-to-reel printing. In gravure printing, the master image is engraved or etched into the gravure cylinder to give microscopic ink-cavity wells. The cylinder is inked, the doctor-blade is cleaned, and the ink image is then transferred to the substrate with pressure applied by impression rollers. In offset printing, the master image is laser-written or lithographed onto a plate cylinder to give a pattern with surface-energy (hydrophilic–hydrophobic) contrast. This plate cylinder is then wetted by water and an oil-based ink from rollers that self-organize onto the hydrophilic and hydrophobic areas, respectively, and the ink-image is then transferred to an intermediate rubber offset cylinder and finally to the substrate with pressure applied by impression rollers. In flexographic printing, the master image is laser-written, lithographed, or molded onto a rubber plate cylinder to create a relief image. This relief image is then inked by a porous anilox metering roller and the ink image transferred to the substrate with pressure applied by impression rollers.

In screen printing, the master image is prepared as a stencil supported on a fine mesh of steel or polymer fibers by lithography or laser writing. The ink is then transferred through this stencil onto the substrate by forcing with a rubber squeegee blade.

In xerographic printing, a photoconductive drum is first charged and the image formed by selective discharge by reflected illumination from a master, laser, or light-emitting diode. The latent image is then bathed in charged solid toner particles that become electrostatically attracted to the image, and the toner image is transferred to the substrate by opposite electrostatic charging with pressure simultaneously applied by rollers. The

particles are then fused together and onto the substrate by heat. The requirement for solid fusible particles places restrictions on the materials that can be deposited.

Different from the well-established printing on paper and other absorbent substrates, printing on nonabsorbent substrates in printed electronics faces new challenges in the fixation of the deposited droplet and liquid shapes. Because these are not pinned by absorption into the substrate, the liquid can flow under surface-tension, capillarity, and Marangoni effects and can break up or migrate away from where the liquid droplet was first deposited. Furthermore, the footprint of the liquid droplets is highly dependent on the surface tension of the solvent. These challenges are common also to the IJP of other materials on impervious substrates.

At present, the printing of polymer light-emitting displays using IJP has reached a significant level of development in industrial research laboratories, where the fixation challenges are overcome by printing into pre-defined wells with appropriate wetting and dewetting properties at the bottom and sidewalls of the well, respectively. By contrast, the "free-form" printing of transistor circuits and solar cell films is in the early stages of exploration.

On the other hand, the development of nanometal inks and printing of metal tracks are being more widely and intensively pursued for various applications. Metal tracks are ubiquitous in numerous devices because they are needed to carry current as bus lines for large-area lighting panels and solar cells, as interconnects and signal lines for transistor and other circuits, and as the antennae in radiofrequency identification tags.[2–11] Other applications of metal films include chemobiomolecular sensors,[12,13] microelectromechanical systems,[14] functional nanostructures,[15] and catalysts,[16] for example, for carbon nanotube and graphene growths.

2.1.2 Printable Metal Systems

For the formation of very low-resolution and thick metal structures such as contact pads and bus lines, the well-established electrically conductive adhesives (ECAs) used in thick-film technology based on screen printing with micron-sized metal particles often suffice.[17] ECA technologies using metal pastes of Ag, Au, and Pt have been available for some time primarily for screen printing. These metal pastes comprise micron-sized metal flakes protected by dispersants such as long-chain thiols or carboxylic acids and dispersed at a volume fraction of >50% in a polymer binder and diluted in an organic solvent for deposition. The dried films can have σ_{dc} up to a few 10^4 S cm^{-1} that is primarily limited by poor contacts between the metal flakes.

Silver occupies a predominant position as the conductive filler in these technologies because of its favorable tradeoff between cost and performance. Gold is considerably more inert to oxidation and can reach higher conductivity because of the absence of surface oxides. However, Au is also considerably more expensive than Ag. Furthermore, Au can form a low-temperature eutectic with Si, which precludes its use in Si semiconductors. However, in the domain of plastic electronics electrochemical stability may be sufficiently important to tilt the balance in its favor as the material of choice, e.g., as source and drain electrodes. One cm³ of Au costs \$700, compared with Ag at \$6 and research-quantity polymer semiconductors of a few hundred dollars. Other metals such as Pt and Pd may also be of interest for specific applications.

For finer metal structures, IJP using nanometal inks based on nanometer-sized metallic particles offers greater promise, and has thus attracted the interest of numerous materials companies.[18] Current IJP technologies can deliver metal nanoparticle (NP) tracks that are as narrow as 50 μm and with a controlled height below several μm. These metal NPs have a metallic core encased in a stabilizing dispersant shell comprising either of molecules or polymers. The core itself is often substantially but not fully crystalline, so they are best called NPs rather than nanocrystals. However, they require a subsequent heat-treatment step at moderately low temperatures below 350°C to sinter the nanoparticulate film to a continuous film with bulk-like metallic properties. During this critical step, the dispersant shell is volatilized to allow neighboring metal NPs to come into contact and sinter together. There is a large volume contraction at this step that must be carefully managed to avoid loss of integrity due to micro-cracking, particularly in thin and narrow tracks.

The nanometal ink systems so far have been dominated by Ag NPs, which have been commercially produced for some time in the 50 nm diameter size range and protected by polymeric dispersants.[19–21] These are soluble in relatively high-surface-tension polar solvents, such as water and alcohols (e.g., ethylene glycol), and can therefore be deposited readily by printing. The dispersant is critical to achieving good dispersibility of the metal NPs and protecting them from premature coalescence. The most common polymeric dispersants here are poly(vinyl alcohol) (PVA) and poly(vinylpyrrolidone) (PVP). These encapsulate the metal NP cores as they form by reduction of metal salts to give a layer thickness of ca. 5 nm. These Ag NPs can be annealed to metallic films with a DC conductivity σ_{dc} that reaches $\approx 6 \times 10^4$ S cm⁻¹ after >1 h at 150°C, and $\approx 2 \times 10^5$ S cm⁻¹ after 5 min at 250°C. Annealing temperatures in excess of 300°C are required to reach one-half of the bulk σ_{dc} of Ag (6×10^5 S cm⁻¹).

There is a considerable motivation to develop alternative dispersants based on molecular self-assembled monolayers (SAMs) for smaller NPs in the 3–10 nm diameter range, and also for other metals such as Au. SAM dispersants are expected not only to give better control of NP characteristics than polymer dispersants, but also more scope for application-specific tailorability. In particular, Au NPs protected by thiol SAMs have a long history spanning over two decades and have spawned a lot of literature.[22,23] They are now commercially available. These are dispersible in aromatic hydrocarbon solvents such as toluene, xylene, and decalin and can be annealed to give $\sigma_{dc} \approx 2 \times 10^5$ S cm⁻¹ at temperatures above 220°C. Aromatic hydrocarbon solvents, however, usually have low surface tensions in the 30 mN m⁻¹ range, which results in severe spreading of the droplet footprint on the substrate.

Although the properties of such NPs have been intensively investigated for various potential applications,[15,24,25] their use in printable metallization is a much more recent endeavor.[26] Detailed investigations of the properties (optical, electrical, and mechanical) of these metal NP films, and of the insulator-to-metal transformation during sintering, have begun only recently. The relations of various physicochemical properties such as solvent dispersibility, processing characteristics, and coalescence temperature are beginning to be understood. Therefore, it is timely to outline here the fundamental aspects of metal NP behavior with a view to providing design guidelines and identifying possibilities for the further development of nanometal inks for print metallization.

2.1.3 Comparison of Nanometal and ECA Characteristics

In contrast to ECAs, metal NP films are initially semi-insulating because of the dispersant layer that acts as a tunneling barrier between the nanosized metal particles. During heat treatment, this shell is disrupted and volatilized, which then allows the neighboring NPs to sinter into a network of interconnected metallic paths. When the network exceeds the percolation limit, σ_{dc} shows a sharp increase.[11,27] Further annealing can bring σ_{dc} to within a factor of two or three of the bulk σ_{dc}. The final morphology of the film, however, is still nanocrystalline, different from bulk metal films made by evaporation or sputtering that show large columnar grains. ECAs do not show similar marked conductivity improvement with sintering because the micronsized metal flakes necessitate a high volume fraction of the polymer binder.

Metal NPs have several advantages over metal pastes for high-end applications: (1) the attainable conductivity in the sintered metal NP films is almost tenfold that of metal pastes, (2) the mechanical strength of the sintered NP films may be better than metal flakes dispersed in a polymer binder, (3) the achievable thickness control for thin film technology (<1 μm thick) is also better because of the smaller size of NPs, and (4) the sintered metal NP films are intrinsically insoluble, and thus compatible with subsequent solvent deposition of overlayers, whereas ECAs require thermal or UV crosslinking to achieve insolubility.

2.1.4 Chapter Organization

This chapter is organized as follows. In Section 2.1, we have summarized various printing technologies and sketched a comparison between the traditional electrically conductive adhesives developed for thick-film technology and the new nanometal inks for high-end thin-film technology. In Section 2.2, we develop geometric models to elucidate various aspects of the packing fraction and porosity of metal NPs and their films and their optimization. In Section 2.3, we outline and compare three phenomenological theories of the thermodynamic size effect on the melting temperature in metal NPs to show that the percolative insulator-to-metal (T_p) transformation is not related to this size-dependent melting. Section 2.4 outlines the systematics of the T_p transformation to show that they point to a cold-sintering mechanism. Section 2.5, discusses a simple and reliable theory of the optical properties of metal NPs in both dilute media and in solid films and shows that the nanostructure details of the T_p transformation in these films can be extracted from the optical plasmon spectrum. In Section 2.6, the methods of synthesis of metal NPs and the recent exciting developments of metal NPs with high dispersibility in water and polar-solvents but low T_p are reviewed. In Section 2.7, we outline some pertinent considerations for inkjet printing and show that high-quality metal tracks and dots can be printed at practical speeds using hydrophilic inks.

2.2 Geometrical Aspects

2.2.1 Estimating the Number of Surface and Bulk Atoms: The Spherical Approximation

Because of the small metal core radius r, NPs have a large surface-area-to-volume ratio. Figure 2.2 shows the total number of atoms in the bulk (open circles) and on the surface (open squares) computed for perfect or truncated octahedrons, which are the idealized shapes for face-centered-cubic (fcc) metals with a magic number of atoms.[23,28] In the spherical approximation of the NP, the number of atoms n in the core of effective radius r is given by

$$n = \frac{\frac{4}{3}\pi r^3}{v_{at}}, \qquad (2.1)$$

where v_{at} is the atomic volume ($=1.7 \times 10^{-23}$ cm^3 for Au and Ag). This line is superimposed on the data in Figure 2.2. Thus, for r of 0.6–100 nm, n increases as r^3 from 55 to 2×10^8 per NP for both Au and Ag.

In this approximation, the number of surface atoms n_{surf} is given by

$$n_{surf} = \frac{4}{3}\pi\left(r^3 - (r - \delta_{at})^3\right)v_{at}^{-1}, \qquad (2.2)$$

where δ_{at} is the thickness of an atomic layer. This is also plotted in Figure 2.2. For an fcc lattice, δ_{at} is taken to be one-third of the body diagonal, i.e., $\delta_{at} = \sqrt{8/3}\, r_{met}$, where r_{met} is the metallic radius. For both Au and Ag, $\delta_{at} = 2.6$ Å. The fraction of atoms on the surface (open squares for the magic-number octahedrons and lines for the spherical approximation) are also shown in the inset of Figure 2.2.

It is clear from both these plots that the spherical approximation is sufficiently accurate to predict the actual n, n_{surf}, and the surface atom fraction down to the smallest practical NP size. These equations show that small NPs with $r < 2$ nm have more than one-third of the atoms residing on the surface. NPs with $r < 0.75$ nm have less than 100 atoms in total and are more appropriately called clusters, as their electronic structure makes a transition towards discrete molecular character. The 55-atom magic-number NP has $r = 0.6$ nm with 76% of the atoms on the surface. Such clusters are no longer metallic because there are not enough atoms below the surface (which is covalently bonded to the SAM) to constitute a metallic core with states at the Fermi level.[24]

2.2.2 Estimating the Fill Fraction and Porosity: The Contracted Random Loose Packing Model

The as-deposited metal NP film has the NPs in nominal van der Waals contact. As a result, a significant volume fraction is occupied by the inter-particle void and also the dispersant layer itself depending on the ratio of the thickness of the dispersant layer t to r. This causes a low initial metal fill fraction, which contributes to occluded porosity and low final metal fill fraction in the sintered film. The porosity arises from volume loss due to the elimination of the dispersant shell. Because the t/r ratio is not small, the volume loss is severe. The hardness and Young's modulus of sintered NP films are thus significantly lower than that of bulk films with similar grain sizes.[19] The final achievable electrical and thermal conductivity are also reduced by a factor of two typically. The porosity causes severe micro-cracking, which coalesces to macroscopic cracks at elevated temperatures when diffusion becomes significant.[20,21]

In the following, we derive analytical approximations of the metal fill fraction before and after sintering. We consider that

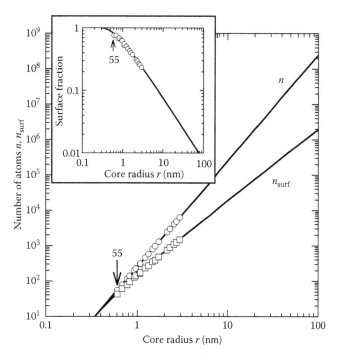

FIGURE 2.2 Number of atoms n and surface atoms n_{surf} as a function of core radius for Au NPs. Symbols give the results for magic clusters. Lines give the theoretical results from the spherical approximation. Inset: Fraction of total atoms residing on the surface.

as-deposited NPs initially adopt random-closed packing (rcp)[29] assuming they have a nonadhesive dispersant shell that allows them to settle to the local energy minimum. Transmission electron microscopy (TEM) images of thin NP films suggests such a packing.[11,27] The metal fill fraction v_f (i.e., volume fraction of the metal cores) in the film is given by geometry in the spherical approximation to be

$$v_f = F\left(1 + \frac{t}{r}\right)^{-3}, \qquad (2.3)$$

where

F is the packing factor for the NPs
t/r is the shell thickness-to-core-radius ratio

For an rcp of monodispersed spheres, $F = 0.634$.[30] For comparison, $F = 0.740$ for fcc.

Figure 2.3a shows as-deposited v_f decreases rapidly with increasing t/r ratio. For $t/r = 0.40$, the as-deposited film contains only 23 vol% metal, which decreases markedly to 8 vol% at $t/r = 1$. This demonstrates clearly the immense sensitivity of the properties of the film to t/r. In order to have high v_f, it is crucial to keep t/r low. For the assumed rcp packing, 36 vol% is the interparticle void and the balance is occupied by the dispersant shell.

To model the sintered v_f, we model the sintered film as a contracted random loose packed (rlp) structure. We assume that as the ligand shell volatilizes, the metal core becomes free to explore the newly formed cavity and contact a neighboring core. Upon contact, a metallic neck forms that fixes the position of the core and frustrates the attainment of a more closely packed rcp. Therefore, the final structure is random loose packed. Although individual cores can move sideways, on average they settle perpendicular to the film plane because the film can macroscopically contract only in the thickness direction. Therefore, we can assume the mean lateral positions of the cores do not change. This frustrated contraction in the film plane leads to

severe porosity, which may coalesce into micro-cracks.[11,19] We further assume that the sintering does not cause bulk melting or deformation of the cores. Prolonged annealing at elevated temperatures can cause void elimination and further densification, however, this almost invariably leaves macroscopic voids that break up the film.[20,21]

To compute the final fill fraction of this contracted rlp structure, we follow the change in fill fraction of the initial rcp structure as the dispersant shell evaporates. We first observe that the local particle arrangement in rcp is to a very good approximation the arithmetic mean of the twelve-coordinated fcc or hcp close-packed (cp) and the six-coordinated non-close-packed simple cubic (sc) structures. The rcp radial distribution function is the mean of the cp and sc distribution functions,[29,31] and the rcp fill fraction (0.634) is to a good approximation the mean (within 0.3%) of the cp (0.740) and sc (0.524) fill fractions. Therefore, we theoretically investigated the local contraction of fcc and sc structures as a model for the contraction of the rcp structure.

The fcc and sc structures were contracted along high-symmetry directions (i.e., [111] and [100] for fcc and [100] and [110] for sc) by allowing the shells to evaporate and the cores to touch in this 1D contraction model without changing the lateral positions. A schematic illustration of this model is given in the inset of Figure 2.3. We obtained the following analytical results for the contractions:

For fcc in the (111) direction,

$$v_f = \frac{\pi}{3\sqrt{3}\left(1 + \frac{t}{r}\right)^2 \sqrt{1 - \frac{1}{3}\left(1 + \frac{t}{r}\right)^2}} \quad \text{for } \frac{t}{r} \leq \frac{1}{2} \qquad (2.4a)$$

$$v_f = \frac{2\pi}{3\sqrt{3}\left(1 + \frac{t}{r}\right)^2} \quad \text{for } \frac{t}{r} > \frac{1}{2} \qquad (2.4b)$$

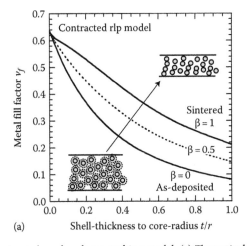

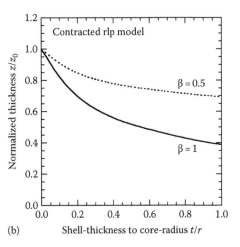

(a)   (b)

FIGURE 2.3 Contracted random loose packing model. (a) Theoretical metal fill fraction before ($\beta = 0$) and after sintering as a function of the shell-thickness to core-radius ratio. Inset: Schematic of the model. (b) Theoretical normalized film thickness after sintering. β is a transformation parameter.

For fcc in the (100) direction,

$$\nu_f = \frac{\pi}{3\sqrt{2}\left(1+\dfrac{t}{r}\right)^2 \sqrt{2-\left(1+\dfrac{t}{r}\right)^2}} \quad \text{for } \frac{t}{r} \le \sqrt{\frac{3}{2}}-1 \qquad (2.4c)$$

$$\nu_f = \frac{\pi}{3\left(1+\dfrac{t}{r}\right)^2} \quad \text{for } \frac{t}{r} > \sqrt{\frac{3}{2}}-1 \qquad (2.4d)$$

For sc in the (110) direction,

$$\nu_f = \frac{\pi}{6\left(1+\dfrac{t}{r}\right)^2 \sqrt{2-\left(1+\dfrac{t}{r}\right)^2}} \quad \text{for } \frac{t}{r} \le \sqrt{\frac{3}{2}}-1 \qquad (2.4e)$$

$$\nu_f = \frac{\pi}{3\sqrt{2}\left(1+\dfrac{t}{r}\right)^2} \quad \text{for } \frac{t}{r} > \sqrt{\frac{3}{2}}-1 \qquad (2.4f)$$

For sc in the (100) direction,

$$\nu_f = \frac{\pi}{6\left(1+\dfrac{t}{r}\right)^2} \qquad (2.4g)$$

The results for sc and fcc were then separately weighted by the degeneracy for the different directions.[32] The rcp → contracted rlp transformation was then computed as the arithmetic mean of the sc and fcc contractions and was smoothed using a polynomial function to remove kink artifacts. This contracted rlp model, therefore, has zero free parameters: the final ν_f is completely determined by the initial t/r ratio. We then introduce a transformation parameter β that gives the fractional change from the initial as-deposited rcp state (β = 0) to the final contracted rlp state (β = 1) to accommodate any incomplete transformation.

The results are plotted in Figure 2.3a for ν_f as a function of t/r. Figure 2.4 gives the corresponding plot of normalized film thickness z/z_o as a function of t/r. This is useful for obtaining the film thickness contraction at-a-glance given by $1 - z/z_o$ for design purposes. For $t/r = 0$ (i.e., no shell), ν_f is always 0.634 and $z/z_o = 1$ as expected, since this film cannot densify. Both the computed initial and final ν_f values decline very quickly with t/r. For $t/r = 0.3$, which corresponds to the usual case of $t = 1\,nm$ and $r = 3\,nm$, the initial ν_f is 0.29 and the final ν_f is 0.48 with $z/z_o = 0.62$ for β = 1.0. The porosity fraction is given by $(1 - \nu_f)$. The final value is 0.52, which is not too different from the random loose packing ratio of 0.55.[30] It is clear from this analysis that significant porosity must be retained in these NP films that have been sintered past the coalescence point, but well below complete melting.

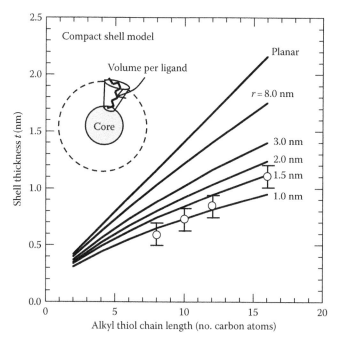

FIGURE 2.4 Compact shell model of the dispersant self-assembled monolayer. Theoretical shell thickness as a function of the alkyl thiol chain length for different nanoparticle radii r. Symbols are data for Au NPs with $r = 1–1.5\,nm$. Inset: Schematic of the model.

Data from the literature and our own work broadly support these conclusions. For polymer-protected Ag NPs with $r = 25\,nm$ and $t = 5\,nm$ (i.e., $t/r = 0.20$), the film thickness contraction is 13%,[19] which suggests β = 0.5. In our work, for SAM-protected Au NPs with $r = 1.4\,nm$ and $t = 0.8\,nm$ (i.e., $t/r = 0.57$), the thickness contraction is 37%, which suggests β = 0.75. For much larger $r = 6.2\,nm$ and $t = 0.8\,nm$ ($t/r = 0.13$), the thickness contraction is 25%, which gives β = 1.0. For SAM-protected Ag NPs with $r = 6.5\,nm$ and $t = 0.8\,nm$ (i.e., $t/r = 0.12$), we have measured the thickness contraction to also be 25%, and hence β = 1.0. Therefore, the contracted rlp model gives a good prediction of the thickness contraction and hence of the final porosity of these films over a wide range of t/r values between 0.1–0.5 and for different metal NPs. The pooled results from these measurements suggests β = 0.9 ± 0.15 for SAM dispersants and lower (ca. 0.5) for polymeric dispersant, which is not surprising considering that polymeric dispersants are difficult to volatilize completely.

The model shows that to achieve high densification with porosity <0.45, it is essential to keep the t/r ratio <0.3. This means the key is to employ the minimum dispersant shell thickness to disperse the maximum NP size possible. There is no penalty in the use of large NPs: the thermodynamic size and surface-area-to-volume ratio effect often asserted in the literature is not in fact relevant to the sintering of these metal NP films (see Sections 2.4 and 2.5). The NPs coalesce and the film sinters because of dispersant shell volatilization rather than size-dependent melting or surface melting.

The final porosity retained suggests that the maximum electrical (and thermal) conductivity in the sintered metal NP films is limited to at most about one-half to one-third of the bulk conductivity for t/r in the region of 0.2–0.3, which is in accord with experiments. Reducing the t/r ratio further should improve this final conductivity, but this approach is limited by the reduced NP dispersibility. Furthermore, the nanocrystalline morphology of the sintered film can also contribute to increased electron scattering, as the granular domains are much finer than the mean free path of the conduction electrons in the bulk metal, which is ca. 40 nm in Au.[33]

For SAM dispersants, t depends on the length of the ligand molecules. For ω-functionalized alkyl chains such as alkyl thiols, this t is related to the packing of the SAM into the shell. For a compact shell model that assumes that the SAM molecules are compacted into the shell at a fixed head group density of 4.5×10^{14} cm^{-2} with volume conservation, the t dependence on the number of carbon atoms for different core radius r can be computed from simple geometry as shown in Figure 2.4. This model essentially draws on the well-established idea of group contribution additivity of van der Waals volumes.[34] The plot reveals a fundamental feature of these NPs. For small r below a few nm, the shell volume expansion away from the core surface is significant. Therefore, the dispersant molecule can be accommodated by a sub-linear increase of the shell thickness with dispersant chain length. Experimental data (shown in symbols) are in accord with this model. The practical lower limit of t is ca. 0.7 nm for a sparse monolayer of short-chain dispersants.[27] Therefore, for $t/r < 0.3$, r should preferably be >2 nm.

2.2.3 A Model for Matrix Dilution Effect: Upper and Lower Bounds of the Average Inter-Core Gap Distance

For some applications, it may be desirable to incorporate a compatible polymer binder into the metal NP film to improve film cohesion, resistance to micro-cracking, and substrate adhesion.[11] "Unfilled" metal NP films develop severe micro-cracking as a result of large local volume contraction when the dispersant shell is eliminated and becomes progressively worse at elevated temperatures with increased atom diffusivity.[20,21] This is fatal for thin metal tracks because the micro-cracks can break line continuity. Recently, it was found that micro-cracking resistance can be markedly improved by incorporating a compatible polymer binder matrix at a controlled volume fraction of ca. 20–30 vol%.[11] The improvement is compared at both optical and sub-optical length scales in Figure 2.5.

Compatibilization can readily be achieved in the case of water- and polar-solvent soluble NPs by employing a polyelectrolyte matrix having the same charge polarity as the ligand shell of the NP.[11,35] The presence of this polymer matrix, however, opens up an additional inter-particle gap d_{gap} due to the matrix dilution effect. This gap is expected to increase with core radius r for a homogeneous dispersion. This can be seen readily in a simple scaling argument: in a homogeneously distributed system with a given core-to-matrix volume fraction, all linear dimensions

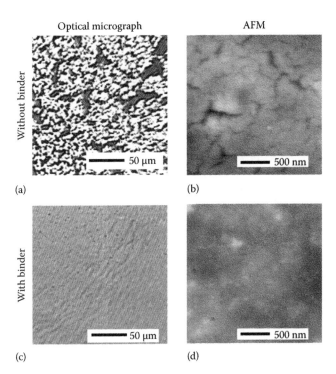

Optical micrograph   AFM

Without binder — 50 µm; 500 nm

(a)   (b)

With binder — 50 µm; 500 nm

(c)   (d)

FIGURE 2.5 Surface morphology of a 0.5 µm thick Au NP film (3.3 nm diameter) heated to 280°C. (a) and (b) are images of the pure NP film without polymeric matrix, (c) and (d) are images of the NP film with 30 vol% poly(3,4-ethylenedioxythiophene):poly(styrenesulfonic acid) as compatibilizing matrix.

including the average gap distance between the cores d_{gap} must scale together with r.

We can thus define an average gap-to-core radius ratio d_{gap}/r and derive an estimate of its upper and lower bounds as follows. We obtain the upper bound by considering that the polymer binder preferentially forms a uniform coating around the cores without filling up the inter-core voids; and we obtain the lower bound by considering the polymer binder preferentially fills up the inter-core voids first. For a purely geometrical continuum model that disregards the discreteness of the molecules, we obtain

$$\max\left(\sqrt[3]{0.634\left(\frac{v_{matrix}}{v_{core}}+1\right)}-1,0\right) < \frac{d_{gap}}{r} < \sqrt[3]{\frac{v_{matrix}}{v_{core}}+1}-1 \quad (2.5)$$

where

v_{matrix} is the volume fraction of the polymer binder matrix
$v_{core} = 1 - v_{matrix}$ is the volume fraction of the metal core
$\max(x, 0)$ denotes the larger of x and 0

We have neglected the shell because we assume it is completely volatilized. For a $v_{matrix} = 0.3$, $0 < d_{gap}/r < 0.13$. For $r = 2$ nm, the upper bound gives $d_{gap} = 2.6$ Å; but for $r = 5$ nm, this $d_{gap} = 6.5$ Å. Therefore, it is very clear from this model that the conductivity of sintered metal NP films becomes less tolerant of the presence of the polymer binder for large r. The average gap between

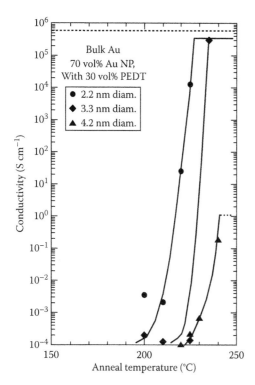

FIGURE 2.6 dc Electrical conductivity of film as a function of heat-treatment temperature for different Au NP diameters dispersed at 70 vol% into a poly(3,4-ethylenedioxythiophene):poly(styrenesulfonic acid) matrix. Average heating rate, 1°C min⁻¹ in N₂. (From Hosteller, M.J. et al., *Langmuir*, 14, 17, 1998. With permission.)

the metal cores can increase rapidly with core size at a constant polymer matrix volume fraction. This eventually breaks up the connectivity of the NPs.

Thus, only a limited amount of polymer matrix (ca. 30 vol%) can be incorporated without degrading σ_{dc}. This prediction has been qualitatively verified for a model Au NP–PEDT system, where PEDT is poly(3,4-ethylenedioxythiophene)–(polystyrene-sulfonic acid), as shown in Figure 2.6. For $r < 1.65$ nm, the ultimate σ_{dc} is 3×10^5 S cm⁻¹, practically independent of the polymer matrix fraction up to 0.3. For $r = 2.1$ nm, the ultimate σ_{dc} drops significantly. This phenomenon may find a useful application as a robust means to printed resistors with controlled resistivity spanning many orders of magnitude.[11]

At the other extreme, the metal core volume fraction needs to exceed a critical threshold v_c for percolation to occur. Site percolation thresholds on simple cubic and fcc lattices are 0.31 and 0.20, respectively.[36–38] The available data for NPs in polymer matrices suggests $v_c \approx 0.26$.[11] Just above the percolation threshold, σ_{dc} is expected to scale as $(v - v_c)^p$ where p is the critical exponent,[36,37] which is ca. 2 for the these materials.

2.2.4 Summary: What Is the Optimal Nanoparticle Core Size?

From the above discussions, the optimal core radius for nano-metal inks depends on several factors. The particles themselves

achieve metallicity for $r > 0.8$ nm. The porosity effect practically requires $r > 2$ nm depending on the dispersant shell thickness. The matrix dilution effect, if present, sets an upper limit for $r < 4$ nm for a matrix volume fraction of 0.2. If the metal NPs are formulated with much higher dispersant and binder content (>25 weight% binder, corresponding to >70 vol% binder in some commercial preparations), a long heat-treatment time at high temperatures will be required to "burn off" the binder in air before satisfactory sintering can be achieved. Unfortunately, this will also be associated with large volume contraction, which causes stress and micro-cracking of the sintered film.

2.3 Thermodynamic Size-Dependent Melting Temperature

2.3.1 Pawlow Equation

It is well known that nanosized metal particles have depressed melting points that depend on radius r.[39–45] Several phenomenological thermodynamic models have been proposed in the literature. One of the earliest and most popular equations is the Pawlow equation formulated for the equilibrium between a solid NP, a liquid NP, and the vapor phase:

$$\frac{T_m}{T_m^o} = 1 - \frac{2}{\rho_s L_o r}\left(\gamma_s - \gamma_l\left(\frac{\rho_s}{\rho_l}\right)^{2/3}\right), \quad (2.6)$$

where

T_m is the melting temperature of the NP
T_m^o is the melting temperature of the bulk metal
ρ_s and ρ_l are the solid and liquid densities, respectively
γ_s and γ_l are the solid and liquid surface energies
L_o is the enthalpy of fusion

All these parameters can be independently measured or obtained phenomenologically by fitting. However, this equation severely overestimates T_m at small r. The correction appears to require second-order terms involving several less well-known parameters.[41]

2.3.2 Hanszen–Wronski Equation

Another popular equation is the Hanszen–Wronski equation formulated for the equilibrium between a solid NP core with a liquid overlayer and the liquid NP. The key difference between this theory and the Pawlow theory is the assumption of the existence of a liquid overlayer enveloping the solid core below its melting temperature:

$$\frac{T_m}{T_m^o} = 1 - \frac{2}{\rho_s L_o}\left(\frac{\gamma_{sl}}{r - \delta_l} - \frac{\gamma_l}{r}\left(1 - \frac{\rho_s}{\rho_l}\right)\right), \quad (2.7)$$

where

ρ_{sl} is the solid–liquid interface energy
δ_l is the thickness of the liquid overlayer

2.3.3 Sambles Equation

A third popular equation is the Sambles equation formulated from the same theory as the Hanszen–Wronski equation but with a different approximation:

$$\ln\frac{T_m}{T_m^o} = -\frac{2}{\rho_s L_o r_s^*}\left(\frac{\rho_l}{\rho_s}\right)^{1/3}\left(\gamma_l\left(1-\frac{\rho_s}{\rho_l}\right) + \gamma_{sl}\left(1-\frac{\delta_l}{r_s^*}\left(\frac{\rho_s}{\rho_l}\right)^{1/3}\right)^{-1}\right),$$

(2.8)

where r_s^* is the hypothetical radius of the NP in the completely solid form.

2.3.4 Comparing the Models with Experiment

We compare the predictions of the Pawlow (dotted line), Sambles (dot-dash line), and Hanszen–Wronski (solid line) equations against the known experimental data for Au, Ag, Sn, and Pb NPs in Figure 2.7.

All three models predict that the melting transition is monotonically depressed for small NPs. All the equations also neglect the possible temperature and radius dependences of ρ_s, ρ_l, and γ_s, which therefore are approximated by their values at T_m^o. Furthermore, both the Hanszen–Wronski and Sambles equations require γ_{sl} and δ_l, which are seldom known to sufficient accuracy and thus have to be treated as fitting parameters. As a result, the Hanszen–Wronski and Sambles equations contain two fitting parameters each, while the Pawlow equation has none.

The results clearly show that both the Hanszen–Wronski and Sambles equations provide a superior description of the data than the Pawlow equation. Of the two, the Hanszen–Wronski equation seems to be more desirable because it is less cumbersome.

The fitted values of δ and γ_{sl} are given in Table 2.1. In all cases, the value of γ_{sl} is within a factor of 2 of the difference $\gamma_s - \gamma_l$, which is thus reasonable. For Au, it is, however, larger than the well-established experimental value for a macroscopic Au surface (0.13 J m^{-2}), which may suggest a possible size effect for this parameter. The best fit δ_l is 4 Å for both equations. Numerous molecular dynamic simulations for Au NPs[44,46,47] have found

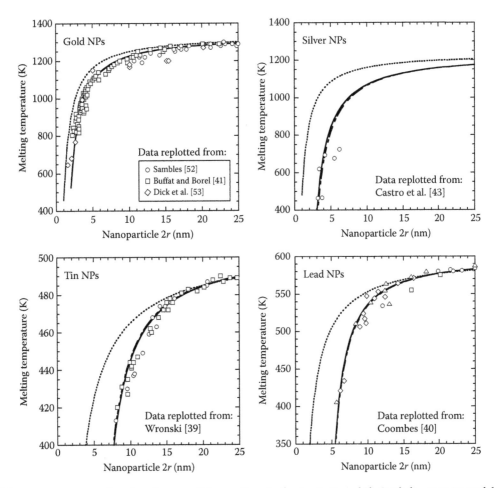

FIGURE 2.7 Melting temperature as a function of nanoparticle core diameter for Au, Ag Sn and Pb. Symbols = experimental data. Dashed line = Pawlow equation; solid line = Hanszen–Wronski equation; dot-dashed line = Sambles equation. Fit parameters for the latter two equations are given in Table 2.1.

TABLE 2.1   Parameters Used in the Fitting of the Melting Point Depression Curves of the Metal NPs in Figure 2.7

	T_m^o (K)	ρ_s (g cm^{-3})	ρ_l (g cm^{-3})	γ_s (J m^{-2})	γ_l (J m^{-2})	γ_{sl} (J m^{-2})	L_o (J g^{-1})	δ_l (Å)
Au	1337	18.4	17.4	1.38	1.14	0.30	63.7	4
Ag	1235	9.82	9.32	1.14	0.95	0.35	104.5	8
Sn	505	7.18	6.98	0.66 (estm.)	0.58	0.080	59.2	22
Pb	601	11.04	10.66	0.53 (estm.)	0.47	0.055	23.0	19

the presence of a liquid layer 1–2 atomic layers (i.e., 2.6–5.2 Å) thick at the surface of Au NPs emerging at ca. 0.8 T_m for the medium-sized NPs.[44,47] Therefore, the presence of this liquid-like layer is also reasonable. For Ag, the fits give a slightly larger δ_l of 8 Å.

For Sn and Pb NPs, however, the fitted δ_l values are much larger, ca. 20 Å. The calorimetry of Sn NPs has found a size-dependent melting enthalpy, which suggests the presence of a disordered overlayer 20 Å thick[45]:

$$\frac{L}{L_o} = \left(1 - \frac{\delta_l}{r}\right)^3 \qquad (2.9)$$

where

L is the enthalpy of fusion of the NPs
L_o is that of the bulk metal

Therefore, the Hanszen–Wronski and Sambles δ_l value also appears to be reasonable for this material.

These results provide strong support for the existence of a disordered liquid-like overlayer at the surfaces of these metal NPs. It should be noted, however, that these δ_l values obtained by fitting are strongly dominated by the T_m depression of the smaller NPs and so give the disordered layer in these NPs. It should be emphasized that the NPs used in these types of measurements are free of dispersion shells and/or oxide shells.

2.3.5 Is Size-Dependent Melting Relevant?

The data and fits show that significant melting point depression of >50°C occurs only for $r < 6$ nm. For Au and Ag, T_m falls below 300°C only for $r < 1$ and 1.7 nm, respectively. For a practical r of 1.7 nm, $T_m = 950$ K for Au and 550 K for Ag, which are much higher than the actual temperatures needed to sinter them into highly conductive films. Therefore, despite repeated assertions in the literature, the size-depression of T_m is not in fact the mechanism that is responsible for the low coalescence temperature of the Au and Ag metal NP films.

This is further conclusively demonstrated by the lack of dependence of the plots of conductivity vs. heat-treatment temperature on r from 1.1 to 6.2 nm for Au NPs, shown in Figure 2.8. Therefore, there is no size-dependent effect on the sintering of these NPs. The coalescence originates from a cold-sintering mechanism that operates upon elimination of the protective dispersant layer.

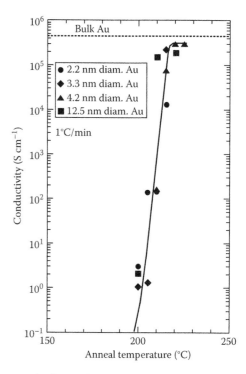

FIGURE 2.8   dc Electrical conductivity of film as a function of heat-treatment temperature for different Au NP diameters. Average heating rate, 1°C min^{-1} in N$_2$.

2.4 Insulator-to-Metal Transformation

2.4.1 Insulator-to-Metal Transformation Temperature (T_p)

Although metal NPs with $r > 0.8$ nm are expected to be metallic, as-deposited metal NP films do not in general show high DC conductivity because the metallic NPs are insulated by the dispersant shell. This leads to hopping conduction across a tunneling barrier.[48] For example, the as-deposited Au NP films with $t \approx 1$ nm typically give $\sigma_{dc} \approx 10^{-7}$–$10^{-6}$ S cm^{-1} and are thus in the semi-insulator regime.[22,23,49,50]

Over a narrow temperature range of 10°C or so, as shown in Figure 2.8, the film conductive rises by over 10 orders of magnitude to reach near bulk conductivity. This insulator-to-metal transformation is essentially an irreversible morphological transition as the dispersant shell volatilizes and the metal cores make contact to form a percolating network.

The transformation is best represented on a temperature-time-transformation diagram. However, for the purpose of comparing NP films, it is useful to define an operational figure-of-merit transformation temperature as the midpoint temperature (T_p) on the log σ_{dc}–linear T plot for the jump in σ_{dc}, at some specified ramp rate, such as 1°C min^{-1}.[11] This is a percolative transition. For SAM-protected Au NPs, this transformation is fairly sharp and can be measured quite precisely (±5°C) for films more than 100 nm thick and deposited on lithographically patterned four-in-line probes by injecting the current through the outer probes and measuring the voltage difference between the inner probes. Four-probe measurements are necessary to measure the mΩ resistances encountered after transformation.

We found that T_p measured at 1°C min^{-1} provides a useful guide to the temperature required for short anneals of a few min to cross the 10^4 S cm^{-1} threshold.

2.4.2 Optical Spectroscopy of the Transformation

Metal NPs have characteristic plasmon features that provide information on the state of the NPs in the film. The plasmon excitation is a cooperative oscillation of the conduction electrons in the NP, which thus probes the electronic quality (scattering) within the NP and their inter-particle coupling in the film. This is very useful for determining the key stages in the transformation of isolated NPs to the final percolated metal film across T_p. For SAM-protected Au NPs, this involves[11,51] (1) relaxation of the metal core (which decreases electron scattering), (2) desorption and thinning of the dispersant shell, and finally (3) core–core coalescence of neighboring NPs, as schematically illustrated in Figure 2.9.

These stages are characterized respectively by (i) sharpening and intensification of the plasmon band at $T_{anneal} \approx$ 100°C–140°C, (ii) red-shifting and broadening of the band, and (iii) partial disappearance of this band into an emerging a long-wavelength Drude tail, as shown in Figure 2.10b. The

onset of the optical Drude tail also corresponds to the T_p measured electrically. These stages can be quantitatively modeled to extract details of the transformation parameters. The final NP metal film still retains a nanodomain structure, as evidenced by vestiges of the plasmon band at 600–800 nm. This provides further evidence that T_p does not involve any bulk melting of the NPs. At this stage, the ligand shell is substantially volatilized. If the NPs are dispersed and isolated in a high volume fraction of a polymer matrix, only stages (i) and (ii) can occur, as shown in Figure 2.10a.

We emphasize here that a "shiny metallic" appearance (i.e., golden sheen in the case of Au NP films) is not by itself evidence for the presence of a percolated metallic conduction path. Both calculations and experiments have shown that as-deposited Au NP films with thin dispersant shells can give a golden appearance even though they are in the semi-insulating state.

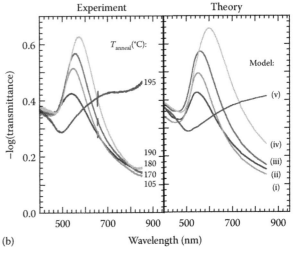

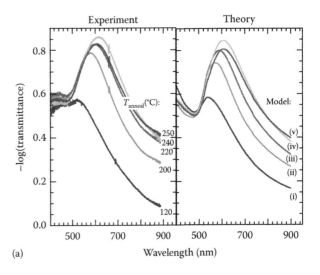

FIGURE 2.10 Experimental and theoretical optical spectra of thin films of Au NPs heated sequentially across the T_p transition. (a) Au NPs dispersed at 50 vol% in a poly(3,4-ethylenedioxythiophene):poly(styrenesulfonic acid) matrix. (b) Au NPs in neat film.

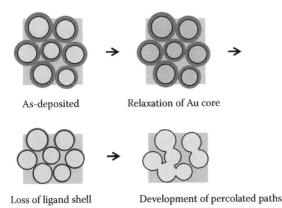

FIGURE 2.9 Schematic of the key stages in the percolative insulator-to-metal transformation of metal NP films.

2.4.3 Vibrational Spectroscopy of the Ligand Shell during Transformation

Fourier-transform vibrational spectroscopy (FTIR) can directly measure changes in the molecular bonding within the ligand shell as it eliminates during the T_p transformation and so reveals details that complement optical spectroscopy. The following markers are useful for tracking this elimination process: stretching ($\nu\ CH_2$) and bending ($\delta\ CH_2$) vibrations of the alkyl chains in the SAM, and the S–CH$_2$ bending ($\delta\ S–CH_2$) of the headgroup.[27] The heat-treatment dependence of the FTIR spectra for two model SAM shells (BT = butanethiol; MHA$_{acid}$ = mercaptohexanoic acid) are shown in Figure 2.11a and b, respectively. It is clear that the SAM shell disappears abruptly though not completely over a small temperature at T_p. From the integrated band intensities of these vibrations, the fraction of SAM

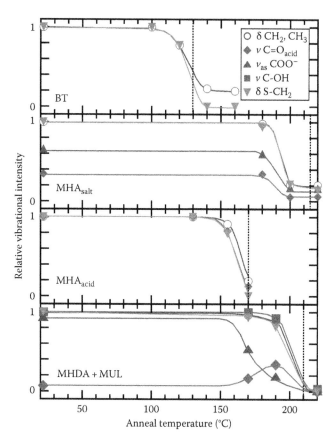

FIGURE 2.12 Relative vibrational intensity of key molecular bands for a film of SAM-protected Au NPs sequentially heated across T_p, for different SAMs. BT = butanethiol, MHA$_{salt}$ = mercaptohexanoate salt, MHA$_{acid}$ = mercaptohexanoic acid, MHDA + MUL = mixture of mercaptohexadecanoate and mercaptoundecanol. Au NP radius = 1.5 nm. Samples were held for 10 min at each temperature step in N$_2$.

remaining in the shell can be quantified as a function of heat treatment, as shown in Figure 2.12. The key conclusion is that T_p (marked by a vertical line) occurs when ca. 80% of the initial dense monolayer shell is removed for the same Au NP r = 1.5 nm, but a variety of SAM shells (MHA$_{salt}$ = mercaptohexanoate salt; MHDA + MUL = mixed mercaptodecanoic acid and mercaptoundecanol). The δ S–CH$_2$ band frequencies (ca. at 1215 and 1265 cm^{-1}) do not change as their intensities decrease during the loss of the ligand shell. This rules out the separate occurrence of Au–S bond scission to give trapped disulfides and/or partial fragmentation intermediates within the shell well before their eventual elimination.[27] The data thus favors a model in which the Au–S bonds are dissociated with near simultaneous fragmentation and/or elimination of the molecule. In all cases, FTIR indicates ligand shell residues, even well above T_p. The occlusion of residual organics in the coalesced NP metal films appears to be a universal feature.

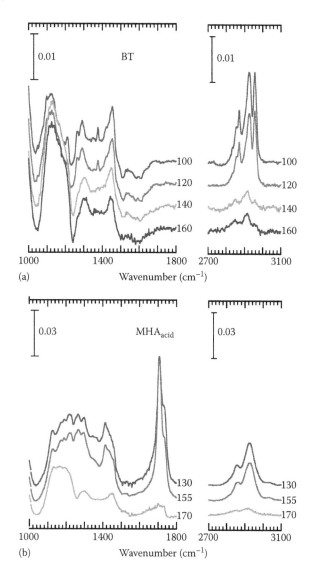

FIGURE 2.11 FTIR spectra of a film of SAM-protected Au NPs sequentially heated across T_p. (a) SAM is butanethiol (T_p = 130°C). (b) SAM is mercaptohexanoic acid (T_p = 170°C). Au NP radius = 1.5 nm. Samples were held for 10 min at each temperature step in N$_2$.

2.4.4 Differential Scanning Calorimetry of the Transformation

Differential scanning calorimetry measures the enthalpy of the transformation. A large exothermic heat flow has been observed during NP coalescence, which could be interpreted as 0.4–0.5 J m^{-2} of the NP surface area.[27] This is on the same order of magnitude as the surface energy of Au (1.5 J m^{-2}).[41,43,52,53] Therefore, the exotherm is assigned to the heat released by sintering and neck formation between adjacent NPs. It cannot be assigned to bond-making and bond-breaking with the ligand shell because the expected enthalpy of reaction, i.e., $2Au_{met} - SR \rightarrow 2Au_{met} + RSSR$,[28,54,55] is only 16 kJ mol^{-1} exothermic and negligible in comparison with the experimental value of 70–85 kJ mol^{-1} when normalized to the number of thiol SAM molecules.

2.4.5 Systematics of T_p with Core Size and Ligand Shell

2.4.5.1 Lack of Size Effect

For different Au NP r between 1.1 and 6.2 nm, the T_p is 210°C independent of r, as shown previously in Figure 2.8. Over this size range, the expected T_m decreases from 970°C to 400°C (Figure 2.7). It is evident that T_p is not related to T_m. The bottleneck step is the elimination of the ligand shell and not the melting of the metal nanocore or even of the surface melting of the core.

2.4.5.2 Strong Ligand-Shell Effect

T_p depends strongly on the properties of the ligand shell (Table 2.2). For alkylthiol SAM-protected Au NPs in which the dominant intermolecular interaction between the ligand molecules is van der Waals interaction, T_p shows a monotonic increase with chain length, from 130°C for butanethiol to 180°C for decanethiol SAMs.[6,27] However, Au NPs protected by short alkyl thiols tend to be unstable, perhaps due to atmospheric oxidation of the thiol and its desorption from the shell. This desorption temperature (from FTIR) appears to be higher than for corresponding thiols on flat Au surfaces,[56–58] perhaps because of slightly stronger bonding on the NP surface.[59] A similar trend is found in Au NPs protected by ω-functionalized alkylthiol SAMs in which the dominant intermolecular interaction is hydrogen bonding via –COOH tail groups. T_p increases from 125°C for mercaptopropionic acid to 170°C for mercaptohexanoic acid and 210°C for mercaptoundecanoic acid.[27] For the corresponding salts of these acids in which the dominant

interaction is ionic via –COO$^-$ M$^+$, the T_p is practically invariant at ca. 210°C. Differences in packing density and conformation of the SAM ligands can be ruled out.[27] Therefore, it is clear that T_p varies strongly with the intramolecular interaction within the shell. When the Au core is protected by a shell of a mixed monolayer, T_p tracks the T_p characteristic of the more robust of the two SAMs, rather than the arithmetic mean.[27] For example, Au NPs protected by a 1:10 mol mixture of mercaptohexanoate salt (T_p = 210°C) and butanethiol (T_p = 130°C) shows a T_p of 200°C. During heat treatment, the butanethiol desorbs over 120°C–160°C followed by an overlapping loss of both the acid and the salt forms of the mercaptohexanoate ligand over 140°C–200°C. Therefore, the presence of small quantities of a robust SAM molecule in the shell can elevate T_p significantly.[27]

2.4.5.3 Sparse Shells

When the Au NPs are protected with a sparse monolayer through selective desorption of a labile second component, the T_p of the NPs can be substantially lowered to 150°C without sacrificing the stability of the NP dispersion.[27] Thus, this provides a method to obtain metal NPs with the desired dispersion characteristics and stability in solution but favorably depressed T_p for applications on a wide variety of plastics and other substrates. In this method, a compact shell is first assembled using a mixture of a mercaptopropanoic acid and a highly water-soluble mercaptopropanol. The mercaptopropanol SAM is subsequently desorbed during subsequent purification cycles in water to give Au NPs that are effectively protected by only a quarter of a monolayer of SAM primarily of the mercaptopropanoate salt.[27] These NPs are soluble to 200 mg mL^{-1} in ethylene glycol and also in water and show a remarkably low T_p of 145°C. The approach also works with Ag NPs to also give similarly low T_p of 155°C. Both these NPs can be annealed to give $\sigma_{dc} > 10^5$ S cm^{-1} for short times just slightly above their T_p. These T_p values are the lowest known so far for Au and Ag NPs that are water- and polar-solvent-soluble.

2.4.5.4 Surface Melting?

For Au NPs suspended in a vacuum, molecular dynamic simulations have found enhanced mobility and loss of correlation of the surface atoms in a distinct "surface-melting" step well below T_m.[44,46,47] For SAM-protected NPs, the covalent bonding of the surface atoms to the ligand molecules may have a significant stabilizing effect on the surface atoms. Whether surface melting still occurs is unclear. Nevertheless, based on empirical observations that T_p varies strongly with the tail group for a given head group of the SAM molecule, surface melting, if it occurs, is not the bottleneck step in the T_p transformation of the film.

2.5 Modeling the Optical Plasmon Resonance

The most distinctive spectroscopic feature of metal NPs is the presence of a plasmon resonance band in the optical spectrum that is not present in continuous thin films. Plasmons are

TABLE 2.2 Dependence of T_p (at 1°C min^{-1}) on the Dispersant Shell for 3 nm Diam. Au NPs

Protection Monolayer	T_p (°C)
Butanethiol	130
Decanethiol	180
Mercaptopropionic acid	125
Mercaptohexanoic acid	170
Mercaptoundecanoic acid	210
Mercaptopropanoate, partial sodium salt	210
Mercaptohexanoate, partial sodium salt	210
Mercaptoundecanoate, partial sodium salt	210

coherent longitudinal oscillations of the free-electron density. The absorption appears as a structureless broad band whose position, intensity, and shape depend on the NP and its surroundings. This band is readily measured in standard UV-Vis spectroscopy and provides useful characterization of the NPs. It turns out that the plasmon bandshape for NPs with $r < 20$ nm, for which the point dipole is an adequate approximation, can be accurately predicted using the simple first-order polarization theory.[60–64]

Although numerous studies have been presented already at various levels of approximations,[63–72] it is useful to cover this again as a first step in the treatment of the optical spectra of their thin films.

2.5.1 Dipole Theory of the Plasmon Resonance Band

2.5.1.1 Metal Core Dielectric Function

The complex dielectric function of metal NPs can be obtained from the dielectric function of the bulk metal in a simple way by correcting for the effects of electron confinement on the scattering frequency. The dielectric function of the bulk metal can be written as the sum of a bound response and a free-electron response[61]:

$$\tilde{\varepsilon}_{bulk} = \tilde{\varepsilon}_{bound} + \tilde{\varepsilon}_{free}, \tag{2.10}$$

where

$\tilde{\varepsilon}_{bound}$ is the dielectric function of the bound electrons
$\tilde{\varepsilon}_{free}$ is the contribution from the free electrons

The free electron contribution can be written explicitly by the Drude theory to give free-electron-like metals:

$$\tilde{\varepsilon}_{bulk} = \tilde{\varepsilon}_{bound} - \frac{\omega_p^2}{\omega^2 + i\omega\gamma_{bulk}}, \tag{2.11}$$

where

ω_p is the plasma frequency of the metal
ω is the frequency of the incident light
γ_{bulk} is the scattering frequency of bulk metal

For the nanosized metal cores, the free electron contribution is modified by the increased scattering frequency within the core [60,61,70,73]:

$$\tilde{\varepsilon}_{nano} = \tilde{\varepsilon}_{bulk} + \frac{\omega_p^2}{\omega^2 + i\omega\gamma_{bulk}} - \frac{\omega_p^2}{\omega^2 + i\omega\gamma_{nano}}, \tag{2.12}$$

where

$\gamma_{nano} = \gamma_{bulk} + C(v_F/r)$, in which C is an empirical scattering factor or order unity that accounts for electron–electron, electron–phonon, electron–defect interactions in the nanocore
v_F is the bulk Fermi velocity
r is the core radius

This equation predicts that the plasmon resonance generally smears out for $r < 1$ nm due to scattering within the confinement potential, as has been observed experimentally.[74] $\tilde{\varepsilon}_{bulk}$ can be obtained from spectroscopic ellipsometry and are extensively tabulated in data books.[75] For Au, fitting to the Drude theory gives $\omega_p = 1.3 \times 10^{16}$ s^{-1}, $\gamma_{bulk} = 1.64 \times 10^{14}$ s^{-1}, while its v_F can be deduced from electronic heat capacity to be 14.1×10^{14} nm s^{-1}.

The complex dielectric function of the metal NP film or the dilute NP dispersion in solvent can be computed using an appropriate effective medium approximation for the system.[76] Once this dielectric function is known, the optical transmission and reflection spectra can be computed self-consistently using Fresnel transfer matrices.[77]

2.5.1.2 Core-Shell Effect

For dilute dispersions of "guest" particles in a solvent host, the appropriate effective medium approximation (EMA) is the Maxwell–Garnet EMA. For a three-component system of a metal core surrounded by a ligand shell and embedded in a matrix, the dimensionless complex polarizability of the core–shell NP is given by the Van de Hulst to be[60,78]

$$\tilde{\alpha} = 3\frac{(\tilde{\varepsilon}^{shell} - \tilde{\varepsilon}^{matrix})(\tilde{\varepsilon}_{nano} + 2\tilde{\varepsilon}^{shell}) + q^3(2\tilde{\varepsilon}^{shell} + \tilde{\varepsilon}^{matrix})(\tilde{\varepsilon}_{nano} - \tilde{\varepsilon}^{shell})}{(\tilde{\varepsilon}^{shell} + 2\tilde{\varepsilon}^{matrix})(\tilde{\varepsilon}_{nano} + 2\tilde{\varepsilon}^{shell}) + q^3(2\tilde{\varepsilon}^{shell} - 2\tilde{\varepsilon}^{matrix})(\tilde{\varepsilon}_{nano} - \tilde{\varepsilon}^{shell})}, \tag{2.13}$$

where

$q = r/r + t$ is the ratio of the core radius (r) to the sum of the core radius and shell thickness ($r + t$), so q^3 is the volume ratio of the core to the NP
$\tilde{\varepsilon}^{matrix}$ is the dielectric function of the matrix in which the NPs are dispersed
$\tilde{\varepsilon}_r^{shell}$ is that of the ligand shell

The initial average value of q can be obtained experimentally from TEM images.

2.5.1.3 Dilute Dispersions

In the dilute dispersions of the NPs in a nonabsorbing medium, the thickness absorption coefficient can be directly found from

$$z_o^{-1} = \frac{2\pi}{\lambda_o}\sqrt{\varepsilon^{matrix}} f\alpha_2, \tag{2.14}$$

where

α_2 is the imaginary component of the polarizability
f is the combined volume fraction of the core and shell

This can be written as an absorption cross section per NP

$$\sigma_{NP} = \frac{z_o^{-1}}{\rho_{NP}}, \tag{2.15}$$

where ρ_{NP} is the number density of the NPs, which in turn can be converted to molar absorptivity ε, as in the standard practice in solution-state spectroscopy, where "molar" here refers to a concentration of 1 mol of metal atoms (not NPs) per liter of solution

$$\varepsilon = \log e \frac{\sigma_{NP}}{c_{mol}}, \qquad (2.16)$$

where

$\log e = 0.43429$

c_{mol} is the number of moles of the metal atoms per NP

This quantity is thus normalized to the amount of metal atoms in the solution. It has been found both experimentally and theoretically to vary slightly with the NP core size, shell thickness, and medium dielectric effects (see Section 2.5.2).

In the limit of very thin or sparse shells, this simplifies to give

$$\sigma = 9 \frac{2\pi}{\lambda_o} \frac{\varepsilon''_{core} (\varepsilon_{matrix})^{3/2} v_{core}}{(\varepsilon'_{core} + 2\varepsilon_{matrix})^2 + (\varepsilon''_{core})^2}, \qquad (2.17)$$

where

v_{core} is the volume of the core

$\tilde{\varepsilon} = \varepsilon' - i\varepsilon''$

This equation explicitly shows the dependence of the absorption cross section of the NP on the medium.

2.5.1.4 Concentrated Dispersion or Solid Films

2.5.1.4.1 Maxwell–Garnet EMA

We now consider the optical behavior of a concentrated dispersion of these NPs in solid films. The Maxwell–Garnett dielectric function ($\tilde{\varepsilon}^{MG}$) of the composite can then be obtained from the Clausius–Mosotti relation to be

$$\tilde{\varepsilon}^{MG} = \tilde{\varepsilon}^{matrix} \frac{1 + \dfrac{2}{3} f\alpha}{1 - \dfrac{1}{3} f\alpha}, \qquad (2.18)$$

where f is the combined volume fraction of the core and shell, with $f + f_{matrix} = 1$.

For a polydispersed or a multi-component system, this is readily extended to give

$$\tilde{\varepsilon}^{MG} = \tilde{\varepsilon}^{matrix} \frac{1 + \dfrac{2}{3} \sum_j f_j \alpha_j}{1 - \dfrac{1}{3} \sum_j f_j \alpha_j}, \qquad (2.19)$$

where j denotes the NPs of size or type j with $\sum_j f_j + f_{matrix} = 1$.

2.5.1.4.2 Bruggeman EMA

At T_p, the metal cores coalesce with their neighbors to give a percolated network with the development of a macroscopic conductivity. This generates a bi-continuous "smeared-out" morphology for which the Maxwell–Garnett EMA is no longer a suitable description. The Bruggeman (Br) EMA becomes more appropriate.[51] The Br EMA dielectric function ($\bar{\varepsilon}_r^{Br}$) can be computed according to

$$f_{Au} \frac{\varepsilon_r^{Au} - \bar{\varepsilon}_r^{Br}}{\varepsilon_r^{Au} + 2\bar{\varepsilon}_r^{Br}} + (1 - f_{Au}) \frac{\varepsilon_r^m - \bar{\varepsilon}_r^{Br}}{\varepsilon_r^m + 2\bar{\varepsilon}_r^{Br}} = 0, \qquad (2.20)$$

where f_{Au} is the volume fraction of Au and $(1 - f_{Au})$ of the matrix (which may be the polymer matrix or air in the case of "unfilled" NP films). The ligand shell is largely desorbed by this stage and so it can be neglected. It is, however, appropriate to use $\varepsilon_r^{Au} = \varepsilon_{r,nano}^{Au}$ because the nanostructure texture persists even after the percolative transition.

2.5.2 Dilute Spectra in a Dielectric Medium

2.5.2.1 Medium Effects

Figure 2.13 shows the computed plasmon spectra of different r of Au NPs with and without a 0.7 nm thick shell in different solvents with optical refractive indices n between 1.30 (e.g., trifluoroacetic acid) and 1.57 (e.g., trichlorobenzene). In the limit of small NP sizes ($r < 10$ nm), the primary dependence is in intensity rather than plasmon band position or band shape. The plasmon band maximum red shifts by less than 1% from 534 nm (2.32 eV) at $n = 1.3$ to 537 nm (2.31 eV) at $n = 1.6$ for Au NPs with $r = 1.4$ nm and $t = 0.72$ nm. On the other hand, the enhancement in the absorption cross section is ca. 20% from $n = 1.30$–1.57. This enhancement is linear in this range. In the absence of a shell, the calculations show that the solvent exerts an even larger influence on the plasmon spectrum, again mainly in the absorption cross section.

The predicted lack of dependence of the plasmon band maximum (λ_{max}) of Au NPs on n has been confirmed experimentally for Au NPs with $r = 1.1$–2.0 nm in hexane, tetrahydrofuran, and benzene[28] and for $r = 2.6$ nm in hexane and dichlorobenzene, which is shown in Figure 2.14.[68] For larger Au NPs with $r = 8$ nm protected by a polymeric dispersant, a sizeable red shift of ≈ 20 nm from cyclohexane to carbon disulfide was claimed,[65] but the effects of solvent-dependent aggregation does not appear to have been ruled out.

2.5.2.2 Core Radius Effects

Figure 2.13 also shows the plasmon band shape is strongly dependent on r but not the band maximum. The weak dependence of band maximum on r has been verified experimentally for $r = 1.0$–1.6 nm protected by alkylthiol,[66] 1.1–2.2 nm protected by dodecanethiol,[28] 0.6–1.8 nm protected by N-[3-(trimethoxysilyl)propyl]diethylenetriamine,[70] and also for Ag NPs

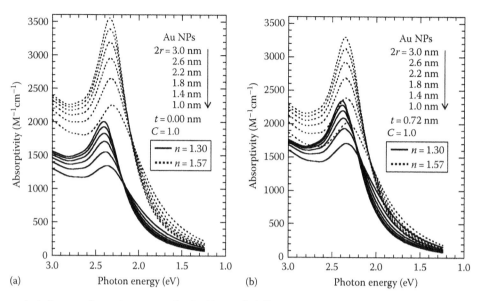

FIGURE 2.13 Theoretical plasmon absorption spectra for Au NPs with different core diameters $2r$ dispersed in dilute media with different refractive indices n. (a) NPs are without shells. (b) NPs have a 0.72 nm thick shell with refractive index of 1.5. The electron scattering factor C used is 1.00.

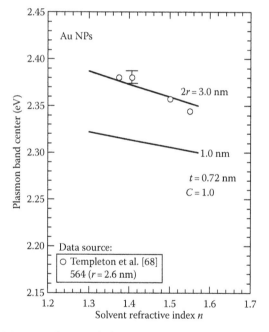

FIGURE 2.14 Theoretical plasmon band absorption maximum as a function of solvent refractive index for 1.0 and 3.0 nm diameter Au NPs dispersed in the solvent. Symbols are experimental data.

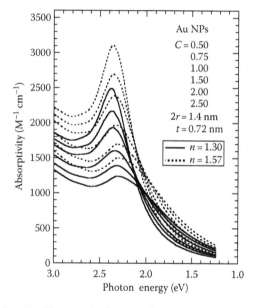

FIGURE 2.15 Theoretical plasmon absorption spectra for 1.4 nm diameter Au NPs with 0.72 nm thick shell with refractive index = 1.5, dispersed in dilute media with refractive index of 1.30 or 1.57, for different electron scattering factor C.

with $r = 2–6$ nm protected with poly(N-vinyl pyrrolidone).[79] The plasmon band red shifts significantly only for $r > 10$ nm due to retardation effects.[63,70,80] The predicted plasmon band sharpening with r has also been experimentally confirmed.[28,66] These remarkable agreements between theory and experiment provide confidence that the simple theory described in Section 2.5.1 is sufficient to describe the optical spectra of small NPs with $r < 5$ nm.

Fitting of the experimental spectral band shape then provides the electron-scattering C factor as the only adjustable parameter. This factor accounts for electron scattering due to core disorder, which itself depends on synthesis conditions and any post-annealing. As C increases, the plasmon band red shifts, broadens, and loses oscillator strength, as shown in Figure 2.15. In general, as-synthesized Au NPs tend to exhibit a large $C \approx 1.5$–2[28,66] that decreases to ≈ 0.75 upon heat-treatment to 130°C.[11,51] In principle, if *C* is used as a global parameter to simultaneously model the NP spectra measured in different solvents, it should be possible to unambiguously recover the mean particle size and distribution of the NPs.

2.5.3 Thin Film Spectra: *In Situ* Measurement of the Nanostructure

In thin films, the plasmon coupling can be used to probe NP interactions and hence the nanostructure of the film. The optical modeling methodology described in Section 2.5.1 has been successfully used to quantitatively analyze the thin film spectra of Au NP films.[51] These analyses reveal the evolution of the effective thickness of the dispersant shell, and of the percolated volume fraction of metal cores, as a function of the heat-treatment temperature (Figure 2.9). The as-deposited films of Au NPs protected by a molecularly thick SAM dispersant shell generally show only a small red shift of plasmon band from the dilute dispersion spectrum, except in the case of Au NPs protected by sparse ionic-protection monolayers, which show a large red shift due to closer NP contacts.[27]

Optical modeling of the spectra recorded after sequential heat-treatment reveals that the effective thickness of the dispersant shell as expected decreases to zero as the heat-treatment temperature crosses T_p. For sintered films of pure SAM-protected NPs, the corresponding volume fraction of percolating Au NPs reaches 90% or higher just beyond this temperature. The presence of a polymer matrix, however, smears out the transformation. For a film with 30 vol% of a polymer filler, the percolated Au fraction is ca. 25%, but this is already sufficient to give σ_{dc} above 10^5 S cm^{-1} without the attendant stress-induced micro-cracking of the films.[51] The agreement between theory and experiment is very remarkable, as shown in Figure 2.10 for two model systems with and without a polymer matrix. There are no free parameters other than the two nanostructure parameters of shell thickness and percolating fraction. Therefore, optical modeling of the plasmon spectra of metal NP thin films provides reliable access to these parameters, which are difficult to obtain otherwise.

2.6 Solution-Phase Preparation of Nanoparticles

In this section, we summarize the main routes of production of metal NPs. These have been prepared by several methods in both gas and solution phases. The gas-phase methods involve either spray pyrolysis or evaporation–condensation. The solution-phase processes are often based on the reduction of a metal precursor salt in the presence of the dispersant to directly obtain the dispersant-stabilized NPs. These often give better control and narrower particle size distribution (PSD) than gas-phase methods and are far more versatile. In particular, the phase-transfer Brust–Schiffrin[81] method can afford high-quality isolable NPs with narrow PSD and controlled diameters. The dispersant shell can be tailored for the introduction of different or mixed ligands.

This survey focuses on solution-phase methods and their synthesis–characteristics relation, such as shape, size, stability, and sintering temperature. Unfortunately, much of the literature to date emphasizes NPs for biochemical applications, for which details on dispersibility, stability, and in particular sintering characteristics are seldom if ever reported. Because the sintering characteristics are sensitive to the presence of nonvolatile materials that may be trapped in-between the NPs, it is vital that the NPs can be purified. The dispersibility of the NPs in solvents determines whether they can be formulated at the desired concentrations typically of >50 mg mL^{-1} into practical ink solvent systems. Through such a survey, the possibilities for further optimization and manipulation of these materials will become apparent.

2.6.1 Silver Nanoparticles

A few reviews have been published recently on the preparation of Ag NPs.[15,82,83] The essential features are summarized in Table 2.3. Traditionally, these are produced by the reduction of Ag$^+$ in a polar solvent in the presence of a soluble polymeric dispersant such as poly(vinyl pyrrolidone) (PVP), polyvinyl alcohol (PVA), or poly(ethylene glycol) (PEG). This gives spherical polymer-protected Ag NPs with a large mean diameter (30–100 nm). These are highly dispersible in water and ethylene glycols at concentrations up to 50 wt%. In the industrially important polyol process, AgNO$_3$ is reduced in ethylene glycol solvent and reductant at 160°C, usually with PVP dispersant to give 100 nm diameter NPs.[84] AgNO$_3$ can also be reduced in dimethyacetamide, dimethylformamide, and acetonitrile.[85–87] In an aqueous medium, Ag$^+$ can be reduced by citrate,[88] sodium borohydride,[82,88–91] PEG and poly(vinyl alcohol),[92] formaldehyde/sorbitol,[93] poly(2,6-dimethyl-1,4-phenylene oxide PPO),[94] tetrathiafulvalene (TTF),[87] PVP,[87] benzylidine-4-phenyl-1,3-dithiol (BPD),[87] or hydroxylamine hydrochloride.[95] Sometimes light,[96,97] microwave,[98] ultrasound,[99] or radiation[100,101] is also applied to speed up the reduction. The PSD of these Ag NPs tends to be broad, depending on the method and conditions. Ag NPs prepared by reducing AgNO$_3$ or AgClO$_4$ in the presence of amine molecular dispersants tend to be fragile.[102] Ag NPs have also been formed in reverse-phase water-in-oil microemulsions using cationic (e.g., cetyltrimethylammonium bromide),[103] anionic (e.g., bis(2-ethylhexyl)sulfosuccinate),[104] or nonionic (e.g., Triton X-100) surfactants.[105–107] The phase-transfer Brust–Schiffrin method, in which the Ag$^+$ is solubilized in the organic phase via an organic-soluble Ag complex and then reduced by NaBH$_4$, is also possible.[102,108,109] This method can afford robust and isolable Ag NPs with finer NP sizes and better PSD and good dispersibility in nonpolar organic solvents.

2.6.2 Gold Nanoparticles

A few reviews on Au NP preparation have also been published recently.[15,110] The essential features are summarized in Table 2.3. Traditionally, these are produced by the reduction of AuCl$_4^-$,

TABLE 2.3 Synthetic Conditions for Metal Nanoparticles

Conditions	NPs	Refs.
Silver: Chemical reduction in aqueous or polar organic solvents		
Substrate: AgNO$_3$	Shape: nanocubes	[84]
Reductant: ethylene glycol	Mean: 100 nm	
Solvent: ethylene glycol (42 mg/mL)	Polydispersity: 8%	
Temperature: 160°C	Solubility: water	
Dispersant: PVP; ratio: 0.6	Coalescence temp.: —	
	Stability: —	
Substrate: AgNO$_3$	Shape: sphere	[88]
Reductant: sodium citrate	Mean: 75 nm	
Solvent: water	Polydispersity: 33%	
Temperature: 100°C	Solubility: water	
Dispersant: sodium citrate; ratio: 1.5	Coalescence temp.: —	
	Stability: —	
Substrate: AgNO$_3$	Shape: sphere	[88]
Reductant: NaBH$_4$	Mean: 12.5 nm	
Solvent: water	Polydispersity: 60%	
Temperature: 22°C	Solubility: water	
Dispersant: SiO$_2$; ratio: 6e-04	Coalescence temp.: —	
	Stability: —	
Substrate: AgNO$_3$	Shape: spheroid	[93]
Reductant: formaldehyde + sorbitol	Mean: 35 nm	
Activator: NaOH	Polydispersity: 43%	
Solvent: water	Solubility: water	
Temperature: 22°C	Coalescence temp.: —	
Dispersant: —; Ratio: —	Stability: stable	
Substrate: AgBF$_4$, AgPF$_6$, AgSbF$_6$, AgSO$_3$CF$_3$, AgClO$_4$, AgNO$_3$	Shape: sphere	[94]
	Mean: 10 nm	
Solvent: mixed methanol and chloroform	Polydispersity: 22%	
	Solubility: chloroform, methanol	
Temperature: 22°C, 6 days		
Dispersant: PPO; ratio: 1.5e-3	Coalescence temp.: —	
	Stability: stable	
Substrate: AgNO$_3$	Shape: —	[87]
Reductant: tetrathiafulvene, PVP	Mean: —	
Solvent: acetonitrile	Polydispersity: —	
Temperature: 22°C, 5 days	Solubility: acetonitrile	
Dispersant: TTF, PVP; ratio: 1	Coalescence temp.: —	
	Stability: stable	
Substrate: AgNO$_3$	Shape: —	[87]
Reductant: benzylydine-4-phenyl-1,3-dithiol	Mean: —	
	Polydispersity: —	
Solvent: DMSO	Solubility: DMSO	
Temperature: 22°C, 5 days	Coalescence temp.: —	
Dispersant: BPD; ratio: 1	Stability: stable	
Substrate: AgNO$_3$	Shape: sphere	[82]
Reductant: NaBH$_4$	Mean: 5 nm	
Solvent: mixture of water and methanol	Polydispersity: 20%	
Temperature: 22°C	Solubility: water	
Dispersant: mercaptosuccinic acid; ratio: 0.5	Coalescence temp.: —	
	Stability: stable	

TABLE 2.3 (continued) Synthetic Conditions for Metal Nanoparticles

Conditions	NPs	Refs.
Silver: Phase-transfer to a nonpolar organic phase, followed by chemical reduction		
Substrate: AgNO$_3$	Shape: sphere	[89]
Reductant: NaBH$_4$	Mean: 2.6 nm	
Solvent: water/toluene	Polydispersity: 70%	
Temperature: 22°C	Solubility: toluene	
Dispersant: dodecanethiol; ratio: 0.3	Coalescence temp.: —	
	Stability: stable	
Substrate: AgNO$_3$, AgClO$_4$	Shape: sphere	[102]
Reductant: NaBH$_4$	Mean: 3 nm	
Solvent: water/toluene	Polydispersity: 20%	
Temperature: 22°C	Solubility: hexane, toluene	
Phase-transfer agent: TOAB		
Dispersant: dodecanethiol; ratio: 0.3	Coalescence temp.: —	
	Stability: stable	
Substrate: AgNO$_3$	Shape: sphere	[109]
Reductant: NaBH$_4$	Mean: 7 nm	
Solvent: water/chloroform	Polydispersity: 7%	
Temperature: 22°C	Solubility: chloroform toluene, hexane	
Phase-transfer agent: TOAB		
Dispersant: dodecanethiol; ratio: 1	Coalescence temp.: —	
	Stability: stable	
Substrate: AgNO$_3$	Shape: sphere	[109]
Reductant: NaBH$_4$	Mean: 14 nm	
Solvent: H$_2$O-MeOH	Polydispersity: 30%	
Temperature: 22°C	Solubility: water, ethylene glycol	
Phase-transfer agent: —		
Dispersant: dodecanethiol; ratio: 0.5	Coalescence temp.: 160°C	
	Stability: stable	
Gold: Chemical reduction in aqueous or polar organic solvents		
Substrate: HAuCl$_4$	Shape: sphere	[111]
Reductant: sodium citrate	Mean: 20 nm	
Solvent: water	Polydispersity: 20%	
Temperature: 100°C	Solubility: water	
Dispersant: sodium citrate; ratio: 1.5	Coalescence temp.: —	
	Stability: —	
Substrate: Citrate Au NPs	Shape: sphere	[113]
Reductant: Citrate, seed growth	Mean: 20 nm	
Solvent: water	Polydispersity: 15%	
Temperature: 22°C	Solubility: water	
Dispersant: (C$_6$H$_5$)$_2$PC$_2$H$_4$SO$_3$Na; ratio: 1	Coalescence temp.: —	
	Stability: unstable	
Substrate: HAuCl$_4$	Shape: sphere	[114]
Reductant: N$_2$H$_4$	Mean: 1—4 nm	
Solvent: water	Polydispersity: 20%	
Temperature: 22°C, 15 h	Solubility: water	
Dispersant: cetyltrimethylammonium chloride or sodium dodecylsulfate, or cetylpyridinium chloride; ratio: 0.25	Coalescence temp.: —	
	Stability: Stable	

TABLE 2.3 (continued) Synthetic Conditions for Metal Nanoparticles

Conditions	NPs	Refs.
Substrate: NaAuCl₄ Reductant: PVA or PEG Solvent: water Temperature: 22°C, 6 days Dispersant: —; Ratio: —	Shape: — Mean: — Polydispersity: — Solubility: water Coalescence temp.: — Stability: stable	[92]
Substrate: HAuCl₄ Reductant: NaBH₄ Solvent: water Temperature: 22°C Dispersant: thiol ionic liquid; ratio: 0.1	Shape: sphere Mean: 3 nm Polydispersity: 15% Solubility: water Coalescence temp.: — Stability: few months	[120]
Substrate: HAuCl₄ Reductant: NaBH₄ Solvent: methanol/acetic acid (6:1) Temperature: 22°C Dispersant:tiopronin; ratio: 0.3	Shape: sphere Mean: 3 nm Polydispersity: 35% Solubility: water Coalescence temp.: — Stability: —	[121]
Substrate: HAuCl₄ Reductant: NaBH₄ Solvent: mixture of water and methanol Temperature: 22°C Dispersant: tripeptide glutathione; ratio: 0.3	Shape: sphere Mean: 1 nm Polydispersity: — Solubility: water Coalescence temp.: — Stability: —	[119]

Substrate: HAuCl₄
Reductant: NaBH₄
Solvent: water
Temperature: 22°C
Dispersant:
  trimethyl(undecylmercapto)
  ammonium; ratio: 0.3 equiv.
	Shape: sphere
	Mean: 5 nm
	Polydispersity: 35%
	Solubility: water
	Coalescence temp.:
	  225°C
	Stability: —
		[50]

Substrate: HAuCl₄
Reductant: NaBH₄
Solvent: water
Temperature: 22°C
Dispersant: cetylpyridinium chloride;
  Phase-transfer dispersant:
  dodecylamine; ratio: 0.3 equiv
	Shape: sphere
	Mean: 15 nm
	Polydispersity: 15%
	Solubility: hexane
	Coalescence temp.: —
	Stability: —
		[118]

Gold: Phase-transfer to a nonpolar organic phase, followed by chemical reduction

Conditions	NPs	Refs.
Substrate: HAuCl₄ Reductant: NaBH₄ Phase-transfer agent: TOAB Solvent: water/toluene Temperature: 22°C Dispersant: dodecanethiol; ratio: 1	Shape: sphere Mean: 2 Polydispersity: 10% Solubility: toluene Coalescence temp.: — Stability: stable	[81]

Substrate: HAuCl₄
Reductant: NaBH₄
Phase-transfer agent: TOAB
Solvent: water/toluene
Temperature: 22°C
Dispersant: dodecylamine, oleylamine;
  ratio: 1
	Shape: sphere
	Mean: 3 nm
	Polydispersity: 25%
	Solubility: toluene,
	  hexane
	Coalescence temp.: —
	Stability: stable
		[122]

TABLE 2.3 (continued) Synthetic Conditions for Metal Nanoparticles

Conditions	NPs	Refs.
Substrate: HAuCl₄ Reductant: NaBH₄ Phase-transfer agent: TOAB Solvent: water/toluene Temperature: 22°C Dispersant: mercaptosuccinic acid; ratio: 0.5	Shape: sphere Mean: 3.5 nm Polydispersity: 20% Solubility: water Coalescence temp.: — Stability: stable	[123]
Substrate: HAuCl₄ Reductant: NaBH₄ Phase-transfer agent: TOAB Solvent: water/toluene Temperature: 22°C Dispersant: PEG-SH; ratio: 0.1	Shape: sphere Mean: 3 nm Polydispersity: 35% Solubility: water Coalescence temp.: 320°C Stability: —	[124]

Substrate: HAuCl₄
Reductant: NaBH₄
Phase-transfer agent: TOAB
Solvent: water/toluene
Temperature: 22°C
Dispersant: ω-functionalized thiol
  mixtures; ratio: 0.5
	Shape: sphere
	Mean: 3.5 nm
	Polydispersity: 30%
	Solubility: water,
	  alcohols
	Coalescence temp.:
	  150°C–210°C
	Stability: stable
		[11,27]

such as by citrate in boiling water.[111] This gives a broad PSD from 15 nm to a few hundred nm in diameter. To obtain a better PSD, a separate nucleation and growth method[112] has been used in which the fine NPs initially formed by citrate reduction (15–20 nm diameter) are subsequently grown by a reduction in sulfonated phosphine dispersant.[24,113] Au NPs have also been prepared by using surfactant molecules (e.g., cetyltrimethylammonium chloride, sodium dodecylsulfate, cetylpyridinium chloride) as dispersants.[114] The citrate-stabilized Au NPs can be ligand exchanged by alkyl thiols and ω-functionalized alkyl thiols,[115] oleates,[116,117] and long-chain alkyl amines[118] to give organic-soluble Au NPs.

The phase-transfer Brust–Schiffrin method is currently the most important and convenient method to prepare high-quality isolable Au NPs.[81] These alkyl thiol-protected Au NPs are synthesized in one pot by transferring the AuCl₄⁻ to an organic phase (usually toluene) tetraoctylammonium bromide phase-transfer agent, then mixing with the desired thiols and reducing with an aqueous reductant (usually NaBH₄). The Au NPs are first formed at the organic–water interface and are dispersed into the organic phase. These are dispersible typically in aromatic hydrocarbon solvents at extremely high concentrations of >50 mg mL⁻¹ without requiring a polymeric dispersant. The PSD is considerably narrower than citrate reduction. Systematic investigations show the mean diameter can be controlled simply by the feed ratio of the thiol to the Au,[28] as shown in Figure 2.16. Bulky thiol groups (e.g., tripeptide glutathione[119] and 4-mercaptobenzoate,[115]) thiol ionic liquids[120] *N*, *N*-trimethyl(undecylmercapto)ammonium,[50] tiopronin, or coenzyme[121] have also been used to improve the stability of

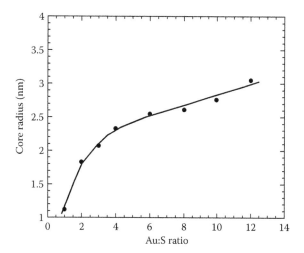

FIGURE 2.16 Au core radius as a function of thiol-to-Au mole feed ratio in the Brust–Schiffrin process.

the Au NPs. These can also be prepared with amines,[122] ω-functionalized alkyl thiols,[11,27,82,123] and thiolated PEG[124,125] as dispersants.

2.6.3 Other Metal Nanoparticles

Pt and Pd NPs can also be produced by reduction of H_2PtCl_6 and H_2PdCl_4, respectively, in water and protected by a SAM molecule such as *p*-aminobenzenesulfonate sodium or mercaptosuccinic acid.[113,126] These can be precipitated and re-dispersed at known concentrations in water. The mean and PSD of these NPs tend to be larger. Reduction of H_2PtCl_6 or K_2PtCl_4 in THF using $NaBH_4$ in the presence of octadecanethiol yields Pt NPs soluble in chloroform and toluene.[127] Alkyl thiol-protected Pt NPs can also be prepared by the Brust–Schiffrin method.[128] Both Pd and Pt NPs can also be prepared by reducing Pd(II) acetate in the presence of a phenanthroline derivative as a molecular dispersant.[113]

2.6.4 Metal NPs with High Polar-Solvent Dispersibility and Low T_p

Au and Ag NPs that are highly dispersible in water and other polar solvents such as alcohols and glycols are important for high-end printed electronics applications. Polar solvents have large surface tension, which acts to self-confine the printed droplets. Nonpolar solvents spread strongly over nonabsorbent surfaces. The use of ionic ω-functionalized SAMs or other water-soluble dispersants such as polyethers tends to produce stabilized NPs with $T_p > 220°C$ but often produces limited dispersibility in water and other polar solvents. Recently, it was found that the use of mixed ω-ionic and ω-hydroxyl alkyl thiol SAMs can produce highly water and polar-solvent dispersible NPs, although the T_p is still high.[11,129] The un-ionized carboxylic acid forms, on the other hand, are typically soluble in alcohol but not water. More recently, it was found that the use of sparse mixed ionic-monolayer protection schematically illustrated in Figure 2.17 can keep the excellent dispersibility of Au and Ag NPs in water and other polar solvents but lower their T_p to 150°C.[27] These nanometal systems, therefore, become compatible with commodity plastic films such as polyesters and with plastic electronics in general. We have extensively tested these materials in inkjet printing and found good jettability and metal track formation characteristics.

2.7 Considerations for Inkjet Printing

2.7.1 Effect of Surface Tension

The required metal NP concentration is determined by the final film thickness and the dimensions of the printed liquid film. If a metal NP track of thickness z and width w is to be printed in a single inkjet pass, the printed liquid column needs to have the same width w. This depends on the droplet contact angle θ on the substrate, droplet volume, and number

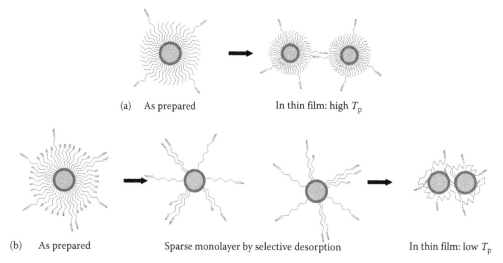

(a) As prepared   In thin film: high T_p

(b) As prepared   Sparse monolayer by selective desorption   In thin film: low T_p

FIGURE 2.17 Schematic illustrating differences between (a) dense mixed ionic monolayer-protected metal NPs, and (b) sparse mixed ionic monolayer-protected metal NPs.

of droplets per unit length. The liquid volume in this column is given by the cross-sectional shape, which is completely determined by θ if its surface tension γ_ℓ dominates over gravity. The presence of NPs at practical concentrations makes a negligible contribution to γ_ℓ.

Due to the small dimension of w, which lies between tens of μm to hundreds of μm, the effect of gravity is indeed negligible. This is quantified by the Bond number Bo = $\rho_\ell g w^2/\gamma_\ell$, which gives 10^{-4}–10^{-2}, where ρ_ℓ is the liquid density and g is the acceleration due to gravity. Thus, the liquid column is a truncated cylinder bounded by θ in contact with the substrate.[130] The cross-sectional area of this truncated cylinder is then given by[130]

$$A = \frac{w^2(\theta - \sin\theta\cos\theta)}{4\sin^2\theta}. \qquad (2.21)$$

The required metal concentration in the ink is then given by

$$c_m = \frac{4\sin^2\theta}{(\theta - \sin\theta\cos\theta)}\frac{\rho_m z}{w}, \qquad (2.22)$$

where ρ_m is the metal density. The required metal volume fraction $\phi_m = n_m/\rho_\ell$ can then be computed and plotted against θ for different z/w between 0.5–5 $\times$ 10^{-3}, as shown in Figure 2.18. This plot reveals that the required concentration of the dispersion for practical IJP increases markedly with decreasing θ. For $w = 100\,\mu$m, $z = 100\,$nm, and $\theta = 90°$, we find $\phi_m = 0.25\,$vol%. If the metal is Au, with $\rho_m = 19.3\,$g cm^{-3}, $n_m = 0.05\,$g cm^{-3}; if Ag, with $\rho_m = 10.5\,$g cm^{-3}, $n_m = 0.026\,$g cm^{-3}.

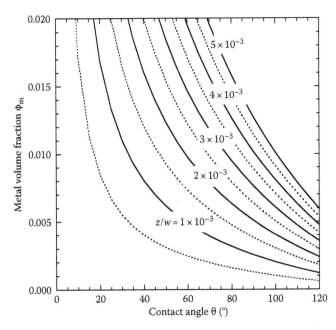

FIGURE 2.18 The required metal NP volume fraction in the dispersion as a function of the droplet contact angle on the substrate for different line thickness-to-width (z/w) ratios.

The relative viscosity of the dispersion if the NPs are noninteracting is given by Einstein's equation

$$\frac{\eta}{\eta_o} = 1 + 2.5\phi_m, \qquad (2.23)$$

where

η is the viscosity of the dispersion
η_o is the viscosity of pure solvent

This equation predicts that the fractional increase in viscosity is only 2.5% for $\phi_m = 1\%$, which is negligible. However, in the presence of interaction, e.g., through the ω-ionic groups in the dispersant shell, the relative viscosity can increase much more quickly.

If θ decreases to 45°, ϕ_m goes up to 0.7 vol%, which now approaches the dispersibility limits and/or the viscosity limit for jettability. This illustrates that it is crucial to achieve high substrate θ, preferably near 90°, for successful free-form IJP, which shows the benefit of using high surface-tension solvents.

The value of θ is related to the solvent and substrate surface tension γ_s according to a simple phenomenological model[34]

$$\gamma_s = \gamma_1\cos\theta + \gamma_{ls}, \qquad (2.24)$$

where $\gamma_{\ell s}$ is the interfacial tension between the liquid and substrate. Fluorocarbon solvents tend to have low γ_ℓ of 10–15 mN m^{-1}, aliphatic hydrocarbons 20–25 mN m^{-1}, aromatic hydrocarbons 25–30 mN m^{-1}, and halogenated or polar solvents above 30 mN m^{-1}. Typical engineering polymer substrates, such as polyethyleneterephthalate, nylons, and polyetheretherketone, have γ_s of 35–45 mN m^{-1},[34] which can be lowered to ca. 20–30 mN m^{-1} by CF$_4$ plasma. This suggests that the solvent should have $\gamma_\ell > 30$ mN m^{-1} to achieve large enough θ. One possible solvent that meets this requirement well is the ethylene glycol–water system with $\gamma_\ell = 47$–72 mN m^{-1}. Furthermore, the edges of the printed liquid column need to be pinned in order to avoid column breakup.[130,131]

2.7.2 Inkjet Droplet Formation and Deposition

Figure 2.19 shows the strobe camera optical images of well-behaved inkjet droplet formation outside the nozzle of a piezoelectric print head. The ink is Au NP dispersed at 50 mg mL^{-1} in a ethylene glycol–water system. We can see that the ink droplet is ejected after ca. 20 μs with a tail. The tail then retracts over the next few μs to a satellite. The satellite then catches up and merges with the primary droplet that flies at ca. 6 m s^{-1} towards the substrate. This is an appropriate droplet flying speed for reliable deposition. The voltage waveform applied on the piezoelectric element determines the IJP droplet characteristics. However, optimization remains somewhat of an art.

Figure 2.20 shows that high-quality Au dots and tracks from these NPs can be printed at a high speed of 1 kHz. The printed structures are continuous and free from the undesirable

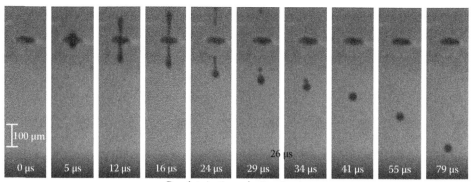

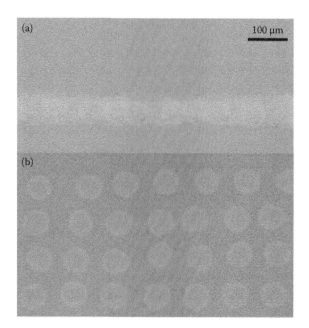

FIGURE 2.19 Strobe-camera optical images of the inkjet droplet formation process.

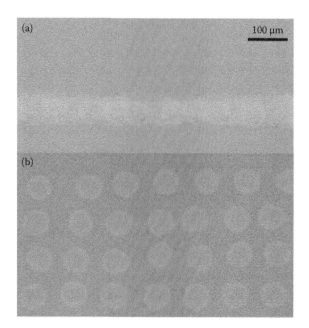

FIGURE 2.20 Optical micrographs of inkjet-printed structures on hydrophilic SiO₂ surface. (a) 40 nm thick Au track. (b) 10 nm thick Au dots.

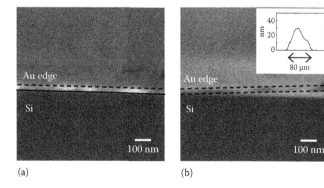

FIGURE 2.21 Cross-sectional scanning electron micrographs of inkjet-printed 30 nm thick Au track on hydrophilic SiO₂ surface. (a) Cleaved edge of track. (b) Oblique view of track.

"coffee stain" effect[132] or the "stack of coin" morphology that is characteristic of extremely slow printing.[6,20,133] Figure 2.21 shows that a relatively uniform cross section can be achieved even for a printed line that is only ca. 30 nm thick. The ink used here is of the hydrophilic type with low T_p, which is eminently suitable for interconnects and FET electrodes for plastic electronics applications.

References

1. H. Kipphan (ed.), *Handbook of Print Media: Technologies and Production Methods* (Springer-Verlag, Berlin, Germany, 2001).
2. H. Sirringhaus, T. Kawase, R.H. Friend, T. Shimoda, M. Inbasekaran, W. Wu, and E.P. Woo, *Science* 290, 2123 (2000).
3. B. Crone, A. Dodabalapur, Y.-Y. Lin, R.W. Filas, Z. Bao, A. LaDuca, R. Sarpeshkar, H.E. Katz, and W. Li, *Nature* 403, 521 (2000).
4. T. Kawase, H. Sirringhaus, R.H. Friend, and T. Shimoda, *Adv. Mater.* 13, 1601 (2001).
5. H.E.A. Huitema, G.H. Gelinck, J.B.P.H. van der Putten, K.E. Kuijk, C.M. Hart, E. Cantatore, P.T. Herwig, A.J.J.M. van der Breemen, and D.M. de Leeuw, *Nature* 414, 599 (2001).
6. D. Huang, F. Liao, S. Molesa, D. Redinger, and V. Subramanian, *J. Electrochem. Soc.* 150, G412 (2003).
7. A.L. Dearden, P.J. Smith, D.Y. Shin, N. Reis, B. Derby, and P. O'Brien, *Macromol. Rapid Commun.* 26, 315 (2005).
8. S. Gamerith, A. Klug, H. Scheiber, U. Scherf, E. Moderegger, and E.J.W. List, *Adv. Funct. Mater.* 17, 3111 (2007).
9. N. Zhao, M. Chiesa, H. Sirringhaus, Y. Li, Y. Wu, and B. Ong, *J. Appl. Phys.* 101, 0645131 (2007).
10. S.H. Ko, J. Chung, H. Pan, C.P. Grigoropoulos, and D. Poulikakos, *Sens. Actuat. A* 134, 161 (2007).
11. S. Sivaramakrishnan, P.J. Chia, Y.C. Yeo, L.L. Chua, and P.K.H. Ho, *Nat. Mater.* 6, 149 (2007).
12. S. Kubitschko, J. Spinke, T. Brückner, S. Pohl, and N. Oranth, *Anal. Biochem.* 253, 112 (1997).
13. E. Chow, J. Herrmann, C.S. Barton, B. Raguse, and L. Wieczorek, *Anal. Chim. Act.* 632, 135 (2008).

14. S.B. Fuller, E.J. Wilhelm, and J.M. Jacobson, *J. Microelectromech. Syst.* 11, 54 (2002).

15. M. Brust and C.J. Kiely, *Colloids Surf. A* 202, 175–186 (2002).

16. H. Hirai, H. Wakabayashi, and M. Komiyama, *Chem. Lett.* 12, 1047 (1983).

17. E.W. Flick, *Adhesives, Sealants and Coatings for the Electronics Industry* (Noyes Publications, New York, 1992).

18. See for example: Cima Nanotech: http://www.cimanotech. com/; Advanced Nano Products: http://www.anapro.com/ english/default.asp; Harima Chemicals: http://www.harima. co.jp/en/index.html; Cabot corporation: http://www.cabot-corp.com/GlobalGateway.aspx; Ulvac Technologies: http:// www.ulvac.com/ (accessed on July 2009).

19. J.R. Greer and R.A. Street, *J. Appl. Phys.* 101, 1035291 (2007).

20. H.H. Lee, K.S. Chou, and K.C. Huang, *Nanotechnology* 16, 2436 (2005).

21. J. Perelaer, A.W.M. de Laat, C.E. Hendriks, and U.S. Schubert, *J. Mater. Chem.* 18, 3209 (2008).

22. M. Brust, C.J. Kiely, D. Schiffrin, and D. Bethell, *Langmuir* 14, 5425 (1998).

23. R.H. Terrill, T.A. Postlethwaite, C.H. Chen, C.D. Poon, A. Terzis, A. Chen, J.E. Hutchison, M.R. Clark, G.D. Wignall, J.D. Londono, R. Superfine, M. Falvo, C.S. Johnson Jr., E.T. Samulski, and R.W. Murray, *J. Am. Chem. Soc.* 117, 12537 (1995).

24. G. Schmid and A. Lehnert, *Angew. Chem. Int. Ed. Engl.* 28, 780 (1989).

25. L.D. Marks, *Rep. Prog. Phys.* 57, 603 (1994).

26. W.P. Wuelfing, F.P. Zamborini, A.C. Templeton, X. Wen, H. Yoon, and R.W. Murray, *Chem. Mater.* 13, 87 (2001).

27. B.T. Anto, S. Sivaramakrishnan, L.L. Chua, and P.K.H. Ho, *Adv. Funct. Mater.* 20, 296 (2010).

28. M.J. Hostetler, J.E. Wingate, C.J. Zhong, J.E. Harris, R.W. Vachet, M.R. Clark, J.D. Londono, S.J. Green, J.J. Stokes, G.D. Wignall, G.L. Glish, M.D. Porter, N.D. Evans, and R.W. Murray, *Langmuir* 14, 17 (1998).

29. R. Zallen, *The Physics of Amorphous Solids* (Wiley-VCH, New York, 1998).

30. C. Song, P. Wang, and H.A. Makse, *Nature* 453, 629 (2008).

31. M.J. Powell, *Powder Technol.* 25, 45 (1980).

32. B.D. Cullity, *Elements of X-ray Diffraction* (Addison-Wesley, Manila, Philippines, 1978).

33. N.W. Ashcroft and D.N. Mermin, *Solid State Physics* (Saunders College, Philadelphia, PA, 1976).

34. D.W. van Krevelen, *Properties of Polymers. Their Correlation with Chemical Structure: Their Numerical Estimation and Prediction from Additive Group Contributions*, 3rd edn. (Elsevier, New York, 1990).

35. P.K.H. Ho, J.S. Kim, N. Tessler, and R.H. Friend, *J. Chem. Phys.* 115, 2709 (2001).

36. G.E. Pike and C.H. Seager, *Phys. Rev. B* 10, 1435 (1974).

37. G.E. Pike and C.H. Seager, *Phys. Rev. B* 10, 1421 (1974).

38. C.D. Lorenz and R.M. Ziff, *J. Phys. A: Math. Gen.* 31, 8147 (1998).

39. C.R.M. Wronski, *Br. J. Appl. Phys.* 18, 1731 (1967).

40. C.J. Coombes, *J. Phys. F: Metal Phys.* 2, 441 (1972).

41. Ph. Buffat and J.P. Borel, *Phys. Rev. A* 13, 2287 (1976).

42. J.-P. Borel, *Surf. Sci.* 106, 1 (1981).

43. T. Castro, R. Reifenberger, E. Choi, and R.P. Andres, *Phys. Rev. B* 42, 8548 (1990).

44. F. Ercolessi, W. Andreoni, and E. Tosatti, *Phys. Rev. Lett.* 66, 911 (1991).

45. S.L. Lai, J.Y. Guo, V. Pertrova, G. Ramanath, and L.H. Allen, *Phys. Rev. Lett.* 77, 99 (1996).

46. S. Arcidiacono, N.R. Bieri, D. Poulikakos, and C.P. Grigoropoulos, *Int. J. Multiphase Flow* 30, 979–994 (2004).

47. L.J. Lewis, P. Jensen, and J.L. Barrat, *Phys. Rev. B* 56, 2248 (1997).

48. B. Abeles, P. Sheng, M.D. Coutts, and Y. Arif, *Adv. Phys.* 24, 407 (1975).

49. M.D. Musick, C.D. Keating, M.H. Keefe, and M.J. Natan, *Chem. Mater.* 9, 1499 (1997).

50. D.E. Cliffel, F.P. Zamborini, S.M. Gross, and R.W. Murray, *Langmuir* 16, 9699 (2000).

51. S. Sivaramakrishnan, B.T. Anto, and P.K.H. Ho, *Appl. Phys. Lett.* 94, 091909 (2009).

52. J.R. Sambles, *Proc. R. Soc. Lond. A* 324, 339 (1971).

53. K. Dick, T. Dhanasekaran, Z. Zhang, and D. Meisel, *J. Am. Chem. Soc.* 124, 2312 (2002).

54. N. Nishida, M. Hara, H. Sasabe, and W. Knoll, *Jpn. J. Appl. Phys.* 35, 5866 (1996).

55. R.G. Nuzzo, B.R. Zegarski, and L.H. Dubois, *J. Am. Chem. Soc.* 109, 733 (1987).

56. E. Delamarche, B. Michel, H. Kang, and Ch. Gerber, *Langmuir* 10, 4103 (1994).

57. D.M. Jaffey and R.J. Madix, *J. Am. Chem. Soc.* 116, 3012 (1994).

58. D.M. Jaffey and R.J. Madix, *Surf. Sci.* 311, 159 (1994).

59. M. Büttner, T. Belser, and P. Oelhafen, *J. Phys. Chem.* 109, 5464 (2005).

60. C.G. Granqvist and O. Hunderi, *Phys. Rev. B* 16, 3513 (1977).

61. U. Kreibig and L. Genzel, *Surf. Sci.* 156, 678 (1985).

62. M. Quinten and U. Kreibig, *Surf. Sci.* 172, 557 (1986).

63. S. Link and M.A. El-sayed, *Int. Rev. Phys. Chem.* 19, 409 (2000).

64. L.M. Liz-Marzan, *Langmuir* 22, 32 (2006).

65. S. Underwood and P. Mulvaney, *Langmuir* 10, 3427 (1994).

66. M.M. Alvarez, J.T. Khoury, T.G. Schaaff, M.N. Shafigullin, I. Vezmar, and R.L. Whetten, *J. Phys. Chem. B* 101, 3706 (1997).

67. S. Link, M.B. Mohamed, and M.A. El-Sayed, *J. Phys. Chem. B* 103, 3073 (1999).

68. A.C. Templeton, J.J. Pietron, R.W. Murray, and P. Mulvaney, *J. Phys. Chem. B* 104, 564 (2000).

69. L.K. Kelly, E. Coronado, L. Zhao, and G.C. Schatz, *J. Phys. Chem. B* 107, 668 (2003).

70. L.B. Scaffardi, N. Pellegri, O. de Sanctis, and J.O. Tocho, *Nanotechnology* 16, 158 (2005).

71. M.M. Miller and A.A. Lazarides, *J. Phys. Chem. B* 109, 21556 (2005).

72. L.B. Scaffardi and J.O. Tocho, *Nanotechnology* 17, 1309 (2006).
73. M.A. El-Sayed, M.B. Mohamed, K.Z. Ismail, and S. Link, *J. Phys. Chem. B* 102, 9370 (1998).
74. U. Kreibig and C.V. Fragstein, *Z. Phys. A* 224, 307 (1969).
75. E.D. Palik (ed.), *Handbook of Optical Constants of Solids* (Academic, New York, 1998).
76. D.E. Aspnes, *Thin Solid Films* 89, 249 (1982).
77. M. Born and E. Wolf, *Principles of Optics: Electromagnetic Theory of Propagation, Interference and Diffraction of Light*, 7th edn. (Cambridge University Press, Cambridge, U.K., 1999).
78. H.C. Van de Hulst, *Light Scattering by Small Particles* (John Wiley & Sons, New York, 1957).
79. L.A. Gómez, C.B. de Araújo, A.M. Brito-Silva, and A. Galembeck, *Appl. Phys. B* 92, 61 (2008).
80. S. Link and M.A. El-Sayed, *J. Phys. Chem. B* 103, 4212 (1999).
81. M. Brust, M. Walker, D. Bethell, D. Schiffrin, and R. Whyman, *J. Chem. Soc. Chem. Commun.* 801 (1994).
82. K. Kimura, H. Yao, and S. Sato, *Synth. React. Inorg. Met.-Org. Chem.* 36, 237 (2006).
83. W. Zhang, X. Qiao, and J. Chen, *Mater. Sci. Eng., B* 142, 1 (2007).
84. Y. Sun and Y. Xia, *Science* 298, 2176 (2002).
85. D.-H. Chen and Y.-W. Huang, *J. Colloid Interface Sci.* 255, 299 (2002).
86. I. Pastoriza-Santos and L.M. Liz-Marzán, *Nano Lett.* 2, 903–905 (2002).
87. X. Wang, H. Hoh, and K. Naka, *Langmuir* 19, 6242 (2003).
88. Z.S. Pillai and P.V. Kamat, *J. Phys. Chem. B* 108, 945–951 (2004).
89. N.K. Chaki, J. Sharma, A.B. Mandle, I.S. Mulla, R. Pasricha, and K. Vijayamohanan, *Phys. Chem. Chem. Phys.* 6, 1304 (2004).
90. S. Chen and K. Kimura, *Chem. Lett.* 1169 (1999).
91. N.H. Kim, J.-Y. Kim, and K.J. Ihn, *J. Nanosci. Nanotechnol.* 7, 3805 (2007).
92. L. Longenberger and G. Mills, *J. Phys. Chem. B* 99, 475 (1995).
93. Y. Yin, Z.-Y. Li, Z. Zhong, B. Gates, Y. Xia, and S. Venkateswaran, *J. Mater. Chem.* 12, 522 (2002).
94. H.S. Kim, J.H. Ryu, B. Jose, B.G. Lee, B.S. Ahn, and Y.S. Kang, *Langmuir* 17, 5817–5820 (2007).
95. N. Leopold and B. Lendl, *J. Phys. Chem. B* 107, 5723–5727 (2003).
96. K. Mallick, M.J. Witcomb, and M.S. Scurrell, *J. Mater. Sci.* 39, 4459 (2004).
97. D.G. Shchukin, I.L. Radtchenko, and G.B. Sukhorukov, *ChemPhysChem* 4, 1101 (2003).
98. S. Komarneni, H. Katsuki, D. Li, and A.S. Bhalla, *J. Phys.: Condens. Matter* 16, S1305–S1312 (2004).
99. Y. Socol, O. Abramson, A. Gedanken, Y. Meshorer, L. Berenstein, and A. Zaban, *Langmuir* 18, 4736–4740 (2002).
100. V. Hornebecq, M. Antonietti, T. Cardinal, and M. Treguer-Delapierre, *Chem. Mater.* 15, 1993–1999 (2003).
101. T. Tsuji, T. Kakitab, and T. Tsuji, *Appl. Surf. Sci.* 206, 314 (2003).
102. J.R. Heath, C.M. Knobler, and D.V. Leff, *J. Phys. Chem. B* 101, 189–197 (1997).
103. S. Thomas, H.S. Mahal, S. Kapoor, and T. Mukherjee, *Res. Chem. Intermed.* 31, 595–603 (2005).
104. A. Taleb, C. Petit, and M.P. Pileni, *Chem. Mater.* 9, 950 (1997).
105. T. Yonezawa, S. Onoue, and N. Kimizuka, *Langmuir* 16, 5218 (2000).
106. J. Eastoe, M.J. Hollambly, and L. Hudson, *Adv. Colloid Interface Sci.* 128, 5 (2006).
107. M.A. López-Quintela, C. Tojo, M.C. Blanco, L.G. Rio, and J.R. Leis, *Curr. Opin. Colloid Interface Sci.* 9, 264 (2004).
108. R.C. Doty, T.R. Tshikhudo, M. Brust, and D.G. Fernig, *Chem. Mater.* 17, 4630 (2005).
109. B.A. Korgel, S. Fullam, S. Connolly, and D. Fitzmaurice, *J. Phys. Chem. B* 102, 8379–8388 (1998).
110. M. Daniel and D. Astruc, *Chem. Rev.* 104, 293 (2004).
111. J. Turkevich, P.C. Stevenson, and J. Hillier, *Discuss. Faraday Soc.* 11, 55 (1951).
112. N.R. Jana, L. Gearheart, and C.J. Murphy, *Langmuir* 17, 6782–6786 (2001).
113. G. Schmid, *Chem. Rev.* 92, 1709 (1992).
114. H. Ishizuka, T. Tano, K. Torigoe, K. Esumi, and K. Meguro, *Colloids Surf.* 63, 337 (1992).
115. M. Giersig and P. Mulvaney, *Langmuir* 9, 3408 (1993).
116. H. Hirai and H. Aizawa, *J. Colloid Interface Sci.* 161, 471 (1993).
117. C.S. Weisbecker, M.V. Merritt, and G.M. Whitesides, *Langmuir* 12, 3763 (1996).
118. A. Swami, A. Kumar, and M. Sastry, *Langmuir* 19, 1168 (2003).
119. T.G. Schaaff, G. Knight, M.N. Shafigullin, R.F. Borkman, and R.L. Whetten, *J. Phys. Chem. B* 102, 10643 (1998).
120. K.-S. Kim, D. Demberelnyamba, and H. Lee, *Langmuir* 20, 556 (2004).
121. A.C. Templeton, S. Chen, S.M. Gross, and R.W. Murray, *Langmuir* 15, 66 (1999).
122. B. Leff, L. Brandt, and J.R. Heath, *Langmuir* 12, 4723 (1996).
123. S. Chen and K. Kimura, *Langmuir* 15, 1075 (1999).
124. W.P. Wuelfing, S.M. Gross, D.T. Miles, and R.W. Murray, *J. Am. Chem. Soc.* 120, 12696 (1998).
125. E.R. Zubarev, J. Xu, A. Sayyad, and J.D. Gibson, *J. Am. Chem. Soc.* 128, 4958 (2006).
126. S. Chen and K. Kimura, *J. Phys. Chem. B* 105, 5397 (2001).
127. C. Yee, M. Scotti, A. Ulman, H. White, M. Rafailovich, and J. Sokolov, *Langmuir* 14, 4314 (1999).
128. J. Yang, J.Y. Lee, T.C. Deivaraj, and H.-P. Too, *Langmuir* 19, 10361 (2003).
129. O. Uzun, Y. Hu, A. Verma, S. Chen, A. Centrone, and F. Stellacci, *Chem. Commun.* 196 (2008).
130. P.C. Duineveld, *J. Fluid Mech.* 477, 175 (2003).
131. S. Schiaffino and A.A. Sonin, *J. Fluid Mech.* 343, 95 (1997).
132. R.D. Deegan, O. Bakajin, T.F. Dupont, G. Huber, S.R. Nagel, and T.A. Witten, *Phys. Rev. E* 62, 756 (2000).
133. D. Soltman and V. Subramanian, *Langmuir* 24, 2224 (2008).

3

Polymer–Clay Nanocomposites

Sabrina Pricl
University of Trieste

Paola Posocco
University of Trieste

Giulio Scocchi
University of Trieste

Maurizio Fermeglia
University of Trieste

3.1 Introduction

The mystery of the *nano-world* has been progressively unraveled in recent years. The nanometer scale is simply a range between micro and molecular dimensions. The sciences in these two-dimensional ranges have been well explored by materials scientists and chemists. Materials science and chemistry are often engaged in research on the nanometer scale, for example, the dimensions of crystal structures. The well-known nanometer-scale technologies developed within materials science and chemistry in the past may not be reasonably regarded as nanotechnology. The real interest in and ultimate goal of nanotechnology is to create revolutionary properties and functions by tailoring materials and designing devices on the nanometer scale. In this respect, polymer–clay nanocomposites (PCNs) are an archetypical example of nanotechnology.

According to the International Union for Pure and Applied Chemistry (IUPAC) (Work et al., 2004), a *composite material* is defined as "a multicomponent material comprising multiple different (non-gaseous) phase domains in which at least one type of phase domain is in a continuous phase." IUPAC also extends its definition to *nanocomposite* materials as those composites "in which at least one of the phases has at least one dimension of the order of nanometers." Based on these definitions, a plethora of systems can be classified among these materials, the dispersions of nanosized objects of different nature—such as metal particles or carbon nanotubes—or intercalated/exfoliated layered minerals in continuous/polymeric phases being prime examples. PCNs fall in the last category of the examples cited above.

The mixing of nanoparticles with polymers to form composite materials has been practiced for decades. For example, the clay-reinforced resin known as Bakelite was introduced in the early 1900s as one of the first mass-produced polymer–nanoparticle composites, and fundamentally transformed the nature of practical household materials. Even before Bakelite, nanocomposites were finding applications in the form of nanoparticle-toughened automobile tires prepared by blending carbon black, zinc oxide, and/or magnesium sulfate particles with vulcanized rubber. Despite these early successes, the broad scientific community was not galvanized by nanocomposites until the early 1990s, when reports by Toyota researchers revealed that adding a clay mineral to nylon produced a fivefold increase in the yield and tensile strength of the material (Kojima et al., 1993; Usuki et al., 1993). Subsequent developments have further contributed to the surging interest in polymer–nanoparticle composites. In particular, the growing availability of the nanoparticles of precise size and shape, such as fullerenes, carbon nanotubes, inorganic nanoparticles, dendrimers, and bio-nanoparticles, and the development of instrumentation to probe small length scales, such as scanning force, laser scanning fluorescence, and electron microscopes, have spurred research aimed at probing the influence of particle size and shape on the properties of polymer–nanoparticle composites.

The subject of hybrids based on layered inorganic compounds such as clay has been tackled for a considerable time, but the area has enjoyed a resurgence of interest and activity owing to the massive industrial exploitation of Nylon6-clay PCNs in the automotive industry by the Toyota Corporation (Okada and Usuki, 2006), and to the exceptional properties exhibited by PCNs (LeBaron et al., 1999; Pinnavaia and Beall, 2000; Sinha Ray and Okamoto, 2003; Utraki, 2004; Zeng et al., 2005; Balazs et al., 2006; Chen et al., 2008). These systems offer a number of material parameters that can be controlled or fine-tuned to achieve a given ultimate property; these include the type of clay, the choice of the clay pretreatment, the selection of the polymeric matrix, and, last but not least, the method by which the polymer is incorporated in the nanocomposite. The last of these is, in turn, dictated by the processing conditions available and whether the end user is an integrated polymer manufacturer or a specialist processor.

As part of this renewed interest in nanocomposites, researchers also began seeking design rules that would allow them to engineer materials that combine the desirable properties of nanoparticles and polymers. The ensuing research revealed a number of key challenges in producing nanocomposites that exhibit a desired behavior. The greatest stumbling block to the large-scale production and commercialization of nanocomposites is the dearth of cost-effective methods for controlling the dispersion of the nanoparticles in polymeric hosts. The nanoscale particles typically aggregate, which negates any benefits associated with the nanoscopic dimension. Another hurdle to the broader use of nanocomposites is the absence of structure–property relationships. Because increased research activity in this area has only spanned the past decade, there are limited property databases for these materials (Ajayan et al., 2003). Thus, greater efforts are needed to correlate the morphology of the mixtures with the macroscopic performance of the materials. Establishing these relationships requires a better understanding of how cooperative interactions between flexible chains and nanoscopic solids can lead to unexpected behavior, like the improved mechanical behavior of clay-reinforced nylon.

3.2 Clay Minerals

3.2.1 Clay Types and Structures

Common clay minerals possess variability in their constitution, as expected for many naturally occurring compounds. Importantly, the composition and purity of given clay can exert an influence on the final properties of the corresponding PCN. These minerals represent a wide class of compounds, also known as layered materials, which can be defined as "crystalline materials wherein the atoms in the layers are cross linked by chemical bonds, while the atoms of adjacent layers interact by physical

forces" (Schoonheydt et al., 1999). Clays are generally classified by structure as allophane, kaolinite, halloysite, smectite, illoite, chlorite, vermiculite, attapulgite-palygorskite-sepiolite, and mixed layered minerals (Grim, 1968). The most used clays for the production of PCNs can be substantially grouped into three main categories:

- *2:1 type*: These minerals belong to the *smectite* family, and present a crystal structure consisting in nanometer thick layers (also called *platelets*) of alumina octahedrons sheets sandwiched between two silica tetrahedron sheets (Figure 3.1a). The stacking of the layers results in a van der Waals gap between the layers. The isomorphic substitution of aluminum (Al) with magnesium (Mg), iron (Fe), lithium (Li) in the octahedron sheets, and/or silicon (Si) with Al in the tetrahedron sheets gives each three-sheet layer an overall negative charge, which is counterbalanced by exchangeable metal cations residing in the interlayer spaces, also called galleries, such as sodium (Na), calcium (Ca), Mg, Fe, and Li.

- *1:1 type*: These clays consist of layers that are made up by alternating Al octahedron and Si tetrahedron sheets, respectively. As no isomorphic substitution occurs, each layer bears no charge. Therefore, except for crystallization and humidity water molecules, cations, and anions are never found in the interlayer galleries, and the layers are held together by a hydrogen bond network between the hydroxyl (–OH) groups in the octahedral sheets and the oxygen atoms of the adjacent tetrahedral layers.

- *Layered silicic acids*: This class of clays is mainly composed of silica tetrahedron sheets characterized by different layer thickness. Basically, their structures are composed on layered silicate networks and interlayer hydrated alkali metal cations.

Smectite clays such as montmorillonite (MMT, Figure 3.1b) are probably the most common types of clay used for nanocomposite formation. The main composition of layered silicates are SiO_2 (30%–70%), Al_2O_3 (10%–40%), and H_2O (5%–10%). The intralayer space of these distinctive materials include the –OH sandwich groups of the octahedral aluminum-hydroxyl sheets and the oxygen of the tetrahedral silicate sheets: the crystal consists of alternating cations planes and negatively charged silicate sheets in the ratio 2:1. Their fundamental building blocks are $Si(O, OH)_4$, silicon–oxygen tetrahedral (*T*-network), and $M(O,OH)_6$ octahedra (*O*-network), with $M = Mg^{2+}$, Al^{3+}, Fe^{2+}, and Fe^{3+} (Figure 3.1c).

Depending on their size, cations may be fully or partially incorporated in the free space between the tetrahedral, producing contracting forces between the layers, in addition to the dispersion forces between the silicon–oxygen tetrahedra in the silica sheets and the electrostatic (Coulombic) forces between the negative charges on opposite layers and the cations in between. Some trivalent Al cations are substituted by Mg^{2+} in MMT; as mentioned above, these isomorphic substitutions result in an overall negative charge on the mineral which

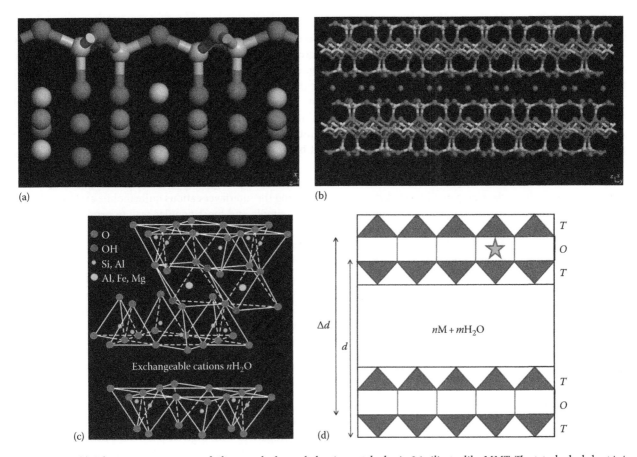

FIGURE 3.1 (a) Schematic arrangement of silica tetrahedra and alumina octahedra in 2:1 silicates like MMT. The tetrahedral sheet is in stick-and-ball representation, with Si atoms in light gray and O atoms in dark gray. The octahedral arrangement is depicted as spheres only, with Al atoms in gray, Mg atoms in light gray, and O atoms in dark gray. (b) Three-dimensional model of MMT showing the relative arrangement of tetrahedral and octahedral sheets and the interlayer cations. The interlayer cations are portrayed as spheres. (c) General crystal structure of layered silicates. (d) Schematic representation of the crystal structures of MMT clay. Symbol legend: square, aluminium–oxygen octahedron (O); square with star, magnesium–oxygen octahedron (O); triangle, silicon–oxygen tetrahedron (T); M = interlayer cations.

is counterbalanced by the presence of external cations (mainly Na^+) in the interlayer galleries. Depending on the valence state of exchangeable ions, the lattice may remain electrically neutral (pyrophyllite and talc groups) or may bear a net negative charge (from 0.25 to 0.6 in smectites up to 1 in micas). In contrast to those in mica, the cations in smectites can readily be exchanged, in particular, for transition-metal ions, and the cation-exchange capacity of smectites attains 0.64–1.50 meq/g. The interlayer space (galleries) of many layered minerals with *T-O-T* structures contain water molecules, which are involved in mineral formation and prevent collapse, a complete sticking of the layers to one another. In natural Na-MMT-type smectite clays, the parameters of the interlayer space are determined by the crystal structure of the aluminosilicate. The general formula of dioctahedral minerals (pyrophyllite, MMT, muscovite, vermiculite, and others) formed by Al octahedra can be represented as $(Si_{8-x}M_y)_4(Al_{4-y})_6(OH)_4O_{20}(H_2O)_w$, and that of trioctahedral minerals based on Mg octahedral (talc, saponite, biotite, hectorite, vermiculite, and others) is $(Si_{8-x}M_x)_4(Mg_{6-y})_6(OH)_4O_{20}(H_2O)_w$. The smectites capable of swelling, such as MMT, HECT, and saponite, are of special interest for intercalation chemistry. Vermiculites

and micas (muscovite and others) swell far less readily. The crystal structures of some layered silicates are shown schematically in Figure 3.1d.

The principal characteristics leading to the exploitation of smectite clays in the preparation of PCNs are

- Their swelling ability, with the consequent potential to host even big molecules as polymers between their layers (*intercalation chemistry*).
- Optimal balance between strength, stiffness, and flexibility. In this respect, each single clay platelet can be considered as a rigid inorganic polymer with a molecular mass (approx. 1.3×10^8) considerably higher than that of typical commercial polymers. Hence, PNCs with very low clay loadings (e.g., 5% weight) can achieve the same final properties of conventional composites.
- High aspect ratio of the individual platelets. The key parameters of layered silicates are the distance from a *T*-network to its analog in one of the neighboring layers, *d*, and the thickness of the interlayer space, Δd. The interlayer spacing in clay minerals depends on the size of

exchangeable cations and the amount of water in the interlayer space. For example, the interlayer spacing is 1.18 nm in anhydrous Na+MMT and increases to 1.25, 1.50–1.55, and even 1.80–1.90 nm upon the incorporation of mono-, bi-, and trimolecular water layers into the interlayer space. The crystals of clay minerals unite in fine flakes, ribbons, or tubes (sometimes of colloidal size), which tend to aggregate, producing secondary porous structures: hexagonal platelets in kaolinite, poorly defined shapes in MMT, thin elongated platelets (*laths*) in hectorite, and others. The longitudinal size of disk-like clay particles depends on the preparation procedure: clays prepared by grinding typically consist of plate-like particles ranging in longitudinal size from 0.1 to 1.0 μm. Since colloidal clay particles have very large aspect ratios, delaminated clays offer extremely large specific surfaces and are nanostructured. This is a necessary condition for good intercalation properties and the enhanced performance of composites Therefore, although layered silicates are not nanoparticles per se, the thickness of the clay layers of the order of 1 nm, the high aspect ratios (typically 100–1500), and the large surface areas (700–800 m²/g) (Theng, 1979; Pinnavaia, 1983; Pinnavaia and Beall, 2000) render the clay platelets truly nanoparticulate objects.

- High natural abundance and, hence, low cost. Clays are ubiquitous in nature, and constitute relatively cheap feedstocks with minimal limitation on supply.

3.2.2 Clay Surface Modification

As mentioned earlier, the interlayer spacing in anhydrous Na-MMT is about 1.0 nm. Each layer is separated from its neighbors by a spacing determined by the interlayer van der Waals forces and forms an interlayer space or gallery. Whereas the Na–O bond length is typically 0.21–0.22 nm, it is increased to 0.36 nm in the galleries. Since the Coulomb electrostatic force is inversely proportional to the square of the distance between the charges, the cation–oxygen bond strength is reduced by ~60%. As a result, the galleries may contain a large amount of water molecules, which, coupled with the clay sheet

surface change and the presence of ions, ultimately contributes to render these substances highly hydrophilic species and, therefore, poorly compatible with a wide range of nonpolar, organic molecules such as polymers. A necessary prerequisite for the successful formation of a PCN is therefore the alteration of the clay polarity to make the clay *organophilic*. Their unique layered structure, coupled to their high intercalation ability, allow these minerals to be easily modified to be compatible with polymers. Further, the relatively low layer charge unbalance (ranging from 0.2 to 0.6) results in weak van der Waals and electrostatic forces between neighboring layers, rendering the interlayer cations quite mobile and interchangeable. A simple and most popular way to produce an organophilic clay from a normally hydrophilic clay is therefore to exchange the interlayer cations with organic cations such as alkylammonium ions. Thus, for example, in MMT the Na+ ions in the clay galleries can be exchanged for dimethyl, bis(hydrogenated tallow) quaternary ammonium chloride (2M2HT, i.e., $(CH_3)_2N(C_{18}H_{37})_2Cl$ see Figure 3.2a) according to the following reaction:

$$Na^+ - MMT + 2M2HT \rightarrow 2M2HT - NH_3^+ - MMT + NaCl$$

Upon this treatment, and depending on the chemical structure of the clay surface organic modifier, the clay becomes more compatible with a given polymer, being it thermoplastic, thermosetting, or elastomer. The resulting modified clay is also often commonly referred to as an *organoclay*. A further, important aspect of clay surface treatment with organic salts is that to accommodate the long, hydrophobic chains within the silicate galleries, the interlayer space must increase with respect to the pristine mineral. For instance, interlayer space Δd in natural MMT is equal to 11.8 Å. After Na+ exchange with 2M2HT, $\Delta d = 24.2$ Å (see Figure 3.2b). This increased interlayer space leads to a twofold advantage in the preparation of a MMT-based PCN: (1) the interlayer binding forces are reduced and (2) the insertion, diffusion, and accommodation of the bulky polymeric chains (or their monomeric precursors) are facilitated.

(a)

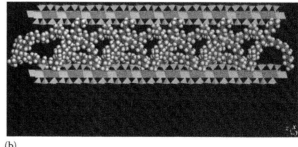

(b)

FIGURE 3.2 (a) Stick-and-ball molecular model of dimethyl, dehydrogenated tallow quaternary ammonium chloride (2M2HT), where tallow stands for a mixture of C18 (~65%), C16 (~30%), and C14 (~5%). Atom color code: gray, C; blue, N; white, H, light green, Cl. (b) 3D molecular model of the organoclay obtained by ion-exchanging interlayer sodium cations with 2M2HT. The 2M2HT molecules are portrayed as CPK spheres, with atom color code as in (a). The MMT sheets are represented as polyhedrons, with the same atom coloring used in Figure 3.1.

3.3 Polymer–Clay Nanocomposites Classification

Nanocomposites are commonly defined as materials consisting of two or more dissimilar materials with well-defined interfaces, at least one of the materials being nanostructured (having structural features ranging in size from 1 to 100 nm) in one, two, or three dimensions. The same refers to the spacing between the networks and layers formed by polymeric and inorganic components. Nanocomposites include materials in which monomer or polymer molecules are incorporated as guests into host lattices. Hosts may be both natural materials and compounds prepared by various synthetic techniques and possessing well-defined intercalation properties. The physicochemical properties of such composites are governed by the distribution of reinforcing constituents (fibrous, dispersion-toughened, or layered structures).

Depending on the strength of the interfacial tension between the polymeric matrix and the layered silicate (modified or not), the then resulting PCNs can be categorized into three types, depending on the extent of the separation of the silicate layers (see Figure 3.3):

1. *Intercalated nanocomposites*: In these PCNs, the polymer chains are inserted between the layers of the clay such that the interlayer spacing d is expanded, but the layers still bear a well-defined relationship to each other.
2. *Exfoliated nanocomposites*: In an exfoliated PCN, the layers of the clay have been completely separated, and the individual mineral sheets are randomly distributed throughout the polymeric matrix.
3. *Conventional composites (microcomposites)*: A third alternative is constituted by the dispersion of whole clay particles (*tactoids*) within the polymer matrix, but this simply represents the use of clay as conventional filler in the formation of a microcomposite.

The synthetic route of choice for making a PCN depends on whether the final material is required in the form of an intercalated or exfoliated hybrid (Pinnavaia and Beall, 2000). Many different factors can exert a control whether a particular PCN can be synthesized as an exfoliated or an intercalated system. Since the presence of either structure can lead to dramatically different characteristics in the ultimate properties of the final clay nanocomposite, it is of utmost importance to be able to understand and control these factors, which include the exchange capacity of the clay, the characteristics of the solvent medium, and the chemical nature of the interlayer cations (*onium ions*). With a thorough modification of the clay surface polarity, onium ions will allow a thermodynamically favorable penetration of the polymer into the intergallery region. The ability of onium ions to assist in clay delamination depends on the nature of its substituents (long vs. short chains, fully apolar vs. partially polar substituents, etc.). The efficiency in loading capacity of the onium ions onto the clay surface also plays a pivotal role, and it should be borne in mind that a commercial organoclay might not have the optimum loading for a given application. The other types of clay surface modifiers can be employed, depending on the nature of the clay surface (e.g., for positively charged clays such as hydrotalcite, negatively charged surfactants can be employed) and the polymer choice, such as ion-dipole interactions, silane coupling agents, and block copolymers.

A customary example of ion-dipole interactions is the intercalation of small molecules into the clay galleries. The entropically driven displacement of the small molecules then provides a route to polymer intercalation. The unfavorable interaction of clay edges with polymers can be overcome by use of silane coupling agents to modify the edges. These can be used in conjunction with the onium-ion-treated organoclay. An alternative approach to compatibilizing clays with polymers is based on the use of block or graft copolymers where one component of the copolymer is compatible with the clay and the other with the polymer matrix. This is similar in concept to the compatibilization of polymer blends. A typical block copolymer would consist of a clay-compatible hydrophilic block and a polymer-compatible hydrophobic block. The block length must be controlled and must not be too long. The high degrees of exfoliation are claimed using this approach.

3.4 Preparation Methods of Polymer–Clay Nanocomposites

The correct selection of the modified clay is essential to ensure the effective penetration of the polymer or its precursor into the interlayer spacing of the clay and result in the desired exfoliated or intercalated product. Substantially, there are three main methods for preparing PCNs:

1. *Intercalation of polymer from solution*. This approach is of interest from various viewpoints. Note, first of all, that it is effective at producing organic–inorganic multilayer composites. Owing to the unusual physical chemistry of the intercalation processes involved, it offers the possibility of improving the physical and mechanical properties of many

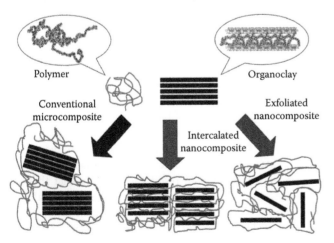

FIGURE 3.3 Schematic representation of the basic types of hybrid nanocomposites from layered silicates.

systems and fabricating electron-conducting materials, e.g., for reversible electrodes. Among the most widespread nanocomposites are polyolefin-MMT, nylon-layered silicates, and epoxy-clay hybrid nanomaterials. This process is based on the choice of a solvent system in which the polymer is soluble and the silicate layers are swellable. The clay is first swollen in a suitable solvent, such as water, chloroform, or toluene. When the polymer and silicate solution are mixed, the polymer chains intercalate and displace the solvent adsorbed within the silicate galleries. Upon solvent removal, the intercalated structure remains, resulting in the corresponding PCN (see Figure 3.4). Generally speaking, for the overall process in which the polymer chains are exchanged with the previously intercalated solvent in the clay galleries, a negative variation of the Gibbs free energy ΔG is required, i.e., $\Delta G = \Delta H - T\Delta S < 0$, where ΔH represents the enthalpic contribution ΔG and $T\Delta S$ is the entropic variation associated to the given process. In this case, the driving force for polymer intercalation into layered silicate from solution is the entropy gained by solvent molecules' release, which compensates for the decrease of entropy of the confined, intercalated chains (Vaia and Giannelis, 1997a). From the application standpoint, however, this method involves the extensive use of solvent. Therefore, exception made for water-soluble polymers (e.g., poly(ethylene oxide), PEO), this procedure is usually environmentally unfriendly and economically prohibitive.

2. *In situ intercalative polymerization method.* The most intriguing type of intracrystalline chemical reaction is the incorporation of monomer molecules into the pores of a host structure, followed by controlled internal transformations into polymer, oligomer, or hybrid-sandwich products (post-intercalative transformations). Monomers intercalated into a clay mineral migrate along its galleries, and, initiated by heat, radiation, or an appropriate agent, polymerization occurs within its layers (see Figure 3.5).

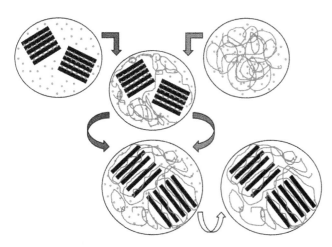

FIGURE 3.4   Schematic view of PCN formation via polymer intercalation from solution. The black bars are MMT layers, the green dots represent solvent molecules, and the red lines symbolize polymer chains.

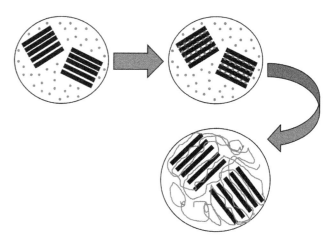

FIGURE 3.5   Schematic view of PCN formation via in situ intercalative polymerization. The black bars are MMT layers, the red dots represent monomeric species, and the red lines symbolize polymer chains.

This approach is often also called "ship-in-the-bottle" polymerization, and the monomer molecules incorporated through displacement reactions form new hydrogen bonds and other types of intercalation compounds with the host. The simplest way of intercalating polymers into inorganic structures is by producing hybrid nanocomposites via a one-step emulsion polymerization of conventional monomers (most frequently, styrene, methyl methacrylate, and acrylonitrile) in the presence of various organophilic minerals. The physical processes underlying intercalation via emulsion polymerization are as follows. The basis of the swelling characteristic of clays in aqueous systems containing monomer micelles 2–10 nm in size allows the micelles to penetrate into swollen MMT layers. At the same time, the monomer drops forming during solution polymerization are very large (10^2–10^4 nm) and are simply adsorbed or bound to the surface of MMT particles. The earliest example of PCN was actually produced *via* this method by Toyota Motors for the synthesis of MMT-nylon6 nanocomposite, which remains the most studied and well-characterized system to date.

3. *Melt intercalation method.* This technique involves annealing (statically or under shear) a mixture of the polymer and organoclay above the polymer softening point. While annealing, the polymer chains diffuse from the bulk polymer melt into the galleries between the silicate layers. A range of nanocomposites with structures from intercalated to exfoliated can be obtained, depending on the degree of the penetration of the macromolecular chains in the interlayer spaces (see Figure 3.6). So far, experimental results point to the fact that the outcome of polymer intercalation depends critically on silicate functionalization and constituent interactions. Undoubtedly, an optimal starting interlayer structure in the organoclay (e.g., a suitable number and size of compatibilizer chains per clay unit area) and the right

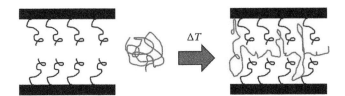

FIGURE 3.6 Schematic representation of PCN formation by melt intercalation. The black bars are MMT layers, the blue lines represent organic surface modifiers (e.g., quaternary ammonium salts), and the red lines symbolize polymer chains.

balance of polar/hydrophobic interactions between the polymer chains and the organoclay are two key issues for a successful melt intercalation process. Contrary to process (a), in which entropy is claimed to be the driving force leading to PCN formation, in melt intercalation the enthalpic component (ΔH) of the Gibbs free energy (ΔG) seems to govern the system thermodynamics (Vaia and Giannelis, 1997b). In fact, in this case the entropy loss associated to the confinement of the polymer melt within the silicate galleries is compensated by the entropy gain of the aliphatic chains of alkylammonium cations associated to the concomitant layer separation, resulting in a net entropy change $\Delta S \approx 0$. The favorable enthalpy is due to the increase in favorable interaction energy between the polymer and the organoclay, due to the formation of weak, hydrogen-like bonds, dipole–dipole, and van der Waals interactions. Overall, this method has great advantages over both processes (a) and (b). In fact, (1) it is environmentally benign as no solvent is employed, and (2) it is totally compatible with current polymer industrial processes such as extrusion or injection molding. Accordingly, the melt intercalation technique has become the standard for the preparation of PCNs.

To date, both thermoset and thermoplastic polymers have been incorporated into nanocomposites, including polyamides (i.e., nylons), polyolefins (e.g., polypropylene, PP), polystyrene (PS), epoxy resins, polyurethanes, polyimides, and poly(ethylene terephthalate, PET). However, the development of compatibilizer chemistry is the key to the expansion of this nanocomposite technology beyond the systems where success has been achieved. Also, polar functionalities such as hydroxyl groups (–OH) can be introduced into the onium salt clay surface modifier to improve the compatibility between the polymer and the mineral via the formation of intermolecular hydrogen bond networks. Similarly, chemistry alterations might be needed for the intercalation/exfoliation of different polymers in different clay minerals. For instance, the intercalation of PP into organomodified MMT is achieved only after maleic anhydride is grafted onto the polymer chain. For the preparation of nanocomposites from high temperature engineering thermoplastics, a major limitation in the use of conventional organoclay is the thermal instability of the alkylammonium species during processing. To overcome this hurdle, imidazolinium or phosphonium salts represent more

stable, valid alternative choices as their use lead to an increase in the degradation temperature of the organoclay from 200°C to 300°C to >300°C. Finally, wholly synthetic organoclays could be employed when exceptional thermal stability is required by the processing conditions.

3.5 Polymer–Clay Nanocomposite Characterization Methods

3.5.1 Wide-Angle X-Ray Diffraction and Transmission Electron Microscopy

The archetypical experimental technique employed to characterize the structure of PCNs is wide-angle x-ray diffraction (WAXD) analysis, mainly due to its ease of use and widespread availability (LeBaron et al., 1999; Biswas and Sinha Ray, 2001). By monitoring the position, shape, and intensity of the basal reflections from the distributed silicate layers, the PCN structure (intercalated or exfoliated) may be identified. For example, in an exfoliated PCN, the extensive layer separation associated with the delamination of the original clay layers in the polymer matrix results in the eventual disappearance of any coherent x-ray diffraction (XRD) from the randomly distributed silicate sheets (compare Figure 3.7a and d). On the contrary, for intercalated PCN structures, the finite layer expansion associated with polymer penetration into the clay galleries results in the appearance of a new basal reflection corresponding to the larger gallery heights (Figure 3.7b). Although WAXD undoubtedly offers a convenient and precise method to determine the interlayer spacing in organoclays (pristine and after modification with, for instance, alkylammonium salts) and in intercalated PCNs (from 1 up to 4 nm), this technique does not allow any speculation about the spatial distribution of the clay platelets or the eventual presence of structural inhomogeneities in the system. Further, some clays (organically modified or not) do not present themselves as well-resolved WAXD peaks; accordingly, peak broadening/shifting and/or intensity decrease may be difficult to detect systematically.

On the other hand, transmission electron microscopy (TEM) allows a qualitative understanding of the internal structure, spatial distribution of the various phases, and views of the defects present in a given PCN by direct observation (see Figure 3.7e through g). However, care must be taken in order to ensure a representative cross section of the sample. The coupling of WAXD and TEM nowadays constitutes the essential tool for evaluating PCN structures (Morgan and Gilman, 2003).

3.5.2 Nuclear Magnetic Resonance

Solid-state ^{1}H and ^{13}C NMR (nuclear magnetic resonance) could be used, in principle, to gain a deeper insight about the morphology, surface chemistry, and, to a limited extent, the dynamics of exfoliated PCNs, although only recently a reliable method for tracking the increase of the interlayer distance d as exfoliation proceeds has been developed (Bourbigot

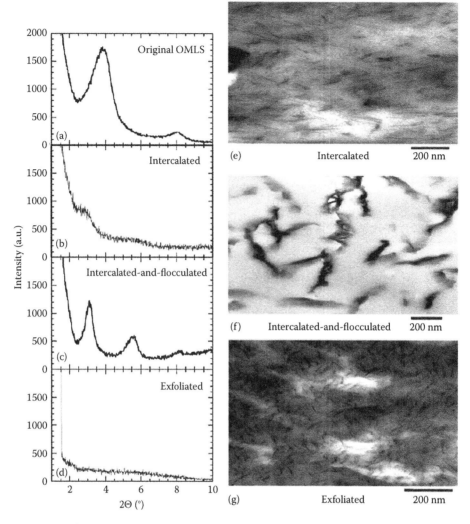

FIGURE 3.7 WAXD patterns (a–d) and TEM images (e–f) of different types of PCNs. (Adapted from Sinha Ray, S. and Okamoto, M., *Prog. Polym. Sci.*, 28, 1539, 2003. With permission.)

et al., 2003). The methodology relies on the measurements of the proton longitudinal relaxation time T_1 under two effects: the paramagnetic character of the silicate (e.g., MMT) that directly reduces T_1 of nearby protons, and spin diffusion, by which the locally enhanced relaxation propagates to more distant protons. The application of this method is actually limited to amorphous PCNs, in which there is no possibility of change in crystallinity, as this would strongly affect the relaxation time.

3.5.3 Differential Scanning Calorimetry

Many of the polymers used to produce PCNs are semi-crystalline. Accordingly, the dispersion of high surface energy clay sheets in such polymer matrices can provide heterogeneous nucleation sites for crystallization. One way to estimate crystallinity in a given PCN is to compare the melting temperature of the actual system with that of a well-known crystalline polymer sample. The fraction of crystalline material

is proportional to the measured enthalpy, after the obvious correction for the mineral mass fraction (Chen and Evans, 2006). Another use of differential scanning calorimetry (DSC) measurements is the estimation of the amount of intercalated polymers by a combination of multiple runs (Chen and Evans, 2005). When a mixture of clay and polymer is run on the DSC twice, the first run allows the polymer to intercalate and the second run only gives the endotherm for melting the excess (or free) polymer since the gallery contents are assumed to behave as they were amorphous. This endotherm is subtracted from the melting enthalpy corresponding to the initial mass of the polymer to provide the amount of amorphous intercalated polymer. Figure 3.8 shows an example of successive DCS runs for a system composed by poly(ethylene glycol) (PEG) and MMT. The first run exhibits an endotherm corresponding to the polymer melting and the superposing, although somewhat delayed, exotherm due to polymer adsorption and intercalation. On the contrary, the second run features only the melting endotherm.

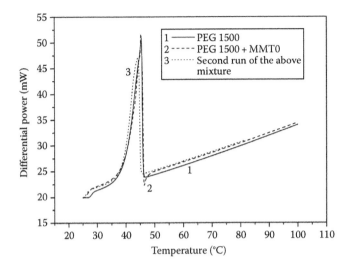

FIGURE 3.8 DSC curves for poly(ethylene glycol) (PEG) and a mixture of PEG and MMT. (From Chen, B. and Evans, J.R.G., *Philos. Mag.*, 85, 1519, 2005. With permission.)

Finally, DSC experiments can be profitably exploited to verify the nature of the thermodynamic driving forces underlying intercalation/exfoliation. As discussed above, DSC observations, for instance, confirmed that the reduction in free energy on intercalation is a compromise between a significant enthalpic change and entropic change for clays with interlayer solvent (e.g., solution intercalation), and primarily from an enthalpic change for PCN produced in the absence of solvent (e.g., melt intercalation).

3.5.4 Computer Simulation Techniques

In current years, computer-based simulation techniques play an ever-increasing role in the design and a priori prediction of new material properties, and in drawing the guidelines for experimental work and characterization (Zeng et al., 2008). In particular, in the complex world of PCNs, computational chemistry and simulation techniques have proved to be extremely useful in addressing the following, topical issues:

- Thermodynamics and kinetics of PCN formation.
- Hierarchical characteristics of the structure and dynamics of PCNs, which span from molecular to meso- up to microscale, with special emphasis on the molecular structures and dynamics at the interface level between the inorganic nanoparticles and the polymeric chain matrices.
- The influence exerted by the presence of the nanoparticles on the flow properties of the pristine polymeric matrices, a critical issue in PCN production and processing.
- The molecular origins lying at the bases of the enhanced macroscopical properties exhibited by PCNs.

Roughly speaking, the most common computational techniques employed so far in the simulation of PCNs can be classified into three main levels:

(i) Fully atomistic simulations, which span length scales of the order of a few nanometers and time scales of the orders of a few nanoseconds, and include validated methods such as molecular dynamics (MD) and Monte Carlo (MC) techniques (Allen and Tildesley, 1989);

(ii) Mesoscale level simulations, in which the length scale is extended up to hundreds of nanometers and the time scale may reach up to hundreds of microseconds. These recent computational techniques substantially include Brownian dynamics (BD) (Carmesin and Kremer, 1988), dissipative particle dynamics (DPD) (Hoogerbrugge and Koelman, 1992), and mean-field density functional theory (MFDFT) (Altevogt et al., 1999);

(iii) Microscale simulations, which basically refer to the well-known micromechanical and finite-element method (FEM) calculations.

Modeling at level (i) usually employs atoms, molecules, or their ensembles as basic constituents, and the techniques of election are MD and MC. MD generates the time evolution of a given system of interacting particles from which the equilibrium (and nonequilibrium) structures, energetic, and thermophysical properties can be estimated by means of the application of the fundamental laws of statistical mechanics. Contrary to MD, which is a deterministic method, the stochastic MC technique uses random numbers to generate a sample population of configuration for a given molecular system, from which, by resorting to the adoption of suitable probabilistic or statistical models, the equilibrium properties can be determined.

Intuitively, given the dimension involved in real PCNs, simulations at this level are mainly directed toward the study and characterization of the thermodynamics of the formation of PCNs, their detailed molecular structures, and the energetic of interactions between the different PCN components (e.g., MMT, onium salt, and polymer). The application of computer-based simulation techniques at level (ii)—or the mesoscale level—aims at filling the gap between detailed atomistic and coarse continuum level, and avoid their shortcomings. In the specific field of PCNs, mesoscale simulation recipes have been employed to study the structural evolution, the microphase structure, and the phase separation of these systems. Of two most popular mesoscale techniques, DPD and DFT, the former is a particle-based method in which the basic unit is no longer a single atom or molecule but a molecular assembly (e.g., an entire particle or a set of linked monomer in a polymer chain). Three main forces act simultaneously upon DPD particles in their motion and interactions, i.e., a conservative force, a dissipative force, and a random force. As these forces are pair wise additive, and the particle momentum is conserved, the system macroscopic behavior directly incorporates Navier–Stokes hydrodynamics.

Finally, level (iii)—or the continuum model level—is known and applied since long, and substantially obeys the fundamental laws of continuity (from mass conservation), equilibrium (from Newton's second law and momentum conservation), conservation of energy (from first law of thermodynamics), and

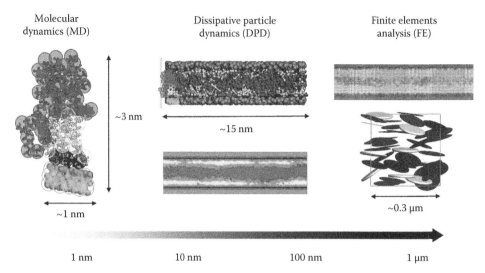

Molecular
dynamics (MD)

Dissipative particle
dynamics (DPD)

Finite elements
analysis (FE)

~3 nm

~15 nm

~1 nm

~0.3 μm

1 nm   10 nm   100 nm   1 μm

FIGURE 3.9   (See color insert following page 22-8.) Concept of multiscale molecular modeling of PCNs involving different simulation techniques at different length and time scales.

conservation of entropy (from second law of thermodynamics). Continuum methods relate to the deformation of a continuous medium to the external forces acting on the medium, and the resulting stresses and strains. Computational approaches at this level range from simple closed-form analytical expressions to micromechanics and complex structural mechanics calculations based on beam and shell theories. The most popular technique so far employed in PCN characterization is FEM, which is a general numerical method for obtaining approximate solutions in space to initial-value and boundary-value problems, including time-dependent processes. It uses preprocessed mesh generation, which enables the model to fully capture the spatial discontinuities of highly inhomogeneous materials, and also to incorporate nonlinear tensile relationships into analysis.

Despite the importance of understanding the molecular structure and nature of PCN materials, their behavior can be homogenized with respect to different aspects which can be observed at different length and time scales. Typically, the macroscopic behavior of a given PCN is usually explained by totally ignoring its discrete atomic and/or molecular structure, and assuming that the material is continuously distributed throughout its volume. In other words, the continuum material is conceived to have an average density and being subjected only to body forces such as gravity and surface forces. Clearly, as we have seen so far, this is not the case for PCNs, where complex structures coexist at different time/length scales. Therefore, the very actual concept of multiscale molecular modeling and simulation comes into play, with the ambitious goal of bridging the models and simulation techniques for PCNs (and for many other complex nanostructure systems as well) across the entire range of length and time scales involved. In this way, computer-based techniques are expected to address first the mesoscopic behavior of PCNs starting from detailed atomistic simulations and then transfer the obtained information to the continuum level. In other terms, the challenge for multiscale modeling is to move, as seamless as possible, from one scale to another so that all the parameters, properties, and topologies obtained at one (lower) scale can be transferred to the next (higher) scale. In the special case of PCNs, the ultimate target is to be able to predict, with a high degree of confidence, their hierarchical structures and behavior, and to capture all the phenomena taking place on length scales that typically span 5–6 orders of magnitudes and time scales encompassing a dozen of orders of magnitude. Figure 3.9 summarizes this concept and gives a graphical, clear view of the multiscale modeling approach to PCNs based on MD [level (i)], DPD [level (ii)], and FEM calculations [level (iii)] (Scocchi et al., 2007a,b).

3.6 Properties of Polymer–Clay Nanocomposites

Nanomaterials additives can provide many property advantages in comparison to both their conventional filler counterparts and base polymers. Properties that have been shown to undergo substantial improvements include

- Mechanical properties (e.g., strength, modulus, and dimensional stability)
- Decreased permeability to gases, water, and hydrocarbons
- Thermal stability and heat distortion temperature
- Flame retardancy and reduced smoke emissions
- Electrical conductivity
- Chemical resistance
- Biodegradability
- Optical clarity in comparison to conventionally filled polymers

In addition, it is important to recognize that nanoparticulate/fibrous loading confers significant property improvements with very low loading levels, traditional nanoparticle additives requiring much higher loadings to achieve similar performance. This, in turn, can result in substantial weight reductions (of obvious importance for various military and aerospace applications) for similar performances, greater strength for similar structural dimensions, and, for barrier applications, increased barrier performance for similar material thickness.

3.6.1 Mechanical Properties

The enhancement in mechanical properties of PCNs can be ascribed to the high rigidity and aspect ratio of the clay nanoparticles, coupled with the good affinity between the polymer and the organoclay. To say, stronger interface interactions significantly reduce the stress concentration point upon repeated distortion which easily occurs in conventional nanocomposites (e.g., those reinforced by glass fibers), and thus leads to weak fatigue strength. As an example, for polyamide–MMT PCNs tensile strength improvements have been reported to be around 40% at room temperature and 20% at 120°C, while for the Young modulus an improvement of 70% and 200% at the same temperatures was reported (Kojima et al., 1993). On the contrary, in the case of the apolar polymers such as polypropylene or polystyrene, only a slight enhancement in tensile stress was observed. This disappointing result was ascribed in part to the lack of an efficient interfacial adhesion between the apolar PP or PS chains and the polar clay surface, and indeed the use of a PP modified by maleic anhydride led to an improvement of the system tensile behavior (Hasegawa et al., 1998).

Generally speaking, however, the enhancement of the mechanical properties of PNCs strongly correlates with the ultimate structure of the material, and many investigations dealt with comparative analyses of results obtained from intercalated and exfoliated PCN structures. When considering the same (or very close) loading values, an exfoliated PCN structure often exhibits both higher elastic modulus and tensile strength, by virtue of the good dispersion and the high moduli of the clay platelets. At the same time, however, an exfoliated PCN can present a lower toughness with respect to the corresponding intercalated system. So, again taking Nylon6 as a proof-of-concept PCN, it has been verified that the impact strength of this polymeric material with the inclusion of 10% (by weight) of nanoclay decreased from approximately 7 to 3.2 kJ/m² and 4.3 kJ/m² for the exfoliated and intercalated PCN, respectively (Dasari et al., 2007). These experimental evidences have been rationalized on the basis of the formation of submicron voids associated with intercalated clay tactoids.

3.6.2 Barrier Properties

The gaseous barrier property improvement that can result from the incorporation of relatively small quantities of nanoclay materials is shown to be substantial. Many data concur to show that oxygen transmission rates in PCNs can be as low as 50% of that of the unmodified polymer. In addition, studies have shown that PCNs have excellent barrier properties against other

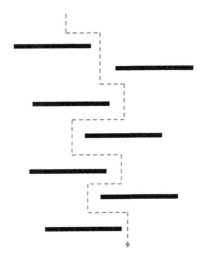

FIGURE 3.10 Sketch of the Nielson labyrinth or tortuous path model according to which a gas or solvent molecule has to follow a zigzag pathway when diffusing through a PCN.

gases (e.g., nitrogen and carbon dioxide), water, hydrocarbons, and other organic solvents such as alcohols, toluene, and chloroform. The main factors contributing to the barrier property enhancement are both the amount of clay incorporated into the polymer matrix and the aspect ratio of the filler particle. In particular, the aspect ratio is shown to have a major effect, with high ratios (and hence high tendencies toward filler incorporation at the nanolevel) quite dramatically enhancing barrier properties.

In general, best barrier effects can be achieved in polymer nanocomposites with fully exfoliated clay minerals, and such evidences can be justified considering the Nielson labyrinth tortuous path model (see Figure 3.10), according to which, once a film of PCN is formed, the sheet-like clay layers orient in parallel with the film surface (Nielsen, 1967). As a result, the diffusing species have to travel a longer way around the impermeable clay platelets than in the corresponding pristine polymer matrix when they traverse an equivalent film thickness.

A further note of interest here is that the improvement of the barrier properties in PCN materials does not involve the chemistry of the systems, as it is practically independent of the nature of the gas/liquid diffusing molecules.

3.6.3 Thermal Stability and Heat Distortion Temperature

Upon heating, polymer molecules start to degrade and, finally decompose, as temperature is increased above a certain critical value, specific for each polymeric species. The thermal stability of a polymer is usually determined via thermogravimetric experiments, in which a weighted sample of polymer is gradually heated and the weight loss upon heating, due to the formation of volatile products (e.g., CO_2, H_2O, NH_3), is recorded. The higher thermal stability experimentally verified for PCNs can be related, in analogy with the barrier properties, by the presence of the dispersed clay platelets, which create a hindered path for the diffusion of the volatile species and assist the formation of char after thermal decomposition. Another important thermal behavior is the heat resistance upon external loading, which is quantified by the so-called heat distortion temperature (HDT). For example, HTD was found to increase from 65°C for pure nylon to 152°C for the corresponding PCN material (Kojima et al., 1993).

3.6.4 Flame Retardancy

The ability of nanoclay incorporation to reduce the flammability of polymeric materials is a battle-horse application of these systems. Indeed, flammability behavior can be restricted in polymers such as PP with as little as 2% nanoclay loading, the effect resulting substantially from very low heat release rates obtained upon the incorporation of nanomaterials. Although conventional microparticle filler incorporation together with the use of flame-retardant agents would ultimately minimize flammability behavior, this is usually accompanied by reduction in various other important properties. With the PCN approach, this is usually achieved while maintaining or even enhancing other properties and characteristics.

The flame retardancy effect is evaluated by quantifying the reduction in the peak of the heat release rate (HHR). As examples, by the addition of 10% organoclay the average HHR of PS decreases by 21% with respect to the pristine polymer (Zheng and Wilkie, 2003) or, more impressively, the HHR of nylon6 made fabric was reduced by 40% with addition of only 5% of organoclay (Bourbigot et al., 2002). The molecular explanation for this characteristic property of PCN finds its roots in the carbonaceous char layers that form when the organic material burns and the structure of the clay minerals. The multilayered clay structures act as excellent insulators and mass transport barriers. Char formation and clay structure thus concur to impede the escape of the decomposed volatiles for the interior of the remaining polymeric matrix. A word of caution, however, must be spent on the fact that the presence of organic surface modifiers has also been shown to be able to catalyze thermal degradation and, hence, act somewhat against the flame retardancy. Therefore, the ultimate flame retardancy property of a given PCN results as a balance between these two counteracting effects.

3.6.5 Electrical Conductivity

Intuitively, being ionic substances clay minerals exhibit peculiar electrical properties. Indeed, although overall a layered clay can be considered as an insulator, the hydrated cationic species present in the interlayer spaces, being quite mobile, can guarantee a notable ionic conductivity of the system. Furthermore, the intercalation of neutral species could affect the hydration shells of these interlayer ions, resulting in a significantly modified ionic mobility and, hence, in altered electrical conductivity and other electrical parameters. As an example, PEO-based PCNs show a remarkably increased ionic conductivity with respect

to the pristine clay materials, and this conductivity increases with increasing temperature. The maximum conductivity in the direction parallel to the clay layer is of the order of 10^{-5}–10^{-4} S/cm (Ruiz-Hitzky et al., 1995). Other parameters that play a pivotal role in these properties are the eventual presence of crystalline phase within the polymeric matrices. In fact, while in conventional polymer/salt systems the ionic conductivity is strongly influenced by the presence of polymeric crystallites, ion-pair formation, and the mobility of the counterions, it is not so in PCNs. Indeed, here the counterions (i.e., the negatively charged clay layers) are substantially immobile and, hence, ion-pairs and anion-complexed cation interactions cannot take place.

3.6.6 Chemical Resistance

Water- or chemicals-laden atmosphere has long been regarded as one of the most damaging environments which polymeric materials can encounter. Thus, the ability to minimize the extent to which water/chemical is adsorbed can be a major advantage. Indeed, nanoclay incorporation can reduce the extent of water/chemical adsorption in a polymer matrix, and this effect again is bound to the clay particle aspect ratio: increasing aspect ratio is found to diminish substantially the amount of water/chemical adsorbed and, hence, water/chemical transmission to the underlying substrate. Thus, application in which contact with water or moist environment is likely could clearly benefit from the use of materials incorporating nanoclay particles.

3.6.7 Biodegradability

In an era in which environmental protection is becoming a must, and recycling and/or biodegradability are two keywords, PCN materials constituted by organoclays and biodegradable polymeric matrices have proven to be outstanding systems by virtue of their improved biodegradability. The first biodegradable PCNs reported were based on organically modified MMT as mineral and poly(ε-caprolactone) (PCL) and poly(lactic acid) (PLA) as matrices. The remarkably improved biodegradability of these PCNs was substantially attributed to the catalytic role of the organoclay in the biodegradation mechanism.

3.6.8 Optical Properties

The presence of filler incorporation at nano-levels has also been shown to have significant effect on the transparency and haze of films. In comparison to conventionally filled polymers, nanoclay incorporation has been shown to significantly enhance transparency and reduce haze. With polymers characterized by a significant amount of crystalline fractions (e.g., polyamides and PET), this effect has been attributed to the modification in the crystallization behavior brought about by the nanoclay particles, the spherulitic domain dimensions being considerably smaller. Similarly, nano-modified polymers have been shown, when employed to coat polymeric transparent materials, to enhance both toughness and hardness of these materials without interfering with light transmission characteristics. An ability to resist high velocity impact combined with substantially improved abrasion resistance has also been verified.

3.7 Applications of Polymer–Clay Nanocomposites

The fact that polymer–clay nanocomposites show concurrent improved performances in various material properties at very low filler content, together with the ease of preparation through simple processes, opens up a new dimension for plastic and composite materials.

Generally speaking, the enhancement of the material properties highlighted above has paved the way for a substantial exploitation of PCN-based systems in industrial applications. For instance, the mechanical property improvements have resulted in major interest in numerous automotive and general/industrial applications. These include potential for utilization and as mirror housing on various vehicle types, door handles, engine covers, and intake manifolds and timing belt covers. More general applications currently being considered include usage as impellers and blades for vacuum cleaners, power tool housing, mower hoods, and covers for portable electronic equipment such as mobile phones, pagers, and so on.

The excellent barrier characteristics exhibited by PCNs have resulted in considerable employment of nanoclay composites for food packaging applications, both flexible and rigid. Specific examples include packaging for processed meats, cheese, confectionery, cereals, and boil-in-the-bag foods, and also extrusion-coating applications in association with paperboard for fruit juice and dairy products, together with co-extrusion process for the manufacture of beer and carbonated drinks bottles. The use of nanocomposite formulations would be expected to enhance considerably the shelf life of many types of foods.

The ability of nanoclay incorporation to reduce solvent transmission through polymers highlights the possibility of the application of PCN systems as both fuel tank and fuel line components for cars. Of further interest for this type of application, the reduced fuel transmission characteristics are accompanied by significant material cost reduction. And, last but not least, the optical clarity coupled with the flexibility and resistance properties have and currently still are opening new avenues for the employment of PCNs in the micro/electronics industry for special applications such as light-emitting devices (LEDs).

Besides these well documented applications, arising by the direct exploitation of the peculiar PCN properties, a list of promising applications of PNCs in important and to-the-edge fields, chosen among the plethora of many other available, will be briefly outlined and discussed.

3.7.1 PCNs as Rheology Modifiers

Rheological modifiers control the flow properties of liquid systems such as paints, inks, emulsions, or pigment suspensions by increasing the medium viscosity, or impart thixotropic flow behavior to liquid system. Several dispersion procedures are used for conventional organoclays; however, they all can be described by two general methods: (1) pre-gel addition and (2) dry addition. The pre-gel method is based on the preparation of a 10%–15% organoclay dispersion in a suitable solvent using a high-speed disperser and a polar activator. The direct addition of organoclays involves adding the organoclay as a dry powder prior to, or during, the grind phase in the manufacturing process. The polar activator is then added and the dispersion continued. The function of polar activators is to disrupt the weak van der Waals forces which tend to hold the clay platelets together. Once these platelets are separated, it allows the organic functional groups to free themselves from the close association with the clay surface, and to solvate in the organic liquid for which they clearly have high affinity.

The rheological properties of a paint system are enhanced by a small addition of organoclays either by pre-gel or dry addition. The gel formation prevents pigment settling and sagging on vertical surfaces, and thus ensures that the proper thickness of the coating is applied. They also guarantee good leveling for the removal of brush marks and storage stability even at high temperatures.

PCNs are also employed in the formulation of printing inks. Here, the role played by these nanocomposites is manifold, ranging from adjusting the consistency of the inks to the desired values, to avoiding pigment sedimentation, providing good color distribution and desired film thickness, reducing the level of mist, and controlling of tack, water pickup, and dot gain control.

Thickening lubricating oils additivated by PCNs can produce especially high temperature-resistant lubricating greases, with enhanced working stability and water resistance. Such greases are typically used for lubrication in foundries, mills, and on high-speed conveyors, as well as in agriculture, automotive, and mining applications.

Finally, the performance of cosmetics is enhanced by the use of PCNs, where they allow good color retention and coverage for nail lacquers, lipsticks, and eye shadows. They have been tested to be nonirritant for both skin and eye contact.

3.7.2 Drug Delivery

The continuous development of new controlled drug delivery systems is driven by the need to maximize therapeutic activity while minimizing negative side effects. One class of drug delivery vehicle that has received more attention in recent years is layered materials that can accommodate polar organic compounds between their layers and form a variety of intercalated compounds. Because the release of drugs in drug-intercalated layered materials is potentially controllable, these new materials have a great potential as a delivery host in the pharmaceutical field. Calcium MMT, for instance, has been used extensively in the treatment of pain, open wounds, colitis, diarrhea, hemorrhoids, stomach ulcers, intestinal problems, acne, anemia, and a variety of other health issues. Not only does MMT cure minor problems such as diarrhea and constipation through local application, but it has also been shown to act on all organs as well.

3.7.3 Wastewater Treatment

The use of PCNs in wastewater treatment has become common in industry today. PCNs exhibit a synergistic effect with many commonly utilized water treatment unit processes, including granular-activated charcoal, reverse osmosis, and air strippers. Although granular-activated carbon is particularly effective at removing a large range of organic molecules from water, it is very poor for removing large molecules such as humic acid and wastewaters containing emulsified oil and grease. Polymer–clay nanocomposites have proven to be the technology of choice for treating oily wastewaters. Humic acid is one of the common contaminants in potable water and is difficult to remove with conventional flocculation techniques commonly used for drinking water treatment, and activated carbon is very ineffective due to its weak interaction with humic acid. If humic acid is not removed from drinking water, subsequent chlorination produces unacceptable levels of trihalomethanes which are known carcinogens.

3.8 Conclusions

Significant progress in the development of polymer–clay nanocomposites has been made over the past 15 years. During these years of intense labor in this field, the advantages and limitations of this (nano)technology have become clear. The data shown in Table 3.1, reporting the expected market size of PCN-based materials in 2009, clearly justify such scientific and technological efforts. However, we have a long way to go before we understand the mechanisms of the enhancement of the major engineering properties of polymers and can tailor the nanostructure of these composites to achieve particular engineering properties. The use of organoclays as rheological modifier is one of the oldest methodologies in industries, and is currently extensively used worldwide. The development of polar activator free PCNs in last 10 years made tremendous impact in the field of paint, ink, and greases. Although the field of nanoclays as drug vehicle for controlled release is one of the born age areas in medicinal application, PCNs have revealed a great potential as compared to polymer and carbon nanotubes for these sophisticated yet fundamental applications. The use of PCNs is still enjoying an expansion in water treatment applications. PCNs operate via partitioning phenomena and have a synergistic effect with activated carbon and other unit processes such as reverse osmosis. They have proven to be superior to any other water treatment

technology in applications where the water to be treated contains substantial amounts of oil, grease, or humic acid. The commercial application of organoclays to trihalomethane control in drinking water has not yet occurred. However, with increasing concerns about the carcinogenic effects of these halogenated substances, the commercialization of this technology could be around the corner.

To date, one of the few disadvantages associated with nanoparticle incorporation has concerned toughness and impact performance. Cleary, this is an issue which would require consideration for application where impact loading events are likely to occur. In addition, further research is necessary to develop a better understanding of formulation/structure/property relationships, better routes to platelet exfoliation and dispersion, and so on. To quote Richard P. Feynman, "there is plenty of space at the bottom" also for further improving the structures and, hence, the performances of these extremely fascinating materials.

References

Ajayan, P. M., L. S. Schadler, and P. V. Braun. 2003. *Nanocomposite Science and Technology*. Weinheim, Germany: Wiley VCH.

Allen, M. P. and D. J. Tildesley. 1989. *Computer Simulation of Liquids*. Oxford, NY: Clarendon Press.

Altevogt, P., O. A. Ever, J. G. E. M. Fraaije, N. M. Maurits, and B. A. C. van Vlimmeren. 1999. The MesoDyn Project: Software for mesoscale chemical engineering. *Journal of Molecular Structure* 463: 139–143.

Balazs, A. C., T. Emrick, and T. P. Russell. 2006. Nanoparticle polymer composites: Where two small worlds meet. *Science* 314: 1107–1110.

Biswas, M. and S. Sinha Ray. 2001. Recent progress in synthesis and evaluation of polymer-montmorillonite nanocomposites. *Advances in Polymer Science* 155: 167–221.

Bourbigot, S., E. Devaux, and X. Flambard. 2002. Flammability of polyamide-6/clay hybrid nanocomposite textiles. *Polymer Degradation and Stability* 75: 397–402.

Bourbigot, S., D. L. VanderHart, J. W. Gilman, W. H. Awad, R. D. Davis, A. B. Morgan, and C. A. Wilkie, 2003. Investigation of nanodispersion in polystyrene-montmorillonite nanocomposites by solid-state NMR. *Journal of Polymer Science. Part B: Polymer Physics* 41: 3188–3213.

Carmesin, I. and K. Kremer. 1988. The bond fluctuation method: A new effective algorithm for the dynamics of polymers in all spatial dimensions. *Macromolecules* 21: 2819–2823.

Chen, B. and J. R. G. Evans. 2005. On the driving force for polymer intercalation in smectite clays. *Philosophical Magazine* 85: 1519–1538.

Chen, B. and J. R. G. Evans. 2006. Poly(ε-caprolactone)-clay nanocomposites: Structure and mechanical properties. *Macromolecules* 39: 747–754.

Chen, B., J. R. G. Evans, H. C. Greenwell, P. Boulet, P. V. Coveney, A. A. Bowden, and A. Whiting. 2008. A critical appraisal of polymer–clay nanocomposites. *Chemical Society Reviews* 37: 568–594.

TABLE 3.1 PCNs' Estimated Market Size by 2009

Technology/Application	Estimated Market Size by 2009[a]
Polymer–clay nanocomposites	Over 1 billion pounds
Packaging	367 million pounds
Automotive	345 million pounds
Building and construction	151 million pounds
Coatings	63 million pounds
Industrial	48 million pounds
Others	67 million pounds

[a] Argonne National Laboratory, Argonne, IL.

Dasari, A., Z.-Zhen Yu, and M. Yiu-Wing. 2007. Transcrystalline regions in the vicinity of nanofillers in polyamide-6. *Macromolecules* 40: 123–130.

Grim, R. E. 1968. *Clay Mineralogy*, 2nd edn. New York: McGraw-Hill Book Company.

Hasegawa, N., M. Kawasumi, M. Kato, A. Usuki, and A. Okada. 1998. Preparation and mechanical properties of polypropylene-clay hybrids using a maleic anhydride modified polypropylene oligomer. *Journal of Applied Polymer Science* 67: 87–92.

Hoogerbrugge, P. J. and J. M. V. A. Koelman. 1992. Simulating microscopic hydrodynamic phenomena with dissipative particle dynamics. *Europhysics Letters* 19: 155–160.

Kojima, Y., A. Usuki, M. Kawasumi, A. Okadam, Y. Fukushima, T. Kurauchi, and O. Kamigaito. 1993. Mechanical properties of nylon 6-clay hybrid. *Journal of Materials Research* 8: 1185–1189.

LeBaron, P. C., Z. Wang, and T. J. Pinnavaia. 1999. Polymer-layered silicate nanocomposites: An overview. *Applied Clay Science* 15: 11–29.

Morgan, A. B. and J. W. Gilman. 2003. Characterization of poly-layered silicates (clay) nanocomposites by transmission electron microscopy and x-ray diffraction: A comparative study. *Journal of Applied Polymer Science* 87: 1329–1338.

Nielsen, L. 1967. Models for the permeability of filled polymer systems. *Journal of Macromolecular Science Part A: Pure and Applied Chemistry* 1: 929–942.

Okada, A. and A. Usuki. 2006. Twenty years of polymer–clay nanocomposites. *Macromolecular and Materials Engineering* 291: 1449–1476.

Pinnavaia, T. J. 1983. Intercalated clay catalysts. *Science* 22: 365–371.

Pinnavaia, T. J. and G. W. Beall, eds. 2000. *Polymer Clay Nanocomposites*, Chichester, U.K.: John Wiley & Sons Ltd.

Ruiz-Hitzky, E., P. Aranda, and B. Casal. 1995. Nanocomposite materials with controlled ion-mobility. *Advanced Materials* 7: 180–184.

Schoonheydt, R. A., T. J. Pinnavaia, G. Lagaly, and N. Gangas. 1999. Pillared clays and pillared layered solids. *Pure and Applied Chemistry* 71: 2367–2371.

Scocchi, G., P. Posocco, M. Fermeglia, and S. Pricl. 2007a. Polymer clay nanocomposites: A multiscale molecular modeling approach. *Journal of Physical Chemistry B* 111: 2143–2151.

Scocchi, G., P. Posocco, A. Danani, S. Pricl, and M. Fermeglia. 2007b. To the nanoscale, and beyond! Multiscale molecular modeling of polymer–clay nanocomposites. *Fluid Phase Equilibria* 261: 366–374.

Sinha Ray, S. and M. Okamoto. 2003. Polymer/layered silicate nanocomposites: A review from preparation to processing. *Progress in Polymer Science* 28: 1539–1641.

Theng, B. K. G. 1979. *Formation and Properties of Clay-Polymer Complexes*. Amsterdam, the Netherlands: Elsevier Scientific Publishing Company.

Usuki, A., M. Kojima, M. Kawasumi, A. Okada, Y. Fukushima, T. Kurauchi, and O. Kamigaito. 1993. Synthesis of Nylon-6 clay hybrid. *Journal of Material Research* 8: 1179–1184.

Utraki, L. A. 2004. *Clay-Containing Polymeric Nanocomposites*. London, U.K.: Rapra Tech. Ltd.

Vaia, R. A. and E. P. Giannelis. 1997a. Lattice of polymer melt intercalation in organically-modified layered silicates. *Macromolecules* 30: 7990–7999.

Vaia, R. A. and E. P. Giannelis. 1997b. Polymer melt intercalation in organically-modified layered silicates: Model predictions and experiment. *Macromolecules* 30: 8000–8009.

Work, W. J., K. Horie, M. Hess, and R. F. T. Stepto. 2004. Definition of terms related to polymer blends, composites, and multiphase polymeric materials. *Pure and Applied Chemistry* 62: 1985–2007.

Zeng, Q. H., A. B. Yu, G. Q. Lu, and D. R. Paul. 2005. Clay-based polymer nanocomposites: Research and commercial development. *Journal of Nanoscience and Nanotechnology* 5: 1574–1592.

Zeng, Q. H., A. B. Yu, and G. Q. Lu. 2008. Multiscale modeling and simulation of polymer nanocomposites. *Progress in Polymer Science* 33:191–269.

Zheng, X. and C. A. Wilkie. 2003. Flame retardancy of polystyrene nanocomposites based on an oligomeric organically-modified clay containing phosphate. *Polymer Degradation and Stability* 79: 551–557.

4

Biofunctionalized TiO$_2$-Based Nanocomposites

Tijana Rajh
Argonne National Laboratory

Nada M. Dimitrijevic
Argonne National Laboratory

Adam Elhofy
Northwestern University

Elena Rozhkova
Argonne National Laboratory

4.1 Introduction

Biology has evolved elegant pathways, molecular structures, and complex machinery at the nanoscale to power sophisticated cell functioning. However, control of molecular processes requires extensive knowledge about the physical and chemical properties of molecules and assemblies that constitute cell machinery. Control and manipulation of biomolecules and supramolecular entities within the living cells are major challenges in understanding the complex functioning of cells. Nanotechnology, on the other hand, has the ability to shape matter on the molecular level and coax the atoms into new combinations resulting in hybrid structures with properties that expend beyond individual entities.

The impact of nanoscience and nanotechnology on cell functioning is critically dependent on the creation of new classes of functionally and physically integrated hybrid materials that incorporate nanoparticles and biologically active molecules (Bard, 1994). These hybrid bioinorganic composites have the ability to integrate with both inorganic materials, via covalent bonding to inorganic supports, and biological entities via multivalent lock-and-key interaction of biomolecules, offering opportunities that impact diverse applications ranging from quantum computation, energy transduction, site-selective catalysis, as well as advanced medical therapies.

Hybrid materials that combine collective properties of crystalline materials and localized properties of molecule-like units are of special interest because they present basic functional units capable of carrying out site-selective redox chemistry. In order to develop new tools for biomedicine and biotechnology that carry out redox chemistry, we focus on the special classes of functional nanomaterials that mimic the exquisite control over energy and electron transfer that occurs in natural energy-transducing processes (Archer, 2004). Thus, in this paper we describe new approaches for creating functionally integrated biomolecule-based hybrid conjugates that combine the physical robustness and chemical reactivities of nanoscale materials with the molecular recognition and selectivity of biology. These systems offer new opportunities to control the electronic properties of functional components, as well as their positioning with biological specificity.

4.2 Nanoparticles as Redox Active Centers

Semiconductor particles that are in the nanometer size regime have attracted significant attention due to their atom-like size-dependent properties (Rajh et al., 1987; Henglein, 1989; Brus, 1991; Bawendi, 1995; Alivisatos, 1996; Micic and Nozik, 1996; Kamat and Meisel, 1997). While most studies have focused on the changes of the electronic properties due to the physical confinement of electrons and holes in potential wells defined by crystallite boundaries, little attention has been paid to the effect of surface and surface reconstruction on the electronic properties of nanoparticles. Recently, new theoretical approaches have shown that pronounced effects on the nanoscale arise from the coupling of surface and core states (Zhou et al., 2003).

The properties of surface atoms that constitute a large fraction of the system are affected by both the surrounding environment and bulk-like behavior of the nanoparticle interior. This can be viewed as a curse or as an opportunity. The coordination sphere of the surface atoms is incomplete. Thus, these sites may trap charges and radical species, thereby reducing their potential for redox active chemistry. On the other hand, the reduced coordination of the surface atoms provides high affinity for coupling with molecules from the surrounding environment and gives the opportunity for creating structures unachievable in other length-scale regimes (Jun et al., 2003).

It is this property of the nanoscale regime that conveys the promise for creating novel electronically coupled hybrid systems that transcend the bio–inorganic interface. Because of the electronic coupling, these systems are expected to produce stable charge separation between two entities as the initial step toward redox active chemistry. When the surface atoms of nanoparticles are coupled to organic ligands, their electronic properties become altered by the properties of organo-metallic complexes formed on the nanoparticle surface. Consequently, the collective orbitals of nanoparticles start mixing with the orbitals of organic ligands covalently linked to the surface atoms of nanoparticles. This creates the opportunity to merge strong nonselective reactivity of semiconductor nanoparticles with molecular recognition and site selectivity of biomolecules.

Metal-oxide semiconductor colloids are of considerable interest because of resistance against photocorrosion, stability, and their photocatalytic properties. Absorption of light having energy greater than the band gap (3.2 eV for anatase TiO_2) results in the promotion of electrons from the valence band to the conduction band of the metal-oxide particle, leaving positively charged holes in the valence band (Figure 4.1). These photogenerated electrons and holes ($E_{cb} \sim -0.5\,V$, $E_{vb} \sim +2.7\,V$ vs. NHE at pH 8, respectively, for TiO_2) can recombine and generate heat, or they can migrate to the surface where they can react with redox species in solution. Surface derivatization strategies have been

developed that enhance the reactive lifetimes, charge separation efficiency, and reaction selectivity of the photogenerated radical pairs in metal-oxide nanoparticles (Rajh et al., 1998, 1999a, 2003a,b; Hara and Mallouk, 2000).

4.2.1 Surface Reconstruction and Adsorption-Induced Restructuring of Metal-Oxide Nanoparticles

The contribution of the surface reconstruction to chemical reactivity of different metal-oxide particles (TiO_2, Fe_2O_3, ZrO_2, and ZnO) having particle sizes ranging from 3 to 40 nm was investigated using x-ray absorption spectroscopies (XANES, EXAFS) (Chen et al., 1997, 2002; Rajh et al., 2001, 2002). When the metal-oxide particles are in the nanocrystalline regime, a large fraction of the atoms that constitute the nanoparticle are located at the surface with significantly altered electrochemical properties. Because of the truncation of the crystal units at the surface and their weaker covalent bonding with solvent species compared to the bonding within the lattice, the energy level of the surface species is found in the mid-gap region, thereby decreasing their reducing/oxidizing abilities (Morison, 1980). In addition, we have found that as the size of nanocrystalline TiO_2 becomes smaller than 20 nm, the surface Ti atoms adjust their coordination environment (Chen et al., 2002; Rajh et al., 2002). Using x-ray absorption spectroscopy, we also found that the change in coordination environment is followed by a compression of the Ti–O bond and a slight extension of Ti–Ti bond to accommodate for the curvature of the nanoparticles (Chen et al., 1997). These undercoordinated defect sites are the source of novel enhanced and selective reactivity of nanoparticles toward bidentate ligand binding. The surface reconstruction distorts the crystalline environment of surface atoms resulting in undercoordinated surface Ti atoms in a square-pyramidal structure. The undercoordinated surface sites (SS) were found to exhibit high affinity for a variety of oxygen-containing ligands and to provide the opportunity for seamless chemical attachments to the nanoparticle surface.

Oxygen-rich enediol ligands were found to restore the octahedral crystalline environment of surface Ti atoms. This, in turn, was found to couple the molecular orbitals of enediol ligands and *d*-orbitals of surface and bulk Ti that constitute the conduction band of TiO_2 nanoparticles altering the effective band gap of nanoparticles. This is manifested by the change of the color of semiconductor nanoparticles (Figure 4.2); while bare TiO_2 nanoparticles are transparent, surface-modified nanoparticles become yellow/red/black, depending on the aromatic character of the ligand used for surface modification (it should be noted that surface modifiers themselves do not have color). The change of a band gap is the consequence of the new hybrid molecular orbitals generated by coupling the orbitals of chelating ligands and the continuum states of metal oxides, which was confirmed by computational studies (Redfern et al., 2003). Hybrid properties arise from the ligand-to-metal charge transfer (CT) interaction between the ligand and surface metal atoms that further

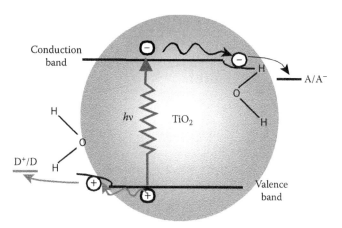

FIGURE 4.1 Salient features of the photoinduced redox reactions at the surface of nanoparticulate TiO_2.

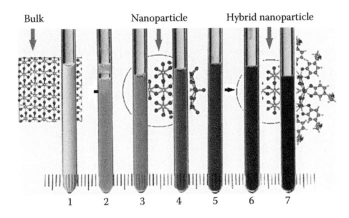

FIGURE 4.2 (See color insert following page 22-8.) Surface-modified 45 Å TiO₂ nanoparticles with different bidentate ligands: (1) bare TiO₂, (2) salicylic acid, (3) dihydroxycyclobutenedione, (4) vitamin C, (5) alizarin, (6) dopamine, and (7) *tert*-butyl catechol. (From Rajh, T. et al., *J. Phys. Chem. B*, 106, 10543, 2002. With permission.)

couple with the semiconductor electronic properties of the core of the nanoparticle. As a result, CT nanocrystallites have the onset of absorption shifted to the red, as compared to that for unmodified nanocrystallites. Depending on the ligand used for surface modification, the optical properties of metal-oxide nanoparticles are tunable throughout the entire visible and near-infrared region (Figure 4.2).

4.2.2 Electron Paramagnetic Resonance (EPR) and Charge Separation

EPR is a spectroscopic technique capable of detecting species containing one or more unpaired electrons, and therefore is the technique of choice for detecting unpaired electrons created during photoinduced charge separation. In our work we use EPR to establish the identity of photogenerated radicals participating in light-induced charge separation as well as the dynamics of the spin interplay between photogenerated charges. It is a powerful, versatile, nondestructive, and nonintrusive analytical method. Unlike many other techniques, EPR yields both structural electronic and dynamical information, even from ongoing chemical or physical processes without influencing the process itself. EPR technique is utilizing the physical interactions of the electron spin (S) and nuclear spin (I) in the external magnetic field (B). These interactions can be described by general Spin–Hamiltonian of the following form:

$$H_S = \beta \vec{S}_e \overline{\overline{g}}_e \vec{B}_0 + \beta \vec{S}_h \overline{\overline{g}}_h \vec{B}_0 + J \vec{S}_e \vec{S}_h + \vec{S}_e \overline{\overline{D}} \vec{S}_h + A \vec{S}_e \vec{I}_n$$

Here, the Zeeman interactions of the electron and holes are described by the first two terms, *J*-electron exchange interaction is described in a third term, zero-field D describes electron spin-spin interactions in a forth term, and *A* represent the nuclear hyperfine tensors that describe electron spin (S) with nuclear spin (I).

4.2.2.1 Zeeman Effects

The energy differences that are studied in EPR spectroscopy are predominately due to the interaction of unpaired electrons in the sample with a magnetic field produced by a magnet in the laboratory. Because the electron has a magnetic moment, it acts like a compass or a bar magnet when placed in a magnetic field, B_0. It will have a state of lowest energy when the moment of the electron, μ, is aligned with the magnetic field and a state of highest energy when μ is aligned against the magnetic field. The two states are labeled by the projection of the electron spin, Ms, on the direction of the magnetic field. Because the electron is a spin 1/2 particle, the parallel state is designated as Ms = −½ and the antiparallel state is Ms = +½.

Because we can change the energy differences between the two spin states by varying the magnetic field strength, we have an alternative means to obtain spectra. We could apply a constant magnetic field and scan the frequency of the electromagnetic radiation as in conventional spectroscopy. Alternatively, we could keep the electromagnetic radiation frequency constant and scan the magnetic field. A peak in the absorption will occur when the magnetic field "tunes" the two spin states so that their energy difference matches the energy of the radiation:

$$h\nu = g\mu_\beta B_0$$

where
- ν is frequency of microwave radiation
- μ_β is Bohr magnetron
- *g*-factor is constant of proportionality

This field is called the "field for resonance." Owing to the limitations of microwave electronics, the latter method offers superior performance. The field for resonance is not a unique "fingerprint" for the identification of a compound because spectra can be acquired at several different frequencies. The g-factor, being independent of the microwave frequency, is much better for that purpose. Notice that the high values of g occur at low magnetic fields and vice versa.

4.2.2.2 Hyperfine Interactions

Measurement of g-factors can give us some useful information; however, it does not tell us much about the molecular structure of our sample. Fortunately, the unpaired electron, which gives us the EPR spectrum, is very sensitive to its local surroundings. The nuclei of the atoms in a molecule or complex often have a magnetic moment, which produces a local magnetic field at the electron. The interaction between the electron and the nuclei is called the *hyperfine interaction*. It gives us a wealth of information about our sample such as the identity and number of atoms which make up a molecule or complex as well as their distances from the unpaired electron.

We used EPR to establish the identity of photogenerated radicals participating in light-induced charge separation as well as dynamics of the spin interplay between photogenerated charges recombination. Using continuous wave (CW) EPR at helium temperatures, we were able to establish the identities of the radical

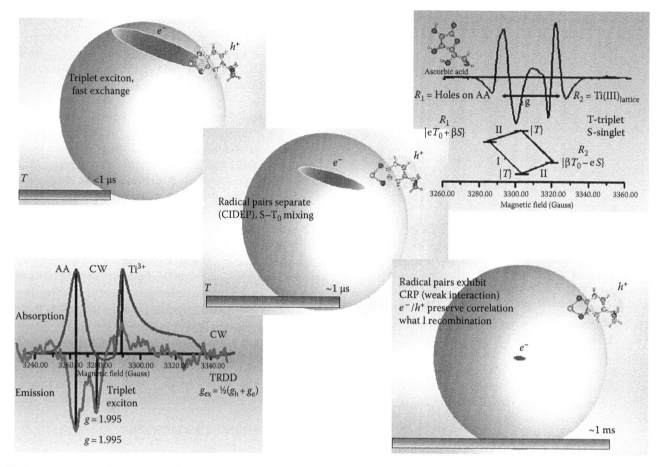

FIGURE 4.3 Spin polarization mechanisms in early stages of photoinduced charge separation in surface-modified TiO_2 nanoparticles showing subsets of electron-hole radical pairs indicative of the range of dynamic properties obtained using EPR. (From Rajh, T. et al., *Chem. Phys. Lett.*, 344, 31, 2001; Thurnauer, M.C. et al., *The Spectrum*, 17, 10, 2004.)

species formed by excitation ($\lambda > 500\,nm$) of the charge transfer complex of surface-modified TiO_2 nanoparticles (Figure 4.3, bottom left, dark curve). Two light-induced reversible signals attributed to the oxidized donor and reduced acceptor in the continuous light CW EPR spectrum were obtained. Photogenerated holes were localized on the enediol ligands used for surface modification ($g = 2.004$) while photogenerated electrons were confined to conduction band and shallow traps of TiO_2 nanoparticles ($g = 1.988$). These photoinduced charges interact, displaying spectral features indicative of the range of dynamic properties (Thurnauer et al. 2004). At earlier times ($<1\,\mu s$ at 4.2 K), a strong exciton-like radical pair in conjunction with a polarized radical pair was observed. Due to large exchange interaction *J*, the environments of holes on organic ligand and electrons in conduction band of TiO_2 are averaged, giving an average signal $g_{ave} = 1.995$ (Figure 4.3, bottom left, gray curve). Excess emission observed in these experiments was consistent with the triplet character of the exciton precursor. However, a subset of surviving electron-hole pairs ($\sim 1\,ms$ at 4.2 K) was found to exhibit very different spin interplay caused by small exchange interaction ($\sim 10^{-4}\,meV$), spin feature of correlated radical pair (CRP). As in natural photosynthesis, these hybrid nanocrystallites through weak interaction mix singlet and triplet states (Figure 4.3, top right). We speculate that this triplet character of the

surviving correlated radical pair may be an indicator and a cause of the long-lived charge separation.

Theoretical modeling of the binding of enediol ligands to the surface Ti atoms has shown that the bidentate dissociative adsorption of enediol ligands to the corner undercoordinated sites is the most favorable binding and leads to the excitation energies that match experimental observations (Redfern et al., 2003; Vega-Arroyo et al., 2007). The investigation of the conformation of surface-bound ligands using FTIR spectroscopy are in agreement with computational studies and confirm that the dominant mechanism of adsorption is bidentate. In addition, we have found that the binding of all investigated enediol ligands follows Langmuir-type adsorption (Rajh et al., 2002). The important aspect of the bidentate binding of ligands for biomedical applications is that the rate of adsorption is a couple of orders of magnitude larger than the rate of desorption. The direct consequence is that formation of the complexes is a favorable event, while pulling the complex apart is a much harder, less favorable process. Therefore, enediol ligands are ideal candidates for the linking of biomolecules to metal-oxide nanoparticles because binding is additionally stabilized due to the bidentate binding to undercoordinated sites. Theoretical studies also indicate that bidentate binding is essential for electronic coupling.

4.2.3 Control of Particle Shape

The ability to systematically manipulate the shape and size of metal-oxide nanoparticles is a major scientific breakthrough in the control of their electronic and chemical properties resulting in new venues for more efficient site-selective chemistry. Recent progress in understanding the surface structure of TiO$_2$ nanoparticles in our laboratory and elsewhere enables the synthesis and control of the SS in differently shaped nanocrystals (Zubavichus et al., 2002; Niederberger et al., 2004; Dimitrijevic et al., 2005). The coordination and structure of Ti surface atoms, and thus, the nature of surface defects are expected to be dependent on the size and shape of the nanoparticles. The question that arises is if the structure of the SS can be used for the control of the crystalline structure of nanomaterials.

We found that the tunable structure of Ti SS can be used for the control of the nanoparticle shape as well as the overall crystalline structure of TiO$_2$ nanostructures. TiO$_2$ nanotubes are formed as a consequence of the presence of undercoordinated SS on thin (9 Å) layered anatase in proton-deficient aqueous systems (Figure 4.4) (Saponjic et al., 2005). Undercoordinated surface titanium sites cause the folding of anatase layers into large curvature nanotube shapes because of their optimal square pyramidal geometry. However, upon charging the undercoordinated sites with protons or lithium ions, the nanotubes unroll into extended sheet-like structures. The exfoliation is accompanied by the compression of the lattice that results in the transformation of the crystalline structure from anatase nanotubes to rutile sheet-like structures. Reversible sheet-to-tube and associated lattice transitions are possible by making a proton-deficient aqueous system of titania nanosheets. Theoretical modeling suggests that reverse transitions are possible since the phase

transition points for anatase to rutile in acid and alkaline solutions lie between the effective surface-to-volume ratios of the nanotubes and nanosheets (Barnard and Zapol, 2004a,b).

These discoveries have opened up a fundamentally different approach to the synthesis of new molecule-nanoparticle systems: the control of nanoparticle shape by changing the surface environment. By taking advantage of the different coordination of Ti atoms on the surface of nanoparticulate TiO$_2$ in different environments (pHs), we have synthesized a library of anisotropic nanoobjects with different aspect ratios as well as nanoparticles and nanocubes (Figure 4.5) (Dimitrijevic et al., 2007). The surface structure of the anisotropic nanocrystallites can have different reconstruction depending on the exposed surfaces, and therefore display diverse chemical reactivity. This, in turn, would allow the site-selective binding of ligands having different geometry and coordination preferences for surface metal atoms.

4.2.4 Effects of Particle Size and Shape on Surface Reconstruction

Axially anisotropic nanoobjects such as nanotubes or nanorods hold a special promise to facilitate the formation of oriented organized structures capable of vectorial electron transport necessary for efficient chemical reactivity. Our initial discovery of the difference between the surface structure of nanostructured and bulk materials extends to the study of localized crystalline structures in anisotropic objects. For this purpose we investigated low temperature electron transfer in differently shaped TiO$_2$ nanoparticles using EPR spectroscopy. These spectra, shown in Figure 4.6, indicate different distribution of electron density in the TiO$_2$ nanoobjects after illumination. Charge separation in nanoparticles that have diameters smaller than exciton radius (30 Å) do not show the existence of lattice electrons, suggesting that charges never separate after strong (excitonic) interaction, and the majority of charges that are formed disappear in fast recombination. Very similar behavior was found for nanotubes that consist of 9 Å layers of anatase TiO$_2$ rolled into tube-like structures. After photoexcitation, only a small fraction of electrons was able to escape excitonic interaction and localize at the surface trapping sites. However, as the size of nanoobjects exceeds the exciton diameter (Bhor radius 15 Å), excitonic interaction is followed by separation of charges, and we start to observe the characteristic EPR spectrum of lattice-trapped electrons. This result suggests that TiO$_2$ nanoparticles that have the radius in the range of exciton size are optimal for carrying out redox chemistry. In this size regime electron hole pairs can separate to experience weak exchange interactions, while still preserve large surface-to-bulk ratio that facilitates reactions with adsorbed species.

We found that the local environment of localized electrons strongly depends on the shape of the nanoobjects. The nanorods show strong localization of electrons at high curvature sites—tips—and display the same Ti(III) environment as high curvature spherical nanoparticles in contrast to faceted nanoparticles

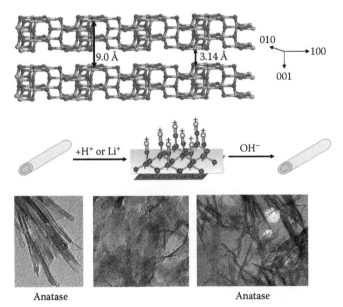

FIGURE 4.4 Schematic presentation of the processes of exfoliation and rolling up of titania nanotubes in conjunction with corresponding low-magnification TEM images. (From Saponjic, Z.V. et al., *Adv. Mater.*, 17, 965, 2005. With permission.)

FIGURE 4.5 TEM images of a library of TiO$_2$ nanoobjects. (From Dimitrijevic, N.M. et al., *J. Phys. Chem. C*, 111, 14597, 2007. With permission.)

that have signals similar to those on single crystal anatase. This leads to the conclusion that the tips of the nanorods have lowest excitation energies even when functionalized with enediol ligands. Two mechanisms can account for the lowering of excitation energy at the tips. One is the simple change of bond

distances at the preferential sites that leads to selective adsorption of bidentate ligands and second, the change of electronic properties of surface Ti atoms at the tips and corner sites.

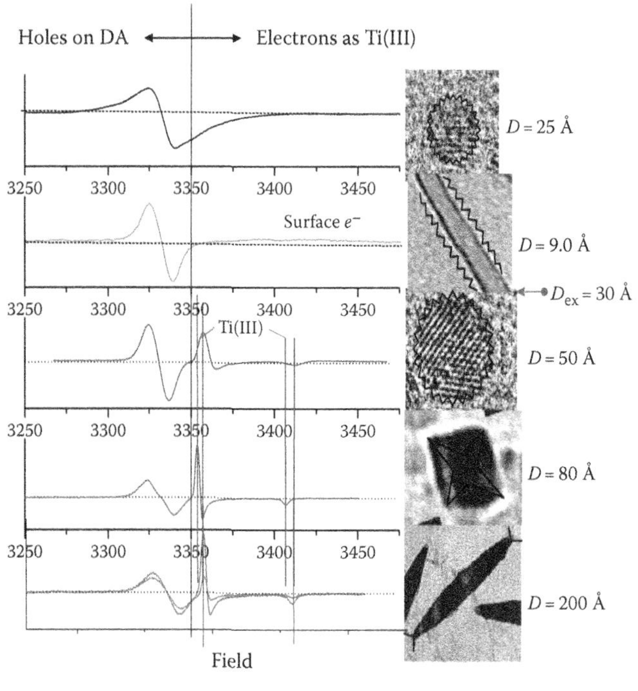

For that purpose we have initiated the study of molecular and electronic structure of anisotropic nanoobjects by using high-resolution transmission electron microscopy (HRTEM) as localized isolated spots that present a small fraction of total atoms within the particles (Rabatic et al., 2006). Figure 4.7 shows the high-resolution image of 010 surface in which strong black features correlate with the positions of Ti atoms in the lattice. It can

FIGURE 4.6 X-band EPR spectra of illuminated TiO$_2$/DA nanocomposites at 4.6 K.

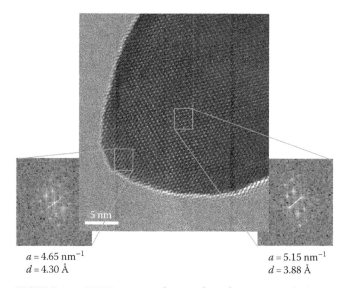

$a = 4.65\ nm^{-1}$
$d = 4.30\ Å$

$a = 5.15\ nm^{-1}$
$d = 3.88\ Å$

FIGURE 4.7 HRTEM image of 010 surface of TiO$_2$ nanorods. (From Rabatic, B.M. et al., *Adv. Mater.*, 18, 1033, 2006. With permission.)

be seen from the image that the Ti–Ti distance in the bulk of nanoparticles is significantly shorter from the Ti–Ti distance within the tip of elongated nanoparticles—nanorods. A detailed inspection of the crystalline fringes reveals that the measured (001) plane separation at the tip is extended compared to the same spacing within the bulk of the nanorod. Specifically, the outermost five atomic layers of the tip are spaced 3.96 ± 0.20 Å along the [100] direction, whereas within the bulk, and nondefect-containing regions of the particles, the plane spacing was measured to be 3.70 ± 0.19 Å. It is within these last five atomic layers where we find the greatest distortion.

The Fourier transform of the image that gives the spacing in the reciprocal space indicates that the Ti–Ti bond along 001 extends by ~0.3 Å at the tips of the nanorods. A further detailed inspection of different orientations in multiple particles revealed that the expansion of the (001) spacing was ranging 0.26–0.28 Å, while contraction of (102) spacing varied from 0.10 to 0.29 Å. This effect is reminiscent of findings for spherical nanoparticles in which the deviation from the octahedral coordination to the undercoordinated square pyramidal symmetry of the surface atoms was accompanied with contraction of a Ti–O bond and elongation of a Ti–Ti distance (Chen et al., 1997). In the case of spherical nanoparticles, all surface atoms were affected by the high curvature of nanoparticles; however, in nanorods, only those located at the high curvature tip are found to change. Due to this change in the coordination of surface Ti atoms, the tips of nanorods are more reactive than the rest of the nanoparticle and, like the surface atoms of spherical nanoparticles, show selective reactivity toward bidentate binding. Once exposed to enediol ligands, the coordination of surface atoms at the tips is restored,

and the bond distance along 001 direction becomes bulk-like. Using HRTEM with FFT analysis, we show that exposing the TiO$_2$ nanorods to dopamine (DA) results not only in the organic molecule binding to the defective tips, but more importantly, in the reorganization of the surface atoms at the tips (Figure 4.8). The reorganization results in a repair of the bond distance along the [100] direction to the associated bulk value and was measured to be 3.76 ± 0.19 Å. This preferential binding of DA to the tips is the result of optimal matching of the enediol groups of DA with the uncoordinated titanium atoms found at the rod tip.

4.2.5 Nature of Trapping Sites

The fact that enediol ligands selectively bind and repair surface trapping sites suggests that binding effects the energetics of the surface state, leaving trapping sites in the particle interior intact. Therefore, by modifying the surface with different enediol ligands, the effects that arise from truncating the crystalline nanoparticles at the surface are selectively removed. We used cyclic voltammetry and spectroelectrochemistry to analyze the changes in the energetics of the surface states upon surface modification (de la Garza et al., 2006). The absorption of light by hybrid systems promotes electrons from the surface-adsorbed ligands to the acceptor states in the TiO$_2$. We have shown that the photogenerated electrons obtained by illumination with visible light can be efficiently collected in nanostructured electrodes and converted into electric current (Figure 4.9a). The wavelength response of the photocurrent correlates with the absorption spectrum of a hybrid system. The spectrochemical studies of excess electrons in nanocrystalline TiO$_2$ injected at different electrode potentials (Boschloo and Fitzmaurice, 1999) suggest the existence of three deep trapping sites (420, 600, and 800 nm) that can be removed by surface modification with enediol ligands. The deepest trapping sites are eliminated at lowest surface coverage, and as the concentration of the ligand increases the shallow trapping sites are removed consecutively (Figure 4.9b). After complete surface modification, excess electrons in nanoparticles display monotonic featureless wavelength dependence. In addition, the injection of excess electrons in enediol-modified electrodes has shown a reversible bleach of the entire effective band gap region ($E > 1.6$ eV) at low chargings (de la Garza et al., 2006).

These results suggest that DA-modified TiO$_2$ behaves as one integrated core-shell system in which the effective band gap relates to the difference of the HOMO of organic ligand and LUMO of TiO$_2$ nanoparticles. Electron accumulation in this system results in the filling of the conduction band states (LUMO) of TiO$_2$ nanoparticles resulting in an apparent increase of the effective band gap of the integrated hybrid nanoparticle and consequent bleaching of the absorption in the visible part of the spectrum (Scheme 4.1). Therefore, enediol ligands couple electronically with nanoparticles and can serve as the conduits of photogenerated charges and are ideal surface modifiers that enable the electronic sensing of attached molecules through the mechanism of light-induced charge separation.

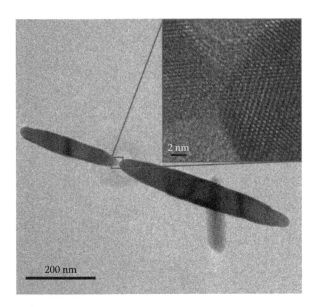

FIGURE 4.8 Transmission electron microscope image of anisotropic TiO$_2$ nanoparticles functionalized with biotinated dopamine and coupled together through avidin. Upper right inset shows a high-resolution image of the coupled nanorods with enhanced contrast seen in the region between the two tips. (From Rabatic, B.M. et al., *Adv. Mater.*, 18, 1033, 2006. With permission.)

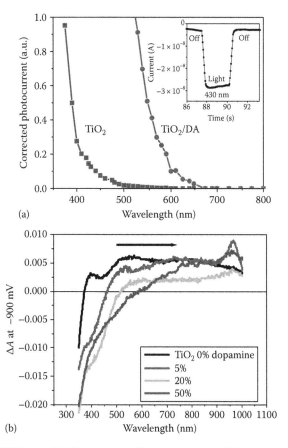

(a)

(b)

FIGURE 4.9 (a) Photocurrent of bare and DA-modified nanocrystalline TiO$_2$ electrode, and (b) the absorption spectra of injected electrons showing healing of surface trapping sites with dopamine. (From de la Garza, L. et al., *J. Phys. Chem. B*, 110, 680, 2006. With permission.)

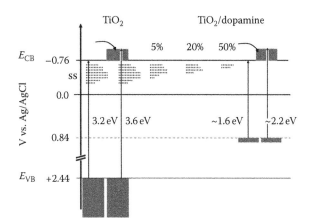

SCHEME 4.1 Salient features of the electronic structure of 4.5 nm TiO$_2$ particles upon surface modification with different concentration of DA studied by spectroelectro-chemistry. The injection of electrons in TiO$_2$ films results in apparent increase of the band gap (3.2–3.6 eV for bare TiO$_2$ and 2.6–2.2 eV in DA-modified TiO$_2$) due to the filling of the conduction band states. Upon surface modification, the deepest surface sites (SS) are removed by using the smallest concentration of dopamine, suggesting that the deepest sites are the most reactive toward bidentate binding. Upon modification of 50% of surface sites, only the shallow SS remain. (From de la Garza, L. et al., *J. Phys. Chem. B*, 110, 680, 2006. With permission.)

4.3 Bioinorganic Nanocomposites and Extended Charge Separation

Analogous to the charge separation in supramolecular triads (Wasielewski et al., 1989) when biomolecules are linked to nanoparticles via conductive enediol ligands, photoinduced electron transfer can extend, ultimately leading to stabilized charge separation. Extended charge separation can be detected electronically or can be used for driving chemical reactions with surrounding redox species allowing a complex light-induced manipulation of biomolecules and their switching functions. Two different biomolecule systems were investigated: DNA oligonucleotides and protein (antibody) systems. These systems were chosen as representatives of the two most important mechanisms of biomolecule interactions: strong site-selective binding through multiple hydrogen bonding (DNA) and molecular recognition interactions through multivalent "lock-and-key" structures (antibody–antigen).

Our experimental strategy for the construction of complex nanostructures is inspired by the strategy that living organisms use for the synthesis of proteins. It involves a three-step process: (1) batch synthesis of each type of building block followed by (2) activation of terminus groups and assembly of hybrid nanocomposites and (3) their integration into templating architectures and/or cell structure. In living organisms, the 20 chemically distinct amino acids are synthesized by independent biochemical pathways and subsequently polymerized as building blocks in a genetically controlled order to form functional proteins. Each protein then folds into a three-dimensional assembly with emergent properties that can be exquisitely specific, depending on the juxtaposition of the different amino acids in the overall structure. We extend this atomic-scale biological design principle to the engineering of nanoscale complex structures.

4.3.1 TiO$_2$–DNA Nanocomposites

Molecular recognition of biomolecules and their site-selective bindings are topics of significant interest because of unique applications in the fields of patterning, genome sequencing, and drug affinity studies (Storhoff and Mirkin, 1999). DNA oligonucleotides hold a special promise because of the exquisite programmability of the nucleic acid-based recognition system. The investigation of oligonucleotide hybridization mechanisms has been mainly focused on optical detection using fluorescence-labeled oligonucleotides with dyes (Pease et al., 1994) or quantum dots (Bruchez et al., 1998) and the enhanced absorption of light by oligonucleotide-modified gold nanoparticles (Taton et al., 2000). It is desirable that after the positioning of hybrid nanoparticles with biological specificity, an activated process changes the biomolecule's redox state and selectively catalyzes a string of consecutive responses. This opens up the possibility of the electronic transduction of DNA sequence and hybridization, as well as the direct detection of oligonucleotide sequence, mismatches, and their possible interactions with DNA binding proteins (Patolsky et al., 2000; Rajh et al., 2004).

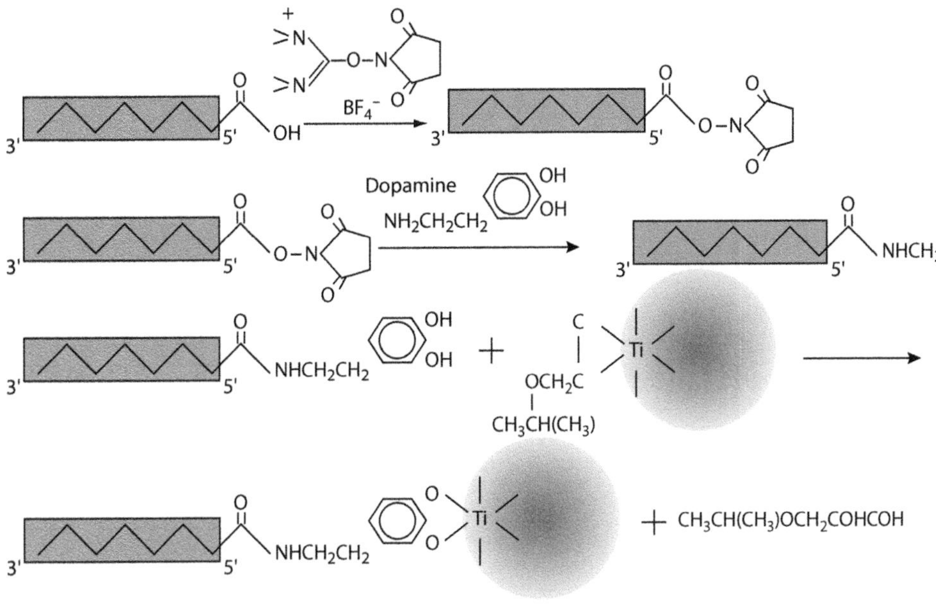

FIGURE 4.10 Schematic presentation of the synthesis of TiO$_2$/DA–DNA nanocomposite.

Nanocrystalline metal-oxide semiconductor particles are photoactive and act as miniaturized photoelectrochemical cells. The absorption of light in DA-modified TiO$_2$ results in the formation of redox reactive species, the holes localized on DA and the electrons delocalized in the conduction band of TiO$_2$ (Rajh et al., 2003a,b). When covalently linked (wired) to electron-donating moieties such as DNA, photoinduced electron transfer can extend further, ultimately leading to stabilized charges that can participate in redox chemistry.

In order to probe this mechanism, we have linked oligonucleotides with different nucleotide sequences to metal-oxide semiconductor particles through a DA linker (Figure 4.10). The synthesis was based on a fact that DA induces the restructuring of the Ti surface atoms, allowing for strong bidentate binding of DA molecules to the nanoparticle surface (Rajh et al., 2002). The appendant amino groups of DA act as point of attachment and could be linked to any molecule that contains carboxyl group. A variety of 20-base pair oligonucleotides terminated with the carboxyl deoxythymidine (CdT) at 5′ end were attached to the appendant amino groups of DA via the intermediate *N*-hydroxy-succinimide ester and chemisorbed on the TiO$_2$ surface (Banwarth et al., 1998; Bonacheva et al., 1999). The number of oligonucleotides bound to each particle was determined from the intensity of the optical absorption of CT complex between DA and TiO$_2$ nanoparticles (Figure 4.11, bottom left). Typically, the change in the absorption after binding was consistent with having between two and three oligonucleotides per particle. Oligonucleotide-modified particles were then hybridized with complementary DNA strands of the same lengths.

The image of TiO$_2$ particles linked to short oligonucleotides (16 base pairs) obtained using atomic force microscopy (AFM) is shown in Figure 4.11. The size of the TiO$_2$ nanoparticle in these experiments was 50 Å, comparable to the size of the oligonucleotide length (40 Å) (see depicted model, Figure 4.11).

The color-coded height profile of the composites clearly shows that composites contain two parts, one with the height of ~25 Å (purple), comparable with the thickness of the DNA oligonucleotide, and the other of 48 Å (orange), comparable with the TiO$_2$ diameter. The height histogram made of 60 images exhibits two maxima, one with the size centered around ~28 Å and the other centered around ~48 Å. The surface area ratio under the deconvoluted curves in the histogram obtained with AFM imaging presents the average number of oligonucleotides per TiO$_2$ particle (throughout this work between 2 and 3).

We have investigated further the role of sequential electron transfer, and electrochemical parameters, and how they affect the distances of charge separation properties in the oligonucleotide-modified TiO$_2$ nanoparticles. We used low-temperature EPR spectroscopy to reveal the identity of the species formed during photoexcitation. EPR spectrum obtained by the illumination of TiO$_2$ nanoparticles modified with DA shows signals attributed to the reduced electron acceptor (Ti^{3+}) and the oxidized electron donor (DA$^+$) (Rajh et al., 1999a). The former signal ($g_\perp = 1.988$, $g_\parallel = 1.958$) is characteristic of a radical in which the unpaired electron occupies the *d*-orbitals of anatase lattice Ti atoms. The latter EPR signal was observed at $g = 2.004$ having a line width of $\Delta H_{pp} = 16$ Gauss. We attribute this signal to C1 centered radical on DA because the signal narrowed to $\Delta H_{pp} = 10$ Gauss upon deuteration of DA (ring D$_2$ or 2,2 D$_2$), indicating the existence of spin density on the pendant CH$_2$–CH$_2$–NH$_2$ side chain of DA (Rajh et al., 2002). The delocalization of photogenerated holes to the pendent side chain of DA suggests DA as a ligand of choice for linking nanoparticles to biomolecules, because it acts as a conduit of photogenerated charges and enables the further extension of charge separation.

The EPR signal obtained after the photoexcitation of single-stranded DNA/TiO$_2$ composite (ssDNA/TiO$_2$) at 10 K was identical

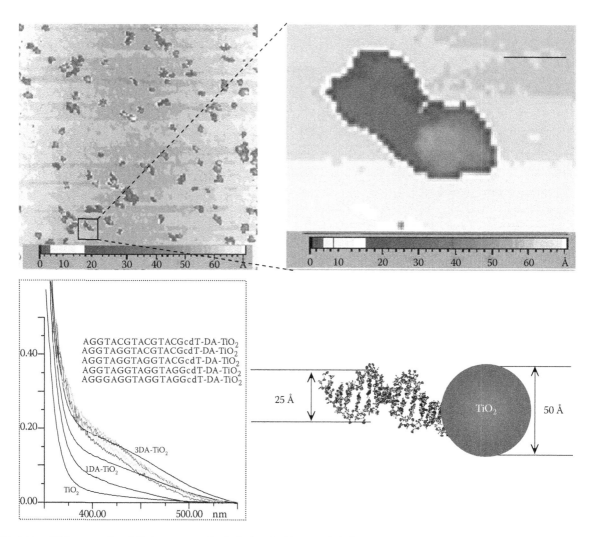

FIGURE 4.11 AFM image of 50 Å TiO_2 nanoparticles linked to double-stranded oligonucleotides. The samples were prepared in 40 mM phosphate buffer and spin-coated onto the freshly cleaved mica substrate. While the two images have different in-plane scale, the height scale is the same for both images. The gray-coded scale of the sample heights is presented on the bottom of each image and presents the scale pointing out of the plane of the page. A model of TiO_2 nanoparticles linked to double-stranded DNA is depicted in the bottom-right corner. The inset on the bottom-left corner presents the absorption spectra of TiO_2 nanoparticles linked to: none, one, two, and three dopamine molecules per particle (smooth lines); and different 16-base pair oligonucleotides end-labeled with dopamine molecules (raster lines) used for obtaining AFM images. (From Rajh, T. et al., *Nano Lett.*, 4, 1017, 2004. With permission.)

to the signal obtained upon the illumination of DA-modified TiO_2 (Figure 4.12a, top curve 1). This result indicates that in the single-stranded TiO_2/DNA nanocomposite charge separation at 10 K terminates at DA and never reaches DNA oligonucleotide. Following the photoexcitation of double-stranded DNA/TiO_2 nanocomposites (ds DNA/TiO_2) at 10 K, however, an EPR signal distinct from the EPR spectrum of DA-modified TiO_2 was observed. This multi-line EPR spectrum, total width of 100 Gauss, resembles the signals of radicals obtained for oxidized DNA at low temperature (Huttermann et al., 1984), but does not duplicate the shapes of signals obtained upon oxidation of the four nucleic bases that constitute DNA (Figure 4.12a, top curve 2).

The main features of the signal obtained upon the illumination of TiO_2/DNA nanocomposite at 10 K could be reasonably simulated assuming hole trapping on CdT (Figure 4.12b, bottom).

The oxidation of CdT as a primary hole trapping site at 10 K is not expected considering that thymidine itself should not be oxidized by photogenerated holes on TiO_2. Therefore, theoretical modeling of the redox properties of thymidine derivatives was carried out (Vega-Arroyo et al., 2003). The computational results indicate that the addition of a carboxyl group to thymine at position 5 (replacing a methyl group) lowers the adiabatic ionization potential of thymine by 0.4 eV (Figure 4.13). Furthermore, it was found that condensation of CdT with an amino group into a peptide bond leads to further lowering of the ionization potential by an additional 0.2 V. Hydrogen bonding and solvation following base pairing of CdT with adenine introduces additional changes in the ionization potential resulting in further reduction of the ionization potential of the CdT···A pair by 0.6 eV. We believe that this last step of base pairing is responsible for our

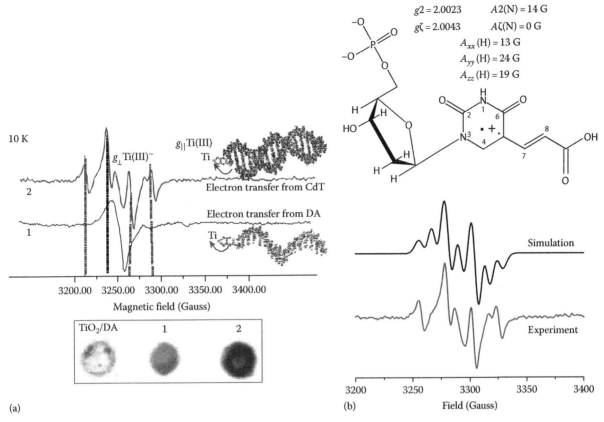

FIGURE 4.12 (a) Charge separation in 50 Å TiO₂ particles coupled to 16-base pair ss DNA, 5′cdTGG-TATATATATATAT 3′ (1), and ds DNA, 5′cdTGGTATATATATATAT 3′hybridized with 5′ATATATATATATACCA (2). Top: EPR spectra recorded at 10 K after illumination with a Xe 300 W lamp. Bottom: Photoreduction of silver ions under ambient conditions. (1) ss DNA/TiO₂ and (2) ds DNA/TiO₂ were spotted onto a glass plate in the presence of silver ions. The samples were illuminated with 100 W white light for 5 min. The dark color indicates metallic silver deposition following accumulation of photogenerated electrons on TiO₂ nanoparticles as a result of extended charge separation. (b) Comparison of experimental and simulated EPR spectra of an oxidized carboxyl deoxythymidine radical (radical structure is presented above). (From Rajh, T. et al., *Nano Lett.*, 4, 1017, 2004. With permission.)

observation that the most efficient electron transfer occurs in double-stranded oligonucleotides attached to TiO₂. As the oxidation potential of CdT decreases upon base pairing, CdT⋯A does not impose a barrier for hole hopping, and the oxidation of double-stranded DNA oligonucleotides becomes efficient. On the other hand, when single-stranded DNA oligonucleotides are linked to TiO₂ nanoparticles, photogenerated holes do not have enough driving force to overcome the barrier imposed by CdT and photogenerated holes localized at DA recombine with electrons on neighboring TiO₂ nanoparticles.

The sequential steps in the hole trapping process were examined by the temperature dependence of the EPR signal. Warming of the sample induces a change in the EPR spectrum indicative of further charge hopping. Upon annealing at 80 K, the EPR signal collapses into a signal with a very weakly resolved hyperfine structure and weak satellite lines. The EPR spectrum of the guanosine radical cation obtained by reaction with radiolytically generated OH radicals matches the 80 K spectrum. This result indicates that sequential hole transfer to guanine in DNA follows the initial localization of photogenerated holes on the terminus CdT. This suggests that hole transfer within DNA

is accompanied with nuclear motion that requires activation energy of ~8 meV (kT at 80 K). Moreover, hole transfer through thymine-bridged (22 tymidines) ds DNA attached to TiO₂ particles was not observed. In this case charge never reaches the DNA oligonucleotides, suggesting that an activation barrier controls the electron transfer processes in the hybrid systems as well as in DNA alone (Henderson et al., 1999; Giese, 2000).

We have also investigated the charge separation in these systems at room temperature by using silver reduction as a messenger of extended charge separation (Figure 4.12, bottom left). Silver ions have a positive deposition potential, and when excess electrons are present in TiO₂ nanoparticles, silver ions act as acceptors for photogenerated electrons and are reduced to their metallic form. As in the photographic process, the amount of silver deposited on the TiO₂ particles, and subsequent color intensity, is proportional to the number of electrons that have survived charge separation. For that purpose, silver ions were added into the solution of DNA/TiO₂ nanocomposites and dried on a glass plate. Illumination of the glass plate resulted in significant silver deposition in the areas where ds DNA/TiO₂ nanocomposites were spotted, indicating that hybridized DNA molecules

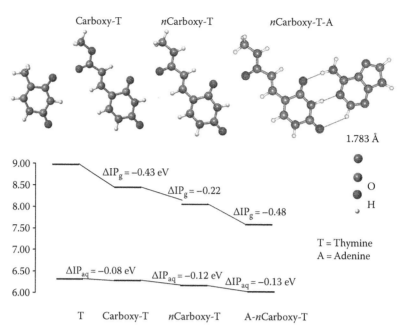

FIGURE 4.13 Theoretical modeling of the structure and ionization potentials of thymine, carboxyl thymine, and carboxyl thymine-adenine pair. The structures and ionization potentials of thymine (T) and carboxyl thymine (cT) and cT⋯A pair in the gas phase and aqueous solution were calculated at the B3LYP/6-31+G**//B3LYP/6-31G** level of theory, including zero-point energy correction. The solvation was treated using the self-consistent reaction field and polarized continuum model. Solvation reduces ionization potentials. The functionalization of thymine by the amino-modified carboxyl group improves its electron-donating properties and the choice of linker strongly influences properties of the base pair. (From Rajh, T. et al., *Nano Lett.*, 4, 1017, 2004. With permission.)

act as efficient electron donors that allow photocatalytic deposition of metallic silver on TiO_2 nanoparticles (Figure 4.12, bottom, spot 2). The deposition of metallic silver in ss DNA/TiO_2 nanocomposites was much less intense, suggesting the existence of the barrier for extended charge separation in agreement with the low temperature EPR results (Figure. 4.12, bottom, spot 1). This small amount of silver deposition may be the result of the thermal mobility of single-stranded DNA at room temperature (Giese, 2000). It should be noted that the deposition of metallic silver was not significant for a sample of TiO_2 linked only to DA under the same conditions (Figure 4.12, bottom).

4.3.2 Sequence Dependence of Charge Separation in TiO_2–DNA Nanocomposites

Charge transfer in DNA and ensuing redox reactions follow thermodynamic and kinetic principles. During the past decade, photochemical methods have been widely used to study charge transfer processes in DNA. Experimental investigations and theoretical treatments of photoinduced charge transfer in DNA have revealed an occurrence of at least two mechanisms: a single-step superexchange mechanism which is strongly distance dependent (Lewis et al., 1997; Bixon and Jortner, 1999), and a multistep hole transfer over several dozen base pairs through either incoherent hole hopping (Bixon et al., 1999; Berlin et al., 2000) or a polaron-like hopping process (Schuster, 2000; Barnett et al., 2001). The latter mechanism explains recent evidence for charge migration over several dozen base pairs that was

obtained from DNA strand cleavage studies (Giese, 2000). Site-selective photooxidation of DNA at guanine-containing sites was attributed to the relative ease of oxidation of G compared to the other nucleobases (Gasper and Schuster, 1997; Kawai et al., 2003b). Greater selectivity was obtained for oxidation at GG or GGG hole trapping sites because their oxidation potential is 0.05–0.1 V more negative compared to single base G (Kurnikov et al., 2002; Lewis and Wasielewski, 2004).

All investigated systems of charge transfer in DNA described so far use electron-acceptor chromophores as light active species that inject holes into DNA. The drawback of this approach is that the study of hole hopping mechanisms is strongly dependent on the efficiency of hole injection into DNA which competes with the fast charge recombination with electrons localized on the neighboring chromophore. Our method of controlling charge transfer reactions in DNA using hybrid nanometer-sized metal-oxide semiconductors allows to separate the contributions of charge recombination from hole migration in DNA. In surface-modified nanoparticles, electron hole pairs that are formed upon illumination are instantaneously separated to relatively large separation distances (~20 Å) before holes are injected into DNA. Fast multi-step electron transfer in surface-engineered nanoparticles increases the initial charge separation distance resulting in a weak coupling of electron hole pairs (10^{-4} meV) and promotes slow charge recombination (Rajh et al., 2001). Moreover, electron hole pairs also exhibit spin multiplicity of excited state precursors, allowing for the triplet character to cause long-lived charge separation. The importance of the effect of charge recombination on the efficiency of charge injection in DNA was outlined previously (Lewis et al., 2003; Liu and Schuster, 2003) and showed that the presence of G/C pairs within three base pairs closest to the chromophore significantly reduced the efficiency of hole injection. This suggests that geminate recombination plays an important role in the efficiency of hole injection, and therefore in determining the mechanism of hole migration in DNA-coupled systems.

Twenty base-pair oligonucleotides having GG accepting sites at different distances from the nanoparticle surface were attached to the appendant chain of DA linked to the TiO$_2$ nanoparticle via an intermediate succinimide ester and hybridized to their complementary strands (Paunesku et al., 2003). Upon illumination,

an electron is excited directly from DA into the conduction band of TiO$_2$. Depending on the redox properties of attached DNA, charge separation will extend further, leading to the long lifetime of the fully separated state. In the case when DNA is oxidizable with DA radicals produced by illumination, photogenerated holes will be transferred to DNA. In the case when DA radicals do not have enough energy to oxidize DNA, the charges will recombine (Figure 4.14a). In this work, charge separation in triad systems was monitored by detecting the ability of electrons localized at TiO$_2$ to escape recombination and reduce electron acceptors in solution. As described in Section 4.3.1, we used silver ions as acceptors because (a) they have favorable redox

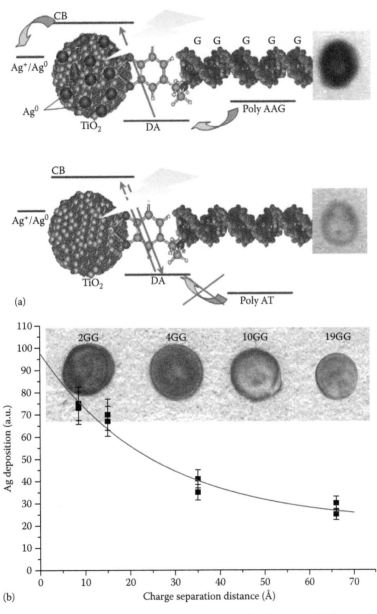

(a)

(b)

FIGURE 4.14 Charge separation in DNA covalently linked to TiO$_2$/DA, monitored by silver deposition. (a) Photogenerated holes are injected into DNA when their oxidation potential is more positive than the one for DNA, allowing the remaining electrons on TiO$_2$ to reduce Ag$^+$ ions. (b) The efficiency of charge transfer of holes generated at the dopamine linker was found to be dependent on the distance of guanine trapping sites (GG) from the nanoparticle surface (position from 5′ end within polyAT run is indicated on respective spots). (From Liu, J.Q. et al., *Chem. Phys.*, 339, 154, 2007. With permission.)

properties, and (b) their reduction can be monitored by a simple photographic approach. The results are presented in Figure 4.14b and show that the efficiency of silver deposition and consequent hole transfer in these hybrid systems depends on the position of guanine pairs in the AT runs in respect to the nanoparticle surface (2GG is ds5′-cdT**GG**ATATATATATATATATA; 4GG is ds5′-cdTAT**GG**ATATATATATATATA; 10GG is ds5′-cdTATATATAT**GG**ATATATATA; 19GG is ds5′-cdTATATATATATATATATA**GG**). Interestingly, single exponential decay describes the distance dependence of the migration of photogenerated holes throughout the entire 70 Å long DNA; there is no evidence for a crossover from strong to weak distance dependence (Berlin et al., 2004; Roginskaya et al., 2004; Lewis et al., 2006). The presence of guanine in the first couple of base pairs leads to efficient charge separation (Φ = 56%) and as the guanine doublet is moved away from the nanoparticle surface, the efficiency of hole transfer decreases. The attenuation coefficient of the exponential decay for a charge transfer through a polyAT bridge was found to be 0.045 Å⁻¹, unexpectedly low for a system that does not contain any guanine hopping sites. We found no evidence for the multiexponential behavior of the rate of migration in these hybrid systems (Berlin et al., 2004), possibly because the separation of charge carriers occurs in surface-modified nanoparticles before the hole is injected into DNA oligonucleotides. Additional results also indicate that the attenuation coefficient varies with the nature of the bridge. The efficiency of hole transfer in a DNA system containing a polyA bridge is even higher, and the attenuation coefficient drops to 0.025 Å⁻¹. These results correlate well with the structural measurements that indicate closer π-stacking, enhanced rigidity, and lower oxidation potential of polyA bridges compared to alternating polyAT bridges (Henderson et al., 1999; Kawai et al., 2003b; Takada et al., 2004; Schuster and Landman, 2004; Bixon and Jortner, 2006; Li et al., 2006). A similar change of attenuation coefficients for long-range distance radical cation migration in DNA was previously obtained in mixed sequences containing repeats of AAGG compared to ATGG repeats by Schuster et al. using electron-acceptor chromophores as light active species (Lewis et al., 2001).

These results show that in a TiO₂/DA/DNA hybrid system, there is a weak, long-range distance dependence of charge separation leading to large separation distances even in the absence of G hole hopping steps (we have examined up to 70 Å). We believe that the low attenuation coefficient obtained in this hybrid system is the consequence of the strong oxidation properties of photogenerated holes in TiO₂, and the initial charge separation mechanism in surface-engineered nanoparticles, and the existence of a surface charge of nanoparticles that cooperatively contribute to electron/hole separation. The prolonged lifetime of weakly interacting electron/hole pairs in appropriately designed nanoparticles enhances the probability of hole migration to distant trapping sites. It should be noted that in the systems that use electron-acceptor chromophores as light active species, in the absence of G hopping sites, the hole is transferred to the G accepting sites <20 Å from the place of its formation (Henderson et al., 1999; Schuster and Landman, 2004).

The small attenuation coefficient (0.045 Å⁻¹) and weak distance dependence determined for hole migration in hybrid systems suggests that a multiple step hole hopping mechanism is the dominant mechanism for hole transfer. As no guanine hopping sites are present, we consider the hopping through adenosine (A) nucleobases that act as shallow trapping sites in the charge migration through DNA. The reactions involved in charge separation and consequent Ag deposition are as follows:

$$TiO_2\text{-}DA \xrightarrow{h<} TiO_2^- \text{-}DA^+ \xrightarrow{k_{cs}} TiO_2^- \text{-}DA \text{-} GG^+ \tag{4.1}$$

with k_{Ag}, k_{Tcr} on the left branch and k_{Ag}, $k_{Gcr} \sim 0$ on the right branch.

where

  k_{Ag} is the rate of Ag deposition
  k_{Trc} is the rate of recombination of photogenerated electrons with the holes on DA
  k_{cs} is the overall rate of charge separation
  k_{Gcr} is the rate of recombination between photogenerated electrons on TiO₂ and holes trapped on a

$$M_{Ag} = \frac{k_{Ag}}{k_{Ag}+k_{Tcr}+k_{cs}} + \frac{k_{cs}}{k_{Ag}+k_{Tcr}+k_{cs}} + \frac{k_{Ag}}{k_{Ag}+k_{Gcr}}$$
$$\cdot 1 - \frac{k_{Tcr}}{k_{Ag}+k_{Tcr}+k_{cs}} = 1 - \frac{1}{a+bN^{-0}} \tag{4.2}$$

GG pair which is negligible due to the large separation distance. The quantum yield of Ag deposition according to the reaction mechanism is where $a = (1 + k_{Ag}/k_{Tcr})$, $b = k_{cs}^{\eta}/k_{Tcr}$, $k_{cs} = k_{cs}^{\eta}N^{-\eta}$, N is the number of hopping sites, and η is the coefficient that defines the hopping mechanism within DNA (Figure 4.15).

Quantum efficiency of charge separation was determined by measuring the number of photons absorbed by TiO₂ films using a phenylglycolic acid actinometer (λ < 400 nm) (Defoin et al., 1986) and the concentration of deposited silver was determined by using the absorption coefficient obtained by the chemical reduction (NaBH₄) of silver under the same experimental conditions. The quantum efficiency of 70% and 56% was obtained for 2GG (number of hopping sites N = 2) in polyA and polyAT DNA runs. Compared to the charge separation yields of 2% in systems that use chromophore as the light active system (Lewis et al., 2001), this is a very high quantum efficiency that most probably reflects slow charge recombination in surface-modified TiO₂ nanoparticles. The measurement of the absolute value of quantum efficiency of silver deposition uniquely determines η = 1.1. This value is indicative of direction-biased consecutive hole hopping through polyA and polyAT bridges to a GG final accepting site. These results indicate that charge separation in this system occurs by sequential electron transfer in which the bridge is populated during hole transfer and forward electron transfer is larger than backward electron transfer. The rate of

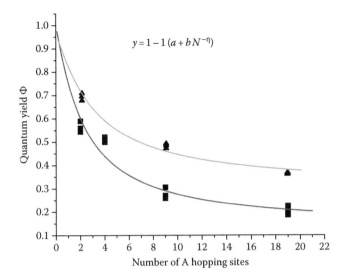

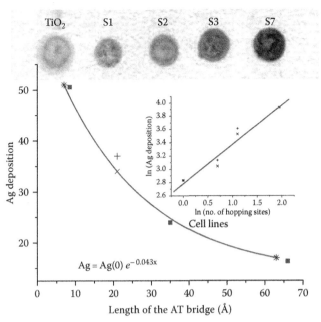

FIGURE 4.15 Fitting of the experimental data (shown in Figure 4.1b) for GG imbedded in polyAT (black, ■) and polyA (gray, ▲) runs assuming the reaction mechanism (1). The figure shows that a set of parameters can uniquely fit experimental data. For polyAT $a = 1.134$, $b = 2.442$, and $\eta = 1.100$ and for polyA $a = 1.337$, $b = 4.980$, and $\eta = 1.100$. These parameters determine k_{cs}^{η}/k_{Tcr} to be 2.442 for polyAT and 4.980 for polyA sequence. (From Liu, J.Q. et al., *Chem. Phys.*, 339, 154, 2007. With permission.)

FIGURE 4.16 Efficiency of charge separation monitored as silver deposition in DNA covalently linked to TiO$_2$/DA depending on the length between 1, 2, 3, and 7 hopping sites (+,×), in conjunction with the normalized data obtained for distance-dependent charge separation in the absence of hopping sites (■). Inset: dependence of silver deposition with the number of hopping sites. (From Liu, J.Q. et al., *Chem. Phys.*, 339, 154, 2007. With permission.)

hole transfer is directly dependent on the rate constant of charge recombination k_{Tcr} between holes generated on DA linkers and electrons localized on TiO$_2$ nanoparticles. Previously it has been determined that the slower component of electron/hole recombination in DA-modified TiO$_2$ is ~1 ns ($k \sim 10^9$ s^{-1}) when TiO$_2$ nanoparticles were deposited into nanocrystalline films (Rajh et al., 2001; Hao et al., 2002). Assuming these values for k_{Tcr}, the rate constant k_{cs} for the polyAT bridge can be obtained from fitting parameter b (Equation 4.2) to provide $k_{cs}^0 = 2 \times 10^9$ s^{-1}. Alternatively, fitting quantum efficiency in hybrid TiO$_2$ systems using polyA bridges leads to the higher value of $k_{cs}^0 = 5 \times 10^9$ s^{-1} indicating that stacking between bases in a polyA sequence is an important parameter in the A-hopping process, as suggested by Majima and coworkers (Kawai et al., 2003b, Kawai 2004).

It is especially informative to investigate the effect of the presence of guanine hopping sites on the efficiency of silver deposition. We investigated the deposition of silver in the hybrid system in which the final accepting site is placed 19 base pairs from the nanoparticle surface, 19GG (S1), while two (S2, ds5′-cdTATATATATGTATATATAGG), three (S3, ds5′-cdTGTATATATGTATATATAGG), and seven (S7, ds5′-cdTGTAGATGTAGATGTAGAGG) guanine hopping sites were introduced at equally spaced distances between the nanoparticle surface and the 19GG accepting site, keeping the distance between the primary and final hole accepting site the same (70 Å). Interestingly, we have found that the dependence of the hole migration efficiency on the length of the polyAT bridge (Figure 4.16) shows the same attenuation coefficient of 0.043 Å^{-1} as the one obtained in the examination of distance dependence of charge separation in the absence of hopping sites (Figure 4.14b), confirming

the A-hopping-assisted mechanism of hole transfer (Berlin et al., 2002). Further, as the number of hole hopping sites is increased (Inset, Figure 4.16), the silver deposition is enhanced in a logarithmic fashion, probably due to enhanced electronic coupling between the primary and final hole accepting sites (Joseph and Schuster, 2006). When G hopping sites are present every 15 Å, the quantum efficiency of charge separation to a final trapping site removed 70 Å from nanoparticle surface (S7) reaches $\Phi = 0.4$. It should be noted that in our measurements, silver deposition reflects the integral number of electrons that escape recombination, not the distance that the hole reaches in the presence of hopping sites usually described using the random walk model (Giese et al., 1999). Our experiments show that as the number of hopping sites increases, the barrier for hole migration decreases leading to more efficient coupling between primary and final hole trapping sites.

It should be noted that under the same experimental conditions, all samples containing single-stranded DNA (ss DNA) show completely different behavior in which the number of polyAT or polyA bridges that separate nanoparticles from GG accepting sites influence the efficiency of silver deposition only slightly. The amount of deposited silver in samples containing ss DNA was an average between the amount deposited when the GG accepting site was placed close to and 19 base pairs from the nanoparticle surface, linearly increasing with the number of guanine bases. This can be explained by the flexibility of single strands which levels out the distance dependence between nanoparticle and hole accepting sites.

The amount of photocatalytic deposition of metallic silver was found to be sensitive to the presence of a single mismatch of nucleic base pairs in a DNA sequence. When one hopping site in a DNA sequence S3 containing two hopping sites is removed creating an AC mismatch (ds5′-cdTGTATATAT*A*TATATATAGG, *A*-AC mismatch), the efficiency of charge separation and consequent silver deposition was decreased 1.5 times. However, when both hopping sites were replaced with AC mismatches (ds5′-cdT*A*TATATAT*A*TATATATAGG, *A*-AC mismatch), the efficiency of silver deposition was even smaller than that in the sequence with no hopping sites, reaching the value of background silver deposition on bare TiO$_2$. These results indicate that the existence of AC mismatches in the DNA make the insulating barrier, preventing hole migration to the final GG accepting site. Introducing a GT mismatch in the same sequence has a smaller effect on hole transfer. One GT mismatch (ds5′-cdTGTATATAT*G*TATATATAGG, *G*-GT mismatch) decreases silver deposition 1.2 times while two mismatches (ds5′-cdT*G*TATATAT*G*TATATATAGG, *G*-GT mismatch) lead to the same effect as if no hopping sites were present. Our work thus shows that light-induced extended charge separation in hybrid systems was found to be a fingerprint of DNA sequence.

4.3.3 Charge Separation in Templated Architectures—Site-Selective DNA Cleavage

The ability to manipulate and control molecules and materials with nanometer resolution is crucial in the field of nanotechnology. DNA is an attractive candidate as a template for arranging nanocomponents with precision in nanometer scale. DNA has already been shown to be a template for the direct growth of metals (Braun et al., 1998; Richter et al., 2000; Richter et al., 2001; Monson and Wooley, 2003; Nakao et al., 2003), semiconductor nanoparticles (Coffer et al., 1996; Dittmer and Simmel, 2004), and conductive polymers (Uemura et al., 2001; Ma et al., 2004; Nickels et al., 2004). However, the difficulty in precise localization and interconnection of nanocomponents impedes further progress toward more complex devices. The highly specific molecular recognition process between DNA strands offers a promising approach for constructing a complex nano-assembly. Gold nanoparticles with attached single-stranded DNA have been assembled into dimers and trimers by sequence-specific DNA hybridization (Alivisatos et al., 1996). Single-wall carbon nanotubes have been precisely addressed onto λ DNA by binding with RecA-modified single-stranded DNA (Keren et al., 2003). However, DNA strands need to be considerably long to form a stable duplex. This hampers the use of double-stranded DNA for the directed assembly of nanoparticles with nanometer scale precision.

We examined the use of peptide nucleic acid (PNA) (Uhlmann et al., 1998), an analog of DNA with an uncharged polyamide backbone, as a building block to assemble nanoparticles along the DNA template (Figure 4.17). Compared to short single-stranded DNA, PNA offers two distinct advantages: greater thermal stability and greater mismatch sensitivity (Uhlmann et al., 1998; Nielsen, 1999; Chakrabarti and Klibanov, 2003). Furthermore, the polyamide backbone is more easily altered than the sugar-phosphate backbone. Motivated by these advantages, we developed a method to assemble TiO$_2$ nanoparticles on double-stranded DNA using the specific hybridization between PNA and DNA.

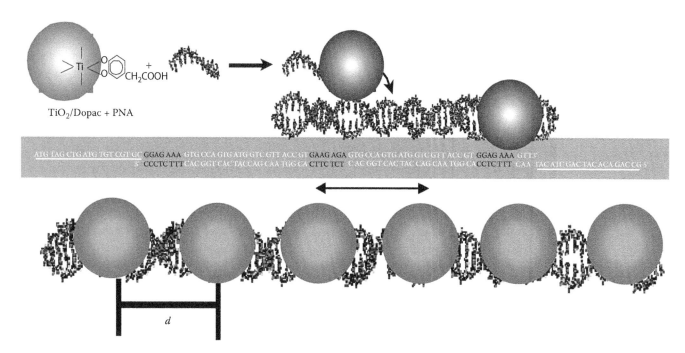

FIGURE 4.17 Schematic presentation of DNA-assisted assembly of nanoparticles. (From Liu, J.Q. et al., *Proc. SPIE*, 6096, 48–57, SPIE, San Jose, CA, 2006. With permission.)

TiO$_2$ nanoparticles anchored to DNA are semiconducting and, in the presence of light, generate a strong oxidizing power that attacks a majority of organic molecules. However, in the same uncontrollable way that laundry bleach attacks all colors in the wash, bare TiO$_2$ nanoparticles uncontrollably oxidize all molecules in their environment. We have found that by using different conductive linkers, one can selectively control oxidation. Through the conductive linkers, it was possible to control the oxidizing power of photogenerated carriers as well as to link biomolecules that introduce site selectivity to the nanoparticle composite and bring the source of oxidation only to the specific biological targets. In this way, when activated by light, the photogenerated holes attack only biological material and cleave chemical bonds within the attached DNA, employing the oxidizing power of photogenerated holes to oxidize nearby nucleobases and leave the rest of the DNA intact. Thus TiO$_2$/DNA nanocomposites present a new vehicle for manipulation useful in biotechnology. These hybrid TiO$_2$/DNA nanocomposites not only retain the intrinsic photocatalytic capacity of TiO$_2$ and the bioactivity of the oligonucleotide DNA (covalently attached to the TiO$_2$ nanoparticle), but also possess the chemically and biologically unique new property of a light-inducible nucleic acid endonuclease. Nanodevices with these extraordinary capabilities could become new tools in biotechnology for the production of building blocks that participate in the synthesis of recombinant DNA, and/or, when introduced into cells, as robust moieties that enable DNA alterations within the cell cultures.

We explore the use of TiO$_2$ nanocomposites as analogs of restriction enzymes with unique specificity that does not exist in current biological approaches. The main goal of this work is to establish the basic mechanisms involved in the process of the interaction of semiconducting nanoparticles with target biomolecules. Initially our research focuses on understanding the effects of sequence and length of anchoring PNA oligonucleotides on the specificity of cleavage of double-stranded target DNA. We attached anchoring PNA oligonucleotides to TiO$_2$ nanoparticles in a similar procedure, as described previously (Rajh et al., 2004). A short 7-mer PNA sequence (TTTCTCC) was synthesized so that its sequence was complementary to λ DNA at designed locations. The *N*-terminal residue of the sequence was neutralized by acetylation, while the *C*-terminus was activated as a carboxylic acid group by using a Merrifield resin. The conjugation of PNA onto TiO$_2$ nanoparticles was preformed by using DA as a bridge linker. In the first step, the carboxylic acid group at the *C*-terminal PNA was converted to a succinimidyl group and reacted with DA with its terminal amino group. DA-labeled PNA binds to the TiO$_2$ nanoparticles through bidentate binding between DA OH groups with undercoordinated Ti surface atoms. The absorption spectrum of the solution of the PNA–TiO$_2$ nanocomposite indicated that each particle was linked to two or three PNA strands.

The assembling of TiO$_2$ nanoparticles onto DNA was performed by the temperature-controlled hybridization of PNA–TiO$_2$ composites with DNA. Briefly, λ DNA was incubated with PNA–TiO$_2$ in TE buffer (10 mM Tris, 1 mM EDTA, pH 8.0) at room temperature for 2 h and then stored at 4°C for at least overnight. The DNA/PNA–TiO$_2$ hybrids were then stretched and fixed on the surface using a moving interface technique (Rajh et al., 2001). First, a freshly cleaved circular mica substrate (Ted Pella, Redding, California) was treated with poly-L-lysine solution for 1 h. The surface then was rinsed with water and dried under a stream of N$_2$. A 1 μL droplet of the hybrid solution, diluted to ~0.1 nM in TE buffer, was deposited onto the edge of a clean glass cover slip, and then carefully placed onto the top of the poly-L-lysine-treated mica surface. The weight of the cover slip forced the solution to separate immediately into a thin layer. The capillary force of the moving interface aligned the hybrids in one direction. After incubation for 10 min, the sample was washed with water to remove any buffer and unbound DNA. The surface was then analyzed with an AFM under ambient conditions.

A topographical AFM image of λ DNA molecules (Figure 4.18a) shows that the DNA was stretched into linear forms by using the capillary force of the moving air/water interface. The average height of DNA is ~0.7 nm, which is consistent with values measured by other groups (Cai, 2001; Woolley and Kelly, 2001). Contour lengths for λ DNA were not measured because they were expected and observed to exceed the range of the scanner (15 μm). The site-specific binding of the PNA–TiO$_2$ nanocomposite to double-stranded target DNA was observed. Representative AFM images are shown in Figure 4.18b. Particles are observed binding on the side of the DNA with the tip of an elongated nanoparticle as a point of attachment. The section analysis demonstrated that the particle size is 5.5 ± 0.5 nm, which is consistent with the size distribution of the TiO$_2$ nanoparticles (Rajh, 1999a). It is also noted that each DNA strand has two particles bound at a consistent contour distance (770 ± 20 nm, 20 nm is the diameter of the tip). The binding of PNA to both an antiparallel site at 5023–5029 bp and a parallel site at 7307–7313 bp should lead to the observed distance of 770 nm. In addition, anchoring of nanoparticles to the complementary runs of the PNA only in the antiparallel positions located at 5023–5029 bp and 6242–6248 bp in λ DNA would lead to the theoretical distance between the two particles of about 420 nm. We have also observed site-specific binding of nanoparticles assembled on λ DNA with separation of 405 nm. These results indicate that the PNA/TiO$_2$ particle adapter complex binds in both parallel and antiparallel fashion onto the complementary runs in target λ DNA.

Our results demonstrate the potential for using the specific recognition properties between PNA/DNA to assemble nanoparticles onto well-defined structures. Previously, we have found that illumination of this assembly results in selective oxidation of DNA at the deepest "thermodynamic traps" (doublets or triplets of guanines) located closest to the nanoparticle surface (Lewis and Wasielewski, 2004). We have also shown previously that TiO$_2$ particles linked to DNA oligonucleotides possess the biologically unique new property of a light-inducible nucleic acid endonuclease (Paunesku et al., 2003). However, these previous experiments employed nanoparticles that were attached to the 5′ terminus of DNA oligonucleotides, limiting their use to the complementary DNA strand. Thus, the use of

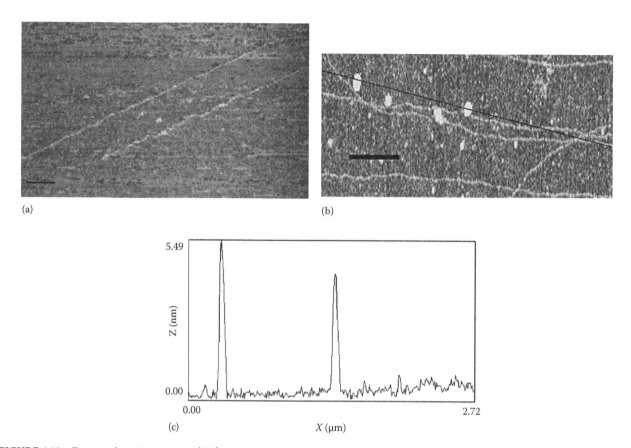

(a)

(b)

(c)

FIGURE 4.18 Topographic AFM images of (a) λ DNA stretched on poly-L-lysine-modified mica surface. (b) DNA/PNA–TiO₂ hybrids stretched on surface and (c) the cross section of the area indicated by the black line in (b). Both images (a) and (b) have a height scale of 5 nm. The scale bars are 500 nm. (From Liu, J.Q. et al., *Proc. SPIE*, 6096, 48–57, SPIE, San Jose, CA, 2006. With permission.)

DNA oligonucleotides as anchoring agents prevented the use of nanoparticles for the manipulation of arbitrary double-stranded target DNA on a desired position for cleavage.

Due to the absence of charge on the PNA backbone, hydrogen bonding between DNA scaffold and PNA adapters is stronger than hydrogen bonding between the two strands of DNA scaffold. As a consequence, PNA binding adapters can easily intercalate into λ DNA by forming DNA/PNA triplex (Uhlmann et al., 1998). Guided by PNA binding adapters, semiconductor particles find unique places for attachments in the target DNA even in equimolar concentrations, suggesting strong binding of PNA adapters to double-stranded DNA at room temperature. Illumination of these composites was found to result in site-specific cleavage of λ DNA, resulting in the formation of DNA fragments with the length that corresponds to the distance between the GGAGAAA sequence in λ DNA runs (Figure 4.19). As it can be observed from AFM image, after cleavage TiO₂ nanoparticles remain attached to the ends of cleaved fragments, suggesting that the points of cleavage are in the close vicinity of TiO₂ surface and most probably involve doublet of guanine in the DNA/PNA triplex.

Herein we demonstrate that nanoparticles linked to PNA adapters preserve light-induced chemistry and are capable of cleaving scaffolding double-stranded DNA in the same manner as restriction enzymes cleave target DNA. These results suggest that the linking of photochemically active nanoparticles to PNA

present ideal approach for developing artificial nanoscale DNA "scissors" that cleave DNA only at recognized segments. Nature utilizes similar principles to protect bacteria from intruding molecules of viral DNA by exploiting proteins as exquisitely precise scissors. Mankind uses these protein-based scalpels as workhorses for diverse applications ranging from mapping of DNA to criminal forensics. In this work we have developed a system that does not require protein for DNA cleavage. Development of artificial robust semiconductor-based DNA "scissors" offers new opportunities for DNA analysis, from DNA sequencing to gene cloning. Semiconductor-based scissors are unique in their ability to cut DNA in rare positions not achievable by using natural restriction enzymes. The unique site specificity of the semiconductor DNA "scissors" is obtained by extending the length of PNA until PNA bound to semiconductor scissors find unique match within the genome. This forms a basis for the synthesis of "rare cutters" nonexistent in the libraries of natural restriction enzymes.

4.3.4 Biology of TiO₂–DNA Nanocomposites

In Sections 4.3.1 through 4.3.3, we have described that TiO₂ nanocomposites retain the intrinsic photocatalytic capacity for carrying out redox functions while retaining the molecular recognition of the oligonucleotide. In this section, we present opportunities to use TiO₂–DNA or TiO₂/PNA nanocomposites

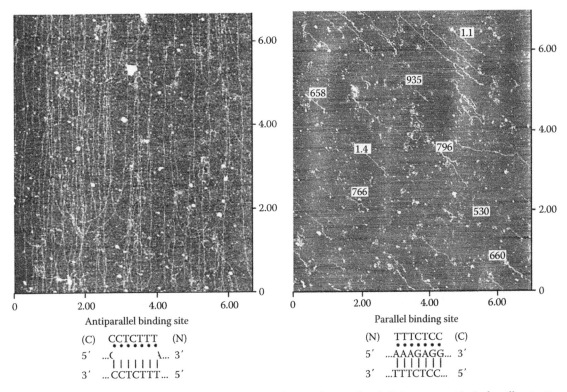

Antiparallel binding site

(C) CCTCTTT (N)
• • • • • •
5′ ...C λ... 3′
| | | | | | |
3′ ...CCTCTTT... 5′

Parallel binding site

(N) TTTCTCC (C)
• • • • • •
5′ ...AAAGAGG... 3′
| | | | | | |
3′ ...TTTCTCC... 5′

FIGURE 4.19 TiO₂ acts as sequence-specific restriction enzyme; left: λ DNA decorated with TiO₂ nanoparticles before illumination and right: TiO₂ decorated λ DNA after illumination when λ DNA is cleaved to the fragments of ~1 Φm length. The sequence of PNA used for attaching TiO₂ nanoparticles to DNA is found in λ DNA genome at eight positions for antiparallel binding and at six positions for parallel binding.

as new vehicles for the manipulation of cell metabolism in vivo and in situ. We have shown that if TiO₂–DNA nanocomposites are present in the excess, they are able to hybridize with long DNA molecules that they would encounter upon entering the cell (Paunesku et al., 2003). We annealed 100 nanocomposites with a 10 kilobase DNA fragment containing a complementary DNA sequence. The hybridization was inspected by atomic force microscopy, which established that the nanocomposites do not clump and that they can withstand incubation at 95°C (which is necessary for annealing and also for polymerase chain reaction (PCR) (Figure 4.20).

We have also shown that TiO₂ oligonucleotide nanocomposites can to be introduced into mammalian cells in vitro by using standard transfection methods. We mapped the location of titanium in cells by detecting titanium-specific Kα x-ray fluorescence induced upon the x-ray excitation of the cells inculcated with TiO₂ nanoparticles. A total of 514 cultured cells from 24 different samples transfected with seven different nanocomposites were inspected for the presence of a titanium signal. Depending on the type of the experiment, 20%–50% of the cells accepted and retained titanium nanoparticles. The acceptance rate was DNA sequence dependent and correlated with the number of complementary sequences in genomic DNA of the investigated cells. The addition of free oligonucleotides generally increased the success of titanium nanocomposite transfection and retention. To confirm that nanocomposites are able to enter the cell nucleus, we isolated nuclei from PC12 cells transfected

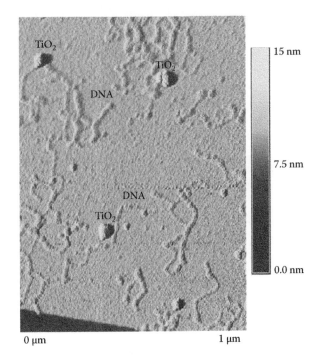

FIGURE 4.20 Atomic force microscopy image of TiO₂–oligonucleotide nanocomposites hybridized with long (phage) DNA. TiO₂–DNA nanocomposites were annealed to long DNA in a HEPES buffer and incubated on freshly cleaved mica. (From Paunesku, T. et al., *Nat. Mater.*, 2, 343, 2003. With permission.)

with TiO$_2$–R18Ss nanocomposite and a "free" R18Sas oligonucleotide. The TiO$_2$–R18Ss oligonucleotide is complementary to the genomic DNA located in the nucleolar region of the nucleus. Scans showing the presence of titanium in 6 out of 13 sampled nuclei demonstrated that the TiO$_2$–DNA nanocomposites, once introduced into mammalian cells, reached the nucleus. The scan presented in Figure 4.21 shows that signals of phosphorus and zinc overlapped with each other, and that the titanium signal shows the highest density in a circular subregion of the nucleus. By its size and shape, this nuclear subregion closely resembles the nucleolus, the subregion of the interphase nucleus where rDNA is located, and would therefore be the most likely nuclear location for the retention of an R18Ss-oligonucleotide-activated TiO$_2$ nanocomposite. Presumably, such retention would be dependent on the hybridization/annealing of R18Ss–TiO$_2$ nanocomposite with the genomic ribosomal 18S rDNA.

As shown in Section 4.3.3, light excitation of TiO$_2$–DNA nanocomposite results in photoinduced endonuclease activity. Photogenerated holes are injected into DNA and are transferred along DNA by a multistep "hopping" charge-transport mechanism. Multiple transfer of the photogenerated holes along the DNA ultimately results in the cleavage of the attached DNA at the point where they accumulate (Von Sonntag, 1987). It is worth noting that the rates of reactions leading to DNA cleavage, that is, the reaction of guanine cation radical with water, are

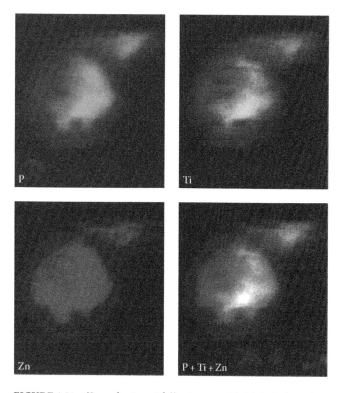

FIGURE 4.21 (See color insert following page 22-8.) A single nucleus containing 3.6×10^6 nanoparticles after incubation of 4×10^6 cells with 80 pmol of R18Ss–TiO$_2$ hybrid nanocomposite. The colocalization of phosphorus (red), titanium (green), and zinc (blue) signals is presented as the overlap of these three colors. (From Paunesku, T. et al., *Nat. Mater.*, 2, 343, 2003. With permission.)

slower than charge recombination in contact ion pairs, which is a process protecting DNA from photochemical damage (Lewis et al., 2001). However, in TiO$_2$ nanocomposites, TiO$_2$ nanoparticles can trap multiple electrons, and this leads to repeated hole creation and accumulation on the DNA molecule attached to the TiO$_2$ nanoparticle.

We annealed nanocomposites TiO$_2$/30 sense and TiO$_2$/50 sense with radiolabeled complementary oligonucleotides. We used the fact that nanocomposites do not enter agarose or polyacrylamide gels during electrophoresis (data not shown), whereas the DNA that is cleaved away from the nanoparticle can enter the gel. To visualize the appearance of the cleaved DNA on the gel during electrophoresis, we radiolabeled the complementary DNA oligonucleotides used for annealing/hybridization with the cleaved DNA. Single-stranded labeled oligonucleotide and double-stranded DNA created by annealing migrate differently after polyacrylamide gel electrophoresis (PAGE). Therefore, it is possible to observe the frequency of cleavage of the DNA away from the TiO$_2$–DNA nanocomposite as an increase in the quantity of the double-stranded DNA in the gel.

The results of a representative experiment are shown in Figure 4.22. As indicated, the three lanes of each sample are obtained for different times of illumination (0, 8, and 16 min, respectively). In both cases, samples were separated into three parts and either left in the dark or illuminated for different times indicated above the figure. Although the amount of single-stranded/not annealed oligonucleotide (TiO$_2$/30 antisense and TiO$_2$/50 antisense) is the same across the three lanes, the amounts of cleaved double-stranded DNA increased with an increase in the illumination time. We would like to note that very similar effects were obtained when TiO$_2$–DNA was hybridized using ^{32}P radioactively labeled oligonucleotides for extended period of times (from 30 to 270 min), indicating that the activation of TiO$_2$ with radiation is equivalent with activation with light (Figure 4.23) (Stojicevic et al.).

To conclude, any molecule synthesized to possess carboxyl groups (such as carboxy-dT-oligonucleotides and long DNA molecules, peptide nucleic acids, and short or long peptides and proteins such as antibodies) can be covalently linked to DA and then stably attached to TiO$_2$ nanoparticles 2–20 nm in size. The experiments shown here are limited to a single type of nanocomposite: DNA oligonucleotides that are attached one, or a few at once to a TiO$_2$ nanoparticle of 45 Å. Nevertheless, even these simple TiO$_2$ oligonucleotide hybrid nanoparticles can perform many chemical and biological tasks, including a light-induced site-specific (within 50 nucleotides) nucleic acid endonuclease activity. Therefore, it would be desirable that these and other types of TiO$_2$–biomolecule nanocomposites can be engineered with novel functional properties to serve as nanodevices for medical biotechnology.

4.3.5 TiO$_2$–Antibody Nanocomposites

Binding of a protein or peptide to TiO$_2$ can enable targeting of the hybrid TiO$_2$–protein conjugate to the specific cell protein or

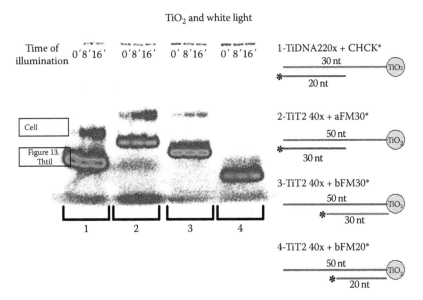

FIGURE 4.22 The cleavage of the double-stranded DNA from the TiO₂ nanoparticle results from the abrogation of the TiO₂–DNA hybrid nanocomposite for different times of illumination. The lanes 1–4 are annealed nanocomposites depicted on the right. 50 nt DNA used as a scaffold fold contains AT-rich base segment as a bridge to TiO₂ nanoparticles that can form a hairpin at room temperature.

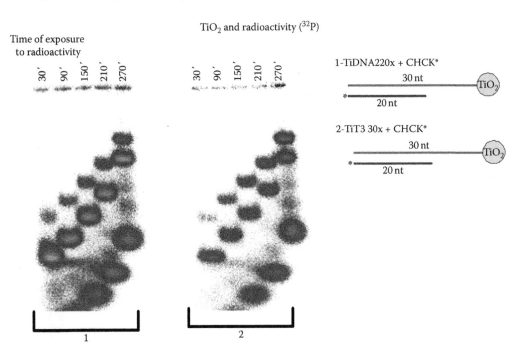

FIGURE 4.23 The cleavage of the double-stranded DNA from the TiO₂ nanoparticles after different times of the exposure to ³²P radioactivity. Note that exposure of 30 nt oligonucleotides hybridized with complementary ³²P-labeled strand do not show DNA cleavage under the same experimental conditions (Stojicevic).

organelle. Of special interest is the binding of monoclonal antibodies or T-cell markers to TiO₂ semiconducting nanoparticles. The reason is that once the conjugated TiO₂-targeting protein binds the target on the cell surface or within the cell matrix, the semiconductor can initiate a series of redox reactions and alter cell functioning. In the same manner, as an oligomer bound to TiO₂ is able to bind target DNA, and after light-induced activation the target DNA was nicked or cleaved, peptide sequence or whole proteins can be used for the lysis of complementary protein structures or whole cells.

Antibodies as cell targeting agents have been used successfully as therapeutics in treating cancer. Antibodies combined with photo-catalytic nanoparticles, such as TiO₂, can form functionalized therapeutics. In this work we created and tested a functionalized TiO₂–Ab conjugate to attack pathogenic cells in psoriasis. We investigated the ability to specifically target and lyse cells associated with psoriases. The strategy was to attach T-cell-specific antibodies to TiO₂ nanoparticles capable of light-induced redox chemistry. The conjugates were allowed to bind T cells and then were activated with either visible or UV light. We use a three-tiered approach. First, the TiO₂ was tested for photocatalytic ability following conjugation to the targeting antibody. Second, the Ab was tested for the ability to bind the target after being conjugated to TiO₂ nanoparticles. Lastly, the conjugates were tested for their ability to specifically lyse targeted cells. It has been shown that in psoriasis T cells migrate to the skin and perpetuate and exacerbate disease symptoms (Schlaak et al., 1994; Gottlieb et al., 1995). The potential therapy then targets T cells specifically by conjugating an antibody specific for T-cell markers to TiO₂. The TiO₂ can then be localized to the T cell with the molecular specificity of markers, where photo-activation would result in the death of T cells while leaving the epidermal cells and Langerhans cells intact.

The first part of this approach requires the conjugation of the targeting protein to TiO₂. The attachment of proteins to photoactive TiO₂ was performed using procedure previously developed for linking PNA oligonucleotides to TiO₂, as described in Section 4.3.3. Similar procedure was also described for linking antibodies to silica nanoparticles using silane chemistry (Zhao et al., 2004). For the targeting protein, we chose antibodies directed against cell surface antigens specific for T cells, namely anti-CD4 antibody (CD4 ab), anti-CD8 (CD8 ab), and anti-CD3 (CD3 ab). The newly formed conjugates were then examined by electron force microscopy that shows that, on average, each antibody was labeled with one nanoparticle (Figure 4.24). The presence of the Ab on TiO₂ was further confirmed by a colorimetric assay using alkaline phosphatase-labeled anti-mouse antibody.

Once it was shown that the antibody could be attached to TiO₂, the functionality of each component was tested. For this purpose, four monoclonal murine antibodies were conjugated to TiO₂. Three of these (α-CD4, α-CD8, and α-CD3) are specific to T-cell epitopes, and a fourth (MOPS), is an isotype-matched negative control antibody. The photocatalytic functionality of TiO₂ in nanoconjugate was tested by the ability of conjugated TiO₂ to deposit metallic Ag upon illumination (Sections 4.3.1

and 4.3.2). When exposed to light, the suspensions of each of the four Ab–TiO₂ nanoconjugates in AgNO₃ rapidly formed silver precipitates. Each batch of nanoconjugates demonstrated similar activities. Unexposed suspensions, TiO₂ suspension, and those lacking TiO₂ formed precipitate much more slowly. The results of a typical silver nitrate experiment are shown in Figure 4.25.

In the next step, we established if the nanoconjugates retain antibodies' biological activity for specific binding and if they are able to bind the target antigen specific of the cell surface. For this purpose, TiO₂ conjugates were incubated with cells expressing the antigens CD4, CD8, and/or CD3 to test the ability of the respective antibody conjugates to bind their targets. Each cell type was incubated with either purified antibodies (CD3, CD4, CD8, or MOPS) or TiO₂–Ab conjugates. A secondary, fluorescently tagged anti-mouse antibody was used for detection in FACS (Figure 4.26). The cells were analyzed by flow cytometric analysis. The experiments show that the α-CD4 antibody and TiO₂–CD4 conjugates bound the 2105 cell lines that express CD4 and CD3 as well as 1942 cell lines that express CD4 and CD8 (both CD4+), but not the 11386 line that does not contain antigen for CD4 expressed on the surface (CD4-). Similarly, both

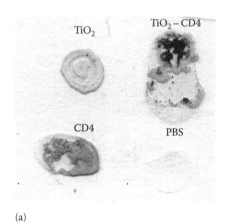

(a)

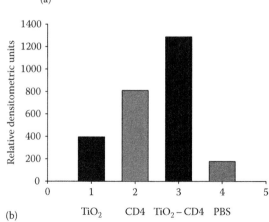

(b)

FIGURE 4.25 TiO₂ conjugate has photo-catalytic functionality. The samples were exposed to UV light for 15 s on a glass slide. The air-dried slide was scanned (a) and the densitometric analysis was performed on the image. The densitometric units are expressed after background subtraction with the treatment listed on the *x*-axis (b).

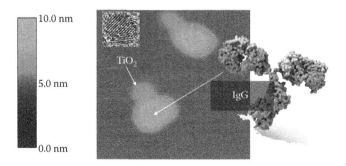

FIGURE 4.24 (See color insert following page 22-8.) Atomic force microscopy image of TiO₂–CD4 nanocomposites. TiO₂–CD4 nanocomposites were dissolved in a HEPES buffer and incubated on freshly cleaved mica.

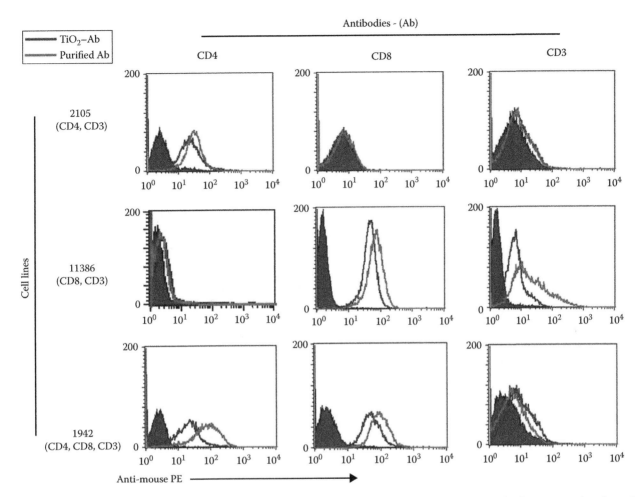

FIGURE 4.26 Antibodies retain targeting ability while linked to TiO$_2$. CD4, CD8, and CD3 are epitopes that mark and are expressed on the surface of T cells. Three different cells lines 1942 (CD4, CD8, and CD3), 2105 (CD4 and CD3), and 11386 (CD8 and CD3) were used to test the ability of either purified antibody of TiO$_2$-linked antibody to target them. Cells were labeled with 5 µg/mL CD4, CD8, or CD3 antibodies and then detected by fluorescently labeled anti-mouse Ab. Antibodies conjugated to TiO$_2$ were used at the same concentration of purified Ab and detected by the same secondary fluorescently labeled antibody. Each cell line was also stained with secondary Ab alone as the unlabeled group in solid black. Flow cytometric analysis was used to examine the data and results are displayed in histogram form with fluorescence on the *x*-axis. Cells lines are indicated on the left and the antibody treatment across the top. Purified antibodies are indicated by gray lines and the TiO$_2$-conjugated Ab are black lines.

α-CD8 and TiO$_2$–CD8 conjugates bound the CD8+ lines 11386 and 1943, but not the CD8- cell line 2105. Although all three cell lines are reported to express CD3, the FACS results show definite binding to either α-CD3 or TiO$_2$–CD3 only for the 11386 cells. It should be noted that the negative control antibody MOPS and the negative control conjugate TiO$_2$–MOPS did not bind cells.

These results have clearly demonstrated the ability to target TiO$_2$ to the cell by using molecular recognition properties of Ab. The important question that remains to be answered at this point is whether TiO$_2$ targeted to the cell of interest is still able to induce cell death after photoactivation. To investigate this outcome, each cell type was incubated with targeting TiO$_2$–Ab conjugate and after photoactivation investigated for cell death using Trypan blue exclusion. The hypothesis is that after photoexcitation, TiO$_2$–Ab nanocomposites will produce either protein damage at the point of binding (antigen site) via oxidation with photogenerated holes or induce oxygen radical-mediated cell damage or programmed

cell death (superoxide radical, singlet oxygen produced via reaction with photogenerated electrons). Both of these would result in targeted cell death (Dimitrijevic et al., 2009).

The most therapeutically useful target in psoriasis is CD4 T cells localized on the skin surface since they have been found to be pathogenic and produce IL-15 which exacerbates disease. We were surprised by the speed and specificity of cell lyse with TiO$_2$–Ab conjugates. When TiO$_2$–CD4 was incubated with the CD4+ cell line within 2 h after 1 min photoactivation, 25% of the cells were killed and additional 30% died 24 h after illumination (see Figure 4.27). Importantly, neither the control Ab, MOPS conjugated to TiO$_2$ nor nonspecific Ab conjugates, like TiO$_2$–CD8 incubated with CD4 cells, caused significant killing. The ability to specifically lyse targeted cells was not limited to TiO$_2$–CD4 on CD4 cell line because TiO$_2$–CD4 was also able to target and lyse CD4+ cell line (positive for CD4, CD8, and CD3), but not those lacking CD4 anti-site (CD4-, such as CD8, CD3 cell line).

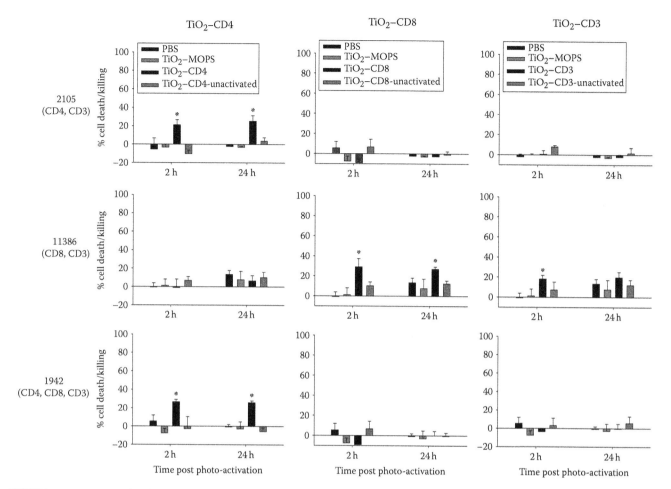

FIGURE 4.27 Functionalized TiO_2 can target and lyse T cells specifically. TiO_2 conjugated to CD4, CD8, CD3, or control Ab (MOPS) were incubated with three different T cell lines as indicated and photo-activated for 1 min. After either 2 or 24 h, an aliquot of cells were enumerated and examined for cells death by Trypan blue exclusion. A plate with similar treatments of each line was also enumerated but the conjugates were not photo-activated (labeled as unactivated TiO_2–Ab). The results are displayed in terms of percent of cell death, on the *y*-axis, at each time point. The cells lines are labeled on the right and the TiO_2 conjugates are labeled across the top. The calculation for percent death at 24 h was calculated with an average of PBS-treated cells as the baseline. *Significant $p < 0.05$ with respect to PBS-treated cells and TiO_2–MOPS-treated cells.

The strategy of conjugating Ab to TiO_2 was functional with other antibodies as well. When TiO_2–CD8 or TiO_2–CD3 was used to target cells, they were both also able to kill their targets specifically. Further, we also demonstrated that the surviving T cells do not oversecrete cytokines (IL-2, IFN-γ, and TNF-α) in response to TiO_2–Ab treatment. These results indicate that cell lysis using TiO_2 composites was not associated with inflammation.

This work demonstrates the power of constructing nanoparticle conjugates which combine the unique qualities of both inorganic and biological components. TiO_2 nanoparticles can be made to harvest a significant part of the light spectrum and convert it into reactive chemical species. In this study, they are successfully linked to antibodies that add the exquisite binding specificity to the composite. Here, we show that the semiconductor properties of materials fabricated on a nanometer scale can be targeted to and destroy the inflammatory cells associated with the pathology of psoriasis.

These nanoconjugates were successfully constructed using several different monoclonal antibodies. Further, different linking chemistries were used for conjugation, with the retention of electron transfer capability. These demonstrate the broad applicability of this technology, and the possibility of optimizing the linkage for maximal performance, as by reducing steric hindrance of the binding site or promoting charge transfer through the organic portion of the nanoconjugate. In each of the FACS experiments showing positive binding to cells, it was observed that there were slightly fewer bindings of nanoconjugates than unconjugated antibody. This is likely due to some degree of steric hindrance induced by the TiO_2 component to either the binding to the cellular epitope and/or binding of the secondary detection antibody. However, the binding of both is comparable, demonstrating that conjugation to the nanoparticles generally left the binding ability of the mAbs intact.

The effect of these nanoconjugates on living cells was notable in its effectiveness, inducibility, and specificity. In a series of viability experiments, around ¼ of CD4+ cells were killed by the TiO_2–CD4 conjugate 2 h after illumination, while no significant killing occurred with the CD4-cell line under the same conditions. Importantly, killing was neither observed for unactivated nanoconjugates nor for the nonspecific TiO$_2$–MOPS nanoconjugate. In a clinical setting, this kind of selectivity and inducibility is of great value, limiting side effects and increasing the safety profile.

The effectiveness of these nanoconjugates lies in their redox potential. However, the biological mechanism by which this electron transfer results in cell death is unknown. Apoptosis was not confirmed, but results are consistent with necrosis. Death occurred soon after treatment, and extended to 24 h after illumination with substantial total cell lysis.

While the subject end goal of this study is the creation of an effective treatment for psoriasis, this application can be easily extended to other disease states. In fact, we view a primary advantage of this technology is its flexibility. We are currently investigating conjugates created using different specificities and classes of targeting molecules. Also, as there are often multiple markers for a disease state, multiple targets can be combined. While steric hindrance limits the creation of multivalent molecular entities, nanoconjugate "cocktails" can be used in a single treatment to target multiple disease-associated epitopes.

4.4 Summary

The long-standing dream of coupling the strong reactivity of semiconductors with the responsiveness of biological materials in order to create an entirely new class of "smart" functional materials was pursued. By welding these seemingly disparate properties, one could unite the robustness of inorganic material and the recognition properties of biological molecules. The major challenge in achieving this dream is bridging the biological molecule to the interface of semiconducting nanoparticles. We have found that the surface of metal-oxide nanocrystalline materials differs from the surface of the bulk materials by the presence of highly reactive under-coordinated surface. This can be viewed as a curse or as an opportunity. The coordination sphere of the surface metal atoms is incomplete and thus traps light-induced charges, but also exhibits high affinity for oxygen-containing ligands and gives the opportunity for chemical modification. Oxygen-rich enediol ligands (readily found in nature as neurotransmitters or antioxidants) form strongly coupled conjugated structures with defect sites and repair the coordination of the surface. As a consequence, the intrinsic properties of a semiconductor change and new hybrid molecular orbitals are generated by mixing the π orbitals of biomolecule and the continuum states of metal oxides. These bio-inorganic conjugates duplicate primary light-induced charge separation mechanism in natural photosynthetic systems resulting in efficient charge separation. Therefore, enediol ligands were used as "conductive leads" or composite "neurotransmitters" that bridge the electronic properties of semiconductors to electroactive biological moieties. Therefore, these unique conjugate hybrid systems (TiO$_2$/enediol) provide means to attach selective functionalities that will lead to the development of highly specialized nanomachines.

We used "conductive linkers" to bridge and initiate the chemical reactions of 5 nm TiO$_2$ nanoparticles with DNA strands. DNA oligonucleotides were linked to metal-oxide nanoparticles through DA—an enediol ligand with amino terminus group—that encapsulates metal-oxide nanoparticles and establishes efficient crosstalk across the biomolecule and metal-oxide interface. The DNA oligonucleotides build in recognition properties to the hybrid system that is used for site-selective reactivities. Metal-oxide nanoparticles, on the other hand, are capable of electronically responding to biomolecule interactions through the mechanism of light-induced charge separation. The frequency of light absorbed by metal oxides was systematically tuned through the chemical modification of nanoparticle surface. Light-induced extended charge separation in these systems was found to be a fingerprint of DNA hybridization and DNA sequence. This approach may lead to a variety of uses such as the development of a new family of DNA-based sensors, electronically tunable site-specific catalysts, new in situ gene therapy procedures, and other medical applications.

A new opportunity of molecular level manipulation of biological molecules was also created by the electronic linking of hybrid semiconductor nanoparticles to DNA molecules. We created artificial nanoscale DNA "scissors" that were designed to bind DNA only at recognized segments. Semiconductor particles are linked to the binding adapters that find unique places for attachments in the target DNA. Fragmentation of target DNA is initiated by light. Nature utilizes similar principles to protect bacteria from intruding the molecules of viral DNA by exploiting proteins as exquisitely precise scissors. Mankind uses them also as workhorses for biotechnology in applications ranging from the cloning of DNA to criminal forensics. However, the use of protein-based DNA scissors is limited to the number of existing naturally expressed proteins. The development of artificial robust semiconductor-based DNA "scissors" offers new opportunities for DNA analysis, from DNA sequencing to gene cloning. Semiconductor-based scissors are unique in their ability to cut DNA in rare positions not achievable by using natural restriction enzymes. Unique site specificity of the semiconductor DNA "scissors" is obtained by extending the "recognizable" segment of binding adapters in a manner that semiconductor scissors find one match only within the genome. This forms a base for the synthesis of "rare cutters" nonexistent in the libraries of natural restriction enzymes (restrictases).

We have also developed a method to control and initiate the chemical reactions of antibodies using novel hybrid nanometer-sized metal-oxide semiconductors. Antibodies were linked to nanoparticles equipped with "conductive leads" that, similar to natural photosynthesis, establish efficient light-induced crosstalk across the biomolecule and metal-oxide interface. We have found that functionally integrated TiO$_2$–antibody complexes retain the photocatalytic properties of nanoparticles and the recognition properties of monoclonal antibodies. Antibodies impart a high degree of specificity to the metal-oxide nanoparticles, "dragging" them only to the specific cells that have antigens on their membrane. In this way, composites are directed to biological targets of interest, such as T lymphocytes. Photoinduced production of reactive oxygenated species (singlet oxygen and superoxide radicals) was found to alter respiratory pathways in mitochondria and causes electron transfer proteins release that triggers programmed cell death.

Acknowledgment

Work at the Center for Nanoscale Materials was supported by the United States Department of Energy, Office of Science, Office of Basic Energy Sciences, under Contract No. DE-AC02-06CH11357.

References

Alivisatos, P.A. 1996. Semiconductor clusters, nanocrystals, and quantum dots. *Science* **271**: 933–937.

Alivisatos, A.P., Johnsson, K.P., Peng, X.G. et al. 1996. Organization of "nanocrystal molecules" using DNA. *Nature* **382**: 609–611.

Archer, M.D. 2004. *Molecular to Global Photosynthesis*, ed. M.D. Archer. Imperial College, London, U.K.

Banwarth, W., Schmidt, D., Stallard, R.L. et al. 1998. Bathophenanthroline-ruthenium(II) complexes as non-radioactive labels for oligonucleotides which can be measured by time-resolved fluorescence techniques. *Helv. Chim. Acta* **71**: 2085.

Bard, A.J. 1994. *Integrated Chemical Systems. A Chemical Approach to Nanotechnology*. Wiley, New York.

Barnard, S. and Zapol, P. 2004a. A model for the phase stability of arbitrary nanoparticles as a function of size and shape. *J. Chem. Phys.* **121**: 4276–4283.

Barnard, S. and Zapol, P. 2004b. Effects of particle morphology and surface hydrogenation on the phase stability of TiO_2. *Phys. Rev. B* **70**: 235403.

Barnett, R.N., Cleveland, C.L., Joy, A. et al. 2001. Charge migration in DNA: Ion-gated transport. *Science* **294**: 567–571.

Bawendi, M.G. 1995. *Prospects of High Efficiency Quantum Boxes Obtained by Direct Epitaxial Growth*, NATO Advanced Science Institutes Series B, Vol. 340, pp. 339–356. Plenum Press, New York.

Berlin, Y.A., Burin, A.L., and Ratner, M.A. 2000. On the long-range charge transfer in DNA. *J. Phys. Chem. A* **104**: 443–445.

Berlin, Y.A., Burin, A.L., and Ratner, M.A. 2002. Elementary steps for charge transport in DNA: Thermal activation vs. tunneling. *Chem. Phys.* **275**: 61–74.

Berlin, Y.A., Kurnikov, I.V., Beratan, D.V. et al. 2004. DNA electron transfer processes: Some theoretical notions. *Top. Curr. Chem.* **237**: 1–36.

Bixon, M. and Jortner, J. 1999. Electron transfer—From isolated molecules to biomolecules. *Adv. Chem. Phys.* **106**: 35–202.

Bixon, M. and Jortner, J. 2006. Shallow traps for thermally induced hole hopping in DNA. *Chem. Phys.* **326**: 252–258.

Bixon, M., Giese, B., Wessely, S. et al. 1999. Long-range charge hopping in DNA. *Proc. Natl. Acad. Sci. U.S.A.* **96**: 11713–11716.

Bonacheva, M., Schibler, L., Lincon, P. et al. 1999. Design of oligonucleotide arrays at interfaces. *Langmuir* **15**: 4317.

Boschloo, G. and Fitzmaurice, D. 1999. Electron accumulation in nanostructured TiO_2 (anatase) electrodes. *J. Phys. Chem. B* **103**: 7860–7868.

Braun, E., Eichen, Y., Sivan, U. et al. 1998. DNA-templated assembly and electrode attachment of a conducting silver wire. *Nature* **391**: 775–778.

Bruchez, Jr., M., Moronne, M., Gin, P. et al. 1998. Semiconductor nanocrystals as fluorescent biological labels. *Science* **281**: 2013.

Brus, L. 1991. Quantum crystallites and nonlinear optics. *Appl. Phys. A-Mater. Sci. Process.* **A53**: 465–474.

Cai, L., Tabata, H., and Kawai, T. 2001. Probing electrical properties of oriented DNA by conducting atomic force microscopy. *Nanotechnology* **12**: 211–216.

Chakrabarti, R. and Klibanov, A.M. 2003. Nanocrystals modified with peptide nucleic acids (PNAs) for selective self-assembly and DNA detection. *J. Am. Chem. Soc.* **125**: 12531–12540.

Chen, L.X., Rajh, T., Wang, Z. et al. 1997. XAFS studies of surface structures of TiO_2 nanoparticles and photo-catalytic reduction of metal ions. *J. Phys. Chem. B* **101**: 10688–10697.

Chen, L.X., Liu, T., Thurnauer, M.C. et al. 2002. Fe_2O_3 nanoparticle structures investigated by X-ray absorption near-edge structure, surface modifications, and model calculations. *J. Phys. Chem. B* **106**: 8539–8546.

Coffer, J.L., Bigham, S.R., Li, X. et al. 1996. Dictation of the shape of mesoscale semiconductor nanoparticle assemblies by plasmid DNA. *Appl. Phys. Lett.* **69**: 3851–3853.

de la Garza, L., Saponjic, Z.V., Dimitrijevic, N.M. et al. 2006. Surface states of titanium dioxide nanoparticles modified with enediol ligands. *J. Phys. Chem. B* **110**: 680–686.

Defoin, A., Defoin-Straatmann, R., Hildenbrand, K. et al. 1986. A new liquid-phase actinometer: Quantum yield and photo-CIDNP study of phenylglyoxylic acid in aqueous-solution. *J. Photochem.* **33**: 237–255.

Dimitrijevic, N.M., Saponjic, Z.V., Rabatic, B.M. et al. 2005. Assembly and charge transfer in hybrid TiO_2 architectures using biotin-avidin as a connector. *J. Am. Chem. Soc. Comm.* **127**: 1344–1345.

Dimitrijevic, N.M., Saponjic, Z.V., Rabatic, B.M. et al. 2007. Effect of size and shape of nanocrystalline TiO_2 on photogenerated charges. An EPR study. *J. Phys. Chem. C* **111**: 14597–14601.

Dimitrijevic, N.M., Rozhkova, E., and Rajh, T. 2009. Dynamics of localized charges in dopamine-modified TiO_2 and their effect on the formation of reactive oxygen species. *J. Am. Chem. Soc.* **131**: 2893–2899.

Dittmer, W.U. and Simmel, F.C. 2004. Chains of semiconductor nanoparticles templated on DNA. *Appl. Phys. Lett.* **85**: 633–635.

Gasper, S.M. and Schuster, G.B. 1997. Intramolecular photoinduced electron transfer to anthraquinones linked to duplex DNA: The effect of gaps and traps on long-range radical cation migration. *J. Am. Chem. Soc.* **119**: 12762–12771.

Giese, B. 2000. Long distance charge transport in DNA: The hopping mechanism *Acc. Chem. Res.* **33**: 631–636.

Giese, B., Wessely, S., Spormann, M. et al. 1999. On the mechanism of long-range electron transfer through DNA. *Angew. Chem. Int. Ed.* **38**: 996–998.

Gottlieb, S.L., Gilleaudeau, P., Johnson, R. et al. 1995. Response of psoriasis to a lymphocyte-selective toxin (DAB(389)IL-2) suggests a primary immune, but not keratinocyte, pathogenic basis. *Nat. Med.* **1**: 442–447.

Hao, E., Anderson, N.A., Asbury, J.B. et al. 2002. Effect of trap states on interfacial electron transfer between molecular absorbates and semiconductor nanoparticles. *J. Phys. Chem. B* **106**: 10191–10198.

Hara, M. and Mallouk, T.E. 2000. Photocatalytic water oxidation by Nafion-stabilized iridium oxide colloids. *Chem. Commun.* **24**: 1903–1904.

Henderson, P.T., Jones, D., Hampikian, G. et al. 1999. Long-distance charge transport in duplex DNA: The phonon-assisted polaron-like hopping mechanism. *Proc. Natl. Acad. Sci. U.S.A.* **96**: 8353–8358.

Henglein, A. 1989. Small-particles research: Physicochemical properties of extremely small colloidal metal and semiconductor particles. *Chem. Rev.* **89**: 1861–1873.

Huttermann, J., Voit, K., Oloff, H. et al. 1984. Specific formation of electron gain and loss centers in x-irradiated oriented fibers of DNA at low-temperatures. *Faraday Discuss. Chem. Soc.* **78**: 135–149.

Joseph, J. and Schuster, G.B. 2006. Emergent functionality of nucleobase radical cations in duplex DNA: Prediction of reactivity using qualitative potential energy landscapes. *J. Am. Chem. Soc.* **128**: 6070–6074.

Jun, Y.-W., Casula, M.F., Sim, J.-H. et al. 2003. Surfactant-assisted elimination of a high energy facet as a means of controlling the shapes of TiO₂ nanocrystals. *J. Am. Chem. Soc.* **125**: 15981–15985.

Kamat, P.V. and Meisel, D. eds., 1997. *Semiconductor Nanoclusters: Physical, Chemical and Catalytic Aspects*, Vol. 103. Elsevier, Amsterdam, the Netherlands.

Kawai, K., Takada, T., Tojo, S. et al. 2003a. Kinetics of weak distance-dependent hole transfer in DNA by adenine-hopping mechanism. *J. Am. Chem. Soc.* **125**: 6842–6843.

Kawai, K., Takada, T., Nagai, T. et al. 2003b. Long-lived charge-separated state leading to DNA damage through hole transfer. *J. Am. Chem. Soc.* **125**: 16198–16199.

Keren, K., Berman, R.S., Buchstab, E. et al. 2003. DNA-templated carbon nanotube field-effect transistor. *Science* **302**: 1380–1382.

Kurnikov, I.V., Tong, G.S.M., Madrid, M. et al. 2002. Hole size and energetics in double helical DNA: Competition between quantum delocalization and solvation localization. *J. Phys. Chem. B* **106**: 7–10.

Lewis, F.D. and Wasielewski, M.R. 2004. Dynamics and equilibrium for single step hole transport processes in duplex DNA. *Top. Curr. Chem.* **236**: 45–65.

Lewis, F.D., Wu, T.F., Zhang, Y.F. et al. 1997. Distance-dependent electron transfer in DNA hairpins. *Science* **277**: 673–676.

Lewis, F.D., Letsinger, R.L., and Wasielewski, M.R. 2001. Dynamics of photoinduced charge transfer and hole transport in synthetic DNA hairpins. *Acc. Chem. Res.* **34**: 159–170.

Lewis, F.D., Liu, J.Q., Zuo, X.B. et al. 2003. Dynamics and energetics of single-step hole transport in DNA hairpins. *J. Am. Chem. Soc.* **125**: 4850–4861.

Lewis, F.D., Zhu, H., Daublain, P. et al. 2006. Crossover from superexchange to hopping as the mechanism for photoinduced charge transfer in DNA hairpin conjugates. *J. Am. Chem. Soc.* **128**: 791–800.

Li, X., Peng, Y., Ren, J. et al. 2006. Effect of DNA flanking sequence on charge transport in short DNA duplexes. *Biochemistry* **45**: 13543–13550.

Liu, C.S. and Schuster, G.B. 2003. Base sequence effects in radical cation migration in duplex DNA: Support for the polaron-like hopping model. *J. Am. Chem. Soc.* **125**: 6098–6102.

Liu, J.Q., de la Garza, L., Zhang, L.G. et al. 2007. Photocatalytic probing of DNA sequence by using TiO₂/dopamine-DNA treads. *Chem. Phys.* **339**: 154–163.

Ma, Y.F., Zhang, J.M., Zhang, G.J. et al. 2004. Polyaniline nanowires on Si surfaces fabricated with DNA templates. *J. Am. Chem. Soc.* **126**: 7097–7101.

Micic, O.I. and Nozik, A.J. 1996. Synthesis and characterization of binary and ternary III-V quantum dots. *J. Lumin.* **70**: 95–107.

Monson, C.F. and Wooley, A.T. 2003. DNA-Templated construction of copper nanowires. *Nano Lett.* **3**: 359–363.

Morison, S.R. 1980. *Electrochemistry at Semiconductor and Oxidized Metal Electrodes*. Plenum Press, New York.

Nakao, H., Shiigi, H., Yamamoto, Y. et al. 2003. Highly ordered assemblies of Au nanoparticles organized on DNA. *Nano Lett.* **3**: 1391–1394.

Nickels, P., Dittmer, W.U., Beyer, S. et al. 2004. Polyaniline nanowire synthesis templated by DNA. *Nanotechnology* 1524–1529.

Niederberger, M., Garnweitner, G., Krumeich, F. et al. 2004. Tailoring the surface and solubility properties of nanocrystalline titania by a nonaqueous in situ functionalization process. *Chem. Mater.* **16**: 1202–1208.

Nielsen, P.E. 1999. Applications of peptide nucleic acids. *Curr. Opin. Mol. Biol.* **10**: 71–75.

Patolsky, F., Lichtenstein, A., and Willner, I. 2000. Amplified microgravimetric quartz-crystal-microbalance assay of DNA using oligonucleotide-functionalized liposomes or biotinylated liposomes. *J. Am. Chem. Soc.* **122**: 418.

Paunesku, T., Rajh, T., Wiederrecht, G. et al. 2003. Biology of TiO₂-oligonucleotide nanocomposite. *Nature Mater.* **2**: 343–346.

Pease, A.C., Solas, D., and Sullivan, E.J. 1994. Light-generated oligonucleotide arrays for rapid DNA-sequence analysis. *Proc. Natl. Acad. Sci. U.S.A.* **91**: 5022.

Rabatic, B.M., Dimitrijevic, N.M., Cook, R.E. et al. 2006. Spatially confined corner defects induce chemical functionality of TiO₂ nanorods. *Adv. Mater.* **18**: 1033–1037.

Rajh, T., Peterson, M.W., Turner, J.A. et al. 1987. Size quantization in small colloidal CdS particles studied with stopped flow spectrometry. *J. Electroanal. Chem. Interfac. Electrochem.* **228**: 55–68.

Rajh, T., Nedeljkovic, J., Chen, L.X. et al. 1998. Photoreduction of Copper on TiO_2 nanoparticles modified with polydentate ligands. *J. Adv. Oxid. Technol.* **3**: 292–298.

Rajh, T., Nedeljkovic, J., Chen, L.X. et al. 1999a. Improving optical and charge separation properties of nanocrystalline TiO_2 by surface modification with vitamin C. *J. Phys. Chem. B* **103**: 3515–3519.

Rajh, T., Thurnauer, M.C., Thiyagarajan, P. et al. 1999b. Structural characterization of self-organized TiO_2 nanoclusters studied by small angle neutron scattering. *J. Phys. Chem. B* **103**: 2172–2177.

Rajh, T., Poluektov, O., Dubinski, A.A. et al. 2001. Spin polarization mechanisms in early stages of photoinduced charge separation in surface-modified TiO_2 nanoparticles. *Chem. Phys. Lett.* **344**: 31–39.

Rajh, T., Chen, L. X., Lukas, K. et al. 2002. Surface restructuring of nanoparticles: An efficient route for ligand-metal oxide crosstalk. *J. Phys. Chem. B* **106**: 10543–10552.

Rajh, T., Makarova, O.V., and Thurnauer, M.C. 2003a. Surface modification of TiO_2: A route for efficient semiconductor assisted photocatalysis. In *Synthesis, Functionalization and Surface Treatment of Nanoparticles*, ed. M.-I. Barton, pp. 147–171. ASP, Stevenson Ranch, CA.

Rajh, T., Poluektov, O.I., and Thurnauer, M.C. 2003b. Charge separation in titanium dioxide nanocrystalline semiconductor revealed by magnetic resonance. In *Chemical Physics of Nanostructured Semiconductors*, Chapter 1, eds. A.I. Kokorin and D.W. Bahnemann, pp. 1–34. NOVA Science Publ. Inc., New York.

Rajh, T., Saponjic, Z., Liu, J. et al. 2004. Charge transfer across the nanocrystalline-DNA interface: Probing DNA recognition. *Nano Lett.* **4**: 1017–1023.

Redfern, P.C., Zapol, P., Curtiss, L.A. et al. 2003. Computational studies of catechol and water interactions with titanium oxide nanoparticles. *J. Phys. Chem. B* **107**: 11419–11427.

Richter, J., Seidel, R., Kirsch, R. et al. 2000. Nanoscale palladium metallization of DNA. *Adv. Mater.* **12**: 507–509.

Richter, J., Mertig, M., Pompe, W. et al. 2001. Low-temperature resistance of DNA-templated nanowires. *Appl. Phys. Lett.* **78**: 536–538.

Roginskaya, M., Bernhard, W.A., and Razskazovskiy, Y. 2004. Diffusion approach to long distance charge migration in DNA: Time-dependent and steady-state analytical solutions for the product yields. *J. Phys. Chem. B* **108**: 2432–2437.

Saponjic, Z.V., Dimitrijevic, N.M., Tiede, D.M. et al. 2005. Shaping nanometer-scale architecture through surface chemistry. *Adv. Mater.* **17**: 965–971.

Schlaak, J.F., Buslau, M., Jochum, W. et al. 1994. T-cells involved in psoriasis-vulgaris belong to the TH1 subset. *J. Invest. Dermatol.* **102**: 145–149.

Schuster, G.B. 2000. Long-range charge transfer in DNA: Transient structural distortions control the distance dependence. *Acc. Chem. Res.* **33**: 253–260.

Schuster, G.B. and Landman, U. 2004. The mechanism of long-distance radical cation transport in duplex DNA: Ion-gated hopping of polaron-like distortions. *Top. Curr. Chem.* **236**: 139–161.

Stojicevic, N., Paunesku, T., Rajh, T. et al. unpublished results.

Storhoff, J.J. and Mirkin, C.A. 1999. Programmed materials synthesis with DNA. *Chem. Rev.* **99**: 1849.

Takada, T., Kawai, K., Cai, X. et al. 2004. Charge separation in DNA via consecutive adenine hopping. *J. Am. Chem. Soc.* **126**: 1125–1129.

Taton, T.A., Mirkin, C.A., and Letsinger, R.L. 2000. Scanometric DNA array detection with nanoparticle probes. *Science* **289**: 1757.

Thurnauer, M.C., Dimitrijevic, N.M., Poluektov, O.G. et al. 2004. Photoinitiated charge separation: From photosynthesis to nanoparticles. *The Spectrum* **17**: 10–15.

Treadway, C.R., Hill, M.G., and Barton, J.K. 2002. Charge transport through a molecular π-stack: Double helical DNA. *Chem. Phys.* **281**: 409–428.

Uemura, S., Shimakawa, T., Kusabuka, K. et al. 2001. Template photopolymerization of dimeric aniline by photocatalytic reaction with Ru(bpy)(3)(2+) in the presence of DNA. *J. Mater. Chem.* **11**: 267–268.

Uhlmann, E., Peyman, A., Breipohl, G. et al. 1998. PNA: Synthetic polyamide nucleic acids with unusual binding properties. *Angew. Chem. Int. Ed.* **37**: 2796–2823.

Vega-Arroyo, M., LeBreton, P.R., Rajh, T. et al. 2003. Theoretical study of the ionization potential of thymine: Effect of adding conjugated functional groups. *Chem. Phys. Lett.* **308**: 54–62.

Vega-Arroyo, M., LeBreton, P.R., Zapol, P. et al. 2007. Theoretical studies of the optical properties of TiO_2-dopamine complex. *Chem. Phys.* **339**: 164–172.

Von Sonntag, C. 1987. *The Chemical Basis of Radiation Biology.* Taylor & Francis, Philadelphia, PA.

Wasielewski, M.R., Niemczyk, M.P., Johnson, D.G. et al. 1989. Ultrafast photoinduced electron-transfer in rigid donor-spacer-acceptor molecules - modification of spacer energetics as a probe for superexchange. *Tetrahedron* **45**: 4785–4806.

Woolley, A.T. and Kelly, R.T. 2001. Deposition and characterization of extended single-stranded DNA molecules on surfaces. *Nano Lett.* **1**: 345–348.

Zhao, X., Hilliard, L.R., Mechery, S.J. et al. 2004. A rapid bioassay for single bacterial cell quantitation using bioconjugated nanoparticles. *Proc. Natl. Acad. Sci. U.S.A.* **101**: 15027–15032.

Zhou, Z., Friesner, R.A., and Brus, L.E. 2003. Electronic structure of 1 to 2 nm diameter silicon core/shell nanocrystals: Surface chemistry, optical spectra, charge transfer and doping. *J. Am. Chem. Soc.* **125**: 15599–15607.

Zubavichus, Y.V., Slovokhotov, Yu.L., Nazeeruddin, M.K. et al. 2002. Structural characterization of solar cell prototypes based on nanocrystalline TiO_2 anatase sensitized with Ru complexes. X-ray diffraction, XPS, and XAFS spectroscopy study. *Chem. Mater.* **14**: 3556–3563.

5

Nanocolorants

Qing Zhang
Shanghai Jiao-Tong University

5.1 Introduction

Colorants are normally understood to include both pigments and dyestuff. Pigments refer mainly to inorganic salts and oxides, such as iron and chromium oxides, which are usually dispersed in crystal or powder form in an application medium. The color properties of the dispersion depends on the particle size and form of the pigment. Pigment colorants tend to be highly durable, heat stable, solvent resistant, lightfast, and migration fast. On the other hand, they also tend to be hard to process and have poor color brilliance and strength. Dyes (also called dyestuff) are conventionally understood to refer to organic molecules dissolved, as molecular chromophores, in the application medium. Examples are azo dyes, coumarin dyes, and perylene dyes. The color imparted by dyestuff to the resulting solution depends on the electronic properties of the chromophore molecule. Dyestuff colorants tend to have excellent brilliance and color strength, and are typically easy to process, but also have poor durability, poor heat and solvent stability, and high migration. Because of the contrasting properties of both types of colorants, much work has been done trying to improve the attributes of each class of colorant.

Nanocolorants were regarded as a new class of colorants that could get out of dilemma between dyestuffs and organic pigments. In nature, nanocolorants are a class of nanocomposites that recombine dye acted as an essential ingredient and suitable polymeric matrix, and its performance target is to integrate excellent chromatic properties and good processibility of dyestuffs and good durability of organic pigments. Nanocolorants are new kind of colorants that can combine the advantages of both pigments and dyes, and will be promisingly applied to photoelectric high-tech fields. The recent progress in the preparations and applications of nanocolorants, and the commentary on them, especially with the focus on discussing the preparation of nanocolorants via miniemulsion polymerization, are summarized.

5.2 Background

5.2.1 Brief Introduction to Colorant [1]

5.2.1.1 What Is a Colorant?

A *colorant* is something added to something else to cause a change in color. Colorants can be dyes, pigments, inks, paint, and other colored chemicals.

5.2.1.2 Dyes

A *dye* can generally be described as a colored substance that has an affinity to the substrate to which it is being applied. The dye is generally applied in an aqueous solution and may require a mordant to improve the fastness of the dye on the fiber. Archaeological evidence shows that, particularly in India and the Middle East, dyeing has been carried out for over 5000 years. The dyes were obtained from animal, vegetable, or mineral origin, with no or very little processing. By far, the greatest source of dyes has been from the plant kingdom, notably roots, berries, bark, leaves, and wood, but only a few have ever been used on a commercial scale. Since the first man-made (synthetic) organic dye, mauveine, was discovered by William Henry Perkin in 1856, thousands of synthetic dyes have since been prepared. Synthetic dyes quickly replaced the traditional natural dyes. They costed less, offered a vast range of new colors, and imparted better properties upon the dyed materials.

5.2.1.2.1 Classification of Dyes

1. Dyes are now classified according to how they are used in the dyeing process:
   - *Acid dyes* are water-soluble anionic dyes that are applied to fibers such as silk, wool, nylon, and modified acrylic fibers using neutral to acid dyebaths. Attachment to the fiber is attributed, at least partly, to salt formation between anionic groups in the dyes and cationic groups in the fiber. Acid dyes are not substantive to cellulosic fibers. Most synthetic food colors fall in this category.
   - *Basic dyes* are water-soluble cationic dyes that are mainly applied to acrylic fibers, but find some use for wool and silk. Usually acetic acid is added to the dyebath to help the uptake of the dye onto the fiber. Basic dyes are also used in the coloration of paper.
   - *Direct or substantive dyeing* is normally carried out in a neutral or slightly alkaline dyebath, at or near boiling point, with the addition of either sodium chloride (NaCl) or sodium sulfate (Na_2SO_4). Direct dyes are used on cotton, paper, leather, wool, silk, and nylon. They are also used as pH indicators and as biological stains.
   - *Mordant dyes* require a mordant, which improves the fastness of the dye against water, light, and perspiration. The choice of mordant is very important as different mordants can change the final color significantly. Most natural dyes are mordant dyes, and there is therefore a large literature base describing dyeing techniques. The most important mordant dyes are the synthetic mordant dyes, or chrome dyes, used for wool; these comprise some 30% of dyes used for wool, and are especially useful for black and navy shades. The mordant, potassium dichromate, is applied as an after-treatment. It is important to note that many mordants, particularly those in the heavy metal category, can be hazardous to health, and extreme care must be taken in using them.
   - *Vat dyes* are essentially insoluble in water and incapable of dyeing fibers directly. However, reduction in alkaline liquor produces the water-soluble alkali metal salt of the dye, which, in this leuco form, has an affinity for the textile fiber. Subsequent oxidation reforms the original insoluble dye. The color of denim is due to indigo, the original vat dye.
   - *Reactive dyes* utilize a chromophore attached to a substituent that is capable of directly reacting with the fiber substrate. The covalent bonds that attach reactive dye to natural fibers make them among the most permanent of dyes. Reactive dyes are by far the best choice for dyeing cotton and other cellulose fibers.
   - *Disperse dyes* were originally developed for the dyeing of cellulose acetate, and are substantially water insoluble. The dyes are finely ground in the presence of a dispersing agent and then sold as a paste, or spray-dried and sold as a powder. Their main use is to dye polyester, but they can also be used to dye nylon, cellulose triacetate, and acrylic fibers.
   - *Azo dyeing* is a technique in which an insoluble azoic dye is produced directly onto or within the fiber. This is achieved by treating a fiber with both diazoic and coupling components. With suitable adjustment of dyebath conditions, the two components react to produce the required insoluble azo dye.
   - *Sulfur dyes* are two-part "developed" dyes used to dye cotton with dark colors. The initial bath imparts a yellow or pale chartreuse color. This is after treated with a sulfur compound in place to produce the dark black we are familiar with in socks.
   - *Food dyes* are another class of dyes that describe the role of dyes rather than their mode of use. Because food dyes are classed as food additives, they are manufactured to a higher standard than some industrial dyes. Food dyes can be direct, mordant, and vat dyes, and their use is strictly controlled by legislation. Many are azoic dyes, although anthraquinone and triphenylmethane compounds are used for colors such as green and blue. Some naturally occurring dyes are also used.
   - *Other important dyes*
     A number of other classes have also been established, including: oxidation bases, for mainly hair and fur; leather dyes, for leather; fluorescent brighteners, for textile fibers and paper; solvent dyes, for wood staining and producing colored lacquers, solvent inks, coloring oils, waxes; and carbene dyes, a recently developed method for coloring multiple substrates.

2. Chemical classification
   By the nature of their chromophore, dyes are divided into
   - *Acridine dyes*, derivates of acridine,

   i.e., Acridine orange, CAS number 494-38-2,
   IUPAC name: *N,N,N′,N′*-tetramethylacridine-3, 6-diamine

- *Anthraquinone dyes*, derivates of 9,10-anthracenedione,

i.e., Alizarin CAS number 72-48-0
IUPAC name: 1,2-Dihydroxyanthraquinone

- *Arylmethane dyes*
  –Diarylmethane dyes, based on diphenyl methane
  i.e., Auramine O, CAS number 2465-27-2

  IUPAC name: bis[4-(dimethylamino)phenyl]metha-niminium chloride

- *Triarylmethane dyes*, derivates of triphenyl methane,

i.e., Methyl blue, CAS number 28983-56-4

- *Azo dyes*, based on azo structure "–N=N–"
  i.e., Fast Yellow AB, CAS number 2706-28-7
  IUPAC name: 2-Amino-5-[(E)-(4-sulfophenyl)diazenyl]-benzenesulfonic acid

- *Cyanine dyes or phthalocyanine dyes*, derivates of phthalocyanine,
  i.e., Phthalocyanine Blue BN

- *Diazonium dyes*, based on diazonium salts

- *Nitro dyes*, based on a –NO$_2$ nitro functional group

- *Nitroso dyes*, based on a –N=O nitroso functional group

- *Quinone-imine dyes*, derivates of quinone
  1,2-Benzoquinone

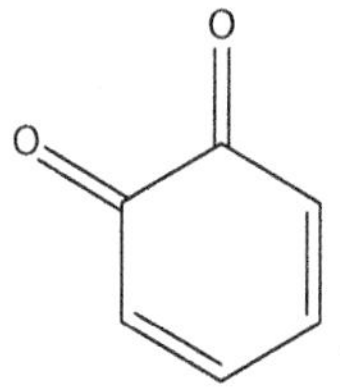

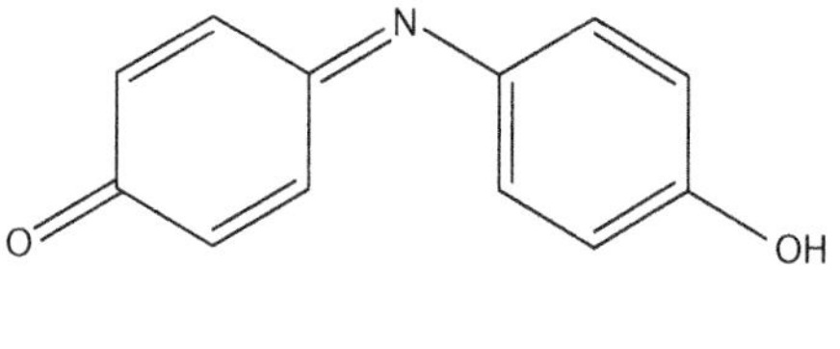

- *Oxazin dyes*, derivates of oxazin,
  i.e., Nile blue, CAS number 3625-57-8

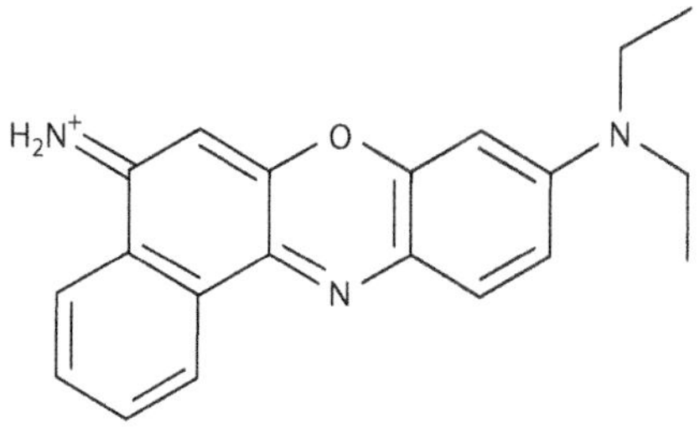

1,4-Benzoquinone

- *Oxazone dyes*, derivates of oxazone
  i.e., Nile red, CAS number 7385-67-3
  IUPAC name: 7-Diethylamino-3,4-benzophenoxazine-2-one

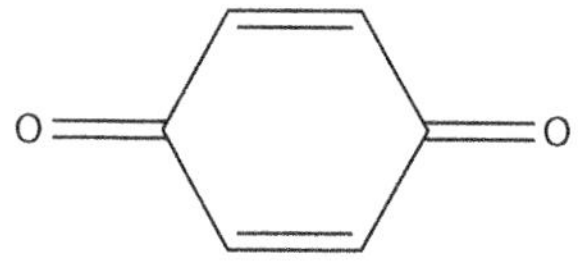

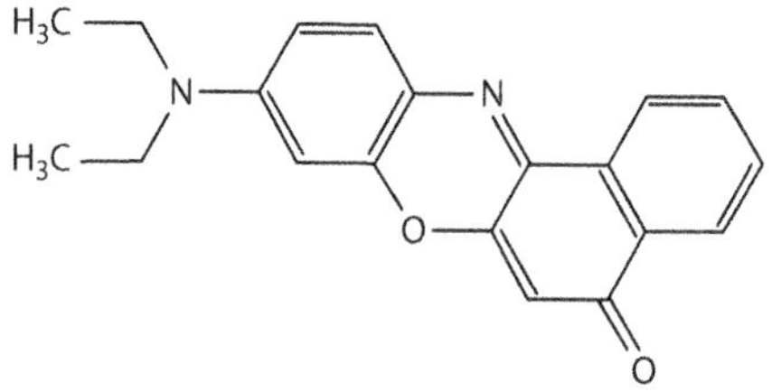

2,3-Dichloro-5,6-dicyano-1,4-benzoquinone

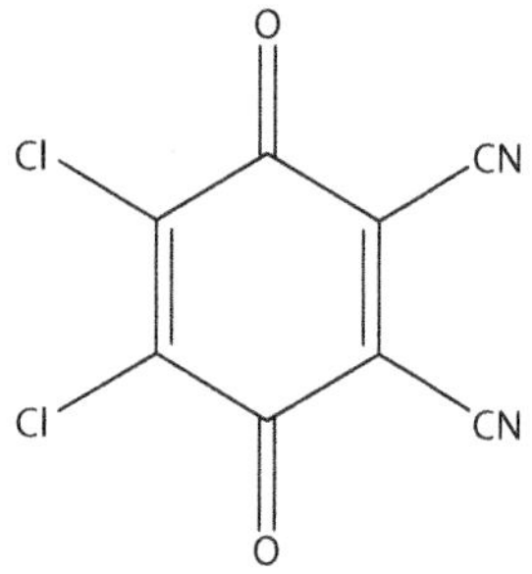

- *Thiazin dyes*, derivates of thiazin
  i.e., Methylene blue, CAS number 61-73-4
  IUPAC name: 3,7-bis(Dimethylamino)-phenothiazin-5-ium chloride

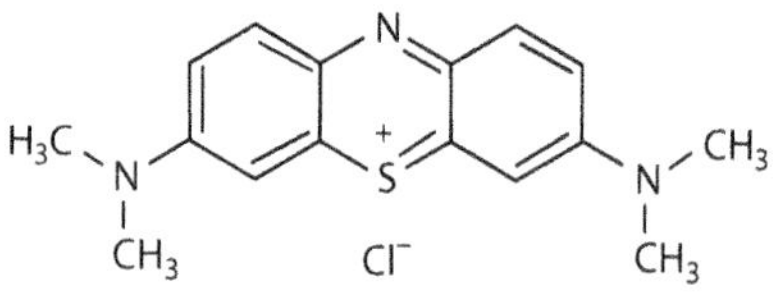

- *Azin dyes*
  Eurhodin dyes, neutral red

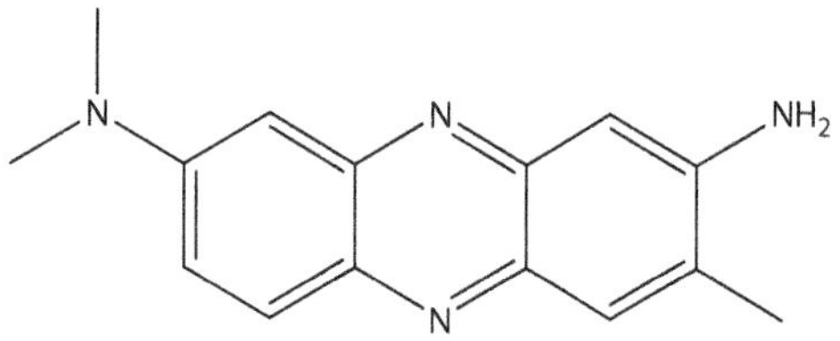

- *Thiazole dyes*, derivates of thiazole

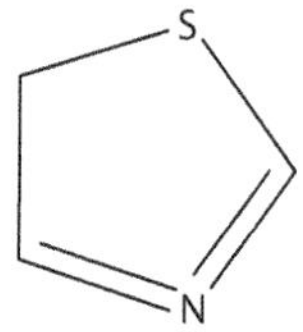

Safranin dyes

i.e., Primuline, CAS number 8064-60-6

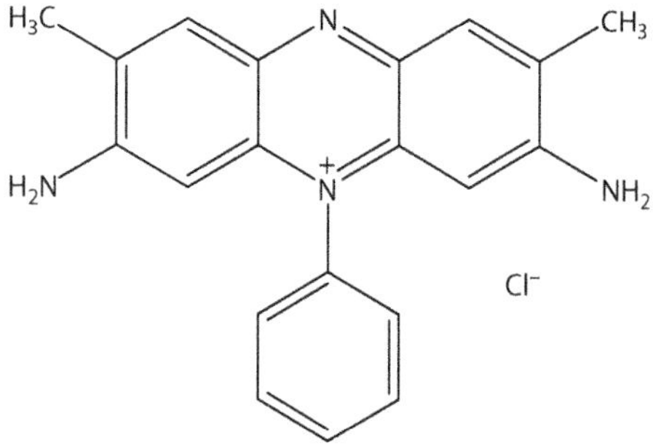

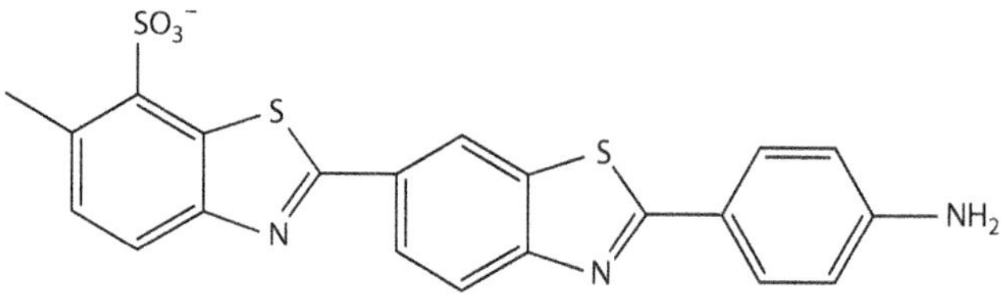

- *Indamins*
  Indophenol dyes, derivates of indophenol

- *Xanthene dyes*, derived from xanthene
  9*H*-xanthene, 10*H*-9-oxaanthracene

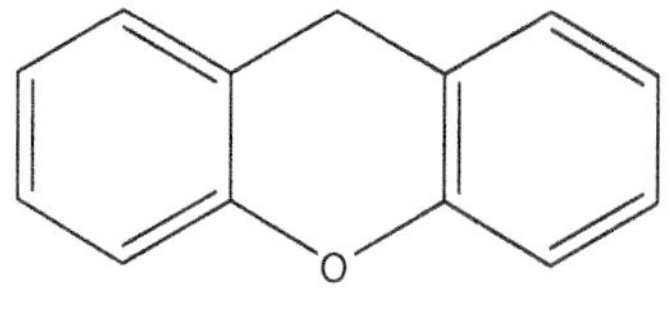

- *Fluorene dyes*, derivates of fluorene

- *Pyronin dyes*
  i.e., Pyronin B

- *Fluorone dyes*, based on fluorone

- *Rhodamine dyes*, derivates of rhodamine rhodamine 123

5.2.1.3 Pigments

A *pigment* is a material that changes the color of light it reflects as a result of selective color absorption. This physical process differs from fluorescence, phosphorescence, and other forms of luminescence, in which the material itself emits light. Both dyes and pigments appear to be colored because they absorb some wavelengths of light preferentially. In contrast with a dye, a pigment generally is insoluble, and has no affinity for the substrate.

Many materials selectively absorb certain wavelengths of light. Materials that humans have chosen and developed for use as pigments usually have special properties that make them ideal for coloring other materials. A pigment must have a high tinting strength relative to the materials it colors. It must be stable in solid form at ambient temperatures.

For industrial applications, as well as in the arts, permanence and stability are desirable properties. Pigments that are not permanent are called fugitive. Fugitive pigments fade over time, or with exposure to light, while some eventually blacken. Pigments

are used for coloring paint, ink, plastic, fabric, cosmetics, food, and other materials. Most pigments used in manufacturing and the visual arts are dry colorants, usually ground into a fine powder. This powder is added to a vehicle (or matrix), a relatively neutral or colorless material that acts as a binder.

A distinction is usually made between a pigment, which is insoluble in the vehicle (resulting in a suspension), and a dye, which either is itself a liquid or is soluble in its vehicle (resulting in a solution). The term biological pigment is used for all colored substances independent of their solubility. A colorant can be both a pigment and a dye, depending on the vehicle it is used in. In some cases, a pigment can be manufactured from a dye by precipitating a soluble dye with a metallic salt. The resulting pigment is called a lake pigment, and based on the salt used, they could be aluminum lake, calcium lake, or barium lake pigments.

5.2.1.3.1 Pigments by Chemical Composition

1. Metallic and carbon
   - Cadmium pigments: Cadmium Yellow, Cadmium Red, cadmium pigments, Cadmium Green, Cadmium Orange
   - Carbon pigments: Carbon Black, Ivory Black, Vine Black, Lamp Black
   - Chromium pigments: Chrome Yellow, Chrome Green
   - Cobalt pigments: Cobalt Violet, Cobalt Blue, Cerulean Blue, Aureolin (Cobalt Yellow)
   - Copper pigments: Han Purple, Egyptian Blue, Paris Green, Verdigris, Viridian
   - Iron oxide pigments: Sanguine, Caput Mortuum, Oxide Red, Red Ochre, Venetian Red, Prussian Blue
   - Clay earth pigments (iron oxides): Yellow Ochre, Raw Sienna, Burnt Sienna Raw Umber, Burnt Umber
   - Lead pigments: Lead White, Cremnitz White, Naples Yellow, Red Lead
   - Mercury pigments: Vermilion
   - Titanium pigments: Titanium Yellow, Titanium Beige, Titanium White, Titanium Black
   - Ultramarine pigments: Ultramarine, Ultramarine Green Shade, French Ultramarine
   - Zinc pigments: Zinc White
2. Biological and organic pigments
   - Biological origins: Alizarin (synthesized), Alizarin Crimson (synthesized), Gamboge, Cochineal Red, Rose Madder, Indigo, Indian Yellow, Tyrian Purple, Gamboge
   - Nonbiological organic: Quinacridone, Magenta, Phthalo Green, Phthalo Blue, Pigment Red 170

5.2.2 What Are Nanocolorants? [2–5]

Generally speaking, any kind of colored complex or mixture contains a dye or a pigment dispersed in nanometer scale could be called a nanocolorant. A nanocolorant can be simply defined as a kind of nano-complex which a dye is dispersed in molecular form and fixed in the cross-linked polymer latex. Nanocolorants are a class of nanocomposites which combine dye as an essential ingredient and suitable polymeric matrix. The performance target is to integrate excellent chromatic properties and good processibility of dyestuffs and good durability of organic pigments. Nanocolorants are new kind of colorants that can combine the advantages of both pigments and dyes, and will be promisingly applied to photoelectric high-tech fields. The recent progress in the preparation and applications of nanocolorants, and the commentary on them, especially with the focus on discussing the preparation of nanocolorants via miniemulsion polymerization, are summarized below.

5.3 Process for Preparing Nanocolorants

5.3.1 Traditional Mechanical Grinding [6]

When the particle size of pigments in ink-jet inks is reduced below 500 nm, the ink compositions achieved extraordinary stability. No dispersant is necessary, and the pigment particles having such a defined size may be incorporated into microemulsions, using a suitable oil (water-immiscible organic compound), at least one amphiphile, and water. Fifteen compositions have been prepared using permutations of three different vehicles and five pigments. The vehicles were based on (a) ethylene glycol phenyl ether (EPH), (b) various invert compositions, and (c) a mixture of xylenes. The pigments were unmodified carbon black (Carbon Black FW18, Degussa), Hostafine Black TS, Hostafine Black T, Hostafine Magenta (HOECHST, Clariant), and Imperon Brilliant Pink (Dystar). Thirteen of the microemulsions are spontaneously formed, providing a success rate of about 86%, in random attempts to make microemulsions from fine particles. The system achieves its stability due to unique size compatibility and adhesion-related "soaking" or submerging of the pigment particles inside the oil phase of the "oil droplet" in the microemulsion.

This is a versatile, broadly enabling, self-sufficient, cost-effective, and yet a relatively simple means for dispersing "off-the-shelf" pigments and colorants without any special treatments, modifications, or use of special dispersants. There are several environmentally benign, nontoxic, noncorrosive microemulsion systems available for extension to the ink technology. These may allow the resolution of material compatibility problems. The microemulsions are formed by mixing a water-immiscible organic solvent, water, and the pigment particles. The pigment particles may be either reduced in size in the microemulsion or prior to mixing with the solvent to form the microemulsion. Preferably, the solvents and amphiphile (preferably, hydrotrope) are mixed first, followed by the addition of water. The particle sizing is typically done after the mixture has been prepared. Ball milling to reduce particle size may be done with zirconia beads for 24 h. Alternatively, a microfluidizer, available from Microfluidics Corporation (Newton, Mass.), may be used. The particles are then filtered through a 5 µm filter, which has been found to result in average particle sizes of less than 500 nm. The concentration of the pigment is generally in the range from about 0.1 to 15 wt% of the ink composition. Here is an example:

A xylene vehicle was prepared by combining 63 g Triton X-100 (a surfactant), 42 g sulfated castor oil, and 36.5 g xylene, providing a total of 141.5 g, which was then diluted to 1000 g with water. A microliquid capsulate suspension was prepared by combining 4.40 g carbon black and 100 g xylene vehicle. The combination was transferred to a ceramic container charged with 100 g of zirconia bead media and the container was rolled until the composition could be filtered through a 5 µm filter (usually for 24 h). Alternatively, fluidization using a microfluidizer for 1 h at 15,000 psi operating pressure accomplishes the same result. This indicates that the average particle size of less than 500 nm has been reached. AEPD (2-amino-2-ethyl-1, 3-propanediol) was added to optimize the performance of the inks. The inks were filtered through 5 µm and filled in pens for use in a DeskJet® 600 printer. Good (acceptable) printing was produced.

5.3.2 Dispersion of Organic Pigments Using Supercritical Carbon Dioxide [7]

In mechanical dispersion methods, shearing forces are usually applied to disperse the aggregated particles into a separate and fine particle, but result in a long processing time and contamination problems when the dispersing apparatus is washed after completion of the dispersion process. In order to reduce these drawbacks, a dispersion method has been proposed as follows: the dispersoid is mixed in a supercritical state and the dispersoid, including the supercritical fluid (SCF), is then rapidly expanded to finely divide the dispersoid; and finally the fine particles are dispersed into a solvent, as illustrated in Figure 5.1.

Under the influence of pressure and temperature, pure substances can assume a gas, liquid, and solid state of matter, except where the equilibrium saturation curve converges such that all three phases co-exit at the triple point. Extension of the liquid–gas phase line ends at the critical point and represents the maximum temperature and pressure at which the liquid and vapor phases coexist in equilibrium, after which gas and liquid have

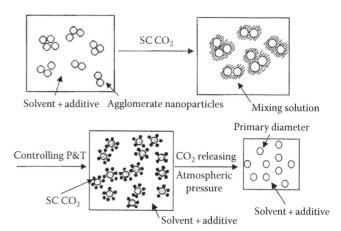

FIGURE 5.1 The concept of dispersion of aggregate pigments using supercritical carbon dioxide. (From Cheng, W.T. et al., *J. Colloid Interface Sci.*, 270, 106, 2004. With permission.)

the same density and appear as a single phase. A fluid is said to be a SCF when its temperature and pressure are in a state above its critical temperature and critical pressure, permitting the gaseous and liquid phases to coexist.

SCFs possess a liquid-like density and solubility and exhibit gas-like transport properties of diffusivity and viscosity. Additionally, SCFs have low surface tension, permeate faster than liquids, and have solubility that gases do not possess. Therefore, high efficiency of dispersion may be achieved using SCF.

In the dispersion method, as illustrated in Figure 5.1, a liquefied carbon dioxide–PGMEA (propylene glycol monomethyl ether acetate)–organic pigment mixture is expressed from the conditions of high pressure to ambient pressure, whereupon the gas undergoes an adiabatic expansion and a fine dispersoid (i.e., a mixture of dispersed pigment with PGMEA) is immediately obtained. It is noted that the cosolvent PGMEA is partially miscible with CO_2 and can be used for preparing an ultrafine color resist to fabricate a color filter in an advanced LCD.

Table 5.1 lists basic information on organic pigments (three kinds) used in this study, for which the chemical structure is shown in Figure 5.2. The pigments were charged into PGMEA, stirred for 20 min using ultrasonic waves (with a Branson 5510 ultrasonic cleaner at 40 kHz), and then transported into a supercritical vessel. Liquid-type carbon dioxide was input for 20 min to reach the supercritical state under conditions of temperature and pressure

TABLE 5.1 The Basic Information on the Examined Organic Pigments

Color	Product Name	Chemical Type	C.I. No.	C.I. Generic Name
Red	Red A3B	Aminoanthraquinone	65300	Pigment red 177
Green	ECG 403	Phthalocyanine	74265	Pigment green 36
Blue	B6700	Phthalocyanine	74160	Pigment blue 15:6

Source: From Cheng, W.T. et al., *J. Colloid Interface Sci*, 270, 106, 2004. With permission.

ranging from 25°C and 1 atm to 55°C and 150 atm. Finally, the dispersoid with CO_2 was released into a collecting bottle under atmospheric pressure. The experimental apparatus of dispersion using SCF is shown in Figure 5.3. This is a batch process, so the flow rates of liquid-type CO_2 and the mixture of PGMEA with organic pigment transported to the dispersion column are unspecified.

As validated by the measurement of laser scattering and UV light, phthalocyanine green 36 pigment with the haloid structure can be dispersed into nanometer-scale particles with mean particle size 93.5 nm, at a concentration of 0.001% (w/w) into the solvent PGMEA using supercritical carbon dioxide under processing conditions of 55°C and 150 atm for 20 min. Additionally, the dispersions of aminoanthraquinone red containing the polarity of amino groups ($-NH_2$) and phthalocyanine blue 15:6 with symmetry benzene and inner hydrogen bonds are difficult to wet and swell by the SCF with PGMEA as cosolvent in the same conditions that disperse phthalocyanine green 36 pigment, and the mean particle sizes are 178.5 and 188.7 nm, respectively, in the dispersing solution.

5.3.3 Vapor Phase Method [8]

An organic pigment contains anthraquinone as principal ingredient such as C.I. Pigment Yellow 23, 108, 147, and 199, and is finely granulated by means of a vapor phase method, including evaporating the pigment in inert gas. More specifically, with a vapor phase method, the organic pigment as raw material is heated to evaporate in a vacuum and the evaporated organic pigment is made to pass through inert gas. As molecules of the evaporated organic pigment continuously collide with inert gas molecules, the former are cooled and aggregated to produce fine particles of the organic pigment. With this method, an organic pigment powder of ultra-fine particles, the primary particles of which are uniform and have a number average particle diameter of 10–50 nm, is produced.

C.I. Pigment Yellow 147 (Figure 5.4), an anthraquinone-type pigment, is placed at an evaporation source arranged in a vacuum vessel, and He is introduced into the vessel as inert gas so that the inner pressure of the vacuum vessel is maintained at 0.7 Torr. Then, the pigment placed at the evaporation source is heated

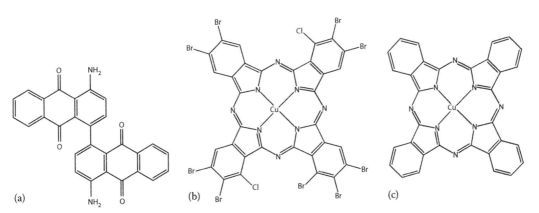

(a)   (b)   (c)

FIGURE 5.2 The chemical structure of organic pigments used in this investigation: (a) red pigment 177, (b) green pigment 36, (c) blue pigment 15:6. (From Cheng, W.T. et al., *J. Colloid Interface Sci*, 270, 106, 2004. With permission.)

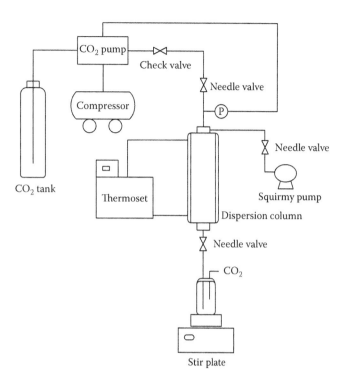

FIGURE 5.3 Schematic diagram of dispersion apparatus using SCF. (From Cheng, W.T. et al., *J. Colloid Interface Sci.*, 270, 106, 2004. With permission.)

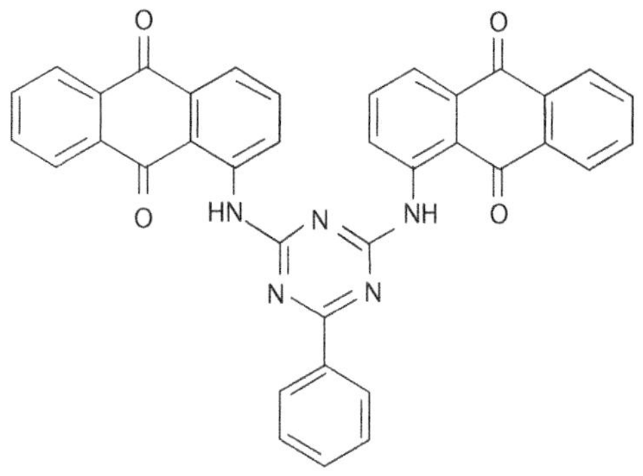

FIGURE 5.4 Chemical structure of C.I. Pigment Yellow 147.

and evaporated by means of a laser. The produced fine particles of the pigment are subsequently deposited to adhere to a rotary substrate arranged over the evaporation source. Thereafter, the deposited fine particles of the pigment are transported by rotating the substrate to a discharge plasma region that is held stably in a discharging state in advance. The fine particles of the pigment produced in this way were observed through a scanning electron microscope to find that the number average particle diameter of primary particles was 35 nm. Since O_2 gas has been introduced near the discharge electrode of the discharge plasma region, plasma is being generated in an O_2 gas atmosphere. Hence, the fine particles of the pigment that have been deposited to adhere to the rotary substrate are made to pass through the region to introduce a hydrophilic group into the surface of the fine pigment particles. The fine particles of the pigment treated by oxygen plasma

on the rotary substrate are scraped off from the latter by means of a blade and collected in a predetermined container.

The particle diameter distribution of the fine particles of the pigment in an aqueous dispersion was observed by means of an electrophoretic light scattering photometer ELS800 (trade name: available from Otsuka Electronics) to find that the volume average particle diameter was 48.5 nm and the particle diameter distribution was in a range between 10 and 200 nm. Then, aqueous ink containing the organic pigment to about 3.5% was prepared by dispersing fine particles of the organic pigment in water and mixing the water with an aqueous liquid medium containing at least glycerol as wetting agent.

5.3.4 Miniemulsion Method

Miniemulsions are specially formulated heterophase systems consisting of stable nanodroplets in a continuous phase. The narrowly size distributed nanodroplets of 50–500 nm can be prepared by shearing a system containing oil, water, a surfactant, and an osmotic pressure agent that is insoluble in the continuous phase. Since each of the nanodroplets can be regarded as a batch reactor, a whole variety of reactions can be carried out starting from miniemulsions clearly extending the profile of classical emulsion polymerization.

5.3.4.1 The Principle of Formation of and Polymerizations in Miniemulsions [9]

In miniemulsion polymerization, the principle of small nanoreactors is realized as demonstrated in Figure 5.5.

In a first step of the miniemulsion process, small stable droplets in a size range between 30 and 500 nm are formed by shearing a system containing the dispersed phase, the continuous phase, a surfactant, and an osmotic pressure agent. In a second step, these droplets are polymerized without changing their identity.

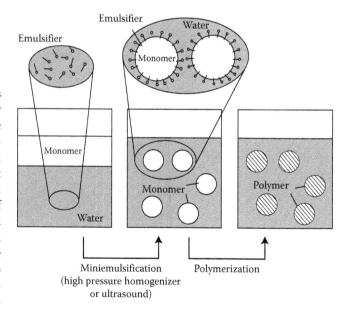

FIGURE 5.5 The principle of miniemulsion polymerization.

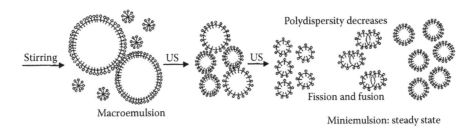

FIGURE 5.6 Scheme for the formation of a miniemulsion by ultrasound. (From Landfester, K., *Top Curr. Chem.*, 227, 75, 2003. With permission.)

Emulsions are understood as dispersed systems with liquid droplets (dispersed phase) in another, nonmiscible liquid (continuous phase). In order to create a stable emulsion of very small droplets, which is, for historical reasons, called a miniemulsion (as proposed by Chou et al. [10]), the droplets must be stabilized against molecular diffusion degradation (Ostwald ripening, a unimolecular process) and against coalescence by collisions (a bimolecular process). Stabilization of emulsions against coalescence can be obtained by the addition of appropriate surfactants, which provide either electrostatic or steric stabilization to the droplets. Mechanical agitation of a heterogeneous fluid containing surfactants always leads to a result of droplets size distribution. Even when the surfactant provides sufficient colloidal stability of droplets, the fate of this size distribution is determined by their different droplet or Laplace pressures, which increase with decreasing droplet sizes resulting in a net mass flux by diffusion between the droplets. If the droplets are not stabilized against diffusional degradation, Ostwald ripening occurs, which is a process where small droplet will disappear leading to an increase of the average droplet size.

In order to obtain emulsification, a premix of the fluid phases containing surface-active agents and further additives is subjected to high energy for homogenization.

Independent of the technique used, the emulsification includes first deformation and disruption of droplets, which increase the specific surface area of the emulsion, and second, the stabilization of this newly formed interface by surfactants.

5.3.4.1.1 Techniques of Miniemulsion Preparation and Homogenization [9]

In high-force dispersion devices, ultrasonication is used today especially for the homogenization of small quantities, whereas rotor–stator dispersers with special rotor geometries, microfluidizers, or high-pressure homogenizers are best for the emulsification of larger quantities. In direct miniemulsions, the droplet size is in turn determined by the amount of oil and water, the oil density, the oil solubility, and the amount of surfactant. It is found for direct miniemulsions that the droplet size is initially a function of the amount of mechanical agitation. The droplets also change rapidly in size throughout sonication in order to approach a pseudo-steady state, assuming a required minimum amount of energy for reaching this state is used. Once this state is reached, it was found that the size of the droplet does not change anymore. At the beginning of homogenization, the

polydispersity of the droplets is still quite high, but by constant fusion and fission processes the polydispersity decreases, and the miniemulsion then reaches a steady state (see Figure 5.6).

The droplet size and size distribution seems to be controlled by a Fokker–Planck type dynamic rate equilibrium of droplet fusion and fission processes. This also means that miniemulsions reach the minimal droplet sizes under the applied conditions (surfactant load, volume fraction, temperature, salinity, and so on), and therefore the resulting nanodroplets are at the critical borderline between stability and instability. This is why miniemulsions directly after homogenization are called "critically stabilized." Practically speaking, miniemulsions potentially make use of the surfactant in the most efficient way possible.

5.3.4.1.2 Surfactant Variation

By varying the relative amount of the surfactant, it was possible to vary the particle size over a wide range [11]. Figure 5.7 shows that, depending on the type of the surfactant, different size ranges can be achieved. Latexes synthesized with ionic surfactants, e.g., sodium dodecyl sulfate (SDS), CTAB, or the C_{12} sulfonium surfactant show about the same size-concentration

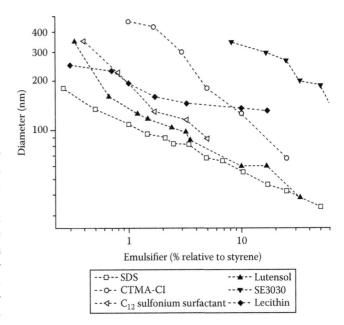

FIGURE 5.7 Variation of the particle size by changing the amount and type of surfactant in a styrene miniemulsion. (From Landfester, K., *Top Curr. Chem.*, 227, 75, 2003. With permission.)

curve, i.e., the efficiency of the surfactants and the size-dependent surface coverage is very similar, independent of the sign of charge. The efficiency of the nonionic surfactants is lower in contrast to the ionic ones, and the whole size-concentration curve is shifted to larger sizes. This is attributed to the lower efficiency of the steric stabilization as compared to electrostatic stabilization and the fact that steric stabilization relies on a more dense surfactant packing to become efficient. As can be derived from the surface tension of the latexes and surfactant titrations, the nonionic particle surfaces are, nevertheless, incompletely covered by surfactant molecules and the latexes show surface tensions well above the values of the saturated surfactant solution where saturated surfactant layers occur. Also, the biosurfactant lecithin can be used for the preparation of stable miniemulsions.

5.3.4.1.3 Characteristics of a Miniemulsion

In some crucial cases, it might be not obvious whether the system represents a miniemulsion or not. Therefore, a short checklist summarizing the characteristics of miniemulsions is provided:

1. Steady-state dispersed miniemulsions are stable against diffusional degradation, but critically stabilized with respect to colloidal stability.
2. The interfacial energy between the oil and water phase in a miniemulsion is significantly larger than zero. The surface coverage of the miniemulsion droplets by surfactant molecules is incomplete.
3. The formation of a miniemulsion requires high mechanical agitation to reach a steady state given by a rate equilibrium of droplet fission and fusion.
4. The stability of miniemulsion droplets against diffusional degradation results from an osmotic pressure in the droplets, which controls the solvent or monomer evaporation. The osmotic pressure is created by the addition of a substance, which has extremely low water solubility, the so-called hydrophobe. This crucial prerequisite is usually not present in microemulsions, but can be added to increase the stability. Such miniemulsions can still undergo structural changes by changing their average droplet number to end up in a situation of zero effective pressure, however, on very long time scales. This secondary growth can be suppressed by an appropriate second dose of surfactant added after homogenization.
5. Polymerization of miniemulsions occurs by droplet nucleation only.
6. During the polymerization, the growth of droplets in miniemulsions can be suppressed. In miniemulsions, the monomer diffusion is balanced by a high osmotic background of the hydrophobe, which makes the influence of the firstly formed polymer chains less important.
7. The amount of surfactant or inherent surface stabilizing groups required to form a polymerizable miniemulsion is comparably small, e.g., with SDS between 0.25% and 25% relative to the monomer phase, which is well below the surfactant amounts required for microemulsions.

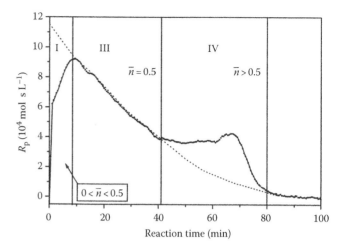

FIGURE 5.8 Calorimetric curve of a typical miniemulsion polymerization consisting of 20% styrene in water, 1.2% SDS (relative to styrene), and KPS as initiator. (From Landfester, K., *Top Curr. Chem.*, 227, 75, 2003. With permission.)

5.3.4.1.4 Radical Polymerizations of Miniemulsions

5.3.4.1.4.1 Mechanisms and Kinetics in Miniemulsion Polymerization

In miniemulsion polymerization, the nucleation of the particles mainly starts in the monomer droplets themselves. Therefore, the stability of droplets is a crucial factor in order to obtain droplet nucleation. The better the droplets are stabilized against Ostwald ripening, the higher is the droplet nucleation. In Figure 5.8, the calorimetric curve of a typical miniemulsion polymerization for 100 nm droplets consisting of styrene as monomer and hexadecane as hydrophobe with initiation from the water phase is shown. Three distinguished intervals can be identified throughout the course of miniemulsion polymerization. According to Harkins' definition for emulsion polymerization, only intervals I and III are found in the miniemulsion process. Additionally, interval IV describes a pronounced gel effect, the occurrence of which depends on the particle size. Similar to microemulsions and some emulsion polymerization recipes, there is no interval II of constant reaction rate. This points to the fact that the diffusion of monomer is in no phase of the reaction the rate-determining step.

The first interval is the interval of particle nucleation (interval I) and describes the process to reach an equilibrium radical concentration within every droplet formed during emulsification. The initiation process becomes more transparent when the rate of polymerization is transferred into the number of active radicals per particle $\bar{n}$, which slowly increases to $\bar{n} \approx 0.5$. Therefore, the start of the polymerization in each miniemulsion droplet is not simultaneous, so that the evolution of conversion in each droplet is different. Every miniemulsion droplet can be perceived as a separate nanoreactor, which does not interact with others. After having reached this averaged radical number, the polymerization kinetics is slowing down again and follows nicely an exponential kinetics as known for interval III in emulsion polymerization or for suspension polymerization. As reasoned by the droplet nucleation mechanism, only the monomer in the droplet is available for polymerization, which is exponentially depleted from the reaction site. The average number of radicals per particle, $\bar{n}$ during interval III, is quite accurately kept at 0.5, implying that the on/off mechanism known from emulsion polymerization upon entry of additional radicals into such small latex particles is strictly valid.

The boost found in interval IV is the typical gel-peak well known also from suspension polymerization, which is due to the viscosity increase inside the particles and the coupled kinetic hindrance of the radical recombination. This is also reflected in a steep rise of $\bar{n}$.

5.3.4.1.4.2 Droplet Size The polymerization kinetics is governed by the droplet size. With increasing shear and increasing concentration of surfactant, the polymerization rate increases.

In steady-state or mechanically equilibrated miniemulsions, the droplet size can be easily varied by varying the amount of surfactant. Depending on the droplet size of the miniemulsions, calorimetric curves with various kinetic features are shown in Figure 5.9. Disregarding the complexity of the kinetics and the existence of the three intervals, the reaction time to reach 95% conversion depends as a rule of thumb about linearly on the particle size and thus varies between 20 and 120 min. Independent of the size of the droplets, interval I (see also Figure 5.8) has a similar duration and takes about 5 min, which again supports the concept that this interval is only influenced by processes in the continuous aqueous phase that do not depend on the droplet size. The maximum reaction speed, however, shows a strong particle size dependence and is proportional to the particle number, i.e., the smaller the particles are, the faster is the reaction.

5.3.4.1.4.3 Initiators For miniemulsion polymerization, the initiator can be either oil- or water-soluble. In the case of an oil-soluble initiator such as lauroyl peroxide (LPO), benzoyl peroxide (BPO), and 2,2ϕ-azoisobutyronitrile (AIBN), the initiator is dissolved in the monomeric phase prior to miniemulsification. Then the reaction starts within the droplets. This is comparable to suspension polymerization where the initiation is carried out in the large droplets. Because of the finite size of the miniemulsions

droplets, radical recombination is here the problem to face. Also a water-soluble initiator such as potassium persulfate (KPS) can be used to start the polymerization from the water phase. The start from the continuous phase is similar to the conventional emulsion polymerization. It was found that the chain length of the resulting polymer is inversely proportional to the square root of the initiator concentration, underlining that the reaction in miniemulsion is rather direct and close to an ideal radical polymerization.

5.3.4.2 Typical Miniemulsion Polymerization Process for Producing Nanocolorants

5.3.4.2.1 Encapsulation of Pigments by Direct Miniemulsification

For the encapsulation of pigments by miniemulsification, two different approaches can be used. In both cases, the pigment/polymer interface as well as the polymer/water interface have to be carefully chemically adjusted in order to obtain encapsulation as a thermodynamically favored system. The design of the interfaces is mainly dictated by the use of two surfactant systems, which govern the interfacial tensions, as well as by the employment of appropriate functional comonomers, initiators, or termination agents. The sum of all the interface energies has to be minimized.

For the successful incorporation of a pigment into the latex particles, both type and amount of surfactant systems have to be adjusted to yield monomer particles, which have the appropriate size and chemistry to incorporate the pigment by its lateral dimension and surface chemistry. For the preparation of the miniemulsions, two steps have to be controlled (see Figure 5.10). First, the already hydrophobic or hydrophobized particulate pigment with a size up to 100 nm has to be dispersed in the monomer phase. Hydrophilic pigments require a hydrophobic surface to be dispersed into the hydrophobic monomer phase, which is usually promoted by a surfactant system 1 with low HLB value. Then, this common mixture is miniemulsified in the water phase employing a surfactant system 2 with high HLB, which has a higher tendency to stabilize the monomer (polymer)/water interface.

Erdem et al. [12] described the encapsulation of TiO$_2$ particles via miniemulsion in two steps. First, TiO$_2$ was dispersed in the monomer using the OLOA 370 (polybutene-succinimide) as stabilizer. Then this phase was dispersed in an aqueous solution to form stable submicron droplets. The presence of TiO$_2$ particles within the droplets limited the droplet size. The complete encapsulation of all of the TiO$_2$ in the colloidal particles was not achieved; the encapsulation of 83% of the TiO$_2$ in 73% of the polymer was reported. Also, the amount of encapsulated material was very low: a TiO$_2$ to styrene weight ratio of 3:97 could not be exceeded.

5.3.4.2.2 Encapsulation of Carbon Black by Co-Miniemulsion [13]

Since carbon black is a rather hydrophobic pigment (depending on the preparation conditions), the encapsulation of carbon black in the latexes by direct dispersion of the pigment powder in the monomer phase prior to emulsification is again a suitable way. Here, the full encapsulation of nonagglomerated carbon particles can be provided by the appropriate choice of the hydrophobe. In this case, the hydrophobe not only acts as the stabilizing agent against Ostwald

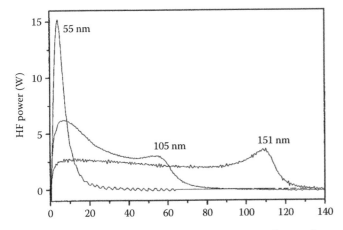

FIGURE 5.9 Calorimetric curves for styrene miniemulsion polymerizations with different SDS contents, e.g., different particle sizes (0.3 rel.% SDS leads to 151 nm, 1.0 rel.% SDS to 105 nm, and 10 rel.% SDS to 55 nm particles). (From Landfester, K., *Top Curr. Chem.*, 227, 75, 2003. With permission.)

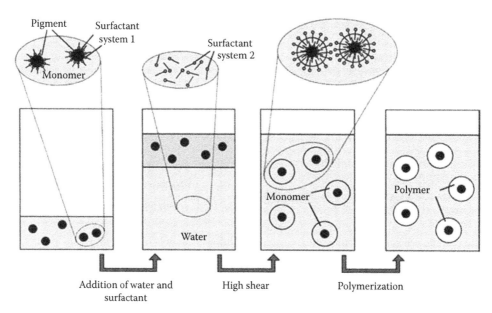

FIGURE 5.10 Principle of encapsulation by miniemulsion polymerization. (From Landfester, K., *Top Curr. Chem.*, 227, 75, 2003. With permission.)

ripening for the miniemulsion process, but also mediates to the monomer phase by partial adsorption. However, this direct dispersion just allows the incorporation of 8 wt% carbon black since the carbon is still highly agglomerated in the monomer. At higher amounts, the carbon cluster broke the miniemulsion, and less-defined systems with encapsulation rates lower than 100%, which also contained pure polymer latexes, were obtained. To increase the amount of encapsulated carbon to up to 80 wt%, another approach was developed, where both monomer and carbon black were independently dispersed in water using SDS as a surfactant and mixed afterward in any ratio between the monomer and carbon. Then this mixture was cosonicated, and the controlled fission/fusion process characteristic for miniemulsification destroyed all aggregates and liquid droplets, and only hybrid particles being composed of carbon black and monomer remain due to their higher stability. This controlled droplet fission and heteroaggregation process can be realized by high-energy ultrasound or high-pressure homogenization. Results showed that this process results in the effective encapsulation of the carbon with practically complete yield: only rather

small hybrid particles, but no free carbon or empty polymer particles, were found. It has to be stated that the hybrid particles with high carbon contents do not possess spherical shape, but adopt the typical fractal structure of carbon clusters, coated with a thin but homogeneous polymer film. The thickness of the monomer film depends on the amount of monomer, and the exchange of monomer between different surface layers is, as in miniemulsion polymerization, suppressed by the presence of an ultra hydrophobe.

Therefore, the process is best described as a polymerization in an adsorbed monomer layer created and stabilized as a miniemulsion ("ad-miniemulsion polymerization"). The process is schematically shown in Figure 5.11.

5.3.4.2.3 Aqueous Dispersions of Polystyrene Latexes Encapsulating a Copper Phthalocyanine Blue Pigment Formulated Using the Miniemulsion Polymerization Technique [14]

As shown in Figure 5.12 organic pigment was first suspended into the monomer phase, and the resulting oily suspension was

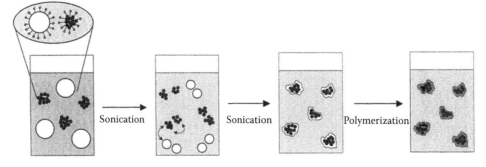

FIGURE 5.11 Principle of co-miniemulsion, where both components have to be independently dispersed in water and mixed afterward. The controlled fission/fusion process in the miniemulsification realized by high-energy ultrasound or high-pressure homogenization destroys all aggregates and liquid droplets, and only hybrid particles being composed of carbon black and monomer remain due to their higher stability. (From Landfester, K., *Top Curr. Chem.*, 227, 75, 2003. With permission.)

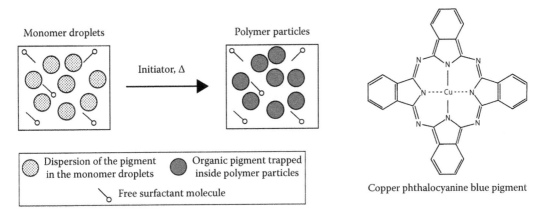

FIGURE 5.12 Schematic representation of the encapsulation reaction of organic pigments through miniemulsion polymerization.

subsequently converted into stable miniemulsion droplets using various types and concentrations of hydrophobe (costabilizer). The pigmented monomer emulsions were finally polymerized using KPS as the initiator. It was shown that the organic pigment could stabilize the miniemulsion droplets, and thus be satisfactorily encapsulated without introducing any other compound in the formulation. In a subsequent approach, the stability of the miniemulsion droplets was improved by using either

hexadecane, hexadecanol, or a polystyrene prepolymer as the hydrophobe. Dynamic light scattering and transmission electron microscopy (TEM) measurements showed that the size and the morphology of the resulting pigmented polymer particles were greatly influenced by the presence of the costabilizer.

It has been showed by TEM that the encapsulation of an organic phthalocyanine blue pigment could be achieved under very simple experimental conditions. Not only did the pigment allow the stabilization of the miniemulsion, but it could also be embedded satisfactorily into the polystyrene latex particles. In order to check the preservation of the identity of the particles during polymerization, and to improve the miniemulsion formulation, the droplet and particle sizes were determined by dynamic light scattering method which appeared to give reliable data with good correlation with TEM observations (Figuer 5.13). Most of the particle sizes of the pigmented lates prepared were greater than 100 nm in diameter. It was confirmed that the addition of a small amount of hydrophobe was helpful in increasing the droplet stability in comparison with polymerizations performed without costabilizer. This allowed the formation of pigmented latexes with different sizes and morphologies.

5.3.4.2.4 Preparation of Dye-Doped Nanocolorants [15]

Anthraquinone-based solvent red, yellow, blue, and green dyes (technical grade, Aolunda Co. Ltd and Bayer Corp.) were used as the essential material, the molecular structure of which is shown in Figure 5.14.

The deionized water (50 g) solution of SDS (based on water 20 mM) and sodium bicarbonate (NaHCO$_3$) was added to styrene (10 g) solution comprising 1.5 g MMA, 1.0 g DVB and

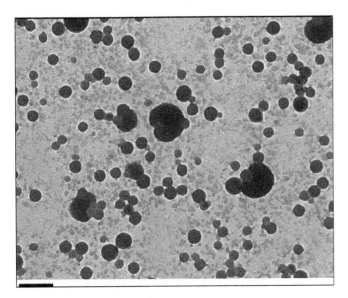

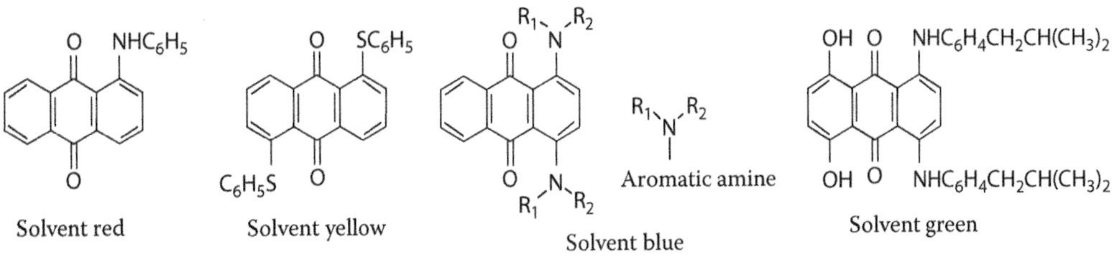

FIGURE 5.13 TEM image of blue pigmented polystyrene latex particles obtained in the presence of a polystyrene prepolymer (run 9). Scale bar: 200 nm.

FIGURE 5.14 Molecular structure of the used dyes.

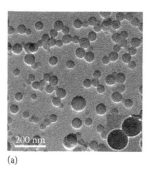 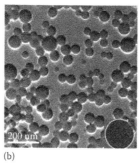 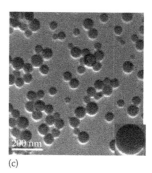

(a)　　　　　　　　　　　　　　(b)　　　　　　　　　　　　　　(c)

FIGURE 5.15 TEM images of nanocolorants: (a) red, (b) yellow, and (c) blue nanocolorant (scale bar 200 nm). Insets: enlarged images by field emission TEM image. Dye/polymer 0.9, 1.2, 1.5 g/12.5 g represented loading of red, yellow, and blue dyes, respectively.

GDMA, 0.8–2.0 g completely dissolved dyes, 0.15 g predissolved polystyrene (M_w 50,000), and reactive light stabilizer (HALS-13, Clariant Corp.), and then quickly stirred at room temperature for 20 min. The resultant macroemulsion was miniemulsified with an ultrasonic homogenizer, operated at 400–500 W for 15 min under ice cooling, and finally obtained a miniemulsion. The miniemulsion was transferred into a flask equipped with an agitator, a thermometer, a reflux condenser, and a nitrogen tube. The system was purged with nitrogen for 10 min and heated to 60°C under nitrogen flow, simultaneously stirring by paddle stirrer at 200 rpm. The reaction was initiated by the injection of a water solution of KPS at 60°C, and then kept polymerization at 65°C for 4 h. The polymerization conversion was determined by gas chromatography until no monomer could be detected.

Typical TEM images obtained demonstrated that the morphology of nanocolorants is homogeneous spherical nanoparticle with a uniform surface, as shown in Figure 5.15.

Under the same dye-content, nanocolorants (a, d, f) exhibited excellent color properties, as seen in Figure 5.16, which showed even brighter lightness and higher saturation than that of the corresponding dyes.

Based on the architecture of nanocolorants originally designed, a set of red, yellow (or green), and blue nanocolorants with monodispersive particle size less than 100 nm were successfully prepared by using a modified miniemulsion polymerization process, which holds the advantages, such as facilitation and availability, low cost, environment friendly, and so on. The obtained nanocolorants exhibited excellent color saturation, lightness, and strong color depth owed to the nanoscale effects of homogeneous nanocolorants. By dissolving primary color dyes in the minidroplet,

the full-tone nanocolorants will be easily obtained according to additive color matching. They really achieved superior migration fastness, light fastness, thermal stability, and good processibility, and their aqueous dispersions could be easily made to high-quality color ink-jet inks, and be used in color resists.

5.4 Summary and Future Perspective

Comparing with the enormous family of classical colorant of dyes and pigments, the monocolorant group is still in its early stages. We have introduced and discussed four special processes for the treatment of colorants in order to achieve mesomeric nanoscale in size. It should be noted that each method has its own limitations.

The mechanical dispersing procedure has low efficiency, consumes a large amount of power, and is very difficult to prevent the ultra-fine nanoparticles from agglomeration. Although much progress has been made for the stabilization of nanocolorant, it still needs further hard work on developing noval dispersant agent and exploring more kinds of colorants. Anyway, most commercialized nanocolorant products are produce via this traditional technology.

The supercritical carbon dioxide processing technology is a novel approach to the dispersion of organic pigment, which is entirely physical dispersing, environmental benign, quick, and easy to handle. But the concentration of the nanocolorant dispersion is rather too low and the investment on supercritical instrument would be very high, thus bringing about more financial problem. Hopefully, it will generate sufficient interest so that future studies can ascertain the nanometer-scale dispersion at high concentrations of organic pigments to manufacture high-resolution color filters for application in advance.

The vapor phase method, or vacuum evaporation process, needs very strict requirements for the instruments and has a similar financial problem like the supercritical carbon dioxide processing technology. The main problem of this process is that it is only appropriate to the dyes that have very high thermal stability, such as anthraquinone derivates. The colorants with low thermal stability such as azo-dyes, the biggest family in colorants, will decompose during the evaporation process.

The miniemulsion method, as discussed more precisely in former context, has attracted more and more research interests

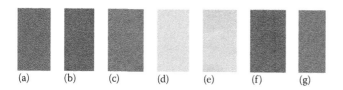

(a)　　(b)　　(c)　　(d)　　(e)　　(f)　　(g)

FIGURE 5.16 **(See color insert following page 22-8.)** Digital photographs of the color films prepared by using nanocolorants and corresponding dyes: (a) red nanocolorant, (b) red dye, (c) uncrosslinked red nanocolorant, (d) yellow nanocolorant, (e) yellow dye, (f) blue nanocolorant, and (g) blue dye.

and commercial attentions. Several problems hinder the applications. Works should be done to obtain better control and a better understanding of the nucleation mechanism in various systems. The commodity of the colorant and monomer in the dispersed phase is crucial in the control of interfacial properties and chromophore concentration in the final product. The contribution of the colorant to monomer conversions and polymer chains growth also needs to be clarified. Work is underway to investigate more in depth the fundamental aspects associated with the miniemulsion formulation and to understand the role of the colorants in the control of the miniemulsion stability and polymerization kinetics. The field of miniemulsion is still on its rise in polymer and material science since there are numerous additional possibilities both for fundamental research and application. Also, a series of research works such as: "High-performance fluorescent particles prepared via miniemulsion polymerization" [17]; "Synthesis and biomedical applications of functionalized fluorescent and magnetic dual reporter nanoparticles as obtained in the miniemulsion process" [18]; "Spirobenzopyran-based photochromic nanohybrids with photoswitchable fluorescence" [18], has been done to show a promising future expectation.

As an end to the section, we present a commercial application of nanocolorant product in digital printing field [19].

Nanocolorants are specialty inks that have been formulated at the nanoparticle level for the digital print environment. For pigment-based inks, nanoparticle formulation offers a range of benefits, including

- Improved jetting reliability for pigments
- Increased color gamut for ink-jet pigments
- Soft fabric hand (feel)
- Simplified processing methods

Pigment-based nanocolorants can be applied to the surface of standard PFP (prepared for print) fabrics via the digital method (Figures 5.17 and 5.18). Unlike dye-based colorants, nanocolorant pigments do not require specialized fabric pretreatment to

FIGURE 5.17 The nanocolorants working with at [TC]² are pigment-based inks supplied by Yuhan-Kimberly along with the Ujet MC2 digital print system.

FIGURE 5.18 Engineered print samples of pigment-based inks.

enable color fixation. Once applied, the colorants are "cured" using dry heat, which activates the binder or cross-linking system in the colorant chemistry. This curing procedure permanently adheres the color to the surface of the cloth, providing fastness to washing and standard care practices.

Yuhan-Kimberly Ltd. created the world's first "Digital Textile Printing" total solution using world-class digital technology [20]. The digital textile printing (DTP) system is completely managed by computers from design process to the printing process, resulting in a simple and fast approach to delicate and complex textile-printed materials (Figures 5.17 and 5.18). And the newest 12-color textile printing system in which all processes have been digitalized, as compared with the conventional textile printing solution, boasts a significant improvement over previous models. The DTP system is being spotlighted as an effective next generation's textile printing system.

Acknowledgments

The author is grateful to Dr. Zheng-Kun Hu, Dr. Min-Zhao Xue, Prof. Yan-Gang Liu, and Miss Qiao-rong Sheng in our research group for the research cooperation and helpful discussion for composing the paragraph.

References

1. http://en.wikipedia.org/wiki/Colorant
2. Barashkov, N.N., Liu, R.H. In: *International Conference on Digital Printing Technologies 2001*, Fort Lauderdale, FL, 2001, pp. 878–880.
3. Clemens, T., Boehm, A.J., Sabine, K.B. In: *The San Francisco Meeting, Polymer Preprints*, San Francisco, CA, Division of Polymer Chemistry, vol. 41(1), 2000, pp. 24–25.
4. Boehm, A.J., Alban, G., Koch, O. In: *The 61st Annual Technical Conference (ANTEC 2003)*, Nashville, TN, 2003, vol. 2, pp. 2419–2422.
5. Boehm, A.J., Sabine, K.B., Peter, R. *Prog. Colloid Polym. Sci.* 1999. 113:121.
6. Gore, M.P. U.S. Patent 6,024,786, Hewlett-Packard Company, Corvallis, OR, CA Issued/Filed Dates: February 15, 2000/ October 30, 1997.

7. Cheng, W.T., Hsu, C.W., Chih, Y.W. *J. Colloid Interface Sci.* 2004. 270:106–112.

8. Tochihara, S., Koike, S., Hiro, M., Shirota, K. U.S. Patent 6,619,791, Assignee: Canon Kabushiki Kaisha, September 16, 2003.

9. Landfester, K. *Top Curr. Chem.* 2003. 227:75–123.

10. Chou, Y.J., El-Aasser, M.S., Vanderhoff, J.W. *J. Dispers. Sci. Technol.* 1980. 1:129–150.

11. Bechthold, N., Tiarks, F., Willert, M., Landfester, K., Antonietti, M. *Macromol. Symp.* 2000. 151:549–555.

12. Erdem, B., Sudol, E.D., Dimonie, V.L., El-Aasser, M.S. *J. Polym. Sci. Polym. Chem. Ed.* 2000. 38:4419–4430.

13. Tiarks, F., Landfester, K., Antonietti, M. *Macromol. Chem. Phys.* 2001. 202:51–60.

14. Lelu, S., Novat, C., Graillat, C., Guyot, A., Bourgeat-lami, E. *Polym. Int.* 2003. 52(4):542–547.

15. Hu, Z., Xue, M., Zhang, Q., Sheng, Q., Liu, Y. *Dyes Pigments* 2008. 76:173–178.

16. Ando, K., Kawaguchi, H. *J. Colloid Interface Sci.* 2005. 285:619–626.

17. Holzapfel, V., Lorenz, M., Weiss, C.K., Schrezenmeier, H., Landfester, K., Mail, V. *J. Phys.: Condens. Matter* 2006. 18:S2581–S2594.

18. Hu, Z., Zhang, Q., Xue, M., Sheng, Q., Liu, Y. *Opt. Mater.* 2008. 30:851–856.

19. http://www.inkdropprinting.com/nanocolorant.htm

20. http://www.qa.dtplink.com/Global/introduce/introduce_introduce.asp

Magnetoelectric Interactions in Multiferroic Nanocomposites

Vladimir M. Petrov
Oakland University

and

Novgorod State University

Gopalan Srinivasan
Oakland University

6.1 Introduction

The magnetoelectric (ME) effect is defined as an induced electric polarization in a material when it is subjected to a magnetic field, or an induced magnetization when in an electric field. The term "magnetoelectric" was first used by Debye and was marked by two major independent discoveries [1]. The first in discovery by Röntgen in 1888 was that a moving dielectric became magnetized when an electric field was applied to it [2]. The second discovery was the realization of the possibility of intrinsic ME behavior of stationary crystals on the basis of symmetry considerations by Pierre Curie in 1894 [3]. The reverse effect of the first event, i.e., the polarization of a moving dielectric in a magnetic field, was observed 17 years after the discovery in 1888. Even though Curie recognized that symmetry was an important issue in ME behavior, it took many decades to demonstrate that the ME response was allowed only in time-asymmetric media. However, after the two discoveries that formed the foundation of ME effect, no work was done until 1958, when Landau and Lifshitz verified the possibility of the ME effect in certain crystals on the basis of the crystal symmetry [4]. In 1959, the symmetry argument was applied by Dzyaloshinskii to an antiferromagnetic Cr_2O_3 [5] to show the violation of time-reversal symmetry for this particular system and suggesting that the ME effect can be seen in Cr_2O_3. In 1960 and 1961, Astrov confirmed this experimentally by measuring the electrically induced ME effect in Cr_2O_3 for the temperature range of 80–330 K [6].

In materials that are magnetoelectric, the induced polarization P is related to the field H by the expression, $P = \alpha H$, where α is the second rank ME-susceptibility tensor and is expressed in the units of s/m in SI units (in Gaussian units $\alpha = 4\pi P/H = 4\pi M/E$ is dimensionless). One generally determines α by measuring P_z for an applied field H along the z-axis. Another parameter of importance is the ME voltage coefficient $\alpha_E = E/H$ which

is related to α by the expression $\alpha = \varepsilon_o\varepsilon_r\alpha_E$, where ε_r is the relative permittivity of the material. Several single-phase materials that are either antiferromagnetic, weak ferromagnetic, or ferrimagnetic, such as Cr_2O_3, $TbFeO_3$, and $Fe_xGa_{2-x}O_3$, show rather weak ME effects [7–12].

6.2 ME Effects in Composites

Composites are of interest for the engineering of materials either with desired properties, or new characteristics that are absent in single-phase materials (for details see [13,14]). Composite materials can be divided into two categories as proposed by van Suchtelen [15]: (a) sum properties and (b) product properties. A sum property of a composite is the weighted sum of the contributions from the constituent phases that is proportional to the volume or weight fractions of these phases. Physical quantities such as density and resistivity are examples of sum properties. For product properties, consider a composite material with two component phases. The first phase has a property $A \rightarrow B$ with a proportionality tensor $dB/dA = X$; and the second phase has a property $B \rightarrow C$ with a proportionality tensor $dC/dB = Y$. Then the composite will have the property $A \rightarrow C$ with a proportionality tensor $dC/dA = Z$, where $Z = Y \cdot Z$; hence, the name "product property." The product property is achieved in a composite but not seen in the individual phases.

An ME composite can be realized with piezomagnetic and piezoelectric components [15–48]. When a magnetic field is applied to a piezomagnetic material, it is strained; the strain in turn causes stress on the piezoelectric material, which becomes electrically polarized as a consequence. The reverse effect is also feasible; but the consequence here will be magnetization of piezomagnetic material. Such ME composites could also be made with magnetostrictive and piezoelectric phases and theories predict ME voltage coefficients as high as 3.2 V/cm Oe [22–25].

Most attempts in the past to realize strong ME effects were on composites of magnetostrictive cobalt ferrite and piezoelectric barium titanate or lead zirconate titanate were unsuccessful [21]. Van den Boomgaard first synthesized composites of $CoFe_2O_4$ and $BaTiO_3$ by two methods: sintering and unidirectional solidification of eutectic melts [16–19]. Both composites developed microcracks due to thermal expansion mismatch and yielded ME coefficients that were a factor 40–60 smaller than calculated values.

Harshe, Dougherty, and Newnham, in their pioneering work on ME composites (a) proposed a theoretical model for multilayer heterostructures with alternating layers of magnetostrictive and piezoelectric phases and (b) fabricated such structures [20,21]. A multilayer structure is expected to be far superior to bulk composites for the following reasons: (a) An essential condition for maximizing ME effects is a large dielectric constant (and piezoelectricity). In bulk composites, the leakage currents due to low-resistivity cobalt ferrite inclusions reduce the overall dielectric constant. A similar decrease in the dielectric constant of piezoelectric layers is also possible in a multilayer composite. But such undesirable effects can easily be eliminated by shorting electrically the magnetostrictive layers. (b) The piezoelectric layer in a layered structure can easily be poled electrically to further enhance the piezoelectricity. They fabricated the ML structures by sintering tape-cast ribbons [21]. The ME coefficients did show an improvement over bulk sintered composites, but were a factor of 2–30 smaller than the theoretical values. This could be due to (a) incompatible structural and thermal properties of the two phases leading to poor coupling at the interface; (b) further weakening of the mechanical bonding due to platinum electrodes at the interface, which makes the composite a *trilayer* structure; (c) possible hexagonal hard magnetic phases at the interface due to high temperature processing; and (d) porosity. But with proper choice for magnetostrictive and piezoelectric phases and a multilayer composite geometry, a much higher α-value has been reported in recent studies [23–25].

A composite with alternate layers of magnetostrictive and piezoelectric phases is called a 2–2 composite since the two phases have mechanical connectivity only in the plane of the layers. Composites of interest in recent years have been $NiFe_2O_4$ (NFO) or $CoFe_2O_4$ (CFO) with lead zirconate titanate (PZT), terfenol-D and PZT, and similar systems [21–33]. Samples of NFO-PZT and CFO-PZT are made by sintering thick films obtained by tape-casting. Thin disks of the samples are polarized with an electric field perpendicularly to its plane. The ME coefficient $\alpha_E = \delta E/\delta H$, where δH is the applied ac magnetic field and δE is the measured ac electric field, and is measured with a setup as shown in Figure 6.1 and for two conditions: (a) transverse or $\alpha_{E,31}$ for bias magnetic H and ac field δH parallel to each other and to the disk plane (1,2) and perpendicular to δE (direction-3) and (b) longitudinal or $\alpha_{E,33}$ for all the three fields parallel to each other and perpendicular to sample plane.

Figure 6.2 shows the static magnetic field dependence of the transverse ME coefficient $\alpha_{E,31}$ and the longitudinal coefficient $\alpha_{E,33}$ for a composite of nickel ferrite (NFO) and lead zirconate

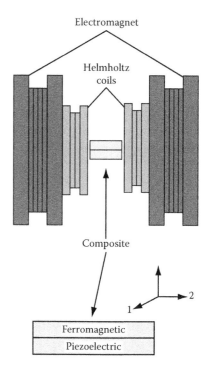

FIGURE 6.1 Schematic diagram showing a setup for the measurements of low-frequency ME effects in a ferromagnetic–piezoelectric composite. The sample is initially poled in an electric field and subjected to a bias magnetic field H and ac magnetic field dH. The ac electric field δE generated due to ME coupling is measured across the piezoelectric layer.

titanate (PZT) [24]. The data at room temperature are for a frequency of 1 kHz and *for unit thickness of the piezoelectric phase*. As H is increased from zero, $\alpha_{E,31}$ increases, reaches a maximum value of 400 mV/cm Oe at 300 Oe, and then drops rapidly to zero above 2 k Oe. We observed a phase difference of 180° between the induced voltages for +H and −H. Similarly, the longitudinal coefficient $\alpha_{E,33}$ for all the fields along direction-3 has a relatively small magnitude compared to the transverse case and peaks at a much higher H. These observations can be understood in terms of the demagnetizing field. The magnitude and the field dependence in Figure 6.2 are related to the variation of magnetostriction λ with H. The coefficients are directly proportional to the piezomagnetic coupling $q = d\lambda/dH$, and the H-dependence of α_E tracks the slope of λ vs. H. The saturation of λ at high field leads to $\alpha_E = 0$. We succeeded in achieving giant ME coefficients on the order of 400–1500 mV/cm Oe predicted by theory [24].

The variation of $\alpha_{E,31}$ with frequency and temperature are also shown in Figure 6.2. Upon increasing the frequency from 20 Hz to 10 kHz, one observes an overall increase of 25% in the ME voltage coefficient, but a substantial fraction of the increase occurs over the 1–10 kHz range (except at 353 K). These variations are most likely due to frequency dependence of the dielectric constant for the constituent phases and the piezoelectric coefficient for PZT. Data on temperature dependence of $\alpha_{E,31}$ at 100 Hz are shown in Figure 6.2. A peak in $\alpha_{E,31}$ is observed at room temperature and it decreases when T is increased from room temperature.

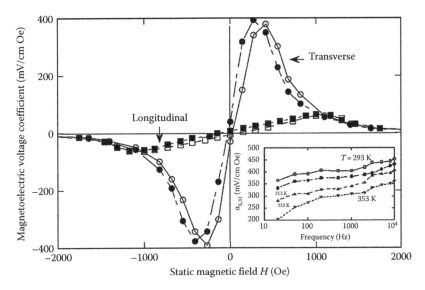

FIGURE 6.2 Representative data on bias field *H* dependence of transverse and longitudinal ME voltage coefficients for a layered sample of NFO and PZT. The inset shows the frequency and temperature dependence of the transverse ME coefficient.

6.3 Device Applications for ME Composites

The ME effect has provided a tool for converting energy from electric to magnetic form or vice versa. In 1948, Tellegen suggested a gyrator using an ME material even before the indication of any ME material [47]. In 1965, O'Dell proposed an ME memory device application [48]. Other possible applications of the ME materials have been proposed by Wood and Austin in 1975 including devices for (a) modulation of amplitudes, polarizations and phases of optical waves, (b) ME data storage and switching, (c) optical diodes, (d) spin-wave generation, (e) amplification, and (f) frequency conversion [47]. Recently magnetoelectric magnetic field sensors have been demonstrated for pico Tesla fields [29].

Other ME phenomena of fundamental and technological interests are the coupling when the electrical or the magnetic subsystem shows a resonant behavior, i.e., electromechanical resonance (EMR) for PZT and ferromagnetic resonance (FMR) for the ferrite. The resonance ME effect is similar in nature to the standard effect, i.e., an induced polarization under the action of an ac magnetic field. But the ac field here is tuned to the electromechanical resonance frequency. As the dynamic magnetostriction is responsible for the electromagnetic coupling, EMR leads to significant increase in the ME voltage coefficients [44,45]. Measurements of ME effects at EMR on ferrite-PZT show a two to three orders of magnitude increase in α_E compared to low frequency values. We developed the theory for ME coupling at EMR at radial and thickness modes for a ferromagnetic–piezoelectric bilayer [44]. There are several device possibilities based on resonance ME effects, such as dual electric and magnetic field tunable microwave and millimeter wave resonators, filters, and phase shifters [49,50].

6.4 Theory of Low-Frequency ME Coupling in a Nanobilayer

The main focus of this chapter is the ME effect in nanobilayers. Nanostructures in the shape of wires, pillars, and films are important for increased functionality in miniature devices. Theories for low frequency ME effects in magnetostrictive-piezoelectric nanostructures were reported recently [46]. The modeling was based on the homogeneous longitudinal strain approach. However, configurational asymmetry of a bilayer implies the presence of a bending strain in the sample in an applied magnetic or electric field. The principal objective of present work is modeling of the ME interaction in a magnetostrictive-piezoelectric nanobilayer on a substrate, taking into account such bending or flexural strains. We concentrate on ME effect in low frequency. Cobalt ferrite (CFO)–barium titanate grown on $SrTiO_3$ substrate is chosen as the model system for numerical estimations. This nanobilayer was recently prepared by Zheng [30]. The ME voltage coefficients α_E have been calculated for field orientations that provide minimum demagnetizing fields and maximum α_E. The effect of substrate clamping has been described in terms of dependence of α_E on substrate dimensions. The ME voltage coefficient is shown to drop with increase in substrate volume. Variations in piezoelectric and piezomagnetic coefficients and permittivity of components due to lattice mismatch are taken into account. We consider here only piezoelectric and ferrite films with thickness much larger than the ferroelectric and ferromagnetic correlation length for both phases. Under these conditions, the size effects may be neglected.

For piezoelectric and magnetostrictive phases and substrate, the following equations can be written for the strain, electric, and magnetic displacements:

$$^P S_i = {}^P s_{ij} \, {}^P T_j + {}^P d_{ki} \, {}^P E_k$$

$$^P D_k = {}^P d_{ki} \, {}^P T_i + {}^P \varepsilon_{kn} \, {}^P E_n$$

$$^m S_i = {}^m s_{ij} \, {}^m T_j + {}^m q_{ki} \, {}^m H_k \qquad (6.1)$$

$$^m B_k = {}^m q_{ki} \, {}^m T_i + {}^m \mu_{kn} \, {}^m H_n$$

$$^s S_i = {}^s s_{ij} \, {}^s T_j$$

where

- S_i and T_j are strain and stress tensor components
- E_k and D_k are the vector components of electric field and electric displacement
- H_k and B_k are the vector components of magnetic field and magnetic induction
- s_{ij}, q_{ki}, and d_{ki} are compliance, piezomagnetic, and piezoelectric coefficients
- ε_{kn} is the permittivity matrix
- μ_{kn} is the permeability matrix
- the superscripts p, m, and s correspond to piezoelectric and piezomagnetic phases and substrate, respectively.

We assume the symmetry of piezoelectric to be ∞m and that of piezomagnetic to be cubic. As shown in Figure 6.3, x_m, x_p, and x_s are the neutral axes that are located at the horizontal mid-plane of piezomagnetic, piezoelectric, and substrate layers, respectively, and separated by distances h_m and h_p. The thickness of the plate is assumed small compared to remaining dimensions.

Assuming that the longitudinal axial strains of each layer are linear functions of the vertical coordinate z_i, it can be shown that

$$^m S_1 = {}^m S_{10} + \frac{z_m}{R_1}$$

$$^P S_1 = {}^P S_{10} + \frac{z_p}{R_1}$$

$$^s S_1 = {}^s S_{10} + \frac{z_s}{R_1}$$

$$^m S_2 = {}^m S_{20} + \frac{z_m}{R_2} \qquad (6.2)$$

$$^P S_2 = {}^P S_{20} + \frac{z_p}{R_2}$$

$$^s S_2 = {}^s S_{20} + \frac{z_s}{R_2}$$

where

- $^i S_{10}$ and $^i S_{20}$ are the centroidal strains along the x- and y-axes at $z_i = 0$
- R_1 and R_2 are the radiuses of curvature

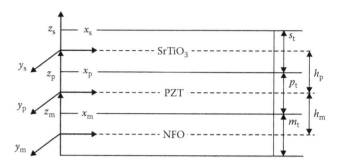

FIGURE 6.3 Schematic diagram showing an NFO PZT bilayer on a strontium titanate substrate.

From geometric considerations, it can be shown that

$$^m S_{10} - {}^P S_{10} = \frac{h_m}{R_1}$$

$$^P S_{10} - {}^s S_{10} = \frac{h_p}{R_1}$$

$$\qquad (6.3)$$

$$^m S_{20} - {}^P S_{20} = \frac{h_m}{R_2}$$

$$^P S_{20} - {}^s S_{20} = \frac{h_p}{R_2}$$

For the transverse field orientation (ac electric field perpendicular to the sample plane and ac magnetic and bias fields in the sample plane) providing the minimum demagnetizing fields, Equation 6.2 can then be rewritten using Equations 6.3 and 6.1 as

$$^m S_{10} + \frac{z_m}{R_1} = {}^m s_{11} \, {}^m T_1 + {}^m s_{12} \, {}^m T_2 + {}^m q_{11} \, {}^m H_1$$

$$^m S_{10} + \frac{(z_p - h_m)}{R_1} = {}^P s_{11} \, {}^P T_1 + {}^P s_{12} \, {}^P T_2 + {}^P d_{31} \, {}^P E_3$$

$$^m S_{10} + \frac{(z_p - h_m - h_p)}{R_1} = {}^s s_{11} \, {}^s T_1 + {}^s s_{12} \, {}^s T_2$$

$$\qquad (6.4)$$

$$^m S_{20} + \frac{z_m}{R_2} = {}^m s_{12} \, {}^m T_1 + {}^m s_{11} \, {}^m T_2 + {}^m q_{12} \, {}^m H_1$$

$$^m S_{20} + \frac{(z_p - h_m)}{R_2} = {}^P S_{12} \, {}^P T_1 + {}^P S_{11} \, {}^P T_2 + {}^P d_{31} \, {}^P E_3$$

$$^m S_{20} + \frac{(z_p - h_m - h_p)}{R_2} = {}^s s_{12} \, {}^s T_1 + {}^s s_{11} \, {}^s T_2$$

The axial forces in the three layers must add up to zero to preserve force equilibrium, that is,

$$\int_{-{}^m t/2}^{{}^m t/2} {}^m T_1 dz_m + \int_{-{}^P t/2}^{{}^P t/2} {}^P T_1 dz_p + \int_{-{}^s t/2}^{{}^s t/2} {}^s T_1 dz_s = 0$$

$$\qquad (6.5)$$

$$\int_{-{}^m t/2}^{{}^m t/2} {}^m T_2 dz_m + \int_{-{}^P t/2}^{{}^P t/2} {}^P T_2 dz_p + \int_{-{}^s t/2}^{{}^s t/2} {}^s T_2 dz_s = 0$$

where ^{m}t, ^{p}t, and ^{s}t are the thicknesses of piezomagnetic, piezoelectric, and substrate layers.

Further, Equation 6.4 should be solved for $^{i}T_j$ and substituted into Equation 6.5. Using Equation 6.5 and taking into account Equations 6.2 through 6.4 and 6.6 enables finding $^{m}S_{10}$ and $^{m}S_{20}$:

$$^{m}S_{10} = s_1 \left[(1-V)\,^{m}Y\,^{m}q_{11}H_1 + V\,^{P}Y\left(^{P}d_{31}E_3 + \frac{h_{m}}{R_1} \right) + V_s\,^{s}Y\frac{h_{m} + h_{p}}{R_1} \right]$$

$$^{m}S_{20} = s_1 \left[(1-V)\,^{m}Y\,^{m}q_{12}H_1 + V\,^{P}Y\left(^{P}d_{31}E_3 + \frac{h_{m}}{R_2} \right) + V_s\,^{s}Y\frac{h_{m} + h_{p}}{R_2} \right]$$

$$(6.6)$$

where

E_3 and H_1 are electric field induced across the piezoelectric layer and applied magnetic field

$s_1 = t\,(^{m}t\,^{m}Y + ^{p}t\,^{P}Y + ^{s}t\,^{s}Y)^{-1}$

$V = ^{P}t/t$

$V_s = ^{s}t/t$

$t = ^{m}t + ^{P}t$

^{m}Y, ^{P}Y, and ^{s}Y are the modules of elasticity of piezomagnetic, piezoelectric components, and substrate, respectively.

To conserve moment equilibrium, the rotating moments of axial forces in the three layers are counteracted by resultant bending moments M_{mj}, M_{pj}, and M_{sj}, induced in piezomagnetic, piezoelectric, and substrate layers. That is,

$$F_{m1}h_{m} + F_{p1}(h_{m} + h_{p}) = M_{m1} + M_{p1} + M_{s1}$$

$$F_{m2}h_{m} + F_{p2}(h_{m} + h_{p}) = M_{m2} + M_{p2} + M_{s2}$$

$$(6.7)$$

where

$$F_{i1} = \int_{-^{i}t/2}^{^{i}t/2} {}^{i}T_1\,dz_1, \quad F_{i2} = \int_{-^{i}t/2}^{^{i}t/2} {}^{i}T_2\,dz_1, \quad M_{i1} = \int_{-^{i}t/2}^{^{i}t/2} z_i\,{}^{i}T_1\,dz_1, \quad \text{and}$$

$$M_{i2} = \int_{-^{i}t/2}^{^{i}t/2} z_i\,{}^{i}T_2\,dz_1$$

Taking into account Equations 6.3, 6.4, and 6.6, the equilibrium condition (6.7) can be solved for R_1 and R_2. The expressions for R_1 and R_2 are not given here because of their complicated nature. The values of these radii of curvature can then be substituted into Equation 6.6 to obtain the centroidal strains. Once the centroidal strains are determined, the axial stress $^{i}T_1$ can be found from Equation 6.4. To obtain the expression for ME voltage coefficient, we use the open circuit condition on the boundary:

$$D_3 = 0. \qquad (6.8)$$

Since electric induction is divergence free and has only one component, it is evident that D_3 is equal to zero for any z. In this case, Equations 6.1 and 6.8 result in the expression for ME voltage coefficient

$$\alpha_{E31} = \frac{E_3}{H_1} = -\frac{^{P}d_{31}\int_{-^{P}t/2}^{^{P}t/2}(^{P}T_1 + ^{P}T_2)dz}{t\,H_1\,^{P}\varepsilon_{33}} \qquad (6.9)$$

where $^{P}T_1$ and $^{P}T_2$ are determined by Equation 6.4 taking into account Equation 6.6.

In case of small flexural strain, the radius of curvature in Equations 6.3, 6.4, and 6.6 must tend to infinity. It is easy to show that expression for ME voltage coefficient reduces in this case to well-known expression, which was obtained with the assumption of homogeneous longitudinal strains [46].

It is well known that the lattice mismatch between the substrate and piezoelectric layers results in variation of piezoelectric coefficients and permittivity. This variation can be found using the Landau–Ginsburg–Devonshire phenomenological thermodynamic theory [51]. According to this approach, the thermodynamic potential G' of a thin film on a thick substrate is defined as [52]

$$G' = G + S_1\sigma_1 + S_2\sigma_2 + S_6\sigma_6 \qquad (6.10)$$

where

G is the elastic Gibbs function for the barium titanate layer without substrate

S_1, S_2, and S_6 are in-plane strains at the film/substrate interface arising from lattice mismatch

σ_1, σ_2, and σ_6 are stress components

In case of a cubic (001) substrate, $S_6 = 0$ and $S_1 = S_2 = S_m$, where the misfit strain $S_m = (b - a_0)/b$ can be calculated using the substrate lattice parameter b and the equivalent cubic cell constant a_0 of the freestanding film. For the barium titanate film on SrTiO$_3$ substrate $a_0 = 0.397\,nm$ and $b = 0.393\,nm$. Using these equations with parameters of the Gibbs function enables determining the dielectric constant and piezoelectric coefficients of PZT: $^{P}\varepsilon_{11}/\varepsilon_0 = 51$, $^{P}d_{31} = -18\,pm/V$ [52]. Similarly, the piezomagnetic parameters of the NFO film should be less compared to bulk NFO. Nevertheless, in the NFO/PZT/STO composite film, the bottom PZT layer acts as a buffer layer and effectively reduces constrains from the STO substrate and compressive strains in the NFO layer are almost released. In what follows, we neglect the contribution of residual strains on NFO parameters arising from lattice mismatch. Substituting the appropriate material property values and the dimensions of the structure into Equation 6.9, we can find the ME voltage coefficient with regard to axial and bending stresses for the bilayer on a substrate.

TABLE 6.1 Material Parameters (Compliance Coefficient s, Piezomagnetic Coupling q, Piezoelectric Coefficient d, and Permittivity ε) for NFO, PZT, and SrTiO$_3$ Used for Theoretical Estimates

Material	s_{11} [10^{-12} m^2/N]	s_{12} [10^{-12} m^2/N]	q_{11} [10^{-12} m/A]	q_{12} [10^{-12} m/A]	d_{31} [10^{-12} m/V]	$\varepsilon_{33}/\varepsilon_0$
PZT	17.3	−7.22	—	—	−175	1750
NFO	6.5	−2.4	−680	125	—	—
SrTiO$_3$	3.3	−0.9	—	—	—	—

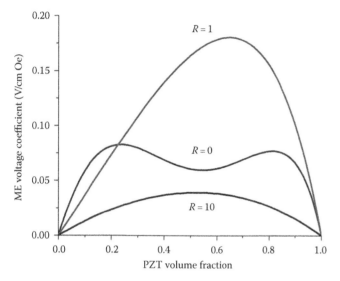

FIGURE 6.4 PZT volume fraction dependence of ME voltage coefficient $\alpha_{E,31}$ as a function of substrate-to-bilayer thickness ratio $R = {}^st/({}^Pt + {}^mt)$.

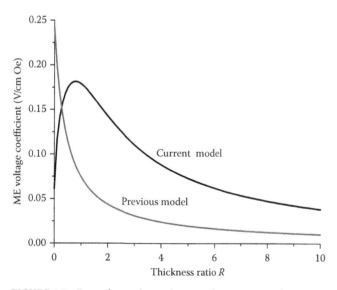

FIGURE 6.5 Dependence of ME voltage coefficient $\alpha_{E,31}$ on substrate-to-bilayer thickness ratio R for PZT volume fraction of $v = 0.6$. The estimates based on present model and the previous model of Ref. [46] are shown.

Next, we apply the model to the case of NFO-PZT on STO substrates. Material parameters used for calculations are listed in Table 6.1. Figure 6.4 shows the PZT volume fraction v dependence of the ME voltage coefficient. For a freestanding bilayer, graph of $\alpha_{E,31}$ vs. v is a double-peaked curve. Such a dependence structure is stipulated by a decrease in the longitudinal strain of PZT due to bending strain in the sample. Figure 6.4 also shows the same dependence for the case when we neglect the flexural strains. That is the case for a freestanding trilayer of NFO-PZT-NFO or PZT-NFO-PZT. This dependence coincides with our earlier model [46]. Our model predicts a decrease in the ME voltage coefficient for the freestanding bilayer compared to the case when neglecting the flexural strains. Placing the bilayer on a substrate of equal thickness gives rise to an increase of ME output due to sign reversal in the contribution of flexural strain to the total strain of PZT in comparison with the freestanding bilayer.

Figure 6.5 shows the ME voltage coefficient as a function of substrate-to-bilayer thickness ratio. Estimates based on this model and our previous model [42] are compared. Increase in the substrate thickness leads to a substantial decrease in ME coupling, as seen in Figure 6.5. However, the rate of change of $\alpha_{E,31}$ is considerably lower than that for the case when neglecting the flexural strains.

6.5 Conclusion

A brief review of ME interactions in layered composites is provided. We then extended the discussion to include low-frequency ME interactions in nanocomposites, with specific focus on a ferrite-piezoelectric bilayer on a dielectric substrate. Expressions have been obtained for the ME voltage coefficients, taking into consideration the substrate clamping effects.

References

1. P. Debye, Bemerkung zu einigen neuen Versuchen uber einen magneto-elektrischen Richteffekt, *Z. Phys.* **36**, 300 (1926).
2. W. C. Rontgen, XLI. Experiments on the electromagnetic action of dielectric polarization, *Phil. Mag.* 19, 385 (1885).
3. P. Curie, Sur la symetrie dans les phenomenes physiques, *J. Phys.* **3**, 393 (1894).
4. L. D. Landau and E. M. Lifshitz, *Electrodynamics of Continuous Media*, Pergamon Press, Oxford, U.K. (1960), p. 119. (Translation of Russian Edition, 1958.)
5. I. E. Dzyaloshinskii, On the magneto-electrical effect in antiferromagnets, *Sov. Phys. JETP* **10**, 628 (1959).
6. D. N. Astrov, Magnetoelectric effect in chromium oxide, *Sov. Phys. JETP* **13**, 729 (1961).
7. G. T. Rado and V. J. Folen, Observation of magnetically induced magnetoelectric effect and evidence for antiferromagnetic domains, *Phys. Rev. Lett.* **7**, 310 (1961); S. Foner and M. Hanabusa, Magnetoelectric effects in Cr$_2$O$_3$ and (Cr$_2$O$_3$)$_{0.8}$ (Al$_2$O$_3$)$_{0.2}$, *J. Appl. Phys.* **34**, 1246 (1963).

8. E. F. Bertaut and M. Mercier, Magnetoelectricity in theory and experiment, *Mater. Res. Bull.* **6**, 907 (1971).

9. G. T. Rado, Linear magnetoelectric effects in gallium iron oxide at low temperatures and in high magnetic fields, *J. Appl. Phys.* **37**, 1403 (1966).

10. E. Kita, S. Takano, A. Tasaki, K. Siratori, K. Kohn, and S. Kimura, Low temperature phase of yttrium iron garnet (YIG) and its first-order magnetoelectric effect, *J. Appl. Phys.* **64**, 5659 (1988).

11. E. Fischer, G. Gorodetsky, and R. M. Hornreich, A new family of magnetoelectric materials: $A_2M_4O_9$ (A = Ta,Nb; M = Mn,Co), *Solid State Commun.* **10**, 1127 (1972).

12. H. Tsujino and K. Kohn, Magnetoelectric effect of $GdMn_2O_5$ single crystal, *Solid State Commun.* **83**, 639 (1992).

13. M. D. Sacks, *Advanced Composite Materials: Processing, Microstructures, Bulk and Interface Properties, Characterization Methods, and Applications Proceedings*, vol. **19**. American Ceramic Society, Westerville, OH (1991).

14. K. V. Logan, Z. A. Munir, and R. M. Spriggs, *Advanced Synthesis and Processing of Composites and Advanced Ceramics II, Ceramic Transactions*, vol. **79**. American Ceramic Society, Westerville, OH (1997).

15. J. van Suchtelen, Product properties: A new application of composite materials, *Philips Res. Rep.* **27**, 28 (1972).

16. J. van den Boomgaard, D. R. Terrell, and R. A. J. Born, An in situ grown eutectic magnetoelectric composite material: Part 1: Composition and unidirectional solidification, *J. Mater. Sci.* **9**, 1705 (1974).

17. A. M. J. G. van Run, D. R. Terrell, and J. H. Scholing, An in situ grown eutectic magnetoelectric composite material: Part 2: Physical properties, *J. Mater. Sci.* **9**, 1710 (1974).

18. J. van den Boomgaard, A. M. J. G. van Run, and J. van Suchtelen, Piezoelectric-piezomagnetic composites with magnetoelectric effect, *Ferroelectrics* **14**, 727 (1976).

19. J. van den Boomgaard and R. A. J. Born, A sintered magnetostrictive composite material $BaTiO_3$-$Ni(Co,Mn)Fe_2O_4$, *J. Mater. Sci.* **13**, 1538 (1978).

20. G. Harshe, J. P. Dougherty, and R. E. Newnham, Theoretical modelling of multilayer magnetoelectric composites, *Int. J. Appl. Electromag. Mater.* **4**, 145 (1993); M. Avellaneda and G. Harshe, Magnetoelectric effect in piezoelectric/ magnetostrictive multilayer (2–2)composites, *J. Intell. Mater. Syst. Struct.* **5**, 501 (1994).

21. G. Harshe, Magnetoelectric effect in piezoelectric-magnetostrictive composites, PhD thesis, The Pennsylvania State University, College Park, PA, 1991.

22. Ce-Wen Nan and D. R. Clarke, Effective properties of ferroelectric and/or ferromagnetic composites: A unified approach and its application, *J. Am. Ceram. Soc.* **80**, 1333 (1997).

23. G. Srinivasan, E. T. Rasmussen, B. J. Levin, and R. Hayes, Magnetoelectric effects in bilayers and multilayers of magnetostrictive and piezoelectric perovskite oxides, *Phys. Rev. B* **65**, 134402 (2002).

24. G. Srinivasan, E. T. Rasmussen, J. Gallegos, R. Srinivasan, Yu. I. Bokhan, and V. M. Laletin, Magnetoelectric bilayer and multilayer structures of magnetostrictive and piezoelectric oxides, *Phys. Rev. B* **64**, 214408 (2001).

25. G. Srinivasan, E. T. Rasmussen, and R. Hayes, Magnetoelectric effects in ferrite-lead zirconate titanate layered composites: Studies on the influence of zinc substitution in ferrites, *Phys. Rev. B* **67**, 014418 (2003).

26. N. A. Spaldin and M. Fiebig, The renaissance of magnetoelectric multiferroics, *Science* **309**, 391 (2005).

27. H. Schmid, *Introduction to Complex Mediums for Optics and Electromagnetics*, Eds. W. S. Weiglhofer and A. Lakhtakia, SPIE Prsee, Bellingham, WA (2003), pp. 167–195.

28. M. Fiebig, Revival of the magnetoelectric effect, *J. Phys. D: Appl. Phys.* **38**, R123 (2005).

29. Ce-Wen Nan, M. I. Bichurin, S. Dong, D. Viehland, and G. Srinivasan, Multiferroic magnetoelectric composites: Historical perspective, status and future directions, *J. Appl. Phys.* **103**, 031101 (2008).

30. H. Zheng, J. Wang, S. E. Lofland, Z. Ma, L. Mohaddes-Ardabili, T. Zhao, L. Salamanca-Riba et al., Multiferroic $BaTiO_3$-$CoFe_2O_4$ nanostructures, *Science* **303**, 661–663 (2004).

31. H. Zheng, J. Wang, L. Mohaddes-Ardabili, M. Wuttig, L. Salamanca-Riba, D. G. Schlom, and R. Ramesh, Three-dimensional heteroepitaxy in self-assembled $BaTiO_3$-$CoFe_2O_4$ nanostructures, *Appl. Phys. Lett.* **85**, 2035 (2004).

32. J. Zhai, Z. Xing, S. Dong, J. Li, and D. Viehland, Detection of pico-Tesla magnetic field using magneto-electric sensors at room temperature, *Appl. Phys. Lett.* **88**, 062510 (2006).

33. J. Zhai, Z. Xing, S. Dong, J. Li, and D. Viehland, Magnetoelectric laminate composites: An overview, *J. Am. Ceram. Soc.* **91**, 351 (2008).

34. Z. P. Xing, J. Y. Zhai, S. X. Dong, J. F. Li, and D. Viehland, Modeling and detection of quasi-static nanotesla magnetic field variations using magnetoelectric laminate sensors, *Meas. Sci. Technol.* **19**, 015206 (2008).

35. S. Priya, R. Bergs, and R. A. Islam, Magnetic field anomaly detector using magnetoelectric composites, *J. Appl. Phys.* **101**, 024108 (2007).

36. R. A. Islam, Y. Ni, A. G. Khachaturyan, and S. Priya, Giant magnetoelectric effect in sintered multilayered composite structures, *J. Appl. Phys.* **104**, 044103 (2008).

37. M. Liu, X. Li, J. Lou, S. Zheng, K. Dui, and N. X. Sun, A modified sol-gel process for multiferroic nanocomposites, *J. Appl. Phys.* **102**, 083911 (2007).

38. Y. Chen, J. Wang, M. Liu, N. X. Sun, C. Vittorai, and V. Harris, Giant magnetoelectric coupling and E-field tenability in a laminate Ni_2MnGa/PMN-PT multiferroic heterostructure, *Appl. Phys. Lett.* **93**, 112502 (2008).

39. J. Zhai, S. Dong, Z. Xing, J. Li, and D. Viehland, Geomagnetic sensor based on magnetoelectric effect, *Appl. Phys. Lett.* **91**, 123513 (2007).

40. J. Ryu, A. V. Carazo, K. Uchino, and H. Kim, Piezoelectric and magnetoelectric properties of lead zirconate titanate/Ni-ferrite particulate composites, *J. Electroceram.* **7**, 17 (2001).

41. K. Mori and M. Wuttig, Magnetoelectric coupling in terfenol-D/polyvinylidenedifluoride composites, *Appl. Phys. Lett.* **81**, 100 (2002).

42. S. Dong, J. F. Li, and D. Viehland, Giant magnetoelectric effect in laminate composites, *Philos. Mag. Lett.* **83**, 769 (2003).

43. S. Dong, J. Zhai, Z. Xing, J. F. Li, and D. Viehland, Extremely low frequency response of magnetoelectric multilayer composite, *Appl. Phys. Lett.* **86**, 102901 (2005).

44. M. I. Bichurin, D. A. Fillipov, V. M. Petrov, U. Laletsin, and G. Srinivasan, Resonance magnetoelectric effects in layered magnetostrictive-piezoelectric composites, *Phys. Rev. B* **68**, 132408 (2003).

45. U. Laletsin, N. Paddubnaya, G. Srinivasan, and C. P. DeVreugd, Frequency dependence of magnetoelectric interactions in layered structures of ferromagnetic alloys and piezoelectric oxides, *Appl. Phys. A* **78**, 33 (2004).

46. V. M. Petrov, G. Srinivasan, M. I. Bichurin, and A. Gupta, Theory of magnetoelectric effects in ferrite-piezoelectric nanocomposites, *Phys. Rev. B* **75**, 224407, (2007).

47. V. E. Wood and A. E. Austin, Possible applications for magnetoelectric materials, *Proceeding of Symposium on Magnetoelectric Interaction Phenomena in Crystals*, Seattle, WA, May 21–24, 1973, eds. A. J. Freeman and H. Schmid, Gordon & Breach Science Publishers, New York (1975), p. 181.

48. T. H. O'Dell, *The Electrodynamics of Magneto-Electric Media*, North-Holland Pub. Co., Amsterdam, the Netherlands (1970).

49. G. Srinivasan and Y. K. Fetisov, Microwave magnetoelectric effects and signal processing devices, *Integr. Ferroelectr.* **83**, 89 (2006).

50. M. I. Bichurin, D. Viehland, and G. Srinivasan, Magnetoelectric interactions in a ferromagnetic-piezoelectric layered structures: Phenomena and devices, *J. Electroceram.* **19**, 243 (2007).

51. N. A. Pertsev, A. G. Zembilgotov, and A. K. Tagantsev, Effect of mechanical boundary conditions on diagrams of epitaxial ferroelectric thin films, *Phys. Rev. Lett.* **80**, 1988 (1998).

52. N. A. Pertsev, V. G. Kukhar, H. Rohlstedt, and R. Waser, Phase diagrams and physical properties of single domain epitaxial Pb $(Zr_{1-x}Ti_x)O_3$ thin films, *Phys. Rev. B* **67**, 054107 (2003).

7

Strain-Induced Disorder in Ferroic Nanocomposites

Anna N. Morozovska
*National Academy of
Science of Ukraine*

Eugene A. Eliseev
*National Academy of
Science of Ukraine*

7.1 Introduction

Ferroic nanomaterials open the way to obtain a variety of new unique magneto-mechanical, electro-mechanical, electronic, and dielectric properties, a lot of which are useful for applications. For example, their ability to store and release energy in well-regulated manners makes them very useful for sensors and actuators, telecommunications, optical and infrared detectors, pyrosensors and thermal imaging, and multidisciplinary fields such as biomedical and environment (Wadhawan 2000, Fiebig 2005, Roduner 2006, Scott 2007).

Ferroelectric, ferromagnetic, and ferroelastic materials are known to belong to the primary ferroics, because the application of an electric, magnetic, or elastic field higher than that of the coercive field leads to the switching of the corresponding order parameter (polarization, magnetization, or elastic strain). Ferroelectric or ferromagnetic domain switching is the phenomenon wherein the ferroic material changes from one spontaneously polarized or magnetized state to another under the applied electric or magnetic field or elastic stress.

The common feature of nanomaterials, such as films and nanocomposites containing nanoparticles with sizes less than 100 nm, is the essential influence of the surface stresses and strains on their properties. Among different ferroic nanomaterials magnetic nanomaterials are the most thoroughly investigated, especially thin magnetic films and their multilayers, while ferroelectric nanomaterials still contain a lot of challenges (Ban et al. 2003, Alpay et al. 2004, Akcay et al. 2008).

Up to now phase transitions in solids attract much scientific and technical interest because of the anomalies in their properties in the vicinity of the phase transition temperature. Recently the possibility to govern the appearance of phase transitions at any arbitrary temperature has been demonstrated in nanosized materials due to the so-called size-driven phase transition. Such transitions were observed in many solids, including ferroelectric, ferromagnetic, and ferroelastic nanomaterials. For instance, it is generally accepted that the ferroelectric and ferromagnetic properties disappear below the critical particle size (Lines and Glass 1977). It is well known that the depolarization of electric fields and demagnetization effects exist in the majority of confined ferroelectric systems (Landau and Lifshits 1980) and cause the disappearance of size-induced ferroelectricity and ferromagnetism in thin films and spherical particles (Ishikava et al. 1988, Tilley 1996).

The cylindrical geometry of particles does not destroy the ferroelectric phase in contrast to the size-induced transition to the paraelectric phase in spherical particles (Uchino et al. 1989, Zhong et al. 1994, Mishra and Pandey 1995, Wang and Smith 1995, Rychetsky and Hudak 1997, Huang et al. 2001, Glinchuk and Morozovska 2003, Fong et al. 2004, Erdem et al. 2006), but sometimes a noticeable enhancement of ferroelectric properties occurs (Mishina et al. 2002, Luo et al. 2003, Morrison et al. 2003a,b, Yadlovker and Berger 2005, Geneste et al. 2006). For instance, Yadlovker and Berger (2005) reported the spontaneous polarization enhancement of up to $0.25–2\,\mu C/cm^2$ and ferroelectric phase conservation in Rochelle salt nanorods. With the help of piezoelectric force microscopy (PFM) Morrison et al. (2003a,b) demonstrated that $PbZr_{0.52}Ti_{0.48}O_3$ (PZT) nanotubes possess a rectangular-shaped local piezoelectric response hysteresis loop with the effective remnant piezoelectric coefficient value compatible with those typical for PZT films. The authors also demonstrated that the ferroelectric properties of the free $BaTiO_3$ nanotubes are perfect. Poyato et al. (2006) with the help of PFM found that nanotube-patterned ("honeycomb") $BaTiO_3$ films of thickness 200–300 nm reveal ferroelectric properties. The inner diameter of the nanotubes ranged from 50 to 100 nm. They also demonstrated the existence of local piezoelectric and oriented ferroelectric responses prior to the application of a dc electric field.

The phenomenological description of ferroelectricity enhancement in cylindrical nanoparticles has been recently proposed by Morozovska et al. (2006, 2007a,b). We proved that the reason for the enhancement of the polar properties and their conservation in ferroelectric nanorods and nanotubes is stress coupled with polarization via an electrostriction effect under the strong decrease of the depolarization field with an increase in particle length. Within the decoupling approximation of electric and elastic problems the tube piezoelectric response was calculated and compared with available experimental data (Kalinin et al. 2006, Morozovska et al. 2007b).

The ability of ferroelectrics to change their sizes in an applied electric field is important for piezoelectric applications (Cao 2005); however, elastic stresses and strains strongly influence the switching behavior of ferroelectric nanomaterials. Inhomogeneous elastic strains are the general feature of nanocomposites and nano-grained ceramics consisting of variously oriented crystalline regions or grains. Therefore, domain switching of a nanocomposite is a typical collective process, wherein an *adequate theoretical description* involves both a statistical approach and consideration of nonlinear correlation effects between neighboring particles or grains. For instance, Chen et al. (1997) studied the evolution of mesostructures (the patterns of internal domain rearrangement) and formulated the incremental nonlinear constitutive relations within the framework of the internal-variable theory. The performed simulation and experimental results were compared and discussed. Li et al. (2005) used a combined theoretical and experimental approach to establish a relation between crystallographic symmetry and the ability of a ferroelectric polycrystalline ceramic to switch. Michael et al. (2007) studied the influence of size effects on the critical temperature, static and dynamic properties of spherical ferroelectric nanoparticles by the modified transverse Ising model and compared theoretical calculations with experimental data. Akdogan and Safari (2007) theoretically considered single-domain, mechanically free and screened $PbTiO_3$ nanocrystals with no depolarization fields. By using experimental data for $PbTiO_3$ particle size–dependent spontaneous polarization, they calculated the Landau coefficients up to the sixth order as a function of particle size in the range <150 nm in a self-consistent manner.

Modern *synthesis methods* of ferroelectric and multiferroic nanoparticles are relatively well-elaborated. As typical examples let us briefly mention the enhanced liquid-precursor-based version of the combined polymerization and pyrolysis preparation route combined with consecutive soft milling used by Erdem et al. (2006) for the preparation of $PbTiO_3$ and $BaTiO_3$ nanoparticles with sizes from 5 to 50 nm; hydrothermal and solvo-thermal methods used by Wei et al. (2007) for the synthesis of high-purity $KTa_{0.3}Nb_{0.7}O_3$ nanoparticles; the single phase sol–gel method used by Fernández-Osorio et al. (2007) for the preparation of free-standing $Pb(Zr_{0.52}Ti_{0.48})O_3$ nanocrystals with an average size of 13 nm; chemical method using a polymer complex of Ba^{2+} and Ti^{4+} with polyvinyl alcohol used by Ram et al. (2007) for the synthesis of crystallite $BaTiO_3$ nanoparticles with 15 nm average size; solid-state chemistry proposed by Mornet et al. (2007) for multiferroic ferroelectric-based nanocomposites design of superparamagnetic nanoparticles around ferroelectric cores; chemical route using polyvinyl alcohol as a surfactant proposed by Verma et al. (2008) for the synthesis of Fe-doped $PbTiO_3$ ferroic nanoparticles with sizes of 19–30 nm. Mornet et al. (2007) pointed out that an approach combining nanotechnology and solid-state chemistry tailors innovative multifunctional nanomaterials such as dense nanocomposites. The sintering of these dense nanocomposites enables a tuning of the ferroic properties and the coexistence between ferroelectricity, piezoelectricity, and superparamagnetism. Mornet et al. (2007) also extended the method to metallic nanoparticles surrounding ferroelectric cores.

As *experimental methods* for studies of temperature- and size-dependent structural changes and phase transitions occurring in ferroic nanoparticles the authors widely use scanning electron microscopy (SEM), transmission electron microscopy (TEM), high-resolution transmission electron microscopy (HR-TEM), x-ray powder diffraction (XRD), Raman spectroscopy, electronic paramagnetic resonance (EPR) and dielectric measurements. In particular Erdem et al. (2006) directly observed the size-driven phase transition at 7 nm for $PbTiO_3$ nanoparticles in comparison with 40 nm for $BaTiO_3$. Corresponding EPR data suggest the gradient shell thickness of about 2 nm at the $PbTiO_3$ particle surfaces and 15 nm shell at the $BaTiO_3$ particle surfaces. Typically Raman spectra reveal that nanoparticles possess a tetragonal or orthorhombic structure in the ordered low-temperature phase and HR-TEM images confirm the crystalline nature of the obtained phase (Fernández-Osorio et al. 2007, Ram et al. 2007). The results of Verma et al. (2008) indicate that the dielectric constant and magnetization value of Fe-doped $PbTiO_3$ nanoparticles depend upon their size and Fe concentration. It is observed that the magnetization is enhanced with reduction in particle size. The largest value of saturation magnetization ($M_s = 41.6 \times 10^{-3}$ emu/g) is observed for 1.2 mol% of Fe. Li et al. (2006) and Kaczmarek et al. (2008) investigated the physical properties of low concentration ferroelectric nematic colloids using calorimetry, optical methods, infrared spectroscopy, capacitance studies, and atomic force microscopy. The resulting homogeneous colloids possess a significantly amplified nematic orientational coupling. Kaczmarek et al. (2008) observed the enhancement of the electro-optic performance in liquid crystals with $Sn_2P_2S_6$ nanoparticles with the average size (45 nm) higher than for $BaTiO_3$ nanoparticles (20 nm).

Despite the technology progress in production of these systems, for instance, using porous matrices (Kumzerov and Vakhrushev 2003, Morrison et al. 2003a,b, Yadlovker and Berger 2005), up to now nothing is known about the superparaelectric phase in ensembles of ferroelectric nanoparticles; also the term "superparaelectric" was used for the description of bulk ferroelectric relaxors (Cross 1987, Glazunov et al. 1995, Li et al. 1997). However, recently Glinchuk et al. (2008b) have predicted the conditions for the appearance of the superparaelectric phase in the ensemble of non-interacting spherical ferroelectric nanoparticles. The superparaelectricity in the ensemble of nanoparticles was defined by an analogy with superparamagnetism that was obtained earlier in small nanoparticles made of magnetic material. Calculations were performed within the Landau–Ginzburg phenomenological approach. The main favorable conditions for the occurrence of superparaelectricity in ferroelectric nanoparticles are a radius smaller than the correlation radius, surface screening of the depolarization field, small Curie–Weiss constant, and high nonlinear coefficients, which guarantee that the barrier of the particle polarization orientation will be smaller than the thermal energy and the particle is of a single domain.

In comparison with the evolved phenomenological description of size-driven phase transitions in ferroic thin films (Ban et al. 2003, Alpay et al. 2004, Bratkovsky and Levanyuk 2005, Akcay et al. 2008) and nanoparticles (Rychetsky and Hudak 1997, Huang et al. 2001, Geneste et al. 2006, Akdogan and Safari 2007, Morozovska et al. 2007a) leading to a relatively comprehensive analytical treatment, we could not find in the literature any simple analytical model of strain-induced collective disorder in ferroic nanocomposites, despite the fact that all available experimental results have proved that the diffused phase transition point temperature dependence, slim and/or smeared hysteresis loops, and broad dispersion of generalized susceptibility temperature maximum are the characteristic features of these nanomaterials (in contrast to square hysteresis loops and sharp maxima in bulk materials). Below we use a concrete example of a strained spherical ferroelectric nanoparticles assembly (e.g., model for porous nanocomposite or fine-grained ceramics) to demonstrate that the ferroelectric disorder originates from an internal elastic strain distribution caused by nanoparticle sizes distribution and interface effects.

This chapter is organized in the following way. The theoretical consideration presented in Section 7.2 consists of a general phenomenological approach for the description of phase transitions and response to external fields in ferroic nanoparticles (Section 7.2.1), detailed tutorial calculations of ferroelectric nanoparticles free energy (Section 7.2.2), depolarization field for a single-domain particle (Section 7.2.3), strain and size-induced effects on phase transition temperature (Section 7.2.4), and the collective effect of strain and depolarization field on the ferroelectric and dielectric response of nanocomposites (Section 7.2.5). Section 7.3 presents the discussion including a brief summary and future perspectives.

7.2 Theoretical Consideration

7.2.1 General Phenomenological Approach for Description of Phase Transitions and Response to External Fields in Nanoferroics

A phenomenological approach for the description of phase transitions and the response to external fields in primary ferroic nanoparticles has been recently proposed by Morozovska et al. (2007c) and Glinchuk et al. (2008a). The intrinsic surface stress, order parameter gradient, and striction as well as depolarization and demagnetization effects were included into the free energy. For the case of primary ferroics, the Landau–Ginzburg–Devonshire expansion of Gibbs free energy bulk part G_V on the multicomponent order parameter η powers (vectors of polarization, magnetization, or strain tensor for ferroelectric, ferromagnetic, or ferroelastic media, respectively) and stress tensor components powers σ_{ij} has the form

$$
G_V = \int_V d^3r \left(\frac{a_{ij}(T)}{2}\eta_i\eta_j + \frac{a_{ijkl}}{4}\eta_i\eta_j\eta_k\eta_l \right.
$$
$$
+ \frac{a_{ijklmn}}{6}\eta_i\eta_j\eta_k\eta_l\eta_m\eta_n + \cdots + \frac{g_{ijkl}}{2}\left(\frac{\partial\eta_i}{\partial x_j}\frac{\partial\eta_k}{\partial x_l}\right)
$$
$$
\left. - \eta_i\left(E_{0i} + \frac{E_i^d}{2}\right) - Q_{ijkl}\sigma_{ij}\eta_k\eta_l - \frac{1}{2}s_{ijkl}\sigma_{ij}\sigma_{kl} \right). \tag{7.1}
$$

The characteristic feature of the nanosized structures' phenomenological description is the surface energy contribution G_S that becomes comparable with the bulk's structure and can exceed it with a decrease in size:

$$
G_S = \int_S d^2r \left(\frac{a_{ij}^S}{2}\eta_i\eta_j + \frac{a_{ij}^S}{4}\eta_i^2\eta_j^2 + \frac{a_{ijk}^S}{6}\eta_i^2\eta_j^2\eta_k^2 - q_{ijkl}^S\sigma_{ij}\eta_k\eta_l \right.
$$
$$
\left. + d_{ijk}^S\sigma_{jk}\eta_i + \mu_{\alpha\beta}s_{\alpha\beta jk}\sigma_{jk} + \frac{v_{ijkl}^S}{2}\sigma_{ij}\sigma_{kl} + \cdots \right). \tag{7.2}
$$

Coefficients $a_{ij}(T) = \alpha_T(T - T_C)$ explicitly depend on temperature T in the framework of the Landau–Ginzburg–Devonshire approach. Coefficients a_{ij}^S of the surface energy expansion may also depend on temperature. High order expansion coefficients a_{ijkl}, a_{ijklmn}, a_{ijkl}^S, and a_{ijklmn}^S are supposed to be temperature independent, and constants, g_{ijkl}, determine the magnitude of gradient energy. Tensors, g_{ijkl} and a_{ijklmn}, are positively defined. The situation with tensor a_{ijkl} depends on the phase transition order: a_{ijkl} is positively defined for the second order phase transition, while it is negatively defined for the first order transition. $\mathbf{E}_0$ is the external field conjugated with the order parameter η. $\mathbf{E}^d$ is the depolarization, demagnetization, or deelastification field that appears due to the nonzero divergence of the order parameter η in confined system (div(η) ≠ 0). It is easy to show that η and $\mathbf{E}^d$ are related to each other via the linear operator $\hat{N}^d[\eta]$ as $\mathbf{E}^d \cong \hat{N}^d[\eta]$. A concrete example of depolarization field calculations will be given in Section 7.2.3.

Coupling terms $Q_{ijkl}\sigma_{ij}\eta_k\eta_l$ and $q_{ijkl}^S\sigma_{ij}\eta_k\eta_l$ determine the influence of mechanical stress on the order parameter for the materials with high symmetry paraphase (paraelectric, paramagnetic, or paraelastic). Here Q_{ijkl} and q_{ijkl}^S are the bulk and surface striction coefficients, respectively; s_{ijkl} are components of elastic compliance tensor (Landau and Lifshitz 1980). d_{ijk}^S is the surface piezoeffect tensor. This tensor arises even in cubic paraelectrics due to the symmetry breaking near the surface (vanishing of inversion center, see e.g., Glinchuk and Morozovska 2004, Bratkovsky and Levanyuk 2005), while in magnetics it exists when there is no inversion of time among the symmetry operations of the material. Tensor v_{jklm}^S is related with the surface excess elastic moduli.

The intrinsic surface stress $\mu_{\alpha\beta}$ under the curved surface of the solid body determines the excess pressure on the surface (Marchenko and Parshin 1980, Marchenko 1981, Shchukin and Bimberg 1999). Intrinsic mechanical stress under the curved surface is determined by the tensor of intrinsic surface stress $\mu_{\alpha\beta}$:

$$n_k\sigma_{kj}\big|_S = -\frac{\mu_{\alpha\alpha}}{R_\alpha}n_j, \qquad (7.3)$$

where

  R_α is the main curvature of the surface free of facets and
    edges in a continuum media approximation
  n_k is the component of the external normal.

If the nanoparticles are clamped to the substrate or porous template, mechanical displacement u_i should be zero in the clamped region: $u_i\big|_{S_{\text{clamped}}} = 0$.

Let us underline that in many experimental papers (e.g., Uchino et al. 1989, Ma et al. 1998, Zhou et al. 2007) the size effects of ferroelectric nanoparticles' phase diagrams are related with the intrinsic surface stress (or surface tension by analogy with liquids). For the case, where a liquid precursor of ferroelectric (e.g., RS, PZT, SBT, or BTO) fills the porous template by capillary effects (Yadlovker and Berger 2005), the uniform stress inside the pores is caused by surface tension. During the following annealing, both the thermal stresses and misfit strain on the nanoparticle-pore interface usually appear. In most of cases, the stress causes the thin strained layer or "shell" on interface. For instance, Luo et al. (2003) and Morrison et al. (2003a,b) reported about the amorphous PZT layer of thickness 5–20 nm that clamped the nanotube crystalline "core." The shell may be partially removed by selective etching. Using the experimental background, the core-and-shell model of spherical ferroelectric nanoparticles proposed earlier by Niepce (1994), Perriat et al. (1994), and Glinchuk and Morozovskaya (2003) was modified by Glinchuk et al. (2008a)

for the description of ferroelectric nanocomposite polar properties. The strong intrinsic surface stress under the curved nanoparticle surface was shown to play an important role in the shift of transition temperature up to the appearance of a new ordered phase that was absent in the bulk ferroic material. Note that the stress is caused by the particle surface clamping by template; i.e., it is related to the surface tension, different thermal expansion coefficients of particle and template, and/or their mismatch strain. Glinchuk et al. (2008a) demonstrated that under the favorable conditions (e.g., radius 5–50 nm and compressive surface stress) the ferroelectric phase appears in incipient ferroelectric nanowires and nanospheres.

Free energy (7.1) is minimal when the order parameter η and relevant stress tensor components σ_{jk} are defined at the nanostructure boundaries. Under such conditions, one should solve the equation of state $\partial G_V/\partial\sigma_{jk} = -u_{jk}$, where u_{jk} is the strain tensor. The equations of state should be solved along with the equations of mechanical equilibrium $\partial\sigma_{ij}(\mathbf{x})/\partial x_i = 0$ and compatibility equations $e_{ikl}e_{jmn}(\partial^2 u_{ln}/\partial x_k\partial x_m) = 0$ (e_{ikl} is the permutation symbol or anti-symmetric Levi–Civita tensor), equivalent to the mechanical displacement vector u_i continuity (Timoshenko and Goodier 1970). For the cases of the clamped system with defined displacement components (or with mixed boundary conditions), one should find the equilibrium state as the minimum of the Helmholtz free energy

$$F_V + F_S \left(F_V = G_V + \int_V d^3r \cdot u_{jk}\sigma_{jk} \text{ and } F_S = G_S + \int_S d^2r \cdot u_\alpha\sigma_{\alpha k}n_k \right)$$

originated from the Legendre transformation of Gibbs energy G (Pertsev et al. 1998).

Euler–Lagrange equations obtained after the Landau–Ginzburg–Devonshire free energy minimization with respect to the order parameter η can be solved by the direct variational method. This leads to the conventional form of the free energy with renormalized coefficients depending on nanoparticle sizes, surface stress, and electrostriction tensor values and so opens the way for the calculations of ferroic properties by algebraic transformations.

7.2.2 Gibbs Energy of Spherical Ferroelectric Nanoparticle

Let us consider the spherical particle or grain (core) with sidewalls covered with thin surface layers (shells) of different structures. For perovskite symmetry the Gibbs energy expansion on averaged polarization $\mathbf{P} = (0, 0, P_3)$ and elastic stress σ_{ij} has the form

$$G_R = 4\pi\int_0^R \rho^2 d\rho \left(a_1 P_3^2 + a_{11}P_3^4 + a_{111}P_3^6 + \frac{g}{2}(\nabla P_3)^2 - P_3\left(E_0 + \frac{E_3^d}{2}\right) \right.$$

$$- Q_{11}\sigma_{33}P_3^2 - Q_{12}(\sigma_{11} + \sigma_{22})P_3^2 - \frac{1}{2}s_{11}(\sigma_{11}^2 + \sigma_{22}^2 + \sigma_{33}^2)$$

$$\left. - s_{12}(\sigma_{11}\sigma_{22} + \sigma_{11}\sigma_{33} + \sigma_{33}\sigma_{22}) - \frac{1}{2}s_{44}(\sigma_{23}^2 + \sigma_{13}^2 + \sigma_{12}^2) \right)$$

$$(7.4)$$

R is the nanoparticle radius. Subscripts 1, 2, and 3 denote Cartesian coordinates x, y, and z, respectively. We use Voigt notation or matrix notation when it is necessary ($xx = 1$, $yy = 2$, $zz = 3$, $zy = 4$, $zx = 5$, and $xy = 6$).

Minimization of the free energy (7.4) on stress components σ_{ij} leads to the following equations of state:

$$\begin{cases} s_{11}\sigma_{11} + s_{12}(\sigma_{22} + \sigma_{33}) + Q_{12}P_3^2 = u_{11}, \\ s_{11}\sigma_{22} + s_{12}(\sigma_{11} + \sigma_{33}) + Q_{12}P_3^2 = u_{22}, \\ s_{11}\sigma_{33} + s_{12}(\sigma_{22} + \sigma_{11}) + Q_{11}P_3^2 = u_{33}, \\ 2u_{23} = s_{44}\sigma_{23}, \quad 2u_{13} = s_{44}\sigma_{13}, \quad 2u_{12} = s_{44}\sigma_{12}. \end{cases} \quad (7.5)$$

Here $u_{ij} = (\partial u_i/\partial x_j + \partial u_j/\partial x_i)/2$ are elastic strain tensor components, u_i is displacement vector components.

Distribution of mechanical displacement u_i should satisfy the conditions of mechanical equilibrium $\partial \sigma_{ij}/\partial x_i = 0$ as well as the appropriate boundary conditions (7.3) on the particle (grain) surface and conditions of continuity at interfaces (if any) in the spherical coordinate system (r, θ, ψ):

$$\begin{cases} \sigma_{rr}(r = R, \theta, \psi) = -p(\theta, \psi) - \dfrac{\mu_{\theta\theta}(\theta,\psi)}{R} - \dfrac{\mu_{\psi\psi}(\theta,\psi)}{R}, \\ \sigma_{r\theta}(r = R) = \dfrac{\partial \mu_{\theta\theta}(\theta,\psi)}{R\partial\theta} + \dfrac{\partial\mu_{\psi\theta}(\theta,\psi)}{R\sin\theta\partial\psi}, \\ \sigma_{r\psi}(r = R) = \dfrac{\partial\mu_{\psi\psi}(\theta,\psi)}{R\sin\theta\partial\psi} + \dfrac{\partial\mu_{\psi\theta}(\theta,\psi)}{R\partial\theta}. \end{cases} \quad (7.6)$$

Here

$p(\theta, \psi)$ is the effective normal pressure caused by ambient, neighboring particles (or polarized grains)

μ_{ij} is the intrinsic surface stress tensor coefficients, which have nontrivial components only on the grain surface ($i, j = \theta, \psi$)

For tutorial purposes of the chapter we neglected the polarization gradient effects. The assumption is rigorous for single-domain particles and *natural* boundary conditions for ferroelectric polarization at the particle surface. The boundary condition $\partial P_3/\partial r = 0$ is called natural (see Korn and Korn 1961 and Tagantsev and Gerra 2006) and corresponds to the case, when one could neglect the surface energy contribution. Mathematically this corresponds to infinite extrapolation length $\lambda = \left(g/a_1^S\right) \to \infty$ in a more general boundary condition:

$$\left(P_3 + \lambda\left(\frac{dP_3}{dr} + \frac{a_{11}^S}{g}P_3^3\right)\right)\Bigg|_{r=R} = 0$$

Without the polarization gradient a homogeneous solution of the elastic problem exists and both the conditions of mechanical equilibrium and compatibility relations are identically valid. Below, two limiting cases of boundary conditions for defined elastic stress and strain are considered.

7.2.2.1 Uniformly Stressed Particles

The situation when a given particle may be regarded stressed (e.g., by surface tension) corresponds to the nanoparticle dipped into elastically "soft" surrounding matrix such as organic polymer, gel, or liquid crystal. For the case, let isotropic effective pressure $p(\theta, \psi) \equiv p_R$ and intrinsic surface stress $\mu_{ij}(\theta, \psi) = \mu_R\delta_{ij}$ (δ_{ij} is the Kronecker symbol) exist at the particle surface. Allowing for Equation 7.6, the solution of Equation 7.5 has the form

$$\sigma_{rr}(r,\theta,\psi) = \sigma_{\theta\theta}(r,\theta,\psi) = \sigma_{\psi\psi}(r,\theta,\psi) = -p_R - \frac{2\mu_R}{R}, \quad \sigma_{r\theta} = \sigma_{r\psi} = 0,$$
$$(7.7a)$$

$$u_{11} = u_{22} = -(s_{11} + 2s_{12})\left(p_R + \frac{2\mu_R}{R}\right) + Q_{12}P_3^2,$$

$$u_{33} = -(s_{11} + 2s_{12})\left(p_R + \frac{2\mu_R}{R}\right) + Q_{11}P_3^2, \quad (7.7b)$$

$$u_{23} = 0, \quad u_{13} = 0, \quad u_{12} = 0.$$

Substituting Equation 7.7a into Equation 7.4 we obtain the Gibbs energy density with renormalized coefficients:

$$g_R = \left(a_1 + (Q_{11} + 2Q_{12})\left(p_R + \frac{2\mu_R}{R}\right)\right)P_3^2 + a_{11}P_3^4$$
$$+ a_{111}P_3^6 - P_3\left(E_0 + \frac{E_3^d}{2}\right). \quad (7.8)$$

Under the absence of external electric field E_0 the average pressure p_R is zero, whereas $p_R = p_R(E_0, R)$ in the general case.

Since the surface stress values are hardly measured for nanoparticles with radii $R = 10^{-8} - 10^{-6}$ m and are strongly dependent on the interface material, let us rewrite the energy (7.8) in terms of the effective strain:

$$u_R(R) = -(s_{11} + 2s_{12})\left(p_R + \frac{2\mu_R}{R}\right) - \left(\frac{2Q_{12} + Q_{11}}{3}\right)P_3^2$$

where $u_R \equiv -(u_{11} + u_{22} + u_{33})/3$ is the average strain (7.7b). After elementary transformations in Equation 7.8 we obtain:

$$g_R = \left(a_1 - \frac{Q_{11} + 2Q_{12}}{s_{11} + 2s_{12}}u_R\right)P_3^2 + \left(a_{11} - \frac{(Q_{11} + 2Q_{12})^2}{s_{11} + 2s_{12}}\right)P_3^4$$
$$- P_3\left(E_0 + \frac{E_3^d}{2}\right) + a_{111}P_3^6. \quad (7.9)$$

7.2.2.2 Uniformly Strained Particles

The situation when a given nanoparticle should be regarded strained corresponds to its clamping by a relatively "rigid" surrounding such as alumina, metallic template, or dense nano-grained ceramic. For the sake of simplicity the effective strain is regarded isotropic: $u_{ij} = u_R \delta_{ij}$. The corresponding solution of Equation 7.5 has the form

$$
\begin{cases}
\sigma_{11} = \sigma_{22} = \dfrac{u_R}{s_{11} + 2s_{12}} + \dfrac{P_3^2}{s_{11} + 2s_{12}} \cdot \dfrac{Q_{11}s_{12} - Q_{12}s_{11}}{s_{11} - s_{12}}, \\[2ex]
\sigma_{33} = \dfrac{u_R}{s_{11} + 2s_{12}} + \dfrac{P_3^2}{s_{11} + 2s_{12}} \cdot \dfrac{2Q_{12}s_{12} - Q_{11}(s_{11} + s_{12})}{s_{11} - s_{12}}, \\[2ex]
\sigma_{11} + \sigma_{22} + \sigma_{33} = \dfrac{3u_R}{s_{11} + 2s_{12}} - \dfrac{P_3^2(2Q_{12} + Q_{11})}{s_{11} + 2s_{12}}, \\[2ex]
\sigma_{23} = 0, \quad \sigma_{13} = 0, \quad \sigma_{12} = 0.
\end{cases} \tag{7.10}
$$

Correspondingly the Helmholtz free energy density acquires the form

$$
f_R = \left(a_1 P_3^2 + a_{11} P_3^4 + a_{111} P_3^6 - P_3 \left(E_0 + \frac{E_3^d}{2} \right) \right.
$$
$$
\left. - \frac{1}{2} Q_{11} \sigma_{33} P_3^2 - \frac{1}{2} Q_{12} (\sigma_{11} + \sigma_{22}) P_3^2 + \frac{u_R}{2} (\sigma_{11} + \sigma_{22} + \sigma_{33}) \right). \tag{7.11}
$$

Substituting solution (7.10) into Equation 7.11, we obtain the energy density with renormalized coefficients:

$$
f_R = \left(\left(a_1 - \frac{Q_{11} + 2Q_{12}}{s_{11} + 2s_{12}} u_R \right) P_3^2 - P_3 \left(E_0 + \frac{E_3^d}{2} \right) \right.
$$
$$
\left. + \left(a_{11} + \frac{Q_{11}^2(s_{11} + s_{12}) + 2Q_{12}^2 s_{11} - 4Q_{11}Q_{12}s_{12}}{2(s_{11} + 2s_{12})(s_{11} - s_{12})} \right) P_3^4 + a_{111} P_3^6 \right). \tag{7.12}
$$

Note that the obtained renormalization of the quadratic term is the same in Equations 7.9 and 7.12, while the renormalization of the quartic ones is different. Also, we neglected the terms proportional to σ_{ij}^4 and $\sigma_{ij}^2 P_3^2$ and stress relaxation and considered the electrostriction effect only.

7.2.3 Depolarization Field Calculations for a Single-Domain Particle

Depolarization electric field $E_d(P_3)$, included into the free energies (7.4), (7.9), and (7.12), appears due to the polarization inhomogeneity in a confined system. The linear operator for $E_d(P_3)$ essentially depends on the system's shape and boundary conditions; at that $E_d(0) \equiv 0$. For most of the cases, it has only integral representations, which reduces to a constant (depolarization

factors) only for the special case of ellipsoidal bodies with homogeneous polarization distribution. In this case of polarization dependence on the x and y coordinates, the simple expression for the electric depolarization field obtained by Kretschmer and Binder (1979) is not valid. For the case when the depolarization field is completely screened by the ambient free charges outside the dielectric particle (short circuit electrical boundary conditions), it is nonzero inside the dielectric particle due to the inhomogeneous polarization distribution (i.e., nonzero div **P** ≠ 0).

Note that open circuit electrical boundary conditions for ferroelectric nanoparticles were considered by Naumov et al. (2004), Ponomareva et al. (2005), and Prosandeev et al. (2008). Partial screening should lead to the appearance of the depolarization field outside the particles. However, it is well known that the decrease of screening ability (e.g., finite conductance of screening layer) leads to an increase of the depolarization field inside the ferroelectrics and causes the domain structure formation, which in turn should decrease the energy of the depolarization field.

Let us suppose that a ferroelectric nanoparticle (or grain) crystalline *core* is single-domain and covered with a thin amorphous (usually dielectric) *shell*. In turn, free screening carriers with surface charge density $\rho_s(\theta)$ adsorbed from the ambient (e.g., air with definite humidity, pores filled with a precursor solution, or inter-grain space) cover the shell. For instance, a thin water layer condensates on the polar oxide surface in the air with humidity 20%–50% (Freund et al. 1999). The surface charges at least partially screen the surrounding medium (usually Si or alumina porous matrix (Mishina et al. 2002 and Yadlovker and Berger 2005) or regular 2D photonic crystal (Luo et al. 2003 and Morrison et al. 2003a,b)) from the nanoparticle electric field.

Let us substitute the real shape of a given nanoparticle (or grain) by an equivalent sphere of radius R. Firstly we calculate the depolarization field for the simplest case of dielectrically isotropic core, shell, and ambient materials (see Figure 7.1 for details).

One can consider zero external field, since equations of electrostatics are linear and the corresponding solution for the

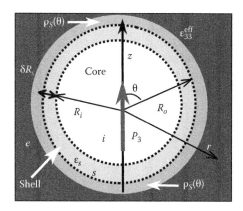

FIGURE 7.1 Ferroelectric nanoparticle core (*i*) covered by dielectric (or magnetic, or surfactant) shell (*s*) and the ambient screening charges $\rho_s(\theta)$ captured by the numerous traps.

Strain-Induced Disorder in Ferroic Nanocomposites   7-7

sphere with shell in the homogeneous external field could be added to the solution found below (see e.g., textbook by Jackson (1962)).

The equations of state relating displacement **D**, electric field **E**, and polarization **P** are

$$\mathbf{D}_i \approx \mathbf{P} + \varepsilon_0 \varepsilon_i \mathbf{E}_i, \quad \mathbf{D}_s = \varepsilon_0 \varepsilon_s \mathbf{E}_s, \quad \mathbf{D}_e = \varepsilon_0 \varepsilon_e \mathbf{E}_e. \quad (7.13)$$

Here, we use the so-called linearized model of ferroelectric nanoparticle core polarization and introduce its isotropic dielectric permittivity $\varepsilon_{11} = \varepsilon_{33} = \varepsilon_i$, where ε_i is called the background (Tagantsev and Gerra 2006) or reference state permittivity (Woo and Zheng 2008). Typically $\varepsilon_i \leq 10$; however, some authors misleadingly put it equal to ferroelectric permittivity $\varepsilon_f \gg 10$ and thus strongly underestimate depolarization field effects. Dielectric shell "*s*" has permittivity ε_s; the ambient medium "*e*" has permittivity ε_e.

Hereinafter, we introduce the potential of electric field $\mathbf{E} = -\nabla\varphi(\mathbf{r})$. In spherical coordinates $\mathbf{r} = \{r, \theta, \varphi\}$ the potential inside each region *i*, *s*, and *e* acquires the form

$$\varphi(r,\theta) = \begin{cases} \varphi_i(r,\theta), & 0 \leq r < R_i, \\ \varphi_s(r,\theta), & R_i \leq r < R_o, \\ \varphi_e(r,\theta), & r \geq R_o. \end{cases} \quad (7.14)$$

The nanoparticle core radius is R_i and shell radius is R_o. The Maxwell equation div **D** = 0 should be supplied with boundary conditions:

$$(\varphi_i - \varphi_s)\big|_{r=R_i} = 0,$$

$$(\mathbf{D}_s - \mathbf{D}_i)\mathbf{e}_r = \left(-\varepsilon_0\varepsilon_s \frac{\partial\varphi_s}{\partial r} + \varepsilon_0\varepsilon_i \frac{\partial\varphi_i}{\partial r} - P_3\cos\theta\right)\Bigg|_{r=R_i} = 0,$$

$$(\varphi_s - \varphi_e)\big|_{r=R_o} = 0, \quad (7.15)$$

$$(\mathbf{D}_e - \mathbf{D}_s)\mathbf{e}_r = \varepsilon_0\left(-\varepsilon_e \frac{\partial\varphi_e}{\partial r} + \varepsilon_s \frac{\partial\varphi_s}{\partial r}\right)\Bigg|_{r=R_o} = \rho\cos\theta$$

Here, the screening charge surface density is $\rho_S(\theta) = \rho \cos \theta$, and $\mathbf{e}_r$ is the outer normal to the spherical surfaces. Two additional conditions should be included, namely, potential at $r = 0$ should be finite, while far from the particle ($r \rightarrow \infty$), the potential vanishes:

$$\varphi_i\big|_{r=0} < \infty, \quad \varphi_e\big|_{r\to\infty} = 0. \quad (7.16)$$

Since we suppose that the polarization inside the sphere is homogeneous, the electrostatic potential inside the particle and screening layer satisfies the Laplace equation $\Delta\varphi = 0$, while the media outside may be semiconducting and its potential should satisfy the equation $\Delta\varphi_e - \varphi_e/R_d^2 = 0$.

The general solution of $\Delta\varphi = 0$, depending only on radius r and polar angle θ, is

$$\varphi(r,\theta) = \sum_{n=0}^{\infty}\left(a_n r^n + \frac{b_n}{r^{n+1}}\right)p_n(\cos\theta) \quad (7.17)$$

where
a_n and b_n are constants
p_n are Legendre polynomials

Since screening charges $\sigma_S(\theta)$ is proportional to $\cos\theta \equiv p_1(\cos\theta)$, one should leave in Equation 7.17, the terms with $n = 1$ only. So we have two constants for each three regions (*i*, *s*, and *e*) to satisfy the boundary conditions (7.15) and (7.16). Taking into account Equation 7.17, one could write

$$\varphi_i(r,\theta) = ar\cos\theta, \quad 0 \leq r < R_i. \quad (7.18a)$$

Potential (7.18a) corresponds to the homogeneous field equal to $-a$:

$$\varphi_s(r,\theta) = \left(cr + \frac{b}{r^2}\right)\cos\theta, \quad R_i \leq r < R_o, \quad (7.18b)$$

$$\varphi_e(r,\theta) = \frac{d}{r^2}\cos\theta, \quad r \geq R_o. \quad (7.18c)$$

Potential (7.18c) corresponds to the field of point dipole with moment ~d. If the external media is a semiconductor, one should modify the solution (7.18c) accordingly:

$$\varphi_e(r,\theta) = d\frac{\exp(-r/R_d)(R_d + r)}{r^2}\cos\theta, \quad r \geq R_o. \quad (7.18d)$$

Boundary conditions (7.15) give the system of four linear equations for constants a, b, c, and d.

For the case of external dielectric media ($R_d \rightarrow \infty$), after the elementary transformations, we obtain that the ambient potential φ_e:

$$\varphi_e(r,\theta) = \frac{1}{\varepsilon_0}\left(\frac{R_o^3\left(3R_i^3\varepsilon_s P_3 + \left(R_i^3(\varepsilon_s - \varepsilon_i) + R_o^3(\varepsilon_i + 2\varepsilon_s)\right)\rho\right)}{R_o^3(\varepsilon_i + 2\varepsilon_s)(\varepsilon_s + 2\varepsilon_e) + 2R_i^3(\varepsilon_s - \varepsilon_i)(\varepsilon_e - \varepsilon_s)}\right)$$

$$\times \frac{\cos\theta}{r^2}, \quad r \geq R_o. \quad (7.19)$$

Since Equation 7.19 can be rewritten as $\varphi_e = \mathbf{P}_{\text{eff}}\mathbf{r}/4\pi\varepsilon_0\varepsilon_e r^3$. So, it has the form of the dipole potential with an effective dipole moment of the particle, and as anticipated, the depolarization electric field outside the particle vanishes as well as the dipole potential, namely,

$$\mathbf{E}_e(\mathbf{r}) = \frac{1}{4\pi\varepsilon_0\varepsilon_e}\left(\frac{3(\mathbf{P}_{\text{eff}}\mathbf{r})\mathbf{r}}{r^5} - \frac{\mathbf{P}_{\text{eff}}}{r^3}\right), \quad r \geq R_o,$$

$$P_{\text{eff}} = \frac{4\pi R_o^3}{3}\frac{3\varepsilon_e\left(3R_i^3\varepsilon_s P_3 + \left(R_i^3(\varepsilon_s - \varepsilon_i) + R_o^3(\varepsilon_i + 2\varepsilon_s)\right)\rho\right)}{R_o^3(\varepsilon_i + 2\varepsilon_s)(\varepsilon_s + 2\varepsilon_e) + 2R_i^3(\varepsilon_s - \varepsilon_i)(\varepsilon_e - \varepsilon_s)}. \quad (7.20)$$

Here, $\mathbf{r} = \{x, y, z\}$ and the effective dipole moment $\mathbf{P}_{\text{eff}} = \{0, 0, P_{\text{eff}}\}$. It should be noted that solution (7.20) is nonzero even in the cases of compensation of either dipole moments $(R_i^3 P_3 + R_o^3 \rho = 0)$ or compensation of surface densities $(P_3 = -\rho)$. The dimensionless ratio

$$d_e(\rho) = \frac{3\varepsilon_e \left(3R_i^3 \varepsilon_s + \left(R_i^3 (\varepsilon_s - \varepsilon_i) + R_o^3 (\varepsilon_i + 2\varepsilon_e) \right) \rho / P_3 \right)}{R_o^3 (\varepsilon_i + 2\varepsilon_s)(\varepsilon_s + 2\varepsilon_e) + 2R_i^3 (\varepsilon_s - \varepsilon_i)(\varepsilon_e - \varepsilon_s)}. \quad (7.21)$$

reflects the degree of polarization screening outside the particle. For the case $P_3 = -\rho$, we obtain that

$$d_e = \frac{3\varepsilon_e (2\varepsilon_s + \varepsilon_i)(R_i^3 - R_o^3)}{R_o^3 (\varepsilon_i + 2\varepsilon_s)(\varepsilon_s + 2\varepsilon_e) + 2R_i^3 (\varepsilon_s - \varepsilon_i)(\varepsilon_e - \varepsilon_s)}$$

The degree of dependence d_e of nanoparticle polarization screening via the shell permittivity ε_s is shown in Figure 7.2 for different background permittivities ε_i.

Note that the depolarization field $\mathbf{E}_i = \{0, 0, E_3^d\}$ inside the nanoparticle is homogeneous:

$$E_3^d = -\frac{\left(R_o^3 (\varepsilon_s + 2\varepsilon_e) - 2R_i^3 (\varepsilon_e - \varepsilon_s) \right) P_3 + 3R_o^3 \varepsilon_s \rho}{\varepsilon_0 \left(R_o^3 (\varepsilon_i + 2\varepsilon_s)(\varepsilon_s + 2\varepsilon_e) + 2R_i^3 (\varepsilon_s - \varepsilon_i)(\varepsilon_e - \varepsilon_s) \right)}, \quad r < R_i.$$

$$(7.22)$$

E_3^d is the true depolarization field that decreases the spontaneous polarization P_3 in comparison with its bulk value P_S.

Solution (7.22) could be simplified in the case of equal dielectric constants of external media and screening shell, that is, $\varepsilon_e = \varepsilon_s$, namely, $E_3^d = -\left((P_3 + \rho)/\varepsilon_0 (\varepsilon_i + 2\varepsilon_e) \right)$. Thus, only for the case $\varepsilon_e = \varepsilon_s$ and $P_3 = -\rho$ the depolarization field (7.22) is absent. Solution (7.22) could be simplified in the case, $R_o - R_i \ll R_i$, namely:

$$E_3^d(R) \approx -\frac{P_3}{\varepsilon_0} \frac{\delta R_s}{R} \frac{2(\varepsilon_e - \varepsilon_s)}{\varepsilon_s (\varepsilon_s + 2\varepsilon_e)}. \quad (7.23)$$

Where the shell thickness $\delta R_s = R_o - R_i$ is introduced and $R_{o,i} \approx R$, under the assumption $\varepsilon_e \gg \varepsilon_s$, one obtains that

$$E_{d3}(R) \approx -\frac{P_3}{\varepsilon_0 \varepsilon_s} \frac{\delta R_s}{R}$$

For dielectric anisotropy media, the problem statement for the depolarization field calculation acquires the form:

$$\varepsilon_{33}^i \frac{\partial^2 \varphi_i}{\partial z^2} + \varepsilon_{11}^i \left(\frac{\partial^2 \varphi_i}{\partial x^2} + \frac{\partial^2 \varphi_i}{\partial y^2} \right) = 0, \quad r < R_i,$$

$$\varepsilon_{33}^s \frac{\partial^2 \varphi_s}{\partial z^2} + \varepsilon_{11}^s \left(\frac{\partial^2 \varphi_s}{\partial x^2} + \frac{\partial^2 \varphi_s}{\partial y^2} \right) = 0, \quad R_i \le r < R_o, \quad (7.24)$$

$$\varepsilon_{33}^e \frac{\partial^2 \varphi_e}{\partial z^2} + \varepsilon_{11}^e \left(\frac{\partial^2 \varphi_e}{\partial x^2} + \frac{\partial^2 \varphi_e}{\partial y^2} \right) = 0, \quad r > R_o.$$

Corresponding boundary conditions

$$(\varphi_i - \varphi_s)\big|_{r=R_i} = 0,$$

$$\left(-\varepsilon_0 \varepsilon_{jk}^s \frac{\partial \varphi_s}{\partial x_k} n_j + \varepsilon_0 \varepsilon_{jk}^i \frac{\partial \varphi_i}{\partial x_k} n_j - P_3 \cos\theta \right)\Bigg|_{r=R_i} = 0,$$

$$(\varphi_s - \varphi_e)\big|_{r=R_o} = 0, \quad (7.25)$$

$$\varepsilon_0 \left(-\varepsilon_{jk}^s \frac{\partial \varphi_s}{\partial x_k} n_j + \varepsilon_{jk}^e \frac{\partial \varphi_e}{\partial x_k} n_j \right)\Bigg|_{r=R_o} = \rho \cos\theta$$

n_j is the Cartesian components of the spherical normal $\mathbf{e}_r$. Additional conditions $\varphi_i\big|_{r=0} < \infty$, $\varphi_e\big|_{r\to\infty} = 0$.

Equations 7.24 could be reduced to the Laplace equations $\Delta\varphi = 0$ by substitution $z \to z/\gamma_k$ in each region, where anisotropy factors $\gamma_k = \sqrt{\varepsilon_{33}^k / \varepsilon_{11}^k}$. However, the boundary conditions (7.25) for the considered spherical particle acquire a cumbersome form either in the spherical or in the Cartesian coordinate systems. The exact analytical solution for the dielectric anisotropy case $\varepsilon_{11} = \varepsilon_{22} \neq \varepsilon_{33}$ is absent allowing for the boundary conditions. For estimations outside the particle, one could use the expressions

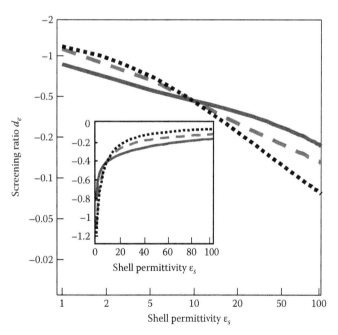

FIGURE 7.2 Degree d_e of nanoparticle spontaneous polarization screening via the shell permittivity ε_s for parameters: $\varepsilon_i = 500$ (dotted curve), $\varepsilon_i = 50$ (dashed curve), and $\varepsilon_i = 5$ (solid curve), $R_i = 10$ nm, $R_i = 12$ nm, ambient permittivity $\varepsilon_e = 10$ and $P_3 = -\rho$. Main plot—log–log scale, inset—linear scale.

$$\mathbf{E}_e(x,y,z) = \frac{1}{4\pi\varepsilon_0\kappa_e}\left(\frac{3(\mathbf{P}_{\text{eff}}\mathbf{R})\mathbf{R}}{R^5} - \frac{\mathbf{P}_{\text{eff}}}{R^3}\right), \quad r \ge R_o,$$

$$P_{\text{eff}} \approx \frac{4\pi R_o^3 P_3}{3}d_e, \tag{7.26}$$

$$d_e = \frac{3\kappa_e\left(3R_i^3\kappa_s + \left(R_i^3(\kappa_s - \varepsilon_i) + R_o^3(\kappa_i + 2\kappa_s)\right)\rho/P_3\right)}{R_o^3(\kappa_i + 2\kappa_s)(\kappa_s + 2\kappa_e) + 2R_i^3(\kappa_s - \kappa_i)(\kappa_e - \kappa_s)}.$$

Here $\mathbf{R} = \{x, y, z/\gamma_e\}$, $\gamma_e = \sqrt{\varepsilon_{33}^e/\varepsilon_{11}^e}$, $R = \sqrt{x^2 + y^2 + (z/\gamma_e)^2}$, and $\mathbf{P}_{\text{eff}} = \{0, 0, P_{\text{eff}}\}$. The dipole moment P_{eff} depends on the effective dielectric constants $\kappa_i = \sqrt{\varepsilon_{11}^i\varepsilon_{33}^i}$, $\kappa_s = \sqrt{\varepsilon_{11}^s\varepsilon_{33}^s}$, and $\kappa_e = \sqrt{\varepsilon_{11}^e\varepsilon_{33}^e}$. Corresponding dimensionless ratio d_e reflects the degree of polarization screening outside the particle.

7.2.4 Strain- and Size-Induced Effects on Phase Transition Temperature

The decrease of paraelectric-ferroelectric phase transition temperature in nanocomposites and nano-grained ceramics should be caused by mechanical stresses and strains (Pertsev et al. 1998, Huang et al. 2001), as well as by the presence of charge defects related with doping (Morozovska and Eliseev 2004) or the depolarization field (Glinchuk and Morozovska 2003).

In the effective media approximation, the simplified expression (7.23) for the depolarization field inside the single-domain particle (or unipolar grain) with polarization P_3 could be rewritten as

$$E_3^d(R) \approx -\frac{P_3}{\varepsilon_0}\frac{2\left(\varepsilon_{33}^{\text{eff}} - \varepsilon_s\right)}{\varepsilon_s 2\left(\varepsilon_{33}^{\text{eff}} + \varepsilon_s\right)}u_d. \tag{7.27}$$

where

$\mathbf{P}$ = vector with components $(0, 0, P_3)$ and P_3 is its component
$u_d = \delta R_s/R \ll 1$ is the small ratio of core and shell sizes (the shell thickness δR_s is much smaller than the grain effective radius R)
ε_s is the shell dielectric permittivity
$\varepsilon_{33}^{\text{eff}}$ is the effective dielectric permittivity of the ambient

Note that the particle splitting into domains could only decrease the depolarization field.

In Section 7.2.2, we consider the free energy coefficients' renormalization caused by the effective strain u_R. Assuming that the contribution of the depolarization field $E_3^d(R)$ and effective elastic strain $u_R(R)$ into the free energy are independent, in what follows we discuss the effects related to the quadratic term renormalization, namely:

$$g_R = \alpha_R(T,R,u)\frac{P_3^2}{2} + \beta_R\frac{P_3^4}{4} + \gamma\frac{P_3^6}{6} - P_3E_0, \tag{7.28a}$$

$$\alpha_R(T,R,u) = \alpha_T\left(T - T_C + \frac{2}{\alpha_T}\frac{Q_{11} + 2Q_{12}}{s_{11} + 2s_{12}}u_R(R)\right.$$
$$\left. + \frac{2}{\alpha_T}\frac{(\varepsilon_{33}^{\text{eff}} - \varepsilon_s)}{\varepsilon_0\varepsilon_s(2\varepsilon_{33}^{\text{eff}} + \varepsilon_s)}u_d(R)\right). \tag{7.28b}$$

Here, parameters T_C, Q_{ij}, and s_{ij} are respectively Curie temperature, electrostriction coefficients, and elastic compliances of the bulk material; α_T is proportional to the inverse Curie constant, core-to-shell sizes ratio $u_d(R)$ is inversely proportional to the particle radius similar to elastic strains $u_R(R)$, despite its purely electric nature. Material coefficient $\gamma = 6a_{111}$ is positive, whereas β_R renormalization is given by Equations 7.9 and 7.12, and its sign determines the ferroelectric–paraelectric phase transition order: $\beta_R < 0$ for the first order phase transitions or $\beta_R > 0$ for the second order ones. Rigorously speaking, the expansion (7.28) is valid inside the single-domain (unipolar) particle.

Note that many authors use the strain dependence $u \sim R^{-1}$ and associate it with effective surface tension (Marchenko 1981, Niepce 1994, Morozovska et al. 2006), whereas we obtained that both surface tension and depolarization effect lead to dependences $u_{R,d} \sim R^{-1}$ within the model of core-and-shell grain and effective media approximation.

Finally, using conventional relation $a_1(T) = \alpha_T(T - T_C)$, where T_C is the Curie temperature, let us estimate the resulting transition temperature change

$$\Delta T = \frac{2}{\alpha_T}\frac{Q_{11} + 2Q_{12}}{s_{11} + 2s_{12}} + \frac{2}{\alpha_T}\frac{\left(\varepsilon_{33}^{\text{eff}} - \varepsilon_s\right)}{\varepsilon_0\varepsilon_s\left(2\varepsilon_{33}^{\text{eff}} + \varepsilon_s\right)} \tag{7.29}$$

for $u_{R,d} = u$ and typical ferroelectric $PbZr_{50}Ti_{50}O_3$ (PZT50/50) material parameters $\alpha_T = 2.66 \cdot 10^5$ m F^{-1} K^{-1}, $a_{11} = 1.898 \cdot 10^8$ m^5 F^{-1} C^{-2}, $a_{111} = 8.016 \cdot 10^8$ m^9 F^{-1} C^{-4}; $Q_{11} = 0.097$ m^4 C^{-2}, $Q_{12} = -0.046$ m^4 C^{-2}, $s_{11} = 10.5 \cdot 10^{-12}$ m^2 N^{-1}, $s_{12} = -3.7 \cdot 10^{-12}$ m^2 N^{-1} and $\varepsilon_s \cong 50$, $\varepsilon_{33}^{\text{eff}} \ge 100$, $\varepsilon_0 = 8.85 \cdot 10^{-12}$ F$\cdot$m^{-1} (Haun et al. 1989). We obtain that $\Delta T = (1 - 2) \cdot 10^4$K and so $u\Delta T \sim (10...10^2)$K in the typical range of strains $u_{R,d} \sim 10^{-3}...10^{-2}$. Really, for reference values of surface stress $\mu = 50$N m^{-1} and radii $R \sim 5$–50nm, the estimation of the effective deformation as $\delta R_s \cong 2\mu(s_{11} + 2s_{12})$ leads to the value $\delta R_s/R \cong 10^{-2} - 10^{-3}$.

7.2.5 Collective Effect of Strain and Depolarization Field on Nanocomposite Ferroelectric and Dielectric Response

Usually the nanoparticles or ceramic grains possess different renormalization of the Curie temperatures (due to the different values of strain $u_R(R)$ and depolarization ratio $u_d(R)$ via size R, pressure p_R, surface stress μ_R, and depolarization field $E_3^d(R)$ distributions) and the polarization orientation in a given sample. Let us assume that all polarization orientations are equivalent without pre-poling. If the uniform electric field $\mathbf{E} = (0, 0, E_0)$ is applied across a rather thick nanocomposite sample, among all possible polarization orientations only P_3 would contribute into the measured effective values of polarization $P_3^{\text{eff}} = \langle\langle P_3 \rangle\rangle$ and dielectric susceptibility $\chi_{33}^{\text{eff}} = \langle\langle dP_3/dE_0 \rangle\rangle$, where the brackets mean averaging over the sample. In the case of thermodynamic equilibrium the spatial averaging should be substituted by the statistical averaging on the effective strain $u_{\min} \leq u \leq u_{\max}$ that includes both elastic strain u_R's and depolarization effect u_d's contributions with the appropriate distribution function $f_R(u)$. So, the dynamic response of the strained nanocomposite should be calculated from the relaxation equations:

$$
\begin{cases}
P_3^{\text{eff}}(E_0) = \displaystyle\int_{u_R^{\min}}^{u_R^{\max}} du\, f_R(u) P_3(E_0, u), \\[2mm]
\Gamma \dfrac{\partial}{\partial t} P_3 + \alpha_R(T, u_R) P_3 + \beta_R P_3^3 + \gamma P_3^5 = E_0(t), \\[2mm]
\alpha_R(T, R, u) = \alpha_T(T - T_C + u \cdot \Delta T), \\[2mm]
\chi_{33}^{\text{eff}}(E_0) = \displaystyle\int_{u_R^{\min}}^{u_R^{\max}} \dfrac{du\, f_R(u)}{\alpha_R(T, u) + 3\beta_R P_3^2 + 5\gamma P_3^4}, \\[2mm]
\left(\dfrac{1}{\chi_{33}}\right)^{\text{eff}} = \displaystyle\int_{u_R^{\min}}^{u_R^{\max}} du\, f_R(u)\left(\alpha_R(T, u) + 3\beta_R P_3^2 + 5\gamma P_3^4\right).
\end{cases}
\tag{7.30}
$$

where Γ is the Landau–Khalatnikov kinetic coefficient, ΔT is given by Equation 7.29 for the particular case $u \cong u_{R,d}(R)$ that is considered hereinafter.

In the general case, Equation 7.30 should be analyzed numerically. Quasi-static ferroelectric hysteresis loops $P_3^{\text{eff}}(E_0)$ shown in Figure 7.3 were simulated for bell-shaped distribution functions $f_R(u)$ with different width and average values.

The transition of the square hysteresis loop $P_3(E_0, u_{\min})$ into the sloped and/or slim one $P_3^{\text{eff}}(E_0)$ is the most evident sequence of the strain distribution. The effect is the most pronounced for the nanocomposites (or ceramics) with compressed or strained particles (or grains) and wide distribution functions (compare Figure 7.3a through f) and resemble the well-known "square-to-slim" transition of the hysteresis loops in disordered ferroelectrics. The physical explanation of the effect is rather simple. When some compressed particles are in the paraelectric phase (dotted loops $P_3(E_0, u_{\max})$ in plots b, c, e, and f), while the other ones are in the ferroelectric phase (dashed curves $P_3(E_0, u_{\min})$ in plots b, c, e, and f), the effective response $P_3^{\text{eff}}(E_0)$ corresponds to the mixed phase with a more thinner and tilted loop than that of the ferroelectric phase. The loop $P_3^{\text{eff}}(E_0)$ becomes much more slim and tilted under the increase of the "paraelectric" particles

ratio (see Figure 7.3c). However, the effective response $P_3^{\text{eff}}(E_0)$ is tilted even in the tensile nanocomposite allowing for the mixing of more "hard" particles with high coercive fields and "soft" particles with low coercive fields (see Figure 7.3g through i). The proposed interpretation correlates with the conventional Cross model proposed for the behavior of the polar regions in the disordered ferroelectrics.

In the simplest case of uniform strain distribution function $f_R(u) = 1/(u_{\max} - u_{\min})$, one can derive analytical expressions for the remnant polarization $P_3^{\text{eff}}(E_0 = 0, T)$ and inverse susceptibility $\left(\chi_{33}^{-1}(0, T)\right)^{\text{eff}}$ temperature dependencies. In the case of the second order phase transitions:

$$
P_3^{\text{eff}}(0, T) = \begin{cases}
2\sqrt{\dfrac{\alpha_T}{\beta_R}} \dfrac{(T_C - u_{\min}\Delta T - T)^{3/2} - (T_C - u_{\max}\Delta T - T)^{3/2}}{3\Delta T(u_{\max} - u_{\min})} \\
\qquad \text{at } T \leq (T_C - u_{\max}\Delta T) \\[2mm]
2\sqrt{\dfrac{\alpha_T}{\beta_R}} \dfrac{(T_C - u_{\min}\Delta T - T)^{3/2}}{3\Delta T(u_{\max} - u_{\min})} \\
\qquad \text{at } (T_C - u_{\max}\Delta T) \leq T \leq (T_C - u_{\min}\Delta T) \\[2mm]
0 \quad \text{at } T \geq T_C - u_{\min}\Delta T
\end{cases}
\tag{7.31}
$$

$$
\left(\chi_{33}^{-1}(0, T)\right)^{\text{eff}} = \begin{cases}
2\alpha_T\left(T_C - \Delta T \dfrac{u_{\max} + u_{\min}}{2} - T\right) \\
\qquad \text{at } T \leq (T_C - u_{\max}\Delta T) \\[2mm]
\alpha_T\left(T - T_C + \Delta T \dfrac{u_{\max} + u_{\min}}{2}\right) \\
\quad + \dfrac{3\alpha_T(T - T_C + u_R^{\min}\Delta T)^2}{2\Delta T(u_{\max} - u_{\min})} \\
\qquad \text{at } (T_C - u_{\max}\Delta T) \leq T \leq (T_C - u_{\min}\Delta T) \\[2mm]
\alpha_T\left(T - T_C + \Delta T \dfrac{u_{\max} + u_{\min}}{2}\right) \\
\qquad \text{at } T \geq (T_C - u_{\min}\Delta T)
\end{cases}
\tag{7.32}
$$

The remnant polarization $P_3^{\text{eff}}(T)$ and linear susceptibility $\chi_{33}^{\text{eff}}(T)$ temperature dependencies are shown in Figure 7.4 for the bell-shaped (solid curves) and uniform (dashed curves) distribution functions of strain. Dotted curves correspond to the bulk material.

The main feature of Figure 7.4 is the strain-induced smearing of the order parameter $P_3^{\text{eff}}(T)$ temperature dependence in the vicinity of the ferroelectric–paraelectric phase transition in comparison with the bulk material (compare solid, dashed,

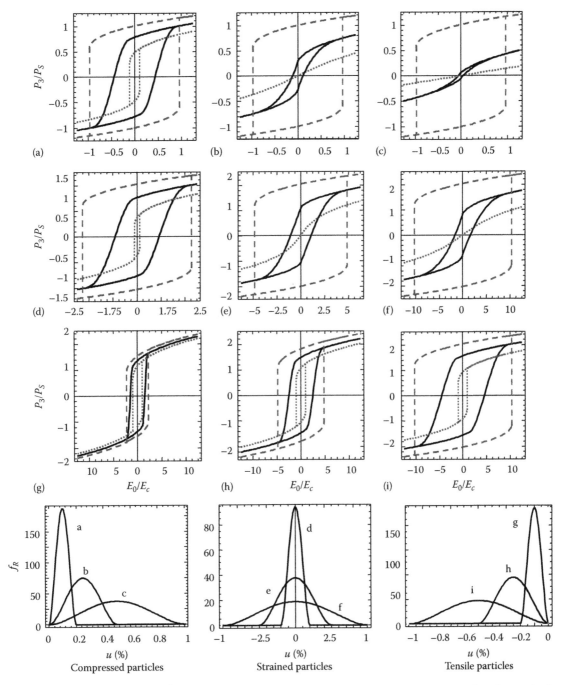

FIGURE 7.3 Quasi-static hysteresis loops $P_3^{eff}(E_0)$ (solid curves), $P_3(E_0, u_{min})$ (dashed curves), and $P_3(E_0, u_{max})$ (dotted curves) of compressed (a, b, and c), strained (d, e, and f) and tensile (g, h, and i) particles calculated at $\Delta T = 10^4$K and PZT (50/50) material parameters. Corresponding distribution functions are shown in the bottom plots by letters a, b, c, d, e, f, g, h, and i. P_S is the bulk material spontaneous polarization, E_c is thermodynamic coercive field of bulk material.

and dotted curves in plot (b)). It is worth to underline that the difference between the bell-shaped (solid curves) and uniform (dashed curves) distribution functions causes noticeable deviation of the $P_3^{eff}(T)$ temperature behavior only in vicinity of the phase transition (compare dashed and solid curves in upper inset). In particular, the phase transition smears in stronger for the uniform distribution function, since it is weakly localized in comparison with the bell-shaped function.

It is clear from Figure 7.4a and c that the maxima of the dielectric susceptibility do not coincide with the disappearance of the corresponding order parameters (see vertical line with arrows). The dielectric susceptibility maxima smear and dispersion increases with the distribution function's half-width increase. The calculated polar and dielectric properties dispersion in the nanocomposites is similar to the one typically observed in the disordered and relaxor ferroelectrics (Cross 1987, Viehland et al. 1992).

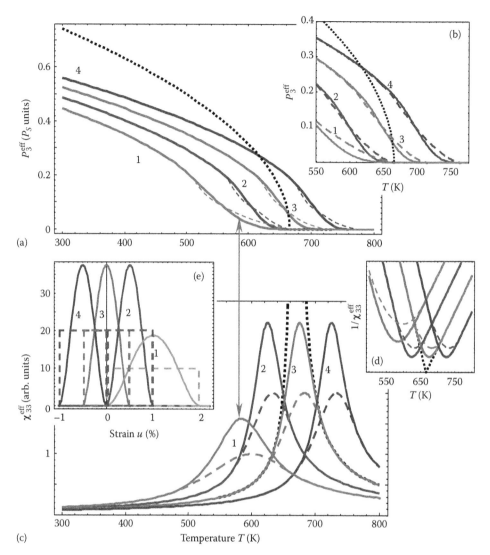

FIGURE 7.4 Remnant polarization (a and b) and linear direct (c) and inverse (d) susceptibility temperature dependencies for bell-shaped (solid curves) and uniform (dashed curves) distribution functions of strain. Corresponding distribution functions (1, 2, 3, and 4) are shown in plot (e). Material parameters correspond to PZT (50/50), $\Delta T = 10^4$K. Dotted curves correspond to the bulk material.

7.3 Discussion

Disorder in nanocomposites with magnetic properties could be considered in the same way as it has been demonstrated for ferroelectric nanoparticles. Elastic strains and demagnetization effects should smear magnetic, dielectric, and magnetoelectric susceptibility temperature maxima and increase their values in the paramagnetic phase. Strains via electrostriction and magnetostriction mechanisms may induce strong built-in fields, which essentially increase generalized susceptibilities and tunability, even in the absence of external fields.

To *summarize*, we have demonstrated that the dense ferroelectric nanocomposites should exhibit disorder properties originating from internal elastic strains and depolarization effects. As anticipated, performed calculations showed that the wider is the distribution function of effective strains (e.g., originated from the distribution of particles sizes and their crystallographic

orientations) the higher is the ferroelectric disorder. In particular, ferroelectric ceramics and porous materials with compressed and strained ferroelectric particles should manifest polar and dielectric properties typical for disordered and relaxor ferroelectrics. Detailed derivation of the evident analytical expressions for effective remnant polarization, dielectric susceptibility, ferroelectric transition temperature dependence on the strain distribution, nanoparticle core and shell radius, dielectric permittivity, intrinsic surface stress, and electrostriction coefficient was presented for tutorial purposes.

Future perspectives of ferroic nanocomposites' novel applications have stimulated the intensive investigations of their magnetic, ferroelectric, and dielectric properties anomalies during the last decade. Theoretical study of the cylindrical and spherical ferroic nanoparticles became a hot topic, because of their properties' new behavior, which is absent in the bulk materials. For instance, room temperature ferromagnetism has been observed in spherical nanoparticles of nonmagnetic oxides. Keeping in mind the similarity of the ferroic properties, one could expect the appearance of ferroelectricity in highly polarizable paraelectric nanoparticles induced by spatial confinement. The theory predicts optimal sizes for ferroelectricity appearance in incipient ferroelectric nanoparticles. The experimental justification of the theoretical forecast is extremely desirable, since the technology roadmap indicates that 2010 should be the year of 3D capacitor nanostructures.

Acknowledgments

Research sponsored by National Academy of Sciences of Ukraine, Ministry of Science and Education of Ukraine and National Science Foundation (Materials World Network, DMR-0908718).

References

Akcay, G., S.P. Alpay, G.A. Rossetti, and J.F. Scott. 2008. Influence of mechanical boundary conditions on the electrocaloric properties of ferroelectric thin films. *J. Appl. Phys.* **103**: 024104-1–024104-7.

Akdogan, E.K. and A. Safari. 2007. Thermodynamic theory of intrinsic finite-size effects in PbTiO$_3$ nanocrystals. I. Nanoparticle size-dependent tetragonal phase stability. *J. Appl. Phys.* **101**: 064114-1–064114-8.

Alpay, S.P., I.B. Misirlioglu, A. Sharma, and Z.-G. Ban. 2004. Structural characteristics of ferroelectric phase transformations in single-domain epitaxial films. *J. Appl. Phys.* **95**: 8118–8123.

Ban, Z.-G., S.P. Alpay, and J.V. Mantese. 2003. Fundamentals of graded ferroic materials and devices. *Phys. Rev. B.* **67**: 184104-1–184104-6.

Bratkovsky, A.M. and A.P. Levanyuk. 2005. Smearing of phase transition due to a surface effect or a bulk inhomogeneity in ferroelectric nanostructures. *Phys. Rev. Lett.* **94**: 107601-1–107601-4.

Cao, W. 2005. The strain limits on switching. *Nat. Mater.* **4**: 727–728.

Chen, X., D.-N. Fang, and K.-C. Hwang. 1997. A mesoscopic model of the constitutive behaviour of monocrystalline ferroelectrics. *Smart Mater. Struct.* **6**: 145–151.

Cross, L.E. 1987. Relaxor ferroelectrics. *Ferroelectrics* **76**: 241–245.

Erdem, E., H.-C. Semmelhack, R. Bottcher et al. 2006. Study of the tetragonal-to-cubic phase transition in PbTiO$_3$ nanopowders. *J. Phys.: Condens. Matter* **18**: 3861–3874.

Fernández-Osorio, A.L., A. Vázquez-Olmos, E. Mata-Zamora, and J.M. Saniger. 2007. Preparation of free-standing Pb (Zr$_{0.52}$Ti$_{0.48}$) O$_3$ nanoparticles by sol–gel method. *J. Sol-Gel Sci. Technol.* **42**: 145–149.

Fiebig, M. 2005. Revival of the magnetoelectric effect. *J. Phys. D: Appl. Phys.* **38**: R123–R152.

Fong, D.D., G.B. Stephenson, S.K. Streiffer et al. 2004. Ferroelectricity in ultrathin perovskite films. *Science* **304**: 1650–1653.

Freund, J., J. Halbritter, and J.K.H. Horber. 1999. How dry are dried samples? Water adsorption measured by STM. *Microsc. Res. Tech.* **44**: 327–338.

Geneste, G., E. Bousquest, J. Junquera, and P. Chosez. 2006. Finite-size effects in BaTiO$_3$ nanowires. *Appl. Phys. Lett.* **88**: 112906-1–112906-3.

Glazunov, A.E., A.J. Bell, and A.K. Tagantsev. 1995. Relaxors as superparaelectrics with distributions of the local transition temperature. *J. Phys.: Condens. Matter* **7**: 4145–4168.

Glinchuk, M.D. and A.N. Morozovska. 2003. Effect of surface tension and depolarization field on ferroelectric nanomaterials properties. *Phys. Status Solidi (b)* **238**: 81–91.

Glinchuk, M.D. and A.N. Morozovskaya. 2003. Radiospectroscopic and dielectric spectra of nanomaterials. *Phys. Solid State* **45**: 1586–1595.

Glinchuk, M.D. and A.N. Morozovska. 2004. The internal electric field originating from the mismatch effect and its influence on ferroelectric thin film properties. *J. Phys.: Condens. Matter* **16**: 3517–3531.

Glinchuk, M.D., E.A. Eliseev, A.N. Morozovska, and R. Blinc. 2008a. Giant magnetoelectric effect induced by intrinsic surface stress in ferroic nanorods. *Phys. Rev. B* **77**: 024106-1–024106-11.

Glinchuk, M.D., E.A. Eliseev, and A.N. Morozovska. 2008b. Superparaelectric phase in the ensemble of noninteracting ferroelectric nanoparticles. *Phys. Rev. B* **78**: 134107-1–134107-9.

Haun, M.J., E. Furman, S.J. Jang, and L.E. Cross. 1989. Thermodynamic theory of the lead zirconate-titanate solid solution system, part V: Theoretical calculations. *Ferroelectrics* **99**: 63–86.

Huang, H., C.Q. Sun, Zh. Tianshu, and P. Hing. 2001. Grain-size effect on ferroelectric Pb(Zr$_{1-x}$Ti$_x$)O$_3$ solid solutions induced by surface bond contraction. *Phys. Rev. B* **63**: 184112-1–184112-9.

Ishikava, K., K. Yoshikava, and N. Okada. 1988. Size effect on the ferroelectric phase transition in PbTiO$_3$ ultrafine particles. *Phys. Rev. B* **37**: 5852–5857.

Jackson, J.D. 1962. *Classical Electrodynamics*. New York: John Wiley & Sons.

Kaczmarek, M., O. Buchnev, and I. Nandhakumar. 2008. Ferroelectric nanoparticles in low refractive index liquid crystals for strong electro-optic response. *Appl. Phys. Lett.* **92**: 103307-1–103307-3.

Kalinin, S.V., E.A. Eliseev, and A.N. Morozovska. 2006. Materials contrast in piezoresponse force microscopy. *Appl. Phys. Lett.* **88**: 232904-1–232904-3.

Korn, G.A. and T.M. Korn. 1961. *Mathematical Handbook for Scientists and Engineers*. New York: McGraw-Hill.

Kretschmer, R. and K. Binder. 1979. Surface effects on phase transition in ferroelectrics and dipolar magnets. *Phys. Rev. B* **20**: 1065–1076.

Kumzerov, Y. and S. Vakhrushev. 2003. Nanostructures within porous materials. In *Encyclopedia of Nanoscience and Nanotechnology*, H.S. Halwa, (ed.), vol. **10**, Stevenson Ranch, CA: American Scientific Publishers, pp. 1–39.

Landau, L.D. and E.M. Lifshits. 1980. *Electrodynamics of Continuous Media*. Oxford, U.K.: Butterworth Heinemann.

Li, Sh., J.A. Eastman, R.E. Newnham, and L.E. Cross. 1997. Diffuse phase transition in ferroelectrics with mesoscopic heterogeneity: Mean-field theory. *Phys. Rev. B* **55**: 12067–12078.

Li, J.Y., R.C. Rogan, E. Ustundag, and K. Bhattacharya. 2005. Domain switching in polycrystalline ferroelectric ceramics. *Nat. Mater.* **4**: 776–781.

Li, F., O. Buchnev, C.I. Cheon et al. 2006. Orientational coupling amplification in ferroelectric nematic colloids. *Phys. Rev. Lett.* **97**: 147801-1–147801-4.

Lines, M.E. and A.M. Glass. 1977. *Principles and Applications of Ferroelectrics and Related Phenomena.* Oxford, U.K.: Clarendon Press.

Luo, Y., I. Szafraniak, N.D. Zakharov et al. 2003. Nanoshell tubes of ferroelectric lead zirconate titanate and barium titanate. *Appl. Phys. Lett.* **83**: 440–442.

Ma, W., M. Zhang, and Z. Lu. 1998. A study of size effects in $PbTiO_3$ nanocrystals by Raman spectroscopy. *Phys. Status Solidi (a)* **166**: 811–815.

Marchenko, V.I. 1981. Possible structures and phase transitions on the surface of crystals. *Pis'ma Zh. Eksp. Teor. Fiz.* **33**: 397–401 [*JETP Lett.* **33**: 381–385].

Marchenko, V.I. and A.Ya. Parshin. 1980. About elastic properties of the surface of crystals. *Zh. Eksp. Teor. Fiz.* **79**: 257–260 [*Sov. Phys. JETP* **52**: 129–132].

Michael, Th., S. Trimper, and J.M. Wesselinowa. 2007. Size effects on static and dynamic properties of ferroelectric nanoparticles. *Phys. Rev. B* **76**: 094107-1–094107-7.

Mishina, E.D., K.A. Vorotilov, V.A. Vasil'ev, A.S. Sigov, N. Ohta, and S. Nakabayashi. 2002. Porous silicon-based ferroelectric nanostructures. *J. Exp. Theor. Phys.* **95**: 502–504.

Mishra, S.K. and D. Pandey. 1995. Effect of particle size on the ferroelectric behaviour of tetragonal and rhombohedral $Pb(Zr_xTi_{1-x})O_3$ ceramics and powders. *J. Phys.: Condens. Matter* **7**: 9287–9303.

Mornet, S., C. Elissalde, O. Bidault et al. 2007. Ferroelectric-based nanocomposites: Toward multifunctional materials. *Chem. Mater.* **19**: 987–992.

Morozovska, A.N. and E.A. Eliseev. 2004. Modeling of dielectric hysteresis loops in ferroelectric semiconductors with charged defects. *J. Phys.: Condens. Matter* **16**: 8937–8956.

Morozovska, A.N., E.A. Eliseev, and M.D. Glinchuk. 2006. Ferroelectricity enhancement in confined nanorods: Direct variational method. *Phys. Rev. B* **73**: 214106-1–214106-13.

Morozovska, A.N., E.A. Eliseev, and M.D. Glinchuk. 2007a. Size effects and depolarization field influence on the phase diagrams of cylindrical ferroelectric nanoparticles. *Physica B* **387**: 358–366.

Morozovska, A.N., M.D. Glinchuk, and E.A. Eliseev. 2007b. Ferroelectricity enhancement in ferroelectric nanotubes. *Phase Transit.* **80**: 71–77.

Morozovska, A.N., M.D. Glinchuk, and E.A. Eliseev. 2007c. Phase transitions induced by confinement of ferroic nanoparticles. *Phys. Rev. B* **76**: 014102-1–014102-13.

Morrison, F.D., L. Ramsay, and J.F. Scott. 2003a. High aspect ratio piezoelectric strontium-bismuth-tantalate nanotubes. *J. Phys.: Condens. Matter* **15**: L527–L532.

Morrison, F.D., Y. Luo, I. Szafraniak et al. 2003b. Ferroelectric nanotubes. *Rev. Adv. Mater. Sci.* **4**: 114–122.

Naumov, I.I., L. Bellaiche, and H. Fu. 2004. Unusual phase transitions in ferroelectric nanodisks and nanorods. *Nature* **432**: 737–740.

Niepce, J.C. 1994. Permittivity of fine grained $BaTiO_3$. *Electroceramics* **4**: 29–37.

Perriat, P., J.C. Niepce, and G. Gaboche. 1994. Thermodynamic consideration of the grain size dependence of materials properties. *J. Thermal Anal.* **41**: 635–649.

Pertsev, N.A., A.G. Zembilgotov, and A.K. Tagantsev. 1998. Effect of mechanical boundary conditions on phase diagrams of epitaxial ferroelectric thin films. *Phys. Rev. Lett.* **80**: 1988–1991.

Ponomareva, I., I.I. Naumov, I. Kornev, H. Fu, and L. Bellaiche. 2005. Atomistic treatment of depolarizing energy and field in ferroelectric nanostructures. *Phys. Rev. B* **72**: 140102(R)-1–140102(R)-4.

Poyato, R., B.D. Huey, and N.P. Padture. 2006. Local piezoelectric and ferroelectric responses in nanotube-patterned thin films of $BaTiO_3$ synthesized hydrothermally at 200°C. *J. Mater. Res.* **21**: 547–551.

Prosandeev, S., I. Ponomareva, I. Naumov, I. Kornev, and L. Bellaiche. 2008. Original properties of dipole vortices in zero-dimensional ferroelectrics. *J. Phys.: Condens. Matter* **20**: 193201-1–193201-14.

Ram, S., A. Jana, and T.K. Kundu. 2007. Ferroelectric $BaTiO_3$ phase of orthorhombic crystal structure contained in nanoparticles. *J. Appl. Phys.* **102**: 054107-1–054107-8.

Roduner, E. 2006. *Nanoscopic Materials. Size-Dependent Phenomena.* Cambridge, U.K.: Royal Society of Chemistry Publishing.

Rychetsky, I. and O. Hudak. 1997. The ferroelectric phase transition in small spherical particles. *J. Phys.: Condens. Matter* **9**: 4955–4965.

Scott, J.F. 2007. Data storage: Multiferroic memories. *Nat. Mater.* **6**: 256–257.

Shchukin, V.A. and D. Bimberg. 1999. Spontaneous ordering of nanostructures on crystal surfaces. *Rev. Mod. Phys.* **71**: 1125–1171.

Tagantsev, A.K. and G. Gerra. 2006. Interface-induced phenomena in polarization response of ferroelectric thin films. *J. Appl. Phys.* **100**: 051607-1–051607-28.

Tilley, D.R. 1996. Finite-size effects on phase transitions in ferroelectrics. In: *Ferroelectric Thin Films*, C. Paz de Araujo, J.F. Scott, and G.W. Teylor (eds.), Amsterdam, the Netherlands: Gordon and Breach, pp. 11–45.

Timoshenko, S.P. and J.N. Goodier. 1970. *Theory of Elasticity.* New York: McGraw-Hill.

Uchino, K., E. Sadanaga, and T. Hirose. 1989. Dependence of the crystal structure on particle size in barium titanate. *J. Am. Ceram. Soc.* **72**: 1555–1558.

Verma, K.C., R.K. Kotnala, and N.S. Negi. 2008. Improved dielectric and ferromagnetic properties in Fe-doped PbTiO$_3$ nanoparticles at room temperature. *Appl. Phys. Lett.* **92**: 1529021-1–1529021-3.

Viehland, D., J.F. Li, S.J. Jang, L.E. Cross, and M. Wutting. 1992. Glassy polarization behaviour of relaxor ferroelectrics. *Phys. Rev. B.* **46**: 8013–8017.

Wadhawan, V.K. 2000. *Introduction to Ferroic Materials.* New York: Gordon and Breach Science Publishers.

Wang, C.L. and S.R.P. Smith. 1995. Landau theory of the size-driven phase transition in ferroelectrics. *J. Phys.: Condens. Matter* **7**: 7163–7171.

Wei, N., D.-M. Zhang, X.-Y. Han et al. 2007. Synthesis and mechanism of ferroelectric potassium tantalate niobate nanoparticles by the solvothermal and hydrothermal processes. *J. Am. Ceram. Soc.* **90**: 1434–1437.

Woo, C.H. and Y. Zheng. 2008. Depolarization in modeling nanoscale ferroelectrics using the Landau free energy functional. *Appl. Phys. A: Mater. Sci. Process.* **91**: 59–63.

Yadlovker, D. and S. Berger. 2005. Uniform orientation and size of ferroelectric domains. *Phys. Rev. B* **71**: 184112.

Zhong, W.L., Y.G. Wang, P.L. Zhang, and D.B. Qu. 1994. Phenomenological study of the size effect on phase transition in ferroelectric particles. *Phys. Rev. B* **50**: 698–703.

Zhou, Z.H., X.S. Gao, J. Wang, K. Fujihara, S. Ramakrishna, and V. Nagarajan. 2007. Giant strain in PbZr$_{0.2}$Ti$_{0.8}$O$_3$ nanowires. *Appl. Phys. Lett.* **90**: 052902-1–052902-3.

8

Smart Composite Systems with Nanopositioning

Kougen Ma
University of Hawaii at Manoa

Mehrdad N. Ghasemi-Nejhad
University of Hawaii at Manoa

8.1 Introduction

The development of biological engineering, nanoscale science and engineering, and space engineering requires nanopositioning capability with precision on the order of tens of nanometers or better. In general, nanopositioning applications require at least 100 times better precision than most machine tools can fabricate and 10 times greater accuracy than most machine shops can measure. For instance, atomic scale manipulation and testing of materials during concomitant high-resolution imaging require 1–5 mm travel range with 1 nm resolution. For an application linking two geosynchronous orbit satellites spaced at 120° around the orbital plane, the total line of sight tracking angular error misalignment budget is about 500 nrad (1/10 arc seconds). Table 8.1 lists the precision requirements for various industries and products from 1 mm to 1 Å. Nanopositioning techniques have great demand in (a) optics, photonics, and measuring technology (image stabilization, interferometry, and mirror positioning); (b) precision mechanics and mechanical engineering (as an additional layer to smart structures to compensate for structural deformation and microengraving systems); (c) semiconductor and microelectronics (nanometrology, wafer and mask positioning, alignment, microlithography, and inspection systems); and (d) life science, medicine, and biology (scanning microscopy, gene manipulation, and patch clamp).

8.1.1 Actuators for Nanopositioning

Traditional actuation approaches for nanopositioning include leadscrews (Physik Instrumente 2008), inchworm (Zhang and Zhu 1997, Suleman et al. 2001, Kim et al. 2002, Yeh et al. 2002,

Mrad et al. 2003, Doherty and Ghasemi-Nejhad 2005, Kim et al. 2005), voice coil actuators (Wang et al. 2008), piezoelectric transducers (Hemsel and Wallaschek 2000, Borodin et al. 2004, Spanner 2006, Mori et al. 2007, Physik Instrumente 2008), magnetic bearing mechanisms (Micromega Dynamics 2008), and flexure reduction mechanisms (Culpepper and Anderson 2004).

A leadscrew is a screw specialized for the purpose of translating rotational to linear motion. The mechanical advantage of a leadscrew is determined by the screw pitch or lead. A leadscrew nut and screw mate with rubbing surfaces, and consequently they have a relatively high friction and stiction compared with mechanical parts that mate with rolling surfaces and bearings. Their efficiency is typically only between 25% and 70%, with higher pitch screws tending to be more efficient. The high internal friction means that leadscrew systems are not usually capable of continuous operation at high speeds, as they will overheat, and also it is typically self-locking. Backlash can be reduced with the use of a second nut to create a static loading force known as preload.

Inchworm actuators are devices that achieve long-range motion by rapidly repeating a clamp-extend-clamp cycle and can act as both coarse and fine positioning elements. The inchworms typically use two actuators for two clamp sections and a third for the extender section. Their main advantage is their high position accuracy, even in open loop systems. Their disadvantages are the narrow tolerances required, the need for temperature compensation, and the fact that the wear of the clamping actuators may lead to early failure (Hemsel and Wallaschek 2000).

For applications that require translation of up to 1 mm with precision down to the subnanometer range, piezoelectric transducers specifically engineered for nanopositioning offer

TABLE 8.1   Precision Requirement for Various Products

	Activity	Typical Required Precision (μm)	Approximate Comparative Dimension
From micropositioning to nanopositioning	Automobile chassis assembly	1000	Thickness of a dime
	Electronic PCB assembly	100	Thickness of paper
	Typical machining tolerances	10	Diameter of small cell
	Read/write heads in disk drives	1	Wavelength of light
	Scanning interferometers	0.100 (100 nm)	Critical dimension of newest computer chips
	Biological cell micromanipulation		
	Adaptive optics		
	Optical fiber alignment		
	Microlithography	0.010 (10 nm)	A few widths of a molecule
	Hubble space telescope		
	Integrated circuits inspection	0.001 (1 nm)	A few diameters of atoms
	Thin-film surface coating	0.0001 (1 Å)	A single atomic layer

Source: Newscale Technologies, http://www.newscaletech.com/nanopositioning.html, 2008.

the optimum mix of performance, cost, size, speed, static, and dynamic accuracy and reliability.

Nanopositioning devices employing piezoelectric materials capitalize on the inverse piezoelectric effect by applying a voltage to the materials to generate motion via material expansion. Most such devices use a polarized ferroelectric ceramic material made from lead, zirconium, and titanium and are constructed of many layers sandwiched between electrodes to reduce the voltage requirement. The benefits of piezoelectric-based nanopositioning devices include unlimited resolution (less than 1 nm), fast expansion and response (microsecond), maintenance-free, inherent vacuum-compatibility, high efficiency, high throughput, and dynamic accuracy. Piezoceramic positioning devices can have bandwidths of tens of kilohertz or more, and they lack the responsiveness-limiting inertia of leadscrews and other conventional mechanisms.

Figure 8.1 describes an inchworm technology (Doherty and Ghasemi-Nejhad 2005). The device consists of a piezoelectric (PZT) actuator, the housing, and two clamping devices. The application and removal of voltage causes the ceramic piezoelectric stack to expand or contract relative to its base. Each end of the piezo-stack is attached to a block. The clamping devices are synchronized such that one locks/releases one end-block while the other releases/locks the other end-block as shown in Figure 8.1. In operation, the inchworm pushes or pulls the load toward the desired load position. In the last cycle of the operation, the final position is achieved through the application of a fraction of the applied voltage to the piezo-stack to precisely position the load and achieve precision positioning. Once positioned, the outer clamp (e.g., Clamp B in Figure 8.1) may be locked, removing the PZT stack from the load path, and the voltage can be removed from the actuator to achieve power-off/hold capability for this inchworm actuator. To achieve vibration suppression (VS), the piezoelectric should be placed in the load path. Hence, in Figure 8.1, Clamp A should be clamped and Clamp B should be unclamped, then vibrate the piezo-stack to actively suppress the vibration. Once, the vibration is suppressed, again Clamp B should be clamped and Clamped A should be unclamped to remove the piezo-stack from the load path to achieve power-off/hold capability. Other issues in the development of an inchworm mechanism include the load-carrying capability and low power consumption. It should be noted that, in practice, a fiberglass cloth should be placed between the piezo-stack and the blocks to provide electrical insulation and prevent electrical short circuit. Doherty and Ghasemi-Nejhad (2005) explained this inchworm mechanism, in detail, for its use in an active composite strut.

Piezoelectric-based nanopositioning development involves many aspects such as stroke range, driving type (open loop or closed loop), bandwidth, sensors, as well as controller design and implementation.

A coarse/fine long stroke assembly consisting of a piezoelectric transducer and a motorized positioner in series is always possible. A highly specialized hybrid controller reads the stage position from an integrated, nanometer-class linear encoder and continuously coordinates both the piezoelectric and servomotor drives in a way that provides the best possible overall performance with rapid pull-in, nanometer-scale bidirectional repeatability, and inherent axial stiffness that are unobtainable

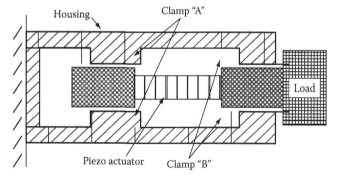

FIGURE 8.1 Schematic of an inchworm technology. Length adjustment to move load to the right: (1) clamp "A," unclamp "B," actuate piezoelectric element; (2) clamp "B," unclamp "A," remove potential from piezoelectric element; (3) repeat steps (1) and (2) as necessary until load moved to desired position.

from conventional architectures. The result is a fast, long-stroke system with extraordinary repeatability as well as resolution.

8.1.2 Sensors for Nanopositioning

Linear variable differential transformers (LVDT), linear encoders, strain gauges, and capacitive sensors are usually employed as sensors for nanopositioning.

A LVDT consists of a hollow cylindrical shaft inside which the position cylinder slides. The outer cylinder contains a primary coil and two secondary coils on either side. Position cylinder movements are detected as changes in the induced voltage in the secondary coils. The analog output is quite sensitive, easily providing submicron resolutions, and it has the advantage of being noncontact. LVDTs provide absolute position metrology, meaning no initialization is necessary on power-up to determine a zero position. Since an LVDT operates on electromagnetic coupling principles in a friction-free structure, it can measure infinitesimally small changes in the core position. This infinite resolution capability is limited only by the noise in an LVDT signal conditioner and the output display's resolution. These same factors also give an LVDT its outstanding repeatability. They generally cannot be used for resolutions better than a few nanometers (Macro Sensors 2003).

Linear encoders (Nyce 2003) are familiar devices for position feedback, and their resolution enters the nanometer realm. Traditional linear encoders use optical or magnetic scales and read head elements. Optical units function by observing the first-order diffraction between a scale, composed of finely pitched lines or facets, and a similar moving reticle. By nanopositioning standards, the period of the scale is often fairly gross, typically several to several dozen microns for moiré scale or holographic encoders.

The position is determined by counting fringe transitions and interpolating between adjacent peaks. Signal-to-noise levels limit even the best encoder systems to the nanometer region, but travel ranges of hundreds of millimeters are possible. Guidance remains a concern: stiction effects limit the effective minimum incremental motion of even the best mechanical bearing systems in long travel stages. Conversely, stability issues, top speed, size, cost, and stiffness limit the air-bearing systems' applicability. For these reasons, stack-based nanopositioning systems are generally guided by flexures, which are stictionless yet highly stable and stiff, but with travels limited to the millimeter region and below.

In addition, encoders suffer from cyclic or subdivisional errors, both optical and electronic. This can lead to repeatable results whose inaccuracy varies quasi-sinusoidally with position. Encoders most commonly provide relative (not absolute) position metrology, meaning the coordinate system must be re-established at each initialization by seeking a reference switch of some sort. The day-to-day repeatability of such a system is only as good as the reference sensing technique.

Cost-effective strain gauges (Window 1992) use films whose electrical characteristics change with strain. These devices are usually attached to the piezoelectric stack itself, or possibly to a structural element of the stage. Consequently, they serve as drivetrain-input metrology rather than direct-motion metrology of the moving platform of the stage. Nevertheless, these are popular and effective position-metrology elements in closed-loop systems. They can provide high sensitivity, are compact, and are adequate for positioning to subnanometer levels. Like LVDTs, they provide absolute position measurement, require no initialization, and can be acceptably stable if good signal conditioning approaches are implemented. Good implementations incorporate multiple sensors and bridge circuits to compensate for thermal changes. But care must be taken when designing them into a mechanism. If there are elastic or frictional elements in the path between the point of motion and the point of measurement, errors will result. Physically small sensors measure only a highly localized region's strain, from which the overall mechanism's motion must be inferred. And these sensors cannot be configured to compensate for orthogonal (parasitic) errors in multi-axis configurations—only parallel metrology of the actual moving platform can provide this valuable capability.

Interferometry (Hariharan 2003) can also be used to measure sample positions and provide a feedback control signal. Such systems offer extremely high accuracy and are noncontact, although bulky specialized optics must be mounted onto the moving and stationary elements of the motion system and safety provisions for beam blockages and eye safety must be considered. Interferometers can be deployed in vacuum systems by feeding laser light through a window or fiber. Some interferometers can provide subnanometer resolutions, though cyclic inaccuracy can be noted in some units.

Because interferometers use air as their working fluid, they can be sensitive to environmental factors (temperature, barometric pressure, humidity, trace gas concentration, and acoustic perturbations). Interferometry tends to be costly, and drift can be problematic, particularly for lower-cost units. Available interferometers are also relative position devices that must be initialized at a repeatable zero position determined by some other metrology incorporated into the motion system.

Capacitive sensors (Baxter 1996) have emerged as the default choice for many nanopositioning applications. They are extremely accurate, ultrahigh-resolution devices for determining the absolute position over ranges of hundreds of microns or even millimeters. The device's positioning motions vary the distance between two nanomachined capacitor plates, providing a sensitive and drift-free positional feedback signal when stimulated by a precision AC carrier. No "home switch" is needed, so day-to-day repeatability is superb.

The best examples of these sensors offer high bandwidth with high precision. They are noncontact and are usually packaged to measure drive-train output (direct metrology)—eliminating errors that are otherwise imposed by intervening mechanisms. The physical characteristics of capacitive sensors make them ideal for multiaxis parallel kinematics approaches, where a single monolithic moving platform is actuated (and measured) simultaneously in several degrees of freedom. This allows active compensation of parasitic errors, which cannot be achieved with stacks of discrete axes. In addition, parallel kinematics facilitates more compact packaging, greater mechanical rigidity, higher throughput, and greater dynamic accuracy in real-time applications. Meanwhile, the single-frequency stimulation of capacitance sensors is inherently resistant to external noise sources (Jordan et al. 2007).

This chapter mainly focuses on nanopositioning systems using coarse/fine linear actuator and smart composite structures. They are an integration of smart structure technology, advanced electromechanical technology, and advanced control technology. Both systems possess simultaneous precision positioning and vibration suppression (SPPVS) capabilities. Section 8.2 describes nanopositioning systems with multifunctional linear actuators. Section 8.3 discusses nanopositioning systems with the smart composite technique and both single-input-single-output (SISO) and multi-input-multi-output (MIMO) systems are discussed. An application is shown in Section 8.4. The chapter ends with the conclusions followed by the acknowledgments.

8.2 Nanopositioning Systems with Multifunctional Linear Actuators

8.2.1 Multifunctional Linear Actuators

As mentioned earlier, nanopositioning goes hybrid, which means that a piezoelectric stack actuator could be placed on top of a motorized positioner to construct a coarse/fine long-travel assembly. This assembly combines the piezoelectric properties of unlimited resolution and very fast response with the long travel ranges and high holding forces of a servomotor/ballscrew arrangement. Figure 8.2 schematically illustrates a multifunctional smart actuator developed in this chapter. This smart actuator consists of a precision and high-speed linear motor, a motor-driven precision lead-screw, a piezoelectric stack translator, a high-resolution encoder, a load cell, a high-resolution linear displacement sensor, a composite housing, and joints at its two ends. The assembly of the linear motor and the lead screw provides a large stroke motion and coarse positioning. The piezoelectric stack actuator is connected to the lead screw in series and provides VS capability and/or fine positioning. The encoder is mounted on the rotational shaft of the linear motor to measure the linear motion of the motor and screw assembly. The linear displacement sensor is appended to the motor and screw assembly and the piezoelectric stack translator to measure the entire motion of the actuator. In the case of using the piezoelectric stack actuator only for VS, the linear displacement sensor may be eliminated. The load cell senses the load transmitting from one actuator end to another. The composite housing houses all parts of the actuator.

Since the repeatability and accuracy of such an actuator is dependent on the repeatability and accuracy of the linear motor, a specialized control strategy has to be designed to ensure hybrid nanopositioning capability. This controller compensates the nonlinearities, such as friction and backlash, of the motor-lead screw assembly in addition to controlling the piezoelectric stack. Therefore, the linear actuators can provide super-precision, large stroke, and simultaneous precision positioning and/or VS capabilities over a wide frequency range of interest.

8.2.2 Mathematical Model

Figure 8.3 shows the schematic of a linear motor and its drive circuit. The motor torque, $T(t)$, is proportional to the armature current, $I(t)$, by an electromotive force constant K_t, that is, $T(t) = K_t I(t)$. The back electromotive force, $E(t)$, is proportional to the rotational velocity of the motor shaft, $\dot{\theta}(t)$, by an electromotive force constant K_θ, that is, $E(t) = K_\theta \dot{\theta}(t)$. Based on Newton's law and Kirchhoff's law, the following equations can be obtained:

$$J\ddot{\theta}(t) + B\dot{\theta}(t) = K_t I(t) \tag{8.1}$$

$$L\dot{I}(t) + RI(t) = u_d(t) - K_\theta \dot{\theta}(t) \tag{8.2}$$

where

J is the moment of inertia of the motor rotor
B is the damping of the motor and its payload
L and R are the electric inductance and resistance of the motor and its drive circuit
u_d is the input voltage

The translation of the motor is as follows:

$$y_a(t) = K_g \theta(t) \tag{8.3}$$

in which K_g is a constant between the motor rotational angle and the translation displacement of the linear motor assembly.

In view of nonlinearity due to the existing friction and backlash of the motor, Equation 8.2 can be rewritten in the form

$$L\dot{I}(t) + RI(t) = f u_d(t) - K_\theta \dot{\theta}(t) \tag{8.4}$$

where f is a nonlinear function that describes the existing nonlinearity. In general, f could be a dead zone or a hysteresis for the linear motor assembly.

The displacement and input voltage relation of the piezoelectric stack actuator can be written as follows:

$$y_p(t) = K_p V_p(t) \tag{8.5}$$

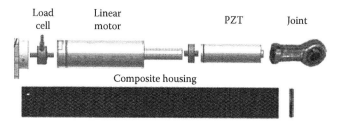

FIGURE 8.2 Schematic of linear actuator.

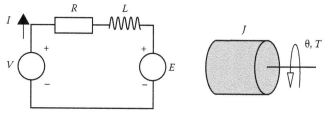

FIGURE 8.3 Linear motor and its drive circuit.

where

K_p is a constant
V_p is the input voltage
y_p is the displacement of the piezoelectric stack actuator

8.2.3 Nonlinear Compensation Control

Figure 8.4 is a measured nonlinear relation between the input voltage of the motor and the linear velocity of the linear motion assembly, showing that the nonlinearity is mainly a dead zone. For simplification, the dead zone is represented by the dash lines in Figure 8.4 and is described by the following equation:

$$\dot{y}_a(t) = \begin{cases} m[u_d(t) - a] & u_d \geq a \\ 0 & -a < u_d < a \\ m[u_d(t) + a] & u_d \leq -a \end{cases} \quad (8.6)$$

where

$\dot{y}_a(t)$ is the motor linear velocity
$u_d(t)$ is the signal to the motor
$m = 3.53$ mm/s/V is the slope of the linear region
$a = 2$ V, $2a$ is the dead zone size

This existing dead zone will cause a large steady-state error. This occurs since the actuator does not respond to the input voltages that fall into the dead zone, and hence the actuator positioning error remains the same as its previous value. Therefore, the dead zone nonlinearity has to be compensated to provide a good trajectory tracking performance. The inverse dead zone technique (Tao and Kokotovic 1994) is a good approach for compensating a dead zone. This technique is used to design a compensator that behaves as the inverse of a dead zone, and then combine the compensator with the controlled plant that has the dead zone behavior. It is easy to imagine that this combination leads to the cancellation of the dead zone effect and the combined system behaves linearly. With regard to the dead zone in Equation 8.6, the inverse dead zone can be written as following:

$$u_d(t) = \begin{cases} \dfrac{e(t)}{m} + a & e \geq 0 \\ \dfrac{e(t)}{m} - a & e < 0 \end{cases} \quad (8.7)$$

where e is the input of the compensator. Combining Equation 8.7 with Equation 8.6 yields $e = \dot{y}_a$.

In this work, an adaptive fuzzy logic control (AFLC) (Ma and Ghasemi-Nejhad 2008), shown in Figure 8.5, is introduced to compensate the dead zone, where m' is the desired slope of the relationship between the input of the AFLC (e) and the motor linear velocity ($\dot{y}_a$). Distinguished from the direct estimation of the dead zone size (Wang et al. 2004), this AFLC scheme consists of a traditional fuzzy logic controller (FLC) and an adaptive input scaling, where the adaptive input scaling dynamically adjusts the input to the traditional fuzzy logic controller.

The fuzzy logic controller with singleton fuzzifier, product-inference rule, center average defuzzifier, and Gaussian membership functions is defined as follows (Wang 1994):

$$u_d(t) = \frac{\sum_{i=1}^{M} f_i(v) V_i}{\sum_{i=1}^{M} f_i(v)} \quad (8.8)$$

$$f_i(v, \sigma_i, b_i) = \exp\left(\frac{-(v - b_i)^2}{2\sigma_i^2}\right) \quad (8.9)$$

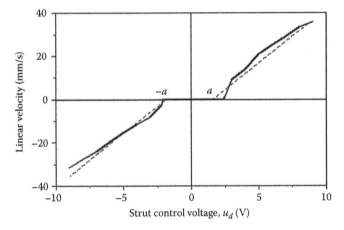

FIGURE 8.4 Linear velocity vs. input voltage of linear motor assembly.

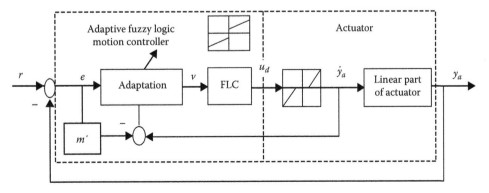

FIGURE 8.5 Adaptive fuzzy logic control with dead zone compensator.

where

f_i are the Gaussian membership functions

v is the input of the FLC

σ_i and b_i are the standard deviations and centers of the Gaussian membership functions, respectively

V_i are the controller output when $v = b_i$

u_d is the FLC output, in general terms

From Equations 8.8 and 8.9, the FLC output can be calculated based on the input (i.e., v) and the selected σ_i, b_i, and V_i. Nevertheless, the existing dead zone in the actuator will cause a residual steady-state error if the fuzzy logic controller output is less than the dead zone size, since the control signal would not be enough to drive the active actuator and the actuator positioning error would remain the same as its previous value. Therefore, the values of σ_i, b_i, and V_i must be selected with care to ensure that the fuzzy logic controller can achieve precision positioning control of the actuator.

To achieve an inverse dead zone, an AFLC strategy is developed in this work. The AFLC is a combination of an innovative adaptive input scaling and the traditional FLC. The basic idea behind the AFLC is as follows: the adaptive input scaling adjusts the input to the traditional FLC such that the FLC with the adjusted input can still provide a control signal greater than the dead zone while the steady-state error is so small that the output of the traditional FLC with the unadjusted input is less than the dead zone. To accomplish this task, the adaptive input scaling must have a large scale for small inputs and small scale for large inputs to avoid the dead zone effects. That is, the adaptive input scaling adjusts its output by multiplying the input by a scale that is inversely proportional to the magnitude of the input. The following defined adaptive input scaling meets the following characteristics:

$$v(t) = \frac{|\varepsilon_1(t)| + |e(t)|}{\varepsilon_2 + |e(t)|} e(t) = \Theta(t)e(t) \qquad (8.10)$$

where

e is the input of the adaptive input scaling that equals to the positioning error of the actuator

ε_1 is an adjustable value

ε_2 is a very small positive value to avoid the denominator being zero

$\Theta(t)$ is the scale of the adaptive input scaling

When the absolute value of e approaches zero, $\Theta(t)$ approaches $\varepsilon_1/\varepsilon_2$, which is a large value since ε_2 has a very small value. $\Theta(t)$ is 1 when the absolute value of e approaches infinity. Therefore, the adaptive input scaling practically adjusts the range of the input to the traditional FLC.

The investigation has demonstrated that the AFLC, consisting of Equations 8.8 through 8.10, can realize an inverse dead zone and the size of the inverse dead zone depends on the value of ε_1 in Equation 8.10 (Ma and Ghasemi-Nejhad 2008). In the case

where the dead zone size is unknown or is varying, the following adaptation for ε_1 is established and utilized:

$$\varepsilon_1(t) = \mu \int_0^t [\dot{y}_a(t) - m'e(t)]e(t)dt \qquad (8.11)$$

where μ is the convergence factor that determines the convergent rate of the adaptation. This process considers both the motor positioning error and the linearity between the motor linear velocity and the input of the AFLC. When $|\dot{y}_a(t) - m'e(t)|$ or $|e(t)|$ in Equation 8.11 is less than a given threshold, the adaptation for ε_1 stops and ε_1 does not change any longer.

The FLC has eleven fuzzy sets, and the Gaussian membership function is employed for each fuzzy set. The selected values for the FLC in Equations 8.8 and 8.9 are as follows:

$$[b_i] = \begin{bmatrix} -5 & -4 & -3 & -2 & -1 & 0 & 1 & 2 & 3 & 4 & 5 \end{bmatrix}\text{mm} \qquad (8.12)$$

$$[\sigma_i^2] = \begin{bmatrix} 1 & 1.5 & 0.5 & 0.5 & 0.5 & 1 & 0.5 & 0.5 & 0.5 & 1.5 & 1 \end{bmatrix} \qquad (8.13)$$

$$[V_i] = \begin{bmatrix} -10 & -8 & -6 & -4 & -2 & 0 & 2 & 4 & 6 & 8 & 10 \end{bmatrix}\text{V} \qquad (8.14)$$

and $\varepsilon_2 = 0.0005$ in Equation 8.10.

Figures 8.6 and 8.7 illustrate the relation between the AFLC input, the linear motor velocity, and the adaptation of ε_1, respectively. In this study, the thresholds for $|\dot{y}_a(t) - m'e(t)|$ and $|e(t)|$ in Equation 8.11 are 0.01 and 0.005, respectively. The desired motion of the motor assembly is a square wave with ±2 mm in amplitude and 0.2 Hz in frequency. The relationship between the velocity and the input of the AFLC demonstrates that the dead zone is narrowed from 4 to about 2 when $\mu = 0$ (i.e., $\varepsilon_1 = 0$) and demonstrates that when $\mu = 10$, the dead zone is compensated completely and the relationship between the velocity and the

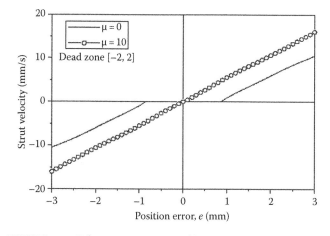

FIGURE 8.6 Velocity vs. AFLC input (e).

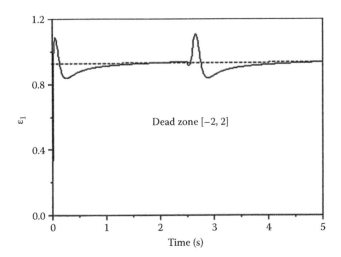

FIGURE 8.7   Time history of ε_1.

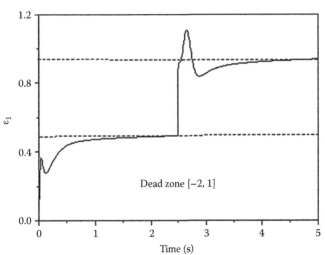

FIGURE 8.9   Time history of ε_1.

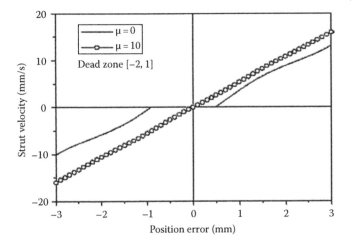

FIGURE 8.8   Velocity vs. AFLC input (e).

AFLC input is linear. The time history of ε_1 for $\mu = 10$ finally converges to 0.93.

Figure 8.8 demonstrates the relationship between the linear actuator velocity and the input of the AFLC for an asymmetric dead zone in which the dead zone is 1 in the positive motion direction and 2 in the negative motion direction. The AFLC can decrease the dead zone size to about 50% of its original value when $\mu = 0$ and completely compensate it when $\mu = 10$. The time history of ε_1, shown in Figure 8.9, depicts that ε_1 converges for both positive and negative motion directions of the linear actuator; however, the final values are different.

8.2.4 Experimental Results

The experimental results are illustrated in Figures 8.10 through 8.12. Figure 8.10 demonstrates the linear actuator displacement and error, showing that it takes 0.2 s to move the linear actuator from −3 to 3 mm and the average steady-state error is 1.8 μm, resulting in a 0.06% average relative steady-state error. The relationship between the input of the AFLC and the velocity of the linear actuator is similar to that in Figure 8.6. For comparison,

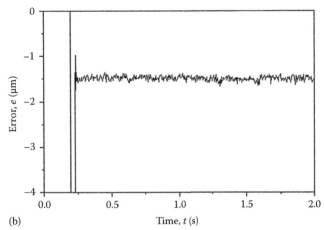

FIGURE 8.10   Displacement and error (AFLC): (a) displacement, (b) error.

the results from a PID/velocity feedforward controller with various proportional gains K_p are given in Figure 8.11, depicting that the best proportional gain is 10.25, and with this best proportional gain, the linear actuator displacement error is still about three times larger than that in AFLC.

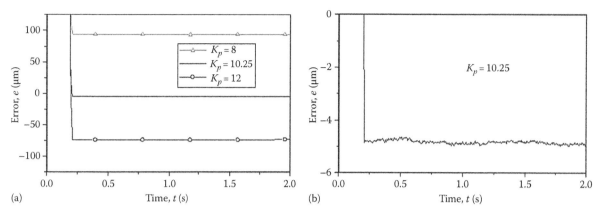

(a)

(b)

FIGURE 8.11 Displacement error (PID): (a) error, (b) error (zoom in around zero of (a)).

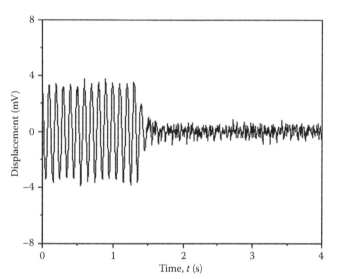

FIGURE 8.12 Vibration reduction using piezoelectric stack actuator.

The linear motor itself can only achieve micropositioning. To achieve nanopositioning and provide VS capability, the piezoelectric stack actuator can be employed. Figure 8.12 illustrates the vibration reduction of the linear actuator, depicting that the vibration is fully reduced and the remaining is the measured noise by employing an adaptive feedforward control (AFC) scheme (Ma 2003, Ma and Ghasemi-Nejhad 2006). This linear actuator has been used for flexible manipulators (Ma and Ghasemi-Nejhad 2005c, 2008).

8.3 Nanopositioning Systems with Piezoelectric Patch Actuators

Smart structures with piezoelectric patch actuators can also be utilized for nanopositioning. This section will discuss two such structures. The first is a SISO smart composite nanopositioning and the second one is a MIMO smart composite nanopositioning.

8.3.1 SISO Smart Composite Nanopositioning

Figure 8.13a gives the smart composite structure with a piezoelectric actuator in pair. It consists of a composite panel, two large-size CCC piezoelectric ceramic patches (CCC 2002) from Continuum Control Corporation, and one small-size ACX piezoelectric ceramic patch (ACX 2002) from Active Control eXperts. The composite panel has four laminae of W3F 282 42″ F593 graphite plain weave fabric prepreg with epoxy resin and 0.254 mm thick each. The composite panel with 0.94 mm in thickness, 55 mm in width, and 235 mm in length was cured in an autoclave. The two bonded CCC PZT patches, which are single patches with the dimensions of 0.33 mm in thickness, 55 mm in width, and 135 mm in length, are collocated back-to-back. One of the CCC PZT patches is used as an actuator, and the other CCC PZT patch could be used as a sensor. The ACX PZT patch with 0.508 mm in thickness, 38.1 mm in width, and 58.8 mm in length is bonded at the tip of the active composite panel (ACP) as an exciter to vibrate the ACP.

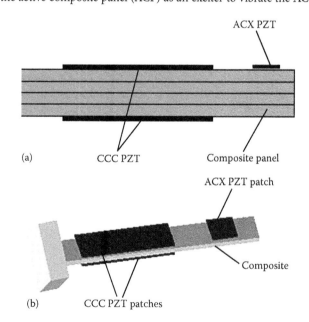

FIGURE 8.13 Smart composite nanopositioning system: (a) 2-D view, (b) 3-D view.

Both CCC and ACX patches employ PZT-5A piezoelectric material and perform as an extension motor. The panel is clamped at one end, as shown in Figure 8.13b, to achieve nanopositioning under a vibrational environment. The first two modal frequencies and modal damping ratios of the smart composite nanopositioning structure are 14.6 Hz, 0.018 and 97.2 Hz, 0.03, respectively.

8.3.1.1 Adaptive Control Schemes

Figures 8.14 through 8.16 show the schematics of the three adaptive control algorithms, namely AFC, adaptive internal model

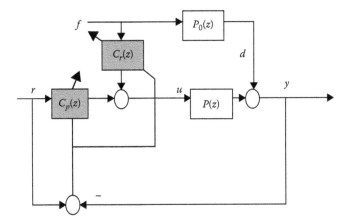

FIGURE 8.14 Adaptive feedforward control.

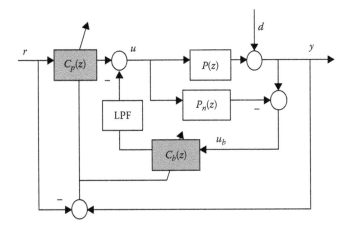

FIGURE 8.15 Adaptive internal model control.

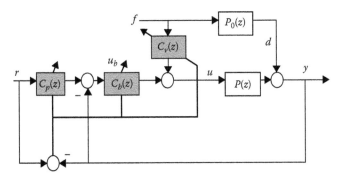

FIGURE 8.16 Hybrid adaptive control.

control (AIMC), and hybrid adaptive control (HAC) for SPPVS. In these diagrams, $P_0(z)$ and $P(z)$ represent the impulse transfer functions of the disturbance path and control path of a controlled structure. f, y, and r are the external disturbance, the displacement of the ACP, and command, respectively. $C_v(z)$ and $C_p(z)$ are the impulse transfer functions of two AFCs and $C_b(z)$ is the impulse transfer function of an adaptive feedback controller. $E(z) = Y(z) - R(z)$ is defined as the tracking error between the command and the displacement at the tip of the smart composite nanopositioning system.

Figure 8.14 shows the diagram of AFC. It consists of two feedforward controllers, where one of them is related to the external disturbance and the other is related to the command. It is obvious that this scheme strongly depends on the external disturbance information. If the disturbance can be known correctly and completely or be observed and constructed, then this scheme can perform well; otherwise, a good performance is hardly achieved. The optimal controllers and adaptive algorithm of this scheme are listed in Table 8.2. The optimal controllers indicate that the tracking error will be zero, if $C_v(z) = -P^{-1}(z)P_0(z)$ and $C_p(z) = P^{-1}(z)$. The filtered-x LMS algorithm is employed to update the two feedforward controllers, as listed in Table 8.2. The details of this scheme can be found in Ma and Ghasemi-Nejhad (2002a, 2006).

Compared with feedforward control, feedback control can provide better performance in the case of unknown external disturbance. To achieve high precision, internal model control has been considered in structural VS and high precision mechanical systems to enhance the vibration reduction and disturbance rejection. Figure 8.15 shows the block-diagram of a two-degree-of-freedom internal model control in which $P_n(z)$ is the nominal impulse response function of the control path $P(z)$. Suppose that the nominal model $P_n(z)$ equals to $P(z)$, the signal u_b is then exactly the same as the response due to the external disturbance, that is, the unknown external disturbance response d can be reconstructed or observed by introducing an internal model $P_n(z)$. The optimal controllers are listed in Table 8.2. The optimal controllers indicate that the tracking error will be zero, if $C_b(z) = C_p(z) = P_n^{-1}(z)$. The filtered-x LMS algorithm is employed to update the two controllers, as listed in Table 8.2. The details of this scheme can be found in Ma and Ghasemi-Nejhad (2003). Figure 8.16 shows the HAC scheme, which takes advantage of both feedforward control and feedback control. It consists of two feedforward controllers and a feedback controller. The two feedforward controllers are related to the external disturbance and the command, respectively to achieve positioning and VS individually, and the feedback controller takes care of both positioning and VS. The optimal controllers and adaptive algorithm of this scheme are also listed in Table 8.2. The optimal controllers indicate that the tracking error will be zero, if $C_v(z) = -P^{-1}(z)P_0(z)$ and $C_p(z) = 1 + C_b^{-1}(z)P^{-1}(z)$. Compared with the optimal controller of the AFC, $C_p(z)$ in HAC is not only related to the inverse of the controlled path, but also to the introduced feedback controller, giving more

TABLE 8.2 Three Adaptive Control Algorithms

Adaptive Feedforward Control	Adaptive Internal Model Control	Hybrid Adaptive Control
$Y(z) = [P_0(z) + P(z)C_v(z)]F(z)$ $+ P(z)C_p(z)R(z)$	$Y(z) = \dfrac{1 - P_n(z)C_b(z)}{1 + [P(z) - P_n(z)]C_b(z)} D(z)$ $+ \dfrac{P(z)C_p(z)}{1 + [P(z) - P_n(z)]C_b(z)} R(z)$	$Y(z) = \dfrac{P_0(z) + P(z)C_v(z)}{1 + P(z)C_b(z)} F(z)$ $+ \dfrac{P(z)C_b(z)C_p(z)}{1 + P(z)C_b(z)} R(z)$
$D(z) = P_0(z)F(z)$ $E(z) = Y(z) - R(z)$ $= [P_0(z) + P(z)C_v(z)]F(z)$ $+ [P(z)C_p(z) - 1]R(z)$	$E(z) = Y(z) - R(z)$ $= \dfrac{1 - P_n(z)C_b(z)}{1 + [P(z) - P_n(z)]C_b(z)} D(z)$ $- \left[1 - \dfrac{P(Z)C_p(z)}{1 + [P(z) - P_n(z)]C_b(z)} \right] R(z)$	$D(z) = P_0(z)F(z)$ $E(z) = Y(z) - R(z)$ $= \dfrac{P_0(z) + P(z)C_v(z)}{1 + P(z)C_b(z)} F(z)$ $- \dfrac{1 + P(z)C_b(z)[1 - C_p(z)]}{1 + P(z)C_b(z)} R(z)$
Optimal solutions: $C_v(z) = -P^{-1}(z)P_0(z)$ $C_p(z) = P^{-1}(z)$	Optimal solutions: $C_b(z) = C_p(z) = P_n^{-1}(z)$	Optimal solutions: $C_v(z) = -P^{-1}(z)P_0(z)$ $C_p(z) = 1 + C_b^{-1}(z)P^{-1}(z)$
Objective: $J(k) = E^2(z,k)$		
Adaptive algorithm:	Adaptive algorithm:	Adaptive algorithm:
$C_p(z,k+1) = C_p(z,k) + \mu_p \dfrac{\partial J(k)}{\partial C_p(z,k)}$	$C_p(z,k+1) = C_p(z,k) + \mu_p \dfrac{\partial J(k)}{\partial C_p(z,k)}$	$C_p(z,k+1) = C_p(z,k) + \mu_p \dfrac{\partial J(k)}{\partial C_p(z,k)}$
$C_v(z,k+1) = C_v(z,k) + \mu_v \dfrac{\partial J(k)}{\partial C_v(z,k)}$	$C_b(z,k+1) = C_b(z,k) + \mu_b \dfrac{\partial J(k)}{\partial C_b(z,k)}$	$C_v(z,k+1) = C_v(z,k) + \mu_v \dfrac{\partial J(k)}{\partial C_v(z,k)}$
$\dfrac{\partial J}{\partial C_v(z,k)} = 2E(z,k)F_{xv}(z,k)$	$U_b(z,k) = \dfrac{D(z,k) + [P(z) - P_n(z)]C_b(z,k)R(z,k)}{1 + [P(z) - P_n(z)]C_b(z,k)}$	$C_b(z,k+1) = C_b(z,k) + \mu_b \dfrac{\partial J(k)}{\partial C_b(z,k)}$
$F_{xv}(z,k) = P(z,k)F(z,k)$	$\dfrac{\partial J}{\partial C_b(z,k)} = -2E(z,k)F_{xb}(z,k)$	$\dfrac{\partial J}{\partial C_v(z,k)} = 2E(z,k)F_{xv}(z,k)$
$\dfrac{\partial J}{\partial C_p(z,k)} = 2E(z,k)F_{xp}(z,k)$	$F_{xb}(z,k) = \dfrac{P(z)}{1 + [P(z) - P_n(z)]C_b(z,k)} U_b(z,k)$	$F_{xv}(z,k) = \dfrac{P(z)}{1 + P(z)C_b(z,k)} F(z,k)$
$F_{xp}(z,k) = P(z,k)R(z,k)$	$\dfrac{\partial J}{\partial C_p(z,k)} = 2E(z,k)F_{xp}(z,k)$	$\dfrac{\partial J}{\partial C_p(z,k)} = 2E(z,k)F_{xp}(z,k)$
	$F_{xp}(z,k) = \dfrac{P(z)}{1 + [P(z) - P_n(z)]C_b(z,k)} R(z,k)$	$F_{xp}(z,k) = \dfrac{P(z)C_b(z,k)}{1 + P(z)C_b(z,k)} R(z,k)$
		$\dfrac{\partial J}{\partial C_b(z,k)} = 2E(z,k)F_{xb}(z,k)$
		$U_b(z,k) = \dfrac{C_p(z,k)R(z,k) - [P_0(z) + P(z)C_v(z,k)]F(z,k)}{1 + P(z)C_b(z,k)}$
		$F_{xb}(z,k) = \dfrac{P(z)}{1 + P(z)C_b(z,k)} U_b(z,k)$

flexibility to the design of the feedforward controller in HAC. The filtered-x LMS algorithm is employed to update the three controllers, as listed in Table 8.2. The details of this scheme can be found in Ma and Ghasemi-Nejhad (2002b, 2005a). A frequency-weighted HAC scheme can be found in Ma and Ghasemi-Nejhad (2004).

8.3.1.2 Experiments and Comparison

The experimental system consists of a dSPACE real time system with two PowerPC processors, two 601B-4 amplifiers from Trek, Inc., a MT600-800 laser displacement sensor from MTI Instruments, Inc., the ACP, a SR785 2-channel dynamic signal analyzer, and an Agilent 33250 function/arbitrary waveform generator. The disturbance comes from the SR785 2-channel dynamic signal analyzer and is applied to the ACX PZT patch through the 601B-4 amplifier to excite the smart composite nanopositioning system. One CCC PZT patch is used as an actuator for the purpose of precision positioning and VS. The AFC, AIMC, and HAC approaches are employed and applied individually.

The command is prescribed as follows:

$$r(t) = Sat[\sin(0.2\pi t)] \quad (V) \tag{8.15}$$

and

$$Sat[x] = \begin{cases} a & x \geq a \\ -a & x \leq -a \end{cases}, \tag{8.16}$$

where a represents the required position and equals 0.2 V (equivalent to 0.508 mm) in the following experiments. Two harmonic disturbances with the first two natural frequencies of the panel are applied to the ACX PZT patch, respectively. The sampling

frequency is 10,000 Hz. To assess the effectiveness of the controllers, the following relative error δ is defined:

$$\delta = \left| \frac{e(t)}{a} \right| \times 100\%. \tag{8.17}$$

Figure 8.17 illustrates the results using the AFC. Figure 8.17a shows the displacement at the ACP tip for VS only with the harmonic disturbance at the first natural frequency of the structure. The amplitude of the displacement decreases from uncontrolled 0.7 V to controlled 0.053 V after the control works for 20 s and to controlled 0.01 V after the control works for 40 s. By changing the frequency of the harmonic disturbance to the second natural frequency of the ACP, the displacement decreases from uncontrolled 0.18 V to controlled 0.005 V after the control works for 3 s. In this case, the displacement tends to be suppressed totally along with the controller action, but the convergent process is slow, especially for the first mode. Figure 8.17b displays the displacement at the panel tip in the case of SPPVS. It depicts that the displacement at the panel tip follows the command properly.

Figure 8.18 illustrates the results using the AIMC. The uncontrolled vibration amplitude of the displacement at the panel tip is

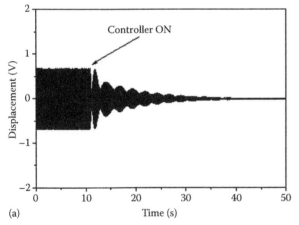

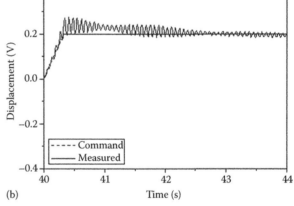

(a)   Time (s)   (b)   Time (s)

FIGURE 8.17 Displacement at the panel tip (AFC): (a) VS, (b) VS + PP.

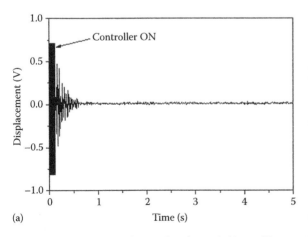

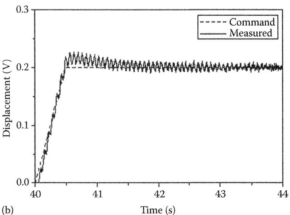

(a)   Time (s)   (b)   Time (s)

FIGURE 8.18 Displacement at the panel tip (AIMC): (a) VS, (b) VS + PP.

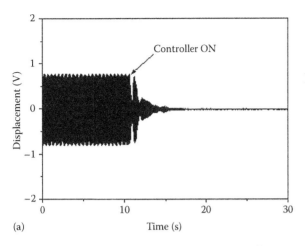
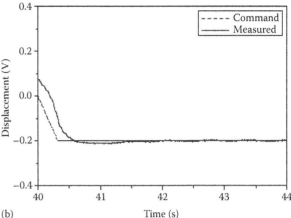

(a) Time (s)   (b) Time (s)

FIGURE 8.19 Displacement at the panel tip (HAC): (a) VS, (b) VS + PP.

0.76 V for the harmonic disturbance at the first natural frequency of the ACP. For VS only, the vibration amplitude is attenuated to 0.03 V after the controller works for 0.7 s. Changing the frequency of the harmonic disturbance to the second natural frequency, the displacement decreases from uncontrolled 0.18 V to controlled 0.02 V after the control works for 0.1 s. Figure 8.18b displays the displacement at the panel tip in the case of SPPVS. It depicts that the displacement at the panel tip follows the command well.

The HAC is applied to the panel. Figure 8.19 shows the displacement at the panel tip for VS only with the harmonic disturbance at the first natural frequency of the panel. The amplitude of the displacement decreases almost to zero from uncontrolled 0.7 V for the harmonic vibration at the first natural frequency of the panel after the controller works for 5 s. By changing the frequency of the harmonic disturbance to the second natural frequency of the panel, the displacement decreases from uncontrolled 0.18 V to controlled 0.01 V after the controller works for 3 s, the remaining signals are measurement noises. Figure 8.19b displays the displacement at the panel tip in the case of SPPVS. It depicts that the displacement at the panel tip follows the command fully.

Figure 8.20 illustrates the relative errors of the displacements at the panel tip, demonstrating that the steady-state relative errors are 3.3%, 1.2%, and 0.7% for the AFC, AIMC, and HAC,

TABLE 8.3 Comparison of Experimental Results

	First Modal Frequency			Second Modal Frequency		
	VS		VS + PP	VS		VS + PP
	UY/Y	ST	RE	UY/Y	ST	RE
AFC	0.70/0.01	40	3.3%	0.18/0.01	3	4.5%
AIMC	0.76/0.03	0.7	1.2%	0.18/0.02	0.1	1.3%
HAC	0.78/0.01	7	0.7%	0.18/0.01	3	0.7%

UY: uncontrolled displacement (V), Y: controlled displacement (V), ST: settling time (s), RE: relative error

respectively in the case of the harmonic disturbance at the first modal frequency of the panel, and these values are 4.5%, 1.3%, and 0.7%, respectively, in the case of the harmonic disturbance at the second modal frequency of the panel. Also, it can be seen that the relative error diminishes along with the control action, showing the adaptation of the three adaptive control schemes. Table 8.3 summarizes the experimental results, indicating that for VS, the AFC and HAC can eventually suppress the entire vibration; the HAC generates a faster reduction and the AIMC can provide the fastest vibration reduction, but some small residual vibration still remains. For SPPVS, the AFC, AIMC, and HAC sequentially produce more accurate positioning, and the HAC makes the best tradeoff between the fastness of the positioning and VS as well as the accuracy of the position.

8.3.2 MIMO Smart Composite Nanopositioning

8.3.2.1 Configuration and Dynamics of the Smart Composite Plate

Figure 8.21 shows the cantilevered smart composite plate (SCP) with four surface-mounted piezoelectric patches. Two of the piezoelectric patches are on one side of the plate and the other two are on the opposite side. The composite has four woven plies of W3F 282 42′ F593 graphite plain weave fabric pre-preg with epoxy resin. Its size is 230 mm × 235 mm × 1 mm. The piezoelectric patches are of active fiber composite type, employ PZT-5A piezoelectric material, and perform as extension motors. Their sizes

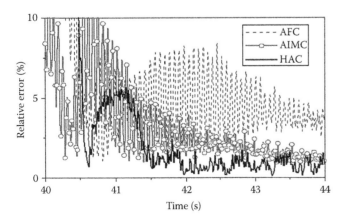

FIGURE 8.20 Relative errors (VS + PP).

FIGURE 8.21 Smart composite plate and sensors.

are 55 mm × 135 mm × 0.25 mm and are placed back-to-back above and below the plate central line at their respective centers, resulting in a distance between the bottom edge of the top piezoelectric patch and the top edge of the bottom piezoelectric patch of 60 mm. In this work, each two back-to-back piezoelectric patches form an actuator pair working in tandem, so that the plate has two pairs of actuators to provide precision positioning and VS. With this actuation configuration, various types of positioning, upon commands, can be achieved, such as the bending or torsional positioning.

ANSYS (Ansys 2006) finite element analysis (FEA) is performed. The detailed parameters of the piezoelectric and composite materials can be found in Doherty and Ghasemi-Nejhad (2005) and Ghasemi-Nejhad et al. (2005). The first two modes of the plate are illustrated in Figure 8.22. The first mode is a bending mode (see Figure 8.22a) and the second mode is a torsional mode (see Figure 8.22b).

Two laser displacement sensors (LDSs) with a sensitivity of 2.54 mm/V are employed to measure the displacements at the upper and lower corners of the free end of the smart composite plate, that is, points A and B in Figure 8.21. The measured frequency response curves are shown in Figure 8.23, where H_{ij} is the frequency response between the ith LDS and the jth

piezoelectric pair actuator, depicting that the first two natural frequencies of the plate are 20.5 and 38 Hz, and their damping ratios are both 0.008. At the first natural frequency (i.e., 20.5 Hz), all phases of H_{11}, H_{12}, H_{21}, and H_{22} are the same, indicating that the plate bends at this frequency; however, at the second natural frequency (i.e., 38 Hz), the phases of H_{11} and H_{12} are opposite, so are H_{21} and H_{22}, indicating that the second mode is torsional. Therefore, the experimental results have good agreement with the results from ANSYS FEA.

Based on the measured frequency responses, a transfer function estimation process is performed to obtain dynamic features of the plate and the obtained transfer functions are as follows:

$$H_{11}(s) = \frac{0.018s^4 + 0.128s^3 + 1.852 \times 10^3 s^2 + 5.174 \times 10^3 s + 3.923 \times 10^7}{s^4 + 5.64s^3 + 7.357 \times 10^4 s^2 + 1.766 \times 10^5 s + 9.436 \times 10^8}$$

$$H_{12}(s) = \frac{150s^2 + 1.923 \times 10^3 s + 1.54 \times 10^7}{s^4 + 5.64s^3 + 7.357 \times 10^4 s^2 + 1.766 \times 10^5 s + 9.436 \times 10^8}$$

$$H_{21}(s) = \frac{120s^2 + 4.152 \times 10^3 s + 1.437 \times 10^7}{s^4 + 5.64s^3 + 7.357 \times 10^4 s^2 + 1.766 \times 10^5 s + 9.436 \times 10^8}$$

$$H_{22}(s) = \frac{0.012s^4 + 0.047s^3 + 1.328 \times 10^3 s^2 + 3.103 \times 10^3 s + 2.935 \times 10^7}{s^4 + 5.64s^3 + 7.357 \times 10^4 s^2 + 1.766 \times 10^5 s + 9.436 \times 10^8}$$

(8.18)

where s is the Laplace variable. These estimated frequency responses are also illustrated in Figure 8.23, demonstrating that the estimated and measured frequency responses match very well.

Due to the very low damping ratios, the plate will vibrate severely when it is moved from one position to another or it is subjected to external disturbances.

8.3.2.2 Adaptive Filtered-x Algorithm with Multireferences and Multichannels (AFAMRMC)

The adaptive filtering-x algorithm has been used in active vibration and noise control (Ma 2003, Peng et al. 2003, Ma and Ghasemi-Nejhad 2006). In most cases, the adaptive filtering-x algorithm only deals with single reference. However, for the case of SPPVS of smart structures here, multiple references have to be considered. Therefore, an adaptive filtering-x algorithm with

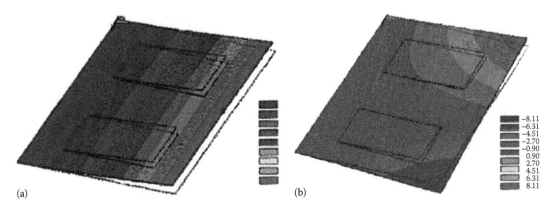

(a)   (b)

FIGURE 8.22 **(See color insert following page 22-8.)** Modal shapes of smart composite plate (ANSYS FEA): (a) first mode, (b) second mode.

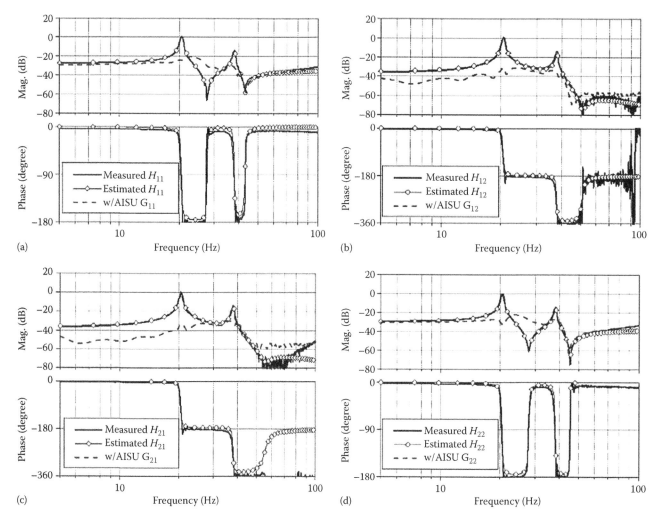

FIGURE 8.23   Bode plots.

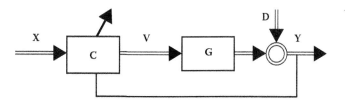

FIGURE 8.24   Adaptive control system with multireferences and multichannels.

multireferences and multichannels (AFAMRMC) is presented here and shown in Figure 8.24.

In Figure 8.24, the AFC **C** has N references $\mathbf{X} = [x_1 \quad x_2 \quad \ldots \quad x_N]^T$ and M control signals $\mathbf{V} = [v_1 \quad v_2 \quad \ldots \quad v_M]^T$ and the controlled plant **G** has L measure signals $\mathbf{Y} = [y_1 \quad y_2 \quad \ldots \quad y_L]^T$; $\mathbf{D} = [d_1 \quad d_2 \quad \ldots \quad d_L]^T$ is the measured response vector due to external disturbances only. Letting the dynamics of the controlled plant **G** and the controller **C** be described by using FIR filters, Figure 8.24 then yields

$$y_l(k) = d_l(k) + \sum_{m=1}^{M} G_{lm}^T V_m(k), \quad l = 1, 2, \ldots, L \qquad (8.19)$$

where

$G_{lm} = [g_{lm}^1 \quad g_{lm}^2 \quad \ldots \quad g_{lm}^{\bar{m}}]^T$ is the coefficient vector of the FIR filter that describes the dynamics of the controlled plant channel formed by the lth sensor and the mth actuator

$\bar{m}$ is the length of that FIR filter

$$V_m(k) = [v_m(k) \quad v_m(k-1) \quad \ldots \quad v_m(k-\bar{m}+1)]^T$$

$$v_m(k) = \sum_{n=1}^{N} C_{mn}^T(k) X_n(k), \quad m = 1, 2, \ldots, M \qquad (8.20)$$

is the control signal to the mth actuator; $X_n(k) = [x_n(k) \quad x_n(k-1) \quad \ldots \quad x_n(k-\bar{n}+1)]^T$; $C_{mn}(k) = [c_{mn}^1(k) \quad c_{mn}^2(k) \quad \ldots \quad c_{mn}^{\bar{n}}(k)]^T$ is the coefficient vector of the FIR filter that describers the dynamics of the controller related to the mth actuator and the nth reference signal x_n, $\bar{n}$ is the length of the FIR controller filter. Assuming **D** varies slowly, so does **C**. Equation 8.20 can be rewritten as the following:

$$y_l(k) = d_l(k) + \sum_{n=1}^{N} \sum_{m=1}^{M} F_{lm}^n(k) C_{mn}(k), \quad l = 1, 2, \ldots, L \quad (8.21)$$

where $F_{lm}^n(k) = [f_{lm}^n(k) \quad f_{lm}^n(k-1) \quad \ldots \quad f_{lm}^n(k - \bar{n} + 1)]$,

$$f_{lm}^n(k) = G_{lm}^T \bar{X}_n(k) \tag{8.22}$$

is called the filtered-x signal, $\bar{X}_n(k) = [x_n(k) \quad x_n(k-1) \quad \ldots \quad x_n(k - \bar{m} + 1)]^T$. Equation 8.21 can also be rewritten in the following matrix form:

$$\mathbf{Y}(k) = \mathbf{D}(k) + \sum_{n=1}^{N} \mathbf{F}_n(k)\mathbf{C}_n(k) \tag{8.23}$$

where

$$\mathbf{F}_n(k) = \begin{bmatrix} F_{11}^n(k) & F_{12}^n(k) & \cdots & F_{1M}^n(k) \\ F_{21}^n(k) & F_{22}^n(k) & \cdots & F_{2M}^n(k) \\ \vdots & \vdots & \vdots & \vdots \\ F_{L1}^n(k) & F_{L2}^n(k) & \cdots & F_{LM}^n(k) \end{bmatrix} \tag{8.24}$$

$$\mathbf{C}_n(k) = [C_{1n}(k) \quad C_{2n}(k) \quad \ldots \quad C_{Mn}(k)]^T \tag{8.25}$$

Defining the objective function as shown in Equation 8.26 and using the steepest descent method (Elliott 2001) yield the adaptive filtering-x algorithm as shown in Equation 8.27:

$$J(k) = \frac{1}{2}\mathbf{Y}^T(k)\mathbf{Y}(k) \tag{8.26}$$

$$\mathbf{C}_n(k+1) = \mathbf{C}_n(k) + \mu \frac{\partial J(k)}{\partial \mathbf{C}_n} \tag{8.27}$$

where μ is the convergence factor that rules the adaptation rate of the AFAMRMC; and the larger the μ is, the faster the adaptation rate is. According to Equations 8.23 and 8.26, Equation 8.27 can be rewritten as the following:

$$\mathbf{C}_n(k+1) = \mathbf{C}_n(k) + 2\mu\mathbf{F}_n^T(k)\mathbf{Y}(k), \quad n = 1, 2, \ldots, N \tag{8.28}$$

It has been proven that the above adaptive algorithm is stable if the convergence factor μ is selected in the range of $0 < \mu < \lambda_{max}^{-1}$, where λ_{max}^{-1} is the maximum eigenvalue of the filtered autocorrelation matrix, given by $E[\mathbf{F}_n^T(k)\mathbf{F}_n(k)]$ (Ma 2003). That is, if λ_{max} is small, large μ can be used and then the adaptation can converge fast. Since the eigenvalues of the autocorrelation matrix are rarely known, the convergence factor is usually selected manually.

8.3.2.3 AFAMRMC for Adaptive Input Shaping and Control

The algorithm developed in the last section can directly be used in many applications such as SPPVS of structures, vibration, and noise control systems with multiple references and multiple channels, etc. However, for a low-damped plant, some difficulties may occur if the AFAMRMC is directly employed due to

the fact that the impulse response of the control channel of a low-damped plant is large in amplitude and of long duration. From Equation 8.22, it can be seen that the filtered-x signal $f_{lm}^n(k)$ comes from the convolution of the impulse response of the control channel G_{lm} and the reference signal $\bar{X}_n(k)$. Therefore, the large and long-time impulse response of the control channel results in a large filtered-x signal and large calculation complexity of calculating the filtered-x signal, since a high-order FIR filter is needed to model this control channel. A large filtered-x signal further causes a large trace of the autocorrelation matrix of the filtered-x signal vector, that is, $E[\mathbf{F}_n^T(k)\mathbf{F}_n(k)]$, then a large maximum eigenvalue of $E[\mathbf{F}_n^T(k)\mathbf{F}_n(k)]$, and hence, a very small available convergence factor and a very slow convergence rate. In addition, the low damping causes a response oscillation when a control signal is applied to the plant.

To raise the convergence rate of the AFAMRMC and reduce the response oscillation induced by controllers, AFAMRMC with a so-called adaptive input shaping technology is introduced here. Figure 8.25 explains this control concept where an adaptive input shaping unit (AISU) W has been integrated with H, which is the controlled plant, and C remains the controller for the new combined plant of W and H. It is expected that the AISU can raise the damping of the combination of the AISU and the controlled plant.

In general, the AFAMRMC developed earlier can be directly employed to design the AISU by assuming that the AISU is in the form of a FIR filter and $\mathbf{D} = 0$, $\mathbf{G} = \mathbf{H}$, and $\mathbf{C} = \mathbf{W}$ in Figure 8.24. As mentioned earlier, the main purpose of the AISU is to increase the damping of the combined W and H and to alleviate the response oscillation in a given frequency range. Reference $\mathbf{X}$ can therefore be chosen as a swept-sinusoidal or random signal in the frequency range of interest, and then applied to all actuators at the same time, resulting in an AFAMRMC with multi-inputs and multi-outputs. However, it might be better to apply the reference $\mathbf{X}$ to an actuator once a time to achieve better vibration alleviation, resulting in a single-input and multi-output AFAMRMC; that is, for a control system with M actuators, M runs of AFAMRMC should be performed to obtain the entire AISU.

The smart composite plate, considered in this study, has two inputs and two outputs. Therefore, the relationship between its inputs and outputs can be written as the following:

$$\begin{Bmatrix} Y_1(z) \\ Y_2(z) \end{Bmatrix} = \begin{bmatrix} H_{11}(z) & H_{12}(z) \\ H_{21}(z) & H_{22}(z) \end{bmatrix} \begin{Bmatrix} U_1(z) \\ U_2(z) \end{Bmatrix} = H(z) \begin{Bmatrix} U_1(z) \\ U_2(z) \end{Bmatrix} \tag{8.29}$$

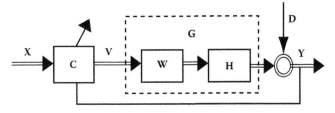

FIGURE 8.25 Adaptive control with adaptive input shaping unit (AISU).

where

$Y_1(z)$ is the z-transform of y_1

z is a complex variable, the same to other variables

For simplification, assuming that the AISU is in the following form

$$\begin{Bmatrix} U_1(z) \\ U_2(z) \end{Bmatrix} = \begin{bmatrix} 1 & W_A(z) \\ W_B(z) & 1 \end{bmatrix} \begin{Bmatrix} V_1(z) \\ V_2(z) \end{Bmatrix} \qquad (8.30)$$

in which W_B and W_B are the two elements of the AISU that need to be determined, substituting Equation 8.30 into Equation 8.29 yields

$$\begin{Bmatrix} Y_1(z) \\ Y_2(z) \end{Bmatrix} = \begin{bmatrix} H_{11}(z) + H_{12}(z)W_B(z) & H_{11}(z)W_A(z) + H_{12}(z) \\ H_{21}(z) + H_{22}(z)W_B(z) & H_{21}(z)W_A(z) + H_{22}(z) \end{bmatrix} \begin{Bmatrix} V_1(z) \\ V_2(z) \end{Bmatrix}$$

$$= \begin{bmatrix} G_{11}(z) & G_{12}(z) \\ G_{21}(z) & G_{22}(z) \end{bmatrix} \begin{Bmatrix} V_1(z) \\ V_2(z) \end{Bmatrix} = G(z) \begin{Bmatrix} V_1(z) \\ V_2(z) \end{Bmatrix} \qquad (8.31)$$

The AISU design for the smart composite plate therefore becomes: applying a swept sinusoidal signal to V_1 and V_2, individually and adaptively adjusting W_B for V_1 or W_A for V_2 to minimize the outputs Y_1 and Y_2. However, a special consideration is added to the AISU of the smart composite plate. The AISU is also expected to weaken the coupling degree of the control channels (i.e., control channel decoupling) while it functions to reduce the response oscillation of the plate. That is, the AISU is expected to make $G(z)$ in Equation 8.31 a diagonally dominant matrix and then the controlled system may be treated as two SISO systems, resulting in significant ease of controller design. Based on this consideration, the AISU design for the smart composite plate becomes the following two steps: first, applying a swept sinusoidal signal to V_1 only and adaptively adjusting W_B to minimize the output Y_2 and second, applying a swept sinusoidal signal to V_2 only and adaptively adjusting W_A to minimize the output Y_1. Then, W_A and W_B are used in Equation 8.31 and the result in Figure 8.25 to design the controller.

It should be noted that the control channel decoupling is different from decoupling the bending and torsional vibration modes of the smart plate. Control channel decoupling is used to convert a MIMO system into multiple SISO subsystems so that the controller design can be simplified. Even after the control channel decoupling, both bending and torsional vibration modes of the smart plate still exist in every SISO decoupled subsystem. Therefore, the AISU introduced here does not decouple the bending and torsional vibration modes. Instead, it controls both bending and torsional vibration modes concurrently while performing the control channel decoupling.

The AFAMRMC developed earlier can directly be employed to adapt W_A and W_B. For W_A, $H_{12}(z)$ and $H_{11}(z)$ form the primary and the secondary channel, respectively; and for W_B, $H_{21}(z)$ and

$H_{22}(z)$ form the primary and secondary channel, respectively. Let $\bar{y}_1(k)$ be $y_1(k)$ due to $v_2(k)$ only, $\bar{y}_2(k)$ be $y_2(k)$ due to $v_1(k)$ only, and W_A and W_B be FIR filters with coefficients $\mathbf{W}_A(k)$ and $\mathbf{W}_B(k)$, respectively, the objective functions $J_1(k)$ and $J_2(k)$ for adapting $\mathbf{W}_A(k)$ and $\mathbf{W}_B(k)$ as well as the adaptations will then be in the following forms:

$$J_1(k) = \frac{1}{2}\bar{y}_1^2(k) \qquad (8.32)$$

$$\mathbf{W}_A(k+1) = \mathbf{W}_A(k) + \mu_A \frac{\partial J_1(k)}{\partial \mathbf{W}_A} \qquad (8.33)$$

$$J_2(k) = \frac{1}{2}\bar{y}_2^2(k) \qquad (8.34)$$

$$\mathbf{W}_B(k+1) = \mathbf{W}_B(k) + \mu_B \frac{\partial J_2(k)}{\partial \mathbf{W}_B} \qquad (8.35)$$

where μ_A and μ_B are two convergence factors.

After W_A and W_B are obtained, the AFAMRMC can be employed to achieve SPPVS of the smart composite plate, as illustrated in Figure 8.26, where there are two reference vectors $\mathbf{X}$ and $\mathbf{R}$ for the adaptive vibration controller $\mathbf{C}_V$ and adaptive positioning controller $\mathbf{C}_P$, respectively. $\mathbf{P}_0$ represents the primary channel (disturbance channel) and $\mathbf{G}$ represents the secondary channel (control channel) including the AISU and the controlled plant. $\mathbf{X}$, $\mathbf{R}$, $\mathbf{D}$, and $\mathbf{Y}$ are the external disturbances, the commands, or the desired positions of the plate, the plate vibrations due to the external disturbances, and the displacements of the plate, respectively.

Figure 8.26 yields

$$\mathbf{Y}(z) = \mathbf{P}_0(z)X(z) + \mathbf{G}(z)[\mathbf{C}_V(z)X(z) + \mathbf{C}_P(z)\mathbf{R}(z)] \quad (8.36)$$

where

$\mathbf{C}_V(z)$, $\mathbf{C}_P(z)$, $\mathbf{G}(z)$, and $\mathbf{P}_0(z)$ are z-domain transfer functions

$X(z)$ and $\mathbf{R}(z)$ are the z-transforms of X and $\mathbf{R}$, respectively

This equation indicates that the output of the controlled structure relies on the external disturbances (causing structural vibration), the commands (prescribing the required positions),

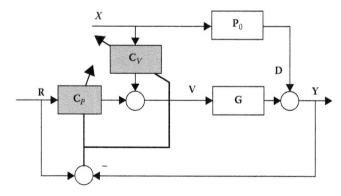

FIGURE 8.26 Adaptive control of positioning and vibration.

the dynamics of the plate, and the controllers. Defining the positioning error $\mathbf{E}(t) = \mathbf{Y}(t) - \mathbf{R}(t)$, then

$$\mathbf{E}(z) = \mathbf{Y}(z) - \mathbf{R}(z) = \mathbf{P}_0(z)X(z) + \mathbf{G}(z)\mathbf{C}_V(z)X(z)$$

$$+ [\mathbf{G}(z)\mathbf{C}_P(z) - \mathbf{I}]\mathbf{R}(z) \qquad (8.37)$$

According to Equation 8.37, it can be easily observed that the positioning error will be zero if the vibration/positioning controllers can be adaptively adjusted to $\mathbf{C}_V(z) = -\mathbf{G}^{-1}(z)\mathbf{C}_P(z)\mathbf{P}_0(z)$ and $\mathbf{C}_P(z) = \mathbf{G}^{-1}(z)$.

The objective functions $J_c(k)$ for adapting $\mathbf{C}_V$ and $\mathbf{C}_P$ as well as the adaptations are as follows:

$$J_c(k) = \frac{1}{2}\mathbf{E}^T(k)\mathbf{E}(k) \qquad (8.38)$$

$$\mathbf{C}_P(k+1) = \mathbf{C}_P(k) + \mu_P \frac{\partial J_c(k)}{\partial \mathbf{C}_P} \qquad (8.39)$$

$$\mathbf{C}_V(k+1) = \mathbf{C}_V(k) + \mu_V \frac{\partial J_c(k)}{\partial \mathbf{C}_V} \qquad (8.40)$$

In this adaptation, the information of the command $\mathbf{R}$ and the disturbance X are needed. $\mathbf{R}$ is always known because it is the desired position, but X sometimes is not available. However, based on the vibration features of a controlled structure, some methods may be employed to reconstruct a signal used as X in this control strategy. For instance, a sinusoidal signal can be used as a reference for suppressing a harmonic vibration with the same frequency or for suppressing a random vibration of which the dominant frequency is close to the frequency of the sinusoidal signal (Peng et al. 2003) or an internal model (Vaudrey et al. 2003, Ma and Ghasemi-Nejhad 2006) or disturbance observer (Bohn et al. 2003) can be employed to observe the vibration response due to unknown disturbances, and then the vibration response due to unknown disturbances can be used as X in this control strategy.

To implement the adaptive input shaping and adaptive control, the dynamics of the controlled plant has to be known or identified. Previous studies have demonstrated that the adaptive algorithm is stable if the phase difference between the controlled plant and its estimated model is less than 90° (Elliott 2001, Vaudrey et al. 2003).

8.3.2.4 Experiments

The experiments are conducted for the SPPVS of the smart composite plate. The sampling rate of the control system is 1 kHz.

The lengths of the FIR filters $\mathbf{W}_A(k)$ and $\mathbf{W}_B(k)$ in the AISU are 200. The convergence factors μ_A in Equation 8.33 and μ_B in Equation 8.35 are both 0.0001, and the reference signal for the AISU is a swept-sinusoidal signal from 0 to 100 Hz. The final values of $\mathbf{W}_A(k)$ and $\mathbf{W}_B(k)$ for the AISU are shown in Figure 8.27. The amplitude-frequency responses for the combination of the smart plate and the AISU, $\mathbf{G}$, is shown in Figure 8.23 by dash lines, indicating that (1) the amplitudes at the first two natural frequencies are dramatically reduced after introducing the

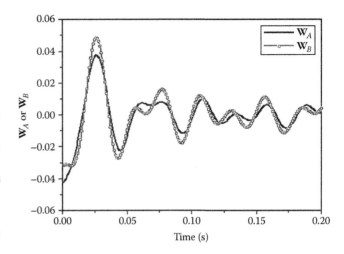

FIGURE 8.27 Impulse responses of $\mathbf{W}_A$ and $\mathbf{W}_B$.

AISU, with 27.7 dB reduction for the first mode and 13.15 dB reduction for the second mode on average and (2) at the low frequency range, the diagonal elements of G and H, that is, $G_{11}(z)$ and $G_{22}(z)$ in Equation 8.31 as well as $H_{11}(z)$ and $H_{22}(z)$ in Equation 8.29, are almost the same, but the differences of the off-diagonal elements, that is, $G_{12}(z)$ and $G_{21}(z)$ in Equation 8.31 as well as $H_{12}(z)$ and $H_{21}(z)$ in Equation 8.29, are large. These characteristics demonstrate that the AISU has concurrently reduced the bending and torsional vibration modes and weakened the coupling degree between the control channels.

After introducing the AISU, the multichannel adaptive control strategy can be applied to the positioning and vibration controls of the smart composite plate. Once again, it should be noted that the controlled plant G here is the combination of the smart composite plate H and the AISU W. The FIR filter lengths of adaptive vibration and positioning controllers are set as 100 and the convergence factors μ_P in Equation 8.39 and μ_V in Equation 8.40 are selected as -0.001.

First, the VS capability of the adaptive vibration controller with or without the AISU is investigated. The external disturbance is sinusoidal at the first modal frequency of the smart composite plate (i.e., 20.5 Hz). Figure 8.28 shows the vibration responses of the plate from the two laser displacement sensors (LDSs), where both the adaptive vibration controllers start at 10 s, demonstrating that the adaptive vibration controller with the AISU performs much better than the adaptive vibration controller without the AISU. It takes about 16 s for the adaptive vibration controller with the AISU to suppress the vibration amplitude from 0.62 V for the LDS1 and 0.5 V for the LDS2 to a tolerance of 0.005 V. However, for the adaptive vibration controller without the AISU, the vibration amplitude is still larger than that tolerance after the controller has worked for 50 s.

Second, the positioning control capability of the adaptive positioning controller with or without the AISU is investigated. The two measurement points A and B of the LDSs in Figure 8.21 are expected to move in the same direction and the desired profile at each measurement point is a square wave at 0.1 Hz with amplitude of 0.2 V (which is corresponding to 0.51 mm), that is,

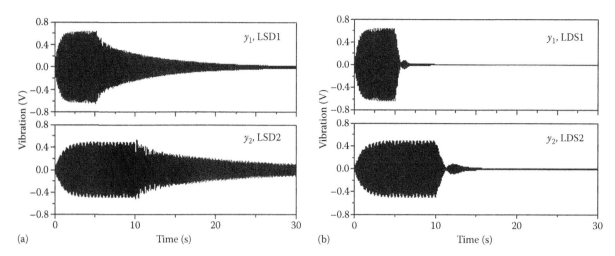

FIGURE 8.28 Vibration suppression (sinusoidal external disturbance at 20.5 Hz). (a) Without AISU; (b) with AISU.

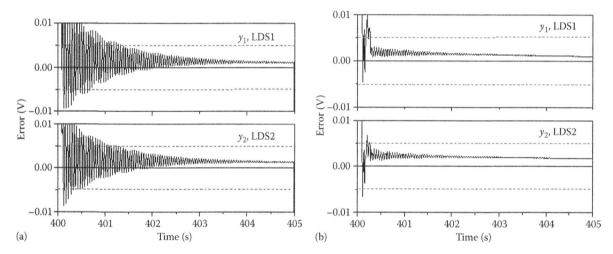

FIGURE 8.29 Positioning errors (positioning control only, no external disturbance): (a) without AISU, (b) with AISU.

R commands the smart plate be in a bending positioning. Figure 8.29 shows the positioning errors of the two measurement points in the bending positioning for the adaptive positioning controller with or without the AISU. For a 0.005 V positioning error tolerance, the maximum settling times of LDS1 and LDS2 for the adaptive positioning controllers with and without the AISU are 0.3 and 1.36 s, respectively. Nevertheless, the steady-state errors for both controllers are 0.0013 V on average. In addition, the adaptive positioning controller without the AISU causes larger free vibration of the plate when the command changes.

Third, the simultaneous precision positioning and vibration control capabilities of the adaptive controller with or without the AISU is investigated. Figure 8.30 shows the positioning errors of the smart composite plate subject to a sinusoidal external disturbance at its first natural frequency of 20.5 Hz. The adaptive controller with the AISU controls the plate positions into the 0.005 V positioning error tolerance within 1.34 s, but for the adaptive controller without the AISU, the positioning errors are larger than 0.1 V even after 5 s and die out very slowly. The steady-state errors for the adaptive controller with the

AISU are 0.0014 V on average. The average residual vibration amplitudes for the adaptive controllers with and without the AISU are 0.0003 and 0.09 V, respectively, demonstrating that the adaptive controller with the AISU can effectively reduce the vibration due to the motion of the plate as well as the external disturbance.

Tables 8.4 and 8.5 summarize the results of the VS and the bending positioning, respectively. In general, the adaptive controller with or without the AISU produces the similar positioning steady-state errors and rise times, but the adaptive controller with the AISU can provide smaller settling times, overshoots, and residual vibrations than the adaptive controller without the AISU does. The excellent performance of the adaptive controller with the AISU is attributed to the damping enhancement of the AISU to the combined plant consisting of the AISU and the controlled smart composite plate, and this enhancement then results in the alleviation of the free vibration, due to the motion of the plate and the fast convergence of the adaptive controller.

A torsional positioning is also tested. In the torsional positioning, the two measurement points A and B in Figure 8.21 are

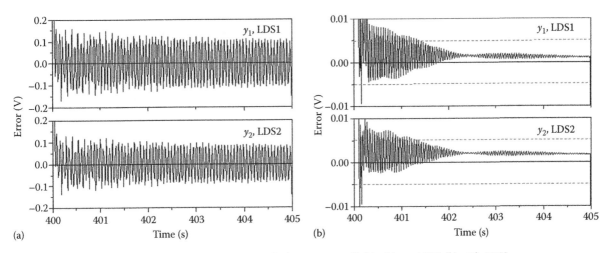

FIGURE 8.30 Positioning errors (simultaneous positioning and vibration control): (a) without AISU, (b) with AISU.

TABLE 8.4 Vibration Suppression at the First Modal Frequency, Bending

Vibration Suppression	Adaptive Vibration Controller without AISU		Adaptive Vibration Controller with AISU	
	LDS1	LDS2	LDS1	LDS2
Time needed to suppress the vibration to 0.005 V tolerance (s)	>50	>50	15	17

TABLE 8.5 Simultaneous Positioning and Vibration Control (in Bending Positioning)

Bending Positioning		Adaptive Vibration Controller without AISU		Adaptive Vibration Controller with AISU	
		LDS1	LDS2	LDS1	LDS2
Positioning control	Steady-state error (V)	0.0012	0.0014	0.0011	0.0016
	Settling time (s)	1.36	1.34	0.3	0.28
	Rise time (s)	0.08	0.1	0.09	0.09
	Overshoot	4.5%	4.5%	2.3%	3.1%
Positioning and vibration controls (vibration at first modal frequency, bending)	Steady-state error (V)	>0.005	>0.005	0.001	0.0017
	Settling time (s)	>5	>5	1.34	1.35
	Rise time (s)	0.07	0.07	0.08	0.08
	Overshoot	70%	85%	5.5%	4.5%
	Residual vibration (V)	0.1	0.085	0.0003	0.0003

TABLE 8.6 Simultaneous Positioning and Vibration Control (in Torsional Positioning)

Torsional Positioning		Adaptive Vibration Controller without AISU		Adaptive Vibration Controller with AISU	
		LDS1	LDS2	LDS1	LDS2
Positioning control	Steady-state error (V)	0.0015	0.0015	0.0016	0.0012
	Settling time (s)	0.36	0.43	0.26	0.23
	Rise time (s)	0.11	0.094	0.087	0.1
	Overshoot	2%	2%	5.7%	0.7%
Positioning and vibration controls (vibration at first modal frequency, bending)	Steady-state error (V)	>0.005	>0.005	0.0016	0.001
	Settling time (s)	>5	>5	0.26	0.21
	Rise time (s)	0.09	0.09	0.09	0.09
	Overshoot	5.5%	5%	5.5%	0.35%
	Residual vibration (V)	0.0054	0.007	0.0003	0.0004
Positioning and vibration controls (vibration at second modal frequency, torsion)	Steady-state error (V)	0.0015	0.001	0.0015	0.001
	Settling time (s)	0.29	0.22	0.26	0.25
	Rise time (s)	0.09	0.09	0.09	0.1
	Overshoot	8%	7.8%	5%	1.5%
	Residual vibration (V)	0.0004	0.0005	0	0

expected to move in opposite directions, and the desired profile remains the square wave used in the bending positioning case. Table 8.6 summarizes the experimental results. Basically, the results are similar to those in the bending positioning. However, the overshoots and residual vibrations are better than those in the bending positioning, even for the adaptive controller without the AISU. This is attributed to the fact that the torsional positioning of the plate does not induce the bending vibration of the plate and the amplitude of the plate at the torsional modal frequency is much less than that at the bending modal frequency (see Figure 8.23).

The above investigations are based on the assumption that the external disturbance X is known. However, X is not always available. An internal model (Ma and Ghasemi-Nejhad 2006) is employed here to reconstruct the uncontrolled vibration of the smart composite plate. Next, this reconstructed uncontrolled

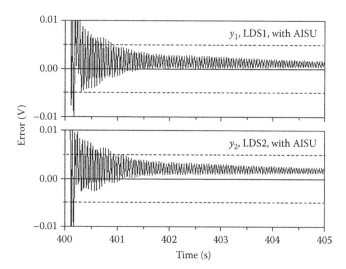

FIGURE 8.31 Positioning errors (simultaneous positioning and vibration control, disturbance unknown).

vibration response is used as the reference signal *X* for VS. The use of the internal model results in the change of the control type from feedforward to feedback (Vaudery et al. 2003). Figure 8.31 shows the positioning errors of the two measurement points (A and B in Figure 8.21) of the smart composite plate for the bending positioning with the external disturbance, depicting that for the 0.005 V positioning error tolerance, the settling times of LDS1 and LDS2 for the adaptive positioning/vibration controller with the AISU are 0.92 and 1.08 s, respectively, and the steady-state errors are about 0.001 and 0.002 V, respectively. The difference of the residual vibrations between Figures 8.30b and 8.19 are insignificant.

To test the effects of the dynamic model errors of the smart plate on the control performance, a dynamic model with 2.5% discrepancy in the first two modal frequencies of the smart plate is employed in the adaptive algorithm. Figure 8.32 shows the

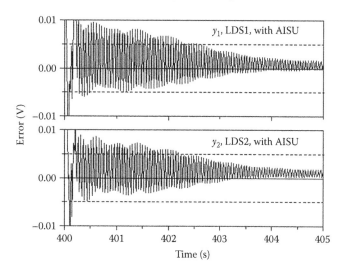

FIGURE 8.32 Positioning errors (simultaneous positioning and vibration control, disturbance unknown, considering a discrepancy in model).

positioning errors using the internal model technique and the inaccurate dynamic model in the adaptive algorithm. For the 0.005 V positioning error tolerance, the settling times of LDS1 and LDS2 for the adaptive positioning/vibration controllers with the AISU are 2.61 and 2.25 s, respectively and the steady-state errors remains 0.001 and 0.002 V, respectively. The VS is not as fast as that in Figure 8.31; however, the positioning errors are still satisfactory. In addition, these investigations indicate that the AISU can help reduce the effects of the dynamic model discrepancy of the smart composite plate on the control performance by decreasing the phase lag between the dynamic model used in the adaptive algorithm and the real plate. For instance, the phase lag is 70° at 20.5 Hz if no AISU is used, and it becomes 65° with the AISU.

8.4 Applications

As mentioned in Section 8.1, the nanopositioning technology developed has board applications. Figure 8.33 shows two applications. The first one is a smart platform with SPPVS capabilities (Ghasemi-Nejhad and Doherty 2002, Ma and Ghasemi-Nejhad

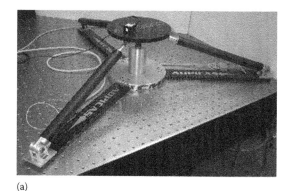

(a)

(b)

FIGURE 8.33 Nanopositioning applications: (a) smart platform, (b) satellite thrust vector control.

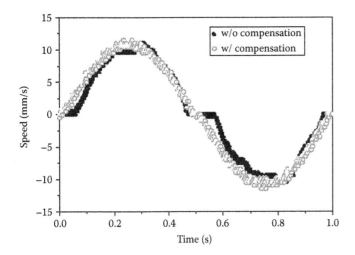

FIGURE 8.34 Actuator velocity history.

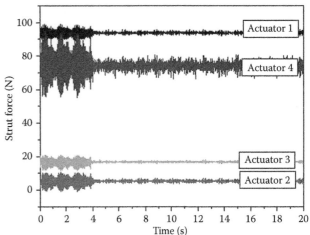

FIGURE 8.36 Loads time history.

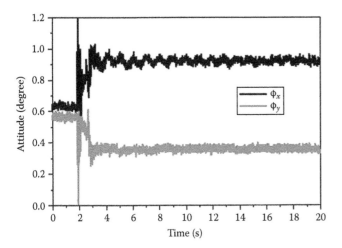

FIGURE 8.35 Satellite rotations.

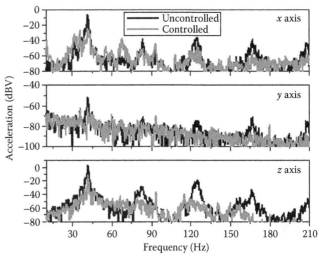

FIGURE 8.37 Satellite vibrations.

2005b) and the second one is satellite thrust vector control utilizing the smart platform (Ma and Ghasemi-Nejhad 2007). The smart platform integrates the linear actuator, the smart composites, sensors, and controllers.

Figure 8.34 shows the actuator velocity history for the cases with or without nonlinearity compensation. Without the compensation, the actuator velocity history is discontinuous, especially at the zero, minimum, and maximum velocities. With the compensation, the velocity history continuity is improved significantly and the velocity curve becomes much smoother. Figure 8.35 shows the satellite attitudes change about 0.3° while the actuators drive the platform to move 2° about the two body axes (ϕ_x and ϕ_y). Figure 8.36 shows the loads measured by the load cells in the actuators. Without the piezoelectric stack actuators, the dynamic forces (peak-peak) in the three actuators and the central supports are 11.16, 11.45, 9.37, and 39.4 N, respectively and with the piezoelectric stack actuators, these dynamic forces reduce to 2.15, 1.78, 1.38, and 7.02 N, respectively. Figure 8.37 is the acceleration spectra for the cases of un-acting and acting piezoelectric stack actuators and demonstrates significant VS at the dominant frequencies.

8.5 Conclusions

Nanopositioning systems and applications are discussed including multifunctional linear actuator systems and smart composites with piezoelectric patch systems. The multifunctional linear actuator nanopositioning system has super-precision, large stroke, and SPPVS capabilities. It is an integration of advanced electro-mechanical technology, smart materials technology, sensing technology, and control technology. The advanced control strategy greatly compensates the hysteresis characteristics such as backlash and/or dead zone and enables the excellent performance of the actuator. The smart composites nanopositioning systems utilize the advanced features of piezoelectric patches to achieve SPPVS functions and several control strategies such as AFC, AIMC, and HAC for SISO systems and adaptive input shaping control for MIMO systems. Bending and/or torsional motions are demonstrated under vibrational environments. All these nanopositioning systems are experimentally tested and excellent results are demonstrated. The applications of these nanopositioning systems are also explored in the precision positioning and VS of a smart composite platform and thrust vector control of satellites.

Acknowledgments

The authors acknowledge the financial support of the Office of Naval Research for the Adaptive Damping and Positioning using Intelligent Composite Active Structures (ADPICAS) project under the government grant numbers of N00014-00-1-0692 and N00014-05-1-0586, and the program officer Dr. Kam W. Ng.

References

ACX. 2002. *QuickPack*® Piezoelectric Actuator. Cambridge, MA: Active Control eXperts, Inc.

ANSYS. 2006. *ANSYS User's Manual*. Canonsburg, PA.

Baxter, L.K. 1996. *Capacitive Sensors: Design and Applications*. Piscataway, NJ: IEEE Press.

Bohn, C., A. Cortabarria, V. Haertel, and K. Kowalczyk. 2003. Disturbance-observer-based active control of engine-induced vibrations in automotive vehicles. *Proc. SPIE Conf. Smart Mater. Struct.* 5049:565–576.

Borodin, S., J.-D. Kim, H.-J Kim, P. Vasiljev, and S.-J. Yoon. 2004. Nano-positioning system using linear ultrasonic motor with shaking beam. *J. Electroceram.* 12(3):169–173.

CCC. 2002. *Product Data*. Billerica, MA: Continuum Photonics, Inc.

Culpepper, M.L. and G. Anderson. 2004. Design of a low-cost nano-manipulator which utilizes a monolithic, spatial compliant mechanism. *Precision Eng.* 28(4):469–482.

Doherty, K.M. and M.N. Ghasemi-Nejhad. 2005. Performance of an active composite strut for an intelligent composite modified Stewart platform for thrust vector control. *J. Intell. Mater. Syst. Struct.* 16(4):335–354.

Elliott, S. 2001. *Signal Processing for Active Control*. London, U.K.: Academic Press.

Ghasemi-Nejhad, M.N. and K.M. Doherty. 2002. Modified Stewart platform for spacecraft thruster vector control. In *Proceedings of the Adaptive Structures, Aerospace Division, ASME International Mechanical Engineering Congress and Exposition*, New Orleans, IMECE2002-39032.

Ghasemi-Nejhad, M.N., R. Russ, and S. Pourjalali. 2005. Manufacturing and testing of active composite panels with embedded piezoelectric sensors and actuators. *J. Intell. Mater. Syst. Struct.* 16(4):319–334.

Hariharan, P. 2003. *Optical Interferometry*, 2nd edn. San Diego, CA: Academic Press.

Hemsel, T. and J. Wallaschek. 2000. Survey of the present state of the art of piezoelectric linear motors. *Ultrasonics* 38(1–8):37–40.

Jordan, S., B. Lula, and S. Vorndran. 2007. Nanopositioning: Racing forward. http://www.photonics.com/content/handbook/2007/positioning/85568.aspx

Kim, J., J.-D. Kim, and S.-B Choi. 2002. A hybrid inchworm linear motor. *Mechatronics* 12:525–542.

Kim, S.-H., I.-H. Hwang, K.-W. Jo et al. 2005. High-resolution inchworm linear motor based on electrostatic twisting microactuators. *J. Micromech. Microeng.* 15:1674–1682.

Ma, K. 2003. Vibration control of smart structures with bonded PZT patches: Novel adaptive filtering algorithm and hybrid control scheme. *Smart Mater. Struct.* 12:473–482.

Ma, K. and M.N. Ghasemi-Nejhad. 2002a. Development of smart composite structural systems with simultaneous precision positioning and vibration control. *Proc. JSME/ASME Int. Conf. Mater. Process.* 1:396–340.

Ma, K. and M.N. Ghasemi-Nejhad. 2002b. Hybrid adaptive control of intelligent composite structures with simultaneous precision positioning and vibration suppression. *Proc. SPIE Model. Signal Process. Control* 4693:13–24.

Ma, K. and M.N. Ghasemi-Nejhad. 2003. Adaptive internal model control of intelligent composite structures with simultaneous precision positioning and vibration suppression. *Proc. SPIE Model. Signal Process. Control* 5049:252–263.

Ma, K. and M.N. Ghasemi-Nejhad. 2004. Frequency-weighted hybrid adaptive control for simultaneous precision positioning and vibration suppression of intelligent composite structures. *Smart Mater. Struct.* 13(5):1143–1154.

Ma, K. and M.N. Ghasemi-Nejhad. 2005a. Adaptive simultaneous precision positioning and vibration control of intelligent structures. *J. Intell. Mater. Syst. Struct.* 16:163–174.

Ma, K. and M.N. Ghasemi-Nejhad. 2005b. Precision positioning of a parallel manipulator for spacecraft thrust vector control. *AIAA J. Guid. Control Dyn.* 28(1):185–188.

Ma, K. and M.N. Ghasemi-Nejhad. 2005c. Simultaneous precision positioning and vibration suppression of reciprocating flexible manipulators. *Smart Struct. Syst.* 1(1):13–28.

Ma, K. and M.N. Ghasemi-Nejhad. 2006. Adaptive precision positioning of smart structures subject to external disturbances. *Intl. J. Mechatronics* 16:623–630.

Ma, K. and M.N. Ghasemi-Nejhad. 2007. Smart composite platform for satellite thrust vector control and vibration suppression. In *Progress in Smart Materials and Structures Research*, P.L. Reece, editor, pp. 151–202. New York: Nova Science Publishers, Inc.

Ma, K. and M.N. Ghasemi-Nejhad. 2008. Adaptive control of flexible active composite manipulators driven by piezoelectric patches and active struts with dead zones. *IEEE Trans. Control Syst. Technol.* 16(5):897–907.

Macro Sensors. 2003. LDVT basic, *Technical Bulletin 0103*. Pennsauken, NJ: Macro Sensors.

Micromega Dynamics. 2008. http://www.micromega-dynamics.com

Mori, S., M. Furuya, A. Naganawa, Y. Shibuya, G. Obinata, and K. Ouchi. 2007. Nano-motion actuator with large working distance for precise track following. *Microsyst. Technol.* 13(8–10):873–881.

Mrad, R.B., A. Abhari, and J. Zu. 2003. A control methodology for an inchworm piezomotor. *J. Mech. Syst. Signal Process.* 17:457–471.

Newscale Technologies. 2008. http://www.newscaletech.com/nanopositioning.html

Nyce, D.S. 2003. *Linear Position Sensors: Theory and Application.* Hoboken, NJ: John Wiley & Sons Inc.

Peng, F.J., M. Gu, and H.-J. Niemann. 2003. Sinusoidal reference strategy for adaptive feedforward vibration control: Numerical simulation and experimental study. *J. Sound Vib.* 265:1047–1061.

Physik Instrumente GmbH. 2008. Micropositioning, nanopositioning and nanoautomation Karlsruhe, Germany.

Spanner, K. 2006. Survey of the various operating principles of ultrasonic piezomotors. *Proc. of 2006 Actuator.*

Suleman, S.B., D. Waechter, R. Blacow et al. 2001. Flexural brake mechanism for inchworm actuator. *Proceedings of the Canada-US CanSmart Workshop on Smart Materials and Structures.* Montreal, Canada, October, 2001, pp. 125–136.

Tao, G. and P.V. Kokotovic. 1994. Adaptive control of plants with unknown dead zones. *IEEE Trans. Autom. Control* 39:59–68.

Thompson, S. and J. Loughlan. 2000. Control of the post-buckling response in thin composite plates using smart technology. *Thin-Wall. Struct.* 36:231–263.

Vaudrey, M.A., W.T. Baumann, and W.R. Saunders. 2003. Stability and operating constraints of adaptive LMS-based feedback control. *Automatica* 39:595–605.

Wang, L.X. 1994. *Adaptive Fuzzy Systems and Control: Design and Stability Analysis.* Englewood Cliffs, NJ: Prentice Hall.

Wang, X.S., C.Y. Su, and H. Hong. 2004. Robust adaptive control of a class of nonlinear systems with unknown dead zone. *Automatica* 40:407–413.

Wang, Y.-C., C.-J. Lin, C.-J. Chen, and H.-C. Liou. 2008. Stage equipped with single actuator for nano-positioning in large travel range. *Key Eng. Mater.* 364–366:768–772.

Window, A.L. 1992. *Strain Gauge Technology,* 2nd edn. London, U.K.: Elsevier Applied Science.

Yeh, R., S. Hollar, and K.S.J. Pister. 2002. Single mask, large force, and large displacement electrostatic linear inchworm motors. *J. Microelectromech. Syst.* 11:330–336.

Zhang, B. and Z. Zhu. 1997. Developing a linear piezomotor with nanometer resolution and high stiffness. *IEEE/ASME Trans. Mechatron.* 2:22–29.

II

Nanoporous and Nanocage Materials

9
Nanoporous Materials

Zheng-Ming Wang
National Institute of Advanced Industrial Science and Technology

9.1 Introduction

There is no a group of material other than nanoporous materials that has attracted remarkable concerns and found tremendous importance widespread in both fundamental research and industrial applications and with a time span from the old times through the present age (Yang 2003). Nanoporous materials abounding both naturally and artificially are indispensable, particularly in the contemporary fields of nanoscience and nanotechnology, which deal with nanomaterials endowed with intriguing size-determined physical and chemical properties. Nanomaterials are those having at least one spatial dimension in the size range of 1–1000 nm and exhibiting peculiar properties as compared with those of the bulk owing to the tunable quantum size effect (Ozin and Arsenault 2005). Sometimes nanoporous materials are considered as a subset of nanomaterials. However, in contrast to the generally meant nanomaterials whose target is size scaling of the bulk materials, the major point aimed at nanoporous materials is the pore, the empty inverse of the bulk materials rather than the bulk materials themselves, which surround or include the pores. Nanoporous materials display their functions by the play of pores such as adsorbing, accommodating, sieving, and separating molecules, which makes them possible to be widely used for applications as adsorbents, catalysts, catalyst supports, and for biological molecular isolation and purification. Therefore, nanoporous materials are different from the general nanomaterials in the sense that the functions of nanoporous materials are also greatly influenced by the properties of nanopores—their sizes, shapes, and amounts—besides the surface and bulk properties of the framework materials. Consequently, unique terminologies and classifications of nanoporous materials are derived. On the other hand, the role of nanoporous materials as the host to accommodate guest molecules brings about a novel nanospace science in which the guest molecules confined inside the nanopore space show extraordinary properties in melting, ordering, light absorption/emitting, reacting (via unique routes), and so forth, differing from those of their bulk counterparts (Kaneko 2000, Hupp and Poeppelmeier 2005, Ozin and Arsenault 2005).

In this chapter, a deep insight into the definition and background of pores and nanopores will be discussed in Section 9.2 and the classifications, typical kinds, and properties of nanoporous materials will be summarized in Section 9.3. The main target for the research of nanoporous materials is how to determine not only their porous properties but also their bulk and surface properties. The typical approaches of these techniques will be introduced in Section 9.4. Finally, a future prospective of nanoporous materials and the related research will be outlined in Section 9.5.

9.2 Definition of Pores and Nanoporous Material

The classic meaning of pores is just voids (holes) in a solid material with various kinds of shapes such as cylinders, balls, slits, hexagons, and so on, which either exist separately or connect to form a network of higher dimensions (Figure 9.1). They can be straight, curved, twisted, different sizes within the same channel, open in one or two ends, or completely closed. The scientific concept of pores can be traced back to some concepts in colloid science, such as one type of dispersion (gas–solid interface), solid foam (Polarz and Smarsly 2002), and pores are currently defined according to phenomena in gas–solid interfaces. The international scientific community (by the International Union of Pure and Applied

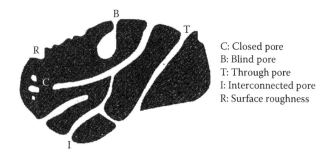

FIGURE 9.1 Schematic description of various types of pores. (From Rouquerol, F. et al., *Adsorption by Powders & Porous Solids*, Academic Press, San Diego, CA, 1999. With permission.)

C: Closed pore
B: Blind pore
T: Through pore
I: Interconnected pore
R: Surface roughness

Chemistry, IUPAC) reached a consensuses on the classification of pores in the context of physisorption according to their sizes (Sing et al. 1985):

1. Macropores: pores with widths exceeding about 50 nm
2. Mesopores: pores with widths between 2 and 50 nm
3. Micropores: pores with widths not exceeding about 2 nm

There is no a strict definition regarding pore width. It is naturally the diameter of a cylindrical pore, the slit width of a slit-shaped pore, but it is not well defined for pores without homogenous openings such as squares, hexagons, ellipses, and so forth.

The classifications of pores are based on the adsorbate–adsorbent (porous material) interaction in the physisorption of adsorptive molecules (Figure 9.2). With respect to chemisorption, where adsorbate molecules are chemically (irreversibly) bonded on the surface of a porous material, physisorption is a general phenomenon with a relatively low degree of specificity, zero activation energy, and reversibility whose total interaction energy, E_0, is determined by a van der Waals-type interaction and other types of weak interactions. For details, E_0 at very low surface coverage can be expressed in the form of the sum (Rouquerol et al. 1999)

$$E_0 = E_D + E_R + E_P + E_{F\mu} + E_{\dot{F}Q} \qquad (9.1)$$

in which E_D and E_R are the contributions from the van der Waals-type interaction, i.e., dispersion and repulsion energies, respectively and the terms E_P, $E_{F\mu}$, and $E_{\dot{F}Q}$ are the contributions of the polarization, field-dipole, and field gradient-quadrupole energies, respectively. E_D and E_R are called nonspecific

TABLE 9.1 Nonspecific and Specific Adsorption Energy

Kinds of Interaction	Subsystem		Adsorption Potential
	Adsorbent (Solid)	Adsorbate	
E_D long distance attraction		Atom-atom	$E_D(r) = -C/r^6$
E_R Shot range repulsion		Atom-atom	$E_R(r) = B/r^m$ $(m = 12)$
E_P	Field F	Field-atom (ion)	$E_P = -(1/2)\alpha F^2$
$E_{F\mu}$		Field-dipole	$E_{F\mu} = -F\mu\cos\theta$ μ: dipole moment
$E_{\dot{F}Q}$	Field gradient $\dot{F}$	Field-quadruple	$E_{\dot{F}Q} = -\dot{F}Q$ Q: quadruple moment

energies and are universal for all kinds of adsorbents and adsorbates whether they are polar or not. By contrast, E_P, $E_{F\mu}$, and $E_{\dot{F}Q}$ are specific energies whose plays require at least one polar term within adsorbent and adsorbate. Table 9.1 lists the kinds of nonspecific and specific adsorption energies. The sum of E_D and E_R is the famous Lennard-Jones type potential $\varepsilon(r)$, which is the total interaction energy always exerted between an adsorbate molecule and an atom in the bulk of porous materials:

$$\varepsilon(r) = \frac{B}{r^{12}} - \frac{C}{r^6} \qquad (9.2)$$

Here, B is a constant and C is a parameter related to the polarizabilities and magnetic susceptibilities of both adsorbate and adsorbent atoms. According to the principle of additivity of the pairwise interactions, the integration of $\varepsilon(r)$ among whole atoms

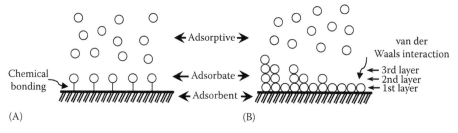

Chemical bonding — Adsorptive — Adsorbate — Adsorbent — van der Waals interaction — 3rd layer — 2nd layer — 1st layer

(A)   (B)

FIGURE 9.2 Description of adsorption terminology. (A) Chemisorption. (B) Physisorption.

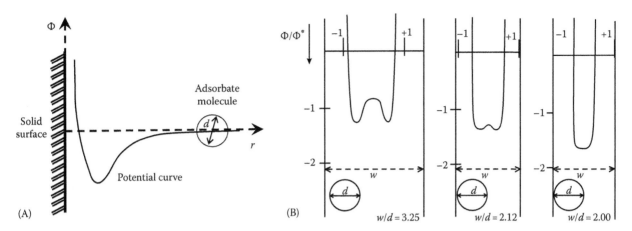

FIGURE 9.3 (A) Potential curve on a flat solid surface and (B) deepening of potential "well" with decreasing micropore sizes. (From Rouquerol, F. et al., *Adsorption by Powders & Porous Solids*, Academic Press, San Diego, CA, 1999. With permission.)

in the bulk of the solid porous material gives the total potential energy, Φ

$$\Phi = \sum \varepsilon_j(r_j), \qquad (9.3)$$

which is the total potential energy that an adsorbate molecule receives in front of a solid surface. As shown in Figure 9.3A, the potential energy curve shows an energy minimum that is the characteristic position at which an adsorbate molecule is most energetically stable. In a porous system, the influence of potential energy from the opposite walls becomes significant or negligible depending on the pore width. It is from the potential energy curve that the adsorption mechanism in a pore system can be roughly understood as follows:

1. In a macropore, the pore wall functions as a flat surface without the influence of the wall on the other side. At first, adsorbate molecules form a monolayer at the position of potential minimum on the pore wall of either side and further adsorption leads to the formation of multilayers beyond the monolayer in which adsorbate molecules receive a comparatively small influence from the pore wall (either from that on the other side). Generally, the attraction effect from the solid surface is limited only within several layers of adsorbate molecules.
2. In a mesopore, the effect of the opposite wall becomes important. After the formation of adsorbate layers with 2–3 molecular thickness on both walls, further adsorption induces attractions between adsorbate molecules on both walls, leading to a sudden condensation of liquid-like adsorbate molecules inside the pores. This phenomenon occurs at a gas phase pressure much lower than its vapor pressure and is generally called capillary condensation.
3. In a micropore, the effect of the opposite wall becomes much more significant. The potential curves from both sides of pores overlap each other due to the close proximity of the pore walls, which produces a deep potential

"well" with an enhanced strength of adsorbate–adsorbent interaction (Figure 9.3B) (Everett and Powl 1976). The enhanced potential field triggers volume filling (micropore filling) of adsorbate molecules from very low gas phase pressure.

Thus, the adsorption mechanism in a pore system determines the classification of pores. It is noteworthy, however, that the boundary pore widths for dividing macropores, mesopores, and micropores are not absolute because potential energies depend not only on pore width but also on pore shape, the nature of porous material, and the size and nature of adsorbate molecules (Sing et al. 1985). Researchers also propose further dividing micropores into the following two categories:

- Ultramicropores: pore widths less than 0.6 nm
- Supermicropores: pore widths between 0.6 and 2 nm

In ultramicropores, adsorption experiences a primary process, which involves the entry of individual adsorbate molecules into very narrow pores, whereas in supermicropores, cooperative adsorption takes place, which involves the interaction between adsorbate molecules inside pores. However, further research is needed for reaching an international consensus on the classification (Sing et al. 1985, Kaneko 1994).

On the other hand, there is not yet a general agreement by an international scientific community like IUPAC with regard to the definition of nanopores. They are generally considered as pores with widths between 1 and 100 nm (Lu and Zhao 2004). Some researchers extend the margin up to 1000 nm or below 1 nm. Nanoporous materials are considered to have not only nanosized pores but also sufficient porosities (the ratio of volume occupied by pores over that by the bulk material), generally surpassing 0.4. Consequently, nanoporous materials actually encompass one part of microporous materials, all mesoporous materials, and one part of macroporous materials. That means that any kind of adsorption mechanism can occur in a nanoporous material.

9.3 Classification, Kinds, and Properties of Nanoporous Materials

Nanoporous materials can be classified either by the properties of pores or by the properties of bulk materials (network or framework materials besides pores) that include or surround pores.

According to pore sizes, nanoporous materials are classified into microporous materials, mesoporous materials, and macroporous materials, which contain only micropores, mesopores, and macropores, respectively (pore classifications as mentioned in Section 9.2). Table 9.2 shows the typical examples of microporous, mesoporous, and macroporous materials. Most of the nanoporous materials don't contain only one type of pore but contain pore systems hybridized with micropores, mesopores, or macropores. Figure 9.4 outlines the pore size distributions of typical nanoporous materials.

When considering the formation structure of pores, nanoporous materials can be classified into two categories: those with pores inside particles (PIP) and those with pores outside particles (POP), i.e., pores being the interstitial spaces between particles (Figure 9.5) (Kaneko 1994). The typical examples for the former materials are zeolite (Van Bekkum et al. 2001), activated carbon (AC) (Marsh and Rodriguez-Reinoso 2006), metal organic framework materials (MOFs) (Eddaoudi et al. 2002, Kitagawa and Matsuda 2007), and so forth. The typical examples for the later materials are various nanoporous oxide materials such as silica and alumina. Nanoporous materials can also be divided according to the regularity of their pore structures. Zeolites and the group materials of mobile crystalline material

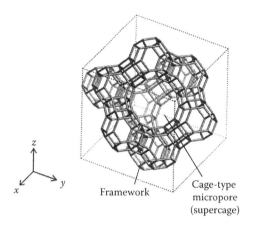

(A) Faujasite-type zeolite

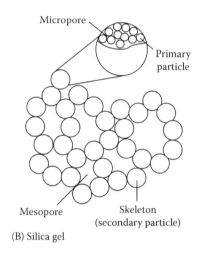

(B) Silica gel

FIGURE 9.5 Typical nanoporous materials with (A) PIP and (B) POP pore structures.

41 (MCM-41) (Kresge et al. 1992) are the typical examples of nanoporous materials with regular pore structures, while pores in AC, silica, alumina, and so forth are not a structure showing any long distance periodicity.

Traditionally, lots of nanoporous materials are named by the species of network or framework materials (the base bulk materials forming pores), for example, silica, alumina, aluminosilicate, carbonaceous adsorbents, and so forth. According to the kinds of network materials, nanoporous materials can be greatly classified into organic and inorganic materials, respectively. As shown in Figure 9.6, the representative organic materials are nanoporous polymeric materials, MOFs, and so forth. Inorganic materials actually cover a large group in nanoporous materials. They can be divided further according to the kinds of bulk materials to (1) inorganic oxide-type materials (including hybrid oxide-type materials), (2) binary/multi compound-type materials other than oxides, (3) carbonaceous materials, and (4) single element (silicon, other metal species, etc.) materials other than carbon (Polarz and Smarsly 2002). The typical ones of inorganic nanoporous materials are also shown in Figure 9.6. Nanoporous materials can also be differentiated by whether the network

TABLE 9.2 Examples of Microporous, Mesoporous, and Macroporous Materials

Microporous materials	Zeolite-like materials, activated carbon fibers (ACF), templated carbon from zeolites, metal organic frameworks (MOFs)
Mesoporous materials	Mesoporous oxides (silica, alumina, zirconia), MCM-41, CMK-1, polymeric materials
Macroporous materials	Porous glass, nanoporous silicon
Hybrid porous materials	Activated carbon (AC), silica gel, pillared clay, nanoporous oxide-bridged carbon nanosheet composites

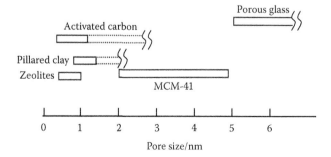

FIGURE 9.4 Pore size distributions of several nanoporous materials.

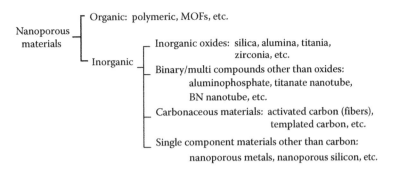

FIGURE 9.6   Classifications of nanoporous materials.

materials are crystalline or not. For example, zeolites are a group of nanoporous materials with both crystalline frameworks and pore regularities (Van Bekkum et al. 2001). MCM-41 has only pore regularities without regular arrangement in a framework structure. Some periodic mesoporous organosilica (PMO) materials present pore regularities with partially regular arrangements in the framework structure (Inagaki et al. 1999). AC possesses neither pore regularity nor regularity in the network structure (Marsh and Rodriguez-Reinoso 2006). Sometimes, nanoporous materials are also characterized by the morphology of the network materials. Pillared clay (Pinnavaia 1983, 1995) and silica-bridged carbon nanosheet composite materials (Chu et al. 2005) are typical examples of nanoporous layered materials. These materials form pores by introducing external pillars or bridges in between layered hosts. Therefore, their pores are an intermediate case of PIP and POP and they present porosity no matter how the host layers are ordered (regularly arranged or not). Table 9.3 summarizes the typical types of nanoporous materials and their properties according to classifications based on network materials.

9.4 Characterization of Nanoporous Materials

9.4.1 Gas Adsorption Method

Nanoporous materials are often characterized by the adsorption of nitrogen or other gases at low temperatures. The purpose is to obtain an adsorption isotherm, which is the relationship between the adsorption amount and equilibrium gas pressure, and then to obtain information on pore types and porosities (specific surface area, pore volume, and pore size) based on the adsorption isotherm. In this section, descriptions on measurement and types of adsorption isotherms and the typical methods of calculating porosities based on adsorption isotherms will be given.

9.4.1.1 Adsorption Isotherms

Adsorption isotherms describe the relationship between the adsorption amount of an adsorptive molecule and the relative equilibrium gas pressure p/p_0 (the ratio of the equilibrium gas pressure p over the saturated vapor pressure p_0 of the adsorptive at the measurement temperature) at a constant temperature. The measurements of adsorption isotherms are generally carried out either by the volumetric method, which measures the difference of gas pressure due to adsorption, or by gravimetric methods that directly measure the weight change due to adsorption. Volumetric and gravimetric adsorption apparatuses are generally equipped with a vacuum system consisting of vacuum lines and a pumping system, gas reservoirs, and gas inlets for storing and introducing gas sources, pressure sensors for measuring the equilibrium pressure and pressure differences or a balance for measuring the weight change of the sample (Rouquerol et al. 1999). The diagrams and descriptions of typical volumetric and gravimetric apparatuses are shown in Figure 9.7. Table 9.4 summarizes the typical gas molecules applied to probe porosities of nanoporous materials.

Adsorption isotherms are very important because they are the first-hand data from which information on porosities of nanoporous materials can be judged. Corresponding to different adsorption mechanisms due to different pore structures, IUPAC recommends six types of adsorption isotherms (Figure 9.8), which have been further extended and refined recently (Figure 9.9) (Rouquerol et al. 1999):

1. Type I: the adsorption isotherm typical of microporous materials. It is characteristic of a sharp uprising in adsorption at a very low p/p_0 range due to the micropore filling mechanism and a long plateau, indicative of a relatively small amount of multi-layer adsorption on the external (open) surface. Generally, there is no adsorption hysteresis for this type of adsorption isotherm. Type I isotherms can be further refined to I_a and I_b types, which differentiate ultramicroporous materials with pores of molecular dimensions in which primary micropore filling takes place at a very low p/p_0 range, and supermicroporous materials with wider micropores in which co-operative micropore filling occurs over a range of a comparatively higher p/p_0 range, respectively.
2. Type II: the adsorption isotherm typical of nonporous, macroporous, and, sometimes, even microporous materials. It generally describes the monolayer–multi-layer adsorption on open and stable external surfaces of

TABLE 9.3 Typical Types and Properties of Nanoporous Materials

	Name	Framework Species (Framework Regularity)	Pore Shape (PIP or POP) (Pore Regularity)	Pore Width (nm)	Surface Area; Porosity
Organic	Polymeric	C, H, etc. (irregular)	Random (PIP) (irregular)	15–55	400–1000 m²/g; >0.6
	Metal organic frameworks (MOFs) or Porous coordination polymers (PCPs) and the similar materials	C, H, N, O, metal ions (regular)	Various types (PIP) (regular)	Several nm	300–5000 m²/g; 0.3–0.7
Inorganic	Zeolite-like materials including various types of aluminosilicate such as Linde-4A/5A, faujasite, mordenite, ZSM-5, silicalite, various kinds of metal-containing silicalite such as titanium silicalite (TS-1, 2), aluminophosphates (ALPO), silicoaluminophosphates (SAPO), cloverite, etc.	Si, Al, O, P, Ga, Ti, V, B, Zn, etc. (regular)	Channel, cage (PIP) (regular)	0.3–1.3	200–800 m²/g; 0.3–0.7
	Silica gel	Si, O (irregular)	Random (POP) (irregular)	2–100	200–750 m²/g; 0.3–0.6
	M41S-like materials including mobile crystalline material MCM-41 and MCM-48, SBA series materials, periodic mesoporous organosilica (PMO) materials, etc.	Si, O (PMO: organic elements) (irregular or partially regular)	Cylinder (PIP) (regular)	2–5	700–1000 m²/g; 0.3–0.6
	Porous glass	Si, O (irregular)	Random (POP) (irregular)	Several 10–10⁴	200 m²/g; >0.3
	Alumina	Al, O (irregular)	Random (POP) (irregular)	1.5–10 or greater	100–300 m²/g; 0.3–0.6
	Titania	Ti, O (irregular)	Random (POP) (irregular)	2–3 (micropores also)	50–400; >0.3
	Zirconia	Zr, O (irregular)	Random (POP) (irregular)	>2 (micropores also)	200–300 m²/g; >0.3
	Inorganic nanotube compounds including titanate nanotube, BN nanotube, etc.	Inorganic species (Ti, B, N), O, H (regular)	Cylinder (PIP)	>3	100–500 m²/g; >0.3
	Activated carbon (AC), Activated carbon fiber (ACF)	C (irregular)	Slit (PIP) (irregular)	0.3–1.3 (mesopore, macropore also)	–3000 m²/g; 0.3–0.6
	Carbon aerogel	C (irregular)	Random (POP) (irregular)	1.5–30	150–800 m²/g; 0.3–0.6
	Novel nanocarbons including carbon nanohorn, carbon nanotube	C (irregular)	Random (POP) (irregular)	Micropore to mesopore	300–1000 m²/g; >0.3
	Mesoporous carbon from MCM-41 template including CMK-1, -3 and SNU-1, -2	C (irregular)	Channel (PIP) (regular)	2–4	1100–2200 m²/g
	Microporous carbon from zeolite template	C (irregular)	Open cage (PIP) (regular)	1–2	1000–4000 m²/g
	Pillared clay	Si, Al, O, etc. (regular or irregular)	Slit-based (PIP or POP) (irregular)	2–4	100–1100 m²/g; 0.3–0.6
	Oxide-bridged carbon nanosheet composite	C, Si, O, etc. (irregular or regular)	Slit-based (PIP or POP) (irregular)	1–4	500–1100 m²/g; 0.3–0.6
	Single element nanoporous materials	Metal atom (Pd, Si, etc.)	Random	Micropore–20	60–150 m²/g; 0.1–07

materials. Type IIa is characteristic of a smooth, non-stepwise increase in adsorption with p/p_0 (without adsorption hysteresis) due to energetic heterogeneity in the adsorbate–adsorbent interaction. The stronger the strength of the adsorbate–adsorbent interaction is, the sharper the uprising in adsorption at low p/p_0. Type IIb is typical of nanoporous materials aggregated from plate-like particles and characteristic of an evident hysteresis loop.

3. Type III: the adsorption isotherm typical of nonporous and macroporous materials, but with weak interaction energy between adsorbate and adsorbent in comparison with the case of Type II.

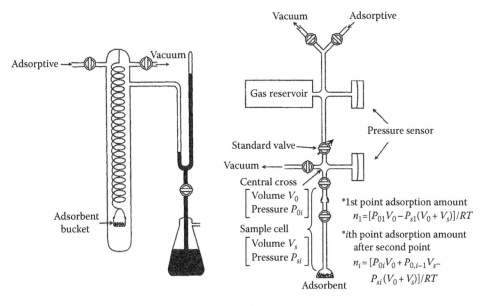

FIGURE 9.7 Measurement methods of adsorption isotherms. (A) Gravimetric method. (B) Volumetric method. (From Rouquerol, F. et al., *Adsorption by Powders & Porous Solids*, Academic Press, San Diego, CA, 1999. With permission.)

TABLE 9.4 Typical Gas Molecules for Probing Porosities of Nanoporous Materials

Probe Molecules	Measurement Temperature	Applicability
Nitrogen	Liquid nitrogen (77 K)	Most popularly used
Argon	77 K or liquid argon (87 K)	Especially good for materials with polar surface (ex. zeolite)
Krypton	77 K	Materials with small specific surface area
Helium	Liquid helium (4.2 K)	Ultramicropores
Carbon dioxide	273–303 K	Microporous materials, especially with slow adsorption rate at low measurement temperature

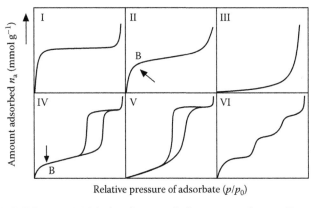

FIGURE 9.8 IUPAC classification of adsorption isotherms. (From Marsh, H. and Rodriguez-Reinoso, F., *Activated Carbon*, Elsevier, San Diego, CA, 2006. With permission.)

4. Type IV: the adsorption isotherm typical of mesoporous materials. It is characteristic of a gradual increase in adsorption at a low p/p_0 range due to monolayer–multilayer adsorption followed by a great uprising in adsorption at a medium p/p_0 range, together with an evident adsorption hysteresis, due to capillary condensation. From the differences in the shape of adsorption hysteresis loops, Type IV isotherms are further divided into Type IVa with a relatively narrow hysteresis loop, Type IVb with a broad hysteresis loop, and Type IVc without an evident hysteresis loop (the typical case for MCM-41).

5. Type V: the adsorption isotherm typical of nanoporous materials with weak adsorbate–adsorbent interaction. It is similar to Type VI with an evident adsorption hysteresis, but characteristic of a small adsorption at the low p/p_0 range due to weak adsorbate–adsorbent interactions. Nanoporous materials showing this type of isotherm unnecessarily contain only mesopores but sometimes also contain micropores with the steep increase in adsorption at a higher p/p_0 range depending on the pore size and properties of adsorbents and adsorbates (for example, H_2O).

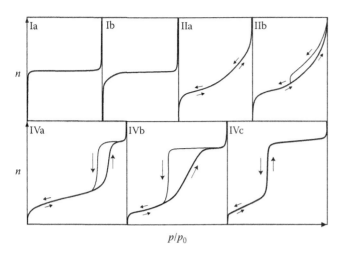

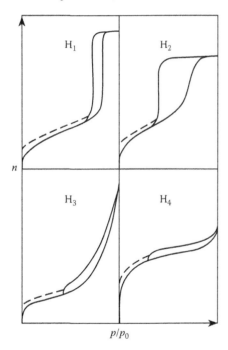

FIGURE 9.9 Subdivision of adsorption isotherms. (From Rouquerol, F. et al., *Adsorption by Powders & Porous Solids*, Academic Press, San Diego, CA, 1999. With permission.)

FIGURE 9.10 IUPAC classification of hysteresis loops. (From Rouquerol, F. et al., *Adsorption by Powders & Porous Solids*, Academic Press, San Diego, CA, 1999. With permission.)

6. Type VI: the adsorption isotherm typical of adsorption phenomena on nonporous surfaces but not on nanoporous materials. The stepwise plateaus are characteristic of layer-by-layer adsorption of nonpolar molecules such as argon, krypton, on uniform surfaces such as graphite surfaces, and are associated with two-dimensional phase changes in the monolayer from which useful information concerning the surface uniformity and adsorbate structure can be obtained.

The presence of a hysteresis loop at medium p/p_0 is a significant characteristic of mesoporous materials because their shapes have a close relationship with mesopore structures. Four typical types of hysteresis loops (Figure 9.10) were classified by IUPAC based on their shapes (Sing et al. 1985):

1. Type H_1: the hysteresis loop associated with pores consisting of agglomerates or compacts of uniform spherical particles with regular arrays, whose pore size distribution are normally narrow.
2. Type H_2: the hysteresis loop associated with pores with ink-bottle shapes or more complicated network structures.
3. Type H_3: the hysteresis loop associated with slit-shaped pores formed from aggregates of plate-like particles.
4. Type H_4: the hysteresis loop associated with slit-shaped pores with narrower sizes.

In many cases, hysteresis loops extend to very low pressure ranges (described by the dashed lines in Figure 9.10), which are attributed either to activation adsorption (irreversible chemisorption), and/or to the changes in adsorbent structures (swelling, etc.), and/or to the physisorption in pores with very narrow sizes in which diffusion is a rate-determining process.

9.4.1.2 Calculations of Pore Parameters

Pore parameters (specific surface area, pore volume, pore size distribution) are calculated by analyzing adsorption isotherms using various theoretical relationships. The typical theories applied to adsorption analysis are the Brunauer–Emmett–Taylor (BET) equation based on the monolayer–multilayer adsorption mechanism in a flat surface or in macropores, which is widely used for the calculation of specific surface area; the Kelvin equation based on capillary condensation in mesopores, which is the basis of various computation algorithms, such as the Barrett–Joyner–Halenda (BJH), Cranston–Inkley (CI), and Dollimore–Heal (DH) methods for mesopore size distribution; and the Dubinin–Radeshkevich (DR) equation based on the micropore filling mechanism for the calculation of pore volumes of micropores. There are also semi-empirical methods such as t and α_s comparison plot methods, which are very valuable and are widely used for computing specific surface area, pore volume, and pore size. Recently, some advanced techniques such as density functional theory (DFT) and simulation methods were begun to apply for pore size distribution calculation. Table 9.5 summarizes the typical determination methods of nanoporous parameters and their theoretical basis, equations, applicability, and parameters. Table 9.6 lists the typical determination methods of nanopore size and nanopore size distribution.

9.4.2 Other Characterization Methods

Besides porous parameters, bulk structures, morphologies, and surface properties are also important characteristics of

TABLE 9.5 Typical Determination Methods of Nanoporous Parameters

Methods	Theoretical Basis	Equations	Applicability	Parameters
Brunauer–Emmett–Taylor (BET) equations	Multilayer adsorption	$w = (V_m Cp)/\{(p_0 - p)[1 + (C - 1)(p/p_0)]\}$ w: adsorption amount p: equilibrium pressure p_0: saturated pressure V_m: monolayer adsorption amount C: constants associated with adsorbate–adsorbent interaction energy $\rightarrow$ BET plot: $x/[w(1 - x)] = 1/(V_m C) + [(C - 1)/(V_m C)]x$ where $x = p/p_0$ Specific surface area (S): Calculated from V_m using molecular sectional area	Nonporous materials, macroporous materials, relative comparisons among materials with various types of pores	Specific surface area (S), C value (a parameter related to adsorption energy)
Kelvin equation	Capillary condensation	$\ln(p/p_0) = -(2\gamma V_m/RT)(\cos\theta/r_m)$ γ: surface tension of adsorbates θ: contact angle V_m: molar volume of adsorbates r_m: curvature radii of meniscus R: ideal gas constant T: temperature	Mesoporous materials	Basis for mesopore calculations (specific surface area, pore volume, and pore size distribution)
Dubinin–Radeshkevich (DR) equation	Volume filling	$w = w_0 \exp[-(A/\beta E_0)^2]$ where $A = RT \ln (p_0/p)$ (adsorption potential) w: adsorption amount w_0: saturated adsorption amount β: affinity coefficient E_0: characteristic heat of adsorption $\rightarrow$ DR plot: $\ln(w/w_0) = -(RT/\beta E_0)^2 \ln^2(p_0/p)$	Microporous materials	Micropore volume (w_0), characteristic heat of adsorption (E_0) at $w/w_0 = 1/e$
t-plot	Comparison plot	$t = (n/n_m) \times 0.354$ (nm) n: adsorption amount n_m: monolayer adsorption capacity · 0.354 nm and t is the statistical thickness of a close-packed monolayer of N_2 molecules · t is a function of p/p_0 and generally obtained from standard data on nonporous materials with similar surface properties $\rightarrow$ t-plot: relationship of t and n	External surface, macroporous materials, mesoporous materials	External and internal specific surface areas, pore volume
α_s-plot	Comparison plot	$\alpha_s = n/n_{0.4}$ n: adsorption amount $n_{0.4}$: adsorption amount at $p/p_0 = 0.4$ α_s is also a function of p/p_0 and generally obtained from standard data on nonporous materials $\rightarrow$ α_s-plot: relationship of α_s and n	External surface and all nanoporous materials	External and internal specific surface areas, pore volume

nanoporous materials that are analyzed by various physico-chemical tools. Table 9.7 summarizes the principles, the obtained information, and various characteristics of these methods.

9.5 Conclusion Remarks

In the above sections, a deep insight into the definitions of pores and nanopores, the classification and typical kinds of nanoporous materials, and their characterization methods were reviewed. From these discussions, the characteristics and specialties of the target "pore" and the relevant nanoporous materials with respective to the generally meant nanomaterials can be comprehended. In the past decades, nanoporous materials brought about a huge impact on both fundamental research and industrial applications in various fields of science and technology such as adsorption, separation, catalysis, sensing, and other functional areas. In recent years, novel nanoporous materials keep appearing one after another and

TABLE 9.6   Typical Determination Method of Nanopore Size and Nanopore Size Distribution

Methods	Theoretical Basis	Characteristics	Applicability
Average method	—	$w = 2V/A$ (slit-shaped) $w = 4V/A$ (cylindrical) w: pore width V: pore volume A: specific surface area	All nanoporous materials
Mercury intrusion method	Washburn equation	$\Delta p_r = -2\gamma \cos\theta$ Δp: excess pressure γ: surface tension θ: contact angle	Macropores (width greater than 50 nm)
Barrett–Joyner–Halenda (BJH) method, Cranston–Inkley (CI) method, Dollimore–Heal (DH) method	Capillary condensation (Kelvin equation)	Based on strict or simplified computation programming, using statistical thickness either by calculation or by standard adsorption data	Mesopore (2–50 nm)
Holvath–Kawazoe (HK) method	Adsorption potential field in micropores	Calculation of average adsorption potential as a function of pore width	Micropores (<2 nm)
Density functional theory (DFT) method	Mean field density functional theory	• Determining adsorption isotherms for individual pores of width w, $V(p/p_0, w)$, by local or nonlocal mean-field molar density calculation • Determining pore size distribution $F(w)$ by curve fitting of experimental isotherm data, $V(p/p_0)$, using the following generalized adsorption isotherm (GAI) equation which relates experimental data with those of individual pores: $V(P/P_0) = \int V(p/p_0, w) \times F(w) \times dw$	All nanopores
Grand Canonical Mont-Carlo simulation (GCMC) method	Statistical mechanics	• Determining adsorption isotherms for individual pores of width w, $V(p/p_0, w)$, by GCMC simulation • Determining pore size distribution $F(w)$ by curve fitting of experimental isotherm data, $V(p/p_0)$, using the GAI equation	Micropores, one part of mesopores (< ~10 nm)

in turn the kinds of nanoporous materials are being updated year after year.

Two issues can be considered for the challenge and trends of research on nanoporous materials in the future. One issue is from the viewpoint of pores: (1) challenging in controlling pore size precisely and liberally, which contributes to the sieving properties and selectivity of nanoporous materials, and (2) challenging for approaching the limitation of pore (void) fractions, which contributes to the maximum volume for accommodating adsorbate molecules. With this respect, the MOFs materials are the hopeful ones. However, the fragile structures of the likewise materials are a challenge for their future applications. Furthermore, it seems that ordered pore structure is not always necessary for realizing functionalities although it contributes to the exact understanding of molecular behaviors

within a well-defined pore system (Rolinson 2003). Another issue is from the viewpoint of the framework materials, which includes the nanoporous realization of functional materials, such as precious metals, silicon, and various ceramic materials, and the functioning of nanoporous materials by tailored composition/modification and by controlling the thickness of pore walls. One prospective material in this respect can be the graphene/carbon nanosheets and their functional nanoporous composite materials.

On the other hand, applying advanced techniques and improvements in theories such as DFT makes it possible to understand pore systems deeply and more precisely. Further progress in both nanoporous materials and their characterization methods will certainly step up the outcome in the fields of nanoscience and nanotechnology in the future.

TABLE 9.7 Other Methods for Characterization of Nanoporous Materials

Methods	Principles	Obtained Information	Others
Methods using x-ray			
Powder x-ray diffraction method (XRD)	Diffraction of scattering x-ray described by Bragg's equation	Crystal structure, phases, crystalline size	
Small angle x-ray scattering (SAXS)	Diffraction of scattering x-ray with small angles (usually < 10°)	Structure of ultrafine particles, pore structure (especially closed pores)	
X-ray photoelectron spectroscopy (XPS) or Electron spectroscopy for chemical analysis (ESCA)	Measurement of kinetic energy of photoelectrons (emitted core electrons) ejected by x-ray	Components and bonding structures of surface atoms	Sensitive to surface atoms within thickness < ~5 nm, resolution: 0.4–1 eV
Extended x-ray absorption fine structure (EXAFS)	Analysis of the fine oscillating structure at energies just above the absorption edge	Local structure (coordination species, numbers and distances) of the central excited atom	Applicable to amorphous materials
X-ray absorption near edge structure (XANES)	X-ray absorption peaks around absorption edges	Formal valence, coordination environment, geometrical distortions	
Methods using electron beams			
Transmission electron microscope (TEM)	Transmission of electrons through thin species of samples	Atomic array, deficiency, morphology	Resolution: ~0.5 nm
Scanning electron microscope (SEM)	Viewing images through secondary electrons	Morphology, particle size and shape, surface structure	Resolution: ~10 nm
Atomic force microscopy (AFM)	Scanning through surface while maintaining inter-atomic forces constant	Surface structure	
Electron diffraction (ED)	Diffraction pattern of transmitted electrons	Crystal structure, crystal axes	
Low energy electron diffraction (LEED)	Diffraction pattern of low energy (20–200 eV) electrons	Surface structure of crystalline materials	Sensitivity: surface 1%
Electron probe micro analyzer (EPMA) including wavelength dispersive x-ray spectrometer (WDX) and energy dispersive x-ray Spectrometer (EDX)	Analysis of fluorescent x-ray produced by bombardment of electron to samples	Surface element distribution	Sensitivity: 0.01–0.1% Resolution: 10 nm–1 μm
Electron energy-loss spectroscopy (EELS)	Vibration spectra from energy loss of incident electrons due to inelastic scattering	Bonding structure of adsorbate atoms/molecules	Sensitivity: surface 1%, higher than IR and RAMAN
Auger electron spectroscopy (AES)	Measurement of kinetic energy of Auger electrons	Surface component, chemical structure of surface	Sensitive to surface atoms within thickness < ~5 nm,
Methods using infra-red (IR) and ultraviolet-visible (UV) light			
IR spectrometry	Absorption and reflection of IR	Bulk structure, surface functional groups	
RAMAN spectrometry	Raman scattering of laser light	Bulk structure	Strong scattering especially at surface
UV-vis spectrometry	UV-vis absorption	Electronic structure of bulk material and adsorbates, valence and coordination states	
Ultraviolet photoelectron spectroscopy (UPS)	Measurement of kinetic energy of photoelectrons emitted from valence-region molecular orbit by ultraviolet photons	Valence state and bonding structure of surface atoms and adsorbates	Sensitive to surface atoms within thickness < ~1 nm, Resolution: several mV to 50 mV
Magnetic resonance methods			
Nuclear magnetic resonance (NMR)	Absorption of electromagnetic wave due to transition of nucleus spins between Zeeman states	Coordination structure of bulk materials, structure of surface organic groups	Applied species: ^{1}H, ^{13}C, ^{27}Al, ^{29}Si, ^{31}P, etc.
Electron spin resonance (ESR)	Absorption of electromagnetic wave due to transition of electron spins between Zeeman states	Electronic states of diamagnetic ions	
Other methods			
Acid–base titration	Neutralization of acids and bases	Surface functional groups	
Temperature-programmed desorption (TPD)	Thermal stability of surface groups or adsorbed species	Surface functional groups, adsorbates	

References

Chu Y.-H., Z.-M. Wang, M. Yamagishi, H. Kanoh, T. Hirotsu, and Y.-X. Zhang. 2005. Synthesis of nanoporous graphite-derived carbon-silica composites by a mechanochemical intercalation approach. *Langmuir* 21: 2545–2551.

Eddaoudi M. E., J. Kim, N. Rosi, D. Vodak, J. Wachter, M. O'Keeffe, and O. M. Yaghi. 2002. Systematic design of pore size and functionality in isoreticular MOFs and their application in methane storage. *Nature* 295: 469–472.

Everett D. H. and J. C. Powl. 1976. Adsorption in slit-like and cylindrical micropores in the Henry's law region. A model for the microporosity of carbons. *Journal of the Chemical Society: Faraday Transactions* 1 72: 619–636.

Hupp J. T. and K. R. Poeppelmeier. 2005. Better living through nanopore chemistry. *Science* 309: 2008–2009.

Inagaki S., S. Guan, Y. Fukushima, T. Ohsuna, and O. Terasaki. 1999. Novel mesoporous materials with a uniform distribution of organic groups and inorganic oxide in their frameworks. *Journal of the American Chemical Society* 121(41): 9611–9614.

Kaneko K. 1994. Determination of pore size and pore size distribution. 1. Adsorbents and catalysis. *Journal of Membrane Science* 96: 59–89.

Kaneko K. 2000. Specific intermolecular structures of gases confined in carbon nanospace. *Carbon* 38: 287–303.

Kitagawa S. and R. Matsuda. 2007. Chemistry of coordination space of porous coordination polymers. *Coordination Chemistry Reviews* 251: 2490–2509.

Kresge C. T., M. E. Leonowicz, W. J. Roth, J. C. Vartuli, and J. C. Beck. 1992. Ordered mesoporous molecular sieves synthesized by a liquid-crystal template mechanism. *Nature* 359: 710–712.

Lu G. Q. and X. S. Zhao. 2004. Nanoporous materials: An overview. In *Nanoporous Materials: Science and Engineering, Series on Chemical Engineering*, Vol. 4, Eds. G. Q. Lu and X. S. Zhao, pp. 1–12. London, U.K.: Imperial College Press.

Marsh H. and F. Rodriguez-Reinoso. 2006. *Activated Carbon*. San Diego, CA: Elsevier.

Ozin G. A. and A. C. Arsenault. 2005. *Nanochemistry: A Chemical Approach to Nanomaterials*. Cambridge, U.K.: The Royal Society of Chemistry.

Pinnavaia T. J. 1983. Intercalated clay catalysts. *Science* 220: 365–371.

Pinnavaia T. J. 1995. Nanoporous layered materials. In *Materials Chemistry—An Emerging Discipline/Advances in Chemistry Series*, Eds. L. V. Interrante, L. A. Caspar, and A. B. Ellis, pp. 283–300. Washington, DC: American Chemical Society.

Polarz S. and B. Smarsly. 2002. Nanoporous materials. *Journal of Nanoscience and Nanotechnology* 2(6): 581–612.

Rolison D. R. 2003. Catalytic nanoarchitectures—The importance of nothing and the unimportance of periodicity. *Science* 299: 1698–1701.

Rouquerol F., J. Rouquerol, and K. Sing. 1999. *Adsorption by Powders & Porous Solids*. San Diego, CA: Academic Press.

Sing K. S. W., D. H. Everett, R. A. W. Haul, L. Moscou, R. A. Pierotti, and T. Siemieniewska. 1985. Reporting physisorption data for gas/solid systems with special reference to the determination of surface area and porosity. *Pure and Applied Chemistry* 57: 603–618.

Van Bekkum H., E. M. Flanigen, P. A. Jacobs, and J. C. Jansen. 2001. *Introduction to Zeolite Science and Practice*. Amsterdam, the Netherlands: Elsevier.

Yang R. T. 2003. *Adsorbents: Fundamentals and Applications*. Hoboken, NJ: John Wiley & Sons Inc.

10

Ordered Nanoporous Structure

Jun Shen
Tongji University

Bin Chu
Wuhan University

Yuan Liu
University of Shanghai for Science and Technology

10.1 Introduction of the Nanoporous Materials

According to the classification made by International Union of Pure and Applied Chemistry (IUPAC) [1], porous materials can be divided into three main categories depending on their pore size, that is, the microporous materials with pore size under 2 nm, the mesoporous materials with pore size between 2 and 50 nm, and the macroporous materials with pore size greater than 50 nm. The commonly referred nanoporous materials are mainly included in the mesoporous range. Therefore, in this chapter, the structure of mesoporous materials is mainly discussed.

In 1992, Kresge et al. and Beck et al. of the United States Mobile Corporation [2,3] reported for the first time a series of mesoporous silicon/aluminosilicate materials (designated M41S) and then their new discoveries of series SBA-n, MSU, MAS, JLU, HMS, and FSM-16 have attracted considerable attention. Such novel mesoporous materials not only break through the pore-size constraint of zeolite molecular sieves but also have uniform channel and ordered arrangement. The pore size can be varied from around 2 nm to over 50 nm in a continuous controllable manner. The high specific surface area, good thermal and hydrothermal stability and larger wall size, has dramatically expanded the range of crystallographically defined pore sizes of molecular sieves from the micropore (lower than 2 nm) to mesopore (lager than 2 nm) regime. As molecular sieves such as zeolite have been researched and applied for a relatively long time, there are quite a number of monographs on them. We will not discuss them in this chapter.

Since ordered nanoporous materials have a uniform and continuous controllable channel structure of nanometer-size and possess the obvious effects of quantum confinement, small size, surface, macroscopic quantum tunneling, and dielectric confinement, they display a variety of novel natures that are very different from the bulk materials. The research on ordered nanoporous materials is significantly valuable in fields ranging from catalysis, membrane-based separations technology to molecular engineering. Ever since ordered mesoporous materials were discovered at the Mobile Corporation in 1992, they have attracted considerable attention in physics, chemistry, and materials all over the world and have developed rapidly becoming one of the key research focuses with an interdisciplinary approach [4–10].

10.2 Fundamental Characteristics of Ordered Mesoporous Materials

Ordered mesoporous materials would have the following characteristics:

1. They are porous materials induced by the supramolecular structure generated from an organic surfactant and self-assembled by sol–gel technology;
2. They have narrow pore size distributions and mesoporous channels are arranged periodically. The pore size and its orientation are generally determined by the type and the nature of the surfactant template;
3. The mesoporous materials can be used as nanometer-scale reaction devices or adsorption agents because of their narrow pore size distribution, high internal surface area, uniform channel alignment, and diversified chemical composition;

4. With the pore size larger than the zeolite molecular sieves, the mesoporous materials can complete large molecular catalysis and separation of large biological macromolecules that the zeolite molecular sieves cannot do;

5. They are also the model materials used in investigating mesopore absorption and can help researchers to study the fantastic property of nanomaterials from the microscopic aspects, such as the small size effect, quantum effect, and surface effect;

6. In microelectronics and optical application, they are also favorable host materials. We can investigate the optical nature exploiting its oriented channel orientation and high specific surface area.

The hydrothermal stability and catalytic activity of mesoporous materials are not so good compared to microporous materials [11]. It is generally believed that the distinction derived from the difference in the manner that the microporous molecular sieves and the atoms of mesoporous materials are linked. Essentially, it is the distinction between the crystalline nature and the amorphous structure. The ordering of the microporous materials is on the atom-scale; that is, every atom has a fixed site in space and can be described by a crystal cell, whereas the ordering of the mesoporous materials is a kind of space orderliness that is irrelevant to atoms. The cell parameters describe only the space orderliness on the mesoscopic scale. The linking manner between atoms is similar to the amorphous materials. Such structures can be characterized as long-range regularity and short-range irregularity structures.

10.3 Structure of Ordered Mesoporous Materials

The synthesis of ordered mesoporous materials was begun as early as the 1970s [12]. Japanese scientists also started the research in the early 1990s [3]. But it is generally believed that the real start in developing such mesoporous materials was when the scientists of Mobile Oil Corporation disclosed the synthesis of the silica/alumina ordered mesoporous molecular sieves (designated as M41S) using CTAB [2,3]. The theory of synthesizing this kind of ordered mesoporous materials broke through the traditional process in which a single solvent molecule or ion works as the template during the synthesis process of the microporous zeolite molecular sieves, and using an orderly supramolecular structure that is self-organized by a kind of cationic quaternary ammonium surfactant as the template successfully synthesized the ordered mesoporous materials of the M41S series with high specific surface area, regular arrays of uniform channels, and controllable pore size (with pore size approximately between 1.6 and 4 nm). It was regarded as a milestone in the developing history of molecular sieves. Figure 10.1 shows the structural schematic of mesoporous M41S materials [116].

The conception of the template derived itself from the synthesis of the molecular sieves firstly [11]. A great deal of organic ammonium molecules were introduced into the zeolite synthesis

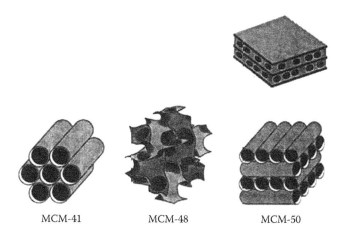

| MCM-41 | MCM-48 | MCM-50 |

FIGURE 10.1 Structural schematic of mesoporous M41S materials. (From Vartuli, J.C. et al., *Advanced Catalysts and Nanostructured Materials: Modern Synthetic Methods*, Moser, W.R. (ed.), Academic Press, New York, 1996. With permission.)

systems, and thus a large amount of zeolite molecular sieves with novel structures was obtained. These organic ammonium molecules played a role of structure directing during the synthesis process, the so-called structure-directing agent. In the synthesis systems of zeolite molecular sieves, inorganic units surround the solvent molecules or ions (such as the tetrapropylammonium ions (TPA$^+$) [14] in the ZSM-5 system, shown in Figure 10.2) that are directed by a single molecule or ion as the structure-directing reagent and condense to form a stable framework. The meaning of the template changed a lot after the discovery of the M41S series of products, resulting in the *de novo* realization of the function of templates. In the synthesis systems of mesoporous materials, a single molecule or ion cannot support a relatively open framework structure that is directed by the supramolecular structure self-assembled by surfactants or macromolecular materials.

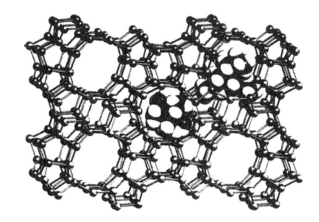

FIGURE 10.2 Schematic drawing of tetrapropylammonium ions (TPA$^+$) that work as a template for ZSM-5, showing how the tetrapropylammonium ion is located in the channel intersections. (From van Bekkum, H. et al., *Stud. Surf. Sci. Catal.*, 137, 207, 2001. With permission.)

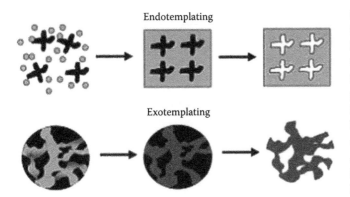

Endotemplating

Exotemplating

FIGURE 10.3 Schematic drawing of the two approaches (endotemplating and exotemplating) to synthesize porous and high-surface-area materials. (From Schüth, F., *Angew. Chem. Int. Ed.*, 42, 3604, 2003. With permission.)

When templates are used in the synthesis of porous solids, two different modes in which the templates can act can be distinguished [15]: the endotemplate and the exotemplate (Figure 10.3). The endotemplate is a template species, which is an isolated precursor and is included as an isolated entity into a growing solid particle; it can be molecules or supramolecular surfactant aggregates. And the exotemplate is a structure providing a scaffold with voids, such as colloidal crystals. If this scaffold is removed after filling of the voids, a porous solid with reverse structure is generated. The two modes both can be used to synthesize ordered mesoporous materials.

Endotemplating: This is the most normal approach to synthesize mesoporous materials. It is different from the structure-directed by metal ion or organic small molecule to yield the zeolite structure. The synthesis of mesoporous materials use the supramolecular structure that is self-assembled by surfactants or macromolecular materials as the template. Two stages in the synthesis process of mesoporous materials can be distinguished:

1. Growth of organic–inorganic phases (mesostructured): amphiphilic surfactant and condensable inorganic monomer or oligomer self-organized into inorganic–organic liquid crystal structure with the crystal lattice on a length scale of nanometers.
2. Generation of mesoporous materials: after removal of organic templates by calcination or other physical or chemical methods, a scaffold with voids can be obtained.

Exotemplating: In this method, using porous materials with three-dimensional continuous channel structures as templates, after filling of the voids, a mesoporous solid that is a negative of the exotemplate can be obtained. This mode is mainly used to synthesize non-silica mesoporous materials, such as mesoporous carbon, metal, and metal oxide.

Among the modes of synthesizing mesoporous materials, endotemplating is well investigated. Almost all the inorganic mesoporous materials possess an amorphous framework only presenting some crystalline structure on the macroscale, except for some organic and inorganic mesoporous skeletons [16–18] with stable microstructure and paracrystalline wall. Until now, most of the synthesized mesostructures have counterparts among known lyotropic liquid crystals, besides a few without the corresponding structure (at least in the same surfactant and water binary system they have none). The mainly common structures of ordered mesoporous materials are as follows: 1-dimensional lamellar structure (*P2*), 2-dimensional hexagonal structure (*P6mm*), 3-dimensional body-centered cubic structure *Ia3d*, 3-dimensional body-centered cubic structure *Im3m*, 3-dimensional simple cubic structure *Pm3n*, 3-dimensional hexagonal (*P6₃/mmc*), and symbiotic structure of 3-dimensional face-centered cubic (*Fm3m*). Table 10.1 lists typical mesoporous solids with different structures and their structural characteristics.

10.4 Synthesizing Mechanism of Ordered Mesoporous Materials

At the time when the first M41S materials were synthesized, people did a lot of research on the synthesizing mechanism of the mesoporous materials and found it to be completely different from that of the traditional microporous molecular sieves. The change was in the synthesis of the traditional molecular sieves using the concept of templates wherein the template generally refers to single molecules such as metal ions, solvent molecules, organic ammonium salt, and so on, whereas mesoporous molecular sieves are synthesized using self-assembled molecular aggregates or supramolecular assemblies as templates. A number of representative synthesis mechanisms have been proposed to explain the formation of mesoporous materials. It is generally believed that no one can explain all the synthesis processes with one mechanism.

In 1992, the Mobile Corporation reported the invention of M41S, proposed a liquid crystal templating mechanism (LCT) [2,3], and a cooperative mechanism (or cooperative templating mechanism) [26,31,117] (see Figure 10.4). The liquid crystal templating mechanism is based on the truth that the high resolution transmission electron microscopy images and x-ray diffraction results of the MCM-41 are strikingly similar to those obtained from lyotropic liquid-crystal phases that are produced in surfactant–water mixtures. In this mechanism, surfactant liquid crystal phases are believed to serve as templates for the formation of the MCM-41 structure. A liquid crystal phase of surfactant is present prior to the addition of inorganic reagents (Figure 10.4, pathway 1). The surfactants (organic templates) with hydrophilic or hydrophobic aggregates form spherical micelles firstly then cylindrical micelles in the water system. When the surfactant is concentrated enough, a liquid crystal structure with regular hexagonal array will be produced. The inorganic monomer molecule or oligomer which dissolved inside the solvent will precipitate in the voids among the micelle rods by the attraction of the hydrophilic surfaces of the micelles. Then the inorganic monomer molecule or oligomer will polymerize and solidify to create inorganic walls. The main suggestion that the liquid crystal templating mechanism proposes is that the liquid crystal

TABLE 10.1 Structure Categories of Mesoporous Materials and Their Structural Characteristics

Typical Materials	Space Group	Crystal System	Structural Characteristics of Channel	Characteristics of the Diffraction (XRD Diffraction Peak, the Diffraction Conditions)
MSU-n, HMS, KIT-1		(Nearly hexagonal)	(Poorly ordered, mostly 1-dimensional)	1 or 2 wider diffraction peaks
MCM-50			1-D lamellar	$\dfrac{1}{d_{001}} = \dfrac{1}{a}$; 001, 002, 003, 004...
MCM-41, SBA-3,15, FSM-16, TMS-1	$P6mm(20)$	Hexagonal	2-D (straight channel)	$\dfrac{1}{d^2} = \dfrac{h^2 + k^2 + l^2}{a^2} + \dfrac{l^2}{c^2}$; 100, 110, 200, 210...
SBA-8, KSW-2	$C2mm(12)$	Tetragonal		$\dfrac{1}{d_{hk}^2} = \dfrac{h^2}{a^2} + \dfrac{h^2}{b^2}$; $h + k = 2n$ (11, 20, 22, 31, 40, ...)
SBA-2, 7, 12 FDU-1	$P6_3/mmc$ 194	Hexagonal	3-D (caged channel, cavity)	$\dfrac{1}{d^2} = \dfrac{4}{3} \cdot \dfrac{h^2 + hk + k^2}{a^2} + \dfrac{l^2}{c^2}$; $hhl: l = 2n$
SBA-1, SBA-6	$Pm\bar{3}n$ (223)	Cubic		$\dfrac{1}{d^2} = \dfrac{h^2 + k^2 + l^2}{a^2}$; $hhl: l = 2n$ (the first peak (110) was not observed) 110, 200, 210, 211, 220, 310, 222, 320, 321, 400...
SBA-16	$Im\bar{3}n$ (229)			110; 200, 211, 220, 310, 222, 321...$h + k + l = 2n$
FDU-2	$Fd\bar{3}m$ (227)			111; 200, 311, 222, $h + k = 2n$, $h + l = 2n$, $k + l = 2n$, $0kl; k + l = 4n$
FDU-12	$Fm\bar{3}m$ (225)			$h + k = 2n$, $h + l = 2n$, $k + l = 2n$ (111; 200, 220, 311, 222, 400...)
SBA-11	$Pm\bar{3}m$ (221)			No restriction on light extinction
SBA-2, 7, 12 FDU-1	$Fm\bar{3}m$ (225) $P6_3/mmc$ 194	Symbiont of hexagonal and cubic structure		$\dfrac{1}{d^2} = \dfrac{4}{3} \cdot \dfrac{h^2 + hk + k^2}{a^2} + \dfrac{l^2}{c^2}$; $hhl: l = 2n$
SBA-16	$Im\bar{3}n$ (229)	Cubic	3-D cross-channel	$h + k + l = 2n$ (110; 200, 211, 220; 310, 222, 321, 400...)
MCM-48 FDU-5	$Ia\bar{3}d$ (230)			$h + k + l = 2n$, $hlk: 2h + l = 4n$ (210; 220, 321, 400, 420, 332, 422, 431, 440, 532...)
HOM-7	$Pn\bar{3}m$ (224)			$0lk: k + l = 2n$ (110; 111, 200, 211, 220, 221, 310, 311, 222...)
CMK-1, HUM-1	$I4_1/a$ (88)	Tetragonal		110, 211, 220...

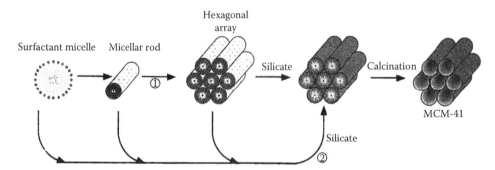

FIGURE 10.4 Schematic drawing of the mechanism for the formation of MCM-41 proposed by Mobile Corporation [2]: (1) for liquid crystal templating mechanism and (2) for cooperative mechanism. (From Beck, J.S. et al., *J. Am. Chem. Soc.*, 114 (27), 10834, 1992. With permission.)

phases or micelles formed by surfactants serve as mesoporous structure-directing agents. Additional experimental evidence for this mechanism is the dependence on the alkyl chain length and the influence of auxiliary organic macromolecules (such as mesitylene) on the pore sizes of MCM-41. This mechanism is simple and easy to be understood, moreover, we can exploit some concepts in liquid chemistry to interpret an amount of experimental phenomena during the synthesis procedure, such as the

principle of phase transition in which the reaction temperature, surfactant concentration, etc. will influence the resulting solids.

However, as the studies developed further, the researchers found that the liquid crystal templating mechanism was so simple that the inconsistency appeared only when using it to interpret some experimental phenomena discovered later. For example, to form liquid crystal phases in the aqueous solution requires high surfactant concentrations (such as, for hexagonal phases of CTAB > 28%, and for cubic phases > 80%). In fact, MCM-41 and MCM-48 can form at very low concentrations (approximate 2% CTAB for MCM-41 and 6%–9% for MCM-48). Furthermore, the hexagonal liquid crystal phases were not detected in the solvent of the surfactant serving as the template and the reaction mixture gel (or sol). All the facts described above indicate that another cooperative mechanism (Figure 10.4, pathway 2) also formulated by the Mobile Corporation is more reasonable under such synthesis conditions.

Like the liquid crystal templating mechanism, the cooperative mechanism also takes the liquid crystal prepared with surfactant or the micelles as the templates of the MCM-41 structure but the difference is believed to be in that the surfactant liquid crystal is present after the addition of inorganic reagents. The surfactant mesophase is created by the interaction between micelles and inorganic species. Such interaction manifests itself as the condensation of inorganic species accelerated by micelles and promotion of the condensation to the formation of the liquid-crystal-like ordered structure. The condensation of inorganic species accelerated by micelles is caused predominantly by complex interface interplays (e.g., electrostatic interactions, hydrogen bonding, or coordination role, etc.) between organic phases and inorganic phases and then results in the condensation of inorganic species at the interface. This mechanism is useful to explain miscellaneous phenomena occurring in the synthesis of mesoporous materials, for example, the synthesis of new products that are different from the known liquid crystal structure [4], the syntheses of mesoporous materials at surfactant concentrations lower than 5% by weight [19,20], and phase transition during the synthesis procedure [21], etc.

After that, Davis proposed the silicate rod assembly mechanism [23] in which he suggested that randomly ordered rod-like organic micelles interact with the silicate species to yield approximately two or three monolayers of silica encapsulation around the external surfaces of the micelles. Subsequently, these composite species spontaneously assemble into the long-range ordered structure (hexagonal packing) characteristic of MCM-41. With further heating, the silicate species in the interstitial spaces of the ordered organic–inorganic composite phase continue to condense (Figure 10.5).

The mechanism presented by Davis cannot be used to explain the long channels of MCM-41 because there are no micelles that are so long in the solution. In fact, besides rod-like micelles, there are spherical micelles in reaction agents. If rod-like micelles encapsulated by the silica around the external surfaces spontaneously assemble into an ordered structure, there should be other species

besides the hexagonal MCM-41. The mechanism also lacks generalization; that is, it cannot rationalize the formation of the cubic phase referred to as MCM-48 and the lamellar phase known as MCM-50. We can assume that some small rods form the MCM-48 with the same length crossing each other. However, the lengths of the micelles are not homogeneous in the surfactant solution. No lamellar micelles that are necessary for the formation of MCM-50 exist in the surfactant solution with low concentrations. There is no doubt that this mechanism can work well in some specific systems and interpret some experimental phenomena successfully.

Firouzi et al. proposed silicatropic liquid crystals' mechanism based on a great deal of synthesis and results obtained by nuclear magnetic resonance as well as their study on inorganic–organic surfactant liquid crystal created under different reaction systems [24]. They believed that the silicate anions ion-exchanged with the surfactant halide counterions (Br^- or Cl^-) pre-absorbed on the surfactant cation head to form a "silicatropic liquid crystal" (SLC) phase that involved silicate-encrusted cylindrical micelles (Figure 10.6). The SLC phase exhibited behavior very similar to that of typical lyotropic systems, except that (1) the surfactant concentrations were much lower and (2) the silicate counterions were reactive (they can condense further with other silicate ions). Heating of the SLC phase caused the silicates to condense irreversibly into the MCM-41. Furthermore, Firouzi et al. also emphasized the matching between the silicate species and the surfactant cation head. They found that in addition to the charge balance requirement (i.e., electrostatic interaction) there was preferential bonding of the ammonium head group to multi-charged D4R (double four-ring, $[Si_{18}O_{20}]^{8-}$) silicate anions under the high pH conditions. The interaction was so strong that an alkyltrimethylammonium surfactant solution could force a silicate solution that did not contain D4R oligomers to re-equilibrate and form D4R species. They suggested that this behavior came from the closely matched projected areas of a D4R anion and an ammonium head group ($0.098\,nm^2$ vs. $0.094\,nm^2$) and the correct distribution of charges on the projected surfaces [19,24].

Cooperative organization mechanism formulated by Stucky et al. is the most generally accepted synthesis mechanism of mesoporous compounds. They proposed that ordered channel arrays were obtained by cooperative interactions among inorganic and organic species. As for the mesopore silica system, the silicate polyanions interact with surfactant cations; the condensation of the silicate at the interface and the hydrophilic/hydrophobic interplay among long surfactant chains result in the approaching of the long surfactant chains. The matching of charge density between the inorganic and organic species

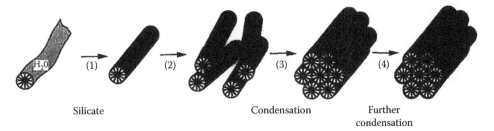

FIGURE 10.5 Assembly of silicate-encapsulated rods. (Adapted from Chen, C.Y. et al., *Micropor. Mater.*, 2, 27, 1993. With permission.)

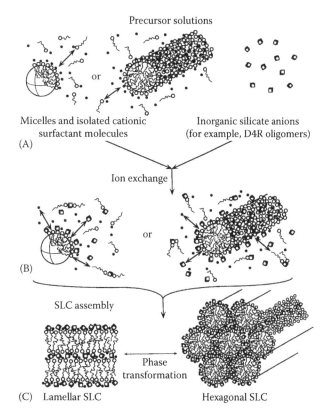

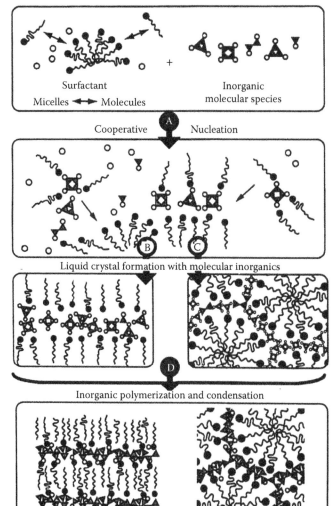

FIGURE 10.6 Formation of a silicatropic liquid crystal phase. (From Firouzi, A. et al., *Science*, 267, 1138, 1995. With permission.)

governs the arrangement of the surfactant. And the pre-organized ordering of the surfactant (for example, rod-like micelles) is not necessary. The charge density of the inorganic layer will be varied during the reaction process, and the interplay between overall inorganic and organic species, hydrophobic interaction among organic species, and condensation among inorganic species will all influence the final products.

On the basis of cooperative mechanism, Huo et al. [25,31] generalized the mechanism mentioned above and proposed a generalized liquid crystal templating mechanism thereby extending the liquid crystal templating mechanism to the general synthesis of non-silica mesoporous materials (Figure 10.7). The generalized liquid crystal templating mechanism suggests that liquid crystal formed by cooperative nucleation of surfactant molecules and inorganic species develops into a mesostructured crystal. This mechanism should work for the synthesis of non-siliceous mesoporous materials as well. There are three main types of cooperative templating: (1) cooperative charge matched templating based on electrostatic interactions, (2) ligand-assisted templating based on covalent bonding interaction, and (3) neutral templating based on hydrogen bonding.

In addition to the above-mentioned mechanisms, there are a variety of synthesis mechanisms relating to ordered mesoporous materials, such as charge density matching mechanism (mechanism of transition from lamellar to hexagonal phases) proposed by Monnier et al. [26] and Stucky et al. [117], silicate layer puckering mechanism proposed by Steel et al. [27], and real liquid

FIGURE 10.7 Cooperative templating of the generalized LCT mechanism. (From Huo, Q. et al., *Chem. Mater.*, 6, 1176, 1994. With permission.)

crystal templating proposed by Attard, Antonietti and coworkers [13,28,29]. These mechanisms can elucidate the synthesis of mesoporous materials under some special conditions but cannot be generalized.

10.5 Pathways to the Synthesis of Ordered Mesoporous Materials

Classical synthesis of mesoporous materials can be divided into two stages: (1) self-assembly of the amphiphilic surfactant molecules and polymerizable inorganic monomer molecules or oligomers (silica source) into organic–inorganic liquid crystal phase under certain conditions, and (2) the removal of surfactant by means of physical or chemical approaches (mostly thermal treatment at high temperature) and a void inorganic scaffolding with mesoporous channels can be obtained.

The three main ingredients in the synthesis procedure of mesoporous materials are inorganic precursors (silica source) that form inorganic network, template (surfactants, etc.) that direct definitively the formation of mesostructured phases, and reaction medium (solvent). There is intense interplay among the three ingredients, and the interaction between the organic templates and inorganic species is a key factor for the synthesis of mesoporous materials and takes the main role of directing in the whole synthesis procedure. In different synthesis systems, there are different interface interactions according to different charge factors between surfactants and inorganic species.

10.5.1 Inorganic–Organic Interaction

Four pathways to the synthesis of normal mesoporous materials are depicted in Figure 10.8 [30]. Cationic or anionic inorganic oligomers can polymerize further in aqueous solution under certain pH conditions. At first, Stucky and coworkers and Huo et al. generalized the pathways to the synthesis of mesoporous materials into four types by synthesizing and observing the synthesis procedure [19,31] as S^+I^-, S^-I^+, $S^+X^-I^+$, and $S^-M^+I^-$. They found that during the synthesis procedure of silicate mesoporous materials in alkali systems since the silica species are oligomeric silicate polyanions, the ordered mesostructured phases can be obtained by organizing inorganic species with negative charges I^- into ordered arrays using cationic surfactant S^+. This route was termed pathway S^+I^-. From this, they proposed four pathways, by such deducing, researchers synthesized different novel mesoporous materials using the interaction between inorganic and organic species. For example, by pathway S^+I^-, the reaction between the aluminum polycation Keggin and anionic surfactant such as alkyl sulfate leads to the formation of mesoporous aluminum oxide materials [32]; by pathway $S^+X^-I^+$, utilizing the cationic surfactant as template the mesoporous silica under extremely acidic conditions can be formed, where S^+ is the quaternary ammonium surfactant such as CTAB, X^- is the haloid ions such as Cl^-, and I^+ is silica species with positive charge in acidic systems [21]. Corresponding to this is pathway $S^-M^+I^-$, where, generally speaking, M^+ is metal cations [31].

After that, Pinnavaia and coworkers [8,22,33–36] used nonionic surfactants such as primary amine and polyethylene oxide surfactants to prepare silicates with cylindrical nanopores referred to as HMS and MSU-n under neutral condition. Contrary to the pathways proposed by Stucky and coworkers where the whole organization process is driven by electrostatic interactions between surfactants and silicate ions, in the presence of neutral surfactants hydrogen bonding becomes the predominant factor, it is called pathway S^0I^0 and N^0I^0. But, because of weak repulsive interactions between the template and the inorganic framework, the resultant mesoporous materials have relatively poor ordering. The syntheses of SBA-15 series of mesoporous materials in a strongly acidic system are prototypic examples of pathway $S^0(IX)^0$ depicted in Figure 10.9 [37,38].

In addition to electrostatic interactions between surfactants and inorganic species and the hydrogen bonding mentioned

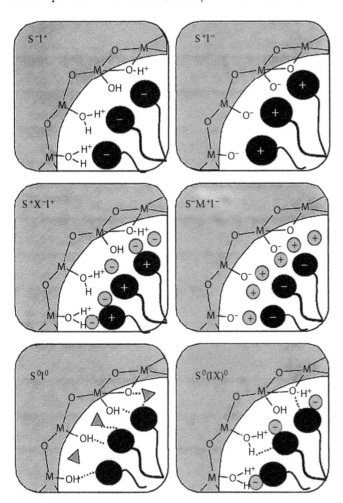

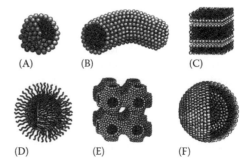

FIGURE 10.9 Schematic diagram of micellar structures. (A) Spherical micelles, (B) cylindrical micelles, (C) planar bilayer, (D) reverse micelles, (E) bicontinuous phase, (F) liposomes. (From Evans, D.F. and Wennerström, H. (eds.), *The Colloidal Domain: Where Physics, Chemistry, Biology and Technology Meet*, VCH Publishers, Weinheim, Germany, 1994; Garcia, C., et al., *Angew. Chem. Int. Ed.*, 42, 1526, 2003. With permission.)

FIGURE 10.8 Schematic representation of the different types of silica–surfactant interfaces. S represents the surfactant molecule and I, the inorganic framework. M^+ and X^- represent the corresponding counterions. Solvent molecules are not shown, except for the I^0S^0 case (triangles); dashed lines correspond to H-bonding interactions. (From Soler-illia, G. et al., *Chem. Rev.*, 102, 4093, 2002. With permission.)

TABLE 10.2 Interactions between Different Types of Surfactants and Inorganic Species

Surfactants	Type of Inorganic Species	Type of Interactions
S^+	$I^- \rightarrow S^+I^-$	Electrostatic interaction
S^-	$I^+ \rightarrow S^-I^+$	Electrostatic interaction
S^+	$I^+ \rightarrow S^+I^+$	Electrostatic interaction
S^-	$I^- \rightarrow S^-I^-$	Electrostatic interaction
S^0	$I^0 \rightarrow S^0I^0$	Hydrogen bonding
N^0	$I^0 \rightarrow N^0I^0$	Hydrogen bonding
S	$I \rightarrow SI$	Covalent bonding

above, there are several interactions that can assist in forming the mesoscopic structure such as chemical bonding interactions (coordination bonding [39–41] or covalent bonding), van der Waals interactions, and dipolar interactions between molecules, as shown in Table 10.2.

10.5.2 Interactions among Surfactants

Interaction between inorganic and organic species is a key aspect of synthesis of mesoporous materials. But to synthesize mesoporous materials, especially high-quality ordered mesoporous materials, the interactions among surfactants are very important. In actual experimental process, many factors can influence the ultimate structure of the mesoporous materials, but the ultimate structure is generated by the self-assembly of organic species. In other words, the mesoscopic structure of the resulting materials is determined by the interactions among some organic species. Table 10.3 gives some mesoscopic structures of mesoporous materials synthesized in the presence of surfactants. From it we can obtain that, different surfactants can synthesize mesoporous materials with different structures, and the same surfactant can prepare mesoporous materials with different structures by modulating reaction conditions.

Israelachvili et al. and Israelachvili gave a model to explain and predict the morphology of liquid crystals of surfactants, and proposed the concept called molecular surfactant packing factor g [42,43] that can characterize the geometry of the mesophase products as a useful molecular structure-directing index. $g = V/a_0l$, where V is the total volume of the surfactant chain plus any cosolvent organic molecules between the chains, a_0 is the effective head group area at the micelle surface, and l is the kinetic length of the surfactant tail. During the research of synthesizing and phase transition of the mesoporous materials, Huo et al. [21] explored firstly the use of the molecular surfactant packing parameter to explain and predict the product structure. Although the calculation of the parameter g is very simple, it can describe which kind of liquid crystal we can get under certain synthesis conditions, and assist us to control synthesis conditions to obtain the mesophase products we want in the syntheses of mesoporous materials, and explain well the phenomena that we observed. Different micelles' structures are depicted schematically in Figure 10.9. The nature of surfactant aggregates and the

TABLE 10.3 Mesostructured Phases Synthesized in the Presence of Surfactants

Mesophase	Code	Surfactants	References
$P6mm$	MCM-41	C_nTMA^+, ($n = 12$–18)	[2,51]
		$C_{16\text{-}n\text{-}16}$ ($n = 4, 6, 8, 10$)	[21]
		CTAB and $C_{20\text{-}3\text{-}1}$	[21]
		$[(CH_3)_3N^+C_{12}H_{24}OC_6H_4OC][2Br^-]$	[62]
		CPBr, $C_nH_{2n+1}N(CH_3)^2$	[63]
	SBA-3	C_nTMA^+, ($n = 14$–18)	[21]
		$C_{16}H_{33}N(CH_3)(C_2H_4OH)_2{}^+$	[21]
		$C_{16}H_{33}N(C_2H_5)^{3+}$ + additive	[21]
		$C_{16\text{-}n\text{-}16}$ ($n = 3, 4, 6, 7, 8, 10, 12$)	[21]
	SBA-15	P123, P103, P85, P65	[32,54]
		B50–1500	[8]
$Pm3n$	SBA-1	$C_{16}H_{33}N(C_2H_5)^{2+}$	[51,33]
	SBA-6	$C_{16}H_{33}N(C_2H_5)^{3+}$	
$Im3m$	SBA-16	F127	[29]
	FDU-1	B50–6600	[32]
		F108	[64]
$Ia3d$	MCM-48	C_nTMA^+, (C_nTAB, C_nTACl, C_nTAOH, ($n = 14$–20))	[2,20,51]
		$C_{16}H_{33}N(CH_2)$ (C_6H_5) + (CDBA cetyldimethyl benzylammonium)	[21]
		$C_nTAB + C_{12}(EO)_m$ ($n = 12, 14, 16, 18$; $m = 3, 4$)	[21]
		CTAB + poly (ethyleneglycol) monoctylphenyl ether (OP-10)	[65]
		CTAB + polyoxyethylene	[21]
		CTAB + $C_{12}NH_2$	[66]
		CTAB + $C_nH_{2n+1}COONa$ ($n = 11, 13, 15, 17$)	[19]
		1-hexadecyl-3-methylimidazolium chloride ($C_{16}mimCl$)	[67]
$Ia3d$	FDU-5	P123 + MPTS (3-mercaptopropyltrimethoxysilane)	[68,90]
		P103, P123, +NaI	[69]
	HOM-5	P123 + TMB	[70]
$Pm3m$	SBA-11	Brij 56($C_{18}EO_{10}$)	[115]
$P6_3/mmc$	SBA-2	$C_{12\text{-}3\text{-}1}$, $C_{14\text{-}3\text{-}1}$, $C_{16\text{-}3\text{-}1}$, $C_{16\text{-}6\text{-}1}$, $C_{18\text{-}3\text{-}1}$, $C_{18\text{-}6\text{-}1}$, $C_{20\text{-}3\text{-}1}$,	[20,31]
	SBA-12	Brij 76($C_{18}EO_{10}$)	[29]
		F127 (film and fiber)	[71,72]
		Omega-hydroxyalkylammonium	[73]
$Fd3m$	FDU-2	$C_{18\text{-}2\text{-}3\text{-}1}$	[76]
	HOM-11	P123 + TMB, L121, PF68 + TMB	[77]
Cmm	SAB-8	$[(CH_3)_3N^+C_{12}H_{24}OC_6H_4C_6H_4OC_{12}H_{24}$ $N^+(CH_3)_3][2Br^-]$	[62]
		P123, Brij 58, F127(film)	[74,75]

morphology of the corresponding mesophases are related to g as shown in Table 10.4.

For most of the simple surfactants, the value of the parameter g can be estimated, but for the block copolymer, we cannot define g as for simple surfactants. The coiling or curling of the organic hydrophobic long-chain of PEO-based surfactants results in the

TABLE 10.4 g Parameter of Different Micellar Structures

$g = V/a_0 l$	Structures	Typical Surfactant	Typical Mesophase
$g < 1/3$	Spherical micelles	Single chain lipids with a large polar head	$Pm3n$, $Fm3m$, $Im3m$
$g = 1/3 – 1/2$	Cylindrical micelles	Single chain lipids with a small polar head	$P3mm$ (MCM-41)
$g = 1/2 – 2/3$	3-D cylindrical micelles	Single chain lipids with a small polar head	$Ia3d$, $Pn3m$
$g = 1$	Bilayer (membranes)	Double-chain lipids and a small polar head	Lamellar (MCM-50)
$g > 1$	Inverse cylindrical micelles or inverse spherical micelles	Double-chain lipids and a small polar head	$Fd3m$

difficulty of the definition of l. At the same time, hydrophilicity of PPO and PEO surfactants vary with temperature, time, and ionic strength in solutions, as well as the volume of water mixed in PEO; chains cannot be calculated resulting in difficulty with the quantifiability of the V value. In this case, people adopt the ratio of hydrophilic chains to hydrophobic chains to discuss phase transition of this kind of surfactants, to explain and predict the mesoporous structure templated by them. For example, using triblock copolymer (EOPOEO) with similar PO chain length (50–60) as template, varying the length of the PEO can obtain different structures [45,46]. Increasing the PEO chain length is equivalent to increasing the polar head group area (a_0) and results in decrease of the g value. PEO surfactants with short-chains lead to the formation of lamellar phases, while medium-chain PEO surfactants give two-dimensional hexagonal structure, and long-chain PEO surfactants result in cubic ($Im3m$) phases.

10.5.3 Other Influence Factors

Different from microporous materials and crystals, mesoporous materials are a kind of materials with the nature of long-range order and short-range disorder. That they have no stable structure in the microscopic regime is their defect, but at the same time, it provides mesoporous materials more choices of framework; that is to say, a wide variety of materials can be utilized to form mesoporous frameworks; therefore, it expands their popularity in applications. Moreover, diverse physicochemical properties of different materials will also influence the formation and phase transition of mesoporous materials. On the whole, there are a number of factors with different influences on the synthesis of mesoporous materials, including: (1) inorganic species that form walls (metal [19,47–51], metal oxide [33,52–57], semiconductor [58–61], organic silicon [62–78], and organics [79–84], etc.) and the tendency by which it promotes the crystallization of walls; (2) concentration of surfactant (molecules, micelles, and liquid crystals) [85,86]; (3) composition of reaction precursors (molar ratio of surfactant to inorganic species [87], ionic intensity [97], pH [28,31,89], type of solvent, etc.); (4) reaction conditions (temperature, reaction time, and additives [88, 90,91]); (5) synthesis approach adopted (hydrothermal, solvent-thermal, supercritical synthesis, solvent evaporation [92–95], etc.); (6) controlling

of product morphology (bulk material, film, fiber, or powder); (7) post-synthesis treatment (expanding the pore size, stabilizing, phase transitions, and removing of surfactant by calcinations or extractions). All these factors associated with each other will have effects on the structure and property of the products in every stage of synthesis.

10.6 Typical Ordered Mesoporous Materials

According to chemical compositions, ordered mesoporous materials can be divided into silica-based and non-silica-based mesoporous materials. Much progress has been made on the research of silica-based mesoporous materials and a variety of reports on this is available.

Mesoporous materials can be mainly divided into several classes as follows: the first is the MCM (Mobil composition of matter) series (produced by the Mobil Corporation), including MCM-41 (hexagonal, $P6mm$), MCM-48 (cubic, $Ia\bar{3}d$), and MCM-50 (lamellar); the second is the SBA-n (Santa Barbara Amorphous) series (reported by the University of California at Santa Barbara, USA) silica-based products including: SBA-1 (cubic, $Pm3n$), SBA-2 (3-D hexagonal, $P6_3/mmc$), SBA-3 (2-D hexagonal, $P6mm$), and SBA-15 (2-D hexagonal, $P6mm$); the third is the MSU series (Michigan State University), in which MSU-X (MSU-1, MSU-2, and MSU-3) has hexagonal mesoporous structure, lower degree of ordering, and only a broad peak in the low angel region on the XRD pattern. MSU-V and MSU-G possess multilamellar vesicles structure, and HMS (hexagonal mesoporous silica) has less ordered hexagonal structure. APMs (acid-prepared mesostructures), the early results of Stucky and coworkers [7,21], were prepared under acidic conditions, and at that time it was expanded for the synthesis techniques (alkaline medium) of the MCM series of materials. Subsequently, they developed their unique SBA-n series. FSM-16 (folded sheet material) is hexagonally mesoporous molecular sieves prepared by Yanagisawa et al. and Inagaki et al. [13,96]. Because of the complexity of the preparation that required adding templates to pre-synthesis layered polysilicate kanemite to modify the morphology and get hexagonal mesoporous materials, FSM-16 did not attract enough attention, but thereafter the success of the MCM preparation opened up a new research region.

The study on non-silica periodic mesoporous materials started later and the first report on the synthesis of ordered mesoporous transition metal oxide with a stable phase was published in 1995. Non-silica mesoporous materials mainly include transition metal oxide, phosphate, and sulfide. Since these materials usually have variable valences, they may find applications in new fields and exhibit prospects that silica-based material cannot reach, they are attracting increasing attention. Compared with silica-based mesoporous materials, their shortcoming is the tendency to structural collapse of the framework after calcination because of their low stability, and low specific surface area as well as channel volume. Besides, their synthesis mechanisms are still elusive.

10.6.1 Transition-Metal Oxide

Oxides of transition metals have some advantages over aluminosilicate materials for use in electromagnetics, photoelectronics, and catalysis, especially in fields such as solid catalysis, photocatalysis, shape-selective separations, mini electromagnetic devices, photochromic materials, electrode materials, storing of information, etc. because transition metal atoms can exist in various oxidation states. However, syntheses of ordered mesoporous structures of transition metal oxide mesoporous materials can be much more complicated and difficult than oxides of main group metals because of the multitude of different coordination numbers and oxidation states.

Ordered mesoporous manganese oxides have been the focus of intense interest in recent years [98] because of their outstanding cation-exchange capacity, molecule adsorption property as well as electrochemical performance, and magnetic property. Their applications in redox catalysis, molecular sieves, cathode materials for lithium cells, and magnetic materials attract a great deal of attention. Tian et al. [54] have recently reported a new family of semiconducting manganese oxide mesoporous structures (MOMS) of both hexagonal and cubic phases. The walls of the mesopores are composed of microcrystallites of dense phases of Mn_2O_3 and Mn_3O_4, with MnO_6 octahedra as the primary building blocks, so that they have very high thermal stability (reaching 1000°C). This is the highest thermal stability of non-silica ordered mesoporous materials reported up-to-date.

Mesoporous zirconia is believed to be the most potential catalysis and catalyst support. Larsen et al. [99] utilized lauryl sulfate (sodium form) as template after calcination at 575°C to obtain ordered mesoporous zirconia with the tetragonal phase. They also implemented n-butane isomerization at low temperature and fragmentation experiments that showed good catalyst activity. Pure zirconia exhibits weakly acidic catalyst activity, in order to obtain strongly acidic catalyst activity anion or metallic ion should be doped in. Wong and Ying [100] improved the acid strength of the zirconia network and the thermal stability by introducing the phosphate groups.

Due to its nontoxic property, high catalyst activity, strongly oxidizing capacity, and good stability, nano-scale titanium

dioxide is one of the most common photocatalyst. Recently, Stone and coworkers [101] studied the photocatalytic activity of mesoporous titania and niobia firstly. It can be presumed that the use of mesoporous titania in photocatalytic applications will attract more attention as further synthesizing of ordered mesoporous titania with high stability and specific surface area.

10.6.2 Other Non-Silica-Based Mesoporous Materials

The synthesis researches on metal sulfates are frequently reported [102]. Semiconducting mesoporous materials with superlattices directed by neutral template, e.g., CdS, SnS, and ZnS [103], have exhibited prominent performance on the applications of phosphor, electroluminescent devices, adsorption, and transducer, therefore attracted much attention. Microporous aluminophosphate (AlPOn) is a kind of crystalline material generally used as a catalyst, and the research focus today is on increasing its pore size thereby increasing the specific surface area and improving its thermal stability. Recently, Kimura et al. [104] reported the preparation of hexagonal mesostructured alkyltrimethylammonium–aluminophosphate materials by using hexadecyltrimethylammonium and dococyltrimethylammonium chlorides as surfactants and by utilizing 1, 3, 5-triisopropylbenzene as an auxiliary organic additive. Calcination of these materials at 600°C yielded thermally stable mesoporous aluminophosphate materials with large surface areas of above $700 \, m^2 \, g^{-1}$ and with pore diameters in the range of 1.8–3.9 nm. The adsorption property were also studied and reported. Besides aluminophosphate, there are also reports on the syntheses and studies of niobotungstate, galloaluminophosphate, vanadium aluminophosphate, cobalt aluminophosphate, etc. with high specific surface areas and good anion-exchange capabilities [33,105].

10.6.3 Ordered Mesoporous Films

Ordered mesoporous films could find possible applications in various functional materials for the features of periodic channel with uniform pore sizes, high specific surface area, and easy manipulation. Hitherto, the study on the synthesis and formation mechanism of ordered mesoporous films has been one of the key focuses on mesoporous film researches. The synthesis approaches of mesoporous films include sol–gel, template assembly, hydrothermal, and solvent thermal, as well as various physical dip-coating methods.

Yang et al. [9,10] reported the synthesis of oriented mesoporous silica films grown at the mica–water interface under acidic condition as well as at the interface between air and water, and proposed a possible mechanism for film growth. Based on the similarity to biomineralization, Brinker et al. [6,106–108] prepared highly ordered silica and organosilica hybrid mesoporous films by employing the dip-coating method. In general, oriented aligning of films is with one-dimensional pore channels aligned parallel to the substrate surface, so that the pores do not facilitate the transport across the film. Therefore, it limited their applications in substance separations and biological sensors. Kuroda and coworkers [109,110] improved the application ability of the mesoporous films by the application of a high magnetic field (>10 T), which is perpendicular to the substrate resulting in most of the one-dimensional mesochannels perpendicular to the substrate during the synthesis process. The key aspect of this synthesis approach that worked well is enough magnetic energy given by a domain of lyotropic liquid crystals formed by the surfactants, resulting in enough coercive force to make mesochannels orient perpendicular to the substrates in magnetic fields.

In order to provide easy accessibility to the substrate, besides the synthesis of mesoporous films with channels perpendicular to the substrate surface, mesoporous films with three-dimensional channels also can be synthesized. Zhao et al. [111] reported the synthesis of a mesoporous silica film with 3D large-scale sponge structure by the self-assembly of inorganic salts and copolymers during an acid-catalyzed process. The Stucky group [112] synthesized mesoporous silica films with 3D hexagonal structure ($P6_3/mmc$) by utilizing double quaternary ammonium salt as the template, its C axis with oriented aligned channels perpendicular to the interface of films, which provides accessibility for the transport of matter in the direction perpendicular to the film.

Other researchers have also implemented the syntheses of mesoporous films with a variety of non-silica materials and investigated their properties. By evaporation-induced self-assembly, Crepaldi et al. [113] prepared mesoporous zirconia thin films. Kuroda et al. [52] synthesized a mesoporous Pt–Ru alloy via the evaporation-mediated direct templating and investigated its electrochemical property. Xue [114] synthesized mesoporous manganese oxide by employing electrodeposition and studied its performance as a supercapacitor.

10.7 Summary

In this chapter, we have tried to cover most aspects of the synthesizing mechanism as well as pathways and to some extent the introduction of typical ordered mesoporous materials. Since the discovery of MCM-41 in 1992, scientific interest in the field of ordered mesoporous materials is blooming very fast. More and more researchers are devoted to further elucidation of the mechanism of formation of a variety of nanoporous structures. We could foresee that such novel materials will become commercially available for advanced industrial applications in the near future.

References

1. Everett D. H. *Pure Appl. Chem.* 1972, 31: 578–638.
2. Beck J. S., Vartuli J. C., Roth W. J. et al. *J. Am. Chem. Soc.* 1992, 114(27): 10834–10843.
3. Kresge L. T., Leonomicz M. E., Roth W. J., Vartuli J. C., Beck J. S. *Nature* 1992, 359: 710–712.
4. Huo Q., Leon R., Petroff P. M., Stucky G. D. *Science* 1995, 268: 1324–1327.
5. Lin H. P., Mou C. Y. *Science* 1996, 273: 765–768.
6. Lu Y. F., Ganguli R., Drewien C. A. et al. *Nature* 1997, 389: 364–368.
7. Schacht S., Huo Q., Voigt-Martin I. G., Stucky G. D., Schüth F. *Science* 1996, 273: 768–771.
8. Tanev P. T., Pinnavaia T. J. *Science* 1995, 267: 865–867.
9. Yang H., Coombs N., Sololov I., Ozin G. A. *Nature* 1996, 381: 589–592.
10. Yang H., Kuperman A., Coombs N., Mamiche-Afara S., Ozin G. A. *Nature* 1996, 379: 703–705.
11. Barrer R. M., Denny P. J. *J. Chem. Soc.* 1961, 2: 971–982.
12. Di Renzo F., Cambon H., Durartre R. *Micropor. Mater.* 1997, 10: 283–286.
13. Yanagisawa T., Shimizu T., Kuroda K., Kato C. *Bull. Chem. Soc. Jpn.* 1990, 63: 988–992.
14. van Bekkum H., Flanigen E. M., Jacobs P. A., Jansen J. C. *Stud. Surf. Sci. Catal.* 2001, 137: 663–668.
15. Schüth F. *Angew. Chem. Int. Ed.* 2003, 42: 3604–3622.
16. Wang X.-S., Ma M., Sun D., Parkin S., Zhou H.-C. *J. Am. Chem. Soc.* 2006, 128: 16474–16475.
17. Zou X., Conradsson T., Klingstedt M., Dadachov M. S., O'Keeffe M. *Nature* 2005, 437: 716–719.
18. Armatas G. S., Kanatzidis M. G. *Nature* 2006, 441: 1122–1125.
19. Firouzi A., Atef F., Oertli A. G., Stucky G. D., Chemlka B. F. *J. Am. Chem. Soc.* 1997, 119: 3596–3610.
20. Chen X., Ding G., Chen H., Li Q. *Sci. China Ser. B-Chem.* 1997, 40: 278–285.
21. Huo Q., Margolese D., Stcuky G. *Chem. Mater.* 1996, 8: 1147–1160.
22. Tanev P. T., Pinnavaia T. J. *Science* 1996, 271: 1267–1269.
23. Chen C. Y., Burkett S. L., Li H. X., Davis M. E. *Micropor. Mater.* 1993, 2: 27–34.
24. Firouzi A., Kumar D., Bull L. M., Besier T., Huo Q., Walker S. A., Stucky G. D., Chmelka B. F. *Science* 1995, 267: 1138–1143.
25. Huo Q., Margolese D., Ciesla U., Feng P., Gier T., Sieger P., Leon R., Petroff P., Schüth F., Stucky G. *Nature* 1994, 368: 317–321.
26. Monnier A., Schuth F., Huo Q. et al. *Science* 1993, 261: 1299–1303.
27. Steel A., Carr S. W., Anderson M. W. *J. Chem. Soc. Chem. Commun.* 1994, 1571–1572.
28. Goltner C., Henke S., Weissenberger M., Antonietti M. *Angew. Chem-Int. Edit.* 1998, 37: 613–616.
29. Clark J., Macquarrie D. *Chem. Commun.* 1998, 853.
30. Soler-illia G., Sanchez C., Lebeau B., Patarin J. *Chem. Rev.* 2002, 102: 4093–4138.
31. Huo Q., Margolese D. I., Ciesla U., Feng P., Gier T. E., Sieger P., Leon R., Petroff P. M., Schuth F., Stucky G. D. *Chem. Mater.* 1994, 6: 1176–1191.
32. Holland B. T., Isbester P. K., Blanford C. F., Munson E. J., Stein A. *J. Am. Chem. Soc.* 1997, 119: 6796.
33. Bagshaw S. A., Prouzet E., Pinnavaia T. J. *Science* 1995, 269: 1242.

34. Tanev P. T., Chibwe M., Pinnavaia T. J. *Nature* 1994, 369: 321–323.

35. Ulagappan N., Rao C. N. R. *Chem. Commun.* 1996, 1685.

36. Attard G. S., Glyde J. C., Goltner C. G. *Nature* 1995, 378: 366.

37. Zhao D., Feng J., Huo Q., Melosh N., Fredrickson G. H., Chmelka B. F., Stucky G. D. *Science* 1998, 279: 548.

38. Voegtlin A. C., Ruch F., Guth J. L., Patarin J., Huve L. *Micropor. Mater.* 1997, 9: 95.

39. Antonelli D. M., Ying J. Y. *Angew. Chem. Int. Ed.* 1996, 35: 426.

40. Antonelli D. M., Nakahira A., Ying J. Y. *Inorg. Chem.* 1996, 35: 3126.

41. Antonelli D. M., Ying J. Y. *Chem. Mater.* 1996, 8: 874.

42. Israelachvili J., Mitchell D., Ninham B. J. *Chem. Soc.* 1976, 72: 1525.

43. Israelachvili J. *Faraday Trans 2*, 2nd edn., Academic Press, London, U.K., 1991.

44. Evans D. F., Wennerström H. Eds. *The Colloidal Domain: Where Physics, Chemistry, Biology and Technology Meet*, VCH Publishers, Weinheim, Germany, 1994.

45. Flodstrom K., Alfredsson V. *Micropor. Mesopor. Mater.* 2003, 59(2–3): 167–176.

46. Kipkemboi P., Fogden A., Alfredsson V., Flodstrom K. *Langmuir* 2001, 17: 5398.

47. Armatas G. S., Kanatzidis M. G. *Science* 2006, 313: 817–820.

48. Sun D., Riley A. E., Cadby A. J., Richman E. K., Korlann S. D., Tolbert S. H. *Nature* 2006, 441: 1126–1130.

49. Attard G. S., Bartlett P. N., Coleman N. R. B., Elliott J. M., Owen J. R., Wang J. H. *Science* 1997, 278: 838.

50. Sakamoto Y., Kaneda M., Terasaki O., Zhao D. Y., Kim J. M., Stucky G., Shim H. J., Ryoo R. *Nature* 2000, 408: 449–453.

51. Yamauchi Y., Ohsuna T., Kuroda K. *Chem. Mater.* 2007, 19: 1335–1342.

52. Vaudry F., Khodabandeh S., Davis M. *Chem. Mater.* 1996, 8: 1451.

53. Yada M., Takenaka H., Machida M., Kijima T. J. *Chem. Soc-Dalton. Trans.* 1998, 1547.

54. Tian Z., Tong W., Wang J., Duan N., Krishnan V., Suib S. *Science* 1997, 276: 926.

55. Srivastava D., Perkas N., Gedanken A., Felner I. *J. Phys. Chem. B* 2002, 106: 1878.

56. Katou T., Lu D., Kondo J., Domen K. *J. Mater. Chem.* 2002, 12: 1480.

57. Wan Y., Yang H., Zhao D. *Acc. Chem. Res.* 2006, 39: 423–432.

58. Maclachlan M. J., Coombs N., Ozin G. A. *Nature* 1999, 397: 681–684.

59. Trikalitis P. N., Rangan K. K., Bakas T., Kanatzidis M. G. *Nature* 2001, 410: 671–675.

60. Trikalitis P. N., Rangan K. K., Kanatzidis M. G. *J. Am. Chem. Soc.* 2002, 124: 2604–2613.

61. Ding N., Takabayashi Y., Solari P. L., Prassides K., Pcionek R. J., Kanatzidis M. G. *Chem. Mater.* 2006, 18: 4690–4699.

62. Lim M., Blanford C., Stein A. *J. Am. Chem. Soc.* 1997, 119: 4090.

63. Lim M., Stein A. *Chem. Mater.* 1999, 11: 3285.

64. Fowler C., Burkett S., Mann S. *Chem. Commun.* 1997, 1769.

65. Lim M., Blanford C., Stein A. *Chem. Mater.* 1998, 10: 467.

66. Burkett S., Sims S., Mann S. *Chem. Commun.* 1996, 1367.

67. Macquarrie D. *Chem. Commun.* 1996, 1961.

68. Yamamoto K., Nohara Y., Tatsumi T. *Chem. Lett.* 2001, 30:648.

69. Fukuoka A., Sakamoto Y., Guan S., Inagaki S., Sugimoto N., Fukushima Y., Hirahara K., Iijima S., Ichikawa M. *J. Am. Chem. Soc.* 2001, 123: 3373.

70. Burleigh M., Dai S., Hagaman E., Lin J. *Chem. Mater.* 2001, 13: 2537.

71. Asefa T., Maclachlan M., Coombs N., Ozin G. *Nature* 1999, 402: 867.

72. Inagki S., Guan S., Fukushima Y., Ohsuna T., Terasaki O. *J. Am. Chem. Soc.* 1999, 121: 9611.

73. Melde B., Holland B., Blanford C., Stein A. *Chem. Mater.* 1999, 11: 3302.

74. Inagki S., Guan S., Ohsuna T., Terasaki O. *Nature* 2002, 416: 304–307.

75. Hoffmann F., Cornelius M., Morell J. et al. *Angew. Chem. Int. Ed.* 2006, 45: 3216–3251.

76. Xia Y. D., Mokaya R. *J. Phys. Chem. B* 2006, 110: 3889–3894.

77. Hatton B., Landskron K., Whitnall W., Perovic D., Ozin G. A. *Acc. Chem. Res.* 2005, 38: 305–312.

78. Zhou X., Qian S., Hao N., Wang X., Yu C., Wang L., Zhao D., Lu G. Q. *Chem. Mater.* 2007, 19: 1870–1876.

79. Meng Y., Gu D., Zhang F. Q., Shi Y. F., Yang H. F., Li Z., Yu C. Z., Tu B., Zhao D. Y. *Angew. Chem. Int. Ed.* 2005, 44: 7053.

80. Zhang F. Q., Meng Y., Gu D., Yan Y., Yu C. Z., Tu B., Zhao D. Y. *J. Am. Chem. Soc.* 2005, 127: 13508.

81. Rzayev J., Hillmyer M. A. *J. Am. Chem. Soc.* 2005, 127: 13373.

82. Tanaka S., Nishiyama N., Egashira Y., Ueyama K. *Chem. Commun.* 2005, 2125.

83. Meng Y., Gu D., Zhang F. et al. *Chem. Mater.* 2006, 18: 4447–4464.

84. Lu Y. F. *Angew. Chem. Int. Ed.* 2006, 45: 7664–7667.

85. Grudzien R. M., Grabicka B. E., Kozak M. et al. *New J. Chem.* 2006, 30: 1071–1076.

86. Kim T. W., Ryoo R., Kruk M. et al. *J. Phys. Chem. B* 2004, 108: 11480–11489.

87. Vartuli J., Schmitt K., Kresge C. et al. *Chem. Mater.* 1994, 6: 2317.

88. Ryoo R., Jun S. *J. Phys. Chem. B* 1997, 101: 317.

89. Ryoo R., Kim J. *J. Chem. Soc.-Chem. Commun.* 1995, 711.

90. Che S. N., Garcia-Bennett A. E., Liu X. Y., Hodgkins R. P., Wright P. A., Zhao D. Y., Terasaki O., Tatsumi T. *Angew. Chem. Int. Ed.* 2003, 42: 3930–3934.

91. Liu X., Tian B., Yu C., Gao F., Xie S., Tu B., Che R., Peng L.-M., Zhao D. *Angew. Chem. Int. Ed.* 2002, 41: 3876–3878.

92. Doshi D. A., Gibaud A., Goletto V., Lu M. C., Gerung H., Ocko B., Han S. M., Brinker C. J. *J. Am. Chem. Soc.* 2003, 125: 11646–11655.

93. Brinker C. J., Dunphy D. R. *Curr. Opinion Coll. Interface Sci.* 2006, 11(2–3): 126–132.

94. Fakcari P., Costacurta S., Mattei G. et al. *J. Am. Chem. Soc.* 2005, 127(11): 3838–3846.

95. Miyata H., Suzuki T., Fukuoka A., Sawada T., Watanabe M., Noma T., Takada K., Mukaide T., Kuroda K. *Nat. Mater.* 2004, 3: 651.

96. Inagaki S., Fukushima Y., Kuroda K. *Chem. Commum.* 1993, 8: 680–681.

97. Antonelli D. M., Ying J. Y., *Angew. Chem. Int. Ed. Engl.* 1995, 34: 2014–2017.

98. Brock S. L., Duea N., Tian Z. R. et al. *Chem. Mater.* 1998, 10: 2619–2628.

99. Larsen G., Lotero E., Nabity M. et al. *J. Catal.* 1996, 164: 246–248.

100. Wong M. S., Ying Y. *Chem. Mater.* 1998, 10: 2067–2077.

101. Victor F., Stone J., Davis R. *J. Chem. Mater.* 1998, 10: 1468–1474.

102. Braun P. V., Osenar P., Stupp S. I. *Nature* 1996, 380: 325–328.

103. Sooklal K., Cucallum B. S., Agnel S. M. *J. Phys. Chem.* 1996, 100: 4551–4555.

104. Kimura T., Sugahara Y., Kuarode K. *Micropor. Mesopor. Mater.* 1998, 22: 115–126.

105. Stein A., Fendorf M., Jarvie T. P. et al. *Chem. Mater.* 1995, 7: 304–313.

106. Sellinger A., Weiss P. M., Nguyen A., Lu Y. F., Assink R. A., Gong W. L., Brinker C. J. *Nature* 1998, 394: 256–260.

107. Lu Y. F., Fan H. Y., Doke N., Loy D. A., Assink R. A., LaVan D. A., Brinker C. J. *J. Am. Chem. Soc.* 2000, 122: 5258–5261.

108. Brinker C. J., Lu Y., Sellinger A., Fan H. *Adv. Mater.* 1999, 11: 576–585.

109. Yamauchi Y., Sawada M., Sugiyama A., Osaka T., Sakka Y., Kuroda K. *J. Mater. Chem.* 2006, 16: 3693–3700.

110. Yamauchi Y., Sawada M., Noma T., Ito H., Furumi S., Sakka Y., Kuroda K. *J. Mater. Chem.* 2005, 15 (11): 1137–1140.

111. Zhao D., Yang P., Melosh N., Chmelka B. F., Stucky G. D. *Adv. Mater.* 1998, 10(16): 1380–1385.

112. Alberius P. C. A., Frindell K. L., Hayward R. C., Kramer E. J., Stucky G. D., Chmelka B. F. *Chem. Mater.* 2002, 14: 3284–3294.

113. Crepaldi E. L., Soler-Illia G., Grosso D., Albouy P. A., Sanchez C. *Chem. Commun.* 2001, 17: 1582–1583.

114. Xue T., Xu C. L., Zhao D. D., Li H. L. *J. Power Sources* 2007, 164: 953–958.

115. Zhao D., Huo Q., Feng J., Chemlka B., Stucky G. *J. Am. Chem. Soc.* 1998, 120: 6024.

116. Vartuli J. C., Kresge C. T., Roth W. J., McCullen S. B., Beck J. S., Schmitt K. D., Leonowicz M. E., Lutner J. D., Sheppard E. W. In *Advanced Catalysts and Nanostructured Materials: Modern Synthetic Methods* pp. 1–19, Moser W. R. Ed., Acadamic Press, New York, 1996.

117. Stucky G., Monnier A., Schüth F. et al. *Mol. Cryst. Liq. Cryst.* 1994, 240: 187–200.

118. Garcia, C., Zhang, Y., DiSalvo, F., and Wiesner, U., *Angew. Chem. Int. Ed.* 2003, 42: 1526–1530.

11

Giant Nanomembrane

Hirohmi Watanabe
*The Institute of Physical
and Chemical Research*

Toyoki Kunitake
*The Institute of Physical
and Chemical Research*

and

Nanomembrane Technologies

11.1 What Is "Giant Nanomembrane"

Membrane, which is self-supporting by definition, separates two spaces of either liquid or solid states, and controls transport of materials and information across them. Its importance in various industrial applications need not be emphasized anew. In the biological world, ion channels and other transport functions are essential for maintenance of the living state, and such functions are supported by biological membranes that consist of lipid bilayers and embedded protein molecules with a membrane thickness of 5–10 nm. The nanoscopic thickness of the biological membrane is the basis of the existence of complex molecular organizations on and in the membrane architecture. Such exquisite organizations have not been achieved with purely artificial materials. On the other hand, macroscopic size is, of course, desired for most industrial membranes. Their ideal features would include defect-free, uniform morphology, macroscopic robustness, efficient permeability, and tailor-made selectivity. The combination of complex molecular architecture as in biological membrane and macroscopic, robust morphology of industrial membrane constitutes a research target of far-reaching significance. An approach toward such an ideal membrane is to develop nanometer-thick (molecular size) membranes with macroscopic mechanical stability and to equip them with well-designed functional units. Polymeric materials and inorganic materials have been used most frequently to fabricate practically interesting membranes, but their thickness basically remained in the micrometer range, and they involved little organized membrane architectures.

Over the last decade, the development of nanometer-thick, freestanding membranes has been accelerated. These efforts often aim at the creation of nano-precision architectures while maintaining macroscopic usefulness. They may be called "giant nanomembrane," when they possess both nanometer thickness and macroscopic size. Thus, it is appropriate to define "giant nanomembranes" as having the following three basic features. First, its thickness is in the range of 1–100 nm. The self-supporting (freestanding) property is the second feature that is required for a membrane to be able to physically separate two spaces. Third, the "giant" nanomembrane should be characterized by aspect ratios of size and thickness greater than 10^6, that is, if the membrane thickness is 10 nm, its size must be greater than 1×1 cm. Two related structural features are included in these requirements. One is macroscopic mechanical strength (robustness), and the other is the uniform and defect-free membrane texture over a large area.

11.2 Approaches toward Fabrication of Nanomembranes

Interests in ultrathin membranes with macroscopic size have a long history. Gold foil is a representative case. A thicker gold leaf is beaten to a larger gold foil of less-than 100 nm thickness, and has been used for gilded ornaments and other purposes. More recently, novel functional materials such as carbon nanotube, graphene, and metallic silicon are used as starting materials to build macroscopic nanomembranes. For instance, a molecular complex of single-walled carbon nanotube and conductive polymer, poly(3-hexylthiophene), was converted to a centimeter-size, freestanding membrane of 250 nm thickness (Gu and Swager 2008), and macroscopic, freestanding nanomembranes of gold and other metals were prepared from the corresponding nanoparticles upon self-assembly at interface and subsequent fusion (Xia and Wang 2008).

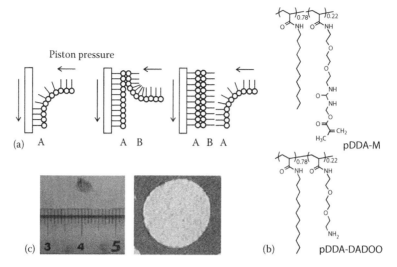

FIGURE 11.1 (a) Illustration of Langmuir–Blodgett technique, (b) chemical structure for the assembly, and (c) obtained nanosheets. (From Endo, H. et al., *Macromolecules*, 39, 5559, 2006. With permission.)

In the case of organic polymers, physical properties arising from extreme thinness of molecular membranes are a target of intensive research from an interest in quasi-2-dimensional behavior of polymer chain. One of the major fabrication techniques is the Langmuir–Blodgett approach in which polymer solutions are spread on water to form very thin films that can be subsequently transferred onto solid substrates. As another example, a 1 cm² film of poly(styrene) with thickness of 29–184 nm is prepared by spin coating and transferred from water surface onto a 3 mm hole for the measurement of glass transition temperature as a function of the film thickness (Forrest et al. 1997). Similar film formation (50 nm thickness) is reported with Nylon 13, 13 polymer (Wang and Porter 1995). These films, as composed of the linear polymer, however lack macroscopic robustness, though they are self-supporting enough for certain measurements. Molecular membranes on water surface were also formed from cross-linked polymers. The monolayer of an amphiphilic siloxane copolymer was cross-linked with poly(vinyl sulfate) and is transferred onto a solid substrate by the conventional Langmuir–Blodgett technique (Mallwitz and Goedel 2001). Most recently, extensive research has been conducted for polymer nanosheet by Miyashita and coworkers at Tohoku University. They found that poly(*N*-dodecyl acrylamide) is especially useful to obtain robust Langmuir–Blodgett multilayers that are freestanding at submillimeter sizes (Figure 11.1) (Endo et al. 2006).

A second molecular approach is electrostatic layer-by-layer assembly, as first demonstrated by Decher (1997). Its schematic representation is shown in Figure 11.2. This method is composed of adsorption of polymer ions in solution onto an oppositely charged solid substrate. Since the adsorption produces excess charges on the surface, alternate adsorption of positive and negative polymer ions becomes feasible, and repeated cycles of adsorption give rise to thin films of the two polymer ions. A freestanding hexagonal sheet (10 mm size and 5 nm thick) of alternate polyions (Huck et al. 2000), as well as a freestanding ultrathin film that incorporates magnetite nanoparticles have

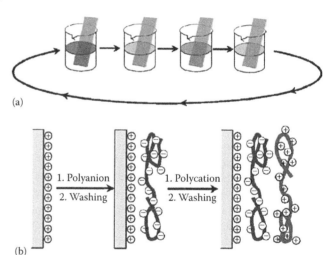

FIGURE 11.2 (a) Schematic representation of the electrostatic alternative assembly process, and (b) the simplified molecular picture of the first cycle; steps 1 and 3 represent the adsorption process of polycation and polyanion, respectively, and steps 2 and 4 are washing steps. (Adapted from Watanabe et al., *Bull. Chem. Soc. Japan*, 80, 433, 2007a. With permission.)

been successfully prepared (Mamedov and Kotov 2000). It is inferred from a literature summary of Jiang and Tsukruk that the alternate layer-by-layer assembly could give self-supporting nanomembranes with thickness of sub-100 nm but with rather small sizes (Jiang and Tsukruk 2006).

11.3 Fabrication of Giant Nanomembranes from Highly Cross-Linked Materials

A recent, general approach for the fabrication of giant nanomembranes is the use of highly cross-linking materials as precursor. The maintenance of macroscopic robustness in spite of nanometer thickness is the key to the fabrication of stable "giant nanomembranes." It is possible to enhance stress responsiveness of materials by simply reducing their thicknesses, since mechanical stress is more readily released in thin materials. High-density cross-linking has been a general approach toward the preparation of hard, strong materials, and such hard matters are converted to soft, flexible materials by reducing its thickness while maintaining satisfactory mechanical strength.

11.3.1 Nanomembrane of Metal Oxides

Common oxide ceramics are made of high-density oxide networks (cross-linking), and are a typical hard matter in the macroscopic size. Therefore, oxide ceramics are a qualified candidate for the fabrication of robust nanomembranes. In the sol-gel method, soluble precursor materials are converted to solid ceramic materials by heating or through reaction with water. This process can be applied to the preparation of freestanding nanomembranes of metal oxides by employing proper sacrificial/underlayers. Certain photoresist polymers are useful as sacrificial layer that facilitates the detachment of nanomembranes from substrate by dissolution in an appropriate solvent. A poly(vinyl alcohol) (PVA) layer as underlayer provides a hydroxyl-rich surface, and apparently promotes the formation of uniform metal oxide layer. As for example, an ultrathin titania layer is formed from titanium *n*-butoxide by spin-coating onto a PVA/photoresist layer on Si wafer. This specimen is isolated as a freestanding, defect-free nanomembrane in the range of 10–200 nm thickness (Figures 11.3 and 11.4) (Hashizume and Kunitake 2003). This fabrication procedure has been extended to obtain robust nanomembranes of Al_2O_3, NbO_5, ZrO_2, SiO_2, and La_2O_3 (Hashizume and Kunitake 2006). The combination of chemical inertness and physical softness allows interesting applications for these nanomembranes.

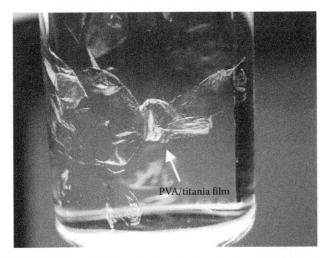

FIGURE 11.3 A PVA/titania film in ethanol. The vial diameter is 27 mm. (Adapted from Watanabe et al., *Bull. Chem. Soc. Japan*, 80, 433, 2007a. With permission.)

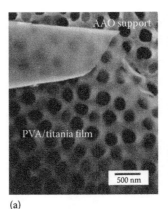 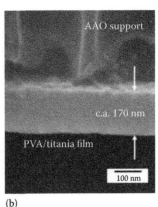

(a) (b)

FIGURE 11.4 SEM images of a PVA/titania film on an anodized aluminum oxide (AAO) support: (a) top view, and (b) cross-sectional view. (Adapted from Watanabe et al., *Bull. Chem. Soc. Japan*, 80, 433, 2007a. With permission.)

11.3.2 Organic Nanomembranes

The fabrication of robust, freestanding nanomembranes is not limited to ceramic components. Highly cross-linked materials are abundantly known for organic polymer components. Thermosetting resins are a representative class of such materials, being insoluble and infusible upon cross-linking by thermal treatment. Epoxy resin is a typical thermosetting resin and has attractive practical properties such as superior adhesiveness, dimensional stability, chemical resistance, and electrical inertness. Spin coating of an aged mixture of the two resin precursors gives uniform nanolayers with thicknesses in the range of 10–100 nm. Large, flexible 20 nm films with a size of over 5 cm² are formed without any traces of cracks and other defects on the surface (Watanabe and Kunitake 2007). This nanomembrane is robust enough to be isolated in solution or in air (Figure 11.5). Other thermosetting resins, melamine resin, urethane resin, and phthalic resin, give similar nanomembranes (Watanabe et al. 2007b). Their mechanical properties appear to reflect those of the corresponding macroscopic (thick) resins. The urethane nanomembrane shows plastic deformation unlike other nanomembranes, when air pressure is applied in the bulge test (see Figure 11.6).

Some of the cross-linkable acrylate monomers are also useful as precursors of nanomembranes (Watanabe et al. 2008). When a dilute solution of acrylic precursors (pentaerythritol tetraacrylate [PETA] or bisphenol A-functionalized acrylic oligomer [Kayarad R-280] (Figure 11.7)) and a photochemical radical initiator is spin-coated on a suitable underlayer and subjected to UV irradiation, efficient polymerization proceeds and freestanding nanomembranes that behave essentially identical to nanomembranes of macroscopic thermosetting resins are formed.

11.3.3 Nanomembrane of Organic and Inorganic Hybrids

Robust giant nanomembranes are obtainable as organic and inorganic hybrids. These hybrid materials may combine advantages of organic polymers (structural versatility, flexibility, and

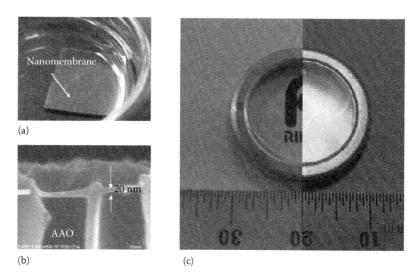

FIGURE 11.5 (a) An epoxy nanomembrane in ethanol, (b) SEM image of an epoxy nanomembrane on AAO support, and (c) combined photograph of a freestanding epoxy nanomembrane on a wire frame, with lighting (left) and without lighting (right).

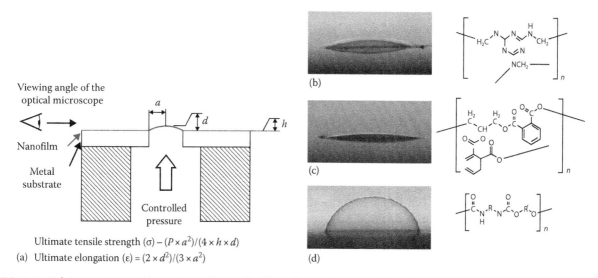

FIGURE 11.6 Bulging experiment: (a) experimental setup, (b–d) lateral views of various deflected nanomembranes under overpressure. (Adapted from Watanabe et al., *Bull. Chem. Soc. Japan*, 80, 433, 2007a. With permission.)

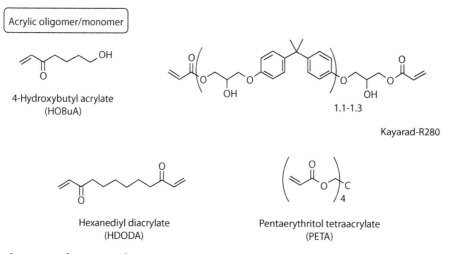

FIGURE 11.7 Chemical structure of various acrylic monomers.

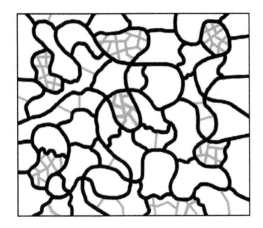

FIGURE 11.8 Schematic illustration of the interpenetrated network of organic (in black) and inorganic (in gray) components. (Adapted from Watanabe et al., *Bull. Chem. Soc. Japan*, 80, 433, 2007a. With permission.)

light weight) and inorganic glasses (thermal and mechanical stability). In fact, it is realized for nanomembranes of zirconium oxide and cross-linked vinyl polymer (Vendamme et al. 2006). Simultaneous radical polymerization and sol-gel reaction during the spin-coating process gives freestanding nanomembranes that contain two types of network structures, as schematically shown in Figure 11.8. A 35 nm thick nanomembrane of this kind is shown by SEM observation to be a uniform and defect-free layer (constant thickness ±10%). TEM observation indicates formation of a smooth amorphous surface at low (11.2 mol%) zirconium oxide (ZrO_2) fractions, but the presence of regular ZrO_2 lattices with domain sizes of 5–10 nm is noted at the ZrO_2 content of more than 20%. When the ZrO_2 component is replaced with SiO_2, the resulting hybrid nanomembranes possess a smooth, amorphous surface, and the domain formation of the inorganic component is not found even at a high SiO_2 fraction of 35 mol% (Vendamme et al. 2007).

In these examples of hybridization, the organic and inorganic components are not covalently bound. Entanglement and/or interpenetration of the two molecular networks provide physical hybridization and maintain the integrity of nanomembrane. The two network structures may be covalently connected to give chemical hybridization. This new system is prepared from a epoxy resin precursor that contains amino-group derivatized SiO_2 precursor (3-aminopropyl triethoxysilane) and epoxy-group containing oligomer (poly[(*o*-creyl glycidyl ether)-*co*-formaldehyde]) (Watanabe et al. 2009). The physical behavior of this nanomembrane is essentially identical to that of nanomembrane from the conventional epoxy resin.

11.4 Physical Properties of Nanomembranes

11.4.1 Macroscopic Behavior and Microscopic Morphology

The freestanding nanomembranes of macroscopic size as described above are all transparent and flexible. Their robustness depends on the nature of the components. The organic and hybrid nanomembranes are quite robust, while nanomembranes of metal oxides are rather fragile. A typical macroscopic behavior is illustrated by the example of a hybrid nanomembrane (thickness ca. 30 nm) of acrylate-zirconia. This freestanding nanomembrane with a size of 16 cm² floating in ethanol (Figure 11.9a) is readily sucked into a micropipette with a tip diameter of 320 μm by folding into small shapes. Figure 11.9b is a camera view of the aspiration process. It is able to pass through a 30,000 times smaller opening of micropipette thanks to its extreme thinness and flexibility. It is again released into the solvent without damage. Soon after the release, the hybrid nanomembrane adopts a folded morphology but can easily regain its original film shape by gentle manipulation with a spatula. Similar characteristic behavior is seen for all the available organic nanomembranes, and must be a common feature of robust nanomembranes.

The microscopic appearance of nanomembranes is essentially identical. As typically shown for an acrylate-zirconia nanomembrane, the whole membrane area is uniform and does not display any defect. The membrane thickness is, in most cases, uniform as seen from the cross-sectional SEM and TEM views of the membrane (Figure 11.10). X-ray diffraction patterns also indicate the membrane texture is generally amorphous for metal oxides and polymers, except for the case of domain formation in the ZrO_2 acrylate hybrid membrane.

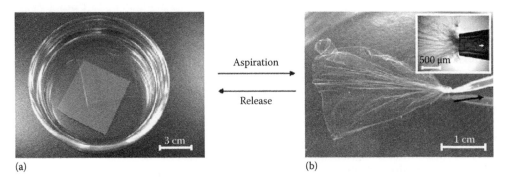

FIGURE 11.9 (a) An IPN hybrid nanofilm detached from the substrate and floating in ethanol and (b) the aspiration process of a 16 cm² nanofilm into a micropipette with a tip diameter of 320 mm (inset; close-up micrograph around the tip during the aspiration process). (Adapted from Watanabe et al., *Bull. Chem. Soc. Japan*, 80, 433, 2007a. With permission.)

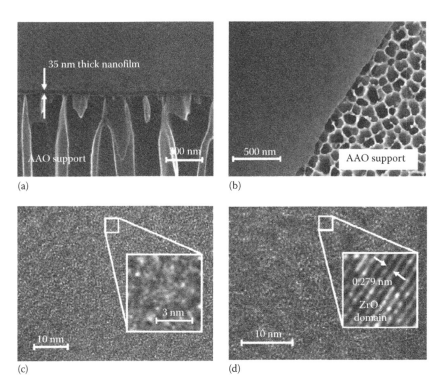

(a)   (b)

(c)   (d)

FIGURE 11.10 Microscopic characterization of freestanding hybrid nanofilms: (a) SEM side-view image of a nanomembrane on AAO support. (b) SEM top-view image of a nanomembrane. (c) TEM image of a low-ZrO_2 nanomembrane on a copper grid. (d) TEM image of a high-ZrO_2 nanomembrane showing the presence of an ordered domain dispersed in an amorphous matrix. (Adapted from Watanabe et al., *Bull. Chem. Soc. Japan*, 80, 433, 2007a. With permission.)

11.4.2 Measurement of Mechanical Properties

The estimation of some mechanical properties (tensile strength and ultimate elongation) of ultrathin membranes has become possible by the two recently developed techniques, i.e., bulge test and the strain-induced elastic buckling instability for mechanical measurements (SIEBIMM) technique. Goedel and Heger have introduced the bulge test for measuring the mechanical property of nanofilms (Goedel and Heger 1998). The experimental setup is shown in Figure 11.6a. The nanomembrane is attached on a metal plate with a circular hole, and air pressure is applied to the nanomembrane from below through the hole. Applied pressure is monitored by a digital manometer, and the membrane deflection is observed with an optical microscope. This behavior is compared for nanomembranes of thermosetting resins in Figure 11.6b through d. The nanomembrane deflects as the applied air pressure is increased and finally breaks. The extent of deflection varies with the kind of thermosetting resins and the nanomembrane of urethane resin gives the largest deflection. The tensile strength and ultimate elongation are estimated from balloon shape and applied pressure at the break.

An ordered wrinkle structure is formed as a result of structural instability by strain-induced elastic buckling, and this phenomenon has been widely used for the determination of mechanical properties of nanofilms (Bowden et al. 1998). Buckling instability that occurs in a bilayer of a stiff, thin layer and a relatively soft, thick substrate layer (poly(dimethylsiloxane); PDMS employed

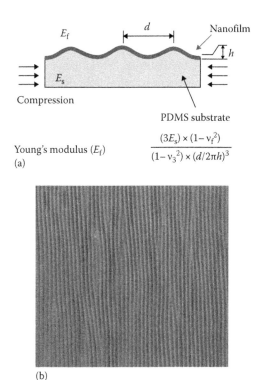

Young's modulus (E_f)

(a)

$$\frac{(3E_s) \times (1 - v_f^2)}{(1 - v_3^2) \times (d/2\pi h)^3}$$

(b)

FIGURE 11.11 SIEBIMM experiment: (a) schematic illustration of strain-induced elastic buckling instability and (b) microscopic picture of an obtained buckling texture of nanomembrane. (Adapted from Watanabe et al., *Bull. Chem. Soc. Japan*, 80, 433, 2007a. With permission.)

TABLE 11.1   Mechanical Properties of Various Nanomembranes

Material	Thickness (nm)[a]	Ultimate Tensile Strength (Pa)[b]	Ultimate Elongation (%)[b]	Young's Modulus (Pa)[c]
Epoxy-amine	24 ± 2	2.2×10^7	0.2	3.5×10^8
Melamine	25 ± 2	1.8×10^7	1.6	1.4×10^9
Phthalic	19 ± 1	1.2×10^7	3.7	1.2×10^9
Urethane	20 ± 3	$>1.0 \times 10^{7d}$	$>33.6^d$	2.8×10^9
Acrylate (R280)	22 ± 2	1.6×10^8	0.25	8.0×10^8
Acrylate (PETA)	28 ± 3	7.8×10^7	1.8	3.9×10^9
Acrylate-ZrO$_2$	30 ± 3	1.1×10^8	2.6	3.5×10^8
Acrylate-SiO$_2$	60 ± 6	4.5×10^7	1.3	1.1×10^9
Epoxy-SiO$_2$	48 ± 2	3.3×10^7	2.6	6.8×10^9
(PAH-PSS)$_9$[e]	35 ± 2	—	—	1.5×10^9

[a] Determined by SEM observation.

[b] Determined by the bulging test.

[c] Determined by the buckling measurement.

[d] Precise value not obtained due to plastic deformation.

[e] From the separate study (Jiang and Tsukruk et al. 2006).

most often) as shown in Figure 11.11a. When compressive force is applied, the bilayer produces highly periodic wrinkles, and the elastic modulus can be calculated from the pitch of the periodic buckling wavelength (Figure 11.11b). In practice, the PDMS composite is compressed by a small vise, and the wrinkle is monitored by an optical microscope and atomic force microscopy, and Young's modulus is determined from the pitch.

Typical data of the mechanical property are shown in Table 11.1.

11.4.3 Cross-Linking Density and Membrane Robustness

The high cross-linking density appears indispensable for maintaining the robustness of giant nanomembranes. It is useful to assess its significance, though qualitatively, on the basis of the available data. Physical properties arising from the extreme thinness of polymer membranes have been considered to be important in relation to quasi-2-dimensional behavior of polymer chain, in the case of poly(styrene) [PS]. However, this linear polymer is not a satisfactory material for giant nanomembranes. It is possible to obtain freestanding 100 nm thick PS (Mw; ca 100,000) film with 2×2 cm size by spin-coating, which is then transferred onto PDMS substrate in the SIEBIMM experiment. This PS nanofilm readily cracks during 2%–3% compression (commonly used pressure for this measurement). In contrast, the cross-linked nanomembranes never show such cracking behavior even at greater compression (5%–7%). Apparently, the chain entanglement of PS alone does not give membrane robustness that characterizes highly cross-linked nanomembranes.

The effective cross-linking is a second factor to affect the membrane robustness. The nanomembrane composed of a copolymer of 4-hydroxybutyl acrylate (HOBuA) and hexanediyl diacrylate (HDODA) (Figure 11.7) can be isolated in ethanol, but it is much swollen, and a robust, large-sized nanomembrane

is obtainable in the presence of the zirconia component. On the other hand, the zirconia nanomembrane alone is freestanding, but it is rather fragile. Satisfactory macroscopic robustness is attained when thermosetting resins are used as nanomembrane materials, probably due to enhanced cross-linking density. Its importance is supported even in the case of acrylate polymers, and robust nanomembranes are available from pentaerythritol tetraacrylate (PETA) that would lead to higher cross-linking density. The structural rigidity of precursor materials may also contribute to the macroscopic robustness. In the case of acrylic monomers, Kayarad-R280 provides a robust nanomembrane, in spite of its functional group density, which is not greater than that of HDODA. This difference appears to arise from the presence of rigid molecular component in Kayarad-R280.

11.4.4 Electrical Property

The electrical property of the cross-linked nanomembrane appears unchanged from that of the macroscopic counterpart. For a 30 nm thick epoxy (PCGF–PEI) nanomembrane that is transferred onto a p-type silicon wafer, an output leakage current of approximately 90 µA at 0.5 V is obtained, corresponding to a resistivity value of 0.5×10^{11} $\Omega \cdot$ cm. This highly insulating character is additional evidence of the defect-free nature, and is not lost upon detachment from the substrate. A conventional bisphenol-A-type epoxy resin gives a similar range of electrical resistivity values (10^{10}–10^{12} $\Omega \cdot$ cm).

11.4.5 Thermal and Chemical Stabilities

An outstanding example of thermal and chemical stabilities is given by a giant nanomembrane of covalently hybridized epoxy resin and silica as prepared from PCGF and aminopropyl-triethoxysilane (APS). There is no indication of thermal decomposition for the nanomembrane when it is heated on Si wafer up to 250°C in thermal desorption spectroscopy. SEM observation

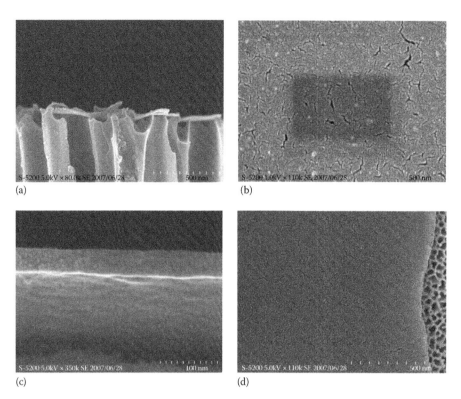

FIGURE 11.12 Cross-sectional and top view of SEM images of PCGF-APS1/1 nanomembranes: (a,b) after thermal treatment at 600°C for 30 min, and (c,d) after HF treatment. (Adapted from Watanabe et al., *Bull. Chem. Soc. Japan*, 80, 433, 2007a. With permission.)

gives a similar result. When the nanomembrane is kept at 600°C for 30 min on AAO substrate, its thickness is reduced from 53.7 to 24.2 nm, and becomes fragile enough to be damaged by electron irradiation for SEM measurement (1.0 kV). Nevertheless, the film morphology is maintained apart from shrinkage and the resulting cleavage. Since the organic component must be removed completely by the heating process, the remaining nanomembrane should be consisted of the inorganic part alone. This unchanged morphology suggests that the organic PCGF and inorganic APS components are dispersed homogeneously in the membrane (Figure 11.12). The PCGF-APS nanomembrane shows superior chemical stability. It does not show any swelling and shrinkage in solvents such as THF, chloroform, DMF, and cyclohexanone. When it is immersed in aqueous HF solution (1%) for 30 min, the film thickness is slightly reduced from 54 to 43 nm (20% shrinkage), but there are no other morphological changes.

11.5 Functional Potentials

The most significant characteristics of giant nanomembranes are the combination of nanometer thickness and macroscopic size, in addition to the variety of precursors and the uniformity and robustness of membranes. Unique industrial potentials arise from these factors. For example, giant nanomembranes may be made of ion-conducting materials and dielectric materials, and these properties will be much benefited by being nanometer thin. Highly efficient fuel cells can be constructed by employing

such nanomembranes, and dielectric membranes will find many uses in electronics. Freestanding insulating nanomembranes could be useful in large-area electronics like displays.

The basic features of giant nanomembranes are most desired in the case of permselective membrane. The outstanding selectivity and efficiency of biomembranes have been an ideal target of artificial membranes. Since those unique features owe much to the intricate molecular organization of biomembranes, it is hoped that analogous molecular organizations may be constructed in nanometer-thick artificial nanomembranes. The extreme thinness of the current giant nanomembrane may provide useful scaffolds for membrane architecture. Related to this potential is the artificial design of highly efficient permselective membrane by taking advantage of giant nanomembranes. Most of the emerging environmental technologies like production of clean water, recovery of rare materials, separation of H_2 and CO_2 and fuel cell membrane, desperately need novel features of membrane systems, as discussed with giant nanomembranes.

References

Bowden, N., Brittain, S., Evans, A. G., Hutchinson, J. W., and Whitesides, G. M. 1998. Spontaneous formation of ordered structures in thin films of metals supported on an elastomeric polymer. *Nature* 393: 146–149.

Decher, G. 1997. Fuzzy nanoassemblies: Toward layered polymeric multicomposites. *Science* 277: 1232–1237.

Endo, H., Kado, Y., Mitusishi, M., and Miyashita T. 2006. Fabrication of free-standing hybrid nanosheets organized with polymer Langmuir-Blodgett films and gold nanoparticles. *Macromolecules* 39: 5559–5563.

Forrest, J. A., Dalnoki-Veress, K., and Dutcher, J. R. 1997. Interface and chain confinement effects on the glass transition temperature of thin polymer films. *Phys. Rev. E* 56: 5705–5716.

Goedel, W. A. and Heger, R. 1998. Elastomeric suspended membranes generated via Langmuir-Blodgett transfer. *Langmuir* 14: 3470–3474.

Gu, H. and Swager, T. M. 2008. Fabrication of free-standing, conductive, and transparent carbon nanotube films. *Adv. Mater.* 20: 4433–4437.

Hashizume, M. and Kunitake, T. 2003. Preparation of self-supporting ultrathin films of titania by spin coating. *Langmuir* 19: 10172–10178.

Hashizume, M. and Kunitake, T. 2006. Preparations of self-supporting nanofilms of metal oxides by casting processes. *Soft Matter* 2: 135–140.

Huck, W. T. S., Stroock, A. D., and Whitesides, G. M. 2000. Synthesis of geometrically well defined, molecularly thin polymer films. *Angew. Chem. Int. Ed.* 39: 1058–1061.

Jiang, C. and Tsukruk, V. V. 2006. Freestanding nanostructures via layer-by-layer assembly. *Adv. Mater.* 18: 829–840.

Mallwitz, F. and Goedel, W. A. 2001. Physically cross-linked ultrathin elastomeric membranes. *Angew. Chem. Int. Ed.* 40: 2645–2647.

Mamedov, A. and Kotov, N. A. 2000. Free-standing layer-by-layer assembled films of magnetite nanoparticles. *Langmuir* 16: 5530–5533.

Vendamme, R., Onoue, S., Nakao, A., and Kunitake, T. 2006. Robust free-standing nanomembranes of organic/inorganic interpenetrating networks. *Nat. Mater.* 5: 494–501.

Vendamme, R., Ohzono, T., Nakao, A., Shimomura, M., and Kunitake, T. 2007. Synthesis and micromechanical properties of flexible, self-supporting polymer-SiO$_2$ nanofilms. *Langmuir* 23: 2792–2799.

Wang, L. H. and Porter, R. S. 1995. Thin films of nylon 13,13 casted on water: Morphology and properties. *J. Polym. Sci. B* 33: 785–790.

Watanabe, H. and Kunitake, T. 2007. A large, freestanding, 20 nm thick nanomembrane based on an epoxy resin. *Adv. Mater.* 19: 909–912.

Watanabe, H., Vendamme, R., and Kunitake, T. 2007a. Development of fabrication of giant nanomembrane. *Bull. Chem. Soc.* 80: 433–440.

Watanabe, H., Ohzono, T., and Kunitake, T. 2007b. Fabrication of large, robust nanomembranes from diverse, cross-linked polymeric materials. *Macromolecules* 40: 1369–1371.

Watanabe, H., Ohzono, T., and Kunitake, T. 2008. Fabrication of large nanomembranes by radical polymerization of multifunctional acrylate monomers. *Polym. J.* 40: 379–382.

Watanabe, H., Muto, E., Ohzono, T., Nakao, A., and Kunitake, T. 2009. Giant nanomembrane of covalently-hybridized epoxy resin and silica. *J. Mater. Chem.* 19: 2425–2431.

Xia, H. and Wang, D. 2008. Fabrication of macroscopic freestanding films of metallic nanoparticle monolayers by interfacial self-assembly. *Adv. Mater.* 20: 4253–4256.

12

Graphitic Foams

Juan Matos
Venezuelan Institute for Scientific Research

Eduardo B. Barros
Universidade Federal do Ceará

Josue Mendes Filho
Universidade Federal do Ceará

Antonio G. Souza Filho
Universidade Federal do Ceará

12.1 Introduction

The science of carbon materials has received an increasing attention in the recent years because of the novel magnetic and electronic properties of carbon nanostructures such as carbon nanotubes, fullerenes, graphene, and carbon nanofoams (Esquinazi et al., 2003, 2004; Makarova, 2004). Within the different forms of carbon materials, carbon foams are very interesting systems, presenting unique properties. Ford et al. first discovered carbon foams in 1964 as insulating foams with a vitreous microstructure (Ford, 1964). In fact, carbon foam is a name associated with a large family of porous structures synthesized from different carbon-based starting materials and with a wide range of structural properties and applications. The structural variation in the carbon foam family can happen in different levels. In the macro structural level, carbon foams that are characterized either by a two-dimensional distribution of open cells, forming a 2D network of ligaments and junctions, or by a three-dimensional distribution of cells, forming a porous medium composed of a 3D network of ligaments and junctions, can be produced (Ford, 1964; Matos and Laine, 1998; Klett, 2000; Klett et al., 2000, 2004). In the microstructural level, the different carbon foams (both the 2D and 3D networks) can be characterized by the different sizes of the cells and by their spatial distribution. The cell sizes range from nanometers to several micrometers, while the distribution of the cells can have different regular patterns or be at random (Matos and Laine, 1998; Matos et al., 2004; Matos et al., 2005; Klett, 2000; Klett et al., 2000; Klett et al., 2004). In a more fundamental level, the carbon foams can be characterized from the point of view of their crystalline structure (or their lack of it, when this is the case) since the solid structure in the ligaments and junctions can be composed of either amorphous (vitreous) carbon or by crystalline graphitic structures that will greatly affect their electrical and thermal management properties. This structural versatility enables a series of different and unique properties such that these materials are very promising for both technological applications and basic studies. In this chapter, we intend to discuss some of the interesting properties of carbon foams. For clarity, this chapter is developed as follows. In Section 12.2, we present the basic steps involved in the synthesis of carbon foams and discuss the possible final structures as results of the appropriate synthesis method. In Section 12.3, we discuss the structural properties of some of the different types of carbon foams. In Sections 12.4 and 12.5, we discuss some general properties of two different types of carbon foams and, how their properties can be modified and analyzed. Section 12.6 follows with some concluding remarks.

12.2 Synthesis Methods

Carbon foams have been prepared by several different methods, using different precursors, and the properties of the resulting material will be strongly dependent on both the production method and the carbon source. Most of the carbon foam preparation methods are based on two basic steps: pyrolysis and carbonization.

In order to start the process the precursor, which is the carbon source, needs to be in a liquid state, which can be obtained by different methods depending on the precursor, such as heating the precursor up until it starts to melt or by dissolving the precursor in a proper solvent. After this preparatory step, the temperature and pressure can be increased at an inert atmosphere so that the lower molecular weight molecules in the precursor start volatilizing and thus forming bubbles. This second part of the process is characterized by the pyrolysis (polymerization) of the precursor, which becomes increasingly denser and less viscous with the release of the volatile matter, and it is at this point that the porous structure of the foam starts to form. The viscosity keeps increasing with the temperature to a point when the material cannot be further melted, and at this point, the foaming is finished. After this, the process is finalized with the carbonization of the resulting foam at high temperatures in order to obtain a pure carbonaceous structure.

Although these basic steps are recurrent in the foaming procedure, some of these steps can be modified, or other steps can be included depending on the precursor and the intended final material. In some cases, an oxidative stabilization was used in order to prevent the porous structure of the foam from melting during the subsequent steps (Kearns, 1999).

12.2.1 Mesophase Pitch-Based Carbon Foams

The highly graphitic carbon foams developed at the Oak Ridge National laboratory (ORNL), produced using mesophase pitch as a precursor, are characterized by having a high strength together with high electrical and thermal conductivities due to the interconnected graphitic ligament network (Klett, 2000; Klett et al., 2000, 2004). In order to obtain this crystalline graphitic structure, the carbon foams were further processed through a graphitization process, done at higher temperatures, of the order of 2800°C, also under an inert environment (Klett, 2000; Klett et al., 2000, 2004).

12.2.2 Saccharose-Based Carbon Nanofoams

Carbon foams have also been prepared by other methods such as controlled pyrolysis of saccharose (Matos and Laine, 1998; Matos et al., 2004, 2005), coal (Calvo et al., 2008), and recently by microwave irradiation of biomass (Wang et al., 2008). Biomass is an interesting carbonaceous precursor because it is cheap, eco-friendly, and renewable and there is abundant natural sources such as wood or agro-industrial wastes whose chemical composition contain a variety of organic polymers that include cellulose, hemicellulose, and lignin. Thermally controlled degradation or pyrolysis of biomass or products derived like saccharose, glucose, fructose, and so on, allows for the synthesis of carbon foams with different shapes and dimensions.

12.3 Carbon Foam Structure

12.3.1 Random Networks

Some carbon foams are random network structures that are cellular structures characterized by cells having random areas and number of sides. They constitute a large class of materials

ubiquitous in nature and include polycrystalline structures observed in metals, ceramics, polymers, Langmuir monolayers, magnet froths, flame cells, biological tissues, soap froths, etc. (Weaire and River, 1984; Fradkov et al., 1985, 1993; Stavans and Glazier, 1989; Stine et al., 1990; Noever, 1991; Weaire et al., 1991; Stavans, 1993; Huang et al., 1995; Kamal et al., 1997). These foams differ from molecular networks such as fullerenes and nanotubes that are very ordered. Another type of carbon network is that corresponding to the microporous structure of activated carbons (Bansal et al., 1998). It is assumed to be constituted by polyhexagonal sheets of carbon, forming micro- and mesopores having pore widths between 10 and 500 Å (Laine and Yunes, 1990; Stoeckli, 1990). Three carbonaceous network structures can be considered according to the dimension: the molecular (fullerenes and nanotubes), the micro (microporous carbon), and the macro (macroporous carbon). Although, in the case of carbon materials, synthetic routes for both molecular-networks and micro-networks have been extensively investigated,[*] few studies exist regarding the synthesis and characterization of macro-networks (Matos and Laine, 1998; Matos et al., 2004, 2005). Particularly, the study of the topological and textural characteristics of carbon films and carbon foams is of considerable research interest since they are closely related both with the physical and chemical properties of these materials (Schwan et al., 1997). Carbon films can be considered as two-dimensional (2D) random cellular networks, and therefore they can be treated as random partitions of the plane by cells, as in the case of bidimensional (2D) polygons that have three or more sides (Weaire and River, 1984; Huang et al., 1995; Kamal et al., 1997; Miri and Rivier, 2001). Although the topology of cellular networks imposes constraints on the possible configuration of the cells, it is assumed for most natural and simulated systems that an infinite space-filling two-dimensional network with trigonal vertices obeys the Euler topological relation (Peshkin et al., 1991; Dubertret et al., 1998a,b):

$$\langle n \rangle = \sum_n \left[n \cdot \rho(n) \right] = 6 \tag{12.1}$$

for the first moment of the cell-side distribution, where $\rho(n)$ is the probability of occurrence of a cell with n sides, which corresponds to the trivial constraint on probabilities, given by

$$\sum_n \left[\rho(n) \right] = 1 \tag{12.2}$$

Detailed description of some of the concepts involved in the topological definition of these kind of random cellular structures

---

[*] Important information on the synthetic routes for both molecular-networks and micro-networks of carbon materials can be found in the works of Bansal et al. (1998); Stoeckli (1990); Laine and Yunes (1990); Kroto et al. (1985); Mochida et al. (1997); Osipov and Reznikov (2002); Pablo et al. (2001); Zhang et al. (2001); Cao et al. (2001); Diehl et al. (2002); Laine et al. (1991); Govor et al. (2000); Dunne et al. (1997); Ng et al. (2002); Imamura et al. (1999); Pan et al. (2002); and Rouzaud et al. (1989).

TABLE 12.1 Comparison of Peak Value in Cell-Side Distribution ρ(*n*) from Various Works

System	Peak *n*-Value	ρ(5)	ρ(6)	References
Carbon foam	5	0.334	0.206	Matos and Laine (1998)
K-doped Carbon foam	5	0.353	0.233	Matos et al. (2004)
Polymer polygrain	5	0.392	0.25	Huang et al. (1995)
Flame cells	5.4	0.36	0.34	Noever (1991)
Soap froths	5.5	0.3	0.3	Stavans and Glazier (1989)
Langmuir monolayer	6	0.225	0.345	Stine et al. (1990)
Polycrystalline metal	6	0.267	0.4	Fradkov et al. (1993)
Polycrystalline SCN	6	—	—	Fradkov et al. (1993); Stavans (1993)
Magnet froth	6	0.25	0.55	Weaire and River (1984)

may be found elsewhere (Fradkov et al., 1985; Rivier, 1985; Stavans and Glazier, 1989; Glazier et al., 1990). Some examples exist in the literature regarding other cell structural systems where the hexagon appears as the most abundant polygon as can be seen in Table 12.1.

12.3.2 Quasi-Regular Structures

Carbon foams can also be developed with a more or less regular structure, where the size and distribution of the pores is more or less homogeneous around the sample. One example of such structures is the carbon foam obtained from mesophase pitch, which is characterized by a regular assembly of spherical cells surrounded by graphitic material forming a regular tetrahedron of graphitic material with a cell associated with each of its vertices. The spherical cells formed in the process described above intersect with each other forming pores of different sizes. Despite this difference in pore size, the pore shapes are fairly uniform with a narrow pore size distribution forming a quasi-regular porous structure. The relation between the size of the tetrahedron and the mean radius of the cells determines the porosity of the medium. Figure 12.1 shows an illustrative diagram of three junctions calculated for different porosities (10%, 40%, 70%, and 95%), defined as the ratio between the cell volume to the total volume (Mukhopadhyay et al., 2003). These tetrahedral junctions are then interconnected to form the carbon foam. Each junction is connected through ligaments to four other junctions in the directions of a regular tetrahedron. In the illustration of the foam with 95% porosity shown in Figure 12.1, the ligaments

are easily seen as the four ends connected to the center of the junction. For porosities less than 70%, the ligaments are not easily observed, because they become thick and are more like the junction itself. It is clear that the difference between the properties of the ligaments and the junction should become less evident for lower porosity foams.

12.4 Structural and Vibrational Properties of Graphitic Foams

In this section, we aim to better understand the correlation between the structural and the vibrational properties in graphitic foams. The highly ordered mesophase pitch-based graphitic foam is used here to exemplify the vibrational properties of carbon foams. This material was chosen due to the fact that its graphitic structure allows for a more precise evaluation and measurement of these properties. However, most of the features presented in this section will be common to all carbon foams, although it is clear that for highly disordered carbon foams some of these features will become increasingly less important.

To understand the vibrational properties of graphitic materials such as the foams described in this chapter it is necessary to have a basic comprehension of the overall vibrational properties of graphite. Raman spectroscopy is one of the most used techniques to probe the vibrational properties of materials. Briefly, the Raman scattering technique is based on the inelastic scattering of light when the radiation interacts with the matter. Due to this interaction, some photons come out

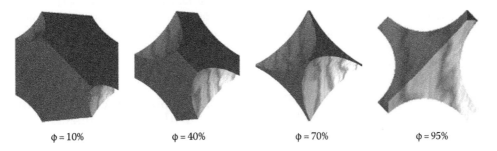

φ = 10%   φ = 40%   φ = 70%   φ = 95%

FIGURE 12.1 Illustrative diagrams of possible junctions of a graphitic foam for different porosities (φ). (From Mukhopadhyay, S.M. et al., *J. Appl. Phys.*, 93, 838, 2003.)

from the sample with higher or lower energy than that of the incoming photons. The gain or loss in the photon energy is due to the atomic vibrations of the material that in turn depends on its symmetry properties and the Raman spectrum carries out information about the material's structure. When the radiation gains energy (the energy contained in vibrations are transferred to the photons) we have a process called anti-Stokes. The Stokes process occurs when energy is transferred from the radiation to the vibrations.

In the case of graphite and its derivatives, the Raman spectra also brings structural and electronic information due to the strong electron–phonon coupling between some of the vibrational modes and electrons near the Fermi level. This strong electron–phonon coupling gives rise to one of the most interesting aspects of the Raman spectra of graphitic systems, which is the observation of Raman features that are associated with a double resonance process (Thomsen, 2000; Saito et al., 2002).

The unique Raman spectra of graphite mainly consist of three strong features. The G band, with a characteristic frequency of approximately $1582\,cm^{-1}$ is observed in single crystal graphite and corresponds to the in-plane vibrations of the nearest neighbor atoms in the graphene (2D graphite) layers. The other two commonly observed features, known as the D- and G'-bands, both originate from a $q \neq 0$ phonon near the Brillouin zone boundary and are associated with a double resonance process (Thomsen, 2000). Since the double resonance process is intrinsically dependent on the electronic properties of the material, it is clear that the behavior of these Raman peaks can be used to probe some of these properties. Furthermore, the double resonance Raman peaks can also be used to probe some of the structural information. For instance, it has been shown that the process responsible for the D-band depends on the presence of lattice defects (Thomsen, 2000; Cançado et al., 2002). Therefore, it is expected that the intensity of the D-band relative to the G-band can be used to probe the distribution of defects in any sp^2 carbon-based system. The other double resonance Raman peak, the G'-band originates from a second-order scattering process involving two phonons with opposite momenta, $+q$ and $-q$. The phonon that is involved in the G'-band process is the same as that of the D-band, however, since this is a two-phonon process, its intensity does not depend on the presence of lattice defects and thus the G'- band can be observed in highly crystalline graphitic materials. The frequency of the G'-band is given by $\omega_{G'} \sim 2\omega_D$, where ω_D denotes the D-band frequency (Koenig and Tuinstra, 1970; Vidano et al., 1981). Although the G'-band cannot be associated with the presence of defects, its line shape can also be used to probe structural information about the graphitic material. This can be obtained by observing that the G'-band of an isolated graphene layer (2D graphite) is different from that of bulk graphite, where the…ABAB… stacking of the graphene layers causes a splitting in this peak. The dispersion of the single peak G'-band feature associated with 2D graphite can be obtained from the work of Cançado et al. (2002) on turbostratic graphite and is found to be $106\,cm^{-1}eV^{-1}$. For 3D graphite, the frequency of the strongest peak and an approximate value for its dispersion can be obtained from the work of Matthews et al. (1999)

on highly oriented pyrolytic graphite (HOPG). For an excitation of 2.41 eV, the frequency of the strong G'-band peak was found to be approximately $2726\,cm^{-1}$ with a dispersion of $\sim 94\,cm^{-1}/eV^{-1}$. The study of the dispersion of the double peak structure of 3D graphite is still in an early stage and the physical reason for the difference in the dispersion for 2D and 3D graphite is still unexplained in the literature. Only a few values for the frequency and the dispersion of the weaker peak on the G' band of 3D graphite could be obtained from the literature (Barros et al., 2005). It is important to stress that although the D band and the G' band are known to originate from the same phonon and using a similar double resonance process, the D band relative intensity and the difference in line shape of the G' band for 2D and 3D graphite are completely uncorrelated effects. The D band intensity is associated with in-plane defects that are able to scatter the photon-excited electrons between different points of the Brillouin zone allowing the double resonance process to have a strong scattering cross-section. On the other hand, the difference in the G' line shape for well-stacked and poorly stacked graphite is, in principle, independent from the presence of in-plane defects, such as vacancies, dislocations and in-plane crystallite boundaries. Instead, the line shape of the G' band is determined by the interaction between different graphene planes that account for the appearance of extra electronic bands that resonantly enhance phonons of slightly different wave vectors and thus contribute to the G' band on slightly different energies. The total G' band line shape is thus a composition of these different double resonance features.

In the case of the graphitic foam, they were observed to be composed of two intermixed graphitic regions, one where the planes are well aligned and the structural behavior follows that of highly aligned pyrolytic graphite (HOPG), and other regions where the planes are not well aligned, and therefore, the interaction between the graphitic planes is small, leading to properties that are similar to that of isolated graphene (2D-graphite) layers. Raman spectroscopy was found to provide a powerful tool to study the structural properties of these carbon-based systems. The Raman spectra analysis was successful in differentiating between the contributions from the two limiting cases of interplane interaction, highly aligned graphitic regions where the planes show stacking order and the structural behavior follows that of HOPG (3D), and poorly aligned graphitic regions where the decrease in the interplane interaction causes the structural behavior to approximate that of isolated graphene (2D) layers. For that reason, the graphitic foam material could be used to provide important information on the contrasting properties of 2D and 3D graphite, such as the thermal conductivity and the vibrational properties.

12.4.1 Probing Structural Information with Raman: Defects

To allow for a better comprehension of how Raman spectroscopy can be used to probe the presence of in-plane lattice defects, a few experiments were conducted on graphitic foams which are

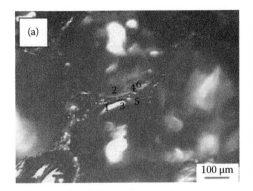
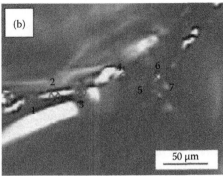

FIGURE 12.2 (a) Optical image of the graphitic foam showing an overall view of the porous structure of the foam. The numbers 1–7 indicate the center of the laser spot where the Raman spectra were taken. (b) Higher magnification image of the studied region.

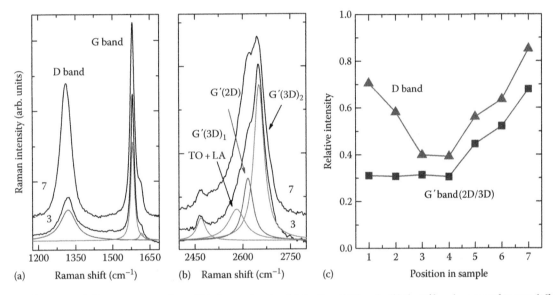

FIGURE 12.3 Characteristic Raman spectra of graphitic foams showing (a) the D and G bands, (b) the G′ band measured in two different points of the sample (labeled 3 and 7 in Figure 12.1). The Raman spectra in (a) and (b) were normalized to the intensity of the G band and the G′(3D)$_2$ peaks, respectively. (c) Ratio between the of G′(2D) and G′(3D) contributions (squares) and relative intensities of the D band (triangles) for the different laser spots indicated by numbers 1–7 in Figure 12.2a,b. (From Barros, E.B. et al., *Vibr. Spectroscop.*, 45, 122, 2007. With permission.)

here described. Figure 12.2a shows an optical image of a selected region of a graphitic foam sample. This picture was taken at low magnification (10×) and it shows the porous structure of the graphitic foam, giving an overall image of the region where the experiment was performed. Throughout the process of cutting the sample into small cubes, several junctions and ligaments were sectioned so that they could be observed from the outer surface of the sample. By studying these sectioned areas, it is possible to probe the properties of the core of the graphitic structure instead of being limited to studying only its surface. Figure 12.2b show a higher magnification image of the sectioned region of the graphitic foam that was studied, this image was obtained using the same 50× objective lens that was used to focus the laser beam. It should also be mentioned that the process of cutting the graphitic foam is likely to increase the level of disorder in the ligaments and junctions. However, the difference in the D-band intensity observed for the various spots (see

Figures 12.2 and 12.3) is likely to be mainly due to differences in the original disorder within the graphitic structure.

Figure 12.3b shows the Raman spectra of the G′-band. The intensity was normalized relative to the intensity of the peak at 2650 cm^{-1}. Based on previous studies of the G′-band of PPP (Cançado et al., 2002) and HOPG (Matthews et al., 1999), the peak at approximately 2616 cm^{-1}, G′(2D), could be assigned to contributions from 2D graphite present in the sample, while the two peaks at 2580 and 2650 cm^{-1}, G′(3D)$_1$ and G′(3D)$_2$, comes from the contribution of highly aligned 3D graphitic structures. It is clear that areas with different interplanar interactions will give rise to slightly different G′ band line shapes, as seen in Figure 12.3b and thus the G′(2D) and G′(3D) peaks observed here should be interpreted as the two limiting cases, while the intermediate cases with the interplanar distance between that of ideal 3D graphite and of 2D turbostratic graphite give rise to a line broadening of these peaks.

To show how the properties of the graphitic foams change along its structure, a series of Raman experiments were performed spanning the regions of the junction and of the ligament. Raman spectra were taken at 7 selected spots, numbered 1–7 in Figure 12.2a,b, and the relative intensities of the D band (triangles) and G′ band (squares) features for the different laser spots are plotted in Figure 12.3c. Each number in the horizontal axis correspond to one of the spots numbered 1–7 in Figure 12.2a,b. The G′ band intensity ratio shown in Figure 12.3c represents the ratio between the intensities of the contribution of the G′(2D) peak to the contribution of the G′(3D) peak. The ratio between the intensities for the G′(2D) and G′(3D) peaks is very low in the region of the ligament (spots 1, 2, and 3), which indicates that the concentration of 2D graphite in this region is low (about 0.3) in comparison to the concentration of 2D graphite in the junction, where this ratio is of about 0.8, and it does not change significantly within this region. However, as the laser beam is focused closer and within the junction the concentration of 2D graphite is seen to increase gradually to its maximum, which occurs at the edge of the junction (spot 7). This technique can only give qualitative information on the distribution of the poorly aligned graphitic structures within the sample and not on the absolute concentration of this disordered structure with respect to the highly aligned structures. The triangles in Figure 12.3c show the evolution of the ratio between the D-band and the G-band intensities as the laser spot is moved across this region of the sample. Note that in spot 3, which is in the middle of the ligament, the relative D-band intensity is very weak and that as the laser is focused closer and within the junction the D-band intensity

gradually gets stronger, as seen for spots 4–7. This behavior follows the same pattern observed for the G′(2D) band to G′(3D) band intensity ratio that indicates that the defects in this region originate from the 2D graphite regions. However, for spots 1 and 2 the D-band intensity becomes stronger while the G′(2D)-band intensity does not change. This suggests that the strong D-band peak observed in spots 1, 2, and 3 originates from defective 3D graphite in this region and that the density of defects increases as the laser is focused away from the center of the ligament.

To better understand this correlation and also to probe other interesting properties of the Raman spectra of graphitic foams, another area was selected for analysis, shown in Figure 12.4a and b (where Figure 12.4b is a higher magnification image of the square area in Figure 12.4a). This area was chosen since there were some optically visible features that could be tracked and used for positioning. The dark squares in Figure 12.4b delimit 5 × 5 μm areas of the graphitic foam that were studied using Raman spectroscopy. To produce a map of the different Raman features as a function of position, we divided the areas in a 21 × 21 square mesh, each point of the mesh separated from one another by 500 nm and Raman spectra were obtained for each point of the mesh. Although the movement within the mesh was achieved with the help of an automated stage with a high reproducibility, the position of the center of the square varied by a small amount each time the experiment was performed. To compensate for this variability, we used a 2D spline algorithm to smoothen the maps and make the correlation between different experiments more accurate. The laser power on the sample was limited to 1 mW to prevent overheating of the sample. The leftmost area is

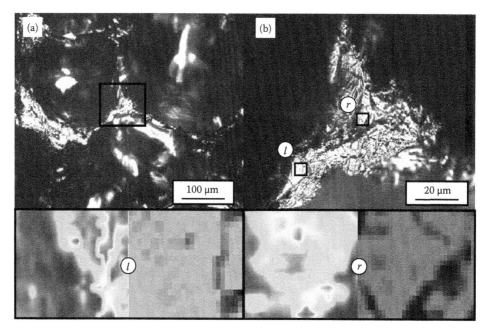

FIGURE 12.4 (See color insert following page 22-8.) (a) Optical image of the graphitic foam sample showing the porous structure of the foam. (b) High magnification image of the area delimited by the square in (a). The small *l* and *r* squares in (b) delimit the 5 × 5 μm areas from which the Raman spectra were obtained. The lower panels show a comparison between the morphological details of the areas *l* and *r* together with the I_D/I_G mapping shown in Figure 12.5. (From Barros, E.B. et al., *Vibr. Spectroscop.*, 45, 122, 2007. With permission.)

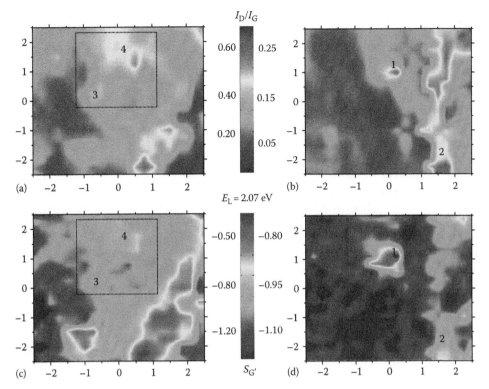

FIGURE 12.5 **(See color insert following page 22-8.)** (a,b) Maps of the spatial distribution of the ID/IG ratio for two different regions of the sample, each with 25 μm² squared area. The Raman spectra were obtained in a 21 × 21 mesh and smoothened to form the map shown here using a 2D spline algorithm. (c,d) Mapping of the Skewness of the G′ band for the same two regions of the graphitic foam as (a) and (b), respectively. (From Barros, E.B. et al., *Vibr. Spectroscop.*, 45, 122, 2007. With permission.)

closer to the ligaments, while the square to the right is close to the center of the junction. Figure 12.5a and b shows the mapping of the D band to G band integrated intensity ratio as a function of sample position for a laser excitation energy of 2.07 eV for the left and right square regions in Figure 12.4b, respectively. The areas in bright red represent highly defective regions, while the areas in blue are regions with high crystallinity (low I_D/I_G values). Different color scales were used in each map in order to enhance the contrast and clarify the visualization. It is interesting to note that the different scales reflect the fact that the mean I_D/I_G ratio is larger at the junctions (Figure 12.4b) than in the region closer to the ligaments, which is in agreement with the initial results shown in Figure 12.3a. Also, it should be noted that the density of defects in these regions of the sample is not continuously distributed. Instead, there are clusters of defective regions that remain even after the process of graphitization. The behavior observed in Figures 12.2a and b, where a correlation between the presence of defects in the graphene plane with the high intensity of the lower frequency G′ peak originating from poorly stacked 2D graphite in the sample could be clearly seen, can be reproduced in these regions. However, in order to accomplish this comparison in a more detailed level, it is necessary to produce a detailed map of the 2D to 3D graphite content in the same region of the sample and compare it with the I_D/I_G ratio map. A simple procedure was developed (Barros et al., 2006) applying the concept of skewness of a distribution in order to

evaluate the line shape of the G′ band and qualitatively obtain the relative amount of 2D graphite in the sample. The skewness is related to the third moment of a distribution and gives the asymmetry of the curve with respect to the mean value. The main advantage of using this concept for evaluating the line shape of the G′ band is that it is independent of fitting parameters and is a computationally cheap. There are several different ways of defining a quantitative measure of the skewness of a distribution. In view of Ref. (Barros et al., 2006) the following equation is used to define the skewness of the G′ band, $S(G')$:

$$S = \frac{1}{A} \int d\omega \frac{(\omega - \bar{\omega})^3}{\sigma^2} I(\omega), \qquad (12.3)$$

where

$I(\omega)$ is the intensity of the G band at a frequency ω
the values A, $\bar{\omega}$, and σ are, respectively, the integrated area, the mean frequency, and the standard deviation of the G′ band

These values are related to the zeroth, first, and second moments of the G′ line shape, respectively. Changes in the relative intensities of different peaks contributing to the G′ band should be correlated with changes in its skewness and thus it is clear that the skewness should carry information about the ratio between highly and poorly aligned graphite within the studied area. Since the lowest frequency peak of the G′ band of 3D graphite is always weaker than the highest frequency peak, the skewness of the 3D graphite G′ is always negative. Therefore, a negative skewness indicates that the 3D graphite phase is dominating the G′ band line shape. The contribution from poorly stacked graphite to the G′ band shows up between the two G′(3D) peaks and tends to increase the value of the skewness towards zero. In this sense, an increase in the value of the skewness can be interpreted as an enhancement in the contribution from the poorly stacked graphite and thus the red areas in Figure 12.5c and d can be interpreted as areas where the density of the 2D graphite is high compared to the density of the well stacked 3D graphite. It should be mentioned that the mapping of the $I_{G'}/I_G$ ratio does not show any distinguishable features, indicating that the G′ band intensity cannot be associated with the presence of defects, only its line shape.

It can be seen from these maps that the areas of Figure 12.5a and b that are characterized by a large I_D/I_G ratio (red areas) are also characterized by a high density of 2D graphite (also shown in red), in such a way that the two sets of maps, (a)–(c) and (b)–(d), showing information of two different peaks, have a clear similarity. As explained above, this correlation is not associated directly to the similar double resonance processes involved for the D band and G′ band Raman features. Instead, this result can be interpreted as a correlation between the in-plane crystallite size (L_a) with the stacking heights of the crystallite (L_c). It can be understood that the presence of defects breaks the translational commensurability of the…ABAB… stacked layers, causing the layers to become well separated and thus to loose the inter-layer site correlation. This interpretation is corroborated by the fact that the frequency of the component of the G′ band that originates from the 2D graphite areas in the graphitic foam contribution is nearly two times the frequency of the D band, as shown in Figure 12.6.

This indicates that the strong D band and the middle frequency peak, which is being associated with the presence of 2D graphite originate from phonons with the same frequency and thus are expected to arise from the same region of the graphitic foam.

However, it should be mentioned that a detailed comparison between the I_D/I_G ratio and the skewness of the G′ band in Figure 12.4 shows that in some areas the correlation between the two maps is stronger than in others. The difference between the two maps originate from the fact that not all the defects in the system are associated with the misalignment of the…ABAB… layers and that highly aligned graphitic areas can also contain a large concentration of defects, while misaligned areas can be composed of defect-free graphene layers. Thus, this correlation is related to the graphitization process and to the fact that graphene planes with a high concentration of defects had a lower chance of forming well stacked crystallites.

12.4.2 Dependence of the Raman Spectra of Graphitic Foams on the Laser Excitation Energy

One of the main problems in using the Raman spectra for the structural characterization of the graphitic foams lies in the fact that the double resonance process is intrinsically connected with the excitation energy. For this reason, both the intensity and the line shape of the Raman features are strongly dependent on the laser used in the experiment. In order to better analyze the samples it is thus important to have a good knowledge of this dependence and make the necessary corrections.

In order to improve the understanding of the dependence of the Raman spectra of graphitic foams on the laser excitation energy (E_L), the mapping of the D band relative intensity was also performed using different laser excitation energies (Barros et al., 2006). The defective features which are distinguishable both in the optical image and on the I_D/I_G map can be used as markers in order to make a good comparison between maps obtained at different laser energies. Four of these distinguishing features are labeled in Figure 12.5 with numbers 1–4 in order of increasing magnitude for the I_D/I_G ratio. Figure 12.7 shows a series of I_D/I_G ratio maps of the same box region of the graphitic foam corresponding to different laser energies (E_L). For comparison, all the maps in this series are plotted using the same scale as that used in Figure 12.5b. It can be seen from this color map that I_D/I_G ratio is strongly dependent on the excitation energy, decreasing with increasing laser energy. To better quantify this dependence, Figure 12.8 shows the dependence of the mean I_D/I_G ratio associated with a $2 \times 2\,\mu m$ area identified with each of the labeled features in Figure 12.5 as a function of E_L. The I_D/I_G ratio plotted in this figure was best fitted to a linear equation that can be written in the form $I_D/I_G = i_0(1 - \alpha E_L)$. The values of i_0 and α that gave the best fit are shown in Table 12.1 and the curves obtained using this fitting procedure are shown as solid lines in Figure 12.8. It is interesting

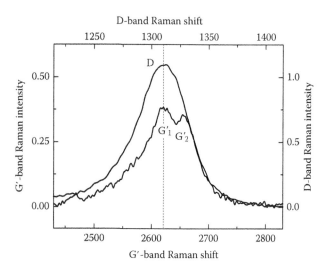

FIGURE 12.6 D and G′ band for the same defective region of the graphitic foam. The top and right scales give information on the frequency and intensity of the D band, respectively, while the bottom and left axes are for the G-band. The intensities were normalized to the G band intensity and the frequency scale of the G′ band is doubled. The dashed line highlights the fact that the lower frequency peak in the G′ band has a frequency of twice the frequency of the observed D band.

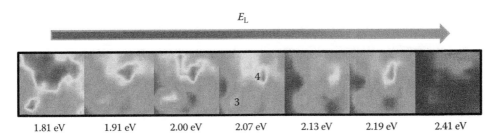

E_{L}

| 1.81 eV | 1.91 eV | 2.00 eV | 2.07 eV | 2.13 eV | 2.19 eV | 2.41 eV |

FIGURE 12.7 **(See color insert following page 22-8.)** Mapping of the $I_{\mathrm{D}}/I_{\mathrm{G}}$ ratio for a selected region of the sample, indicated with a square in Figure 12.5a, obtained using different laser excitation energies. All the maps are normalized with the same parameters to emphasize the dependence of the $I_{\mathrm{D}}/I_{\mathrm{G}}$ ratio on the laser excitation energy.

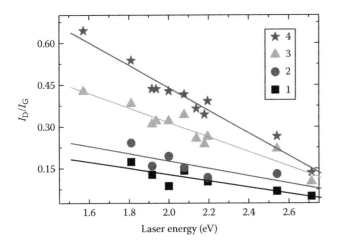

FIGURE 12.8 Dependence of the $I_{\mathrm{D}}/I_{\mathrm{G}}$ ratio for the four spots in Figure 12.3 on the laser excitation energy. The solid lines show the $i_0(1 - \alpha E_{\mathrm{L}})$ best fit curves. The values for i_0 and α for each curve are shown in Table 12.2. (From Barros, E.B. et al., *Vibr. Spectroscop.*, 45, 122, 2007. With permission.)

to see that for all regions of the sample the obtained value for α was of approximately 0.31 eV^{-1}. Thus it seems that this value is independent of the crystallite size or the density of defects. The only fitting parameter that changes for different crystallite sizes is the parameter i_0, which seems to be intrinsically related to the crystallite size or the density of defects. It can thus be concluded that, in first approximation, the laser energy dependence of the D band relative intensity is independent from the density of defects and that a careful parametrization and calibration of the experiment will, in principle, allow for a quantitative measurement of the mean crystallite size in the studied area of the graphitic foam sample using the obtained Raman spectra. However, it should be stressed that this can only be done after taking into consideration the effect of the laser excitation energy. This is of great importance when comparing measurements that were performed using different laser excitation energies. The laser energy dependence of the $I_{\mathrm{D}}/I_{\mathrm{G}}$ ratio has been also investigated for other graphitic materials, such as graphite nanocrystals with different crystallite sizes. Through a series of careful experiments, Cançado et al. (2006) obtained an accurate empirical relationship that associates the

relative D band intensity ($I_{\mathrm{D}}/I_{\mathrm{G}}$ ratio) to the crystallite size and the laser excitation energy for these materials. For the graphite nanocrystals, the $I_{\mathrm{D}}/I_{\mathrm{G}}$ ratio was observed to scale with E_{L}^{-4} and with the inverse of the crystallite size. It is possible to tentatively compare the laser energy dependence obtained for the $I_{\mathrm{D}}/I_{\mathrm{G}}$ ratio in graphitic foams with that obtained for graphite nanocrystals by fitting the behavior in Figure 12.8 with an allometric function written as $I_{\mathrm{D}}/I_{\mathrm{G}} = aE_{\mathrm{L}}^{-b}$. The best-fit parameters obtained by using this procedure are also shown in Table 12.1. It can be seen that the value of b varies between 1.75 and 2.62 for the graphitic foams, which is very different from the $b = 4$ value measured by Cançado et al. on the graphite nanocrystals (Cançado et al., 2006). The different behavior with laser energy observed between the graphitic foams and the graphite nanocrystals come from intrinsic differences between the two materials. This effect can be related to the fact that the graphitic foam material is composed of a mixture between the 2D and 3D phases and that the microscopic foam structure implies constraints on the crystallite orientation and on the curvature of the graphene planes where the strain effects should be important. A careful analysis of this effect is still needed and would be of good use for the understanding of this system. To complete the analysis of the Raman spectra of graphitic foams with the laser excitation energy it is also necessary to investigate the behavior of the G′ band for the different laser lines. The relative intensity of the G′ band ($I_{\mathrm{G'}}/I_{\mathrm{G}}$) was obtained for the same four regions described earlier and the behavior was found to be similar to that of the D band. Indicating that this behavior is in some way connected to the double resonance process involved in the Raman intensity of the D band and G′ band, and thus not directly connected with the defects. The line shape of the G′ band was found to vary with the changing excitation energy. In order to account for that the spectrum of each of the four areas labeled from 1–4 in Figure 12.5 was fitted to four Lorentzians, one for the TO + LA feature and three for the three contributions to the G′ band (as is shown in Figure 12.2). The TO + LA feature can also be decomposed into two Lorentzians (Tan et al., 1998).

However, for simplicity, we chose to fit the TO + LA peak with a single peak. The best-fit frequencies for the region 3 are shown as solid symbols in Figure 12.9a. It can be seen that the peak frequencies could be well fitted to linear behaviors. The best-fit

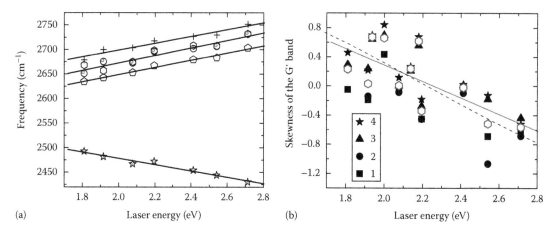

FIGURE 12.9 (a) Dependence of the frequencies of the three G′ band components together with the TO + LA combination mode with the laser excitation energy (E_L). Open symbols indicate the value of $2\omega_D$, where ω_D is the frequency of the D band. (b) Dependence of the skewness of the spectral region of the G′ band on the excitation laser energy (E_L). The solid symbols represent four different studied areas. The open symbols show the mean value for each excitation energy, while the solid line shows the best fit of these data points to a linear dependence on E_L. The dashed line shows the calculated effect of the TO + LA peak on the skewness of the G′ band line shape.

TABLE 12.2 Values i_0 and α Are the Best-fit Parameters for the Dependence of the I_D/I_G Ratio for the Four Different Spots Labeled 1–4 in Figure 12.4 to an $i_0(1 - \alpha E_L)$ Curve and for aE_L^{-b}

	I_0	A	I_0	a	B	A	S_0	β
1	0.35 ± 0.06	0.31 ± 0.02	0.31 ± 0.02	1.7 ± 0.2	2.0 ± 0.2	0.49 ± 0.04	1.5 ± 0.8	0.51 ± 0.05
2	0.44 ± 0.06	0.30 ± 0.05	0.47 ± 0.03	1.0 ± 0.2	1.8 ± 0.3	0.70 ± 0.05	2.1 ± 0.9	0.49 ± 0.03
3	0.83 ± 0.06	0.31 ± 0.02	0.78 ± 0.02	0.7 ± 0.4	2.2 ± 0.8	1.19 ± 0.05	2.3 ± 0.7	0.43 ± 0.02
4	1.25 ± 0.06	0.32 ± 0.01	1.15 ± 0.02	0.7 ± 0.4	2.7 ± 0.8	1.66 ± 0.04	2.7 ± 0.7	0.42 ± 0.01

Note: The values I_0 and A are the best-fit parameters for curves $I_0(1-0.31E_L)$ and AE_L^{-2}, where α and b where considered to be constant and equal to 0.31 eV^{-1} and 2, respectively, for all regions. The values of S_0 and β are the best-fit parameters obtained for the linear dependence of the G′ band skewness on the laser excitation energy.

parameters for the laser dependence of the G′ band features are shown in Table 12.2. The value obtained for the dispersion of the TO + LA peak (~69 cm^{-1}) is much larger than previously obtained in highly oriented pyrolytic graphite (~38 cm^{-1}) [Tan et al., 1998]. This may be due to the fact that this peak was fitted to a single Lorentzian peak, instead of two.

For completeness, an analysis was performed on the dependence of the skewness of the G′ band on the focusing position in the sample and on the laser excitation energy by following the behavior of the four areas labeled 1–4 in Figure 12.5 for all the different laser excitation energies. The skewness values obtained for these regions are shown as solid black symbols in Figure 12.9b. It can be seen that, although there is a large variation for the skewness values, the total skewness seems to decrease with increasing laser excitation energy. However, this laser energy dependence is likely to be a contribution from the TO + LA peak to the line shape of the G′ band. It is possible to make a simple evaluation of the role of the TO + LA peak to the calculated skewness of the G′ band. This can be accomplished by considering that the skewness of a combination of two peaks can be obtained to a first approximation by the following equation:

$$S = a_0 S_0 + a_1 S_1 + \frac{a_0 a_1}{\left(\sigma^2 + a_0 a_1 \Delta\omega(E_L)^2\right)^{3/2}} (a_0^2 - a_1^2)\Delta\omega(E_L)^3 \quad (12.4)$$

where the skewness of each individual curve is S_0 and S_1. Here $a_0 = A_0/(A_0 + A_1)$, $a_1 = A_1/(A_0 + A_1)$, $\sigma^2 = a_0\sigma_0^2 + a_1\sigma_1^2$, and $\Delta\omega(E_L) = \omega_0(E_L) - \omega_1(E_L)$. The values of ω_0 and ω_1 are the mean frequencies of the two peaks, which will depend on the laser energy, as shown in Figure 12.9a. σ_0^2 and σ_1^2 are the standard deviation of the peaks, which are assumed to be independent of the laser energy, and A_0 and A_1 are the integrated areas of each of the peaks. The values $\sigma = 50$ cm^{-1}, $a_0 = 0.96$, $a_1 = 0.04$ and $\Delta\omega(E_L) = 94–141E_L$ cm^{-1}, obtained from the fitting of the G′ band line shape, can then be used to estimate the maximum effect of the contribution of the TO + LA peak to the skewness of the spectral region of the G′ band as a function of the laser excitation energy. The resulting dependence is shown as a dashed line in Figure 12.8b. It can be clearly seen that, although Equation 12.4 does not show a linear dependence explicitly within this laser energy region, the skewness dependence on E_L can be well approximated by a linear equation. It is thus clear that the G′ band dependence on the laser excitation energy follows a linear behavior that originates from the contribution of the TO + LA feature to the lowest energy side of the spectra and this dependence can be fitted to a linear equation written as $S = S_0(1 - \beta E_L)$. The best fit coefficients obtained for the four studied regions are shown in Table 12.2.

12.5 Magnetic Properties of Carbon Foam Macro-Networks

Magnetism is a property that is conventionally associated with transition metals and transition metal complexes with unpaired d electrons. In one allotropic form, carbon exists as graphite, which is structurally highly regular and does not contain any unpaired electrons. However, some structurally highly irregular carbon materials with high surface areas have been reported to display magnetic properties (Makarova, 2004). Magnetic properties of carbon materials depend on its allotropic modification and this is important because these solid carbon samples may potentially be employed as optoelectronic and photovoltaic devices (Harris, 2003). Particularly in the case of graphite or activated carbon, the magnetic properties are governed by circular currents driven on and between the graphene layers. Recent studies have shown the influence of defects and dislocations in graphene layers in carbon nanofoams (Rode et al., 2004) and physisorption in microporous carbon (Sato et al., 2007) on the magnetic behavior of carbon materials. We verified the magnetic behavior of carbon foams (Matos et al., 2006, 2009). Samples were synthesized by the controlled pyrolysis of saccharose previously dissolved at 80°C in water (Matos and Laine, 1998). Samples were recrystallized at 25°C and thermally stabilized under vacuum (200 mbar) at 110°C by 1 h and finally carbonized for 1 h under N_2 flow at 450°C. Carbon films and foams were obtained simultaneously and scanning electron microscopy (SEM) confirmed the highly irregular topology of the samples. The presence of unpaired electrons in the sample was established by electron paramagnetic resonance (EPR) spectra of both samples at ambient temperature. Figure 12.10 shows the SEM image of the sample. The random topological structure (shape of polygonal cells with n-sides) of this carbon foam and its lack of structural regularity can be seen. The presence of a hyperbolic curvature in the surface of this carbon foam must be noted. This can be associated to the presence of topological and bonding defects in graphene layers as suggested by Rode and co-workers (Rode et al., 2004). The presence of free electrons or radicals is readily observable in the EPR spectra where the measured Lande factor or g factor is 2.0075, which is very close to that of a free electron (ge = 2.002319). The line width (4.37 against 4.58 Gauss) is similar to radicals or free electrons (between 3 and 15 Gauss).

The magnetization of the samples can be measured using SQUID analysis and two basic data sets can be obtained for the carbon foam sample. The first one describes magnetization as a function of applied field at different temperatures, and the second data set describes magnetization as a function

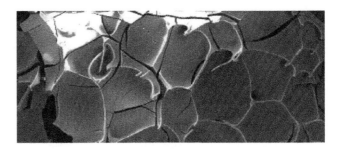

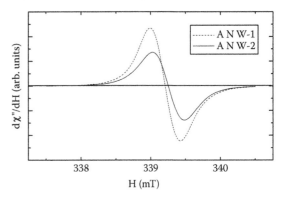

FIGURE 12.10 SEM image of carbon foam (upper panel) and its electron spin resonance spectrum (lower panel, from Matos et al., *Open Mater. Sci. J.*, 3, 28, 2009).

of temperature at some constant applied field. In case of foam, the reproducibility of magnetic data was verified. Figure 12.11 shows the relationship between magnetization and applied magnetic field at various temperatures for the sample shown in the SEM image in the temperature ranges from 1.8 to 300 K and the field ranges from 0 to 5 T. From these plots, it is readily observable that the magnetization is the result of both paramagnetic and diamagnetic components. At low temperatures, 1.8–10 K, the paramagnetic component is significant in relation to the diamagnetic component and the carbon foam samples are standard paramagnets at low temperature. As the temperature is varied from 10 to 300 K, the diamagnetic component outweighs the paramagnetic component, and what we observe is a standard diamagnetic response. Figure 12.11b illustrates the relationship between magnetism and temperature, plotted as χ/gm (T) in 1 kOe. The inverse relationship between χ and T, which is indicative of a paramagnet, can be readily seen for the foam sample. The data in Figure 12.11b has been corrected for the intrinsic background of the sample holder, but no corrections have been made for either diamagnetism, due to the variable nature of the structure and composition of the material. It must also be noted, that the magnitude of the magnetization vector varies on the basis of the synthetic methods used to prepare the samples. This clearly reflects the variable number of free electrons or radicals in the samples based on the synthetic approach utilized and the network structure plays an important role on the magnetic behavior of carbon foams. In conclusion we confirmed that our samples were paramagnetic at low temperatures (i.e., 1.8–10 K) and diamagnetic at temperatures higher than 10 K.

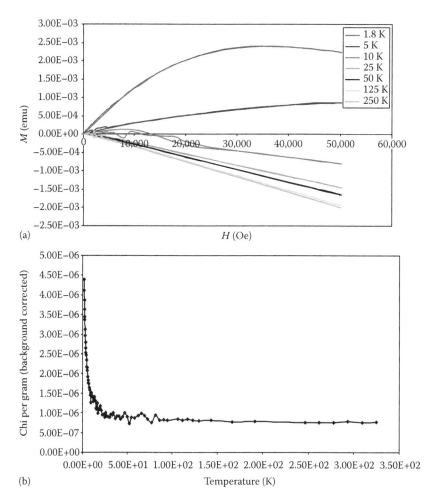

FIGURE 12.11 (a) Magnetization as a function of H at various temperatures. (b) χ/gm as a function of temperature for $H = 1\,kOe$.

Results showed that saccharose could be used as a precursor for the synthesis of carbon materials for magnetic applications at low temperatures. An important issue of this kind of foam preparation is the raw material that is abundant and cheap. Also, preparation is relatively easy to perform and eco-friendly.

12.6 Concluding Remarks

We have discussed in this chapter the basic properties of carbon materials named graphitic foams. It should be remarked that these carbon foam materials are characterized by having high strength, high electrical and thermal conductivities, and an interesting magnetic behavior at low temperatures for the case of graphitic carbon foams obtained by the controlled pyrolysis of biomass and its derivatives. It is clear that these materials are attractive for developing both basic science and carbon-based technology. There is an expectation that these materials will be important for many applications, but much research should continue in order to optimize the relationship between morphology and physical properties.

Acknowledgments

The authors would like to acknowledge Prof. Mildred S. Dresselhaus, Toshiaki Enoki, David Tomanek, and A. Jorio for stimulating discussions related to carbon foams and Prof. J. Klett from the Oak Ridge National Laboratory for providing the graphitic foam samples. Grants from Brazilian agencies CNPq, CAPES, and FUNCAP are acknowledged.

References

Bansal, R.C., J.-B. Bonnet, and F. Stoeckli. 1998. *Active Carbon.* New York: Marcel Dekker.

Barros, E.B., N.S. Demir, A.G. Souza Filho et al. 2005. Raman spectroscopy of graphitic foams. In: *Materials for Space Applications*, Eds. Chipara, M., Edwards, D.L., Benson, R.S., and Phillips, S., Material Research Society Symposium Proceedings, 851, Warrendale, PA.

Barros, E.B., H. Son, G.G. Samsonidze et al. 2006. *Phys. Rev. B* **76**, 035444.

Barros, E.B., A.G. Souza Filho, H. Son, and M.S. Dresselhaus. 2007. *Vibr. Spectr.* **45**, 122.

Calvo, M., R. García, and S.R. Moinelo. 2008. *Energy Fuels* **22**, 3376–3383.

Cançado, L.G., M.A. Pimenta, R. Saito et al. 2002. *Phys. Rev. B* **66**, 035415.

Cançado, L.G., K. Takai, T. Enoki et al. 2006. *Appl. Phys. Lett.* **88**, 163106.

Cao, A., X. Zhang, C. Xu et al. 2001. *Appl. Surf. Sci.* **181**, 234.

Diehl, M.R., S.N. Yaliraki, R.A. Beckman et al. 2002. *Angew. Chem. Int. Ed.* **41**, 353.

Dubertret, B., T. Aste, H.M. Ohlenbusch, and N. Rivier. 1998a. *Phys. Rev. E* **58**, 6368.

Dubertret, B., T. Aste, and M.A. Peshkin. 1998b. *J. Phys. A* **31**, 879.

Dunne, L.J., P.F. Nolan, J. Munn et al. 1997. *J. Phys.: Condens. Matter* **9**, 10661.

Esquinazi, P., D. Spermann, R. Hohne et al. 2003. *Phys. Rev. Lett.* **91**, 227201.

Esquinazi, P., R. Hohne, K.-H. Han et al. 2004. *Carbon* **42**, 1213.

Ford, W. 1964. U.S. Patent 3,121,050.

Fradkov, V.E., A.S. Kravchenko, and L.S. Shvindlerman. 1985. *Scr. Metall.* **19**, 1291.

Fradkov, V.E., M.E. Glicksman, M. Palmer et al. 1993. *Physica D* **66**, 50.

Glazier, J.A., M.P. Anderson, and G.S. Grest. 1990. *Philos. Mag. B* **62**, 615.

Govor, L.V., M. Goldbach, I.A. Bashmakov et al. 2000. *J. Phys. Rev. B* **62**, 2201.

Harris, P.J.F. 2003. *Chem. Phys. Carbon* **28**, 1.

Huang, T., M.R. Kamal, and A.D. Rey. 1995. *J. Mater. Sci. Lett.* **14**, 220.

Imamura, R., K.M.J. Ozaki, and A. Oya. 1999. *Carbon* **37**, 997.

Kamal, M.R., T. Huang, and A.D. Rey. 1997. *J. Mater. Sci.* **32**, 4085.

Kearns, K.M. 1999. U.S. Patent 5,868,974.

Klett, J. 2000. Process for making carbon foam. U.S. Patent 6,033,506.

Klett, J., R. Hardy, E. Romine et al. 2000. *Carbon* **38**, 953.

Klett, J.W., A.D. Mcmillan, N.C. Gallego, and C.A. Walls. 2004. *J. Mater. Sci.* 39, 3659.

Koenig, F. and J.L. Tuinstra. 1970. *J. Chem. Phys.* **53**, 1126.

Kroto, H.W., J.R. Heath, S.C. O'brian et al. 1985. *Nature* **318**, 162.

Laine, J. and S. Yunes. 1990. *Carbon* **30**, 601.

Laine, J., S. Simoni, and R. Calles. 1991. *Chem. Eng. Commum.* **99**, 15.

Makarova, T.L. 2004. *Semiconductors* **38**, 615.

Matos, J. and J.J. Laine. 1998. *Mater. Sci. Lett.* **17**, 649–651.

Matos, J., M. Labady, A. Albornoz, and J.L. Brito. 2004. *J. Mater. Sci.* **39**, 3705–3716.

Matos, J., M. Labady, A. Albornoz, and J.L. Brito. 2005. *J. Mol. Catal. A: Chem.* **228**, 198–194.

Matos, J., C.P. Landee, P. Silva et al. 2006. In: *World Conference on Carbon*, Aberdeen, U.K.

Matos, J., T. Dudo, C.P. Landee et al. 2009. *Open Mater. Sci. J.* **3**, 28–32.

Matthews, M.J., M.A. Pimenta, G. Dresselhaus et al. 1999. *Phys. Rev. B* **59**, R6585.

Miri, M.F. and N. Rivier. 2001. *Europhys. Lett.* **54**, 112.

Mochida, I., M. Egashira, Y. Korai, and K. Yokogawa. 1997. *Carbon* **35**, 1707.

Mukhopadhyay, S.M., R.V. Pulikollu, E. Ripberger, and A.K. Roy. 2003. *J. Appl. Phys.* **93**, 838.

Ng, H.T., M.L. Foo, A. Frang et al. 2002. *Langmuir* **18**, 1.

Noever, D.A. 1991. *Phys. Rev. A* **44**, 968.

Osipov, E.V. and V.A. Reznikov. 2002. *Carbon* **40**, 955.

Pablo, P.J. De, C. Gómez-Navarro, A. Gil et al. 2001. *Appl. Phys. Lett.* **79**, 2979.

Pan, L., T. Hayashida, and Y. Nakayama. 2002. *J. Mater. Res.* **17**, 145.

Peshkin, M.A., K.J. Strandburg, and N. Rivier. 1991. *Phys. Rev. Lett.* **67**, 1803.

Rivier, N. 1985. *Philos. Mag. B.* **52**, 795.

Rode, A.V., G. Gamaly, A.G. Christy et al. 2004. *Phys. Rev. B* **70**, 054407.

Rouzaud, J.N. and A. Oberlin. 1989. *Carbon.* **27**, 517.

Saito, R., A. Jorio, A.G. Souza Filho et al. 2002. *Phys. Rev. Lett.* **88**, 027401.

Sato, H., N. Kawatsu, T. Enoki et al. 2007. *Carbon* **45**, 203.

Schwan, J., S. Ulrich, T. Theel et al. 1997. *Appl. Phys.* **82**, 6024.

Stavans, J. 1993. *Rep. Prog. Phys.* **56**, 733.

Stavans, J. and J.A. Glazier. 1989. *Phys. Rev. Lett.* **62**, 1318.

Stine, K.J., S.A.R. Auseo, B.G. Moore et al. 1990. *Phys. Rev. A* **41**, 6884.

Stoeckli, F. 1990. *Carbon* **28**, 1.

Tan, P.H., Y.M. Deng, and Q. Zhao. 1998. *Phys. Rev. B* **58**, 5435.

Thomsen, C. 2000. *Phys. Rev. Lett.* **85**, 5214;4542.

Vidano, R.P., D.B. Fishbach, L.J. Willis, and T.M. Loehr. 1981. *Solid State Commun.* **39**, 341.

Wang, C., D. Ma, and X. Bao. 2008. *J. Phys. Chem. C.* **112**, 17596.

Weaire, D. and N. River. 1984. *Contemp. Phys.* **25**, 59.

Weaire, D., F. Bolton, P. Molho, and J.A. Glazier. 1991. *J. Phys.: Condens. Matter.* **3**, 21012.

Zhang, Y., A. Chang, J. Cao et al. 2001. *Appl. Phys. Lett.* **79**, 3155.

13

Arrayed Nanoporous Silicon Pillars

Xin Jian Li
Zhengzhou University

13.1 Introduction

Si-based nanosystems have attracted much attention because of their unique nanostructures and unusual physical properties that might be applied in areas as diverse as optoelectronics, single electron devices, sensors, detectors, and cold cathodes for field-emission displays (Canham 1990, Bisi et al. 2000, Birner et al. 2001, Bettotti et al. 2002, Shang et al. 2002, Fujita et al. 2003, Scheible and Blick 2004, Garguilo et al. 2005, Heitmann et al. 2005, Kanechika et al. 2005, Shao et al. 2005). The fabrication of the high-performance field-effect transistors based on *n*-type Si nanowires (Zheng et al. 2004) and piezoelectric nanogenerators based on zinc oxide nanowire arrays (Wang and Song 2006) are two of the most exciting examples to demonstrate the importance of Si-based nanostructures in developing optoelectronic nanodevices. In the past decade, various interesting Si nanostructures, such as porous Si (PS) (Canham 1990, Cullis et al. 1997, Bisi et al. 2000, Birner et al. 2001, Bettotti et al. 2002), arrays of nanocones (Shang et al. 2002), nanopillars (Scheible and Blick 2004), nanorods (Cluzel et al. 2006), and nanowire *p–n* junction diodes (Peng et al. 2005), etc., were prepared and their unique optical or electrical properties were observed. Nevertheless, considering the role played by Si in modern microelectronics and its potential application in

future nanoelectronics, the space for preparing novel Si nanostructures with enhanced physical properties should still be tremendous.

In this chapter, we present a concise review of the recent research on a novel Si micron/nanometer structural composite system, the silicon nanoporous pillar array (Si-NPA). The contents include the preparation, characterization and structure control of Si-NPA, the optical and humidity sensing properties of Si-NPA, the detection of low-concentration biomolecules by Ag/Si-NPA, the enhanced field emission from carbon nanotube (CNT)/Si-NPA, and the optical and electrical properties of the CdS/Si-NPA nanoheterojunction array. A brief summary is given at the end of the chapter.

13.2 Preparation and Characterization of Si-NPA

13.2.1 Preparation

Si-NPA was prepared by hydrothermally etching single crystal Si (*sc*-Si) wafers in the solution of hydrofluoric acid (HF) containing ferric nitrate (Fe(NO$_3$)$_3$) (Xu and Li 2008a). The initial silicon wafers were heavily doped (111) oriented *p*-type *sc*-Si wafers. The structural features of Si-NPA, such as the porosity, surface morphology, and elemental components, as well as the size and size distribution of the pores and Si nanocrystallites (*nc*-Si), could be tuned by changing the preparation conditions or by using *sc*-Si wafers with different doping levels. The preparation conditions include the components and concentrations of the hydrothermal solution, the etching temperature, pressure, and time, etc. For preparing Si-NPA with well-separated porous silicon pillars, the resistivity of the initial *sc*-Si wafers should not be bigger than 2.0 Ω cm.

13.2.2 Structural and Morphological Characterization

Si-NPA has been proved to be a micron–nanometer structural composite system characterized by its regular hierarchical structure (Xu and Li 2008a). The morphological characteristics of Si-NPA obtained by a field-emission scanning electron microscope (FE-SEM) are illustrated in Figure 13.1a and b. Obviously, a regular array composed of large quantities of well-separated pillars is formed on the sample surface. The silicon pillars are perpendicular to the surface, with a height of ~4.1 μm and a top-to-top distance of ~3.6 μm. Between the pillar layer and the *sc*-Si substrate is a transitional porous layer with clear upper and lower boundaries. From the amplified FE-SEM image of the transitional porous layer (Figure 13.1c), the average pore size in this layer was estimated to be ~8.2 nm. By cleaving the pillar layer from the sample, the fine structure of the pillars was studied by a transmission electron microscope (TEM) and a high resolution TEM (HRTEM). Figure 13.1d is

the TEM image of an individual pillar, where large quantities of pores were observed. From the locally magnified TEM images, it was found that the average pore size varies gradually from the pillar top to root, ~40 nm at the top and ~15 nm at the root. Figure 13.1e presents the typical HRTEM image of pore walls, where uniformly distributed crystal zones dispersed in a kind of amorphous matrix were observed. These crystal zones were proved to be the cross-sections of *nc*-Si, and their size distribution is depicted in Figure 13.1f. The *nc*-Si sized from ~1.95 to ~4.4 nm, with an average size of ~3.4 nm and a planar density of ~2.3 × 10^{12} cm^{-2}.

The elemental components of the sample surface as well as their chemical valence states were analyzed by the experiments of room-temperature x-ray photoelectron spectroscopy (XPS). On the XPS spectrum given in Figure 13.2, the visible three peaks were determined to correspond to the binding energy of Si 2p, Si 2s, and O 1s, respectively. This directly proves that the elements presented on the surface of Si-NPA are Si and O. Further analysis discloses that the Si 2p peak can be broken into three sub-peaks (the inset of Figure 13.2) located at ~99.7, ~101.7, and ~103.1 eV, representing the valence states of Si^0, Si^{2+}, and Si^{4+}, respectively (Chen et al. 2005, Xu and Li 2008a). This indicates that the Si atoms present on the surface of Si-NPA have three different chemical valence states. On the other hand, the sharp O 1s peak located at ~533.3 eV is found to originate from O atoms in the valence state of O^{2-} (Cheng et al. 2003). Combining with the structural characterization given in Figure 13.1e, the Si atoms with Si^0 valence state should be those constructing *nc*-Si, while those with Si^{2+} and Si^{4+} valence states should be Si atoms

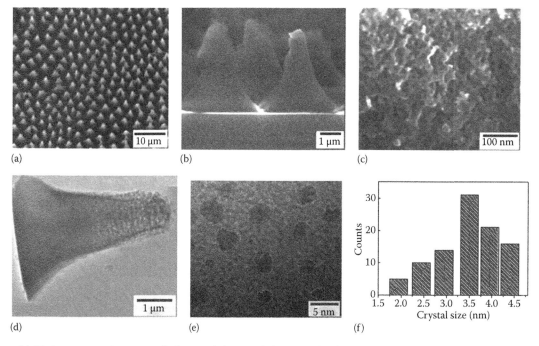

(a)   (b)   (c)

(d)   (e)   (f)

FIGURE 13.1 (a) FE-SEM image of Si-NPA with the sample being titled at an angle of 45°, (b) cross-sectional FE-SEM image of Si-NPA, (c) the amplified image to illustrate the finer pores located in the transitional porous layer beneath the pillar layer, (d) TEM image of an individual nanoporous Si pillar, (e) HRTEM image of pore walls, and (f) the size distribution of *nc*-Si in Si-NPA. (From Xu, H.J. and Li, X.J., *Opt. Express*, 16(5), 2933, 2008a.)

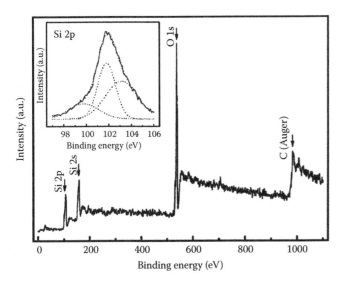

FIGURE 13.2 Room temperature XPS result of Si-NPA. Inset: Si 2p peak as well as its fitting results. (From Xu, H.J. and Li, X.J., *Opt. Express*, 16(5), 2933, 2008a.)

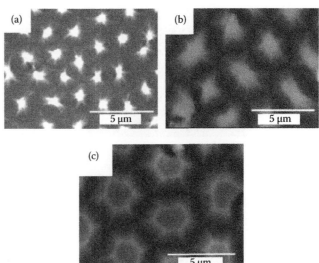

FIGURE 13.3 Morphology of Si-NPA prepared with HF concentration of (a) 7.62 mol/L, (b) 13.33 mol/L, and (c) 15.23 mol/L.

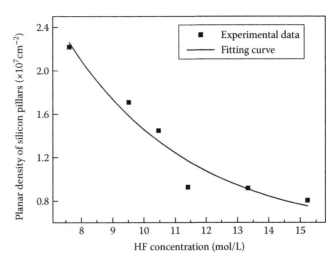

FIGURE 13.4 Plot of planar pillar density of Si-NPA with HF concentration.

constructing the amorphous matrix in the form of Si monoxide and dioxide. Judged from the area ratio between the peaks corresponding to Si^{2+} and Si^{4+}, the quantities of Si dioxide and monoxide are almost equal. The result infers that a large amount of SiO_x ($x = 1, 2$) has been natively formed in Si-NPA, which is most probably the SiO_x matrix that encapsulates the *nc*-Si.

Based on the structural and compositional analysis of Si-NPA, it could be concluded that Si-NPA is a micron/nanometer structural composite system characterized by its hierarchical structure, that is, a regular array composed of micron-sized, quasi-identical Si pillars, nanopores densely distributed on the Si pillars and in the transitional porous layer beneath, and *nc*-Si constructing the pore walls and dispersed in the SiO_x matrix.

13.2.3 Morphology Control

The shape and the planar density of the nanoporous silicon pillars of Si-NPA could be modulated through changing the preparing conditions, such as the solution concentrations and the etching temperature, etc.

By solidifying the solution concentration of Fe^{3+} at 0.04 mol/L and the etching temperature at 140°C, the morphology evolution of Si-NPA with HF concentration was studied by changing its concentration from 7.62 to 19.99 mol/L. It is verified that the samples with well-separated silicon pillars could be obtained only when the HF concentration is lower than 15.23 mol/L, or serious connection among the porous pillars would occur. Form the FE-SEM images given in Figure 13.3a through c, two obvious changes could be found. One is the shape of the pillars, changing from blunt cone-like (Figure 13.3a and b) to crater-pit-like (Figure 13.3c). The other is the planar pillar density, changed from ~2.22×10^7 (HF concentration: 7.62 mol/L) to 8.05×10^6 cm^{-2} (HF concentration: 15.23 mol/L). The evolution

curve of the pillar planar density with HF concentration is plotted in Figure 13.4. Obviously, the pillar planar density of Si-NPA decreases exponentially with the HF concentration of the solution.

The shape and the planar density of the nanoporous Si pillars could also be tuned by varying the solution concentration of Fe^{3+}. By solidifying the HF concentration at 11.42 mol/L and the etching temperature at 140°C, it was determined that 0.17 mol/L is the upper limit of Fe^{3+} concentration for preparing Si-NPA with well-separated pillars. The evolution of the morphology and the pillar planar density with Fe^{3+} concentration are demonstrated in Figures 13.5 and 13.6, respectively. The planar pillar density initially drops exponentially with Fe^{3+} concentration and then approached to a constant. With Fe^{3+} concentration changed from 0.01 to 0.17 mol/L, the planar

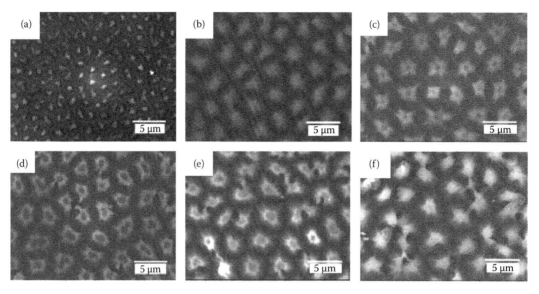

FIGURE 13.5 Morphologies of Si-NPA prepared with Fe^{3+} concentration of (a) 0.01 mol/L, (b) 0.04 mol/L, (c) 0.08 mol/L, (d) 0.14 mol/L, (e) 0.17 mol/L, and (f) 0.25 mol/L.

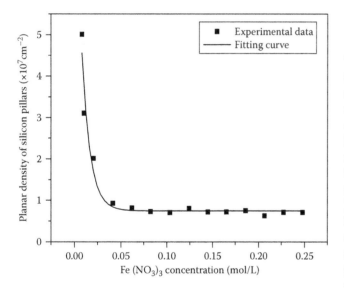

FIGURE 13.6 Plot of planar pillar density of Si-NPA with Fe^{3+} concentration.

13.2.4 Preparation of Nanocomposites Based on Si-NPA

The unique structure and chemical and physical properties of Si-NPA makes it an ideal substrate or template for preparing or assembling various Si-based functional nanosystems with regularly patterned structures. In the past several years, we have demonstrated the preparation of CdS and ZnS/Si-NPA nanoheterojunction array by the heterogeneous chemical reaction method (Xu and Li 2007a,b, 2008b, Xu et al. 2008), ZnO/Si-NPA by vapor-liquid-solid method (Yao et al. 2007a,b), CNT/Si-NPA nest array by thermal chemical vapor deposition method (Li and Jiang 2007, Jiang et al. 2007, 2008), Fe_3O_4, TiO_2, WO_3, and $BaTiO_3$/Si-NPA by spin-coating method (Wang and Li 2005a,b), and noble metal (Au, Ag, and Cu)/Si-NPA by immersion plating method (Fu and Li 2005, 2007). It is found that these nanocomposites could usually exhibit enhanced physical properties and might have potential applications in the fields of optoelectronics, field-emission flat displays, gas/humidity sensors, biomolecule detectors, and nanostructured solar cells.

13.3 Optical Properties of Si-NPA

13.3.1 Light Absorption

The realization of high antireflection is important for improving the efficiency of Si-based solar cells, and presently it is usually realized by constructing some kinds of surface periodical structures (Hadobas et al. 2000). Si-NPA might be an enhanced light absorber because of the coexistence of the numerous nanopores and the periodical pillar array.

The integral diffuse reflection spectrum of Si-NPA at a wavelength range of 240–2400 nm is shown in Figure 13.7. Obviously,

pillar density decreased from ~3.1×10^7 to 7.05×10^6 cm^{-2}. Obviously, both the evolution trends of the pillar planar density and the pillar shape are similar to that by changing HF concentration, that is, the higher the Fe^{3+}/HF concentration, the lower the planar pillar density.

The related experiments disclosed that the morphology control of Si-NPA could also be realized by varying the etching temperature. During the temperature range 40°C–200°C, it is found that the higher the etching temperature, the lower the planar pillar density. With temperature increasing, the shape of the nanoporous Si pillars would also change from corn-like to crater-pit-like.

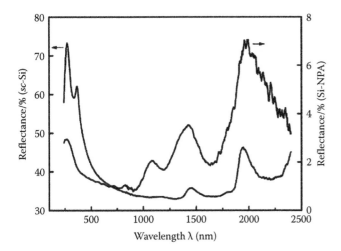

FIGURE 13.7 The integral diffuse reflection spectrum taken in the wavelength range of 240 ~ 2400 nm of Si-NPA, together with that of *sc*-Si for comparison. (From Xu, H.J. and Li, X.J., *Opt. Express*, 16(5), 2933, 2008a.)

the profile of the reflectance spectrum of Si-NPA is similar to that of *sc*-Si, because the peak positions for both of them are predominantly decided by the lattice structure of crystal Si. But the strength of the reflectance peaks of Si-NPA is greatly reduced compared with the counter part of *sc*-Si. Even compared with traditional PS, a sponge-like silicon nanostructure with strong light-absorption capability, the reflectance of Si-NPA is rather low. For example, in the wavelength range of 400–1000 nm, a band concerned mostly in the field of solar cells, the average reflectivity of Si-NPA is less than 2.0%, while that for PS is above 5.0% (Strehlke et al. 1999, Lipinski et al. 2003, Xu and Li 2008a). Clearly, the strength of all the reflection peaks of *sc*-Si is reduced correspondingly for Si-NPA, but their reduction amplitudes are different. For example, the strength of the strong reflection peaks locating at ~270 and ~1900 nm reduces from ~74% to ~2.9% and ~42% to ~6.1%, respectively. The broadband low reflectance might be directly due to the complex hierarchical structure of Si-NPA. When a branch of incident light arrives at the surface of Si-NPA, two kinds of reflection processes would occur. One is the multiple reflection of light amongst the micron-sized porous pillars, and the other is the reciprocating motion of light within the nanopores locating both on the silicon pillars and in the transitional porous layer. It is easy to deduce that the multiplicity of the character sizes of the hierarchical structure, ~3.6 μm for top–top distance between the pillars, ~15–40 and ~8.2 nm for the pores on the pillars and in the transitional porous layer, would be helpful for the capture of incident photons with different wavelengths. Furthermore, the existence of large quantities of *nc*-Si as well as their wide size distribution would surely increase the absorption probability of the captured photons. Just as has been proved both experimentally and theoretically (Wolkin et al. 1999), the band-gap of *nc*-Si decreases with its size. Therefore, the unique hierarchical structure of Si-NPA leads to an integral increment of light absorption.

13.3.2 Photoluminescence

Si nanostructures with visible photoluminescence (PL) have been intensively studied for their potential applications in future Si-based optoelectronics. Si-NPA also exhibits strong and stable PL at room temperature (Xu and Li 2008a). Figure 13.8a presents the PL spectra of Si-NPA obtained by changing the excitation wavelength (EW) from 340 to 430 nm. Clearly, each PL spectrum has three emission bands, one in blue range and the other two in red range. Fitting each spectrum by the Gauss–Newton method, the peak positions of the three emission bands could be determined correspondingly. For example, the PL spectrum under the irradiation of 340 nm ultraviolet light could be decomposed into three separate emission bands, peaked at ~420, ~640, and ~705 nm, respectively. The evolution of the PL peak positions and intensities with EW could be seen more clearly in Figure 13.8b and c. In the applied EW scale, the peak intensities of the two red PL bands decrease with EW, while the peak positions and the full width at half maximum of the peaks remain almost unchanged. Similar to the two red PL bands, the peak intensity of the blue band also decreases with EW, but its peak position increases with EW. When the EW was changed from 340 to 430 nm, a redshift of ~60 nm (from ~420 to ~480 nm) of the peak position was observed.

The PL excitation (PLE) spectra of Si-NPA were measured by setting the monitoring wavelength at 420, 640, and 705 nm, respectively (Figure 13.9). It was found that all the three PLE spectra have an effective excitation band peaked at ~365 nm, showing no dependence on the monitoring emission wavelength. This indicates that the excitation process for the three emission bands might be an identical one. Deduced from the microstructure of Si-NPA and the PL mechanism put forward for various Si nanostructures (Duan et al. 1994, Li et al. 1998, Qin and Li 2003), the PL excitation process in Si-NPA should be attributed to a band–band transition of electrons in *nc*-Si, as is illustrated in Figure 13.10. Therefore, the observation of the three emission bands could be only attributed to three different radiative recombination paths.

The strong dependence between the emission and excitation wavelengths for the blue emission in Si-NPA indicates that this is a band–band recombination process obeying the quantum confinement (QC) model (Wolkin et al. 1999, Islam and Kumar 2003). According to QC model, the emission wavelength of *nc*-Si depends on its grain size, that is, the bigger the size, the longer the emission wavelength. Theoretical calculations demonstrated that with the size of *nc*-Si being tuned from 0.8 to 4.3 nm, its band-gap would vary from ~5.0 to 1.6 eV (~248.5– 776.5 nm) (Kanemitsu et al. 1993, Bruno et al. 2007). Applying the estimation method to Si-NPA, the band-gap corresponding to the smallest *nc*-Si (~1.95 nm, Figure 13.1f) is evaluated to be ~2.95 eV (~421 nm). This gap energy is well in accordance with the peak energy of the blue emission band of Si-NPA. Therefore, the blue emission from Si-NPA could be described as a two-step process illustrated by procedure (a) and (b) in Figure 13.10. When Si-NPA is irradiated by a given light, only the carriers

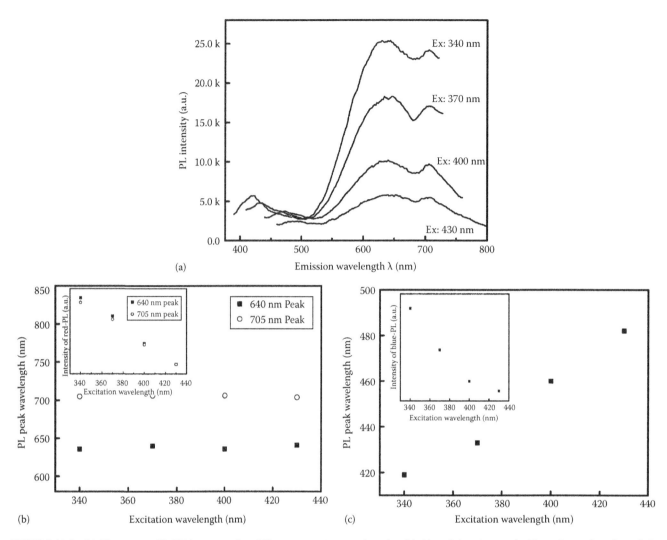

FIGURE 13.8 (a) PL spectra of Si-NPA measured at different excitation wavelengths. (b), (c) and their insets, the PL peak wavelengths and the PL intensities as a function of the excitation wavelength for the two red emission bands and the blue emission band, respectively. (From Xu, H.J. and Li, X.J., *Opt. Express*, 16(5), 2933, 2008a.)

confined in the *nc*-Si whose size is bigger than a critical size could be effectively excited to a high energy level. The excited carriers would fall down to the excitation states with relatively low energy levels through some non-radiative recombination process (procedure (a)), and then transit back to the ground state accompanied with photon emissions (procedure (b)). This model well explained the phenomenon of the redshift of the blue emission band with EW.

Different from the evolution behavior of the blue PL band, the peak positions of the two red PL bands remain almost unchanged when irradiated with different EW. This result directly denied the possibility to attribute the origin of the two red PL bands to the pure QC model. In fact, similar evolutions for the peak positions of the red PL bands have been observed and studied in anodized PS and *nc*-Si-embedded silicon oxide, and it has been demonstrated that the corresponding PL mechanism could be well explained by a QC/luminescence center (QCLC) model (Duan et al. 1994, Cooke et al. 1996, Qin and Li 2003). According to

the QCLC model, the excitation process also occurred through band–band transition in *nc*-Si, but the emission process mainly occurred through the luminescence centers (LCs) locating at the surface of *nc*-Si or in the SiO$_x$ layer encapsulating *nc*-Si. Among the electrons excited to the high energy levels of *nc*-Si, excluding the electrons participating in the procedure of the blue PL, the others would be responsible for the red PL. These carriers would firstly transit to the high-energy states of the LCs via diffusion or tunneling process (procedure (c) in Figure 13.10), then fall down to the low-energy states of the LCs through some non-radiative recombination process (procedure (d)), and finally transit back to the ground state accompanied with photon emissions (procedures (e) and (f)). Because the energy levels of the LCs are generally independent of the size of *nc*-Si, the peak positions of the two red emission bands remain unchanged when irradiated with different wavelengths. According to the QCLC model, there should be two kinds of LCs to explain the origin of the two separate red emission bands observed in Si-NPA. In fact, it has been demonstrated

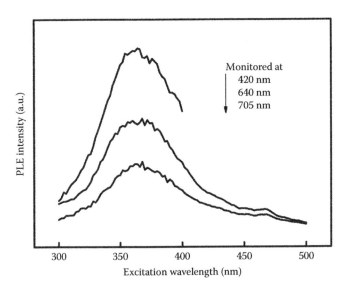

FIGURE 13.9 The PLE spectra of Si-NPA taken under three different monitored emission wavelengths. (From Xu, H.J. and Li, X.J., *Opt. Express*, 16(5), 2933, 2008a.)

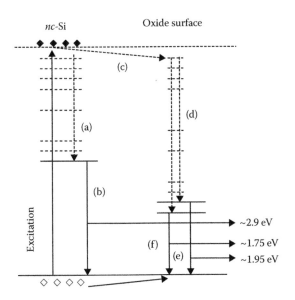

FIGURE 13.10 The schematic diagram to illustrate the PL mechanism of Si-NPA: (a) the non-radiative recombination for the excited carriers descending from the initial high energy level to the bottom of the conduction band of the *nc*-Si, (b) the radiative recombination for the excited carriers transiting back to the valance band from the bottom of the conduction band of the *nc*-Si, (c) the transition for the excited carriers from the initial exciting states to the high-energy states of the LCs locating at the surface of *nc*-Si via diffusion or tunnelong process, (d) the non-radiative recombination for the excited carriers descending from the high- to the low-energy states of the LCs, (e) and (f) the radiative recombination for the excited carriers transiting back to the valance band from the low-energy states of the LCs. (From Xu, H.J. and Li, X.J., *Opt. Express*, 16(5), 2933, 2008a.)

that the PL peak wavelength of a high-quality SiO_x thin film could change from visible to near-IR with x variation (Carius et al. 1981). For Si-NPA, the XPS data (the inset in Figure 13.2) have shown

that except Si^{4+}, large quantities of Si atoms are in the valence state of Si^{2+}, and this directly confirmed that the x value of the natively formed SiO_x layer deviates from 2. Considering the complexity of the hierarchical structure of Si-NPA, especially the size evolution of the nanopores with the depth from the sample surface, it is reasonable to deduce that two kinds of SiO_x films with different x values might have been formed because of the existence of oxidation degree. This might lead to the formation of two kinds of LCs and therefore two red emission bands were observed. The clarification on the origin of the three emission bands is important for controlling the position and the intensity of the PL bands according to different device requirements. These results indicate that Si-NPA might be a good candidate both as a functional Si nanostructure and as a template to assemble Si-based nanocomposites for future optoelectronic nanodevices.

13.4 Capacitive Humidity Sensing Properties of Si-NPA

An ideal humidity sensing material should usually possess the properties of high sensitivity, short response/recovery time, wide operating temperature range, small humidity hysteresis and long-term stability (Traversa 1995, Yuk and Troczynski 2003). For humidity sensors based on porous materials, a suitable control on the average pore size and size distribution is important (Golonka et al. 1997, Wang and Virkar 2004). In the past decade, PS as a humidity sensing material has aroused much attention because of its tunable nanoporous structures and the high sensitivity (Rittersma et al. 2000, Baratto et al. 2001, Foll et al. 2002, Fürjes et al. 2003, Francia et al. 2005). But the relatively long recovery time and large hysteresis brought by its integral sponge-like structure make its practical application being baffled (Foucaran et al. 2000). Si-NPA might be an ideal candidate to overcome these difficulties, because the channel network constructs by the valleys around the Si pillars might have provided an effective pathway for vapor transport.

13.4.1 Capacitive Humidity Sensitivity

The humidity sensing properties of Si-NPA was studied by constructing coplanar interdigital electrodes illustrated in Figure 13.11 (Li et al. 2008). The humidity environments were provided by encapsulating a series of standard saturated salt solutions (LiCl, $MgCl_2$, $Mg(NO_3)_2$, $CuCl_2$, NaCl, KCl, and KNO_3) in conical flasks with stoppers, with the room-temperature relative humidity (RH) being ~11.3%, 33.1%, 54.4%, 75.5%, 85.1%, and 94.6%, respectively. The capacitance response to humidity was measured using a TH2818 automatic component analyzer/precision LCR multi-frequency meter at different frequencies.

In order to describe the capacitance variation of Si-NPA with RH more accurately, define the device sensitivity S as

$$S = \frac{C_{RH} - C_{11}}{C_{11}} \times 100\%,$$

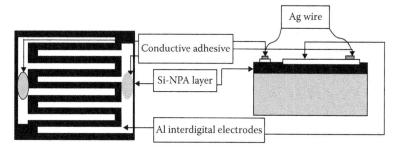

FIGURE 13.11 A schematic diagram to illustrate the structure of Si-NPA humidity sensor. (From Li, L.Y. et al., *Thin Solid Films*, 517(2), 948, 2008.)

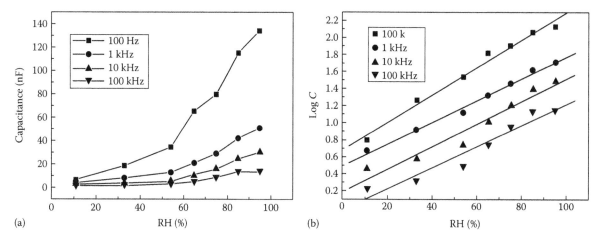

FIGURE 13.12 (a) Experimental capacitance–RH curves of Si-NPA humidity sensor measured at different frequencies, and (b) the transformed logarithmic capacitance–RH curves corresponding to the data presented by (a). (From Li, L.Y. et al., *Thin Solid Films*, 517(2), 948, 2008.)

where C_{11} and C_{RH} represent the capacitances at RH = 11.3% and at a certain measuring RH level, respectively. For testing the frequency dependence of the device, the RH-capacitance response was measured with four frequencies, 100 Hz, 1 kHz, 10 kHz, and 100 kHz, respectively, and the capacitance–RH curves are depicted in Figure 13.12a. The capacitance increases monotonically with RH, indicating that Si-NPA is a suitable humidity sensing material. The sensitivity S at the four measuring frequencies was calculated to be ~2048%, 1190.6%, 959.8%, and 733.1%, respectively. The lower the measuring frequency applied, the higher the device sensitivity obtained. Such sensitivity could fully satisfy the practical requirement. As for the nonlinearity of the capacitance–RH curve, one of the shortcomings mostly met in capacitive humidity sensors, it could be solved through the function processing method (Traversa 1995, Sundaram et al. 2004). The transferred logarithmic capacitance *versus* RH curves for Si-NPA are depicted in Figure 13.12b. Clearly, the curve measured with 1 kHz is mostly approximated to a straight line. So considering the sensitivity and the linearity of the logarithmic response curve synthetically, 1 kHz might be the most suitable signal frequency for Si-NPA humidity sensors.

13.4.2 Response and Recovery Time

The response time for RH increasing processes and the recovery time for RH decreasing processes for a sensor are usually defined as the time taken to achieve 90% of its total capacitance variation. By alternating the RH between 11.3% and 94.6% and *vice versa*, the response and recovery curves for Si-NPA were measured, as shown in Figure 13.13a. Through a simple calculation, the response and recovery time were evaluated to be ~11 and ~2 s, respectively. For characterizing the response and recovery speed of Si-NPA under different RH alternating processes, two dynamic processes were performed by continuously changing the RH from 11.3% to 94.6% and 94.6% to 11.3%. From the results presented in Figure 13.13b, it was found that both the response and recovery time for the first several humidity switchovers were very short, but they were obviously prolonged in the succeeding switchovers. The evolution trend disclosed that for Si-NPA humidity sensors the vapor absorption is easy to occur in low humidity environment, while the vapor desorption is easy to occur in high humidity environment. It should be especially noted that both in RH increasing and decreasing processes, the response and recovery time corresponding to the switchover between 33.1% and 54.4% RH are much longer than that for other humidity switchovers. This might be caused by the transition of the adsorption mode of the water molecules from monolayer to multilayer physisorption, just as the phenomenon of the abrupt increment of capacitance observed in Figure 13.12a (Li et al. 2008).

That the recovery time is shorter than the response time has been proved to be an important feature for the humidity or gas

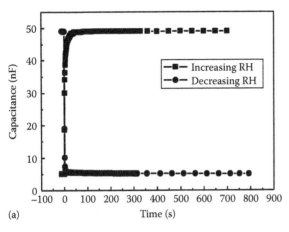

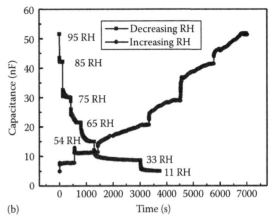

(a)   (b)

FIGURE 13.13 (a) Static and (b) dynamic response/recovery time plots of Si-NPA humidity sensor measured with a frequency of 1 kHz. (From Li, L.Y. et al., *Thin Solid Films*, 517(2), 948, 2008.)

sensors based on Si-NPA (Wang and Li 2005a,b, Xu et al. 2005a, Jiang et al. 2007, Li et al. 2007a), because for most humidity sensors prepared by other materials, the recovery time is usually much longer than the response time. Therefore, the sensing process of Si-NPA might have been dominated by the physisorption rather than chemisorption, because the physisorption dominated by van der Waals interaction is much weaker than the chemisorption dominated by the chemical bonding interaction. In fact, in the instances such as the microsensors based on the strontium titanate–niobate film (Li et al. 1999) and the mesoporous silica aerogel thin film (Wang and Wu 2006), it was also reported that the recovery time was shorter than the response time. But both the times were longer than the counterparts of Si-NPA sensors. So we tend to attribute the high response and recovery speed of Si-NPA sensors to its unique surface morphology and structures. As demonstrated in Figure 13.1, the nanoporous silicon pillars that act as the main bodies for humidity sensing are regularly arrayed and are perpendicular to the surface, while the valleys around the pillars are well connected and form a well-defined channel network. Such a geometrical configuration makes Si-NPA an effective three-dimensional humidity sensing system. Therefore, the humidity balance between Si-NPA and the detected environment could be quickly built up through the effective pathway provided by the channel network, which would definitely result in a fast response and recovery speed for Si-NPA sensors.

13.4.3 Humidity Hysteresis

The relatively big humidity hysteresis has long been a serious problem for practical humidity sensors. The hysteresis for Si-NPA sensor was studied by measuring its capacitance dependence on cyclic humidity changes, and the results were shown in Figure 13.14 (Li et al. 2008). It was found that humidity hystereses were observed in all the three humidity cycles measured with 100 Hz, 1 kHz, and 10 kHz. A simple comparison disclosed that the hysteresis observed under 1 kHz was the smallest. As has been analyzed by other groups (Li et al. 2000), the hysteresis for a porous humidity sensing material should be mainly attributed to the different

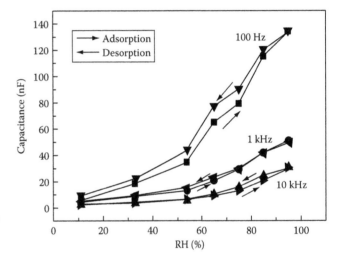

FIGURE 13.14 Humidity hysteresis characteristic of Si-NPA humidity sensor measured at 100 Hz, 1 kHz, and 10 kHz, respectively. (From Li, L.Y. et al., *Thin Solid Films*, 517(2), 948, 2008.)

conditions required for the occurrence of the capillary condensation effect that occurred in the adsorption and desorption processes. If the critical pore radius estimated by Kelvin equation for capillary condensation in the adsorption process was r_k, then that in the desorption process would be $2r_k$. Therefore, the difference on the critical radius required for capillary condensation determines the unavoidability of the hysteresis. But our experiments showed that the hysteresis was also affected by other measuring factors, such as the measuring frequency (Li et al. 2008).

In summary, humidity sensors based on Si-NPA could exhibit excellent humidity sensing properties including high sensitivity, relatively short response/recovery times, and small hysteresis. In fact, similar advantages have also been proved in the sensors based on metal oxides and carbon nanotube/Si-NPA nanocomposite systems (Wang and Li 2005a,b, Jiang et al. 2007, Li et al. 2007a, Dong et al. 2009). These results indicate that Si-NPA might be a promising humidity sensing material or a template for preparing nanocomposite-sensing materials for practical applications.

13.5 Detection of Biomolecules by Ag/Si-NPA

Since the observation of surface-enhanced Raman scattering (SERS) phenomenon (Fleischmann et al. 1974), great progress has been achieved in detecting low-concentration biomolecules by the SERS technique (Kneipp et al. 1997, Nie and Emory 1997, Xu et al. 1999). In the past several years, the SERS detection of various biomolecules or species has been reported, such as proteins (Driskell et al. 2007), enzymes (Ruan et al. 2006), viruses (Wabuyele and Vo-Dinh 2005), bacteria (Jarvis et al. 2006), and nucleotides (Kneipp et al. 1998), etc. The most exciting idea was that the SERS technique might also be used as an effective tool in the DNA rapid sequencing process at single-molecular level, because it has been demonstrated that the SERS peaks of nucleotide bases could be easily distinguished out from the accompanied noise brought by sugar or phosphate groups on DNA backbones (Green et al. 2006).

The quality of the SERS signal depends highly upon the metal species and the surface microstructures of the active substrate. Among various SERS active materials, the Ag-based substrates, such as Ag colloids, electrochemically roughened Ag electrodes, and Ag films, were the most popularly used ones (Nie and Emory 1997, Juan et al. 2000, Chattopadhyay et al. 2005, Baia et al. 2006, Fang et al. 2008). Once Ag is chosen as the active metal, its microstructure would be the most important factor for obtaining strong, stable, and reliable SERS signals. In the past decade, PS was intensively used as a template for preparing Ag-based SERS-active substrates and many exciting results were obtained, mainly because of its large specific area, open and controllable nanoporous structure, strong capability in reducing Ag+ from its salt solutions without any reducing agent, and the technical compatibility in realizing one-chip devices (Chan et al. 2003, Lin et al. 2004, Kalkan and Fonash 2006, Panarin et al. 2007). This indicates that developing novel Ag nanostructures on Si substrate might be a promising direction for obtaining high-quality SERS active substrates.

Considering the unique morphological and structural properties of Si-NPA (Xu and Li 2008a) and the predominance of Ag nanostructures in SERS detection of low-concentration biomolecules, here we will try to develop a patterned Ag structure on Si-NPA and Ag/Si-NPA, by a simple immersion plating method. For demonstrating the SERS effect of Ag/Si-NPA, adenine was chosen as the detected biomaterial.

13.5.1 Preparation and Characterization

Ag/Si-NPA is prepared by a simple immersion plating method. After hydrothermal etching, the Si-NPA samples used as templates were aged in air for one day and then immersed in 0.01 mol/L AgNO$_3$ solution for 5 min. Then the immersed samples were washed with deionized water and dried in nitrogen flow at room temperature. Considering the similarity between the nanoporous structures of Si-NPA and PS, the chemical reactions occurred in the immersion plating of Ag on Si-NPA should be the same with that on PS (Andsager et al. 1993, Tsuboi et al. 1998).

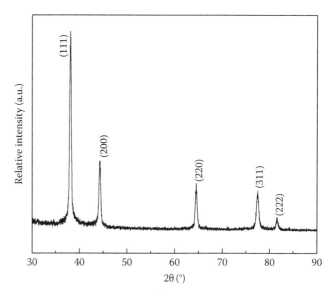

FIGURE 13.15 XRD pattern of Ag/Si-NPA prepared by immersion plating method.

The XRD pattern measured for Ag/Si-NPA is shown in Figure 13.15, from which the Ag deposited on Si-NPA was determined to be of body-centered cubic structure. Figure 13.16a presents the surface morphology of Ag/Si-NPA, with the insets being the images taken with the samples being tilted at an angle of 30°. Figure 13.16b is a locally amplified image. Clearly, two kinds of Ag structures were formed on Si-NPA. One was the continuous Ag film covered completely over the porous Si pillars and composed of Ag nanocrystallites (nc-Ag), whose size increased gradually with its location from the pillar top to root (~40–100 nm). The other was the quasi-regular, interconnected Ag network composed of submicron Ag crystallites (sub-mc-Ag) through forming large quantities of loop-chains at the valleys surrounding the porous Si pillars. All the sub-mc-Ag was in ellipsoid-like shape, with a length of ~770 nm and a width of ~550 nm. Obviously, the growing of Ag on Si-NPA obeyed the Volmer–Weber mode because of the weak interaction between Ag and Si atoms (Oskam et al. 1998). The co-presentation of the nc-Ag on the pillars and the sub-mc-Ag at the valleys surrounding the pillars exhibits high regularity that indicates that the growing of Ag on Si-NPA was of a kind of site-selective mode. Considering the fact that the Si-NPA used for Ag immersion plating was one-day aged and the surface variation caused by oxidation, the site-selective growth of Ag might originate from the speed difference of the Ag atoms reduced from ions because the top sites were surely easy to be oxidized than the valley sites.

13.5.2 Detection of Adenine by SERS Technique

Water solutions of adenine with two different concentrations, 10^{-4} and 10^{-6} mol/L, were prepared for SERS detection. After being washed with KCl solution to remove the possible impurities adsorbed on it, Ag/Si-NPA samples were immersed into

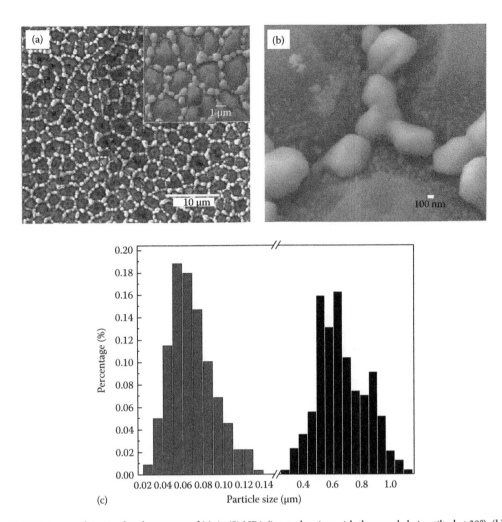

FIGURE 13.16 FE-SEM images showing the plane views of (a) Ag/Si-NPA (inset: the view with the sample being tilted at 30°), (b) locally amplified photograph of Ag/Si-NPA with the sample being tilted at 30°, and (c) the size-distribution of *nc*-Ag and sub-*mc*-Ag deposited on Si-NPA.

adenine solutions, washed with deionized water, and dried in nitrogen flow at room temperature. The surface enhanced Raman scattering spectra were tested by a micro-Raman spectroscope with a laser wavelength of 532 nm.

To determine the SERS activity of Ag/Si-NPA to adenine, three related Raman spectra were firstly measured and the results were depicted in Figure 13.17. Curve (a) was the spectrum for Ag/Si-NPA that was firstly immersed in 0.1 mol/L KCl solution for 20 min and then washed with deionized water. Here only one strong Raman peak appeared, locating at ~520 cm^{-1} and corresponding to the first-order optical phonon scattering of Si substrate. Curve (b) was the spectrum for Si-NPA that was also firstly dealt by 0.1 mol/L KCl solution, then immersed in 10^{-4} mol/L adenine solution for 30 min, and finally washed with deionized water. Here, in addition to the peak that appeared in curve (a), a wide weak peak locating at ~960 cm^{-1} was observed. The newly appeared peak was determined to be from the Si horizontal optics double phonon scattering process. Obviously, the silicon-related Raman peaks of Si-NPA were highly weakened after silver deposition. Curve (b) definitely indicated that

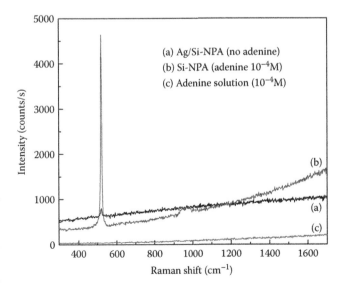

FIGURE 13.17 Raman spectra of (a) Ag/Si-NPA (without adenine), (b) Si-NPA immersed into 10^{-4} mol/L adenine for 30 min, and (c) the solution of adenine with a concentration of 10^{-4} mol/L.

Si-NPA itself is not an effective SERS active substrate for adenine detection, because no adenine-related Raman peak was observed in the spectrum for Si-NPA with adenine adsorption. Curve (c) was the spectrum for 10^{-4} mol/L adenine solution, in which no obvious Raman peak appeared. This indicated that the direct detection of adenine in its low-concentration solution by Raman spectroscopy is also ineffective.

Figure 13.18 shows the Raman spectra of adenine adsorbed on Ag/Si-NPA (spectrum (a)) and that of solid adenine (spectrum (b)) for comparison. The sample corresponding to spectrum (a) was prepared by firstly immersing Ag/Si-NPA in the solution of KCl and then in the solution of adenine with a concentration of 10^{-4} mol/L. For the convenience of comparison, the intensity of spectrum (b) was reduced to 1/5 of its original data. Definitely, almost all the bands that appeared in spectrum (a) could be found correspondingly in spectrum (b), and were highly in accordance with those determined to be from adenine by the SERS technique by the other groups (Giese and McNaughton 2002, Li and Fang 2007). The typical bands in spectrum (a) peaked at ~733, ~1333, and ~1459 cm^{-1} were determined to correspond to those of adenine peaked at ~721, ~1331, and ~1459 cm^{-1} in spectrum (b), respectively. Considering the concentration of the adenine solution used in preparing adenine-adsorbed Ag/Si-NPA was low, the signal-to-noise ratio exhibited by spectrum (a) seems satisfactory. Furthermore, it should be noted that compared with spectrum (b), the peak positions for most bands in spectrum (a) shifted correspondingly, accompanied with the disappearance of the bands that peaked at ~535 and ~1418 cm^{-1}, and the appearance of the band that peaked at ~687 cm^{-1}. These spectrum variations should be attributed to the environmental difference of the adenine molecules in the two different samples that could be explained by the enhancement mechanism about the SERS effect (Akemann et al. 1997, Campion and Kambhampati 1998, Moskovits 2005).

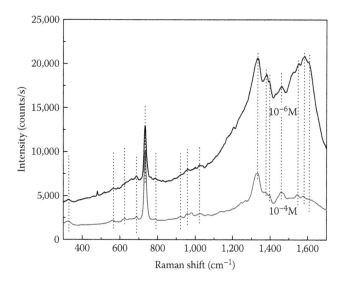

FIGURE 13.19 The SERS spectra for adenine adsorbed on Ag/Si-NPA through immersing in 10^{-4} and 10^{-6} mol/L adenine solution, respectively.

Figure 13.19 shows the SERS spectrum of the neutral aqueous solution of adenine with a concentration of 10^{-6} mol/L adsorbed on Ag/Si-NPA. It was found that the peak positions of the characteristic bands for adenine were highly consistent with those for the 10^{-4} mol/L solution, except that the spectrum shape at high wave number was slightly changed because of the elevation of the noise baseline. Judged from the range of the wave number, the fall-down of the signal-to-noise ratio of the SERS peaks might originate from the partial decomposition and carbonization of adenine molecules under laser irradiation, because the Raman bands for amorphous carbon have been determined to be around ~1350 and ~1580 cm^{-1} (Casiraghi et al. 2005). For further clarifying the problem, the SERS detection for the adenine solution with high-concentration was carried out with a purposively prolonged detecting time. It was found that with the prolonged detecting time, a similar phenomenon to that observed with low concentration solutions also appeared. Furthermore, this spectrum variation would gradually become dominant with the decrement of the adenine concentration of the detected solutions, from 10^{-4} to 10^{-5}, 10^{-6}, 10^{-7}, 10^{-8}, and 10^{-9} mol/L. In the present experiments, a clear adenine SERS spectrum could only be obtained with the solution concentration of adenine being lower than 10^{-6} mol/L. This limit was in accordance with that obtained by Ishikawa et al., who achieved the single-molecule SERS imaging and spectroscopy of adenine adsorbed on colloidal Ag nanoparticles using 10^{-6} mol/L adenine solution (Ishikawa et al. 2002). So, Ag/Si-NPA is a SERS active substrate with significant enhancement effect, and the distinct Raman scattering enhancement of adenine indicates that the direct detection and identification of single native nucleotides should be possible.

In addition to the physical and chemical characteristics of the metal species, the enhancement capability of a SERS-active substrate depends also highly upon the size, shape, arrangement mode of the metal particles, and the spacing sites formed between

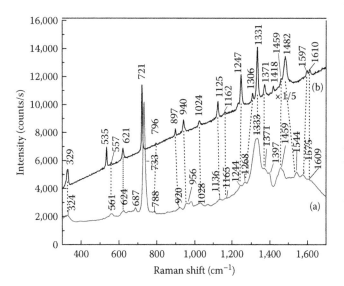

FIGURE 13.18 (a) The SERS spectrum for adenine adsorbed on Ag/Si-NPA through immersing in 10^{-4} mol/L adenine solution, and (b) the Raman spectrum of solid adenine.

them. In the case of Ag/Si-NPA, two kinds of silver structures on Si-NPA are formed, the continuous silver film composed of *nc*-Ag sized from ~40 to ~100 nm and the interconnected silver network composed of ellipsoid-like shape sub-*mc*-Ag specified by a length of ~770 nm and a width of ~550 nm. The two typical kinds of silver particles are located on the nanoporous silicon pillars and the valleys surrounding them, respectively. The great difference between the two kinds of silver particles decided the roles that they played in the SERS process. Theoretically, the occurrence of the SERS effect originates from the electromagnetic field and chemical enhancements, with the former being mainly due to the resonant excitation of localized surface plasmons (LSPs) and the latter due to the interaction between the molecules and the metal surface. Generally, the electromagnetic field enhancement is the dominant factor in most Ag-based SERS systems. In the present experiment, however, the LSPs were not excited effectively in the sub-*mc*-Ag because its size is bigger than or similar to the excitation light's wavelength (Kreibig and Volmer 1995). On the other hand, it has been proved that the spacing sites formed between the metallic particles are the hot sites where the optical field intensity is much higher than that at the other sites, which results in obtaining a strong Raman signal for the locally adsorbed biomolecules (Wang et al. 2005). The optimal spacing for the observation of an effective SERS phenomenon is sub-10 nm. Therefore, although it was difficult to excite the LSPs from the sub-*mc*-Ag themselves, the large quantities of the spacing sites formed between the silver particles (either sub-*mc*-Ag or *nc*-Ag) with sub-10 nm spacing will only be SERS-active sites that introduce big SERS effects. Furthermore, it has been pointed out that the optimal size of silver particles to produce an obvious SERS effect is in the range of 80–100 nm (Nie and Emory 1997).

Based on this judgment, the *nc*-Ag formed in Ag/Si-NPA that is sized about 40–100 nm would also bring in a tremendous SERS effect. Therefore, together with the SERS contribution from the hot sites between the particles with sub-10 nm spacing, the SERS effect in Ag/Si-NPA is surely strong.

13.5.3 Stability of SERS Detection

Stability is another key factor for the practical application of a SERS active substrate. It has been disclosed that the SERS activity of the electrochemically roughened silver electrodes and the silver foil roughened with nitric acid etching could only last for a few days and about a week, respectively, and the silver and gold colloids after the addition of the analytes would tend to aggregate and exhibit poor reproducibility of the SERS signals (Norrod et al. 1997, Dou et al. 1999, Hu et al. 2002). The SERS stability of Ag/Si-NPA was also studied by measuring its Raman spectra after the adenine adsorption for different storage times.

Figure 13.20 depicts the SERS spectra measured after the sample was stored in air for 0, 32, 50, 64, 72, and 132 days, respectively. Here it could be found that the Raman peaks corresponding to the adenine molecules adsorbed on Ag/Si-NPA for all the stored samples exhibit no obvious shape variation. This indicates that the SERS activity of Ag/Si-NPA substrate maintains even after being stored for 132 days. It should be noted that with increasing time, the relative intensity of the Raman peaks firstly increased (32 days), then kept stable a level for the fresh sample (50 and 64 days), and finally slightly decreased (72 and 132 days). This trend was shown more clearly in the inset of Figure 13.20. Compared with the initial measurement, the intensity of the band peaked at ~733 cm^{-1} increasing over two

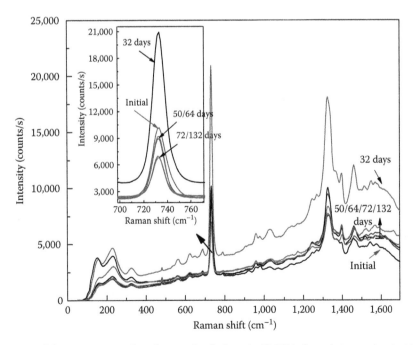

FIGURE 13.20 Time evolution of the SERS spectra for adenine adsorbed on Ag/Si-NPA through immersing in 10^{-4} mol/L adenine solution (inset: those for the SERS peak around at ~733 cm^{-1}).

times after 32 days, then decreased to nearly the same intensity as that for the initial sample after 50–64 days, and finally decreased slightly after 72–132 days. This indicates that at least after 132 days, Ag/Si-NPA still kept the SERS activity to the adsorbed adenine molecules without obvious reduction of the SERS intensity. Therefore, the SERS activity of Ag/Si-NPA is much more stable than that of the usually used Ag-based substrates. The slight instability of the Ag/Si-NPA active substrate is probably caused by the analyte desorption from the SERS active sites or the activity decay of the metal particles in the scattering process, as has been clarified by other groups (Chattopadhyay et al. 2005). However, the SERS intensity increased after the samples were stored for 32 days. It may be caused by the unique structure of Ag/Si-NPA. The adenine molecules adsorbed on the inactive sites, such as on sub-*mc*-Ag, might gradually transfer to the sub-10 nm spacing sites formed between the silver particles or *nc*-Ag by Brownian motion. Because these spacing sites play the role of SERS active spots, they lead to the great enhancement of the total SERS intensity. These results indicated that Ag/Si-NPA might be a promising active substrate for practical SERS detection of low-concentration biomolecules.

13.6 Field Emission from CNT/Si-NPA

Since the synthesis of carbon nanotube (CNT) (Iijima 1991), high expectation has been put on its potential applications as electron-emission cathodes for large area panel displays, because CNT seems to satisfy all the technical requirements of an ideal cathode, such as small tip radius of curvature, high aspect ratio, relatively low work function, high thermal conductivity, and chemical stability (de Heer et al. 1995, Rinzler et al. 1995, Fan et al. 1999, Huang et al. 2005). In the past several years, field emission (FE) has been demonstrated in various CNT films (Suh et al. 2002, Ahn et al. 2003, Zhang et al. 2004, Chen et al. 2005, Uh et al. 2005, Lee et al. 2006b), but the realization of large-acreage FE with high emission current density (ECD) and low turn-on field is still a challenge (Chen et al. 2005, Jeong et al. 2005, Jung et al. 2005, Saurakhiya et al. 2005, Uh et al. 2005, Hahn et al. 2006). The unique porous structure and the regular pillar array of Si-NPA (Xu and Li 2008a) might make it an ideal template for fabricating silicon-based composite FE cathodes. Here we will show that an enhanced FE could be obtained through growing CNT on Si-NPA.

13.6.1 Preparation and Characterization

CNT/Si-NPA is prepared by growing CNTs on Si-NPA substrate using thermal chemical vapor deposition (CVD) method (Li and Jiang 2007). The surface morphology and the Raman spectrum of the sample are presented in Figures 13.21 and 13.22, respectively. Through analyzing the positions of the three Raman peaks, 1329, 1588, and 2659 cm⁻¹, respectively, the entangled line-shaped matters growing on Si-NPA were determined to be CNTs (Farrari and Robertson 2000). When directly judged from the density distribution of CNTs, they grew densely on the pillar

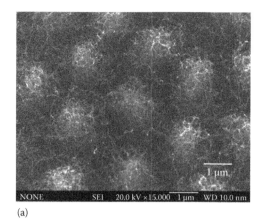

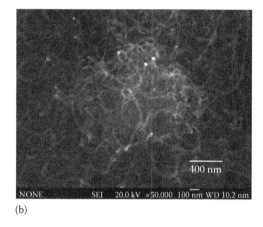

FIGURE 13.21 The surface morphology of CNT/Si-NPA obtained by FE-SEM. (a) The CNT nest array of CNT/Si-NPA titled at an angle of 30° and (b) the plan-view image of an individual CNT nest. (From Li, X.J. and Jiang, W.F., *Nanotechnology*, 18, 065203, 2007. With permission.)

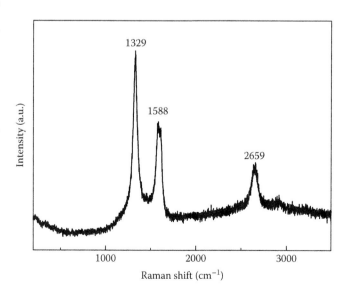

FIGURE 13.22 Experimental micro-Raman spectrum of CNT/Si-NPA. (From Li, X.J. and Jiang, W.F., *Nanotechnology*, 18, 065203, 2007. With permission.)

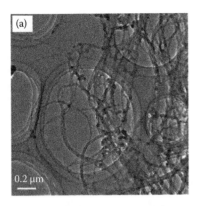
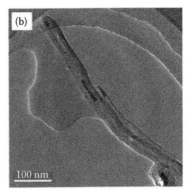
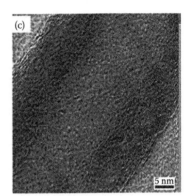

FIGURE 13.23   TEM images of (a) a cluster of CNTs, (b) an individual CNT with closed end, and (c) HRTEM image of a CNT with clearly observed multi-walls. (From Li, X.J. and Jiang, W.F., *Nanotechnology*, 18, 065203, 2007. With permission.)

but sparsely at the valleys surrounding the pillar to form a kind of nest-shaped assemblage of CNTs, and integrally construct a regularly patterned nest array. Deducing from the locations of the CNTs on Si-NPA, its growth might be of a site-selective mode in which the microstructure and surface chemical properties of the Si-NPA must have played key roles (Li and Jiang 2007).

By fetching the upper layer from the CNT/Si-NPA, the structural characteristics of the grown CNTs were studied by TEM and HRTEM. Figure 13.23a is a TEM image of a cluster of CNTs. The typical CNT length is evaluated to be of the magnitude of several microns. It is easy to find that there were numerous small grains encapsulated in the carbon tubes. Just as what has been discussed in similar experiments (Zhang et al. 2002), these grains should be iron particles obtained from the reduction of the decomposition product of ferrocene that plays the role of a catalyst in the growing process of CNTs. Figure 13.23b disclosed that the CNTs have closed ends and all the iron particles were encapsulated in the CNTs and were dispersed at the ends or the intermediate sections of the tubes. The HRTEM image (Figure 13.23c) disclosed that all the CNTs are multi-walled CNTs with an average diameter of ~40 nm, and the wall of the tubes contained 35 graphitic layers with a 10° tilt towards the tube axis.

13.6.2 Enhanced Field Emission from CNT/Si-NPA

The FE measurements of CNT/Si-NPA were performed with a diode structure, with the actual emission area being 1.0 cm × 1.5 cm. Defining the applied electric field to extract a current density of 2 μA/cm² as the turn-on field, the measured turn-on field for CNT/Si-NPA was 0.56 V/μm. Figure 13.24 depicted the curve of the ECD *versus* the applied electric field, and the corresponding F–N curve was depicted in the inset of it. Clearly, all the dots on the F–N curve could be well fitted by a single straight line. According to the F–N law (Fowler and Nordheim 1928), the electron emission from CNT/Si-NPA could be determined to be cold FE. Obviously, the ECD of NACNT/Si-NPA increased monotonically with the applied field, and the curve shape was rather smooth. This indicated that the emitters of CNT/Si-NPA were kept in a normal emission state during the whole FE measuring

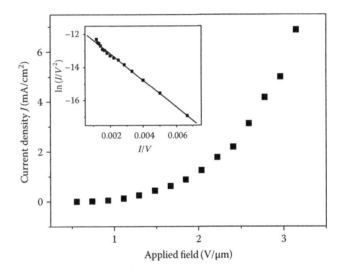

FIGURE 13.24   The experimental curve of the FE current density *versus* the applied electric field of CNT/Si-NPA. Inset: the corresponding F–N curve and its fitting line. (From Li, X.J. and Jiang, W.F., *Nanotechnology*, 18, 065203, 2007. With permission.)

process, and the increment of the current density was mainly caused by the strengthening of the emission capability of each emitter with field enhancement. At an electric field of 3.1 V/μm, an ECD of 6.8 mA/cm² was achieved. Considering that the emission area of CNT/Si-NPA was 1.5 cm², this is a rather strong performance among carbon-related cathodes. Matsumoto et al. and Sato et al. have tried to grow CNTs on silicon protrusions formed by gas etching or photolithography methods for lowering the turn-on field and increasing the ECD (Matsumoto et al. 2001, Sato et al. 2005). They demonstrated that with the emission areas being 0.14 and 0.05 cm², the turn-on fields as low as 1.67 and 1.8 V/μm, and the ECDs as high as 0.36 mA/cm² (at ~3.3 V/μm) and 1 mA/cm² (at 3.3 V/μm) could be achieved. Their results indicated that for CNT/Si system, the emission performance could be effectively improved through introducing silicon protrusions on the substrate. On the other hand, CNT films grown on various suitably treated substrates with an ECD higher than 1 mA/cm² were also reported, although with an actual emission area of only several square millimeters. For examples, ECDs of ~6 A/cm² (at 7.4 V/μm) and ~25 mA/cm^2 (at 4 V/μm) were demonstrated by the CNT films grown on Fe/Al/TiN/Si and Ti substrates, respectively, with the emission areas being 10^{-4} and 0.04 cm^2, respectively (Hahn et al. 2006, Lee et al. 2006a). However, when the actual emission area reaches the magnitude of a square centimeter, the ECD would drastically drop down to the scale of several tens to several hundreds of microamperes. For example, for the CNT films grown on SiO$_2$/Si (0.8 cm^2), ITO-coated soda lime glass (64 cm^2), and the glass modified by organic functional groups (1 cm^2) the corresponding ECDs were only 42 μA/cm^2 (at 6 V/μm), 62 μA/cm^2 (at 2 V/μm), and 160 μA/cm^2 (at 4.8 V/μm), respectively (Jung et al. 2005, Uh et al. 2005, Lee et al. 2006b). Based on these comparisons, it could be concluded that the field electron emission from CNT/Si-NPA has been greatly enhanced, and the technical parameters such as turn-on field and ECD has almost satisfied the typical technical requirements for flat panel display operation (Amaratunga and Silva 1996).

The field-enhancement factor β is one of the most important technical parameters to judge the performances of a cold field emitter, and usually, it is calculated from the work function of cathode material and the slope coefficient of the fit line of the F–N curve. In the case of CNT/Si-NPA, if we choose the work function as 5.0 eV, which is the mostly adopted value of CNTs (Suh et al. 2002), the β of CNT/Si-NPA was calculated to be ~25,000. Such a β value is much higher than the reported typical values of CNT cathodes, such as 400–1200 of CNTs on silicon and glass substrates, and 2600–3500 of highly ordered CNT arrays on porous aluminum oxide (Xu and Brandes 1999, Suh et al. 2002). Therefore, judging from the actual emission area, the turn-on field, the current density, and the calculated field enhancement factor, CNT/Si-NPA was an ideal candidate cathode. Combining with the morphological and structural analysis on CNT/Si-NPA, the reasons leading to the excellent FE performances might be concisely concluded as follows. (1) The geometrical predominance of CNT. Just as what have been demonstrated both theoretically and experimentally (Miline et al. 2003), both the small tip radius of curvature and the high aspect ratio of CNT would be helpful for the emission of electrons from the emitter bulk. (2) The formation of the regular nest array of CNTs. The array of the silicon pillars in Si-NPA template brought regular surface undulation, and therefore controlled the site-selection growth of CNTs as well as the final space-distribution of CNT emitters. Such a geometrical configuration of CNT/Si-NPA would largely decrease the electrostatic shield among the emitters and might partly bring the great increment of the field-enhancement factor β. (3) The formation of the numerous iron particles encapsulated in CNTs. As has been demonstrated by other groups very recently (Xu et al. 2005, Lee et al. 2006b), the presentation of the residual iron catalyst particles in CNTs could significantly increase the field enhancement factor β and lower the work function of CNTs, and finally lead to the enhancement of the emission capability of CNTs specified by a relatively low turn-on field and high ECD. Although much effort is still needed to clarify the roles of iron particles played in the emission process of CNT/Si-NPA, the very low turn-on field (0.56 V/μm) and the very big field-enhancement factor β (~25,000) should strongly depend on the presentation of the iron particles. Furthermore, the presentation of large quantities of iron particles in CNTs would lead to the increment of the electric conductivity of CNT/Si-NPA. All these factors might play important roles for achieving such a high ECD in CNT/Si-NPA. These results indicated that CNT/Si-NPA might be an ideal candidate cathode for cold field emissions and might find potential applications in vacuum microelectronics and field-emission displays.

13.7 CdS/Si Nanoheterojunction Array Based on Si-NPA

Semiconductor nanoheterostructures have attracted much attention in recent years because they could exhibit electrical or light-emitting properties different from their traditional counterparts (Prabhakaran et al. 2003, Tzolov et al. 2004, Oh et al. 2005, Vasa et al. 2005, Xiang et al. 2006, Li et al. 2007b, Nakamura et al. 2007, Shen et al. 2007). Technically, the nanoheterostructures were usually prepared by growing one kind of semiconductor onto the other, with the latter being pre-treated to form some kinds of nanostructures and acting as both a compositional material and a well-established nano-template. Considering that crystal silicon has been a dominant electronic material, the exploration of Si-based nanoheterostructures might be of key importance in developing future optoelectronic nanodevices.

In the past several years, mainly due to the properties of tunable PL, broadband light transparency, and high electrical conductibility of cadmium sulfide (CdS) the probe on CdS/Si nanoheterostructure arrays has aroused much interest, mainly aimed at the potential applications in the field of solar cells, light-emitting diodes, light detectors, and laser sources (Dobson et al. 2001, Duan et al. 2003, Agarwal et al. 2005, Greytak et al. 2005). Here we present the research on the photoluminescent and electrical properties of a CdS/Si nanoheterostructure array based on Si-NPA. We show that three-primary-color PL and excellent rectification performances could be achieved in CdS/Si-NPA.

13.7.1 Preparation and Characterization

The CdS/Si nanoheterojunction array was prepared by growing CdS nanocrystallites (*nc*-CdS) onto Si-NPA through a heterogeneous reaction process (Xu and Li 2007b). The freshly prepared Si-NPA was firstly immersed in CdCl$_2$ alcohol solution and then exposed to H$_2$S flow to grow *nc*-CdS. The chemical reaction obeys the chemical reaction formula CdCl$_2$ (s) + H$_2$S (g) = CdS (s) + 2HCl (g). For obtaining properly crystallized and stabilized *nc*-CdS, an annealing treatment was performed in a highly pure nitrogen atmosphere. The as-prepared CdS/Si heterostructure was named CdS/Si-NPA.

The XRD spectra for Si-NPA and CdS/Si-NPA are depicted in Figure 13.25. A broadened Si (111) diffraction peak was observed for Si-NPA and a group of diffraction peaks corresponding to wurtzite-CdS were observed for CdS/Si-NPA. This directly

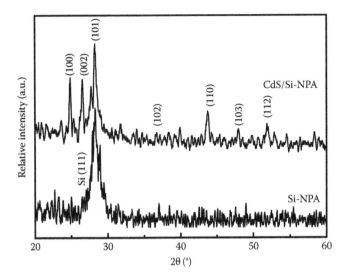

FIGURE 13.25 XRD patterns of CdS/Si-NPA and Si-NPA. (From Xu, H.J. and Li, X.J., *Appl. Phys. Lett.*, 91, 201912, 2007b.)

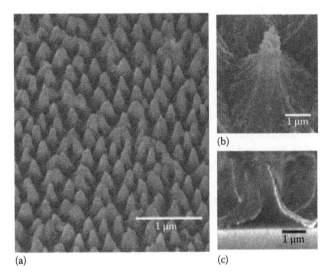

FIGURE 13.26 FE-SEM images of (a) CdS/Si-NPA, (b) an individual pillar of CdS/Si-NPA, and (c) cross-sectional image of CdS/Si-NPA. The images of (a) and (b) were taken with the samples being titled at an angle of 45°. (From Xu, H.J. and Li, X.J., *Appl. Phys. Lett.*, 91, 201912, 2007b.)

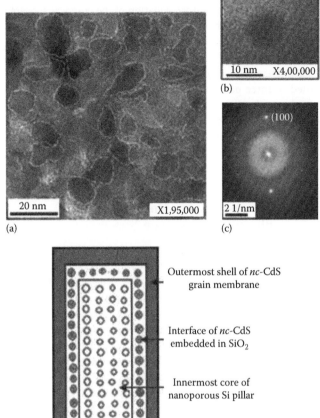

FIGURE 13.27 HRTEM images of (a) the interface formed between *nc*-CdS membrane and Si-NPA substrate and (b) an individual crystalline zone of *nc*-CdS, (c) the FFT pattern corresponding to the field given in (b), and (d) the schematic representation of one individual pillar of CdS/Si-NPA. (From Xu, H.J. and Li, X.J., *Appl. Phys. Lett.*, 91, 201912, 2007b.)

proved the growth of crystalline CdS on Si-NPA. The morphology of CdS/Si-NPA given in Figure 13.26 disclosed that both the pillars and the valleys around the pillars of Si-NPA were covered with CdS particles and a continuous grain membrane of CdS was formed on the Si-NPA. Judging from the cross-sectional image of CdS/Si-NPA (Figure 13.26c), the thickness of the CdS membrane was estimated to be ~80 nm and the interface between the CdS membrane and the substrate could be distinguished. Clearly, even after the growth of CdS, the pillars were still well-separated and the characteristic of the pillar array of the Si-NPA was kept. To clarify the composition and microstructure of the interface, high HRTEM observation was performed and the image is presented in Figure 13.27. In the image with a

relatively large field of vision (Figure 13.27a), numerous crystalline zones surrounded by amorphous districts were observed. The size of the crystalline zones scaled from ~4 to ~15 nm, with an average size of ~9 nm. To determine the phases of the crystalline zones as well as the neighboring amorphous districts, a typical area shown in Figure 13.27b was selected to carry out the electron diffraction measurement with two-dimensional fast Fourier transform (FFT) mode. Figure 13.27c shows that a set of diffraction spots with hexagonal symmetry and a diffraction ring were observed, and they were determined to correspond to the diffraction from the (100) crystal plane of wurtzite-CdS and the diffraction from amorphous SiO_2, respectively. Therefore, the interface between CdS and Si-NPA could be described as a two-phase system with large quantities of *nc*-CdS embedded in an amorphous SiO_2 matrix. Such an interface might have been formed through the CdS growing into the nanopores located at the upper bed of the silicon pillars. Considering that the growth of *nc*-CdS was performed in a reducing atmosphere, the detected amorphous SiO$_2$ should be formed during the preparing process of Si-NPA, that is, the native oxidation of the pore walls located at the upper bed of the pillars (Xu and Li 2007a).

Based on the above analysis, the structure of the CdS/Si-NPA could be illustrated by the diagrammatic sketch given in Figure 13.27d. Here each individual pillar of the CdS/Si-NPA is composed of three parts, the outmost shell of continuous *nc*-CdS membrane (region I), the interface layer composed by large quantities of *nc*-CdS embedded in amorphous SiO$_2$ matrix (region II), and the innermost core of nanoporous Si pillar (region III). Therefore, the CdS/Si-NPA could be viewed as a complex and unique nanoheterostructure.

13.7.2 Three-Primary-Color Photoluminescence

Figure 13.28 presents the PL spectra of CdS/Si-NPA measured at room temperature, together with those of Si-NPA and annealed Si-NPA for comparison. Here the annealing conditions taken for the Si-NPA were the same as that for the CdS/Si-NPA. All the PL spectra were measured under the irradiation of 370 nm ultraviolet. Three distinct PL bands were observed in the CdS/Si-NPA, which peaked at ~420, ~520, and ~745 nm, respectively, together with an additional weak PL band centered at ~470 nm. It should be noted that the peak positions for the three strong PL bands were approximated to the three-primary-color emission bands, the blue, green, and red bands in display techniques. This indicates that the CdS/Si-NPA might be a potential light source for white light emission.

As an essential requirement for practical solid-state white lighting devices, both the peak positions and the relative intensities of the emission bands should be adjustable. To realize the

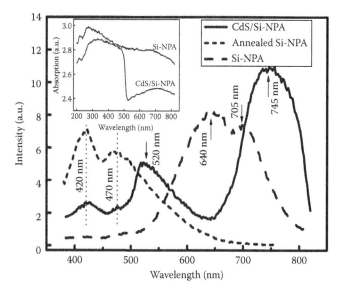

FIGURE 13.28 Room-temperature PL spectra of CdS/Si-NPA, Si-NPA, and annealed Si-NPA, excited by 370 nm ultraviolet. Inset: the absorption spectra of CdS/Si-NPA and annealed Si-NPA measured in the wavelength scale of 200–820 nm. (From Xu, H.J. and Li, X.J., *Appl. Phys. Lett.*, 91, 201912, 2007b.)

adjustability, the origins of these emission bands must be clarified. Clearly, the PL spectrum of the CdS/Si-NPA was different from both the Si-NPA and the annealed Si-NPA that exhibited two red PL bands centered at ~640 and ~705 nm, and two blue PL bands centered at ~420 and ~470 nm, respectively (Figure 13.28). The two red PL bands of the Si-NPA were not observed in the PL spectra of the annealed Si-NPA and the CdS/Si-NPA, but two blue PL bands were observed simultaneously. Considering that both the latter two samples have experienced an annealing process under identical conditions, it was reasonable to deduce that the emitters contributing to the two red PL bands of the Si-NPA have disappeared during the annealing process, accompanied with the formation of the emitters contributing to the two blue PL bands observed in the annealed Si-NPA and the CdS/Si-NPA. Combined with the structure of the CdS/Si-NPA demonstrated by Figure 13.27d, it could be presumed that the two blue PL bands of the CdS/Si-NPA originated from the Si-NPA substrate beneath the outermost CdS shell and the CdS/Si-NPA interface, the same origin as that for the annealed Si-NPA. The peak intensities of the two blue PL bands from the CdS/Si-NPA were a little lower than those from the annealed Si-NPA, and this should be due to the intensity weakening of the excitation ultraviolet when it passed through the covered *nc*-CdS shell.

The green and red PL bands observed in the CdS/Si-NPA could be directly attributed to the *nc*-CdS or CdS/Si-NPA nanoheterostructure as an integral system. Judging from their peak positions and profiles, the green band was only observed in the CdS/Si-NPA and the red band was obviously different from that observed in the Si-NPA. In fact, the green emission band peaking at ~520 nm has been observed by many other groups studying the PL properties of nanostructured CdS, and the origin was attributed to the band edge transition in the *nc*-CdS (Ge and Li 2004, Liu et al. 2004, Wang et al. 2006). The idea was further proved by the light absorption experiment carried out for the CdS/Si-NPA, as shown by the spectrum given in the inset of Figure 13.28. Here the absorption edge was determined to be at ~516 nm (~2.40 eV), almost identical with the PL peak position of the green band of CdS/Si-NPA, ~520 nm (~2.39 eV). Therefore, it could be concluded that the green PL from the CdS/Si-NPA originates from the intrinsic radiative recombination in the *nc*-CdS. As for the red PL band centered at ~745 nm, it should be related with the surface defect states of the *nc*-CdS. A similar broad PL band centered at ~758 nm has been observed in CdS nanobelts and the origin was determined to be the transition of the electrons from the defect states to the valence band (Wang et al. 2006). Further investigations also have disclosed that the peak positions for the PL originated from the defect states of sulfur vacancies (V_s^+) located in the wavelength range of ~550 to ~800 nm (Sun et al. 2000, Ge and Li 2004, Liu et al. 2004). The experimental data of x-ray energy dispersed spectrum for the CdS/Si-NPA (Xu and Li 2007a) indicates that the atom ratio between Cd and S was about 1.00: 0.90, which infers that large quantities of V_s^+ might have been formed on the surface of the *nc*-CdS. Therefore, it is reasonable to attribute the red

PL band from the CdS/Si-NPA to the transition of the excited electrons from the surface states of V_s^+ to the valence band of *nc*-CdS. The clarification on the PL origins of the CdS/Si-NPA provides a base for realizing PL adjustability. The realization of three-primary-color PL in the CdS/Si-NPA might have provided another candidate path for solid-state lighting.

13.7.3 Electrical Properties

To realize the electroluminescence (EL) in the CdS/Si-NPA and improve the EL efficiency thereafter, a study on its electrical properties and the transport mechanism is necessary. As has been specified above (Figures 13.26 and 13.27), an individual pillar of the CdS/Si-NPA was composed of three parts, the outmost shell of the *nc*-CdS, the interface with the *nc*-CdS embedded in the nanopores, and the innermost core of the nanoporous Si pillar. The room-temperature Hall effect experiments disclosed that the Si-NPA and the *nc*-CdS layer deposited on the Si-NPA behaved as *p*- and *n*-type semiconductors, respectively. This indicated that a unique CdS/Si nanoheterojunction array was formed.

For carrying out electrical measurements, the electrodes designed as per the illustration in Figure 13.29 were prepared through depositing In/Ag alloy and Ag on the front and back side of the CdS/Si-NPA, respectively. The electrical detections were performed at room temperature. By taking the bias with the Ag electrode as being positive to the In/Ag electrode in the forward direction, the measured dark current density–voltage (*J–V*) curve for the CdS/Si-NPA is depicted in Figure 13.30. Here an obvious rectification effect was observed, with an onset voltage of ~1 V at a current density of ~1.6 mA/cm², a forward current density of ~170 mA/cm² at 4.5 V, a leakage current density of ~8 × 10⁻² mA/cm² at a reverse bias of 6 V, and a breakdown voltage of ~8 V. The forward-to-reverse rectifying ratio was calculated to be ~215 at ±4.5 V. All the parameters satisfy or approach the technical requirements for practical applications. Because the contacts formed between the In/Ag and *nc*-CdS, and *sc*-Si and Ag were both proved to be ohmic (Gokarna et al. 2000), the rectification effect observed in the CdS/Si-NPA might originate from the *nc*-CdS/Si-NPA or the Si-NPA/*sc*-Si. Figure 13.31 presents the experimental dark *J–V* curve of the Si-NPA/*sc*-Si. Clearly, across the measuring range, *J* varied linearly with *V* and

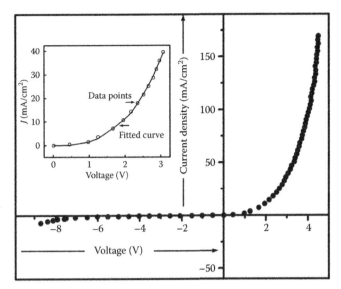

FIGURE 13.30 *J–V* characteristic of CdS/Si-NPA. The inset shows a fitted *J–V* curve via the trap-limited model $J \propto V^m$ with $m = 4$ and the proportional constant being 0.4. (From Xu, H.J. and Li, X.J., *Appl. Phys. Lett.*, 93, 172105, 2008b.)

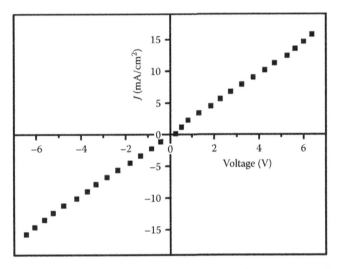

FIGURE 13.31 *J–V* characteristic of Si-NPA/*sc*-Si. (From Xu, H.J. and Li, X.J., *Appl. Phys. Lett.*, 91, 201912, 2007b.)

no rectification phenomenon was observed. Therefore, *nc*-CdS/Si-NPA is the only contributor to the rectification effect that occurred in the CdS/Si-NPA.

Theoretically, the dark *J–V* relation for a heterojunction could be described as (Maruska et al. 1992)

$$J = J_0 \left[\exp\left(\frac{q(V - IR_s)}{nkT} \right) - 1 \right], \tag{13.1}$$

Here

J_0 is the reverse saturation current density
I is the current
R_s is the serial resistance
n is the ideality factor of the diode

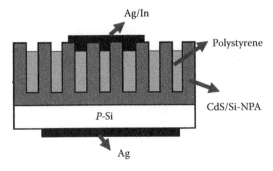

FIGURE 13.29 The diagrammatic sketch of the electrode configuration on CdS/Si-NPA for electrical measurements. (From Xu, H.J. and Li, X.J., *Appl. Phys. Lett.*, 93, 172105, 2008b.)

Clearly, J is decided by three parameters, J_0, n, and R_s that are determined by the potential barrier of the heterojunction, the carrier transport process, and the resistance to the holes transporting across the interfacial space–charge region because of carrier depletion, respectively (Anderson et al. 1991). Compared with the heterojunctions of CdS, ZnS, and SnO_2: Sb/Si based on PS (Gokarna et al. 2000, Deshmukh et al. 2001, Dimova-Malinovska and Nikolaeva 2003, Elhouichet et al. 2005), the J for CdS/Si-NPA is much bigger, and according to formula (13.1) this would require big J_0 and small n and R_s. But the J_0 measured for the CdS/Si-NPA is very small, only $\sim 8 \times 10^{-2}$ mA/cm^2 at a reverse bias of 6 V. As for n, the parameter that describes the deviation degree of a heterojunction from an ideal diode, it could be calculated through the following equation transformed from formula (13.1):

$$n = \frac{q}{kT} \frac{\partial V}{\partial (\ln J)}, \tag{13.2}$$

with $\partial V / \partial (\ln J)$ being the inverse slope of $\ln(J)$–V curve. By substituting the experimental J–V data and the slope coefficient of $\ln(J)$–V into formula (13.2), the value of n for CdS/Si-NPA was calculated to be ~ 23.3. This indicated that in the CdS/Si-NPA, both the values of J_0 and n would actually go against obtaining a big J. Therefore, the big J realized in the CdS/Si-NPA could only be due to its small R_s. In fact, Si-NPA as the functional substrate for developing the CdS/Si-NPA possesses a very large specific surface area. This would surely enlarge the contact interface between the nc-CdS and Si-NPA and therefore result in the great reduction of R_s.

To clarify the charge transport mechanism across the CdS/Si-NPA, a simulation on the J–V curve of the CdS/Si-NPA (Figure 13.30) was carried out and the result is shown in the inset of the figure. It was found that the curve could be well fitted by $J \propto V^m$, a rule deduced from the trap-limited model for the charge transport across a heterojunction (Schlamp et al. 1997). This indicated that the carrier transport process in the CdS/Si-NPA was controlled by the interface defect states acting as carrier traps, just as that in ZnS/Si heterojunctions (Hazdra et al. 1995). According to the results on the CdS/Si nanoheterostructure by other groups (Lei et al. 2005), the interface defect states in the CdS/Si-NPA most probably come from the sulfur vacancies (V_S^+) located at the interface between the nc-CdS layer and the Si-NPA and this deduction was proved by the x-ray photoelectron spectra measured at room temperature (Xu and Li 2007a). It is these V_S^+ states that determine the transport process and mechanism of the CdS/Si-NPA.

In summary, the J–V curve of the CdS/Si-NPA was measured and an obvious rectification phenomenon was observed with the parameters satisfying or approaching the technical requirements for device applications. Theoretical analysis has shown that the electron transport across CdS/Si-NPA was dominated by the trap-limited model, and this might be very helpful for further optimizing the device performance in fabricating light-emitting diodes.

13.8 Summary

Si-NPA is an Si hierarchical structure possessing simultaneously the following three advantages: the material importance inherited from sc-Si, the porous structure and physical properties of traditional PS, and the regularly patterned morphology that is useful in preparing various functional nanosystems. Both the Si-NPA and Si-NPA-based nanocomposite systems could exhibit enhanced physical properties that might be applied in future nanodevices, such as optical devices, field emitters, sensors, SERS detectors for biomolecules, arrayed rectifiers, etc. Therefore, there are still many scientific or technical problems to be probed or solved. For example, the formation mechanism of the Si-NPA is to be clarified, and the carrier implantation efficiency for promoting the electroluminescence strength of the CdS/Si-NPA is to be greatly improved. Anyway, the presently achieved results strongly indicate that the Si-NPA is one of the most promising material systems worthy of being concerned.

Acknowledgments

The related work was partially supported by the National Natural Science Foundation of China (NSFC, 19904011 and 10574112). The author would like to thank all the graduate students who have contributed to the related research, especially Dr. H. J. Xu, W. F. Jiang, L. Y. Li, and F. Feng.

References

Agarwal, R., Barrelet, C. J., Lieber, C. M. 2005. Lasing in single cadmium sulfide nanowire optical cavities. *Nano Lett.* 5(5): 917–920.

Ahn, K. S., Kim, J. S., Kim, C. O., Hong, J. P. 2003. Non-reactive rf treatment of multiwall carbon nanotube with inert argon plasma for enhanced field emission. *Carbon* 41(13): 2481–2485.

Akemann, W., Otto, A., Schober, H. R. 1997. Raman scattering by bulk phonons in microcrystalline silver and copper via electronic surface excitations. *Phys. Rev. Lett.* 79(25): 5050–5053.

Amaratunga, G. A. J., Silva, S. R. 1996. Nitrogen containing hydrogenated amorphous carbon for thin-film field emission cathodes. *Appl. Phys. Lett.* 68(18): 2529–2531.

Anderson, R. C., Muller, R. S., Tobias, C. W. 1991. Investigations of the electrical properties of porous silicon. *J. Electrochem. Soc.* 138(11): 3406–3411.

Andsager, D., Hilliard, J., Hetrick, J. M., AbuHassan, L. H., Plish, M., Nayfeh, M. H. 1993. Quenching of porous silicon photoluminescence by deposition of metal adsorbates. *J. Appl. Phys.* 74(7): 4783.

Baia, M., Baia, L., Astilean, S. 2006. Toward rapid and inexpensive identification of bulk carbon nanotubes. *Appl. Phys. Lett.* 88(14): 143121.

Baratto, C., Faglia, G., Comini, E. et al. 2001. A novel porous silicon sensor for detection of sub-ppm NO_2 concentrations. *Sens. Actuat. B* 77(1–2): 62–66.

Bettotti, P., Cazzanelli, M., Negro, L. D. et al. 2002. Silicon nanostructures for photonics. *J. Phys.: Condens. Matter* 14: 8253–8281.

Birner, A., Wehrspohn, R. B., Gosele, U. M., Busch, K. 2001. Silicon-based photonic crystals. *Adv. Mater.* 13(6): 377–388.

Bisi, O., Ossicini, S., Pavesi, L. 2000. Porous silicon: Quantum sponge structure for silicon based optoelectronics. *Surf. Sci. Rep.* 38: 1–126.

Bruno, M., Palummo, M., Marini, A., Sole, R. D., Ossicini, S. 2007. From Si nanowires to porous silicon: The role of excitonic effects. *Phys. Rev. Lett.* 98: 036807.

Campion, A., Kambhampati, P. 1998. Surface-enhanced Raman scattering. *Chem. Soc. Rev.* 27: 241–250.

Canham, L. T. 1990. Silicon quantum wire array fabrication by electrochemical dissolution of wafers. *Appl. Phys. Lett.* 57(10): 1046–1048.

Carius, R., Fischer, R., Holzenkampfer, E., Stuke, J. 1981. Photoluminescence in the amorphous system SiO_x. *J. Appl. Phys.* 52(6): 4241–4243.

Casiraghi, C., Ferrari, A. C., Robertson, J. 2005. Raman spectroscopy of hydrogenated amorphous carbons. *Phys. Rev. B* 72(8): 085401.

Chan, S., Kwon, S., Koo, T. W., Lee, L. P., Berlin, A. A. 2003. Surface-enhanced Raman scattering of small molecules from silver-coated silicon nanopores. *Adv. Mater.* 15(9): 1595–1598.

Chattopadhyay, S., Lo, C.-H., Hsu, C.-H., Chen, L.-C., Chen, K.-H. 2005. Surface-enhanced Raman spectroscopy using self-assembled silver nanoparticles on silicon nanotips. *Chem. Mater.* 17(3): 553–559.

Chen, Z., Engelsen, D., Bachmann, P. K. et al. 2005. High emission current density microwave-plasma-grown carbon nanotube arrays by postdepositional radio-frequency oxygen plasma treatment. *Appl. Phys. Lett.* 87(24): 243104.

Cheng, X., Feng, Z. -D., Luo, G. -F. 2003. Effect of potential steps on porous silicon formation. *Electrochim. Acta* 48: 497–501.

Cluzel, B., Pauc, N., Calvo, V., Charvolin, T., Hadji, E. 2006. Nanobox array for silicon-on-insulator luminescence enhancement at room temperature. *Appl. Phys. Lett.* 88: 33120.

Cooke, D. W., Bennett, B. L., Famum, E. H. et al. 1996. SiO_x luminescence from light-emitting porous silicon: Support for the quantum confinement/luminescence center model. *Appl. Phys. Lett.* 68: 1663–1665.

Cullis, A. G., Canham, L. T., Calcott, P. D. J. 1997. The structural and luminescence properties of porous silicon. *J. Appl. Phys.* 82: 909–965.

Deshmukh, N. V., Bhave, T. M., Ethiraj, A. S. et al. 2001. Photoluminescence and *I–V* characteristics of a CdS-nanoparticles-porous-silicon heterojunction. *Nanotechnology* 12(3): 290–294.

Dimova-Malinovska, M., Nikolaeva, M. 2003. Transport mechanisms and energy band diagram in ZnO/ porous Si light-emitting diodes. *Vacuum* 69: 227.

Dobson, K. D., Visoly-Fisher, I., Hodes, G., Cahen, D. 2001. Stabilizing CdTe/CdS solar cells with Cu-containing contacts to p-CdTe. *Adv. Mater.* 13(19): 1495–1499.

Dong, Y. F., Li, L. Y., Jiang, W. F., Wang, H. Y., Li, X. J. 2009. Capacitive humidity-sensing properties of electron-beam-evaporated nanophased WO_3 film on silicon nanoporous pillar array. *Physica E* 41(4): 711–714.

Dou, X., Jung, Y. M., Cao, Z., Ozaki, Y. 1999. Surface-enhanced Raman scattering of biological molecules on metal colloid II: Effects of aggregation of gold colloid and comparison of effects of pH of glycine solutions between gold and silver colloids. *Appl. Spectrosc.* 53(11): 1440–1447.

Driskell, J. D., Uhlenkamp, J. M., Lipert, R. J., Porter, M. D. 2007. Surface-enhanced Raman scattering immunoassays using a rotated capture substrate. *Anal. Chem.* 79(11): 4141–4148.

Duan, J. Q., Song, H. Z., Yao, G. Q., Zhang, L. Z., Zhang, B. R., Qin, G. G. 1994. Variation of double-peak structure in photoluminescence spectra from porous silicon with excitation wavelength. *Superlat. Microstruct.* 16: 55–58.

Duan, X., Huang, Y., Agarwal, R., Lieber, C. M. 2003. Single-nanowire electrically driven lasers. *Nature (London)* 421(6920): 241–245.

Elhouichet, H., Moadhen, A., Oueslati, M., Romdhane, S., Roger, J. A., Bouchriha, H. 2005. Structural, optical and electrical properties of porous silicon impregnated with SnO_2: Sb. *Phys. Status Solidi (c)* 2(9): 3349–3353.

Fan, S., Chapline, M. G., Franklin, N. R., Tombler, T. W., Cassell, A. M., Dai, H. 1999. Self-oriented regular arrays of carbon nanotubes and their field emission properties. *Science* 283(5401): 512–514.

Fang, J. X., Yi, Y., Ding, B. J., Song, X. P. 2008. A route to increase the enhancement factor of surface enhanced Raman scattering (SERS) via a high density Ag flower-like pattern. *Appl. Phys. Lett.* 92(13): 131115.

Ferrari, A. C., Robertson, J. 2000. Interpretation of Raman spectra of disordered and amorphous carbon. *Phys. Rev. B* 61(20): 14095–14107.

Fleischmann, M., Hendra, P. J., McQuillan, A. J. 1974. Raman spectra of pyridine adsorbed at a silver electrode. *Chem. Phys. Lett.* 26(2): 163–166.

Foll, H., Christophersen, M., Carstensen, J., Hasse, G. 2002. Formation and application of porous silicon. *Mater. Sci. Eng. Rep.* 39(4): 93–140.

Foucaran, A., Sorli, B., Garcia, M., Pascal-Delannoy, F., Giani, A., Boyer, A. 2000. Porous silicon layer coupled with thermoelectric cooler: A humidity sensor. *Sens. Actuat. B* 79(3): 189–193.

Fowler, R. H., Nordheim, L. W. 1928. Electron emission in intense electric fields. *Proc. R. Soc. Lond. Ser. A* 119: 173–181.

Francia, G. D., Castaldo, A., Massera, E., Nasti, I., Quercia, L., Rea, I. 2005. A very sensitive porous silicon based humidity sensor. *Sens. Actuat. B* 111–112: 135–139.

Fu, X. N., Li, X. J. 2005. Preparation, structural characterization and nitrogen adsorption properties of a self-supported nanostructured gold film. *Acta Phys. Sin.* 54(11): 5257–5261.

Fu, X. N., Li, X. J. 2007. Enhanced field emission of composite Au nanostructured network on Si nanoporous pillar array. *Chin. Phys. Lett.* 24(8): 2335–2337.

Fujita, S., Uchida, K., Yasuda, S., Ohba, R., Tanamoto, T. 2003. Novel random number generators based on Si nanodevices for mobile communication security systems. *Nanotechnology* 3: 309–312.

Fürjes, P., Kovács, A., Dücso, Cs., Ádám, M., Müller, B., Mescheder, U. 2003. Porous silicon-based humidity sensor with interdigital electrodes and internal heaters. *Sens. Actuat. B* 95(1–3): 140–144.

Garguilo, J. M., Koeck, F. A. M., Nemanich, R. J., Xiao, X. C., Carlisle, J. A., Auciello, O. 2005. Thermionic field emission from nanocrystalline diamond-coated silicon tip arrays. *Phys. Rev. B* 72(1): 165404.

Ge, J., Li, Y. 2004. Selective atmospheric pressure chemical vapor deposition route to CdS arrays, nanowires, and nanocombs. *Adv. Funct. Mater.* 14(2): 157–162.

Giese, B., McNaughton, D. 2002. Surface-enhanced Raman spectroscopic and density functional theory study of adenine adsorption to silver surfaces. *J. Phys. Chem. B* 106(1): 101–112.

Gokarna, A., Bhoraskar, S. V., Pavaskar, N. R., Sathaye, S. D. 2000. Optoelectronic characterisation of porous silicon/CdS and ZnS systems. *Phys. Status Solidi (a)* 182(1): 175–179.

Golonka, L. J., Licznerski, B. W., Nitsch, K., Teterycz, H. 1997. Thick film humidity sensors. *Meas. Sci. Technol.* 8: 92–98.

Green, M., Liu, F. M., Cohen, L., Köllensperger, P., Cass, T. 2006. SERS platforms for high density DNA arrays. *Faraday Discuss.* 132: 269–280.

Greytak, A. B., Barrelet, C. J., Li, Y., Lieber, C. M. 2005. Semiconductor nanowire laser and nanowire waveguide electro-optic modulators. *Appl. Phys. Lett.* 87(15): 151103.

Hadobas, K., Kirsch, S., Carl, A., Acet, M., Wassermann, E. F. 2000. Reflection properties of nanostructure-arrayed silicon surfaces. *Nanotechnology* 11: 161–164.

Hahn, J., Jung, S. M., Jung, H. Y., Heo, S. B., Shin, J. H., Suh, J. S. 2006. Fabrication of clean carbon nanotube field emitters. *Appl. Phys. Lett.* 88(11): 113101.

Hazdra, P., Reeve, D. J., Sands, D. 1995. Origin of the defect states at. ZnS/Si interfaces. *Appl. Phys. A* 61(6): 637–641.

de Heer, W. A., Châtelain, A., Ugarte, D. 1995. A carbon nanotube field-emission electron source. *Science* 270(5239): 1179–1180.

Heitmann, J., Muller, F., Zacharias, M., Gosele, U. 2005. Silicon nanocrystals: Size matters. *Adv. Mater.* 17(7): 795–803.

Hu, J., Zhao, B., Xu, W., Fan, Y., Li, B., Ozaki, Y. 2002. Simple method for preparing controllably aggregated silver particle films used as surface-enhanced Raman scattering active substrates. *Langmuir* 18(18): 6839–6844.

Huang, H., Liu, C., Wu, Y., Fan, S. 2005. Aligned carbon nanotube composite films for thermal management. *Adv. Mater.* 17(13): 1652–1656.

Iijima, S. 1991. Synthesis of carbon nanotubes. *Nature* 354: 56–58.

Ishikawa, M., Maruyama, Y., Ye, J. Y., Futamata, M. 2002. Single-molecule imaging and spectroscopy of adenine and an analog of adenine using surface-enhanced Raman scattering and fluorescence. *J. Lumin.* 98(1–4): 81–89.

Islam, Md. N., Kumar, S. 2003. Influence of surface states on the photoluminescence from silicon nanostructures. *J. Appl. Phys.* 93: 1753–1759.

Jarvis, R. M., Brooker, A., Goodacre, R. 2006. Surface-enhanced Raman scattering for the rapid discrimination of bacteria. *Faraday Discuss.* 132: 281–292.

Jeong, T., Heo, J., Lee, J. et al. 2005. Improvement of field emission characteristics of carbon nanotubes through metal layer intermediation. *Appl. Phys. Lett.* 87(6): 063112.

Jiang, W. F., Xiao, S. H., Feng, C. Y., Li, H. Y., Li, X. J. 2007. Resistive humidity sensitivity of arrayed multi-wall carbon nanotube nests grown on arrayed nanoporous silicon pillars. *Sens. Actuat. B* 125: 651–655.

Jiang, W. F., Li, L. Y., Xiao, S. H., Dong, Y. F., Li, X. J. 2008. Study on the vacuum breakdown in field emission of a nest array of multi-walled carbon nanotube/silicon nanoporous pillar array. *Microelectron. J.* 39: 763–767.

Juan, F. A., Mark, S. W., Isabel, L. T., Juan, C. O., Juan, I. M. 2000. Complete analysis of the surface-enhanced Raman scattering of pyrazine on the silver electrode on the basis of a resonant charge transfer mechanism involving three states. *J. Chem. Phys.* 112(17): 7669.

Jung, M.-S., Ko, Y. K., Jung, D.-H. et al. 2005. Electrical and field-emission properties of chemically anchored single-walled carbon nanotube patterns. *Appl. Phys. Lett.* 87(1): 013114.

Kalkan, A. K., Fonash, S. J. 2006. Laser-activated surface-enhanced Raman scattering substrates capable of single molecule detection. *Appl. Phys. Lett.* 89(23): 233103.

Kanechika, M., Sugimoto, N., Mitsushima, Y. 2005. Field-emission characteristics of a silicon tip defined by oxygen precipitate. *J. Appl. Phys.* 98: 054907.

Kanemitsu, Y., Uto, H., Masumoto, Y., Matsumoto, T., Futagi, T., Mimura, H. 1993. Microstructure and optical properties of free-standing porous silicon films: Size dependence of absorption spectra in Si nanometer-sized crystallites. *Phys. Rev. B* 48: 2827–2830.

Kneipp, K., Wang, Y., Kneipp, H. et al. 1997. Single molecule detection using surface-enhanced Raman scattering (SERS). *Phys. Rev. Lett.* 78(9): 1667–1670.

Kneipp, K., Kneipp, H., Kartha, V. B. et al. 1998. Detection and identification of a single DNA base molecule using surface-enhanced Raman scattering (SERS). *Phys. Rev. E* 57(6): 6281–6284.

Kreibig, U., Volmer, M. 1995. *Optical Properties of Metal Clusters, in Springer Series in Material Science*, Vol. 25 (Berlin, Germany, Springer).

Lee, H. J., Lee, Y. D., Cho, W. S. 2006a. Field-emission enhancement from change of printed carbon nanotube morphology by an elastomer. *Appl. Phys. Lett.* 88(9): 093115.

Lee, Y. H., Kim, D. H., Kim, D. H, Ju, B. K. 2006b. Magnetic catalyst residues and their influence on the field electron emission characteristics of low temperature grown carbon nanotubes. *Appl. Phys. Lett.* 89(8): 083113.

Lei, Y., Chim, W. K., Sun, H. P., Wilde, G. 2005. Highly ordered CdS nanoparticle arrays on silicon substrates and photoluminescence properties. *Appl. Phys. Lett.* 86(10): 103106.

Li, J., Fang, Y. 2007. An investigation of the surface enhanced Raman scattering (SERS) from a new substrate of silver-modified silver electrode by magnetron sputtering. *Spectrochim. Acta Part A* 66(4–5): 994–1000.

Li, X. J., Jiang, W. F. 2007. Enhanced field emission from a nest array of multi-walled carbon nanotubes grown on a silicon nanoporous pillar array. *Nanotechnology* 18: 065203.

Li, P., Wang, G. Z., Ma, Y. R., Fang, R. C. 1998. Origin of the blue and red photoluminescence from aged porous silicon. *Phys. Rev. B* 58: 4057–4065.

Li, G. Q., Lai, P. T., Zeng, S. H., Huang, M. Q., Li, B. 1999. A new thin-film humidity and thermal micro-sensor with $Al/SrNb_xTi_{1-x}O_3/SiO_2/Si$ structure. *Sens. Actuat.* 75(1): 70–74.

Li, G. Q., Lai, P. T., Huang, M. Q., Zeng, S. H., Li, B., Cheng, Y. C. 2000. A humidity-sensing model for metal–insulator–semiconductor capacitors with porous ceramic film. *J. Appl. Phys.* 87(12): 8716.

Li, X. J., Chen, S. J., Feng, C. Y. 2007a. Characterization of silicon nanoporous pillar array as room-temperature capacitive ethanol gas sensor. *Sens. Actuat. B* 123: 461–465.

Li, Y. Q., Tang, J. X., Wang, H., Zapien, J. A., Shan, Y. Y., Lee, S. T. 2007b. Heteroepitaxial growth and optical properties of ZnS nanowire arrays on CdS nanoribbons. *Appl. Phys. Lett.* 90(9): 093127.

Li, L. Y., Dong, Y. F., Jiang, W. F., Ji, H. F., Li, X. J. 2008. High-performance capacitive humidity sensor based on silicon nanoporous pillar array. *Thin Solid Films* 517(2): 948–951.

Lin, H. H., Mock, J., Smith, D., Gao, T., Sailor, M. J. 2004. Surface-enhanced Raman scattering from silver-plated porous silicon. *J. Phys. Chem. B* 108(31): 11654–11659.

Lipinski, M., Bastide, S., Panek, P., Levy-Clement, C. 2003. Porous silicon antireflection coating by electrochemical and chemical methods for silicon solar cells manufacturing. *Phys. Status Solidi A* 197: 512–517.

Liu, W., Jia, C., Jin, C., Yao, L., Cai, W., Li, X. 2004. Growth mechanism and photoluminescence of CdS nanobelts on Si substrate. *J. Cryst. Growth.* 269(2-4): 304–309.

Maruska, H. P., Namavar, F., Kalkhoran, N. M. 1992. Current injection mechanism for porous-silicon transparent surface light-emitting diodes. *Appl. Phys. Lett.* 61(11): 1338–1340.

Matsumoto, K., Kinosita, S., Gotoh, Y., Uchiyama, T., Manalis, S., Quate, C. 2001. Ultralow biased field emitter using single-wall carbon nanotube directly grown onto silicon tip by thermal chemical vapor deposition. *Appl. Phys. Lett.* 78(4): 539–540.

Milne, W. I., Teoa, K. B. K., Chhowallaa, M. et al. 2003. Electrical and field emission investigation of individual carbon nanotubes from plasma enhanced chemical vapour deposition. *Diam. Relat. Mater.* 12(3-7): 422–428.

Moskovits, M. 2005. Surface-enhanced Raman spectroscopy: A brief retrospective. *J. Raman Spectrosc.* 36(6–7): 485–496.

Nakamura, A., Ohashi, T., Yamamoto, K. et al. 2007. Full-color electroluminescence from ZnO-based heterojunction diodes. *Appl. Phys. Lett.* 90(9): 093512.

Nie, S., Emory, S. R. 1997. Probing single molecules and single nanoparticles by surface-enhanced Raman scattering. *Science* 275: 1102–1106.

Norrod, K. L., Sudnik, L. M., Rousell, D., Rowlen, K. L. 1997. Quantitative comparison of five SERS substrates: Sensitivity and limit of detection. *Appl. Spectrosc.* 51(7): 994–1001.

Oh, D. C., Suzuki, T., Kim, J. J. et al. 2005. Capacitance-voltage characteristics of ZnO/GaN heterostructures. *Appl. Phys. Lett.* 87(16): 162104.

Oskam, G., Long, J. G., Natarajan, A., Searson, P. C. 1998. Electrochemical deposition of metals onto silicon. *J. Phys. D: Appl. Phys.* 31(16): 1927–1949.

Panarin, A. Y., Terekhov, S. N., Khodasevich, I. A., Turpin, P.-Y. 2007. Silver-coated nanoporous silicon as SERS-active substrate for investigation of tetrapyrrolic molecules. *Proc. SPIE* 6728: 672828.

Peng, K., Xu, Y., Wu, Y., Yan, Y., Lee, S. -T., Zhu, J. 2005. Aligned single-crystalline Si nanowire arrays for photovoltaic applications. *Small* 1: 1062–1067.

Prabhakaran, K., Meneau, F., Sankar, G. et al. 2003. Luminescent nanoring structures on silicon. *Adv. Mater.* 15(18): 1522–1526.

Qin, G. G., Li, Y. J. 2003. Photoluminescence mechanism model for oxidized porous silicon and nanoscale-silicon-particle-embedded silicon oxide. *Phys. Rev. B* 68: 085309.

Rinzler, A. G., Hafner, J. H., Nikolaev, P. et al. 1995. Unraveling nanotubes: Field emission from an atomic wire. *Science* 269(5230): 1550–1553.

Rittersma, Z. M., Splinter, A., Bödecker, A., Benecke, W. 2000. A novel surface-micromachined capacitive porous silicon humidity sensor. *Sens. Actuat. B* 68(1–3): 210–217.

Ruan, C., Wang, W., Gu, B. 2006. Detection of alkaline phosphatase using surface-enhanced Raman spectroscopy. *Anal. Chem.* 78(10): 3379–3384.

Sato, H., Hata, K., Miyake, H., Hiramatsu, K. 2005. Selective growth of carbon nanotubes on silicon protrusions. *J. Vac. Sci. Technol. B* 23(2): 754–758.

Saurakhiya, N., Zhu, Y. W., Cheong, F. C., Ong, C. K., Wee, A. T. S., Lin, J. Y., Sow, C. H. 2005. Pulsed laser deposition-assisted patterning of aligned carbon nanotubes modified by focused laser beam for efficient field emission. *Carbon* 43(10): 2128–2133.

Scheible, D. V., Blick, R. H. 2004. Silicon nanopillars for mechanical single-electron transport. *Appl. Phys. Lett.* 84(23): 4632–4634.

Schlamp, M. C., Peng, X., Alivisatos, A. P. 1997. Improved efficiencies in light emitting diodes made with CdSe(CdS) core/shell type nanocrystals and a semiconducting polymer. *J. Appl. Phys.* 82(11): 5837–5842.

Shang, N. G., Meng, F. Y., Au, F. C. K. et al. 2002. Fabrication and field emission of high-density silicon cone arrays. *Adv. Mater.* 14: 1308–1311.

Shao, M. W., Shan, Y. Y., Wong, N. B., Lee, S. T. 2005. Silicon nanowire sensors for bioanalytical applications: Glucose and hydrogen peroxide detection. *Adv. Funct. Mater.* 15(9): 1478–1482.

Shen, G., Ye, C., Golberg, D., Hu, J., Bando, Y. 2007. Structure and cathodoluminescence of hierarchical Zn3P2/ZnS nanotube/nanowire heterostructures. *Appl. Phys. Lett.* 90(7): 073115.

Strehlke, S., Bastide, S, Levy-Clement, C. 1999. Optimization of porous silicon reflectance for silicon photovoltaic cells. *Sol. Energy Mater. Sol. Cell* 58: 399–409.

Suh, J. S., Jeong, K. S., Lee, J. S. 2002. Study of the field-screening effect of highly ordered carbon nanotube arrays. *Appl. Phys. Lett.* 80(13): 2392–2394.

Sun, L., Fu, X., Wang, M., Liu, C., Liao, C., Yan, C. 2000. Synthesis of CdS nanocrystal within copolymer. *J. Lumin.* 87–89: 538–541.

Sundaram, R., Raj, E. S., Nagaraja, K. S. 2004. Microwave assisted synthesis, characterization and humidity dependent electrical conductivity studies of perovskite oxides, $Sm_{1-x}Sr_xCrO_3$ ($0 \leq x \leq 0.1$). *Sens. Actuat. B* 99(2–3): 350–354.

Traversa, E. 1995. Ceramic sensors for humidity detection: The state-of-the-art and future developments. *Sens. Actuat. B* 23(2): 135–156.

Tsuboi, T., Sakka, T., Ogata, Y. H. 1998. Metal deposition into a porous silicon layer by immersion plating: Influence of halogen ions. *J. Appl. Phys.* 83(8): 4501.

Tzolov, M., Chang, B., Yin, A., Straus, D., Xu, J. M., Brown, G. 2004. Electronic transport in a controllably grown carbon nanotube-silicon heterojunction array. *Phys. Rev. Lett.* 92(7): 075505.

Uh, H. S., Ko, S. W., Lee, J. D. 2005. Growth and field emission properties of carbon nanotubes on rapid thermal annealed Ni catalyst using PECVD. *Diam. Relat. Mater.* 14: 850–854.

Vasa, P., Singh, B. P., Ayyub, P. 2005. Coherence properties of the photoluminescence from CdS–ZnO nanocomposite thin films. *J. Phys.: Condens. Matter* 17(1): 189–197.

Wabuyele, M. B., Vo-Dinh, T. 2005. Detection of human immunodeficiency virus type 1 DNA sequence using plasmonics nanoprobes. *Anal. Chem.* 77(23): 7810–7814.

Wang, W. S., Virkar, A. V. 2004. A conductimetric humidity sensor based on proton conducting perovskite oxides. *Sens. Actuat. B* 98(2–3): 282–290.

Wang, H. Y., Li, X. J. 2005a. Capacitive humidity-sensing properties of Si-NPA and Fe_3O_4/Si-NPA. *Acta Phys. Sin.* 54(5): 2220–2205.

Wang, H. Y., Li, X. J. 2005b. Structural and capacitive humidity sensing properties of nanocrystal magnetite/silicon nanoporous pillar array. *Sens. Actuat. B* 110(2): 260–263.

Wang, Z. L., Song, J. 2006. Piezoelectric nanogenerators based on zinc oxide nanowire arrays. *Science* 312(5771): 242–246.

Wang, C. T., Wu, C. L. 2006. Electrical sensing properties of silica aerogel thin films to humidity. *Thin Solid Films* 496(2): 658–664.

Wang, H., Levin, C. S., Halas, N. J. 2005. Nanosphere arrays with controlled sub-10-nm gaps as surface-enhanced Raman spectroscopy substrates. *J. Am. Chem. Soc.* 127(43): 14992–14994.

Wang, Z. Q., Gong, J. F., Duan, J. H. et al. 2006. Direct synthesis and characterization of CdS nanobelts. *Appl. Phys. Lett.* 89(3): 033102.

Wolkin, M. V., Jorne, J., Faucher, P. M., Allan, G., Delerue, C. 1999. Electronic states and luminescence in porous silicon quantum dots: The role of oxygen. *Phys. Rev. Lett.* 82: 197–200.

Xiang, J., Lu, W., Hu, Y., Wu, Y., Yan, H., Lieber, C. M. 2006. Ge/Si nanowire heterostructures as high-performance field-effect transistors. *Nature (London)*, 441(7092): 489–493.

Xu, X., Brandes, G. R. 1999. A method for fabricating large-area, patterned, carbon nanotube field emitters. *Appl. Phys. Lett.* 74(17): 2549–2551.

Xu, H. J., Li, X. J. 2007a. Preparation, structural and photoluminescent properties of CdS/silicon nanoporous pillar array. *J. Phys.: Condens. Matter* 19: 056003.

Xu, H. J., Li, X. J. 2007b. Three-primary-color photoluminescence from CdS/Si nanoheterostructure grown on silicon nanoporous pillar array. *Appl. Phys. Lett.* 91: 201912.

Xu, H. J., Li, X. J. 2008a. Silicon nanoporous pillar array: A silicon hierarchical structure with high light absorption and triple-band photoluminescence. *Opt. Express* 16(5): 2933–2941.

Xu, H. J., Li, X. J. 2008b. Rectification effect and electron transport property of CdS/Si nanoheterostructure based on silicon nanoporous pillar array. *Appl. Phys. Lett.* 93: 172105.

Xu, H. X., Bjerneld, E. J., Käll, M., Börjesson, L. 1999. Spectroscopy of single hemoglobin molecules by surface enhanced Raman scattering. *Phys. Rev. Lett.* 83(21): 4357–4360.

Xu, Y. Y., Li, X. J., He, J. T., Hu, X., Wang, H. Y. 2005a. Capacitive humidity sensing properties of hydrothermally-etched silicon nano-porous pillar array. *Sens. Actuat. B* 105(2): 219–222.

Xu, Z., Bai, X. D., Wang, E. G., Wang, Z. L. 2005b. Field emission of individual carbon nanotube with in situ tip image and real work function. *Appl. Phys. Lett.* 87(16): 163106.

Xu, H. J., Jia, H. S., Yao, Z. T., Li, X. J. 2008. Photoluminescence and *I–V* characteristics of Zns grown on silicon nanoporous pillar array. *J. Mater. Res.* 23(1): 121–126.

Yao, Z. T., Sun, X. R., Xu, H. J., Jiang, W. F., Xiao, X. H., Li, X. J. 2007a. The structure and photoluminescence properties of ZnO/silicon nanoporous pillar array. *Acta Phys. Sin.* 56(10): 6098–6103.

Yao, Z. T., Sun, X. R., Xu, H. J., Li, X. J. 2007b. Preparation, structural and electrical properties of zinc oxide grown on silicon nanoporous pillar array. *Chin. Phys.* 16(10): 3108–3113.

Yuk, J., Troczynski, T. 2003. Sol–gel $BaTiO_3$ thin film for humidity sensors. *Sens. Actuat. B* 94(3): 290–293.

Zhang, X., Cao, A., Wei, B., Li, Y., Wei, J., Xu, C., Wu, D. 2002. Rapid growth of well-aligned carbon nanotube arrays. *Chem. Phys. Lett.* 362(3–4): 285–290.

Zhang, J., Feng, T., Yu, W., Liu, X., Wang, X., Li, Q. 2004. Enhancement of field emission from hydrogen plasma processed carbon nanotubes. *Diam. Relat. Mater.* 13(1): 54–59.

Zheng, G., Lu, W., Jin, S., Lieber, C. M. 2004. Synthesis and fabrication of high-performance *n*-type silicon nanowire transistors. *Adv. Mater.* 16: 1890–1893.

Nanoporous Anodic Oxides

Martin S. Bojinov
University of Chemical Technology and Metallurgy

14.1 Introduction

Nanostructured materials can be regarded as artificially synthesized materials that have constituent grain structures or morphological features modulated on a length scale less than 100 nm. The principal impetus toward fabricating nanostructured materials lies in the promise of achieving unique properties and superior performance due to their inherent nano-architectures.

Nanoporous materials are of great interest due to their high surface-to-volume ratios and size-dependent properties. There are several reasons to expect oriented nanotubular structures to exhibit superior performance in a range of devices. One reason is their potentially high surface areas as compared to conventional materials. The surface area to volume ratio of an array of nanopores increases as the diameter of these pores decreases. The porosity of the nanostructured architecture results in a large surface area available for interfacial reactions. Another reason is the possibility of vectorial charge transport through the nanotubular layer. Furthermore, for example, due to light scattering within a porous structure, incident photons would be more effectively absorbed than on a flat electrode.

The discovery of carbon nanotubes with their variety of interesting properties has stimulated the quest for the synthesis of nanotubular and nanocolumnar structures of other substances and chemical compounds. Several recent studies have indicated that nanoporous oxides with such a structure, formed on valve metals such as Al, Ti, Zr, Nb, Ta, and W, have improved properties compared to any other form of these oxides for application in catalysis, photocatalysis, sensing, photoelectrolysis, and photovoltaics. Such nanoporous materials consisting of nanotube arrays have been produced by a variety of methods including template deposition, sol–gel transcription using organo-gelators as templates, seeded growth, and hydrothermal processes. Of these nanoporous oxides fabrication routes, one of the most promising is the method of anodic oxidation in suitable electrolyte solutions, by which the dimensions of the nanopores can be accurately controlled. Uniform nanoporous oxides on Ti, for example, that consist of titania nanotube arrays of various pore sizes, depths, and wall thicknesses can be grown by precise tailoring of the electrochemical conditions. In recent literature, there is a lot of evidence of the unique properties such material architectures possess, making them of considerable scientific interest as well as practical importance.

In this chapter, the main aspects of the kinetics of formation, properties and applications of nanoporous oxides formed by anodic oxidation are briefly reviewed and discussed. First, a necessary background on the electrochemical processes of anodic oxide formation on valve metals is given and the main concepts related to the structure and morphology of the obtained layers are outlined. Second, the role of the key parameters of the anodic oxidation process (current density, potential, electrolyte composition, temperature, and substrate composition) on the main characteristics of the obtained nanoporous oxides is presented in some detail. Next, the level of understanding on the mechanism of growth of nanoporous and nanotubular oxide structures on valve metals is comprehensively reviewed. A special emphasis is put on the strong and weak points of the proposed approaches and the degree of predictability of the process kinetics they are poised to achieve. Further, the methods to determine the main structural characteristics of the oxides are concisely described, more attention being paid to the electronic properties of the nanoporous oxide structures that are extremely important for a range of prospective applications. In the following section, basic information on the processes and devices in which oxides on valve metals with nanoporous and nanotubular morphology can be applied is given. In Section 14.4, an attempt is made to summarize and rationalize these findings in order to stress the main unresolved questions and trace the path toward further research and development in this area.

14.2 Background

Metals such as Al, Zr, Ti, Nb, Ta, and W, which belong to a group known as valve metals, usually have their surfaces covered by a thin oxide film spontaneously formed in air or in electrolytes at open circuit. This film constitutes a barrier between metal and medium. Typical values of the initial thickness of these oxide films are in the range 2–5 nm, when formed in air at room temperature, but the thickness of the films can be increased by anodic oxidation. The history of anodic oxidation of the most common of these materials, aluminum, dates back to the beginning of the last century (Diggle et al. 1969, Thompson et al. 1987, Lohrengel 1993). Anodic treatments of aluminum were intensively investigated to obtain protective and decorative films on its surface. More recently, applications of nanoporous alumina with a huge and well-ordered surface area and a relatively narrow pore size distribution have been exploited (Menon 2004). Several attempts to fabricate inorganic membranes have been reported. Presently, nanoporous anodic oxides on aluminum and other valve metals are considered as prominent template materials for the synthesis of nanowires or nanotubes with monodisperse, controllable diameter, and high aspect ratios.

14.2.1 Basics of Anodic Oxidation of Metals

For anodic films on valve metals, the thickness is determined by the applied potential and may be estimated from the anodization ratio a_r, which typically lies in the range 1.0–2.0 nm V^{-1} (Table 14.1).

TABLE 14.1 Anodizing Ratios for Anodic Film Formation on Valve Metals

Metal	Anodizing Ratio a_r (nm V^{-1})
Ta	1.6
Nb	2.2
Ti	2.5
Zr	2.0–2.7
W	1.8
Al–15% H_2SO_4	1.0
Al–2% $(COOH)_2$	1.2
Al–4% H_3PO_4	1.2
Al–3% H_2CrO_4	1.25
Al–non-dissolving electrolytes	1.3–1.37

Note: In the case of aluminum anodized in sulfuric, oxalic, phosphoric, and chromic acid, anodizing ratios of the barrier layer formed beneath the porous alumina are indicated.

The maximum attainable thickness in the barrier-type oxide films was reported to be less than 1 μm, corresponding to breakdown voltages in the range of 500–700 V. Above the limiting voltages, the dielectric breakdown of the films occurs (Lohrengel 1993).

The growth of such anodic films, commonly irreversible, occurs with a fixed stoichiometry under an electrical field strength $\vec{E}$ of 1–10 MV cm^{-1}. The current density passing across the anodic film during oxidation can be in general written as

$$i = i_{ion} + i_{elec} = i_a + i_c + i_{elec} \qquad (14.1)$$

where i_a, i_c, and i_{elec} are the anion-contributing, cation-contributing, and electron-contributing current densities, respectively. Since the electronic conductivity in anodic oxides is considered rather low, the ionic current density $i_{ion} = i_a + i_c$ is the predominant mode to transport the charges. The relationship between the ionic current, i_{ion}, and the electric field, $\vec{E}$, can be expressed in terms of the Guntherschultze–Betz equation

$$i_{ion} = A \exp(B\vec{E}) \qquad (14.2)$$

where A and B are temperature- and metal-dependent parameters. As an example, for aluminum oxide, A and B are in the range of 1×10^{-16} to 3×10^{-2} mA cm^{-2} and 1×10^{-7} to 5.1×10^{-6} cm V^{-1}, respectively. Based on the Guntherschultze–Betz equation, the rate-limiting steps of the film formation are determined by the ionic transport either at the metal/oxide interface, within the bulk oxide, or at the oxide/electrolyte interface. Nowadays, it is generally accepted that the oxides simultaneously grow at both interfaces, for example, at the oxide/electrolyte interface by metal cation transport and at the metal/oxide interface by oxygen ion transport. The transport numbers of cations and anions were reported as for example, 0.45 and 0.55, respectively, for oxidation of Al at a current density of 5 mA cm^{-2}.

14.2.2 Specifics of Porous Oxide Formation on Metals

Depending on several factors, in particular the ability of the electrolyte to dissolve the formed oxide, two types of anodic oxide films can be produced (Diggle et al. 1969). Barrier type films can be formed in electrolytes in which the solubility of the oxide is negligible. Porous type films can be created in electrolytes in which the oxide is soluble, such as sulfuric, phosphoric, chromic and oxalic acid solutions for Al, and electrolytes containing hydrofluoric acid of alkali fluorides for the other valve metals (Ti, Zr, Nb, Ta). As the pore diameter of such oxides is in the nanometer range (10–100 nm), they can be recognized as nanoporous structures.

As shown in Figure 14.1, both the barrier-type and the porous-type films consist of an inner oxide of high purity material and an outer oxide layer that is usually thought to have incorporated electrolyte anions (Thompson et al. 1987, Thompson 1997, Menon 2004). The inner oxide is adjacent to the

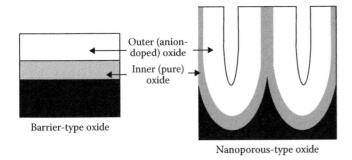

Barrier-type oxide

Nanoporous-type oxide

FIGURE 14.1 A scheme of a barrier and nanoporous oxides on a valve metal.

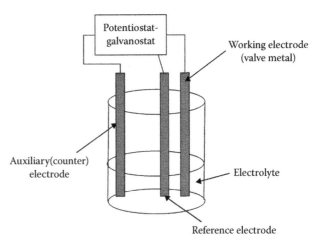

FIGURE 14.2 A scheme of an electrochemical setup for the anodic oxidation of valve metals.

oxide/metal interface, while the outer oxide is adjacent to the electrolyte/oxide interface. The degree of incorporation of electrolyte species in the outer oxide layer of barrier-type alumina strongly depends on the type of electrolyte, the concentration of adsorbed anions, and the faradaic efficiency of film growth.

On the other hand, the thickness of the nanoporous oxide films is time-dependent, and films much thicker than barrier-type oxides can be obtained. Anodizing time, current density, temperature, the nature and concentration of the electrolyte are the most important parameters in determining the film thickness of porous oxides. For instance, in the case of alumina, thick, compact, and hard porous films are formed at low temperatures (0°C–5°C, so-called hard anodizing conditions), whereas thin, soft, and non-protective films are produced at high temperatures (60°C–75°C), so-called soft anodizing conditions). As the temperature increases, the corresponding current density also increases. This does not mean that a higher current density increases the film thickness since the rate of oxide film dissolution at the electrolyte/oxide interface increases as well. The thickness of the thin barrier layer at the bottom of the porous structure is only dependent on the anodizing voltage, regardless of the anodizing time. However, electrolyte effects on the anodizing ratio in the barrier films have to be considered. Comparing the anodizing ratio of the barrier-type oxide formed in non-dissolving electrolytes (Table 14.1), the electrolyte effect can be ascribed to the dissolution of the already formed oxide in acidic electrolytes.

14.3 State of the Art

14.3.1 Overview of Nanoporous Oxide Formation

The anodic oxidation of valve metals is usually carried out using a two- or three-electrode electrochemical cell with the respective valve metal as anode and a cathode (most frequently Pt) at a constant current density or cell voltage, or the combination of both (Figure 14.2) (Menon 2004). Anodization experiments are commonly conducted with magnetic agitation of the electrolyte that reduces the thickness of the diffusion layer at the anode/electrolyte interface and ensures uniform local current density and temperature over the anode surface.

In the case of the formation of nanoporous films, two types of growth curves are observed depending on the method of film preparation (galvanostatic, i.e., at constant current density, Figure 14.3a, and potentiostatic, i.e., at constant voltage in the cell, Figure 14.3b) (Thompson 1997). In the galvanostatic regime, there are two distinct parts to the voltage *vs.* time profiles separated by a maximum in the voltage: the first linear increase (1) and the quasi-exponential drop (2). During part 1, oxide film formation is controlled by the kinetic rate of the reaction dictated by the composition of the electrolyte and anodization current. Thus, for a given current and electrolyte, the thickness of the oxide film increases linearly and hence the voltage drop through it. During part 2 of anodization, an array of pores develops on the barrier oxide and the pore depths increase. The anodic voltage drop is dictated by the parallel resistances posed by the oxide in the porous and nonporous areas. The control of the parallel resistance is dominated by the smaller of the two resistances, which, in this case, is the pore resistance. Since, during part 2 anodization the pore depth increases, that is, the thickness of the oxide film decreases, the resistance and the anodic voltage drop decreases. Finally, a dynamic steady state is reached in region 3, in which the porous oxide grows at a constant rate.

The transient of the potentiostatic current density reflects the formation of barrier-type or porous type oxides at constant voltage (see Figure 14.3b). In the beginning of the oxide formation, both transients have qualitatively similar behavior. However, for the barrier film formation, the current density i_b decays according to a power-type law. Eventually, the barrier film current is dominated by an ionic current i_{ion} through the growing layer. On the other hand, in a film-dissolving electrolyte, the current density decreases rapidly at first (region 1 in Figure 14.3b). Then, it passes through minimum value (region 2 in Figure 14.3b), increases subsequently to reach a maximum value (region 3), slowly decreasing thereafter to reach a steady-state value (region 4 in Figure 14.3b). The overall current density i_{porous} can be considered as the sum of i_b and a hypothetic current density i_{cp}, which means the current density responsible for the creation of

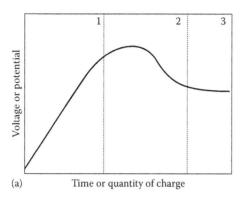 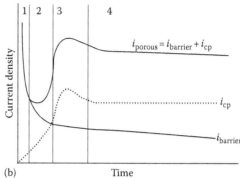

FIGURE 14.3 Schematic diagram of the initial growth of a nanoporous oxide at (a) constant current density, (b) constant voltage.

pores. The current density i_b is determined by the applied potential in terms of the anodizing ratio, while i_{cp} depends on the composition of the electrolyte and the temperature as well as on the applied potential.

The pore formation mechanism is displayed schematically in Figure 14.4, corresponding to the four stages of the current vs. time relationship in Figure 14.3b. At the beginning of the anodization, the barrier film covers the entire surface of the metal (region 1). Due to the non-uniform dissolution process, the electric field is focused locally on fluctuations of the surface (region 2). This presumably leads to a field-enhanced dissolution of the formed oxide and thus to the growth of pores (region 3). Since some pores begin to stop growing due to competition among the pores, the current decreases again and a steady state of stable pore growth is reached (region 4). However, it is very often observed that during the stable pore growth, the current

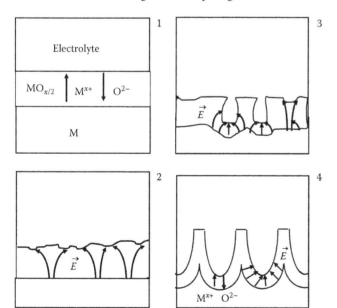

FIGURE 14.4 Schematic diagram of the pore formation at the beginning of the anodization. Region 1: formation of barrier oxide on the entire area; region 2: local field distributions caused by surface fluctuations; region 3: creation of pores by field-enhanced or/and temperature-enhanced dissolution; region 4: stable pore growth.

density continues to decrease slightly. This is supposed to be due to diffusion limitations in the long pore channels.

14.3.2 Influence of Process Parameters on the Key Properties of Nanoporous Oxides

14.3.2.1 Applied Potential

The key factor controlling the tube diameter and length is the anodization potential (Masuda et al. 1997, Sulka and Parkoła 2006, Inguanta et al. 2007, Belwalkara et al. 2008). For example, in the case of TiO_2 layers on Ti, a wide variety of nanopore diameters can be achieved (Mor et al. 2003, Macak et al. 2005a,b, Macak and Schmuki 2006, Prakasam et al. 2007). For anodization experiments carried out in $1\,M\ H_3PO_4 + 0.3\%\ HF$ it has been shown that the pore diameter can be grown in the range of 15–120 nm in the potential range between 1 and 25 V (Mor et al. 2006a,b). Particularly, self-organized nanoporous structures are obtained even at potentials as low as 1 V, although the morphology shows rather a web like structure rather than a clear nanotubular morphology. Recently, for mixed glycerol–water electrolytes containing NH_4F, the tube diameter range was further extended from 20 up to 300 nm in the potential range between 2 and 40 V (Prakasam et al. 2007). This level of pore diameter control bears significant potential for applications where this parameter needs to be adjusted for specific use, for example, when a defined size for embedding of electroactive species is desired.

In non-aqueous electrolytes, typically higher voltages are reported to grow tubes of a given diameter. These higher values are presently attributed to the typically very significant voltage drop that occurs in such electrolytes. It appears that in general the tube diameter can be described by $d = kU$, where k is rather close to $2a_r$ (a_r being the anodization ratio for the corresponding barrier anodic oxide, see Table 14.1) and U is the applied potential.

14.3.2.2 Composition of the Electrolyte

For all the valve metals except Al, the presence of fluoride ions in the anodization electrolyte is critical to the achievement of nanoporous structures. In any given F^- containing electrolyte, there is an optimal window of anodization voltages wherein nanotubular structures are obtained. In an aqueous electrolyte containing 0.5 wt % HF, random pores are obtained at a wide range of anodization potentials from 3 to 75 V. At low anodizing voltages, a porous film with sponge-like morphology is obtained. On increasing the voltage, the surface becomes nodular with an appearance similar to particulate films. On further increasing the voltage, the particulate appearance is lost, with the emergence of discrete, hollow cylindrical tube-like structures. The nanotubular structure is lost at anodizing voltages greater than 40 V and once again, a spongelike randomly porous structure is obtained. In more dilute HF solutions, larger anodization voltages are required to produce the tubular structure (Macak et al. 2006).

At a constant temperature, anodization potential and fluoride ion concentration, the electrolyte pH is the principal factor determining the thickness of the nanoporous layers in aqueous electrolytes (Mor et al. 2006a). Dilution of the HF-based electrolytes is not sufficient to create conditions favorable for deep nanopore growth because of the pH variation during the oxidation process. However, when the fluoride-ion containing electrolyte is buffered, the pH variation is reduced and control of the pore depth is achieved. For titanium oxide, the optimal pH range for the formation of the micron-long nanotubes is between pH 3 and 5. Lower pH forms shorter but cleaner nanotubular structures, whereas higher pH values result in longer nanotube arrays that suffer from unwanted precipitates. Self-organized nanotube formation is not found to occur in alkaline solutions.

With organic electrolytes, the exchange of oxygen with the metal substrate is more difficult in comparison to water and results in a reduced tendency to form oxide. Thus, the reduction in water content is thought to allow for thinner and/or more defective barrier layers through which ionic transport is supposed to be enhanced (Yoriya et al. 2007). On the other hand, incorporation of organic components from the electrolyte into the oxide is presumed to lower the relative permittivity of the film and increase its dielectric breakdown potential, which would allow for a larger voltage window for nanopore growth. Solvents such as formamide are highly polar, with dielectric constants much greater than that of water. The higher polarity of the solvent allows HF to be easily dissolved and makes it chemically available at the oxide/electrolyte interface. In such electrolytes, the pore length and aspect ratio have been found to increase when the associated cations are larger (Shankar et al. 2007). The thinnest nanoporous layers have been obtained when the cation is H^+ and the longest nanotubes were obtained when the cation is tetrabutylammonium (Bu_4N^+). This behavior could be attributed to the presence of a thinner interfacial oxide layer (due to a presumed inhibiting effect of quaternary ammonium cations) that promotes faster migration of the oxide-metal interface deeper into the oxide, a decrease in the rate of chemical dissolution and higher conductivity of the electrolytes containing larger cations.

Glycerol is one of the most viscous solvents in common use. The high resistance of glycerol constrains the anodic oxidation current densities in electrolytes based on this solvent to small values, whereas the large viscosity of glycerol causes diffusion of reactants to be slow. Therefore, nanoporous-layer growth rates in glycerol are generally small. However, nanotube arrays on Ti fabricated in glycerol exhibit high aspect ratios and very smooth sidewall profiles due to the suppression of local concentration fluctuations (Mor et al. 2006a). Conversely, anodic oxidation in ethylene glycol is characterized by rapid nanoporous layer growth rates that are approximately five times the rate of nanotube formation in amide-based electrolytes and an order of magnitude greater than that in aqueous solutions. This rapid growth rate does not degrade the porous structure of the nanotubes. Nanoporous layers up to 360 μm in length, long range order manifested in hexagonal close-packing and very high aspect ratios (~2200) have been produced on Ti by careful selection of the anodic oxidation parameters (Shankar et al. 2007). Very recently, the growth of anodic nanotube arrays in ionic liquids has also been reported with promising results (Paramasivam et al. 2008).

14.3.2.3 Temperature

From the above discussion, it appears that chemical dissolution and field-assisted electrochemical dissolution processes are the two crucial factors in the growth of systems of nanopores. Varying the electrolyte temperature can change the rate of both dissolution processes (Mor et al. 2006a, Sulka and Parkoła 2007). For nanoporous titanium oxide, the pore diameter remains essentially the same in the temperature range 5°C–50°C, whereas the wall thickness changes by approximately a factor of four and the tube-length changes approximately twice. As the wall thickness increases with decreasing anodization temperature, the voids in the interpore areas fill; as the tubes become more interconnected, the discrete nanotube-like structures approach in appearance a more dense nanoporous array.

14.3.2.4 Composition of the Substrate—Nanoporous Oxides on Valve Metal Alloys

The growth of nanoporous layers on a range of alloys would lead to a significant increase of the potential functionality of the material (by e.g., incorporation of doping species in the oxide structure). It also could result in the formation and growth of nanoporous layers as surface coatings on various technical alloys. Using the same approach as for the pure valve metals, nanoporous layers have been grown on intermetallic compounds (TiN and TiAl), binary alloys (Ti-Nb, Ti-Zr), or complex alloys with biomedical application (Ti-6%Al-7%Nb and Ti-29%Nb-13%Ta-4.6%Zr). For Ti-Nb, synergistic effects on the growth morphologies of the nanoporous oxides have been found and the range of achievable diameters and lengths of TiO_2-based nanotube arrays can be significantly expanded (nanoporous layer thickness from 0.5 to 8 μm, and the diameter from 30 to 120 nm) (Feng et al. 2007). The morphology of the tubes differs significantly from that of the nanostructures grown on pure Ti or Nb substrates: only considerably shorter tubes grow on Ti whereas irregular porous structures grow on Nb in similar conditions. For anodic porous oxides formed on TiZr alloys, the morphological character is reported to be between those of titanium oxide and zirconium oxide (Yasuda et al. 2007). The nanotubes have a straight and smooth morphology, a diameter from 15 to 470 nm and a length up to 21 μm (i.e., an extended structural flexibility compared with layers formed on the pure metals is once more observed). On more complicated technical alloys such as Ti29Nb13Ta4.6Zr and Ti-28Zr-8N, oxide nanotube arrays with two discrete sizes and geometries have been reported (Tsuchiya et al. 2006, Feng et al. 2008). However, on alloys containing reactive elements such as V, the ordered porous structure is not readily achieved. Thus, it can be tentatively concluded that the alloys that would form nanoporous oxide structures must contain essentially only valve metal components. However, using potentiostatic anodization combined with ultrasound, nanotube arrays were grown recently on a Ti8Mn alloy that bears a large potential as a catalytic and photocatalytic material (Mohapatra et al. 2007).

14.3.3 More Detailed View of Process Mechanisms

In this section, a more detailed overview of the existing models for the initiation and growth of the nanopores, as well as their subsequent organization in well-ordered nanopore or nanotube arrays, will be presented. Generally, three types of pore-generation models are considered—models that attribute the pore generation and propagation to the field-assisted dissolution of metal cations at the barrier oxide film/electrolyte interface, models that consider local breakdown/thinning of the initial barrier film, and approaches that take into account the growth-induced plasticity of the oxide that results in a flow of film material from the barrier layer toward the regions of pore growth. The three concepts will be treated consecutively in the following.

14.3.3.1 Electrochemical Dissolution Assisted by a Non-Homogeneous Field Strength

As mentioned already above, with the onset of anodic oxidation, a thin barrier oxide forms on the metal surface (Figure 14.5a). Small pit-like pore nuclei are proposed to originate in this oxide layer due to the localized field-enhanced dissolution of this oxide (Figure 14.5b), or also to preferential growth of the barrier film that follows the original structure of the metal substrate (Mor et al. 2006a). The barrier layer at the bottom of these

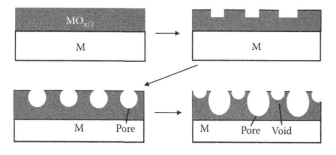

FIGURE 14.5 A simplified scheme of the field-assisted dissolution model of nanoporous layer growth.

nuclei becomes thinner, which, in turn, increases the electric field strength and leads to further pore growth (Figure 14.5c). The pore entrance is not affected by electric field-assisted dissolution and hence remains relatively narrow, while the electric field distribution in the curved bottom surface of the pore causes pore widening and deepening, the result being a pore with a scallop shape. Due to the relatively low ion mobility and relatively high solubility of the oxide in the appropriate electrolyte, only pores having thin walls are supposed to be formed. Thus it is assumed that un-anodized metallic portions can initially exist between the pores, and during the progress of the anodic oxidation, the electric field in these protruded metallic regions increases, enhancing the field-assisted oxide growth and oxide dissolution. The result is that well-defined inter-pore voids start forming simultaneously with the pores (Figure 14.5d).

Thereafter, both voids and tubes grow in a dynamic steady state, forming eventually nanotube arrays. The nanotube length increases until the field-assisted dissolution rate becomes equal to the chemical dissolution rate at the top surface of the nanotubes. Thus the chemical dissolution reduces the thickness of the oxide layer (barrier layer) keeping the electrochemical processes (field-assisted oxidation and dissolution) processes active. The rate of the electrochemical processes is determined by the potential as well as the concentration of the electrolyte, whereas the chemical dissolution rate by the anion concentration and solution pH.

14.3.3.2 Local Breakdown/Thinning of the Barrier Layer

The potentiostatic growth phases of the nanoporous oxide according to this model are shown schematically in Figure 14.6 (Taveira et al. 2005, Macak et al. 2007a,b, Yasuda et al. 2007, Hildebrand et al. 2008). Once again, in the very first stage of the anodization, a thin and compact barrier-type oxide layer is formed (Figure 14.6a). This layer has been demonstrated to contain some statistically distributed "breakdown sites," that is, locations where next to the compact oxide localized accelerated dissolution occurs. In the next stage, such pore nucleation events are apparent on almost the entire surface (Figure 14.6b). On certain locations, undermined patches of the initial compact oxide remain and have been detected by microscopic observations (Figure 14.6c). Just after this stage, the pores have a random appearance and pronounced dissolution at the pore bottoms takes place resulting in pore deepening (Figure 14.6d). Subsequently, the nanoporous structure starts to readily convert into a nanotubular structure by a natural selection mechanism (Figure 14.6e).

Several complications of the process are found to arise when anodic oxidation is conducted in the galvanostatic mode (Figure 14.7) (Taveira et al. 2006). First, as usual, a compact layer of oxide is formed, leading to a fast increase of the potential–stage 1 of the schematic E–t curve (Figure 14.7). However, due to the presence of F$^-$ ions, breakdown/thinning of the barrier film takes place, causing the potential to drop to a minimum value (regions 1–2 of the E–t curve). A porous oxide layer grows at the breakdown/thinning spots, the voltage increases slowly but gradually because of the thickening of that layer. However, if instabilities occur, a detrimental process can be triggered leading to a

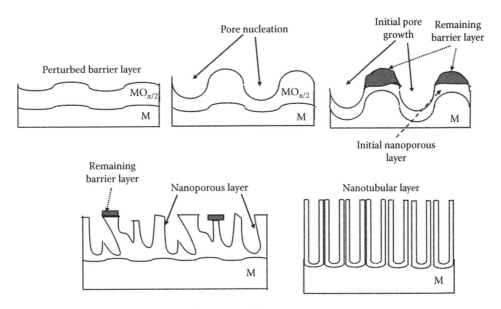

FIGURE 14.6 A simplified scheme of the barrier film breakdown model.

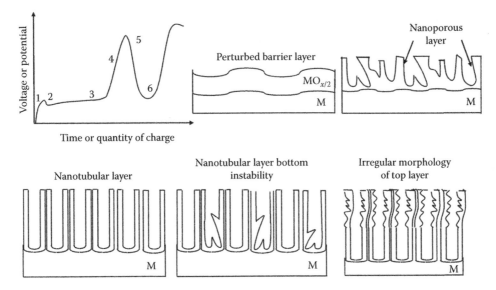

FIGURE 14.7 Possible complications during nanoporous layer growth in the galvanostatic regime.

steep rise in potential. The important role of the local pH at the nascent pore tip for establishing a certain porous morphology has been already stressed above (Section 14.3.2.2). It has been deemed plausible that the instabilities (regions 3–4 in the *E–t* curve) are associated to a significant acidification in some pores. This would lead to an enhanced local current density and a rapid growth of the layer resulting in a fast increase in its resistance and therefore a higher potential. At the peak point of the oscillations, the local conditions (e.g., pH, potential, stress) are such that it is probable that a partial detachment of the layer occurs. This leads to an easy electrolyte access to the surface and therefore to a voltage drop (regions 5–6 of the *E–t* curve).

Experiments with electrolyte stirring have been recently interpreted so that the rate-limiting factor of nanoporous layer growth on Ti is associated with a transport step of a diffusional nature (Yasuda et al. 2007). It has been assumed that diffusion can be separated in two parts: a part within the tubes and another in the adjacent electrolyte layer. By stirring, mainly the part within the electrolyte can be affected. Due to the change in the diffusion layer thickness, the profile will be influenced thus changing the concentration at the pore entrance. In the proposed diffusion model, the diffusing ions are consumed or produced only at the bottom of the nanopore, and the top dissolution is neglected. The corresponding ionic fluxes in the diffusion layer and nanoporous layer are given by the Fick equations

$$J = -D\frac{\partial c}{\partial x} \quad (-x_0 \le x \le 0)$$

$$J = -PD\frac{\partial c}{\partial x} \quad (0 \le x \le \delta)$$

(14.3)

where P, D, c, x_0, and δ are the porosity, diffusion coefficient and concentration of the ionic species, thickness of the diffusion layer, and length of the nanopores, respectively. The anodic current is obtained as

$$i = nFJ = \frac{PnFDc_0}{\delta + Px_0} \tag{14.4}$$

where n is the number of electrons transported in the reaction. Therefore, when the reaction is controlled by a diffusion of ionic species in the electrolyte, the inverse of the current is found to depend linearly on the nanopore length

$$\frac{1}{i} = \frac{1}{PnFDc_0}(\delta + Px_0) \tag{14.5}$$

This relationship has been verified for nanoporous oxides anodically grown on Ti and Ti-Zr alloys. For larger nanopore lengths, deviations from the simple relationship are observed presumably due to the increase in porosity.

Additional proof for the local barrier layer breakdown/thinning model comes from very recent *in situ* ellipsometric measurements (Joo et al. 2008). The optical models used to simulate the experimental two-parameter ellipsometric curves during anodic oxidation of Ti at a constant potential in a fluoride-containing solution at the stages of pore formation and nanotube growth are shown in Figure 14.8.

Using these optical models, the experimental ellipsometric curves have been successfully simulated, which lends further credibility to the local barrier-layer breakdown concept. However, there are several arguments against the mass transport explanation of stable nanopore growth offered above (Mor et al. 2006a):

1. A problem with the mass transport explanation as applied to the anodization process, especially in electrolytes based on organic solvents such as ethylene glycol, is the strong dependence of the nanotube growth rate on anodization voltage.
2. When the process is diffusion controlled, the diffusion coefficient is inversely proportional to the viscosity, that is, the growth rates in more viscous solutions should be lower. However, much higher growth rates have been obtained in the much more viscous ethylene glycol based electrolytes than, for example, in formamide based electrolytes. Also, the use of higher viscosity formamide solutions of bulkier cations such as Bu_4N^+ has been found to result in higher nanotube growth rates compared to lower viscosity solutions of smaller cations such as NH_4^+. Also, when the concentration of tetrabutylammonium fluoride is increased, clearly no new sources of oxygen bearing anionic species were added to the electrolyte. The viscosity of the solution increased and the growth rate increased as well.

14.3.3.3 Stress Development, Plasticity, and Flow of Oxide Film Material

The formation of porous oxide structure on a barrier oxide could be analyzed by taking into account the stress generation/relief (Jessensky et al. 1998, Choi et al. 2005) and surface instability of this film subject to small perturbations (Raja et al. 2005). The instability of the barrier layer during anodic oxidation of valve metals can be visualized to occur because of two competing processes, that is, surface energy acting as a stabilizing force and increase in strain energy due to electrostriction and electrostatic and recrystallization stresses trying to destabilize the surface. The electrostriction stress σ_{er} can be expressed as

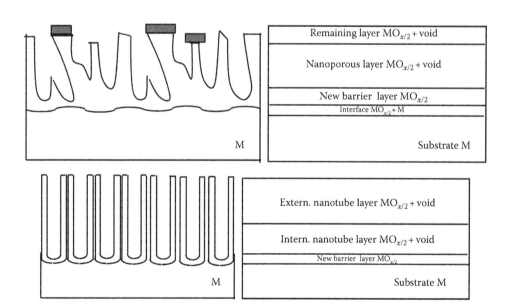

FIGURE 14.8 Optical models of the layers in the stages of pore formation and nanotube growth.

$$\sigma_{er} = \gamma_{11} Y \vec{E}^2 \tag{14.6}$$

where

Y is the Young's modulus
γ_{11} is the electrostriction coefficient in the direction of field
$\vec{E}$ is the electric field strength

In turn, the electrostatic stress σ_{es}, which is compressive in nature, can be expressed as

$$\sigma_{es} = \frac{\varepsilon \varepsilon_0 \vec{E}^2}{2} \tag{14.7}$$

where

ε_0 is the permittivity of free space
ε is the dielectric constant of barrier layer

Volume expansion due to formation of the oxide induces further compressive stress that can be represented as

$$\sigma_{vol} = -\frac{(\partial v/v) Y (1-v)}{(1+v)(1-2v)} \tag{14.8}$$

where v is the Poisson's ratio. The total strain energy density is then

$$U_s = \frac{\sigma^2}{2Y} = \frac{(\sigma_{er}^2 + \sigma_{es}^2 + \sigma_{vol}^2)}{2Y} \tag{14.9}$$

For a surface perturbed with, for example, a sinusoidal wave, the energy density varies across the crests and valleys on the surface. The valley shows high stress concentration effect and thus higher strain energy. The heterogeneity (energy gradient along crest and valley) in strain energy density along the surface acts as a driving force for instability. Below a critical wavelength of perturbation λ, the surface becomes unstable. According to the authors (Raja et al. 2005), this wavelength is determined by the ratio of surface energy γ to strain energy

$$\lambda \leq \frac{8\pi \gamma Y}{3\sigma^2} \tag{14.10}$$

With increase in strain energy during anodization, the stability of the barrier layer is supposed to be affected. For example, it is possible that the dielectric constant of the barrier layer will not be uniform because of non-stoichiometry of the film, since dielectric constants were observed to increase with temperature, voltage, and crystallinity of the respective material. Similarly, electrostriction coefficient also could vary because of the variation in chemistry of the oxide. The variations in dielectric constant and electrostriction coefficient could increase the strain energy and cause instability.

Figure 14.9a through c schematically illustrates how perturbation analysis can explain a nanoporous oxide-layer formation (Raja et al. 2005). Because of curvature (or equivalently, when

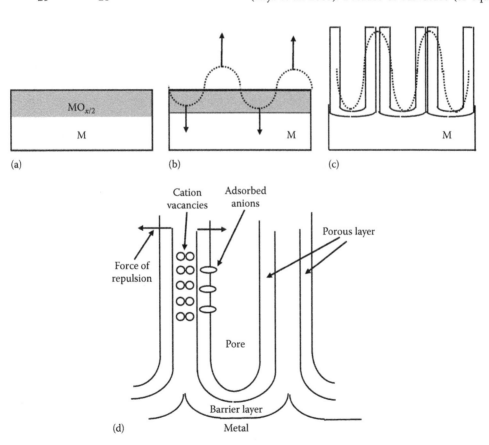

FIGURE 14.9 (a–c) A simplified scheme of the pore initiation and growth according to the surface perturbation model; (d) a more detailed view of the proposed mechanism for pore detachment and formation of the nanotubular structure.

the surface is sinusoidally perturbed) and increased stress concentration at the valleys, the field strength would be larger there than at the crests. It is likely that the surface energy could be reduced when aggressive anions are present in the environment that could increase the critical wavelength of the perturbation for sustained instability. Due to the difference in chemical potentials, these anions could be preferentially adsorbed on the valleys. In order to maintain electroneutrality, more hydroxonium ions could migrate to these sites and lead to dissolution of cations due to local acidity increase. Therefore, valleys dissolve preferentially and crests grow at the expense of the valleys leading to a nanoporous structure. However, this mechanism does not explain the individually separated nano-tubular structure observed during the anodization of a range of valve metals in acidified fluoride solutions. As mentioned above, the walls of anodized valve metals are considered to have inner and outer layers. When cations dissolve into solution, cation vacancies are generated in the film lattice. Assuming that the cation vacancies are transported radially, rows of vacancies would reach the center of the inter-connected nanopore walls from two neighbor pores, as schematically illustrated in Figure 14.9d. To maintain neutrality, oxygen vacancies could be generated. If the dissolution rate is much higher than the generation of oxygen vacancies at the metal/barrier oxide interface that is responsible for the growth of the barrier layer inward, the repelling forces of the cation vacancies could cause separation of the neighboring pores and form individual nanotubes.

The mechanism described above is to a certain extent analogous to the so-called flow model of growth of nanoporous oxides, that has been proposed recently on the basis of experiments involving tracer studies of nanoporous alumina and titania (Garcia-Vergara et al. 2006a,b, Garcia-Vergara et al. 2007, LeClere et al. 2008). In Figure 14.10, the three possible mechanisms of pore generation are considered. In the first two diagrams, Figure 14.10a and b, the formation of pores by dissolution is depicted. The dissolution is assumed to be field-assisted, as above, although the influence of the field is not essential to the model. In the second case, Figure 14.10c, field-assisted ejection of cation species accompanies field-assisted

dissolution of film material. The proportion of the cation species that are ejected to the electrolyte is assumed the same as for a barrier film, with pores formed by field-assisted dissolution of film material at the pore base. In the third case, field-assisted ejection of cationic species is accompanied by flow of film material in the barrier layer region of the porous film (Figure 14.10d). The pores originate from the displacement of film material by flow toward the cell walls, such that the film thickness exceeds the thickness of oxidized metal by a factor *f* related to the pore volume in the film.

According to this mechanism (LeClere et al. 2008), the constant thickness of the barrier film is maintained by the flow of oxide from the barrier layer toward the cell wall, driven by compressive stresses from electrostriction and possibly through volume expansion due to oxidation, as outlined already above. The flow is facilitated by the plasticity of the barrier layer due to participation of most of the film constituents in ionic transport. Evidence of flow in anodic oxides is available from their ability to accommodate the expansion of oxygen bubbles in films, with bubbles having estimated gas pressures of the order of 100 MPa. Similar stress levels have been estimated for electrostriction, which are sufficient to deform oxides. The compressive stresses act along the direction of the field, the pressure facilitating the flow of anodic oxide around the scalloped metal/oxide interface. The flat film/electrolyte interface is then unstable in response to local perturbations of the electric field that can stabilize pores embrya. Pore filling is prevented by the absence of growth of new oxide at the film surface, while increased stresses from electrostriction assist stabilization of the pores. The sites of the initial pores depend upon the topography of the original metal substrate surface. Some incipient pores stop growing, while others develop into the major pores of the film.

For generation of major pores by field-assisted flow, the thickness of the film relative to that of the oxidized metal, F_m, can be estimated as

$$F_m = \frac{d_p}{d_M} = \frac{kr_{PB}}{1-P} g_{eff} \tag{14.11}$$

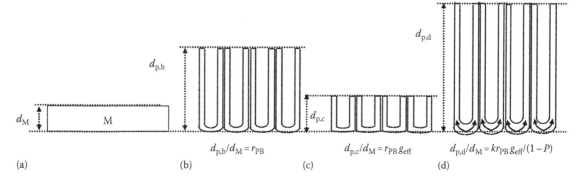

FIGURE 14.10 The possible mechanisms of pore generation as considered in the flow model of nanopore growth and the associated ratio of the porous film thickness vs. substrate metal thickness in each case. (a) metal substrate, (b) field-assisted film dissolution, (c) field-assisted dissolution + field-assisted cation ejection, (d) field assisted cation ejection + flow model.

where

> k is a factor representing the influence of incorporated contaminant species on the volume of the anodic oxide
>
> r_{PB} is the Pilling–Bedworth ratio of the oxide
>
> g_{eff} is the efficiency of film growth, assuming losses due only to field-assisted ejection
>
> P is the porosity generated by flow of film material, that is, the ratio of the pore volume to the total volume of the film

Calculations of F_m for nanoporous anodic oxides using this model have shown a fair comparison with experimentally estimated values, notwithstanding the fact that both incorporation of electrolyte species and the proportion of cations ejected in the solution have not been taken into account.

14.3.4 Characterization of Nanoporous Oxides

14.3.4.1 Basic Characteristics—Thickness and Porosity

A simple method to estimate the porosity of anodic oxides is to measure voltage–time curves during re-anodizing of the nanoporous layer (formed, for example, in an acidic solution) in a neutral buffer solution, for example, consisting of 0.5 M boric acid and 0.05 M sodium tetraborate at 20°C, as shown schematically in Figure 14.11 (Ono and Masuko 2003, Ono et al. 2004, 2005). Such a measurement is based on the fact that the anodic barrier film growth is assumed to take place both at the oxide/metal interface by the inward migration of oxygen and the oxide/electrolyte interface by the outward cation migration (Section 14.2.1). The porosity P of the layer requires prior knowledge of the transference numbers of cations (t_+) and anions (t_- in the growing oxide and makes use of the equation

$$P = \frac{\beta t_+}{1 - \beta t_-}, \quad \beta = \frac{m_2}{m_1} \tag{14.12}$$

where

> m_1 is the slope of the voltage–time curve during re-anodizing of the metallic specimen with already formed porous layer
>
> m_2 is the slope of the voltage–time curve during the growth of barrier film by direct anodizing of a metal substrate

This method for porosity measurement is well established and called the "pore-filling" technique.

Another method to determine the porosity of the oxide films is based on the general equation for the terms contributing to the current density during anodic oxidation (Bocchetta et al. 2002, Sunseri et al. 2006). As mentioned already in Section 14.2.1, the current density measured during the growth of a porous layer is the sum of the ionic current density, i_{ion}, the oxidation current of the metal, and electronic current density, i_{el}, due to redox processes occurring at the oxide/electrolyte interface, such as oxygen evolution. Moreover, i_{ion} is considered as the sum of the formation current density, i_{form}, indicating the contribution to form new oxide, and the dissolution current density, i_{diss}, related to the field-assisted dissolution of cations into the electrolyte at the pore bottom

$$i = i_{ion} + i_{el} = i_{form} + i_{diss} + i_{el} \tag{14.13}$$

By weight measurements and applying Faraday's law, it is possible to evaluate each term of that equation. The average i_{ion} and i_{form} are evaluated as follows:

$$i_{ion} = \frac{\Delta m_M z F}{S_m A_M \Delta t}$$
$$i_{form} = \frac{m_{form} z F}{S_m M_{MO} \Delta t} \tag{14.14}$$

where

> S_m is the apparent surface area of the sample
>
> Δm_M is the amount of metal consumed during anodizing
>
> m_{form} is the weight of the porous layer
>
> A_M and M_{MO} are the atomic weight of the metal and molecular weight of oxide, respectively

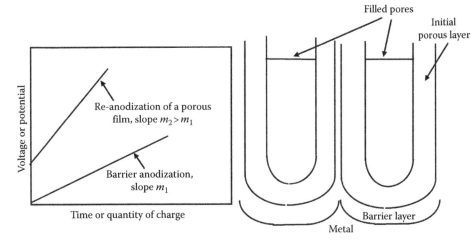

FIGURE 14.11 Schematic representation of "pore filling" technique.

From the above equations, the average dissolution current can be calculated as

$$i_{diss} = i_{ion} - i_{form} \tag{14.15}$$

If the barrier layer thickness does not change during the steady-state formation of a porous layer, the rate of the displacement of the metal/oxide interface toward the metal must be equal to the rate of displacement of the pore bottom toward the oxide. Therefore, assuming that the pores are cylindrical and extend throughout the porous layer thickness, the porosity can be estimated from the ratio of the dissolution and total ionic current densities

$$P = \frac{i_{diss}}{i_{ion}} \tag{14.16}$$

Values calculated using this equation and the associated gravimetric method have been found to be in good agreement with the porous layer thickness and porosity data obtained using the most common method of estimating these parameters—the combination between scanning electron microscopic (SEM) observations of the oxide surface and cross section. A summary of the parameters that can be estimated using such observations as depending on the formation potential is schematically shown in Figure 14.12 (Macak et al. 2007a,b), in which the film formation voltage has been denoted as U_f.

Another independent method to determine the porosity of the obtained layers relies on their optical properties (Mor et al. 2005, 2006a). Using transmittance *vs.* wavelength of incident light spectra for nanoporous oxides, the thickness of the film can be estimated using the relation

$$d = \frac{\lambda_1 \lambda_2}{2\left[\lambda_2 n(\lambda_1) - \lambda_1 n(\lambda_2)\right]} \tag{14.17}$$

where λ_1 and λ_2 are the wavelengths corresponding to two adjacent maxima or minima in the transmittance spectrum and $n(\lambda_1)$ and $n(\lambda_2)$ are the refractive indices at λ_1 and λ_2, respectively. The refractive indices of a nanoporous oxide deposited on a transparent substrate can be estimated from the transmittance spectrum in the range 380–1100 nm employing Manifacier's envelope method

$$n(\lambda) = \sqrt{S + \sqrt{S^2 - n_S^2(\lambda)n_0^2(\lambda)}}$$

$$S = \frac{\left[n_S^2(\lambda) + n_0^2(\lambda)\right]}{2} + 2n_0 n_S \frac{T_{max}(\lambda) - T_{min}(\lambda)}{T_{max}(\lambda)T_{min}(\lambda)} \tag{14.18}$$

where

$\quad n_0$ and n_S are the refractive indices of air and transparent substrate

$\quad T_{max}$ is the maximum envelope

$\quad T_{min}$ is the minimum envelope

From the transmittance spectrum, the refractive index of the substrate is calculated as a function of wavelength using the relationship

$$n_S(\lambda) = \frac{1}{T_S(\lambda)} + \sqrt{\frac{1}{T_S^2(\lambda)} - 1} \tag{14.19}$$

The porosity of the oxide can then be estimated from the equation

$$P = 1 - \frac{n^2 - 1}{n_d^2 - 1} \tag{14.20}$$

where n and n_d are the refractive indices of the nanoporous structure and dense barrier films, respectively. The obtained porosity values for nanoporous titanium oxides have been found to be in good agreement with those estimated by SEM.

14.3.4.2 Electrical and Photoelectrical Properties

The electrical and photoelectrical properties of nanoporous oxides, which are usually regarded as large bandgap semiconductors, have been extensively studied in the recent literature due to the importance of these properties in prospective catalyst, photocatalyst, and sensor applications. The two main characterization methods employed have been electrochemical impedance spectroscopy (EIS), and photocurrent spectroscopic/transient techniques in electrolyte solutions that are essentially inert toward the formed nanoporous layers (Muñoz 2007, Muñoz et al. 2007, Tsuchiya et al. 2007, Taveira et al. 2008). From EIS measurements, space charge layer and surface state capacitances as depending on the applied potential were extracted and interpreted using both the classical Mott–Schottky theory of the semiconductor/electrolyte interface and the amorphous Schottky barrier model (Muñoz 2007). Voltammetric measurements with suitable redox couples added to these electrolyte solutions have been also employed.

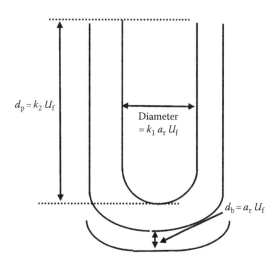

FIGURE 14.12 Schematic representation of the dependence of the nanopore diameter, the barrier layer thickness at the bottom of the nanopores and the nanoporous layer thickness on the applied potential. All values are evaluated from SEM images.

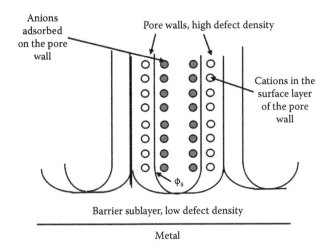

FIGURE 14.13 Surface charges of nanotube walls and the related surface potential drops.

The electrochemical behavior of tubular oxides can be explained considering a structure conformed by the tubes and a base oxide with different semiconducting properties (Muñoz 2007). Here it is expected that a higher concentration of defects (a significant surface charge and an associated potential drop, as shown in Figure 14.13) with a consequent higher electronic conductivity is present in the wall of tubes than in the underlying layer, whose behavior can be considered similar to the compact barrier oxide.

Figure 14.14 presents an idealized band scheme of the porous oxide depicting the different layers in depletion and accumulation conditions (Muñoz 2007). According to this scheme and assuming that the system is under electronic equilibrium, the introduction of the pore walls with a higher donor concentration will shift the energy levels of the compact layer toward higher energies. This is reflected in a positive shift of the observed flat band potential, that is, that of the barrier sublayer. In the situation where the porous oxide is at potentials higher than the flat band state (depletion conditions), the hydrogen evolution (Red$_1$/Ox$_1$ couple) is controlled by the Schottky barrier formed in the barrier sublayer and the reduction reaction of a redox couple

can take place via the injection of holes transported through the band states (cf. Figure 14.14a). This concept is able to explain the experimentally observed similarity of the hydrogen evolution currents for nanoporous oxides of different thicknesses. On the other hand, the mechanism of transport of holes is different for each type of nanoporous oxides (with shallow or deep pores), as reflected in the deviation of the oxidation current in the presence of the redox couple Red$_2$/Ox$_2$.

In the case of a downward band bending (accumulation conditions, Figure 14.14b), hydrogen evolution takes place by the ejection of electrons from the conduction band edge of the underlying compact oxide. Due to the relative position of the bands, the porous oxide layer presents a deeper band bending than the compact one. Therefore, the reduction of the oxidized form of a couple is thought to be controlled by charge ejection at the barrier sublayer. This implies that practically no differences under the different porous layers are expected for the cathodic current density in the accumulation regime, as observed.

The influence of a large amount of surface states of the pore walls on the base oxide has been demonstrated by their charging under illumination. Also, different distributions of energy states promoted by the distinct structures of nanoporous oxides give rise to different charging and recombination kinetics reflected in the capacitance values and trends as deduced from EIS measurements. The presence of the nanoporous structures also brings about new charge transfer paths, increasing the observed photocurrents.

The absorption coefficient α and the bandgap E_g of a semiconductor material, such as the nanoporous oxide, are related through

$$(\alpha h\nu)^s = h\nu - E_g \tag{14.21}$$

where
  ν is the frequency
  h is the Planck's constant
  $s = 0.5$ for indirect bandgap material

The optical bandgap of nanoporous titanium oxide has been found to exhibit a slight blue shift with respect to that of a dense

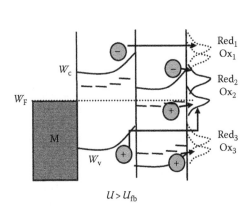

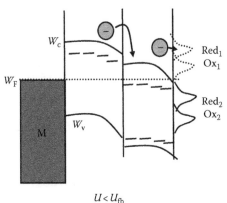

FIGURE 14.14 Energy diagrams of the double oxide structure at potentials more negative and more positive than the flat band potential showing possible charge transfer paths.

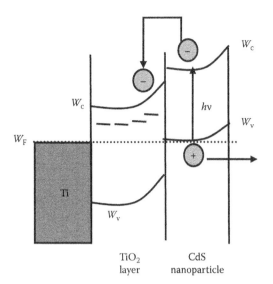

FIGURE 14.15 Schematic diagram illustrating charge injection from excited CdS nanoparticle into TiO_2 nanoporous oxide. CB and VB refer to the energy levels of the conduction and valence bands, respectively, of CdS and TiO_2 nanotube wall.

oxide, which might be due to a quantization effect in the nanoporous film where the pore wall thickness is of the range of 10 nm (Mor et al. 2005, 2006a,b). A band tail has been also observed. The degree of lattice distortion is likely to be relatively higher for nanotube array films, thus causing aggregation of vacancies acting as trap states along the nanotube walls leading to a lower band-to-band transition energy.

The UV–Vis spectra of TiO_2 nanoporous films electrochemically doped with nitrogen and fluorine have revealed that fluorine incorporation resulted in no discernable change in the onset of optical absorption, whereas N-doping leads to a slightly higher optical absorption in the visible wavelength range, the optical absorption being a function of both film thickness and nitrogen concentration. Bandgap shifting has been also attempted by the modification of nanoporous anodic oxides via electrodeposition of metallic or narrow-bandgap semiconductor nanoparticles (Chen et al. 2006, Mor et al. 2006a). Such sandwich electrodes are advantageous since electron injection may be optimized through confinement effects, and a sensitizer is well approximated by a narrow bandgap semiconductor material such as CdS. Since the conduction band of bulk CdS is found to be approximately 0.5 V more negative than that of TiO_2, the coupling of the semiconductors has proven to have a beneficial role in improving charge separation as shown in Figure 14.15; excited electrons from the CdS nanoparticles can quickly transfer to the TiO_2 ordered nanoporous structure, arriving at the photocurrent collector.

14.3.5 Applications of Nanoporous Oxides

In this section, a range of examples for the prospective applications of nanoporous oxides in various technology domains are presented and briefly discussed. By no means exhaustive, this list gives a general idea of the widespread possibilities that these nanostructural materials offer, as well as of the main problems and challenges during the practical implementation of such technologies to produce novel devices and techniques.

14.3.5.1 Templates for Deposition of Various Nanomaterials and Nanostructures

Self-organized nanoporous oxides, mainly alumina and titania, have been used to fabricate a variety of nanomaterials (Shingubara 2003, Piao et al. 2005, Mor et al. 2006a,b). These methods could be classified as follows: etching of semiconductor substrate using a nanoporous film as a mask, pattern transfer using porous oxide as a template, deposition of functional materials in the nanoporous oxide template by electrochemical (both galvanic and currentless) and sol–gel techniques, and deposition of functional materials by chemical vapor deposition.

Pattern transfer of nanoporous structure to semiconductor substrates is a promising technique for applications such as photonic band materials, field emitter arrays and quantum dot arrays (Menon 2004). A modern method to achieve this is to deposit the nanoporous oxide directly on the semiconductor (for example, Si) substrate. Pattern transfer of anodic oxide nanopore arrays to metallic hole arrays using a replica was also proposed.

Numerous studies have been conducted on filling of conductive materials in nanoporous oxides by electroplating (Luo et al. 2004, Ohgai et al. 2004, 2005, Dalchiele et al. 2007, Drury et al. 2007, Sharma et al. 2007). An important condition is that prior to electroplating, the barrier layer that forms under the nanoporous structure should be thinned to less than about 10 nm (Menon 2004, Mor et al. 2006a). Chemical etching of the anodic film (pore widening treatment), or stepwise lowering of the anodic voltage have been used for that purpose. An optimum combination of anodization voltage step-down and chemical etching can result in a membrane in which the barrier layer is sufficiently thin to enable electrodeposition, and yet mechanically robust enough to be easily handled. Concerning the electrodeposition process, alternating current or pulsed-current electroplating has been widely used since the impedance of the barrier layer at the nanopore bottom is too large to afford for direct current electroplating. The electrodeposited nanowires include Au, Ag, Pt, Cu, Ni, Cu-Co, Ni-Fe, Fe-Co-Ni, Fe-Pt, Co-Pt, Ni-Pt, polyaniline, as well as Ag-polyaniline composites and magnetic coordination polymers such as Prussian blue.

A simplified sequence of processes used to fabricate a freestanding platinum nanowire array electrode is shown schematically in Figure 14.16 (Piao et al. 2005). In order to prepare an ordered nanowire array, a thin platinum layer (30 nm) was sputtered onto the bottom side of the through-hole alumina film and this film served as the working electrode in a conventional three-electrode cell for electrochemical deposition of platinum. After electrochemical deposition, a further platinum layer of about 5 μm thick was deposited via evaporation. Finally, the resulting platinum film, with nanowire arrays/porous alumina membrane composite, was immersed into 5% NaOH solution to remove the alumina layer.

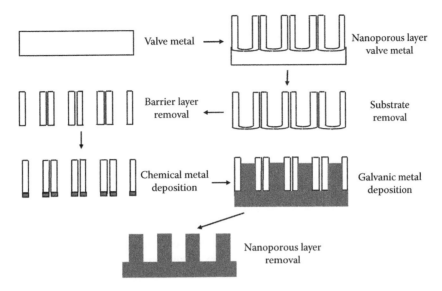

FIGURE 14.16 Schematic diagram of the method used to prepare ordered metal nanowire arrays.

In order to be able to deposit inorganic compound nanowires, a "paired cell" using nanoporous anodic alumina membrane has been proposed (Figure 14.17a) (Piao et al. 2005). According to this method, two solutions, A and B, of the same volume and proper concentration are poured into each half-cell as shown in Figure 14.17b. The cells are then left for a certain time at room temperature in order for A and B to meet inside the alumina nanopores and form A_mB_n nanowires. In this experimental

configuration, each pore of the membrane is supposed to serve as a reaction vessel. Superionic conductor or semiconductor compounds such as AgI, Cu_2S, Ag_2S, CuS, and CdS have been successfully synthesized using this method.

14.3.5.2 Catalysis, Photocatalysis, and Solar Cells

Highly ordered nanoporous titanium oxide arrays, oriented perpendicular to the substrate, have been applied as the working electrodes in a liquid junction dye-sensitized solar cell (DSC) under both front-side (Figure 14.18a) and backside (Figure 14.18b) illumination (Mor et al. 2006a, 2008, Ong et al. 2007). Non-transparent oxide nanotube arrays have been produced by potentiostatic anodic oxidation of a valve metal foil. The resulting nanotubes remain attached to the metal substrate, which is opaque, and can only be subjected to backside illumination. Backside illumination is not optimal in DSC since the Pt counter electrode partially reflects light, while certain components of the electrolyte absorb photons in the visible range. Only transparent nanoporous oxide arrays produced on conducting glass substrates (e.g., fluorine doped SnO_2, FTO) permit use of the front-side illumination geometry. The FTO also serves as the negative electrode.

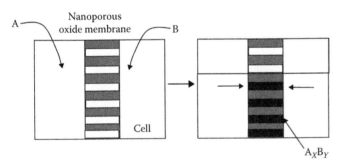

FIGURE 14.17 Schematic diagram for the fabrication of nanowires of the A_XB_Y compound using a paired cell with a nanoporous anodic oxide as the reaction vessel.

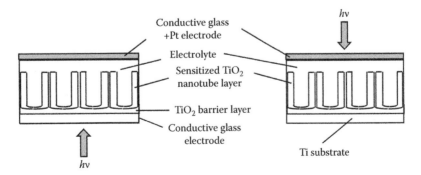

FIGURE 14.18 Front-side (left) and rear-side (right) illuminated dye solar cell with incorporated nanoporous oxide structure.

Extensive research is also carried out to prepare and optimize the negative and positive catalytic electrodes for photo-splitting of water using nanoporous oxides (Mukherjee et al. 2003, de Tacconi et al. 2006, Xie 2006, Guo et al. 2007, Zlamal et al. 2007, Kim et al. 2008a,b, Varghese and Grimes 2008, Watcharenwong et al. 2008). In order to optimize their properties with regard to the enlargement of the spectral window of light absorption, binary nanoporous arrays composed of mixed oxides such as n-Ti(Fe)O$_2$, n-Ti(Mn)O$_2$, and n-Ti(Cu)O$_2$ have been produced on conductive glass substrates. In addition, TiO$_2$ nanoporous arrays loaded with WO$_3$ have been prepared (Mor et al. 2006a, 2008). Vertically oriented p-type Ti(Cu)O$_2$ nanoporous array films have been synthesized by anodic oxidation of copper rich (60%–74%)–Ti films, and n-type Ti(Cu)O$_2$ nanotube array films by anodization of copper poor (11%–24%) Ti films, co-sputtered onto FTO coated glass. The p-type Cu-Ti-O nanotube array films exhibit external quantum efficiencies up to 11%, with a spectral photo-response indicating that the complete visible spectrum contributes significantly to the photocurrent generation. Water-splitting photo-electrochemical p-n-junction diodes have been very recently fabricated using p-type Cu-Ti-O nanotube array films in combination with n-type TiO$_2$ nanotube array films (Figure 14.19). With the glass substrates oriented back-to-back, light is incident upon the UV absorbing n-TiO$_2$ side, with the visible light passing to the p-Cu-Ti-O side. Photocatalytic reactions are powered only by the incident light to generate fuel with oxygen evolved from the n-TiO$_2$ side and hydrogen from the p-Cu-Ti-O side.

Recent studies report also the prospective use of noble metal- and noble metal oxide-doped TiO$_2$ and WO$_3$ nanoporous oxide arrays as fuel cell and supercapacitor electrodes (Yong-gang and Xiao-gang 2004, Bo et al. 2006, Xie et al. 2008). For example, a system consisting of a TiO$_2$ nanotubular layer loaded with Au nanoparticles has been demonstrated to be a very effective

catalyst for the electrocatalytical reduction of oxygen (Macak et al. 2007a,b). The nanotubular support has been reported to enhance significantly the activity of the TiO$_2$/Au system in comparison with flat TiO$_2$ layer. On the other hand, catalytic oxidation of methanol has been explored on a system consisting of bimetallic Pt/Ru nanoparticles embedded in nanoporous TiO$_2$ and WO$_3$ matrices (Macak et al. 2005a,b, Barczuk et al. 2006). These matrices provide a high surface area thus significantly enhancing the electrocatalytic activity of Pt/Ru for methanol oxidation (relative to the performance of Pt/Ru at the same loading but immobilized on a conventional compact oxide matrix). Moreover, annealed to anatase, the TiO$_2$ nanoporous support exhibits even higher enhancement effect during electrooxidation of methanol than when employed in the original amorphous state. Concerning WO$_3$, while tungsten oxide supports obtained by deposition are found to enhance reactivity of Pt/Ru through the formation of hydrogen bronzes, the rigid nanoporous WO$_3$ matrix fabricated via anodic oxidation provide high surface area, a large degree of ordering, and is expected to protect the nanoparticles against agglomeration. Nanoporous oxides are also likely to be more stable than the respective electrodeposited hydrated oxides during prolonged electrolysis.

14.3.5.3 Humidity and Gas Sensors

Ceramic sensor operation is based on either electronic or ionic conductivity. The ability of, for example, alumina to sense humidity is based upon ionic conduction; the presence of an adsorbed layer of water at the surface reduces the total sensor impedance due to the increase in the ionic conductivity, as well as capacitance due to the high dielectric constant of water. Porous alumina is usually preferred for sensing applications due to the large surface area available for water adsorption (Dickey et al. 2002, Varghese and Grimes 2003, Varghese et al. 2002, 2003a,b).

According to the contemporary views, capillary condensation occurs inside pores that are sufficiently small (i.e., nanopores) at lower humidity levels, as illustrated in Figure 14.20. This condensed water reduces considerably the impedance of the nanoporous layer. No condensation occurs inside pores of radius greater than a certain value at a particular humidity

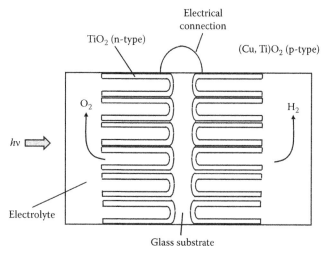

FIGURE 14.19 Illustration of photo-electrochemical diode for water splitting comprised of n-type TiO$_2$ and p-type Cu-Ti-O nanotube array films, with their FTO glass substrates connected through an ohmic contact.

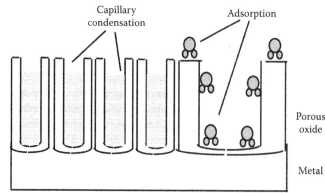

FIGURE 14.20 Schematic representation of water vapor interaction with mesoporous and nanoporous valve metal oxide.

level. In such pores, as in the case of a plane surface, it is the number of physisorbed layers that determines the impedance. The results of recent studies on uniform nanoporous oxide films show that an easily built and highly reproducible wide-range humidity sensor can be achieved with nanodimensional pores of a narrow size distribution made by anodic oxidation of Al in the manner described in the present chapter. In addition, the use of alumina for room temperature sensing of ammonia has been demonstrated, with all sensors showing a completely reversible response upon changing ammonia partial pressures. However, higher operating temperatures dramatically reduce sensor sensitivity, indicating that the ammonia molecules are weakly attached to the nanoporous alumina surfaces. The results indicate that physisorption is responsible for the sensing action.

On the other hand, it has been reported that nanoporous TiO_2 arrays are probably the material that responds to a change in hydrogen concentration with the largest drop in impedance, that is, largest sensitivity found so far (Varghese et al. 2003a,b, Mor et al. 2004, 2006b). Hydrogen molecules dissociate at defects on the titania surface, and can subsequently diffuse into the titania lattice, acting as electron donors. It is believed that chemisorption of the dissociated hydrogen on the titania surface is the underlying sensing mechanism. During chemisorption, hydrogen acts as a surface state, and a partial charge transfer from hydrogen to the titania conduction band takes place. This is supposed to create an accumulation layer on the nanoporous surface, enhancing its electrical conductance. Upon removal of hydrogen, it is presumed that an electron transfer back to the hydrogen molecule takes place, which subsequently desorbs, restoring the original electrical resistance of the material. Evaporation of a thin Pd metal layer on the surface of the nanoporous TiO_2 array has been reported to increase the room-temperature sensitivity of the device considerably, and self-cleaning of the contaminated surface with UV irradiation that adds to the versatility of such sensor devices has also been described.

14.3.5.4 Biochemical and Biological Applications

Nano-patterned enzymatic electrodes utilizing porous alumina as a sensor's substrate for biochemical detection have been recently developed (Tarkistov 2004). For example, the enzyme penicillinaze was immobilized onto the porous alumina to detect penicillin concentration in the solution. A quasi-linear dependence of the imaginary part of the impedance on penicillin concentration has been found for concentrations down to 10^{-5} M, which is thought to meet the sensitivity requirements.

Further, nanoporous anodic oxides on Nb (Sieber et al. 2005, Choi et al. 2006, 2007, Tzvetkov et al. 2007, 2008) have been used as immobilization matrices of, for example, Cytochrome C and horseradish peroxidase (HRP) for their large surface areas, narrow pore size distributions, and good biocompatibility (Choi et al. 2008, Rho et al. 2008). Niobium oxide matrices with different structural features were templated with the surfactants and the selectivity of these hosts to specific protein characteristics was determined. It was observed that proteins could be readily assembled onto the nanoporous films with detectable retention

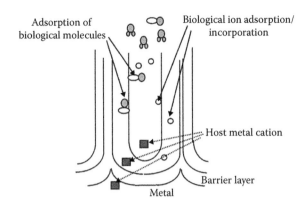

FIGURE 14.21 A simplified scheme of molecular level events at the surface of metal implants.

of bioactivity. The Nb_2O_5 matrix with a tailored pore size and counterpoised surface charge to that of the proteins allowed for a maximum adsorption capacity of biomolecules. Furthermore, the immobilized HRP onto Nb_2O_5 derived electrode presented good bioactivity and thus an amperometric biosensor for the response of hydrogen peroxide in the range from 0.1 mM to 0.1 M has been fabricated and tested.

As titanium and its alloys are widely used for orthopedic and dental implants due to their superior mechanical properties, excellent corrosion resistance, and good biocompatibility, there have been numerous attempts to tailor the oxide properties for a better tissue-materials interaction (Popat et al. 2007b). Figure 14.21 shows some of the molecular interaction events that are supposed to happen when a metallic implant is placed in the body. Oxygen diffuses from the surface oxide into the bulk metal, and metal ions can diffuse from the bulk into the surface oxide as well. Biological ions can also be adsorbed onto the surface oxide. Interactions of biological molecules (protein, enzymes, etc.) with the implant surface can cause transient or permanent changes in the conformation and thus function of these molecules.

In view of these considerations, it is not surprising that barrier and nanoporous TiO_2 morphologies have shown distinct differences in behavior with regard to biological cell-materials interactions (Oh et al. 2005, Oh and Jin 2006, Kim et al. 2008a,b). Materials that adsorb more attachment proteins are expected to provide more sites for osteoblast precursor bonding to the implant, which then would lead to faster bone in-growth and implant stabilization. Surface topography plays an important role as it gives focal adhesion points for the proteins to get attached and thus help in the cell adhesion process. The open microstructure of oxide nanotube arrays also increases the surface energy, which also influences more osteoblast interactions. Significantly, larger levels of cell attachment were noticed on nanoporous surfaces with higher surface energy. From these results, it has been concluded that nanoporous anodized titania surface is more osteoconductive than the control-Ti surface, a result that can have direct impact toward reducing the healing time through faster bone cell attachment, growth, and differentiation *in vivo* for load bearing dental and orthopedic Ti implants.

In this sequence of applications, the use of nanoporous alumina and titania membranes as drug eluding coatings for implantable devices is also worth mentioning (Popat et al. 2007a). The nanotubes can be loaded with various amounts of drugs and their release can be controlled by varying the tube length, diameter, and wall thickness. It has been shown that the release rates of, for example, bovine serum albumin can be controlled by varying the amount of protein loaded into the nanotubes. Further, by changing the nanopore diameter, wall thickness, and length, the release kinetics can be altered for each specific drug to achieve a sustained release.

14.4 Discussion

In this section, based on the information gathered and systematized in Section 14.3, an attempt to summarize and critically discuss the major findings concerning the mechanism of growth of nanoporous oxides and therefore the means of controlling their main characteristics and properties is attempted. First, a summary of the author's view on modeling the kinetics of growth of barrier oxide films on valve metals in electrolytes that dissolve the formed oxide is presented, with an emphasis put on the processes of electrochemical and chemical dissolution at the barrier layer/electrolyte interface. Next, the effect of stress and strain generated in anodic oxides on the pore initiation is considered in connection to the mechanism of the barrier layer growth and dissolution. Finally, an attempt is made to reconcile the different views on stable nanopore/nanotube growth existing in the literature with the prospect of finding a deterministic way to optimize the surface characteristics, electrical, photoelectrical, and catalytic properties of oxide nanopore arrays for a range of possible applications.

14.4.1 Barrier Film Growth and Dissolution

To account for both the film growth and underlying valve metal dissolution under potentiostatic conditions in aggressive electrolytes, such as fluoride-containing electrolytes for W, Nb, Ta, Ti, Zr, etc., a quantitative model approach has been recently proposed (Bojinov et al. 2003, Karastoyanov and Bojinov 2008). It has been found to reproduce successfully both the steady-state current *vs.* potential curves and impedance spectra in a range of potentials. This approach, which treats the processes of film growth and metal dissolution through the film as sequences of generation, transport and annihilation of point defects (cation and anion vacancies) and features a defect recombination reaction at the film/electrolyte interface, is briefly described below. A scheme of the processes that are taken into account in the model is presented in Figure 14.22.

The oxide that spontaneously forms on a range of valve metals at open circuit in an aqueous solution can be regarded as mixed valence oxides in which $M(X)$ and $M(Y)$ positions, $X < Y$, coexist with the cation sublattice, together with a certain concentration of $M(Y)$ cation vacancies. A significant concentration of oxygen vacancies is assumed to exist in the anion sublattice as well.

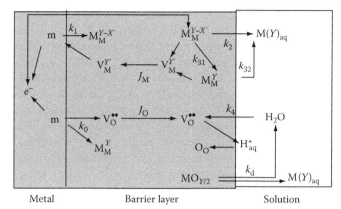

Metal   Barrier layer   Solution

FIGURE 14.22 A scheme of the processes of barrier layer formation and dissolution.

$M(X)$ cation vacancies are neglected for simplicity. The growth of the barrier film proceeds at the metal/film interface with oxidation of the metal

$$m \rightarrow M_M^Y + \frac{Y}{2}V_O^{\bullet\bullet} + Ye_m \qquad (14.22)$$

where e_m is an electron in the metal substrate.

The oxygen vacancy is transported through the barrier film by high-field assisted migration and reacts with absorbed water at the film/solution interface achieving film growth

$$\frac{Y}{2}V_O^{\bullet\bullet} + \frac{Y}{2}H_2O \xrightarrow{k_4} \frac{Y}{2}O_O + YH^+ \qquad (14.23)$$

In order for the barrier oxide thickness to be invariant with time in the steady state, as found experimentally, a chemical film dissolution reaction is included:

$$MO_{Y/2} + YH^+ \xrightarrow{k_d} M_{aq}^{Y+} + \frac{Y}{2}H_2O \qquad (14.24)$$

In fluoride-containing electrolytes, fluoro-complexes of M_{aq}^{Y+} are probably formed, which is expected to increase the solubility of the film and thus the steady-state current density, governed by chemical dissolution of the oxide. On the other hand, $M(X)$ at the F/S interface undergoes either oxidative dissolution as M_{aq}^{Y+} or transforms into a passivating species that dissolves isovalently:

$$M_M^{(Y-X)\prime} \xrightarrow{k_2} M_{aq}^{Y+} + (Y-X)e' + V_M^{Y\prime} \qquad (14.25)$$

$$M_M^{(Y-X)\prime} \xrightarrow{k_{31}} M_M^{Y\cdot} + (Y-X)e'$$

$$M_M^{Y\cdot} \xrightarrow{k_{32}} M_{aq}^{Y+} + V_M^{Y\prime} \qquad (14.26)$$

In the above equations, e' represents an electronic defect in the barrier layer of oxide. If the concentration of hypovalent cations in the lattice is very small and can be neglected (such as in the case of Al_2O_3), the above sequence of reactions reduces to

$$M_M^{Y^*} \xrightarrow{k_3} M_{aq}^{Y+} + V_M^{Y\prime} \qquad (14.27)$$

The cation vacancies, generated by the above processes, are transported through the film via high-field assisted migration; recombine with oxygen vacancies to recreate the perfect lattice according to the reaction

$$V_O^{\bullet\bullet} + V_M^{Y\prime} \rightarrow null \qquad (14.28)$$

The high-field migration equations for the transport of cation and anion vacancies are

$$J_O = -\frac{D_O}{2a}c_O(L)e^{\frac{2Fa\vec{E}_0}{RT}}, \quad J_M = c_M(0)\frac{D_M}{2a}e^{\frac{YFa\vec{E}_0}{RT}} \qquad (14.29)$$

The total defect transport current density is $i = -2FJ_O + YFJ_M$.

The electric field strength in the metal/oxide/electrolyte system $\vec{E}$ is given by the sum of all the potential drops in the system divided by the oxide film thickness

$$\vec{E} = \frac{\phi_{M/F} + \vec{E}_b L + \phi_{F/S}}{L} \qquad (14.30)$$

where

$\phi_{M/F} = (1 - \alpha_{F/S})E - \vec{E}L$

$\phi_{F/S} = q_n L_{F/S}/\varepsilon\varepsilon_0$

$\alpha_{F/S}$ is the part of the applied potential that is consumed at the film/electrolyte interface

q_n is a negative surface charge due to the accumulation of cation vacancies at the film/solution interface

$\vec{E}_b$ is the field strength in the film bulk

$L_{F/S}$ is the width of the cation vacancy accumulation zone

Further, the steady-state potential drop at the metal/film interface has been demonstrated to not depend on the applied potential. In other words, the reactions at this interface are not considered rate limiting, rather, their rate is adjusted by the transport processes of point defects in the passive film. The expressions of the instantaneous partial currents due to the transport of oxygen and cation vacancies acquire the form:

$$I_O = \frac{FD_0}{a}c_O(L)\exp\left\{\frac{2Fa}{RTL}\left[(1-\alpha)E + \frac{q_n L_{F/S}}{\varepsilon\varepsilon_0}\right]\right\},$$
$$I_M = \frac{YFD_0}{2a}c_M(0)\exp\left\{\frac{YFa}{RTL}\left[(1-\alpha)E + \frac{q_n L_{F/S}}{\varepsilon\varepsilon_0}\right]\right\} \qquad (14.31)$$

At the film/solution interface, cation vacancies are (1) generated with a current density $I_{M,F/S}$, (2) transported via high-field migration with a current density I_M, and (3) react with the oxygen vacancies according to the recombination reaction at a rate of $I_O S q_n$, where S is a recombination cross-section. Then,

$$\frac{dq_n}{dt} = I_{M,F/S} - I_M - I_O S q_n = I_O S\left(\frac{I_{M,F/S} - I_M}{I_O S} - q_n\right). \qquad (14.32)$$

According to the dissolution scheme described by reactions (14.25) and (14.26), the current density due to cation vacancies at the film/solution interface is given by:

$$\frac{I_M}{YF} = J_M^{Y\prime} = k_2\gamma_X + k_{32}\gamma_Y^* \qquad (14.33)$$

γ_X and γ_Y^* being the fractions of the F/S interface occupied by $M(X)$ and $M(Y)^*$, referred to the cation sublattice only. The material balances in the outermost cation layer are correspondingly:

$$\frac{\beta d\gamma_X}{dt} = \frac{I_M}{YF} - k_2\gamma_X - k_{31}\gamma_X \qquad (14.34)$$

$$\frac{\beta d\gamma_Y^*}{dt} = k_{31}\gamma_X - k_{32}\gamma_Y^* \qquad (14.35)$$

In the steady state, Equations 14.34 and 14.35 become

$$\frac{\bar{I}_{M,F/S}}{YF} - \bar{k}_2\bar{\gamma}_X - \bar{k}_{31}\bar{\gamma}_X = 0 \qquad (14.36)$$

$$k_{32}\bar{\gamma}_X - \bar{k}_{31}\bar{\gamma}_Y^* = 0 \qquad (14.37)$$

with, obviously

$$\bar{\gamma}_X + \bar{\gamma}_Y^* + \bar{\gamma}_Y = 1 \qquad (14.38)$$

$\bar{\gamma}_Y$ being the surface fraction occupied by regular $M(Y)$ sites, at which the chemical dissolution of the oxide to balance its growth is assumed to proceed. The partial current density due to oxygen vacancies $\bar{I}_O$ is assumed to flow only on the fraction $\bar{\gamma}_Y$.

$$\bar{I}_O = 2Fk_d c_{H^+}^n \bar{\gamma}_Y = 2Fk_d'\bar{\gamma}_Y = k_{MO_{Y/2}}\bar{\gamma}_Y \qquad (14.39)$$

where n is the reaction order of the chemical dissolution reaction, with respect to H^+.

Equations 14.33, 14.37, and 14.38 can be used to calculate $\bar{I}_{M,F/S}$:

$$\bar{I}_{M,F/S} = \frac{YFk_{32}(\bar{k}_2 + \bar{k}_{31})}{\bar{k}_{31} + \bar{k}_{32}}(1 - \bar{\gamma}_Y) = \bar{k}_M(1 - \bar{\gamma}_Y) \qquad (14.40)$$

In turn, $\bar{I}_M$ is proposed to be proportional to the surface fraction $(1 - \bar{\gamma}_Y)$ occupied by $M(Y)^*$, on which the electrochemical dissolution of the metal cations proceeds. If we define $\bar{\gamma}_Y$ as a function of the relative reaction rate for the formation of $M(Y)$ centers

$$\bar{\gamma}_Y = \frac{k_{MO_{Y/2}}}{\bar{k}_M + k_{MO_{Y/2}}} \qquad (14.41)$$

then

$$\bar{I}_{\mathrm{M}} = \frac{\bar{k}_{\mathrm{M}}^2}{\bar{k}_{\mathrm{M}} + k_{\mathrm{MO}_{Y/2}}} \tag{14.42}$$

$$\bar{I}_{\mathrm{O}} = \frac{k_{\mathrm{MO}_{Y/2}}^2}{\bar{k}_{\mathrm{M}} + k_{\mathrm{MO}_{Y/2}}} \tag{14.43}$$

The main message of this treatment in view of the subsequent description of the initiation of nanopores on the barrier layer surface is that there are two types of sites at the interface. On the first type of sites, the $M(Y)$ sites, barrier film formation and chemical dissolution of the oxide proceed. In fact, on these sites, it is expected that film growth will prevail over dissolution since growth is influenced by the applied potential, whereas dissolution is not. On the other hand, the $M(Y)^*$ sites are the sites at which ejection of cations stimulated by the applied potential takes place, most probably further accelerated by the reactive adsorption of fluoride. These sites are considered as the likely places for nanopore nucleation. However, for the stable pore growth, it is clear that other factors do contribute that are most probably related to the mechanical stresses and associated strain accumulated in the barrier oxide during its formation.

14.4.2 Nanopore Nucleation and Growth

The growth of an oxide film via vacancy generation and transport has been demonstrated to initiates compressive stresses at the metal/film interface regardless of the exact growth mechanism (Tzvetkov et al. 2008). Two important stress relief mechanisms are considered. First, the oxidation-induced strain can be compensated by a relatively large flow of cations from the metal into the oxide. A second mechanism of stress generation and relief can be proposed by considering the fact that the stoichiometry of the barrier layer varies from the metal/film to the film/solution interface. Due to the higher oxidation state of the cation at the film/electrolyte interface when compared to that at the metal/film interface, the internal part of the layer is probably subjected to larger compressive stresses than the external part. Therefore, the rate of cation transport through the film in its outer part and the rate of the reaction that controls cation ejection will be higher.

The net result of the division of the surface into zones of growth/chemical dissolution and zones of electrochemical cation ejection could be the perturbation of the surface, as shown in Figure 14.23 (Raja et al. 2005, Tzvetkov et al. 2008). In fact, what is assumed is that the adsorption of a water molecule on a surface cation site could lead to further film growth according to the reaction

$$M_{\mathrm{M}} + (Y/2)\mathrm{H}_2\mathrm{O} \rightarrow \mathrm{MO}_{Y/2,\mathrm{pore\ wall}} + \mathrm{V}_{\mathrm{M}}^{Y'} + Y\mathrm{H}^+ \tag{14.44}$$

which is supposed to proceed preferentially on the normal $M(Y)$ sites and can be regarded as displacement of metal cations

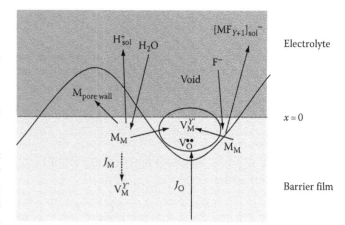

FIGURE 14.23 Schematic of pore nucleation due to perturbation of the surface and spatial differentiation of film growth and dissolution reactions.

toward the electrolyte (Figure 14.23), assisted by the faster rate of cation transport in the outer part of the oxide. This reaction will leave a cation vacancy behind. On the other hand, a fluoride ion adsorbed on a surface cation would lead to the direct abstraction of the cation to the electrolyte with the net result of formation of a fluoride complex and another cation vacancy. This process, that is, the field-assisted cation ejection, is proposed to proceed on the $M(Y)^*$ sites on the barrier layer surface.

It is obvious from the schematic shown in Figure 14.23 that it is less probable that the oxygen vacancies will reach the top of the pore walls than the bottom of the valleys, so the growth could proceed preferably by (14.44). Thus, both oxygen and cation vacancies take part in the film formation and dissolution processes at the film surface. As the ionic conduction in anodic layers on most valve metals is owed rather to J_{O} than to J_{M} (the transference numbers of oxygen are usually greater), the increased amount of the $\mathrm{V}_{\mathrm{M}}^{Y'}$ generated during metal dissolution will be accumulated at the film/solution interface, forming a negative surface charge that attracts the oxygen vacancies. As the oxide lattice at the interface is far from perfect, the combination between the two types of defects is more likely to produce a vacancy pair according to the reaction

$$\mathrm{V}_{\mathrm{O}}^{\bullet\bullet} + \mathrm{V}_{\mathrm{M}}^{Y'} \rightarrow \left[\mathrm{V}_{\mathrm{O}}\mathrm{V}_{\mathrm{Nb}}\right]^{(Y-2)'} \tag{14.45}$$

which we could visualize as a void formation, that is, a local recess of the layer at its interface with the electrolyte. This is a viable explanation of how the pores initiate and grow into the barrier layer.

If we include in this picture the surface stress perturbation to define the zones of vacancy recombination into voids in a regular manner, nanopore arrays can form as a result of pore nucleation. Subsequently, the mechanisms of pore growth into nanotubes as proposed above (Sections 14.3.3.2 through 14.3.3.3) could be well applicable (Yasuda and Schmuki 2007). Within the frames of these mechanisms, pore growth initially proceeds in a wormlike manner, establishing a pH-gradient between the top and the bottom of the pore. In the next step, a natural selective mechanism owed to the greater oxidation area around the deeper pores helps the self-organization process to form well-ordered nanostructures. The accumulation of vacancies and/or vacancy pairs at the opposite walls of a nanotube could lead to an increase of the electronic conductivity of the nanopore array due to the fact that these point defects can act as donors or acceptors of electrons, and also could generate repulsion forces between neighboring pores so that the nanopores are detached from each other and form separated nanotubes, as experimentally observed.

14.5 Summary and Outlook

In this chapter, we have reviewed the fabrication, properties, and selected applications of highly ordered nanoporous oxide films formed by anodic oxidation of valve metals, with focus on Al and Ti. To introduce the subject, a short overview of the main features of the anodic oxidation process has been presented in the first part of the chapter. In the second part, the influence of different operational parameters (potential, current density, electrolyte composition and concentration, temperature, substrate pretreatment, etc.) on the characteristics of the obtained nanoporous oxides has been given. It has been emphasized that to produce mechanically robust nanoporous oxide structures with high aspect ratio, careful control of the experimental conditions is paramount, including the electrolyte concentration, pH, viscosity, temperature, as well as the applied potential and the duration of the anodic oxidation. By doing so, large variations in the main characteristics of the nanoporous oxides, namely, pore (nanotube) diameters, nanotube walls and pore (nanotube) lengths, could be achieved. For prospective applications, these structures could be modified by the deposition of metallic or semiconductor nanoparticles, or doped with nitrogen, carbon, and fluorine using appropriate substrates or electrolyte compositions. In the following part of the chapter, the three main concepts underlying the possible mechanism of formation and ordering of the nanoporous structures have been overviewed in some detail, emphasizing both the common grounds and the distinct features of these approaches, namely, the field-assisted dissolution model, the film breakdown/thinning model, and the flow model. A possible bridge between these model concepts and modern theories of the growth of barrier oxide films on valve metals that is oriented toward an explanation of the nucleation of the nanopore arrays is presented in Section 14.4. Further, a range of prospective applications of the nanoporous oxide structures formed by anodic oxidation of valve metals has been critically reviewed, by putting an emphasis on the diversity of possible uses of such materials as templates for the synthesis of metallic and inorganic nanowires microelectronic devices, in catalysis, photocatalysis, sensor development, and biochemical and biological fields. The anodically formed nanostructures have been found to be especially promising in photocatalysis, in which field a hydrogen evolution rate higher than any reported hydrogen

generation rate for any oxide material system by photoelectrolysis has been obtained, and also in hydrogen sensor development, presenting the largest known sensitivity of any material, to any gas, at any temperature.

In summary, although still rather new as material concept, the photolysis, charge transport, photocatalytic, and gas-sensing properties of highly ordered valve metal oxide nanotube arrays are quite remarkable. As such, it is believed that such materials warrant extended in-depth studies, comparable to the efforts that have been spent in recent years investigating the properties of carbon nanotubes. In view of the prospective use of such nanostructures in various fields of industrial endeavor, there are several problems that merit special attention: the range of characteristics (diameter and length) of the nanopore and nanotube arrays formed via anodic oxidation is still limited, a method to control nanohole diameters in the nanometer range for observing the quantum confinement effect is yet to be devised, fabrication procedures for integrating porous nanohole membranes and nanotube arrays on semiconductor substrates, while at the same time maintaining their mechanical robustness, is yet to be firmly established. The full potential of nanoporous anodic oxides in nano-sciences and technologies can only be realized through persistent efforts at solving these problems, which calls for a quantitative approach to the modeling of the formation mechanism of these structures that would enable the prediction of their valuable properties in a more deterministic way.

References

Barczuk, P.J., Macak, J.M., Tsuchiya, H. et al. 2006. Enhancement of the electrocatalytic oxidation of methanol at Pt/Ru nanoparticles immobilized in different WO_3 matrices. *Electrochem. Solid State Lett.* 9:E13–E16.

Belwalkara, A., Grasinga, E., Van Geertruyden, W. et al. 2008. Effect of processing parameters on pore structure and thickness of anodic aluminum oxide (AAO) tubular membranes. *J. Membr. Sci.* 319:192–198.

Bo, G., Xiao-gang, Z., Changzou, Y. et al. 2006. Amorphous $Ru_{1-y}Cr_yO_2$ loaded on TiO_2 nanotubes for electrochemical capacitors. *Electrochim. Acta* 52:1028–1032.

Bocchetta, P., Sunseri, C., Bottino, A. et al. 2002. Asymmetric alumina membranes electrochemically formed in oxalic acid solution. *J. Appl. Electrochem.* 32:977–985.

Bojinov, M., Cattarin, S., Musiani, M. et al. 2003. Evidence of coupling between film growth and metal dissolution in passivation processes. *Electrochim. Acta* 48:4107–4117.

Chen, S., Paulose, M., Ruan, C. et al. 2006. Electrochemically synthesized CdS nanoparticle-modified TiO_2 nanotube-array photoelectrodes: Preparation, characterization, and application to photoelectrochemical cells. *J. Photochem. Photobiol. A: Chem.* 177:177–184.

Choi, J., Wehrspohn, R.B., Gösele, U. 2005. Mechanism of guided self-organization producing quasi-monodomain porous alumina. *Electrochim. Acta* 50:2591–2595.

Choi, J., Lim, J.H., Lee, J. et al. 2006. Porous niobium oxide films prepared by anodization in HF/H₃PO₄. *Electrochim. Acta* 51:5502–5507.

Choi, J., Lim, J.H., Lee, J. et al. 2007. Porous niobium oxide films prepared by anodization–annealing–anodization. *Nanotechnology* 18:055603.

Choi, J., Lim, J.H., Rho, S. 2008. Nanoporous niobium oxide for label-free detection of DNA hybridization events. *Talanta* 74:1056–1059.

Dalchiele, E.A., Marotti, R.E., Cortes, A. et al. 2007. Silver nanowires electrodeposited into nanoporous templates: Study of the influence of sizes on crystallinity and structural properties. *Physica E* 37:184–188.

de Tacconi, N.R., Chenthamarakshan, C.R., Yogeeswaran, G. et al. 2006. Nanoporous TiO₂ and WO₃ films by anodization of titanium and tungsten substrates: Influence of process variables on morphology and photoelectrochemical response. *J. Phys. Chem. B* 110:25347–25355.

Dickey, E.C., Varghese, O.M., Ong, K. et al. 2002. Room temperature ammonia and humidity sensing using highly ordered nanoporous alumina films. *Sensors* 2:91–110.

Diggle, J.W., Downie, T.C., Goulding, C.W. 1969. Anodic oxide films on aluminum. *Chem. Rev.* 69:365–405.

Drury, A., Chaure, S., Kröll, M. et al. 2007. Fabrication and characterization of silver/polyaniline composite nanowires in porous anodic alumina. *Chem. Mater.* 19:4252–4258.

Feng, X.J., Macak, J.M., Schmuki, P. 2007. Flexible self-organization of two size-scales oxide nanotubes on Ti45Nb alloy. *Electrochem. Commun.* 9:2403–2407.

Feng, X.J., Macak, J.M., Albu, S.P. et al. 2008. Electrochemical formation of self-organized anodic nanotube coating on Ti–28Zr–8Nb biomedical alloy surface. *Acta Biomater.* 4:318–323.

Garcia-Vergara, S.J., Iglesias-Rubianes, L., Blanco-Pinzon, C.E. et al. 2006a. Mechanical instability and pore generation in anodic alumina. *Proc. R. Soc. A* 462:2345–2358.

Garcia-Vergara, S.J., Skeldon, P., Thompson, G.E. et al. 2006b. A flow model of porous anodic film growth on aluminium. *Electrochim. Acta* 52:681–687.

Garcia-Vergara, S.J., Skeldon, P., Thompson, G.E. et al. 2007. Compositional evidence for flow in anodic films on aluminum under high electric fields. *J. Electrochem. Soc.* 154:C540–C545.

Guo, Y., Quan, X., Lu, N. et al. 2007. High photocatalytic capability of self-assembled nanoporous wo₃ with preferential orientation of (002) planes. *Environ. Sci. Technol.* 41:4422–4427.

Hildebrand, H., Marten-Jahns, U., Macak, J.M. et al. 2008. Mechanistic aspects and growth of large diameter self-organized TiO₂ nanotubes. *J. Electroanal. Chem.* 621:254–266.

Inguanta, R., Butera, M., Sunseri, C. 2007. Fabrication of metal nano-structures using anodic alumina membranes grown in phosphoric acid solution: Tailoring template morphology. *Appl. Surf. Sci.* 253:5447–5456.

Jessensky, O., Muller, F., Gösele, U. 1998. Self-organized formation of hexagonal pore structures in anodic alumina. *J. Electrochem. Soc.* 145:3735–3740.

Joo, S., Muto, I., Hara, N. 2008. In Situ ellipsometric analysis of growth processes of anodic TiO₂ nanotube films. *J. Electrochem. Soc.* 155:C154–C161.

Karastoyanov, V., Bojinov, M. 2008. Anodic oxidation of tungsten in sulphuric acid solution-Influence of hydrofluoric acid addition. *Mater. Chem. Phys.* 112:702–710.

Kim, E.Y., Park, J.H., Han G.Y. 2008a. Design of TiO₂ nanotube array-based water-splitting reactor for hydrogen generation. *J. Power Sources* 184:284–287.

Kim, S.E., Lim, J.H., Lee, S.C. et al. 2008b. Anodically nanostructured titanium oxides for implant applications. *Electrochim. Acta* 53:4846–4851.

LeClere, D.J., Velota, A., Skeldon, P. et al. 2008. Tracer investigation of pore formation in anodic titania. *J. Electrochem. Soc.* 155:C487–C494.

Lohrengel, M.M. 1993. Thin anodic oxide layers on aluminium and other valve metals: high field regime. *Mater. Sci. Eng. R* 11:243–294.

Luo, H., Chen, X., Zhou, P. et al. 2004. Prussian blue nanowires fabricated by electrodeposition in porous anodic aluminum oxide. *J. Electrochem. Soc.* 151:C567–C570.

Macak, J.M., Schmuki, P. 2006. Anodic growth of self-organized anodic TiO₂ nanotubes in viscous electrolytes. *Electrochim. Acta* 52:1258–1264.

Macak, J.M., Barczuk, P.J., Tsuchiya, H. et al. 2005a. Self-organized nanotubular TiO₂ matrix as support for dispersed Pt/Ru nanoparticles: Enhancement of the electrocatalytic oxidation of methanol. *Electrochem. Commun.* 7:1417–1422.

Macak, J.M., Sirotna, K., Schmuki, P. 2005b. Self-organized porous titanium oxide prepared in Na₂SO₄/NaF electrolytes. *Electrochim. Acta* 53:3679–3685.

Macak, J.M., Taveira, L.V., Tsuchiya, H. et al. 2006. Influence of different fluoride containing electrolytes on the formation of self-organized titania nanotubes by Ti anodization. *J. Electroceram.* 16:29–34.

Macak, J.M., Tsuchiya, H., Ghicov, A. et al. 2007a. TiO₂ nanotubes: Self-organized electrochemical formation, properties and applications. *Curr. Opin. Solid State Mater. Sci.* 11:3–18.

Macak, J.M., Schmidt-Stein, F., Schmuki, P. 2007b. Efficient oxygen reduction on layers of ordered TiO₂ nanotubes loaded with Au nanoparticles. *Electrochem. Commun.* 9:1783–1787.

Masuda, H., Hasegawa, F., Ono, S. 1997. Self-ordering of cell arrangement of anodic porous alumina formed in sulfuric acid solution. *J. Electrochem. Soc.* 144:L127–L130.

Menon, L. 2004. Nanoarrays from porous alumina. *The Dekker Encyclopedia of Nanoscience and Nanotechnology*. Marcel-Dekker Publishers: New York, 2221–2235.

Mohapatra, S.K., Raja, K.S., Misra, M. 2007. Synthesis of self-organized mixed oxide nanotubes by sonoelectrochemical anodization of Ti–8Mn alloy. *Electrochim. Acta* 52:590–597.

Mor, G.K., Varghese, O.M., Paulose, M. et al. 2003. Fabrication of tapered, conical-shaped titania nanotubes. *J. Mater. Res.* 18:2588–2593.

Mor, G.K., Carvalho, M.A., Varghese, O.M. et al. 2004. A room-temperature TiO_2-nanotube hydrogen sensor able to self-clean photoactively from environmental contamination. *J. Mater. Res.* 19:628–634.

Mor, G.K., Varghese, O.M., Paulose, M. et al. 2005. Transparent highly ordered TiO_2 nanotube arrays via anodization of titanium thin films. *Adv. Func. Mater.* 15:1291–1296.

Mor, G.K., Varghese, O.M., Paulose, M. et al. 2006a. A review on highly ordered, vertically oriented TiO_2 nanotube arrays: Fabrication, material properties, and solar energy applications. *Sol. Energy Mater. Sol. Cells* 90:2011–2075.

Mor, G.K., Varghese, O.M., Paulose, M. et al. 2006b. Fabrication of hydrogen sensors with transparent titanium oxide nanotube-array thin films as sensing elements. *Thin Solid Films* 496:42–48.

Mor, G.K., Varghese, O.K., Wilke, R.H.T. et al. 2008. P-type Cu–Ti–O nanotube arrays and their use in self-biased heterojunction photoelectrochemical diodes for hydrogen generation. *Nano Lett.* 8:1906–1911.

Mukherjee, N., Paulose, M., Varghese, O.M. et al. 2003. Fabrication of nanoporous tungsten oxide by galvanostatic anodization. *J. Mater. Res.* 18:2296–2299.

Muñoz, A.G. 2007. Semiconducting properties of self-organized TiO_2 nanotubes. *Electrochim. Acta* 52:4167–4176.

Muñoz, A.G., Chen, Q., Schmuki, P. 2007. Interfacial properties of self-organized TiO_2 nanotubes studied by impedance spectroscopy. *J. Solid State Electrochem.* 11:1077–1084.

Oh, S.-H., Jin, S. 2006. Titanium oxide nanotubes with controlled morphology for enhanced bone growth. *Mater. Sci. Eng. C* 26:1301–1306.

Oh, S.-H., Finõnes, R.R., Daraio, Ch. et al. 2005. Growth of nanoscale hydroxyapatite using chemically treated titanium oxide nanotubes. *Biomater.* 26:4938–4943.

Ohgai, T., Hoffer, X., Gravier, L. et al. 2004. Electrochemical surface modification of aluminium sheets for application to nano-electronic devices: Anodization aluminium and electrodeposition of cobalt–copper. *J. Appl. Electrochem.* 34:1007–1012.

Ohgai, T., Gravier, L., Hoffer, X. et al. 2005. CdTe semiconductor nanowires and NiFe ferro-magnetic metal nanowires electrodeposited into cylindrical nano-pores on the surface of anodized aluminum. *J. Appl. Electrochem.* 35:479–485.

Ong, K.G., Varghese, O.K., Mor, G.K. et al. 2007. Application of finite-difference time domain to dye-sensitized solar cells: The effect of nanotube-array negative electrode dimensions on light absorption. *Sol. Energy Mater. Sol. Cells* 91:250–257.

Ono, S., Masuko. N. 2003. Evaluation of pore diameter of anodic porous films formed on aluminum. *Surf. Coat. Technol.* 169–170:139–142.

Ono, S., Saito, M., Ishiguro, M. et al. 2004. Controlling factor of self-ordering of anodic porous alumina. *J. Electrochem. Soc.* 151:B473–B478.

Ono, S., Saito, M., Asoh, H. 2005. Self-ordering of anodic porous alumina formed in organic acid electrolytes. *Electrochim. Acta* 51:827–833.

Paramasivam, I., Macak, J.M., Selvam, T. et al. 2008. Electrochemical synthesis of self-organized TiO_2 nanotubular structures using an ionic liquid (BMIM-BF_4). *Electrochim. Acta* 54:643–648.

Piao, Y., Lima, H., Chang, J.Y. et al. 2005. Nanostructured materials prepared by use of ordered porous alumina membranes. *Electrochim. Acta* 50:2997–3013.

Popat, K.C., Eltgroth, M., LaTempa, T.J. et al. 2007a. Titania nanotubes: A novel platform for drug-eluting coatings for medical implants? *Small* 3:1878–1881.

Popat, K.C., Leoni, L., Grimes, C.A. et al. 2007b. Influence of engineered titania nanotubular surfaces on bone cells. *Biomater.* 28:3188–3197.

Prakasam, H.E., Shankar, K., Paulose, M. et al. 2007. A new benchmark for TiO_2 nanotube array growth by anodization. *J. Phys. Chem. C* 111:7235–7241.

Raja, K.S., Misra, M., Paramguru, K. 2005. Formation of self-ordered nano-tubular structure of anodic oxide layer on titanium. *Electrochim. Acta* 51:154–165.

Rho, S., Jahng, D., Lim, J.H. et al. 2008. Electrochemical DNA biosensors based on thin gold films sputtered on capacitive nanoporous niobium oxide. *Biosens. Bioelectron.* 23:852–856.

Shankar, K., Mor, G.K., Fitzgerald, A. et al. 2007. Cation effect on the electrochemical formation of very high aspect ratio TiO_2 nanotube arrays in formamide-water mixtures. *J. Phys. Chem. C Lett.* 111:21–26.

Sharma, G., Pishko, M.V., Grimes, C.A. 2007. Fabrication of metallic nanowire arrays by electrodeposition into nanoporous alumina membranes: effect of barrier layer. *J. Mater. Sci.* 42:4738–4744.

Shingubara, S. 2003. Fabrication of nanomaterials using porous alumina templates. *J. Nanopart. Res.* 5:17–30.

Sieber, I., Hildebrand, H., Friedrich, A. et al. 2005. Formation of self-organized niobium porous oxide on niobium. *Electrochem. Commun.* 7:97–100.

Sulka, G.D., Parkoła, K.G. 2006. Anodising potential influence on well-ordered nanostructures formed by anodisation of aluminium in sulphuric acid. *Thin Solid Films* 515:338–345.

Sulka, G.D., Parkoła, K.G. 2007. Temperature influence on well-ordered nanopore structures grown by anodization of aluminium in sulphuric acid. *Electrochim. Acta* 52:1880–1888.

Sunseri, C., Spadaro, C., Piazza, S. et al. 2006. Porosity of anodic alumina membranes from electrochemical measurements. *J. Solid State Electrochem.* 10:416–421.

Tarkistov, P. 2004. Electrochemical synthesis and impedance characterization of nano-patterned biosensor substrate. *Biosens. Bioelectron.* 19:1445–1456.

Taveira, L.V., Macak, J.M., Tsuchiya, H. et al. 2005. Initiation and growth of self-organized TiO_2 nanotubes anodically formed in $NH_4F/(NH_4)_2SO_4$ electrolytes. *J. Electrochem. Soc.* 152:B405–B410.

Taveira, L.V., Macak, J.M., Sirotna, K. et al. 2006. Voltage oscillations and morphology during the galvanostatic formation of self-organized TiO$_2$ nanotubes. *J. Electrochem. Soc.* 153:B137–B143.

Taveira, L.V., Sagüés, A.A., Macak, J.M. et al. 2008. Impedance behavior of TiO$_2$ nanotubes formed by anodization in NaF electrolytes. *J. Electrochem. Soc.* 155:C293–C302.

Thompson, G.E. 1997. Porous anodic alumina: fabrication, characterisation and applications. *Thin Solid Films* 297:192–201.

Thompson, G.E., Xu, Y., Skeldon, P. et al. 1987. Anodic oxidation of aluminum. *Philos. Mag. B* 55:651–667.

Tsuchiya, H., Macak, J.M., Ghicov, A. et al. 2006. Nanotube oxide coating on Ti–29Nb–13Ta–4.6Zr alloy prepared by self-organizing anodization. *Electrochim. Acta* 52:94–101.

Tsuchiya, H., Macak, J.M., Ghicov, A. et al. 2007. Characterization of electronic properties of TiO$_2$ nanotube films. *Corros. Sci.* 49:203–210.

Tzvetkov, B., Bojinov, M., Girginov, A. et al. 2007. An electrochemical and surface analytical study of the formation of nanoporous oxides on niobium. *Electrochim. Acta* 52:7724–7731.

Tzvetkov, B., Bojinov, M., Girginov, A. 2008. Nanoporous oxide formation by anodic oxidation of Nb in sulphate–fluoride electrolytes. *J. Solid State Electrochem.* DOI 10.1007/s10008-008-0651-y.

Varghese, O.M., Grimes, C.A. 2003. Metal oxide nanoarchitectures for environmental sensing. *J. Nanosci. Nanotech.* 3:277–293.

Varghese, O.K., Grimes, C.A. 2008. Appropriate strategies for determining the photoconversion efficiency of water photoelectrolysis cells: A review with examples using titania nanotube array photoanodes. *Sol. Energy Mater. Sol. Cells* 92:374–384.

Varghese, O.M., Gong, D., Paulose, M. et al. 2002. Highly ordered nanoporous alumina films: Effect of pore size and uniformity on sensing performance. *J. Mater. Res.* 17:1162–1171.

Varghese, O.M., Gong, D., Paulose, M. et al. 2003a. Extreme changes in the electrical resistance of titania nanotubes with hydrogen exposure. *Adv. Mater.* 15:624–627.

Varghese, O.M., Gong, D., Paulose, M. et al. 2003b. Hydrogen sensing using titania nanotubes. *Sens. Actuat. B* 93:338–344.

Watcharenwong, A., Chanmanee, W., de Tacconi, N.R. et al. 2008. Anodic growth of nanoporous WO$_3$ films: Morphology, photoelectrochemical response and photocatalytic activity for methylene blue and hexavalent chrome conversion. *J. Electroanal. Chem.* 612:112–120.

Xie, Y. 2006. Photoelectrochemical application of nanotubular titania photoanode. *Electrochim. Acta* 51:3399–3406.

Xie, Y., Zhou, L., Huang, Ch. et al. 2008. Fabrication of nickel oxide-embedded titania nanotube array for redox capacitance application. *Electrochim. Acta* 53:3643–3649.

Yasuda, K., Schmuki, P. 2007. Control of morphology and composition of self-organized zirconium titanate nanotubes formed in (NH$_4$)$_2$SO$_4$/NH$_4$F electrolytes. *Electrochim. Acta* 52:4053–4061.

Yasuda, K., Macak, J.M., Berger, S. et al. 2007. Mechanistic aspects of the self-organization process for oxide nanotube formation on valve metals. *J. Electrochem. Soc.* 154:C472–C478.

Yong-gang, W., Xiao-gang, Z. 2004. Preparation and electrochemical capacitance of RuO$_2$/TiO$_2$ nanotubes composites. *Electrochim. Acta* 49:1957–1962.

Yoriya, S., Paulose, M., Varghese, O.M. et al. 2007. Fabrication of vertically oriented TiO$_2$ nanotube arrays using dimethyl sulfoxide electrolytes. *J. Phys. Chem. C* 111:13770–13776.

Zlamal, M., Macak, J.M., Schmuki, P. 2007. Electrochemically assisted photocatalysis on self-organized TiO$_2$ nanotubes. *Electrochem. Commun.* 9:2822–2826.

15

Metal Oxide Nanohole Array

Tsuyoshi Hamaguchi
Toyota Central R&D Labs, Inc.

Masayoshi Uno
University of Fukui

Shinsuke Yamanaka
Osaka University

15.1 Introduction

15.1.1 Background

Since Bill Clinton, who was then the president of United States, announced the National Nanotechnology Initiative on January 21, 2000, the research of nanotechnology has spread quickly around the world. In the material science field, nanostructured material has also been paid much attention. Developments in the field of nanomaterials have produced very significant and interesting results in all the areas investigated. Many topics like nanotubes,[1–5] porous materials,[6–9] nanoparticles,[10–14] and other nanostructured materials[15–18] are focused. The synthesis of nanostructured materials can be separated by two approaches: one is the top-down approach from the macroscale to the nanoscale materials and the other is the bottom-up approach from atom- or ion-scale to nanoscale materials. Since top-down approach has size limitation, the bottom-up approach, especially chemical approach, has been paid attention. In this chapter, we focus on the synthesis process of the metal oxide nanostructured materials by a kind of bottom-up approach, "solution processing."

15.1.2 Soft Solution Process

In the 1970s, Rouxel and Linvage first proposed the term "chemie douce," called soft chemistry now.[19] Soft chemistry is the synthesis chemistry that uses reactions that occur at mild conditions. The examples of reactions that occur at low temperature are hydrothermal reaction, hydrolysis reaction, dehydration reaction, intercalation reaction, ion-exchange reaction, and so on. Soft chemistry consists of these reactions as they progress by themselves or with small stimulations. The solution process involved in soft chemistry is called soft solution processing.

Soft solution processing, proposed by Yoshimura et al., is the "low temperature fabrication methods that use aqueous solutions to create shaped, sized, located, and oriented materials directory."[20,21] It is a concept that has extended from soft chemistry. Soft solution processing seems to give results that are similar to those of other processes that use fluid and/or beam/vacuum processing. It is worth noting that the total energy consumption in this solution system should be the lowest among all these processing routes. More energy is needed to create melts, vapor, gas, or plasma than to form aqueous solution at the same temperature. Species in aqueous solutions are hydrated (or created some complex agents), and thus they have only a small ΔG for the reaction, and relatively high activation energies are necessary for defeating the hydration energies of ions. In the case of ceramics, anions must be oxidized at the same time that cations are reduced. Therefore, some particular activation processes are required to accelerate the kinetics for synthesizing crystallized single- or multicomponent ceramic materials from the solution.

15.1.3 Nanostructured Material

15.1.3.1 Self-Assembly Phenomena

In order to synthesize nanostructured materials from solution directory, self-assembly phenomena are commonly applied. Anodic alumina[6,7] is a typical example of self-assembled materials. Figure 15.1 shows the schematic diagram of anodic alumina. The geometry of anodic alumina may be schematically represented as a honeycomb structure, which is characterized by a close packed array of columnar hexagonal cells, each containing a central pore normal to the substrate. It is well known that the cell or pore size of the anodic alumina is controlled by the synthesis condition.

Recently, Masuda et al. reported that a highly ordered honeycomb structure with an almost ideally arranged hexagonal could

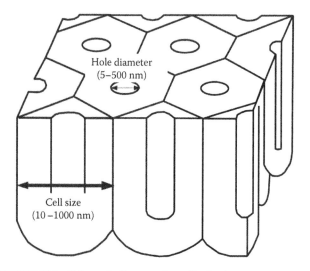

FIGURE 15.1 Schematic diagram of anodic alumina.

be obtained over relatively large areas under a specific anodizing condition in phosphoric acid,[22] oxalic acid,[23] and sulfuric acid.[24] The condition of the self-assembling of the cell arrangement was characterized by a longer anodizing period with appropriate anodizing potential. These highly ordered anodic alumina were used as a template for highly ordered porous materials that consisted of other materials.

15.1.3.2 Nanoparticles

A homogeneous nucleation system is commonly used as the synthesis of metal oxide nanoparticles.[10,11] The homogeneous nucleation system is actualized in the solution with no nuclei of synthesis material. A synthesis from solution can be divided in four steps: precursor formation, nucleation, growth, and aging.[25] The process can be illustrated as a plot of the precursor concentration against time, as shown in Figure 15.2.

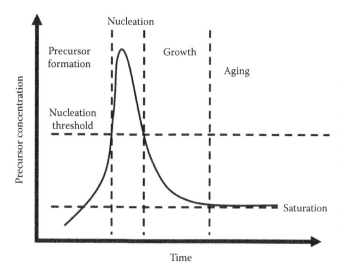

FIGURE 15.2 Precursor concentration against time for the nucleation, growth, and aging of nanoparticles prepared by solution phase reaction.

The driving force of the crystal growth and nucleation from the solution is described as follows:

$$\Delta\mu = RT\ln\frac{C}{C_s} = RT\ln(1+\sigma) = RT\ln S \tag{15.1}$$

where

C_s means saturation concentration
C means the concentration of the precursor molecules
σ is called as the relative supersaturation ($\sigma = (C - C_s)/C_s$)

Before nucleation can occur, zero-charge precursor molecules need to be formed with a specific water/hydroxyl content to allow condensation to occur. The free enthalpy of nucleation is

$$\Delta G = -nRT\ln S + A\gamma \tag{15.2}$$

where

A is the surface area of the solid ($A = n^{2/3}(36\pi v^2)^{1/3}$)
n is the number of precursor
γ is the interfacial tension or energy

The surface tension is usually positive, and the solution is usually supersaturation ($S > 1$). The activation energy required for nucleation corresponds to that of the maximum change in ΔG as a function of n. The number of precursor molecules n^* in critical nuclei is given by $d(\Delta G)/dn = 0$.

$$\Delta G^* = \frac{n^*}{2}RT\ln S = \frac{16\pi\gamma^3 v^2}{3(RT\ln S)^2} \quad \left[n^* = \frac{32\pi\gamma^3 v^2}{3(RT\ln S)^3}\right] \tag{15.3}$$

The radius of the critical nuclei r^* is given as follows:

$$r^* = \frac{3n^* v}{4\pi} \frac{2\gamma v}{RT\ln S} \tag{15.4}$$

The size of the critical nuclei is smaller for higher supersaturation and smaller surface tension. If a cluster whose radius is smaller than the critical nuclei is formed, it dissolves in the solution. The buildup of precursor molecules results in a supersaturation, and once the precursor concentration is above the nucleation threshold, nucleation can occur until the precursor concentration is again below the nucleation threshold. Further growth from the supersaturated solution is possible until the saturation concentration of the solid is reached. After nucleation and growth, the average particle size and size distribution may change by aging. Two dominant processes are aggregation and coarsening, whose rates depend strongly on the experimental parameters. The detailed interaction between the rates of precursor formation, nucleation, growth, and aging determines the final particle size and size distribution.

15.1.3.3 Thin films

A secondly nucleation system is used as the synthesis of the metal oxide thin films. The liquid phase deposition (LPD) method is a typical example of the synthesis process of the metal oxide thin films.[26–31] The LPD method is a novel wet process, which has been developed for the preparation of metal oxide thin films. In this process, it is possible to produce thin metal oxide films or hydroxide films directly on a substrate that is immersed in the treatment solution for deposition. Metal oxide or hydroxide thin films are formed by the means of both a hydrolysis equilibrium reaction of a metal–fluoro complex ion and an F^--consuming reaction of boric acid or aluminum metal that acts as a scavenger for F^- in the treatment solution for deposition. A hydrolysis reaction of metal–fluoro complex ions $[MF_x^{(x-2n)-}]$ is presumed.

$$MF_x^{(x-2n)-} + nH_2O \rightleftarrows MO_n + xF^- + 2nH^+ \qquad (15.5)$$

The equilibrium reaction (Equation 15.5) is shifted to the right-hand side by the addition of boric acid or aluminum metal, which readily reacts with the F^- ions to form stable complex ions (Equations 15.6 and 15.7)

$$H_3BO_3 + 3H^+ + 4F^- \rightleftarrows BF_4^- + 3H_2O \qquad (15.6)$$

$$Al^{3+} + 3H^+ + 6F^- \rightleftarrows H_3AlF_6 \qquad (15.7)$$

The LPD method is a very simple process and does not require any special equipment such as a vacuum system. It can, moreover, be applied readily to the preparation of thin films on various types of substrates with large surface areas and complex morphologies, since the LPD method is performed in a homogeneous aqueous solution. The precursor molecules are formed slowly in the LPD process because of slow reaction, as described in Equation 15.6 or Equation 15.7, and thus it is impossible to actualize high supersaturation concentration. The homogeneous nucleation reaction does not occur because of its low supersaturation concentration; however, nucleation reaction, which requires a lower driving force than homogeneous nucleation reaction, occurs, resulting in the formation of the metal oxide thin film. The synthesis processes of the metal oxide thin film, such as the LPD or solgel methods, are suitable for the template of the nanostructured materials. For example, Aoi et al. proposed the synthesis process of the three-dimensional, ordered macroporous titanium oxide by LPD method[32], and Imai et al. proposed the synthesis process of the titania nanotubes by LPD method.[33]

In the next section, a synthesis process of new metal oxide nanostructured materials called metal oxide nanohole array[34–37] is described. The anodic alumina was used as a template for the metal oxide nanohole array, and LPD was used as a template method.

15.2 Metal Oxide Nanohole Array

15.2.1 Formation Mechanism of the Metal Oxide Nanohole Array

Figure 15.3 shows a field emission scanning electron microscope (FE-SEM) photograph of the surface structure of anodic alumina used as a starting material. Anodic alumina in Figure 15.3

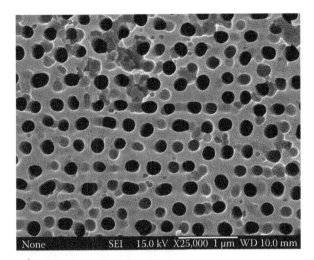

FIGURE 15.3 FE-SEM photograph of the surface structure of the anodic alumina. (Courtesy of Whatman, Inc., Clifton, NJ.)

has many pores whose mean diameter and cell size are approximately 200 and 250 nm, respectively.

Ammonium hexafluorotitanate $((NH_4)_2TiF_6)$ was dissolved at a concentration of 0.10 M in distilled water as a treatment solution. Anodic alumina was prepared at 200 V in phosphoric acid. The anodic alumina was immersed in the treatment solution and the reaction was carried out at 293 K for 2.0 h. The sample was then removed from the treatment solution, washed with distilled water and acetone, and dried at room temperature. Figure 15.4 shows a FE-SEM photograph of the surface structure of titania nanohole array synthesized by LPD method. Titania nanohole array had many tubes whose mean inner and outside diameters were about 250 and 300 nm, respectively. These tubes consisted of the titania nanoparticles.

We derived the following formation mechanism as shown in Figure 15.5 on the basis of experimental results. In this figure, the variation of the surface structure and cross-sectional structure is shown. In the cross-sectional view, side A means the top

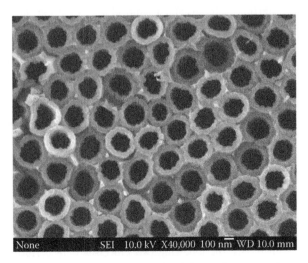

FIGURE 15.4 FE-SEM photograph of the surface area of titania nanohole array synthesized by LPD method.

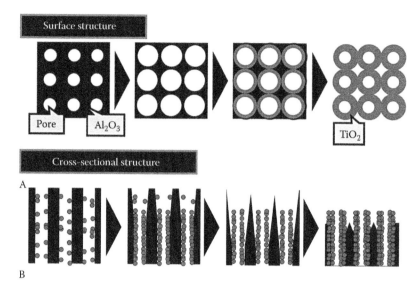

FIGURE 15.5 Schematic diagram of the formation mechanism of the oxide nanohole array.

and side B means the center of the sample. At the first stage, a few titania particles began to deposit on the wall of anodic alumina and the hole of the anodic alumina began to be widened by the dissolution. At the second stage, the holes of the anodic alumina became wide and many titania particles began to form titania thin films. However, since side A of alumina consisted of soft alumina-like gel,[38] the holes were widening from the end side and the titania particles dropped off and were not able to form thin films near side A. At the next stage, some titania tubes and some remaining alumina, which formed some scaffolds, were observed. At the last stage, no alumina was observed from the surface side because the anodic alumina dissolution had proceeded further. Furthermore, more titania particles had deposited, so the hole diameter of the tubes decreased and the wall thickness of the tubes increased. Generally, the formation process of thin films by template process requires at least two processes: a deposition process of new materials on the surface of the starting materials and a removal process of the starting materials. In the formation process of the oxide nanohole array, two other reactions occurred at the same time: the dissolution of the anodic alumina and the deposition of the metal oxide. Additionally, the alumina ions formed by the dissolution reaction of the anodic alumina act as F^- scavengers. Therefore, only one process was required at the formation of the metal oxide nanohole arrays. The detail structure of the metal oxide nanohole array was described in the previous report.[35]

15.2.2 Other Nanostructured Materials

The hydrolysis reaction described as Equation 15.5 occurred in the treatment solution during the synthesis process of the oxide nanohole array. In order to synthesize other nanostructured materials, the following two processes were performed. In one of the processes, the reaction temperature is increased. Figure 15.6 shows a FE-SEM photograph of the surface structure

FIGURE 15.6 FE-SEM photograph of the surface structure of the titania nanorod array.

of titania nanorod array. There are many arranged rods that do not have any holes. The diameter of the rod was about 350 nm. Since the deposition rate of the titania became faster because of the increased reaction temperature, the hole was filled with titania powder. This structure was obtained not only in the case of titania, but also in other cases such as tin oxide and zinc oxide.

The other process is the addition of the boric acid or aluminum ions in the treatment solution. Boric acid and aluminum ions are called the F^- scavengers at the LPD method because of Equations 15.6 and 15.7. Since boric acid and aluminum ions expend the hydrofluoric acid by the formation of BF_4^- or H_3AlF_6 in the treatment solutions, Equation 15.5 shifts to the right side. Figure 15.7 shows FE-SEM photographs of the surface structure of titania nanohole array synthesized by the addition of the boric acid to the treatment solution. The similar surface structure to anodic alumina was observed in the Figure 15.7.

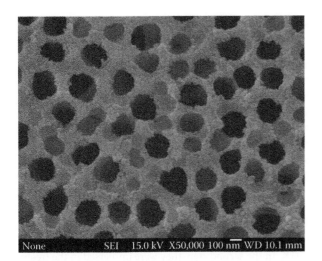

FIGURE 15.7 FE-SEM photograph of the surface structure of the network-type titania nanohole array.

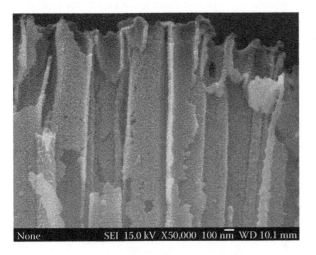

FIGURE 15.8 FE-SEM photograph of the cross-sectional structure of the network-type titania nanohole array.

Oxide nanohole arrays like these were named network-type nanohole arrays. Figure 15.8 shows an FE-SEM photograph of the cross-sectional structure of the network-type titania nanohole array. Many tubes connected by the titania powder on the top of the sample can be seen in Figure 15.8, and there is no anodic alumina between each tube near the surface. In other words, the anodic alumina was dissolved after titania film coated on the surface of anodic alumina. This result suggested that the titania film synthesized by LPD method had many small pores. Titania nanorod array and network-type titania nanohole array were synthesized in almost the same condition, in which the deposition amount of titania was increased; however, the surface structure of these samples was fundamentally different. We derived the following formation mechanism on the basis of experimental results, as shown Figure 15.9. The dissolution rate of anodic alumina at 40°C was faster than that at 20°C. Dissolving anodic alumina increased the Al ions in the treatment solution, and the Al concentration in the hole of the anodic alumina became apparently higher than that at the other area in the treatment solution. The deposition of the titania powder occurred in the hole selectivity. On the other hand, when the boric acid was added, there were many scavenger ions in the treatment solution. Therefore, the deposition of titania powder occurred in the entire treatment solution. According to previous reports,[27] the addition of boric acid decreases the induction period for deposition. The deposition of titania powder with boric acid did not need the dissolution of the anodic alumina, so the deposition of titania powder starts before dissolution of the anodic alumina. Additionally, the pH of the treatment solution with boric acid does not change dramatically; therefore, it is considered that the dissolution rate of anodic alumina in the treatment solution is almost the same in the treatment solution without boric acid. After anodic alumina is coated with titania powder, its dissolution starts. It follows that each tube was connected. This phenomenon occurred not only when the boric acid was added in the treatment solution, but also when the Al

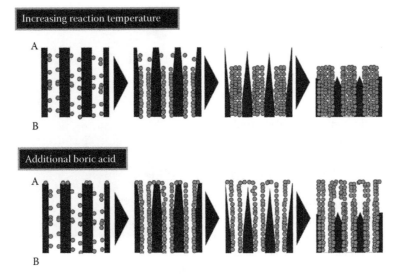

FIGURE 15.9 Schematic diagram of the formation mechanism of the various nanostructured materials.

concentration in the treatment solution was high, such as in the case that the titania nanohole array was synthesized from the same treatment solution used several times.

15.3 Application of Oxide Nanohole Array

15.3.1 Li-Ion Battery

The nanostructured metal oxide has received much attention as electric devices, sensors, and catalysis. Adachi et al. reported that the titania nanotube applied to the photocatalysis.[39] Shimizu et al. reported that the porous titania applied to gas sensor electrode.[8] The porous titania was synthesized by the anodic oxidation of Ti metal. The porous titania sensor showed a quick response to H_2 in dry air with high sensitivity caused by its structure. Ide et al. reported the increase in the intensity of electrochemiluminescence from cells consisting of titania nanohole array film.[40] Hamaguchi et al. reported that titania nanohole array or tin oxide nanohole array applied to photocatalysis or gas sensor electrode.[35,36] In this chapter, the application example of the titania nanohole array to the anode of Li-ion battery is introduced.

The demand for high-energy-density rechargeable batteries has stimulated the search for a new range of electrode materials and the development of new methods to upgrade old materials.[41,42] The lithium-ion battery is considered to be a good system for the commercialization of large secondary batteries. Many systems have been developed for lithium-ion batteries with cobalt, manganese, or nickel compounds as the positive electrode and graphite as the negative electrode.[43–47] To conquer these drawbacks, another class of materials, called transition metal oxides, has been evaluated.

Titania has been found to be one of the good candidates as a lithium host because it is a high-capacity material with low cost and is not harmful.[48,49] The redox reaction of anatase titania and Li is shown in the following equation:

$$TiO_2 + xLi^+ + xe^- \rightarrow Li_xTiO_2 \quad (0 < x < 1) \qquad (15.8)$$

The theoretical capacity of anatase TiO_2 is 336 mAh/g, and the initial open-circuit voltage is more than 3 V. A flat discharge curve is seen around 1.8 V.

There has been an interest in three-dimensional electrodes for the new designs of a lithium battery because of their great potential to achieve an improvement in both energy and power densities. Three-dimensional electrodes in lithium batteries typically have active surfaces exposed to the electrolyte in three dimensions.[50–52] Thus, the available volume of the active material can be much more effectively utilized as compared with planar thin-film electrodes. More importantly, within the constraints of how uniformly the three-dimensional structure is created, the thicker the electrode, the larger will be the energy density without sacrificing power density.

15.3.2 Li-Ion Charge–Discharge Properties of Titania Nanohole Array

The electrode type of the oxide nanohole array was prepared by the two-step LPD method. The purpose of employing the two steps of the LPD methods as stated above was to close up the through-holes in this metal oxide nanohole array. If the through-holes exist, lithium ion does not intercalate or deintercalate into the nanohole array and the ions and the electrons exchange directly to the current collector through the holes because the electrical conductivity of the current collector is higher than that of the nanohole arrays. In the first step, a metal fluoride was dissolved in distilled water to make a 0.10 M solution. An anodic alumina disk (anodisc, Whatman Ltd., Clifton, New Jersey) as the starting material and the F^- scavenger were immersed in the treatment solution for 1 h at 293 K. In the second step, they were immersed in the treatment solution (the concentration was same as in the first step) for 30 min at 323 K. The sample was then removed from the treatment solution, washed with distilled water and acetone, and dried at room

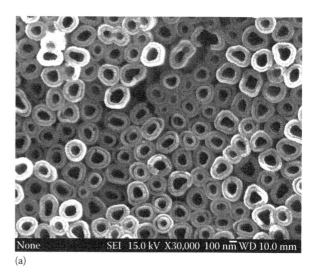

(a)

(b)

FIGURE 15.10 FE-SEM photographs of the surface structure of the titania nanohole array electrode (a) topside (b) downside.

temperature. Heat treatments of the sample were carried out in air for 1 h at temperatures ranging from 773 to 1173 K.

Figure 15.10 shows the titania nanohole array electrode synthesized by the two-step LPD method. In Figure 15.10a, it can be observed that many tubes with double-layer structure; the wall thickness of each layer is 45 nm. The hole size of these tubes is about 200 nm. As the downside of titania nanohole array electrode (Figure 15.10b) has no holes, this titania nanohole array electrode could be applied to the lithium ion battery.

Figure 15.11 shows the XRD patterns for the titania nanohole array electrode sintered at 773 and 1173 K. For the titania nanohole array sintered at 773 K, diffraction peaks other than that of anatase titania were not observed. The diffraction pattern for titania nanohole array electrode sintered at 1173 K shows the peak of the anatase titania and γ-Al$_2$O$_3$.

The following part of this study demonstrates the discharge properties of titania nanohole array electrode. Table 15.1 shows the charge–discharge properties of the sample between 3 and 1 V. To confirm the suitability of the electrolysis cell, the charge–discharge properties of TiO$_2$ powder were measured. The discharge capacity of TiO$_2$ powder was about 169 mAh/g. According to the previous report [48], the theoretical discharge capacity of titania powder between 3 and 1 V was about 168 mAh/g, so the measured value agreed with the theoretical value. This result demonstrated that the design of the cell is reasonable. The discharge capacity of titania nanohole array sintered at 773 K was about 148 mAh/g; therefore, only 88% TiO$_2$ in the titania nanohole array electrode had reacted. The discharge capacity of crashed titania nanohole

array sintered at 773 K was about 168 mAh/g. When the discharge measurement of a sample with the high electrical resistivity was performed, the electrode consisted of the sample and the electrical conducting material such as carbon. In these measurements, crashed titania nanohole array sintered at 773 K was in contact with the electrical conducting material but titania nanohole array sintered at 773 K reacted only by themselves. As the electrical resistivity of the titania nanohole array electrode was very high because of the existence of alumina, and it was impossible for the titania nanohole array electrode to contain the electrical conducting material, the flow of current was insufficient for the reaction and all of titania could not react with Li. Figure 15.12 shows the FE-SEM photographs of the titania nanohole array electrode after measurement. Figure 15.12a and b demonstrates almost the same structure as that before the measurement. This result demonstrates that the nanostructure of the oxide nanohole array electrode was not destroyed by the charge–discharge of Li ion.

According to the above discussion, the high electrical resistivity of the titania nanohole array electrode led to its low discharge

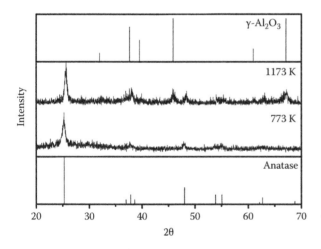

FIGURE 15.11 X-ray diffraction patterns of the titania nanohole array electrode sintered at 773 and 1173 K.

TABLE 15.1 Discharge Capacity of Any Sample

Sample Name	Discharge Capacity (mAh/g)
TiO$_2$ powder	169
Titania nanohole array sintered at 773 K	148
Crashed titania nanohole array sintered at 773 K	168

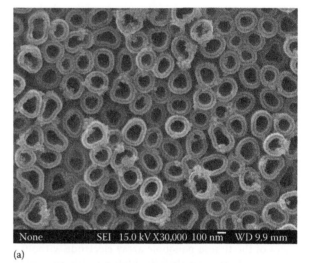

(a)

(b)

FIGURE 15.12 FE-SEM photographs of titania nanohole array electrode after measurement (a) topside (b) downside.

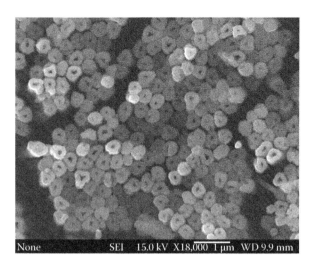

FIGURE 15.13 FE-SEM photograph of the surface structure of the titania nanohole array electrode with ITO.

capacity. The titania nanohole array electrode with low resistivity was prepared by the two-step LPD method and the solgel method. In the first step, the anodic alumina was immersed for 1 h at 293 K in the treatment solution of ammonium hexafluorotitanate at a concentration of 0.10 mol/dm³ distilled water. ITO was deposited on the oxide nanohole array synthesized by the solgel method after the first steps of the LPD method. The second steps of LPD method was curried after these reactions. Figure 15.13 shows the titania nanohole array electrode with ITO.

In this figure, there were many tubes without double-layer wall, resulting from the fact that ITO was deposited between two walls. The content of titania was almost the same value as that in the sample without ITO. The discharge capacity of the sample was about 270 mAh/g. The reason for the fact that the capacity of the titania nanohole array electrode with ITO being larger than the theoretical capacity is as follows. The features of the titania nanohole array electrode without ITO are its high electrical resistivity, high aspect ratio, and high specific surface area. The reason for using a material with high specific surface area as an electrode was that such a material was expected to have a electrical double-layer capacity. Since all of the titania powder in the titania nanohole array did not react, the electrical double layer was formed only in a part of the titania nanohole array. As the titania nanohole array electrode with ITO has low electrical resistivity, almost all of the titania nanohole array electrode reacted and the electrical double layer was formed in the entire sample. This presumption derives from the electrical capacity to be larger than the theoretical capacity of titania powder.

15.4 Summary

The purpose of this chapter is "the development of the synthesis process of nanostructured materials applied to the chemical solution process carried out at room temperature." Focusing on this purpose, each chapter is looked back on.

The new synthesis process of the nanostructured materials named the oxide nanohole array was demonstrated. The LPD method is applied as the template method of the structure of the anodic alumina. The notable features of this process were described as follows. The oxide nanohole array was synthesized by the template method. Normally, the template process of the anodic alumina required three processes: (1) filling the pore of the anodic alumina, the dissolution of the anodic alumina, and the formation of p-type mold; (2) other materials filling in the p-type mold; and (3) the dissolution of the p-type mold. In the formation process of the oxide nanohole arrays, two other reactions—the dissolution of the anodic alumina and the deposition reaction of the other materials—occurred at the same time. Therefore, the one-step template process was actualized. Additionally, the formation process of the oxide nanohole array was carried out around room temperature. The structure of the oxide nanohole array was easy to control because it is very easy to control the structure of the anodic alumina and the deposition rate of the objective substance.

The application possibility of the oxide nanohole array to anodic electrode of the lithium ion secondary batteries was estimated. The titania nanohole array electrode with ITO had higher electrical capacitance than the theoretical capacitance of titania. These results were caused by the high specific surface area of the titania nanohole array electrode. The titania nanohole array electrode with ITO is not suitable for practical use because of its small theoretical capacitance. However, if the synthesis of metal oxide nanohole arrays, which have a large theoretical capacitance, succeeds, they will have a high commercial potential. The weak points of the oxide nanohole array were its brittleness and high electrical resistivity. It is found that the applications that did not require these two weak points, such as photocatalyst, had a high potential for practical use.

References

1. P. Ajayan and S. Iijima, Smallest carbon nanotube, *Nature*, 358 (1992) 23.
2. S. Iijima and T. Ichihashi, Single-wall carbon nanotubes of 1-nm diameter, *Nature*, 363 (1993) 603.
3. A. Thess, R. Lee, P. Nikolaev, H. Dai, P. Petit et al., Crystalline ropes of metallic carbon nanotubes, *Science*, 273 (1996) 483.
4. O. Varghese, D. Gong, M. Paulose, C. Grimes, and E. Dickey, Crystallization and high-temperature structure stability of titanium oxide nanotube array, *J. Mater. Res.*, 18 (2003) 156.
5. T. Kasuga, M. Hiramatsu, A. Hoson et al., Titania nanotubes prepared by chemical processing, *Adv. Mater.*, 11 (1999) 1307.
6. H. Masuda, K. Nishio, and N. Baba, Fabrication of porous TiO₂ films using two-step replication of microstructure of anodic alumina, *Jpn. J. Appl. Phys.*, 31 (1992) L1775.
7. H. Masuda and K. Fukuda, Ordered metal nanohole arrays made by a 2-step replication of honeycomb structures of anodic alumina, *Science*, 268 (1995) 5216.

8. Y. Shimizu, N. Kuwano, H. Takeo et al., High H_2 sensing performance of anodically oxidized TiO_2 film contacted with Pd, *Sens. Actuat. B*, 83 (2002) 195.

9. Y. Ishikawa and Y. Matsumoto, Electrodeposition of TiO_2 photocatalyst into nanopores of hard alumite, *Electrochim. Acta*, 46 (2001) 2819.

10. G. Oskam and F. Poot, Synthesis of ZnO and TiO_2 nanoparticles, *J. Sol-Gel Sci. Technol.*, 37 (2006) 157.

11. G. Oskam, Metal oxide nanoparticles; synthesis, characterization and application, *J. Sol-Gel Sci. Technol.*, 37 (2006) 161.

12. G. K. Chuah, S. Jaenicke, and B. K. Pong, The preparation of high-surface-area zirconia-2. Influence of precipitating agent and digestion on the morphology and microstructure of hydrous zirconia, *J. Catal.*, 175 (1998) 80.

13. K. Sue, M. Suzuki, K. Arai et al., Size-controlled synthesis of metal oxide nanoparticles with a flow-through supercritical water method, *Green Chem.*, 8 (2006) 634.

14. X. Zhang, H. Liang, and F. Gan, Novel anion exchange method for exact antimony doping control of stannic oxide nanocrystal powder, *J. Am. Ceram. Soc.*, 89 (2006) 792.

15. J. Wu, H. Zhang, N. Du et al., General solution route for nanoplates of hexagonal oxide or hydroxide, *J. Phys. Chem. B*, 110 (2006) 11196.

16. S. Yamabi and H. Imai, Crystal phase control for titanium dioxide films by direct deposition in aqueous solutions, *Chem, Mater.*, 14 (2002) 609.

17. S. Yamabi and H. Imai, Growth conditions for wurtzite zinc oxide films in aqueous solutions, *J. Mater. Chem.*, 12 (2002) 3773.

18. A.B. Ellis, M. J. Geselbracht, B. J. Johnson, G. C. Linsensky, and W. R. Robinson, *Teaching General Chemistry: A Materials Science Companion*, American Chemical Society, Washington, DC (1993).

19. B. Gerand, G. Nowogrocki, J. Guenot et al., Structural study of a new hexagonal form of tungsten trioxide, *J. Solid State Chem.*, 29 (1979) 429.

20. M. Yoshimura, Importance of soft solution processing for advanced inorganic materials, *J. Mater. Res.*, 13 (1998) 796.

21. M. Yoshimura and J. Livage, Soft processing for advanced inorganic materials, *MRS Bull.*, 25 (2000) 51.

22. H. Masuda, K. Yada, and A. Osaka, Self-ordering of cell configuration of anodic porous alumina with large-size pores in phosphoric acid solution, *Jpn. J. Appl. Phys.*, 37 (1998) L1340.

23. H. Masuda, M. Yotsuya, M. Asano et al., Self-repair of ordered pattern of nanometer dimensions based on self-compensation properties of anodic porous alumina, *Appl. Phys. Lett.*, 78 (2001) 826.

24. H. Masuda, F. Hasegawa, and S. Ono, Self-ordering of cell arrangement of anodic porous alumina formed in sulfuric acid solution, *J. Electrochem. Soc.*, 144 (1997) L127.

25. J. P. Jolivet, *Metal Oxide Chemistry and Synthesis—From Solution to Solid State*, John Wiley & Sons, West Sussex,U. K. (2003).

26. A. Hishinuma, T. Goda, M. Kitaoka et al., Formation of silicon dioxide films in acidic solutions, *J. Surf. Sci.*, 48–49 (1991) 405–408.

27. S. Deki, Y. Aoi, and Y. Asaoka, Monitoring the growth of titanium oxide thin films by the liquid-phase deposition method with a quartz crystal microbalance, *J. Mater. Chem.*, 7 (1997) 733.

28. S. Deki, Y. Aoi, J. Okibe et al., Preparation and characterization of iron oxyhydroxide and iron oxide thin films by liquid-phase deposition, *J. Mater. Chem.*, 7 (1997) 1769.

29. H. Yu Yu Ko, M. Mizuhata, A. Kajinami et al., Fabrication and characterization of Pt nanoparticles dispersed in Nb_2O_5 composite films by liquid phase deposition, *J. Mater. Chem.*, 12 (2002) 1495–1499.

30. K. Tsukuma, T. Akiyama, N. Yamada et al., Liquid phase deposition of a film of silica with and organic functional group, *J. Non-Cryst. Solids*, 231 (1998) 161–168.

31. K. Tsukuma, T. Akiyama, and H. Imai, Liquid phase deposition film of tin oxide, *J. Non-Cryst. Solids*, 210 (1997) 48–54.

32. Y. Aoi, S. Kobayashi, E. Kamijo et al., Fabrication of three-dimensional ordered macroporous titanium oxide by the liquid-phase deposition method using colloidal template, *J. Mater. Sci.*, 40 (2005) 5561–5563.

33. H. Imai, Y. Takei, K. Shimizu et al., Direct preparation of anatase TiO_2 nanotubes in porous alumina membranes, *J. Mater. Chem.*, 9 (1999) 2971–2973.

34. S. Yamanaka, T. Hamaguchi, H. Muta et al., Fabrication of oxide nanohole arrays by a liquid phase deposition method, *J. Alloys Compd.*, 373 (2004) 312–315.

35. T. Hamaguchi, M. Uno, K. Kurosaki et al., Study on the formation process of titania nanohole arrays, *J. Alloys Compd.*, 386 (2005) 265–269.

36. T. Hamaguchi, M. Uno, and S. Yamanaka, Photocatalytic activity of titania nanohole arrays, *J. Photochem. Photobiol. A*, 173 (2005) 99–105.

37. T. Hamaguchi, N. Yabuki, M. Uno et al., Synthesis and H_2 gas sensing properties of tin oxide nanohole arrays with various electrodes, *Sens. Actuat. B*, 83 (2002) 209–215.

38. G. E. Thompson, R. C. Furneaux, G. C. Wod et al., Nucleation and growth of porous anodic films on aluminum, *Nature*, 272 (1978) 433.

39. M. Adachi, Y. Murata, M. Harada et al., Formation of titania nanotubes with high photo-catalytic activity, *Chem. Lett.*, 29 (2000) 942–943.

40. K. Ide, M. Fujimoto, T. Kado et al., Increase in intensity of electrochemiluminescence from cell consisting of TiO_2 nanohole array film, *J. Electrochem. Soc.*, 155, (2008) B645–B649.

41. K. M. Abraham, Directions in secondary lithium battery research and development, *Electrochim. Acta*, 38 (1994) 1233.

42. Y. Toyoguchi, Y. Moriwaka, T. Sotomura et al., Progress in materials applications for new-generation secondary batteries, *IEEE Trans. Electr. Insul.*, 26 (1991) 1044.

43. P. Fragnud, R. Nagarajan, D. M. Schleich et al., Thin-film cathodes for secondary lithium batteries, *J. Power Sour.*, 54 (1994) 362.

44. S. Yamada, M. Fuliwara, and M. Kanda, Synthesis and properties of $LiNiO_2$ as cathode material for secondary batteries, *J. Power Sour.*, 54 (1994) 209.

45. Y. Gao and J. R. Dahn, The high temperature phase diagram of $Li_{1+x}Mn_{2-x}O_4$ and its implications, *J. Electrochem. Soc.*, 143 (1996) 100.

46. M. Ugaji, M. Hibino, and T. Kudo, Evaluation of a new-type of vanadium-oxide from peroxo-polyvanadate as a cathode material for rechargeable lithium batteries, *J. Electrochem. Soc.*, 142 (1995) 3664.

47. C. Natarajan, K. Setoguchi, and G. Nogami, Preparation of a nanocrystalline titanium dioxide negative electrode for the rechargeable lithium ion battery, *Electrochim. Acta*, 43 (1998) 3371.

48. S.Y. Huang, L. Kavan, I. Exnar et al., Rocking chair lithium battery based on nanocrystalline TiO_2 (anatase), *J. Electrochem. Soc.*, 142 (1995) L142.

49. F. Bonino, L. Busani, M. Lazzari et al., Anatase as a cathode material in lithium-organic electrolyte rechargeable batteries, *J. Power Sour.*, 6 (1981) 261.

50. D. Li, H. Zhou, and I. Honma, Design and synthesis of self-ordered mesoporous nanocomposite through controlled in-situ crystallization, *Nat. Mater.*, 3 (2004) 65.

51. H. Zhou, S. Zhu, M. Hibino et al., Lithium storage in ordered mesoporous carbon (CMK-3) with high reversible specific energy capacity and good cycling performance, *Adv. Mater.*, 15 (2003) 2107.

52. S. Zhu, H. Zhou, T. Miyoshi et al., Self-assembly of the mesoporous electrode material $Li_3Fe_2(PO_4)_3$ using a cationic surfactant as the template, *Adv. Mater.*, 16 (2004) 2012.

16

From Silicon to Carbon Clathrates: Nanocage Materials

Patrice Mélinon
Université de Lyon

Alfonso San Miguel
Université de Lyon

16.1 Introduction

The synthesis of materials having unusual properties is one of the major challenges in crystal engineering. In this quest, one of the more salient parameters that must be considered is the anisotropy inside the crystal lattice. This anisotropy can be accomplished by the mixing between different atoms, molecules, bonds, and/or through the arrangement of the elementary building blocks of the structures themselves. The structural complexity inevitably increases and this seems to be the price to pay to increase material potentialities in different applications. The YBaCuO family or other type-II cuprate superconductors constitute one of the best examples of the increase in complexity with respect to the classical (type-I) pure elements superconductors such as niobium. Likewise, the technological hard magnetic cermets are comprised of the engineered combination of several pure elements. In these two examples, we can consider how performances were improved through the increase in complexity driven by the combination of basic elements. Another promising route in the development of new materials is the structural architecture based on the assembling of "cells" or "cages" rather than atoms. This approach introduces novel features that are related to the topology of the "cells" themselves. Infinite crystalline frameworks can be then be constructed by the regular assemblage of the elementary cells or cages. These building blocks can then form framework structures exhibiting cavities, channels, or windows. This fact is of prime importance since these cavities can accommodate guest molecules, which act as a template for the whole symmetry of the lattice. We obtain then two interpenetrating networks in which unusual properties can be brought directly from the nature of the interaction between the two subnets rather than from the host network bonding or other short range order interactions.

The recent development of nanomaterials from fullerenes to opals going through nanotubes can respond through intercalation to this kind of scheme. Nevertheless, our interest will be focused on a particular and original family of systems in which the host structures constitute a subset of the extensively studied and huge family of zeolites: clathrates. Clathrates have a particular position since the elementary building blocks are made of atomic polyhedra (the nanocages), which organize in a crystalline structure by sharing their faces and fill all the available space, without other voids that are the ones at the interior of the nanocages. These voids have the possibility of guest elements, leading to an original intercalation scheme. Among the broad family of materials having clathrate structures, we will further focus on covalent-like clathrate structures based on elemental group 14 (C, Si, Ge, Sn) host structures.

The interest on these nanocage-based materials has been propelled by their potentialities in different domains such as optoelectronic engineering, integrated batteries, thermoelectric power, hard materials, or superconductivity.

In the following sections, we review the main properties of covalent Si clathrates. The connection between clathrate physics and other fields (cluster physics, surface physics, amorphous structure, etc.) will be also briefly included. In the first part, we discuss the properties of the clathrates without an interpenetrating lattice. In the second part, we focus on the case of endohedrally doped clathrates and some of their relevant properties.

16.2 Silicon Clathrate

16.2.1 A Piece of History

Interestingly, the original ancient Greek meaning of the word *klethra*, from which the word clathrate originates, refers to a particular tree, namely the alder. Let us tell you about this history, as it allows to bring some kind of poetry to the atomic scale. It appears that its second and more modern meaning referring to "cage" finds its origin in the Virgil's *Aeneid* version of the myth of the *Metamorphosis of the Heliades*. "Clathrate" is, in its modern meaning, the Greek roots for "bar" and for "close" and is also present in the Latin roots for "furnished with a lattice." In the Greek mythology, the Heliades are the daughters of Helios, who represents the sun. In Virgil's version of the myth, at the death of their brother Phaëthon, the Heliades were converted to alder (*klethra*) thickets, which encaged the dead Phaëton in the oracular island by growing around its shores. Curiously, the only possible natural clathrate could have been produced by the impact of a meteorite (El Goresy et al. 2003), which is nothing but Heliade, one of the daughters of the sun.

The mythological etymology of clathrate is not the only aspect of these materials related to the ancient Greek civilization. In fact, the nanocage structure of clathrates derives from one of the atomic polyhedra, which fascinated the intellectuals of this ancient civilization: regular dodecahedra that have been recognized long time ago by Plato and Euclid. According to Plato's description in the *Timaeus*, the dodecahedron corresponds to the universe because the zodiac has 12 signs (corresponding to the 12 faces of the dodecahedron). In our "modern" world (in reference to the ancient Greek world), the dodecahedron and its dual* form the icosahedron (the elemental piece of the clathrate architecture) are at the basis of the old problem of "space filling by polyhedra" discussed since two centuries and partially solved by Fedorov in 1885 (Fedorov 1885).

Experimentally, the first determination of a "clathrate "lattice formed by host H_2O molecules and with guest atoms or molecules inside the cages was recognized by Claussen (Claussen 1951). H_2O clathrate lattices are commonly named gas hydrates as the guest species contained in the nanocages are common

gases such as noble gases or methane. In 1958, Frank and Kasper (Frank 1959) made a thorough topological study of these particular crystalline structures. Following their work, the clathrate structure of germanium and silicon was identified (Cros et al. 1965) on crystals obtained from the thermal decomposition of zintl phases of Si or Ge and alkali metals by Cros et al. (1965). Clathrate was recognized as the dual Frank Kasper phase. The duality was extensively studied by Kléman (1989). In the past 30 years, a large number of covalent clathrate species have been synthesized. Recent reviews on clathrates have been undertaken by Bobev and Sevov (2000), San-Miguel and Toulemonde (2005), and Beekman and Nolas (2008). The interest was renewed at the beginning of the last decade since the discovery of the superconductivity in the clathrate compound $Na_2Ba_6Si_{46}$ by Kawaji et al. (1995). The origin of the amazing clathrate structure can be understood from the following remark cited by Brus (1997):

> Materials scientists seek to understand how to create new substances. One strategy is to explore thermodynamic metastability, that is, to search for materials seemingly unfavored by their higher energy, yet permitted to exist by barriers that prevent a transformation to lower energy forms.

In other words, the kinetics pathway is of prime importance. All the clathrates with Si, Ge, Sn, and C are expected to be metastable phases. However, while Si, Ge and Sn clathrates have been successfully synthesized, C clathrates, which are extremely promising materials, appear to resist all synthesis attempts so far.

16.2.2 Synthesis

Till now, clathrate structure has not been clearly identified in natural materials. The synthetic path is presently the only way to produce silicon clathrates. Two methods are commonly reported for the Si clathrate synthesis, which we will discuss separately. Both methods are based on the thermal treatment of silicides containing the species that should be intercalated in the clathrate structure and differ on the pressure conditions needed for the synthesis.

16.2.2.1 Low-Pressure Synthesis

The silicon clathrate samples are synthesized in two steps (the same applies for Ge and Sn clathrates): the synthesis of the precursor (silicide MSi) followed by its pyrolysis. The silicides MSi (M being an alkaline) are prepared by treating stoichiometric mixtures of Si and M elements. The mixtures are loaded in an inert metallic container that is subsequently subjected to arcmelting under inert atmosphere. The reaction vessels are heated at moderate temperatures (700–1000 K) for a long period (2–20 days). The following reaction takes place:

$$M + Si \rightarrow MSi \tag{16.1}$$

This method lead to the first group-14 clathrate synthesis and has been developed by the group of Bordeaux (France) (Cros et al. 1965). In the pyrolysis step, silicide is then transferred in

---

* Polyhedra are duals when the vertices of one correspond to the faces of the other.

a vacuum chamber and heated at moderate temperature (700–1000 K). For the clathrate I phase, silicide is heated under argon pressure. Clathrate II is prepared by heating the silicide in vacuum. During heating, the samples lose alkali atoms. The stoichiometry can be adjusted by incorporating metallic alkaline vapor:

$$M_x@\text{Si-34} + X(\text{vapor}) \rightarrow M_{x'}@\text{Si-34}x' + x \qquad (16.2)$$

Heating in free alkaline vapor permits the synthesis of $M_x@$Si-34 (34 atoms in the primitive cell) with x down to one. Gryko (Gryko et al. 2000) washed $M_x@$Si-34 (M = Na) in concentrated hydrochloric acid. Subsequently, the samples are dried and degassed at 700 K in vacuum over several days. The process is repeated several times. The low reported value of x is about 0.02 close to the guest-free clathrate form even if it contained about 5% of diamond-type silicon impurities. Ammar, Cros, and coworkers (Ammar et al. 2004) with additional reactions with iodine obtained $x \approx 0.0058$ and traces of Na_8Si_{46}. More recently, Na_4Ge_9 zintl phase was used to synthesize pure Ge-34 clathrate (Guloy et al. 2006).

The global concentration of alkaline is determined by x-ray and/or by the flame emission technique. The purity of the clathrate is checked by Rietveld analysis of the diffraction data.

This method is suitable for the production of $M_x@$Si-34 and $M_8@$Si-46, where M is an alkaline atom. The quantity of obtained products is compatible with any chemical or physical characterization. Of particular interest was the inclusion of Ba atoms in the structure, $Na_xBa_y@$Si-46, reported by Yamanaka et al. (1995) using both Zintl phases (2:1 molar mixture) NaSi and $BaSi_2$. As we will discuss later, this clathrate showed superconductivity and led to research interest on the study of clathrate materials as well as to the development of the clathrate high-pressure synthesis.

16.2.2.2 High-Pressure Synthesis

This method was developed by Yamanaka and collaborators (Fukuoka et al. 2000) who were the first to prepare the $Ba_8@$Si-46 clathrate superconductor. Starting from the $BaSi_2$ Zintl phase mixed with silicon powder and placed in an h-BN cell, the synthesis occurs at approximately 1000 K under high pressure (1–5 GPa). The sample is quenched at room temperature before the pressure is slowly released. The synthesis relies on two principles:

1. The clathrate stoichiometry needed in the precursor mixture should allow for the complete filling of all cages.
2. The atomic volume of the products should be smaller than the volume of the reactants (Le Chatelier principle).

This technique is versatile but the quantity of sample that can be prepared in a run is limited to the press capacity to several hundreds of milligrams. Figure 16.1 shows as an example a belt-type apparatus used for the high-pressure synthesis of clathrates. In this particular case, the maximum sample volume is limited to less than 100 mm³.

FIGURE 16.1 View of the different parts of a belt apparatus used for the high pressure synthesis of silicon clathrates. (A) is one of the two tungsten carbide pistons that will confine the sample from the two sides of the central hole of the "belt" which central part is also made of tungsten carbide (B). The reactants are mixed in a glove-box and placed in a crucible (usually in *h-BN*) which is itself inserted in a cylindrical graphite resistive furnace. These elements (sample, crucible and furnace) are introduced in a hole practiced in the pyrophyllite gasket shown in (C) which is itself introduced in the cavity of the "belt" and closed in both sides with pyrophyllite (D) and teflon (E) gaskets. Stainless steel truncated cones (F) are used to establish from each side of the belt the electrical contact between the furnace and the pistons where the electrical current will be injected. The completed set-up is then inserted between the jaws of the 4-columns hydraulic press (G).

Using this technique, Reny et al. (2000) also prepared iodine compounds $I_8@$Si-46. Other alkali earths (Sr, Ca) have been incorporated with this method (Toulemonde et al. 2006) as well as Te atoms (Jaussaud et al. 2004).

16.2.3 Si Clathrate Structure

An important difference between the two types of Si clathrates considered here is that the type II clathrate has been obtained nearly pure, i.e., without guest atoms inside the nanocages. This is not possible in the type I clathrate. We will then start our discussion with the type II clathrate, which in addition is the one with the smaller number of atoms in the unit cell.

16.2.3.1 Clathrate II

Crystallographic data deduced from Rietveld refinement are reported in Table 16.1. In Figure 16.2 is shown the associated x-ray powder diffraction pattern. The relative intensity of the diffraction peaks strongly depends on the purity of the material, especially on the presence of residual guest atoms inside the cage. This lattice can be considered as being built by the coalescence of

TABLE 16.1   Crystallographic Data for Clathrates I and II

Name Notation	Diamond Si-2	Clathrate II Na_x@Si-34[a]	Clathrate I Na_8@Si-46
Space group	$Fd\bar{3}m$	$Fd\bar{3}m$ origin at center $\bar{3}m$	$Pm\bar{3}m$ origin at $4\bar{3}m$
Lattice constant (Å)[b]	5.4286	14.6428(8)	10.1983(2)
[c]		14.61963(1)	10.19648(2)
Position	x, y, z	x, y, z	x, y, z
[b]	(Si) $x = y = z = 1/8$	$x = y = z = 1/8$	(Si) $x = 1/4$ $y = 0$ $z = 1/2$
[c]	(Si) $x = y = z = 1/8$	$x = y = z = 1/8$	(Si) $x = 1/4$ $y = 0$ $z = 1/2$
[b]		(Si) $x = y = z = 0.7827(2)$	(Si) $x = y = z = 0.1847(2)$
[c]		(Si) $x = y = z = 0.7818(1)$	(Si) $x = y = z = 0.1851(1)$
[b]		(Si) $x = y = 0.8169(1)$ $z = 0.6288(2)$	(Si) $x = 0$ $y = 0.3088(2)$ $z = 0.1173(2)$
[c]		(Si) $x = y = 0.8168(1)$ $z = 0.6288(1)$	(Si) $x = 0$ $y = 0.3077(2)$ $z = 0.1175(2)$
		(Na) $x = y = z = 0$	(Na) $x = y = z = 0$
		(Na) $x = y = z = 3/8$	(Na) $x = 1/4$ $y = 1/2$ $z = 0$
Number of positions, Wyckoff notation, point symmetry	8, a, $4\bar{3}m$	(Si) 8, a, $4\bar{3}m$ (Si) 32, e, 3 m (Si) 96, g, m (Na) 16, c, $\bar{3}m$ (Na) 8, b, $4\bar{3}m$	(Si) 6, 2, 42 m (Si) 16, i, 3 (Si) 24, k, m (Na) 2, a, m3 (na) 6, d, $\bar{4}2m$

[a] $x = 1/4$ (Reny et al. 1998) and $x = 1$ (Ramachandran et al. 1999).
[b] From Reny et al. (1998).
[c] From Ramachandran et al. (1999).

two Si_{28} and four Si_{20} per unit cell (Figure 16.2). It belongs to the same space group than the common diamond structure $Fd3m$. Using the crystallographic notation, clathrate II is labeled Si-34 since we have $1/4(2 \times 28 + 4 \times 20) = 34$ atoms in the primitive cell. Such a structure is obtained by templating one Si atom in the Si_5 basic sp^3 tetrahedron with Si_{28} cage, this latter having T_d point group symmetry. Si_{28} has four hexagons and shares these hexagons with its four Si_{28} neighboring cages. The space filling needs additional silicon atoms in a tetrahedral symmetry forming Si_{20} cages. Since 85.7% of the membered rings are pentagons, the electronic properties will be sensitive to the frustration effect (see Section 16.2.6.3). As we will discuss later, the mean hybridization, $n = 2.86$, in the type II clathrate can be easily estimated from Equation 16.6. The three inequivalent silicon sites (96 g, 8a, and 32e) and their relative populations have been confirmed by magic angle spinning ^{29}Si NMR spectra (Gryko et al. 2000). The NMR spectra exhibit three sharp peaks with well-resolved resonances at 0.3 (96 g), 93 (32e), and 50.4 (8a) ppm, respectively.* The local order has been also studied by EXAFS spectroscopy at the Si K-edge. Figure 16.2 displays the pseudoradial distribution function deduced from EXAFS performed at the Si K threshold for $Na_{0.25}$@Si-34 (clathrate II), Na_8@Si-46 (clathrate I), and Si-2 lattices, respectively. Simulation within the FEFF code (Ankudinov et al. 1998) framework is also done. First, second neighbors are nearly the same in all the structures showing the similarities at the short-range order around a silicon atom (tetrahedral arrangement). The disappearance of the peak attributed to the third

neighbors (see Figure 16.2) in the clathrates lattices is the short-range order signature of the cage-like structure with respect to other sp^3 lattices.

16.2.3.2 Clathrate I

This lattice is built from the coalescence of two Si_{20} and six Si_{24} per unit cell (Figure 16.3). It belongs to the primitive cubic $Pm3m$ space group with $1/4(6 \times 24 + 2 \times 20) = 46$ atoms. This lattice called Si-46 is formed by 88.9% of pentagons with a mean hybridization $n = 2.81$. As we will discuss later, this lattice cannot be observed without the incorporation of guest atoms inside the cages. The expected crystallographic data deduced from Rietveld refinement are reported in Table 16.1. The 14-sided Si_{24} cages are arranged as three mutually perpendicular interlocking columns with the pentagonal dodecahedra lying between them on a bcc (body centered cubic) lattice.

16.2.4 Origin of the Clathrate Structure

The clathrate structures appear to correspond to a natural solution of the geometrical space-filling problem. This fact and others demand that these structures are discussed in multidisciplinary fields, which will be briefly included in this section.

16.2.4.1 Fivefold Symmetry and Lattices

The clathrate structures are characterized by the importance of dodecahedra (Si_{20}) and more generally of fivefold membered rings. It is well known that fivefold symmetry is forbidden due to long-range order constraints since it is not compatible with

---

* Measured from a tetramethylsilane standard.

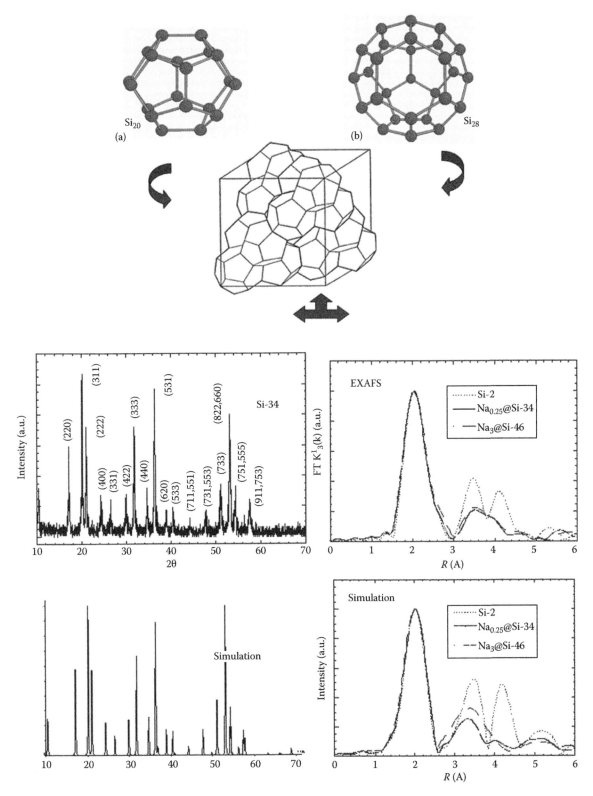

FIGURE 16.2 (top) (a) Si_{20} and (b) Si_{28} cages and drawing showing the arrangement of the both polyhedra in the Si-34 structure. (left panel) x-ray powder diffraction pattern of $Na_{0.25}@Si-34$ and calculated spectrum assuming the crystallographic data displayed in Table 16.1. (right panel) EXAFS Si-K edge, pseudo radial distribution function of Si-2 (diamond structure), $Na_{0.25}@Si-34$ and $Na_8@Si-46$. The temperature was 20 K. Simulation (lower panels) were done using the FEFF code. The parameters are given in Table 16.1. For this purpose, the Debye temperature were $\theta_D = 545$ K, $\theta_D = 430$ K and $\theta_D = 395$ K for Si-2, $Na_{0.25}@Si-34$ and $Na_8@Si-46$ respectively. (From San-Miguel, A. et al., *Phys. Rev. B*, 65, 054109, 2002. With permission.)

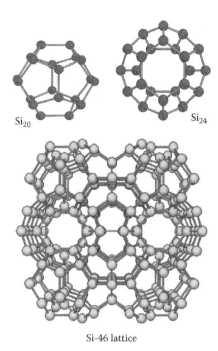

Si$_{20}$ Si$_{24}$

Si-46 lattice

FIGURE 16.3 The type-I Si-46 clathrate structure (below) can be viewed as the association of Si$_{20}$ and Si$_{24}$ cages (above). (From San-Miguel, A. et al., *Phys. Rev. B*, 65, 054109, 2002. With permission.)

the translation symmetry. Kepler argued that pentagons do not allow a complete filling of 2D space. The same holds good for icosahedra when considering 3D space. Twenty tetrahedra combine naturally to form an icosahedron. However, they cannot be perfectly packed. This geometric mismatch is called "geometric frustration." In a "flat" 3D space, the filling of space by polyhedra needs a combination of two effects: a distortion of the dodecahedron and/or the combination with other polyhedra having a point symmetry compatible with the translation symmetry (Si$_{28}$ or Si$_{24}$).

16.2.4.2 Bubble Soap and Beyond

Everyone is familiar with the spherical shape of a bubble soap caused by surface tension. The tension causes the bubble to form a sphere, as a sphere has the smallest possible surface area for a given volume. When two bubbles merge, the same physical principles apply, and the bubbles will adopt the shape with the smallest possible surface area. This introduces a flatness of the wall between the two bubbles. Due to the Laplace Young internal pressure, the curvature of the wall depends on the relative radius of the two bubbles. The merging of three bubbles introduces additional walls. The driving force for the merging of soap bubbles obeys the complex mathematical problem of minimal partitioning of space into equal volumes, as they will always find the smallest surface area between points or edges. Since the magic angle is found to be 120°, hexagon will be the natural polyhedron between bubbles. In 1887 Lord Kelvin stated that the cubooctahedron was the best candidate. Nevertheless, real foams do not follow Kelvin's conjecture not just because it was concerned with bubbles of equal size but also because it was

thought that in practice such a macroscopic system could not be expected to find its global minimum. In 1993, Weaire and Phelan (1994) overthrew Kelvin's partition of space criterion with a new proposition, even though, there is no proof that the Weaire–Phelan partition scheme is optimal. This solution associates two sorts of polyhedra: one is a docahedron formed by 12 (distorted) pentagons and another is 14-sided with two opposite hexagonal faces and 12 pentagonal faces. The network is the so called "clathrate of type I." It has been established (Mutoh 2004) that both polyhedra are the polyhedra of minimal volume circumscribed about a sphere (called isoperimetric quotient). We should also consider that the clathrate structure needs a slight distortion of the initial polyhedra (see Figure 16.4).

The truncated octahedron mentioned by Kelvin is less "spherical" than the dodecahedron or the icosahedron (Figure 16.4). However, it is the best solution taking into account its capability for space-filling. These forms correspond to the Wulff's form (Fonseca 1991) mentioned in face-centered cubic (fcc) crystal (the first Brillouin zone). In fact, the general sphere packing problem requires the densest packing of equal spheres in the 3D space. Physicists argue that fcc or hexagonal closed packed (hcp) dense packing is the solution with a density of 0.74 (Kepler conjecture). Mathematicians argue that this is an unsolved problem and propose an upper value of 0.7784, which is higher than the fcc packing (Conway and Sloane 1993). Among other solutions, the icosahedron and its dual form (pentagonal decahedron) play a crucial role. While the AA_{12}(cub) (12 atoms assembled around one central atom in the fcc structure) bonding units appear in the fcc crystal structures, the crystal structures that include AA_{12}(icos) (12 atoms assembled around one central atom in the icosahedral structure) icosahedral units are generally much more complicated. However, there is a simple and naive argument to explain the similarity between an fcc and an icosahedral structure. Starting with an fcc structure, the first coordination polyhedron is a cuboctahedron depicted on Figure 16.5. In a solar system analogy, let us consider the central atom as the sun and the neighboring atoms the "planets." It is possible to achieve any permutation of the 12 planets by placing them around the sun in such a way they never overlap (Conway and Sloane 1993). If the "planets" move in the directions indicated by the arrows (see Figure 16.5), they can be brought continuously into the icosahedral arrangement. This continuous deformation presents some analogies with the martensitic metals or the silicon-pressure-induced transformation from the four-coordinated diamond phase to the six-coordinated β-tin phase.

16.2.4.3 Topology and Electronic Structure

The elementary building blocks of the clathrate structures, the so called "cages" or "nanocages" are polyhedra classified following the type of polygons (regular or not) at their surface and their point group symmetry. A 3D crystal needs polyhedra filling totally the space without overlap. Of course, such polyhedra share common faces. For regular polyhedra, only the cube satisfies the criterion. For nonregular polyhedra, the answer is

Polyhedron	QF	Shape	SPC	Polyhedron	QF	Shape	SPC
Sphere	1	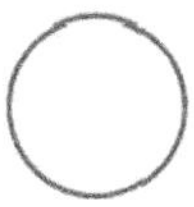	No	Pentagonal dodecahedron Si_{20}	0.7547	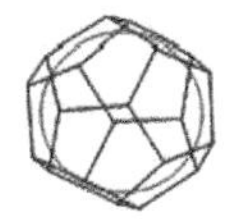	No
Tetrahedron	0.3023	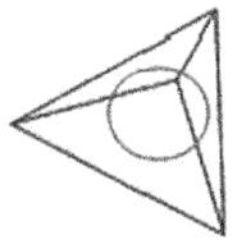	No	Tetrakai-decahedron Si_{24}	0.7856	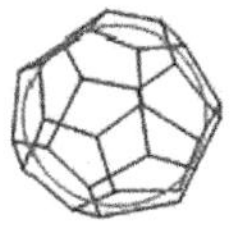	No
Cube	0.5236	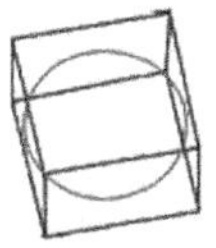	Yes	Hexakai-decahedron Si_{28}	0.8121		No
Truncated octahedron (Kelvin)	0.757	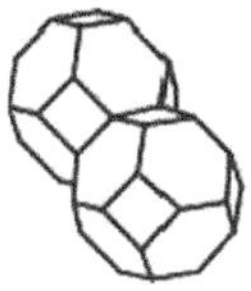	Yes	Weaire Phelan (clathrate l) polyhedra are distorted	0.764	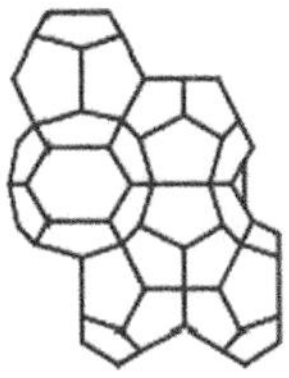	Yes
Undistorted clathrate I	0.7779	$6Si_{24} + 2Si_{20}$	No	Undistorted clathrate II	0.7738	$2Si_{28} + 4Si_{20}$	No

FIGURE 16.4 Polyhedron, quality factor QF (isoperimetric quotient) and shape of selected polyhedra. SPC indicates space-filling cells. Note that the hypothetic undistorted clathrate I (regular Si_{20} and semiregular Si_{24} polyhedra) has the better QF but does not fill space since both cages have to be distorted. The true clathrate with distorted polyhedra named Weaire Phelan structure is better that the classical truncated octahedron conjectured by Kelvin. All these polytypes are a particular case of the bubble soap problem since in their initial works, Kelvin, Weaire and Phelan discussed polyhedra with a surface curvature. The clathrate I is the so called flat limit approximation.

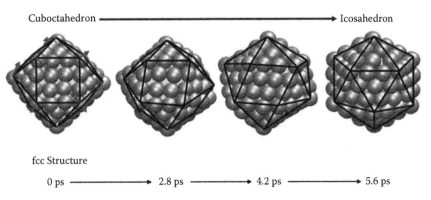

Cuboctahedron ⟶ Icosahedron

fcc Structure

0 ps ⟶ 2.8 ps ⟶ 4.2 ps ⟶ 5.6 ps

FIGURE 16.5 Transformation of a cuboctahedron of the fcc lattice into an icosahedron view perpendicular to a 100 facet obtained from molecular dynamics simulation. The simultaneous rotation of triangular faces {111} about their normals reconstructs the cuboctahedron into the icosahedron. The value of the rotation angle is 22.24°. The times for each snapshot from left to right are 0, 2.8, 4.2, and 5.6 ps, respectively. (After Chen, F. and Johnson, R., *Appl. Phys. Lett.*, 92, 023112, 2008. With permission.)

given by the Schläfli's (Sivardière 1995) theorem, which cannot be solved practically:

$$S + F - A - C + 1 = 0 \qquad (16.3)$$

where S, F, A and C are the number of vertices, edges, faces, and cells, respectively. As a matter of fact, the nonregular polyhedra are still simply the Dirichlet regions of the Bravais lattice: in other words, the Wigner Seitz cells (named Voronoï cells) defined in the 230 crystallographic groups. Among their common features, all these polyhedra do not present pentagons since fivefold axis are not compatible with translation symmetry. The set of solutions for 3D crystalline pavement can be increased significantly if we allow for the stacking of two or more types of polyhedra. Andreini (1905) provided the solution when two kinds of polyhedra (one regular and another semiregular) are used. Consequently, the available possibilities to form 3D framework structures are probably infinite but the problem of determining the nature of possible polyhedra with odd polygons allowing for such nonoverlapping space filling remains nowadays unsolved, although it is possible to converge toward the Andreini's pavement by a weak deformation of two polyhedra. This is an important point: translational symmetry may require not only even-membered rings but also distortion of angle and bond lengths. Assuming that polyhedra are distorted, we are not aware of a general method to assess the complete solution for space filling, but the set is limited taking new assumptions into account:

1. The polyhedra are convex and obey the Euler's theorem: all the atoms are located on a surface of the polyhedra leading the core empty. These polyhedra are defined as "cages."
2. The hybridization of the atoms making the polygons are close to the sp^3 basis: each atom belongs to four different "cages," the coordination inside a single cage being three.
3. We minimize the set of polyhedra: we use two polyhedra for a single structure and no more.

Since the hybridization is close to sp^3, the corresponding dihedral angle $\theta = 109, 47°$ gives the type of n-gons forming the cage. The ideal case corresponds to $n = 5$ with $\theta = 108°$. The translational symmetry needs even-membered rings, the closest being $n = 4$ ($\theta = 90°$) and $n = 6$ ($\theta = 120°$). Since the coordination for a single polyhedron is 3(2), Euler's theorem gives rise to the relationship

$$2F_4 + F_5 = 12 \qquad (16.4)$$

for any F_6, F_n being the number of n-gon in the "cage." Since F_5 favors sp^3 hybridization, the ratio between F_5 and F_n should be maximum. The presence of squares (F_4) leading to a pure p bonding is not discussed since the dihedral angle is significantly far away the ideal value: thus $F_5 = 12$. Moreover, translational symmetry needs at least two or more even membered n-gon in the cage. Pi orbital vector analysis (POAV) (Haddon 1986) gives the mean hybridization versus the mean dihedral angle $\theta_{\pi\sigma}$ ($\theta_{\pi\sigma} = 90°$ for sp^2 and $\theta_{\pi\sigma} = 109, 47°$ for sp^3, respectively).

$$n = \frac{2}{\left(1 - 3\cos\theta_{\pi\sigma}^2\right)} \qquad (16.5)$$

After calculation and introducing the F_6/F_5 ratio,

$$n = 3\frac{(1 - 12\pi/3^{3/2}20)}{\left(1 - 12\pi/3^{3/2}(20 + 24F_6/F_5)\right)} \qquad (16.6)$$

For silicon, which prefers sp^3 hybridization, we overlook until $F_6 = 2, 3, 4$. This set leads to the fullerene family and holds for carbon since sp mixing ranges between sp^2 and sp^3. For silicon,

the corresponding fullerenes will be Si_{24}, Si_{26}, and Si_{28}, respectively. The topology give us new conditions for filling the 3D Euclidean space. Among these fullerenes, we do not consider Si_{26}, as it needs to be associated to two other polyhedra (Si_{20} and Si_{24}) (Benedek and Colombo 1996) for space filling and then do not satisfy our criterion (3). Finally, we obtain two networks satisfying all the assumptions. The first lattice is a combination of Si_{20} and Si_{24} and the second one combines Si_{20} and Si_{28}. They are labeled clathrate (I) and (II), respectively. Both are observed for silicon. In summary, clathrates are formed by the stacking of two types of polyhedra sharing common faces. In addition, a slight distortion is needed for satisfying the Andreini's rule.

16.2.4.4 Topology: A Little Bit More

After Riemmann's work, the world was never flat again despite the first opinion of Euclid: briefly, the plane is an Euclidean surface, the sphere belongs to the spherical geometry, and a surface composed of saddle points corresponds to an hyperbolic geometry (Figure 16.6). The graphite surface formed by hexagons is flat. Likewise, the diamond structure is well described in 3D Euclidean (flat) space with a collection of (armchair) hexagons. Using the general properties of Euler's law (Terrones et al. 2002) and applying for surfaces combined with the Gauss–Bonnet theorem, it holds in a simple case that

$$N_5 - N_7 = 12(1 - g) \qquad (16.7)$$

where

  g called "genus" is the number of handles or holes in the structure

  N_7 is the number of heptagons of the polyhedron

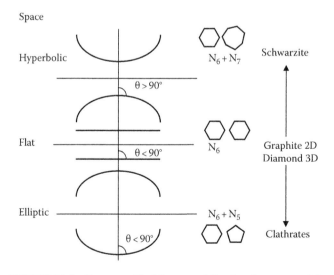

FIGURE 16.6 Curvature K of the space following the nature of the polygons for the pavement of the plan. Euclidean flat space corresponds to a set of pure hexagons $K = 0$, elliptic space $K > 0$ corresponds to a set of hexagons and pentagons, hyperbolic space corresponds to a set of hexagons and heptagons. Since the genius is much more when $K < 0$, the network will be strongly tortuous.

The complexity of the space increases with g. For example, $g = 0$ in a sphere and all topologically equivalent closed surfaces can be mapped continuously onto a sphere and one in the case of torus topology. Since hexagons do not participate in the curvature, N_6 does not appear in the formula. The mapping of a surface with pentagons and hexagons induces a positive curvature K where the mapping of the surface with heptagons and hexagons induces a negative curvature K. From the Riemann's analysis, the two cases correspond to spherical (or elliptical) and hyperbolic geometries, respectively (see Figure 16.6).

Considering the case $N_7 = 0$, with $g = 0$ (sphere), then

$$N_5 = 12 \qquad (16.8)$$

This is the Si_{20} dodecahedron. Taking $N_6 = 4$ leads to Si_{28}, both polyhedra form the clathrate structure in fcc structure with 34 atoms per primitive cell. Now considering the case $N_5 = 0$. The most primitive space is defined with a genius $g = 2$, and

$$N_7 = 12 \qquad (16.9)$$

This is a structure with 12 heptagons instead of 12 pentagons. We can define a very simple architecture with $N_6 = 0$ called schwarzite (in honor of the mathematician Karl Hermann Amandus Schwarz known for his work in complex analysis). Table 16.2 reports the crystallographic data of the most simple schwarzite structure (Terrones and Terrones 2003, Spagnolatti

et al. 2003). This is the dual form of the so-called "amorphon" formed by the pavement of the 3D space with distorted dodecahedra. The schwarzite is a complex structure where all the atoms are threefold coordinated like in graphene (a sheet of graphite) (see Figure 16.7). This structure is characterized by a very low density: the "filling-space" criterion is not achieved in hyperbolic geometry. Since the sp^2 hybridization is not reported in silicon, the silicon schwarzite has never been observed. Some authors argue that complex carbon forms including vitreous carbon are derived from the schwarzite networks.

TABLE 16.2 Crystallographic Data for Carbon Schwarzite Compared to Carbon Diamond

Name Notation	Diamond C-2	Schwarzite
Space group	$Fd\bar{3}m$	$Fd3$ origin at center $\bar{3}$
Lattice constant (Å)[a]	3.5669	14.96
Position	x, y, z	x, y, z
a	(C) $x = y = z = 1/8$	(C) $x = y = z = 0.8$[b]
b		(C) $x = 0.927$ $y = 0.698$ $z = 0.768$
b		(C) $x = 0.826$ $y = 0.885$ $z = 0.018$
Number of positions, Wyckoff notation, point symmetry	8, a, $\bar{4}3m$	32, (e), 3 96, (g), 1 96, (g), 1

[a] From Donohue (1974).
[b] From Spagnolatti et al. (2003).

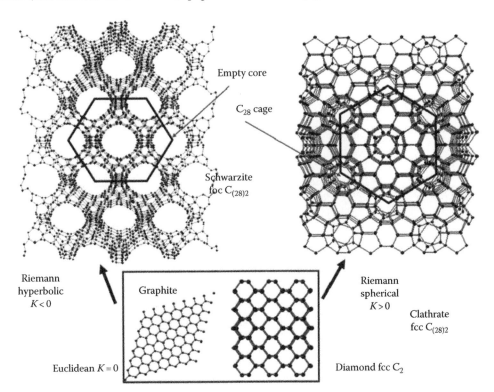

FIGURE 16.7 Diamond, clathrate and schwarzite observed along (110) direction. The graphite is the reference for a 2D flat surface. The typical hexagonal symmetry observed in the fcc structure is shown by the hexagon. Note that schwarzite have an empty core with a low density.

16.2.4.5 Crystallography

In 16.2.4.(2,3) sections, we have given examples on the basis of a topological criterion: *the maximum space filling with a minimum set of polyhedra*. Even if this is an interesting point of view to understand the clathrate structure, it will be a wrong analysis to conclude that the driving force for building a clathrate lattice obeys this golden rule. We address now several examples showing that the clathrate lattice can be built without a direct correlation with the "space filling minimum" criterion. What is then the real reason for clathrate formation? It has been evoked in the case of micelles, which lead also to the formation of clathrate structures. The organization should be adopted in the liquid phase, where the solvation of the guest species through encaging is favored by the entropy and enthalpy balance.

16.2.4.6 Laves Phase versus Clathrate II

The Laves phase structures (see Figure 16.8) are adopted by a large number of binary intermetallic compounds of composition AB_2. These phases belong to the class of tetrahedrally close-packed alloys (space group *Fd3m n°227* structure type Cu_2Mg). Two atoms are in position 8a and 16d, respectively (lattice parameter 7.048 Å). The first coordination polyhedron corresponds to a deformed icosahedron and the second to a polyhedron formed by triangular faces with a fivefold symmetry plus four hexagons. These polyhedra mimic the clathrate structure: they are the dual forms of the dodecahedron Si_{20} and Si_{28} as reported in clathrate II.

16.2.4.7 β-Tungsten Phase versus Clathrate I

β-Tungsten (see Figure 16.8) is recognized as a Frank–Kasper tetrahedrally close-packed type structure. The metastable β-tungsten polytype has a cubic A15 lattice (space group *O3h-Pm3n n°223* structure type Cr_3Si). Two atoms are in position 2a and 6c, respectively (lattice parameter 5.05 Å). The first coordination polyhedron corresponds to an icosahedron, the second to a polyhedron formed by triangular faces with a fivefold symmetry plus two hexagons. We just remark that these polyhedra are the dual forms of the dodecahedron Si_{20} and Si_{24} as reported in clathrate I. Such phase has been extensively studied due to the superconductivity at low temperature ($T_c = 3$ K) while superconductivity in common bcc α-tungsten is only 0.01 K.

16.2.4.8 Tetrahedra Stacking

Clathrates can be seen as pure sp^3 lattices, which differ from the diamond lattice by the stacking mode of its tetrahedral units. In the clathrate, the tetrahedra are mainly stacked in eclipsed mode while diamond is formed by stacking in the staggered mode (Cros et al. 1965, O'Keeffe 2006). We recall here that combining eclipsed and staggered modes together leads to the hexagonal "diamond" structure. Figure 16.9 shows the natural appearance of the pentagonal rings within the eclipsed mode. We underline that neither topological arguments nor polyhedra formation are required at this stage of our discussion.

16.2.5 Elementary Cages: A Piece of Molecular Physics

Before reviewing the physical properties of the clathrate lattices, it is worthwhile to define the properties of the individual cages that constitute the elemental building blocks of these structures. Both type I and type II clathrates have a cubic symmetry while containing an important proportion of Si_{20} cages with an icosahedral symmetry. Cubic and icosahedral symmetry point groups are not compatible, and this will give rise to some

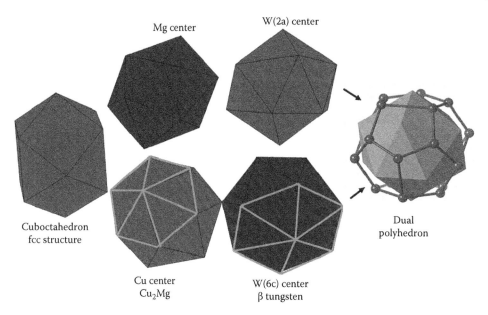

FIGURE 16.8 First coordination polyhedra in Cu_2Mg Laves and β tungsten phase. The hexagonal faces are underlined. The central atom is also mentioned. The dual transformation of an icosahedron leading to a dodecahedron is displayed in the right panel. The first polyhedron in the fcc structure is also displayed (cuboctahedron).

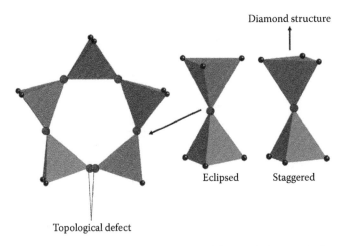

FIGURE 16.9 Formation of a pentagonal face with five tetrahedra in eclipsed mode. (From O'Keeffe, M., *Mater. Res. Bull.*, 41, 911, 2006. With permission.)

features, which characterize the clathrate phases. In particular, the symmetry breakdown involves a confinement effect, in other words, the whole clathrate lattice keeps the memory of the elementary cages that constitute the network. We briefly discuss the expected properties of these individual cages. The correlation between the cage itself and the lattice was already underlined by Saito (Saito and Oshini 1986).

16.2.5.1 Si$_{20}$

Si$_{20}$ is not a stable molecule, the ground state for such number of Si atoms corresponding to two Si$_{10}$ clusters (Sun et al. 2002). Free Si$_{20}$ is one of the regular Archimedean polyhedron dual of the icosahedron. Since the icosahedron shape is very close to a sphere we can use the spherical harmonics approach to describe its electronic structure. The first symmetries of the spherical harmonics in the icosahedral group are the following (Fowler and Woolrich 1986):

$$\Gamma_{20} = A_g + T_{1u} + H_g + (T_{2u} + G_u) + \cdots \qquad (16.10)$$

Since A_g, T_{1u}, H_g and $T_{2u} + G_u$ correspond to s, p, d, and f shells, the Hückel scheme gives rise to a HOMO (highest occupied molecular orbital) in the neutral Si$_{20}$ cluster, which has an open shell, while the closed-shell structure occurs for Si$_{20}^{2+}$. In this way, Si$_{20}^{2+}$ has a fivefold degenerate HOMO and a fourfold degenerate LUMO (lowest unoccupied molecular orbital). Furthermore, degeneracy favors the Jahn–Teller distortion. This has been pointed out by Adams et al. and Parasuk et al. (Parasuk and Almöf 1991) for carbon and by Saito et al. for silicon (Saito and Oshini 1986). One can ensure the stability of the cage by hydrogen-terminated bonds. This molecule, Si$_{20}$H$_{20}$, is stable in the LDA-DFT framework. Relaxation using conjugated gradient indicates a very small distortion with respect to the I_h symmetry. The HOMO and LUMO in the Si$_{20}$H$_{20}$ cluster are fivefold (H_u) and singlet (A_g) degenerate. Connetable (2003) found a HOMO–LUMO separation to be around 2.9 eV, which is higher than in the clathrate phases as we

will see. This gap opening with respect to the clathrate structure could be attributed to the confinement effect.

16.2.5.2 Si$_{24}$

The Si$_{24}$ cage has a low point group symmetry D_{6d}. The HOMO (B_1)–LUMO (A_1) separation in Si$_{24}$H$_{24}$ (D_{6d} is calculated to be 2.8 eV (Connetable 2003), a value close to the one in other cages Si$_{20}$H$_{20}$ and Si$_{28}$H$_{28}$).

16.2.5.3 Si$_{28}$

The Si$_{28}$ cage has a threefold degenerated HOMO (T_u) with a small separation from a singlet A_g state (LUMO). This molecule has four electrons shared between p-like state and a s-like one. In analogy with the free silicon atom, hybrid orbitals can be built up from linear combinations of states. In other words, Si$_{28}$ can accommodate four additional electrons to completely fill the states and leave a large gap between those states and the new LUMO (LUMO + 1 in free Si$_{28}$). On the other side, Si$_{28}$H$_{28}$ (T_d symmetry) has a filled HOMO level (A_1) with a HOMO–LUMO (A_1 symmetry) separation of about 2.9 eV. The geometry optimization gives rise to the Si coordinates observed in the clathrate. This provides a proof for the correlation existing between the Si$_{28}$H$_{28}$ cage and the clathrate lattice, analogous to the observed correlation between SiH$_4$ silane and the diamond structure.

16.2.6 Electronic Structure: The Case of Si-34

16.2.6.1 Cohesive Energy

No thermodynamic data about sublimation energy are reported for the clathrates. The cohesive energy is estimated within the local density approximation (LDA) (Kohn and Sham 1965) of the density functional theory (DFT). The difference in energy between Si-34 and Si-2 is of 0.06 eV per bond. The cohesive energy is just above the hexagonal lattice (wurtzite), which itself lies above the diamond structure. Clathrate is nevertheless remarkably stable as compared to the β-tin structure, which lies at 0.17 eV above the ground state. Data about cohesive energies are summarized in Table 16.3. The slight difference between the clathrate and the diamond structure is due to the balance between the tetrahedral stacking mode (eclipsed versus staggered), frustration (odd membered rings), and lattice distortions. We examine now all these features separately.

TABLE 16.3 Cohesive Energy of Diamond, Hexagonal Diamond, Si-34, and β-Tin Structures

	Si-2	2H-Si	Si-34	β-Tin
E_{coh} (eV)[a]	5.412 (4.63)	—	5.357 (4.58)	—
E_{coh} (eV)[a]	5.952 (4.63)	—	5.875 (4.57)	—
E_{coh} (eV)[b]	–(4.63)	—	(4.56)	—
E_{coh} (eV)[c]	4.67	4.654	—	4.4
E_{coh} (eV)[d]	4.63	—	—	—

[a] From Dong et al. (1999).
[b] From San-Miguel et al. (1999).
[c] From Yin and Cohen (1980).
[d] From Kittel (2004).

16.2.6.2 Stacking Mode

Wurtzite differs from diamond only in the stacking mode: staggered in diamond and staggered-eclipsed mode in wurtzite. At the ground state level, the difference of cohesive energy (0.01 eV) corresponds to half the cost of energy between a staggered mode and an eclipsed mode. In a tight-binding scheme, the difference arises from the balance between the more negative covalent bond energy and a more positive on-site energy in the wurtzite (Sutton 1994). Since half of the tetrahedra are in eclipsed mode, the energy cost per eclipsed mode will be about 2×0.01 eV $= 0.02$ eV. This difference is low as compared to the clathrate energy cost (0.06 eV).

16.2.6.3 Frustration

The concept of "frustration" has been treated by many authors and from different points of view. This term was first introduced by Toulouse (1989) in the framework of antiferromagnetic systems where exchange interactions between nearest neighbors cannot be simultaneously satisfied for an odd number of atoms. This has also been recognized in Na$_3$ trimer and identified due to a Jahn–Teller effect (Broyer et al. 1986, Delacrétaz et al. 1986). Briefly, the frustration appears as long as the number of atoms into a ring has an odd parity. Let us remember that clathrate is mainly built up of fivefold rings while Si diamond is built up of sixfold rings. Since the atom–atom (or electron–electron) pairing is not possible, the system forms an incomplete basis set involving the breakdown of the symmetry in the energy diagram. The breakdown of the so-called "bonding/antibonding" states has been pointed out by Weaire and Thorpe (Weaire and Thorpe 1971) and Friedel (Friedel and Lannoo 1973). It is nevertheless possible in odd cycles to form complete bonding s-like or p-like state sets while antibonding s-like or p-like state basis remain incomplete. In other words, antibonding states contain one bonding node in odd-membered rings. In fact, this effect might favor cohesive energy. This is well illustrated while looking for the allowed energies in the Hückel approximation for a ring formed by monovalent atoms:

$$E = \alpha + 2\beta \cos\left(\frac{m\pi}{N}\right) \quad (16.11)$$

where
N is the number of atoms ($N > 3$) in the ring
$m = 1, 2, ..., N$
α and β are the classical on-site and hopping integrals

For even values, the antibonding state is the mirror of the bonding state (i.e., the same absolute energy) while odd values give a dissymmetry between bonding and antibonding states: the deeper bonding states lie at higher energy than the higher antibonding state. In this way, we expect a lack in cohesive energy.

16.2.6.4 Distortion

We can characterize the distorted tetrahedron in clathrate by three terms: the mean bond distance, the mean dihedral angle, and the spread around the mean value of these two quantities (see Table 16.4). Amorphous silicon (a-Si) presents a large spread of

TABLE 16.4 Mean Distance and Dihedral Angle and Spread

—	Si-2	Si-34	Si-46	a-Si[a]
d_{Si-Si}	2.35	2.38	2.38	2.35
σ_{Si-Si}	0	0.022	0.024	0.074
$\theta_{\pi\sigma}$	109.47	109.4	109.9	108.4
$\sigma_{\theta\pi\sigma}$	0	2.8	2.5	10.8

[a] From Fortner and Lanin (1989).

geometrical parameters compared to the clathrate. The difference between cohesive energy between a-Si and the crystal one (Si-2) is estimated to be 0.18 eV/at (Roorda et al. 1989). This value is higher than those found in clathrates. However, this energy takes also into account the mean reduction of the coordination number owing to the dangling bonds. The calculation of the energy cost gained by a distortion is rather straightforward. We can still estimate its amplitude within the hybrid covalent energy V_h (Harrison 1980).

$$\frac{\delta V_h}{V_h} = \frac{-2\Delta d}{d} = -2\lambda\Delta\theta^2 \quad (16.12)$$

where
Δd is the mean interatomic distance deviation
$\Delta\theta$ is the spread of the dihedral angle
λ a defined parameter (Harrison 1980)

In a crude model, the cohesive energy is roughly proportional to the hybrid covalent energy and its raise for the clathrate structure with respect to diamond is found to be $\Delta E = 0.05$ eV. This value is very close to the previously referred DFT calculations (0.06 eV). Consequently, the distortion brings the main contribution to the corresponding enthalpy difference $\Delta H_{(clathrate-diamond)}$. For a-Si this hand-built model gives for the clathrate $\Delta E = 0.4$ eV and overestimates by a factor 2 the enthalpy difference $\Delta H_{(clathrate-diamond)} = 0.2$ eV. This shift can be explained since in this discussion we consider the spread around the mean value. We also question the role of the sixfold defects in the clathrate since the hexagonal rings have a planar structure close to the benzenic aromatic rings. This configuration is expected to be strongly unfavorable for silicon introducing a considerable energy strain. Using DFT-LDA framework, we calculate this energy by comparing the total energy of Si$_6$H$_{12}$ having the clathrate configuration (aromatic ring) with the diamond phase (armchair). The total energy difference between both structures is about 1 eV in favor of the armchair structure. We note that complete relaxation of the distorted planar structure using a conjugated gradient scheme gives rise to the continuous transformation from the aromatic ring toward the armchair structure without energy barrier. The strain energy gained by atom in the sixfold ring is about 0.17 eV (Table 16.5). The contribution for the clathrate lattice is about 0.02 eV/atom. Nevertheless, the strain energy gained in the clathrate plays an important role in the clathrate stability and ensures a decrease of the stability compared to the diamond phase. Contrary to the amorphous silicon where the strain energy is distributed among the atoms, the

TABLE 16.5 Total DFT Energy Calculated for Three Isomers Si_6H_{12} (C_6H_{12} respectively)

	Si_6H_{12}[a]	Si_6H_{12}[b]	Si_6H_{12}[c]	C_6H_{12}[a]	C_6H_{12}[b]	C_6H_{12}[c]
Total energy in eV	−828.32	−828.98	−829.32	−1117.02	−1117.59	−1118.75

[a] Planar structure as observed in clathrate.
[b] Planar structure after geometry optimization.
[c] Relaxed armchair structure corresponding to the expected ground state.

strain energy in the clathrate has two contributions the first one (0.05 eV/atom) corresponds to the mean strain energy distributed among the atoms and the second one (0.02 eV/atom) finds its origin in the localization of the strain energy around a sixfold ring (0.17 eV/atom in the cycle). These arguments provide a plausible origin for the clathrate phase transition toward the diamond phase at a temperature lower than the one observed in the a-Si. This excess of strain energy localized in the planar hexagons could act as embryo for the crystallization in the diamond phase.

16.2.6.5 e-DOS in Si-34 Lattice

The theoretical electron density of states (e-DOS) (Mélinon et al. 1998) is displayed Figure 16.10 in comparison with the Si diamond phase. The present calculations are first performed within

the LDA to the DFT. Since LDA underestimated the band gap, we perform further quasiparticle energy calculation providing a good description of the measured band gap within 0.2 eV.

The band gap in Si-34 is much larger (1.9 eV) than the one in Si-2 (1.2 eV). This last value is in good agreement with the experimental data (Gryko et al. 2000). The angular momentum decomposition of the DOS is obtained by projecting the wave functions on a spherical harmonic basis centered on each Si atom. In that way, the e-DOS can be decomposed in three regions: s-like p-like and sp-like. The mean value of the p-band is almost the same in both phases (Si-34 and Si-2). Nevertheless, the p-band presents a strong steepening just below the top of the valence band in the clathrate. Moreover, the bandwidth of the p-like DOS decreases by 2 eV from the Si-2 to Si-34 phase. The

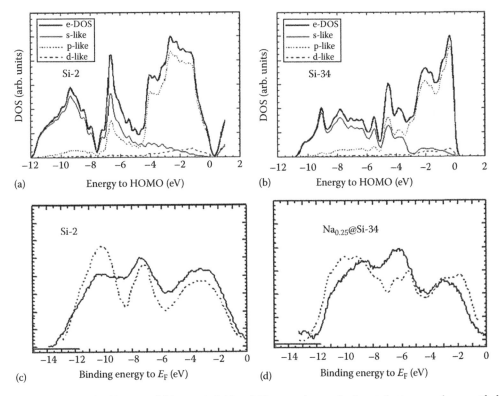

FIGURE 16.10 Theoretical e-DOS for (a) Si-2 and (b) Si-34 (solid lines). The s- and p-resolved contributions are shown with dotted and dashed lines, respectively. The angular momentum decomposition of the e-DOS is obtained by projecting the wave functions on a spherical harmonics basis centered on each Si atom. The e-DOS have been aligned at the valence band top edge. The calculations are performed within the local density approximation to the DFT. The kinetics cut-off energy was fixed at 20 Ry. A 0.1 eV Gaussian broadening has been used. The bandgap was calculated within the quasiparticle energy calculation. The Fermi level was located at 0.6 and 0.95 eV above the VB top edge for intrinsic Si-2 and Si-34, respectively. (c) and (d) valence band for Si-2 and $Na_{0.25}$@Si-34 (solid lines), respectively. The dotted lines are calculated from the e-DOS taking into account the effect of the cross-section modulation for 3s and 3p (XPS data Al K_α *x-rays*) and an additional broadening for a best fit with experimental spectra. (From Mélinon, P. et al., *Phys. Rev. B*, 58, 12590, 1998. With permission.)

s-band is more affected and is significantly shifted toward low energy in Si-34. This effect has been reported by Saito in Si-46 phase (Saito and Oshiyama 1995). Finally, the total width of the valence band (11.1 eV) is significantly narrower than that of the Si-2 lattice (12.17 eV). These features are attributed to the already discussed frustration phenomenon. p-band is less affected than s-band. The reduction of the e-DOS width is related to a less efficient mixing between s- and p-like. The frustration involves an incomplete antibonding state giving rise to a strong molecular character in the clathrate. Finally, we report the experimental valence band (VB) DOS for both phases. Figure 16.10 displays the VB-DOS for Si-2 (Figure 16.10a) and Si-34 (Figure 16.10b), respectively. VB-DOS are obtained by x-ray spectroscopy near the Fermi level. The experimental spectra are corrected for the inelastic scattering tail and reflect the true e-DOS after modulating by 3s and 3p cross-sections. For a better comparison, we have simulated theoretical VB spectra by broadening the e-DOS (Figure 16.10) after the modulation of the 3s and 3p yielding. The agreement is rather good and corroborates most of the predictions cited in the previous section.

16.2.6.6 p-DOS in Si-34 Lattice

Total vibrational density of states (p-DOS) has been studied both experimentally and theoretically. Figure 16.11 displays the phonon density of states (p-DOS) obtained from neutron scattering experiments (Mélinon et al. 1999). Both spectra for Si-2 and Si-34 present similar features with three regions attributed to acoustic modes (AM), acoustic-optical modes OM + AM, and optical modes OM. The redshift of the highest mode frequency at **k** = 0 is correlated to the features observed in the whole spectrum. Optical modes in Si-34 are shifted toward low energy while acoustic band is shifted toward higher energies. The band narrowing is quite similar to the one reported for the e-DOS. This underlines the correlation between electronic and vibrational features for covalent bonded systems. The softening of the optical modes can be easily understood in terms of the electronic structure. Two effects could be invoked: the bond and angle spreading and frustration. To examine the first one, let us consider the case of amorphous silicon. Experimental and theoretical vibrational densities of states for amorphous silicon (Kamitakahara et al. 1987) show that the vibrational density of states is quite insensitive to their

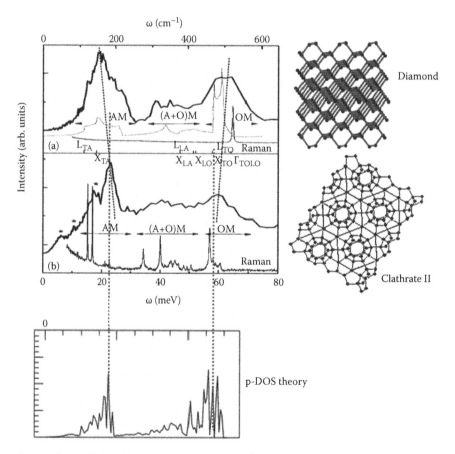

FIGURE 16.11 Phonon density of states deduced from neutron scattering and Raman scattering spectra of (a) silicon diamond and (b) Na$_i$@Si-34 clathrate phase. (*) on curve (b) mark the peaks corresponding to the sodium p-DOS residual contribution. Note the decoupling between sodium states and silicon p-DOS. Calculated phonon frequencies of selected acoustic (AM) and optical (OM) modes as well as frequencies calculated at the high-symmetry points G, X, and L by Giannozzi et al. (1995) in the Si-diamond phase are indicated. In addition, the Si-2 diamond p-DOS calculated by these authors is also reported in dashed line (a). Bottom part: theoretical value of the *p-DOS* (Connétable 2003) in the Si-34 clathrate. (From Mélinon, P. et al., *Phys. Rev. B*, 59, 10099, 1999. With permission.)

fine details of the local structure. In particular, no extra vibrational modes are reported. We have then to conclude that the distortion in distances and angles cannot entirely explain the redshift. It remains then the frustration effect that avoids a complete set of optical modes within fivefold rings. This leads to a softening of the stretching modes in the clathrate.

The "hardening" of the pseudoacoustic band (the modes are blueshifted) is not correlated to the frustration but rather to the topology. Each atom is located at the node of four polyhedra. The bond bending mode will be difficult to excite since it needs the deformation of the four polyhedra at the same time. Consequently, the bond-bending force constants α (Alben et al. 1975) will be greater than in diamond.

16.2.6.7 Si-34 Lattice under Pressure

It is well known that Si-2 phase is the most stable at standard thermodynamic conditions and undergoes a phase transition toward the β-tin structure at 11.5 GPa (Yin and Cohen 1980). This transformation is well recognized by crystallographic arguments. Diamond structure can be described in the well-known cubic system $Fd\bar{3}m$ or in the tetragonal system (I_{41}/amd). In this representation, we have the following structural parameters (Libotte and Gaspard 2000) $a = b$ and $c/a = \sqrt{2}$. In this way, the β-tin structure is just obtained by varying the c/a ratio. This transformation is first order with an important atomic volume reduction. The Si-34 structure is metastable and undergoes a phase transition toward the diamond structure at normal pressure and high temperature and changes into β-tin structure at room temperature and high pressure. The most fascinating result is the transition pressure. San-Miguel et al. (1999) reported that both the Si-2 and the Si-34 undergo the phase transformation at the same pressure (11 ± 0.5 GPa) towards the β-tin structure. Similar features have been reported by Ramachandran et al. (2000). Thirty years ago, Bundy and Kasper (1970) also reported that clathrate collapsed to metallic state of relative low resistivity at pressures over 10 GPa. The obtained bulk modulus of Si-34 was only $8\% \pm 5\%$ smaller than the one of Si-2, in good agreement with our cohesive energy arguments.

16.3 Clathrates and Doping

The nanocages of the clathrate structure offer the possibility of endohedral intercalation with charge transfer, i.e., nonsubstitutional doping. A_x@Si-34 compounds were first prepared and extensively studied by Hagenmuller and coworkers (Cros et al. 1965, 1970). Cros reported Cs_8@Si-34 compound with a lattice parameter (14.64 Å) close to Na_8@Si-34 (14.62 Å). Ramachandran et al. (2000) and Bobev and Sedov (2000) studied Na_4Cs_2@Si-34 and Rb_2Na_4@Si-34, respectively. Many clathrate doping schemes have been developed in the last few years.

16.3.1 Outline of the Doping Problem

Silicon in its diamond structure can be doped by substitution. For example, phosphorus has five valence electrons and when it sits at a substitutional site in the silicon lattice, only four of these

electrons make up the bond with the neighboring Si atoms, the latter being "free." Clathrates can also be doped by substitution like the diamond phase. However, the cage-like structure allows the incorporation of a guest atom X inside the cage. This new doping called endohedral renews some interest since it offers unusual features that question the physical meaning of the so-called word "doping" face to intercalation schemes. The specific features emerging from clathrate doping can be classified as follows:

1. A large flexibility in the nature and the strength of the coupling between the guest atom and the host cage following the valence and the size of the guest atom.
2. The doping factor: in diamond, large doping values lead to mechanical stress due to the local modification of the lattice parameter. Additionally, the dopant bonding can also create misfits. In fact, the relevant limit for technical applications comes from the so-called "pinning defects" that trap free carriers. These limitations are filtered out in clathrates. In fact, endohedral doping allows for a weak interaction of the dopant with the host lattice itself and a high density of dopants reaching unusual values up to 10% can be obtained. These values largely overpass the greatest doping free carriers density reported in classical semiconductors.
3. The coupling between the electronic levels of the guest atom and the molecular levels of the cage depends on their own local symmetry, i.e., with the cage considered at a molecular level.
4. Local distortions. We can expect a Jahn–Teller effect to lift degeneracy. Also the dimerization due to the covalent bonding between two guest atoms belonging to two abutting cages has been evoked. Both effects break the symmetry of the lattice by the displacement of one sublattice with respect to the other.

16.3.2 Doping in Isolated Cages

16.3.2.1 Endohedrally Doped Si_{20}

Using first principles calculations, Sun et al. (2002) reported the stability of endohedrally doped Si_{20} cage even though the HOMO state remains partially filled (no bandgap). These authors found that divalent metals (Ba, Sr, Ca) transfer two electrons while tetravalent Zr atoms transfer four electrons. In this case the HOMO state (T_g) is completely filled and opens a gap. Connétable et al. (2001) studied the incorporation of xenon, iodine, and tin in $(SiH)_{20}$ within the DFT framework. A gap opening is observed with a strong modification of the unoccupied orbitals near the LUMO level while the HOMO region remains weakly affected. This picture is amazing when we compare it with the classical "hybridization" scheme. 5p orbitals in iodine and xenon lie at lower energy than HOMO. The hybridization should pull up this level at higher energy beyond HOMO, leading to a gap reduction. The observed effect is just the opposite. The gap opening is explained by symmetry. In fact, in the icosahedral group, silicon s- and p-orbitals span the irreducible representations (IR) as follows:

$$IR[Si_{3s}] = A_g + G_g + H_g + T_{1u} + T_{2u} + G_u \qquad (16.13)$$

$$IR[Si_{3p}] = A_g + T_{1g} + T_{2g} + 2G_g + 3H_g + 2T_u + 2T_{1u} + 2G_u + 2H_u \qquad (16.14)$$

$$IR[Si_{3d}] = 2A_g + 2T_{1g} + 2T_{2g} + 4G_g + 4H_g + A_u + 3T_{1u}$$
$$+ 3T_{2u} + 4G_u + 3H_u \qquad (16.15)$$

The common symmetries between the two representations are

$$A_g + T_{1u} \qquad (16.16)$$

Such symmetries are the same than those observed in $(SiH)_{20}$ LUMO (A_g) and LUMO+1 (T_{1u}) while HOMO and HOMO-1 have H_u and G_u symmetries, respectively. This result explains the DFT conclusions, since whatever the nature of the dopant, the coupling is only possible near the LUMO region. As a consequence, the occupied states are weakly affected and the hybridization between the guest atom and the cage pulls up the LUMO at higher energy explaining the gap opening. Another aspect of endohedral doping is the stability of the guest atom in the geometrically centered position. Using first principles calculations Tournus et al. (2004) studied the position of Na atom inside the $(SiH)_{20}$. The minima just coincides with the center of the cage keeping the whole symmetry of the $(SiH)_{20}$.

16.3.2.2 Endohedrally Doped Si$_{28}$

As Si_{20} molecule, Si_{28} needs four electrons to fill the HOMO state. Gong et al. (Gong and Zheng 1995) studied the incorporation of tetravalent atoms (C, Si, Ge, Ti, Zr) without success owing to a poor charge transfer between the guest atom and the Si_{28} cage (see Figure 16.12a). Tsumuraya et al. (1995) reported a severe reconstruction of Ba@Si$_{28}$, the barium atom remaining inside the cage. Besides, Jackson et al. (Jackson 1993) reported successful incorporation of these elements in the C$_{28}$ cage. Bonding between Si_{28} and the endohedral atom is ionic like with a small amount of electronic charge transferred. This is due to the large cage radius compared to Si_{20} or C$_{28}$ in agreement with the low elastic coupling factor, K, calculated in the harmonic oscillator framework (see Figure 16.11, Section 16.2.6.6). This corroborates the description of vibrations of guest atoms in the clathrate cages as Einstein rattlers. Increasing the guest–host coupling needs a displacement of the guest atom from the center as observed in the EXAFS spectra (see the section below). This displacement is also predicted for C$_{28}$ cage. It was found than an off-center relaxation is observed for the Ti@C$_{28}$ cage while Zr was found to be stable at the cage center (Jackson et al. 1993). The displacement

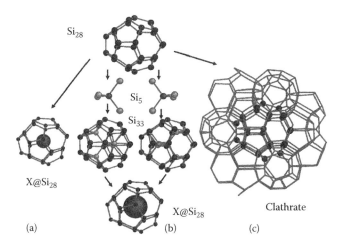

FIGURE 16.12 (a) HOMO–LUMO separation through encaging tetravalent atom. Because of the large Si_{28} size, the electron transfer is not enough to achieve a complete HOMO–LUMO separation. (b) HOMO–LUMO separation through encaging Si_5 molecule (left $Si_{33}b$ right $Si_{33}a$). This is equivalent to a giant super tetravalent atom (bottom side). The cluster corresponds to the magic number Si_{33}. (c) HOMO–LUMO separation through the sharing of common faces (type II clathrate).

is also reported in sodium inside $Si_{28}H_{28}$ where the potential is found to be very flat (Demkov et al. 1994, Tournus et al. 2004). An alternative route consists of encaging a tetravalent superatom using Si_5 T_d molecule (see Figure 16.12b). This molecule has four valence electrons in the twofold HOMO with a large equivalent radius (0.344 nm). The stuffed cage corresponds to the net formula Si_{33} with a filled HOMO gap that ensures a low reactivity as pointed out by Elkind et al. (1987). Since the main parameter is the transfer of four electron charge, several positions of the Si_5 molecule inside the cage are still possible, leading to numerous isomers with equivalent cohesive energies. Numerous isomeric forms have been already reported: Kaxiras (1990), Patterson et al. (Patterson and Messmer 1990), Rothlisberger et al. (1994), Ramakrishna et al. (1994), Mukherjee et al. (1996), and Mélinon et al. (1997). In this way, the symmetry of the Si_5 molecule does not play a significant role, and can be considered as a supertetravalent atom (Mélinon and Masenelli 2008). One of them is derived from the clathrate lattice ($Si_{33}a$), another to the surface reconstruction scheme ($Si_{33}b$). The first one ($Si_{33}a$) has four Si atoms fivefold-coordinated while the second one ($Si_{33}b$) has four atoms sevenfold coordinated. These four atoms belong to the second-neighbor shells. The super atom can be also replaced by four hydrogen atoms ($Si_{28}H_4$). However, HOMO–LUMO separation is weak enough compared to the template molecule SiH_4. Finally, incorporating endohedral atoms (iodine, xenon) induces a gap opening (0.5 eV) lower than those observed in the $(SiH)_{20}$ molecule. This is of prime importance that encaged atoms contribute to the HOMO–LUMO separation in isolated cages while encaged atom in the clathrate cages provides an additional net hybridization and/or an electron transfer toward the host lattice.

16.3.3 Metal Insulator Transition

In a crude model, the potential energy V_{pot} for the "free" electron is given by the hydrogenoid model (Sutton 1994):

$$V_{pot} = \frac{e^2 m^*}{4\pi\epsilon_0 \epsilon^2 m_e a_0} \qquad (16.17)$$

where

    m^* is the effective mass
    ϵ is the dielectric constant (ϵ =12 in Si-2)
    a_0 the Bohr radius (0.529)

For silicon with phosphorus, this equation holds (V_{pot} = 0.045 eV). At low temperature, electrons are localized just below the bottom of the conduction band, the system being an insulator. Nevertheless, if we increase the donor concentration, the hydrogen orbit of neighboring atoms can overlap leading to the well-known insulator–metal transition. This Mott transition occurs for a critical donor concentration n_c:

$$n_c^{1/3} a_0 m_c \epsilon > 0.24 m^* \qquad (16.18)$$

For silicon-doped with phosphorus, a value of $n_c = 3.8 \times 10^{18}$ cm^{-3} is obtained. This concentration is very low compared to that of the silicon atoms. Now, let us consider the clathrate structure. Clathrates are not usually doped by substitution. Since the cage-like structure allows the incorporation of a guest atom, the doping is endohedral-like. The question is the following: applying the same crude approach, what is the critical donor concentration? For simplicity, we consider the case of a sodium donor. Using the random phase approximation in RPA-GW formalism, we find the macroscopic dielectric constant of Si-34 (including nonlocal effects) to be $\epsilon_M = 9.42$.* This is ~22% smaller than the dielectric constant value 12.2 of Si-2 as calculated using the same formalism. Moreover, as mentioned by Mott (1973), the 3s electron of the sodium atom is attracted by the force $e^2/\epsilon_{eff} r^2$ to the surface of the polyhedron (the cage being empty), ϵ_{eff} is a parameter ranging between 1 and ϵ_M. Moreover, the mean effective masses for one electron averaged over the different crystallographic directions are $m_e^* = 0.73^*$ and $m_e^* = 0.26$ (Sze 1981) for clathrate and silicon diamond, respectively. Thus the critical donor concentration is crudely approximated by

$$n_{c(clat)} \approx n_{c(dia)} \left(\frac{\epsilon}{\epsilon_{eff}}\right)^3 \left(\frac{m^*_{(clat)}}{m^*_{(dia)}}\right)^3 \qquad (16.19)$$

$n_{c(clat)} \sim 1.4 \times 10^{21}$ atoms/cm^3 with $\epsilon_{eff} \sim 1/2\epsilon_M$. This concentration is not so far from the number of sodium atoms $n = 1.9 \times 10^{21}$ atoms/cm^3 in Na$_2$@Si-34 when the system becomes a conductor.

Increasing the alkaline atom size (from sodium to cesium) pulls up the effective dielectric constant and then reduces the number of alkaline atoms per cell for the insulator/metal transition. This is the most amazing result in such a lattice: the metallization in clathrate needs a high concentration of guest atoms and the concept of impurity is ruled out. The host lattice is not perturbed by the guest atoms since they are located inside the cages. In addition, we can incorporate different atoms with different valence states. The spread in number of valence electrons and the different extension of the outer orbitals following the guest atom are new parameters and open a wide variety of doping possibilities with unusual properties.

16.3.4 Hard Materials

San-Miguel et al. (2002) reported the pressure stability and compressibility of endohedrally doped silicon clathrates M$_8$@Si-46 with M = Na, Ba, and I. Combining experimental results and ab initio calculations, they found that endohedral doping of silicon clathrates does not alter significantly the compressibility of the empty clathrate or even can decrease it to values approaching that of the diamond Si-2 phase. Further, they showed that selective intercalation can prevent the collapse of the clathrate structure up to very high pressures. Table 16.6 reports some selected bulk modulus estimated in various compounds.

A direct correlation between a high bulk modulus value and hardness is not straightforward. Nevertheless in the particular case of strong covalent systems, many studies converge to a strong correlation between hardness and bulk modulus or even better with the shear modulus.

16.3.5 X$_8$@Si-46 under Pressure

The experimental (P, V) data points can be well fitted with the Murnaghan equation of state giving bulk modulus values that are summarized in Table 16.6 (San-Miguel and Toulemonde 2005). A Murnaghan fit of the energy versus volume data points

TABLE 16.6 Bulk Modulus B_0 in GPa of Silicon Clathrates Compared to the One of the Diamond Structure

Compound	Experiment	Theory	Difference from Si-2(%)
Xe$_8$@Si-46	Not synthesized	85[a]	12
Si-46	Not synthesized	87[a]	10
Si-34	90 ± 5[b]	87.5[b]	9.5
Ba$_8$@Si-46	93 ± 5[a]	—	—
I$_8$@Si-46	95 ± 5[a]	91[a]	6
Te$_8$@Si-46	Not synthesized	95[a]	2
Sn$_8$@Si-46	Not synthesized	97[a]	0
Si-2 (diamond)	97.88[c]	97[b]	0

[a] From San-Miguel et al. (2002).
[b] From San-Miguel et al. (1999).
[c] From Hellwege and Hellwege (1966).

---

* This value is calculated for Si-46 lattice (Saito and Oshiyama 1995). This is consistent with calculations performed on Si-34 lattice by X. Blase et al., private communication.

for the iodine clathrate yields $B_0 = 91$ GPa, which is 4% larger than the one of the undoped Si-46 clathrate, despite the 2.4% volume increase of the unit cell under iodine intercalation. The large iodine binding energy and the increase of B_0 point out again to a mixing of the iodine and silicon orbitals to form iono-covalent bonds.

The main instability observed in intercalated silicon clathrates is the observation of an isostructural homothetic volume collapse at a pressure that depends on the nature of the doping atom (San-Miguel et al. 2005). For the type-I Ba intercalated clathrate for instance, we observe a strong but progressive modification in the slope of the volume reduction with pressure from 13 to 18 GPa, but the cubic structure is preserved at least up to 30 GPa. Only Na_8@Si-46, as Si-34, seems to avoid this scheme transforming directly into the silicon stable phase at the corresponding pressure. The size of the atom appears to be dependent on the pressure stability of silicon clathrates. Recent works point out to an extensive rehybridization of silicon with a loss of covalency to explain the homothetic volume collapse (Tse et al. 2007).

In addition to the volume collapse transition, at least two other types of pressure-induced modifications have been observed. First of all, at pressures well above the volume collapse, an irreversible amorphization is observed (San-Miguel et al. 2002). Secondly at pressures preceding the volume collapse, different types of elastic or phonon anomalies have been described having few or nonobservable effects on the structure compressibility (Tse et al. 2002, Kume et al. 2003, Machon et al. 2009).

Doping has then a dramatic effect on the pressure stability of the clathrate structure. In particular, it allows in some cases to push the limit of cage-structure stability up to pressures at least three times higher than the stability limit of the empty silicon clathrate.

16.3.6 Superconductivity

Kawaji et al. (1995) reported the occurrence of superconductivity in barium compounds ($T_c = 8$ K in Ba_8@Si-46). In the conventional Bardeen–Cooper–Schrieffer (BCS) theory for phonon-mediated superconductivity, the critical temperature for superconductivity is given by

$$T_c = 1.13\Theta_D \exp\left(\frac{-1}{N(E_F)V}\right) \quad (16.20)$$

where

Θ_D is the Debye frequency

V is the average electron pairing interaction

This crudest form of BCS theory neglects the electron–electron Coulomb repulsion μ^*. A more versatile formulation developed by McMillan (1968) takes into account μ^*

$$T_c = \frac{\hbar\omega_{ph}}{1.2}\exp\left[\frac{-1.04(1+\lambda_{ph})}{\lambda_{ph} - \mu^*(1+0.62\lambda_{ph})}\right] \quad (16.21)$$

λ_{ph} is the electron–phonon coupling constant. It is worthwhile to compare the superconductivity in clathrates and those observed in A_3@C_{60} compounds. Nevertheless, some striking differences exist. In A_3@C_{60} compounds, molecular orbitals on neighboring molecular sites mix to form a band without overlapping. The Fermi level is located in the t_{1u}-derived band where the bandwidth does not exceed 0.5 eV. The phonon modes are well separated in A_3@C_{60} compounds and span a wide frequency. In sp^3-doped clathrates, the electronic band width is much greater than the expected intramolecular modes and then electron–phonon scattering occurs close to the Fermi level.

Partial substitution of Ba in Ba_8@Si-46 by Sr and Ca (Toulemonde et al. 2006) leads to a reduction of the critical superconductivity temperature, which scales perfectly with the evolution of the cage size. In fact, the more or less important injection of d-electrons in the conduction band through the guest–host hybridization appears to be the main factor ruling the value of T_c. Multigap superconductivity has been proposed for Ba_8Si-46 (Lortz et al. 2008).

16.3.7 Thermoelectric Properties

Thermoelectricity is related to the coupled thermal and electric behavior. The strength of a thermoelectric behavior is quantified by evaluating its figure of merit. In particular, one defines the conventional dimensionless factor

$$ZT = \left(\frac{S^2\sigma}{\kappa}\right)T$$

where

S is the Seebeck coefficient

σ the electrical conductivity

κ the thermal conductivity

Conventional gas-phase refrigeration devices have figures of merit much higher than 1, whereas the best known solid-state devices show values lower than 1. Basically, in simple metals, σ and κ are strongly correlated. Several authors (see Tse et al. (2000)) have suggested doped clathrates as possible candidates for thermoelectric applications. Doped clathrates display metallic behavior with possible large σ values. The existence of localized Einstein-like modes (see Figure 16.11) more or less coupled with the host lattice is a reducing factor for the thermal conductivity (Tse et al. 2000). The observation of the p-DOS spectra suggests that sodium compounds are the best candidates. The experimental thermal conductivity (Tse et al. 2000) κ in Na_8@Si-46 is found almost constant as a function of the temperature ($\kappa = 5 \pm 1.5$ Wm/K) and the behavior is close to an amorphous-like material. This thermal conductivity is much smaller than the one found in the diamond phase ($\kappa = 150$ Wm/K). This ensures the role of the localized modes and the potentiality of such compounds. Tse et al. (2000) have calculated the Seebeck coefficient for $Na_x@Si$-34 and $Na_8@Si$-46. For this last one, they found a positive Seebeck coefficient increasing with temperature ($S \sim 20\,\mu V/K$ at $T = 300\,K$). For another compound, S increases with the dopant concentration. A large negative value is obtained for $Na_8@Si$-34 but such a compound is a poor electrical conductor and consequently a bad candidate for thermoelectric applications.

An alternative to the use of smaller guest atoms is the use of larger cages and consequently most of the work on the development of new clathrate thermoelectric materials concentrated on Ge or Sn clathrate frameworks.

16.3.8 Optoelectronic Properties

Several authors mentioned the clathrate potentiality for applications in optoelectronic devices. First of all, the wide band gap opening (around 1.9 eV) (Adams et al. 1994, Mélinon et al. 1998, Gryko et al. 2000, Connétable et al. 2003) ensures electronic transition in the visible region. More recently, Connetable et al. (2003) mentioned a large gap using guest atoms belonging to columns IVA, VA, and VIA. The larger gap, however indirect, is calculated for the $Sn_8@Si$-46 compound in the green region. In $X_8@Si$-46, the gap is found to be indirect while in $X_4@Si$-34 it is found direct. Even in the case of direct gap, the matrix elements of the dipolar transitions are not allowed, leading to a main limitation. We can nevertheless expect that the observed flatness of the electronic bands could promote allowed transitions by disorder-induced breakdown symmetry or phonon-assisted transitions. Besides these fundamental limitations, the major barrier for the development of a clathrate-based optoelectronics is at present the production of clathrate thin films. In fact, the available samples are currently in small powder form. Munetoh et al. (2001) reported a molecular dynamics study of the epitaxial clathrate growth. They claimed that crystal growth processes of clathrates during solid phase epitaxy are energetically favorable.

16.3.9 Clathrates under Temperature

Clathrate I can be identified to the A15 structure (Cr_3Si type or β-tungsten, $Pm\bar{3}n$). In such lattice, the Frank–Kasper lines (the lines crossing the centre of the hexagons of the Si_{24} cages) never intersect. Clathrate II can be identified to the cubic Friauf–Laves structure of $MgCu_2$ ($Fd\bar{3}m$). In such a lattice, the lines intersect. Jund et al. (1997) studied the melting of both structures. The salient parameter was the connectivity of the disclination lines. As mentioned above, this connectivity is higher in Friauf–Lave structure than in A15 structure. The increase of the thermal disorder due to the heating contributes to the lowering of the connectivity. The connectivity remains in solid phase just beneath the first-order transition toward liquid phase (Jund et al. 1997). The critical transition is significantly lower in Friauf–Lave than in A15 structure. Since the empty clathrate I lattice exists, we cannot check this assumption. As a matter of fact, clathrates undergo a phase

transition toward diamond structure at below $T = 750$–780 K, a value much lower than the one commonly reported in a-Si films ($T \sim 950\,K$) (Roorda et al. 1989). We can postulate that clathrates become amorphous-like just before to crystallize in diamond phase. Since a-Si contains fivefold defects, such defects could act at the embryo of the fivefold/sixfold transition. a-Si is thermodynamically unstable in contact with Si-2 but does not crystallize owing to the kinetic Gibbs barrier. The low temperature of transition in clathrate suggests a large reduction of the Gibbs free energy difference between clathrate and Si-2 as compared to a-Si and Si-2. However, the phase transition observed in the clathrate ranges above the temperature where structural relaxation in a-Si is observed. This indicates that the thermal energy is high enough to induce structural relaxation in clathrate before phase transition. The origin of this effect will be discussed later.

16.4 Carbon Clathrate: The Holy Grail

The position of dodecahedrane ($C_{20}H_{20}$) [can be considered] as the structurally most complex, symmetric, and aesthetically appealing member of the C,H, convex polyhedra ($n = 20$)

This sentence is found in the introduction of the paper from Paquette and collaborators (Ternansky et al. 1982) who first synthesized the carbon dodecahedra achieved in 23 steps starting from the cyclopentadienide anion. In the same way, the carbon clathrate is the "holy grail" for the physicists since the expected properties are clearly unusual. We mention mechanical properties close to diamond. In fact, the ideal tensile strength in C-46 is larger by at least 25% than the ideal tensile strength of diamond in its <111> direction (Blase et al. 2004). Carbon clathrate compounds are also expected to be superconductors with a very high T_c (>100 K) (Connétable et al. 2003). This is also a wide band gap material (Gryko et al. 2000). Contrary to diamond, where n-type material has proved difficult to manufacture, n-doping in carbon clathrate could be achieved by incorporating donor atoms inside the cages (Bernasconi et al. 2000). Unfortunately, type-I or type-II carbon clathrate has not been yet synthesized. As observed in other elements of the column IVA, carbon clathrates are metastable but lie not so far from the stable graphite phase (Benedek et al. 1995). In particular they are expected to be more stable than fullerites. The question of its synthesis can then be raised. Generally, two ways are investigated: the direct chemical way with the decomposition of a precursor and the synthesis under high pressure and high temperature. Unfortunately, the precursors do not exist so long and owing to the particular phase diagram of carbon, the synthesis under high pressure and high temperature needs unusual values ($P > 10$–50 GPa and $T \gg 1000\,K$) with severe conditions for the experimental devices (Rey et al. 2008). Moreover, the clathrates are in competition with other complex carbon forms including onions, nanotubes, foams, and schwartzites (Spagnolatti et al. 2003) (see Section 16.2.4.4).

16.5 Other Clathrates

16.5.1 Column IVA

Isomorph clathrates lattices can be synthesized from all the elements belonging to column IVA. Germanium clathrates have been investigated (Dong et al. 2000, Guloy et al. 2006). Gallmeier et al. (1969) reported the synthesis of $K_8@Ge-46$ and $K_8@Sn-46$. Crystallographic data of such compounds are given by Bobev and Sevov (2000). Moriguchi et al. (2000) studied the energetics and electronic states in hypothetic Si–Ge clathrates. The band gap is predicted to range between 1.2 and 2.0 eV following the ratio Ge/Si. These authors suggested the application of these clathrates in optoelectronics since the band gap could be easily tuned.

16.5.2 Compounds

Crude arguments tell us that a large variety of clathrates can be synthesized taking into account the octet rule. Elements belonging to the column IIIA will be tetrahedrally coordinated if the guest atoms providing the lacking electrons. For example, clathrates with divalent guest atoms provide 16e. The octet rule will be checked for the formula $X_8@(A_{16/46}B_{30/46})-46$, X being a divalent atom, A and B belong to the columns IIIA and IVA, respectively. Likewise, the elements of the column VA provide one electron per atom. The octet rule will be checked taking an acceptor atom belonging to the column VII, for example. Possible clathrates with the following formula $X_8@(A_{8/46}B_{38/46})-46$ are expected to be stable (X, A, and B belonging to the columns VA, IVA and VII, respectively).

16.5.3 Compounds from Columns IVA and VA

Mixed clathrates including elements from columns IVA and VA together have been already reported. Eisenmann et al. (1986) reported $A_8B_{16}@C_{30}$ with A = Sr, Ba, B = Al, Ga and C = Si, Ge, Sn. Chu et al. (1982) prepared a germanium arsenide iodine $I_8@(As_{8/46}Ge_{38/46})-46$.

16.5.4 Compounds from Columns IVA and IIIA

Kuhl et al. (1995) prepared $Ba_8@(In_{16/46}Ge_{30/46})-46$. Chakoumakos et al. (2000, 2001), Nolas and Kendziora (2000) reported $Sr_8@(Ga_{16/46}Ge_{30/46})-46$ and $Eu_8@(Ga_{16/46}Ge_{30/46})-46$. More recently, Sales et al. (2001) studied magnetic, electrical, and thermal transport in $Eu_8@(Ga_{16/46}Ge_{30/46})-46$, $Sr_8@(Ga_{16/46}Ge_{30/46})-46$ and $Ba_8@(Ga_{16/46}Ge_{30/46})-46$. Thermoelectric measurements in $Sr_8@(Ga_{16/46}Ge_{30/46})-46$ were reported by Meng et al. (2001).

16.5.5 Compounds from Column IVA and Metals

Cordier and Woll (1991) reported mixed clathrates $Ba_8@Si^*-46$ with $Si^* \equiv (T_xSi_{1-x})_{6/46}Si_{40/46}$ and $Ba_8@Ge^*-46$, $Ge^* \equiv (T_xGe_{>1-x})_{6/46}Ge_{40/46}$ with T = Ni, Pd, Pt, Cu, Ag, and Au. Kuhl et al. (1995) reported $Ba_8@Ge^*-46$ with $Ge^* \equiv T_{8/46}Ge_{38/46}$, T = Zn, Cd.

16.5.6 Ionocovalent Compounds

The most popular ionocovalent clathrate is the one prepared from SiO_2 (clathrasil). Among them, the melanophlogite (clathrate I) is a natural mineral, which consists of polyhedral cages of SiO_2. This material is only 0.1 eV/molecule above α-quartz despite a large spread in Si–O–Si angles ((143.8°–180°) instead of 143.7° in α-quartz) (Demkov et al. 1995).

16.6 Conclusion

The properties of clathrate compounds arise from their particular topology and from their particular host–guest relationship. In fact, these unusual materials present amazing properties related to the coupling strength between their two interpenetrated lattices. This coupling can be discussed in terms of a more or less efficient charge transfer rather than in terms of chemical bonding. The study of the clathrate phase is a challenge from a theoretical point of view since the electron–electron correlations are large and poorly described by first-principles calculations. Clathrates constitute equally challenging materials from the point of view of applications as optical devices, thermoelectric materials, or superconducting devices. Clathrates constitute a bridge between fullerene science, zeolite world, and cluster physics. Finally, we should note that the most probable candidate for what could be the only known natural group 14 clathrate structure ever reported is a carbon clathrate formed in rocks by a meteoritic impact (El Goresy et al. 2003) and putting a bit of poetry on all this: aren't meteorite daughters from the sun as denoted by the etymological/mythical origin of the word "clathrate"?

References

Adams, G.B., Page, J.B., Sankey, O.F., and O'Keeffe, M. 1994. Polymerized C_{60} studied by first-principles molecular dynamics. *Phys. Rev. B* 50: 17471–17479.

Alben, R., Weaire, D., Smith, J.E., and Brodsky, M.H. 1975. Vibrational properties of amorphous Si and Ge. *Phys. Rev. B* 11: 2271–2296.

Ammar, A., Cros, C., Pouchard, M. et al. 2004. On the clathrate form of elemental silicon, Si_{136}: Preparation and characterisation of Na_xSi_{136} ($x \rightarrow 0$). *Solid State Sci.* 6(5): 393–400.

Andreini, A. 1905. Sulle reti di poliedri regolari e semiregolari. *Mem. Soc. Ital. delle Sci* (3)14: 75–129.

Ankudinov, A., Ravel, B., Rehr, J.J., and Conradson, S. 1998. Real space multiple scattering calculation and interpretation of x-ray absorption near edge structure. *Phys. Rev. B* 58: 7565–7578.

Beekman, M. and Nolas, G.S. 2008. Inorganic clathrate-II materials of group 14: Synthetic routes and physical properties. *J. Mater. Chem.* 18: 842–851.

Benedek, G. and Colombo, L. 1996. Hollow diamonds from fullerenes. *Mater. Sci. Forum* 232: 247–274.

Benedek, G., Galvani, E., Sanguinetti, S., and Serra, S. 1995. Hollow diamonds: Stability and elastic properties. *Chem. Phys. Lett.* 244(5–6): 339–344.

Bernasconi, M., Gaito, S., and Benedek, G. 2000. Clathrates as effective p-type and n-type tetrahedral carbon semiconductors. *Phys. Rev. B* 61: 12689–12692.

Blase, X., Gillet, P., San-Miguel, A., and Mélinon, P. 2004. Exceptional ideal strength of carbon clathrates. *Phys. Rev. Lett.* 92: 215505–215508.

Bobev, S. and Sevov, S.C. 2000. Clathrates of group 14 with alkali metals: An exploration. *J. Solid State Chem.* 153: 92–105.

Broyer, M., Delacrétaz, G., Labastie, P. et al. 1986. Spectroscopy of Na$_3$. *Z. Phys. D Atoms, Mol. and Clusters* 3(2): 131–136.

Brus, L. 1997. Chemistry: Metastable dense semiconductor phases. *Science* 276: 373–374.

Bundy, F.P. and Kasper, J.S. 1970. Electrical behaviour of sodium-silicon clathrates at very high pressures. *High Temp. High Press.* 2: 429–436.

Chakoumakos, B.C., Sales, B.C., Mandrus, D., and Nolas, G.S. 2000. Structural disorder and thermal conductivity of the semiconducting clathrate Sr$_8$Ga$_{16}$Ge$_{30}$. *J. Alloys Compd.* 296: 80–86.

Chakoumakos, B.C., Sales, B.C., and Mandrus, D.G. 2001. Structural disorder and magnetism of the semiconducting clathrate Eu$_8$Ga$_{16}$Ge$_{30}$. *J. Alloy. Compd.* 322(1–2): 127–134.

Chen, F. and Johnson, R. 2008. Martensitic transformations in Ag-Au bimetallic core-shell nanoalloys. *Appl. Phys. Lett.* 92: 023112-1–023112-3.

Chu, T.L., Chu, S.S., and Ray, R.L. 1982. Germanium arsenide iodide: A clathrate semiconductor. *J. Appl. Phys.* 53: 7102–7103.

Claussen, W.F. 1951. Suggested structures of water in inert gas hydrates. *J. Chem. Phys.* 19: 259–260.

Connétable, D. 2003. PhD thesis, University of Lyon 1, Lyon, France.

Connétable, D., Timoshevskii, V., Artacho, E., and Blase, X. 2001. Tailoring band gap and hardness by intercalation: An ab initio study of I$_8$@Si-46 and related doped clathrates. *Phys. Rev. Lett.* 87: 206405–206408.

Connétable, D., Timoshevskii, V., Masenelli, B. et al. 2003. Superconductivity in doped *sp*^{3} semiconductors: The case of the clathrates. *Phys. Rev. Lett.* 91: 247001–247004.

Conway, J.H. and Sloane, N.J.A. 1993. *Sphere Packing, Lattices and Groups.* Springer-Verlag, New York.

Cordier, G. and Woll, P. 1991. Neue ternäre intermetallische Verbindungen mit Clathratstruktur: Ba$_8$(T,Si)$_6$Si$_{40}$ und Ba$_6$(T,Ge)$_6$Ge$_{40}$ mit T ≡ Ni, Pd, Pt, Cu, Ag, Au. *J. Less. Common Met.* 169(2): 291–302.

Cros, C., Pouchard, M., and Hagenmuller, P. 1965. Two new phases of the silicon-sodium system, *C. R. Seances Acad. Sci. Ser. A* 260: 4764–4767.

Cros, C., Pouchard, M., and Hagenmuller, P. 1970. Sur une nouvelle famille de clathrates minéraux isotypes des hydrates de gaz et de liquides. Interprétation des résultats obtenus. *J. Solid State Chem.* 2: 570–581.

Delacrétaz, G., Grant, E.R., Whetten, R.L., Wöste, L., and Zwanziger, J.W. 1986. Fractional quantization of molecular pseudorotation in Na$_3$. *Phys. Rev. Lett.* 56: 2598–2601.

Demkov, A., Sankey, O., Schmidt, K., Adams, G., and O'Keefe, M. 1994. Theoretical investigation of alkali-metal doping in Si clathrates. *Phys. Rev. B* 50: 17001–17008.

Demkov, A.A., Ortega, J.O., Sankey, O.F., and Grumbach, M.P. 1995. Electronic structure approach for complex silicas. *Phys. Rev. B* 52: 1618–1630.

Dong, J., Sankey, O.F., and Kern, G.I. 1999. Theoretical study of the vibrational modes and their pressure dependence in the pure clathrate-II silicon framework. *Phys. Rev. B* 60: 950–958.

Dong, J., Sankey, O.F., Ramachandran, G.K., and McMillan, P.F. 2000. Chemical trends of the rattling phonon modes in alloyed germanium clathrates. *J. Appl. Phys.* 87: 7726–7734.

Donohue, J. 1974. *The Structure of the Elements.* Wiley Interscience, New York.

Eisenmann, B., Schäfer, H., and Zagler, R. 1986. Die verbindungen AII8BIII16BIV30 (AII ≡ Sr, Ba; BIII≡ Al, Ga; BIV ≡ Si, Ge, Sn) und ihre käfigstrukturen. *J. Less Common Met.* 118(1): 43–55.

El Goresy, A., Dubrovinsky, L.S., Gillet, P., Mostefaoui, S., Graup, G., Drakopoulos, M., Simionovici, A.S., Swamy, V., and Masaitis, V.L. 2003. A new natural, super-hard, transparent polymorph of carbon from the Popigai impact crater, Russia. *Compd. Rend. Geosci.* 335: 889–898.

Elkind, J.L., Alfonr, J.M., Weiss, F.D., Laaksonen, R.T., and Smalley, R.E. 1987. FT-ICR probes of silicon cluster chemistry: The special behavior of Si$^+$$_{39}$. *J. Chem. Phys.* 87: 2397–2399.

Fedorov, E.S. 1885. The symmetry of regular systems of figures, *Zap. Miner. Obshch.* 21: 1–279.

Fonseca, I. 1991. The Wulff theorem revisited. *Proc. R. Soc. Lond. A* 432: 125–145.

Fortner, J. and Lannin, J.S. 1989. Radial distribution functions of amorphous silicon. *Phys. Rev. B* 39: 5527–5530.

Fowler, P.W. and Woolrich, J. 1986. π-systems in three dimensions. *Chem. Phys. Lett.* 127(11): 78–83.

Frank, F.C. 1959. Complex alloy structures regarded as sphere packings. II. Analysis and classification of representative structures. *Acta Cryst.* 12: 483–499.

Friedel, J. and Lannoo, M. 1973. Covalence in elemental structures and AB compounds. I. Theorems of Leman, Thorpe and Weaire. *J. Phys.* 34(1): 115–121.

Fukuoka, H., Ueno, K., and Yamanaka, S. 2000. High-pressure synthesis and structure of a new silicon clathrate Ba$_{24}$Si$_{100}$. *J. Organomet. Chem.* 611(1–2): 543–546.

Gallmeier, J., Schaefer, H., and Weiss, A. 1969. Cage structure as a common building principle of compounds K$_8$E$_{46}$ (E=silicon, germanium, tin). *Zeitschrift fuer Natur-forschung, Teil B: Anorganische Chemie, Organische Chemie, Biochemie, Biophysik, Biologie* 24(6): 665–667.

Giannozzi, P., de Gironcoli, S., Pavone, P., and Baroni, S. 1995. *Ab initio* calculation of phonon dispersions in semiconductors. *Phys. Rev. B* 43: 7231–7242.

Gong, X. and Zheng, Q. 1995. Electronic structures and stability of Si_{60} and $C_{60}@Si_{60}$ clusters. *Phys. Rev. B* 52: 4756–4759.

Gryko, J., McMillan, P.F., Marzke, R.F. et al. 2000. Low-density framework form of crystalline silicon with a wide optical band gap. *Phys. Rev. B* 62: R7707–R7710.

Guloy, A.M., Ramlau, R., Tang, Z. et al. 2006. Guest-free germanium clathrate. *Nature* 443: 320–323.

Haddon, R.C. 1986. Hybridization and the orientation and alignment of pi-orbitals. *J. Am. Chem. Soc.* 108(11): 2837–2842.

Harrison, W. 1980. *Electronic Structure and the Properties of Solids*. W.H. Freeman, New York.

Hellwege, K.H. and Hellwege, A. 1966. *Crystal and Solid State Physics, Landolt-Bornstein, Numerical Data and Functional Relationships in Science and Technology, New Series, Group III*, Vol 1. Springer-Verlag, Berlin, Germany.

Jackson, K., Kaxiras, E., and Pederson, M.R. 1993. Electronic states of group-IV endohedral atoms in C_{28}. *Phys. Rev. B* 48: 17556–17561.

Jaussaud, N., Toulemonde, P., Pouchard, M. et al. 2004. High pressure synthesis and crystal structure of two forms of a new tellurium-silicon clathrate related to the classical type I. *Solid State Sci.* 6: 401–411.

Jund, P., Caprion, D., Sadoc, J.F., and Jullien, R. 1997. Melting of model structures: Molecular dynamics and Voronoï tessellation. *J. Phys.: Condens. Matter* 9: 4051–4059.

Kamitakahara, W.A., Soukoulis, C.M., Shanks, H.R., Buchenau, V., and Grest, G.S. 1987. Vibrational spectrum of amorphous silicon: Experiment and computer simulation. *Phys. Rev. B* 36: 6539–6542.

Kawaji, H., Horie, H.O., Yamanaka, S. et al. 1995. Superconductivity in the silicon clathrate compound $(Na,Ba)_xSi_{46}$. *Phys. Rev. Lett.* 74: 1427–1429.

Kaxiras, E. 1990. Guest effect of surface reconstruction on stability and reactivity of Si clusters. *Phys. Rev. Lett.* 64: 551–554.

Kittel, C. 2004. *Introduction to Solid State Physics*. Wiley & Sons, New York.

Kléman, M. 1989. Curved crystals, defects and disorder. *Adv. Phys.* 38: 605–667.

Kohn, W. and Sham, L.J. 1965. Self-consistent equations including exchange and correlation effects. *Phys. Rev.* 140: A1133–A1138.

Kuhl, B., Czybulka, A., and Schuster, H.U. 1995. Neue ternäre Käfigverbindungen aus den Systemen Barium-Indium/Zink/Cadmium-Germanium: Zintl-Verbindungen mit Phasen-breite? *Z. Anorg. Allg. Chem.* 621: 1–6.

Kume, T., Fukuoka, H., Koda, T., Sasaki, S., Shimizu, H., and Yamanaka, S. 2003. High-pressure Raman study of Ba doped silicon clathrate. *Phys. Rev. Lett.* 90: 155503–155506.

Libotte, H. and Gaspard, J.P. 2000. Pressure-induced distortion of the β-Sn phase in silicon: Effects of nonhydrostaticity. *Phys. Rev. B* 62: 7110–7115.

Lortz, R., Viennois, R., and Petrovic, A. 2008. Phonon density of states, anharmonicity, electron-phonon coupling, and possible multigap superconductivity in the clathrate superconductors Ba_8Si 46 and $Ba_{24}Si_{100}$: Factors behind large difference in T_c. *Phys. Rev. B* 77: 2245071–22450712.

Machon, D., Toulemonde, P., McMillan, P.F., Amboage-Castro, M., Muñoz, A., Rodríguez-Hernández, P., and San-Miguel, A. 2009. High-pressure phase transformations, pressure-induced amorphization and a polyamorphic transition for Rb_6Si_{46} clathrate composition. *Phys. Rev. B* 79(18): 184101.

McMillan, W.L. 1968. Transition temperature on strong-coupled superconductors. *Phys. Rev. B* 167: 331–344.

Mélinon, P. and Masenelli, B. 2008. Cage like based materials with carbon and silicon. *ECS Trans.* 13(14): 101–104.

Mélinon, P., Kéghélian, P., Prével, B. et al. 1997. Nanostructured silicon films obtained by neutral cluster depositions. *J. Chem. Phys.* 107: 10278–10287.

Mélinon, P., Kéghélian, P., Blase, X. et al. 1998. Electronic signature of the pentagonal rings in silicon clathrate phases: Comparison with cluster-assembled films. *Phys. Rev. B* 58: 12590–12593.

Mélinon, P., Kéghélian, P., Perez, A. et al. 1999. Phonon density of states of silicon clathrates: Characteristic width narrowing effect with respect to the diamond phase. *Phys. Rev. B* 59: 10099–10104.

Meng, J.F., Chandra Shekar, N.V., Badding, J.V., and Nolas, G.S. 2001. Threefold enhancement of the thermoelectric figure of merit for pressure tuned $Sr_8Ga_{16}Ge_{30}$. *J. Appl. Phys.* 89: 1730–1733.

Moriguchi, K., Munetoh, S., and Shintani, A. 2000. First-principles study of $Si_{34-x}Ge_x$ clathrates: Direct wide-gap semiconductors in Si-Ge alloys. *Phys. Rev. B* 62: 7138–7143.

Mott, N.F. 1973. Properties of compounds of type Na_xSi_{46} and Na_xSi_{136}. *J. Solid State Chem.* 6: 348–351.

Mukherjee, S., Seitsonen, A.P., and Nieminen, R.M. 1996. Structure of nanoscale Si_{33} clusters from ab initio electronic structure calculation. *Proceedings of the International Workshop on Clusters and Nanostructured Materials*, P. Jena and S.N. Behera (Eds.), Nova Science Publishers, New York, pp. 165–171.

Munetoh, S., Moriguchi, K., Kamei, K., Shintani, A., and Motooka, T. 2001. Epitaxial growth of a low-density framework form of crystalline silicon: A molecular-dynamics study. *Phys. Rev. Lett.* 86: 4879–4882.

Mutoh, N. 2004. The polyhedra of maximal volume inscribed in the unit sphere and of minimal volume circumscribed about the unit sphere. *Lecture Notes in Computer Science*, 2866. Springer, Berlin/Heidelberg, Germany, pp. 204–214.

Nolas, G.S. and Kendziora, C.A. 2000. Raman scattering study of Ge and Sn compounds with type-I clathrate hydrate crystal structure. *Phys. Rev. B* 62: 7157–7161.

O'Keeffe, M. 2006. Tetrahedral frameworks TX2 with T-X-T angle = 180° rationalization of the structures of MOF-500 and of MIL-100 and MIL-101. *Mater. Res. Bull.* 41(5): 911–915.

Parasuk, V. and Almöf, J. 1991. C_{20}: The smallest fullerene? *Chem. Phys. Lett.* 184: 187–190.

Patterson, C.H. and Messmer, R.P. 1990. Bonding and structures in silicon clusters: A valence-bond interpretation. *Phys. Rev. B* 42: 7530–7555.

Ramakrishna, M.V. and Pan, J. 1994. Chemical reactions of silicon clusters. *J. Chem. Phys.* 101: 8108–8118.

Ramachandran, G.K., McMillan, V., Diefenbacher, J. et al. 1999. 29Si NMR study on the stoichiometry of the silicon clathrate Na_8Si_{46}. *Phys. Rev. B* 60: 12294–12298.

Ramachandran, G.K., McMillan, P.F., Dep, S.K. et al. 2000. High-pressure phase transformation of the silicon clathrate Si_{136}. *J. Phys. Matter* 12: 4013–4020.

Reny, E., Gravereau, P., Cros, C., and Pouchard, M. 1998. Structural characterisations of the Na_xSi_{136} and Na_8Si_{46} silicon clathrates using the Rietveld method. *Mater. Chem.* 8: 2839–2844.

Reny, E., Yamanaka, S., Cros, C., and Pouchard, M. 2000. High pressure synthesis of an iodine doped silicon clathrate compound. *Chem. Commun.* 2505–2506.

Rey, R., Muñoz, A., Rodríguez-Hernández, P., and San-Miguel, A. 2008. First-principles study of lithium-doped carbon clathrates under pressure. *J. Phys.: Condens. Matter* 20: 215218–215225.

Roorda, S., Doorn, S., Sinke, W.C. et al. 1989. Calorimetric evidence for structural relaxation in amorphous silicon. *Phys. Rev. Lett.* 62: 1880–1883.

Röthlisberger, U., Andreoni, W., and Parrinello, M. 1994. Structure of nanoscale silicon clusters. *Phys. Rev. Lett.* 72: 665–668.

Saito, S. and Oshini, S. 1986. *Microclusters*, vol. 4., Sugano, S., Nishina, Y., and Ohnishi, S. (Eds.), Springer Series in Materials Science, Berlin, Germany, p. 263.

Saito, S. and Oshiyama, A. 1995. Electronic structure of Si_{46} and $Na_2Ba_6Si_{46}$. *Phys. Rev. B* 51: 2628–2631.

Sales, B.C., Chakoumakos, B.C., Jin, R., Thompson, J.R., and Mandrus, D. 2001. Structural, magnetic, thermal, and transport properties of $X_8Ga_{16}Ge_{30}$ (X=Eu, Sr, Ba) single crystals. *Phys. Rev. B* 63: 245113–245120.

San-Miguel, A. and Toulemonde, P. 2005. High-pressure properties of group IV clathrates, *High Press. Res.* 25: 159–185.

San-Miguel, A., Kéghélian, P., Blase, X. et al 1999. High pressure behavior of silicon clathrates: A new class of low compressibility materials. *Phys. Rev. Lett.* 83: 5290–5293.

San-Miguel, A., Melinon, P., Connetable, D. et al. 2002. Pressure stability and low compressibility of intercalated cagelike materials: The case of silicon clathrates. *Phys. Rev. B* 65: 054109–054112.

San-Miguel, A., Merlen, A., and Toulemonde, P. 2005. Pressure-induced homothetic volume collapse in silicon clathrates. *Europhys. Lett.* 69: 556–562.

Sivardière, J. 1995. *La Symétrie en Mathématiques, Physique et Chimie*, Presses Universitaires de Grenoble, Grenoble, France, p. 182.

Spagnolatti, I., Bernasconi, M., and Benedek, G. 2003. Electron-phonon interaction in carbon clathrate hex-C40. *Eur. Phys. J. B* 34: 63–67.

Sun, Q., Wang, Q., Briere, T.M. et al. 2002. First-principles calculations of metal stabilized Si_{20} cages. *Phys. Rev. B* 65: 235417–235422.

Sutton, A.P. 1994. *Electronic Structure of Materials*, Science Publications Clarendon press, New York, p. 130.

Sze, S.M. 1981. *Physics of Semiconductors Devices*. Wiley, New York.

Ternansky, R.J., Balogh, D.W., and Paquette, L.A. 1982. Dodecahedrane. *J. Am. Chem. Soc.* 104(16): 4503–4504.

Terrones, T. and Terrones, M. 2003. Curved nanostructured materials. *New J. Phys.* 5: 126.1–126.37.

Terrones, M., Terrones, G., and Terrones, H. 2002. Structure, chirality, and formation of giant icosahedral fullerenes and spherical graphitic onions. *Struct. Chem.* 13: 373–384.

Toulemonde, P., San-Miguel, A., Merlen, A. et al. 2006. High pressure synthesis and properties of intercalated silicon clathrates. *J. Phys. Chem. Solids* 67: 1117–1121.

Toulouse, G. 1989. How frustration set in. *Phys. Today* 42(12): 97.

Tournus, F., Masenelli, B., Mélinon, P. et al. 2004. Guest displacement in silicon clathrates. *Phys. Rev. B* 69: 035208–035213. Available at: <http://apps.isiknowledge.com/CitingArticles.do? product=WOS&SID=P1aHaFLLgjk@OckMpia&searchmode=CitingArticles&parentQid=10&recid=133264339&parentDoc=34&db id = WOS>

Tse, J.S., Uehara, K., Rousseau, R., Ker, A., Ratcliffe, C.I., White, M.A., and Mackay, G. 2000. Structural principles and amorphouslike thermal conductivity of Na-doped Si clathrates. *Phys. Rev. Lett.* 85: 114–117.

Tse, J.S., Desgreniers, S., Li, Z., Ferguson, M.R., and Kawazoe, Y. 2002. Structural stability and phase transitions in K_8Si_{46} clathrate under high pressure. *Phys. Rev. Lett.* 89: 195507–195510.

Tse, J.S., Flacau, R., Desgreniers, S., Iitaka, T., and Jiang, J. 2007. Electron density topology of high-pressure Ba_8Si_{46} from a combined Rietveld and maximum-entropy analysis. *Phys. Rev. B* 76: 1741091–1741098.

Tsumuraya, K., Nagano, T., Eguch, H., and Takenaka, H. 1995. Optimized structures of Si_{28} and $Ba@Si_{28}$ clusters: Ab initio study. *Int. J. Quantum Chem.* 91: 328–332.

Weaire, D. and Phelan, R. 1994. A counterexample to Kelvin's conjecture on minimal surfaces. *Philos. Mag. Lett.* 69: 107–110.

Weaire, D. and Thorpe, M. F. 1971. Electronic properties of an amorphous solid. I. A simple tight-binding theory. *Phys. Rev. B* 4: 2508–2520.

Yamanaka, S., Horie, H.O., Nakano, H., and Ishikawa, M. 1995. Preparation of barium-containing silicon clathrate compound. *Fullerene Sci. Technol.* 3(1): 21–29.

Yin, M.T. and Cohen, M.L. 1980. Microscopic theory of the phase transformation and lattice dynamics of Si. *Phys. Rev. Lett.* 45: 1004–1007.

III

Nanolayers

Self-Assembled Monolayers

Frank Hagelberg
East Tennessee State University

17.1 Self-Assembled Monolayers: An Overview

Self-assembled monolayers (SAMs) are ordered monomolecular films of an organic species that organize spontaneously on a substrate (Ulman 1996, Chaki 2001, Schreiber 2004, Love et al. 2005). The great and growing attention paid to these composites is motivated by their interest as prototypical organic–inorganic interfaces, but chiefly by the wide variety of their applications, most of them in the field of nanotechnology. SAMs are nanosystems, as they have at least one nanoscopic dimension. In the most thoroughly studied case of organic molecules attached to a planar surface, this dimension is the thickness of the film, which is usually 1–3 nm. Metal, metal oxide, and semiconductor surfaces have been used as SAM substrates. However, SAMs can assemble on surfaces of any shape or size, and many currently active projects related to SAM research involve monolayers deposited on nanoparticles (e.g., Doellefeld et al. 2002, Berry and Curtis 2004, Antolini et al. 2007), and thus deal with objects confined to the nanoscale in all three dimensions.

The high interest in SAMs is motivated by the fact that they are expedient tools for manipulating and functionalizing surfaces and allow control of the physical and chemical nature of the interface between the substrate and its environment. Thus, the conductivity, the electrostatic and optical characteristics, wettability or corrosion resistance of the surface, among many other features, may be sensitively modified by SAM engineering. Increasingly, SAMs also play a role in nanoparticle design. Covering quantum dots or magnetic nanocrystals with monolayers is currently discussed as a strategy for making these toxic nanosystems biocompatible (see Section 17.5). Superstructures of nanoparticles may be created by suitably tailored SAMs attached to the particle surfaces (Section 17.2.3).

The history of research on SAMs reaches back into the middle of the last century. Zisman et al. are credited with pioneering accomplishments in identifying and characterizing SAMs (Bigelow et al. 1946, Shafrin and Zisman 1949). Shafrin and Zisman (1949) fabricated an oriented monomolecular layer of alkyl amines adsorbed from water to a platinum surface. Nuzzo and Allara (1983) were the first to demonstrate the stability of SAMs composed of alkanethiolates ($S\text{-}(CH_2)_{n-1}\text{-}CH_3$) in contact with gold, as prepared by adsorption of di-*n*-alkyl disulfides from dilute solutions. Early research on thiol monolayers demonstrated the highly specific interaction between gold and sulfur to produce organic disulfide (Nuzzo and Allara 1983), sulfide (Troughton et al. 1988), and thiol (Nuzzo et al 1987) monolayers on gold.

Although SAM research has explored numerous combinations of substrates and adsorbates, fundamental as well as applied work on SAMs has centered on systems consisting of thiol molecules in contact with noble metal surfaces. In our survey, we will therefore preferentially refer to these composites. The predominance of these systems may be attributed to the relative simplicity of their components, the ease of their fabrication, and to their stability that is rooted in the high affinity of noble metal surfaces for thiols. Further, a large number of actual and potential applications have been identified for these composites.

In contrast to Langmuir–Blodgett films (Peterson 1990), where the cohesion between the surface adsorbate layer is due to *physisorption*, SAMs *chemisorb* to their substrates. A constituent molecule of a SAM may be subdivided into three segments, whose SAM-forming interactions are characterized by clearly separated energy scales. The adsorbates chemisorb to the substrate through their head groups, for instance the S–H subunit in the case of alkanethiol (SH-$(CH_2)_{n-1}$-CH_3) layers. This interaction is a strong chemical bonding effect that attaches the molecule to a specific site of the substrate. Often the nature of the bond is covalent. Thus, the Au–S bond that anchors alkanethiols in gold surfaces is predominantly covalent with polar admixtures. For some systems, in contrast, such as carboxylic acids on AgO/Ag (Ulman 1991), ionic bonds have been reported. The corresponding adsorption energies are typically in the order of a few eV. For thiolates on gold, they have been determined to fall into the range between about 1.7 and 2.0 eV (Dubois et al. 1990).

The head group connects to an alkyl chain (see Figure 17.1 for a schematic by representation), which interacts with neighboring chains through van der Waals forces. These processes are exothermic with typical energies in the 0.1 eV regime. Chemical modification of the alkyl chain may introduce additional, e.g., polar, interactions to superpose the van der Waals component (Kitaigorodtskii 1973). Interchain effects determine the SAM structure beyond a critical level of molecular surface coverage. At sufficiently low surface density of adsorbates on metal substrates, alkanethiols tend to bind to the surface, stabilizing in a horizontal, in-plane position. With increasing density, however, adjacent chains begin to interact and, as a result of this process, to lift off the surface, adopting an out-of-plane phase, where the nature and strength of the interaction determines the tilt angle with respect to the surface normal (Figure 17.1).

The alkyl chain terminates in an end group that may be, in the simplest case, a methyl unit, but also might be a complex biological molecule chosen to functionalize the surface. As established by Fourier transform infrared spectroscopy (FTIR, Nuzzo et al. 1990b), methyl end groups are thermally disordered at room

temperature. A typical energy scale for end-group processes is given by kT, with T as absolute temperature.

While SAMs are generally labeled as highly ordered (Ulman 1991), they exhibit, by the standard of inorganic solids, a substantial propensity for defects. In the first place, these are induced by the substrate that has been chosen as polycrystalline gold in the majority of SAM studies performed so far. The defects in this class are attributed largely to step edges and grain boundaries as the causes of local disorder in the monolayer structure. Further, the adsorption of alkanethiol on Au(111) has been shown to lift the complex reconstruction of the clean gold substrate at sufficiently high coverage for most investigated adsorbate species (Maksymovych et al. 2006, Mazzarello et al. 2007). In this process, gold vacancies are created that have been observed to nucleate into islands (Poirier 1997). Further defects include the boundaries between zones of different monolayer orientation or substrate impurities.

Since the first description of alkanethiol monolayers on gold substrates was given (Nuzzo and Allara 1983), a multiplicity of different SAM systems and structures have been identified. At the same time, the paradigmatic configuration, alkanethiol in contact with Au(111), has been the subject of a vivid, often controversial, and ongoing discussion. From experimental observation (Camillone et al. 1993, Fenter et al. 1994, Poirier and Tarlov 1994), the order adopted by thiol adsorbates on Au(111) deviates from the hexagonal symmetry of the substrate, as they stabilize in a $(2\sqrt{3} \times 3)$ rather than $(\sqrt{3} \times \sqrt{3})$ superlattice. This behavior is ascribed to a dimerization, or pairing effect between the sulfur head groups of the adsorbates. For alkanethiolate, a recent interpretation of the pairing phenomenon is supported by both experimental and computational evidence (Maksymovych et al. 2006, Mazzarello et al. 2007, Cossaro et al. 2008). This model involves the formation of an *adatom geometry*, where two adjacent S atoms are joined by an intermediate Au atom that is elevated from the Au surface (Cossaro et al. 2008). This process lifts the reconstruction of the pristine substrate.

The present understanding of structural features of alkanethiol on gold will be summarized in Section 17.2.1 of this survey. We also will highlight the structural properties of a very different SAM type, involving organosilicon derivatives on hydroxylated surfaces that tend to form vertically as well as laterally stabilized three-dimensional polymers (Section 17.2.2). In particular, we will focus on the structures of the arguably most novel SAMs, namely those attaching to nanoparticles, where the SAM geometry is critically determined by the high substrate curvature (Section 17.2.3). Importantly, an array of nanoparticles can be designed by covering them with SAMs. Two- and three-dimensional superlattices of nanocrystals may be formed as they are capped with alkanethiol ligands. The particles are linked by interlocking chains, and the chain length represents a parameter that allows control of the order and dimensionality of the resulting superstructure (Martin et al. 2000).

Studying SAM evolution under kinetic and thermodynamic viewpoints is essential for understanding their mechanisms of self-assembly. SAM preparation from the gas phase has allowed

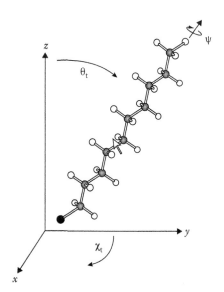

FIGURE 17.1 A chemisorbed alkyl chain, stabilized in an out-of-plane SAM phase. Angles θ, χ, ψ refer to the tilt of the molecule with respect to the surface normal, the tilt orientation, and the twist (the rotation angle about the axis of the molecule), respectively. (From Schreiber, F., *Prog. Surf. Sci.*, 65, 151, 2000. With permission.)

investigating their formation as a function of coverage for the reference system alkanethiol on Au(111) (Barrena et al. 2003, Schreiber 2004). A general consensus has been reached regarding the extremes of low and high coverage, where an in-plane phase, with alkyl chains oriented parallel to the surface, and an out-of-plane phase, involving a tilt angle of 30° with respect to the surface normal, have been identified, respectively. As the condition for self-assembly, a physisorbed precursor state, preceding the final chemisorbed configuration, has been found (Lavrich et al. 1998). This state is essential for both the self-organization of the molecules on the surface and for the emergence of covalent bonding between the substrate and the monolayer as the root of SAM stability. These matters are surveyed in Section 17.3.

Section 17.4 focuses on nanotechnological applications of SAMs. Much attention has been paid to recently developed SAM-based methodologies to immobilize biological molecules on monolayers. Different procedures to achieve this goal include attaching suitably modified biomolecules on a metal surface (Case et al. 2003) or connecting them to the end group of a conventional monolayer (Sigal et al. 1996). Among the most promising applications of SAMs are those in the field of biological and chemical sensing. Microcantilevers coated with monolayers display small but measurable bending responses as well as shifts in their resonance frequencies upon adsorption of minute traces of the species of interest (Section 17.4.1.1). Nanoparticles may be strategically loaded with SAMs such that they are linked by the targeted agent and arranged in arrays whose properties differ significantly from those of the free nanoparticle ensemble (see Section 17.4.1.2).

The basic functionality of *active materials* based on SAMs has recently been demonstrated (Lahann et al. 2003). These are exemplified by the concept of *switchable surfaces* (Section 17.4.2). The monolayer responds to an external stimulus, namely the electrochemical surface potential. This parameter controls conformational changes that are coupled with alterations of the chemical nature of the SAM, which can be varied in this manner from hydrophilic to hydrophobic.

Nanolithography is a broad basin of SAM applications, taking *microcontact* to *nanocontact printing*, as outlined in Section 17.4.3. A variety of *positive printing techniques* involving SAMs have been developed over the last two decades. Nanocontact printing may be performed in a parallel mode where monolayers are transferred from a stamp to the printing substrate (Li et al. 2002). Serial nanoprinting, in contrast, transports molecules in temporal succession to selected surface sites, as may be accomplished by use of an atomic force microscope (AFM) tip (Piner et al. 1999). *Negative nanoprinting*, in contrast, proceeds by selective reduction of a preexisting SAM that may be realized by AFM or scanning tunneling microscopy (STM), or by more traditional means of patterning, involving photon, ion, or electron impact.

The high versatility of the SAM concept is documented by its relevance for widely diverging branches of nanotechnology, such as nanobioengineering and molecular electronics. In Section 17.4.4, a summary of SAM-related activities in the latter field is attempted. Various configurations have been explored in testing

the conductivity features of SAMs. Thus, alkanethiol monolayers or bilayers have been sandwiched between metal electrodes, realizing a metal insulator–metal tunnel junction (Ratner et al. 1998). Various conduction mechanisms have been observed for different choices of monolayer materials. For the prototype of alkanethiol enclosed between gold surfaces, however, electron tunneling has been secured as the main conduction mode.

Novel classes of monolayers have been proposed in the area of nonlinear optics (see Section 17.4.5). Materials suitable for this application have to comply with a set of very specific criteria, and various experimentally tested SAMs have been demonstrated to satisfy them. These are mostly chromophores of high density and high polarizability. A large second-order nonlinear susceptibility has been reported for SAMs composed of hemicyanine (Naraoka et al. 2002). Specific techniques of materials engineering, such as the fabrications of *ionically self-assembled monolayers* (ISAMs), are employed to design SAMs with enhanced nonlinear optical properties (Heflin et al. 1999).

Prior to surveying, in Sections 17.2 through 17.4, the SAM features of major relevance to nanoscience and nanotechnology, we focus on two procedural aspects of SAM research, namely their preparation (Section 17.1.1) and their characterization (Section 17.1.2).

17.1.1 Preparation of SAMs

Among the advantages of SAMs is the relative ease of their manufacturing. As self-assembly is based on the strong adhesion of a selected molecular groups to a substrate, SAMS may be fabricated by direct deposition from the liquid or the gas phase. Care must be taken in preparing the substrate that determines critically the SAM formation mechanism and thereby controls crucial parameters of the resulting phase, such as packing density, stability, long-range order, and the number of defects. Thus, the nature and the roughness of the surface, as well as its cleanliness and crystallinity are major concerns related to the substrate preparation. For alkanethiol on gold, evaporated gold films have often been used. These form preferentially terraces with (111) orientation since Au(111) is the most stable gold surface. The cleaning of the metal surfaces involved is often performed by a combination of ion sputtering and annealing in an UHV environment. An important criterion for the clean Au(111) surface is the appearance of the $(22 \times \sqrt{3})$ or "herringbone" reconstruction (Barth et al. 1990).

Both liquid and gas phase deposition are conceptually simple. Monolayers will assemble on a sufficiently cleaned substrate dipped into a solution that contains the adsorbate species. A frequently used procedure to fabricate alkanethiol monolayers on gold consists of exposing the substrate to ethanolic solution with alkanethiol at micro- to millimolar concentrations. Assembly from the gas phase is more elaborate than from the liquid phase (Poirier and Pylant 1996). Here, deposition usually proceeds under UHV conditions, which implies the advantage that SAM formation can be studied using a variety of surface analysis instruments while the assembly is taking place.

Strong attention is paid to a very different class of SAM composites, involving monolayers deposited on metal colloids (Freeman et al. 1995) or nanoparticles (Daniel and Astruc 2004). The synthesis of these complexes is more intricate than that of their planar counterparts. Again, systems composed of alkanethiol adsorbates in contact with gold substrates have been studied most thoroughly and may be considered as reference units. An important step toward the controlled fabrication of gold nanoparticles was done with the introduction of the Brust–Schiffrin method (Brust et al. 1994, 1995). Here, a thiol component contributes to the reduction of gold precursor salts such as $AuCl_4$, resulting in the formation of gold nanoclusters surrounded by thiol polymers that anchor in their surfaces. The size of the nanoparticles is governed by the ratio of thiol to gold, as a larger ratio leads to smaller average core sizes (Daniel and Astruc 2004). The gold particles obtained by this procedure range in extension from 1.5 and 5.2 nm. Fabricating gold nanoclusters with larger diameters is achieved by the use of surfactants with higher desorption rates than thiols. By the means of ligand-exchange techniques, these weakly bound adsorbates can be replaced by thiol molecules after the particle has attained its terminal size. Also, the functionalities of SAM-covered nanoparticles may be changed by chemical reactions that modify the thiol chain end groups (Templeton et al. 1998b). In this context, it has to be taken into account that the adsorption structures of thiols on nanoparticles differ from those on plane surfaces (Section 17.2.3), corresponding to different reactivities for both configurations. The curved substrate geometry allows certain reactions to proceed that are excluded from occurring in the plane surface case by steric constraints. An example for this group of reactions is given by bimolecular nucleophilic substitutions (S_N2 reactions, Templeton et al. 1998a).

17.1.2 Characterization of SAMs

The whole arsenal of surface characterization methods has been employed to study the chemical and physical properties of SAMs, and giving a complete survey of these techniques and their specific application to SAMs is beyond the scope of this chapter. In what follows, we name some experimental procedures that have proven to be efficient tools of SAM analysis and comment briefly on them. We do not aim at completeness, but attempt to list the most customary experimental procedures used to specify SAM properties.

Microscopy provides the most direct structural information about SAMs, and has been widely used for their characterization. Both STM and AFM have been instrumental in determining SAM architectures and provided detailed information on geometric features, pertaining to regular SAM equilibrium arrangements as well as irregularities such as defects, or to structural changes during SAM growth. Increasingly, *scanning force microscopy* (SFM), providing surface scans by the means of the AFM technique, is utilized for the study of SAMs (e.g., Munuera and Ocal 2006). It should be noted that these procedures are local by nature and thus do not allow for conclusions related to

the surface as a whole, in contrast to the diffraction and spectroscopic methods mentioned below. The STM procedure is limited with respect to out-of-plane phases of hydrocarbon chains, as the tunneling current through these chains strongly diminishes with their length and can hardly be monitored beyond an extension of 12 segments (Schreiber 2000). A strength of the STM method, however, is that it allows to gain information about the electronic structure of the considered system, as may be obtained by comparing the measured tunneling current with simulations based on first principles theory. The *conductive scanning force microscopy* (C-SFM) method combines the virtues of STM and SFM as is it optimized to map simultaneously the topography and the current distribution of a sample.

Various *diffraction techniques* have been employed to monitor the two-dimensional structure of SAMs, most importantly *grazing incidence X-ray diffraction* (GIXD) (Marra et al. 1979, Feidenhans'l 1989), *He atom diffraction* (Hudson 1992), and *low-energy electron diffraction* (LEED) (Somorjai and van Hove 1979). These methods provide information about the reciprocal lattice of the considered system. Observation of interfering waves scattered from a large number of centers allows reconstructing the charge density of the studied material. Due to the different characteristic energy scales involved, 10 keV for GIXD versus 10 meV for He atom diffraction, the first two of the three procedures indicated above may be viewed as complementary (Fenter et al. 1998). While x-rays penetrate the monolayer and thus probe the adsorbate–substrate interface, the He atoms are scattered by the monolayer surface. The two techniques are thus sensitive to different SAM regions. The LEED method is of lower resolution than GIXD. A further advantage of the latter over the former procedure is that x-ray data may be interpreted by use of elementary or "kinematic" scattering theory. Extracting information beyond the unit cell size and symmetry from LEED measurement, in contrast, requires inclusion of multiple scattering events (Lueth 1993).

The electron density profile along the surface normal can be derived from *x-ray reflexivity* (XR) data (Tidswell et al. 1990, Tolan 1999), as provided by x-ray measurement that does not involve any surface parallel momentum transfer component. X-ray reflexivity results may be used to obtain detailed information about thin film thickness and roughness.

A broad variety of *spectroscopy-based methods* have been utilized for SAM characterization, and we will not attempt a comparative discussion of these procedures. As diffraction, spectroscopy yields spatial averages of the measured quantities. With reference to monolayers, these averages include all adsorbed molecules, while the information from the diffraction techniques stems exclusively from molecules arranged in crystalline and thus ordered domains (Schreiber 2000).

In *infrared* (Nuzzo et al. 1990a) and *Raman* (Cipriani et al. 1974) *spectroscopy*, observation of molecular vibrational modes leads to conclusions about the adsorbate structure. For a vibration to be infrared active, it must be associated with a change in the respective dipole moment, as realized, for instance, by the asymmetric stretch of the CO_2 molecule. Raman activity, on the other hand, depends on a variation of the molecular polarizability, as in the symmetric stretch of CO_2. For polyatomic species that possess a center of inversion, the two spectroscopic methods are complementary, as Raman active bands are usually not infrared active, and vice versa. Without a symmetry constraint imposed on the molecular structure, vibrational bands are often accessible by both techniques.

An application of Raman scattering with particular relevance to surface studies is *surface-enhanced Raman spectroscopy* (SERS) (Fleischmann et al. 1974), which refers to a dramatic enhancement of the Raman scattering intensity from molecules deposited on solid substrates. As enhancement factors as high as 10^{15} were reported, the method allows, in principle, for detection at the single-molecule level (Nie and Emory 1997). While the SERS-generating mechanism is still controversial, theory (Garcia-Vidat and Pendry 1996) and experiment (Kim et al. 1989) point at the critical role of localized surface plasmons that are excited by the incoming radiation and greatly increase the electric field strength felt by the adsorbate. It should be noted that perfectly flat surfaces are not SERS active. In keeping with this constraint, the SERS technique is useful for analyzing systems that involve roughened or nanostructured substrates.

Various instruments of thin film analysis probe the electronic structure of the adsorbates. Information about the chemical environment of the head group and thus the salient bonding features of the atoms comprised in this group is available form *X-ray photoelectron spectroscopy* (XPS) (Tillman et al. 1989), where the energy of an electron ejected by interaction with x-rays is determined. More specifically, XPS exploits the fact that the energy of this electron equals the difference between the primary x-ray energy and the electron binding energy. *Auger electron spectroscopy* (AES) (Grant 2003) is sensitive to film composition in terms chemical elements. A primary electron beam of 2–3 keV ionizes the most strongly bound electronic shells of the sample. An outer shell electron fills the vacancy, giving rise to the loss of an *Auger electron* from an even more weakly bound shell. The energy of this electron provides a fingerprint for the ejecting atom.

A synthesis of vibrational and electronic spectroscopy is provided by the *high-resolution electron energy loss spectroscopy* (HREELS) (Lueth 1993). The physical basis of this technique is inelastic electron scattering, and specifically the excitation of atomic and molecular vibrations in the studied surface. The energy of the highly monochromatic electron beam is typically between 1 and 10 eV. Several spectroscopic techniques based on nonlinear optics have been used to analyze SAMs: most prominently sum frequency generation (SFG, Yeganeh et al. 1995) and second harmonic generation (SHG, Dannenberger et al. 1999). Among numerous additional spectroscopy-based methods that have been employed to determine SAM properties, we name the *near-edge x-ray adsorption fine structure* (NEXAFS, Stöhr 1992) technique.

It might seem surprising to encounter *nuclear magnetic resonance* (NMR) among the techniques that have been found useful for characterizing SAMs. NMR, being used routinely for the study of condensed matter systems, is not one of the traditional surface analysis methods. As recently demonstrated, however, NMR is an excellent tool for investigating monolayers on colloidal and other high-area materials (Reven and Dickinson 1998). In particular, this method has been applied to SAMs on nanostructured substrates, such as alkanethiols on gold nanoparticles (Badia et al. 1996, 1997). NMR measurement employing signals from ^1H, ^{13}C, and ^2H provides information about the adsorbate–substrate interaction, thermal properties, and about the impact of hydrogen bonding groups on the chain dynamics of alkanethiol SAMs.

Various frequently used SAM characterization procedures do not fit under the categories mentioned above. *Ellipsometry* (Azzam and Bashara 1977, Wassermann et al. 1989) is a common optical method to determine film thickness. Polarized light reflected from the studied SAM is analyzed with respect to the phase shift between its surface parallel and surface perpendicular components, as well as its amplitude as compared to the incident beam. In conjunction with the refractive index of the organic layer, these data yield the film thickness. If the latter exceeds 50 Å, ellipsometry is capable of providing both, the thickness and the refractive index of the film.

Essential information about monolayer composition, stability, as well as its thermal properties has been obtained from *temperature-programmed desorption* (TPD, also known as *thermal desorption spectroscopy*, TDS), where the sample is heated in vacuum while the residual gas is analyzed by mass spectrometry (Dubois et al. 1990, Wetterer et al. 1998, Rzeznicka et al. 2005).

A further class of methods for monitoring SAMs we name *electrochemical techniques*. *Cyclic voltammetry* (Bandyopadhyay et al. 1997) and *impedance measurements* (Bandyopadhyay and Vijayamohanan 1998) shed light on basic SAM characteristics such as the redox properties of the adsorbate molecules, the distribution of defects, or the level of coverage. Voltammetry allows to draw conclusions with respect to some of these features by comparing the SAM capacitance with that of the pure substrate.

A large multiplicity of experimental tools has been employed to characterize SAMs, and the above compilation is far from exhaustive. Among the powerful methods that are not easily included in any of the indicated rubrics are *x-ray standing waves* (XSW, Zegenhagen 1993, 2004, Jackson et al. 2000) or *quartz crystal microbalance* (QCM, Shimazu et al. 1994) measurement.

17.2 Systems and Structures

Our survey of SAM systems will focus largely on the two most frequently investigated SAM types. These are alkanethiol molecules on gold substrates and organosilicon derivatives on hydroxylated surfaces. We will use the former class, which usually displays a higher degree of long-range order than the latter, to exemplify the essential structural properties of SAMs. Referring to SAM structure, we distinguish between two different geometric components that constitute the architecture of a given SAM system. These are (a) the two-dimensional order as defined by the distribution of the molecular head groups over the surface and (b) the geometry of the molecular chain and the terminating group. Figure 17.1 illustrates the latter for the case of an alkyl chain anchoring in a SAM substrate. The fundamental structural parameters for this adsorbate component are obviously the chain length and the tilt angle included by the chain with the surface normal. Besides the chemical nature of the substrate and the adsorbate, other variables of critical impact on the resulting SAM structure are the temperature and the level of coverage (or adsorbate surface density). The latter ranges between the extremes of full coverage, corresponding to the highest achievable packing density (the case of surface saturation) and sufficiently low coverage for the mutual interaction between the adsorbing molecules to be deemed negligible.

17.2.1 Alkanethiol Molecules on Au(111)

We first comment on the case of full coverage. For longer chains, involving C_nH_{2n} backbones with typically 10 or 12 C atoms and tilt angles close to 30°, elementary geometric arguments suggest that the adsorbates arrange on the gold surface in a $(\sqrt{3} \times \sqrt{3})$ R30° lattice. As established by helium diffraction (Camillone et al. 1993) and STM measurement (Poirier and Tarlov 1994) for methyl terminated chains, the order of alkanethiol on Au(111) is more properly described by a $(2\sqrt{3} \times 3)$ superlattice (more frequently referred to in the literature as $c(4 \times 2)$ structure), as illustrated in Figure 17.2. This observation implies that the molecular pattern deviates from the hexagonal symmetry of the substrate.

While interpretations based on twist angle differences between neighboring adsorbates have failed to account for the prevalence of the $(2\sqrt{3} \times 3)$ superlattice, SH head group dimerization has been proposed as the cause of the symmetry breaking (Fenter et al. 1994). The latter effect is thus attributed to internal molecular degrees of freedom. A fit based on the GIXD data by

Fenter et al. (1994) yielded a nearest neighbor S–S distance of only 2.2 Å. Direct evidence for pairing between adjacent thiol head groups on the Au(111) surface was provided by a HREELS investigation on octadecanethiol on Au(111), where a S–S stretch mode was detected at 530 cm⁻¹ (Kluth et al. 1999). As shown in Figure 17.3, the naïve model of uniformly distributed molecules (Figure 17.3a) is replaced by a modified scheme that proposes a regular alternation between adsorbates attaching to threefold coordinated hollow sites and to bridge sites of the substrate (Figure 17.3b).

We point out that HREELS spectroscopy was able to detect the S–S stretch only after annealing the studied SAM to 375 K (Kluth et al. 1999). This was ascribed to an activation barrier related to the formation of gauche defect, as shown in Figure 17.3b. Obviously, head group dimerization between two neighboring molecules requires different orientations of their S–C bonds.

Chemical intuition is not easily reconciled with the proposal of S–S dimerization of adjacent S–H head groups on the gold surface, since sulfur tends to have two dangling bonds, or valences. Instead of *dimerization*, *pairing* may be the more adequate term to be used in this context. The pairing mechanism of alkanethiol head groups on Au(111) is a surface-assisted effect (Zhou et al. 2007, see below), and bonding arguments

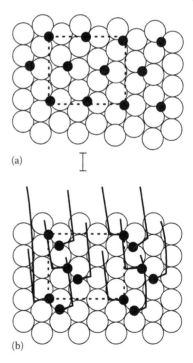

(a)

(b)

FIGURE 17.3 Two-dimensional schematic of adsorption sites for octadecanethiol molecules. (a) The reference case of uniform adsorbate distribution. All thiol head groups attach to threefold hollow sites, with a nearest neighbor distance of 5.0 Å. (b) Revised model derived from STM (Poirier and Tarlov 1994), HREELS (Kluth et al. 1999), and force field modeling studies, accommodating a S–S dimerization effect. A gauche defect at the S–C bond is indicated. The dashed rectangle refers to the unit cell of the modified superlattice. (From Kluth, G.J. et al., *Phys. Rev. B*, 59, R10449, 1999. With permission.)

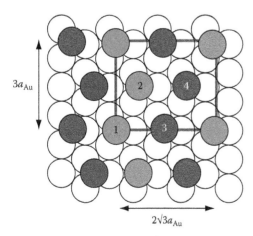

FIGURE 17.2 Representation of the $(2\sqrt{3} \times 3)$ superlattice formed by long-chain alkanethiol adsorbates on the Au(111) surface. The open circles denote gold atoms, while the gray and black circles refer to two nonequivalent alkanethiol adsorbates. (From Schreiber, F., *J. Phys.: Condens. Matter*, 16, R881, 2004. With permission.)

suggested by gas phase chemistry are to be applied with great caution in this case.

The equilibrium structures reported for alkanethiol on Au(111) are the product of the characteristic interactions between the head groups and the substrate, as well as between the adsorbate chains, including the terminating group. The impact of the latter component is of higher importance for long than for short chains. Conversely, SAMs composed of alkanethiols with short chains ($n < 8$) are more strongly determined by head group substrate than by interchain effects. Marked deviations from the geometric pattern prevailing for longer chains have been reported for n-alkyl species as small ethanethiol (Porter et al. 1987), and more recently for butanethiol (Kang and Rowntree 1996).

Several recent research initiatives related to thiols and thiolates on Au(111) have focused on the influence exerted by the long-range order of the substrate on the monolayer. Modeling the Au(111) surface as a flat surface is a problematic premise that has to be justified in every individual case. The clean Au(111) surface reconstructs into a complex structure described by a $(22 \times \sqrt{3})$ unit cell. This pattern forms since the lower coordinated surface atoms prefer a closer equilibrium spacing than the regularly coordinated atoms in deeper layers. The surface thus tends toward an increase of its atom number density. In this process, some Au atoms move from their original *fcc* to *hcp* locations, and some come to occupy places between these two position types. These are incommensurate zones, or *soliton lines*, which are elevated relative to the regions of regular stacking, forming ridges that separate *fcc* from *hcp* regions. The *herringbone* profile arises from long-range elastic forces that cause the emergence of alternating domains, involving a periodic reorientation of the direction of closest packing by 60°, and with a repetition length of about 15 nm.

Substantial experimental as well as computational work has dealt with the question of the interaction between thiolate monolayers and the reconstructed Au(111) surface (Maksymovych et al. 2006, Mazzarello et al. 2007, Wang et al. 2007, Cossaro et al. 2008). Supported by DFT calculations, Wang et al. (2007) described a process of reconstruction lifting by thiolate monolayers that was based on a mechanism proposed by Mazzarello et al. (2007). At a critical level of coverage, a novel adsorption geometry was shown to be maximally stable. In this structure, a surface Au atom is raised to occupy a bridge position between the S atoms of two adjacent thiolate molecules. This *adatom* geometry removes the Au(111) reconstruction. As the coverage threshold that makes the flat substrate more favorable than the reconstructed one, the authors determined a small fraction, namely 7.8%, corresponding to 1 adsorbate molecule per 13 Au atoms, with reference to the non-reconstructed geometry.

17.2.1.1 The Case of CH₃SH

The extreme case of $n = 0$ is realized for methanethiol adsorbates. As self-assembly is understood to be caused by the interaction between alkane chains, no such effect is expected for the simple CH_3SH unit. At the same time, the simplicity of this unit makes it ideally suited to study elementary processes related to the substrate–adsorbate interaction at the molecular level. Further, since the influence of the van der Waals force is strongly deemphasized for methanethiol as compared to thiols with longer chains, CH_3SH monolayers are much more readily accessible by modeling from first principles than $CH_3(CH_2)_{n-1}SH$ monolayers with $n > 1$. While experimental and theoretical endeavors to clarify the basic processes involved in the adsorption and mutual interaction of methanethiol molecules on metal substrates are still on the way (Rzeznicka et al. 2005, Zhou and Hagelberg 2006, Zhou et al. 2007, Maksymovych and Dougherty 2008, Nenchev et al. 2009), essential information has been obtained about structural, dynamical, and bonding properties of this system at the quantum chemical level.

In most of these efforts, modeling was performed by the use of density functional theory (DFT). An early effort (Grönbeck et al. 2000) arrived at the conclusion that the S–H bond of the methanethiol head group should be cleaved once the molecules adsorb on the Au(111) surface, yielding methylthiolate (CH_3S) as dissociation product. This proposition was used as a premise for further DFT simulation of methanethiol on Au(111) (e.g., Yourdshahyan and Rappe 2002). However, a recent combined TPD, AES, and low-temperature scanning tunneling microscopy (LT-STM) measurement (Rzeznicka et al. 2005) did not yield any evidence for S–H cleavage. From observation of dimethylsulfide traces at elevated temperature following ion bombardment of the surface, however, it was concluded that scission of this bond occurs on Au defect sites. This model was confirmed by DFT computation (Zhou and Hagelberg 2006), where non-dissociative adsorption was found to be thermodynamically stable.

First principles theory has also recently commented on the problem of sulfur head group dimerization (Zhou et al. 2007). Monomer and dimer patterns were clearly distinguished CH_3SH on Au(111) at various levels of coverage. The dimerized structures were found to be of higher stability than the monomers. In the former case, adjacent S atoms turned out to be separated by about 3.5 Å, while the corresponding distance was approximately 5.0 Å for the latter. The separation between the interacting molecules in the dimerized phase is sizably larger than the bond length proposed for longer alkane chains in Fenter et al. (1994). It excludes a chemical bonding effect as the physical cause of the pairing phenomenon. Neither can, as argued in Zhou et al. (2007), the van der Waals force be the agent that generates the dimer phase. Instead, thiol dimerization is a surface-mediated effect, modified by electrostatic repulsion since the adsorbed molecules are not electrically neutral: Electron transfer proceeds from the metal substrate to the adsorbate, endowing each molecule with a small negative charge.

Self-assembly of methanethiol on the intact Au(111) reconstruction was observed by Maksymovych and Dougherty (2008). As shown by STM investigation at low temperature

($T < 60\,\text{K}$), the preferred adsorption geometry is characterized by an adsorbate domain orientation parallel to the soliton lines. As the salient feature of the resulting methanethiol superstructure, its closest packed direction aligns itself with that of the Au(111) reconstruction.

Very recently, Nenchev et al. presented a combined experimental and theoretical study of methanethiol adsorption on the Au(111) surface as a function of temperature (Nenchev et al. 2009) in the range between 90 and 300 K. At saturation coverage, two SAM structures were detected. A regular arrangement of adsorbate rows that form on top of the intact Au(111) reconstruction is observed at temperatures below 125 K. These rows are seen to line up with the closest packed substrate direction. While this finding agrees qualitatively with the respective result of Maksymovych and Dougherty (2008), controversy still exists about the surface sites occupied by the adsorbates. Deposition at 160 K gives rise to the formation of a new ordered phase characterized by a closely packed arrangement, displaying the hexagonal symmetry of the substrate. The Au(111) reconstruction is not present any longer in this phase. This is in keeping with a dissociative bonding model, involving thermally produced thiolate and subsequent adatom formation (see above, Maksymovych et al. 2006). Eventually, the observed hexagonal phase is established as a consequence of the increased diffusion of gold atoms at the temperature of the experiment, 160 K (Fenter 1998).

The recent results on methanethiol in contact with Au(111) demonstrate that the presence of an alkane chain is not a necessary condition for the emergence of monolayers with long-range order. From present theoretical assessment (Zhou et al. 2007), the van der Waals force, while strongly influencing the geometry as well as stability of alkanethiols with longer chains, is not the organizing agent that drives the self-assembly of methanethiol. A surface-assisted interaction is thus responsible for the latter effect.

17.2.2 Organosilicon Derivatives on Hydroxylated Surfaces

In this category, the most extensively studied SAM systems are composed of alkylsilane derivatives in contact with oxidized silicon (SiO_2) surfaces or analogous hydroxylated substrates such as SnO_2 or TiO_2 (Helmy and Fadeev 2002). The adsorbate molecules are of the form $RSiX_3$, R_2SiX_2, or R_3SiX, where R denotes a hydrocarbon backbone, and X stands for alkoxy or chloride. An example for a popular SAM forming molecule of this type is given by octadecyltrichlorosilane, or *OTS* (CH_3-$(CH_2)_{17}$-SiX_3). In what follows, we will focus to the latter adsorbate as a model species. The reaction of chloro- and alkoxysilanes with various types of Si-based surfaces has been investigated in great detail (e.g., Kallury et al. 1988). As the molecule adsorbs to the substrate, Si–Cl bonds react with OH groups attached to the surface, as well as trace water (van Roosmalen and Mol 1979). When the Cl components are split off, a network based on the Si–O–Si motifs emerges, connecting

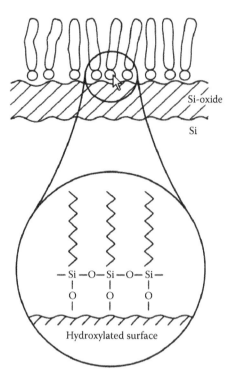

FIGURE 17.4 Cross-linked architecture formed by *n*-alkyltrichlorosilane on a hydroxylated SiO_2 surface. This scheme, however, does not give rise to global crystalline order (see text). (From Schreiber, F., *Prog. Surf. Sci.*, 65, 151, 2000. With permission.)

the adsorbates to each other and to the surface. The resulting structure is illustrated in Figure 17.4.

This three-dimensional polymer is vertically as well as laterally stabilized by chemical bonding, in contrast to alkanethiol on Au(111), where self-assembly is achieved by laterally acting van der Waals forces. Directional constraints rule out cross-linking by Si–O bonds as the construction principle of globally ordered SAMs at full coverage (Stevens 1999). As the O–Si–O bond angle is less than 180°, the distance between next neighbor O atoms is less than 3.2 Å, assuming an Si–O bond length of 1.6 Å. This length interval, however, is too narrow to allow for parallel, upright hydrocarbon chains. For the out-of-plane phase of OTS on oxidized silicon, a film thickness of about 25 Å was reported (Tidswell et al. 1990, Wassermann et al. 1989), from which a tilt angle of 20° was inferred. This is lower than the tilt angle found for alkanethiol on Au(111) (30°, see above), corresponding to a smaller interchain spacing than is typical for SAMs based on thiols.

17.2.3 SAMs on Nanoparticles

The study of metal nanoparticles is largely motivated by the surface and quantum size effects characteristic for these species. The dimensional reduction of the respective bulk materials gives rise to novel phenomena, among them the emergence of superparamagnetism, fluorescence yields in excess of 80%, lowered melting points, altered surface reactivities, or electromagnetic energy localization as observed in plasmon spectra (Love et al.

2005, Vericat et al. 2008). These effects are both of systematic and technological interest. As a basic structural feature, nanoparticles exhibit a high surface-to-bulk ratio. While 0.2% of the Au atoms of a gold particle with a diameter of 1000 nm are located on the surface, this fraction increases to 88% for a 1.3 nm particle (Love et al. 2005).

Exploring SAMs on metal nanoparticle substrates involves studying adsorption of the SAM-forming molecules on surfaces with high curvature. These do not need to be spherical. For instance, gold nanocrystals with diameters larger than about 0.8 nm tend to stabilize as truncated octahedra or as cuboctahedra (Love et al. 2005). Eight (111) faces are here truncated by six (100) faces, implying that the nature of the substrate varies over the surface of the nanoparticle. More importantly, a large fraction of the metal atoms inhabit boundary regions that separate different surface segments from each other. Assuming a gold cluster that is loaded with alkanethiol chains over the whole of its surface, it is obvious that a molecule attached to an edge has a higher lateral mobility than a molecule occupying a site near the center of a face.

With increasing particle radius, the surface curvature diminishes, and the adsorption conditions become comparable to those realized for planar substrates. A high degree of conformational order in the alkanethiolate chains, similar to that found in the corresponding planar SAMs, was reported by Badia et al. (1997), who investigated gold nanoparticles of about 30 Å in diameter with octadecanethiolate monolayers, using various characterization techniques. From molecular dynamics simulation (Luedtke and Landman 1996, 1998) of thiolate chains on gold nanocrystallites ranging in size from 140 to 1289 Au atoms, the formation of compact monolayers on the (111) and (100) faces of the substrate was obtained, with substantially higher packing densities than found on a flat Au surface. The molecules tend, at lower temperature, to organize into preferentially oriented molecular bundles, with approximately parallel adsorbate backbone within each of the bundles, and zones of disorder at the corners and vertices.

17.2.3.1 Superlattices Formed by Nanocrystals

Once the problem of fabricating and characterizing nanoparticles coated with SAMs is solved, novel structures can be built from these composites. SAM-loaded metal nanoparticles may form monolayers themselves, and they also may arrange as thin films or three-dimensional superlattices (Wang 1998, Gutierrez-Wing et al. 2000, Wang et al. 2000). In all of these cases, the organic ligands are the agents that facilitate the self-assembly of the nanocrystals.

The separation of closely packed adjacent spherical nanoparticles (see Figure 17.5a) with adsorbed alkyl chains has been determined to increase linearly with the chain length (Martin et al. 2000). The interparticle distance was shown to increase with the number of carbon atoms contained in the alkyl backbone at about half the rate expected from high performance liquid chromatography (HPLC) studies of the capped nanoclusters in solution. This suggests that the bonding mechanism

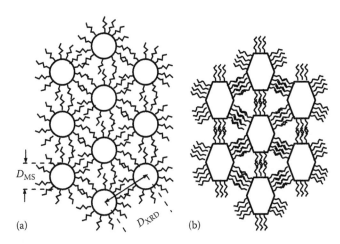

FIGURE 17.5 Two types of self-assembled superlattices formed by SAM-coated nanocrystals, involving (a) spherical and (b) faceted substrate particles. (From Wang, Z.L., *Adv. Mater.*, 10, 13, 1998. With permission.)

operative in the arrays of functionalized nanoparticles involves interlocking alkyl chains. Martin et al. (2000) demonstrated by transmission electron microscopy (TEM) studies that the interparticle spacing of the nanocrystal superlattice can be controlled by capping with alkanethiols within a certain regime of chain lengths. Utilization of thiols with very short backbones was shown to result in precipitation, and nanoparticles capped with substantially longer chains (involving C_{16} and C_{18}) displayed a wide size distribution and did not exhibit crystalline order. Intermediate chain lengths (C_6 and C_{14}), however, led to the formation of ordered architectures. Adsorbates with C_6 backbones were shown to build exclusively three-dimensional structures, while two-dimensional lattices emerged from choosing the species with C_8 to C_{14} backbones, implying that not only the interparticle distance, but also the dimensionality of the resulting superlattice can be controlled by manipulating the alkanethiol chain length.

17.3 Growth and Kinetics

The various systems and structures highlighted in the preceding section are the result of self-assembly, as determined by the growth behavior of the considered monolayer. In an attempt to understand this behavior, one has to ask for the phases traversed by the system as it develops from an initial condition of extremely low molecular surface density to the final stage of saturation coverage. While the former situation is governed by the kinetics of individual molecules interacting with the substrate, the latter is strongly impacted by the geometric constraints valid for the considered combination of adsorbate and substrate, such as closest packing conditions. In the context of SAM evolution, many questions arise that pose experimental as well as theoretical challenges of higher complexity than those associated with SAM equilibrium structures. These concern mostly the forces that drive self-assembly, the various phases adopted by SAMs as a function of control parameters such as temperature, the level of coverage, and the dependence of the specific modes of SAM growth on the physical and chemical properties of the chosen substrate–adsorbate combination.

Due to the intricacy of the subject matter, the studies presented on SAM growth phenomena are not as numerous as those on structural aspects of SAMs at equilibrium. The majority of the publications devoted to questions of SAM self-assembly mechanisms refers to the prototype alkanethiol on Au(111). An agreement has been reached about the phase observed at low coverage: In this regime, alkanethiol tends to form an in-plane, or "lying-down" geometry, with alkyl chains oriented parallel to the substrate. The highest possible coverage in this phase is readily estimated from surface filling arguments. For decanethiol on Au(111), it amounts to 27% of the out-of-plane saturation coverage (Schreiber 2004). This fraction decreases with increasing chain length. The well-characterized out-of-plane phase, involving a tilt angle of about 30° with respect to the surface normal and formation of a $(2\sqrt{3} \times 3)$ superlattice (see Section 17.2.1) is realized at a level of coverage larger or equal to 0.5. The intermittent phase, mediating between the limiting case of the in-plane and the conventional out-of-plane orientation, is a subject of lively discussion (Barrena et al. 2003). In this regime, molecular arrangements with a tilt angle of 50° have been reported (Barrena et al. 2003) as well as new structures that involve mixed in-plane regions often coexisting with zones of pure in-plane phases. In view of this multiplicity of phases it is obvious that no simple unique description of SAM growth can be given. The diagram shown in Figure 17.6, displaying the phases of decanethiol as a function of the level of coverage as well as temperature, clarifies that any path leading from low to high SAM coverage has to cross several phase boundaries.

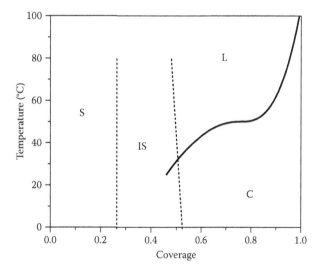

FIGURE 17.6 Various phases of decanethiol on Au(111) in a landscape described by temperature and the level of coverage as independent variables. (From Schreiber, F., *J. Phys.: Condens. Matter*, 16, R881, 2004.) Symbols denote the in-plane (or "striped") phase (S), a domain of intermediate structures (IS), a liquid (L), and an ordered out-of-plane (C) phase.

For this reason, the time behavior of SAM formation is not adequately captured by the most elementary scenario of Langmuir kinetics, which leads to an exponential law of the form $\sigma(t) = 1 - \exp(-t/\tau)$ for the level of coverage σ as a function of time, where τ is a time constant. As the growth of the system does not proceed within a single uniform phase, it cannot be entirely defined by the number of unoccupied sites. As one assembles the SAM from the gas phase and thus enhances the coverage level steadily from an initial level of zero, one moves from an initial in-plane structure through the intermediate regime to enter, depending on the temperature chosen, a liquid or the crystalline $(2\sqrt{3} \times 3)$ phase (Poirier and Pylant 1996). These various domains are governed by differing time constants, such that SAM growth proceeds in multiple time scales.

Evidently, transitions between neighboring SAM phases do not occur instantly, but within a characteristic time interval required for altering the respective SAM architecture. The initial in-plane phase has proven to be of high stability, requiring substantial adsorbate flux for conversion into the final out-of-plane phase and temperatures above 200 K (Love et al. 2005). It is thus plausible that the kinetics of alkanethiol assembly is sensitively dependent on the molecule chain length. Chain–substrate interaction as well as steric effects inhibit the transition from in-plane to out-of-plane structures more efficiently for longer than for shorter chains. The out-of-plane phase of decanethiol on Au(111) has been reported to grow by a factor of 500 more slowly than the initial in-plane phase (Schreiber 1998).

For a deeper understanding of the initial interaction of the adsorbate with the substrate, the existence of a *physisorbed* precursor that eventually converts into the well-characterized *chemisorbed* state is essential. The physisorption of alkanethiol on Au(111) is understood as a manifestation of the van der Waals force (Lavrich et al. 1998, Schreiber 2000). The molecule is here already oriented parallel to the surface, albeit retaining a mobile thiol head group, as presented schematically in Figure 17.7. The two equilibrium states are separated by the potential barrier shown in the figure whose effective height has been estimated to be about 0.3 eV for $SH(CH_2)_{n-1}CH_3$, independent of n. The physisorbed state makes it possible for the adsorbate to remain close to the surface for periods that are long as compared to the times typical for an elastic or weakly inelastic scattering event between the molecule and the surface. This preliminary state is instrumental for the self-organization of the adsorbate system since the molecule–substrate interaction is less dominant here than in the chemisorbed equilibrium state and thus the intermolecular effect more emphasized. Also, the precursor state greatly enhances the probability for the eventual formation of a chemical bond between the substrate and the adsorbate.

Growth from the gas phase provides experimental conditions that facilitate detailed studies of SAM assembly as a function of coverage. For most practical purposes, however, deposition from the liquid phase is preferred (see above), as the simplicity of this fabrication route is among the major advantages of

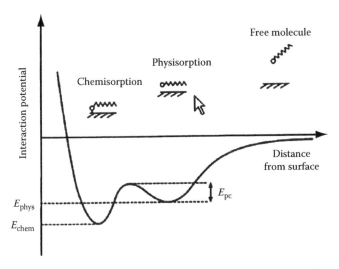

FIGURE 17.7 Schematic potential curve for the interaction of a single decanethiol molecule with the clean Au(111) surface. (From Schreiber, F., *Prog. Surf. Sci.*, 65, 151, 2000. With permission.). Minima for chemisorbed decanethiol and its physisorbed precursor are indicated along with sketches of the respective adsorption geometry. The chemisorption well segment of the curve only refers to the chemical bond between the surface and the head group, excluding the chain–surface interaction.

SAMs. The time dependence of assembly from solution has been shown to obey, to a good approximation, the basic exponential law that follows from the assumption of Langmuir kinetics (e.g., Dannenberger et al. 1999). For usual concentrations of the solution from which the SAM is grown, a coverage level of 80%–90%, corresponding to the chemisorbed out-of-plane phase, is typically reached within several minutes. Subsequently, the onset of long-term growth behavior is observed. A temporal succession of well-separated growth steps was obtained by sum frequency generation (SFG) measurement on docosanethiol ($SH–(CH_2)_{21}–CH_3$) deposited on Au(111) (Himmelhaus et al. 2000). By analysis of vibrational intensities as a function of the immersion time, it was concluded that the fastest step, chemisorption of the S head group, is followed by a much slower process that involves straightening the hydrocarbon chains. This step was found to proceed more slowly than the initial one by a factor of three to four while being 35–70 times faster than the final stage of SAM evolution, which was dominated by reorientation of the end groups.

The role played by the in-plane phase under conditions of SAM growth from the liquid phase is still controversial (Xu et al. 1998, Munuera and Ocal 2006). This phase has been detected by STM and AFM investigation. The use of very dilute solutions provided time windows sufficiently wide to identify in-plane structures in the beginning stages of SAM assembly. A recent scanning force microscopy study (Munuera and Ocal 2006) identified an initial growth stage where the in-plane configuration coexists with islands formed by out-of-plane molecules. An intermediate disordered phase leads to the final out-of-plane configuration at high coverage.

17.4 Applications of SAMs

An abundance of SAM applications in nanoscience and nanotechnology has been proposed, as SAM engineering has proved to be an extremely versatile tool of surface modification and functionalization. We highlight five major categories of practical current interest in SAMs: nanobiotechnology (Section 17.4.1), flexible interfaces responding to external stimuli (Section 17.4.2), nanolithography (Section 17.4.3), molecular electronics (Section 17.4.4), and nonlinear optics (Section 17.4.5).

17.4.1 Nanobiomolecular Applications

Numerous biological applications have been identified for SAMs (Love et al. 2005), related to immobilizing biomolecular species on surfaces. The main strategies for achieving this goal involve the direct deposition of appropriately selected or designed biomolecules on substrates and the attachment of the targeted molecule to suitably modified SAM end groups.

The first of these two methodologies is based on grafting a biological molecule into a conventional SAM. The studied molecular species may be tailored specifically to adhere tightly to the metal substrate, as described in Case et al. (2003), where a metalloprotein incorporating a thiol group terminated by COOH was attached to the gold surface. This species was embedded into a traditional SAM composed of alkanethiols with C_{18} backbones (see Figure 17.8) by the process of *nanografting*. Here, the AFM technique is applied to identify a sufficiently flat SAM region. Subsequently, the AFM tip is used to stimulate exchange between the biomolecules and those in the monolayer throughout the region of interest. AFM imaging of the selected area at low force makes it possible to visualize the exchanged molecules.

The second procedure involves a suitably modified end group to activate a mechanism of lock–key recognition. The adsorbate thus terminates in a receptor group chosen to bind selectively to a targeted biological species that is to be immobilized on the

FIGURE 17.8 (See color insert following page 22-8.) Model of six three-helix metalloproteins grafted into a SAM with C18 backbones. (From Case M. et al., *Nano. Lett.*, 3, 425, 2003. With permission.) For clarity of presentation, the alkanethiol molecules are shown as oriented perpendicularly with respect to the surface, and not with their tilt angle of 30°.

SAM. For instance, Sigal et al. (1996) generated a SAM that binds selectively proteins whose primary sequence terminates with a stretch of six histidines. A tetravalent chelate of Ni(II) was chosen as the alkanethiol end group, and the histidine-tagged proteins were shown to bind to the SAM by histidine interaction with the two vacant sites on the Ni(II) ions of the chelate.

17.4.1.1 Sensing by SAM-Coated Microcantilevers

The design of micromechanical systems used for the purpose of chemical and biological sensing is an area of rapid current development (Raiteri et al. 2001, Ziegler 2004). This method is based on observing the response of a mechanical *transducer* upon adsorption of the targeted species. This element may be a thin silicon cantilever with a clean or monolayer-coated gold film attached on one side. As a result of a heat transfer, mass changes, or adsorption-induced surface stress, the cantilever undergoes bending by a small margin in the nanometer regime. Deflections related to this deformation may be monitored by the use of optical equipment, such as a laser diode in conjunction with a position sensitive photodetector. Figure 17.9 illustrates the mechanic effect of alkanethiol adsorption on a thin gold-coated silicon layer (Lavrik et al. 2004). The resulting SAM causes compressive strain that curves the cantilever. This behavior is ascribed to the charge transfer from the Au substrate to the chemisorbed thiols and the corresponding electrostatic repulsion between the molecules.

Wu et al. (2001) investigated single-stranded DNA (ssDNA) immobilized on a gold-coated silicon nitride substrate that was functionalized by a $SH(CH_2)_6$ layer. In this case, the net negative charge of the ssDNS backbones generates a repulsive intermolecular force. Further, under the experimental conditions, the highly flexible ssDNA units were forced to occupy a space smaller than their natural size. The combination of

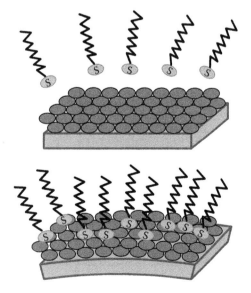

FIGURE 17.9 Bending response of a microcantilever composed of a silicon layer coated with gold film upon formation of an alkanethiol SAM on the Au substrate. (From Lavrik, N.V. et al., *Rev. Sci. Instrum.*, 75, 2229, 2004. With permission.)

both effects was found to cause the cantilever to bend. The release of surface strain and thus a decrease of the cantilever deformation was observed after the addition of the complementary DNA strands, followed by hybridization between the strands. This phenomenon was interpreted as largely entropic in nature: the emerging double-stranded DNS (dsDNA) molecules are more compact and rigid than their single-stranded counterparts and thus occupy less lateral space on the surface. This effect may be exploited to design biological sensors of high accuracy and efficiency by suitable calibration of the cantilever response.

Explosives immobilized by SAMs consisting of 4-mercapto-benzoic acid in contact with Au films are detectable with sensitivities in the parts-per-trillion (ppt) range (Pinnaduwage et al. 2003). This was accomplished by measuring the change in the resonance frequency as well as the bending response of microcantilevers that were coated with the SAM-covered thin Au layer upon adsorption of minute traces of the respective agent.

17.4.1.2 Biological Sensing by Use of SAM-Coated Nanoparticles

Efficient procedures for assembling nanoparticles (see Section 17.2.3) in a well-controlled, reversible way by functionalizing them with biological species have been described (Mirkin et al. 1996). Thus, in experiments on DNA/nanoparticle hybrid materials (Mirkin et al. 1996), short DNA segments with typically 20 or fewer bases (DNA oligonucleotides) were attached to thiol groups on the surfaces of 13 nm Au particles. When oligonucleotide duplexes with "sticky ends" that are complementary to the base sequences on the SAMs were added to the solution, the DNA fragments immobilized on the Au particles and the duplexes were found to interlock, leading to the formation of larger aggregates. Disintegration and reintegration of the resulting structure can be controlled by temperature.

This effect may be exploited for the purpose of colorimetric sensing (Elghanian et al. 1997), which involves detecting a biological agent that links the nanoparticles by measuring the light extinction spectra in the nanoparticle ensemble. If the next neighbor distances substantially exceed the average particle diameter, the aggregate appears red. As the distances shrink to less than this average, the color changes to blue. This behavior is ascribed to the surface plasmon resonance of the Au particles and provides the basis for a simple yet efficient optical sensing technique.

17.4.2 Switchable Interfaces

A major goal of interfacial engineering is the design of two-dimensional systems that react to external stimuli, responding to their environment by changes of their properties. A recent example for the implementation of this notion was given by Lahann et al. (2003), who demonstrated external control of wetting, and associated properties, by use of a (16-mercapto) hexadecanoic acid (MHA) SAM on Au(111) (Lahann et al. 2003). As an essential prerequisite for the mechanism proposed by the authors, this molecule contains hydrophobic as well as hydrophilic segments, namely an alkyl chain and a carboxylate end group, respectively. A switchable interface was introduced by devising a prescription to confer these contrasting properties to the surface upon a change of the electrochemical potential of the substrate.

The elementary steps involved in this procedure are illustrated in Figure 17.10. Initially, the substrate is covered by a MHA derivative, namely MHA (2-chlorophenyl) diphenylmethyl ester (MHAE). As a consequence of their expansive head groups, the SAM formed by these molecules (shown in the left image of Figure 17.10) displays a lower level of coverage than typically achieved in the case of MHA adsorption. In the following hydrolysis step, however, MHAE is reduced to MHA, which maintains the low concentration as well as the upright orientation of the precursor species.

On account of the low coverage level of the resulting SAM, it has a high degree of flexibility to undergo conformational changes. Manipulating the electrochemical potential of the Au substrate, it is possible to control the SAM structure and, by the same token, the chemical nature of the interface. As demonstrated by sum frequency generation (SFG), as well as wetting measurement (Lahann et al. 2003), the molecules bend down, attaching their negatively charged end groups to the surface, if the Au(111) substrate acquires a sufficient net positive charge. The standing-up phase is recovered as the substrate polarity is changed. The former configuration gives rise to a hydrophobic, the latter to a hydrophilic interface. The substrate potential thus represents a simple control parameter for the definition of SAM architecture and associated macroscopic properties, such as wettability, adhesion, friction, and biocompatibility.

17.4.3 Nanolithography

Nanolithography involves the fabrication of nanometer-scale structures, and specifically the design of surface or interface patterns with at least one lateral dimension ranging between 1 Å and approximately 100 nm. Nanolithographic methods are used in many diverse areas of current technology, among them manufacturing miniature circuits on electronic chips, engineering nanoelectromechanical systems, or submicron machinery that translates electrical signals into mechanical responses. Another use of these procedures involves optical applications, where the availability of minute relief features to be employed as diffraction gratings, waveguides, or microlens arrays is relevant.

With their vertical dimension, perpendicular to the surface, in the order of nanometers, the films provided by SAMs are well suited for the purpose of patterning at the nanoscale. Thus, SAMs have been used as resists for pattern transfer (Kumar et al. 1995) and as templates to pattern proteins and other biosystems (Bernard et al. 1998). Earlier, micrometer-sized structures have been created from SAMs by the use of techniques such as photolithography (Huang et al. 1994, Wollman et al. 1996), microcontact printing (Abbott et al. 1994, Kumar et al. 1995), microwriting (Abbott et al. 1994, Xia and Whitesides 1995), and micromachining (Xia and Whitesides 1995).

Among these methodologies for the use of SAMs in the design of lateral structures, microcontact printing (μCP), as introduced by the Whitesides group (Tien et al. 1998), is probably the most frequently used one. In analogy to conventional printing, μCP involves "ink" attached to a "stamp." The latter is a template usually cast from a photolithographically fabricated resist pattern. The stamp consists of a polymer, such as polydimethylsiloxane (PDMS), which exhibits elastic in contrast to plastic deformation over a wide range of strain. The

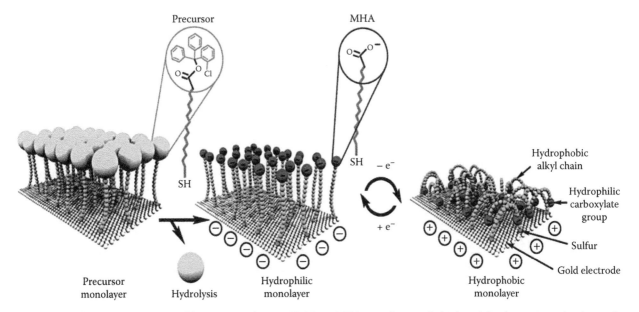

FIGURE 17.10 Schematic presentation of the transition from a MHAE to a MHA monolayer on Au(111), and the alternative molecular conformations of MHA as a function of the Au substrate potential. (From Lahann, J. et al., *Science*, 299, 371, 2003. With permission.)

utilization of PDMS therefore makes it possible to alternate features sizes by applying pressure to the stamp. The raised portions of the stamp are coated with a SAM precursor that plays the role of the ink. Microprinting proceeds by bringing the PDMS stamp in direct contact with a substrate such as Au(111) for several seconds. As a result, monolayers are transferred to the substrate areas in contact with the stamp.

The smallest features achievable by means of this simple and robust scheme are in the order of 50 nm, spreading over an area of about 10 μm² (Tien et al. 1998). Higher resolution is accomplished by use of hardened PDMS stamps whose raised segments are sharpened, i.e., V-shaped (Li et al. 2002). Further, species of high molecular weight, dendrimers and proteins, were employed as SAM-forming adsorbates in order to reduce diffusion. As was demonstrated by AFM imaging, this procedure allows for *nanocontact printing*, leading to the formation of sharply defined arrays on the substrate at the 40 nm scale over an area of 3 mm².

Serial nanoprinting techniques have been developed that complement the *parallel*, or single-step, methods outlined in the preceding paragraphs. We mention *dip-pen nanolithography* (DPN), which was developed by the Mirkin group (Piner et al. 1999), as an alternative procedure of transporting molecules to selected substrate sites. It involves the use of an AFM device in air and is based on the observation that, under this experimental condition, water condenses in the space between the tip and the substrate, as shown in Figure 17.11. As Piner et al. were able to demonstrate, molecules flow from the tip to the sample by capillary action when an AFM tip coated with octadecanethiol (ODT) is brought into contact with a sample surface. The size of the water meniscus forming between the tip and the surface, as determined by the relative humidity, controls various parameters that impact the printed profile, namely the rate of adsorbate transfer, the effective tip–substrate contact area, and the DPN resolution. Thus, DPN is instrumental to creating well-defined adsorbate islands whose distribution over the substrate is unconstrained by any rigidly patterned stamp.

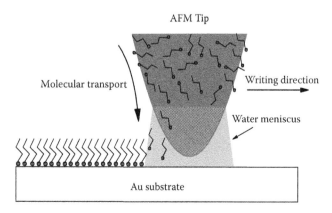

AFM Tip

Molecular transport

Writing direction

Water meniscus

Au substrate

FIGURE 17.11 Basic operation of "dip-pen" nanolithography (DPN). Transfer of molecules proceeds from an AFM tip to the substrate through a volume of condensed water.

A recent extension of the method is *polymer pen lithography* (Huo et al. 2008), which employs arrays of soft elastomeric tips, driven by a piezoelectric scanner rather than a single AFM tip to implement "direct writing" to the substrate. This scheme combines the feature size control of DPN with the large-area capability of contact printing.

The methods described so far in this subsection are *positive printing* techniques, associated with the addition of SAM-forming molecules to a substrate. The complementary methodology involves *negative printing*, where a preexisting SAM is reduced to generate the desired spatial arrangement. This approach takes advantage of the strong and localized AFM or STM tip–surface interaction. *Scanning probe lithography* (SPL) has been applied successfully in various configurations (Liu et al. 2000). *Nanoshaving* utilizes an AFM tip that exerts high local pressure on the chosen contact regions, corresponding to a high shear force that results in the displacement of adsorbate molecules. STM-based variations of scanning probe lithography operate without physical contact between the microscope tip and the substrate. Here, the selection of a targeted SAM area is followed by increasing the tunneling current beyond a certain threshold to cause electron-induced diffusion or electron-induced evaporation, corresponding to displacement of metal atoms or desorption of adsorbate molecules, respectively.

Beyond the procedures indicated above, SAM patterning has been achieved by more traditional methods, such as the impact of photons (Calvert 1993) as well as ion (Ada et al. 1995) and electron (Lercel et al. 1994) beams. These types of irradiation do not necessarily remove the adsorbed molecules in the targeted SAM area, but might as well be employed for proper modification of these molecules. Felgenhauer et al. (2001) investigated a biphenylthiol SAM patterned by exposure to a 300 eV electron beam and observed a tightly connected network of the biphenyl moieties, bridging defects and reducing the occurrence of transient pinholes induced by thermal fluctuations. Subsequent copper deposition proceeded selectively to the nonirradiated surface segments. For alkanethiol SAMs, in contrast, electron irradiation causes the degradation of the alkane chains, counteracting the blocking behavior of the native SAM and favoring the irradiated portions of the surface as copper deposition areas. Thus, as nanoprinting media, alkanethiol and biphenyl-based thiols display inverse responses to electron irradiation, acting as positive and negative resist for copper deposition, respectively. Nanoprinting on SAMs is a lively area of continuing research (e.g., Huo et al. 2008, and references therein).

17.4.4 Molecular Electronics

The extension of electronic circuitry to molecular dimensions has been pursued under many different viewpoints (Tour et al. 1998). The overarching aim of these efforts is the fabrication of both passive and active electronic elements, wires as well as transistors, from molecules as fundamental building blocks. SAMs are a natural focus of this initiative as they are usually composed of organic molecules with excellent electric contact to a metal substrate that might be employed as an electrode. Different schemes of integrating SAMs into electronic circuits have been explored. Evidently, the monolayer may be sandwiched between two metal plates instead of adsorbed to a single surface. If this configuration is realized, the metal substrates assume the function of electrodes. To ensure equally tight bonding to both of them, alkanedithiol (SH-$(CH_2)_n$-SH) may be employed as monolayer substance instead of alkanethiol (SH-$(CH_2)_{n-1}$-CH_3) (Wang et al. 2005).

A circuit element comprising a SAM between two metal electrodes may be understood as a metal–insulator–metal (M–I–M) tunnel junction, since the SAM layer, when properly manufactured, represents a single van der Waals crystal with a HOMO–LUMO gap of approximately 8 eV (Ratner et al. 1998). Here, HOMO refers to the highest occupied and LUMO to the lowest unoccupied molecular orbital of alkanethiol. For the investigated case of a defect-free, large-energy-gap SAM, electron tunneling could be identified unambiguously as the main intrinsic conduction mechanism. This conclusion was based on current-voltage ($I(V)$) measurements spanning a sufficiently wide temperature range, namely 80–300 K. No significant temperature dependence of the $I(V)$ characteristics was found when the voltage was varied between 0.0 and 1.0 V, a behavior indicative of tunneling as opposed to thermionic emission or hopping conduction. The latter mechanisms have been established, for example, in case of SAMs consisting of 4-thioacetylbiphenyl SAMs (Zhou et al. 1997) and 1,4-phenelyene diisocyanide (Chen et al. 1999). When, however, the electrode Fermi levels lie within a large HOMO–LUMO gap, as is realized for short alkanethiol sandwiched Au surfaces, conduction is expected to be dominated by tunneling. Thus, the conduction behavior of a SAM-based electronic element depends sensitively on the monolayer species.

Electron tunneling through alkanethiol monolayers has recently also been studied with the help of conductive scanning force microscopy (SFM, Munuera et al. 2007), where the role of the second electrode is taken by the microscope tip. Further, tunneling junctions designed by coating Hg drops with alkanethiols have been proven to be functional (Rampi et al. 1998, Rampi and Whitesides 2002). Using micromanipulators, York et al. brought two Hg drops in contact with each other after having covered one or both of them with alkanethiol monolayers (York et al. 2003). In this fashion, a comparison between the conductive properties of bilayers and monolayers was made possible. As a result, the tunneling current recorded for bilayers of octanethiol or nonanethiol turned out to be two times as large as that observed for hexadecanethiol or octadecanethiol monolayers. This finding was rationalized in terms of weak electronic coupling associated with noncovalent molecule–electrode interface that is part of the monolayer, but not the bilayer system. From these data, the presence or absence of a sulfur–metal contact is of critical impact on the current through the SAM. This and related research on the conduction properties of SAMs (e.g., Haag et al. 1999, Holmlin et al. 2001) lay the ground for their future use as ingredients of nanoelectronic devices.

17.4.5 Nonlinear Optics

The design of efficient nonlinear optical media is among the primary concerns of current electro-optical technology. In particular, much effort is invested into the search for suitable alternatives to traditional optical materials with nonlinear properties. These are chiefly inorganic single crystals with high second- or third-order susceptibilities. However, these materials tend to be costly, fragile, and difficult to fabricate and incorporate into commercial equipment (Bierlein and Vanherzeele 1989, Dagani 1996). Suitably selected and prepared polymeric materials, in contrast, offer a multitude of desirable features, such as large nonlinear optical susceptibilities as well as thermal and mechanical stability, and combine low cost with ease of manufacturing (Garito et al. 1994).

SAMs formed by chromophores of high density and high polarizability have proven to be extremely efficient media for nonlinear optics applications (Heflin et al. 1999, Naraoka et al. 2002). The polarizability criterion can be satisfied by choosing a species that contains a conjugated electron system, such an aromatic ring, coupling a donor–acceptor pair. An example for such an arrangement is given by hemicyanine, which exhibits strong intramolecular charge transfer (CT) from the alkylamino group to the pyridinium ion group. In order to anchor the hemicyanine molecule in a gold substrate, it was joined in a recent experiment (Naraoka et al. 2002) with an organosulfur group, to yield hemicyanine disulfide. From second harmonic generation (SHG) experiment performed by observing reflection from hemicyanine SAM, a large second-order nonlinear susceptibility was obtained.

The optimal design SAM-forming materials with respect to their nonlinear optical response is among the challenges of current molecular engineering. SAM-based materials with very substantial second-order nonlinear susceptibilities have been fabricated by use of *ionically self-assembled monolayers* (ISAMs, Heflin et al. 1999). These are manufactured by dipping a negatively charged substrate into a solution containing water-soluble cationic polymer molecules. The procedure is repeated with anionic instead of cationic polymer molecules. Electrostatic attraction orients the anions and attaches them to the cations that are bound to the substrate. The resulting composites are stable, robust, uniform organic layers that display a marked deviation from centrosymmetry, favoring nonlinear optical effects.

The list of SAM applications as presented in this section could be complemented by adding numerous further actual or potential operation modes of SAMs, most of them similar to what has been described under Sections 17.4.1 through 17.4.5. In particular, we name corrosion inhibition and fabrication of protective coatings (Scherer et al. 1997), lubrication and adhesion studies (Carpick and Salmeron 1997), manufacturing of tunable surfaces by SAM end-group modification (Barrena et al. 1999), and the investigation of dynamic interactions involving spatially fixed molecules (Bracco and Scoles 2003) as areas of interest for present or future SAM engineering.

17.5 Summary and Future Perspectives

An extensive body of information has accrued over the past two decades on the paradigmatic SAM system, namely alkanethiol adsorbates anchored in planar gold surfaces. Experiment and theory have collaborated to elucidate both the microscopic and macroscopic aspects of these composites. Pertaining to molecular characteristics of SAMs, adsorbate formation has been analyzed in terms of kinetics and thermodynamics, geometric properties, intermolecular as well as molecule–surface interactions. The organic film properties of SAMs have been studied along various avenues of investigation, yielding information on structural and energetic film parameters such thickness, stability, and persistence besides chemical features such as wettability and affinity toward forming bonds with external groups. Much information on the practical applications of SAMs is available, related to creating, modifying, and functionalizing these systems.

Although the rapid rate and daunting number of publications related to SAMs testify to a tremendous amount of research and engineering done on this subject, the field of SAMs is still wide open to scientific inquiry. Continuing challenges include the extension from the model system alkanethiol on a plane metal surface to more complex composites, involving nonplanar substrates and adsorbates that deviate from the linear chain prototype. Among many materials that have not yet been explored as SAM substrates, or only to a relatively small extent, are various semiconductors of commercial interest, such as InP or GaN. Mixed substrates covered by different SAMs with differing functionalities seem to be worthwhile of future developmental work as well as SAMs that resist heat and strain to a higher extent than those presently existing.

The research on novel substances for the design of SAMs is to be supplemented by studies on a wider range of physical and chemical SAM properties than before. Little is known, for instance, about magnetic phenomena pertaining to SAMs, and their nonlinear optical characteristics are presently a topic of lively discussion and investigation.

Nanotechnological innovation is expected to benefit from further advances in SAM-related research. Within this effort, developing *dynamic SAM systems* that react to external stimuli will have a high priority. Some exemplary implementations of this concept have been mentioned above, namely switchable surfaces (Section 17.4.2), or alkanethiols attached to gold nanoparticles and functionalized with single DNA strands (Section 17.4.1.2), which are designed to respond to the changes in the electric potential and to the presence of complementary DNA segments, respectively. A rich diversity of SAMs that are optimized to adapt or react to their physical and chemical environment are conceivable.

Further progress in patterning SAMs at the nanometer scale will be interesting for various areas of nanoscience and technology and might be especially valuable for the advancement of molecular electronics (Section 17.4.4). Strategies of manipulating the local topography and composition of SAMs could provide essential tools for tailoring electronic junctions at the molecular level.

Recent concepts that have emerged during the past 10 years call for some novel applications of SAMs. Among them is the concern about nanotoxicity. Thus, semiconductor nanocrystals (*quantum dots*) are of high interest for medical imaging and diagnostics, since they display increased intensity of fluorescent light emission and fluoresce with longer lifetimes as compared with conventional materials. They are, however, toxic and require addition of biocompatible interfaces. Functionalizing quantum dots with SAMs in response to this challenge is an ongoing research effort (Doellefeld et al. 2002, Antolini et al. 2007). Similarly, the use of magnetic nanoparticles for various clinical purposes has been proposed (Berry and Curtis 2004), such as improving the contrast in magnetic resonance imaging (MRI) or the quality of site-specific drug delivery. The endeavors of making these compounds biocompatible are confronted with the problem that conventional, thiol-based SAMs in general do not form stable complexes with magnetic materials. An exception to this rule, however, are iron–platinum alloys, as was recently demonstrated (Xu et al. 2004, and references therein). Quantum dots and magnetic nanoscrystals provide two examples for nanoparticles of high recent interest as SAM substrates; more can be found in Love et al. (2005).

The considerable number and the broad diversity of currently active projects related to SAMs reflect the relevance of these systems for widely separated sectors of nanoscience, spanning a range that extends from nanolithography to nanobiotechnology. At the same time, novel applications of SAMs and novel features even of the well-established SAM prototypes continue to be discovered. While, during the past two decades, SAMs have acquired the status of a mainstay of nanoscience and nanotechnology, many questions for both their fundamental understanding and their practical use still remain to be answered.

References

Abbott, N.L., Kumar, A., Whitesides, G.M. 1994. Using micromachining, molecular self-assembly, and wet etching to fabricate 0.1–1-µm-scale structures of gold and silicon. *Chem. Mater.* 6: 596–602.

Ada, E.T., Hanley, L., Etchin, S. et al. 1995. Ion beam modification and patterning of organosilane self-assembled monolayers. *J. Vac. Sci. Technol. B* 13: 2189–2196.

Antolini, F., Di Luccio T., Laera, F. et al. 2007. Direct synthesis of II–VI compound nanocrystals in polymer matrix. *Phys. Stat. Sol. B* 244: 2768–2781.

Azzam, R.M.A. and Bashara, N.M. 1977. *Ellipsometry and Polarized Light.* Amsterdam, the Netherlands: North-Holland Publishing Company.

Badia, A., Gao, W., Singh, S., Demers, L., Cuccia, L., Reven, L. 1996. Structure and chain dynamics of alkanethiol-capped gold colloids. *Langmuir* 12: 1262–1269.

Badia, A., Cuccia, L., Demers, L., Morin, F.G., Lennox, R.B. 1997. Structure and dynamics in alkanethiolate monolayers self-assembled on gold nanoparticles: A DSC, FT-IR, and deuterium NMR study. *J. Am. Chem. Soc.* 119: 2682–2692.

Bandyopadhyay, K., Vijayamohanan, K. 1998. Formation of a self-assembled monolayer of diphenyl diselenide on polycrystalline gold. *Langmuir* 14: 625–629.

Bandyopadhyay, K., Sastry, M., Paul, V. 1997. Formation of a redox active self-assembled monolayer: Naphtho[1,8-*cd*]-1,2-dithiol on gold. *Langmuir* 13: 866–869.

Barrena, E., Kopta, S., Ogletree, D.F., Charych, D.H., Salmeron, M. 1999. Relationship between friction and molecular structure: Alkylsilane lubricant films under pressure. *Phys. Rev. Lett.* 82: 2880–2883.

Barrena, E., Palacios-Lido, E., Munuera, C. 2003. The role of intermolecular and molecule-substrate interactions in the stability of alkanethiol nonsaturated phases on Au(111). *J. Am. Chem. Soc.* 126: 386–395.

Barth, J.V., Brune, H., Ertl, G. 1990. Scanning tunneling microscopy observations on the reconstructed Au(111) surface: Atomic structure, long-range superstructure, rotational domains, and surface defects. *Phys. Rev. B* 42: 9307–9318.

Bernard, A., Delamarche, E., Schmid, H. et al. 1998. Printing patterns of proteins. *Langmuir* 14: 2225–2229.

Berry, C.C., Curtis, A.S.G. 2004. Functionalisation of magnetic nanoparticles for applications in biomedicine. *J. Phys. D: Appl. Phys.* 36: R198–R206.

Bierlein, H., Vanherzeele, J.D. 1989. Potassium titanyl phosphate: Properties and new applications. *J. Opt. Soc. Am. B* 6: 622–633.

Bigelow, W.C., Pickett, D.L., Zisman, W.A. 1946. Oleo-phobic monolayer. *J. Colloid Sci.* 1: 513–538.

Bracco, G., Scoles, G. 2003. Study of the interaction potential between He and a self-assembled monolayer of decanthiol. *J. Chem. Phys.* 119: 6277–6281.

Brust, M., Walker, M., Bethell, D., Schiffrin, D., Whyman, R. 1994. Synthesis of thiol derivatised nanoparticles in a two-phase liquid-liquid system. *J. Chem. Soc. Chem. Commun.* 7: 801–802.

Brust, M., Fink, J., Bethell, D., Schiffrin, D., Kiely, C. 1995. Synthesis and reactions of functionalized gold nanoparticles. *J. Chem. Soc. Chem. Commun.* 16: 1655–1656.

Calvert, J.M. 1993. Lithographic patterning of self-assembled films. *J. Vac. Sci. Technol. B* 11: 2155–2163.

Camillone, N., Chidsey, C.E.D., Liu, G.Y. 1993. Superlattice structure at the surface of a monolayer of octadecanethiol self-assembled on Au(111). *J. Chem. Phys.* 98: 3503–3511.

Carpick, R.W., Salmeron, M. 1997. Scratching the surface: Fundamental investigations of tribology with atomic force microscopy. *Chem. Rev.* 97: 1163–1194.

Case, M., McLendon, G.L., Hu, Y. et al. 2003. Using nano-grafting to achieve directed assembly of de novo designed metalloproteins on gold. *Nano Lett.* 3: 425–429.

Chaki, N.K. 2001 Applications of self-assembled monolayers in materials chemistry. *Proc. Indian Acad. Sci.* 113, 659–670.

Chen, J., Reed, M.A., Rawlett, A.M. et al. 1999. Large on-off ratios and negative differential resistance in a molecular electronic device. *Science* 286: 1550–1552.

Cipriani, J., Racins, S., Dupeyrat, R. et al. 1974. Raman scattering of Langmuir-Blodgett barium stearate layers using a total reflection method. *Opt. Commun.* 11: 70–73.

Cossaro, A., Mazzarello, R., Rousseau, R. 2008. X-ray diffraction and computation yield the structure of alkanethiols on gold(111). *Science* 321: 943–946.

Dagani, R. 1996. Devices based on electro-optic polymers begin to enter marketplace. *Chem. Eng. News* 74: 22–27.

Daniel, M.C., Astruc, D. 2004. Gold nanoparticles: Assembly, supramolecular chemistry, quantum-size-related properties, and applications toward biology, catalysis, and nano-technology. *Chem. Rev.* 104: 293–346.

Dannenberger, O., Buck, M., Grunze, M. 1999. Self-assembly of *n*-alkanethiols: A kinetic study by second harmonic generation. *J. Phys. Chem. B* 103: 2202–2213.

Doellefeld, H., Hoppe, K., Kolny, J. et al. 2002. Investigations on the stability of thiol stabilized semiconductor nanoparticles. *Phys. Chem. Chem. Phys.* 4: 4747–4753.

Dubois, L.H., Zegarski, B.R., Nuzzo, R.G. 1990. Fundamental studies of microscopic wetting. 2. Interaction of secondary adsorbates with chemically textured organic monolayers. *J. Chem. Phys.* 112: 570–579.

Elghanian, R., Storhoff, J.J., Mucic, R.C. et al. 1997. Selective colorimetric detection of polynucleotides based on the distance-dependent optical properties of gold nanoparticles. *Science* 277: 1078–1081.

Feidenhans'l, R. 1989. Surface structure determination by x-ray diffraction. *Surf. Sci. Rep.* 10: 105–188.

Felgenhauer, T., Yan, C., Geyer, W. 2001. Electrode modification by electron-induced patterning of aromatic self-assembled monolayers. *Appl. Phys. Lett.* 79: 3323–3325.

Fenter, P. 1998. X-ray and He atom diffraction studies of self-assembled monolayers. In *Thin Films: Self Assembled Monolayers of Thiols*, A. Ulman, ed., Vol. 24, pp. 111–147. London, U.K.: Academic Press.

Fenter, P., Eberhardt, A., Eisenberger, P. 1994. Self-assembly of *n*-alkyl thiols as disulfides on Au(111). *Science* 266: 1216–1218.

Fenter, P., Schreiber, F., Berman, L. et al. 1998. On the structure and evolution of the buried S/Au interface in self-assembled monolayers: X-ray standing wave results. *Surf. Sci.* 412–413: 213–235.

Fleischmann, M., Hendra, P.J., McQuillan, A.J. 1974. Raman spectra of pyridine adsorbed at a silver electrode. *Chem. Phys. Lett.* 26: 163–166.

Freeman, R.J., Grabar, K.C., Allison, K.J. et al. 1995. Self-assembled metal colloid monolayers: An approach to SERS substrates. *Science* 267: 1629–1632.

Garcia-Vidat, F.J., Pendry, J.B. 1996. Collective theory for surface enhanced Raman scattering. *Phys. Rev. Lett.* 77: 1163–1166.

Garito, A., Shi, R.F., Wu, M. 1994. Nonlinear optics of organic and polymer materials. *Phys. Today* May: 51–57.

Grant, J.T. 2003. *Surface Analysis by Auger and X-ray Photoelectron Spectroscopy*. Chichester, U.K.: IM Publications.

Grönbeck, H., Curioni, A., Andreoni, W. 2000. Thiols and disulfides on the Au(111) surface: The head group–gold interaction. *J. Am. Chem. Soc.* 122: 3839–3842.

Gutierrez-Wing, C., Santiago, P., Ascencio, J.A. et al. 2000. Self-assembling of gold nanoparticles in one, two, and three dimensions. *Appl. Phys. A* 71: 237–243.

Haag, R., Rampi, M.A., Holmlin, E. 1999. Electrical breakdown of aliphatic and aromatic self-assembled monolayers used as nanometer thick organic dielectrics. *J. Am. Chem. Soc.* 121: 7895–7906.

Heflin, J.R., Figura, C., Marciu, D. 1999. Thickness dependence of second-harmonic generation in thin films fabricated from ionically self-assembled monolayers. *Appl. Phys. Lett.* 74: 495–497.

Helmy, R., Fadeev, A.Y. 2002. Self-assembled monolayers supported on TiO_2: Comparison of $C_{18}H_{37}SiX_3$ (X = H, Cl, OCH_3), $C_{18}H_{37}Si(CH_3)_2Cl$, and $C_{18}H_{37}PO(OH)_2$. *Langmuir* 18: 8924–8928.

Himmelhaus, M., Eisert, F., Buck, M. 2000. Self-assembly of *n*-alkanethiol monolayers. A study by IR-visible sum frequency spectroscopy. *J. Phys. Chem. B* 104: 576–584.

Holmlin, R.E., Haag, R., Chabinyc, M.L. 2001. Electron transport through thin films in metal–insulator–metal junctions based on self-assembled monolayers. *J. Am. Chem. Soc.* 123: 5075–5085.

Huang, J.Y., Dahlgren, D.A., Hemminger, J.C. 1994. Photopatterning of self-assembled alkanethiolate monolayers on golds. A simple monolayer photoresist utilizing aqueous chemistry. *Langmuir* 10: 626–628.

Hudson, J.B. 1992. *Surface Science—An Introduction*. Boston, MA: Butterworth-Heinemann.

Huo, F., Zheng, Z., Zheng, G. et al. 2008. Polymer pen lithography. *Science* 312: 1658–1660.

Jackson, G.J., Woodruff, D.P., Jones, R.G. et al. 2000. Following local adsorption sites through a surface chemical reaction: CH_3SH on $Cu(111)$. *Phys. Rev. Lett.* 84: 119–122.

Kallury, K.M.R., Krull, U.J., Thompson, M. 1988. X-ray photoelectron spectroscopy of silica surfaces treated with polyfunctional silanes. *Anal. Chem.* 60: 169–172.

Kang, J., Rowntree, P.A. 1996. Molecularly resolved surface superstructures of self-assembled butanethiol monolayers on gold. *Langmuir* 12: 2813–2819.

Kim, J.H., Cotton, T.M., Uphaus, R.A. 1989. Surface-enhanced resonance Raman scattering from Langmuir-Blodgett monolayers: Surface coverage–intensity relationships. *J. Phys. Chem.* 93: 3713–3720.

Kitaigorodtskii, A.I. 1973. *Molecular Crystals and Molecules*. New York: Academic Press.

Kluth, G.J., Carraro, C., Maboudian, R. 1999. Direct observation of sulfur dimers in alkanethiol self-assembled monolayers on Au(111). *Phys. Rev. B* 59: R10449–R10452.

Kumar, A., Abbott, N.L., Kim E. 1995. Patterned self-assembled monolayers and mesoscale phenomena. *Acc. Chem. Res.* 28: 219–226.

Lahann, J., Samir Mitragotri, S., Tran, T. et al. 2003. A reversibly switching surface. *Science* 299: 371–374.

Lavrich, D.J., Wetterer, S.M., Bernasek, S.L. et al. 1998. Physisorption and chemisorption of alkanethiols and alkyl sulfides on Au(111). *J. Phys. Chem. B* 102: 3456–3465.

Lavrik, N.V., Sepaniak, M.J., Datskos, P.G. 2004. Cantilever transducers as a platform for chemical and biological sensors. *Rev. Sci. Instrum.* 75: 2229–2253.

Lercel, M.J., Redinbo, G.F., Pardo, F.D. et al. 1994. Electron beam lithography with monolayers of alkylthiols and alkylsiloxanes. *J. Vac. Sci. Technol. B* 12: 3663–3667.

Li, H.W., Muir, B., Fichet, G. 2002. Nanocontact printing: A route to sub-50-nm-scale chemical and biological patterning. *Langmuir* 19: 1963–1965.

Liu, G.Y., Xu, S., Qian, Y. 2000. Nanofabrication of self-assembled monolayers using scanning probe lithography. *Acc. Chem. Res.* 33: 457–466.

Love, J.C., Estroff, L.A., Kriebel, J.K. 2005. Self-assembled monolayers of thiolates on metals as a form of nanotechnology. *Chem. Rev.* 105: 1103–1169.

Luedtke, W.D., Landman, U. 1996. Structure, dynamics, and thermodynamics of passivated gold nanocrystallites and their assemblies. *J. Phys. Chem.* 100: 13323–13329.

Luedtke, W.D., Landman, U. 1998. Structure and thermodynamics of self-assembled monolayers on gold nanocrystallites. *J. Phys. Chem. B* 102: 6566–6572.

Lueth, H. 1993. *Surfaces and Interfaces of Solids*. Berlin, Germany: Springer.

Maksymovych, P., Dougherty, D.B. 2008. Molecular self-assembly guided by surface reconstruction: CH_3SH monolayer on the Au(111) surface. *Surf. Sci.* 602: 2017–2024.

Maksymovych, P., Sorescu, D.C., Yates, J.T. 2006. Gold-adatom-mediated bonding in self-assembled short-chain alkanethiolate species on the Au(111) surface. *Phys. Rev. Lett.* 97: 146103–146106.

Marra, W.C., Eisenberger, P., Cho, A.Y. 1979. X-ray total-external-reflection–Bragg diffraction: A structural study of the GaAs–Al interface. *J. Appl. Phys.* 50: 6927–6933.

Martin, J.E., Wilcoxon, J.P., Odinek, J. et al. 2000. Control of the interparticle spacing in gold nanoparticle superlattices. *J. Phys. Chem. B* 104: 9475–9486.

Mazzarello, P., Cossaro, A., Verdini, A. et al. 2007. Structure of a CH_3S monolayer on Au(111) solved by the interplay between molecular dynamics calculations and diffraction measurements. *Phys. Rev. Lett.* 98: 016102–016105.

Mirkin, C.A., Letsinger, R.L., Mucic, R.C., Storhoff, J.J. 1996. A DNA-based method for rationally assembling nanoparticles into macroscopic materials. *Nature* 382: 607–608.

Munuera, C., Ocal, C. 2006. Real time scanning force microscopy observation of a structural phase transition in self-assembled alkanethiols. *J. Chem. Phys.* 124: 206102–206106.

Munuera, C., Barrena, E., Ocal, C. 2007. Scanning force microscopy three-dimensional modes applied to conductivity measurements through linear-chain organic SAMs. *Nanotechnology* 18: 125505–125511.

Naraoka, R., Kaise G., Kajikawa, K. 2002. Nonlinear optical property of hemicyanine self-assembled monolayers on gold and its adsorption kinetics probed by optical second-harmonic generation and surface plasmon resonance spectroscopy. *Chem. Phys. Lett.* 362: 26–30.

Nenchev, G., Diaconescu, B., Hagelberg, F. et al. 2009. Self-assembly of methanethiol on the reconstructed Au(111) surface. *Phys. Rev. B* 80: 081401(R).

Nie, S., Emory, S.R. 1997. Probing single molecules and single nanoparticles by surface-enhanced Raman scattering. *Science* 275: 1102–1106.

Nuzzo, R.G., Allara, D.L. 1983. Adsorption of bifunctional disulfides on gold surfaces. *J. Am. Chem. Soc.* 105: 4481–4483.

Nuzzo, R.G., Fusco, F.A., Allara, D.L. 1987. Spontaneously organized molecular assemblie 3. Preparation and properties of solution adsorbed monolayers of organic disulfides on gold surfaces. *J. Am. Chem. Soc.* 109: 2358–2368.

Nuzzo, R.G., Dubois, D.H., Allara, D.L. 1990a. Fundamental studies of microscopic wetting on organic surfaces. 1. Formation and structural characterization of a self-consistent series of polyfunctional organic monolayers. *J. Am. Chem. Soc.* 108: 558–569.

Nuzzo, R.G., Korenic, E.M., Dubois, L.H. 1990b. Studies of the temperature dependent phase behavior of long chain n-alkyl thiol monolayers on gold, *J. Chem. Phys.* 93: 767–773.

Peterson, I.R. 1990. Langmuir–Blodgett films. *J. Phys. D* 23: 379–395.

Piner, R.D., Zhu, J., Xu, F. et al. 1999. "Dip-Pen" nanolithography. *Science* 283: 661–663.

Pinnaduwage, L.A., Boiadjiev, V., Hawk, J.E. et al. 2003. Sensitive detection of plastic explosives with self-assembled monolayer-coated microcantilevers. *Appl. Phys. Lett.* 83: 1471–1473.

Poirier, G.E. 1997. Characterization of organosulfur molecular monolayers on Au(111) using scanning tunneling microscopy. *Chem. Rev.* 97: 1117–1128.

Poirier, G.E., Pylant, E.D. 1996. The self-assembly mechanism of alkanethiols on Au(111). *Science* 272: 1145–1148.

Poirier, G.E., Tarlov, M.J. 1994. The c(4×2) superlattice of n-alkanethiol monolayers self-assembled on Au(111). *Langmuir* 10: 2853–2856.

Porter, M.D., Bright, T.B., Allara, D.L. et al. 1987. Spontaneously organized molecular assemblies: 4. Structural characterization of n-alkyl monolayers on gold by optical ellipsometry, infrared spectroscopy and electrochemistry. *J. Am. Chem. Soc.* 109: 3559–3568.

Raiteri, R., Grattarola, M., Berger, A. 2001. Micromechanics senses biomolecules. *Sens. Actuat. B* 79: 115–126.

Rampi, M.A., Whitesides, G. 2002. A versatile experimental approach for understanding electron transport through organic materials. *Chem. Phys.* 281: 373–391.

Rampi, M., Schueller, O.J.A., Whitesides, G.M. 1998. Alkanethiol self-assembled monolayers as the dielectric of capacitors with nanoscale thickness. *Appl. Phys. Lett.* 72: 1781–1783.

Ratner, M.A., Davis, B., Kemp, M. et al. 1998. Molecular wires: Charge transport, mechanisms and control. In *The Annals of the New York Academy of Sciences*, A. Aviram and M. Ratner, eds., Vol. 852, pp. 22–37. New York: The New York Academy of Sciences.

Reven, L., Dickinson, L. 1998. NMR spectroscopy of self-assembled monolayers. In *Thin Films: Self Assembled Monolayers of Thiols*, A. Ulman, ed., Vol. 24, pp. 111–147. London, U.K.: Academic Press.

Rzeznicka, I., Lee, J., Maksymovych, P. 2005. Nondissociative chemisorptions of short chain alkanethiols on Au(111). *J. Phys.: Chem. B* 109:15992–15996.

Scherer, J., Vogt, M.R., Magnussen, O.M., Behm, R.J. 1997. Corrosion of alkanethiol-covered Cu(100) surfaces in hydrochloric acid solution studied by in-situ scanning tunneling microscopy. *Langmuir* 13: 7045–7051.

Schreiber, F. 2000. Structure and growth of self-assembling monolayers. *Prog. Surf. Sci.* 65: 151–256.

Schreiber, F. 2004. Self-assembled monolayers: From 'simple' model systems to biofunctionalized interface. *J. Phys.: Condens. Matter* 16: R881–R900.

Schreiber, F., Eberhardt, A., Leung, T. Y. B. 1998. Adsorption mechanisms, structures, and growth regimes of an archetypal self-assembling system: Decanethiol on Au(111). *Phys. Rev. B* 57: 12476–12481.

Shafrin, E.G., Zisman, W.A. 1949. Hydrophobic monolayers adsorbed from aqueous solution. *J. Colloid Sci.* 4: 571–589.

Shimazu, K., Yagi, I., Sato, Y. 1994. Electrochemical quartz crystal microbalance studies of self-assembled monolayers of 11-ferrocenyl-1-undecanethiol: Structure-dependent ion-pairing and solvent uptake. *J. Electroanal. Chem.* 372: 117–124.

Sigal, G.B., Bamdad, C., Barberis, A. et al. 1996. A self-assembled monolayer for the binding and study of histidine-tagged proteins by surface plasmon resonance. *Anal. Chem.* 68: 490–497.

Somorjai, G.A., van Hove, M.A. 1979 *Adsorbed Monolayers on Solid Surface*. Berlin, Germany: Springer.

Stevens, M.J. 1999. Thoughts on the structure of alkylsilane monolayers. *Langmuir* 15: 2773–2778.

Stöhr, J. 1992. *NEXAFS Spectroscopy*. New York: Springer.

Templeton, A.C., Hostetler, M.J., Kraft, C.T. et al. 1998a. Reactivity of monolayer-protected gold cluster molecules: Steric effects. *J. Am. Chem. Soc.* 120: 1906–1911.

Templeton, A.C., Hostetler, M.J., Warmoth, E.K. 1998b. Gateway reactions to diverse, polyfunctional monolayer-protected gold clusters. *J. Am. Chem. Soc.* 120: 4845–4849.

Tidswell, I.M., Ocko, G.M., Pershan, P.S. et al. 1990. X-ray specular reflection studies of silicon coated by organic monolayers (alkylsiloxanes). *Phys. Rev. B* 41: 1111–1128.

Tien, J., Xia, Y., Whitesides, G.M. 1998. Microcontact printing of SAMS. In *Self Thin Films: Assembled Monolayers of Thiols*, A. Ulman, ed., Vol. 24, pp. 227–254. London, U.K.: Academic Press.

Tillman, N., Ulman, A., Elman, J. 1989. Oxidation of a sulfide group in a self-assembled monolayer. *Langmuir* 5: 1020–1026.

Tolan, M. 1999. *X-ray Scattering from Soft-Matter Thin Films* (*Springer Tracts in Modern Physics*, Vol. 41). Heidelberg, Germany: Springer.

Tour, J.M., Reinerth, W.M., Jones, L. et al. 1998. Recent advances in molecular scale electronics. *Ann. NY Acad. Sci.* 852: 197–204.

Troughton, E.B., Bain, C.D., Whitesides, G.M. et al. 1988. Monolayer films prepared by the spontaneous self-assembly of symmetrical and unsymmetrical dialkyl sulfides from solution onto gold substrates: Structure, properties, and reactivity of constituent functional groups. *Langmuir* 4: 365–385.

Ulman, A. 1991. *An Introduction to Ultrathin Organic Films.* Boston, MA: Academic Press.

Ulman, A. 1996. Formation and structure of self-assembled monolayers. *Chem. Rev.* 96: 1533–1554.

van Roosmalen, A.J., Mol, J.C. 1979. An infrared study of the silica gel surface. 2. Hydration and dehydration. *J. Phys. Chem.* 83: 2485–2488.

Vericat, C., Benitez, G.A., Grumelli, D.E. 2008. Thiol-capped gold: From planar to irregular surfaces. *J. Phys.: Condens. Matter* 20: 184004–184012.

Wang, Z.L. 1998. Structural analysis of self-assembling nanocrystal superlattices. *Adv. Mater.* 10: 13–30.

Wang, Z.L., Dai, Z.R., Sun, S.H. 2000. Polyhedral shapes of cobalt nanocrystals and their effect on ordered nanocrystal assembly. *Adv. Mater.* 12: 1944–1946.

Wang, W., Lee, T., Reed, M.A. 2005. Electron tunneling in self-assembled monolayers. *Rep. Prog. Phys.* 68: 523–544.

Wang, Y., Hush, N.S., Reimers, J.R. 2007. Formation of gold-methanethiyl self-assembled monolayers. *J. Am. Chem. Soc.* 129: 14532–14533.

Wassermann, S.R., Whitesides, G.M., Tidswell, I.M. et al. 1989. The structure of self-assembled monolayers of alkylsiloxanes on silicon: a comparison of results from ellipsometry and low-angle x-ray reflectivity. *J. Am. Chem. Soc.* 111: 5852–5861.

Wetterer, S.M., Lavrich, D.J., Cummings, T. 1998. Energetics and kinetics of the physisorption of hydrocarbons on Au(111). *J. Phys. Chem. B* 102: 9266–9275.

Wollman, E.W., Kang, D., Frisbie, C.D., Lorkovic, I.M. 1996. Photosensitive self-assembled monolayers on gold: Photochemistry of surface-confined aryl azide and cyclopentadienylmanganese tricarbonyl. *J. Am. Chem. Soc.* 116: 4395–4404.

Wu, G., Ji, H., Hansen, K. et al. 2001. Origin of nanomechanical cantilever motion generated from biomolecular interactions. *PNAS* 98: 1560–1564.

Xia, Y., Whitesides, G.M. 1995. Use of controlled reactive spreading of liquid alkanethiol on the surface of gold to modify the size of features produced by microcontact printing. *J. Am. Chem. Soc.* 117: 3274–3275.

Xu, S., Cruchon-Dupeyrat, S., Garno, J.C. et al. 1998. *In situ* studies of thiol self-assembly on gold from solution using atomic force microscopy. *J. Chem. Phys.* 108: 5002–5012.

Xu, C.J., Xu, K.M., Gu, H.W. et al. 2004. Nitrilotriacetic acid-modified magnetic nanoparticles as a general agent to bind histidine-tagged proteins. *J. Am. Chem. Soc.* 126: 3392–3393.

Yeganeh, M.S., Dougal, S.M., Polizzotti, R.S., Rabinowitz, P. 1995. Interfacial atomic structure of a self-assembled alkyl thiol monolayer/Au(111): A sum-frequency generation study. *Phys. Rev. Lett.* 74: 1811–1814.

York, R., Nguyen, P.T., Slowinski, K. 2003. Long-range electron transfer through monolayers and bilayers of alkanethiols in electrochemically controlled Hg–Hg tunneling junctions. *J. Am. Chem. Soc.* 125: 5948–5953.

Yourdshahyan, Y., Rappe, A. 2002. Structure and energetic of alkanethiol adsorption on the Au(111) surface. *J. Chem. Phys.* 117: 825–835.

Zegenhagen, J. 1993. X-ray standing waves. *Surf. Sci. Rep.* 18: 199–271.

Zegenhagen, J. 2004. X-ray standing waves imaging. *Surf. Sci.* 554: 77–79.

Zhou, J.G., Hagelberg, F. 2006. Do methanethiol adsorbates on the Au(111) surface dissociate? *Phys. Rev. Lett.* 97: 045505-1–045505-4.

Zhou, C., Deshpande, M.R., Reed, M.A. et al. 1997. Nanoscale metal/self-assembled monolayer/metal heterostructures. *Appl. Phys. Lett.* 71: 611–613.

Zhou, J.G., Williams, Q.L., Hagelberg, F. 2007. Head group dimerization in methanethiol monolayers on the Au(111) surface: A density functional theory study. *Phys. Rev. B* 76: 75408–75413.

Ziegler, C. 2004. Cantilever-based biosensors. *Anal. Bioanal. Chem.* 379: 946–959.

Graphene and Boron Nitride Single Layers

Thomas Greber
Universität Zürich

This chapter deals with single layers of carbon (graphene) and hexagonal boron nitride on transition metal surfaces. The transition metal substrates take the role of the support and allow, due to their catalytic activity, the growth of perfect layers by means of chemical vapor deposition. The layers are sp^2-hybridized honeycomb networks with strong in plane σ and weaker π bonds to the substrate and to the adsorbates. This hierarchy in bond strength causes anisotropic elastic properties, where the sp^2 layers are stiff in plane and soft out of plane. A corrugation of these layers imposes a third hierarchy level in bond energies, with lateral bonding to molecular objects with sizes between 1 and 5 nm. These extra bond energies are in the range of thermal energies $k_B T$ at room temperature and are particularly interesting for nanotechnology. The concomitant template function will be discussed in Section 18.2.6.2. The peculiar bond hierarchy also imposes intercalation as another property of sp^2 layer systems (Section 18.2.5). Last but not least, sp^2 layer systems are particularly robust, i.e., they survive immersion into liquids (Widmer et al., 2007), which is a promise for sp^2 layers being useful outside ultrahigh vacuum.

The chapter shortly recalls the synthesis, describes the atomic and electronic structures, and discusses properties like intercalation and the use of sp^2 layers on metals as tunneling junctions or as templates. The sections are divided into subsections along the sketch in Figure 18.1, i.e., for flat and corrugated layers. Since there are flat and corrugated layers for graphene (g) as well as hexagonal boron nitride (h-BN), the similarities and differences between C-C and B-N are discussed in every subsection. The

chapter ends with an appendix that summarizes the basics of atomic and electronic structures of honeycomb lattices.

Of course, the chapter does not cover all aspects of sp^2 single layers. Topics like free-standing layers (Geim and Novoselov, 2007), edge structures of ribbons (Enoki et al., 2007; Cervantes-Sodi et al., 2008), topological defects (Coraux et al., 2008), or mechanical and chemical properties are not covered.

18.1 Single Layer Systems

A single layer of an adsorbate strongly influences the physical and chemical properties of a surface. Sticking and bonding of atoms and molecules may change by orders of magnitude, as well as the charge transport properties across and parallel to the interface.

There are many single layer systems like graphite/graphene (Shelton et al., 1974; N'Diaye et al., 2006), hexagonal boron nitride (Paffett et al., 1990; Auwärter et al., 1999), boron carbides (Yanagisawa et al., 2004), molybdenum disulfide (Helveg et al., 2000), sodium chlorides (Bennewitz et al., 1999; Pivetta et al., 2005), aluminum oxide (Nilius et al., 2008), and copper nitride (Burkstrand et al., 1976; Ruggiero et al., 2007) to name a few. In order to decide whether single layers are "dielectric" or "metallic," the electronic structure at the Fermi level has to be studied, where a metallic layer introduces new bands at the Fermi energy, while a dielectric layer does not.

Figure 18.1 shows a schematic view of the single layer systems, with adsorbed molecules. The beneath conduction

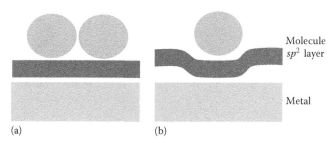

Molecule
sp^2 layer

Metal

(a)   (b)

FIGURE 18.1 Schematic side view of a single layer on top of a metal. The layer changes the properties of adsorbed molecules as compared to the pristine surface. (a) Flat layer, where molecule–molecule interactions are present (see Figures 18.4 and 18.18). (b) Corrugated layer, which is a template on which single molecules can be laterally isolated (see Figure 18.7).

electrons of the substrate still have tunneling contact to the molecule, though the bonding is much weaker compared to the bare metal. If the single layer is flat (Figure 18.1a), adsorbed molecules may touch, while in corrugated layer systems (Figure 18.1b), molecules may be laterally isolated on the surface. The corrugation (see Section 18.2.2.2) results from the epitaxial stress due to the lattice misfit between the substrate and the sp^2 layer. This results in superstructures with nanometer periodicities and corrugations in the 0.1 nm range. For developments in nanotechnology, it is useful to have single layer systems that are inert, remain clean at ambient conditions, and are stable up to high temperatures. In this field, graphene and hexagonal boron nitride are outstanding examples. These two intimately related model systems are discussed in more detail. Both are sp^2-hybridized layers with about the same lattice constant, but, while on most transition metals, graphene is metallic, h-BN is an insulator.

18.2 sp^2 Single Layers

In this chapter, we restrict ourselves to the sp^2-hybridized single layer systems of graphene and hexagonal boron nitride. Except for some considerations in Appendix 18.A we always deal with layers on transition metal supports. The comparison between graphene and boron nitride is particularly helpful for the understanding of their functionalities. They are siblings with similarities and differences. Both form a sp^2 honeycomb network with similar lattice constant (0.25 nm), with strong σ in plane bonds and soft π bonds to the substrate and the adsorbates. Depending on the substrate, both form flat or corrugated overlayers. The corrugation is a consequence of the lattice mismatch and the anisotropic bonding of the atoms in the honeycomb with the substrate, and leads to the peculiar functionality of molecular trapping (Dil et al., 2008). Graphene and h-BN differ in their atomic structures: In the case of h-BN, two different atoms constitute the honeycomb unit cell, while the two carbon atoms are equivalent within the honeycomb lattice. This causes most graphene overlayers to be metallic, while the h-BN layers are insulators.

It is furthermore the reason for an "inverted topography" if corrugation occurs (see Figure 18.8).

The theoretical description of sp^2 single layers (Wallace, 1947) is easier than the experimental realization (Novoselov et al., 2005). The first experimental reports on sp^2 layer production on supports dates back to early times of surface science (Karu and Beer, 1966) and has sometimes been considered to be an annoyance since they poison catalysts (Shelton et al., 1974). In the field of ultrathin epitaxial films of graphene and hexagonal boron nitride on solid surfaces, the work of the Oshima group has to be highlighted (Oshima and Nagashima, 1997). They studied the production of sp^2 layer systems systematically and carefully characterized their electronic and vibronic structures. Today this work is the starting point for numerous ongoing investigations, also because of the rise of graphene (Geim and Novoselov, 2007), i.e., the discovery that devices containing free-standing single layer graphene may be realized. Therefore, there is a strong demand for production routes for these materials that have, somehow, to start on a solid surface.

18.2.1 Synthesis

There are several roads to the production of sp^2 single layers. Here the chemical vapor deposition (CVD) processes that base on the reactive adsorption of precursor molecules from the gas phase onto the substrate, and the segregation method, where the constituents diffuse from the bulk to the surface, are briefly summarized.

The sp^2 layers may also be grown on substrates without C_{3v} symmetry as, e.g., present for the (111) surfaces of face-centered cubic *fcc* crystals. Examples are the growth of h-BN on Mo(110) (Allan et al., 2007) or on Pd(110) (Corso et al., 2005).

Although this chapter deals with single sp^2 layers only, it could be desirable to grow multilayers. There is, e.g., one report on the growth of graphene on h-BN/Ni(111) (Nagashima et al., 1996), and for the segregation approach it seems to be easier to grow multilayers, as it was shown for silicon carbide (SiC) substrates (Forbeaux et al., 1998), since this method has not to overcome the low sticking probability of precursor gases on complete sp^2 layers.

18.2.1.1 Chemical Vapor Deposition

CVD processes comprise the adsorption and reaction of the educts, i.e., precursor molecules on the surface where a new material shall grow. The process often involves the cracking or decomposition of the precursor molecules and a partial release of products into the gas phase.

The first h-BN single layers have been synthesized on Ru(0001) and Pt(111) surfaces by CVD of benzene-like borazine $(HBNH)_3$ (Paffett et al., 1990). The process comprises the hydrogen abstraction from the borazine molecules, the assembly of hexagonal boron nitride, and the desorption of H_2 gas. The h-BN growth rate drops after the formation of the first layer by several orders of magnitude. This has the practical benefit that it is easy to prepare single layers. In Figure 18.2, the growth of h-BN

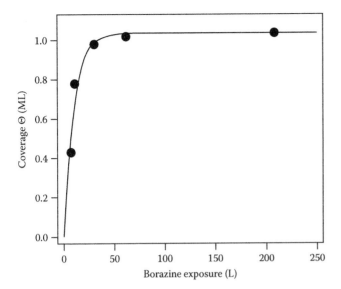

FIGURE 18.2 Growth of a h-BN layer on Ni(111): coverage as a function of borazine exposure. 1 Langmuir (L) = 10^{-6} Torr·s. Note the drop of growth rate after the completion of the first layer. (Courtesy of W. Auwärter.)

on Ni(111) is shown as a function of the exposure to the precursor molecules. Clearly, the growth rate drops by more than two orders of magnitude after the completion of the first layer. This behavior is quite general for sp^2 layer systems, and similar growth behavior is expected for graphene layers. The drop in growth rate is presumably due to the much lower sticking probability of borazine after the completion of the first h-BN layer, and the much lower catalytic activity of h-BN compared to clean transition metals. However, not much is known on the details of the growth process of h-BN. From the study of h-BN island morphologies on Ni(111), it was suggested that the borazine BN six-ring is opened during the self-assembly process (Auwärter et al., 2003; 2004). For h-BN layer formation, also trichloroborazine (ClBNH)$_3$ (Auwärter et al., 2004), and diborane (B$_2$H$_6$) ammonia (NH$_3$) gas mixtures (Desrosiers et al., 1997) were successfully used.

For the formation of graphene layers many different precursors have been used, as, e.g., ethene (C$_2$H$_4$) (N'Diaye et al., 2008), acetylene (C$_2$H$_2$) (Nagashima et al., 1996), or propylene (C$_3$H$_6$) (Shikin et al., 2000). It turns out that almost all hydrocarbons react in nonoxidizing environments on transition metal surfaces to graphene, though there are also growth conditions where diamond films grow (Gsell et al., 2008).

18.2.1.2 Segregation

An alternative way to the above-mentioned CVD processes is the use of segregation. Here the educts are diluted in the bulk of the substrate. If the substrate is heated, they start to diffuse and eventually meet the surface, where they have a larger binding energy than in the bulk. At the surface, they may then react with the new material. If the substrate temperature exceeds the stability limit of the sp^2 layer, back dissolution into the bulk or

desorption may occur. Well-known examples are, e.g., the formation of graphene on Ni(111) (Shelton et al., 1974; Eizenberg and Blakely, 1979), Ru(0001) (Marchini et al., 2007), SiC (Forbeaux et al., 1998) or the formation of BC$_3$ layers on NbB$_2$ (Yanagisawa et al., 2004).

18.2.2 Atomic Structure

The atomic structure of the sp^2 layer systems is fairly well understood. It bases on a strong sp^2-hybridized in plane bonded honeycomb lattice and a, relative to these σ bonds, weak π bonding to the substrate. The π (p_z) bonding depends on the registry to the substrate atoms, where the layer-substrate ($p_z - d_{z^2}$) hybridization causes a tendency for lateral lock-in of the overlayer atoms to the substrate atoms. For systems, where substrates have the same symmetry (C_{3v}), as the sp^2 honeycombs, the lattice mismatch M is defined as

$$M = \frac{a_{ovl} - a_{sub}}{a_{sub}} \qquad (18.1)$$

where

a_{ovl} is the lattice constant of the overlayer

a_{sub} that of the substrate

Often the sign of the lattice mismatch is not indicated, and the absolute value of the difference between the substrate and the overlayer lattice constant is used. Then it has to be explicitly said whether the mismatch induces compressive (+) or tensile (−) stress in the overlayer, or vice versa, tensile (+) or compressive (−) stress in the substrate. The mismatch plays a key role in the epitaxy of the sp^2 layers. Of course, the bonding to the substrate also depends on the substrate type, and in turn may lead to the formation of corrugated superstructures with a peculiar template function.

Figure 18.3 shows the calculated BN bond strength on different transition metals (TM) (Laskowski et al., 2008). The calculations were performed for (1 × 1) unit cells, where the h-BN was strained to the lattice constant of the corresponding substrate, with nitrogen in on-top position. The BN bond energy is determined from the difference of the energy of the (1 × 1) h-BN/TM system and the strained h-BN + TM system. The trend indicates the importance of the d-band occupancy of the substrate and it is, e.g., similar to the dissociative adsorption energies of ammonia NH$_3$ on transition metals (Bligaard et al., 2004). Since the lock-in energy, i.e., the energy that has to be paid when the nitrogen atoms are moved laterally away from the on-top sites, is expected to scale with the adsorption energy, Figure 18.3 also rationalizes, e.g., why h-BN/Pd(111) is less corrugated than h-BN/Rh(111).

This section is divided into two parts, where we distinguish flat and corrugated layers. Flat means that the same types of atoms have the same height above the substrate, which has only to be expected for (1 × 1) unit cells like that in the h-BN/Ni(111) system. Corrugation occurs due to lattice mismatch and lock-in energy gain, i.e., due to the formation of dislocations.

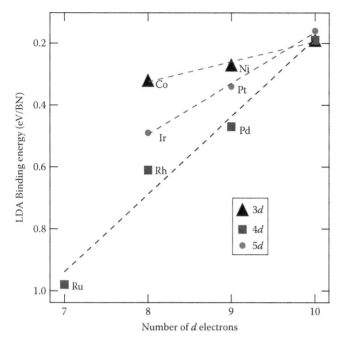

FIGURE 18.3 Calculated BN binding energies of h-BN on various transition metals. The h-BN is strained to the lattice constant of the substrate. The bond strength correlates with the d-band occupancy, where the bonding is strongest for the $4d$ metals. The dashed lines are guides to the eyes. Results within the LDA are shown. (Data from Laskowski, R. et al., *Phys. Rev. B*, 78(4), 045409, 2008.)

18.2.2.1 Flat Layers: Domain Boundaries

The Ni(111) substrate has C_{3v} symmetry and a very small lattice mismatch of +0.4% to the sp^2 layers. On Ni(111) h-BN forms perfect single layers. Figure 18.4 shows scanning tunneling microscopy (STM) images from h-BN/Ni(111). The layers appear to be defect free and flat. The STM may resolve the nitrogen and the boron sublattices, where a Tersoff Haman calculation indicated

that the nitrogen atoms map brighter than the boron atoms (Grad et al., 2003). The production process of the h-BN layers (see Section 18.2.1) also leads to the formation of larger terraces compared to uncovered Ni(111), where widths of 200 nm are easily obtained. The stability of the (111) facet was also observed in experiments with stepped Ni(755) (Rokuta et al., 1999) and Ni(223) surfaces, where the miscut of these surfaces relative to the [111] direction leads to large (111) facets and step bunches.

For the case of the h-BN/Ni(111) system the mismatch M is +0.4%, i.e., the h-BN is laterally weakly compressed. This small mismatch leads to (1 × 1) unit cells and atomically flat layers. The atomic structure of h-BN/Ni(111) is well understood, and there is good agreement between experiment (Gamo et al., 1997; Auwärtar et al., 1999; Muntwiler et al., 2001) and theory (Grad et al., 2003; Che and Cheng, 2005; Huda and Kleinman, 2006). Figure 18.5 shows six (1 × 1) configurations for h-BN within the Ni(111) unit cell. The nomenclature of the structure (B,N)=(fcc,top) indicates that boron is sitting on the fcc site, i.e., on a site where no atom is found in the second nickel layer, and where the nitrogen atom sits on top of the atom in the first nickel layer. Theory found two stable structures (B,N)=(fcc,top) and (hcp,top) only. In both cases, nitrogen is on top (Grad et al., 2003). The (fcc,top) structure has the lowest energy, which was consistent with the published experimental structure determinations. The calculated energy difference between the structure with boron on fcc and on hcp differs only by 9 meV, which is reasonable since it indicates interaction of the layer with the second nickel layer. The 9 meV are, however, one order of magnitude smaller than the thermal energies k_BT during the synthesis, and from this viewpoint it is not clear why pure (B,N)=(fcc,top) single domain can be grown. The inspection of Figure 18.5 shows that the (fcc,top) structure is a translation of the (top,hcp) or (hcp,fcc). A transformation from (fcc,top) to (hcp,top) would involve a rotation, or a permutation of B and N and a translation. The growth of the h-BN layers on Ni(111) proceeds via the

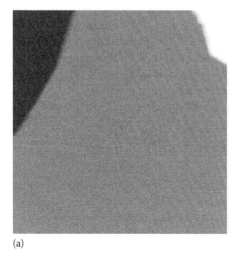

(a)

(b)

FIGURE 18.4 Constant current STM images of a flat layer of hexagonal boron nitride on Ni(111). (a) 30 × 30 nm. The gray scales indicate three terraces with different heights. Large defect-free terraces form. (b) 3 × 3 nm. The nitrogen (bright) and boron (gray) sublattices are resolved with different topographical contrast. (From Auwärtar, W. et al., *Surf. Sci.*, 429(1–3), 229, 1999. With permission.)

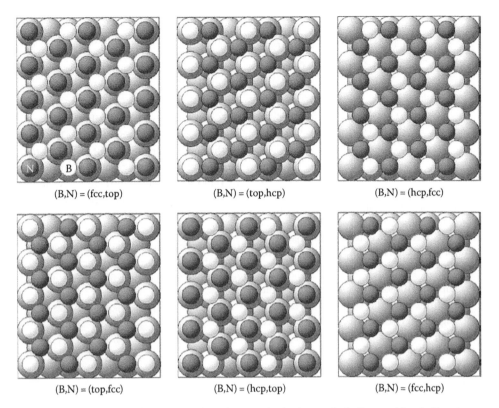

FIGURE 18.5 Possible (1 × 1) configurations for one-monolayer h-BN/Ni(111). The top, fcc hollow, and hcp hollow sites are considered for the position of the boron and nitrogen atoms, respectively. The (B,N)=(fcc,top) registry has the strongest bond, while calculations predict 9 meV lower binding energy for the (B,N)=(hcp,top) configuration. (From Grad, G.B. et al., *Phys. Rev. B*, 68(8), 085404, 2003. With permission.)

nucleation and growth of triangular islands that are separated by distances between 10 and 100 nm (Auwärtar et al., 2003). Therefore, if (fcc,top) and (hcp,top) nucleation seeds form, two domains are expected since the change of orientation of islands with sizes larger than 10 nm costs too much energy (Auwärtar et al., 2003). The concomitant domain boundaries (see Figure 18.16a) have interesting functionalities. They act as collectors for intercalating atoms, and clusters like to grow on these lines. Also, it is expected that such domain boundaries are model systems for sp^2 edges. Unfortunately, the (fcc,top):(hcp,top) ratio is not yet under full experimental control. But it is, e.g., known that the omission of the oxygen treatment of the Ni(111) surface before h-BN growth causes two domain systems (Auwärtar et al., 2003).

The vertical position of the nitrogen and the boron atoms is not the same within the (1 × 1) unit cell. This local corrugation or buckling, where nitrogen is the outermost atom, and boron sits closer to the first Ni plane, was taken as an indication of the compressive stress on the h-BN in the h-BN/Ni(111) system (Rokuta et al., 1997; Auwärtar et al., 1999). The comprehensive density functional theory study of Laskowski et al. (2008) indicated however that this buckling, i.e., height difference between boron and nitrogen, persists also for systems with tensile stress in the overlayer, where the substrate lattice constant is larger than that of the h-BN. B is closer to the first substrate plane for all investigated cases. This is a consequence of the bonding to the substrate, where the boron atoms are attracted and the nitrogen atoms are repelled from the surface (Laskowski et al., 2008).

For the corresponding graphene Ni system, the same structure, i.e., the (C_A,C_B)=(fcc,top), which is equivalent to the (top,fcc) configuration has been singled out against the (fcc,hcp) structure (Gamo et al., 1997). There are no reports on (top,fcc)/(top,hcp) domain boundaries, as observed for the h-BN/Ni(111) case (Auwärtar et al., 2003). For g/Ir(111), which belongs to the corrugated layer systems, dislocation lines that are terminated by heptagon-pentagon defects were found (Coraux et al., 2008). Of course, these kinds of defects were less likely for BN, since this would imply energetically unfavorable N-N or B-B bonds in the BN network.

18.2.2.2 Corrugated Layers: Moiré and Dislocation Networks

When the lattice mismatch M (Equation 18.1) of the laterally rigid sp^2 networks exceeds a critical value, superstructures with large lattice constants are formed. If the lattices of the overlayer and the substrate are rigid and parallel, the superstructure lattice constant gets $a_{ovl}/|M|$, where a_{ovl} is the 1 × 1 lattice constant of graphene or h-BN (≈0.25 nm).

For rigid sp^2 layers, i.e., if there would be no lateral lock-in energy available, we expect flat floating layers, reminiscent to incommensurate moiré patterns without a directional lock-in of the structures (see Figure 18.6a). For the case of h-BN/Pd(111) such a tendency to form moiré type patterns, also without a

Moiré

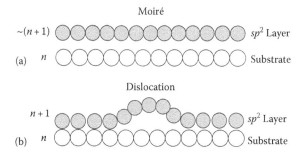

FIGURE 18.6 Schematic view of a moiré and a dislocation in an overlayer system. (a) In a moiré pattern ~$(n + 1)$ sp^2 units fit on n substrate units, and the directions of the substrate and the adsorbate lattice do not necessarily coincide. (b) In a dislocation network $n + 1$ sp^2 units coincide with n substrate units. Here the lock-in energy is large for on-top and weak or repulsive for bridge sites, where the dislocation evolves.

preferential lock-in to a substrate direction, was found (Morscher et al., 2006). These h-BN films have the signature of the electronic structure of flat single layers (see Section 18.2.4). The lock-in energy is the energy that an epitaxial system gains on top of the average adhesion energy if the overlayer locks into preferred bonding sites. It involves the formation of commensurate coincidence lattices between the substrate and the adsorbate layer with dislocations (see Figure 18.6b). The lock-in energy is expected to be proportional to the bond energy shown in Figure 18.3. For the case of h-BN/Rh(111), the lattice mismatch (Equation 18.1) is −6.7%, and 13×13 BN units coincide with 12×12 Rh units (Corso et al., 2004; Bunk et al., 2007). With the room temperature lattice constants of h-BN and Rh this leads to a residual compression of the 13 h-BN units by 0.9%. Figure 18.7 shows a relief-view of this superstructure coined "h-BN nanomesh" (Corso et al.,

2004). It has a super cell with a lattice constant of 3.2 nm, i.e., (12×12) Rh (111) units and displays the peculiar topography of a mesh with "wires" and "holes" or "pores." It turned out that this structure is a corrugated single layer of hexagonal boron nitride with two electronically distinct regions that are related to the topography (Berner et al., 2007; Laskowski et al., 2007). This structure also has the ability of trapping single molecules in its holes (see Section 18.2.6.2). The accompanying variation of the local coordination of the substrate and the adsorbate atoms divides the unit cells into regions with different registries. In reminiscence to Figure 18.5, the notation (B,N)=(top,hcp) refers to the local configuration, where a B atom sits on top of the substrate atom in the first layer and N on top of the hexagonal close packed (hcp) site that is on top of the substrate atom in the second layer. Again three regions can be distinguished with atoms in (fcc,top), (top,hcp), and (hcp,fcc) configurations (see Figure 18.8). A force field theory approach indicated that the

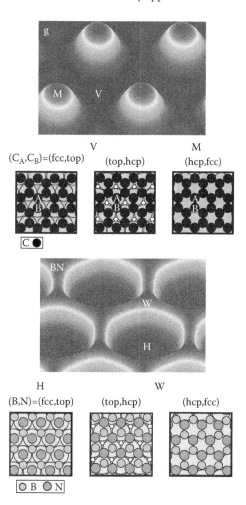

FIGURE 18.8 Views of the height-modulated graphene (g) and h-BN nanomesh (BN) on Ru(0001), M and V denote mounds (hcp,fcc) and valleys of the graphene, H and W holes (fcc,top) and wires of the h-BN nanomesh. The six ball model panels illustrate the three different regions [(fcc,top), (top,hcp), and (hcp,fcc)], which can be distinguished in both systems. (From Brugger, T. et al., *Phys. Rev. B*, 79(4), 045407, 2009. With permission.)

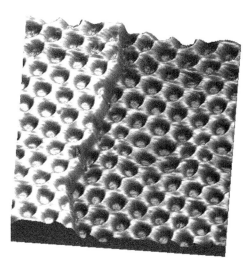

FIGURE 18.7 Relief view of a constant current STM image of a corrugated layer boron nitride (nanomesh) (30×30 nm^2, $I_t = 2.5$ nA, $V_s = -1$ V). This nanostructure with 3.2 ± 0.1 nm periodicity consists of two distinct areas: the wires, which are $1.2 + 0.2$ nm broad and the holes with a diameter of 2.0 ± 0.2 nm. The corrugation is 0.07 ± 0.02 nm. (From Greber, T. et al., *Surf. Sci.*, 603, 1373, 2009. With permission.)

(B,N)=(fcc,top) sites correspond to the tightly bound regions of the h-BN layers, i.e., the holes, while the weakly bound or even repelled regions correspond to the wires (Laskowski et al., 2007). The corrugation, i.e., height difference of the h-BN layer from the top of the substrate, in accordance between experiment and theory is about 0.05 to 0.1 nm. This is sufficient to produce a distinct functionality (see Section 18.2.6.2) (Berner et al., 2007). For h-BN/Ru(0001), a structure very similar to that of h-BN/Rh(111) was found (Goriachko et al., 2007).

For the graphene case the situation is related, though not identical. The difference lies in the fact that the base in the 1 × 1 unit cell of free-standing graphene consists of two identical carbon atoms C_A and C_B, while that of h-BN does not. C_A and C_B become only distinguishable by the local coordination to the substrate. In g/Ru the local (fcc,top) and (top,hcp) coordination leads to close contact between the (C_A, C_B) atoms and the substrate (Wang et al., 2008) while (B,N) is strongly interacting only in the (fcc,top) coordination (Laskowski et al., 2008). As a result, twice as many atoms are bound in strongly interacting regions in g/Ru when compared to h-BN/Ru. In reminiscence to morphological terms the strongly bound regions of g/Ru(0001) are called valley (V) and the weakly bound regions with the (C_A, C_B) atoms on (hcp,fcc) sites are called mounds (M) or hills. The fact that (top,hcp) leads to strong bonding for graphene but weak bonding for h-BN gives rise to an inverted topography of the two layers: a connected network of strongly bound regions for graphene (valleys) and a connected network of weakly bound regions for h-BN (wires). Also for the graphene case "moiré"-type superstructures were found, where g/Ir(111) is the best studied so far (N'Diaye et al., 2006).

18.2.3 Electronic Structure I: Work Function

18.2.3.1 Flat Layers: Vertical Polarization

The work function, i.e., the minimum energy required to remove an electron from a solid, is material dependent. There are excellent reviews on the topic (Hölzl et al., 1979; Kiejna and Wojciechowski, 1981). For our purpose, where we want to discuss the electric fields near the surface, it is sufficient to recall the Helmholtz equation that relates the work function Φ of a flat surface with vertical electric dipoles:

$$\Phi = \frac{e}{\epsilon_0} N_a \cdot p \qquad (18.2)$$

where

    e is the elementary charge
    ϵ_0 is the permittivity
    N_a is the areal density of the dipoles p

If the dipoles that are caused by the leaking of the electron wave functions into the vacuum are assigned to the atoms, we get, e.g., for Ni(111) with a work function of 5.2 eV and an in plane lattice constant of 0.25 nm a dipole of 0.7 Debye (1 Debye = $3.34 \cdot 10^{-30}$ Cm).

It has to be said that this is the classical view of the work function and quantum mechanical effects beyond the uncertainty principle may be incorporated empirically into Equation 18.2 in using an effective dipole.

Now the influence of overlayers shall be analyzed. If a single layer of a medium is placed on top of the substrate, the electric fields in the surface dipole polarize the layer, i.e., decreases the work function by $\Delta \Phi_s = (e/\epsilon_0) N_a \cdot p_{ind}$ by screening. To first order the induced dipole p_{ind} is proportional to the electric field perpendicular to the surface $E_\perp : p_{ind} = \alpha \cdot E_\perp$, where α is the polarizability.

In Figure 18.9, it is shown how a polarizable medium screens out the electric field and decreases the work function. These considerations apply for dielectrics and metals, if they are not in contact with the substrate. The strong vertical distance dependence of the electric field in the surface dipole layer involves a correlation with the screening-induced work function shift $\Delta \Phi_s$. In the case of contact of a metallic overlayer with the Fermi level of the substrate, no vertical dipoles in the sense of Figure 18.9 are induced, although the work function may change. However, as we know from the Smoluchowski effect (Smoluchowski, 1941) a corrugation induces a nonuniform surface charge density and thus lateral electric fields (see Section 18.2.3.2).

For a dielectric, as it is h-BN, the screening causes, e.g., in the case of h-BN/Ni(111) a work function lowering of 1.6 eV (Grad et al., 2003), which corresponds to an induced dipole of 0.3 Debye. In this case, charge gets displaced, but not transferred from the sp^2 layer to the substrate. If the medium is metallic, as it is for most graphene cases, we expect a charge transfer that aligns the chemical potentials of the substrate and the overlayer. A strong interaction also alters the chemical potentials and in turn influences the charge transfer. In any case, the charge redistribution changes the work function, i.e., the surface dipole. For weakly bound graphene theory predicts a correlation between the work function of the substrate and the doping level (Giovannetti et al., 2008).

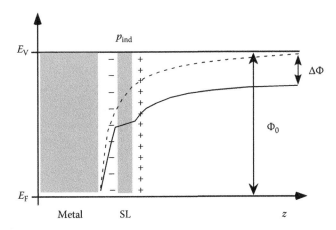

FIGURE 18.9 Schematic view of the effect of a thin insulated, polarizable single layer on the work function. The layer gets polarized and accordingly reduces the surface dipole. The dashed line is the electrostatic potential without single layer, while the solid line shows the effect of the single layer.

18.2.3.2 Corrugated Layers: Lateral Electric Fields, Dipole Rings

If the surface is flat, the lateral electric fields will only vary for ionic components in the two sublattices of the sp^2 networks. For the case of h-BN, the different electronegativities of the two elements cause a local charge transfer from the boron to the nitrogen atoms. By means of density functional theory calculations it has been found that for a free-standing h-BN sheet about 0.56 electrons are transferred from B to N (B,N)=(0.56,−0.56)e^- (Grad et al., 2003). This yields a sizable Madelung contribution to the lattice stability of 2.4 eV per 1 × 1 unit cell (see Appendix 18.A.2). If the h-BN sits on a metal, this Madelung energy is reduced by a factor of 1/2 due to the screening of the charges in the half space taken by the metallic substrate. For h-BN on Ni(111) the ionicity slightly increases (B,N)=(0.65,−0.59) e^-, where the net charge displacement of 0.06 e^- toward the substrate is consistent with the above-mentioned work function decrease due to polarization (Grad et al., 2003). It is expected that the relatively strong local ionicity of the h-BN surface has an influence on the diffusion of atoms with the size of the BN bond length of 0.15 nm. If the surface is not flat, but corrugated, this causes lateral electric fields on the length scale of the corrugation, which is particularly interesting for trapping atoms or molecules (Dil et al., 2008). Lateral fields may occur in dielectric or above metallic overlayers. Figure 18.10 shows the lateral and vertical electrostatic potential in a corrugated single layer nanostructure. For the case of a dielectric (Figure 18.10a), a different distance of the layer from the surface imposes a different screening and lateral potential variations, also within the dielectric. For the case of a metal (Figure 18.10b), a corrugation causes lateral potential variations, reminiscent to the Smoluchowski effect (Smoluchowski, 1941), but not within the layer. Also, it was found for g/Ru(0001) that the metallic case causes smaller lateral potential variations (Brugger et al., 2009).

For dielectrics or insulators, the key for the understanding of the lateral electrostatic potential variation came from the σ band splitting (see Section 18.2.4.2). This splitting, which is in the order of 1 eV, is also reflected in N1s core level x-ray photoelectron spectra (Preobrajenski et al., 2007), where

the peak assignment is in line with the σ band assignment (Berner et al., 2007; Goriachko et al., 2007). Without influence of the substrate the energy of the band, which reflects the in plane sp^2 bonds, is referred to the vacuum level. This means that the sum of the work function and the binding energy as referred to the Fermi level, is a constant. Vacuum level alignment arises for physisorbed systems, as, e.g., for noble gases (Wandelt, 1984; Janssens et al., 1994), or as it was proposed for h-BN films on transition metals (Nagashima et al., 1995). The σ band splitting causes the conceptual problem of aligning the vacuum level and the Fermi level with two different work functions. The *local* work function, or the electrostatic potential near the surface may, however, be different. This electrostatic potential with respect to the Fermi level can be measured with photoemission of adsorbed Xe (PAX) (Kuppers et al., 1979; Wandelt, 1984). Xenon does not bond strongly to the substrate and thus the core level binding energies are a measure for the potential energy difference between the site of the Xe core (Xe has a van der Waals radius of about 0.2 nm) and the Fermi level. Recently, the method of photoemission from adsorbed Xe was successfully applied to explore the electrostatic energy landscape of h-BN/Rh(111) (Dil et al., 2008). In accordance with density functional theory calculations it was found that the electrostatic potential at the Xe cores in the holes of the h-BN/Rh(111) nanomesh is 0.3 eV lower than that on the wires. This has implications for the functionality, since these sizable potential gradients polarize molecules, and it may be used as an electrostatic trap for molecules or negative ions. The peculiar electrostatic landscape has been rationalized with dipole rings, where in plane dipoles, sitting on the rim of the corrugations, produce the electrostatic potential well. For in plane dipoles that sit on a ring, the electrostatic potential energy in the center of the ring becomes

$$\Delta E_{\text{pot}} = \frac{e}{4\pi\epsilon_0} \frac{P}{R^2} \qquad (18.3)$$

where

   e is the elementary charge
   R is the radius of the hole
   $P = \Sigma \,|\mathbf{p}_i|$ is the sum of the absolute values of the dipoles on the ring

For $R = 1$ nm and $\Delta E_{\text{pot}} = 0.3$ eV, P gets 10 Debye, which is equivalent to 5.4 water molecules with the hydrogen atoms pointing to the center of the holes. The strong lateral electric fields in the BN nanomesh may be exploited for trapping molecules, or negatively charged particles. It can also act as an array of electrostatic nanolenses for slow charged particles that approach or leave the surface. Figure 18.11 shows the electrostatic potential in a dipole ring and the square of the related electric fields. The square of the gradient of the

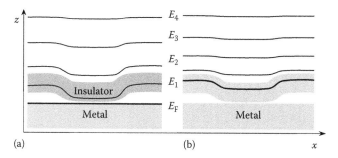

(a)   (b)   *x*

FIGURE 18.10 Contour plot of the electrostatic potential at the surface of a corrugated single layer on a flat metal. The energies continuously increase from the Fermi level E_{F} toward the vacuum level E_{V} at $z = \infty$, with $E_{\text{F}} < E_1 < E_2 < E_3 < E_4 < E_{\text{V}}$. (a) For a single layer dielectric and (b) for a single layer metal.

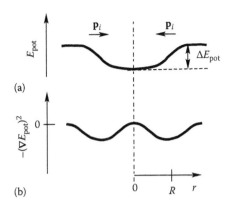

FIGURE 18.11 Schematic drawing of a dipole ring where in plane dipoles $\mathbf{p}_i$ are sitting on a ring with radius R. (a) The electrostatic potential energy E_{pot} and (b) the polarization energy, which is proportional to $(\nabla E_{pot})^2$. (From Greber, T. et al., *Surf. Sci.*, 603, 1373, 2009. With permission.)

electrostatic potential, or the electric field, is expected to be proportional to the polarization induced bond energy E_{pol}:

$$E_{pol} = \mathbf{E}_{\parallel} \cdot \mathbf{p}_{ind} = \alpha \cdot \mathbf{E}_{\parallel}^2 \qquad (18.4)$$

where

$\mathbf{E}_{\parallel}$ is the lateral electric field
$\mathbf{p}_{ind}$ is the induced dipole
α is the polarizability

Equation 18.4 is a first order approximation and neglects the weakening of the dipole ring field due to the induced dipole. For

the case of h-BN/Rh(111) the origin of the in plane electrostatic fields is not the Smoluchowski effect, where the delocalization of the electrons at steps forms in plane dipoles (Smoluchowski, 1941). It is due to the contact of differently bonded boron nitride with different local work functions since the screening depends on the vertical displacement of the dielectric from the metal, i.e., on the corrugation (see Section 18.2.3.1).

The concept of dipole rings is not restricted to the above-mentioned polarization of corrugated dielectrics. As it was recently shown for the case of g/Ru(0001), lateral electric fields also occur above corrugated graphene, where the dipoles are created due to lateral polarization like in the Smoluchowski effect (Brugger et al., 2009).

18.2.4 Electronic Structure II: Band Structure, Fermisurfaces

The band structure of sp^2 layers on transition metals has the same signature as free-standing single layers. The binding energy difference between the in plane σ bands and the out of plane π bands does not remain constant, when the layers come into contact with a substrate. This is a consequence of the bonding via the π orbitals, while for the σ bands vacuum level alignment is observed.

18.2.4.1 Flat Layers: The Generic Case

The electronic band structure of h-BN on Ni(111) is similar to that of a free-standing layer of h-BN, with σ and π bands that have the generic structure of a sp^2 honeycomb lattice (see

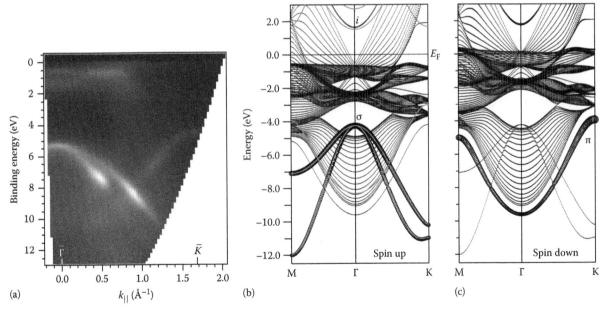

FIGURE 18.12 (a) Experimental grayscale dispersion plot on the ΓK azimuth from angle-resolved He I_α photoemission. The gray scale reproduces the intensity (white maximum). Bottom: Spin-resolved theoretical band structure of h-BN/Ni(111) along MΓ and ΓK in the hexagonal reciprocal unit cell (Brillouin zone). (b) Spin up, the radius of the circles is proportional to the partial $p_x = p_y$ charge density centered on the nitrogen sites (σ bands). Thick gray lines show the bands of a buckled free-standing h-BN monolayer, which have been aligned to the h-BN/Ni(111) σ band at Γ. (c) Spin down, the radius of the circles is proportional to the partial p_z charge density on the nitrogen sites (π bands). (From Grad, G.B. et al., *Phys. Rev. B*, 68(8), 085404, 2003. With permission.)

Appendix 18.A). In Figure 18.12, angle-resolved photoemission data from h-BN/Ni(111) along the ΓK azimuth are compared with density functional theory calculations. In the bottom panel, the band structure calculations are shown for h-BN/Ni(111) and free-standing h-BN. The bandwidth of the σ bands increases less than 1% by the adsorption of the h-BN. The widths of the spin split π bands increase in the spin average by 4%, where the width is larger for the minority spins. This band structure with σ and π bands is generic for all sp^2 systems. Though, the σ and π band do not shift equally in energy upon adsorption of the sp^2 layer. The shift between the σ and π bands of about 1.3 eV indicates that the interaction between the Ni substrate and the h-BN π and σ bands is not the same. Also other substrates like Pt(111) and Pd(111) cause a similar h-BN band structure, although h-BN/Pt(111) and h-BN/Pd(111) are systems with large negative lattice mismatch (Oshima and Nagashima, 1997; Morscher et al., 2006). For these "flat" cases, the bonding to the substrates is rather independent of the positions within the unit cell. In the section on corrugated layers (Section 18.2.4.2), we will see that this changes when substrates with mismatch and strong lock-in energy are used. For Ni(111), Pd(111), and Pt(111) it was found empirically that the σ bands have, within ±100 meV, the same binding energy, if they are referred to the vacuum level (Nagashima et al., 1995). This vacuum alignment confirms that the σ bands do not strongly interact with the underlying metal. The binding energy of the π band, on the other hand, varies and indicates the π bonding to the substrate.

For grapheme, the electronic structure is also well understood (Oshima and Nagashima, 1997). The π bands are of particular interest since they are decisive for the issue on whether the overlayer is metallic or not. On substrates the ideal case of free-standing graphene may generally not be realized since the site selective p_z interaction with the substrate is causing a symmetry breaking between the two carbon sublattices C_A and C_B. This opens a gap at the Dirac point. The Dirac point coincides with the K point of the Brillouin zone of graphene, where the band structure is described by cones on which electrons behave quasi relativistic (see Appendix 18.A).

The magnitude of the gap is a measure for the anisotropy of the interaction of the C_A and C_B carbon atoms with the substrate. The position of the Fermi level with respect to the Dirac energy is decisive on whether we deal with p-type or n-type graphene (or h-BN). Since the center of the gap of h-BN lies below the Fermi level, this means that, e.g., h-BN/Ni(111) is n-type. Figure 18.13 shows the π band electronic structure of sp^2 honeycomb lattices near the Fermi energy and around the K point. If the Fermi energy lies above the Dirac energy E_{Dirac}, which is the energy of the center of the gap, the majority of the charge carriers move like electrons (n-type). Correspondingly, if the Fermi energy lies below the Dirac point the charge carriers move like holes (p-type). This n-type, p-type picture is analogous to the semiconductors, where the Fermi level takes, depending on its position, the role of the acceptors (p-type) and the donors (n-type), respectively. Nagashima et al. first measured the π gaps at K for the case of one and two layers of graphene on TaC(111), where they found

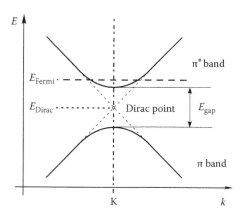

FIGURE 18.13 Schematic diagram showing the π band electronic structure of sp^2 honeycomb lattices around the K point. For free-standing graphene, the π and π^* bands lie on cones intersecting at the Dirac point. Interaction with a support causes a shift of the Fermi energy and, if it is anisotropic, a gap opens. Here ($E_{Fermi} > E_{Dirac}$), the case for n-type conduction is shown (for details see text).

n-type behavior and a gap of 1.3 eV for the single layer and about 0.3 ± 0.1 eV for the double layer (Nagashima et al., 1994).

18.2.4.2 Corrugated Layers: σ Band Splitting

The superstructures as described in Section 18.2.2.2 are also reflected in the electronic structure. Figure 18.14 shows valence band photoemission spectra for three different h-BN single layer systems in normal emission. We would like to draw the attention on the σ bands at about 5 eV binding energy (see Table 18.1). Figure 18.14 shows valence band photoemission data for h-BN/Ni(111), h-BN/Pd(111), and h-BN/Rh(111). Hexagonal boron nitride has three σ bands and one π band that are occupied. Along Γ, i.e., perpendicular to the sp^2 plane, the two low binding energy σ bands are degenerate, while the third is not accessible for He I_α radiation. Accordingly, the σ bands of h-BN Ni(111) and of h-BN/Pd(111) show one single peak. Those of h-BN/Rh(111) are weak and split in a σ_α contribution and a σ_β contribution (Goriachko et al., 2007). In measuring the spectra

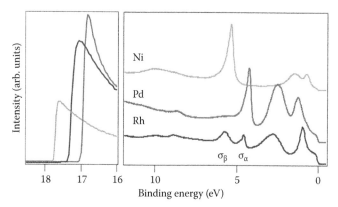

FIGURE 18.14 He I_α normal emission photoemission spectra of h-BN/Ni(111), h-BN/Pd(111), and h-BN/Rh(111). While h-BN/Ni(111) and h-BN/Pd(111) show no sizable σ band splitting, the splitting into σ_α and σ_β for h-BN/Rh(111) is about 1 eV. (From Greber, T. et al., *Surf. Sci.*, 603, 1373, 2009. With permission.)

TABLE 18.1 Experimental Values (in eV) for the Photoemission Binding Energies Referred to the Fermi Level E_B^F of the σ Bands for h-BN Single Layers on Three Different Substrates Along Γ, and the Corresponding Work Functions Φ

Substrate	$E_B^F\sigma_\alpha$	$E_B^F\sigma_\beta$	Φ	$E_B^V\sigma_\alpha$	References
Ni(111)	5.3	5.70	3.5	8.8	Greber et al. (2002)
Rh(111)	4.57		4.15	8.7	Goriachko et al. (2007)
Pd(111)	4.61		4.26	8.9	Greber et al. (2009)

Note: The binding energies with respect to the vacuum level E_B^V are determined by $E_B^F + \Phi$.

away from the Γ point it is seen that both, the σ_α and σ_β bands split in two components each, as it is known from angular resolved measurements of h-BN single layers with σ_α bands only (Nagashima et al., 1995) (see Figure 18.12).

The σ band splitting indicates two electronically different regions within the h-BN/Rh(111) unit cell. They are related to strongly bound and weakly bound h-BN. Later on, h-BN/Ru(0001) (Goriachko et al., 2007) was found to be very similar to h-BN/Rh(111). Theoretical efforts (Laskowski et al., 2007) and atomically resolved low temperature STM (Berner et al., 2007) showed that the h-BN/Rh(111) nanomesh is a corrugated single sheet of h-BN on Rh(111) (see Section 18.2.2.2). The peculiar structure arises from the site dependence of the interaction

with the substrate atoms that causes the corrugation of the h-BN sheet, with a height difference between strongly bound regions and weakly bound regions of about 0.05 to 0.1 nm.

Interestingly, no σ band splitting could be found for the g/Ru(0001) system, although these graphene layers are also strongly corrugated (Brugger et al., 2009). This is likely related to the inverted topography (see Figure 18.8) and the metallicity of graphene on ruthenium. The metallicity can be investigated locally by means of scanning tunneling spectroscopy. It has, however, to be said that tunneling spectroscopy is most sensitive to the density of states at the Γ point but the graphene conduction electrons reside around the K point. This experimental limitation does not exist in angular resolved photoemission, where on the other hand the sub-nanometer resolution of scanning probes is lost. Though with photoemission, it is possible to determine the average Fermi surface of the sp^2 layers. The Fermi surface indicates the doping level, i.e., the number of electrons in the conduction band. In the case of graphene, cuts across Dirac cones (see Figure 18.13) are measured. If the Fermi surface map is compared to another constant energy surface near the Fermi energy, it can be decided whether the Dirac point lies below or above the Fermi energy, i.e., whether we deal with *n*- or *p*-type graphene. In Figure 18.15, the Fermi surface map as measured for g/Ru(0001) and h-BN/Ru(0001) is compared. For the case of g/Ru(0001), extra intensity at the K points indicates Dirac cones. The cross sections

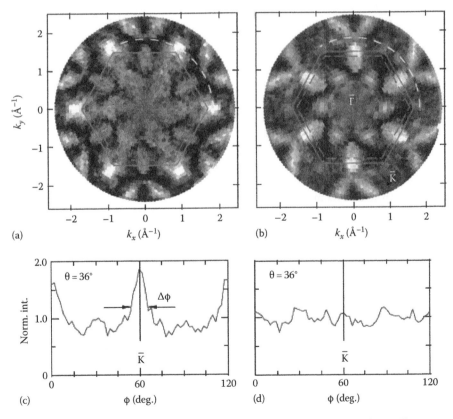

FIGURE 18.15 Fermi surface maps. (a) g/Ru(0001). (b) h-BN/Ru(0001). The hexagons indicate the surface Brillouin zones of Ru(0001) (dashed), graphite (solid), and h-BN (solid). (c) and (d) show the normalized intensities of azimuthal cuts along the dashed sectors in (a) and (b), respectively. (From Brugger, T. et al., *Phys. Rev. B,* 79(4), 045407, 2009. With permission.)

of which correspond to 5% of the area of the Brillouin zone, or 0.1 e^-. The cross section shown in Figure 18.15c does not resolve a single cone with a bimodal distribution around K, as it is, e.g., the case for g/SiC (Ohta et al., 2006), but a single peak from which an average cone diameter at the Fermi energy is determined. The measurement of the band structure along ΓK indicates that these 0.1 e^- are donated from the substrate to the graphene. For the case of h-BN/Ru(0001), the Fermi surface shows no peak at K which is expected for an insulating or semiconducting sp^2 layer (Greber et al., 2009). All features that can be seen in this Fermi surface map stem from the underlying Ru(0001) substrate.

18.2.5 Sticking and Intercalation

Graphite is well known for its ability to intercalate atoms (Dresselhaus and Dresselhaus, 1981) and there are also reports on intercalation in h-BN (Budak and Bozkurt, 2004). Intercalation is the reversible inclusion of guest atoms or molecules between other host molecules. For the host material graphite or h-BN intercalation occurs between the honeycomb sheets, where the bonding in the sheet is strong and the bonding between the sheets is relatively weak. Here, we deal with single sp^2 layers on transition metals and use the term "intercalation" also for irreversible intercalation, as it is observed if metal atoms slip below the sp^2 layer. In the case of g/Ni(111) it has, e.g., been found that Cu, Ag, and Au intercalate irreversibly, although they form no graphite intercalation compounds (Shikin et al., 2000).

Intercalation is preceded by sticking (adsorption) and diffusion of the intercalating species. In the following, the model case of cobalt on h-BN/Ni(111) is discussed in more detail. Figure 18.16 shows the growth of cobalt on h-BN/Ni(111). On flat terraces, as shown in Figure 18.16a, three different patterns are observed: (1) three-dimensional (3D) clusters, whose heights scale with the lateral diameter; (2) triangular, two-dimensional (2D) islands with a constant apparent height; and (3) line patterns (Auwärtar et al., 2002). Often these line patterns are found to be connected to 2D islands, and 3D islands tend to nucleate on such lines. A careful analysis relates the lines with domain boundaries, where (B,N)=(fcc,top) and (B,N)=(hcp,top) domains touch

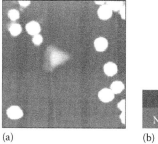 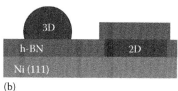

(a)   (b)

FIGURE 18.16 Cobalt on h-BN/Ni(111). (a) STM image, showing triangular 2D intercalated Co islands, and circular 3D clusters and a defect line connecting to a intercalated island (27 × 27 nm). (b) Schematic side view of the situation, showing the topology of the 3D and 2D agglomerates. (Data from Auwärtar, W. et al., *Surf. Sci.*, 511, 379, 2002. With permission.)

(Auwärtar et al., 2003) (see Section 18.2.2.1). The 2D islands are irreversibly intercalated Co below the h-BN layer, while the 3D clusters remain on top of the h-BN. This results, e.g., in the property that the 3D islands may be removed, cluster by cluster, with a STM manipulation procedure (Auwärtar et al., 2002).

The sticking coefficient, i.e., the probability that an impinging atom sticks on the surface, is determined by x-ray photoelectron spectroscopy (XPS). While the sample was exposed to a constant flux of about 3.5 monolayers of Co per hour, the measurement of the Co uptake on the sample gives a measure for the sticking probability (Auwärtar et al., 2002). Figure 18.17a shows the temperature dependence of the sticking coefficient of cobalt atoms that are evaporated by sublimation onto h-BN/Ni(111). Clearly, the sticking is not unity, as it is commonly assumed for a metal that is evaporated onto a metal. Also, the sticking is strongly temperature dependent. This means that at higher temperature more Co atoms scatter back into the vacuum and that the bond energy of the individual Co atoms must be fairly small. The solid line in Figure 18.17a indicates the result of an extended Kisliuk model (Kisliuk, 1957) for the sticking of Co on h-BN/Ni(111). This model predicts about 30% of the Co atoms not to thermalize on the surface and to directly scatter back, and says that the activation energy for diffusion is about 190 meV smaller than the desorption energy (Auwärtar et al., 2002).

Also the intercalation is thermally activated. In Figure 18.17b the intercalated amount of Co as compared to the total amount

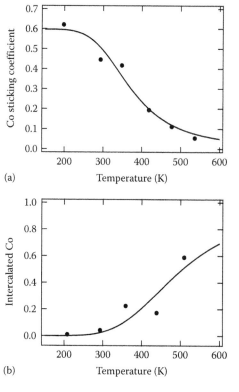

(a)

(b)

FIGURE 18.17 Temperature dependence of (a) sticking coefficient of Co from a sublimation source at about 1400 K and (b) intercalation of Co below h-BN on Ni(111). (Data from Auwärtar, W. et al., *Surf. Sci.*, 511, 379, 2002. With permission.)

of Co on the surface is shown for low Co coverages. The fact that at low substrate temperatures almost no Co slips below the h-BN indicates that the intercalation must be thermally activated. The number N^{2D} of intercalated atoms may be modeled with a simple ansatz like

$$N^{2D} = k_2 \exp - \frac{E_A^{2D}}{k_B T} \tag{18.5}$$

where

k_2 is the reaction constant that comprises the sticking and the diffusion of Co to the intercalation site

E_A^{2D} is the thermal activation energy for intercalation at this site

$k_B T$ is the thermal energy

For the 3D clusters, the number N^{3D} of atoms in the clusters is assumed to be constant:

$$N^{3D} = k_3 \tag{18.6}$$

where k_3 is the reaction constant that comprises the sticking and the diffusion of Co to the 3D cluster nucleation sites. The backreaction, i.e., the dissolution of atoms from 2D islands or 3D clusters, is neglected since there the Co bond energies are much larger. The solid line in Figure 18.17b shows a fit with $E_A^{2D} = 0.24 \pm 0.1$ eV and a ratio $k_2/k_3 = 1/250$. In k_2/k_3, the temperature dependence of the sticking and the diffusion cancel. A temperature dependence of the 3D nucleation site density is not included in this kinetic modeling because the data do not allow the extraction of more than two independent parameters. The small value of k_3/k_2 is an indication that a Co atom finds an intercalation site more easily than a 3D cluster. This is consistent with the assignment that the defect lines (see Figure 18.16a) act as collectors for the intercalation since the probability for a diffusing particle to hit a line is much larger than hitting a point like 3D cluster.

18.2.6 Functionality: Tunneling Junctions, Templates

The functionality of a material is given by a set of particular capabilities, like e.g. selective adsorption and desorption of certain molecules. For sp^2 layers it backs on properties like high thermal stability and, compared to clean transition metal surfaces, low reactivity. The functionality may feature a change in electron transport perpendicular and parallel to the surface. In a metal sp^2-layer metal hetero-junction the layer changes the phase matching of the electrons in the two electrodes. This is particularly interesting in view of spintronic applications where in a magnetic hetero-junction, the resistance for the two spin components is not the same, and magnetoresistance or spin filtering is expected (Karpan et al., 2007). If the layers are insulating or dielectric, as it is the case for h-BN they act as atomically sharp tunneling junctions. The corrugated sp^2 layers bear, on top of the potential of the flat layers, the possibility of using them as templates for molecular architectures.

18.2.6.1 Tunneling Junctions

An insulating single layer on a metal acts as a tunneling junction between the substrate and the adsorbate. It decouples adsorbates from the metal. This decoupling also influences the time scale on which the electronic system equilibrates. The electronic equilibration time scale increases and eventually becomes comparable to that of molecular motion. This opens the doors for physics beyond the Born-Oppenheimer approximation, where it is assumed that the electronic system is always in equilibrium with the given molecular coordinates. Such an example is the C_{60}/h-BN/Ni(111) junction (Muntwiler et al., 2005). Figure 18.18a shows three substrate terraces separated by the Ni(111) step height of 0.2 nm. The whole surface is wetted by a hexagonally closed packed C_{60} layer with a 4×4 superstructure, with one C_{60} on 16 h-BN/Ni(111) unit cells. The electronic structure of the C_{60} layer shows a distinct temperature dependence

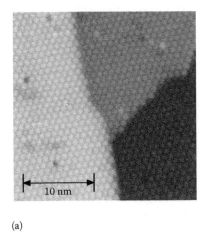

(a)

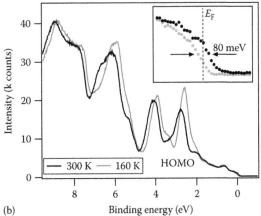

(b)

FIGURE 18.18 C_{60} on h-BN/Ni(111). (a) STM shows that a monolayer of C_{60} wets the substrate and forms a regular 4×4 structure. (b) The valence band photoemission spectrum is dominated by molecular orbitals of C_{60}. The intensity and energy position of the orbitals are strongly temperature dependent. Note the shift of the photoemission leading edge of 80 meV, which corresponds to the transfer of about half an electron onto the molecules in going from 150 K to room temperature. (From Muntwiler, M. et al., *Phys. Rev. B*, 71(12), 121402, 2005. With permission.)

that is in line with the onset of molecular libration. The freezing of the molecular rotation at low temperature is well known from C_{60} in the bulk of fullerene (Heiney et al., 1991) and C_{60} at the surface of fullerene (Goldoni et al., 1996). The C_{60} molecular orbitals as measured with photoemission shift by about 200 meV in going from 100 K to room temperature. The shift is parallel to the work function, which indicates a vacuum level alignment of the molecular orbitals of C_{60}. It signals a significant charge transfer from the substrate to the adsorbate. As can be seen in the inset of Figure 18.18b it causes an upshift of the leading edge in photoemission. This upshift translates into an average charge transfer of about 0.5 e^- onto the lowest unoccupied molecular orbital (LUMO) of the C_{60} cage. The LUMO occupancy therefore changes by about a factor of 7 in going from 100 K to room temperature, which indicates a molecular switch function. The normal emission photoemission intensity of the highest occupied molecular orbital (HOMO), as a function of the substrate temperature, shows the phase transition between 100 K and room temperature. The fact that the phase transition is visible in the normal emission intensity gives a direct hint that the molecular motion which leads to the phase transition must be of rocking type (cartwheel) since azimuthal rotation of the C_{60} molecules would not alter this intensity. The experiments also show that C_{60} is indeed weakly bound to the substrate since it desorbs at temperatures of ~500 K, which is only 9% higher than the desorption temperature from bulk C_{60}. If we translate the work function shift in the phase transition into a pyroelectric coefficient $p_i = \partial P_S / \partial T$, where P_S is the polarization, we get an extraordinary high value of $-2000\,\mu C/m^2\,K$, which is larger than that of the best bulk materials. The correlation of the molecular orientation with the charge on the LUMO can be rationalized with the shape of the LUMO of the C_{60}. The LUMO wave function is localized on the pentagons of the C_{60} cage and, at low temperature, C_{60} does not expose the pentagons toward the h-BN/Ni(111). Therefore, the charge transfer is triggered by the onset of rocking motion where the LUMO orbitals get a larger overlap with the Fermi sea of the nickel metal. In turn, this will increase electron tunneling to the LUMO. In an adiabatic picture, however, the back tunneling probability is as large as the forth tunneling probability, and the magnitude of the observed charge transfer could not be explained. It was, therefore, argued that the magnitude of the effect is a hint for nonadiabatic processes, i.e., that the back tunneling rate gets lower due to electron self-trapping. This self-trapping can only be efficient, if the electron tunneling rate is low, i.e., if the electron resides for times that the molecule needs to change its coordinates.

18.2.6.2 Templates

Templates are structures that are able to host objects in a regular way. In nanoscience, they play a key role and act as a scaffold or construction lot for supramolecular self-assembly that allow massive parallel production processes on the nanometer scale. Therefore, the understanding of the template function is of paramount importance. It requires the exploration of the atomic

structure that defines the template unit cell geometry, and the electronic structure that imposes the bond energy landscape for the host atoms or molecules. On surfaces, the bond energy landscape has a deep valley perpendicular to the surface that is periodically corrugated parallel to the surface. The perpendicular valley is responsible for the adsorption and the parallel corrugation governs surface diffusion. If the surface shall act as a template, and, e.g., impose a lateral ordering on a length scale larger than the (1 × 1) unit cell, it has to rely on reconstruction and the formation of superstructures. For the case of h-BN/Rh(111), the superstructure has a size of 12 × 12 Rh(111) unit cells on top of which 13 × 13 h-BN unit cells coincide. It is a corrugated sp^2 layer, which displays a particular template function for molecular objects with a size that corresponds with the nanostructure. The super cell divides into different regions, the wires, where the layer is weakly bound to the substrate and the "holes" or "pores," where h-BN is tightly bound to the substrate. The holes have a diameter of 2 nm and are separated by the lattice constant of 3.2 nm (see Section 18.2.2.2).

Figure 18.19 documents the template function of h-BN/Rh(111) for three different molecules at room temperature. For C_{60} with a van der Waals diameter of about 1 nm it is found that h-BN is wetted, and 12 C_{60} molecules sit in one Rh (12 × 12) unit cell. It can also be seen that the ordering is not absolutely perfect. There are super cells with one C_{60} missing and such where one extra C_{60} is "coralled" in the holes of the h-BN nanomesh. If we take a molecule that has the size of the holes of 2 nm, here naphthalocyanine (Nc) ($C_{48}H_{26}N_8$), we observe that the molecules self-assemble in the holes. Apparently, the trapping potential is larger than the molecule–molecule interaction that would lead to the formation of Nc islands with touching molecules. If we take a molecule with an intermediate size, i.e., copper phthalocyanine (Cu-Pc) ($C_{32}H_{16}CuN_8$) with a van der Waals diameter of 1.5 nm, it is seen that they also assemble in the nanomesh holes. There is a significant additional feature compared to the Nc case, i.e., Cu-Pc does not sit in the center of the holes but likes to bind at the rims. This observation gives an important hint to the bond energy landscape within this superstructure.

In Section 18.2.3.2, it was shown that corrugated sp^2 layer systems have relatively large lateral electric fields due to dipole rings. These fields may polarize molecules and provide an additional bond energy. Figure 18.11 shows that this bonding scales with α (Equation 18.4) $\cdot E_\parallel^2$, where $E_\parallel$ is the lateral electric field and α is the polarizability of the molecule. $E_\parallel^2$ is largest at the rims of the h-BN nanomesh holes or the g/Ru(0001) mounds. This feature in the bond energy landscape is in line with the observation that molecules like Cu-Pc like to sit at the corrugation rims of the sp^2 layers. A quantitative measure for the adsorption energy can be obtained from thermal desorption spectroscopy (TDS), which was performed for xenon.

Figure 18.20 shows a comparison of Xe/h-BN/Rh(111) and Xe/g/Ru(0001) TDS, where the surfaces are heated with a constant heating rate $\beta = dT/dt$, and where the remaining Xe on the surface was monitored with photoemission from adsorbed

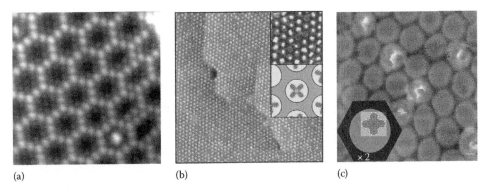

FIGURE 18.19 Room temperature STM images of molecules trapped in h-BN/Rh(111) nanomesh. (a) C_{60}: Individual molecules are imaged throughout this region, following closely the topography ($15 \times 15\,nm$). The positions in the hole centers are occupied by either zero or one C_{60} molecule; at two places, large protrusions may represent additional corralled molecules. (From Corso, M. et al., *Science*, 303, 217, 2004. With permission.) (b) Naphthalocyanine (Nc) ($C_{48}H_{26}N_8$): Site-selective adsorption ($120 \times 120\,nm$). The inset ($19 \times 19\,nm$) on the top right shows an enlargement. The inset on the right is a schematic representation of the molecule in the h-BN nanomesh. (From Berner, S. et al., *Angew. Chemie-Int. Ed.*, 46, 5115, 2007. With permission.) (c) Copper phthalocyanine (Cu-Pc) ($C_{32}H_{16}CuN_8$) molecules trapped at the rim of the holes ($18 \times 18\,nm$). At the given tunneling conditions, the mesh wires map dark, the holes map gray, and Cu-Pc are imaged as bright objects. The inset shows a magnified model of the trapped Cu-Pc molecule. (From Dil, H. et al., *Science*, 319, 1824, 2008. With permission.)

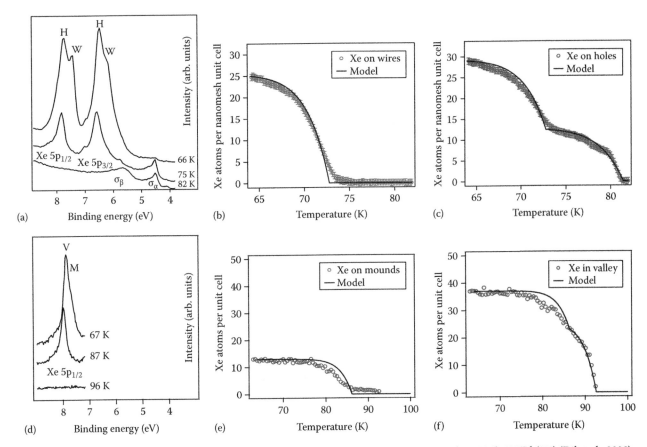

FIGURE 18.20 Temperature dependence of normal emission valence band photoemission spectra from Xe/h-BN/Rh(111) (Dil et al., 2008) and Xe/g/Ru(0001) (Brugger et al., 2009). (a) Energy distribution curves extracted for three different temperatures (66, 75, and 82 K) for Xe/h-BN/Rh(111). (b) Spectral weight of the Xe on the wires as a function of temperature. (c) Spectral weight of the Xe in the holes as a function of temperature. (d) Energy distribution curves extracted for three different temperatures (67, 87, and 96 K) for Xe/g/Ru(0001). (e) Spectral weight of the Xe on the mounds as a function of temperature. (f) Spectral weight of the Xe in the valleys as a function of temperature. The solid lines in (b), (c), (e), and (f) are fits obtained from a zero-order desorption model.

xenon (PAX). Since the Xe core level energy is sensitive to the local electrostatic potential it also indicates where it is sitting in the sp^2 super cell. This allows a TDS experiment where the Xe bond energy on different sites in the super cell is inferred (Dil et al., 2008; Brugger et al., 2009).

The desorption data for Xe/h-BN/Rh(111) on the wires and in the holes are shown in Figure 18.20b and c, respectively, and for Xe/g/Ru(0001) on the mounds and the valley in Figure 18.20e and f. The data of the weakly bound sp^2 layer regions (wires and mounds) are well described with a zero order desorption model where dN molecules desorb on increasing temperature dT:

$$-dN = \frac{\nu}{\beta} \cdot \exp\left(-\frac{E_d}{k_B T}\right) \cdot dT \qquad (18.7)$$

where

ν is the attempt frequency in the order of 10^{12} Hz
β is the heating rate
E_d is the desorption energy
$k_B T$ is the thermal energy

The data are fitted to the integrals of Equation 18.7 where the initial coverage N_1 is taken from the intensity of the photoemission peaks, the sizes of the Xe atoms, and the super cells. From this it is, e.g., found that 25 Xe atoms cover the wires in the h-BN/Rh(111) unit cell. The desorption energies of 169, 181, 222, and 249 meV for Xe/Xe (Kerner et al., 2005), Xe^W/h-BN/Rh(111), Xe^M/g/Ru(0001), and for Xe/graphite (Ulbricht et al., 2006) indicate that they are similar, but have a trend to increase when going to more metallic substrates. The initial coverages and desorption energies are summarized in Table 18.2. For the strongly bound regions (hole and valley), this single desorption energy picture does not hold. For Xe/h-BN/Rh(111), it turned out that 12 Xe atoms in the holes have to be described with a larger bond energy. Correspondingly different phases C and R have been introduced, where C stands for coexistence and R for ring or rim (Dil et al., 2008). Twelve Xe atoms fit on the rim of the h-BN/Rh(111) nanomesh holes, and thus a dipole ring–induced extra bond energy of 13% was inferred. For g/Ru(0001), the situation is less clear-cut since a two phase fit fits the data less well than those of the h-BN/Rh(111) case. This may be due to the different topography of the two systems (see Figure 18.8). The fact that the extra bond energy in Xe/g/Ru(0001) is 4% only is rationalized

with the lower local work function difference, i.e., the smaller Xe core level energy splitting in Xe/g/Ru(0001) (240 meV) than in Xe/h-BN/Rh(111) (310 meV).

Appendix 18.A: Atomic and Electronic Structures in Real and Reciprocal Space

Graphene and hexagonal boron nitride layers are isoelectronic. Both the (C,C) and the (B,N) building blocks have 12 electrons and both have a honeycomb network structure with fairly strong bonds of 6.3 and 4.0 eV for C=C and B=N, respectively. The electronic configurations of the constituent atoms are shown in Figure 18.21.

18.A.1 sp^2 Hybridization

The strong bonding within the sp^2 network foots on the hybridization of the $2s$ and the $2p$ valence orbitals. The $2s$ and $2p$ energies in atomic boron, carbon, and nitrogen are close in energy (in the order of 10 eV) and have similar spacial extension. If the atoms are assembled into molecules, where the overlap between adjacent bonding orbitals is maximized, the linear combination of the s and p orbitals provides higher overlap than that of two $2p$ orbitals. Figure 18.22 shows a s and a p_x wave function and their sp_x hybrid, which is a coherent sum of the wave function amplitudes. The phase in the p wave function depends on the direction and produces the lobe along the x direction, which allows a large overlap with the wave functions of the neighboring atoms.

The hybrid orbitals are obtained as a linear combination of atomic orbitals. For the case of carbon sp^3, sp^2, and sp^1 hybrids may be formed, where CH_4 (methane), C_2H_4 (ethene), and C_2H_2 (acetylene) are the simplest hydrocarbons representing these hybrids. The tetrahedral symmetry of sp^3 and the planar symmetry of sp^2 are also reflected in the two allotropes diamond

5	6	7
B	C	N
$1s^2 2s^2 2p^1$	$1s^2 2s^2 2p^2$	$1s^2 2s^2 2p^3$

FIGURE 18.21 Boron, carbon, and nitrogen are neighbors in the periodic table. Accordingly, h-BN and graphene are isoelectronic and are constituted by 12 electrons per unit cell.

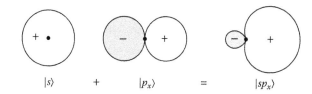

FIGURE 18.22 The coherent sum of a $|s\rangle$ and $|p_x\rangle$ wave function leads to a directional $|sp^2\rangle$ hybrid. The phases of the wave functions are marked with + and −, where negative amplitudes are shaded.

TABLE 18.2 Experimentally Determined Xe Desorption Energies E_d and Xe Atoms per Unit Cell at Full Coverages N_1 for h-BN/Rh(111) (Dil et al., 2008) and g/Ru(0001) (Brugger et al., 2009)

	h-BN/Rh(111)			g/Ru(0001)		
Phase	C^W	C^H	R^H	C^M	C^V	R^V
E_d (meV)	181	184	208	222	222	234
N_1	25	17	12	13	14	23

Note: For all fits an attempt frequency ν of 1.2×10^{12} Hz has been used.

and graphite, where graphite is the thermodynamically stable phase. For boron nitride, accordingly cubic and hexagonal boron nitrides are found, where here the cubic form is thermodynamically more stable at room temperature and 1 bar pressure.

The *sp* hybrid orbitals are obtained in combining *s* and *p* orbitals. This corresponds to a new base for the assembly of atoms into molecules. In the *sp²* hybrid orbitals, the index 2 indicates that two *p* orbitals are mixed with the *s* orbital. The *2s* orbital is combined with $2p_x$ and $2p_y$ orbitals, into 3 twofold spin degenerate orbitals that form the σ bonds:

$$\left| sp_1^2 \right\rangle = \sqrt{\frac{1}{3}} |s\rangle + \sqrt{\frac{2}{3}} |p_x\rangle \tag{18.8}$$

$$\left| sp_2^2 \right\rangle = \sqrt{\frac{1}{3}} |s\rangle - \sqrt{\frac{1}{6}} |p_x\rangle + \sqrt{\frac{1}{2}} |p_y\rangle \tag{18.9}$$

$$\left| sp_3^2 \right\rangle = \sqrt{\frac{1}{3}} |s\rangle - \sqrt{\frac{1}{6}} |p_x\rangle - \sqrt{\frac{1}{2}} |p_y\rangle \tag{18.10}$$

These three σ orbitals contain a mixture of 1/3 of an s electron and 2/3 of a p electron. It can easily be seen that the three orbitals lie in a plane and point in directions separated by angles of 120°. $\left| sp_1^2 \right\rangle$ points into the direction $\left[\sqrt{2/3}, 0, 0 \right]$, $\left| sp_2^2 \right\rangle$ into the direction $\left[-\sqrt{1/6}, \sqrt{1/2}, 0 \right]$, and from the scalar product between the two directions, we get the angle of 120°. In the case of the *sp²* hybridization, the fourth orbital has pure *p* character and forms π bonds.

$$\left| sp_4^2 \right\rangle = |p_z\rangle \tag{18.11}$$

$\left| sp_4^2 \right\rangle$ is perpendicular to the *sp²* σ bonding plane. For the case of graphene, these p_z orbitals on the honeycomb *sp²* network are responsible for the spectacular electronic properties of the conduction electrons in the π bands because they are occupied with one electron.

18.A.2 Electronic Band Structure

The *sp²* hybridization determines the atomic structure of both, hexagonal boron nitride and graphene sheets. They form a two-dimensional honeycomb structure as shown in Figure 18.23. The lattice can be described as superposition of two coupled sublattices *A* and *B* (see Figure 18.23a). In the case of graphene both sublattices are occupied by one carbon atom (C_A, C_B), while in the case of h-BN one sublattice is occupied by boron atoms and the other sublattice by nitrogen atoms (B,N). The interference of the electrons between these lattices causes the peculiar electronic structure of *sp²* layer networks. The 12 electrons in the unit cell are filled into 4 *1s* core levels, and into the 16 *sp²* hybrids that form three σ bonding, one π bonding,

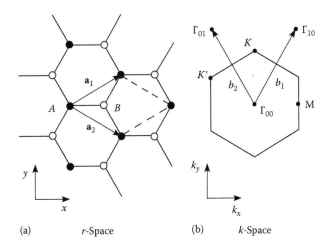

(a)   *r*-Space   (b)   *k*-Space

FIGURE 18.23 Honeycomb structure like that of a single layer h-BN or graphene. (a) Real space: The *sp²* hybridization causes the formation of two coupled sublattices *A* and *B* with lattice vectors $\mathbf{a}_1$ and $\mathbf{a}_2$ and lattice constant $a = |\mathbf{a}_1| = |\mathbf{a}_2|$. (b) Reciprocal space: Brillouin zone. The high symmetry points Γ, *M*, and *K* are marked, where the reciprocal distance Γ*K* is $2\pi/\sqrt{3}a$. The distinction of *K* and *K'* is possible if system is threefold symmetric (trigonal), but not sixfold symmetric (hexagonal).

one π* antibonding, and 3σ* antibonding bands. In the bonding bands, the two adjacent *sp²* hybrids are in phase, while in the antibonding case they are not. The atomic orbitals sp_1^2, sp_2^2, and sp_3^2 constitute the in plane σ bands, while the $sp_4^2 = p_z$ orbitals form the π bands. From this, it can be seen that for flat layers the σ and π electrons do not interfere, i.e., may be treated independently. In Figure 18.23, the real space lattice and the corresponding Brillouin zone in reciprocal space are shown. In *k*-space the reciprocal lattice constant is $2\pi/a$, where *a* is the lattice constant of graphene or h-BN, of about 0.25 nm. For the case of graphene, where the two sublattices are indistinguishable, this leads to a gap-less semiconductor with Dirac points at the *K* points. In h-BN, the distinguishability of boron and nitrogen leads to an insulator, where the π valence band is mainly constituted by the nitrogen sublattice, and the conduction band by the boron sublattice.

The different electronegativities of boron and nitrogen lead for free-standing h-BN to 0.56 *e⁻* transferred from B to N (Grad et al., 2003). This ionicity produces a Madelung energy E_{Mad}:

$$E_{\text{Mad}} = \alpha_{\text{Mad}} \cdot \frac{1}{4\pi\epsilon_0} \cdot \frac{q^2}{a} \tag{18.12}$$

with

$\alpha_{\text{Mad}} = 1.336$ for the honeycomb lattice (Rozenbaum, 1996)
q is the displaced charge (in the above case 0.56 *e⁻*)
a the lattice constant

It has to be noted that this Madelung energy applies for the free-standing case. If the ionic honeycomb layer sits on top of a metal, the energy in Equation 18.12 reduces by a factor of 1/2.

18.A.2.1 π Bands

It is instructive to recall the basic statements within the framework of the "tight binding" scheme of Wallace that he developed for the band theory of graphite (Wallace, 1947). Also the description of tight binding calculations of molecules and solids of Saito and Dresselhaus is recommended (Saito et al., 1998) and the most recent review of Castro Neto et al. (Castro Neto et al., 2009). Tight binding means that we approximate the wave functions as superpositions of atomic p_z wave functions on sublattices A and B, respectively. Furthermore, they are Bloch functions with the periodicity of the lattice. The essential physics lies in the interference between the two sublattices. It is just another beautiful example for quantum physics with two interfering systems. In order to solve the Schrödinger equation with this ansatz, we have to solve the secular equation:

$$\begin{vmatrix} H_{AA} - E & H_{AB} \\ H_{BA} & H_{BB} - E \end{vmatrix} = 0 \qquad (18.13)$$

where the matrix elements H_{AA} and H_{BB} describe the energies on the sublattices A and B and, most importantly, H_{AB} the hybridization energy due to interference or hopping of the π electrons between the two lattices.

The solutions for the energies E_- (bonding) and E_+ (antibonding) are

$$E_\pm = \frac{1}{2}\left(H_{AA} + H_{BB} \pm \sqrt{(H_{AA} - H_{BB})^2 + 4|H_{AB}|^2}\right) \qquad (18.14)$$

Here we show the simplest result that explains the essential physics. It is the degeneracy of the bonding π band and the antibonding π* band at the K point of the Brillouin zone, if the two sublattices are indistinguishable. This means that at K the square root term in Equation 18.14 has to vanish.

The tight binding ansatz delivers values for H_{AA}, H_{BB}, and H_{AB}:

$$H_{AA} = E_A \qquad (18.15)$$

$$H_{BB} = E_B \qquad (18.16)$$

where E_A and E_B are the unperturbed energies of the atoms on sublattices A and B. The square of the interference term H_{AB} is k-dependent and gets

$$|H_{AB}|^2 = \gamma_{AB}^2\left(1 + 4\cos(k_x a)\cos\left(k_y a/\sqrt{3}\right) + 4\cos^2\left(k_y a/\sqrt{3}\right)\right) \qquad (18.17)$$

where

k_x and k_y are coordinates in k-space pointing along x and y, respectively

a is the lattice constant (see Figure 18.23)

γ_{AB} describes the hybridization between the sublattices and is proportional to the electron hopping rate between two adjacent sites on sublattice A and sublattice B. Basically it determines the π band width, which turns out to be $3\gamma_{AB}$. If hopping within the sublattices, e.g., between two adjacent A-sites, is allowed this leads to k-dependent corrections in H_{AA} and H_{BB} and to a symmetry breaking between the π and π* bands (Wallace, 1947). If the phase of the electron wave functions is considered, the hexagonal symmetry is broken and a trigonal symmetry, i.e., two distinct K points, K and K', have to be considered (McClure, 1956; Castro Neto et al., 2009). For equivalent sublattices, i.e., $H_{AA} = H_{BB}$ the π band structure is given by the hopping between the two sublattices, i.e., by H_{AB} and H_{BA}, respectively. From Equation 18.17, it is seen that H_{AB} vanishes at the K point, i.e., for $\mathbf{k} = (k_x, k_y) = \left(0, \frac{2\pi}{\sqrt{3}a}\right)$. The fact that the electrons in the two sublattices do not interfere if they are at the K point of the Brillouin zone is a direct consequence of the symmetry of the crystal and does not change if hopping between non-nearest neighbors is included in the model. It should also be mentioned that these results are expected for the electronic structure of any system that forms a honeycomb lattice.

In Figure 18.24, the generic tight binding band structure of a sp^2-hybridized lattice with two sublattices is shown for hopping γ_{AB} between the sublattices A and B, only. Figure 18.24a presents the case of graphene, i.e., equivalent sublattices A and B. The two p_z electrons from the atoms C_A and C_B fill the π

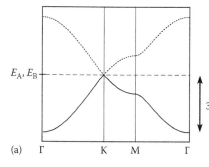

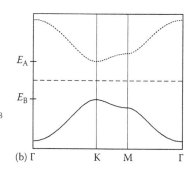

(a) Γ   K   M   Γ   (b) Γ   K   M   Γ

FIGURE 18.24 Tight binding π and π* band structures of honeycomb lattices as shown in Figure 18.23 along ΓKMΓ. The hopping rate $\gamma_{AB}/\hbar$ is kept constant for both cases. (a) Case for two undistinguishable sublattices A and B with $E_A = E_B$. Note the emergence of a gapless semiconductor: If each sublattice contributes one electron, the Fermi surface is constituted by the Dirac point at K. (b) Case for two distinguishable sublattices A and B with $E_A > E_B$. Note that at K, a gap with the magnitude $E_g = E_A - E_B$ opens, and with an even number of π electrons the system is an insulator.

band. The highest energy corresponds to the Fermi energy and lies at the K point. This means that the Fermi surface of this system consists of points at the K points. The peculiarity that the resulting Fermi surface encloses no volume leads to the notion that graphene is a gapless semiconductor. The band structure in the vicinity of the K point is most interesting. The dispersion $E(k)$ of the electrons is linear, i.e., $(\partial E/\partial k)_K = v_F =$ const.. The linear dispersion resembles that of massless photons, or relativistic particles with $E \gg m_0 c^2$, where m_0 is the rest mass. It constitutes so called Dirac cones at the K point of the Brillouin zone (see Figure 18.24a). From Equation 18.17 we get with $\partial E/\partial k$ at the K point the Fermi velocity $v_F = (a \cdot \gamma_{AB})/\hbar$, which is for a π bandwidth of 6 eV or hopping rate of $1/10$ fs about $c/300$. In brackets a seeming contradiction has to be clarified: The second derivative of $E(k)$ at the K point is zero. With the relation for the second derivative of $\partial^2 E/\partial k^2 = \hbar^2/m^*$, the effective mass m^* of the electrons and the holes is infinite. That is, electrons and holes at the K point may not be accelerated, as it is the case for photons. However, since the Fermi velocity is not zero, electrons need not be accelerated in order to be transported.

In Figure 18.24b the result for a lattice with two inequivalent sublattices that corresponds to the case of h-BN is shown. For the sake of simplicity, the same hybridization γ_{AB} has been chosen. The symmetry between the A and B lattices is broken, if the energies of the unperturbed atoms E_A and E_B are not the same (see Equation 18.14), which is obviously the case for boron and nitrogen. The band structure is similar, but at the K point a gap with the magnitude of $|E_A - E_B|$ opens, and no Dirac physics is expected.

18.A.2.2 σ Bands

The σ bands form the strong bonds between the atoms in the $x - y$ plane. For flat layers, they are orthogonal to the π bands and can be treated independently. The case is more involved than that of the π bands since here the three atomic orbitals (s, p_x and p_y) on the two sublattices give rise to six bands. Also, the overlap between the different atomic orbitals gets larger, and the s–p mixing is a function of the k vector. With the tight binding ansatz similar to Equations 18.14 through 18.17 three σ bands were derived (Saito et al., 1998). Essentially the secular equation is now the determinant of a 6×6 matrix leading to six bands. At Γ the lowest lying band, σ_0, has s character and the two remaining σ bands, σ_1 and σ_2 are degenerate and mainly p_x and p_y derived.

It is interesting to note that also for the σ bands the band structure forms cones, if the base of the honeycomb lattice is homonuclear (graphene). They are reminiscent to the Dirac cone, where the π and π* bands touch. At K and a binding energy of about 13 eV, the σ_0 and σ_1 bands touch.

For heteronuclear bases in the honeycomb (h-BN) also a gap opens, as it is observed for the π and π* bands. Of course, the conical band touching is less important since both involved σ bands remain fully occupied. The measurement of this gap, nevertheless, would open a way to distinguish the σ bonding and the π bonding to the substrate.

Figure 18.25 shows a state-of-the-art density functional theory (DFT) band structure calculation for a single layer graphene and h-BN, respectively. These calculations consider a much larger basis set than the 2s and three 2p orbitals as it is done in

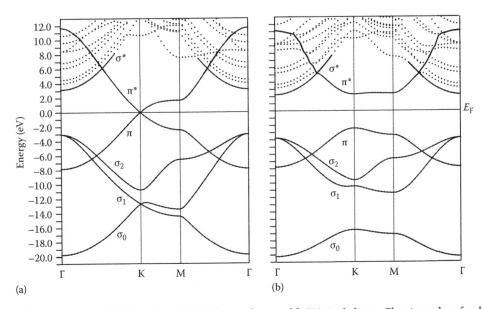

FIGURE 18.25 DFT band structure calculation along ΓKMΓ for graphene and h-BN single layers. The eigenvalues for discrete k-points are shown as dots. The solid lines connect the eigenvalues in the Brillouin zone where the band assignment from the tight binding ansatz with a 2s and 2p basis set holds: Three σ bands, σ_0, σ_1, and σ_2, and the two π bands, π and π*, are shown. The three σ* antibonding bands are more difficult to resolve since they are also mixing in the DFT calculation with 3s and 3p contributions. (a) Graphene. Note the Dirac cone at K at the Fermi level E_F, where the π and π* bands touch. For the σ_0 and σ_{1-} bands, a similar conical band touching occurs at about −13 eV. (b) h-BN. Note the band narrowing and the opening of a gap at the K point for the π–π* and σ_0–σ_1 band cones. (Calculations courtesy of Peter Blaha.)

the tight binding picture. Accordingly, more bands are found. The dots in Figure 18.25 are the energy eigenvalues for given *k* vectors. The three σ, the π and π* bands reproduce the tight binding result (Saito et al., 1998). At higher energies, also 3*s* and 3*p* orbitals contribute eigenvalues, and the identification of the antibonding bands becomes difficult.

Acknowledgments

Most material presented in this chapter was obtained with Jürg Osterwalder and thanks to the empathy and work of our students Wilhelm Auwärter, Matthias Muntwiler, Martina Corso, and Thomas Brugger. It is also a big pleasure to acknowledge Peter Blaha, who started in an early stage of our endeavor to contribute significantly with theory to the understanding of *sp*² single layers.

References

Allan, M. P., Berner, S., Corso, M., Greber, T., and Osterwalder, J. (2007). Tunable self-assembly of one-dimensional nano-structures with orthogonal directions. *Nanoscale Research Letters*, 2(2):94–99.

Auwärter, W., Kreutz, T. J., Greber, T., and Osterwalder, J. (1999). XPD and STM investigation of hexagonal boron nitride on Ni(111). *Surface Science*, 429(1–3):229–236.

Auwärter, W., Muntwiler, M., Greber, T., and Osterwalder, J. (2002). Co on *h*-BN/Ni(111): From island to island-chain formation and Co intercalation. *Surface Science*, 511:379–386.

Auwärter, W., Muntwiler, M., Osterwalder, J., and Greber, T. (2003). Defect lines and two-domain structure of hexagonal boron nitride films on Ni(111). *Surface Science*, 545(1–2):L735–L740.

Auwärter, W., Suter, H. U., Sachdev, H., and Greber, T. (2004). Synthesis of one monolayer of hexagonal boron nitride on Ni(111) from B-trichloroborazine (ClBNH)₃. *Chemistry of Materials*, 16(2):343–345.

Bennewitz, R., Barwich, V., Bammerlin, M., Loppacher, C., Guggisberg, R., Baratoff, A., Meyer, E., and Guntherodt, H. J. (1999). Ultrathin films of NaCl on Cu(111): A leed and dynamic force microscopy study. *Surface Science*, 438(1–3):289–296.

Berner, S., Corso, M., Widmer, R., Groening, O., Laskowski, R., Blaha, P., Schwarz, K. et al., (2007). Boron nitride nanomesh: Functionality from a corrugated monolayer. *Angewandte Chemie-International Edition*, 46(27):5115–5119.

Bligaard, T., Norskov, J. K., Dahl, S., Matthiesen, J., Christensen, C. H., and Sehested, J. (2004). The Bronsted-Evans-Polanyi relation and the volcano curve in heterogeneous catalysis. *Journal of Catalysis*, 224(1):206–217.

Brugger, T., Gunther, S., Wang, B., Dil, J. H., Bocquet, M. L., Osterwalder, J., Wintterlin, J., and Greber, T. (2009). Comparison of electronic structure and template function of single-layer graphene and a hexagonal boron nitride nanomesh on Ru(0001). *Physical Review B*, 79(4):045407.

Budak, E. and Bozkurt, C. (2004). The effect of transition metals on the structure of h-BN intercalation compounds. *Journal of Solid State Chemistry*, 177(4–5):1768–1770.

Bunk, O., Corso, M., Martoccia, D., Herger, R., Willmott, P. R., Patterson, B. D., Osterwalder, J., van der Veen, I., and Greber, T. (2007). Surface x-ray diffraction study of boron-nitride nanomesh in air. *Surface Science*, 601(2):L7–L10.

Burkstrand, J. M., Kleiman, G. G., Tibbetts, G. G., and Tracy, J. C. (1976). Study of n-Cu(100) system. *Journal of Vacuum Science Technology*, 13(1):291–295.

Castro Neto, A., Guinea, F., Peres, N., Novoselov, K., and Geim, A. K. (2009). The electronic properties of graphene. *Reviews of Modern Physics*, 81:109.

Cervantes-Sodi, F., Csanyi, G., Piscanec, S., and Ferrari, A. C. (2008). Edge-functionalized and substitutionally doped graphene nanoribbons: Electronic and spin properties. *Physical Review B*, 77(16) 165427-1–165427-13.

Che, J. G. and Cheng, H. P. (2005). First-principles investigation of a monolayer of C-60 on h-BN/Ni(111). *Physical Review B*, 72(11):115436.

Coraux, J., N'Diaye, A. T., Busse, C., and Michely, T. (2008). Structural coherency of graphene on Ir(111). *Nano Letters*, 8(2):565–570.

Corso, M., Auwärter, W., Muntwiler, M., Tamai, A., Greber, T., and Osterwalder, J. (2004). Boron nitride nanomesh. *Science*, 303:217–220.

Corso, M., Greber, T., and Osterwalder, J. (2005). h-BN on Pd(110): A tunable system for self-assembled nanostructures? *Surface Science*, 577(2–3):L78–L84.

Desrosiers, R. M., Greve, D. W., and Gellman, A. J. (1997). Nucleation of boron nitride thin films on Ni(100). *Surface Science*, 382(1–3):35–48.

Dil, H., Lobo-Checa, J., Laskowski, R., Blaha, P., Berner, S., Osterwalder, J., and Greber, T. (2008). Surface trapping of atoms and molecules with dipole rings. *Science*, 319(5871):1824–1826.

Dresselhaus, M. S. and Dresselhaus, G. (1981). Intercalation compounds of graphite. *Advances in Physics*, 30(2):139–326.

Eizenberg, M. and Blakely, J. M. (1979). Carbon monolayer phase condensation on Ni(111). *Surface Science*, 82(1):228–236.

Enoki, T., Kobayashi, Y., and Fukui, K.-I. (2007). Electronic structures of graphene edges and nanographene. *International Reviews in Physical Chemistry*, 26(4):609–645.

Forbeaux, I., Themlin, J. M., and Debever, J. M. (1998). Heteroepitaxial graphite on 6H-SiC(0001): Interface formation through conduction-band electronic structure. *Physical Review B*, 58(24):16396–16406.

Gamo, Y., Nagashima, A., Wakabayashi, M., Terai, M., and Oshima, C. (1997). Atomic structure of monolayer graphite formed on Ni(111). *Surface Science*, 374(1–3):61–64.

Geim, A. K. and Novoselov, K. S. (2007). The rise of graphene. *Nature Materials*, 6(3):183–191.

Giovannetti, G., Khomyakov, P. A., Brocks, G., Karpan, V. M., van den Brink, J., and Kelly, P. J. (2008). Doping graphene with metal contacts. *Physical Review Letters*, 101(2):026803.

Goldoni, A., Cepek, C., and Modesti, S. (1996). First-order orientational-disordering transition on the (111) surface of C-60. *Physical Review B*, 54(4):2890–2895.

Goriachko, A., He, Y. B., Knapp, M., Over, H., Corso, M., Brugger, T., Berner, S., Osterwalder, J., and Greber, T. (2007). Self-assembly of a hexagonal boron nitride nanomesh on Ru(0001). *Langmuir*, 23(6):2928–2931.

Grad, G. B., Blaha, P., Schwarz, K., Auwärtar, W., and Greber, T. (2003). Density functional theory investigation of the geometric and spintronic structure of h-BN/Ni(111) in view of photoemission and STM experiments. *Physical Review B*, 68(8):085404.

Greber, T., Auwärtar, W., Hoesch, M., Grad, G., Blaha, P., and Osterwalder, J. (2002). The Fermi surface in a magnetic metal-insulator interface. *Surface Review and Letters*, 9(2):1243–1250.

Greber, T., Corso, M., and Osterwalder, J. (2009). Fermi surfaces of single layer dielectrics on transition metals. *Surface Science*, 603:1373–1377.

Gsell, S., Berner, S., Brugger, T., Schreck, M., Brescia, R., Fischer, M., Greber, T., Osterwalder, J., and Stritzker, B. (2008). Comparative electron diffraction study of the diamond nucleation layer on Ir(001). *Diamond and Related Materials*, 17(7–10):1029–1034.

Heiney, P. A., Fischer, J. E., McGhie, A. R., Romanow, W. J., Denenstein, A. M., McCauley, J. P., Smith, A. B., and Cox, D. E. (1991). Orientational ordering transition in solid C_{60}. *Physical Review Letters*, 66(22):2911–2914.

Helveg, S., Lauritsen, J. V., Laegsgaard, E., Stensgaard, I., Norskov, J. K., Clausen, B. S., Topsoe, H., and Besenbacher, F. (2000). Atomic-scale structure of single-layer MoS_2 nanoclusters. *Physical Review Letters*, 84(5):951–954.

Hölzl, J., Schulte, F., and Wagner, H. (1979). *Work Function of Metals*, Volume 85 of *Springer Tracts in Modern Physics*. Springer, Berlin, Heidelberg, Germany/New York, Solid Surface Physics Edition.

Huda, M. N. and Kleinman, L. (2006). h-BN monolayer adsorption on the Ni(111) surface: A density functional study. *Physical Review B*, 74(7):075418.

Janssens, T. V. W., Castro, G. R., Wandelt, K., and Niemantsverdriet, J. W. (1994). Surface-potential around potassium promoter atoms on Rh(111) measured with photoemission of adsorbed Xe, Kr, and Ar. *Physical Review B*, 49(20):14599–14609.

Karpan, V. M., Giovannetti, G., Khomyakov, P. A., Talanana, M., Starikov, A. A., Zwierzycki, M., van den Brink, J., Brocks, G., and Kelly, P. J. (2007). Graphite and graphene as perfect spin filters. *Physical Review Letters*, 99(17):176602.

Karu, A. E. and Beer, M. (1966). Pyrolytic formation of highly crystalline graphite films. *Journal of Applied Physics*, 37(5):2179.

Kerner, G., Stein, O., Lilach, Y., and Asscher, M. (2005). Sublimative desorption of xenon from Ru(100). *Physical Review B*, 71(20):205414.

Kiejna, A. and Wojciechowski, K. F. (1981). Work function of metals - relation between theory and experiment. *Progress in Surface Science*, 11(4):293–338.

Kisliuk, P. (1957). The sticking probabilities of gases chemisorbed on the surfaces of solids. *Journal of Physics and Chemistry of Solids*, 3(1–2):95–101.

Kuppers, J., Wandelt, K., and Ertl, G. (1979). Influence of the local surface-structure on the 5p photoemission of adsorbed xenon. *Physical Review Letters*, 43(13):928–931.

Laskowski, R., Blaha, P., Gallauner, T., and Schwarz, K. (2007). Single-layer model of the hexagonal boron nitride nanomesh on the Rh(111) surface. *Physical Review Letters*, 98(10):106802.

Laskowski, R., Blaha, P., and Schwarz, K. (2008). Bonding of hexagonal BN to transition metal surfaces: An ab initio density-functional theory study. *Physical Review B*, 78(4):045409.

Marchini, S., Gunther, S., and Wintterlin, J. (2007). Scanning tunneling microscopy of graphene on Ru(0001). *Physical Review B*, 76(7):075429.

McClure, J. W. (1956). Diamagnetism of graphite. *Physical Review*, 104(3):666–671.

Morscher, M., Corso, M., Greber, T., and Osterwalder, J. (2006). Formation of single layer h-BN on Pd(111). *Surface Science*, 600(16):3280–3284.

Muntwiler, M., Auwärtar, W., Baumberger, F., Hoesch, M., Greber, T., and Osterwalder, J. (2001). Determining adsorbate structures from substrate emission x-ray photoelectron diffraction. *Surface Science*, 472(1–2):125–132.

Muntwiler, M., Auwärtar, W., Seitsonen, A. P., Osterwalder, J., and Greber, T. (2005). Rocking-motion-induced charging of C-60 on h-BN/Ni(111). *Physical Review B*, 71(12):121402.

Nagashima, A., Itoh, H., Ichinokawa, T., Oshima, C., and Otani, S. (1994). Change in the electronic states of graphite overlayers depending on thickness. *Physical Review B*, 50(7):4756–4763.

Nagashima, A., Tejima, N., Gamou, Y., Kawai, T., and Oshima, C. (1995). Electronic-structure of monolayer hexagonal boron-nitride physisorbed on metal-surfaces. *Physical Review Letters*, 75(21):3918–3921.

Nagashima, A., Gamou, Y., Terai, M., Wakabayashi, M., and Oshima, C. (1996). Electronic states of the heteroepitaxial double-layer system: Graphite/monolayer hexagonal boron nitride/Ni(111). *Physical Review B*, 54(19):13491–13494.

N'Diaye, A. T., Bleikamp, S., Feibelman, P. J., and Michely, T. (2006). Two-dimensional Ir cluster lattice on a graphene moire on Ir(111). *Physical Review Letters*, 97(21):215501.

N'Diaye, A. T., Coraux, J., Plasa, T. N., Busse, C., and Michely, T. (2008). Structure of epitaxial graphene on Ir(111). *New Journal of Physics*, 10:043033.

Nilius, N., Ganduglia-Pirovano, M. V., Brazdova, V., Kulawik, M., Sauer, J., and Freund, H. J. (2008). Counting electrons transferred through a thin alumina film into Au chains. *Physical Review Letters*, 100(9):096802.

Novoselov, K. S., Geim, A. K., Morozov, S. V., Jiang, D., Katsnelson, M. I., Grigorieva, I. V., Dubonos, S. V., and Firsov, A. A. (2005). Two-dimensional gas of massless Dirac fermions in graphene. *Nature*, 438(7065):197–200.

Ohta, T., Bostwick, A., Seyller, T., Horn, K., and Rotenberg, E. (2006). Controlling the electronic structure of bilayer graphene. *Science*, 313(5789):951–954.

Oshima, C. and Nagashima, A. (1997). Ultra-thin epitaxial films of graphite and hexagonal boron nitride on solid surfaces. *Journal of Physics-Condensed Matter*, 9(1):1–20.

Paffett, M. T., Simonson, R. J., Papin, P., and Paine, R. T. (1990). Borazine adsorption and decomposition at Pt(111) and Ru(001) surfaces. *Surface Science*, 232(3):286–296.

Pivetta, M., Patthey, F., Stengel, M., Baldereschi, A., and Schneider, W. D. (2005). Local work function moire pattern on ultra-thin ionic films: NaCl on Ag(100). *Physical Review B*, 72(11):115404.

Preobrajenski, A. B., Vinogradov, A. S., Ng, M. L., Cavar, E., Westerstrom, R., Mikkelsen, A., Lundgren, E., and Martensson, N. (2007). Influence of chemical interaction at the lattice-mismatched h-BN/Rh(111) and h-BN/Pt(111) interfaces on the overlayer morphology. *Physical Review B*, 75(24):245412.

Rokuta, E., Hasegawa, Y., Suzuki, K., Gamou, Y., Oshima, C., and Nagashima, A. (1997). Phonon dispersion of an epitaxial monolayer film of hexagonal boron nitride on Ni(111). *Physical Review Letters*, 79(23):4609–4612.

Rokuta, E., Hasegawa, Y., Itoh, A., Yamashita, K., Tanaka, T., Otani, S., and Oshima, C. (1999). Vibrational spectra of the monolayer films of hexagonal boron nitride and graphite on faceted Ni(755). *Surface Science*, 428:97–101.

Rozenbaum, V. M. (1996). Coulomb interactions in two-dimensional lattice structures. *Physical Review B*, 53(10): 6240–6255.

Ruggiero, C. D., Choi, T., and Gupta, J. A. (2007). Tunneling spectroscopy of ultrathin insulating films: CuN on Cu(100). *Applied Physics Letters*, 91(25):253106.

Saito, R., Dresselhaus, G., and Dresselhaus, M. (1998). *Physical Properties of Carbon Nanotubes*. Imperial College Press, Covent Garden, London, U.K.

Shelton, J. C., Patil, H. R., and Blakely, J. M. (1974). Equilibrium segregation of carbon to a nickel (111) surface - surface phase-transition. *Surface Science*, 43(2):493–520.

Shikin, A. M., Prudnikova, G. V., Adamchuk, V. K., Moresco, F., and Rieder, K. H. (2000). Surface intercalation of gold underneath a graphite monolayer on Ni(111) studied by angle-resolved photoemission and high-resolution electron-energy-loss spectroscopy. *Physical Review B*, 62(19):13202–13208.

Smoluchowski, R. (1941). Anisotropy of the electronic work function of metals. *Physical Review*, 60(9):661–674.

Ulbricht, H., Zacharia, R., Cindir, N., and Hertel, T. (2006). Thermal desorption of gases and solvents from graphite and carbon nanotube surfaces. *Carbon*, 44(14):2931–2942.

Wallace, P. R. (1947). The band theory of graphite. *Physical Review*, 71(9):622–634.

Wandelt, K. (1984). Surface characterization by photoemission of adsorbed xenon (pax). *Journal of Vacuum Science Technology A-Vacuum Surfaces and Films*, 2(2):802–807.

Wang, X. R., Ouyang, Y. J., Li, X. L., Wang, H. L., Guo, J., and Dai, H. J. (2008). Room-temperature all-semiconducting sub-10-nm graphene nanoribbon field-effect transistors. *Physical Review Letters*, 100(20):206803.

Widmer, R., Berner, S., Groning, O., Brugger, T., Osterwalder, E., and Greber, T. (2007). Electrolytic in situ STM investigation of h-BN-nanomesh. *Electrochemistry Communications*, 9(10):2484–2488.

Yanagisawa, H., Tanaka, T., Ishida, Y., Matsue, M., Rokuta, E., Otani, S., and Oshima, C. (2004). Phonon dispersion curves of a BC3 honeycomb epitaxial sheet. *Physical Review Letters*, 93(17):177003.

19

Epitaxial Graphene

Walt A. de Heer
Georgia Institute of Technology

Xiaosong Wu
Georgia Institute of Technology

Claire Berger
*Georgia Institute of Technology
and National Center for
Scientific Research*

19.1 Introduction

Graphene is a two-dimensional (2D) sheet of carbon atoms in which each carbon atom is bound to its three neighbors to form the honeycomb structure shown in Figure 19.1a. Graphene occurs naturally as the basic building sheet of graphite. In graphite, each graphene layer is arranged such that half of the atoms lie directly over the center of a hexagon in the lower graphene sheet, and half of the atoms lie directly over an atom in the lower layer, seen in Figure 19.1b. This stacking of the graphene layers is called the Bernal stacking. This particular arrangement of the graphene layers in graphite greatly affects the electronic properties of graphite; other types of stacking give different electronic properties.

The covalent chemical bonds between the carbon atoms in a graphene sheet, known as sp² bonds, are among the strongest in nature. They are in fact even stronger than the carbon bonds in diamond: diamond reverts to graphite when heated. This exceptional property of the sp² bond, which results in the exceptional stability of graphitic systems, is very important for electronic applications and especially for nanoelectronics in which the structures are subjected to extremely large thermal and electrical stresses. Moreover, the low temperature electronic mobility of graphite is of the order of 10^6 cm²/Vs, exceeding silicon by about three orders of magnitude. The very strong bonds, combined with the low mass of the carbon atom causes a very high sound velocity resulting in a large thermal conductivity, which is advantageous for graphitic electronics. The most important graphene properties actually emerge from the unique band structure of this material as explained in detail below.

In contrast to the very strong bonds between the carbon atoms in a graphene sheet, the inter-sheet van der Waals bonding in graphite is more than 100 times weaker and ranks among the weakest bonds. This very weak interlayer bond causes the layers to become essentially decoupled from each other, which produces large anisotropies in the electronic and thermal properties. For example, at low temperatures in-plane electrical and thermal conductivity of single crystal graphite is several thousand times greater than perpendicular to the planes (Kelly 1981). It is because of the very weak interlayer bonds that microscopically thin layers of graphite are easily mechanically cleaved from a larger crystal by rubbing it against a surface, for example paper, causing the familiar pencil trace that consists of microscopic graphite flakes. In fact graphite derives its name from this property (*graphein* is Greek for *to draw*).

The recent surge in scientific enquiry into graphene gives the impression that this is a new material. This is in fact not so, free-standing graphene has been first produced in 1962 (Boehm 62) and on many surfaces since then. It has been known and described for at least half a century. The chemical bonding and structure has been described since the 1930s while the electronic band structure of graphene had been first calculated by Wallace (Wallace 1947) in 1949. These calculations have been refined ever since, in the wake of new experimental results. In fact, the word *graphene* was officially adopted in 1994, after it had already been used interchangeably with monolayer graphite in surface science to denote graphene sheets on various metallic, metal carbide, and silicon carbide surfaces. It was known in the surface science community that the bonding of a single graphene layer on metal surface depends sensitively on the metal surface. In these experiments, graphene is grown on carefully prepared, contamination-free metal surface by heating them to high temperatures in a carbon containing gas (methane, ethylene, acetylene, etc.). Depending on the surface, the graphene layer

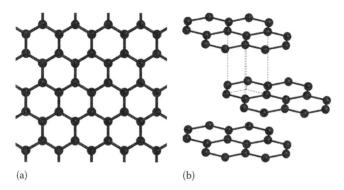

(a)   (b)

FIGURE 19.1 (a) Crystal structure of graphene, in which carbon atoms arrange in a honeycomb structure. (b) AB stacked graphite. One set of carbon atoms are over the center of the hexagons in the adjacent layers.

may grow epitaxially (e.g., on a Ni surface (Rosei et al. 1983)), whereby the graphene atoms registers with the metal atoms on the surface, or the graphene crystallites may have various orientations or even be randomly oriented (as, e.g., on Pt surface (Land et al. 1992)). In many cases, the bonding to the surface is so weak that the properties of pristine graphene are retained.

It is also noteworthy that several recently discovered graphitic nanostructures, like, for example, carbon nanotubes, are directly related to graphene. In fact, carbon nanotubes have been traditionally referred to as rolled up graphene sheets (Iijima 1991), and the carbon nanotube properties are most easily described and understood in terms of those of graphene (Dekker 1999, Special-Issue 2000). Indeed, essential properties of graphene were well understood for decades.

Epitaxial graphene layers can be grown on silicon carbide surfaces by heating single crystals to high temperatures (>1100°C) in vacuum or in other inert atmospheres (Charrier et al. 2002). First reports of single and multilayer graphene on silicon carbide substrates date back to the 1970s (Van Bommel et al. 1975). In this process silicon sublimes, leaving a carbon-rich surface, which reconstructs into a layer composed of one or more epitaxial graphene sheets. The properties of epitaxial graphene sheets on single crystal silicon carbide had been of interest for several reasons. First of all, graphene growth on silicon carbide can occur during various silicon carbide crystal growth processing steps (e.g., molecular beam epitaxy) where it affects the silicon carbide growth; methods have been developed to inhibit the effect of graphene coating. On the other hand, a thin graphitic layer on the silicon carbide had been considered as a method to electrically contact the silicon carbide for silicon carbide–based electronics (Lu et al. 2003).

Researchers at the Georgia Institute of Technology (GIT) pioneered graphene electronics: they were the first to propose and to actively pursue epitaxial graphene for graphene-based electronics. The initial steps were taken in 2001, after preliminary calculations revealed the potential of patterned graphene for electronics in general (Nakada et al. 1996, Wakabayashi et al. 1999, de Heer 2001). In 2002, graphene layers were grown on single crystal silicon carbide and microelectronics methods

to pattern epitaxial graphene were developed. The first patent (de Heer et al. 2006) was issued in 2006 based on an application filed in 2003. In 2004, the GIT team published the first scientific paper describing both the vision of the epitaxial graphene electronics as well as providing the earliest results of the potential of epitaxial graphene electronics properties. This includes indications that the material could be gated (Berger et al. 2004), meaning that an electrostatic potential applied to the graphene layers, thereby altering the electron density, could significantly modulate the conductivity.

Soon after, epitaxial graphene was selected as the most promising platform for graphene-based electronics. In fact, epitaxial graphene has appeared on the International Semiconductor Technology Roadmap as a potential successor of silicon for post-CMOS electronics (ITRS 2007). Below we will outline developments along these lines.

There has been a parallel development in another form of graphene called exfoliated graphene. In 2004, researchers at Manchester University developed a simple method to exfoliate graphite in order to produce thin graphitic layers on conducting (degenerately doped) silicon wafer supplied with a 300 nm thick silicon oxide dielectric layer (Novoselov et al. 2004). The process involves repeatedly "cleaving" a small graphite crystal using Scotch tape, resulting in successively thinner graphite flakes on the tape. Pressing the graphite-flake-coated tape against the oxidized silicon wafer transfers some of the exfoliated flakes to the wafer so that the surface is sparsely covered with microscopically small flakes of varying thicknesses. Most importantly, an optical microscopic method was also developed by which flakes as thin as a monolayer could be reliably identified. The first publication (Novoselov et al. 2004) showed an electric field effect, i.e., that the ultrathin graphite flakes on oxidized silicon carbide surfaces can be gated. The paper demonstrated that these ultrathin graphite flakes could be made and that they essentially had the electronic properties of graphite, which is distinct from those of graphene. It was further suggested and demonstrated in following papers (Novoselov et al. 2005, Zhang et al. 2005) that single layer graphene could be produced by this method. The monolayer graphene was next shown to be distinct from those of graphite and consistent with predictions. This development opened a new direction in 2D electron gas physics, one based on graphene rather than the more familiar 2D electron gases produced at the inversion layer of gallium arsenide/gallium aluminum arsenide interfaces, which had been the focus of the 2D electron gas community for several decades (Beenakker and Vanhouten 1991).

As is clear from various factors (i.e., the method of production, variations in the flake thickness, the microscopic size of the flakes, variations in the surface condition of the substrate), exfoliated graphene samples are suited only for purposes of research into the properties of graphene and ultrathin graphite, and are not considered to have applications potential (Geim and Novoselov 2007). Despite these drawbacks, research into the 2D electronic properties of exfoliated graphene has blossomed into one of the hottest new topics in condensed matter physics.

This recent great interest in graphene can be traced to the confluence of three circumstances:

1. The development of epitaxial graphene, with its promise as a new electronics platform.
2. The development of exfoliated graphene on oxidized silicon substrates for basic research.
3. The theoretical appeal of the simple but unusual electronic structure of a single isolated graphene sheet.

The last point needs to be expanded. Two-dimensional electron gases have been studied for many years resulting in a deep understanding of their properties. The fields of 2D electron gas physics as well as the field of mesoscopic physics (which primarily concerns the properties of very small, i.e., mesoscopic, 2D electron gas structures like quantum wires and quantum dots) are of considerable importance and are well-understood (Beenakker and Vanhouten 1991). The basic electronic structure of graphene is in principle understood and several of its properties have been verified, for example in carbon nanotubes. The fact that the electronic properties of graphene are fundamentally distinct from those of "normal" electron gases provides a bonanza for experimental and theoretical physics. It should be mentioned that theoretical research into the properties of epitaxial graphene are far less advanced. This material is more complex due to the physics at the substrate/graphene interface, which up to recently has been improperly downplayed in most graphene studies.

Here we review some properties, applications and potential of epitaxial graphene grown on silicon carbide substrates.

19.2 The Electronic Structure of Graphene

The electronic structure of graphene is relatively easy to comprehend and it forms the foundation of its electronic properties. Graphene is not a metal: it is essentially a giant organic molecule and technically a semi-metal. The electrons follow the so-called pi-bonds that connect neighboring atoms when they travel from one atom to the next. In contrast to metals (e.g., copper), the precise geometry of the carbon atoms is essential. In further contrast to metals, where the electrons behave, more or less, as if they were essentially free electrons (that is with a bare electron charge and mass), the electrons in graphene interact with the lattice in such a way that they appear to be massless (see below). Moreover, graphene electrons come in two "flavors," which relates to the direction of their "pseudo-spin." The pseudo-spin is an additional quantum number that characterizes graphene electrons. It actually is a label to differentiate bonding and antibonding orbitals that have identical wave numbers and energies. This combination of properties provides graphene with its unique electronic properties. Below we provide more detail.

The four carbon valence electrons cause the chemical bonds between neighboring carbon atoms. The orbitals of these electrons arrange in such a way to produce three covalent sigma bonds with neighboring atoms in the plane. The electrons participating

in these bonds give the graphene sheet its exceptional strength. While electrons in the sigma bonds do not participate in the electronic transport they provide the rigid scaffolding (i.e., the rigid hexagonal honeycomb structure) of the graphitic system.

All of the electronic properties result from the fourth valence electron of each carbon atom. We here provide a simplified picture of the fourth electron. For simplicity, we will refer to these electrons as the electrons and ignore the sigma bonded electrons. For each carbon atom, the electron is centered on the carbon atom and oscillates up and down through the graphene plane. This up and down motion produces a quantum mechanical p_z orbital. This orbital consists of two lobes (charge density clouds): one above and one below the graphene sheet, shown in Figure 19.2a. The motion of the p_z electrons on two neighboring atoms can be in phase or out of phase with each other. Moreover, the orbitals on neighboring atoms slightly overlap. The energy of a pair of neighboring p_z orbitals decreases when their relative motion is in phase with each other, and it increases when the motion is out of phase. Hence, when the motion is in phase, the pair is said to be in a bonding (attractive) configuration, and when they are 180° out of phase, they are in an antibonding (repulsive) configuration. In general, the interaction strength can have any intermediate value depending on the relative phase.

Due to the non-negligible overlap of the p_z orbitals on neighboring atoms, the electrons readily "hop" from one atom to its neighbors. In fact each electron hops at a rate of about $\nu = 10^{15}$ times per second between neighbors. This hopping is so rapid that it is impossible to associate a specific electron with a specific carbon atom. Moreover, a classical description of the electron fails and a quantum mechanical picture is required.

In the quantum mechanical picture, each electron is described as a wave that extends over the entire surface of the graphene sheet. This electron wave has a specific phase and amplitude at each atom. (The amplitudes at each atom are minute: summing the magnitude of the squares of the amplitudes $\Psi(i)$ of a specific electron wave at each atom in the sheet produces unity: $\Sigma|\Psi(i)|^2 = 1$). The total energy of the electron wave is found by summing the energy at each bond, thereby by weighting the amplitude of the electron wave at each atom with a factor that depends on the relative phase of the electron ϕ_{ij} wave at neighboring atoms i, j: $E = -\gamma \Sigma |\Psi(i)| |\Psi(j)| \cos(\phi_{ij})$, where $\gamma \approx 3$ eV represents the energy due to the overlap of neighboring p_z orbitals. Two extreme cases of electron waves can be easily visualized: the first one is when $\phi = 0$ (all in phase) and $|\Psi| = 1/\sqrt{N}$ at each atom. In this case the total energy of the wave is $E_- = -3\gamma$. In the other case, $\phi = \pi$ so that the wave is exactly out of phase at each atom and $E_+ = +3\gamma$. Waves with all intermediate values can be constructed. Those waves for which the total energy is exactly 0 are special as these define the Fermi surface for neutral graphene as explained below.

As demonstrated in the two extreme examples above for neighboring p_z orbitals, the electron waves come in two varieties: π and π^* representing bonding and anti-bonding waves respectively. While for a finite graphene sheet the electron waves are standing waves, in an infinite sheet they are more aptly represented as traveling waves.

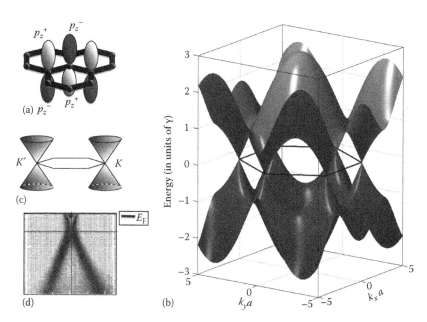

FIGURE 19.2 (a) p_z orbits of carbon atoms are slightly overlapped in graphene. Depending on the relative phase of electron waves in neighboring atoms, the pairing of p_z orbits can change from bonding to antibonding. (b) The band structure of graphene calculated from tight binding model. The valence band meets the conduction band at Dirac points. (c) In the vicinity of Dirac points, the energy of electron waves is linear in momentum. (d) Band structure of epitaxial graphene measured by ARPES shows a linear dispersion and a Dirac point.

A traveling electron wave in graphene is much like a light wave (photon). It has a wavelength, a direction of propagation, energy, and momentum. For simplicity and following quantum mechanical convention, we will refer to the traveling waves as electrons. As a matter of fact in quantum mechanics, an electron is a wave (actually a wave function) that is spread out over space and has an amplitude and phase everywhere on the graphene sheet. It is specifically not a small, localized charged object. Indeed, whereas the spread-out nature of a photon appears more natural, in fact when photons are absorbed, they behave like discrete particles. This particle/wave dichotomy is inherent in the quantum mechanical description of matter.

The momentum p of the electron in a π (or π^*) state is given by $p = hk/2\pi$, where h is Planck's constant and k is the wave number which is a vector of magnitude $2\pi/\lambda$ that points in the propagation direction, and where λ is the wavelength. The energy of a π or a π^* electron, with wave number k propagating in the x direction (perpendicular to a hexagon edge) is $E = \pm h\nu(1 + 2\cos(kd/2))$; ($-$ for π; $+$ for π^*) where $d = 0.26$ nm is the graphene lattice constant. This equation describes the π and π^* energy bands as first found by Wallace (Wallace 1947) in 1947 (Figure 19.2). The two energy bands cross when $k = K = +4\pi/3d$ and $k = K' = -4\pi/3d$ where $E = 0$. These energy bands have been directly measured by angle resolved photoemission by researches at Berkeley (Ohta et al. 2006, Rollings et al. 2006) (Figure 19.2d).

Electrons are Fermions (as a result of their half-integer spin which can be either up or down). Consequently, as for all Fermions, no two electron waves (of similar spin) in graphene can have exactly the same momentum and energy. This property (the Heisenberg exclusion principle) dictates the electronic structure of the graphene as follows.

The electronic structure is found by filling up in energy the π and π^* bands with all of the electrons (the fourth valence electron of each carbon atom) in the graphene sheet consistent with the exclusion principle. Hence, starting with the electron with the least energy at the bottom of the π band, electrons are added one by one, until all N electrons are included. In that situation the graphene sheet is neutral (since one electron was added for each carbon atom). This procedure fills the bands up to a maximum energy (Fermi energy), which corresponds exactly to the energy of the K and K' points in neutral graphene. If the sheet is negatively charged, the additional electrons cause the Fermi level to rise to higher energies; if it is positively charged, the Fermi level is lowered.

Note in Figure 19.2c that near the K and K' points the energy E of the electron waves varies linearly with momentum (i.e., with k): $E = \pm c^*p^*$ where p^* is the momentum measured from the K point and $c^* \approx 10^6$ cm/s. It turns out that taking all directions in the plane into account, this relation still holds: $E = \pm c^*|p^*|$. This is a very interesting situation that is reminiscent of the energy-momentum relation (dispersion relation) of a photon: $E = c|p|$ where p is the momentum of the photon: $p = hk/2\pi$ and k is the wave number. In this case, c is the speed of light. Hence, the dispersion relation of graphene resembles that of a photon, which implies that the velocity of electrons in graphene (its Fermi velocity) is $c^* \approx 10^6$ m/s, independent of its energy, unlike the velocity of electrons in copper for example, that changes with its energy. This property, that the electron velocity is independent of its energy, provides it with the massless quality of a photon; however, the speed of these particles is a factor of 300 smaller than the speed of light. We should point out that in most materials the band structure gives rise to unusual effective masses that range from 0 to ∞. These "effective masses" reflect the property of electrons in a material and not at all the actual masses of the particles involved.

Examining the graphene dispersion relation more carefully reveals that near the Fermi level it is described by two pairs of cones (Figure 19.2c), one pair with its apexes touching the *K* point and the other pair touching the *K*′ point. In neutral graphene, the electrons fill the π and π* bands up to the *K* and *K*′ points (that is the bottom cones), while above the *K* and *K*′ points (that is the top cones) are empty. The fact that there are bottom, filled, cones and top, unfilled, cones and that the electrons are Fermions, indicates that the electrons in graphene actually resemble massless Fermions, like neutrinos, rather than photons. This fact was first pointed out by Ando et al. (Ando et al. 1998), 10 years ago, when he examined the properties of graphene in the context of his well-known investigations of carbon nanotubes. The analogy with neutrinos prompted him to identify the dynamics of electrons in graphene as determined by the Weyl's Hamiltonian (Ando et al. 1998). This property has been rediscovered and colorfully renamed so that the electrons in graphene are now considered to obey the "massless Dirac Hamiltonian." In the same vein, the cones that define the band structure are referred to as Dirac cones and the *K* and *K*′ points as "Dirac points."

Knowing the band structure of graphene allows one to predict some of its electronic transport properties. An electric field *E* applied to the electrons forces the electrons to move. Specifically, only electrons near the Fermi level (charge carriers) can move by shifting from occupied levels to unoccupied levels. Consequently, the current density *J* is given by $J = \sigma E$, where the conductivity $\sigma = e^2ND$. Here *e* is the electronic charge, *D* is the electronic diffusivity, proportional to the electron mobility μ, and *N* is the density of states at the Fermi level E_F. The density of states is the number of states per unit energy and per unit area that are available for electrons. Therefore, at the Fermi energy, *N* reflects the number of charge carriers in the material. In graphene, the density of states at E_F is proportional to the circumference of the circle that is defined by the intercept of E_F and the Dirac cones. For negatively doped graphene (electron doped), E_F is above the Dirac point and for positively doped graphene (hole doped) E_F is below the Dirac point. Moreover, note that *N* varies linearly with E_F and that for neutral graphene $E_F = 0$ so that $N = 0$.

The vanishing of the density of states at E_F in neutral graphene is reminiscent of a zero gap semiconductor (graphene is actually a semi-metal). In either case, it might be expected that the conductivity of graphene increases with E_F (i.e., with the degree of electron doping) and that it vanishes at zero doping. It turns out, however that the situation is not so simple: the physics near the Dirac point is complex.

A second important property of graphene is its relative immunity to scattering from impurities and phonons. This property is perhaps one of the most important features of carbon nanotubes and leads to its ballistic conductance at room temperature (Ando et al. 1998, Frank et al. 1998). The origin of this property can be understood, in principle, from the band structure. Specifically when absorbed, a phonon can impart a rather large momentum change to an electron but with very little energy change. Hence, only electrons at the Fermi level can be scattered and they will be scattered only to other points at the Fermi level. This already greatly restricts which electrons can interact with a specific phonon. There is even a further restriction. In order to back scatter an electron, say one going in the $+p_x$ direction, to one going in the $-p_x$ direction, will require not only a phonon with a momentum of $2p_x$, it must also change the electronic wave from one occupying a bonding orbital to one occupying an antibonding orbital. A single phonon cannot perform this feat and so electron-phonon backscattering is inhibited. A similar argument demonstrates that backscattering from long-range scatterers (e.g., charged particles on the surface) is forbidden, however electrons can scatter from defects in the graphene structure.

As mentioned in the introduction, graphite consists of graphene sheets that are Bernal stacked. Even though the layers are only very weakly coupled, the electronic structure especially near the Fermi level is profoundly affected. This is because of the fact that once a second layer is placed over the first one, every second carbon atom in one sheet has an atom in the other sheet directly above it. This causes a lifting of the graphene symmetry in which all atoms are essentially identical. As a consequence, the Dirac cones develop into touching paraboloids and the electrons become massive and many of the distinguishing properties of graphene vanish.

19.3 Properties of Structured Graphene

The electronic structure described above applies for an infinite single graphene sheet and serves as a starting point for patterned and other forms of graphene. Perhaps the best known graphene structure is the carbon nanotube. A simple carbon nanotube consists of a graphene sheet that is rolled in a seamless tube. Its electronic structure is described in many review articles (Special-Issue 2000, Anantram and Leonard 2006) and has traditionally been derived from the band structure of graphene by imposing appropriate boundary conditions. For a nanotube, electronic motion about its axis results in discrete angular momentum states about the axis while motion along the axis is unconfined. Hence, a nanotube is quantum mechanically a one-dimensional conductor or semiconductor. Most significantly, the confinement effect causes (for semiconducting nanotubes) a band gap to open of the order of 1 eV/*D* where *D* is the diameter of the nanotube.

Likewise, quantum confinement effects also affect a graphene ribbon of width *W*. Specifically, whereas the motion along such a ribbon is unbounded, the motion perpendicular to the ribbon is confined to produce standing waves, resulting in normal modes (or states) that resemble the electromagnetic waves in a waveguide. Since the electronic speed is constant (like for photons), this analogy has merit and is useful to understand electronic propagation through graphene ribbons. Narrow graphene ribbons should be semiconducting, and according to theory their band gap depends on the way the ribbon is "cut." For semiconducting ribbons, the gap is inversely proportional to its width and is predicted to be of the order of 1 eV/W where W is the width in nanometers (Nakada et al. 1996, Wakabayashi et al. 1999).

Hence, the electronic properties can be tuned by shape. This property, which is shared with carbon nanotubes, is probably the single most important property of graphene ribbons. However, in contrast to carbon nanotubes, graphene ribbons can be rationally patterned using relatively standard microelectronics methods. This overcomes one fundamental problem of carbon nanotubes, which need to be accurately placed for electronic applications purposes. Complex graphene structures can be patterned to produce differently shaped interconnected graphene devices, thus avoiding problems of interfaces and interconnects. This resolves a second major problem of carbon nanotube–based electronics. It was precisely these considerations that lead to the pioneer investigations in graphene electronics (Berger et al. 2004, de Heer 2006).

19.4 Epitaxial Graphene on Silicon Carbide

Graphene has been grown on metallic substrates (transition metals and carbides) for many years by pyrolysis of carbon containing gases on heated metal surfaces ((Aizawa et al. 1990, Marchini et al. 2007) and references therein). Most studies were motivated by heterogeneous catalysis, because deposit of graphitic layers occurring as a by-reaction were found to be a major reason for catalyst deactivation. This is because a layer of graphite is inert to most chemical reactions.

In some cases (e.g., on Ni(111) (Rosei et al. 1983) or Ru(0001) (Marchini et al. 2007)), the carbon grows epitaxially which means that the graphene registers with the crystal lattice of the substrate. The properties of these epitaxial films have been extensively studied, however since these films are grown on conducting substrates, they have little electronics applications potential which requires insulting substrates. Consequently, this large body of original graphene work remains relatively unknown to surface-science outsiders.

Silicon carbide is a semiconductor with the chemical composition SiC. It forms in a wide variety of crystal structures (polytypes). We will concentrate on two hexagonal (H) phases, namely the 4H and 6H structures. The 4 and 6 refer to the number of SiC bilayers that make up a unit cell of the crystal. We further focus on the (0001) and (000-1) polar faces (i.e., surfaces). The (000-1) face is one where the surface presents Si atoms, whereas the (000-1) face (on the other side of the crystal) presents C atoms. Ultrathin graphitic material can be grown on silicon carbide crystals simply by heating the material in vacuum at high temperatures (>1100°C). The process involves the sublimation of silicon from the surfaces causing them to become carbon rich. The carbon rearranges to form graphene layers. The layers form epitaxially on the silicon carbide surface. This is clearly a growth process that proceeds in a different way than commonly associated with epitaxial growth on surfaces in which the materials

from which the films grow are supplied externally. Nevertheless, the process is relatively simple and has been known for many years (Van Bommel et al. 1975, Forbeaux et al. 1998). It was also known that growth on the Si and on the C terminated surfaces are different. Material on the C face usually consists of several graphene layers while that grown on the Si face has few and even only one layer (Charrier et al. 2002).

However, several aspects were not known and discovered by the GIT team (Berger et al. 2004, 2006, Hass et al. 2006, 2008a,b, Varchon et al. 2007). The graphene grows over steps on the surface to produce continuous films, seen in Figure 19.3a. This is clearly important for applications. Furthermore it was found, that the interface layer is charged due to the Schottky barrier at the interface to produce charge densities of the order of $n = 10^{12}$ electrons/cm² on the graphene layer at the interface (Berger et al. 2004, 2006, Ohta et al. 2006, Rollings et al. 2006, de Heer et al. 2007, Hass et al. 2008a). The other layers are essentially uncharged (Sadowski et al. 2006, Orlita et al. 2008). This intrinsic charging is advantageous for some applications since it enhances the conductivity.

19.4.1 Si-Face Epitaxial Graphene

Graphene grown on the Si face typically consists of one to about five layers. Like bulk graphite it is Bernal stacked. The interface layer is charged (Ohta et al. 2006, Rollings et al. 2006): $n \approx 5 \times 10^{12}$/cm² so that the Fermi energy $E_F \approx 0.3$ eV. The interaction with the substrate has been reported to cause an energy gap at the interface while this gap rapidly closes as more layers are added (Zhou et al. 2007). The graphene/SiC interface is found to be complex, and composed of various layers (Hass et al. 2008a). It is possible to grow graphene monolayers on the Si face (Bostwick et al. 2007, Mallet et al. 2007, Rutter et al. 2007, Zhou et al. 2007, Emtsev et al. 2008, 2009) although up to now, the electronic mobility is found to be relatively low (i.e., comparable to silicon $\mu \leq 1000$ cm²/Vs at room temperature) (Berger et al. 2004, Emtsev et al. 2009). The low mobility, which appears to be related to defects in the structure, will need to be overcome before graphene grown on the Si face can be considered for electronic applications.

19.4.2 C-Face Epitaxial Graphene

Graphene grown on the C-face has exceptional properties. Figure 19.3c shows a large area of graphene free of SiC steps that was achieved by utilizing step bunching of SiC. C-face graphene is typically multilayered from about 5 to about 100 layers, as seen in Figure 19.3d, and it turns out to be very difficult to grow only a single layer. Like for the Si-face, the interface graphene layer is charged with a similar charge density $n \approx 5 \times 10^{12}$/cm². Electrical transport measurements reveal that the interface layer dominates the electronic transport, at least for relatively narrow ribbons (width <1 μm) and that the mobility of that layer exceeds 10^4 cm²/Vs (Berger et al. 2006). Moreover, quantum coherence has been established over micron distances in relatively narrow

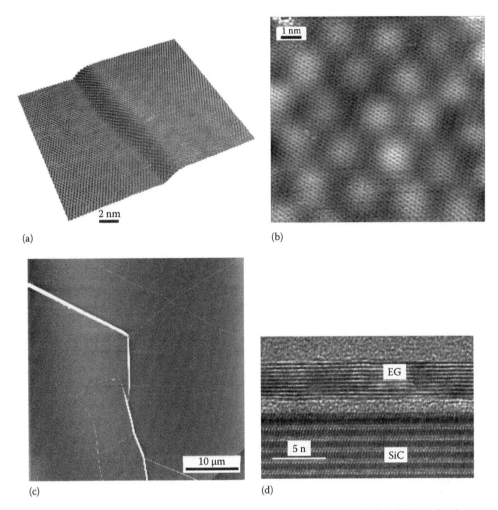

(a)

(b)

(c)

(d)

FIGURE 19.3 (a) STM image of an epitaxial graphene layer over an H-SiC substrate step on the Si-face. (SiC step height 0.25 nm.) (b) STM image of epitaxial graphene on the C-face of 4H-SiC grown in a levitation furnace, in low vacuum. The Moiré pattern arises from a rotational stacking of the graphene layers. (c) AFM image of epitaxial graphene on the C-face of 4H-SiC. The white lines are "puckers" in the graphene sheets. Flat graphene terraces extend over several tens of micrometers. (d) TEM image of the cross section of multilayer epitaxial graphene on the C-face of 4H-SiC.

ribbons (Berger et al. 2006). Both these properties are very important for applications. In particular, high mobility material is advantageous for high speed, and low dissipation electronics, whereas long phase coherence lengths indicate that quantum coherent devices can be realized in epitaxial graphene devices.

The most important feature of the C-face material is that the layers are not Bernal stacked, but rather exhibit rotational disorder (Hass et al. 2008a,b). For this material, the successive layers prefer to be rotated at several specific angles with respect to each other, forming Moiré patterns, shown in Figure 19.3b. The reason for this is not understood, but as a consequence, the electronic structure is the same as that of a single graphene layer. The material behaves like a stack of independent decoupled graphene layers, rather that exhibiting the properties of graphite. This material is therefore multilayered epitaxial graphene (MEG) and it is not simply thin graphite. We believe that this will lead to very important new developments both in fundamental physics and in new device architectures.

The layers above the first one are found to be essentially undoped (Sadowski et al. 2006, Orlita et al. 2008). In fact, infrared absorption measurements in a magnetic field reveal that for these layers $n \le 5 \times 10^9/cm^2$ so that the Dirac point is only a few meV below the Fermi level. Hence, for these layers the Fermi wavelength is of the order of $1\,\mu m$, compared with the first layer where the Fermi wavelength is of the order of 10 nm. This is very important, especially for nanostructured devices. For example, a MEG ribbon that is 100 nm wide, will only conduct though the interface layer and not the top layers, because the wavelength is smaller than the ribbon width only for the interface layer (de Heer et al. 2007, Darancet et al. 2008). Hence, even though the top layers are graphene, they do not conduct which is consistent with observations. On the other hand, in ribbons that are larger than $10\,\mu m$, the top layers will also contribute to the transport. This is also seen in experiments.

The top, essentially undoped layers have been recently probed by infrared spectroscopy in a magnetic field B. In this situation,

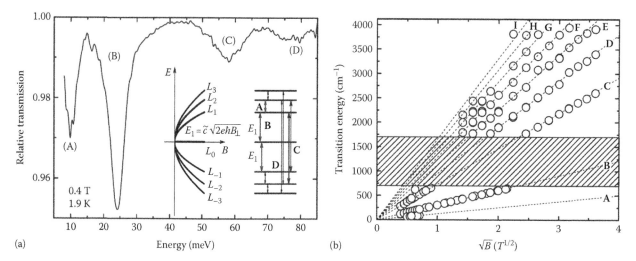

FIGURE 19.4 (a) Infrared transmission spectrum of epitaxial graphene, where transitions occur as a valley. Transitions happen when electrons absorb photon energy and jump from one Landau level to another. As a result, the transition energy corresponds to the distance between two Landau levels. (b) √B dependence of the transition energy, as expected for the Dirac spectrum in graphene.

the electrons bunch in separate energy levels, the Landau levels, which are labeled by a Landau level index N. The energy of the Landau levels in graphene is: $E_N(B) = C \mathrm{sgn}(N)\sqrt{(B|N|)}$. The √B dependence is directly related to the Dirac cone and serves as its signature. The Landau level spectrum has been probed by monitoring the absorption of infrared light (Figure 19.4a). The analysis of these spectra have revealed that the bandstructure of MEG is indeed described by Dirac cones (Figure 19.4b) and more importantly that the mobility of the material at room temperature exceeds 250,000 cm²/Vs. This is in the same range as measured in thoroughly cleaned suspended graphene flakes (Bolotin et al. 2008, Du et al. 2008), but at low temperatures.

19.4.3 Chemically Modified Epitaxial Graphene

One extremely important feature of epitaxial graphene on silicon carbide, as graphene in general, is that it can be chemically modified which greatly enhances its utility. For example, it is possible to convert MEG to semiconducting graphene oxide by a chemical treatment (Wu et al. 2008). Moreover, it is possible to mask the material with a chemically inert coating and to produce openings in the mask where the chemical reactions take place. In this way graphene oxide regions can be patterned on the surface. This process is reminiscent of patterned doping procedures that are employed in silicon processing technology. This is a very promising new direction in graphene processing. Epitaxial graphene also lends itself very well to surface chemistry in order to chemically modify the surfaces. This is primarily so, because the surface is so well organized and well adapted for organic chemistry processing. Chemical modification and functionalization is important from a technological point of view because this will allow the fabrication of precise dielectric coatings that inevitably will be very important for most applications.

Another aspect is the possibility of functionalization of edges of patterned graphene ribbons. Because a carbon atom at

an edge has a bond missing, edges in general are expected to be very reactive, contrary to graphene surfaces which requires strong chemical agents for functionalization. It is envisioned to bind molecules to graphene ribbon edges, for doping of sensor purposes.

19.5 Epitaxial Graphene Electronics

19.5.1 Graphene, Why Bother?

Graphene is a semi-metal (and often referred to, less accurately, as a zero gap semiconductor). That means that in neutral graphene there is no energy gap between occupied and unoccupied states. Semiconducting devices rely on this gap since it is ultimately this gap that allows a semiconductor to be switched from conducting to non-conducting. In fact, for many applications, the larger the gap the better since it insures a large on to off ratio for switching devices. However, as mentioned above, graphene has the interesting property that a band-gap develops in confined structures, of which carbon nanotubes are prototypical. Hence, the electronic properties are structure dependent. Indeed, a graphene ribbon with a width of 1 nm has theoretically a band gap of about 1 eV. An energy gap of this magnitude is very well suited for electronics applications, while the size scale of 1 nm is in the range of anticipated ultimate lithographic methods. Hence, in general, graphene-based electronics is intrinsically nanoscopic and in line with microelectronics developments. Besides this property, graphene, like nanotubes, can sustain very large current densities that will necessarily occur in nanoscopic electronic devices.

But why should one bother with graphene nanostructures since nanotubes can be produced relatively easily and they too have these properties? The reason is twofold. First of all it has been proven to be very difficult to accurately place well defined and shaped carbon nanotubes on a prepared substrate. However,

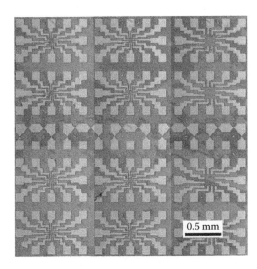

FIGURE 19.5 Picture of patterned epitaxial graphene on a $3 \times 4\,mm^2$ SiC chip. Pd/Au electrical contacts (bright color) connect about a hundred graphene devices.

such problems do not exist in epitaxial graphene. Since epitaxial graphene is a 2D material that can be grown over large areas, devices can be simply patterned using lithographic methods. For example, in Figure 19.5, hundreds of graphene devices are made in a 3-by-4 millimeter SiC substrate. Next, and very importantly, even though carbon nanotubes (like graphene ribbons) are ballistic conductors (Frank et al. 1998, Liang et al. 2001), it is impossible to inject electrons into a nanotube without causing significant heating of the contact. This is due to the nature of the contact between a metal and a carbon nanotube, which conducts by virtue of its π electron bands. This heating not only consumes energy, it also causes large electrical stresses at the contact giving rise to electromigration of the metal atoms that will ultimately cause the contact to fail. Patterned graphene does not suffer from these problems. Since graphene is rationally patterned, the placement problem does not exist. Furthermore, narrow graphene ribbons can be seamlessly connected to their graphene leads. In that case, the π electron system of the leads continues into the ribbon and essentially eliminates contact heating. A further advantage, and probably one that will only become more in the distant future, is that all graphene electronics may be phase coherent (Berger et al. 2006, Heersche et al. 2007, Tombros et al. 2007). From a quantum mechanical point of view, electron wave propagates through the entire structure or ribbons and leads much like microwaves propagate through wave guides. Because of that, the electronic phase is well defined everywhere and can be utilized as a "state variable," that is, it can be used to propagate information.

19.5.2 Ballistic Transport

One of the most remarkable features of carbon nanotubes and graphene is that electrons can travel large distances without scattering. This property already mentioned above, has important consequences for electronics. Not only does this provide for efficient (low power) electronic devices, it also is an important enabling property for very high speed (THz) electronics. In fact it is anticipated that ultrahigh speed electronics will be one of the most important applications of graphene electronics, at least initially.

19.5.3 Gate Doping

It is possible to controllably change the number of electrons in a graphene sheet by the gate doping. A gate is produced by covering the sheet with a metal-coated insulating layer. Applying an electric potential between the graphene and the metal induces charges on the graphene (Berger et al. 2004, Novoselov et al. 2004, 2005, Zhang et al. 2005). If electrons are added, then states in the upper cones will become occupied, so that the Fermi energy is raised. If electrons are removed then electrons are drained from the lower cones and the Fermi energy is lowered. This, in turn affects the conductivity σ, since σ = $|n|e\mu$, where n is the additional number of electrons per unit area (the doping density), e is the electronic charge, and μ is the mobility. Furthermore, the electronic wavelength depends on n: $\lambda_F = \sqrt{(4\pi/n)}$, so that this wavelength can be tuned. Gate doping has been demonstrated in prototype field effect transistors and quantum dot devices (Berger et al. 2004, Novoselov et al. 2004, 2005, Zhang et al. 2005, Kedzierski et al. 2008, Ponomarenko et al. 2008).

19.6 Conclusion

In summary, epitaxial graphene grown on silicon carbide substrates is a complex but promising material for graphene-based electronics because high-quality graphene can be grown to uniformly cover the entire crystal. It can be chemically converted to a semiconductor. The material can also be patterned using standard microelectronics patterning methods, which is attractive for technology.

References

Aizawa, T., Souda, R., Otani, S., Ishizawa, Y., and Oshima, C. 1990. Anomalous bond of monolayer graphite on transition-metal carbide surfaces. *Physical Review Letters* 64: 768–771.

Anantram, M. P. and Leonard, F. 2006. Physics of carbon nanotube electronic devices. *Reports on Progress in Physics* 69: 507–561.

Ando, T., Nakanishi, T., and Saito, R. 1998. Berry's phase and absence of back scattering in carbon nanotubes. *Journal of the Physical Society of Japan* 67: 2857–2862.

Beenakker, C. W. J. and Vanhouten, H. 1991. Quantum transport in semiconductor nanostructures. *Solid State Physics—Advances in Research and Applications* 44: 1–228.

Berger, C., Song, Z. M., Li, T. B. et al. 2004. Ultrathin epitaxial graphite: 2D electron gas properties and a route toward graphene-based nanoelectronics. *Journal of Physical Chemistry B* 108: 19912–19916.

Berger, C., Song, Z. M., Li, X. B. et al. 2006. Electronic confinement and coherence in patterned epitaxial graphene. *Science* 312: 1191–1196.

Boehm, H., Clauss, A., Hofmann, U., Fischer, G. 1962. Dunnste kohlenstoff folien. *Zeitschrift fur Naturforschung B* 17: 150.

Bolotin, K. I., Sikes, K. J., Jiang, Z. et al. 2008. Ultrahigh electron mobility in suspended graphene. *Solid State Communications* 146: 351–355.

Bostwick, A., Ohta, T., Seyller, T., Horn, K., and Rotenberg, E. 2007. Quasiparticle dynamics in graphene. *Nature Physics* 3: 36–40.

Charrier, A., Coati, A., Argunova, T. et al. 2002. Solid-state decomposition of silicon carbide for growing ultra-thin heteroepitaxial graphite films. *Journal of Applied Physics* 92: 2479–2484.

Darancet, P., Olevano, V., and Mayou, D. 2008. Coherent electronic transport through graphene constrictions: Subwavelength regime and optical analogies. *Physical Review Letters* 101: 116896.

Dekker, C. 1999. Carbon nanotubes as molecular quantum wires. *Physics Today* 52: 22–28.

Du, X., Skachko, I., Barker, A., and Andrei, E. Y. 2008. Approaching ballistic transport in suspended graphene. *Nature Nanotechnology* 3: 491–495.

Emtsev, K. V., Speck, F., Seyller, T., Ley, L., and Riley, J. D. 2008. Interaction, growth, and ordering of epitaxial graphene on SiC{0001} surfaces: A comparative photoelectron spectroscopy study. *Physical Review B* 77: 155303.

Emtsev, K. V., Bostwick, A., Horn, K. et al. 2009. Towards wafer-size graphene layers by atmospheric pressure graphitization of silicon carbide. *Natural Materials* 8: 203–207.

Forbeaux, I., Themlin, J. M., and Debever, J. M. 1998. Heteroepitaxial graphite on 6H-SiC(0001): Interface formation through conduction-band electronic structure. *Physical Review B* 58: 16396.

Frank, S., Poncharal, P., Wang, Z. L., and De Heer, W. A. 1998. Carbon nanotube quantum resistors. *Science* 280: 1744–1746.

Geim, A. K. and Novoselov, K. S. 2007. The rise of graphene. *Nature Materials* 6: 183–191.

Hass, J., Feng, R., Li, T. et al. 2006. Highly ordered graphene for two dimensional electronics. *Applied Physics Letters* 89: 143106.

Hass, J., De Heer, W. A., and Conrad, E. H. 2008a. The growth and morphology of epitaxial multilayer graphene. *Journal of Physics: Condensed Matter* 20: 323202.

Hass, J., Varchon, F., Millan-Otoya, J. E. et al. 2008b. Why multilayer graphene on 4H-SiC(000(1)over-bar) behaves like a single sheet of graphene. *Physical Review Letters* 100: 125504.

de Heer, W. A. 2001. Tight-binding calculation of the electronic structure of graphene nanoribbons.

de Heer, W. A., Berger, C., and First, P. N. 2006. Patterned thin films graphite devices and methods for making the same. US Patent 7,015,142 (Provisional filed June 2003, 2004, Issued March 21, 2006).

de Heer, W. A., Berger, C., Wu, X. S. et al. 2007. Epitaxial graphene. *Solid State Communications* 143: 92–100.

Heersche, H. B., Jarillo-Herrero, P., Oostinga, J. B., Vandersypen, L. M. K., and Morpurgo, A. F. 2007. Bipolar supercurrent in graphene. *Nature* 446: 56–59.

Iijima, S. 1991. Helical microtubules of graphitic carbon. *Nature* 354: 56–58.

ITRS 2007. International technology roadmap for semiconductors: Emerging research materials.

Kedzierski, J., Hsu, P. L., Healey, P. et al. 2008. Epitaxial graphene transistors on SiC substrates. *IEEE Transactions on Electron Devices* 55: 2078–2085.

Kelly, B. T. 1981. *Physics of Graphite*. Applied Science Publishers, London, U.K.

Land, T. A., Michely, T., Behm, R. J., Hemminger, J. C., and Comsa, G. 1992. STM investigation of single layer graphite structures produced on Pt(111) by hydrocarbon decomposition. *Surface Science* 264: 261–270.

Liang, W. J., Bockrath, M., Bozovic, D., Hafner, J. H., Tinkham, M., and Park, H. 2001. Fabry-Perot interference in a nanotube electron waveguide. *Nature* 411: 665–669.

Lu, W. J., Mitchel, W. C., Thornton, C. A., Collins, W. E., Landis, G. R., and Smith, S. R. 2003. Ohmic contact behavior of carbon films on SiC. *Journal of the Electrochemical Society* 150: G177–G182.

Mallet, P., Varchon, F., Naud, C., Magaud, L., Berger, C., and Veuillen, J. Y. 2007. Electron states of mono- and bilayer graphene on SiC probed by scanning-tunneling microscopy. *Physical Review B* 76: 041403(R).

Marchini, S., Gunther, S., and Wintterlin, J. 2007. Scanning tunneling microscopy of graphene on Ru(0001). *Physical Review B* 76: 075429.

Nakada, K., Fujita, M., Dresselhaus, G., and Dresselhaus, M. S. 1996. Edge state in graphene ribbons: Nanometer size effect and edge shape dependence. *Physical Review B* 54: 17954–17961.

Novoselov, K. S., Geim, A. K., Morozov, S. V. et al. 2004. Electric field effect in atomically thin carbon films. *Science* 306: 666–669.

Novoselov, K. S., Geim, A. K., Morozov, S. V. et al. 2005. Two-dimensional gas of massless Dirac fermions in graphene. *Nature* 438: 197–200.

Ohta, T., Bostwick, A., Seyller, T., Horn, K., and Rotenberg, E. 2006. Controlling the electronic structure of bilayer graphene. *Science* 313: 951–954.

Orlita, M., Faugeras, C., Plochocka, P. et al. 2008. Approaching the Dirac point in high-mobility multilayer epitaxial graphene. *Physical Review Letters* 101: 267601.

Ponomarenko, L. A., Schedin, F., Katsnelson, M. I. et al. 2008. Chaotic Dirac billiard in graphene quantum dots. *Science* 320: 356–358.

Rollings, E., Gweon, G. H., Zhou, S. Y. et al. 2006. Synthesis and characterization of atomically thin graphite films on a silicon carbide substrate. *Journal of Physics and Chemistry of Solids* 67: 2172–2177.

Rosei, R., Decrescenzi, M., Sette, F., Quaresima, C., Savoia, A., and Perfetti, P. 1983. Structure of graphitic carbon on Ni(111)—A surface extended-energy-loss fine-structure study. *Physical Review B* 28: 1161–1164.

Rutter, G. M., Guisinger, N. P., Crain, J. N. et al. 2007. Imaging the interface of epitaxial graphene with silicon carbide via scanning tunneling microscopy. *Physical Review B* 76: 235416.

Sadowski, M. L., Martinez, G., Potemski, M., Berger, C., and De Heer, W. A. 2006. Landau level spectroscopy of ultrathin graphite layers. *Physical Review Letters* 97: 266405.

Special-Issue 2000. *Physics World* 13: 29–53.

Tombros, N., Jozsa, C., Popinciuc, M., Jonkman, H. T., and Van Wees, B. J. 2007. Electronic spin transport and spin precession in single graphene layers at room temperature. *Nature* 448: 571–574.

Van Bommel, A. J., Crobeen, J. E., and Van Tooren, A. 1975. Leed and Auger electron observations of the SiC (0001) surface. *Surface Science* 48: 463.

Varchon, F., Feng, R., Hass, J. et al. 2007. Electronic structure of epitaxial graphene layers on SiC: Effect of the substrate. *Physical Review Letters* 99: 126805.

Wakabayashi, K., Fujita, M., Ajiki, H., and Sigrist, M. 1999. Electronic and magnetic properties of nanographite ribbons. *Physical Review B* 59: 8271–8282.

Wallace, P. R. 1947. The band theory of graphite. *Physical Review* 71: 622–634.

Wu, X. S., Sprinkle, M., Li, X. B., Ming, F., Berger, C., and De Heer, W. A. 2008. Epitaxial-graphene/graphene-oxide junction: An essential step towards epitaxial graphene electronics. *Physical Review Letters* 101: 026801.

Zhang, Y. B., Tan, Y. W., Stormer, H. L., and Kim, P. 2005. Experimental observation of the quantum Hall effect and Berry's phase in graphene. *Nature* 438: 201–204.

Zhou, S. Y., Gweon, G. H., Fedorov, A. V. et al. 2007. Substrate induced band gap opening in epitaxial graphene. *Nature Materials* 6: 770–775.

20

Electronic Structure
of Graphene Nanoribbons

Juan Jose Palacios
Universidad Autonoma

Joaquin Fernández-Rossier
Universidad de Alicante

Luis Brey
*Instituto de Ciencia de
Materiales de Madrid*

Herb A. Fertig
Indiana University

20.1 Introduction

Finding new phenomena in condensed matter is strongly linked to the discovery, synthesis, or fabrication of new materials and, even more importantly, to the quality of them. Semiconductor heterojunctions, that is, layered structures composed of two or more semiconductors, are a good example of this. The discovery of the integer (Klitzing et al. 1980) and fractional quantum Hall effects (QHEs) (Tsui et al. 1982) was possible due to the expertise developed in the growth of interfaces between properly doped GaAs and AlGaAs semiconductors. The two-dimensional (2D) electron gas confined at the interface of these two semiconductors was hiding surprising physics, unveiled only when the growth techniques were mastered and the mobility of the 2D electrons was sufficiently high. Nowadays, a novel material, graphene, is responsible for the renewed interest in 2D electronic systems. The term graphene refers to a one-atom thick layer of carbon atoms arranged in a 2D honeycomb lattice. It can be considered as the starting material for other low-dimensional carbon-based systems. For instance, it can be wrapped to form fullerenes, rolled into one-dimensional (1D) nanotubes, or stacked to form graphite. Theoretically, graphene has been studied for almost 60 years, but it was only recently that it was fully appreciated that graphene also provides an excellent playground for "condensed-matter quantum electrodynamics." This has

spurred the interest of this material beyond the limits of the condensed-matter physics community.

Interestingly, graphene had been considered an academic system, presumed to be unstable with respect to the formation of other carbon allotropes. It was argued by Peierls long ago (Peierls 1935) that strictly 2D crystals were thermodynamically unstable. Thermal fluctuations should destroy long-range order at any finite temperature. A little more than 4 years ago graphene was, however, isolated and its electronic structure revealed by electrical measurements (Novoselov et al. 2004). In hindsight, one can justify the existence of graphene arguing that, in reality, it is a metastable state where the appearance of dislocations or other defects is energetically too costly and, in practice, presents an infinite lifetime. Another point of view is that of attributing the stability to the inherent corrugation of the sheet, as observed in experiments (Meyer et al. 2007).

Earlier attempts to isolate graphene were based on chemical exfoliation. Bulk graphite was intercalated so that graphene planes would become separated by layers of atoms or molecules. In some cases, a large separation would be achieved such that the resulting compounds could be considered as isolated graphene layers embedded in a bulk matrix. Removal of the intercalated species would result in a material similar to graphite, but without interplane order. Graphene can also be grown by thermal decomposition of hydrocarbons on metallic substrates (Parga et al. 2008) or by controlled segregation of C from various substrates (Berger et al. 2004, Parga et al. 2008). Unless graphene can be removed after growth, the presence of the substrate partially narrows down the scientific interest of this type of graphene, since their intrinsic properties are shadowed or completely masked by the substrate. Few-layer graphene obtained by epitaxial growth can, on the other hand, be of interest in electronic applications due to the high mobility of its carriers.

By far, the most popular way of obtaining graphene is by micromechanical cleavage of bulk graphite, as pioneered by A. Geim's group (Novoselov et al. 2004). The technique consists of drawing with graphite and repeatedly peeling with adhesive tape until the thinnest flakes are found. The main problem is that of actually finding the single-atom-thick flakes, which are rare and hidden among the thicker ones. Scanning-probe techniques are not efficient in this search due to the typical small exploration areas. Luckily, graphene becomes visible with an optical microscope if placed on top of a Si wafer with the appropriate thickness of SiO$_2$. Graphene has probably always been there, but nobody noticed it.

It is the exceptional electronic quality exhibited by isolated graphene that makes it worth considering when it comes to searching for new physics. Graphene quality clearly reveals itself in the very high mobilities μ, which exceed 15,000 cm^2/V s even under ambient conditions. Furthermore, the observed mobilities depend weakly on temperature T. This means that at 300 K, μ is still limited by impurity scattering and, therefore, can be improved significantly, perhaps, even up to ≈100,000 cm^2/V s. This translates into ballistic transport on long scales (up to ≈0.3 μm at 300 K). An additional indication of the extreme electronic quality of graphene is the fact that the QHE can be observed even at room temperature. This extends the previous temperature range for the observation of this effect by a factor of 10 (Novoselov et al. 2007).

An equally important reason for the interest in graphene is the unique nature of its charge carriers. In condensed matter physics, the Schrödinger equation describes the electronic properties of materials. Graphene is no exception, but its charge carriers mimic massless particles and can be described, maybe more naturally, by an effective Dirac equation, at least at long wavelengths. Although there is nothing particularly relativistic about electrons moving in a lattice of carbon atoms, their interaction with the potential of the honeycomb lattice gives rise to a new type of quasiparticles that, at low energies, are accurately described by a Dirac equation with an effective speed of light $\upsilon_F \approx 10^6$ m/s. The "relativistic" description of electrons on honeycomb lattices has been known theoretically for many years, and the experimental discovery of graphene now provides a way to probe quantum electrodynamics phenomena on a simple piece of carbon.

Unfortunately, despite the high quality and mobility of graphene, it is not so convenient for standard electronic applications that are nowadays mostly based on semiconducting materials. Graphene is a zero-gap semiconductor and this limits the range of applicability of this material for electronics. The absence of an energy gap, however, may be circumvented by proper atomic-scale engineering. For instance, recent developments in the fabrication of graphene nanoribbons with top-down techniques

(Han et al. 2007) as well as in the chemical synthesis of these (Li et al. 2008) are blazing the trail in this direction. One might expect that the lateral confinement of the carriers in these systems may open a gap in the spectrum of graphene. This would allow for fabrication of devices based on the same principles as today's, but with the advantages of the high mobilities and the intrinsic nanometer scale of graphene. This naive expectation is true; however, somewhat unexpectedly the physics found in graphene nanoribbons is far more interesting and richer than this as we show in this chapter.

20.2 Theoretical Background

20.2.1 Electronic Structure of Bulk Graphene

Bulk graphene crystallizes in a bipartite honeycomb lattice with two atoms per unit cell, usually denoted A and B, located at (0,0) and at $a(0,1/\sqrt{3})$, where a is the lattice parameter of the triangular sublattice. The primitive translation vectors of the lattice are **a** = $a(1, 0)$ and **b** = $a(-1/2, \sqrt{3}/2)$, and the primitive reciprocal lattice vectors are given by **a*** = $(2\pi/a) (1, 1/\sqrt{3})$ and **b*** = $(2\pi/a) (0, 2/\sqrt{3})$.

Usually one only considers the low-energy physics that takes place in the subspace expanded by the single p_z orbital (the one perpendicular to the graphene plane). The first neighbors tight-binding Hamiltonian for electrons in graphene has the form

$$\hat{H}_0 = -t \sum_{\langle i,j \rangle} (\hat{c}_i^+ \hat{c}_j + h.c.) \tag{20.1}$$

where

$\hat{c}_i (\hat{c}_i^+)$ annihilates (creates) an electron on site i
t is the nearest neighbor hopping energy, and the sum runs only over first neighbors

The energy bands derived from this Hamiltonian have the form

$$\varepsilon_\pm(k) = \pm t \sqrt{1 + 4\cos\left(\frac{ak_x}{2}\right)\cos\left(\frac{\sqrt{3}ak_y}{2}\right) + 4\cos^2\left(\frac{ak_x}{2}\right)} \tag{20.2}$$

where the minus sign applies to the occupied band and the plus sign to the empty band. These are shown in Figure 20.1. As can be seen, the density of states (DOS) vanishes at the six corners of the Brillouin zone, only two of which are inequivalent. For neutral graphene, the Fermi energy E_F lies precisely at these points. We take these to be **K** = $(2\pi/a)$ $(1/3, 1/\sqrt{3})$ and **K**′ = $(2\pi/a)$ $(-1/3, 1/\sqrt{3})$ (see Figure 20.1). These are the locations of the famous Dirac cones of graphene. Close to one of the Dirac points, the dispersion relation is linear:

$$\varepsilon_\pm(\mathbf{k}) = \pm \upsilon_F |\mathbf{k}| \tag{20.3}$$

where

k is the momentum measured relative to the Dirac points
υ_F represents the Fermi velocity, given by $\upsilon_F = \sqrt{3}/2 ta$

FIGURE 20.1 Band structure of bulk graphene obtained within a first-neighbor tight-binding model.

20.2.2 Basic Graphene-Nanoribbon Geometries

Graphene nanoribbons are obtained by "cutting" graphene in the form of a quasi-1D wire along a given crystallographic direction. Because a large portion of the carbon atoms lies on the edges, the electronic and magnetic properties of a graphene nanoribbon can be somewhat different from that of bulk graphene. There are two basic shapes for graphene edges: armchair and zigzag. Theoretical works have shown that graphene systems such as clusters, or nanoribbons with zigzag edges have localized edge states at the Fermi energy (Fujita et al. 1996, Nakada et al. 1996, Brey and Fertig 2006b, Pereira et al. 2006, Fernández-Rossier and Palacios 2007).

Armchair-terminated graphene systems, on the other hand, do not show zero-energy states (the Fermi energy is usually set to zero). Edges along generic crystallographic directions are composed of armchair and zigzag sections, and zero-energy edge states can survive provided that three or four contiguous zigzag sites can be found (Nakada et al. 1996). In this review, we focus on the more general cases of armchair and zigzag-terminated nanoribbons.

The atomic structure of the armchair and zigzag nanoribbons are shown in Figure 20.2. The left one is the zigzag graphene nanoribbon and it has zigzag-shaped edges. The armchair graphene nanoribbon, right panel of Figure 20.2, has armchair-shaped edges. In both cases, the dangling σ-bonds at the edges are assumed to be passivated by hydrogen atoms. In the zigzag nanoribbon the atoms at each edge belong to the same sublattice A on the top edge and B on the bottom edge. The unit cell contains A-type atoms that alternate along the unit cell with B-type atoms. Zigzag nanoribbons may have axial symmetry (as the one shown in Figure 20.2a) or may not (constructed from these by adding or removing a single zigzag chain). The width of zigzag nanoribbons is approximately given by $W = (N/4)\sqrt{3}a$, where N is the number of atoms in the unit cell (this expression gives a value that is actually larger than the true width, but approaches to it as $N \rightarrow \infty$). In armchair-terminated nanoribbons, on the other hand, the edges consist of a line of A–B dimers. The width is related to the number of atoms in the unit cell N through the expression $W = (N/4)a$. This expression gives a number that is larger than the actual width of the ribbons that do not have axial symmetry (as the one shown in Figure 20.2b), but is smaller than the width for those that have it (constructed from these by adding or removing an A–B dimer from the unit cell). Again, the approximate expression for the width becomes exact as $N \rightarrow \infty$ for both types of armchair ribbons.

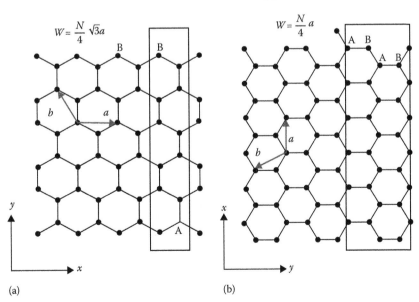

$$W = \frac{N}{4}\sqrt{3}a$$

$$W = \frac{N}{4}a$$

(a)

(b)

FIGURE 20.2 The lattice structure of a zigzag- (a) and an armchair- (b) terminated graphene nanoribbon. The primitive lattice vectors are denoted by **a** and **b**. Atoms enclosed in the vertical rectangles represent the unit cell of the nanoribbons. The approximate width of the nanoribbons W as a function of the number of atoms N in the unit cell is also indicated. Note that the coordinate axes in Figure 20.2b is rotated 90°, with respect to Figure 20.2a.

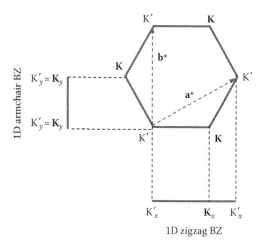

FIGURE 20.3 The first Brillouin zone of 2D graphene. The **a*** and **b*** are the primitive reciprocal lattice vectors. The vertices of the hexagon, **K** and K′, are the Dirac points. We also plot the 1D Brillouin zone of zigzag (horizontal) and armchair (vertical) graphene nanoribbons. In the case of zigzag nanoribbons, the projection of the 2D Brillouin zone on the 1D Brillouin zone preserves the existence of the Dirac points K_x and K'_x. However, in the case of armchair-terminated nanoribbons, the Dirac points are projected on just one: $K_y = K'_y$.

The zigzag nanoribbons are periodic only in the x direction with a primitive translation vector $a(1, 0)$. The Brillouin zone is 1D with a reciprocal lattice vector, $\mathbf{a}^*_{zz} = 2\pi/a(1, 0)$, see Figure 20.3. In the armchair-terminated nanoribbons, the system is periodic in the y direction and the primitive translation vector is $a(0, \sqrt{3})$. Therefore, the 1D reciprocal lattice vector is $\mathbf{a}^*_{ac} = (2\pi/a\sqrt{3})(0, 1)$.

20.3 Electronic Structure of Graphene Nanoribbons: Non-Interacting Theory

20.3.1 Tight-Binding Approximation

As we are interested in the basic relationship between the electronic structure and the edge shape, the hopping parameters between the nearest-neighbor sites are all set to the same value t for simplicity. For the same reason, in this section we neglect effects due to electron–electron interactions. In Figure 20.4, we plot an example of the band structure of a nanoribbon with zigzag edges. The finite width of the ribbon produces confinement of the electronic states near the Dirac points. The two bands of dispersionless localized surface states that occur between K_x and K'_x in Figure 20.4 are also affected by the finite width; they admix and the two bands are slightly offset from zero. In the zigzag termination, the nanoribbon does not mix the Dirac points (see Figure 20.3). This is the reason why, as seen in Figure 20.4, confined states emerge from the two independent Dirac points, K_x and K'_x.

The electronic properties of armchair nanoribbons depend strongly on their width. In Figure 20.5a and b, we plot two examples of band structures of armchair-terminated graphene

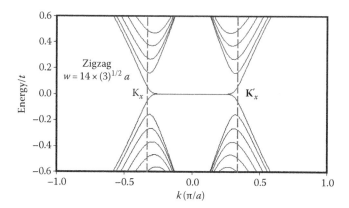

FIGURE 20.4 Energy bands for zigzag-terminated graphene nanoribbons with 56 atoms in the unit cell. κ is the wave vector parallel to the nanoribbon edge and takes values in the Brillouin zone $(-\pi/a, \pi/a)$. The dispersionless states correspond to confined edge states. Note that this Brillouin zone is shifted π/a with respect to the zigzag Brillouin zone of Figure 20.3.

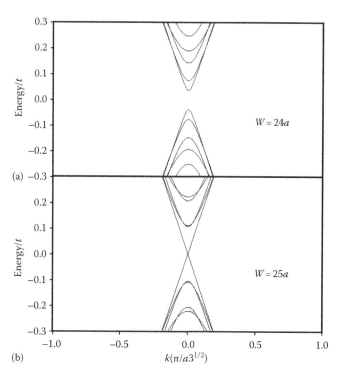

FIGURE 20.5 Energy bands for armchair-terminated graphene nanoribbons, as the one shown in Figure 20.3b, with (a) $W = 24a$ and (b) $W = 25a$ atoms in the unit cell. κ is the wave vector parallel to the nanoribbon edge and takes values in the Brillouin zone $\left(-\pi/\sqrt{3}a, \pi/\sqrt{3}a\right)$. Note that this Brillouin is shifted by half Brillouin zone with respect to the armchair Brillouin zone of Figure 20.3.

nanoribbons. One sees that in Figure 20.5b there is a Dirac point leading to metallic behavior, whereas in Figure 20.5a there is a gap leading to a band insulator. In general, we find that armchair nanoribbons without axial symmetry that are formed by a number of armchair chains $N_y = 3M + 1$ with M an integer ≥ 0 are metallic, whereas all the other cases are insulators. In the case of nanoribbons with axial symmetry, we find that those having a maximum cross section of $W = (3M - 1)a$ with M positive integer are metallic, being semiconducting otherwise. We will specify the type of armchair nanoribbon, that is, with or without axial symmetry, when necessary.

In the armchair nanoribbons, both Dirac points **K** and **K′** are projected in the same point of the 1D Brillouin zone, and the confined states emerge only from a Dirac point. The energy of the confined states also behave in a discontinuous way with respect to the width of the ribbon. In Figure 20.7, we plot the energy of the lowest energy confined states at the center of the Brillouin zone as a function of the nanoribbon width.

20.3.2 Dirac Equation

The low energy, long wavelength electronic properties of graphene can be appropriately described in the $\mathbf{k} \cdot \mathbf{p}$ approximation (DiVincenzo and Mele 1984, Ando 2005). Wavefunctions can be expressed in terms of envelope functions $\Phi(r) = [\psi_A(r), \psi_B(r)]$ and $\Phi'(\mathbf{r}) = [\psi'_A(\mathbf{r}), \psi'_B(\mathbf{r})]$, which multiply the states at the **K** and **K′** points respectively. The envelope functions satisfy the low-energy Schrödinger equations, $H\Phi = \varepsilon\Phi$ and $H'\Phi' = \varepsilon\Phi'$, with

$$H = \begin{pmatrix} 0 & k_x - ik_y \\ k_x + ik_y & 0 \end{pmatrix} \quad (20.4)$$

and

$$H' = \begin{pmatrix} 0 & -k_x - ik_y \\ -k_x + ik_y & 0 \end{pmatrix} \quad (20.5)$$

Note that $\mathbf{k}$ denotes the separation in reciprocal space of the wavefunction from the **K** (**K′**) point in the Hamiltonian H (H').

The bulk solutions of the $\mathbf{k} \cdot \mathbf{p}$ Hamiltonian retain their valley index as a good quantum number and the wavefunctions with energies $\varepsilon = \pm \upsilon_F |k|$ are $[e^{i\mathbf{k}\mathbf{r}}, \pm e^{i\mathbf{k}\mathbf{r}} e^{i\theta_k}]$ for the **K** valley, and $[e^{i\mathbf{k}\mathbf{r}}, \mp e^{i\mathbf{k}\mathbf{r}} e^{-i\theta_k}]$ for the **K′** valley. Here $\theta_k = \arctan k_x/k_y$.

20.3.2.1 Zigzag Nanoribbons

The dependence of the electronic states on the width of the nanoribbon may be understood in terms of eigenstates of the Dirac Hamiltonian with appropriate boundary conditions (Brey and Fertig 2006a,b), setting the wavefunction to zero on the A sublattice on one edge and on the B sublattice for the other. We can understand the lines of vanishing wavefunction to be lattice sites that would lie just beyond the edges if bonds had not been cut from them.

We assume that the edges of the nanoribbons are parallel to the x-axis and our wave functions exist in the space $0 < x < W$. Translational invariance in the x direction guarantees the wavefunctions can be written in the form $\psi_\mu(')(\mathbf{r}) = e^{ik_x x} \phi_\mu(')(y)$. To find wavefunctions for a system with edges, we make the

replacement $k_y \to -i\partial_y$ in the Hamiltonians (20.4) and (20.5) and, acting on the spinor state twice with the Hamiltonian, one easily finds for the **K** (**K′**) valley that the wavefunctions obey

$$(-\partial_y^2 + k_x^2)\phi_A(') = \tilde{\varepsilon}^2 \phi_A(')$$
$$(-\partial_y^2 + k_x^2)\phi_B(') = \tilde{\varepsilon}^2 \phi_B(') \quad (20.6)$$

with $\tilde{\varepsilon} = \varepsilon / \upsilon_F$. It is easy to see if one solves the equations for ϕ_B and ϕ'_A, the remaining wavefunctions are determined by

$$\tilde{\varepsilon}\phi_B = (\partial_y + k_x)\phi_A$$
$$\tilde{\varepsilon}\phi'_A = (-\partial_y - k_x)\phi'_B \quad (20.7)$$

For the zigzag nanoribbon, we meet the boundary condition for each type of wavefunction separately:

$$\phi_B(y = 0) = \phi'_B(y = 0) = \phi_A(y = W) = \phi'_A(y = W) = 0 \quad (20.8)$$

Solutions near different Dirac cones can satisfy these boundary conditions separately. In the following equations we present the solution for the **K** valley. The most general solution of Equations 20.6 and 20.7 with the boundary conditions Equation 20.8 has the form

$$\phi_A(y) = \pm\sinh(z(y - W)) \quad (20.9)$$

$$\phi_B(y) = \sinh(zy) \quad (20.10)$$

where z is obtained from the transcendental equation

$$\tanh(zW) = \frac{z}{k_x} \quad (20.11)$$

and the eigenenergies get the form $\varepsilon = \pm\sqrt{k_x^2 - z^2}$.

Equation 20.11 supports solutions with real values of $z \equiv k$ for $k_x > k_x^c = 1/W$, which correspond to the edge states. These have energies $\pm\sqrt{k_x^2 - k^2}$ and are linear combinations of states localized on the left and right edges of the ribbon. For large values of k_x, $k \to -k_x$ and the surface states become decoupled. For $k_x < 0$, there are no states with real z that can meet the boundary conditions, so surface states are absent. For values of k_x in the range $0 < k_x < k_x^c$, the surface states are so strongly admixed that they are indistinguishable from confined states.

For pure imaginary $z = ik_n$, the transcendental equation becomes

$$k_x = \frac{k_n}{\tan(k_n W)} \quad (20.12)$$

and for each solution k_n, there are two confined states with energies $\tilde{\varepsilon} = \pm\sqrt{k_n^2 + k_x^2}$ and wavefunctions

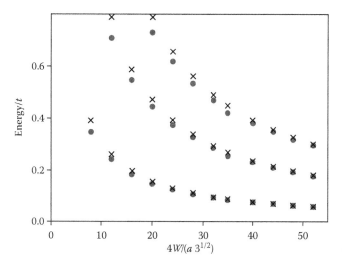

FIGURE 20.6 Calculated confined state energies at a Dirac point versus the nanoribbon width in a zigzag nanoribbon. The dots are tight-binding results and the crosses are the results of the $\mathbf{k} \cdot \mathbf{p}$ approximation.

$$\begin{pmatrix} \phi_A \\ \phi_B \end{pmatrix} = \begin{pmatrix} \sin(k_n(y-W)) \\ \pm \sin(k_n y) \end{pmatrix} \qquad (20.13)$$

Here, the index n indicates the number of nodes of the confined wavefunction. Interestingly, for values of k_x larger than k_x^c, Equation 20.12 does not support nodeless solutions, indicating the existence of surface states in this region of reciprocal space. The critical value k_x^c is the momentum where the lowest energy solution of the transcendental Equation 20.11 changes from pure real to pure imaginary and the energy is equal to $\pm \upsilon_F |k_x^c|$.

For the $\mathbf{K}'$, the solutions have the same form as Equation 20.9, but the values of z are obtained from the transcendental equation

$$\tanh(zW) = -\frac{z}{k_x} \qquad (20.14)$$

and the eigenenergies get the form $\varepsilon = \mp \sqrt{k_x^2 - z^2}$.

In order to analyze the accuracy of the $\mathbf{k} \cdot \mathbf{p}$ approximation for describing the electronic properties of carbon nanoribbons, in Figure 20.6 we plot the energies at the Dirac point of the three lowest confined states of a zigzag nanoribbon as a function of its width, both from the tight-binding approach and from our solutions to the Dirac equation. It is apparent that the two approaches match quite well, even for rather small widths (~35 Å).

20.3.2.2 Armchair Nanoribbons

As in the case of the zigzag nanoribbons, the electronic properties of armchair nanoribbons may be understood in terms of eigenstates of the Dirac Hamiltonian with the correct boundary conditions. In what follows, we consider armchair nanoribbons without axial symmetry. In Figure 20.2b, one can see that the termination consists of a line of A–B dimers, so it is natural to have the wavefunction amplitude vanish on both sublattices at $x = 0$ and $x = W + a/2$. To do this we must admix valleys and require

$$\phi_\mu(x = 0) = \phi_\mu'(x = 0)$$

$$\phi_\mu(x = W + a/2) = \phi_\mu'(x = W + a/2)e^{i\Delta K \left(W + \frac{a}{2} \right)}$$

with $\Delta K = 4\pi/3a$. The $a/2$ offset in the boundary condition on the right is necessary because the two upmost atoms in the ribbon unit cell are displaced $W + a/2$ away from the two lowest atoms (see Figure 20.2b). With these boundary conditions, the general solutions of the Dirac equation are plane waves:

$$\phi_B(x) = e^{ik_n x} \quad \text{and} \quad \phi_B'(x) = e^{-ik_n x} \qquad (20.15)$$

The wavefunctions on the A sublattice may be obtained via Equation 20.7. The wave vector k_n satisfies the condition

$$e^{2ik_n(W+a/2)} = e^{i\Delta k(W+a/2)} \qquad (20.16)$$

so that

$$\left(k_n - \frac{2\pi}{3a} \right)(2W+a) = 2\pi n \qquad (20.17)$$

with n an integer. Thus for armchair nanoribbons, the allowed values of k_n are

$$k_n = \frac{2\pi n}{2W+a} + \frac{2\pi}{3a} \qquad (20.18)$$

with energies $\pm \sqrt{k_n^2 + k_y^2}$. Note that this is in contrast to the zigzag nanoribbon for which the allowed values of k_n depend on the momentum along the nanoribbon. For a width $(3M + 1)a$,

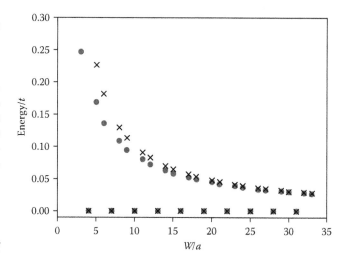

FIGURE 20.7 Calculated lowest energy confined states at the center of the Brillouin zone versus the nanoribbon width for an armchair nanoribbon. The dots correspond to the tight-binding results and the crosses are the results of the $\mathbf{k} \cdot \mathbf{p}$ approximation. The $\mathbf{k} \cdot \mathbf{p}$ results are slightly shifted to the right for clarity. Note that for $W = 25a$, the $\mathbf{k} \cdot \mathbf{p}$ results are doubly degenerate.

the allowed values of k_n, $k_n = (2\pi/3a)((2M + 1 + n)/(2M + 1))$ create doubly degenerate states for $|2M + 1 + n| \geq 0$, and allow a zero-energy state when $k_x \to 0$. Nanoribbons of widths that are not of this form have nondegenerate states and do not include a zero-energy mode. Thus, these nanoribbons are band insulators. The quality of the $\mathbf{k} \cdot \mathbf{p}$ approximation for describing the electronic states of armchair nanoribbons is reflected in Figure 20.7, where the energies of the confined states obtained by diagonalizing the tight binding Hamiltonian and by solving the low-energy Hamiltonian are compared. The quantitative agreement is apparent for all but the narrowest ribbons, where one does not expect the $\mathbf{k} \cdot \mathbf{p}$ approximation to work well.

20.4 Electronic Structure of Graphene Nanoribbons: Electron–Electron Interactions

20.4.1 Modeling the Interactions

In this section, we explore how the electronic structure of graphene nanoribbons can be affected by Coulomb interactions. Taking electron–electron interactions into consideration is a major theoretical problem that usually needs to be tackled through various approximations. Before we present results for different representative cases, we summarize the basics behind the different approximations considered. To this aim, an atomistic point of view is taken throughout.

The simplest way to include the electronic repulsion is within a mean field Hubbard model, which only considers Coulomb repulsion between two electrons at the same atomic site:

$$\hat{H} = \hat{H}_0 + U \sum_i \left(\hat{n}_{i\uparrow} \langle \hat{n}_{i\downarrow} \rangle + \hat{n}_{i\downarrow} \langle \hat{n}_{i\uparrow} \rangle \right) - U \sum_i \langle \hat{n}_{i\downarrow} \rangle \langle \hat{n}_{i\uparrow} \rangle \quad (20.19)$$

where $\hat{n}_{i\sigma} = \hat{c}_{i\sigma}^+ \hat{c}_{i\sigma}$ are the number operators. This Hamiltonian must be solved self-consistently. After an initial guess, the occupation numbers $\langle \hat{n}_{i\sigma} \rangle$ are updated in each iteration. When the system is finite (e.g., a ribbon with periodic boundary conditions), the occupation numbers are obtained for a given number of electrons, N_e, from the occupied eigenvectors after diagonalizing the Hamiltonian. Alternatively, they can be obtained from the Green's function $\hat{G}$:

$$\langle \hat{n}_{i\sigma} \rangle = -\frac{1}{\pi} \int_{-\infty}^{E_F} \text{Im} \left[\langle i\sigma | \hat{G}(E) | i\sigma \rangle \right] dE \quad (20.20)$$

where $\hat{G}(E) = (E\hat{1} - \hat{H} + i\hat{0})^{(-1)}$ and $\hat{1}$ and $\hat{0}$ are the unit and zero matrices, respectively. Equation 20.20 is convenient when the natural variable is E_F instead of the electronic charge N_e, which is usually the case when the ribbons are infinite.

When feasible, density functional theory (DFT) (Hohenberg and Kohn 1964) can be used to check to what extent the results obtained within the Hubbard approximation are robust.

As we will see, this is the case in most situations. Standard implementations of DFT as those found in commercial codes such as GAUSSIAN (Frisch et al. 2003) or CRYSTAL (Saunders et al. 2003) consist in solving the self-consistent Kohn–Sham equations (Kohn and Sham 1965) for appropriate choices of the exchange-correlation potential and basis set. It turns out that, at least at a qualitative level, most of the results obtained for graphene nanoribbons do not depend on the choice of functional and basis set.

20.4.2 Armchair Nanoribbons

The electronic structure of armchair nanoribbons is unaffected by the interactions when these are considered in the Hubbard approximation. These only introduce an overall trivial shift of value $U/2$ since $\langle \hat{n}_{i\uparrow} \rangle = \langle \hat{n}_{i\downarrow} \rangle = 1/2$ for all the atoms. Figure 20.8a shows the DFT bands of an axially symmetric $W = 8a$ nanoribbon calculated with the help of the CRYSTAL03 code (Saunders et al. 2003), using a hybrid functional (B3LYP (Frisch et al. 2003, Saunders et al. 2003)) and a minimal basis set. One can see that electron–electron interactions, as approximated by DFT, modify the electronic structure of the metallic nanoribbons, opening a gap in the spectrum at $k = 0$ (see Figure 20.8b). The difference is related to the fact that the DFT Hamiltonian includes next-to-near neighbor interactions, absent in the Hubbard model. These are responsible for the opening of the gap. On the basis of this result, armchair metallic nanoribbons as obtained from the tight-binding approximation with or without Hubbard-like interactions can be considered as mathematical objects to be looked at with academic interest only. The electronic structure of nominally semiconducting armchair ribbons is, on the other hand, qualitatively unaffected by the interactions treated at the DFT level.

20.4.3 Zigzag Nanoribbons

As shown in Sections 20.3.1 and 20.3.2, the single-particle band structure of zigzag nanoribbons features two almost degenerate quasi-flat bands at the Fermi energy associated to edge states. Partially-filled flat bands or, more generally, degenerate states are expected to be strongly affected by Coulomb interactions. Electrons can gain exchange energy by occupying as many orbitals as possible and aligning their spins while not paying kinetic energy. In atoms, this is known as Hund's rule (Kittel 1976), in solids, Stoner's criterion (Buschow and de Boer 2004). In other words, interactions are expected to give rise to magnetism in zigzag nanoribbons.

Figure 20.9b shows the spin-degenerate non-interacting energy bands for a nanoribbon with $W = 4.5\sqrt{3}a$ (Figure 20.9a shows the unit cell). Solid (dashed) lines represent full (empty) states. At zero temperature, the lower (valence) band is full and the upper (conduction) band is empty. Figure 20.9c shows the mean field interacting bands as obtained from the Hubbard model, shifted rigidly by $-U/2$, for $U = 2$ eV. This value of U reproduces fairly well the bands resulting from more sophisticated calculations such as DFT (see below). As a result of the interactions,

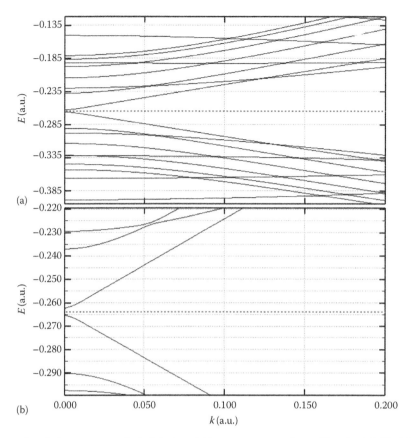

FIGURE 20.8 (a) DFT energy bands (in atomic units) of a $W = 8a$ axially symmetric armchair ribbon (only half of the Brillouin zone is shown). The Fermi energy is denoted by the dotted line. (b) Blow-up of the energy region near the Fermi energy where a gap can be appreciated for $k = 0$.

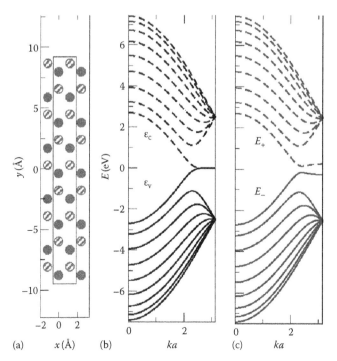

FIGURE 20.9 (a) $W = 4.5\sqrt{3}\,a$ zigzag nanoribbon unit cell. Bands for $U = 0$ (b) and $U = 2$ eV (c). Only half of the Brillouin zone is shown.

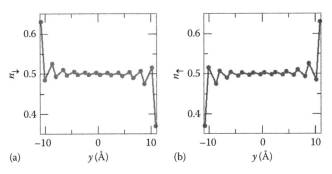

FIGURE 20.10　Spin-resolved occupation numbers for the nanoribbon in Figure 20.9 for $U = 2\,eV$.

local moments of opposite signs form on the edges, with a total spin $S = 0$, and a gap opens at the Fermi energy. The average spin-resolved charges along a unit cell are shown in Figure 20.10a and b. Spin up (down) electrons pile at the top (bottom) edge of the ribbon and leave a charge deficit in the opposite side. Therefore, the edges have local magnetization with opposite sign. For a given spin, there is an excess of electrons in one edge that are missing in the other. The total electronic charge turns out to be the same in all the atoms.

Figure 20.11 shows the fully degenerate spin up and spin down bands obtained from a DFT calculation using the B3LYP (Saunders et al. 2003) functional and a minimal basis set. As can be appreciated, the bands are identical, reflecting the antiferromagnetic nature of the ground state. Apart from a slight loss of the particle–hole symmetry, the DFT bands are similar to the ones obtained with the Hubbard model for interaction values in the range $U = 1$–$3\,eV$.

20.5 Defects and Magnetism in Graphene Nanoribbons

In this section, we explore how the electronic structure of graphene nanoribbons is affected by defects. Among the many possible defects one can find on a graphene lattice, we select to study

here single-atom vacancies, voids, and notches. Incidentally, hydrogen-adsorbed atoms have a similar effect on the electronic structure as single-atom vacancies. Hydrogen adsorption on graphene is a major topic nowadays and likely to play a fundamental role on the functionalization of graphene for electronic applications. This is the reason why we mainly focus on these types of defects here.

20.5.1 Preliminary Considerations

Defects in metals are expected to increase the localization of the carriers and therefore decrease their mobility. The most intriguing consequence of the existence of defects in bulk graphene is that the conductivity at charge neutrality is actually *larger* ($\approx 4e^2/h$) than its theoretical value (Novoselov et al. 2004). The presence of defects in nanoribbons, on the other hand, affect the transport properties of both electrons and holes in the usual way, namely, reducing the conductance. In bipartite lattices such as graphene, however, the presence of certain defects such as vacancies can lead to the formation of local magnetic moments, which is a preliminary condition for the existence of magnetic order. This is intimately related to the appearance of midgap states and how they are affected by electron–electron interactions.

The existence of zero-energy states in bipartite lattices (as that of graphene) for some types of disorder was proved by Inui et al. (Inui et al. 1994). Within the first-neighbor tight-binding model, a sufficient condition for the existence of midgap states is the existence of a finite sublattice imbalance, $N_I \equiv N_A - N_B$, where N_A and N_B are the number of atoms belonging to each sublattice or missing from each sublattice in an otherwise balanced system. Thus, whereas ideal graphene has $N_I = 0$ and no midgap states, defective graphene can present finite sublattice imbalance and $|N_I|$ midgap states.

We model vacancies and voids by removing atoms, actually, by removing the representing p_z orbitals in the tight-binding model. We ignore the lattice distortion and we assume that the on-site energy is the same for edge and bulk atoms. The single-orbital Hamiltonian implicitly assumes full hydrogen passivation of the

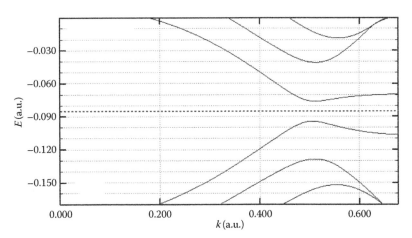

FIGURE 20.11　DFT energy bands (in atomic units) of a $W = 4.5\sqrt{3}\,a$ zigzag ribbon. Only half of the Brillouin zone is shown.

sp_2 dangling bonds of the atoms without full coordination. This assumption, which might not be completely realistic in the case of actual vacancies (Lehtinen et al. 2004), does not invalidate our model; for one can consider an alternative physical realization. The chemisorption of a hydrogen atom on top of a bulk graphene atom (Yazyev and Helm 2007) effectively removes a p_z orbital from the low-energy Hamiltonian. In our one-orbital model, there is no difference between these two possibilities.

We characterize voids by the number and type, A or B, of atoms removed from the otherwise perfect structure. We will label voids as one would do for chemical compounds, $A_{N_A}B_{N_B}$. Voids will be unbalanced when they are created by removing N_A and N_B atoms such that $N_I = N_A - N_B \neq 0$. The sublattice imbalance N_I, which can be either positive or negative, can be interpreted as an imbalance "charge." This quantity is central to our discussion, although the exact chemical "formula" of the void is also important since it gives an idea of the size and shape of the void. In the case of ribbons, the voids can be close to the edges, thus becoming notches. For a single void characterized by N_I, $|N_I|$ zero-energy states appear in the non-interacting spectrum with weight on only one sublattice (Inui et al. 1994). When the graphene structure presents a gap E_g, as in armchair ribbons, these states are normalizable and localized around the void (in contrast to 2D graphene where the zero-energy states are not normalizable (Pereira et al. 2006)). Figure 20.12 shows various examples of voids with different sublattice imbalances N_I.

Let us consider now two voids, characterized *locally* by $N_I(1)$ and $N_I(2)$, sufficiently separated so that they do not affect each other. From the previous discussion, we know that the single particle spectrum has, at least,

$$N_Z^{min} = |N_I(1) + N_I(2)| \tag{20.21}$$

midgap states. Within the non-interacting model, midgap states are 100% sublattice polarized (Inui et al. 1994). The non-interacting Hamiltonian has finite matrix elements between states that have weight on different sublattices. Hence, the mechanism for midgap state annihilation is hybridization of midgap states localized in different sublattices. This annihilation occurs as bonding–antibonding pairs of midgap states form, resulting in a shift in their energy and in a loss of the sublattice polarization. If $N_I(1)$ and $N_I(2)$ have the same sign, there are $|N_I(1)| + |N_I(2)|$ midgap states, regardless of the distance. If they have different signs, for example, $N_I(1) + N_I(2) = 0$, the hybridization apparently warrants the annihilation of all

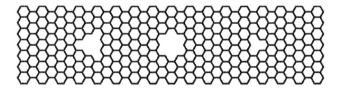

FIGURE 20.12 Examples of voids with different sublattice imbalance in the middle of a graphene ribbon. From left to right, the associated imbalance charges are $N_I = -2$, 0, and 1 (monovacancy).

midgap states. For large distances, however, this annihilation vanishes and midgap states remain.

In the general case, one can conclude that the minimum number of zero-energy states will be given by

$$N_Z^{min} = \sum_{\alpha, \beta} |N_I(\alpha) + N_I(\beta)| \tag{20.22}$$

where the integer indices α and β run over voids with the same imbalance sign, respectively. In practice, within an arbitrarily small energy interval $|E| \to 0$, the number of zero-energy states can be as large as

$$N_Z^{max} = \sum_{\alpha} |N_I(\alpha)| + \sum_{\beta} |N_I(\beta)| \tag{20.23}$$

In the general case, N_Z will be a number between N_Z^{min} and N_Z^{max}.

20.5.2 Non-Interacting Theory

When a defect is placed in a nanoribbon, it loses its translational invariance. In this case, one can resort to a partition method for the evaluation of the (local) DOS. We split the system in three sectors, the left and right electrodes and the central region where the defect(s) is (are) located. The Hamiltonian reads

$$\hat{H}_0 = \hat{H}_L + \hat{H}_R + \hat{H}_C + \hat{V}_{LC} + \hat{V}_{RC} \tag{20.24}$$

where $\hat{H}_{(L,R,C)}$ are the Hamiltonians of the left, right electrodes, and the central region, respectively. $\hat{V}_{(R,L)C}$ describes the hopping between the central region and the electrodes. The first step is the determination of the Green's function of the semi-infinite electrodes, $\hat{g}_{L,R}$. This requires the solution of a self-consistent Dyson equation (Muñoz-Rojas et al. 2006). Once this is done, the Green's function of the central region reads

$$\hat{G}_C(E) \equiv [E\hat{1} - \hat{H}_C - \hat{\Sigma}_L(E) - \hat{\Sigma}_R(E)]^{-1} \tag{20.25}$$

with

$$\hat{\Sigma}_{L,R}(E) = \hat{V}_{LC,RC}\hat{g}_{L,R}(E)\hat{V}_{CL,CR}^+ \tag{20.26}$$

The simplest defective structure one can consider is a perfect semiconducting graphene ribbon from which a single atom, A or B, is removed. We consider an axially symmetric semi-conducting armchair nanoribbon. As described in Section 20.3, armchair nanoribbons are semiconducting whenever $W \neq (3M - 1)a$ for M integer. In agreement with the result in Inui et al. (1994), a zero-energy state appears in the spectrum. For neutral graphene, this state is half filled. In other words, a spin unpaired electron occupies the midgap state. The spatially resolved DOS at zero energy, which is nothing but the modulus square of the wave function associated with the zero-energy

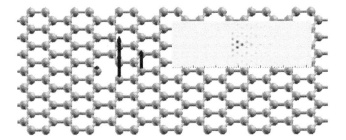

FIGURE 20.13 Magnetic moments on lattice sites around a single vacancy. Inset: Probability density of the zero-energy state built with the help of Gaussian functions located on lattice sites.

state $|\phi_0|^2$, is shown in the inset of Figure 20.13 for a nanoribbon of $W = 7a$. The state is localized in the neighborhood of the vacancy. The shape of the midgap state is also peculiar: It has a clear directionality. Importantly, the *integrated* charge, including both mid-gap and band states below the Fermi energy, yields a homogeneous charge distribution: There is one electron per atom in every atom, even in the presence of the vacancy. Hence, the localized midgap state does not imply charge localization, yet there is a finite spin density.

We now consider two vacancies. If the imbalance numbers N_I are the same, for example, two A-type vacancies, the global structure has $2|N_I|$ zero-energy states. The non-interacting Hamiltonian does not couple sites on the same sub-lattices so that the zero-energy states associated with the same-sign vacancies cannot interact, regardless of the distance separating them. On the contrary, for imbalance numbers of different sign, the hybridization depends on the distance D and, given the directional character of the midgap states in ribbons, on the relative orientation. In Figure 20.14, we show the DOS for a system with two vacancies A and B ($N_I = \pm 1$, respectively). They are aligned along the ribbon axis and placed at a distance of $6.35a$ away from each other for the two possible spatial orderings, A + B (head to head) and B + A

(tail to tail), as shown in the insets. Due to the high directional character of the associated zero-energy states, the coupling is not invariant against the interchange of positions and the zero-energy states hybridize differently, depending on the spatial ordering. In one case, the twofold zero-energy peak clearly splits into two above and below the Fermi energy. In the other, the splitting is much smaller (not visible in this scale). For one relative orientation, the wave functions overlap and the degeneracy is strongly removed. For the other, the wave functions do not couple at this distance and the degeneracy is practically unaffected. In the lower panel of Figure 20.14 we show a logarithmic plot of the energy splitting Δ as a function of the distance for the two cases. The splitting decays exponentially in both, reflecting the localized character of the vacancy states.

As a final example and in order to stress the fact that there is nothing in the previous discussion specific to vacancies in the bulk of the ribbon, we compute the non-interacting DOS for an A_6B_4 void plus an AB_2 void placed on the edges (i.e., notches) with $N_I = 2$ and $N_I = -1$, respectively. Removing just one atom to create a notch with $N_I = -1$ would have given the same charge as the AB_2 defect, but it would be chemically very unstable and we ignore that possibility. The notches are located on opposite edges, although the results are the same for notches on the same edge. A single doubly-degenerate state appears at zero energy for the $N_I = 2$ notch (see Figure 20.15). When the second notch is added in close proximity, only a single zero-energy state remains, according to the total sublattice imbalance of the system $N_I = 2 - 1 = 1$.

20.5.3 Electron–Electron Interactions and Magnetism

The existence of degenerate states at the Fermi level, as in the cases where the global imbalance charge N_I is different from zero, make it imperative to consider electron–electron interactions in

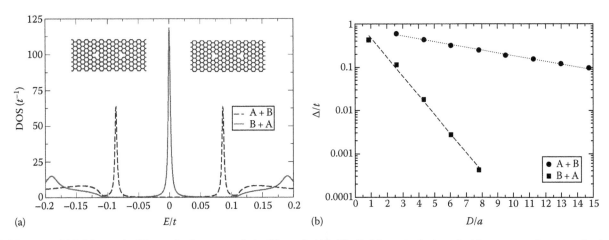

(a) E/t (b) D/a

FIGURE 20.14 (a) DOS near the Dirac point for an armchair ribbon of width $W = 7a$ (shown in the insets) with two vacancies presenting imbalance charge of different sign and same modulus ($N_I = \pm 1$). Solid lines correspond to the left inset and dashed lines correspond to the right inset. A finite broadening has been added to the delta functions for visibility's sake. The finite, but small, energy splitting in the latter case is not visible at this scale. (b) Bonding–antibonding energy splitting as a function of the distance between vacancies for the two different cases shown in (a). A linear fit on the logarithmic scale has been added to confirm the localized nature of the vacancy states.

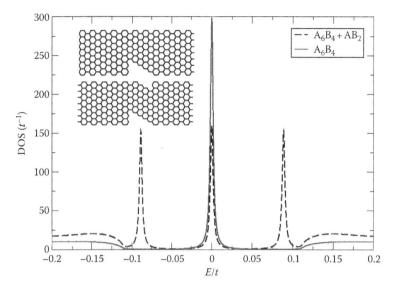

FIGURE 20.15 DOS for a single notch with imbalance charge $N_I = 2$ (upper inset, solid line). The same notch with an additional notch nearby of charge $N_I = -1$ (lower inset, dashed line).

the theoretical description. As in the case of zigzag nanoribbons presented in Section 20.4.3, the simplest way to do this is with a Hubbard model. We now verify, within this model, whether the physical picture anticipated in the non-interacting model holds. Because of the particle–hole symmetry, midgap states are half filled for neutral graphene, and the appearance of magnetic order is expected in analogy with Hund's rule in atomic magnetism. Importantly, a theorem by Lieb (1989), valid for the exact ground state of the Hubbard model and neutral bipartite lattices, states that the total spin S of the ground state is given by $2S = |N_A - N_B| = |N_I|$. Lieb's theorem provides a rigorous connection between vacancies in the graphene lattice and the emergence of magnetism. Incidentally, this was known long ago in the context of chemical studies of hydrocarbons as the Longuet-Higgins conjecture (Longuet-Higgins 1950). As a result, sublattice unbalanced neutral graphene will always present a finite total magnetic moment.

The numerical calculations are done in this case for periodic systems with a unit cell of width W, length $L = N_x \left(\sqrt{3}/4 \right) a$, where N_x is the number of carbon atoms along an armchair chain, and with periodic boundary conditions along the x direction. We consider unit cells as long as 25 nm and the typical number of atoms in a self-consistent calculation is 1000. Alternatively, the calculations can be done combining Equations 20.20 and 20.25 with identical results. The self-consistent solution of structures with single atom vacancies have one unpaired electron in the zero-energy state, that is, $S = 1/2$. Our mean field calculation for $U = 2$ eV shows a spin splitting of the midgap state Δ_S and a smaller spin splitting δ of the conduction and valence band states (see Figure 20.16). The spin degeneracy is thus broken, with only one of the spin states occupied, the other being empty. This results in a finite magnetization density $m_i = (n_{i\uparrow} - n_{i\downarrow})/2$, localized around the vacancy, as shown in Figure 20.13. Interestingly, although the magnetization resides mostly on the majority

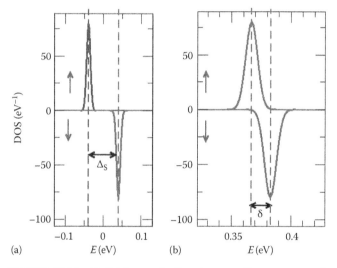

FIGURE 20.16 (a) DOS near the Fermi energy (here set to zero) showing the spin-split midgap state. (b) Zoom of the spin-split conduction band minimum. For clarity, we substitute the delta functions by Gaussian functions with a finite broadening.

sublattice, interactions induce some reversed magnetization in the other sublattice (not appreciated in the figure).

Whereas, according to Lieb's theorem, the total magnetic moment $S = \sum_i \langle m_i \rangle$ is given by the sublattice imbalance, the degree of localization of the spin texture is not. In order to quantify it, one can define the standard deviation

$$\sigma_1 = \sqrt{\sum_i \langle m_i \rangle^2} \qquad (20.27)$$

In this definition, σ_1 is not normalized as usual by the total number of atoms of the sample since σ_1 characterizes a localized object.

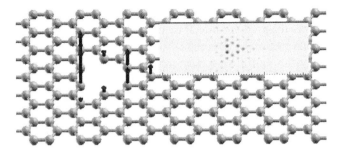

FIGURE 20.17 Magnetic moments on lattice sites around a triangular void with $N_I = 2$. Inset: Probability density of the zero-energy states built with the help of Gaussian functions placed on lattice sites.

For finite U, the graphene lattice responds with a staggered magnetization to the presence of defects, and σ_1 is thus an appropriate magnitude to measure the magnitude of that response.

A less trivial example of the confirmation of Lieb's theorem to predict the total magnetic moment of the ground state is that of nanoribbon with a void larger than a monovacancy ($N_I > 1$), for example, a void with $N_I = 2$ as the A_3B_1. Figure 20.17 shows the magnetization profile for this case. For the chosen value of $U = 2$ eV, the staggered magnetization is barely visible in this scale. In agreement with Lieb's theorem, the ground state has a spin $S = 1$, made out of local moments localized, mostly, on the triangle boundaries. The inset shows the sum of the probability densities associated with the two zero-energy states.

We finally study the interaction between two magnetic defects with local sublattice imbalance $N_I = \pm 1$ (vacancies). The Lieb's theorem warrants that, when the sign of the sublattice imbalance is the same for the two defects, the total spin of the ground state is the sum of the spin of the individual defects. Hence, they are coupled *ferromagnetically*. In contrast, if the two defects have opposite sublattice imbalances so that the global sublattice imbalance is zero, Lieb's theorem warrants that the total spin is zero. Our calculations show that this can happen in two different scenarios: the local magnetization might be zero everywhere or the two defects could be magnetized along opposite directions, that is, could be coupled *antiferromagnetically*.

In Figure 20.18, we plot the normalized standard deviation of the two magnetic moments σ_2 for a ribbon with $W = 7a$ as a function of the vacancies separation. We normalize the computed σ_2 to the one corresponding to two independent single-defect magnetic textures, $\sqrt{2}\sigma_1$. When the defects are sufficiently far away, σ_2 must tend to $\sqrt{2}\sigma_1$, that is, the normalized σ_2 must tend to 1. We show results for both monatomic vacancies lying on the same sublattice (A + A, open circles), whose ground state total spin is $S = 1$, and on different sublattices (A + B, full circles), whose ground state spin is $S = 0$. The two curves are calculated with $U = 2$ eV. At large distances, the two defects become decoupled as expected. At short distances, the behavior of the magnetic texture is radically different for both A + A and A + B structures. In the former case, σ_2 is enhanced, indicating the localization of the magnetic texture in a smaller region. Since the total spin is 1, local moments survive even when the two defects are very close. As the separation between defects increases, they

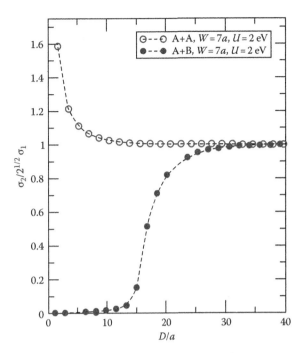

FIGURE 20.18 Normalized standard deviation of the magnetization as a function of the distance between two monovacancies for two cases: Vacancies in the same sublattice (open circles) and vacancies in different sublattices (full circles). The nanoribbon width is $W = 7a$ and the interaction strength is $U = 2$ eV.

become independent from each other and $\sigma_2 = \sqrt{2}\sigma_1$. When this happens, the energy gap between $S = 1$ and $S = 0$ should vanish. In contrast to the A + A case, the *local* magnetization of the A + B structure progressively vanishes below a *minimal distance* D_c. This can be easily understood since at small separations, the two electrons accompanying the zero-energy states of the two vacancies occupy the lowest-energy bonding state (see Figure 20.14) in a singlet configuration. In other words, there is a *maximal density* of defects above which zero-energy states hybridize and local magnetic moments vanish. The critical density depends, of course, on the energy scales of the problem, the single particle gap E_g, controlled by the ribbon width, and the on-site repulsion U.

20.6 An Application of Graphene Nanoribbons: Field-Effect Transistors

The electronic versatility exhibited by graphene nanoribbons makes them ideal candidates for all imaginable electronic applications. One important aspect in semiconductor microelectronics is the electrical tunability of the resistance of a semiconductor through the field effect. In this section, we explore the possible application of graphene nanoribbons as field-effect transistors. We consider here only the case of armchair nanoribbons to avoid unnecessary complications coming from the magnetic edges in zigzag nanoribbons.

20.6.1 Electrostatics and Gating

First we address the problem of how the Fermi energy is actually tuned and how this can be described theoretically. In other words, how graphene nanoribbons are driven out of charge neutrality and how this changes their electronic properties. Experimentally this is done through a capacitive coupling to a metal gate. Figure 20.19 shows a graphene ribbon of width W on top of an insulating slab of thickness d that lies above a metallic gate. Application of a gate voltage V_G injects carriers in the graphene layer accompanied by a corresponding change in the metal.

20.6.1.1 Long-Range Interactions

While the magnetic properties of graphene nanoribbons are correctly accounted for by the Hubbard model, the theoretical description of a gate voltage and the deviations from charge neutrality induced thereby require consideration of the true long-range nature of the Coulomb interactions. These can be added to the one orbital Hamiltonian:

$$\hat{H} = \hat{H}_0 + \hat{U}_C \qquad (20.28)$$

The second term describes now the Coulomb interaction between the extra charges in the system:

$$\hat{U}_C = e \int \delta \hat{n}(\vec{r}) \phi_{ext}(\vec{r}) d\vec{r} + \frac{e^2}{2\epsilon} \int \frac{\delta \hat{n}(\vec{r}) \delta \hat{n}(\vec{r}')}{|\vec{r} - \vec{r}'|} d\vec{r}\, d\vec{r}' \qquad (20.29)$$

where $\delta \hat{n}(\vec{r})$ is the operator describing the departure of the local electronic density from charge neutrality:

$$\delta \hat{n}(r) \simeq \sum_{i\sigma} |\phi_i(r)|^2 (\hat{c}_{i\sigma}^+ \hat{c}_{i\sigma} - n_0) \qquad (20.30)$$

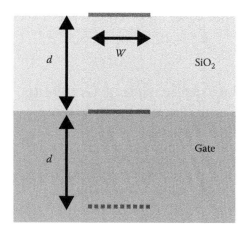

FIGURE 20.19 Scheme of a field-effect transistor with a graphene nanoribbon. This is placed on top of an insulating substrate that in turn is located on top of a metal plate. The voltage gate is applied between the metal and the nanoribbon.

Here the sum runs over all the atoms in the lattice and $\phi_i(r)$ denotes the p_z atomic orbital centered around the atom i. As in Section 20.2.1, $\hat{c}_{i\sigma}^+ (\hat{c}_{i\sigma})$ is the second quantization operator that creates (annihilates) one electron with spin σ in the orbital $\phi_i(r)$, localized around atom i. In Equation 20.30, $n_0 = 1/2$ is the number of electrons per site and per spin in neutral graphene.

The electrostatic potential created by the extra charges in the metallic gate is denoted by $\phi_{ext}(\vec{r})$. Considered as a whole, the graphene and the metal layer form a neutral system. Therefore, the extra carriers in one side are missing in the other. Since the charges in the metallic gate move as to cancel the electric field inside the metal, they depend on the density distribution in the graphene electrode, which in turn depends on $\phi_{ext}(\vec{r})$. This self-consistent problem is solved using the image method: for a given charge density profile in the graphene, $e\langle\delta\hat{n}(\vec{r})\rangle$, the potential created by the corresponding extra charges in the metal is given by a distribution of fictitious image charges inside the metal (see Figure 20.19): $-e\delta\hat{n}_{im}(\vec{r}) = -e\delta\hat{n}(\vec{r} - 2\vec{d})$ with $\vec{d} = (0, 0, d)$.

$$V_{ext}(\vec{r}) = -\frac{e}{\epsilon} \int \frac{\langle \delta_n(\vec{r}') \rangle}{|\vec{r} - \vec{r}' + 2\vec{d}|} d\vec{r}' \qquad (20.31)$$

It must be pointed out that the image charges are a mathematical construct that provide a simple way to obtain a solution of the electrostatic problem, which automatically satisfies that the surface of the metallic gate is a constant potential surface, as expected for a metal. With this method, the potential evaluated at the metal gate ($z = 0$) is exactly zero. Therefore, the potential difference between the metal–insulator interface and the graphene layer is the potential evaluated at the graphene layer.

20.6.1.2 Hartree Approximation

The electronic repulsion is now treated in the Hartree approximation where $\hat{H}_0 + \hat{U}_C$ become a one-body Hamiltonian that can be represented in the basis of localized p_z atomic orbitals. The electrons feel the electrostatic potentials created by both the gate and themselves:

$$\hat{U}_C \simeq e \int \delta\hat{n}(\vec{r})(V_{ext}(\vec{r}) + V_{SC}(\vec{r})) d\vec{r} \qquad (20.32)$$

The argument of the integral, $V_T = V_{ext} + V_{SC}$, can be written as

$$V_T(\vec{r}) = \frac{e}{\epsilon} \int \langle \delta_n(\vec{r}') \rangle \mathcal{K}(|\vec{r} - \vec{r}'|, d) d\vec{r}' \qquad (20.33)$$

where

$$\mathcal{K}(|\vec{r} - \vec{r}'|, d) = \frac{1}{|\vec{r} - \vec{r}'|} - \frac{1}{|\vec{r} - \vec{r}' + 2\vec{d}|} \qquad (20.34)$$

The term $\hat{U}_C$ introduces site-dependent shifts in the diagonal matrix elements of $\hat{H}_0$:

$$U_i = eV_i = \frac{1}{\epsilon} \sum_j q_j \upsilon_{ij} \qquad (20.35)$$

where q_j is the average excess charge in site j and $\upsilon_{ij} = \int d\vec{r} d\vec{r}' \, |\phi_i(\vec{r})|^2 |\phi_j(\vec{r}')|^2 \, \mathcal{K}(|\vec{r} - \vec{r}'|, d)$ is the potential created by an unit charge located at site j on site i. At this point, we adopt an approximation that permits to evaluate υ_{ij} without a detailed model of $\phi_i(\vec{r})$. Since the atomic orbitals are highly localized for $|\vec{R}_i - \vec{R}_j| > a$, we can approximate υ_{ij} by the first term in the multipolar expansion:

$$\upsilon_{ij} \simeq \frac{1}{|\vec{R}_i - \vec{R}_j|} - \frac{1}{|\vec{R}_i - \vec{R}_j - 2\vec{d}|}.$$

We have verified that the potential in a given site is quite independent of the approximation adopted to evaluate the $i = j$ contribution, for which the monopolar approximation fails. Without sacrificing accuracy, we adopt the simplest strategy of removing that term from the sum. As the Hubbard Hamiltonian of Section 20.4.1, the Hamiltonian in Equation 20.28 is now solved self-consistently.

20.6.1.3 Self-Consistent Electronic Structure of Armchair Ribbons

In Figure 20.20 we show both the self-consistent (upper panels) and the non-interacting (lower panels) bands for electrically doped ribbons in two cases: semiconducting and metallic ($N_y = 9$ and $N_y = 10$ respectively). Notice that the non-interacting bands

have electron–hole symmetry. In these figures, we plot the self-consistent chemical potential in the upper panels and the naive Fermi energy obtained upon integration of the non-interacting DOS. The electrical injection of extra carrier results in a large shift of the bands, due to the electrostatic interactions. The inhomogeneity of the electronic density and the electrostatic potential result in a moderate change of the shape of the bands. Notice for instance that for the $N_y = 9$ ribbon the chemical potential intersects two self-consistent bands (see Figure 20.20a) whereas the naive Fermi energy intersects three non-interacting bands (see Figure 20.20c).

In Figure 20.21 we show the self-consistent density profile of q_i and U_i for the semiconducting ribbon $N_y = 9$ with two different values of the average density n_{2D}. We see how the density of carriers is larger on the edges than in the middle of the ribbon as expected for a conducting system. As a result, the electrostatic potential has an inverted U shape. Superimposed to these overall trends, both q_i and U_i feature oscillations arising from quantum mechanical effects. Expectedly, the average U_i is much larger in the high density than in the low density. In the high density case, it is apparent that the potential is flat in the inner part of the ribbon, very much like in a metal. Very similar trends are obtained for ribbons with different widths.

20.6.1.4 Density vs. V_G

In Figure 20.22a, we plot the gate voltage vs. n_{2D} in graphene nanoribbons for different widths. The common feature in all the curves is the linear relation between V_G and n_{2D} that reflects the dominance of the classical electrostatic contribution over

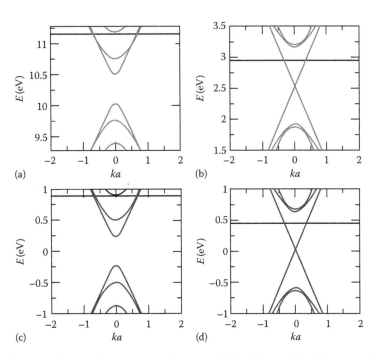

FIGURE 20.20 Energy bands for $N_y = 9$ (left panels) and $N_y = 10$ (right panels). The Hartree (bare) bands are shown above (below). The corresponding densities are $n_{2D} = 7.6 \times 10^{12}$ cm^{-2} ($N_y = 9$) and $n_{2D} = 1.7 \times 10^{12}$ cm^{-2} ($N_y = 10$). Notice that the non-interacting and Hartree bands are shifted with respect to each other. The horizontal lines indicate the Fermi energy.

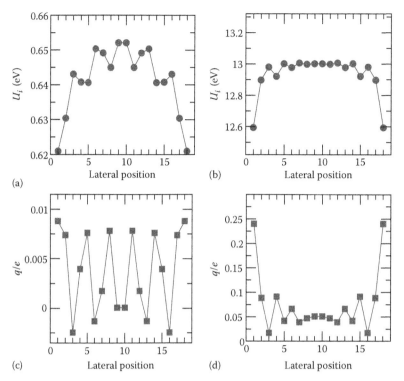

FIGURE 20.21 Potential (a),(b) and density (c),(d) profiles, as a function of the position across the ribbons, for $N_y = 9$ ribbon with $n_{2D} = 7.6 \times 10^{11}$ cm^{-2} (a, c) and $n_{2D} = 7.6 \times 10^{12}$ cm^{-2} (b, d). The higher density results correspond to the bands shown in Figure 20.20a.

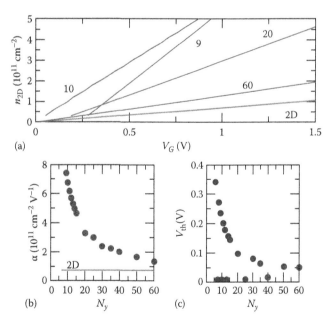

FIGURE 20.22 (a) Gate voltage versus average density for armchair ribbons of different widths. (b) Inverse capacitance, α, as a function of width (measured in terms of N_y). (c) Threshold voltage as a function of the ribbon width N_y. As a rule of thumb, $N_y = 4$ corresponds to 1 nm.

quantum effects. This is also apparent from Figure 20.20, where the shift of the bands is much larger than the position of the Fermi energy with respect to the bottom of the conduction bands. In addition, we see that some of the curves do not intersect at $V_G = 0$ for $n_{2D} = 0$ in the case of semiconducting ribbons.

This is clearly the case of $N_y = 9$ and $N_y = 20$. In contrast, the curve $N_y = 10$ extrapolates to zero. We denote the threshold gate potential as V_{th}. Notice that it corresponds to the change in chemical potential of the ribbon when a single electron is added. Since the chemical potential for a semiconductor system lies in the middle of the gap, and the chemical potential when a single electron is added is the lowest energy state of the conduction band, we must have $eV_{th} = E_g/2$ where E_g is the gap of the semiconducting ribbon. We fit all the curves in Figure 20.22 according to

$$n_{2D}(N_y) = \alpha(N_y)[V_G - V_{th}(N_y)] \qquad (20.36)$$

In Figure 20.22b and c, we plot α and V_{th} as a function of N_y. The wider ribbon considered has $N_y = 60$ that corresponds to $W \approx 14.5$ nm. In Figure 20.22b we see how α rapidly decreases towards the 2D value $\alpha_{2D} = e^2 d/\epsilon = 0.7 \cdot 10^{11}$ cm^{-2} V^{-1} as W increases. We have verified that α scales inversely proportional to N_y. Interestingly, $\alpha(N)$ approaches to the 1D case very quickly, even if $W \ll d$. The evolution of V_{th} as a function of N reflects two facts: On one side, two out of three ribbons are semiconducting. On the other, it reflects the fact that the gap decreases as W^{-1} for these. Notice that ribbons of $W = 20$ nm ($N \simeq 80$) still present a gap and, according to Figure 20.22b, their capacitance is almost that of 2D graphene. It must be pointed out that Equation 20.36 takes into account the capacitive coupling to the gate electrode but leaves out the capacitive coupling to source and drain electrodes normally present in the system.

20.6.1.5 Ideal Conductance vs. V_G

We finally study the number of occupied bands $\mathcal{N}$ in a given ribbon as a function of V_G. This is related to the ideal conductance as a function of V_G through the Landauer formula (Datta 1995) for perfectly transmitting ribbons:

$$G = \frac{2e^2}{h} \mathcal{N} \qquad (20.37)$$

The value of conductance so obtained can be considered an upper limit for the real conductance of the nanoribbon when this is not perfect. In an ideal armchair ribbon, the number of channels $\mathcal{N}$ increases one by one as the Fermi energy, with respect to the Dirac point, is increased. Our approach permits to obtain both $\mathcal{N}$ and the quantum shift of the chemical potential as a function of the gate voltage and to plot $\mathcal{N}$ as a function of V_G.

In Figure 20.23a, we plot the conductance $G(V_G)$ for two narrow ribbons ($N = 9$ and $N = 10$) in the small gate regime. It is apparent that the $N = 9$ ribbon only conducts above a threshold gate whereas the $N = 10$ ribbon conducts even for $V_G = 0$. Therefore, semiconducting ribbons can be electrically tuned from insulating to conducting behavior with a gate voltage. In Figure 20.23b, we show the conductance for the same ribbons at higher V_G. The different sizes of the plateaus for $N_y = 9$ and $N_y = 10$ reflect the different structure of the bands, as seen in Figure 20.20. In Figure 23c and d, we show the conductance for wider ribbons ($N_y = 40$ and $N_y = 60$ respectively). Although the shape of the curves is superficially similar for all these ribbons,

the V_G necessary to have a fixed number of bands at the Fermi energy is a decreasing function of W. This is a consequence of quantum confinement: the smaller the ribbon, the larger the sub-band level spacing $\Delta E_n = E_n(k = 0) - E_{n-1}(k = 0)$. A shift of the Fermi energy, relative to the self-consistent bands by an amount ΔE_n, so that a new band is available for transport, implies also an electrostatic overhead, due to the change in density, which accounts for most of the eV_G, as we have discussed above.

On average, the steps in the stepwise curve $\mathcal{N}(V_G)$ increase in size as V_G increases. This trend is more apparent in wider ribbons. In this sense it can be said that $\mathcal{N}$ is sublinear in V_G. In the limit of very wide ribbons, for which the DOS is almost 2D, we can derive a qualitative relation between number of channels and V_G. The number of conducting modes $\mathcal{N}(E_F)$ is proportional to the perimeter of the Fermi circle, which is $2\pi e V_Q$, that is, $\mathcal{N} \propto V_Q$, where V_Q is the quantum capacitance. On the other side, in 2D, V_Q scales as $\sqrt{n_{2D}}$ whereas V_G scales linearly. As a result we have

$$\mathcal{N} \propto V_Q \propto \sqrt{V_G} \qquad (20.38)$$

In Figure 23c and d we have fitted the numerical data to the curve $G = a V_G^b$, where a and b are fitting parameters. We have obtained exponents $b = 0.45$ and $b = 0.47$ for $N_y = 40$ and $N_y = 60$, respectively, in agreement with the qualitative discussion above. In Figure 23c and d, the solid line is the best fit to the equation $G = a\sqrt{V_G}$.

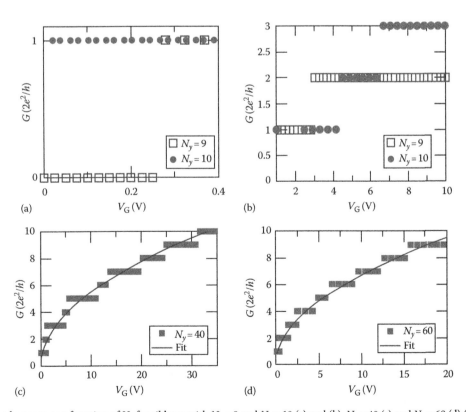

FIGURE 20.23 Conductance as a function of V_G for ribbons with $N_y = 9$ and $N_y = 10$ (a) and (b), $N_y = 40$ (c) and $N_y = 60$ (d) (see text).

20.6.2 Band-Gap Engineering

One of the major difficulties in the use of carbon-based systems for nanoelectronics resides in the complexity of the physics behind the contact with the metallic electrodes. For instance, Schottky barriers in the case of field-effect transistors with carbon nanotubes are a major concern (Palacios et al. 2008b). A way to get around the lack of reproducibility is to avoid the use of metallic electrodes or, at the very least, to minimize their influence. Band-gap engineering through the manipulation of the atomic structure of graphene nanoribbons offers a way out of this problem, as we show in this section.

20.6.2.1 The Device

A single graphene ribbon with armchair edges and a narrow channel in the middle (see Figure 20.24a and b) presents three regions with different band gaps that can be doped differently by a single back gate (see Figure 20.24c). For channel widths with N_y smaller than ≈ 500, the constriction presents a transport gap at room temperature (Son et al. 2006) and can act as the semiconductor active channel does in standard transistors. Since the gap scales inversely with the width of the ribbon, the wide sections of the ribbon or leads can present a zero or vanishingly small gap for large enough widths and behave as metallic electrodes at finite temperature or when they are driven out of the charge neutrality point

by the action of a back gate voltage V_G, as shown in previous section. The same gate voltage controls the on–off state by bringing E_F into the gap of the constriction or out of it (see Figure 20.24c).

Theoretically, this system should behave as an ambipolar transistor where the threshold voltage, V_{th} (the voltage at which the $I(V_G)$ characteristics begins to deviate from an exponential behavior), is determined solely by the width of the narrow channel (see Section 20.6.1). The on-conductance, $G_{on} = dI/dV \mid_{V_G > V_{th}}$, and the subthreshold swing, $S = (d \log I/dV_G)^{-1}$, magnitudes that determine the transistor performance are also controlled by the capacitive couplings and by the *atomic structure of the lead-channel contact*. The ultimate performance limits of graphene ribbon field-effect transistors are established here for two contact models: The first, *square*, where the crystallographic orientation of one edge changes by 90° at the contact (see Figure 20.24a). The second, *tapered*, where the lead narrows down progressively until reaching the channel width. This is achieved in practice by a single change in the crystallographic orientation of one of the edges of 60° (see Figure 20.24b). In this case, the armchair edge is never interrupted, except for the change in the crystallographic orientation. The channel is always placed laterally, that is, on one side of the ribbon. This choice can be justified on the basis that, in order to fabricate a constriction with reduced disorder, it is desirable to leave untouched one edge of the original graphene flake and, for example, etch away only the other edge (constrictions created by

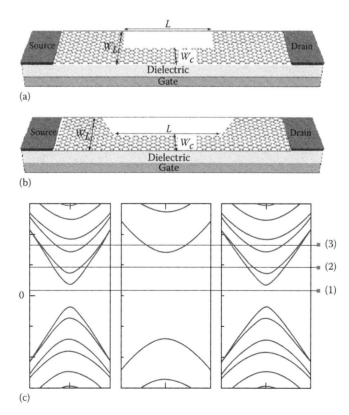

(a)

(b)

(c)

FIGURE 20.24 Schematic view of a graphene ribbon field-effect transistor with a square- (a) and tapered- (b) constriction placed on one side. Example of band structures corresponding to the three regions: left lead, right lead, and constriction (c). The horizontal lines denote different Fermi levels corresponding to different conduction situations: (1) at the charge neutrality point, (2) off state, and (3) on state.

etching both edges have already been fabricated (Han et al. 2007)). Our proposed device could improve the final performance.

20.6.2.2 Electronic Transport

The electronic structure is calculated with the standard one-orbital tight-binding model discussed in Section 20.2.1. For simplicity's sake, we put aside electron–electron interactions in this section, since we are only concerned here with some basic geometrical issues. As we have seen, the presence of zigzag edges in the square contacts might introduce a magnetic contribution when interactions are considered. Electron–electron interactions also determine the relation between V_G and E_F, which has already been studied in the previous section.

We compute the conductance G using the Landauer formalism, which assumes coherent transport across the constriction. The scattering problem is solved using the partition method and the Green's functions introduced in Section 20.5.2. This procedure involves (1) the calculation of the Green's function of the central region (which includes the constriction and part of the leads), (2) the calculation of the self-energies $\hat{\Sigma}_{L,R}(E)$ associated to the semi-infinite ribbon leads by iterative solution of the Dyson equation (Muñoz-Rojas et al. 2006), and (3) the subsequent evaluation of the transmission probability T using the Caroli expression (Caroli et al. 1971):

$$T(E) = Tr(\hat{\Gamma}_L(E)\hat{G}_C\hat{\Gamma}_R(E)\hat{G}_C^+(E)) \qquad (20.39)$$

where Tr denotes the trace and

$$\hat{\Gamma}_{L,R}(E) = i(\hat{\Sigma}_{L,R}(E) - \hat{\Sigma}_{L,R}^+(E)) \qquad (20.40)$$

Within the Landauer formalism, the conductance is given now by $G = (2e^2/h)T(E_F)$, where $T(E_F)$ substitutes the number of bands $\mathcal{N}$ in perfect ribbons (see section 20.6.1.5). In order to simplify the discussion, we assume semi-infinite graphene ribbons on both sides. The role played by the metallic electrodes to which the graphene ribbons are ultimately connected is neglected. One can anticipate that they should not affect the results as long as $W_L \gg W_C$, where W_L is the width of the bulk ribbon and W_C is the width of the narrow channel.

We begin by showing in Figure 20.25a and b the zero-temperature G as a function of the energy for a channel of length $L \approx 10$ nm and $W_C = 1/2$, 1/4, and 1/8 W_L, for $W_L = 5.8$ nm. The Fermi energy is set to zero as usual. For both types of square and tapered contacts we obtain the expected step-wise increase of the conductance as a function of E (in units of the hopping parameter t) associated with the increase in the number of bands crossing E in the channel. The steps come in pairs, reflecting the band structure of semiconducting armchair ribbons (see Figure 20.24c). On top of the steps we obtain strong Fabry–Perot-like oscillations, that is, interference oscillations, as a consequence of the finite reflection at the contacts. The periodicity of these oscillations is consistent with the length and with the dispersion relation in the channel. The amplitude of the oscillations, on the other hand, depends on how abrupt the contact is. For the square contacts, the amplitude of the oscillations is larger than for the tapered ones as expected.

In order to separate the contribution of the scattering at the interfaces from the quantum interference effects, we have computed the transmission of a single interface, for example, the transmission between a semi-infinite lead and a semi-infinite narrower channel (dashed lines in Figure 20.25). In all cases, the transmission steps saturate to their quantum limit very slowly. For both contact types, as the width of the channel decreases, the transmission worsens as a consequence of the increasing mismatch between lead and channel wave functions. The increasing reflection on decreasing W_C/W_L is somewhat damped for the tapered contacts, but still clearly visible. When $W_C/W_L \to 0$, quasilocalized states form in the channel as a result of a significant loss in the transparency of the contacts. In general, one can anticipate that channels with tapered contacts to the leads should perform better than their square counterparts for both G_{on} and S, although the overall performance diminishes as the channel narrows down in both models.

We now turn our attention to the actual room-temperature performance of the proposed devices. Although the

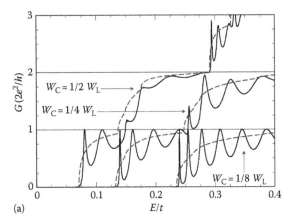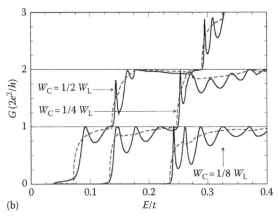

FIGURE 20.25 (a) Conductance for an $L \approx 10$ nm channel and different widths in the case of the square contact model (solid line). The width of the lead is $W_L = 5.8$ nm. The conductance for a single interface is shown in dashed lines for both (a) and (b) cases.

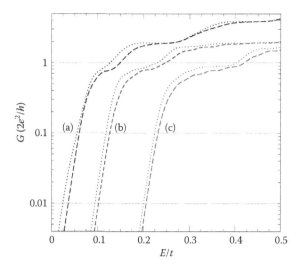

FIGURE 20.26 Logarithmic plot of the room-temperature conductance for the square (dashed lines) and tapered (dotted lines) contact models shown in Figure 20.24a and b for different channel widths $W_C/W_L = 1/2$ (a), 1/4 (b), and 1/8 (c) with $W_L = 5.8$ nm.

channel widths and lengths considered (≤ 10 nm) are still difficult to achieve with present lithographic techniques, the results can be easily extrapolated to more realistic constrictions. The ones considered here behave as intrinsic semiconductors at room temperature since their gaps are in the range $\approx 0.1 - 1$ eV for a typical value of $t = 2.7$ eV, which is much larger than $kT \approx 25$ meV for $T = 300$ K. Figure 20.26 shows a logarithmic plot of $G(E_F)$ for the cases presented previously. If we assume, for simplicity, that the Hartree capacitive coupling is negligible, the horizontal axis also corresponds to $-eV_G/t$, the experimentally controllable quantity. As seen in the previous section, the classical capacitance goes to zero for a vanishing dielectric width. In all cases, the conductance oscillations have disappeared, smeared out by temperature. The combined action of finite reflection at the interfaces, finite temperature, and tunneling prevents G_{on} from reaching the quantum limit $2e^2/h$ for $E_F > eV_{th}$ before the next channel opens up. We can also extract the respective values of S from the logarithmic plots. While, for the wider channels, S is larger for the tapered contacts (83 vs. 66 meV/decade), these contacts outperform the square ones for the narrowest channels (63 vs. 68 meV/decade). Other issues such as the classical capacitance and the overall role of electrostatics need to be considered for a more realistic evaluation of the performance of the graphene nanoribbon field-effect transistors, but this is beyond the scope of this review.

20.7 Summary and Perspectives

Here we have examined a few aspects related to the peculiar electronic properties of graphene nanoribbons and how to possibly make use of them for electronic applications. We have shown how the crystallographic orientation of the ribbons determines largely their electronic properties. While armchair nanoribbons can be metallic or semiconducting in a first-neighbors

tight-binding approximation, zigzag nanoribbons present a large degeneracy at the Fermi level (for a more detailed account see Brey and Fertig 2006b). When electron–electron interactions are taken into account, this picture is strongly modified in the case of zigzag ribbons. The edges develop a finite spin density of magnetism, being antiferromagnetically coupled to each other (for more details see Fernández-Rossier 2008). The electronic structure of armchair ribbons is not strongly modified by interactions, except in the case of the metallic ones where a small gap opens (for more details see Son et al. 2006). The presence of defects such as vacancies and voids in armchair ribbons can induce either ferromagnetic or antiferromagnetic order (a complete study can be found in Ref. (Palacios et al. 2008a)). Finally, we have examined some basic issues regarding the possible use of graphene nanoribbons as field-effect transistors such as changes in their electronic structure when electrically charged by a gate (Fernández-Rossier et al. 2007) and changes in their electronic transport properties when constrictions are introduced, breaking the translational symmetry (Muñoz-Rojas et al. 2008).

One of the major experimental difficulties when it comes to comparing with theory is the precise definition and knowledge of the atomic structure at the edges of the ribbons. As we have seen, the electronic properties depend very much on the crystallographic orientation of the nanoribbon. In particular the existence of zigzag edges is a matter of debate at present. These are very reactive to the presence of adsorbates such as H (Wassmann et al. 2008) and are prone to suffer deformations or Jahn–Teller distortions (distortions to lower the symmetry) to reduce the energy (Koskinen et al. 2008). Until this is verified, the magnetic properties of zigzag edges remain an academic curiosity. This is by no means an extensive review and does not pretend to cover all aspects of this very active field. Ongoing theoretical and experimental work will likely modify some of the conclusions that can be drawn from this review. The reader is invited to do so.

Acknowledgments

We acknowledge Federico Muñoz Rojas for help with some of the figures. This work has been financially supported by MEC-Spain under Grant Nos. MAT2007-65487, MAT2006-03741, and CONSOLIDER CSD2007-00010, by Generalitat Valenciana under Grant No. ACOMP07/054, and by Grant No. DMR-0704033.

References

Ando, T. 2005. Theory of electronic states and transport in carbon nanotubes. *Journal of the Physical Society of Japan*, 74:777.

Berger, C., Song, Z., Li, T. et al. 2004. Ultrathin epitaxial graphite: 2d electron gas properties and a route toward graphene-based nanoelectronics. *The Journal of Physical Chemistry B*, 108(52):19912.

Brey, L. and Fertig, H. A. 2006a. Edge states and the quantized hall effect in graphene. *Physical Review B (Condensed Matter and Materials Physics)*, 73(19):195408.

Brey, L. and Fertig, H. A. 2006b. Electronic states of graphene nanoribbons studied with the Dirac equation. *Physical Review B (Condensed Matter and Materials Physics)*, 73(23):235411.

Buschow, K. H. J. and de Boer, F. R. 2004. *Physics of Magnetism and Magnetic Materials*. Kluwer Academic Publishers, New York.

Caroli, C., Combescot, R., and Dederichs, P. 1971. Direct calculation of the tunneling current. *Journal of Physics C: Solid State Physics*, 4:916.

Datta, S. 1995. *Electronic Transport in Mesoscopic Systems*. Cambridge University Press, Cambridge, U.K.

DiVincenzo, D. P. and Mele, E. J. 1984. Self-consistent effective-mass theory for intralayer screening in graphite intercalation compounds. *Physical Review B*, 29(4):1685–1694.

Fernández-Rossier, J. 2008. Prediction of hidden multiferroic order in graphene zigzag ribbons. *Physical Review B (Condensed Matter and Materials Physics)*, 77(7):075430.

Fernández-Rossier, J. and Palacios, J. J. 2007. Magnetism in graphene nanoislands. *Physical Review Letters*, 99(17):177204.

Fernández-Rossier, J., Palacios, J. J., and Brey, L. 2007. Electronic structure of gated graphene and graphene ribbons. *Physical Review B (Condensed Matter and Materials Physics)*, 75(20):205441.

Frisch, M. J., Trucks, G. W., Schlegel, H. B. et al. 2003 www.gaussian.com.

Fujita, M., Wakabayashi, K., Nakada, K., and Kusakabe, K. 1996. Peculiar localized state at zigzag graphite edge. *Journal of the Physical Society of Japan*, 65:1920.

Han, M. Y., Özyilmaz, B., Zhang, Y., and Kim, P. 2007. Energy band-gap engineering of graphene nanoribbons. *Physical Review Letters*, 98(20):206805.

Hohenberg, P. and Kohn, W. 1964. Inhomogeneous electron gas. *Physical Review*, 136(3B):864–871.

Inui, M., Trugman, S. A., and Abrahams, E. 1994. Unusual properties of midband states in systems with off-diagonal disorder. *Physical Review B*, 49(5):3190–3196.

Kittel, C. 1976. *Introduction to Solid State Physics*. Wiley, New York.

Klitzing, K. V., Dorda, G., and Pepper, M. 1980. New method for high-accuracy determination of the fine-structure constant based on quantized hall resistance. *Physical Review Letters*, 45(6):494–497.

Kohn, W. and Sham, L. J. 1965. Self-consistent equations including exchange and correlation effects. *Physical Review*, 140(4A):1133–1138.

Koskinen, P., Malola, S., and Hakkinen, H. 2008. Self-passivating edge reconstructions of graphene. *Physical Review Letters*, 101(11):115502.

Lehtinen, P. O., Foster, A. S., Ma, Y., Krashenin-nikov, A. V., and Nieminen, R. M. 2004. Irradiation-induced magnetism in graphite: A density functional study. *Physical Review Letters*, 93(18):187202.

Li, X., Wang, X., Zhang, L., Lee, S., and Dai, H. 2008. Chemically derived, ultrasmooth graphene nanoribbon semiconductors. *Science*, 319(5867):1229.

Lieb, E. H. 1989. Two theorems on the Hubbard model. *Physical Review Letters*, 62(10):1201–1204.

Longuet-Higgins, H. C. 1950. Some studies in molecular orbital theory. I. Resonance structures and molecular orbitals in unsaturated hydrocarbons. *The Journal of Chemical Physics*, 18(3):265–274.

Meyer, J. C., Geim, A. K., Katsnelson, M. I., Novoselov, K. S., Booth, T. J., and Roth, S. 2007. The structure of suspended graphene sheets. *Nature*, 446(7131):60.

Muñoz-Rojas, F., Jacob, D., Fernández-Rossier, J., and Palacios, J. J. 2006. Coherent transport in graphene nanoconstrictions. *Physical Review B (Condensed Matter and Materials Physics)*, 74(19):195417.

Muñoz-Rojas, F., Fernández-Rossier, J., Brey, L., and Palacios, J. J. 2008. Performance limits of graphene-ribbon field-effect transistors. *Physical Review B (Condensed Matter and Materials Physics)*, 77(4):045301.

Nakada, K., Fujita, M., Dresselhaus, G., and Dresselhaus, M. S. 1996. Edge state in graphene ribbons: Nanometer size effect and edge shape dependence. *Physical Review B*, 54(24):17954–17961.

Novoselov, K. S., Geim, A. K., Morozov, S. V. et al. 2004. Electric field effect in atomically thin carbon films. *Science*, 306(5696):666.

Novoselov, K. S., Jiang, Z., Zhang, Y. et al. 2007. Room-temperature quantum hall effect in graphene. *Science*, 315(5817):1379.

Palacios, J. J., Fernández-Rossier, J., and Brey, L. 2008a. Vacancy-induced magnetism in graphene and graphene ribbons. *Physical Review B (Condensed Matter and Materials Physics)*, 77(19):195428.

Palacios, J. J., Tarakeshwar, P., and Kim, D. M. 2008b. Metal contacts in carbon nanotube field-effect transistors: Beyond the Schottky barrier paradigm. *Physical Review B (Condensed Matter and Materials Physics)*, 77(11):113403.

Parga, A. L. V. d., Calleja, F., Borca, B. et al. 2008. Periodically rippled graphene: Growth and spatially resolved electronic structure. *Physical Review Letters*, 100(5):056807.

Peierls, R. E. 1935. Quelques proprietes typiques des corpses solides. *Ann. I. H. Poincare*, 5:177–222.

Pereira, V. M., Guinea, F., Santos, J. M. B. L. d., Peres, N. M. R., and Neto, A. H. C. 2006. Disorder induced localized states in graphene. *Physical Review Letters*, 96(3):036801.

Saunders, V. R., Dovesi, R., Roetti, C. et al. 2003. www.crystal.unito.it.

Son, Y.-W., Cohen, M. L., and Louie, S. G. 2006. Energy gaps in graphene nanoribbons. *Physical Review Letters*, 97(21):216803.

Tsui, D. C., Stormer, H. L., and Gossard, A. C. 1982. Two-dimensional magnetotransport in the extreme quantum limit. *Physical Review Letters*, 48(22):1559–1562.

Wallace, P. R. (1947). The band theory of graphite. *Physical Review*, 71(9):622–634.

Wassmann, T., Seitsonen, A. P., Saitta, A. M., Lazzeri, M., and Mauri, F. 2008. Structure, stability, edge states, and aromaticity of graphene ribbons. *Physical Review Letters*, 101(9):096402.

Yazyev, O. V. and Helm, L. 2007. Defect-induced magnetism in graphene. *Physical Review B (Condensed Matter and Materials Physics)*, 75(12):125408.

21

Transport in Graphene Nanostructures

Christoph Stampfer
ETH Zürich

Johannes Güttinger
ETH Zürich

Françoise Molitor
ETH Zürich

Stephan Schnez
ETH Zürich

Eric Schurtenberger
ETH Zürich

Arnhild Jacobsen
ETH Zürich

Sarah Hellmüller
ETH Zürich

Thomas Ihn
ETH Zürich

Klaus Ensslin
ETH Zürich

21.1 Introduction

Graphene is a truly two-dimensional (2-D) crystal consisting of a honeycomb-like hexagonal lattice of carbon atoms, as illustrated in Figure 21.1a. The carbon atoms form strong covalent bonds by three in-plane sp^2 hybridized orbitals, whereas the fourth valence electron of the $2s^2 2p^2$ orbitals of carbon, assigned to the perpendicular $p_{\parallel}$ orbital can move freely in plane forming the so-called π-electron system [sai98]. The delocalized π-electronic states are responsible for electrical conductance and make graphene, in contrast to sp^3-hybridized insulating diamond, to a gapless semiconductor. In particular, the π-electronic system in combination with the hexagonal lattice, where two carbon atoms sit in the unit cell (see A and B in Figure 21.1a, where the dashed lines mark the unit cell), lead to a number of unique electronic properties of graphene [wal47, cas07, gei07]. The electronic band structure can be approximated by a nearest-neighbor tight-binding approach of linear combinations of the perpendicular $p_{\parallel}$ orbitals. According to

Wallace [wal47], this leads to $E\vec{k} = E_0 \pm \left| \sum_{i=1}^{3} \gamma_i \exp(-i\vec{k} \cdot \vec{r}_i) \right|$ (illustrated in Figure 21.1b), where $\gamma_i \approx 2.8\,\text{eV}$ [rei02] are the so-called nearest-neighbor hopping integrals, E_0 is the energy of the bare $p_{\parallel}$ orbital, and $\vec{r}_i$ are the vectors pointing to the three A atoms neighboring each B atom (see arrows in Figure 21.1a). The two freely moving valence electrons (per unit cell) completely fill the (valence) π-band and leave the (conduction) π^*-band unfilled resulting in a point-like Fermi surface (see Figure 21.1b). Therefore, in the near vicinity of the Fermi energy, the band structure of graphene [wal47, mac57] can be linearized, leading to two cone-like structures centered at the two inequivalent (so-called) K and K' points at the corners of the also hexagonal Brillouin zone (see Figure 21.1b and cross sections in Figure 21.2a). Consequently, the dynamics of charge carriers in graphene can be described in the near vicinity of the Fermi energy by a linear dispersion relation, $E = \hbar v_{\text{F}} |\vec{\kappa}|$ (and $\vec{\kappa} = \vec{k} - \vec{K}^{(')}$), where the carriers behave like massless particles with a constant Fermi velocity $v_{\text{F}} \approx 10^6$ m/s, about 300 times smaller than the speed of light [cas06, kat06, kat07]. Additionally, the presence of the two sublattices (A and B in Figure 21.1a), due to two carbon atoms per unit cell, allows to express the wavefunction for a unit cell by $\phi = c_A \phi_A + c_B \phi_B$, where $\phi_{A,B}$ are the $p_{\parallel}$ wave functions at the A, B site and the two component vector (c_A, c_B) forms the so-called pseudospin in graphene. Indeed, in close analogy to neutrino physics [and99], the dynamics of electrons in the near vicinity of the K points can be fully described by the Dirac–Weyl Hamiltonian, $H = v_F \sigma \cdot p$, where σ are the 2-D Pauli spin matrices and p is the momentum operator. Here, the graphene pseudospin takes the role of "real" spin and this analogy actually gives rise to the pseudospin terminology. The pseudospin up (or down) state is related to an A–B symmetric (or antisymmetric) wave function. Since the symmetry of these wave functions is

a function of the $\vec{k}$-vector direction, helicity ($h = \sigma \cdot p/2 \, |p|$) becomes a good quantum number $h_{1,2} = \pm 1/2$ [and98, and99, mce99]. Therefore, electrons (holes) around the K-point have a positive (negative) helicity. This implies that σ has its two eigenvalues either in the direction (+) or opposite (−) to the momentum p (as illustrated in Figure 21.2a). In other words, charge carriers in graphene can be described in terms of 2-D massless Dirac fermions [nov05, zha05, zho06, bos07]. Thus, the carriers can be considered as behaving analogously to relativistic particles but with a strongly reduced velocity allowing for the observation of quantum electrodynamic (QED)–specific phenomena in a solid-state environment [cas07].

This has also major impact on the transport properties of graphene. For example, the helicity as a good (quasi-particle) carrier

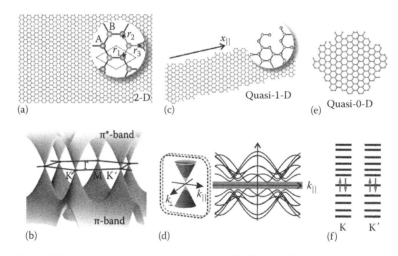

FIGURE 21.1 (a) Illustration of truly 2-D graphene consisting of a honeycomb-like hexagonal lattice with two sublattices A and B (inset). The unit cell is marked by dashed lines. (b) Dispersion relation of graphene. The π- and π*-band touch at the K and K' points. (c) Quasi-1-D nanoribbon with rough edges. At the edges, the A–B symmetry is broken (inset). (d) Schematic illustration of the quasi-1-D band structure of a nanoribbon (localized states not shown). The left panel illustrates the zone-folding approach. (e) Quasi-0-D graphene quantum dot made out of a nanometer-sized graphene flake with badly defined edges. (f) Expected energy levels of an ideal graphene quantum dot with depicted spin and valley (K and K') degeneracy.

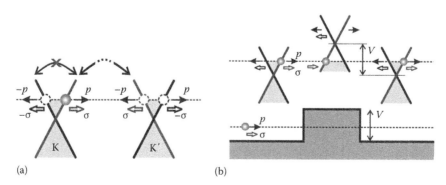

FIGURE 21.2 (a) Schematic illustration of the quasi-particle spectrum in graphene. Long-range disorder, intravalley backscattering is blocked by the pseudospin (denoted by σ) conservation. In contrast, for short-range disorder intervalley (K–K') scattering is possible. (b) The three upper diagrams show the position of the Fermi level and corresponding quasi-particle participating to transport across a potential barrier, which is shown below. The conservation of the pseudospin and the matching of electron states (outside the barrier) and hole states (within the barrier) give rise to Klein tunneling. (Adapted from Katsnelson, M.I. et al., *Nat. Phys.*, 2, 620, 2006. With permission.)

quantum number completely blocks direct backscattering by long-range disorder [and98] (as illustrated in Figure 21.2a). This is due to low-momentum (low-$\vec{k}$) scattering processes, which only involve a single dispersion cone. There, direct backscattering processes would also require the transition from a symmetric to an antisymmetric wavefunction, which is not included in long-range Coulomb scattering (see Figure 21.2a). This leads to unprecedented carrier mobilities at room temperature of up to 20.000 cm²/V s [gei07], making graphene a promising material for high mobility nanoelectronic applications. Pseudo-relativistic Klein tunneling [kle29, dom99, kat06a] and related phenomena make it hard to confine carriers by electrostatic potentials. The so-called Klein paradox [cal99, su93] that refers to the transmission of relativistic particles through high and wide potential barriers is an exotic and counterintuitive consequence of QED [kat06a]. The chiral nature of the quasiparticles in graphene leads to a matching between electron and hole (i.e., QED-positron) wavefunctions across the potential barrier resulting in a high-probability tunneling, as illustrated in Figure 21.2b. Therefore, it has been suggested to use graphene barriers rather for the realization of electronic lenses, e.g., a Veselago lens [ves68]. On the other hand, the roughness of the edge potentials will make it difficult to observe quantized conductance in 1-D graphene structures [lin08]. Furthermore, the cyclotron energy of Dirac fermions scales like $\sqrt{B}$ (where B is the magnetic field), in contrast to the linear behavior for particles with parabolic dispersions [cas07], lead to very different cyclotron energy scales in graphene compared to standard semiconductors. This makes it, for example, possible to observe the quantum Hall effect even at room temperature [nov07]. However, not only the energy scale and B-field dependence differ from the ordinary 2-D electron gas but there is also an anomalous "half-integer" quantum Hall effect with a level at zero energy (i.e., at the so-called Dirac point) that includes both hole and electron states [nov05, zha05]. This is related to the coupling of the pseudospin with orbital motion giving rise to a geometric phase of π [and98a] (the so-called Berry phase [ber84]) accumulated along cyclotron trajectories [nov05, zha05, mik99]. For more comprehensive reviews about properties of bulk graphene, please refer to [cas06, cas07, gei07, gei07a, gus07, kat07, gei08].

In addition to its unique electronic—purely charge based—properties, graphene and carbon materials in general have promising electron spin properties. It is believed that electrons in these materials have exceptionally long spin decoherence times due to weak spin–orbit interactions (light weight of carbon) [hue06, min06] and weak hyperfine coupling due to the low nuclear spin concentration, arising from the ~99% natural abundance of ^{12}C [tra07]. This promises spin decoherence times superior to the GaAs material system in which solid-state spin qubits are most advanced today [pet05, kop06], making graphene in particular interesting as the host material for quantum dots where individual electron spins could be used as spin-qubits [tra07]. Spin-qubits are considered as promising building blocks for future quantum computation or information processing in general. However, in contrast to state-of-the-art semiconductor (e.g., GaAs) quantum dots, the electron

confinement in gapless graphene is rather challenging and needs novel technological approaches for fabricating low-dimensional graphene nanostructures.

In this chapter, we review work on the fabrication and characterization of graphene nanostructures. We discuss quasi-1-D nanoribbons consisting of etched graphene strips with rough edges (see Figure 21.1c). These systems are expected to exhibit a washed out linear 1-D density of states and a band structure similar to carbon nanotubes [sai98] (Figure 21.1d), where an effective transport gap is observed near the charge neutrality point. Furthermore, we focus on quasi-0-D graphene quantum dots (Figure 21.1e) with discrete energy levels, as illustrated in Figure 21.1f. In this context, we discuss the tunability of graphene single-electron transistors, transport via quantum confined excited states, and charge detection by using a nearby nanoribbon. The chapter is organized as follows: We start with Section 21.2 by reviewing the transport properties of 2-D graphene. In Section 21.3, the fabrication of graphene nanodevices is described in detail. Section 21.4 focuses on graphene nanoribbons and Section 21.5 on graphene quantum dots.

21.2 Transport Properties of Bulk Graphene

Electronic transport experiments allow probing electronic properties in the vicinity of the Fermi level. At sufficiently low temperatures, this provides direct insights into the electronic density of states, carrier densities, and mobilities.

In the context of diffusive transport, the zero-temperature conductivity can be expressed according to the Drude model by $\sigma(E_F) = n(E_F)e\,\mu(E_F)$, where n is the carrier density (i.e., the integrated density of states) and μ is the mobility [dat95]. By varying the Fermi energy, for example by applying a global electrostatic potential, we can study the conductivity as a function of E_F obtaining insights into mobility and band structure properties [nov04].

An established method to experimentally characterize the electronic transport properties of quasi-2-D materials is based on the so-called Hall bar geometries, as shown in the inset of Figure 21.3a. By injecting a current from the left most to the right most contact, the longitudinal resistance can be extracted in a four-point measurement geometry by recording the potential drop between two side contacts. Additionally, one can measure the transverse Hall resistance in weak (and strong) magnetic field, which allows to deduce the carrier density n and the mobility μ, independently. A global back gate (separated by $d \approx 300$ nm thick oxide) is used to tune the Fermi level. The influence of an applied back gate voltage V_{bg} on the carrier density in bulk graphene can be described by a simple parallel-plate capacitor model. The induced charge carrier density as function of applied back gate voltage is then given by $n(V_{bg}) = \alpha V_{bg}$, where $\alpha = \varepsilon_0\varepsilon_{ox}/(de) = 7.4 \times 10^{10}$ cm^{-2} V^{-1}. Using this model, the mobility can be extracted from the

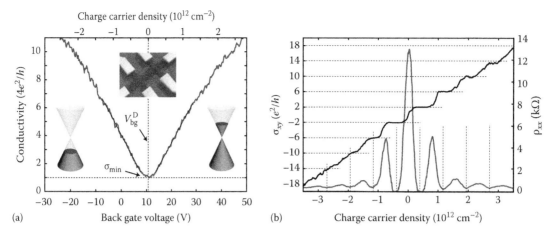

FIGURE 21.3 (a) Conductivity as a function of back gate voltage V_{bg} and carrier density n measured on a graphene hall bar, as illustrated by the central inset. Transport can be tuned from the hole (left) to the electron (right) regime by crossing the so-called Dirac point, V_{bg}^D, with a minimum conductivity of approx. $4e^2/h$. The black lines indicate the fits used for the extraction of the mobility for both regimes. (Source drain current $I_{sd} = 20$ nA rms.) The inset shows a schematic of the measured device. (b) Hall conductivity ρ_{xy} (black curve) and longitudinal resistivity ρ_{xx} (gray curve) plotted as a function of charge carrier density at $B = 8$ T. The typical half-integer quantum Hall plateaus for graphene are observed. The vertical dashed lines indicate the filling factors of the Landau levels obtained theoretically, in good agreement with the experiment ($I_{sd} = 100$ nA rms). (Adapted from Molitor, F. et al., *Phys. Rev. B*, 76, 245426, 2007. With permission.)

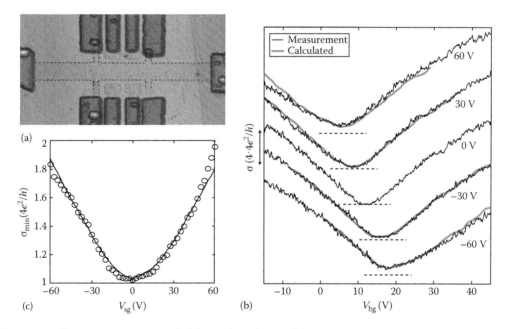

FIGURE 21.4 (a) Scanning force microscope image of a fabricated graphene Hall bar with two graphene side gates on each side of the channel. The channel is approx. $0.72\,\mu$m wide and $2\,\mu$m long. (b) Conductivity as function of back gate voltage, plotted for different side gate voltages. The black lines are measurements and the thicker, gray lines are calculated by an electrostatic model, using the zero side gate voltage measurement. Conductivity values of $4e^2/h$ are indicated by dashed lines, spaced by an offset of $4 \times 4e^2/h$ for clarity. (c) Minimum conductivity as function of side gate voltage. The circles correspond to the minima of the smoothed measured traces, whereas the line is calculated with a model from the zero side gate measurement. (Adapted from Molitor, F. et al., *Phys. Rev. B*, 76, 245426, 2007. With permission.)

four-point longitudinal graphene Hall bar conductivity measurement shown in Figure 21.3a (the corresponding sample is shown in Figure 21.4a). The conductivity minimum at roughly $V_{bg}^D = 10$ V reflects the Dirac (or charge neutrality) point, where the carrier density is lowest. For ideal 2-D graphene, theory actually predicts a vanishing carrier density. The back gate voltage offset from 0 V is due to unintentional p-doping of the graphene sample during fabrication (see [sta07a, hwa07]). By applying a more positive voltage $V_{bg} > V_{bg}^D$ the electron density increases and the conductivity increases linearly in V_{bg}. Similarly, we find for $V_{bg} < V_{bg}^D$ the regime of hole transport, where the conductivity also increases linearly as a function of $|V_{bg}|$.

21.2.1 Mobility

From the linear slopes of the conductivity traces, we can extract the carrier mobilities for both the electron and hole regimes. The presented measurements belong to a characteristic sample with a mobility close to 5.000 cm^2/Vs for both carrier types. For a charge carrier density of $|n| = 10^{12}$ cm^{-2}, this corresponds to a mean free path $\Lambda_m = \hbar\mu\sqrt{\pi n}/|e| \sim$ 60nm, which is consistent with the assumption that transport in the micrometer-sized Hall bar (see Figure 21.3a) can be considered as diffusive [nov04, mol07]. The measured mobility is temperature independent and might be limited by (1) warping and ripples (or frozen phonons) [mey07], (2) strong doping fluctuations [sta07a, mos07], or (3) electron–hole puddles [mar08]. So far mobilities up to 20.000 cm^2/V s have been reported for graphene samples resting on silicon oxide surfaces [gei07]. More recently, measurements on freely suspended graphene Hall bars have exhibited mobilities as high as 200.000 cm^2/V s (at 2 K) [bol08, bol08a]. The mobility in suspended graphene exhibits a strong temperature dependence indicating that phonons (i.e., frozen phonons in non-suspended graphene) may play an important role.

21.2.2 Minimum Conductivity

Another important observation is that graphene's minimum conductivity σ_{min} does not disappear in the limit of vanishing $\langle n \rangle$ but rather tends to a minimum value of the order of the quantum unit e^2/h [two06, tan07]. This quantum-limited conductivity is an intrinsic property of 2-D Dirac fermions, which persists in ideal graphene crystals without any impurities or lattice defects [fra86, lud04, per06, kat06, ost06, zie98, two06]. The experimentally observed conductivity minima σ_{min} depend on sample geometry, disorder (i.e., doping fluctuations and electron-hole puddles), and overall doping of the graphene flake [gei07, tan07]. Several experiments [gei07, tan07, mia07, dan08] on samples with similar geometries show such a minimum conductance value. The presence of the conductivity minimum σ_{min} has major impact on the electrostatic tunability of graphene devices [nov04], making it difficult to fully pinch off currents in 2-D graphene devices.

21.2.3 "Half-Integer" Quantum Hall Effect

The most spectacular transport phenomenon reported so far is the well-understood anomalous "half-integer" quantum Hall effect (QHE) reflecting the consequences of graphene's QED-like energy spectrum. Figure 21.3b shows the QHE of single-layer graphene [nov05, zha05, jia07, jia07a, mol07], which is the relativistic counterpart of the well-known integer QHE observed in semiconductors with a band gap and a parabolic dispersion [kli80]. It manifests itself as a continuous ladder of equidistant plateaus in Hall conductivity σ_{xy}, which persists through the neutrality (Dirac) point. There a charge carrier crossover from electrons to holes takes place. The sequence of plateaus is shifted with respect to the standard QHE ladder by 1/2. The plateaus arise at $\sigma_{xy} = \pm 4e^2/h(N + 1/2)$, where N is the Landau

level index and the factor 4 appears due to twofold valley (K and K′) and twofold spin degeneracy. The unusual sequence is well understood as arising from the QED-like quantization of graphene's electronic energy spectrum in magnetic field B, given by $E_N = \pm v_F\sqrt{2e\hbar B N}$, where the sign $\pm$ refers to electrons and holes, respectively [gor02, zhe02, gus05, per06]. The existence of the quantized level at zero energy $N = 0(V_{bg} = V_{bg}^D)$, which is shared by electrons and holes, is essentially everything one needs to explain the anomalous QHE [nov05, zha05]. This measurement also provides a clear fingerprint for distinguishing between single- and multilayer graphene [gei07].

21.2.4 Graphene Gates

We have seen that electrostatic tunability (e.g., by a global back gate) is crucial to investigate the electronic transport properties as a function of the Fermi energy. Far more insights and tunability can be gained when having access to local gates, which allow to shift the Fermi level locally. For example, p–n junctions [wil07, oez07a], tunable tunnel barriers [sta08], or strong carrier density variations along the graphene sample [mol07] can thus be obtained. Local gate structures have (so far) been either made (1) as local top gates electrodes [wil07, hua07, oez07, liu08] or (2) as lateral in-plane structures (see e.g., Figure 21.4a) [mol07]. Top gates have the advantage of potentially huge lever arms, when making use of thin top gate oxides. However, they have the disadvantage of requiring top gate oxides sitting on the graphene sheet resulting in most cases in strongly reduced carrier mobilities. The rather weak lever arm of in-plane gates can be significantly improved when making the gate electrodes out of the very same graphene sheet used to make the device (e.g., the Hall bar shown in Figure 21.4a). This allows to bring lateral graphene gates very close to the conducting channels since alignment issues are not relevant. So far it has been shown that lateral graphene gates can be placed as close as 20 nm [gue09] to the active device. This provides good local tunability for a large number of graphene nanodevices, such as side-gated nanoribbons [mol09] and Hall bars [mol07], tunnel barriers [sta08, sta08a], quantum dots [pon08, gue08, sch09], and tunable Aharonov–Bohm rings [hue09].

In Figure 21.4a, we show a scanning force microscope (SFM) image of a graphene Hall bar with four lateral graphene gates placed ≈100 nm separated from the Hall bar [mol07]. For simplicity, we discuss here only measurements where the same potential has been applied to all four gates. Figure 21.4b shows conductivity traces as a function of V_{bg}, similar to Figure 21.3a, but for different side-gate voltages (see labels). These measurements reveal clearly (1) the overall relative lever arm between the side gates and the back gate $\alpha_{bg/sg} \approx 0.2$ (see shift of the Dirac point as function of the side-gate voltage), (2) an increase of the minimum conductivity σ_{min} as function of $|V_{sg}|$ (see also Figure 21.4c), and (3) an appearing asymmetry between electron and hole mobilities for large side-gate voltage (compare the traces for ±60 V side gate voltage). Apart from the mobility asymmetry (which might be related to a change in the effective disorder potential [sta09a]), the influence of the graphene side gates can be well understood by a resistor model, where the in-plane gates mainly affect the outermost 70 nm of the graphene flake [mol07].

21.3 Fabrication of Graphene Devices

The state-of-the-art fabrication of graphene devices and in particular of graphene nanodevices has been mostly developed in Manchester by Geim and Novoselov [nov04, nov05] and at Columbia University by Kim and colleagues [zha05]. In Figure 21.5, we summarized the different process steps. The substrate material consists of highly doped silicon (Si^{++}) bulk material covered with around 300 nm of silicon oxide (SiO$_2$), where thickness and roughness of the SiO$_2$ top layer is crucial for the identification and further processing of single-layer graphene samples (a). Before depositing graphene, standard photolithography followed by metallization (usually chrome/gold) and lift-off is used to pattern arrays of reference alignment markers on the substrate in order to reidentify locations (for addressing individual graphene flakes) on the chip (b). For a general overview on micro-fabrication, please refer to [fra04].

21.3.1 Mechanical Exfoliation of Graphite

The deposition of graphene, including also few-layer graphene flakes is based on mechanical exfoliation of (natural) graphite by adhesive tapes [nov04]. In 2005, two groups [nov05, zha05] proved that this simple technique indeed leads to high-quality single-layer graphene flakes resting on the SiO$_2$ surface. Other techniques based either on local exfoliation by scanning tunneling and atomic force microscope tips [zha05a] or direct chemical vapor-based growth of graphene [nan10] are still under development. In Figure 21.6, we show two pictures of this, in principle, rather dirty process. Natural graphite flakes are distributed on a sticky tape followed by folding this tape several times, as shown in Figure 21.6a. By folding the sticky tape covered with graphite, the stacked graphene sheets are pulled apart leading to thinner and thicker few-layer graphene and ultimately to single-layer graphene sheets. In a next step, the Si/SiO$_2$ dies get pressed hard onto the tape (see Figure 21.6b)

in order to transfer the carbon material from the adhesive tape to the SiO$_2$ surface, as illustrated in Figure 21.5d.

21.3.2 Single-Layer Graphene Identification

After exfoliation and deposition, there are graphite flakes with all kinds of layer thicknesses on the sample. The main difficulty is to find the thinnest among them (i.e., the single-layer graphene). Surprisingly, an optical microscope in combination with a carefully selected oxide thickness is sufficient for finding (i.e., detecting) individual graphene flakes (see e.g., Figure 21.7b). This technique is widely used for the identification of graphene flakes because optical microscopy in contrast to AFM and Raman imaging techniques (e.g., [sta07]) provides sufficient throughput for scanning large samples. However, an optical image cannot unambiguously distinguish between single- and bilayer graphene. Therefore, Raman spectroscopy [fer06, gup06, gra07] and SFM techniques are indispensable for fully characterizing graphene flakes.

21.3.2.1 Visibility of Graphene

The optical visibility of graphene on various substrates (including SiO$_2$) can be well understood in terms of a Fresnel-law-based model [abe07, bla07, rod07, cas07a]. Thin graphite flakes are sufficiently transparent to add to an optical path, which changes their color by interference with respect to an empty wafer [nov04, bla07]. For a certain thickness of SiO$_2$, even a single layer was found to give sufficient, albeit faint, contrast to allow the huge image processing power of the human brain to spot in seconds, a few micron-sized graphene crystallites among copious thicker flakes scattered over a millimeter-sized area. The origin of the detectable optical contrast is not only due to an increased optical path but also due to the notable opacity of graphene. The intensity of the reflected light can be described by the Fresnel law, assuming a two- or three-layer interface, as illustrated by the inset in Figure 21.7a.

The dependence of the contrast on the SiO$_2$ thickness and the wavelength of the incident light is shown in Figure 21.7a.

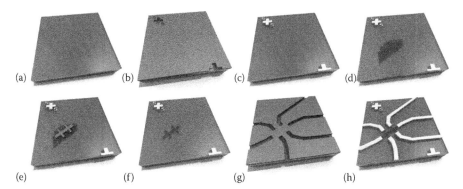

(a)   (b)   (c)   (d)

(e)   (f)   (g)   (h)

FIGURE 21.5 Fabrication process: (a) Starting point is a silicon substrate with a 290 nm thick silicon dioxide on top. (b) With optical lithography orientation numbers for the flakes and markers used for the alignment during the subsequent electron beam lithography (EBL) steps, are written into the photo resist. (c) The structure is then made with evaporation of chromium and gold, followed by a lift-off process. (d) Deposition of graphene on the substrate and identification of single-layer flakes using an optical microscope and Raman spectroscopy. (e and f) Patterning of the flakes by EBL and etching by reactive ion etch (RIE). (g) An additional EBL step is performed for defining the contacts, which are then made by evaporation of chromium and gold, followed by a lift-off process. (h) Final device, ready for bonding.

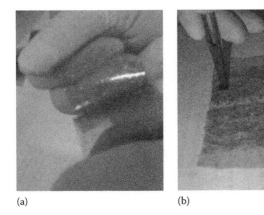

| (a) | (b) |

FIGURE 21.6 (a) Mechanical exfoliation of graphene. The adhesive tape with graphite on top is folded and unfolded several times in order to mechanically cleave multilayer graphite into graphene. (b) Deposition of graphene onto a silicon die. The die is pressed onto the scotch tape and as a result graphene and few-layer graphite flakes may stick randomly to the silicon oxide surface.

For green light (550 nm) at which human eyes are most sensitive, the contrast is maximized by using an oxide thickness of approx. 90 or 290 nm [bla07]. Figure 21.7b shows an example of an optical image of a single-layer graphene flake (proven to be single-layer [sta07a]) on a 297 nm thick SiO_2 layer demonstrating the strong visibility of a monolayer graphene. Since there is a strong dependence on the oxide thickness (see Figure 21.7a), it is necessary to have an accurate and homogeneous thickness over the wafer for providing a homogenous visibility of graphene. In summary, the strong dependence on the substrate material and the light source (including filtering) make it very hard to clearly distinguish between single and few layer graphene. Thus, optical microscopy is good for finding the location of a potential single-layer graphene flake. However, full characterization and clear fingerprint for single-layer graphene has to be provided by an additional technique.

21.3.2.2 Raman Imaging

Raman spectroscopy has been identified to be a very powerful tool to unambiguously distinguish between single- and few-layer graphene [fer06, gup06, gra07]. The 2D nature of graphene has pronounced effects not only on their electronic band structure but also on their vibrational properties. Spectroscopic work on carbon materials has become extremely rewarding as it provides fundamental insight as well as the information used to characterize graphene [pis07, sta07a]. Since the resonant Raman-scattering signal of carbon materials (for a review see [rei03, dre02, fer07b]) is so large that even individual micron-sized graphitic flakes can be investigated, it has become an in situ nondestructive tool for its characterization. Without going into too many details here, the Raman spectra provide a clear fingerprint of single-layer graphene including three main peaks (see e.g., spectra in Figure 21.8a).

One distinguishes three peak families: (1) The D mode peak at 1350 cm⁻¹, which does not originate from a Γ-point Raman-active vibration, is induced by disorder including edges [gra07]. Therefore, the absence of this peak provides a good quality check of the deposited graphene material. The spectra shown in Figure 21.8a where the D-line is clearly present have been taken at the edge of the graphene flake. (2) The high-energy mode (HEM) vibrations in sp^2 bound carbon materials correspond to the Γ-point optical phonon mode at 1582 cm⁻¹. The carbon atoms move tangentially to the graphene sheet and this so-called G-peak is prominent in graphite, nanotubes, and graphene. In the latter, it can be fitted nicely by a single Lorentzian peak. From the actual G-peak position, we gain direct insight into the doping level of the graphene flake [pis07, sta07a, cas07b, fer07a, das08, ber09]. (3) The double-resonant 2D peak [tho00], which is the overtone of the D mode, shifts at about twice the rate of the D mode (i.e., 2679 cm⁻¹). Since no defects are involved in this two-phonon scattering process, the 2D mode is always observed independently of the defect

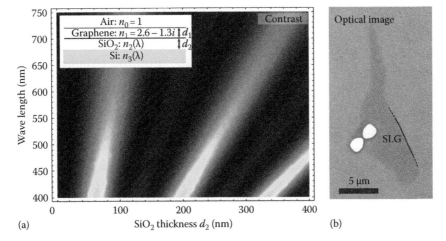

| (a) | (b) |

FIGURE 21.7 (a) Optical contrast of graphene, in dependence of the wavelength of the incident light and the oxide thickness. Bright color reflects high contrast. (b) Optical light microscope image of single-layer graphene (SLG) flake on 297 nm silicon oxide. The image has been taken with a green filter. (Adapted from Stampfer, C. et al., *Appl. Phys. Lett.*, 91, 241907, 2007.)

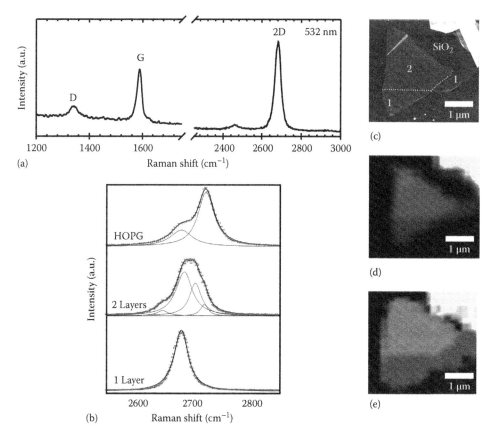

FIGURE 21.8 (a) Raman spectra of single-layer graphene. (b) 2D peaks for single- and double-layer graphene with HOPG as a bulk reference. The dashed lines show the Lorentzian peaks used to fit the data, the solid lines are the fitted results. The single peak position for the single-layer graphene is at 2678.8 ± 1.0 cm^{-1}. The peak position of the two inner most peaks for double-layer graphene are 2683.0 ± 1.5 cm^{-1} and 2701.8 ± 1.0 cm^{-1}. (c) SFM micrograph of a graphitic flake consisting of one double- and two single-layer sections (white dashed line along the boundaries), highlighted in the Raman map (d) showing the integrated intensity of the G line. (e) Raman mapping of the FWHM of the 2D line clearly showing the difference between single- and double-layer regions.

concentration (or edges) [mau02]. Moreover, the line shapes of the 2D peak also carry valuable information about the underlying electronic band structure. This is due to the large-$\vec{k}$ phonons involved in this process, which resonantly connect the two cones at the inequivalent K-points (see Figure 21.1b). The number of possible phonon transitions is now directly linked with the number of electronic bands available. Thus, for single-layer graphene, only one band per valley is present and therefore only one D-phonon transition is possible. This leads to a perfect Lorentzian peak shape, as shown in the lower panel of Figure 21.8b. In contrast, due to the four carbon atoms per unit cell in bilayer graphene, we find two electronic bands per valley resulting in a total of four possible transitions and the 2D line shape indeed is composed of four individual Lorentzian peaks (see center panel in Figure 21.8b). Thus, the 2D line shape provides a strong fingerprint of graphene and is nowadays widely used for identifying graphene. For comparison, the upper panel of Figure 21.8b shows the 2D peak of graphite. In Figure 21.8d and e, we show confocal Raman images of a graphitic flake with regions of single- and bilayer. A corresponding SFM image is shown in Figure 21.8c. The laser-spot size is approx. 400 nm in diameter. The single and bilayer regions can be nicely resolved in both (1) the G-peak intensity map (Figure 21.8d)

where mainly the different amount of carbon material (per spot size) enters and (2) the line width of the 2D-peak (Figure 21.8e) which unambiguously pins down the single-layer region. Indeed, the line width of 33 cm^{-1} for single-layer graphene has also been shown to be doping independent [sta07a] making this quantity a very reliable measure for identifying single-layer graphene.

21.3.3 Etching Graphene Nanostructures

In order to fabricate well-designed graphene (nano) devices, a reliable technique to structure graphene is required. The commonly used technique is based on resist spin coating, electron beam lithography (EBL), development and subsequent etching of the unprotected graphene. This process sequence is illustrated in Figure 21.5d through f. The resist (mainly polymethyl methacrylate (PMMA)) thickness varies between 50 and 100 nm directly limiting on one hand the minimum feature size of the final graphene device and on the other hand the process window for the actual etching step. It has been found that short (5–15 s) mainly physical reactive ion etching (RIE) based on argon and oxygen (80/20) provides good results without influencing the overall quality of the flake [mol07]. With reactive ion etching, the chemical bonds of the etch target are broken up by physical bombardment with argon ions which were created in a plasma. On the bombarded sites, chemical etching (in our case with oxygen) can take place enhanced by the heightened chemical reactiveness of the etchant species due to the plasma. Hence, this process combines the anisotropic etching possibilities of the physical bombardment, with the material selectivity of the chemical etching. The ability of this process to etch up to around five-layer thick flakes facilitates contacting of structures by reducing the risk for shorted contacts due to thicker graphitic regions. In Figure 21.9,

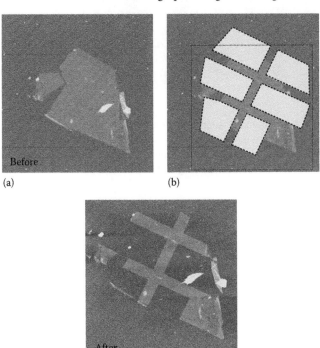

(a)

(b)

(c)

FIGURE 21.9 (a) Scanning force microscope image of a few-layer graphene flake before etching. (b) The very same image with the etch mask on top and (c) shows a scanning force microscope image of the reactive ion etched graphene flake.

SFM images of a flake before (a) and after the RIE etching (c) is shown. In Figure 21.9b, the used etch mask is highlighted.

21.3.4 Contacting Graphene Nanodevices

After etching and removing the residual EBL resist, the SFM images (as shown in Figures 21.9c and 21.10a) are important for proving the quality of the patterned graphene flakes, mainly in terms of contamination. Here the step height (see Figure 21.10b) and the surface roughness of the flake, which should be lower than 0.2 nm rms, are good quality measures for flakes before contacting. Selected graphene (nano) structures are next contacted by an additional EBL step, followed by metallization and lift-off, as illustrated in Figure 21.5f through h. Two layers of EBL resist are used to provide a T-shape resist profile allowing to pattern metal structures down to 70 nm. After development, 2–5 nm chrome (Cr) or titanium (Ti) and 30–50 nm gold (Au) are evaporated for contacting the preselected graphene nanostructures. In Figure 21.10c, we show an optical microscope image of the metallic wiring of graphene nanostructures after successful lift-off process. The SFM image (shown in Figure 21.10d) proves that the alignment between graphene nanostructure and metal electrodes is sufficient for studying transport through these devices. (see also Figure 21.16a). The sample is now ready for packaging and wire bonding, generally both are crucial and important steps for transport studies on nanodevices.

21.4 Graphene Nanoribbons

Graphene nanoribbons (see Figures 21.1c and 21.11), in contrast to 2-D graphene, exhibit an effective energy gap (Figure 21.1d), overcoming the gapless band structure of graphene (Figure 21.1b). Nanoribbons (including constrictions) show an overall semiconducting behavior, which makes these quasi-1-D graphene nanostructures promising candidates for the fabrication of nanoscale graphene transistors [wan08, zha08, mer08], tunnel barriers, and quantum dots (see Section 21.5) [sta08, pon08, sta08a, sch09].

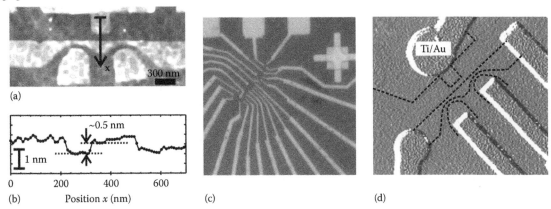

(a)

(b)   Position *x* (nm)

(c)

(d)

FIGURE 21.10 (a) Scanning force microscope image of a graphene nanostructure after RIE etching, where (b) shows a SFM cross section along a path *x* [marked in (a)] averaged over approx. 40 nm perpendicular to the path proving the selective etch process. (c) Light microscope image of the metal electrodes contacting the graphene structure. (d) SFM close-up of (c), see box therein. The minimum feature size is approx. 50 nm. The dashed lines indicate the outline of the graphene areas. (Adapted from Stampfer, C. et al., *Appl. Phys. Lett.*, 92, 012102, 2008.)

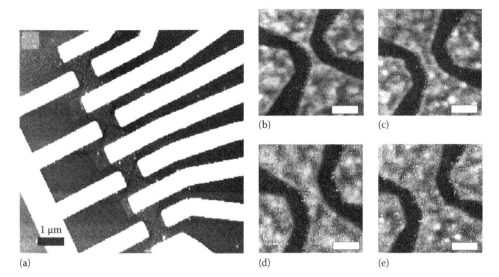

FIGURE 21.11 (a) Scanning force microscope image of constrictions with different widths after metal contacts have been made. The Cr/Au contacts can be seen in white and the graphene flake with the constrictions can be seen underneath. (b–e) Close up SFM images of each single constriction. The gray areas consist of graphene and the black areas are the silicon oxide substrate. The white spots that can be seen on the graphene are most probably dirt assembled during the fabrication process. The white scale bars correspond to a length of 100 nm. (Adapted from Molitor, F. et al., *Phys. Rev. B*, 79, 075426, 2009.)

On the other hand, ideal graphene nanoribbons [bre06, whi07, wak07, wak09] promise interesting quasi-1-D physics with strong relations to single-walled carbon nanotubes [sai98, rei03]. Transport through graphene nanoribbons has been studied by a number of groups [che07, han07, tod09, sta09, mol09, liu09] and an interpretation of the experimentally observed transport and energy gaps in fabricated nanoribbons is discussed below.

In Figure 21.11, we show SFM images of fabricated graphene nanoribbons [mol09] with double graphene side gates. These structures were fabricated by using the technological methods and processes described in Section 21.3. For example, the devices shown in Figure 21.11b through e consist of nanoribbons with length $l = 100$ nm and widths w varying from 30 to 100 nm. The graphene side gates are separated by a gap of 50 nm from the nanoribbon.

In Figure 21.12a, we show a typical low source–drain bias ($V_{bias} = 300 \mu V \ll 4K_B T$) back-gate characteristic of a graphene nanoribbon ($w = 45$ nm; sample not shown) [sta09]. This measurement has been recorded at an electron temperature of ≈2 K. By comparing this measurement with the back-gate characteristics of 2-D graphene (Figure 21.3a), we find manifestations of the (quasi-1-D) nanoribbon size effects in transport properties. Similar to the measurement shown in Figure 21.3a, we can tune transport from the hole (left side; see inset) to the electron regime (right side). In contrast to bulk graphene, where we find a linear dependence of the conductance (i.e., the current) on the back-gate voltage (i.e., carrier density), here we find a sub-linear slope, an overall strongly reduced conductance, and reproducible fluctuations, which may be due to local resonances or attributed to conductance fluctuations [hei03]. In contrast to the conductance

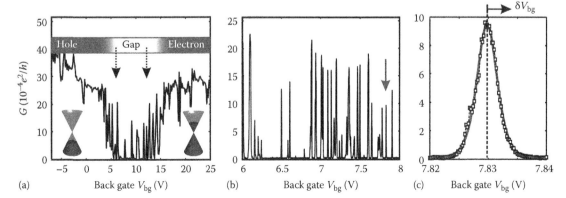

FIGURE 21.12 Transport gap in a graphene nanoribbon. (a) Low bias ($V_{bias} = 300 \mu V$) back gate characteristics (of a nanoribbon with $w = 45$ nm) showing that the regimes of hole and electron transport are separated by the so-called transport gap, delimited by the vertical arrows. (b) High resolution close-up inside the gap displaying a large number of sharp resonances within the gap region. (c) Close-up of a single resonance [see arrow in panel (b)]. (Adapted from Stampfer, C. et al., *Phys. Rev. Lett.*, 102, 056403, 2009.)

minimum ($\sigma_{min} = 4e^2/h$) at the Dirac point observed in 2-D, we find a region of strongly suppressed current (see region 6 V < V_{bg} < 12 V in Figure 21.12a) around the charge neutrality point. This region of suppressed current is the so-called transport gap ΔV_{bg} in back-gate voltage ($\Delta V_{bg} \approx 6$ V). In the following section (Section 21.2.1), we summarize the experimental observation of two different energy scales for the transport gap in both, the back gate and source–drain bias direction. In Section 21.2.2 we describe a (simple) model based on Coulomb blockade in disordered systems to explain the observed transport through fabricated graphene nanoribbons.

21.4.1 Energy Gaps in Nanoribbons

In close analogy to carbon nanotubes, which can be considered as graphene stripes wrapped up to seamless cylinders [rei03], a number of theoretical predictions have been put forward to describe a (clean confinement) energy gap in graphene nanoribbons. Along these lines, zone folding approximations [whi07], π-orbital tight-binding models [per06, dun07], and first-principles calculations [fer07, yan07] predict an energy gap E_g scaling as $E_g = \alpha/w$ with the nanoribbon width w, where α is in the range of 0.2 eV × nm to 1.5 eV × nm, depending on the model and the crystallographic orientation of the graphene nanoribbon [lin08]. However, these theoretical estimates can neither explain the experimentally observed energy gaps of etched nanoribbons (with rough edges) of widths beyond 20 nm, which turn out to be significantly larger than predicted, nor do they explain the large number of resonances experimentally found inside the transport gap (see e.g., Figure 21.12) [che07, han07, tod09, mol09, sta09, liu09]. This has led to the suggestion that localized states (and interactions effects)

due to edge roughness, disorder (e.g., bulk disorder), or bond contractions at the edges [yan07] may dominate the transport gap. Several mechanisms have been proposed to describe the observed gap, including re-normalized lateral confinement [muc09], quasi-1D Anderson localization [muc09], percolation models [ada08], and many-body effects (including quantum dots) [sol07], where substantial edge roughness is required. It has also been shown that moderate amounts of edge roughness can substantially suppress the linear conductance near the charge neutrality point [eva08], giving rise to localized states relevant for both single-particle and many-body descriptions.

In contrast to the prediction of clean energy gaps (predicted for samples without bulk disorder and edge roughness), where transport should be completely pinched-off, transport experiments show a large number of reproducible conductance resonances inside the gap reminiscent of conductance resonances in the Coulomb blockade regime of quantum dots (Figure 21.12a and b). Sequences of resonances with small linewidths indicate strong localization. An example of a particularly narrow resonance is shown in Figure 21.12c (see arrow in Figure 21.12b). The line shape can be well fitted by $I \propto \cosh^{-2}(e\alpha_{bg}\delta V_{bg}/2.5\,k_B T_e)$, where $\alpha_{bg} \approx 0.2$ is the back gate lever arm and $\delta V_{bg} = V_{bg} - V_{peak,bg}$ (see Figure 21.11c) [bee91]. The estimated effective electron temperature, $T_e = 2.1$ K, is close to the base temperature (of 2.0 K) of the cryostat, leading to the conclusion that the peak broadening is mainly limited by temperature rather than by the lifetime of the resonant state.

In Figure 21.13a we show a source–drain current measurement of the nanoribbon as a function of source–drain bias V_{bias} and back-gate voltage V_{bg} (i.e., Fermi energy). Here, regions of strongly suppressed current (white areas) are observed leading

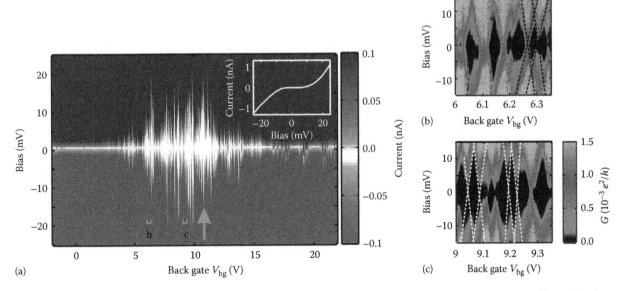

FIGURE 21.13 (a) Source-drain current measurements as function of bias and back gate voltage on a 45 nm wide nanoribbon. The white areas are regions of strongly suppressed current forming the energy gap. The inset shows a typical nonlinear *I–V* characteristic (V_{bg} = 10.63 V, see arrow). (b, c) Differential conductance (*G*) measurements as close-ups of panel (a) at two different back gate regimes (see labels in (a)). These measurements show diamonds with suppressed conductance (highlighted by dashed lines) allowing to extract the charging energy from individual diamonds. (Adapted from Stampfer, C. et al., *Phys. Rev. Lett.*, 102, 056403, 2009.)

to an effective energy gap in bias direction inside the transport gap in back-gate voltage (shown in Figure 21.12a). Highly non-linear I–V characteristics (see e.g., inset in Figure 21.13a) are characteristic for the energy gap in bias direction. This energy gap agrees reasonably well with the observations in Refs. [han07, sol07] of an energy gap of $E_g = 8$ meV for $w = 45$ nm (see also below). The transport gap in source–drain bias voltage corresponding to the energy gap E_g, and the transport gap ΔV_{bg} in back gate voltage are two distinct voltage scales resulting from the experiment. The quantity ΔV_{bg} is measured at constant (nearly zero) V_{bias} (transport window) by varying Fermi energy E_F. Varying the magnitude of the transport window V_{bias} at fixed Fermi energy gives rise to E_g. However, an explicit quantification of both energy scales is rather challenging.

21.4.1.1 Transport Gap in Back Gate

One way of quantifying the size of the transport gap in back gate voltage ΔV_{bg} is shown in Figure 21.14a (and Figure 21.14b). Here, the conductance trace is smoothened over a back gate voltage range large enough to eliminate the (reproducible) resonances without affecting the general shape (compare gray and black trace). The regions of a linear increase of the conductance at both sides of the transport gap are selected by hand, and a linear fit is performed (black bold lines). The gap size in back gate voltage ΔV_{bg} is then defined as the distance between the intersection points of the fitted traces with the $G = 0$ line ($\Delta V_{bg} = 3.4$ V). This is a reasonable approach since the conductance values are much smaller than the minimal conductivity observed for extended graphene systems, which is of the order of $4e^2/h$ (see e.g., Figure 21.3a). Also, different approaches to define the gap have been applied, for example by defining a cut-off current. However, the overall results are the same, even if the details are changed slightly [mol09].

The corresponding energy scale ΔE_F related to ΔV_{bg} can be estimated from $\Delta E_F \approx \hbar v_F \sqrt{2\pi C_g \Delta V_{bg}/|e|}$, (where C_g is the back gate capacitance per area) [sta09]. For the measurements shown in Figures 21.12 and 21.13, we find for example an energy gap

$\Delta E_F \approx 110$–340 meV which is more than one order of magnitude larger than $E_g = 8$ meV (see above and Figure 21.14a). This discrepancy is attributed to different physical situations described by these two energy scales, which will be further discussed below (see Section 21.4.2).

21.4.1.2 Energy Gap in Bias Direction

More insight into the energy gap in bias direction E_g (Figures 21.13a and 21.14b) can be gained by focusing on a smaller back gate voltage range, as shown in Figures 21.13b, c and 21.14c, which are high-resolution differential conductance dI/dV_{bias} close-ups of Figure 21.13a (see labels therein) and Figure 21.14b, respectively. At this scale, transport is dominated by well distinguishable (or intersecting) diamonds of suppressed conductance (see marked areas and dashed lines in the corresponding figures) which indicate that transport is blocked by localized electronic states or quantum dots (see also Ref. [tod09]). The related charging energy E_c which is itself related to the quantum dot size, depends on the Fermi energy in a small back gate voltage range (see e.g., different diamond sizes in Figure 21.13b and c) but also on a large scale (see Figure 21.13a). It has been confirmed that the outline of Figure 21.13a can indeed be obtained by plotting individually extracted charging energies as function of V_{bg} over a large back gate voltage range [sta09].

In the following, the energy gap in bias direction E_g will be defined as the maximum charging energy $E_{c,max}$, which is directly related to the smallest charged island along the nanoribbon ever forming when the density is swept (more details below).

21.4.1.3 Energy Scales as Function of Geometry

In Figure 21.15a we summarize the transport gaps in back gate voltage ΔV_{bg} (determined as shown in Figure 21.14a) as function of w for different graphene samples consisting of different nanoribbons with width w and length l. The error bars in horizontal direction result from the resolution of the SFM scans, while the vertical error bars are determined by applying the procedures to different measurements of the same constriction, and using different ranges for smoothening and fitting. We see

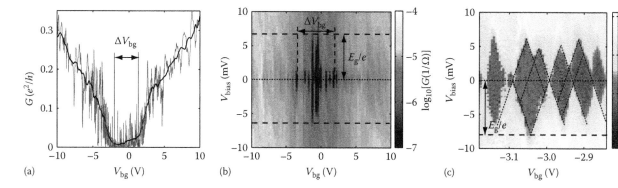

FIGURE 21.14 Transport gap measurements for the constriction with dimensions $w = 85$ nm, $l = 500$ nm (a) Conductance versus back gate voltage V_{bg}, illustrating the procedure used to determine the size of the transport gap in back gate voltage ΔV_{bg}. The measurement is performed with a bias voltage of $300\,\mu$V. The trace form is smoothened over 2.5 V in V_{bg}. The black lines indicate the linear fits used to determine the gap size. (b) Color plot of the conductance as a function of applied back gate and bias voltage. (c) Zoom of the gap measured in (b). (Adapted from Molitor, F. et al., *Phys. Rev. B*, 79, 075426, 2009.)

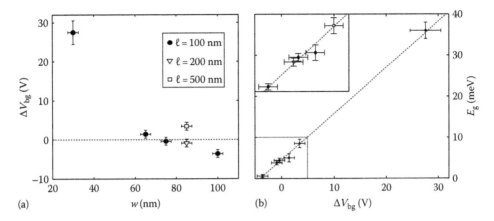

(a) w (nm)   (b) ΔV_{bg} (V)

FIGURE 21.15 (a) ΔV_{bg}, determined by the procedure described in Figure 21.14a, as a function of the constriction width. The dashed line indicates $\Delta V_{bg} = 0$ V. (b) E_g, determined by taking the charging energy of the largest diamond ($E_{c,max}$), as shown in Figure 21.13b, as a function of ΔV_{bg}. The dashed line is a linear fit to the data. The inset shows a zoom of the region -5 V $< \Delta V_{bg} < 5$ V, indicated by the dotted lines in the lower left corner of the figure. (Adapted from Molitor, F. et al., *Phys. Rev. B*, 79, 075426, 2009.)

a decay of the gap size with increasing constriction width w, in good agreement with the observations by Han et al. [han07] for the size of the transport gap in bias direction E_g. For the longer constrictions with $l = 200$ nm and especially $l = 500$ nm, ΔV_{bg} is larger than the value one would expect for a 100 nm long constriction of the same width w. This can be understood by the fact that increasing the width of the constriction or decreasing its length increases the probability of at least one percolating conductive path through the constriction, and therefore decreases the region in energy where transport is governed by localization [mol09]. Please note that for the constrictions with $w = 75$ nm and $l = 100$ nm, $w = 100$ nm and $l = 100$ nm, $w = 85$ nm and $l = 200$ nm, $\Delta V_{bg} < 0$. The value of ΔV_{bg} includes an offset which depends on the conductance value chosen to measure the distance between the fitted lines. With the choice of taking the intersection at $G = 0$, a negative value of ΔV_{bg} means that the intersection point of the two fitted lines lies at a positive conductance value. In these cases, even though the conductance is reduced due to localized states in the constriction, it is never completely suppressed [mol09].

Figure 21.14b displays the size of the energy gap in bias direction E_g, determined from the largest diamond in the gap region, as a function of the gap extension ΔV_{bg} in back gate voltage. The energy gap E_g increases approximately linearly with ΔV_{bg}. The dashed line indicates the result of a linear fit, taking into account the different error bars of the individual data points. The fit gives basically the same result when taking into account only the five lowest values. Even though the slope of the fit relates the back gate voltage to an energy, it does not represent the lever arm of the back gate on the constriction, as the latter can be determined from the diamond measurements in Figure 21.14c. The slope can rather be understood as describing the envelope of the diamond-shaped region of suppressed conductance, as shown in Figures 21.13b, c and 21.14b. The constant proportionality (linear slope) of E_g and ΔV_{bg} suggests that the geometry of the constrictions has no major influence on the relation of these two quantities. The connection between ΔV_{bg} and E_g relates two energy scales characterizing the disorder potential. The quantity E_g is presumably related to the charging energy of the individual islands. The gap size in back gate voltage ΔV_{bg} is a measure of the energy interval in which localized electrons govern transport at the Fermi energy. It may be related to the average amplitude of the potential fluctuations due to disorder [mar08]. However, the reason for this apparently linear relation between these two quantities is not yet understood in detail.

21.4.2 Coulomb Blockade in Nano Constrictions

The experimental data shown above provide indications that the two experimentally observed energy scales E_g and ΔE_F are related to charged islands or quantum dots forming spontaneously along the nanoribbon. Similar observations were reported in [tod09, liu09]. This is supported by the observation (1) of Coulomb diamonds, which vary in size as function of the Fermi energy (Figures 21.13b, c and 21.14c), (2) of a variation of the relative lever arms of individual resonances [sta09] and (3) of local islands charging inside the nanoribbon [tod09, sta09, mol09]. Quantum dots along the nanoribbon can arise in the presence of a quantum confinement energy gap (ΔE_{con}) combined with a strong bulk and/or edge-induced disorder potential Δ_{dis}, as illustrated in Figure 21.16. The confinement energy can be estimated by $\Delta E_{con}(w) = \gamma \pi a_{C-C}/w$, where $\gamma \approx 2.7$ eV and $a_{C-C} = 0.142$ nm [whi07]. This leads to $\Delta E_{con} = 26$ meV for $w = 45$ nm, which by

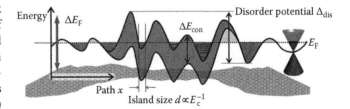

FIGURE 21.16 Schematic illustration of the potential landscape along the graphene nanoribbon allowing the formation of charged islands and quantum dots. For more information see text. (Adapted from Stampfer, C. et al., *Phys. Rev. Lett.*, 102, 056403, 2009. With permission.)

itself can neither explain the observed energy scale ΔE_F, nor the formation of quantum dots in the nanoribbon.

However, by superimposing a disorder potential giving rise to electron–hole puddles near the charge neutrality point [mar08], the confinement gap ensures that Klein tunneling (from puddle to puddle) gets substituted by real tunneling. Within this model, ΔE_F depends on both the confinement energy gap and the disorder potential. An upper bound for the magnitude of the disorder potential can be estimated from our data to be given by ΔE_F. Comparing to Ref. [mar08] where a bulk carrier density fluctuation of the order of $\Delta n \approx \pm 2 \times 10^{11}$ cm^{-2} was reported, we find reasonable agreement as the corresponding variation of the local potential is $\Delta E_F \approx 126$ meV. We can estimate the fraction of overlapping diamonds by summing over all charging energies E_c extracted from Figure 21.13a. This leads to $\Sigma E_c \approx 630$ meV. Comparison with the estimate for ΔE_F gives 45%–82% overlapping diamonds. We expect that this value depends strongly on the length of the nanoribbon in agreement with findings of Ref. [mol09]. The energy gap E_g in bias direction does not tell much about the magnitude of the disorder potential, but it is rather related to the sizes of the charged islands. In particular, the minimum island size is related to the maximum charging energy $E_{c,max}$. By using a simple capacitor disc model, we can estimate the effective charge island diameter by $d = e^2/(4\varepsilon\varepsilon_r E_c) \approx 100$ nm [where $\varepsilon_r = (1 + 4)/2$], which exceeds the nanoribbon width w. Thus, in ribbons of different width the charging energy will scale with w giving the experimentally observed $1/w$ dependence of the energy gap in bias direction, as shown in Figure 21.15 and Refs. [han07, mol09].

21.5 Graphene Quantum Dots

Quantum dots [kou98, kou01] are small man-made structures in a solid, typically with sizes ranging from nanometers to a few microns, consisting of 10^3 – 10^9 atoms where the number of free electrons can be tuned over a wide range. The (quasi-0-D) confinement of the electrons in all three spatial directions results in a quantized energy spectrum (see also Figure 21.1f) and quantum dots are therefore regarded as artificial atoms [kas93].

Quantum dots are not only interesting from a fundamental point of view but are also promising hosts for spin qubits [los98], which have been recognized as potential building blocks for future quantum information technology. In semiconductor material systems (e.g., GaAs), in which quantum dot–based solid-state spin qubits are most advanced today [elz04, pet05, kop06], spin–orbit and hyperfine interactions have been identified as the main processes limiting spin decoherence times. Therefore, alternative materials for quantum dots (such as, e.g., silicon [ang07, sha08], semiconducting carbon nanotubes [bie04, sap06], or nanowires [fas05, sho06]) attract interest as hosts for potential spin-qubits.

Graphene and carbon materials in general are believed to have exceptionally long spin coherence times due to weak spin–orbit interactions (light weight of carbon) [min06, hue06] and the low nuclear spin concentration, arising from the ≈1% natural abundance of ^{13}C suppresses hyperfine coupling [tra07]. In addition, graphene quantum dots allow elucidating the quantum-to-classical

crossover in both regular and chaotic-confined billiard systems with massless particles [ber87, pon08, lib09]. However, confining electrons in graphene is challenging, mainly due to the gapless electronic structure [cas07] and phenomena related to Klein tunneling [dom99, kat06a]. Therefore, split gate techniques (e.g., [ree88]) or SFM-based local oxidation [lüs99] well-known for semiconducting materials, to fabricate quantum dots are hard to apply. However, it has been shown recently that cutting out nanostructures from 2-D graphene allows to structurally confine electrons [sta08, pon08, sta08a, gue08, gue08a, tod09, sch09]. Actually, the functionality of graphene nanodevices can be engineered by designing the structure of the graphene flake. This will be shown with three different examples, discussing (1) a graphene single-electron transistor based on a width-modulated graphene nanoribbon, (2) graphene single-electron transistor device with a nearby graphene nanoribbon acting as charge detector, and (3) by discussing a graphene nanodevice small enough so that quantum confinement effects start to play an important role.

21.5.1 Graphene Single-Electron Transistor

Single-electron transistors (SETs) [kou97], in general, consist of a conducting island connected by tunneling barriers to two conducting (source and drain) leads and (at least one) capacitively coupled gate to tune the potential on the island (see Figure 21.17a). Electronic transport through such devices (from source to drain) can be completely blocked by Coulomb interaction for temperatures and bias voltages lower than the characteristic energy required on adding an electron to the island [kou97, hei03]. This is shown in Figure 21.17b. Since no energy level of the island lies inside the transport window (see dashed horizontal lines), the device is out of resonance and transport (i.e., current) is

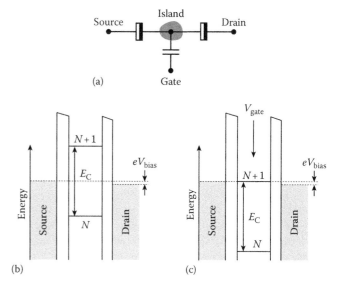

FIGURE 21.17 (a) Schematics of a single-electron transistor (b) Illustration of energy levels in Coulomb blockade regime (off resonance) (c) By aligning an energy level of the island with the transport window the device is in resonance and transport becomes possible.

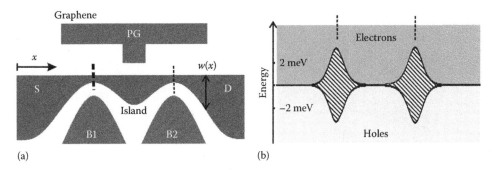

FIGURE 21.18 (a) Schematic illustration of a tunable single-electron transistor device with electrode assignment. (b) Effective energy band structure of the device as depicted in panel a. The tunnel barriers exhibit an effective energy gap in the range of a few meV, depending on the width of the constrictions. For more information on the model see text and Ref. [sta08]. (Adapted from Stampfer, C. et al., *Nano Lett.*, 8, 2378, 2008. With permission.)

suppressed by the so-called Coulomb blockade. Here, temperature and bias (eV_{bias}) are significantly smaller than the charging energy E_c. Moreover, it is assumed that the tunneling resistance of the two tunneling barriers is significantly larger than e^2/h, such that (the island) level broadening can be neglected [ihn04]. By changing the gate potential, we can tune the single-electron transistor into resonance, such that an island level is aligned within the transport window and carriers can resonantly hop from source to drain, as shown in Figure 21.17c.

The following discussion follows mainly the outline of Ref. [sta08, sta08a]. In Figure 21.18a, we show a schematic illustration of a tailored (i.e., etched) graphene flake providing the functionality of a graphene single-electron transistor [sta08, sta08a, gue08]. The width modulated graphene nanostructure consists of a central island, two spatially separated constrictions, which act as effective tunnel barriers and three lateral graphene gates (plunger gate PG and barrier gates B1 and B2) for full electrostatic tunability. It is crucial that narrow graphene constrictions (incl. nanoribbons) have been found to exhibit an effective transport gap [che07, han07, sol07, li08] enabling effective tunable tunneling barrier (see also Section 21.4). Although the transport gap is not a clean energy gap and is likely to be formed by a series of local quantum dots inside the nano constriction [tod09, sta09, mol09, liu09], the overall transport gap behavior can be modeled by the expression $E_g(w) = a/w\, \exp(-bw)$ [sol07], where $a = 1\,\text{eV nm}$ and $b = 0.023\,\text{nm}^{-1}$ are constants extracted from fits to experimental data [han07]. Thus, the exact shape of the effective energy gap, $E_g(x)$ (x is the transport direction, see Figure 21.18a) is given (mainly) by the lateral confinement, that is, by the variation of the width $w(x)$ along the device (see Figure 21.18a). By further assuming that electron–hole symmetry holds in the confined geometry, we can plot the effective conduction band edge at $+E_g(x)/2$, and an effective valence band edge at $-E_g(x)/2$. This is shown in Figure 21.18b and we refer to it as the effective band structure of the width-modulated graphene nanoribbon. Indeed, in this respect we can think of graphene band gap engineering by tailoring the graphene sheet width [han07].

By adjusting the Fermi level (see horizontal line in Figure 21.18b) such that both effective band gaps are within the

transport window, an electrically isolated island can be formed and transport can be tuned at the single-electron level. In Figure 21.18a, we show a (false color) SFM image of a fabricated graphene device consisting of a flake tailored into the shape shown in Figure 21.18a (see also Figure 21.10a). The flake rests on approx. 300 nm SiO_2 and the highly doped Si substrate is used as a back gate to adjust the overall Fermi level. In Figure 21.17b a low bias source-drain back gate characteristic of this device is shown. The transport gap of strongly suppressed current ($-25\,\text{V} < V_{bg} < -12\,\text{V}$) separates the region of hole (left side; see also inset) and electron (right side) transport. This transport characteristic is dominated by the two narrow constrictions, as also seen by direct comparison with Figure 21.12a. Fixing the back gate voltage (i.e., the Fermi level) to a value inside the transport gap, single-electron transport phenomena can be investigated. Before focusing on the significantly smaller energy scale of the Coulomb blockade resonances, we discuss the full tunability on a larger energy scale. Figure 21.20a shows source–drain current measurements as function of applied barrier gate potentials (V_{b1} and V_{b2}) at fixed back gate ($V_{bg} = -15\,\text{V}$; see arrow in Figure 21.19b). The bright areas represent regions of suppressed current, whereas in the corners (dark areas) we find either hole- or electron-dominated transport. Vertical and horizontal stripes of suppressed current are observed, indicating that transport through each of the two constrictions is characterized by a transport gap, which can be individually tuned with the respective barrier gate. For example, keeping $V_{b1} = -20\,\text{V}$ constant and sweeping V_{b2} from -20 to $+5\,\text{V}$ keeps constriction 1 conducting whereas constriction 2 is tuned from large conductance to very low conductance (into the transport gap). The capacitive cross talk from barrier gate 1 (B1) to constriction 2 and from barrier gate 2 (B2) to constriction 1 has been found to be smaller than 2% [sta08a].

Since the back gate voltage has been fixed (see arrow in Figure 21.19b and bullet in Figure 21.20a) such that the Fermi energy in the contacts of the structure lies within the conduction band, it can be indicated by horizontal dashed lines in the four drawings in Figure 21.20b. The four drawings represent energy diagrams corresponding to the four corners of Figure 21.20a, as indicated by the white numbers. For example, in corner 2 transport takes

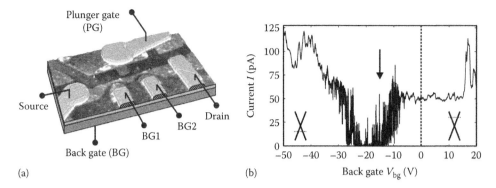

(a)   (b)

FIGURE 21.19 (a) False color scanning force microscope image of an investigated graphene nanodevice. Bright areas mark the etched single layer graphene flake, whereas the elevated structures highlight the metal contacts. The minimum graphene feature size is approximately 50 nm. (b) Low bias source-drain back gate characteristic of the device shown in panel a at 1.7 K and $V_{bias} = 300\,\mu V$ for $V_{b1} = V_{b2} = V_{pg} = 0$ V. The resolved transport gap separates between hole and electron transport as shown by the insets. (Adapted from Stampfer, C. et al., *Int. J. Mod. Phys. A*, 23, 2647, 2009. With permission.)

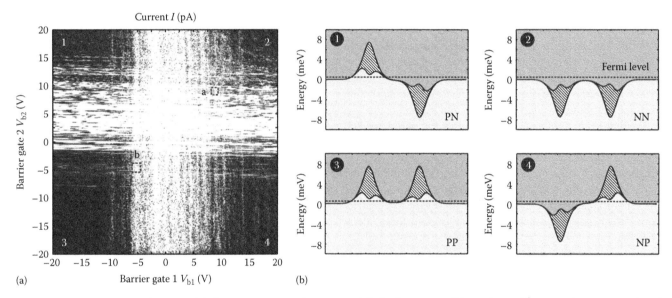

(a)   (b)

FIGURE 21.20 Transport as function of the barrier gate potentials V_{b1}, V_{b2} and the back gate at small bias voltages. (a) Source–drain current plotted as function of V_{b1} and V_{b2} for constant back gate ($V_{bg} = -15$ V; see arrow in Figure 21.18b). Here, both individual gaps can clearly be seen. The white areas correspond to suppressed current. Please note also the gap homogeneity as a function of the back gate. (b) Schematic illustrations of the barrier configurations explaining the different transport regimes shown in panel a. (Adapted from Stampfer, C. et al., *Nano. Lett.* 8, 2378, 2008. With permission.)

place in the conduction band throughout the whole structure. In corner 1 (4) transport occurs in the conduction band in the right (left) part of the structure. The left (right) constriction is traversed via states in the valence band. The situation is even more complex in corner 3, where the Fermi energy cuts both barrier regions in the valence band. Although these situations imply two or even four p–n-like transitions along the structure, no distinctive features are observed in these measurements. This may be a manifestation of the suppression of backscattering due to Klein tunneling [kle29]. These measurements (Figure 21.20a) combined with the heuristic energy diagram model describing our sample (Figure 21.20b) shows and explains the large tunability of the graphene nanodevice, where both tunneling barriers can be individually tuned over a large range including the crossover from hole to electron transport. In order to operate

the device in the single-electron transport configuration, we focus on a regime where both tunneling barriers are active (i.e., lie within the transport window). Thus, we focus, for example, on small barrier gate voltage regimes, as highlighted by the (labeled) dashed boxes in Figure 21.20a.

In Figure 21.21 we show three different close-ups of transport characteristics in two different regimes. We distinguish between the NN (Figure 21.21a) and the PP (Figure 21.21b) regimes, depending on either having the tunnel barriers (according to B1 and B2) shifted down (N) or up (P), as indicated in Figure 21.20b. We observe in all regimes sequences of horizontal and vertical stripes of suppressed current, and current resonances (for more details on different regimes see Refs. [sta08a, sta09a]). Their direction in the $V_{b1} - V_{b2}$ plane indicates that their physical origin has to be found within constriction 1 (vertical stripes)

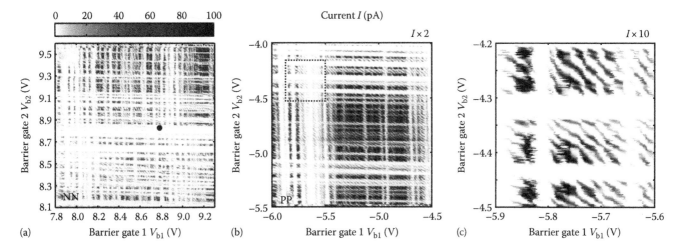

FIGURE 21.21 Source–drain current through the graphene single-electron transistor as function of the barrier gates V_{b1} and V_{b2} for constant bias $V_{bias} = 300\,\mu V$ and back gate $V_{bg} = -15\,V$. (a, b) Close-ups of Figure 21.20a (as indicated therein by labeled boxes), showing transport in the NN (a) and PP (b) regimes. On top of the horizontal and vertical transmission modulations, we observe (diagonal) Coulomb blockade resonances. This is best seen in panel c, which is a close-up of panel b. In panels b and c, the current has been multiplied by factors of 2 and 10, respectively, to meet the color scale shown above panel b. (Adapted from Stampfer, C. et al., *Nano. Lett.* 8, 2378, 2008. With permission.)

or constriction 2 (horizontal stripes). According to Section 21.4, some of these resonances can be understood as Coulomb blockade resonances of isolated charge islands inside the corresponding graphene constriction forming the effective tunneling barrier. A blowup of a small region in Figure 21.20a is shown in Figure 21.21c. The current exhibits even finer resonances which are almost equally well tuned by both constriction gates (see diagonal lines). These resonances are therefore attributed to states localized on the island between the barriers. It will be shown below that these resonances occur in the Coulomb blockade regime of the island. We attribute the deviations from perfectly straight diagonal lines to the presence of rough edges and inhomogeneities within the graphene island which has dimensions (slightly) larger than the elastic mean free path of the electrons [sta08]. This characteristic pattern (Figure 21.21c) can be found within a large $V_{b1} - V_{b2}$ parameter range within the regime where the two barrier gaps cross each other (i.e., the inner bright part of Figure 21.20a).

So far we have mainly focused on the barriers. In the following, we concentrate on the charging of the island itself. We fix

the barrier gate potentials (V_{b1} and V_{b2}) either in the NN regime or in the NP regime in order to study Coulomb blockade. Figure 21.22a shows sharp conductance resonances with a characteristic period of about 20 mV ($V_{b1} = 5.570\,V$ and $V_{b2} = -2.033\,V$ are fixed). Their amplitude is modulated on a much larger voltage scale of about 200 mV by the transparency modulations of the constrictions (cf. Figure 21.21a). These resonances in the narrow graphene constrictions can significantly elevate the background of the Coulomb peaks (see, e.g., black arrow). Figure 21.21b, a close-up of Figure 21.22a (see marked area therein), confirms that transport can also be completely pinched off between Coulomb blockade peaks. Corresponding Coulomb diamond measurements [kou97], that is, measurements of the differential conductance ($G_{diff} = dI/dV_{bias}$) as a function of bias voltage V_{bias} and plunger gate voltage V_{pg} are shown in Figure 21.22a. Within the plunger gate voltage range shown, no charge rearrangements have been observed and the peak positions were stable over more than 10 consecutive plunger gate sweeps. It is found that the Coulomb peaks and the Coulomb diamonds are not very sensitive to the tunneling barrier regime, although in some cases

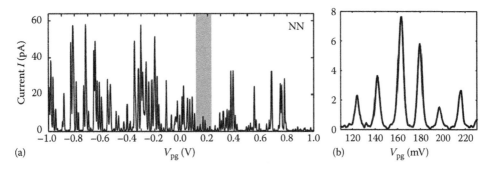

FIGURE 21.22 (a) Coulomb resonances on top and nearby strong transport modulations in the NN regime ($V_{bg} = -15\,V$, $V_{b1} = 8.79\,V$, and $V_{b2} = 8.85\,V$). (b) A close-up highlighting Coulomb peaks (see marked area in panel a). (Adapted from Stampfer, C. et al., *Int. J. Mod. Phys. A*, 23, 2647, 2009. With permission.)

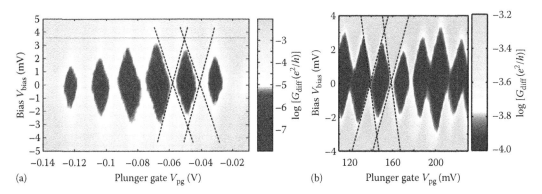

FIGURE 21.23 Coulomb diamonds in differential conductance G_{diff}, represented in a logarithmic color scale plot (dark regions represent low conductance). A dc bias V_{bias} with a small ac modulation (50 μV) is applied symmetrically across the dot and the current through the dot is measured. Coulomb diamonds have been measured in both the NP regime (panel a) and PP regime (panel b). (Adapted from Stampfer, C. et al., *Appl. Phys. Lett.*, 92, 012102, 2008; Stampfer, C. et al., *Nano Lett.*, 8, 2378, 2008.)

p–n-like junctions are present, whereas in other cases a more uniform island might be expected. This can be seen by comparing diamonds measured in different configurations as shown in Figure 21.23a and b, for the NN and NP regime, respectively.

From the extent of all the diamonds in bias direction, we estimate the average charging energy of the graphene single-electron transistor operated in both regimes to be $E_c \approx 3.4$ meV. This charging energy corresponds to a sum-capacitance of the graphene island $C_\Sigma = e^2/E_c \approx 47.3$ aF, whereas the extracted back gate capacitance $C_{\text{bg}} \approx 18$ aF is higher than the purely geometrical parallel plate capacitance of the graphene island $C = \varepsilon_0\varepsilon A/d \approx 7.4$ aF. This is related to the fact that the diameter of the graphene island (A) is approximately the same as the gate oxide thickness d [ihn04, sta08]. The lever arms and the electrostatic couplings of the electrodes to the graphene island do not change significantly between the NN, PP (not shown), and the NP regimes.

Thus, the lever arm of the plunger gate is $\alpha_{\text{pg}} = C_{\text{bg}}/C_\Sigma \approx 0.15$ ($C_{\text{bg}} \approx 6.9$ aF), whereas the electrostatic coupling to the other gates was determined to be $C_{\text{b1}} \approx 5.5$ – 6.0 aF and $C_{\text{b2}} \approx 5.0$ aF. It is found that the island geometry and dot location with respect to the lateral gates stay almost constant [sta08a]. However, the capacitive coupling to the source and drain contacts (i.e., C_S and C_D) changes significantly as a function of the tunneling barrier configuration. This can be seen when comparing the symmetry of the diamonds in the NN and NP regime, as shown in panels a and b of Figure 21.23. While the size and fluctuations of the diamonds remain (almost) constant as function of the regime, the lever arms of the source and drain contacts change strength. In one case (NP regime), we extract $C_S \approx 1.8$ aF and $C_D \approx 9.6$ aF, whereas in the other (NN regime) $C_S \approx 10.1$ aF and $C_D \approx 1.8$ aF, which can be seen from the different slopes of the diamond edges. However, the individual tunnel barriers strongly depend on the local barrier configuration and change also within the NN or the NP region [sta08a].

We now estimate the energy scale of the resonances in the constrictions. The spacing of the constriction resonances in plunger gate is about 200 mV, whereas the spacing of Coulomb peaks is 20 mV. By assuming that the capacitance between the

plunger gate and the localized states in the constrictions leading to the resonances is about three times smaller than C_{pg} (estimated from the geometry of the device), the energy scale of the resonances in the constriction is about 10 mV, in agreement with the measured transport gap in Figure 21.16. Alternatively, this characteristic energy scale can also be estimated by considering that the back gate voltage sweep from −25 to −15 V (around the charge neutrality point) at $V_{\text{bg}}^D = -20$ V, Figure 21.19b translates to a Fermi energy sweep over an energy interval of approximately 120 meV. Near the Dirac point the spacing of the constriction resonances in back gate voltage is found to be of the order of 200 mV, leading again to a characteristic energy scale of 10 meV.

21.5.2 Charge Detection with a Nanoribbon Near a SET

Next we discuss an integrated graphene device consisting of a graphene single-electron transistor with a nearby graphene nanoribbon acting as a quantum point-contact-like charge detector. Charge detection techniques [fie93] have been shown to extend the experimental possibilities with single-electron devices (incl. quantum dots) significantly. They are, powerful for detecting individual charging events and spin-qubit states [elz04, pet05] and molecular states in coupled quantum dots [dic04]. Furthermore, charge detectors have been successfully used to investigate shot noise on the single electron level, and full counting statistics [gus06]. This makes charge detection also interesting for advanced investigation of graphene quantum dots and graphene nanodevices in general.

In the following discussion, we follow mainly the outline of Ref. [gue08]. Figure 21.24a shows a SFM image of an integrated graphene device consisting of a graphene single-electron transistor with a graphene nanoribbon nearby. The single-electron transistor device consists of two 35 nm wide graphene constrictions separating source (S) and drain (D) contacts from the graphene island. The diameter of the graphene island is approximately 200 nm. The constrictions and the island are electrostatically tuned independently again by two barrier gates (B1 and B2)

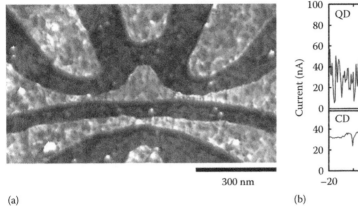

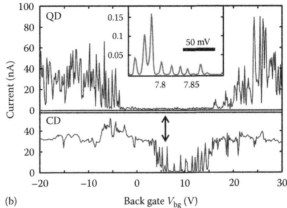

(a)   300 nm   (b)   Back gate V_{bg} (V)

FIGURE 21.24 Nanostructured graphene single-electron transistor with nanoribbon and characteristic transport measurements. (a) Scanning force micrograph of the measured device. The central island is connected to source (S) and drain (D) contacts by two constrictions. The effective diameter of the dot is 200 nm and the constrictions are 35 nm wide. The graphene nanoribbon acts as charge detector. Three lateral gates B1, B2, and PG are used to tune the devices. (b) Back gate characteristics of the single-electron transistor upper panel and the charge detector lower panel. Both measurements were performed at a source–drain bias voltage of $V_{bias,set} = V_{bias,cd} = 500\,\mu V$ and at 1.7 K. The inset shows Coulomb blockade resonances observed inside the transport gap as a function of the back gate voltage over a range of 150 mV. (Adapted from Güttinger, J. et al., *Appl. Phys. Lett.*, 93, 212102, 2008. With permission.)

and a plunger gate (PG), respectively. The highly doped silicon back gate (BG) allows again to adjust the overall Fermi level. In addition, we placed a 45 nm wide graphene nanoribbon approximately 60 nm next to the island, which acts as a charge detector, as shown below.

The individual characterization of both devices is shown in the upper and lower panels of Figure 21.24b. Here we show the source-drain current I_{sd} as a function of the back gate voltage (at a temperature of 1.7 K) of both the single-electron transistor (upper curve) and the charge detector (lower curve). In both cases we observe a transport gap extending roughly from −4 to 15 V and from 4 to 14 V for the single-electron transistor and charge detector, respectively. From high source-drain voltage $V_{bias,set}$ measurements (not shown), we estimate the characteristic energy scale of these effective energy gaps to be about 13 and 8 meV, respectively. This is in reasonable agreement with measurements on graphene nanoribbons, where the transport gap is dominated by the width of the graphene nanostructure (as shown in Figure 21.16). The large-scale current fluctuations are again attributed to resonances in the graphene constrictions. By focusing on a smaller back gate voltage range of 150 mV (see inset of Figure 21.24b), Coulomb blockade resonances of the single-electron transistor are resolved in regions where these resonances are suppressed. From additional Coulomb diamond measurements as function of back gate and plunger gate voltage (not shown here), a charging energy $E_c = 4.3$ meV and a plunger gate and back gate lever arm of $\alpha_{pg,set} = 0.06$ and $\alpha_{pg,set} = 0.34$ can be extracted [gue08]. After having demonstrated the functionality of both devices independently, their joint operation is shown in Figures 21.25 and 21.26, where we demonstrate the functionality and sensitivity of the graphene charge detector.

For these measurements the back gate voltage is set to $V_{bg} = 6.5$ V such that the Fermi level is close to the charge neutrality

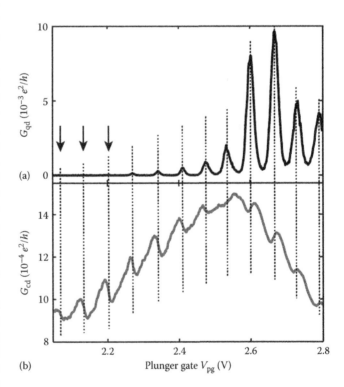

FIGURE 21.25 (a) Single-electron transistor (SET) conductance G_{set} and (b) charge detector (CD) conductance G_{cd} as a function of plunger gate voltage V_{pg} for a fixed V_{bg} voltage, $V_{bg} = 6.5$ V. The vertical arrows in (a) indicate Coulomb blockade resonances that can be hardly measured by conventional means because the current levels are too low. However, they can be detected by the charge detector [see (b)]. Bias on the SET: $V_{bias,set} = 500\,\mu V$. Bias on CD: $V_{bias,cd} = 8.2$ mV. (Adapted from Güttinger, J. et al., *Appl. Phys. Lett.*, 93, 212102, 2008. With permission.)

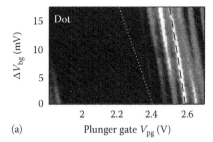
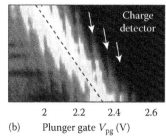
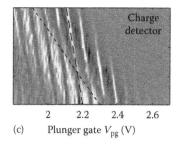

(a)   (b)   (c)

FIGURE 21.26 (a) Conductance G_{set} of the single-electron transistor as a function of plunger gate voltage V_{pg} and back gate voltage V_{bg}. The back gate voltage is converted to a relative scale starting at $V_{bg} = 6.505\,V$ with $\Delta V_{bg} = 0\,V$. For this measurement, a source-drain bias of $V_{bias,set} = 500\,\mu V$ is symmetrically applied. Narrowly spaced periodic lines are Coulomb blockade resonances as marked by the black line with long dashes, while the larger scale features are attributed to a modulation of the transmission through the barrier's white dotted line. (b) Simultaneous measurement of the charge detector conductance G_{cd}. The broad line with increased conductance is less affected by changing the PG voltage compared to the Coulomb blockade resonances in the island, and it is attributed to a local resonance in the charge detector short dashed line. In addition to this broad line, faint lines with a slope corresponding to the Coulomb blockade resonances in the single-electron transistor are observed arrows. (c) Derivative of the charge detector conductance is plotted in (b) with respect to plunger gate voltage dG_{cd}/dV_{pg}, where the lines with short and long dashes indicate the two different lever arms. (Adapted from Güttinger, J. et al., *Appl. Phys. Lett.*, 93, 212102, 2008. With permission.)

point (see arrows in Figure 21.24b) as well as inside the transport gap of the charge detector. We operate the charge detector in a regime where strong resonances are accessible in order to make use of the steep slopes of the conductance as a function of V_{pg}, which is of the order of $4 - 6 \times 10^{-4}\ (e^2/h)/100\ mV$ to detect individual charging events on the single-electron transistor.

Figure 25a shows almost equidistantly spaced $\Delta V_{pg} = 65 \pm 3.8\ mV$ Coulomb blockade resonances as a function of V_{pg} (at $V_{bias,set} = 500\,\mu V$). The strong modulation of the conductance peak amplitudes is due to superimposed resonances in the graphene constrictions defining the island. In Figure 21.25b we plot the simultaneously measured conductance through the charge detector at a bias voltage of $V_{bias,cd} = 8.2\ mV$ for the same V_{pg} range. On top of the peak-shaped charge detector resonance we observe conductance steps that are well aligned (see dotted lines) with single charging events on the island. The conductance step is related to a shift in the charge detector resonance with respect to the plunger gate voltage. From an analysis of more than 300 charging events, an average shift in plunger gate voltage of $\Delta V_{pg,cd} = 34 \pm 6\ mV$ is observed. Despite the fact that in Figure 21.25b the shifts are larger for lower plunger gate voltage, no systematic dependence on the plunger voltage, or on the single-electron transistor conductance or on the individual charge detector resonance can be extracted. Figure 21.26a and b show 2-D plots of a set of traces corresponding to those shown in Figure 21.25a and b taken for different back gate voltages and $V_{bias,cd} = V_{bias,set} = 500\ mV$. Figure 21.26a shows Coulomb blockade resonances in the single-electron transistor conductance following a relative lever arm of plunger gate and back gate of $\alpha_{pg,set}/\alpha_{bg,set} = 0.18$ (see black dashed line). In Figure 21.26b these resonances are observed through charge detection and are marked with arrows.

The charge detector resonance used for detection can be distinguished from the island resonances by its larger width and its different slope given by $\alpha_{pg,cd}/\alpha_{bg,cd} = 0.04$ black dashed line. This reduced slope is due to the larger distance of the charge detector nanoribbon to the plunger gate (350 nm) compared to the

island-plunger gate distance. The modulation of the Coulomb blockade resonances in Figure 21.26a is due to resonances in the tunneling constriction located around 300 nm away from the plunger gate (see Figure 21.24a). This leads to a slope of 0.08 for the peak modulations (see white dotted line). Independent of this modulation we identify single charging events on the island as conductance fringes (see arrows in Figure 21.26b) on top of the up and down slopes of the charge detector resonance. This can be even better seen by numerical differentiation of G_{cd} versus V_{pg}, as shown in Figure 21.26c. Here the sharp conductance changes are due to the fact that charging events in the dot are strongly pronounced, and both relative lever arms to the Coulomb blockade peaks and the constriction resonance in the charge detector are indicated by dashed lines. The detection range can be improved by increasing the bias $V_{bias,cd}$, leading to a broadening of the constriction resonance, as seen by comparing Figure 21.26c with Figure 21.25b. From the measurement shown in Figure 21.25b a nanoribbon conductance change of up to 10% can be extracted for a single charging event. For lower bias voltages, for example, $V_{bias,cd} = 500\,\mu V$ the change in the conductance can be increased to 60% [gue08].

21.5.3 Quantum Confinement Effects

By down-scaling the island of the single-electron transistor device (see schematics in Figure 21.18a) we step into a regime where quantum confinement effects dominate over (classical) Coulomb interaction effects [ihn04]. Thus, for the detection of quantum confinement effects and the identification of individual orbital quantum states in graphene quantum dots, we move to smaller system sizes and lower temperatures. Here we discuss a graphene quantum dot device (see inset of Figure 21.27) consisting of two about 60 and 70 nm wide graphene constrictions which separate the source (S) and drain (D) contacts from the graphene island with an effective diameter of approximately 140 nm. The island can be further tuned by a plunger gate (PG) separated from the island by 100 nm, whereas the overall Fermi level is again adjusted with a highly doped silicon back gate (BG).

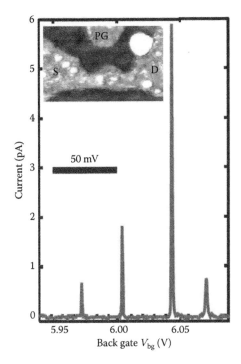

FIGURE 21.27 (inset) Scanning force micrograph of the graphene quantum dot. The quantum dot can be tuned by a nearby plunger gate (PG). The central island is connected to source (S) and drain (D) contacts by two constrictions. The effective diameter of the dot is approximately 140 nm. (main panel) Coulomb blockade measured as function of the back gate at an electronic temperature of 200 mK and a bias voltage of $V_{\text{bias}} = 16 \mu\text{V}$. (Adapted from Schnez, S. et al., *Appl. Phys. Lett.*, 94, 012107, 2009. With permission.)

The following discussion follows mainly the outline of Ref. [sch09]. In Figure 21.27 we show characteristic Coulomb blockade peaks (here as function of applied back gate voltage) measured inside the overall transport gap of this graphene quantum dot device. In contrast to Figure 21.20b, these resonances exhibit a much smaller linewidth (which is related to the significantly lower base temperature of approx. 50 mK) and a significantly larger peak spacing, which is due to the smaller size of the graphene island leading to both (i) larger charging energy and (ii) detectable quantum confinement (see below). The linewidth of these Coulomb blockade peaks has also been used to estimate an upper bound for the electronic temperature, which was found to be around 200 mK [sch09]. For further characterization, the back gate voltage has been set close to the overall charge neutrality point of this graphene device.

Characteristic Coulomb diamond measurements, i.e., differential conductance $G_{\text{diff}} = dI/dV_{\text{bias}}$, as a function of the quantum dot bias voltage V_{bias} and plunger gate voltage V_{pg} are shown in Figure 21.28. Within this plunger gate voltage range, no charge rearrangements were observed and the sample was stable for more than 2 weeks. Here, we extract a typical energy scale of the order of 10 meV. A strong fluctuation of the addition energy is observed over the plunger gate voltage range $-0.1\,\text{V} < V_{\text{pg}} < 1.2\,\text{V}$ (full data range not shown), corresponding to an energy range of around 100 meV, indicating the importance of quantum confinement effects. This is supported by the observation of excited states, which appear in Figure 21.28a as distinct lines of increased conductance running parallel to the edge of the Coulomb diamonds [kou97]. Figure 21.28b showing a close-up of Figure 21.28a allows extracting an excitation energy 1.6 meV as marked

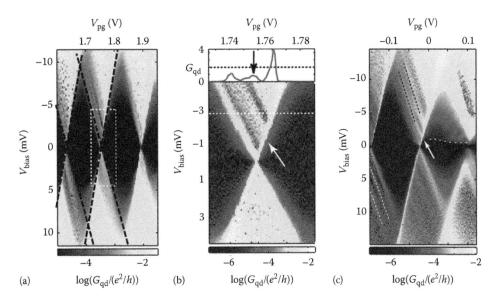

FIGURE 21.28 (a) Differential conductance G logarithmic plots a function of source-drain voltage V_{bias} and plunger gate voltage V_{pg}. (b) The lower panel is a zoom of the framed area in panel a. An excited state is represented by white arrow. A cut along the dashed line at $V_{\text{bias}} = -2.87$ mV is shown in the upper panel here G is measured in units of $10^{-3}\ e^2/h$ and was smoothed over four points. (c) Stability diagram at different plunger gate voltages. Several excited states are visible as shown by dashed lines. In the right diamond, regions of higher conductance in the upper part of the diamond can be seen. This is interpreted as a signature of cotunneling in a graphene quantum dot see arrow. In all measurements shown in Figures 21.27 and 21.28, the back gate voltage was set to $V_{\text{bg}} = 0$ V and the electronic temperature was around 200 mK as deduced from the Coulomb peak width. (Adapted from Schnez, S. et al., *Appl. Phys. Lett.*, 94, 012107, 2009. With permission.)

by the white arrow. A line cut at V_{bias} = 2.78 mV dashed line presented in the upper panel of Figure 21.28b shows the peak of the excited state at finite bias (marked by the arrow). The broadening of the peak significantly exceeds thermal broadening and might be due to the energy-dependent coupling of the excited state to the graphene leads. This would be in good agreement with the current understanding of the underlying mechanism of the effective tunneling barrier formation in graphene nano constrictions (see Section 21.4.2). Figure 21.28c shows two Coulomb diamonds at lower plunger gate voltage, where more than one excited state is observed as a function of increasing energy, as shown by pairs of dashed lines (left diamond). These excitations are found at energies of around 1.6 and 3.3 meV (black dashed lines) and 2.1 and 4.2 meV (white dashed lines), respectively.

The observation of excitations at finite source-drain voltage finds support by the detection of inelastic cotunneling onsets at lower bias. Inside the right Coulomb diamond of Figure 21.28c, we distinguish between regions of suppressed and slightly elevated conductance separated by the dotted line. The edge of this conductance step is aligned with the first excited state outside the diamond at an energy of 1.6 meV as highlighted by an arrow.

In order to further explore the excitation spectrum of this graphene quantum dot, we show the energy shift in nine consecutive Coulomb peaks in a magnetic field applied normal to the plane of the quantum dot in Figure 21.29. The vertical energy axis was obtained by converting plunger gate voltage into energy using the measured lever arm α_{pg} = 0.075. In the constant-interaction model, the ground-state energy of an N-particle quantum dot can be written as the sum of the single-particle energies $\varepsilon_i(B)$ plus an electrostatic charging

energy NE_c. The ground-state energy is tuned by the gate voltage V_g. The experiment was done in the zero-bias regime; hence, we measured the chemical potential of the Nth Coulomb resonance as explained in Ref. [sch09]. Experimentally, the single-particle energy $\varepsilon_N(B)$ of the Nth Coulomb resonance is then determined by $\varepsilon_N(B) = e\alpha_{pg}V_g(N, B) + NE_c$ + const. The constant part and the electrostatic contribution NE_c are subtracted such that consecutive peaks shown as red triangles and blue circles, respectively, touch each other alternatingly. Characteristic lines (see dashed lines in Figure 21.29) linear in B with slopes of around 2.5 meV/T can be seen. This strong B-field dependence cannot be explained by the Zeeman effect, which would result in a slope of $g\mu_B$ = 116 eV/T, assuming a g-factor of g = 2. For higher magnetic fields, the Landau level degeneracy increases and fewer Landau levels are filled. Consequently, the energy spectrum is expected to evolve from single-level fluctuations into a regular pattern. This transition can be seen at around 4 T. Recent theoretical approximations are in reasonable qualitative agreement with our experimental data [sch08, rec09].

21.6 Summary

This work demonstrates the maturity and versatility of transport experiments on graphene nanodevices. We have shown that by downscaling graphene devices they strongly alter their transport properties. In contrast to bulk graphene, where pseudo-relativistic carriers and a semimetallic band structure make it hard to confine carriers or to fully suppress transport, nanostructured graphene enables strongly tunable electronic devices, such as transistors, tunneling barriers, or quantum dots. The fabrication of graphene nanodevices has been discussed in detail and the main characterization methods have been touched. We have strong indications that in graphene nanodevices the formation of a confinement energy gap replaces pseudo-relativistic "Klein" tunneling by "real" tunneling leading to the replacement of disorder-induced electron–hole puddles by fully isolated charged islands. These charged islands or quantum dots are found to dominate transport in graphene nanostructures strongly. In particular, we discussed transport through graphene nanoribbons, single-electron transistors, and quantum dots in detail. In all cases we explained the relevant energy scales related to the transport gaps, charging of the isolated islands, or quantum confinement effects. These insights are important to understand transport in graphene nanostructures in general and may help in designing future graphene nanoelectronic devices.

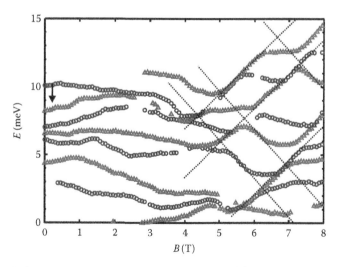

FIGURE 21.29 Experimental energy spectrum of the quantum dot in a perpendicular magnetic field. The typical magnetic field scale at which a significant change is expected is approximately given by one flux quantum Φ_0 = e/h per dot area, i.e., $4\Phi_0/\pi d^2$ = 270 mT and is indicated by the black arrow. Starting around B = 4 T, a regular pattern with characteristic linear slopes evolves; see dashed lines which show the transition from single particle fluctuations to B-field dependence. (From Schnez, S. et al., *Appl. Phys. Lett.*, 94, 012107, 2009. With permission.)

References

[abe07] Abergel, D. S. L., Russell, A., and Fal'ko, V. I., Visibility of graphene flakes on a dielectric substrate, *Appl. Phys. Lett.* **91**, 063125 (2007).

[ada08] Adam, S., Cho, S., Fuhrer, M.S. and Das Sarma, S., Density Inhomogeneity Driven Percolation Metal-Insulator Transition and Dimensional Crossover in Graphene Nanoribbons, *Phys. Rev. Lett.* **101**, 046404 (2008).

[and98] Ando, T. and Nakanishi, T., Impurity scattering in carbon nanotubes – Absence of back scattering, *J. Phys. Soc. Jpn.* **67**, 1704 (1998).

[and98a] Ando, T., Nakanishi, T., and Saito, R., Berry's phase and absence of back scattering in carbon nanotubes, *J. Phys. Soc. Jpn.* **67**, 2857 (1998).

[and99] Ando, T., Nakanishia, T., and Saito, R., Conductance quantization in carbon nanotubes: neutrinos on cylinder surface, *Microelectron. Eng.* **47**, 421 (1999).

[ang07] Angus, S. J., Ferguson, A. J., Dzurak, A. S., and Clark, R. G., Gate-defined quantum dots in intrinsic silicon, *Nano Lett.* **7**, 2051 (2007).

[bee91] Beenakker, C. W. J., Theory of coulomb-blockade oscillations in the conductance of a quantum dot, *Phys. Rev. B* **44**, 1646 (1991).

[ber84] Berry, M. V., Quantal phase factors accompanying adiabatic changes, *Proc. R. Soc. Lond. A* **392**, 45 (1984).

[ber87] Berry, M. V. and Mondragon, R. J., Neutrino billiards: Time-reversal symmetry-breaking without magnetic fields, *Proc. R. Soc. Lond. A* **412**, 53–74 (1987).

[ber09] Berciaud, S., Ryu, S., Brus, L. E., and Heinz, T. F., Probing the intrinsic properties of exfoliated graphene: Raman spectroscopy of free-standing monolayers, *Nano Lett.* **9**, 346 (2009).

[bie04] Biercuk, M. J., Mason, N., and Marcus, C. M., Local gating of carbon nanotubes, *Nano Lett.* **4**, 1 (2004).

[bla07] Blake, P., Novoselov, K. S., Castro Neto, A. H., Jiang, D., Yang, R., Booth, T. J., Geim, A. K., and Hill, E. W., Making graphene visible, *Appl. Phys. Lett.* **91**, 063124 (2007).

[bol08] Bolotin, K. I., Sikes, K. J., Hone, J., Stormer, H. L., and Kim, P., Temperature dependent transport in suspended graphene, *Phys. Rev. Lett.* **101**, 096802 (2008).

[bol08a] Bolotin, K. I., Sikes, K. J., Jiang, Z., Fundenberg, G., Hone, J., Kim, P., and Stormer, H. L., Ultrahigh electron mobility in suspended graphene, *Solid State Commun.* **146**, 351–355 (2008).

[bos07] Bostwick, A., Ohta, T., Seyller, T., Horn, H. K., and Rotenberg, E., Quasiparticle dynamics in graphene, *Nat. Phys.* **3**, 36 (2007).

[bre06] Brey, L. and Fertig, H. A., Electronic states of graphene nanoribbons studied with the Dirac equation, *Phys. Rev. B* **73**, 235411 (2006).

[cal99] Calogeracos, A. and Dombey, N., History and physics of the Klein paradox, *Contemp. Phys.* **40**, 313 (1999).

[cas06] Castro-Neto, A., Guinea, F., and Peres, N. M., Drawing conclusions from graphene, *Phys. World* **19**, 33 (2006).

[cas07] Castro-Neto, A., Guinea, F., Peres, N. M., Novoselov, K. S., and Geim, A. K., The electronic properties of graphene *Rev. Mod. Phys.* **81**, 109 (2009).

[cas07a] Casiraghi, C., Hartschuh, A., Lidorikis, E., Qian, H., Harutyunyan, H., Gokus, T., Novoselov, K. S., and Ferrari, A. C., Rayleigh imaging of graphene and graphene layers, *Nano Lett.* **7**, 2711 (2007).

[cas07b] Casiraghi, C., Pisana, S., Novoselov, K. S., Geim, A. K., and Ferrari, A. C., Raman fingerprint of charged impurities in graphene, *Appl. Phys. Lett.* **91**, 233108 (2007).

[che07] Chen, Z., Lin, Y., Rooks, M., and Avouris, P., Graphene electronics, *Physica E* **40**, 228 (2007).

[dan08] Danneau, R., Wu, F., Craciun, M. F., Russo, S., Tomi, M. Y., Salmilehto, J., Morpurgo, A. F., and Hakonen, P. J., Evanescent wave transport and shot noise in graphene: Ballistic regime and effect of disorder, *J. Low. Temp. Phys.* **153**, 374 (2008), arXiv:0807.0157.

[das08] Das, A., Pisana, S., Chakraborty, B., Piscanec, S., Saha, S. K., Waghmare, U. V., Novoselov, K. S., Krishnamurthy, H. R., Geim, A. K., Sood, A. K., and Ferrari, A. C., Monitoring dopants by Raman scattering in an electrochemically top-gated graphene transistor, *Nat. Nanotech.* **3**, 210 (2008).

[dat95] Datta, S. *Electronic Transport in Mesoscopic Systems*, Cambridge University Press, New York, 1995.

[dic04] DiCarlo, L., Lynch, H., Johnson, A., Childress, L., Crockett, K., Marcus, C., Hanson, M., and Gossard, A. C., Differential charge sensing and charge delocalization in a tunable double quantum dot, *Phys. Rev. Lett.* **92**, 226801 (2004).

[dom99] Dombey, N. and Calogeracos, A., Seventy years of the Klein paradox, *Phys. Rep.* **315**, 41–58 (1999).

[dre02] Dresselhaus, M. S., Dresselhaus, G., Jorio, A., Souza, A. G., and Saito, R., Raman spectroscopy on isolated single wall carbon nanotubes, *Carbon* **40**, 2043 (2002).

[dun07] Dunlycke, D., Areshkin, D. A., and White, C. T., Semiconducting graphene nanostrips with edge disorder, *Appl. Phys. Lett.* **90**, 142104 (2007).

[elz04] Elzerman, J. M., Hanson, R., Willems van Beveren, L. H., Witkamp, B., Vandersypen, L. M. K., and Kouwenhoven, L. P., Single-shot read-out of an individual electron spin in a quantum dot, *Nature* **430**, 431 (2004).

[eva08] Evaldsson, M., Zozoulenko, I.V., Xu, H., and Heinzel, T., Edge-disorder-induced Anderson localization and conduction gap in graphene nanoribbons, *Phys. Rev. B* **78**, 161407(R) (2008).

[fas05] Fasth, C., Fuhrer, A., Bjork, M. T., and Samuelson, L., Tunable double quantum dots in InAs nanowires defined by local gate electrodes, *Nano Lett.* **5**, 1487, (2005).

[fer06] Ferrari, A. C., Meyer, J. C., Scardaci, V., Casiraghi, C., Lazzeri, M., Mauri, F., Piscanec, S., Jiang, Da., Novoselov, K. S., Roth, S., and Geim, A. K., The Raman fingerprint of graphene, *Phys. Rev. Lett.* **97**, 187401 (2006).

[fer07] Fernandez-Rossier, J., Palacios, J. J., and Brey, L., Electronic structure of gated graphene ribbons, *Phys. Rev. B* **75**, 205441 (2007).

[fer07a] Ferrari, A. C., Raman spectroscopy of graphene and graphite: Disorder, electron-phonon coupling, doping and nonadiabatic effects, *Solid State Commun.* **143**, 47 (2007).

[fie93] Field, M., Smith, C. G., Pepper, M., Ritchie, D. A., Frost, J. E. F., Jones, G. A. C., and Hasko, D. G., Measurements of coulomb blockade with a noninvasive voltage probe, *Phys. Rev. Lett.* **70**, 1311 (1993).

[fra86] Fradkin, E., Critical behavior of disorderd degenerate semiconductors, *Phys. Rev. B* **33**, 3263 (1986).

[fra04] Franssila, S., *Introduction to Micro Fabrication*, John Wiley & Sons, Ltd., Sussex, U.K., 2004.

[gei07] Geim, A. K. and Novoselov, K. S., The rise of graphene, *Nat. Mater.* **6**, 183 (2007).

[gei07a] Geim, A. K. and MacDonald, A. H., Graphene: Exploring carbon flatland, *Phys. Today* **60**, 35 (2007).

[gei08] Geim, A. K. and Kim, P., Carbon wonderland, *Sci. Am.* **298**, 68 (2008).

[gor02] Gorbar, E. V., Gusynin, V. P., Miransky, V. A., and Shovkovy, I. A., Magentic field driven metal-insulator phase transition in planar systems, *Phys. Rev. B* **66**, 045108 (2002).

[gra07] Graf, D., Molitor, F., Ensslin, K., Stampfer, C., Jungen, A., Hierold, H., and Wirtz, L., Spatially resolved Raman spectroscopy of single- and few-layer graphene, *Nano Lett.* **7**, 238 (2007).

[gue08] Güttinger, J., Stampfer, C., Hellmüller, S., Molitor, F., Ihn, T., and Ensslin, K., Charge detection in graphene quantum dots, *Appl. Phys. Lett.* **93**, 212102 (2008).

[gue08a] Güttinger, J., Stampfer, C., Molitor, F., Graf, D., Ihn, T., and Ensslin, K., Coulomb oscillations in three-layer graphene nanostructures, *New J. Phys.* **10**, 125029 (2008).

[gue09] Guettinger, J., Stampfer, C., Libisch, F., Frey, T., Burgdörfer, J., Ihn, T., and Ensslin, K., Electron-Hole Crossover in Graphene Quantum Dots, *Phys. Rev. Lett.* **103**, 046810 (2009).

[gup06] Gupta, A., Chen, G., Joshi, P., Tadigadapa, S., Eklund, P.C., Raman scattering from high frequency phonons in supported n-graphene layer films, *Nano Lett.* **6**, 2667 (2006).

[gus05] Gusynin, V. P. and Sharapov, S. G., Unconventional integer quantum effect in graphene, *Phys. Rev. Lett.* **95**, 146801 (2005).

[gus07] Gusynin, V. P., Sharapov, S. G., and Carbotte, J. P., Ac conductivity of graphene: From tight-binding model to 2 + 1-dimension electrodynamics, *Int. J. Mod. Phys. B* **21**, 4611 (2007).

[gus06] Gustavsson, S., Leturcq, R., Simovic, B., Schleser, R., Ihn, T., Studerus, P., Ensslin, K., Driscoll, D. C., and Gossard, A. C., Counting statistics of single electron transport in a quantum dot, *Phys. Rev. Lett.* **96**, 076605 (2006).

[hei03] Heinzel, T., *Mesoscopic Electronics in Solid State Nanostructures*, Wiley-VCH, Weinheim, Germany, 2003.

[han07] Han, M. Y., Özyilmaz, B., Zhang, Y., and Kim, P., Energy band gap engineering of graphene nanoribbons, *Phys. Rev. Lett.* **98**, 206805 (2007).

[hua07] Huard, B., Sulpizio, J. A., Stander, N., Todd, K., Yang, B., and Goldhaber-Gordon, D., Transport measurements across a tunable potential barrier in graphene, *Phys. Rev. Lett.* **98**, 236803 (2007).

[hue06] Huertas-Hernando, D., Guinea, F., and Brataas, A., Spin-orbit coupling in curved graphene, fullerenes, nanotubes, and nanotube caps, *Phys. Rev. B* **74**, 155426 (2006).

[hue09] Huefner, M., Molitor, F., Jacobsen, A., Pioda, A., Stampfer, C., Ensslin, K. and Ihn, T., Investigation of the Aharonov-Bohm effect in a gated graphene ring, *Physica Status Solidi B* **246**, 2756 (2009).

[hwa07] Hwang, E. H., Adam, S., and Das Sarma, S., Carrier transport in two-dimensional graphene layers, *Phys. Rev. Lett.* **98**, 186806 (2007).

[ihn04] Ihn, T., *Electronic Quantum Transport in Mesoscopic Semiconductor Structures* (Springer Tracts in Modern Physics, 192) Springer, Berlin, Germany, p. 102 (2004).

[jia07] Jiang, Z., Zhang, Y., Tan, Y.-W., Stormer, H. L., and Kim, P., Quantum hall effect in graphene, *Solid State Commun.* **143**, 14 (2007).

[jia07a] Jiang, Z., Zhang, Y., Stormer, H. L., and Kim, P., Quantum hall states near the charge neutral Dirac point in graphene, *Phys. Rev. Lett.* **99**, 106802 (2007).

[kas93] Kastner, M. A., Artificial atoms, *Phys. Today* **46**, 24 (1993).

[kat06] Katsnelson, M. I., Zitterbewegung, chirality, and minimal conductivity in graphene, *Eur. J. Phys. B* **51**, 157 (2006).

[kat06a] Katsnelson, M. I., Novoselov, K. S., and Geim, A. K., Chiral tunneling and the Klein paradox in graphene, *Nat. Phys.* **2**, 620 (2006).

[kat07] Katsnelson, M. I. and Novoselov, K. S., Graphene: New bridge between condensed matter physics and quantum electrodynamics, *Solid State Comm.* **143**, 3 (2007).

[kle29] Klein, O., Die reflexion von elektronen an einem potentialsprung nach der relativistischen dynamik von dirac, *Z. Phys.* **53**, 157 (1929).

[kli80] Klitzing, V. K., Dorda, G., and Pepper, M., New method for high-accuracy determination of the fine-structure constant based on quantized hall resistance, *Phys. Rev. Lett.* **45**, 494 (1980).

[kop06] Koppens, F. H. L., Buizert, C., Tielrooij, K. J., Vink, I. T., Nowack, K. C., Meunier, T., Kouwenhoven, L. P., and Vandersypen, L. M. K., Driven coherent oscillations of a single electron spin in a quantum dot, *Nature* **442**, 766 (2006).

[kou97] Kouwenhoven, L. P., Markus, C. M., and McEuen, P. L., Electron transport in quantum dots. *Mesoscopic Electron Transport*, Sohn, L. L., Kouwenhoven, L. P., Schön, G., Eds.; NATO Series; Kluwer, Dordrecht, the Netherlands, 1997.

[kou98] Kouwenhoven, L. P. and Marcus, C. M., Quantum dots, *Phys. World* **11**, 35–39 (1998).

[kou01] Kouwenhoven, L. P., Austing, D. G., and Tarucha, S., Few-electron quantum dots, *Rep. Prog. Phys.* **64**, 701 (2001).

[li08] Li, X., Wang, X., Zhang, L., Lee, S., and Dai, H., Chemically derived, ultrasmooth graphene nanoribbon semiconductors, *Science* **319**, 1229 (2008).

[lib09] Libisch, F., Stampfer, C., and Burgdörfer, J., Graphene quantum dots: Beyond a Dirac billiard, *Phys. Rev. B* **79**, 115423 (2009).

[lin08] Lin, Y.-M., Perebeinos, V., Chen, Z., and Avouris, P., Electrical observation of subband formation in graphene nanoribbons, *Phys. Rev. B* **78**, 161409(R) (2008).

[liu08] Liu, G., Velasco, J., and Lau, C. N., Graphene p-n-p junctions with contactless top gates, *Appl. Phys. Lett.* **92**, 203103 (2008).

[liu09] Liu, X., Oostinga, J. B., Morpurgo, A. F., and Vandersypen, L. M. K., Coulomb blockade in top-gated graphene nanoribbons, *Phys. Rev. B* **80**, 121407 (2009).

[los98] Loss, D. and DiVincenzo, D. P., Quantum computation with quantum dots, *Phys. Rev. A* **57**, 120 (1998).

[lu99] Lu, X., Huang, H., Nemchuk, N., and Ruoff, R., Patterning of highly oriented pyrolytic graphite by oxygen plasma etching, *Appl. Phys. Lett.* **75**, 193 (1999).

[lud04] Ludwig, A. W. W., Fisher, M. P. A., Shankar, R., and Grinstein, G., The integer quantum hall transition: A new approach and exact results, *Phys. Rev. B* **50**, 7526 (1994).

[lüs99] Lüscher, S., Fuhrer, A., Held, R., Heinzel, T., Ensslin, K., and Wegscheider, W., In-plane gate single electron transistor fabricated by scanning probe lithography, *Appl. Phys. Lett.* **75**, 2452 (1999).

[mac57] McClure, J. W., Band structure of graphite and de Haas-van alphen effect, *Phys Rev.* **108**, 612 (1957).

[mar08] Martin, J., Akerman, N., Ulbricht, G., Lohmann, T., Smet, J. H., von Klitzing, K., and Yacoby, A., Observation of electron–hole puddles in graphene using a scanning single-electron transistor, *Nat. Phys.* **4**, 144 (2008).

[mau02] Maultzsch, J., Reich, S., Thomsen, C., Webster, S., Czerw, R., Carroll, D. L., Vieira, S. M. C., Birkett, P. R., and Rego, C. A., Raman characterization of boron-doped multiwalled carbon nanotubes, *Appl. Phys. Lett.* **81**, 2647, (2002).

[mce99] McEuen, P. L., Bockrath, M., Cobden, D. H., Yoon, Y. -G., and Louie, S. Disorder, pseudospins, and backscattering in carbon nanotubes, *Phys. Rev. Lett.* **83**, 5098 (1999).

[mer08] Meric, I., Han, M. Y., Young, A. F., Oezyilmaz, B., Kim, P., and Shepard, K., Current saturation in zero-bandgap, top-gated graphene field-effect transistors, *Nat. Nanotechnol.* **3**, 654 (2008).

[mey07] Meyer, J., Geim, A. K., Katsnelson, M. I., Novoselov, K. S., Booth, T. J., and Roth, S., The structure of suspended graphene sheets, *Nature* **446**, 60 (2007).

[mia07] Miao, F., Wijeratne, S., Zhang, Y., Coskun, U. C., Bao, W., and Lau, C. N., Phase coherent transport of charges in graphene quantum billiard, *Science* **317**, 1530 (2007).

[mik99] Mikitik, G. P. and Sharlai, Y. V., Manifestation of Berry's phase in metal physics, *Phys. Rev. Lett.* **82**, 2147–2150 (1999).

[min06] Min, H., Hill, J. E., Sinitsyn, N. A., Sahu, B. R., Kleinman, L., and MacDonald, A. H., Intrinsic and rashba spinorbit interactions in graphene sheets, *Phys. Rev. B* **74**, 165310 (2006).

[mol07] Molitor, F., Güttinger, J., Stampfer, C., Graf, D., Ihn, T., and Ensslin, K., Local gating of a graphene Hall bar by graphene side gates, *Phys. Rev. B* **76**, 245426 (2007).

[mol09] Molitor, F., Jacobsen, A., Stampfer, C., Güttinger, J., Ihn, T., and Ensslin, K., Transport gap in side-gated graphene constrictions, *Phys. Rev. B* **79**, 075426 (2009).

[mos07] Moser, J., Barreiro, A., and Bachtold, A., Current-induced cleaning of graphene, *Appl. Phys. Lett.* **91**, 163513 (2007).

[muc09] Mucciolo, E.R., Castro Neto, A.H., and Lewenkopf, C.H., Conductance quantization and transport gaps in disordered graphene nanoribbons, *Phys. Rev. B* **79**, 075407 (2009).

[nan10] Nandamuri, G., Roumimov, S., and Solanki, R., Chemical vapor deposition of graphene films, *Nanotechnol.* **21**, 145604 (2010).

[nov04] Novoselov, K. S., Geim, A. K., Morozov, S. V., Jiang, D., Zhang, Y., Dubonos, S. V., Grigorieva, I. V., and Firsov, A. A., Electric field effect in atomically thin carbon films, *Science* **306**, 666 (2004).

[nov05] Novoselov, K. S., Geim, A. K., Morozov, S. V., Jiang, D., Katsnelson, M. I., Grigorieva, I. V., Dubonos, S. V., and Firsov, A. A., Two-dimensional gas of massless Dirac fermions in graphene. *Nature* **438**, 197 (2005).

[nov07] Novoselov, K. S., Jiang, Z., Zhang, Y., Morozov, S. V., Stormer, H. L., Zeitler, U., Maan, J. C., Boebinger, G. S., Kim, P., Geim, A. K., Room-temperature quantum hall effect in graphene, *Science* **315**, 1379 (2007).

[ost06] Ostrovsky, P. M., Gornyi, I. V., and Mirlin, A. D., Electron transport in disordered graphene, *Phys. Rev. B* **74**, 235443 (2006).

[oez07] Oezyilmaz, B., Jarillo-Herrero, P., Efetov, D., and Kim, P., Electronic transport in locally gated graphene nanoconstrictions, *Appl. Phys. Lett.* **91**, 192107 (2007).

[oez07a] Oezyilmaz, B., Jarillo-Herrero, P., Efetov, D., Abanin, D., Levitov, L. S., and Kim, P., Electronic transport and quantum hall effect in bipolar graphene p-n-p junctions, *Phys. Rev. Lett.* **99**, 166804 (2007).

[per06] Peres, N. M. R., Guinea, F., and Castro-Neto, A. H., Electronic properties of disordered two-dimensional carbon, *Phys. Rev. B* **73**, 125411 (2006).

[pet05] Petta, J. R., Johnson, A. C., Taylor, J. M., Laird, E. A., Yacoby, A., Lukin, M. D., Marcus, C. M., Hanson, M. P., and Gossard, A. C., Coherent manipulation of coupled electron spins in semiconductor quantum dots, *Science* **309**, 2180–2184 (2005).

[pis07] Pisana, S., Lazzeri, M., Casiraghi, C., Novoselov, K. S., Geim, A. K., Ferrari, A. C., and Mauri, F., Breakdown of the adiabatic Born–Oppenheimer approximation in graphene, *Nat. Mater.* **6**, 198 (2007).

[pon08] Ponomarenko, L. A., Schedin, F., Katsnelson, M. I., Yang, R., Hill, E. H., Novoselov, K. S., and Geim, A. K., Chaotic dirac billiard in graphene quantum dots, *Science* **320**, 356 (2008).

[ree88] Reed, M. A., Randall, J. N., Aggarwal, R. J., Matyi, R. J., Moore, T. M., and Wetsel, A. E., Observation of discrete electronic states in a zero-dimensional semiconductor nanostructure, *Phys. Rev. Lett.* **60**, 535 (1988).

[rec09] Recher, P., Nilsson, J., Burkard, G., and Trauzettel, B., Bound states and magnetic field-induced valley splitting in gate-tunable graphene quantum dots, *Phys. Rev. B*, **79**, 085407 (2009).

[rei02] Reich, S., Maultzsch, J., Thomsen, C., and Ordejón, P., Tight-binding description of graphene, *Phys. Rev. B* **66**, 035412 (2002).

[rei03] Reich, S., Thomsen, C., and Maultzsch, J., *Carbon Nanotubes*, Wiley-VCH, Weinheim, Germany, 2003.

[rod07] Roddaro, S., Pingue, P., Piazza, V., Pellegrini, V., and Beltram, F., The optical visibility of graphene: Interference colors of ultrathin graphite on SiO2, *Nano Lett.* **7**, 2707 (2007).

[sai98] Saito, R., Dresselhaus, G., and Dresselhaus, M. S., *Physical Properties of Carbon Nanotubes*. Imperial College Press, London, U.K., 1998.

[sap06] Sapmaz, S., Meyer, C., Beliczynski, P., Jarillo-Herrero, P., and Kouwenhoven, L. P., Excited state spectroscopy in carbon nanotube double quantum dots, *Nano Lett.* **6**, 1350 (2006).

[sch08] Schnez, S., Ensslin, K., Sigrist, M., and Ihn, T., Analytical model of the energy spectrum of a graphene quantum dot in a perpendicular magnetic field, *Phys. Rev. B* **78**, 195427 (2008).

[sch09] Schnez, S., Molitor, F., Stampfer, C., Güttinger, J., Shorubalko, I., Ihn, T., and Ensslin, K., Observation of excited states in a graphene quantum dot, *Appl. Phys. Lett.* **94**, 012107 (2009).

[sha08] Shaji, N., Simmons, C. B., Thalakulam, M., Klein, L. J., Qin, H., Luo, H., Savage, D. E., Lagally, M. G., Rimberg, A. J., Joynt, R., Friesen, M., Blick, R. H., Coppersmith, S. N., and Eriksson, M. A., Spin blockade and lifetime-enhanced transport in a few-electron Si/SiGe double quantum dot, *Nat. Phys.* **4**, 540 (2008).

[sho06] Shorubalko, I., Pfund, A., Leturcq, R., Borgström, M. T., Gramm, F., Müller, E., Gini, E., and Ensslin, K., Tunable few electron quantum dots in InAs nanowires, *Nanotechnology* **18**, 044014 (2006).

[sol07] Sols, F., Guinea, F., and Castro-Neto, A. H., Coulomb blockade in graphene nanoribbons, *Phys. Rev. Lett.* **99**, 166803 (2007).

[sta07] Stampfer, C., Bürli, A., Jungen, A., and Hierold, C., Raman imaging for processing and process monitoring for nanotube devices, *Phys. Status Solidi B* **244**, 4341 (2007).

[sta07a] Stampfer, C., Molitor, F., Graf, D., Ensslin, K., Jungen, A., Hierold, C., and Wirtz, L., Raman imaging of doping domains in graphene on SiO2, *Appl. Phys. Lett.* **91**, 241907 (2007).

[sta08] Stampfer, C., Güttinger, J., Molitor, F., Graf, D., Ihn, T., and Ensslin, K., Tunable coulomb blockade in nanostructured graphene, *Appl. Phys. Lett.* **92**, 012102 (2008).

[sta08a] Stampfer, C., Schurtenberger, E., Molitor, F., Güttinger, J., Ihn, T., and Ensslin, K., Tunable graphene single electron transistor, *Nano Lett.* **8**, 2378 (2008).

[sta09] Stampfer, C., Güttinger, J., Hellmüller, S., Molitor, F., Ensslin, K., and Ihn, T., Energy gaps in etched graphene nanoribbons, *Phys. Rev. Lett.* **102**, 056403 (2009).

[sta09a] Stampfer, C., Schurtenberger, E., Molitor, F., Güttinger, J., Ihn, T., and Ensslin, K., Transparency of narrow constrictions in a graphene single electron transistor, *Int. J. Mod. Phys. A* **23**, 2647 (2009).

[su93] Su, R. K., Siu, G. C., and Chou, X., Barrier penetration and Klein paradox, *J. Phys. A* **26**, 1001 (1993).

[tan07] Tan, Y.-W., Zhang, Y., Bolotin, K., Zhao, Y., Adam, S., Hwang, E. H., Das Sarma, S., Stormer, H. L., and Kim, P., Measurement of scattering rate and minimum conductivity in graphene, *Phys. Rev. Lett.* **99**, 246803 (2007).

[tho00] Thomsen, C. and Reich, S., Double resonant Raman scattering in graphite, *Phys. Rev. Lett.* **85**, 5214 (2000).

[tod09] Todd, K., Chou, H.-T., Amasha, S., and Goldhaber-Gordon, D., Quantum dot behavior in graphene nanoconstrictions, *Nano Lett.* **9**, 416 (2009).

[tra07] Trauzettel, B., Bulaev, D. V., Loss, D., and Burkard, G., Spin qubits in graphene quantum dots, *Nat. Phys.* **3**, 192 (2007).

[two06] Tworzydlo, J., Trauzettel, B., Titov, M., Rycerz, A., and Beenakker, C. W. J., Sub-poissonian shot noise in graphene, *Phys. Rev. Lett.* **96**, 246802 (2006).

[ves68] Veselago, V. G., The electrodynamics of substances with simultaneous negative values of e and μ, *Sov. Phys. Usp.* **10**, 509 (1968).

[wak07] Wakabayashi, K., Takane, Y., and Sigrist, M., Perfectly Conducting Channel and Universality Crossover in Disordered Graphene Nanoribbons, *Phys. Rev. Lett.* **99**, 036601 (2007).

[wak09] Wakabayashi, K., Takane, Y., Yamamoto, M., and Sigrist, M., Electronic transport properties of graphene nanoribbons, *New J. Phys.* **11**, 095016 (2009).

[wal47] Wallace, P. R., The band theory of graphite, *Phys Rev.* **71**, 622 (1947).

[wan08] Wang, X., Ouyang, Y., Li, X., Wang, G., Guo, J., and Dai, H., Room-temperature all-semiconducting sub-10-nm graphene nanoribbon field-effect transistors, *Phys. Rev. Lett.* **100**, 206803 (2008).

[whi07] White, C. T., Li, J., Gunlycke, D., and Mintmire, J. W., Hidden one-electron interactions in carbon nanotubes revealed in graphene nanostrips, *Nano Lett.* **7**, 825 (2007).

[wil07] Williams, J. R., DiCarlo, L., and Marcus, C. M., Quantum hall effect in a gate-controlled p-n junction of graphene, *Science* **317**, 938 (2007).

[yan07] Yang, L., Park, C.-H., Son, Y.-W., Cohen, M. L., and Louie, S. G., Quasiparticle energies and band gaps in graphene nanoribbons, *Phys. Rev. Lett.* **99**, 186801 (2007).

[zha05] Zhang, Y., Tan, Y.-W., Stormer, H. L., and Kim, P., Experimental observation of the quantum Hall effect and Berry's phase in graphene, *Nature* **438**, 201–204 (2005).

[zha05a] Zhang, Y., Small, J.P., Pontius, W.V., and Kim, P., Fabrication and electric field dependent transport measurements of mesoscopic graphite devices, *Appl. Phys. Lett.* **86**, 073104 (2005).

[zha08] Zhang, Q., Fang, T., Xing, H., Seabaugh, A., and Jena, D., Graphene nanoribbon tunnel transistors, *IEEE Electron Device Lett.* **29**, 1344 (2008).

[zhe02] Zheng, Y. and Ando, T., Hall conductivity of a two-dimensional graphite system, *Phys. Rev. B* **65**, 245420 (2002).

[zho06] Zhou, S. Y., Gweon, G.-H., Graf, J., Fedorov, A. V., Spataru, C. D., Diehl, R. D., Kopelevich, Y., Lee, D.-H., Louie, S. G., and Lanzara, A., First direct observation of Dirac fermions in graphite, *Nat. Phys.* **2**, 595 (2006).

[zie98] Ziegler, K., Delocalization of 2D Dirac fermions: the role of a broken symmetry, *Phys. Rev. Lett.* **80**, 3113–3116 (1986).

22

Magnetic Graphene Nanostructures

Oleg V. Yazyev

Ecole Polytechnique Fédérale de Lausanne (EPFL)

and

Institut Romand de Recherche Numérique en Physique des Matériaux (IRRMA)

22.1 Introduction

Magnetic materials constitute an essential part of modern technology. These materials are mostly based on the elements belonging to either the *d*- or the *f*-block of the periodic table. Among the periodic table elements only Fe, Co, and Ni exhibit spontaneous ferromagnetism at room temperature, that is, in the absence of an external magnetic field. The magnetic properties of these transition metal elements originate from the partially filled *d*-electron bands. Spontaneous magnetism is not common for the *p*-block elements belonging to the second period of the periodic table, despite the fact that carbon is able to form an extraordinary number of different molecular structures. Magnetic materials based on light elements are nevertheless very promising for technological applications as they may possess a number of precious properties such as low density, low production costs, and biocompatibility.

The first encouraging experimental results were reported in 1991 when magnetism was observed in crystalline *p*-nitrophenyl nitronyl nitroxide (*p*-NPNN) (Takahashi et al. 1991; Tamura et al. 1991) and in a charge-transfer complex of C_{60} and tetrakis(dimethylamino)ethylene (TDAE) (Allemand et al. 1991). These organic materials are made of light elements only (C, H, N, and O). In their crystals, the unpaired electron spins are localized on weakly coupled molecular building blocks. Because of the weak coupling between the localized electron spins, the long-range magnetic order in these materials is realized only below 0.6 and 16 K, respectively, which renders them useless for practical applications. Since 1991, many other similar compounds have been investigated. However, in all cases the temperatures

below which the long-range magnetic order establishes (Curie temperatures and Néel temperatures for the ferromagnetic and antiferromagnetic materials, respectively) were much lower than the room temperature. The second wave of interest in magnetic carbon materials came 10 years later after ferromagnetism with Curie temperature near 500 K was observed in the high-pressure rhombohedral C_{60} (Makarova et al. 2001). Then, room-temperature ferromagnetism was observed in proton-irradiated graphite (Esquinazi et al. 2003). Although the magnetic signal in rhombohedral C_{60} was later attributed to the presence of iron impurities (Makarova et al. 2006), the observation of ferromagnetism of irradiated graphite samples received increasing confirmation by numerous research groups. Recent experimental investigations have revealed that the magnetic order in proton-bombarded graphite has a two-dimensional (i.e., graphene-like) character (Barzola-Quiquia et al. 2007) and originates from the carbon π-electron system rather than from impurities (Ohldag et al. 2007). These results are supported by theoretical calculations, which show the crucial role of defect-induced states in explaining the magnetism of irradiated graphite (Lehtinen et al. 2004; Kumazaki and Hirashima 2007; Yazyev and Helm 2007; Yazyev 2008b).

The experimental achievements mentioned above stimulated further search of novel forms of magnetism in carbon-based systems through both experimental and theoretical approaches (Makarova and Palacio 2006). This search focuses primarily on materials and nanostructures derived from graphene (Figure 22.1a), a truly two-dimensional form of carbon that has been isolated only very recently (Novoselov et al. 2004), but quickly attracted enormous attention in both science and

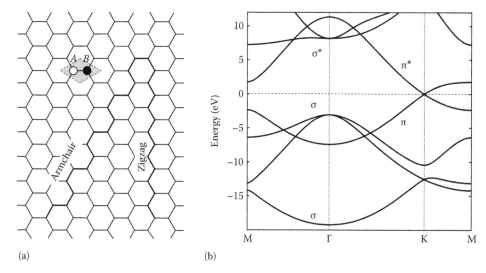

(a)   (b)

FIGURE 22.1 (a) Two-dimensional crystalline lattice of graphene. The shaded area marks the unit cell of graphene containing two carbon atoms. Carbon atoms which belong to sublattice A and sublattice B are shown as empty and filled circles, respectively. The two high-symmetry directions are emphasized. (b) Band structure of graphene obtained from first-principles calculations. The bands are labeled according to their symmetry and binding character. The zero energy corresponds to the Fermi level.

technology (Geim and Novoselov 2007; Katsnelson 2007). This broad class of sp^2 carbon materials involves polyaromatic molecules, fullerenes, graphite, and their further modifications obtained by patterning, chemical treatment, or the implantation of defects and impurities.

The magnetic graphene nanostructures are particularly promising for applications in the field of spintronics, one of the most probable turns of the evolution of electronics industry, which promises information storage, processing, and communication at faster speeds and lower energy consumption. While traditional electronics utilizes only the charge of electrons, spintronics will also make use of the spin of electrons. Magnetic materials based on light p-block elements are expected to have high magnitudes of the spin wave stiffness (Edwards and Katsnelson 2006) and, thus, nanostructures made of these elements would possess higher Curie temperatures or spin correlation lengths (Yazyev and Katsnelson 2008). Several novel physical effects have been predicted for magnetic graphene nanostructures. For instance, the half-metallicity of zigzag graphene nanoribbons triggered by external electric fields (Son et al. 2006b) would allow realizing in practice efficient electric control of spin transport. Finally, the light element materials display weak spin-orbit and hyperfine couplings that are the main channels of relaxation and decoherence of electron spins (Trauzettel et al. 2007; Yazyev 2008a). This makes carbon nanomaterials promising for the transport of spin-polarized currents and for the spin-based quantum information processing.

This tutorial chapter illustrates how the magnetism in graphene nanostructures can be understood and modeled by means of simple theoretical approaches such as the mean-field Hubbard model. Section 22.2 introduces the theoretical model and its specific consequences for describing the electronic structure and magnetic properties of graphene-based materials. Then, the described theory is applied to three important cases:

(1) finite graphene nanofragments, (2) one-dimensional edges and nanoribbons of graphene, and (3) the magnetism in graphene and graphite induced by the presence of point defects. The future perspectives of the field are outlined in Section 22.4.

22.2 Computational Tools

22.2.1 Tight Binding and Mean-Field Hubbard Model

The vast majority of computational studies of magnetic carbon nanostructures is currently preformed using the first-principles electronic structure methods relying on density functional theory. Surprisingly, most of the results obtained by means of first-principles electronic structure calculations can be reproduced using simple model Hamiltonians. Moreover, the simplified models allow deeper understanding of the results obtained. The most important model is the one-orbital mean-field Hubbard model. This model considers only π symmetry electronic states that are formed by the unhybridized p_z atomic orbitals of sp^2 carbon atoms. The π states play the dominant role in low-energy electronic properties of graphene systems (Figure 22.1b). The Hubbard model Hamiltonian can be written as

$$\mathcal{H} = \mathcal{H}_0 + \mathcal{H}'. \tag{22.1}$$

The first term is the nearest-neighbor tight-binding Hamiltonian

$$\mathcal{H}_0 = -t \sum_{\langle i,j \rangle, \sigma} \left[c_{i\sigma}^\dagger c_{j\sigma} + h.c. \right], \tag{22.2}$$

in which the operators $c_{i\sigma}$ and $c_{i\sigma}^\dagger$ annihilate and create an electron with spin σ at site i, respectively. The notation $\langle i,j \rangle$ stands for the pairs of nearest-neighbor atoms and "h.c." is its Hermite conjugate counterpart. The hopping integral $t \approx 2.7$ eV defines the energy scale of the tight-binding Hamiltonian. This physical model is equivalent to the Hückel method familiar to chemists. From the computational point of view, the Hamiltonian matrix is determined solely by the atomic structure: the off-diagonal matrix elements (i,j) and (j,i) are set to $-t$ when atoms i and j are covalently bonded, and to 0 otherwise. In a neutral graphene system, each sp^2 carbon atom contributes one p_z orbital and one π electron. The π electron system is thus said to be half-filled. The spectrum of the eigenvalues of tight-binding Hamiltonian matrix is symmetric with respect to zero energy. This means that in a neutral graphene system for each eigenvalue $\varepsilon < 0$ corresponding to an occupied (binding) state, there is an unoccupied (anti-binding) state with $\varepsilon^* = -\varepsilon$. This property is often called electron-hole symmetry. The states with $\varepsilon = 0$ are called zero-energy states (also referred to as non-binding or midgap states).

The nearest-neighbor tight-binding model has proved to describe accurately the electronic structure of graphene, carbon nanotubes, and other nonmagnetic sp^2 carbon materials. However, electron–electron interactions have to be introduced in some form in order to describe the onset of magnetism. Within the Hubbard model these interactions are introduced through the on-site Coulomb repulsion

$$\mathcal{H}' = U \sum_i n_{i\uparrow} n_{i\downarrow}, \qquad (22.3)$$

where $n_{i\sigma} = c_{i\sigma}^\dagger c_{i\sigma}$ is the spin-resolved electron density at site i; the parameter $U > 0$ defines the magnitude of the on-site Coulomb repulsion. This model considers only short-range Coulomb repulsion, that is, two electrons interact only if they occupy the p_z atomic orbital of the same atom. Despite its apparent simplicity, this term is no longer trivial from the practical point of view. A step toward its simplification is the mean-field approximation

$$\mathcal{H}' = U \sum_i \left(n_{i\uparrow} \langle n_{i\downarrow} \rangle + \langle n_{i\uparrow} \rangle n_{i\downarrow} - \langle n_{i\uparrow} \rangle \langle n_{i\downarrow} \rangle \right) \qquad (22.4)$$

under which a spin-up electron at site i interacts with the average spin-down electron population $\langle n_{i\downarrow} \rangle$ at the same site, and vice versa. An attentive chemist may notice that this mean-field model represents a variation of the unrestricted Hartree–Fock method. From the computational point of view, the electron–electron interaction term concerns only the diagonal elements of the Hamiltonian matrix. The diagonal elements of the spin-up and spin-down block now depend on initially unknown $\langle n_{i\downarrow} \rangle$ and $\langle n_{i\uparrow} \rangle$, respectively. Thus, the problem has to be solved self-consistently. The solution provides the spin densities

$$M_i = \frac{\langle n_{i\uparrow} \rangle - \langle n_{i\downarrow} \rangle}{2} \qquad (22.5)$$

at each atom i and the total spin of the system $S = \sum_i M_i$. For a given graphene structure both local and total spins depend exclusively on the dimensionless parameter U/t. It has been shown that the mean-field Hubbard model is able to reproduce closely the results of first-principles calculations if the parameter U/t is chosen appropriately (Fernández-Rossier and Palacios 2007; Pisani et al. 2007). In particular, the results of density functional theory calculations using a generalized gradient approximation exchange-correlation functional are best reproduced when $U/t \approx 1.3$. The results of the local spin density approximation calculations are best fitted using $U/t \approx 0.9$ (Pisani et al. 2007). On the experimental side, the magnitudes of $U \sim 3.0$–$3.5\,\mathrm{eV}$ have been deduced from the magnetic resonance studies of neutral soliton states in *trans*-polyacetylene, a one-dimensional sp^2 carbon material (Thomann et al. 1985; Kuroda and Shirakawa 1987). This interval corresponds to $U/t \sim 1.1$–1.3 in good agreement with the results of the generalized gradient approximation calculations for graphene. Increasing U/t leads to the enhancement of magnetic moments. The range of meaningful magnitudes is limited by $U/t \approx 2.2$ above which the ideal graphene breaks the spin-spatial symmetry and develops antiferromagnetic order (Kumazaki and Hirashima 2007). In the computational examples considered below a value of $U/t = 1.2$ is used.

22.2.2 Benzenoid Graph Theory and Lieb's Theorem

There are two important statements concerning the nearest-neighbor tight-binding model and the Hubbard model of graphene-based systems: the benzenoid graph theory and Lieb's theorem. The combination of these two simple counting rules yields general predictions on magnetism in finite graphene systems (Wang et al. 2009). The honeycomb lattice of graphene belongs to a broad class of bipartite lattices. A bipartite lattice can be partitioned into two mutually interconnected sublattices A and B. Each atom belonging to sublattice A is connected to the atoms in sublattice B only, and vice versa. The graphene systems whose faces are topological hexagons are called benzenoid systems (or honeycomb systems). Carbon atoms in such systems have either three or two nearest neighbors. The class of benzenoid systems is a subclass of bipartite systems.

The spectrum of tight-binding Hamiltonian of a honeycomb system can be analyzed using a mathematically rigorous approach of the benzenoid graph theory (Fajtlowicz et al. 2005). An important result for us is that this theory is able to predict the number of zero-energy states of the Hamiltonian. The number of such states is equal to the graph's nullity

$$\eta = 2\alpha - N, \qquad (22.6)$$

where

N is the total number of sites
α is the maximum number of nonadjacent sites

The onset of magnetism in the system is determined by the competition of the exchange energy gain and the kinetic energy penalty associated with the spin-polarization of the system (Mohn 2003). The gain in exchange energy is due to the exchange splitting of the electronic states subjected to spin-polarization (Palacios et al., 2008)

$$\Delta_S = \varepsilon_\uparrow - \varepsilon_\downarrow = \frac{U}{2} \sum_i n_i^2, \qquad (22.7)$$

where $\sum_i n_i^2$ is the inverse participation ratio, a measure of the degree of localization of the corresponding electronic state. The kinetic energy penalty is proportional to the energy of this state. Thus, the zero-energy states undergo spin-polarization at any $U > 0$, irrespective of their degree of localization. One can view spin-polarization as one of the mechanisms for escaping an instability associated with the presence of low-energy electrons in the system. Other mechanisms, such as the Peierls distortion, were shown to be inefficient in graphene nanostructures (Pisani et al. 2007).

Although the benzenoid graph theory is able to predict the occurrence of zero-energy states, it is not clear how the electron spins align in these states. The complementary knowledge comes from Lieb's theorem (Lieb 1989), which determines the total spin of a bipartite system described by the Hubbard model. This theorem states that in the case of repulsive electron–electron interactions ($U > 0$), a bipartite system at half-filling has the ground state characterized by the total spin

$$S = \frac{1}{2}\left| N_A - N_B \right|, \qquad (22.8)$$

where N_A and N_B are the numbers of sites in sublattices A and B, respectively. The ground state is unique and the theorem holds in all dimensions without the necessity of a periodic lattice structure.

In Section 22.3, the application of these two simple counting rules will be illustrated on small (~1 nm size) graphene fragments (Wang et al. 2009).

22.3 Magnetism in Graphene Nanostructures

22.3.1 Finite Graphene Fragments

Let us consider the three graphene fragments shown in Figure 22.2. From the point of view of our single-orbital physical models, only the connectivity of the π-electron conjugation network is important. Generally speaking, these π-systems may form only small parts of much more complex molecular systems. Some readers may prefer to make a correspondence between the π-electron networks shown in Figure 22.2 and the corresponding polycyclic aromatic hydrocarbon molecules. In other words, the edges of the graphene fragments are supposed to be passivated by hydrogen atoms. The hexagonal graphene fragment shown in Figure 22.2a is thus equivalent to the coronene molecule. For this fragment, the number of sites belonging to the two sublattices is equal, i.e., $N_A = N_B = 12$. The number of nonadjacent sites can be maximized by selecting all atoms belonging to one of the sublattices, i.e., $\alpha = 12$. Thus, both the number of zero-energy states η and the total spin S are zero. The tight-binding model predicts a wide bandgap of $1.08t$ ($\approx 3.0\,eV$) for this graphene molecule. The mean-field Hubbard model does not show any local magnetic moments.

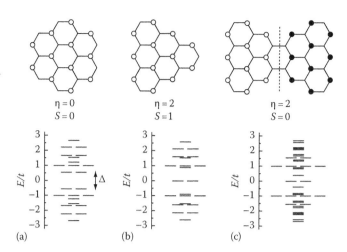

FIGURE 22.2 Atomic structures and tight-binding model spectra of three graphene fragments: (a) coronene, (b) triangulane, and (c) a bow-tie-shaped fragment. Nonadjacent sites are labeled by circles. Empty circles (filled) correspond to sublattice $A(B)$.

Secondly, we consider a triangle-shaped fragment shown in Figure 22.2b. The corresponding hypothetical aromatic hydrocarbon is called triangulane. Contrary to coronene, the two sublattices of triangulane are no longer equivalent: $N_A = 12$ and $N_B = 10$. The unique choice maximizing the number of nonadjacent sites is achieved by selecting the atoms belonging to sublattice A only, i.e., $\alpha = N_A = 12$. Thus, the benzenoid graph theory predicts the presence of two zero-energy states on sublattice A. Lieb's theorem predicts the $S = 1$ (spin-triplet) ground state. The two low-energy electrons populate a pair of zero-energy states according to Hund's rule, i.e., their spins are oriented parallel to each other. The results of the mean-field Hubbard model calculations for this system are shown in Figure 22.3a. One can see that spin-polarization lifts the degeneracy of zero-energy electronic states and opens an energy gap of $0.30t$. Therefore, the system is stabilized by the spin-polarization. Most of the spin-up electron density localized on the atoms in sublattice A originates from the two electrons populating the non-binding states. However, one can notice an appreciable amount of spin-down density on the atoms in sublattice B that is compensated by an equivalent contribution of the spin-up density in sublattice A. These induced magnetic moments are due to the spin-polarization of fully populated states driven by the exchange interaction with the two unpaired electrons. Although the triangulane hydrocarbon itself has never been isolated, the synthesis of its chemical derivatives has been reported by several groups (Allinson et al. 1995; Inoue et al. 2001). Indeed, electron spin resonance studies have verified the spin-triplet ground state of these triangulane derivatives, nature's striking examples of stable non-Kekuléan aromatic molecules. It is remarkable that the number of zero-energy states and, hence, the total spin of the triangle-shaped fragments scale linearly with their linear size (Ezawa 2007; Fernández-Rossier and Palacios 2007; Wang et al. 2008).

The third bow-tie-shaped graphene molecule (Figure 22.2c) is composed of two triangulane fragments sharing a corner

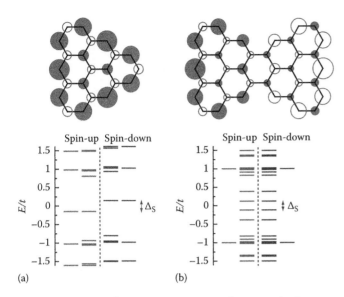

FIGURE 22.3 Local magnetic moments and spin-resolved energy levels obtained through the mean-field Hubbard model calculations of (a) triangulane and (b) the bow-tie-shaped graphene fragment ($U/t = 1.2$). Area of each circle is proportional to the magnitude of the local magnetic moment at each atom. Filled circles represent positive magnitudes and empty circles represent negative magnitudes of the spin density.

hexagon. This molecular structure was first proposed by Eric Clar and named "Clar's goblet" after him (Clar 1972). For this molecule Lieb's theorem predicts the spin-singlet ground state ($N_A = N_B = 19$). The largest set of nonadjacent sites ($\alpha = 20$) involves the atoms belonging to both sublattice A and sublattice B in the left and right parts of the structure as shown in Figure 22.2c. Consequently, there are $\eta = 2 \times 20 - 38 = 2$ zero-energy states in agreement with the tight-binding calculation. These states are spatially segregated in the two triangular parts of the molecule (Wang et al. 2009). In order to satisfy the spin-singlet ground state, the two zero-energy states have to be populated by two unpaired electrons oriented antiparallel to each other. Such a singlet state is said to break the spin-spatial symmetry or, in other words, the system exhibits antiferromagnetic ordering. This result can be verified by the mean-field Hubbard model calculations, as shown in Figure 22.3b. At first sight, this simple structure seems to violate Hund's rule. However, one has to keep in mind that the non-binding states are localized within one of the two graphene sublattices. Thus, there are two electronic sub-bands, each populated by electrons according to Hund's rule. The coupling between the electron spins in these two sub-bands is antiferromagnetic due to the superexchange mechanism.

These two counting rules thus provide an easy way for predicting the occurrence of local magnetic moments in small graphene nanostructures as well as the relative orientation of these moments (Wang et al. 2009). This simple approach allows for designing graphene molecules with predetermined magnetic properties.

22.3.2 Zigzag Edges and Nanoribbons

As one moves on to larger graphene fragments, the application of counting rules may no longer seem to be practical. An alternative approach focuses on the possible effects of the edges of graphene nanostructures. Generally speaking, there are two high-symmetry crystallographic directions in graphene, armchair and zigzag, as shown in Figure 22.1a. Cutting graphene along these directions produces armchair and zigzag edges, respectively. The edge effects can be conveniently modeled by considering one-dimensional periodic strips of graphene that are usually referred to as nanoribbons (Figure 22.4).

The electronic structure of armchair and zigzag nanoribbons shows, however, pronounced differences. Figure 22.4a shows the tight-binding band structure of a ~1.5-nm-width armchair nanoribbon. In this particular case, confining electrons in one direction by introducing a pair of parallel armchair edges opens a gap of $0.26t$. Generally speaking, within the tight-binding model, armchair nanoribbons exhibit either metallic or semiconducting behavior depending on their width (Nakada et al. 1996; Son et al. 2006a). According to the same model, all zigzag nanoribbons are predicted to be metallic and feature a flat band in the low-energy region (Figure 22.4b). The physical phenomenon behind these low-energy states is however different from the one discussed in Section 22.3.1. Strictly speaking, this flat band does not represent zero-energy states, but the states whose energies approach zero with increasing the nanoribbons width. The low-energy states are localized at the edge and decay quickly away from it. The flat band develops in exactly one-third of the one-dimensional Brillouin zone at $k \in [(2/3)(\pi/a);(\pi/a)]$

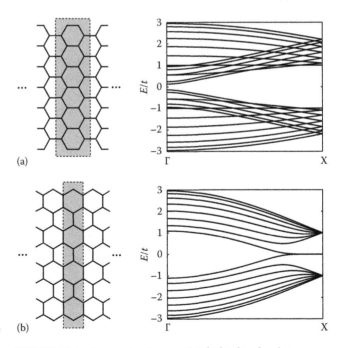

FIGURE 22.4 Atomic structures and tight-binding band structures of (a) armchair and (b) zigzag graphene nanoribbons. Shaded areas represent the unit cells of the graphene nanoribbons.

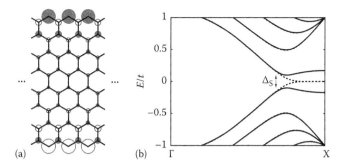

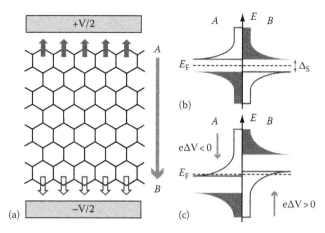

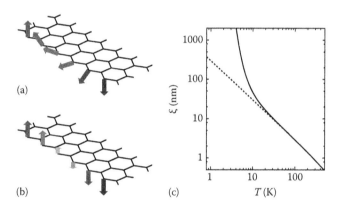

FIGURE 22.5   (a) Local magnetic moments obtained through the mean-field Hubbard model calculation of zigzag graphene nanoribbons ($U/t = 1.2$). Area of each circle is proportional to the magnitude of the local magnetic moment at each atom. Filled circles represent positive magnitudes and empty circles represent negative magnitudes of the spin density. (b) Mean-field Hubbard model band structure (solid lines) compared to the tight-binding band structure (dashed lines) of the zigzag graphene nanoribbon.

FIGURE 22.6   Scheme of the electric field induced half-metallicity of graphene nanoribbons. (a) Electric field is applied across the nanoribbon, from edge A (spin-up, filled arrows) to edge B (spin-down, empty arrows). (b) Schematic representation of the edge- and spin-resolved density of states of the system at zero applied field. (c) Applied electric field closes the bandgap at the Fermi level E_F for spin-down electrons only.

FIGURE 22.7   Schematic representation of the transverse (a) and the longitudinal (b) low-energy spin excitation at zigzag graphene edges. The magnetic moments are shown by arrows. Their directions are represented by orientation of the arrows. The magnitude is illustrated through the arrow lengths. (c) Spin correlation length ξ as a function of temperature when the weak magnetic anisotropy of carbon is taken into account (solid line) and neglected (dashed line). (Reprinted from Yazyev, O.V. and Katsnelson, M.I., *Phys. Rev. Lett.*, 100, 047209, 2008. With permission.)

($a = 0.25\,$nm is the unit cell of the zigzag edge). The presence of low-energy states suggests a possibility of induced magnetic moments. Indeed, the mean-field Hubbard model solution for this system shows localized magnetic moments at the edges (Figure 22.5a). The localized magnetic moments display ferromagnetic ordering along the zigzag edge. According to Lieb's theorem the net magnetic moment of the nanoribbon has to vanish because of the equal number of atoms in sublattices A and B. This condition is satisfied since the orientation of the magnetic moments localized at the opposite edges is antiparallel. The band structure of the mean-field Hubbard model solution is compared to the tight-binding band structure in Figure 22.5b. One can see that spin-polarization splits the low-energy flat band ($\Delta_S = 0.20t$ at $U/t = 1.2$), while leaving the rest of electronic states untouched.

The intriguing magnetic properties of zigzag graphene nanoribbons may find applications in future spintronic devices. An attractive one is the recent prediction of the controlled half-metallicity of zigzag nanoribbons (Son et al. 2006b). The half-metallicity, that is, the coexistence of metallic nature for electrons with one-spin orientation and insulating nature for electrons with the other, can be triggered by means of a transverse external electric field (Figure 22.6a). This physical effect is schematically explained in Figure 22.6b. The direction of applied electric field defines the spin channel with metallic band structure. If realized in practice, this simple device would offer an efficient electric control of spin transport – the Holy Grail of spintronics.

More importantly, however, these novel nanoscale devices also face novel limitations. One of them stems from the one-dimensional character of magnetism at zigzag graphene edges. The Mermin–Wagner theorem excludes long-range order in one-dimensional magnetic systems at any finite temperature (Mermin and Wagner 1966). The range of magnetic order is limited by the spin correlation length ξ (and, thus, the dimensions of the device), which is determined by the transverse

and longitudinal spin fluctuations (Figure 22.7a and b). First-principles calculations have shown that the transverse spin fluctuations (spin waves) play a dominant role in breaking the long-range magnetic order at zigzag graphene edges (Yazyev and Katsnelson 2008). Despites the high spin-wave stiffness, a spin correlation length of ~1 nm at 300 K is predicted for this truly one-dimensional system. Reducing temperature increases ξ, which reaches the micrometer range below 10 K (Figure 22.7b).

It is worth mentioning that at the time this chapter was written no direct proof of edge magnetism in graphene was reported. Exciting magnetic phenomena described in this section are still awaiting experimental confirmations. Nevertheless, the presence of localized low-energy states at the zigzag edges of graphene has been verified by means of scanning tunneling microscopy (STM; Kobayashi et al. 2005, 2006).

22.3.3 Point Defects in Graphene and Graphite

At last, let us discuss the origin of ferromagnetism in proton-irradiated graphite, the first carbon-based material in which the long-range magnetic order has been experimentally observed at sufficiently high temperatures. The three-dimensional crystalline lattice of graphite is composed of weakly coupled graphene layers (Figure 22.8a). Irradiating graphite with high-energy particles, such as protons, results in creation of several types of point defects. The incident particles lose part of their kinetic energy by transferring momentum to carbon atoms in graphite lattice. If the transferred energy is larger than the displacement threshold $T_d \approx 20\,\text{eV}$, the carbon atom may leave its equilibrium position (Banhart 1999). This results in the formation of a pair of point defects: a vacancy, i.e., a one-atom hole in the graphene layer (Figure 22.8b), and an interstitial, the carbon atom trapped between the adjacent graphene layers (Figure 22.8c) (Telling et al. 2003). The slowed-down protons stick to carbon atoms resulting in hydrogen chemisorption defects (Figure 22.8d). From the point of view of one-orbital models, both vacancy and hydrogen chemisorptions defects remove one p_z orbital from the π-electron system of graphene. In the first case, the p_z orbital is eliminated as a part of the knocked-out carbon atom. The chemisorption of hydrogen does not remove the carbon atom itself, but changes its hybridization to the sp^3 state that can not contribute a p_z orbital to the π-electron system. These two types of defects are further referred as p_z-vacancies. On the contrary, the stable form of interstitial carbon atoms, shown in Figure 22.8c is equivalent to a pair of di-vacancies (i.e., the missing pairs of

neighboring p_z orbitals) in the two graphene layers adjacent to the interstitial atom.

The p_z-vacancy defects introduce remarkable changes into the electronic structure of ideal graphene. Let us consider a periodically repeated supercell of graphene composed of $2N$ ($N_A = N_B = N$) carbon atoms. The removal of one atom from sublattice A introduces a zero-energy state in the complementary sublattice ($\alpha = N_B$; hence $\eta = 2N_B - ((N_A - 1) + N_B) = 1$). These non-binding states have been observed in the STM images of graphite as triangle-shaped $\left(\sqrt{3} \times \sqrt{3}\right)R30°$ superstructures localized around defects (Mizes and Foster 1989; Kelly and Halas 1998; Ruffieux et al. 2000). Such defect-induced states are quasi-localized states since they decay as $1/r$ when $r \to \infty$ (Pereira et al. 2006). Lieb's theorem predicts a magnetic moment of $|(N_A - 1) + N_B| = 1\mu_B$ per defect, that is, the presence of defects induces ferromagnetic order in graphene. This result has been widely confirmed by both first-principles (Duplock et al. 2004; Lehtinen et al. 2004; Yazyev and Helm 2007) and mean-field Hubbard model (Kumazaki and Hirashima 2007; Palacios et al. 2008) calculations. The distribution of spin density around a single p_z-vacancy in graphene obtained from the mean-field Hubbard model calculations clearly shows the $\left(\sqrt{3} \times \sqrt{3}\right)R30°$ superstructure (Figure 22.9a).

However, such a system with one defect per supercell does not present a realistic model of disordered graphene because of the following assumptions. First, all p_z-vacancies are located in the same sublattice. Second, the defects form a periodic superlattice. A step toward a more realistic description of disordered graphene may consist in constructing models with defects randomly distributed in a large enough supercell (Yazyev 2008b). Such models allow defects to occupy both sublattices at arbitrary concentrations. Any short-range order in the spatial arrangement of defect is eliminated. Figure 22.9b shows the distribution of spin density in such a model. Let us try to understand this result. From Lieb's theorem the total spin per supercell is

$$S = \frac{1}{2}\left|N_A - N_B\right| = \frac{1}{2}\left|N_B^d - N_A^d\right|,$$

where $N_A^d\left(N_B^d\right)$ is the number of vacancies created in sublattice $A(B)$. That is, electron spins populating the quasi-localized states induced by defects in the same (different) sublattice orient parallel (antiparallel) to each

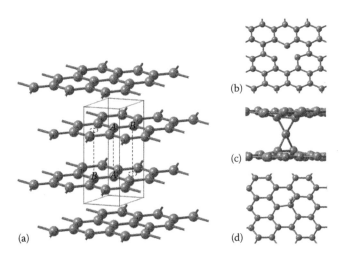

(a)   (b)   (c)   (d)

FIGURE 22.8 (a) Atomic structure of crystalline graphite (*ABA* stacking). Unit cell edges are shown as solid lines. Dashed lines illustrate the nonequivalence of atoms labeled as *A* and *B* inside the unit cell of *ABA* graphite. Atomic structures of vacancy (b), interstitial (c), and hydrogen chemisorption (d) defects in graphite.

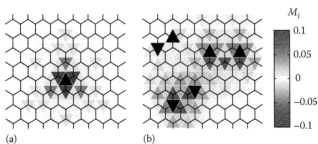

(a)   (b)

FIGURE 22.9 Spin density distribution around a single p_z-vacancy (a) and in a system with random distribution of defects obtained through the mean-field Hubbard model calculations ($U/t = 1.2$). Positions of defects belonging to sublattice $A(B)$ are labeled by ▼ (▲).

other. Moreover, the degeneracy of two quasi-localized states populating complementary sublattices is lifted, i.e., they form a pair of weakly binding and antibinding states (Kumazaki and Hirashima 2007; Palacios et al. 2008). The energy splitting Δ increases with decreasing distance between two defects. Below some critical distance, the gain in exchange energy does not compensate the kinetic energy penalty due to the splitting, and the defect-induced magnetic moments quench (Boukhvalov et al. 2008). The energy splitting argument also explains the nonmagnetic character of the stable form of carbon interstitials in graphite. The competition between these effects is well illustrated in Figure 22.9b. Figure 22.10 shows the sublattice-resolved mean magnetic moments $\langle M^\gamma \rangle$ ($\gamma = A, B$) averaged over many random configurations as a function of defect concentration x. The two panels, Figure 22.10a and b, correspond to the case of defects equally distributed over the two sublattices $\left(N_A^d = N_B^d \right)$ and to the situation when defects belong to sublattice B only ($N_A^d = 0$), respectively (Yazyev 2008b). An important conclusion is that the situation of equivalent sublattices (Figure 22.10a) results in vanishing net magnetic and antiferromagnetic correlations. The situation of defects populating only one sublattice leads to ferromagnetic ordering. In this case, the net magnetic

moment per unit cell $\langle M \rangle = \langle M^A \rangle + \langle M^B \rangle$ is proportional to x in full agreement with Lieb's theorem.

One can argue that the prediction of antiferromagnetic order for the fully disordered graphene fails to explain the observed ferromagnetic order in irradiated graphite. In other words, a mechanism discriminating between the two sublattices of graphene is required to explain the experimental observations. In fact, such a mechanism is intrinsically present in graphite. The lowest-energy ABA stacking order of individual graphene sheets in graphite breaks the equivalence of the two sublattices (Figure 22.8a). Indeed, first-principles calculations show that the chemisorption of hydrogen on sublattice B is 0.16 eV lower in energy than on the complementary sublattice (Yazyev 2008b). This energy difference is more than sufficient to trigger considerable difference in the equilibrium populations of the two sublattices. The energy barrier for the hopping of hydrogen atoms from one carbon atom to another is ~1 eV. Similar discriminating mechanism may also exist for the formation of vacancies. Cross sections for the momentum transfer due to knock-on collisions are likely to be equal for both A and B carbon atoms in graphite. However, the stacking order of graphite has a strong influence on the recombination kinetics of interstitial and vacancy defects. An instantaneous recombination of low energy recoils was found to be significantly more likely for the atoms in position A (Yazyev et al. 2007). All these results indicate that the most probable physical picture of magnetic order in irradiated graphite is ferrimagnetism with the larger magnetic moment induced in sublattice A.

22.4 Conclusions and Perspectives

The numerical examples shown above have illustrated several scenarios of magnetism in graphene nanostructures induced by reduced dimensions and disorder. However, beyond these simple models the field of carbon-based magnetism faces a number of challenges. The most important problems are related to the experimental side of the field. In particular, although the physics of magnetic graphene edges has already attracted a large number of computational and theoretical researchers, no direct experimental evidences have been reported by the time this chapter was written. Further progress in this field will also require novel manufacturing techniques that would allow a precise control of the edge configuration. The area of defect-induced magnetism in graphite demands for detailed studies of defects produced upon irradiation and their relation to the observed ferromagnetic ordering. The limits of saturation magnetization and Curie temperature in irradiated graphite still have to be investigated. On the theory side of this field, an understanding of the magnetic-phase transitions in graphene materials and nanostructures has to be developed. Other important directions of theoretical research include spin transport and the magnetic anisotropy of carbon-based systems.

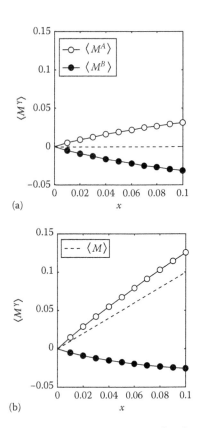

FIGURE 22.10 Average magnetic moments of carbon atoms in A and B sublattices of graphene as a function of p_z-vacancy concentration x. The defects are either distributed equally among sublattices (a) or belong to sublattice B only (b). Average magnetic moments $\langle M \rangle$ per unit cell are shown as dashed lines.

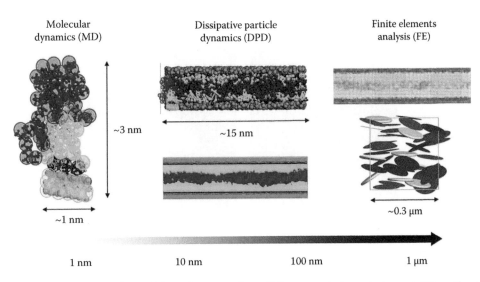

Molecular
dynamics (MD)

Dissipative particle
dynamics (DPD)

Finite elements
analysis (FE)

~3 nm

~1 nm

~15 nm

~0.3 μm

1 nm   10 nm   100 nm   1 μm

FIGURE 3.9   Concept of multiscale molecular modeling of PCNs involving different simulation techniques at different length and time scales.

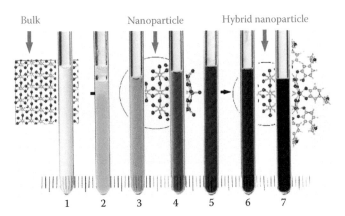

Bulk   Nanoparticle   Hybrid nanoparticle

1   2   3   4   5   6   7

FIGURE 4.2   Surface-modified 45 Å TiO₂ nanoparticles with different bidentate ligands: (1) bare TiO₂, (2) salicylic acid, (3) dihydroxycyclobutene-dione, (4) vitamin C, (5) alizarin, (6) dopamine, and (7) *tert*-butyl catechol. (From Rajh, T. et al., *J. Phys. Chem. B*, 106, 10543, 2002. With permission.)

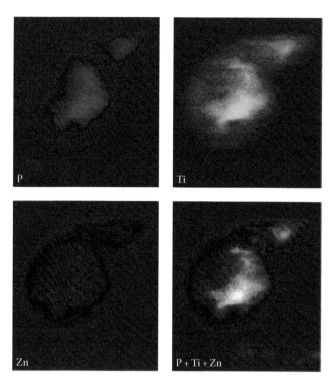

FIGURE 4.21 A single nucleus containing 3.6×10^6 nanoparticles after incubation of 4×10^6 cells with 80 pmol of R18Ss–TiO$_2$ hybrid nanocomposite. The colocalization of phosphorus (red), titanium (green), and zinc (blue) signals is presented as the overlap of these three colors. (From Paunesku, T. et al., *Nat. Mater.*, 2, 343, 2003. With permission.)

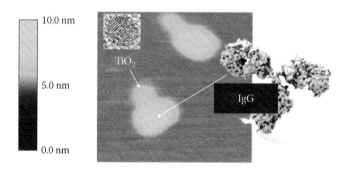

FIGURE 4.24 Atomic force microscopy image of TiO$_2$–CD4 nanocomposites. TiO$_2$–CD4 nanocomposites were dissolved in a HEPES buffer and incubated on freshly cleaved mica.

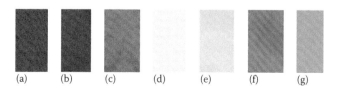

FIGURE 5.16 Digital photographs of the color films prepared by using nanocolorants and corresponding dyes: (a) red nanocolorant, (b) red dye, (c) uncrosslinked red nanocolorant, (d) yellow nanocolorant, (e) yellow dye, (f) blue nanocolorant, and (g) blue dye.

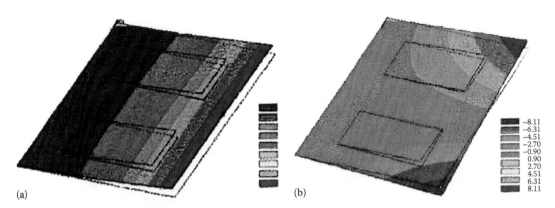

(a)   (b)

                                                                                −8.11
                                      −6.31
−4.51
−2.70
−0.90
0.90
2.70
4.51
6.31
8.11

FIGURE 8.22 Modal shapes of smart composite plate (ANSYS FEA): (a) first mode, (b) second mode.

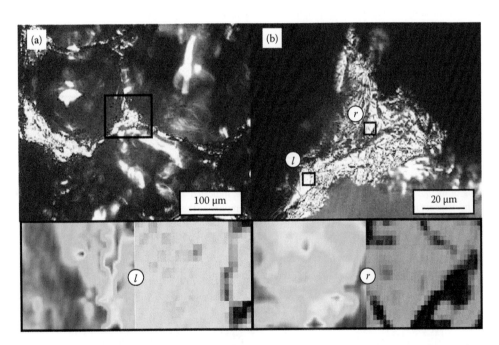

FIGURE 12.4 (a) Optical image of the graphitic foam sample showing the porous structure of the foam. (b) High magnification image of the area delimited by the square in (a). The small l and r squares in (b) delimit the $5 \times 5\,\mu m$ areas from which the Raman spectra were obtained. The lower panels show a comparison between the morphological details of the areas l and r together with the I_D/I_G mapping shown in Figure 12.5. (From Barros, E.B. et al., *Vibr. Spectroscop.*, 45, 122, 2007. With permission.)

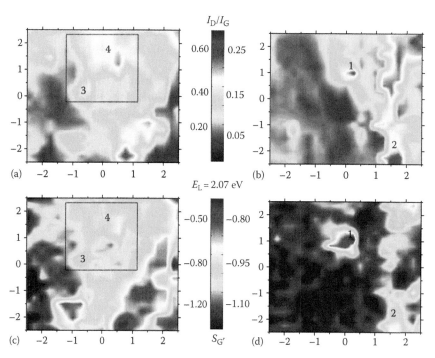

FIGURE 12.5 (a,b) Maps of the spatial distribution of the ID/IG ratio for two different regions of the sample, each with 25 μm² squared area. The Raman spectra were obtained in a 21 × 21 mesh and smoothened to form the map shown here using a 2D spline algorithm. (c,d) Mapping of the Skewness of the G′ band for the same two regions of the graphitic foam as (a) and (b), respectively. (From Barros, E.B. et al., *Vibr. Spectroscop.*, 45, 122, 2007. With permission.)

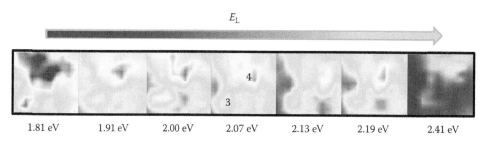

1.81 eV   1.91 eV   2.00 eV   2.07 eV   2.13 eV   2.19 eV   2.41 eV

FIGURE 12.7 Mapping of the I_D/I_G ratio for a selected region of the sample, indicated with a square in Figure 12.5a, obtained using different laser excitation energies. All the maps are normalized with the same parameters to emphasize the dependence of the I_D/I_G ratio on the laser excitation energy.

FIGURE 17.8 Model of six three-helix metalloproteins grafted into a SAM with C18 backbones. (From Case M. et al., *Nano. Lett.*, 3, 425, 2003. With permission.) For clarity of presentation, the alkanethiol molecules are shown as oriented perpendicularly with respect to the surface, and not with their tilt angle of 30°.

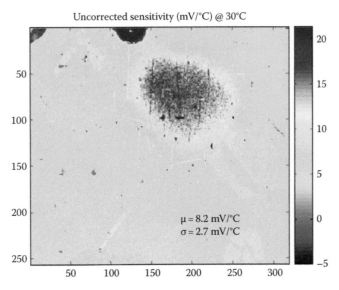

FIGURE 34.29 Measured sensitivity map of the 320 × 256 FPA at 30°C. The average sensitivity is 8.3 mV/°C with a standard deviation σ of 8.3 mV/°C.

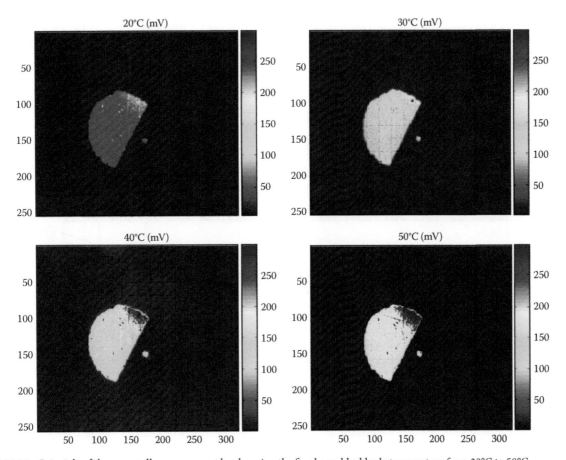

FIGURE 34.32 Principle of the cross-talk measurement by changing the fixed area blackbody temperature from 20°C to 50°C.

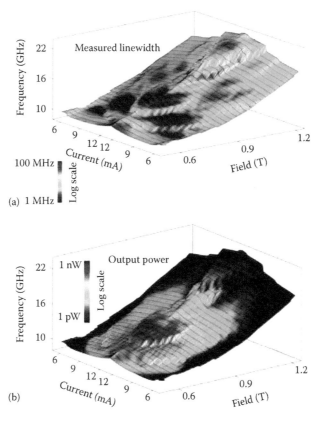

FIGURE 38.11 (a) Plot showing the measured frequency of a spin valve nanocontact STNO as a function of applied current and field. For these plots the device current is increased from 6 to 12 mA and then decreased from 12 to 6 mA to investigate any regions that may be hysteretic with current. The oscillator linewidth is shown by the color map with blue being small linewidths (~1 MHz) and red being large linewidths (~100 MHz). (b) is the same as (a) except the total output power is shown on the color map. (Adapted from Rippard, W.H. et al., *Phys. Rev. B*, 74, 224409, 2006.)

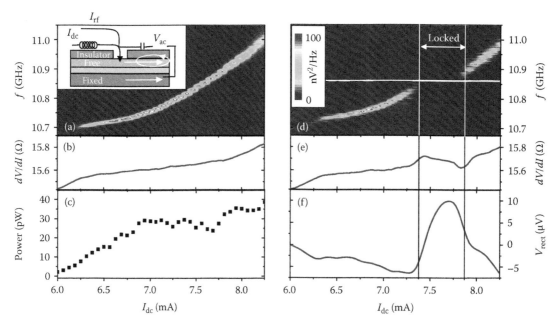

FIGURE 38.19 Measurement of an STNO phase locking to an injected microwave signal. The device geometry is shown in the inset (a); (a), (b) and (c) show the measured frequency, device resistance, and output power as a function of bias current with no injected microwave signal. (d), (e), and (f) show the measured frequency, device resistance, and dc voltage as a function of bias current with an injected microwave signal. (Adapted from Rippard, W.H. et al., *Phys. Rev. Lett.*, 95, 067203, 2005.)

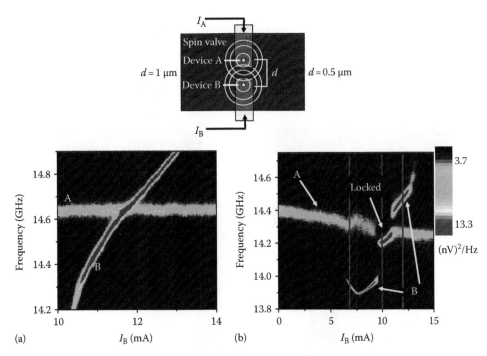

FIGURE 38.20 Demonstration of mutual phase locking of two STNOs (Kaka et al. 2005). (a) The emission spectra of two oscillators as a function of the current through oscillator B when the oscillators are far apart showing no interaction. (b) The emission spectra of two oscillators as a function of the current through oscillator B when the oscillators are close showing strong interactions and phase locking.

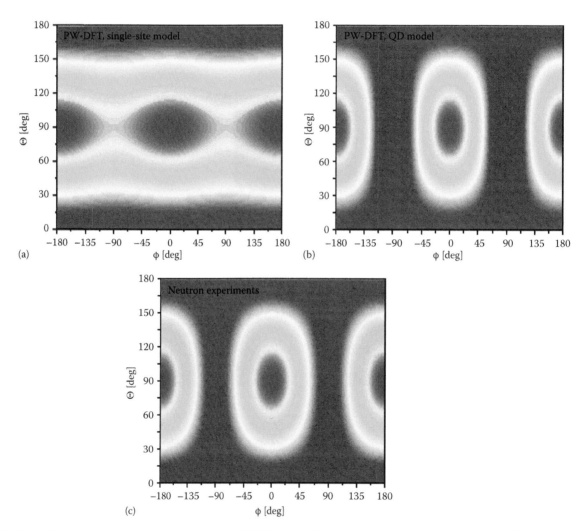

FIGURE 40.21 H_2 orientational potentials deduced from PW-DFT calculations (a,b) and neutron experiments (c). Red/blue denote regions of high/low potential energy.

Acknowledgments

I would like to thank my collaborators L. Helm, E. Kaxiras, M. I. Katsnelson, U. Rothlisberger, I. Tavernelli, and W. L. Wang. I am also grateful to two of them, L. Helm and W. L. Wang, for valuable comments on the manuscript.

References

Allemand, P.-M., Khemani, K. C., Koch, A. et al. 1991. Organic molecular soft ferromagnetism in a fullerene C_{60}. *Science* 253: 301.

Allinson, G., Bushby, R. J., Paillaud, J.-L., and Thornton-Pett, M. 1995. Synthesis of a derivative of triangulene; the First Non-Kékule Polynuclear Aromatic. *J. Chem. Soc., Perkin Trans.* 1: 385.

Banhart, F. 1999. Irradiation effects in carbon nanostructures. *Rep. Prog. Phys.* 62: 1181.

Barzola-Quiquia, J., Esquinazi, P., Rothermel, M., Spemann, D., Butz, T., and Garcia, N. 2007. Experimental evidence for two-dimensional magnetic order in proton bombarded graphite. *Phys. Rev. B* 76: 161403.

Boukhvalov, D. W., Katsnelson, M. I., and Lichtenstein, A. I. 2008. Hydrogen on graphene: Electronic structure, total energy, structural distortions and magnetism from first-principles calculations. *Phys. Rev. B* 77: 035427.

Clar, E. 1972. *The Aromatic Sextet*. London, U.K.: Wiley.

Duplock, E. J., Scheffler, M., and Lindan, P. J. D. 2004. Hallmark of perfect graphene. *Phys. Rev. Lett.* 92: 225502.

Edwards, D. M. and Katsnelson, M. I. 2006. High-temperature ferromagnetism of sp electrons in narrow impurity bands: Application to CaB_6. *J. Phys.: Condens. Matter* 18: 7209.

Esquinazi, P., Spemann, D., Höhne, R., Setzer, A., Han, K.-H., and Butz, T. 2003. Induced magnetic ordering by proton irradiation in graphite. *Phys. Rev. Lett.* 91: 227201.

Ezawa, M. 2007. Metallic graphene nanodisks: Electronic and magnetic properties. *Phys. Rev. B* 76: 245415.

Fajtlowicz, S., John, P. E., and Sachs, H. 2005. On maximum matchings and eigenvalues of benzenoid graphs. *Croat. Chem. Acta* 78: 195–201.

Fernández-Rossier, J. and Palacios, J. J. 2007. Magnetism in graphene nanoislands. *Phys. Rev. Lett.* 99: 177204.

Geim, A. K. and Novoselov, K. S. 2007. The rise of graphene. *Nat. Mater.* 6: 183.

Inoue, J., Fukui, K., Kubo, T. et al. 2001. The first detection of a clar's hydrocarbon, 2,6,10-Tri-tert-Butyltriangulene: A ground-state triplet of non-Kekulé polynuclear benzenoid hydrocarbon. *J. Am. Chem. Soc.* 123: 12702.

Katsnelson, M. I. 2007. Graphene: Carbon in two dimensions. *Mater. Today* 10: 20.

Kelly, K. F. and Halas, N. J. 1998. Determination of a and b site defects on graphite using C_{60}-adsorbed STM tips. *Surf. Sci.* 416: L1085.

Kobayashi, Y., Fukui, K.-i., Enoki, T., Kusakabe, K., and Kaburagi, Y. 2005. Observation of zigzag and armchair edges of graphite using scanning tunneling microscopy and spectroscopy. *Phys. Rev. B* 71: 193406.

Kobayashi, Y., Fukui, K.-i., Enoki, T., and Kusakabe, K. 2006. Edge state on hydrogen-terminated graphite edges investigated by scanning tunneling microscopy. *Phys. Rev. B* 73: 125415.

Kumazaki, H. and Hirashima, D. S. 2007. Nonmagnetic-defect-induced magnetism in graphene. *J. Phys. Soc. Jpn.* 76: 064713.

Kuroda, S. and Shirakawa, H. 1987. Electron-nuclear double-resonance evidence for the soliton wave function in polyacetylene. *Phys. Rev. B* 35: 9380.

Lehtinen, P. O., Foster, A. S., Ma, Y., Krasheninnikov, A. V., and Nieminen, R. M. 2004. Irradiation-induced magnetism in graphite: A density functional study. *Phys. Rev. Lett.* 93: 187202.

Lieb, E. H. 1989. Two theorems on the Hubbard model. *Phys. Rev. Lett.* 62: 1201.

Makarova, T. and Palacio, F. eds. 2006. *Carbon Based Magnetism: An Overview of the Magnetism of Metal Free Carbon-based Compounds and Materials*. Amsterdam, the Netherlands: Elsevier.

Makarova, T. L., Sundqvist, B., Höhne, R. et al. 2001. Magnetic carbon. *Nature* 413: 716.

Makarova, T. L., Sundqvist, B., Höhne, R. et al. 2006. Retraction: Magnetic carbon. *Nature* 440: 707.

Mermin, N. D. and Wagner, H. 1966. Absence of ferromagnetism or antiferromagnetism in one- or two-dimensional isotropic Heisenberg models. *Phys. Rev. Lett.* 17: 1133.

Mizes, H. A. and Foster, J. S. 1989. Long-range electronic perturbations caused by defects using scanning tunneling microscopy. *Science* 244: 559.

Mohn, P. 2003. *Magnetism in the Solid State: An Introduction*. Berlin, Germany: Springer-Verlag.

Nakada, K., Fujita, M., Dresselhaus, G., and Dresselhaus, M. S. 1996. Edge state in graphene ribbons: Nanometer size effect and edge shape dependence. *Phys. Rev. B* 54: 17954.

Novoselov, K. S., Geim, A. K., Morozov, S. V. et al. 2004. Electric field effect in atomically thin carbon films. *Science* 306: 666.

Ohldag, H., Tyliszczak, T., Höhne, R. et al. 2007. π-Electron ferromagnetism in metal-free carbon probed by soft x-ray dichroism. *Phys. Rev. Lett.* 98: 187204.

Palacios, J. J., Fernández-Rossier, J., and Brey, L. 2008. Vacancy-induced magnetism in graphene and graphene ribbons. *Phys. Rev. B* 77: 195428.

Pereira, V. M., Guinea, F., Lopes dos Santos, J. M. B., Peres, N. M. R., and Castro Neto, A. H. 2006. Disorder induced localized states in graphene. *Phys. Rev. Lett.* 96: 036801.

Pisani, L., Chan, J. A., Montanari, B., and Harrison, N. M. 2007. Electronic structure and magnetic properties of graphitic ribbons. *Phys. Rev. B* 75: 064418.

Ruffieux, P., Gröning, O., Schwaller, P., Schlapbach, L., and Gröning, P. 2000. Hydrogen atoms cause long-range electronic effects on graphite. *Phys. Rev. Lett.* 84: 4910.

Son, Y.-W., Cohen, M. L., and Louie, S. G. 2006a. Energy gaps in graphene nanoribbons. *Phys. Rev. Lett.* 97: 216803.

Son, Y.-W., Cohen, M. L., and Louie, S. G. 2006b. Half-metallic graphene nanoribbons. *Nature* 444: 347.

Takahashi, M., Turek, P., Nakazawa, Y. et al. 1991. Discovery of a quasi-1D organic ferromagnet, p-NPNN. *Phys. Rev. Lett.* 67: 746.

Tamura, M., Nakazawa, Y., Shiomi, D. et al. 1991. Bulk ferromagnetism in the β-phase crystal of the p-nitrophenyl nitronyl nitroxide radical. *Chem. Phys. Lett.* 186: 401.

Telling, R. H., Ewels, C. P., El-Barbary, A. A., and Heggie, M. I. 2003. Wigner defects bridge the graphite gap. *Nat. Mater.* 2: 333.

Thomann, H., Dalton, L. R., Grabowski, M., and Clarke, T. C. 1985. Direct observation of Coulomb correlation effects in polyacetylene. *Phys. Rev. B* 31: 3141.

Trauzettel, B., Bulaev, D. V., Loss, D., and Burkard, G. 2007. Spin qubits in graphene quantum dots. *Nat. Phys.* 3: 192.

Wang, W. L., Meng, S., and Kaxiras, E. 2008. Graphene nanoflakes with large spin. *Nano Lett.* 8: 241.

Wang, W. L., Yazyev, O. V., Meng, S., and Kaxiras, E. 2009. Topological frustration in graphene nanoflakes: Magnetic order and spin logic devices. *Phys. Rev. Lett.* 102: 157201.

Yazyev, O. V. 2008a. Hyperfine interactions in graphene and related carbon nanostructures. *Nano Lett.* 8: 1011.

Yazyev, O. V. 2008b. Magnetism in disordered graphene and irradiated graphite. *Phys. Rev. Lett.* 101: 037203.

Yazyev, O. V. and Helm, L. 2007. Defect-induced magnetism in graphene. *Phys. Rev. B* 75: 125408.

Yazyev, O. V. and Katsnelson, M. I. 2008. Magnetic correlations at graphene edges: Basis for novel spintronics devices. *Phys. Rev. Lett.* 100: 047209.

Yazyev, O. V., Tavernelli, I., Rothlisberger, U., and Helm, L. 2007. Early stages of radiation damage in graphite and carbon nanostructures: A first-principles molecular dynamics study. *Phys. Rev. B* 75: 115418.

23

Graphene Quantum Dots

Prabath Hewageegana
University of Kelaniya

Vadym Apalkov
Georgia State University

23.1 Introduction

23.1.1 Graphene: Ultrathin Layer of Graphite

Graphene is a layer of crystalline carbon that just one atom thick (Novoselov et al. 2004). Graphene is a two-dimensional crystal with a honeycomb structure. The crystalline structure of graphene results in a very unique energy dispersion law (Wallace 1947). Namely, the low-energy excitations are described by the Dirac–Weyl equations for massless relativistic particles with a linear dependence on a momentum (Ando 2002). This property of graphene makes it a very interesting academic material, which has been studied theoretically in great detail for the last 50 years. But only in 2004, graphene has been realized experimentally (Novoselov et al. 2004). Since then graphene has been a subject of extensive theoretical and experimental research (Ando et al. 1998, Novoselov et al. 2005, Katsnelson 2007). Now high-quality graphene crystallites with the sizes up to 100 μm can be easily grown (Novoselov et al. 2005). The quality of graphene crystals is very high and is sufficient for many research and commercial applications. The charge carriers in graphene can travel thousands of interatomic distances without scattering. The mobility of the carriers is high (exceeding 15,000 cm²/V s) and has weak dependence on the temperature up to 300 K. Even with the doping, when the electron density exceeds 10^{12} cm⁻², the mobility is still high. The manifestation of the extremely high quality of graphene is observation of the quantum Hall effect in this system (Zhang et al. 2005).

The ballistic transport of charge carriers, i.e., electrons or holes, in graphene suggests that graphene materials could provide the foundation for a new generation of nanoscale devices (Westervelt 2008). With the charge transport over a large distance, the different elements of nanodevices can be easily coherently connected. This property opens a possibility to build coherently connected nanostructures, based on the wave nature of the charge carriers.

Another reason why researches are so interested in graphene is the unique nature of its charge carriers. For both electrons and holes in graphene, the energy dispersion law is linear and the energy of the carrier is proportional to its momentum. This dispersion law is similar to the massless relativistic particles. Therefore, the natural description of the carrier in graphene is based not on the Schrödinger equation but on the Dirac–Weyl relativistic equation (Novoselov et al. 2005). The effective speed of "light," which enters the Dirac–Weyl equation, for electrons and holes in graphene is $v_F \approx 10^6$ m/s (Ando 2002).

23.1.2 Relativistic Nature of Electrons in Graphene: Klein's Paradox

A unit cell of two-dimensional graphene honeycomb lattice contains two carbon atoms, say A and B (Slonczewski and Weiss 1958). The low-energy dynamics of electrons in graphene is described by the tight-binding Hamiltonian with the nearest-neighbor hopping. In the continuum limit, this Hamiltonian generates the band structure with two π bands and the Fermi levels located at two inequivalent points, $K = (2\pi/a_0)(1/3, 1/\sqrt{3})$ and $K' = (2\pi/a_0)(2/3, 0)$, of the first Brillouin zone (see Figure 23.1), where $a_0 = 0.246$ nm is a lattice constant (Saito et al. 1998). Near the points K and K' an electron has a linear Dirac–Weyl dispersion relation of relativistic type. Finally, in the continuum limit, the electron wave function is described by an eight-component spinor, $\psi_{s,k,\alpha}$, where $s = 1/2$ is a spin index, $k = K$, K' is a valley index, and $\alpha = A, B$ is a sublattice index. Without spin–orbit interaction (Kane and Mele 2005), the Hamiltonian of electronic system is described by two 4×4 matrices for each component of electron spin (Ando 2005):

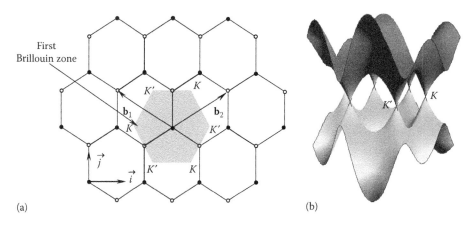

(a)   (b)

FIGURE 23.1 (a) Reciprocal lattice and first Brillouin zone (shaded) of graphene. The vectors $\mathbf{b}_1$ and $\mathbf{b}_2$ are the lattice basis vectors. Here K and K' corresponds to two valleys. (b) Electronic band structure of graphene. Near K and K' points the charge carriers have linear dispersion.

$$H = \frac{\gamma}{\hbar}\begin{pmatrix} 0 & p_x - ip_y & 0 & 0 \\ p_x + ip_y & 0 & 0 & 0 \\ 0 & 0 & 0 & p_x - ip_y \\ 0 & 0 & p_x + ip_y & 0 \end{pmatrix} \quad (23.1)$$

where

(p_x, p_y) is a two-dimensional momentum of an electron

$\gamma = \sqrt{3}a_0\gamma_0/2$ is the band parameter (Saito et al. 1998)

Here $\gamma_0 = 3.03\,\text{eV}$ is the transfer integral between the nearest-neighbor carbon atoms. Within a single valley, the Hamiltonian matrix becomes 2×2 matrix:

$$H = \frac{\gamma}{\hbar}\begin{pmatrix} 0 & p_x - ip_y \\ p_x + ip_y & 0 \end{pmatrix}. \quad (23.2)$$

where two components of the wavefunction correspond to two sublattice indices, A and B. These two components of the wavefunction can be described in terms of a pseudospin. It is convenient to rewrite the Hamiltonian (23.2) in terms of 2×2 Pauli matrices, $\vec{\sigma}$:

$$H = \frac{\gamma}{\hbar}(\vec{\sigma}\vec{p}) \quad (23.3)$$

The Hamiltonians (23.2) and (23.3) result in relativistic energy dispersion law with linear dependence on the momentum

$$E(p) = \pm\frac{\gamma}{\hbar}|p|, \quad (23.4)$$

where "+" and "−" signs correspond to electron and hole states, respectively. The corresponding wavefunctions are given by the following expression:

$$\begin{pmatrix} \psi_1 \\ \psi_2 \end{pmatrix} = \frac{e^{i\vec{p}\vec{r}/\hbar}}{\sqrt{2}}\begin{pmatrix} 1 \\ \pm e^{i\phi_p} \end{pmatrix}, \quad (23.5)$$

where $\tan(\phi_p) = p_y/p_x$. The expression (23.5) clearly shows the chiral nature of the electron states in graphene. For example, at $p_y = 0$ the expression (23.5) becomes

$$\begin{pmatrix} \psi_1 \\ \psi_2 \end{pmatrix}_{p_y=0} = \frac{e^{ip_xx/\hbar}}{\sqrt{2}}\begin{pmatrix} 1 \\ \pm\text{sgn}(p_x) \end{pmatrix}, \quad (23.6)$$

which means that for electron states $(E(p) = +(\gamma/\hbar)p_x)$ the pseudospin part of the wavefunction is $\begin{pmatrix} 1 \\ 1 \end{pmatrix}$ at $p_x > 0$ and $\begin{pmatrix} 1 \\ -1 \end{pmatrix}$ at $p_x < 0$.

The dispersion law (23.4) is shown schematically in Figure 23.2a. For comparison, in Figure 23.2b, a nonrelativistic dispersion law with parabolic dependence on a momentum is shown. The main features of the energy dispersion law (23.4) are the following: linear dependence on the momentum, zero-gap excitations, and chiral nature of the electronic states (see Figure 23.2a).

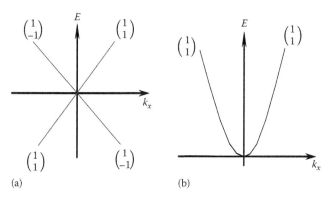

(a)   (b)

FIGURE 23.2 The energy dispersion law is shown (a) for massless relativistic electrons in graphene and (b) for nonrelativistic electrons. The chiral nature of electron states in graphene is illustrated by a pseudospin part of the wavefunction, which depends on the direction of the momentum. For nonrelativistic electrons the pseudospin part of the wavefunction does not depend on the momentum.

There is one important manifestation of this type of energy dispersion law, which is explored below in detail. Namely, the confinement of an electron in graphene becomes quite a challenging task due to so-called Klein's paradox (Klein 1929, Calogeracos and Dombey 1999, Katsnelson et al. 2006). Klein's paradox means that an electron can penetrate through any potential barrier. This is due a gapless nature of the dispersion law (23.4). Indeed, if an electron approaches a potential barrier then due to the gapless nature of the energy dispersion law, there are states, i.e., hole states, inside the barrier. Then the electron penetrates into the barrier and emerges as a hole inside the barrier. It does not matter what the height of the barrier is, there are always real states inside the barrier. It means that for any potential barrier, the electron, which becomes a hole inside the barrier, can freely propagate inside the barrier and finally it can penetrate through the barrier without any losses. The Klein's tunneling is illustrated in Figure 23.3a, where both the electron states outside the barrier and the corresponding hole states inside the barrier are shown.

For comparison, in Figure 23.3b, the tunneling process is illustrated for a standard "nonrelativistic" electron with a finite electron-hole gap. Since in this case the system has a gap, then there are no states inside the barrier and the electron cannot freely propagate inside the barrier. Then the electron transport through the barrier is exponentially suppressed.

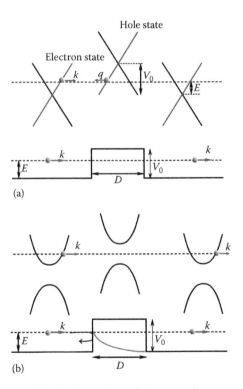

(a)

(b)

FIGURE 23.3 (a) Klein's tunneling of relativistic electron in graphene is shown schematically for a potential barrier. There are freely propagating electronic states both inside and outside the barrier. The electron state outside the barrier becomes the hole state inside the barrier. (b) Tunneling through a potential barrier is show for nonrelativistic electrons. Due to a finite gap in the energy dispersion law there are no freely propagating states inside the barrier.

Based on the Klein's tunneling, we can immediately conclude that an electron in graphene cannot be localized by a confinement potential. Indeed, we can introduce a confinement potential of any strength (even a very strong one), but due to the Klein's tunneling, an electron easily escapes from the confinement well. It means that there are no conventional quantum dots in graphene. Here by quantum dot we mean a finite region within a graphene layer within which an electron is localized, i.e., it stays within this region for an infinite time.

23.1.3 Quantum Dots in "Nonrelativistic" Systems

Transport and optical properties of electronic systems strongly depend on dimensionality of the system (Chakraborty et al. 2002). Reducing the dimensionality of quantum semiconductor structures introduces an additional control to the system. In this context, quantum dots (Chakraborty 1999) represent an ultimate reduction in the dimensionality of nanoscale devices. In quantum dots, electrons are confined in all three directions. As a result, they occupy spectrally sharp energy levels like those found in atoms. In this sense, quantum dots are also called artificial atoms. The typical dimension of quantum dots is between nanometers to a few microns. Due to their extremely small size, quantum dots can be occupied by just a few electrons and this property provides many advantages and opportunities to optimize nanoscale devices. In addition to the nanoscale size of the quantum dots, they show superior transport and optical properties. These properties have beneficial applications in many fields of science, including semiconductor lasers, quantum computers (in a future), and medical applications.

One of the promising applications of quantum dots is related to their unique optical properties. The discrete nature of the energy spectra of quantum dots makes them optically active only at discrete values of the light wavelengths. The values of the wavelengths can be easily controlled by the dot parameters, such as quantum dot's diameter and height. The energy emitted from the quantum dots as a light is close to 100% of the energy put into the system. This exceptionally high efficiency makes quantum dots appealing for their use as light sources and as individual color pixels in color flat panel displays. The unique optical properties of quantum dots have found their applications in the quantum dot photodetectors (Liu et al. 2001), quantum dot lasers (Fafard et al. 1996), and other optical systems. Another important application of quantum dots is related to the energy or information storage, which can be used in the quantum information processing (Loss and DiVincenzo 1998).

The quantum dots created in carbon nanotubes have been already reported in the literature (Buitelaar et al. 2002, Cobden and Nygard 2002, Ke et al. 2003, Moriyama et al. 2005, Ishibashi et al. 2006). Due to broad applications of quantum dots in nanoscale devices, it is quite crucial to understand how to realize the quantum dots in graphene. In the next sections we discuss unique properties of quantum dots in graphene.

23.2 Quantum Dots in Graphene

23.2.1 Two Types of Quantum Dots in Graphene

There are two types of quantum dots in graphene.

1. The first type of quantum dot (type I quantum dot) is just a small piece of graphene (see Figure 23.4). Such quantum dots have been successfully realized experimentally (Ponomarenko et al. 2008), where the quantum dots of different sizes ranging from 15 to 250 nm have been created. For such quantum dots, the usual properties of quantum dots, such as Coulomb blockade, have been observed. Since the motion of an electron is restricted by a piece of graphene, i.e., quantum dot, the electron is localized in quantum dots of type I. Ponomarenko et al. (2008) studied the transport through such quantum dots experimentally. They found that for small quantum dots (diameter is less 100 nm) the main contribution to the formation of the dot energy levels comes from the quantum confinement. The positions of the energy levels in such quantum dots depend on the structure of the boundary of the quantum dot, i.e., on the boundary conditions. In graphene, there are difference types of boundary conditions, e.g., zigzag or armchair (Nakada et al. 1996, Brey and Fertig 2006, Akhmerov and Beenakker 2008). Therefore to find the properties of the quantum dots of type I, the correct boundary condition should be used.

2. The second type of quantum dot (type II quantum dot) is a quantum dot, which in conventional semiconductor systems is realized through a confinement electric potential, e.g., gate potential (see Figure 23.5). In this case in graphene, there is the Klein's tunneling and, strictly speaking, the electrons cannot be localized by such confinement potential. Therefore, based on this statement, we can conclude that there are no standard quantum dots of type II, which can accumulate electrons, in graphene. This statement is not completely true. What we should

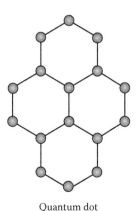

Quantum dot

FIGURE 23.4 The type I quantum dot in graphene is shown schematically. The quantum dot consists of a finite number of carbon atoms.

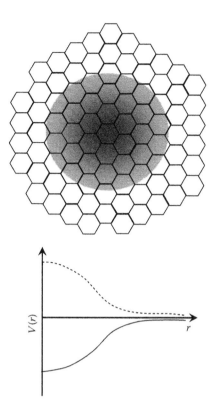

FIGURE 23.5 The type II quantum dot in graphene is shown schematically. The quantum dot is realized through a confinement potential. The confinement potential can be of electron type (solid line) or hole type (dashed line).

discuss in graphene is not a localization, i.e., trapping of an electron for infinitely long time, but a trapping of an electron for a finite long time. Therefore, the following question should be addressed: Can an electron in graphene be trapped within some spatial region for a long enough time? Here, a long enough time means that the time should be long enough for a specific application of graphene quantum dots.

Below we discuss only the type II quantum dots. We show that in the case there are two mechanism of trapping. The first one is due to formation of semiclassical tunneling barrier, i.e., classically forbidden region. Such tunneling barrier exists only if an electron has a transverse momentum, i.e., the electron has a component of the momentum parallel to equipotential lines of the confinement potential. Such transverse momentum creates an effective nonzero electron mass, which results in the classically forbidden regions (Chen et al. 2007, Silvestrov and Efetov 2007). In the case of a cylindrically symmetric quantum dots, the transverse momentum is generated by an angular momentum (Chen et al. 2007). The height and the width of the tunneling barrier are determined by the values of the angular momentum and the slope of the confinement potential. The most efficient trapping occurs in a smooth confinement potential and at large values of electron angular momentum.

The second mechanism of trapping is due to interference of electron waves within the quantum dot region (Hewageegana and Apalkov 2008). This mechanism of trapping is very sensitive to the parameters of the confinement potential.

First, we introduce a main system of equations, which needs to be solved to find the electronic properties of the quantum dots. We assume that the quantum dot is cylindrically symmetric, which means that the corresponding confinement potential, $V(r)$, depends only on the distance from the center of the quantum dot. The confinement potential does not introduce any intervalley mixture. Then it is enough to consider the states of a single valley only. Finally, all electronic states discussed below will have fourfold degeneracy: twofold degeneracy due to valley and twofold degeneracy due to spin. With the cylindrically symmetric confinement potential, the Hamiltonian (23.3) takes the form

$$H = \frac{\gamma}{\hbar}(\vec{\sigma}\vec{p}) + V(r). \qquad (23.7)$$

We can rewrite the Hamiltonian (23.7) in the matrix form (see Equation 23.2):

$$H = \frac{\gamma}{\hbar}\begin{pmatrix} 0 & p_x - ip_y \\ p_x + ip_y & 0 \end{pmatrix} + \begin{pmatrix} V(r) & 0 \\ 0 & V(r) \end{pmatrix}. \qquad (23.8)$$

Then the corresponding Schrödinger equation takes a form

$$H\begin{pmatrix} \psi_1 \\ \psi_2 \end{pmatrix} = E\begin{pmatrix} \psi_1 \\ \psi_2 \end{pmatrix}, \qquad (23.9)$$

where

$\psi_1(\vec{r})$ and $\psi_2(\vec{r})$ are two components of the wavefunction
E is an eigenenergy of a stationary state

Here the two components of the wavefunction correspond to two sublattice indices, A and B. Taking into account the matrix expression (23.8) for the Hamiltonian we obtain

$$V(r)\psi_1 + \frac{\gamma}{\hbar}(p_x - ip_y)\psi_2 = E\psi_1 \qquad (23.10)$$

$$V(r)\psi_2 + \frac{\gamma}{\hbar}(p_x + ip_y)\psi_1 = E\psi_2. \qquad (23.11)$$

It is convenient to rewrite this system of equations in the cylindrical coordinates, r and θ. Then Equations 23.10 and 23.11 take the form

$$V(r)\psi_1 + \gamma e^{-i\theta}\left(-i\frac{\partial}{\partial r} + \frac{1}{r}\frac{\partial}{\partial \theta}\right)\psi_2 = E\psi_1 \qquad (23.12)$$

$$V(r)\psi_2 + \gamma e^{i\theta}\left(-i\frac{\partial}{\partial r} - \frac{1}{r}\frac{\partial}{\partial \theta}\right)\psi_1 = E\psi_2. \qquad (23.13)$$

We are looking for the solution of the system of equations (23.12) and (23.13) in the form

$$\psi_1(r,\theta) = \chi_1(r)e^{i(m-1/2)\theta},$$
$$\psi_2(r,\theta) = \chi_2(r)e^{i(m+1/2)\theta}, \qquad (23.14)$$

where $m = \pm 1/2, \pm 3/2,\ldots$ is an electron angular momentum. Then Equations 23.12 and 23.13 become

$$V(r)\chi_1 - i\gamma\frac{d\chi_2(r)}{dr} - i\gamma\frac{m+1/2}{r}\chi_2(r) = E\chi_1 \qquad (23.15)$$

$$V(r)\chi_2 - i\gamma\frac{d\chi_1(r)}{dr} + i\gamma\frac{m-1/2}{r}\chi_1(r) = E\chi_2. \qquad (23.16)$$

The system of equations (23.15) and (23.16) describe the electronic states of a graphene sheet with confinement potential $V(r)$. This is our main system of equations. Below we use two approaches to solve this system of equations: (1) semiclassical approximation (Chen et al. 2007) and (2) exact numerical solution of eigenvalue problem (Hewageegana and Apalkov 2008, Matulis and Peeters 2008).

23.2.2 Electron Trapping due to Formation of Tunneling Barrier

First, we solve the system of equations (23.15) and (23.16) within the semiclassical approach. This approach can be applied if the electron angular momentum is large enough. We are looking for the solution of the system of equations (23.15) and (23.16) in the form $\chi_1, \chi_2 \propto \exp(iq\rho)$. Then under the condition $m \gg 1$ we obtain

$$[E - V(r)]^2 = \gamma^2\left(\frac{m}{r}\right)^2 + \gamma^2 q^2. \qquad (23.17)$$

This equation shows that now we have a classically forbidden region. Namely, if $|E - V(r)| < \gamma m/\rho$, then there are no solution with the real q. Therefore the transverse momentum m/r introduces a gap (or effective electron mass) in the energy dispersion law. The classical turning points can be found from the condition $q = 0$, i.e.,

$$[E - V(r)]^2 = \gamma^2\left(\frac{m}{r}\right)^2. \qquad (23.18)$$

For a monotonic confinement potential (see Figure 23.6), there are two classical turning points. If r_0 is a solution of equation $E - V(r_0) = 0$, then from Equation 23.18, we can find the positions of the classical turning points (see Figure 23.6):

$$r_1 = r_0 - \Delta r_1 = r_0 - \frac{m/r_0}{F/\gamma + m/r_0^2}, \qquad (23.19)$$

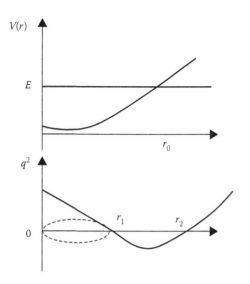

FIGURE 23.6 Schematic illustration of classically forbidden region at large values of angular momentum. The classical regions are determined by the condition $q^2 > 0$. Points r_1 and r_2 are the classical turning points. The electron is trapped inside the quantum dot at $r < r_1$. Tunneling through the classically forbidden region ($r_1 < r < r_2$) determines the escape rate from the quantum dot.

$$r_2 = r_0 + \Delta r_1 = r_0 - \frac{m/r_0}{F/\gamma - m/r_0^2}, \qquad (23.20)$$

where $F = dV(r)/dr\,|_{r=r_0}$ is the slope of the confinement potential. Therefore, within the semiclassical approximation, the electron can freely propagate within the regions $r > r_2$ and $r < r_1$. The coupling between these two regions occurs through the tunneling barrier ($r_1 < r < r_2$). Due to the presence of the tunneling barrier, the electron can be trapped within the region $r < r_1$, i.e., in the quantum dot. Here the size of the quantum dot should be defined by the condition $r < r_1$.

We can estimate the lifetime of the electron in the quantum dot. The tunneling through the classically forbidden region ($r_1 < r < r_2$) determines the escape rate from the quantum dot. We assume that $F \gg \gamma m / r_0^2$, then $\Delta r_1 = \Delta r_2 = m\gamma/Fr_0$. The escape rate from the quantum dot is determined by the tunneling exponent and is given by the following expression (Landau and Lifshitz 1991):

$$R = \exp\left(-\int_{r_0 - \Delta r_1}^{r_0 + \Delta r_1} |q(r)|\,dr\right), \qquad (23.21)$$

where the integral is calculated over the classically forbidden region. Within this region, we can find from Equation 23.17 the following expression for $|q|$:

$$|q(r)| = \sqrt{(m/r_0)^2 - (F/\gamma)^2 (r - r_0)^2}. \qquad (23.22)$$

Then we can find the escape rate from the quantum dot:

$$R = \exp\left(-\int_{r_0 - \Delta r_1}^{r_0 + \Delta r_1} \sqrt{\left(\frac{m}{r_0}\right)^2 - \left(\frac{F}{\gamma}\right)^2 (r - r_0)^2}\,dr\right)$$

$$= \exp\left(-\frac{\pi\gamma m^2}{2Fr_0^2}\right). \qquad (23.23)$$

Therefore, in order to trap an electron, we need to have a smooth confinement potential, i.e., the slope, F, should be small, and large angular momentum, m. For example, for a potential of a form $V(r) = (u/p)r^p$ the escape rate becomes

$$R = \exp\left[-\frac{\pi\gamma m^2}{2ur_0^{p+1}}\right] = \exp\left[-\frac{\pi m^2}{p(E/\epsilon_p)^{p+1/p}}\right], \qquad (23.24)$$

where

$$\epsilon_p = \left(\frac{\gamma^p u}{p}\right)^{1/(p+1)}. \qquad (23.25)$$

The condition of electron trapping within the quantum dot region means that the escape rate is small, i.e., $R \ll 1$, or the tunneling exponent is large. Then from Equation 23.24, we obtain

$$\frac{\pi m^2}{p(E/\epsilon_p)^{p+1/p}} \gg 1. \qquad (23.26)$$

This equation determined the upper limit on the energy of the strongly trapped state at a given angular momentum, m:

$$E < \epsilon_p m^{2p/(p+1)}. \qquad (23.27)$$

For example, for parabolic confinement potential, i.e., $p = 2$, the upper limit is

$$E < \epsilon_2 m^{4/3}. \qquad (23.28)$$

Based on the semiclassical approximation, we can also find the interlevel separation of the trapped states in the quantum dot. The condition of semiclassical quantization in the quantum dot region, i.e., $r < r_1$, takes the form (Landau and Lifshitz 1991)

$$S(E) = \int_0^{r_1} q(r)\,dr$$

$$= \int_0^{r_1} \frac{1}{\gamma} \sqrt{(E - V(r))^2 - (\gamma m/r)^2}\,dr = \pi N, \qquad (23.29)$$

where N is integer and the integral is calculated over the classical region. Then the interlevel separation, ΔE, can be found from the equation

$$\frac{dS}{dE}\Delta E = \pi \qquad (23.30)$$

or

$$\Delta E = \frac{\pi}{dS/dE}. \qquad (23.31)$$

From expression (23.29), we can find the derivative dS/dE in the following form:

$$\frac{dS}{dE} = \int_0^{r_1} \frac{dr}{\gamma} \frac{E - V(r)}{\sqrt{(E-V(r))^2 - \gamma^2 (m/r)^2}}$$

$$= \int_0^{r_1} \frac{dr}{\gamma} \frac{E - (u/p)r^p}{\sqrt{(E-(u/p)r^p)^2 - \gamma^2 (m/r)^2}}, \qquad (23.32)$$

where $V(r) = (u/p)r^p$. It is convenient to introduce a new variable $x = (r/r_1)^2$. Then expression (23.32) takes the form

$$\frac{dS}{dE} = \frac{r_1}{\gamma} \int_0^1 dx \frac{E - (u/p)r_1^p x^p}{\sqrt{(E-(u/p)r_1^p x^p)^2 - \gamma^2 (m/r_1)^2 x^{-2}}}$$

$$= \frac{r_1}{\gamma} \int_0^1 dx \frac{1 - x^p}{\sqrt{(1-x^p)^2 - (\gamma/E)^2 (m/r_1)^2 x^{-2}}}, \qquad (23.33)$$

where we took into account that $E = (u/p)r_0^p$ and $r_1 \approx r_0$. If the energy is large enough, i.e.,

$$E \gg m/r_1, \qquad (23.34)$$

then we can disregard the terms containing the energy in the argument of the integral. Under this condition, the expression (23.33) becomes

$$\frac{dS}{dE} = \alpha \frac{r_1}{\gamma}, \qquad (23.35)$$

where $\alpha \sim 1$ is a constant. Now we can find the interlevel separation:

$$\Delta E = \frac{\pi}{\alpha} \left(\frac{\gamma}{r_1} \right). \qquad (23.36)$$

Substituting expression $r_1 \approx r_0 = (p/uE)^{1/p}$ we obtain the interlevel separation as a function of the energy

$$\Delta E(E) = \alpha \epsilon_p (E/\epsilon_p)^{-1/p}, \qquad (23.37)$$

where ϵ_p is given by expression (23.25) and depends only on the shape of the confinement potential. For example, for parabolic confinement potential, i.e., $p = 2$, we have $\Delta E \propto 1/\sqrt{E}$, which is different from the nonrelativistic electrons with parabolic dispersion law, where the interlevel separation does not depend on the energy.

Based on the expressions for the interlevel separation (23.37) and for the maximum energy of trapped states (23.27), we can estimate the number, $N_{p,m}$, of the trapped states within a quantum dot with a given value of the angular momentum:

$$N_{p,m} = - \int_0^{m^2 p/(p+1)} \frac{dE}{\Delta E} \sim \int_0^{m^2 p/(p+1)} \frac{dE^{1/p}}{E} \sim \frac{p}{p+1} m^2. \qquad (23.38)$$

Therefore, a trapping of an electron by a confinement potential can be achieved in a smooth potential and at large values of angular momentum. The trapping is due to formation of classically forbidden region, i.e., tunneling barrier. The width of the classically forbidden region depends on the transverse momentum, which in the cylindrical geometry becomes the angular momentum. The above semiclassical approach shows that the electron trapping strongly depends on the values of the angular momentum, m. At small m the tunneling barrier is narrow and the electron can easily escape from the quantum dot. Based on this analysis, we can conclude that there are no trapped states at small values of angular momentum, at which the tunneling exponent (see Equation 23.23) becomes small. For example, at $m = \pm 1/2$, the transverse momentum of one of the components of the wavefunction becomes zero and the tunneling barrier disappears. We should not also expect any trapped states in the confinement potential with sharp boundaries, e.g., box-like potential. Indeed, in this case the slope, F, of the confinement potential at the boundary of the quantum dot becomes large and the tunneling exponent becomes small (see Equation 23.23). These statements are not completely true since the semiclassical approach does not work when the tunneling exponent is small. To address the problem of the states with small values of the angular momentum and the confinement potential with sharp boundaries, we need to solve our main system of equations (23.15) and (23.16) exactly.

23.2.3 Electron Trapping due to Interference

From the analysis of electron trapping presented in Section 23.2.2, we can conclude that at small angular momentum and in a confinement potential with sharp boundaries the electron cannot be trapped within the quantum dot region. It happens that in this case there is another mechanism of trapping. This type of trapping is not due to formation of tunneling barrier, but due to interference effects within the quantum dot region.

To study this problem we need to solve the system of equations (23.15) and (23.16) exactly. To simplify a problem, we consider a special profile of the confinement potential, which has sharp boundaries. Namely, the potential has the following form:

$$V(r) = \begin{cases} 0 & \text{if } r < R \\ V_0 & \text{if } r > R \end{cases}, \qquad (23.39)$$

where

$V_0 > 0$ is the strength of the confinement potential

R is the radius of the quantum dot

The confinement potential in Equation 23.39 has a sharp boundary and from the analysis of Section 23.2.2, we can conclude that the width of the classically forbidden region is very narrow and the trapping time of the electron within such quantum dot is small even at large values of angular momentum. This means that the trapping of an electron in such potential should be determined by the behavior of the wavefunctions within the whole region of the quantum dot, i.e., a possible electron trapping is determined by the interference effects within the quantum dot. In this case, we should not expect an exponential (as in Section 23.2.2) but a power law dependence of a trapping time on the parameters of the confinement potential.

There are two main approaches to the problem of trapped states. In the first approach, the trapped states of a quantum dot are considered as resonances, which are revealed as the peaks in the scattering cross section (Goldberger and Watson 1964) or as the first-order poles of the scattering matrix in the complex energy plane. Then the width of the peaks in the scattering cross section or the value of the imaginary part of the complex pole determines the lifetime of the trapped state (Matulis and Peeters 2008).

In the second approach, the resonances, i.e., the trapped states, are defined as the time-independent solutions of the Schrödinger equation with purely outgoing boundary conditions. The stationary solutions with such boundary conditions exist only at complex energies. Therefore the trapped states in this approach have complex energies and are considered as the long-lived states in the decay process (Goldberger and Watson 1964). The real part of the complex energy is associated with the energy of the trapped state. The inverse of the imaginary part of the energy is associated with the lifetime of the decaying state. That is, if E is the complex energy of the trapped state, then the trapping time is $\tau = \hbar/\mathrm{Im}[E]$. Here $\mathrm{Im}[E]$ is the imaginary part of the energy and $\hbar$ is the reduced Planck constant. This approach was originally introduced by Gamow (1928) and the resonances in this approach are called the Gamow's vectors.

Below we are using the second approach to the problem of trapped states. The wavefunction of a trapped state is shown schematically in Figure 23.7. The electron is trapped within the quantum dot region, i.e., at $r < R$. Outside the quantum dot, the wavefunction is oscillating and corresponds to outgoing waves, $\psi \propto \exp(ikr)$.

Therefore, to find the complex energy, E, of the trapped states, we need to solve the system of equations (23.15) and (23.16) with the following boundary conditions: (1) at $r = 0$ both functions χ_1 and χ_2 should be finite and (2) at infinite distance the functions χ_1 and χ_2 should satisfy outgoing boundary conditions, i.e., χ_1, $\chi_2 \propto \exp(ikr)$.

With the confinement potential (23.39), there are two regions $r < R$ and $r > R$. Within each of these regions, the potential is constant. Then by eliminating the function χ_1 (or χ_2) from the system of equations (23.15) and (23.16), we can find that χ_1 and χ_2 satisfy Bessel's differential equations of the orders $|m - 1/2|$ and

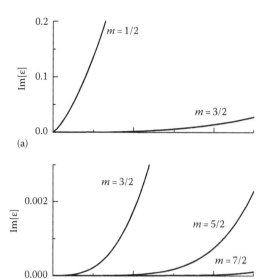

FIGURE 23.7 The imaginary part of the energy of a trapped state is shown as a function of deviation of the parameters of confinement potential from the condition of strong localization. The numbers next to the lines are the angular momenta of the corresponding states. The graph for $m = 3/2$ is the same in figures (a) and (b).

$|m + 1/2|$, respectively. Taking into account that the solutions of equations (23.15) and (23.16) inside the quantum dot should be finite at $r = 0$, we can write the general solution of the system of equations (23.15) and (23.16) inside the quantum dot, where $V = 0$, in the following form:

$$\begin{pmatrix} \chi_1(r) \\ \chi_2(r) \end{pmatrix} = A \begin{pmatrix} J_{|m-1/2|}(\varepsilon r/R) \\ iJ_{|m+1/2|}(\varepsilon r/R) \end{pmatrix}, \qquad (23.40)$$

where J_n is the Bessel function of the nth order and we introduced the dimensionless energy $\varepsilon = RE/\gamma$, which is complex in our approach.

Outside the quantum dot, i.e., at $r > R$, the solution of the corresponding Bessel differential equations should satisfy the outgoing boundary conditions, i.e., $\chi_1 \propto \chi_2 \propto \exp(ikr)$. Then the general solution of the system of equations (23.15) and (23.16) at $r > R$, where $V = V_0$, is described by the Hankel functions of the first kind and has the following form:

$$\begin{pmatrix} \chi_1(r) \\ \chi_2(r) \end{pmatrix} = B \begin{pmatrix} H^{(1)}_{|m-1/2|}\dfrac{(\varepsilon - v_0)r}{R} \\ iH^{(1)}_{|m+1/2|}\dfrac{(\varepsilon - v_0)r}{R} \end{pmatrix}, \qquad (23.41)$$

where $H^{(1)}_n$ is the Hankel function of the first kind. We also introduced the dimensionless confinement potential $v_0 = RV_0/\gamma$.

At the boundary between two regions, i.e., at $r = R$, the two-component wavefunction should be continuous. From this condition, we can find the energy eigenvalue equation in the following form:

$$\frac{H^{(1)}_{|m-1/2|}(\varepsilon - v_0)}{H^{(1)}_{|m+1/2|}(\varepsilon - v_0)} = \frac{J_{|m-1/2|}(\varepsilon)}{J_{|m+1/2|}(\varepsilon)}. \qquad (23.42)$$

The energy, ε, in this equation is complex. The imaginary part of the energy determines the electron trapping time. We can also notice that Equation 23.42 is symmetric with respect to the change of a sign of the angular momentum, $m \to -m$. This means that the energy spectra of an electron with positive and negative angular momenta are identical. Therefore below, we can consider only positive values of m, $m = 1/2, 3/2, \ldots$.

In terms of the complex energy, the condition of strong trapping means that the imaginary part of the energy is small. It happens that in the present geometry of the quantum dot, there is a special state, at which an electron is strongly localized, which means that the imaginary part of the energy of this state is zero and the electron lifetime is *infinitely large*. Such state exists only if the dimensional potential strength, v_0, is a zero of the Bessel function of the order $(m - 1/2)$, i.e., $J_{m-1/2}(v_0) = 0$. Indeed, we can see that in this case, $\varepsilon = v_0$ is a solution of Equation 23.42. Since the potential strength is real, then the imaginary part of the energy of this state is zero.

Therefore a strongly trapped state, i.e., localized state, of an electron in the quantum dot exists only if the confinement potential satisfies the following condition:

$$v_0 = \lambda_{n,i}, \qquad (23.43)$$

where $\lambda_{n,i}$ is the ith zero of the Bessel function of the order $n = 0, 1, 2, \ldots$. Here $n = m - 1/2$. To get an idea about the typical values of the strength of the confinement potential determined by Equation 23.43, we rewrite the expression (23.43) in the original units. Since the unit of the energy is γ/R, then we obtain

$$V_0 = \frac{\gamma}{R} \lambda_{n,i}. \qquad (23.44)$$

For example, for $R = 30$ nm and $\gamma = 645$ meV·nm, the heights of the confinement potential are given by the following expression:

$$V_0 = 21.5 \lambda_{n,i} \, (\text{meV}). \qquad (23.45)$$

Expression (23.45) determines the infinite set of confinement potentials at which the electron can be strongly localized, i.e., imaginary part of the energy is zero. In Table 23.1, a few lowest values of the heights of the confinement potential are shown for different values of angular momentum, $m = n + 1/2$.

Therefore, if we know the spatial size of the quantum dot, R, then we can adjust the height of the confinement potential according to Equation 23.44. In such confinement potential, the electron is strongly localized. Of course, the electron is strongly localized only at one energy, which is equal to the height of the confinement potential.

TABLE 23.1 Heights of the Confinement Potential, V_0, at Which the Electron Can Be Strongly Localized, Is Shown for a Few Lowest Values of n and i

	$i=1$	$i=2$	$i=3$	$i=4$	$i=5$	$i=6$	$i=7$
$n=0$	52	119	187	254	322	389	458
$n=1$	84	152	219	287	356	423	491
$n=2$	110	182	250	319	387	454	523
$n=3$	137	210	281	349	418	486	554
$n=4$	164	239	309	379	449	518	587

Note: The potential satisfies Equation 23.45. The potential strength is in units of meV. The radius of the quantum dot is $R = 30$ nm.

The strong electron localization is due to interference between the electron waves inside the quantum dot, so finally the outgoing wave has zero amplitude. As a result, the condition of the trapping becomes very sensitive to the parameters of the confinement potential. It means that any deviation of the parameters of the confinement potential from the condition (Equation 23.43) of localization introduces an electron escape from the quantum dot.

To study a sensitivity of the trapped states to the parameters of the confinement potential, we assume that the condition of localization is weakly violated, i.e., $\delta_v \equiv v_0 - \lambda_{m-1/2,i}$ is small and nonzero. In this case, the electron is still trapped within the quantum dot but the trapping time is finite. It means that imaginary part of the energy, $\text{Im}[\varepsilon]$, is nonzero but small. To find $\text{Im}[\varepsilon]$, we consider $\delta_v = v_0 - \lambda_{m-1/2,i}$ as a small parameter and from Equation 23.42 obtain the first nonzero correction to $\text{Im}[\varepsilon]$.

It is convenient to introduce the following notation:

$$\delta_\varepsilon \equiv \varepsilon - v_0. \qquad (23.46)$$

Then the energy of the level can be expressed in the following form:

$$\varepsilon = \lambda_{n,i} + \delta_\varepsilon + \delta_v. \qquad (23.47)$$

Taking into account expression (23.47), we rewrite the right and left-hand sides of Equation 23.42 as

$$\frac{J_{m-1/2}(\varepsilon)}{J_{m+1/2}(\varepsilon)} = \frac{J_{m-1/2}(\lambda_{m-1/2,i} + \delta_\varepsilon + \delta_v)}{J_{m+1/2}(\lambda_{m-1/2,i} + \delta_\varepsilon + \delta_v)}, \qquad (23.48)$$

$$\frac{H^{(1)}_{m-1/2}(\varepsilon - v_0)}{H^{(1)}_{m+1/2}(\varepsilon - v_0)} = \frac{H^{(1)}_{m-1/2}(\delta_\varepsilon)}{H^{(1)}_{m+1/2}(\delta_\varepsilon)}. \qquad (23.49)$$

Then the eigenvalue equation (Equation 23.42) becomes

$$\frac{J_{m-1/2}(\lambda_{m,i} + \delta_\varepsilon + \delta_v)}{J_{m+1/2}(\lambda_{m,i} + \delta_\varepsilon + \delta_v)} = \frac{H^{(1)}_{m-1/2}(\delta_\varepsilon)}{H^{(1)}_{m+1/2}(\delta_\varepsilon)}. \qquad (23.50)$$

Now we need to take into account that δ_ε and δ_v are small and find the Taylor series of the right and left-hand sides of Equation 23.50. We need to find the first nonzero imaginary correction to the energy of the state, i.e., we need to find the imaginary part of δ_ε. The left-hand side of Equation 23.50 contains only the Bessel functions. For a real argument, the Taylor series of the Bessel function is always real. Therefore, the left-hand side of Equation 23.50 gives the contribution only to the real terms in Equation 23.50. Keeping only the lowest order corrections in the left-hand side of Equation 23.50, we obtain

$$(\delta_\varepsilon + \delta_\nu)\frac{J'_{m-1/2}(\lambda_{m,i})}{J_{m+1/2}(\lambda_{m,i})} = \frac{H^{(1)}_{m-1/2}(\delta_\varepsilon)}{H^{(1)}_{m+1/2}(\delta_\varepsilon)}. \qquad (23.51)$$

The right-hand side of Equation 23.51 contains the Hankel functions. Even for the real argument, the Taylor series of the Hankel function contains both the real and imaginary terms. The Taylor series of the Hankel function depends on its order, i.e., on the value of $m - 1/2$. The special case is zero order Hankel function, i.e., $m = 1/2$. In this case, the Taylor series expansion contains logarithm and the right-hand side of Equation 23.51 becomes

$$\frac{H^{(1)}_0(\delta_\varepsilon)}{H^{(1)}_1(\delta_\varepsilon)} = -\delta_\varepsilon \ln \delta_\varepsilon + i\frac{\pi}{2}\delta_\varepsilon. \qquad (23.52)$$

Substituting this expression in Equation 23.51 we obtain

$$(\delta_\varepsilon + \delta_\nu)\frac{J'_0(\lambda_{0,i})}{J_1(\lambda_{0,i})} = -\delta_\varepsilon \ln \delta_\varepsilon + i\frac{\pi}{2}\delta_\varepsilon. \qquad (23.53)$$

Taking into account that $J'_0(\lambda_{0,i}) = -J_1(\lambda_{0,i})$ we find the solution of Equation 23.53 in the form

$$\delta_\varepsilon = \frac{\delta_\nu}{\ln \delta_\nu} + i\frac{\pi}{2}\left(\frac{\delta_\nu}{\ln \delta_\nu}\right). \qquad (23.54)$$

Since Im ε = Im δ_ε then the imaginary part of the energy is

$$\mathrm{Im}[\varepsilon] = \frac{\pi}{2}\left(\frac{\delta_\nu}{\ln \delta_\nu}\right). \qquad (23.55)$$

If $m > 1/2$ then the Taylor series of the right-hand side of Equation 23.51 takes the form

$$\frac{H^{(1)}_{m-1/2}(\delta_\varepsilon)}{H^{(1)}_{m+1/2}(\delta_\varepsilon)} = \frac{\delta_\varepsilon}{2m-1} + i\frac{\pi\delta_\varepsilon^{2m}}{2^{2m}[(m-1/2)!]^2}. \qquad (23.56)$$

Taking into account that $J'_{m-1/2}(\lambda_{m,i}) = -J_{m+1/2}(\lambda_{m,i})$ and combining Equations 23.51 and 23.56 we obtain the final equation

$$\frac{2m}{2m-1}\delta_\varepsilon = -\delta_\nu - i\frac{\pi}{2^{2m}[(m-1/2)!]^2}\delta_\varepsilon^{2m}, \qquad (23.57)$$

from which we can find the energy of the trapped state in the following form:

$$\delta_\varepsilon = -\left(1 - \frac{1}{2m}\right)\delta_\nu - i\frac{\pi}{[2^m(m-1/2)!]^2}\left[1 - \frac{1}{2m}\right]^{2m+1}\delta_\nu^{2m}. \qquad (23.58)$$

From this expression, we can find the imaginary part of the energy

$$\mathrm{Im}[\varepsilon] = \frac{\pi}{[2^m(m-1/2)!]^2}\left[1 - \frac{1}{2m}\right]^{2m+1}\delta_\nu^{2m}. \qquad (23.59)$$

Equations 23.55 and 23.59 determine an electron escape rate from the quantum dot if the condition of strong localization is weakly violated, i.e., δ_ν is small. We can see from these expressions that with increasing the angular momentum, m, of the trapped state, the imaginary part of the energy and correspondingly the electron escape rate from the quantum dot becomes less sensitive to the value of δ_ν. In Figure 23.8, the imaginary part of the energy is shown as a function of δ_ν for a few states with the lowest values of the angular momentum. We can see that at the same value of δ_ν the imaginary part is large at small values of angular momentum, m, and is small at large values of m. Therefore, we can conclude that any small deviation of the parameters of confinement potential from the condition of strong localization results in a fast escape of an electron from the states of the quantum dot with small angular momentum.

The violation of the condition of strong localization (Equation 23.43) can be due to the change of the height of the confinement potential or due the fluctuations of the radius of the quantum dot, e.g., for elliptical dot. Here the fluctuation δ_R of the radius of the quantum dot is related to parameter δ_ν through the following expression:

$$\delta_\nu = \frac{V_0}{\gamma}\delta_R = \nu_0\frac{\delta_R}{R}. \qquad (23.60)$$

For example, for the state with $m = 5/2$ and $\nu_0 = 21.1$ we can find from expressions (23.59) and (23.60) that the trapping time is around 10^{-11} s if $\delta_R/R = 5\%$ and the trapping time is 10^{-7} s if $\delta_R/R = 1\%$.

In general, to find the energy of an electron in quantum dot we need to solve Equation 23.42 numerically. The solution of this equation provides both the real and imaginary parts of the energy. From the values of the imaginary part of the energy, we

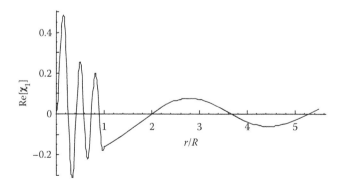

FIGURE 23.8 The first component of the wavefunction, Re[χ_1], is shown schematically as a function of r for a trapped state of the quantum dot. Outside the quantum dot the wavefunction satisfies outgoing boundary condition.

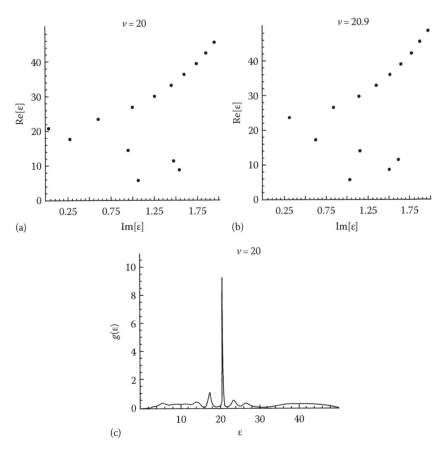

FIGURE 23.9 The energy spectra of an electron in graphene quantum dot is shown for angular momentum $m = 5/2$ in the complex energy plane (a) at $v = 20$ and (b) at $v = 20.9$. (c) The density of states, $g(\varepsilon)$, is shown as a function of ε at $v_0 = 20$. The energy is shown in dimensionless units.

can find the electron escape rate from the corresponding state of the quantum dot. In Figure 23.9, a typical energy spectra of an electron is shown at $m = 5/2$ and two values of the height of the confinement potential. From Equation 23.43 we can find that for $m = 5/2$, i.e., $n = m - 1/2 = 2$, the strong localization is achieved at the following values of the strength of the confinement potential

$$v_0 = 5.1; \ 8.4; \ 11.6; \ 14.8; \ 18.0; \ 21.1; \ ... \qquad (23.61)$$

In Figure 23.9b the confinement potential height, $v_0 = 20.9$, is close to the value 21.1 in the set (23.61). Then in this case the imaginary part of the energy of one of the states is very small (Matulis and Peeters 2008). The electron in this state can be trapped for a long time. For all other states in Figure 23.9b and also for the states in Figure 23.9a, where the strength of the confinement potential is far from any value in the set (23.61), the typical imaginary part of the energy is of the order of 1. The electron in these states is weakly trapped. We can estimate the electron trapping time in these states. In the real units, the imaginary part of the energy is of the order of γ/R. Then the trapping time is $\tau \sim \hbar R/\gamma$. For a quantum dot of size $R = 30$ nm, we obtain $\tau \sim 10^{-13}$ s. This is a relatively short time and the electron in such states should be considered as the weakly trapped.

The strongly trapped states can be also described as sharp peaks in the electron density of states. The density of states is expressed through the real and imaginary parts of the energy and in the dimensionless units has the following form:

$$g(\varepsilon) = \frac{1}{\pi} \sum_j \frac{|\mathrm{Im}(\varepsilon_j)|}{[\varepsilon - \mathrm{Re}(\varepsilon_j)]^2 + |\mathrm{Im}(\varepsilon_j)|^2}. \qquad (23.62)$$

The density of states for $v_0 = 20$ and $m = 5/2$ is shown in Figure 23.9c. The sharp maximum in the density of states corresponds to a highly trapped state with the energy close to v_0.

23.3 Conclusion

A graphene quantum dot, created by a confinement potential, has unique properties. Due to Klein's tunneling, an electron in graphene can tunnel through a confinement potential of any strength. It means that if an electron is placed inside the quantum dot, eventually it tunnels through the confinement potential into the hole state and escapes from the quantum dot. Therefore the electron cannot be localized by the confinement potential of the quantum dot. In this case we need to consider not the electron localization but the electron trapping. Then the states of the quantum dots should be considered as resonances with finite lifetimes. Such resonances are revealed as the peaks in the scattering cross section, where the width of the peak determines the electron escape rate from the quantum dot or the electron lifetime.

The trapped states of the quantum dots can be also described as the decaying states, which can be obtained as the stationary solutions of the Schrödinger equation with outgoing boundary conditions. Such solutions exist only at complex energies. The real part of the energy is the energy of the trapped state and the imaginary part of the energy is proportional to the electron escape rate from the quantum dot. The definition of the trapped states as the stationary solutions of the Schrödinger equation with outgoing boundary conditions can simplify the analysis of the problem. For some type of confinement potential, the analytical expressions for the complex energy of the trapped state can be obtained.

There are two different mechanisms of electron trapping in a graphene quantum dot. The first mechanism is due to formation of a semiclassical trapping barrier. The electrons in graphene are "relativistic" electrons with zero effective mass. Due to a gapless structure of the energy dispersion law, the electron can tunnel through any confinement potential. To suppress the Klein's tunneling, we need to introduce a nonzero electron effective mass. The nonzero effective mass can be induced by the transverse momentum. In the case of cylindrically symmetric confinement potential the transverse momentum is generated by the angular momentum. Finally, at finite and large angular momentum, the electron is trapped inside a quantum dot due to formation a semiclassical tunneling barrier. The height and the width of the barrier depend on the angular momentum and on the slope of the confinement potential at the classical turning points. The most efficient trapping is achieved in a smooth trapping potential and at large values of electron angular momentum. At small angular momentum, this mechanism of trapping is inefficient.

The second mechanism of trapping is due to interference of the electron waves within the quantum dot region. Due to this interference, the outgoing wave is strongly suppressed, which results in electron trapping. This mechanism of trapping is very sensitive to the parameters of the confinement potential and the electron energy. At the same time, it is possible to tune the parameters of the confinement potential so that the electron can be strongly localized, i.e., the trapping time is infinitely large. This strong localization occurs only at one energy.

Comparing two mechanisms of trapping, we can conclude that the advantage of the first mechanism of trapping, which is due to formation of semiclassical tunneling barrier, is that it is much less sensitive to the structure and parameters of the confinement potential than the second mechanism of trapping, which is due to interference. At the same time, only the electrons with large values of the angular momentum can be trapped in the quantum dot due to the first mechanism of trapping. In the second mechanism of trapping, the electrons with both small and large angular momenta can be strongly trapped by the confinement potential, but this trapping occurs only at one value of the electron energy.

Summarizing, the trapping due to formation of semiclassical tunneling barrier occurs in any *smooth* confinement potential and at *large* values of angular momentum. Such trapping occurs within broad energy interval. The trapping due to interference can occur in a potential of any shape and at any value of the angular momentum, but such trapping exists only at one value of electron energy and at special values of the parameters of the confinement potential.

References

Akhmerov, A. R. and C. W. J. Beenakker. 2008. Boundary conditions for Dirac fermions on a terminated honeycomb lattice. *Phys. Rev. B* 77: 085423–085432.

Ando, T. 2002. In *Nano-Physics & Bio-Electronics: A New Odyssey*, T. Chakraborty, F. Peeters, and U. Sivan (Eds.). Chap. 1. Elsevier, Amsterdam, the Netherlands.

Ando, T. 2005. Theory of electronic states and transport in carbon nanotubes *J. Phys. Soc. Jpn.* 74: 777–817.

Ando, T., T. Nakanishi, and R. Saito. 1998. Berry's phase and absence of back scattering in carbon nanotubes. *J. Phys. Soc. Jpn.* 67: 2857–2862.

Brey, L. and H. A. Fertig. 2006. Edge states and the quantized Hall effect in graphene. *Phys. Rev. B* 73: 195408–195412.

Buitelaar, M. R., A. Bachtold, T. Nussbaumer, M. Iqbal, and C. Schonenberger. 2002. Multiwall carbon nanotubes as quantum dots. *Phys. Rev. Lett.* 88: 156801–156804.

Calogeracos, A. and N. Dombey. 1999. History and physics of the Klein paradox. *Contemp. Phys.* 40: 313–321.

Chakraborty, T. 1999. *Quantum Dots*. Elsevier, Amsterdam, the Netherlands.

Chakraborty, T., F. Peeters, and U. Sivan. 2002. *Nano-Physics & Bio-Electronics: A New Odyssey*. Elsevier, Amsterdam, the Netherlands.

Chen, H.-Y., V. M. Apalkov, and T. Chakraborty. 2007. Fock-Darwin states of Dirac electrons in graphene-based artificial atoms. *Phys. Rev. Lett.* 98: 186803–186806.

Cobden, D. H. and J. Nygard. 2002. Shell filling in closed single-wall carbon nanotube quantum dots. *Phys. Rev. Lett.* 89: 046803–046806.

Fafard, K., Hinzer, S. R. et al. 1996. Red-emitting semiconductor quantum dot lasers. *Science* 274: 1350–1353.

Gamow, G. 1928. Zur quantentheorie de atomkernes. *Z. Phys.* 51: 204–212.

Goldberger, M. L. and K. M. Watson. 1964. *Collision Theory*. Wiley, New York.

Hewageegana, P. and V. Apalkov. 2008. Electron localization in graphene quantum dots. *Phys. Rev. B* 77: 245426–245433.

Ishibashi, K., S. Moriyama, D. Tsuya, and T. Fuse. 2006. Quantum-dot nanodevices with carbon nanotubes. *J. Vac. Sci. Technol. A* 24: 1349.

Kane, C. L. and E. J. Mele. 2005. Quantum spin Hall effect in graphene. *Phys. Rev. Lett.* 95: 226801–226804.

Katsnelson, M. I. 2007. Graphene: Carbon in two dimensions. *Mater. Today* 10: 20–27.

Katsnelson, M. I., K. S. Novoselov, and A. K. Geim. 2006. Chiral tunnelling and the Klein paradox in graphene. *Nat. Phys.* 2: 620–625.

Ke, S.-H., H. U. Baranger, and W. Yang. 2003. Addition energies of fullerenes and carbon nanotubes as quantum dots: The role of symmetry. *Phys. Rev. Lett.* 91: 116803–116806.

Klein, O. 1929. Die reflexion von elektronen an einem potential-sprung nach der relativischen dynamik von Dirac. *Z. Phys.* 53: 157–165.

Landau, L. D. and E. M. Lifshitz. 1991. *Quantum Mechanics: Non-Relativistic Theory,* 3rd edition. Pergamon, Oxford, NY.

Liu, H., M. Cao, J. McCaffrey, Z. R. Wasilewski, and S. Fafard. 2001. Quantum dot infrared photodetectors. *Appl. Phys. Lett.* 78: 79–81.

Loss, D. and D. P. DiVincenzo. 1998. Quantum computation with quantum dots. *Phys. Rev. A* 57: 120–126.

Matulis, A. and F. M. Peeters. 2008. Quasibound states of quantum dots in single and bilayer graphene. *Phys. Rev. B* 77: 115423–115429.

Moriyama, S., T. Fuse, M. Suzuki, Y. Aoyagi, and K. Ishibashi. 2005. Four-electron shell structures and an interacting two-electron system in carbon-nanotube quantum dots. *Phys. Rev. Lett.* 94: 186806–186809.

Nakada, K., M. Fujita, G. Dresselhaus, and M. S. Dresselhaus. 1996. Edge state in graphene ribbons: Nanometer size effect and edge shape dependence. *Phys. Rev. B* 54: 17954–17961.

Novoselov, K. S., A. K. Geim, S. V. Morozev et al. 2004. Electric field effect in atomically thin carbon films. *Science* 306: 666–669.

Novoselov, K. S., D. Jiang, F. Schedin et al. 2005. Two-dimensional atomic crystals. *Proc. Natl. Acad. Sci. USA* 102: 10451–10453.

Ponomarenko, L. A., F. Schedin, M. I. Katsnelson et al. 2008. Chaotic Dirac Billiard in graphene quantum dots. *Science* 320: 356–358.

Saito, R., G. Dresselhaus, and M.S. Dresselhaus, *Physical Properties of Carbon Nanotubes.* 1998. Imperial College Press, London, U.K.

Silvestrov, P. G. and K. B. Efetov. 2007. Quantum dots in graphene. *Phys. Rev. Lett.* 98: 016802–016805.

Slonczewski, J. C. and P. R. Weiss. 1958. Band structure of graphite. *Phys. Rev.* 109: 272–279.

Wallace, P. R. 1947. The band theory of graphite. *Phys. Rev.* 71: 622–634.

Westervelt, R. M. 2008. Graphene nanoelectronics. *Science* 320: 324–325.

Zhang, Y., Y.-W. Tan, H. L. Stormer, and P. Kim. 2005. Experimental observation of the quantum Hall effect and Berry's phase in graphene. *Nature* 438: 201–204.

24

Gas Molecules on Graphene

Tim O. Wehling
Hamburg University

and

Radbound University of Nijmegen

Mikhail I. Katsnelson
Radboud University of Nijmegen

Alexander I. Lichtenstein
Hamburg University

24.1 Introduction

The variety of existing carbon allotropes has put this element into the focus of basic as well as applied research for a long time. Its three-dimensional crystallographic forms—graphite and diamond—are known from the ancient times and widely used in industrial applications. Starting in the 1980s, lower dimensional forms of carbon have been discovered, namely, the *zero-dimensional* fullerenes or cage molecules (Kroto et al. 1985) and the *one-dimensional* carbon nanotubes (CNTs) (Iijima 1991), which are now extensively studied due to their remarkable mechanical and electronic properties. At the same time, despite very intensive research in the area, no *two-dimensional* form of carbon has been known until very recently.

Graphene—the arrangement of carbon atoms on a bipartite planar hexagonal lattice (see Figure 24.1)—has been the starting point in all calculations on graphite, CNTs, and fullerenes since the late 1940s (Wallace 1947), but its experimental realization has been postponed till 2004 (Novoselov et al. 2004) when a technique called micromechanical cleavage was employed to obtain the first graphene crystals. The observation of a peculiar electronic spectrum (Novoselov et al. 2005, Zhang et al. 2005), high electron mobility (Novoselov et al. 2004, Morozov et al. 2008), and the possibility of electronic as well as gas-sensing applications (Schedin et al. 2007) initiated enormous interest in this field (for reviews, see Castro Neto et al. 2009, Geim and Novoselov 2007, Katsnelson and Novoselov 2007).

From the electronic point of view, graphene is a two-dimensional zero-gap semiconductor with the energy spectrum shown schematically in Figure 24.1. The corner points K and K' of the Brillouin zone make up the Fermi surface of pristine graphene. Near these Fermi points, the electron dispersion is linear. Thus, graphene's low-energy quasiparticles resemble massless Dirac fermions. This band structure results in a linearly vanishing pseudogap shape density of electronic states around the Fermi level and allows studying the ultra-relativistic physics in this system (Katsnelson and Novoselov 2007). Gas molecules adsorbed on graphene cause a scattering potential acting on the "Dirac" electrons. Consequently, gas molecules on graphene can be used to investigate scattering of quasi-relativistic particles in a condensed matter experiment.

Moreover, graphene proves promising for future applications. According to the semiconductor industry roadmap, the conventional Si-based electronics is expected to encounter fundamental limitations at the spatial scale below 10 nm, thus calling for novel materials that might substitute or complement Si. The main problem of silicon is the enhanced electron scattering for devices structured at a smaller scale. Graphene is atomically thin, but its electrons scatter surprisingly little: at room temperature, some of the first graphene devices exhibited already an electronic mean free path of 0.3 μm (Geim and Novoselov 2007). Nowadays, semiconductor devices are based on the ability of combining gate voltages and chemical doping for locally controlling the density of charge carriers. One drawback of this technology is that dopants are charged impurities that may enhance the electron scattering and decrease the device performance. In graphene, however, the ballistic transport on a submicron scale persists even heavy electrostatic and chemical doping (Novoselov et al. 2004, Schedin et al. 2007). Gas molecule adsorbates provide model systems for studying and optimizing doping and electron scattering of graphene.

In addition to these electronic applications, recent experiments (Schedin et al. 2007, Robinson et al. 2008) have demonstrated graphene's potential for solid-state gas sensors and even the possibility of single molecule detection. In this chapter, we address the microscopic processes responsible for the doping of graphene by gas molecule adsorbates and explain the function

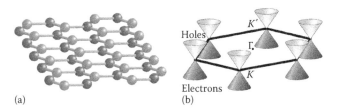

(a)   (b)

FIGURE 24.1 (a) Crystallographic structure of graphene. Atoms from different sublattices (A and B) are marked by different shades of gray. (b) Band structure of graphene in the vicinity of the Fermi level. Conduction band touches the valence band at K and K' points. The six corners of the Brillouin zone decompose into two classes, K and K', which are nonequivalent by reciprocal lattice vector translations. (From Katsnelson, M.I. and Novoselov, K.S., *Solid State Commun.*, 143, 3, 2007. With permission.)

of graphene gas sensors. The article is organized in the following way: In Section 24.2, we give an overview of experiments on gas sensing and chemical doping in graphene and related materials. Current realizations of graphene gas sensors are discussed. We proceed with the theoretical understanding of these experiments in Section 24.3 and explain how atomistic simulations of graphene interacting with gas molecules can be performed. Finally, we give the outlook and conclusions in Section 24.4.

24.2 Experiments on Graphene–Gas Molecule Interaction

Graphene and derived materials can be viewed as large nonpolar molecules and have a strong affinity to adsorb other nonpolar organic molecules. Related applications are of great industrial importance: Microporous graphene-derived structures known as "activated carbons" have an extremely high internal surface area and are widely used in filtering water and gas (Harris 2005). These filtering applications rely on the adsorption itself but do not directly make use of the modifications in the electronic properties of the graphenic filter material.

Since 2000, it has been shown that gas molecule adsorption can strongly influence the electronic behavior of graphenic materials and new applications have been emerging. CNT-based gas sensors that measure changes in electronic transport properties upon gas exposure have been reported (Collins et al. 2000, Kong et al. 2000, Kim et al. 2003). In the setup used in the Refs. (Kong et al. 2000, Kim et al. 2003), single-walled CNTs were placed on a SiO_2/Si substrate and contacted by normal metal electrodes (see Figure 24.2). The CNT-SiO_2/Si system acts like a capacitor with the CNT, and the Si back gate acts as the "plates" that are separated by the SiO_2 dielectric. By applying a gate voltage V_g between the tube and the Si gate, the CNT can be charged and the chemical potential inside the CNT can be tuned. For a semiconducting CNT, a field effect transistor (FET) is created in this way. The current through such CNT–FET turns out to be sensitive to gas exposure. Kong et al. found shifting of the current vs. gate voltage curves in different directions upon NH_3 and NO_2 exposure (see Figure 24.3). This shifting of current vs. V_g curves indicates an adsorbate-induced shifting of the Fermi level, that is, the adsorbates act as dopants.

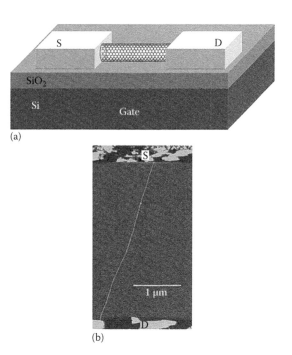

(a)

(b)

FIGURE 24.2 (a) Schematic structure of a back-gated SWNT FET. (b) Atomic force microscopy image of a device with a single nanotube (lying on SiO_2) bridging source (S) and drain (D) electrodes. (From Kim, W. et al., *Nano Lett.*, 3, 193, 2003. With permission.)

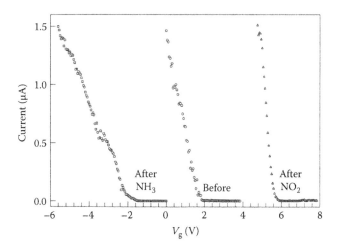

FIGURE 24.3 Chemical gating effects to the semiconducting SWNT. Current vs. gate voltage curves before NO_2 (circles), after NO_2 (triangles), and after NH_3 (squares) exposures. The measurements with NH_3 and NO_2 were carried out successively after sample recovery. (From Kong, J. et al., *Science*, 287, 622, 2000. With permission.)

Defining the sensitivity of such devices as the ratio of the resistance of the tube before (R_{before}) and after (R_{after}) gas exposure, sensitivities up to 10^3 were achieved for 200 ppm NO_2 diluted in Ar, and the typical response times were on the order of 1–10 s (Kong et al. 2000). This result was obtained at room temperature, and heating for 1 h to 200°C recovered the device to its original state. Therefore, CNT gas sensors are very interesting for applications—other solid-state sensors like Cd-doped SnO_2 operate at 250°C and have five times longer response times.

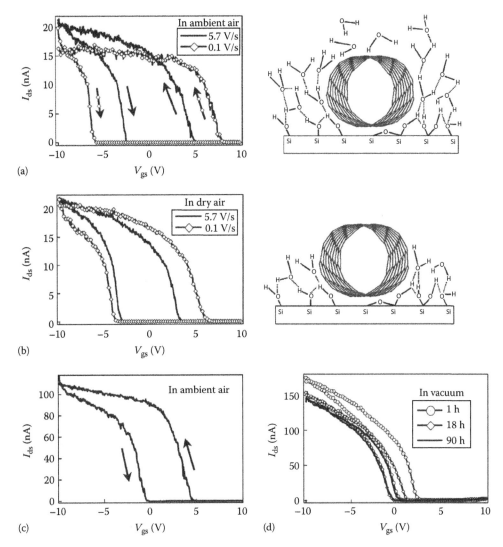

FIGURE 24.4 Source-drain current I_{ds} vs. gate voltage V_{gs} curves recorded for a SWNT FET under two V_{gs} sweep rates (as indicated) in ambient air. Bias voltage $V_{ds} = 10\,mV$. The schematic drawing at the right depicts water molecules' hydrogen bonding to the surface Si-OH groups and physisorbed on a nanotube. (b) Data recorded for the same device as in (a) after placing the device in dry air for 0.5 h. The schematic drawing at the right shows surface Si-OH-bound water molecules. (c) $I_{ds} - V_{gs}$ data for a second SWNT FET in ambient air. Gate-sweep rate = 5.7 V s^{-1}. Bias voltage $V_{ds} = 10\,mV$. (d) Data recorded after the same device as in (c) was pumped in vacuum for various periods of time. (From Kim, W. et al., *Nano Lett.*, 3, 193, 2003. With permission.)

CNT–FETs turned out to be also sensitive to H_2O adsorption (Kim et al. 2003) and exhibit hysteresis in the current vs. gate voltage curves, if water is adsorbed on the CNT (see Figure 24.4). Interestingly, the hysteresis is reduced if the CNTs are coated by the hydrophobic poly(methyl methacrylate) (PMMA). This reduction of hysteresis has been argued to be due to either the PMMA preventing the H_2O molecules from reaching the CNT and adsorbing to it or the PMMA blocking a more complicated interaction mechanism involving the CNT, the H_2O adsorbates, as well as the SiO_2 surface.

Graphene is a two dimensional material and—in contrast to CNTs—allows for experiments in Hall bar geometry. This stimulated investigations on graphene–gas molecule interactions. Already the first experiments on graphene found a strong sensitivity of single- and few-layer graphenes to gas exposure (Novoselov et al. 2004): The resistance R of a graphene multi-layer device changes upon H_2O, NH_3, and C_2H_5OH adsorption. R drops upon exposure to water, while ethanol and ammonia exposure as well as placement in vacuum increase it. Assuming unintentional doping by a film of water on all probes under ambient conditions, these changes in resistance can be attributed to different types of doping by the adsorbates (Novoselov et al. 2004): H_2O causes p-type doping, NH_3 and C_2H_5OH n-type doping.

Subsequent experiments (Schedin et al. 2007) studied the sensitivity of graphene-based devices to active gases in more detail. By combining measurements of the longitudinal and the Hall resistivity (ρ_{xx} and ρ_{xy}, respectively), the chemically induced charge carrier concentrations Δn and their signs were determined. The measurements (Figure 24.5a) show that NO_2

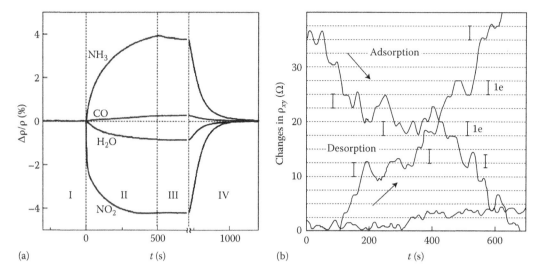

FIGURE 24.5 (a) Response in resistivity ρ of single-layer devices to NO_2, H_2O, NH_3, and CO in concentrations of 1 ppm in He at atmospheric pressure. The positive (negative) sign of changes in ρ were added to indicate electron (hole) doping. Region I—the device is in vacuum prior to its exposure; II—exposure to the diluted chemical; III—evacuation of the experimental setup; and IV—annealing at 150°C. (From Novoselov, K.S. et al., *Science*, 306, 666, 2004. With permission.) (b) Examples of changes in Hall resistivity observed near the neutrality point during adsorption of strongly diluted NO_2 and its desorption in vacuum at 50°C. The bottom curve is a reference—the same device thoroughly annealed and then exposed to pure He. The curves are for a three-layer device in $B = 10$ T. The grid lines correspond to changes in ρ_{xy} caused by adding one electron charge. (From Schedin, F. et al., *Nat. Mater.*, 6, 652, 2007. With permission.)

and H_2O act as the acceptors, while NH_3 and CO are donors. In the limit of strong doping, charge carrier concentrations on the order of 10^{13}e cm^{-2}, that is, 0.005e per graphene unit cell, could be induced by NH_3 and NO_2 adsorption. In the opposite limit of extremely dilute active gases, it is possible to detect NO_2 and NH_3 at a level below 1 ppb—the best sensitivities demonstrated so far for any gas sensors. For NO_2, the adsorption of single molecules could be detected: the height of steps occurring in Hall resistance ρ_{xy} (Figure 24.5b) is peaked around the value corresponding to the removal or addition of one electron to the graphene sample (Schedin et al. 2007). Such steps occur only, when the sample is exposed to NO_2 or annealed after exposure but not, for example, for a clean sample in He flow. Hence, one NO_2 molecule appears to withdraw one electron from the graphene sheet. While this leads to charged impurities on top of the graphene, the charge carrier mobilities are almost unaffected by this NO_2 doping (Schedin et al. 2007). Recently, the NO_2-induced shift of the Fermi level has been measured directly by means for angular resolved photoelectron spectroscopy of epitaxially grown graphene on top of SiC (Zhou et al. 2008).

To date, the search for a large-scale production scheme of graphene needed for industrial applications is still underway. A recent promising route is chemically derived graphene (Gilje et al. 2007), where a graphite oxide solution is spin casted on a SiO_2 substrate and chemically reduced after removal of the solvent. In this way, sensor devices were created, which operate similar to those mentioned in Schedin et al. (2007) and allowed for the detection of adsorbate molecules at ppb concentrations (Robinson et al. 2008).

24.3 Microscopic Theory of Graphene–Gas Molecule Interactions

The experiments discussed in the previous section show that electron transport through graphene-derived FET devices is sensitive to various adsorbates. They suggest that the adsorbates may be grouped into two classes: NO_2, for example, is a strong dopant that is effective for graphene on SiO_2 as well as on SiC substrates and which can be detected in the limit single adsorbates. On the other hand, no NH_3, H_2O, and CO single molecule detection has been achieved. For graphene's interaction with strong dopants like NO_2, the substrate seems to be of minor importance. However, for H_2O, the strong dependence of the hysteresis on the PMMA coating (see Figure 24.4 and Kim et al. 2003) indicates that the substrate is crucial for graphene's sensitivity to this adsorbate. In this section, we review theories on graphene-adsorbate interactions and discuss an experiment investigating the differences in doping strength and the mechanism mentioned above.

The microscopic mechanisms responsible for the sensitivity of graphene to certain gas molecules are likely similar to those for gas molecules on CNTs. Experiments like those in Collins et al. (2000), Kong et al. (2000), and Kim et al. (2003) stimulated numerous theoretical investigations, and a picture based on the so-called charge transfer analysis has been developed to explain the gas molecule adsorption effects on CNTs. Charge transfer analysis means partitioning the electronic charge density among the atoms of the system. To this end, a density functional theory (DFT) calculation is performed, which yields the electronic density and the wave functions. Partitioning schemes using projections of the Kohn–Sham wave functions onto localized atomic orbitals (Löwdin and Mulliken analysis, see, for example, Segall et al. 1996) as well as schemes dealing with the electronic density (Hirschfeld and Bader analysis, see, for example, Meister and Schwarz 1994) have been employed and yield similar results.

DFT calculations for NO_2, H_2O, and NH_3 on nanotubes showed that these molecules are physisorbed on nondefective CNTs, that is, they are weakly bonded by image charges or the induced dipole moments. Mulliken or Löwdin charges have been considered for different adsorbates (Peng and Cho 2000, Chang et al. 2001, Zhao et al. 2002): NO_2 is found to be negatively charged by $-0.1e$ per molecule from the tube, whereas one NH_3 molecule is predicted to be positively charged by $+0.03$ to $+0.04e$ (Chang et al. 2001, Zhao et al. 2002). Similar calculations for NO_2, H_2O, and NH_3 adsorbed on graphene (Leenaerts et al. 2008) showed that NO_2 is charged by $-0.1e$ per molecule, independently of adsorption geometry, whereas H_2O and NH_3 have charge transfer depending strongly on orientation (between 0.0 and $+0.03e$ for NH_3 and from -0.03 to $+0.02e$ for H_2O). The calculated adsorption energies are on the order of room temperature thermal energies and they differ by less than 30 meV with orientation.

For NO_2 on graphene, a reliable prediction of charge transfer is possible, whereas the situation for H_2O and NH_3 molecules is ambiguous. Moreover, the problem in directly relating charge transfer and with doping or electronic transport properties is that all schemes for detecting charge transfer are sensitive to two different types of charge transfer (see Figure 24.6 for illustration). (1) On adsorption, the wave functions of the graphene bands and the orbitals of the molecule change their shape, while the band structure close to the Fermi level remains unchanged.

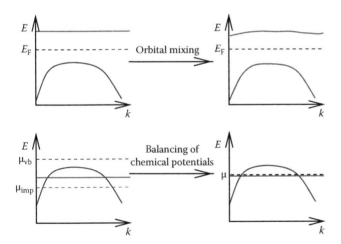

FIGURE 24.6 Schematic band structure diagrams to illustrate two different mechanisms that lead to charge transfer. The dispersive valence band of the host material (graphene) and a flat impurity band are shown. Upper panel: The charge densities of the impurity/adsorbate and the host material (graphene) overlap and the shape of the wave functions is changed, but acceptor or donor levels close to the Fermi level do not occur. Lower panel: An empty molecular orbital of the adsorbate with energy below the valence band maximum of the host is populated with the host electrons. It gives rise to charge transfer and acts as the acceptor.

The change of the involved wave functions can be expected to depend strongly on the adsorbate's orientation. But—independent of orientation—the adsorbates do not change the number charge carriers contributing to the electronic transport. (2) Electrons from an adsorbate impurity band can be transferred to a graphene band or *vice versa*, because, for example, an empty adsorbate orbital has lower energy than the valence band maximum of the host material. This type of charge transfer leads to additional electrons or holes in the graphene bands that contribute to electronic transport. A microscopic theory of adsorbate-induced doping of graphene entirely based on band structure arguments does not encounter this ambiguity.

To develop such a theory, firstly reconsider the experiment (Schedin et al. 2007). Graphene showed sensitivity to *single* (paramagnetic) NO_2 molecules, whereas no single molecule detection for the (diamagnetic) H_2O, NH_3, and CO has been found. This suggests a connection between doping strength and the presence of unpaired electrons in the adsorbate molecule. It is therefore helpful to start with the NO_2 system providing both open-shell single molecules and closed-shell dimers N_2O_4. This model system has been investigated in a combined theoretical and experimental study (Wehling et al. 2008b) that is explained below.

Theoretically, the electronic and structural properties of the graphene NO_2 systems can be addressed by means of DFT. Although van der Waals forces are ill represented in the local density approximation (LDA) as well as in gradient corrected exchange correlation functionals (GGA) resulting in over- and under-bonding, respectively (Meijer and Sprik 1996), applying both functionals yields the upper and lower bounds for the adsorption energies and the related structural properties, and allows for qualitative insights. Such calculations can be carried using a supercell approach and a plane wave-based DFT code like the Vienna *ab initio* simulation package (VASP) (Kresse and Hafner 1994) employed in Ref. (Wehling et al. 2008b). In that periodic scheme, single NO_2 and N_2O_4 adsorbates have been modeled in 3×3 and 4×4 graphene supercells, respectively. Supercells of this size still allow for full relaxation of all atomic coordinates.

Gaseous NO_2 stands in equilibrium with its dimer N_2O_4, allowing for various different adsorption mechanisms on graphene—similar to the case of graphite (Sjovall et al. 1990, Moreh et al. 1996). As the first step in simulating the interaction of these molecules with graphene, structural relaxations are performed. For both, possible adsorption geometries are shown in Figure 24.7 (right). The corresponding adsorption energies in GGA range from 85 to 67 meV (67 to 44 meV) for the monomer (dimer) per molecule. As usual, LDA yields higher adsorption energies—approximately 169–181 meV for the monomer and 112–280 meV for the dimer—and favors the adsorbates by 0.5–1 Å nearer to the sheet than the 3.4–3.9 Å obtained in GGA. Adsorption near defects can cause higher adsorption energies, for example, chemisorption of NO_2 at a vacancy defect yields 1.8 eV. However, the doping effects occurring there turn out to be similar to those on perfect graphene (see below). So, we first focus on adsorption on perfect graphene.

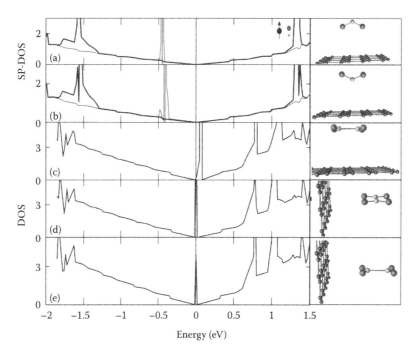

FIGURE 24.7 Left: Spin-polarized DOS of the graphene supercells with adsorbed NO_2, (a,b), and DOS of graphene with N_2O_4, (c–e), in various adsorption geometries. The energy of the Dirac points is defined as $E_D = 0$. In the case of NO_2, the Fermi level E_f of the supercell is below the Dirac point, directly at the energy of the spin down POMO, whereas for N_2O_4 E_f it is directly at the Dirac points. Right: Adsorption geometries obtained with GGA. (From Wehling, T.O. et al., *Nano Lett.*, 8, 173, 2008b. With permission.)

The effect of the adsorbates on the electronic structure can be understood from the spin-polarized electronic density of states (DOS) of the supercells shown in Figure 24.7. The molecular orbitals of NO_2 correspond to flat bands and manifest themselves as peaks in the DOS (see Figure 24.7a and b), which reveal a strong acceptor level at 0.4 eV below the Dirac point in both adsorption geometries. This acceptor state is caused by the partially occupied molecular orbital (POMO) of NO_2, which is split by a Hund-like exchange interaction. The spin-up component of this orbital is approximately 1.5 eV below the Dirac point and fully occupied, as it is also for the case of free NO_2 molecule. The spin down component of the NO_2 POMO is *unoccupied* for free NO_2, but 0.4 eV below the Dirac point in the adsorbed configuration. Hence, it can accept one electron from graphene in the dilute limit that corresponds to the limit of an infinitely large supercell. This is exactly the scenario depicted schematically in Figure 24.6 (lower part). Similar findings have been reported by several authors (Chang et al. 2001, Zhao et al. 2002, Santucci et al. 2003, Leenaerts et al. 2008, Wehling et al. 2008b).

By means of band structure calculations, the hybridization of the NO_2 acceptor bands with the graphene bands has been investigated. It turns out that the acceptor band is localized almost entirely at the adsorbate and no significant mixing with the graphene bands occurs. This weak coupling of graphene and the dopant bands means that no covalent bond is formed upon adsorption, which might be important for achieving doping without significant loss of mobility (Schedin et al. 2007).

In contrast to the paramagnetic monomer, the dimer N_2O_4, has no unpaired electrons and is diamagnetic. The possibility

of doping effects due to the adsorbed dimers can be understood with the DOS depicted in Figure 24.7c through e. Again, the molecular orbitals of the adsorbates are recognizable as sharp peaks in the supercell DOS and, similar to NO_2, the band structure calculations reveal weak coupling of the adsorbate and the graphene bands. One finds that the N_2O_4's highest occupied molecular orbital (HOMO) is in all cases more than 3 eV below the Fermi level and, therefore, does not give rise to any doping. However, the lowest unoccupied molecular orbital (LUMO) is always quite near to the Dirac point, that is, between 1 and 66 meV above it. Those initially empty N_2O_4 LUMOs can be populated by the graphene electrons due to thermal excitations and act consequently as acceptor levels. Thus, both N_2O_4 and NO_2 cause p-type doping of graphene, but the affinity of the open shell monomer to accept electrons from graphene is much stronger than for the closed-shell dimer.

The prediction of the monomer and the dimer causing acceptor levels far below and rather close to the Dirac point, respectively, is well in line with the experimental indications for two different classes of dopants regarding the doping strength and has been examined experimentally in more detail. To this end, electric field effect and Hall measurements at different adsorbate concentrations have been combined (Wehling et al. 2008b). Hall bar devices from monolayer graphene flakes on heavily doped oxidized (300 nm SiO_2) silicon substrate—similar to the FETs used in (Schedin et al. 2007)—were firstly annealed to remove any unintentional doping. Then, the samples were exposed to NO_2 strongly diluted in nitrogen (100 ppm of NO_2). After the exposure, the chamber was evacuated and the samples were

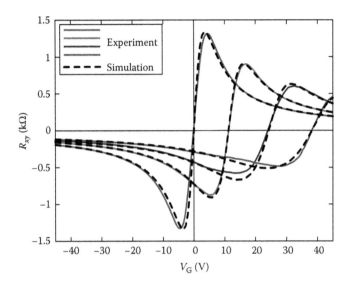

FIGURE 24.8 The Hall resistance R_{xy} as a function of the gate voltage V_G for a graphene sample with different levels of NO_2/N_2O_4 doping. The solid lines are the experimental results with the brown curve corresponding to the highest concentration of adsorbates and the red curve to almost zero doping. The dashed lines are the simulations. They are fitted to the experimental curves by adjusting the dopant concentrations c_1 (NO_2) and c_2 (N_2O_4) for each curve (see text). The simulation close to the red curve corresponds to undoped graphene, $c_1 = c_2 = 0$. (From Wehling, T.O. et al., *Nano Lett.*, 8, 173, 2008b. With permission.)

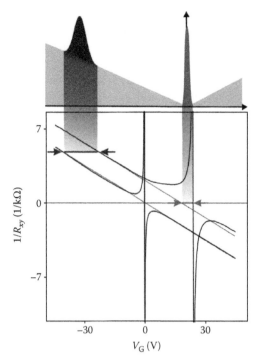

FIGURE 24.9 $1/R_{xy}$ for pristine (left curve) and doped (right curve) graphene samples. Upper panel: the DOS of doped graphene (corresponds to the right curve) with light gray depicting DOS for pure graphene, the dark peak representing the DOS for NO_2, and the midgap peak for N_2O_4. Shifting of the right curve with respect to the undoped one (left) suggests the presence of the low-laying NO_2 peak, and the fact that the electron branch for right curve is shifted with respect to the hole branch indicates the presence of the N_2O_4 peak. (From Wehling, T.O. et al., *Nano Lett.*, 8, 173, 2008b. With permission.)

annealed in a number of annealing cycles while being constantly kept under vacuum. During each annealing cycle, the devices were heated up to 410 K, kept at that temperature for some time, allowing for desorption of some NO_2/N_2O_4 (such that reducing the doping level slightly), and then cooled down to room temperature at which longitudinal R_{xx} and Hall R_{xy} resistances were measured at $B = 1$ T as a function of the gate voltage V_G. This procedure allowed varying the level of doping gradually in the range from 3×10^{12} cm^{-2} down to the practically pristine state with doping as low as 10^{11} cm^{-2}.

For a sample annealed in 16 cycles, Figure 24.8 presents R_{xy} as a function of the gate voltage taken in the fourth, seventh, and thirteenth cycles, plus the curve for the pristine state. These R_{xy} vs. V_G measurements exhibit two characteristic features. Firstly, the curves move toward higher positive gate voltages with increasing NO_2/N_2O_4 doping. Secondly, the transition region, where R_{xy} depends linearly on the gate voltage (corresponding to the presence of both types of carriers), becomes wider, and simultaneously, the maximum R_{xy} achieved becomes lower for a higher amount of NO_2/N_2O_4 on the graphene sample.

This is a clear signature of the two distinct acceptor levels. Consider $1/R_{xy}$ as shown in Figure 24.9. The deep acceptor level causes a solid shift at all V_G, while the acceptor level close to the Dirac point gives rise to an additional shift of the electron branch (straight line at negative $1/R_{xy}$). The curve for doped graphene (blue curve) exhibits these two shifts with respect to the red curve, which corresponds to the undoped graphene. The NO_2-acceptor level shifts the entire doped curve to the right, whereas

the additional shift of the electron branch reflects the presence of the N_2O_4 impurity level near the Dirac point. The latter additional shift in $1/R_{xy}$ displays as broadening of the transition region near the charge neutrality point in the R_{xy} curves that were discussed above.

For a more quantitative analysis, a model based on four types of carriers—electrons and holes in graphene as well as electrons in the NO_2 and N_2O_4 acceptor states—can be considered (see Wehling et al. (2008b) for details) to simulate the Hall resistance as a function of V_G. In this way, it can be shown that the Hall effect measurements are an indirect probe of the impurity DOS N_{imp}. For reasonable agreement of the experimental curves presented in Figure 24.8 with the simulations, $N_{imp}(E)$ has to be peaked around *two* distinct energies, $E_1 \leq -300$ meV and $E_2 \approx -40$ meV. Hence, the good agreement of the simulations and experiment (see Figure 24.8) is well in line with the presence of two distinct impurity levels due to the NO_2 and N_2O_4 as predicted by DFT. The deep acceptor level at $E_1 \leq -300$ meV due to NO_2 is always fully occupied under the experimental conditions. This full occupancy corresponds to the transfer of *one* electron from graphene sheets to NO_2 per adsorbate molecule, as observed in Schedin et al. (2007). The graphene-NO_2 system appears to be well explained in the band theory picture.

An analogue approach to single molecule NH_3 and H_2O adsorption on graphene does not yield any doping at all. DFT/GGA calculations with, for example, one H_2O molecule per 3×3 graphene supercell (Wehling 2006) or per 4×4 supercell (Leenaerts et al. 2008) predicts H_2O to physisorb at 3–4 Å above the graphene sheet with adsorption energy on the order of $\lesssim 50$ meV. However, there are no changes in the DOS and in the band structure close to the Fermi level upon the adsorption of the molecule. The HOMO of H_2O is more than 2.4 eV below the Fermi energy and its LUMO more than 3 eV above it (see also Figure 24.11a). The absence of any additional impurity level close to the Dirac point shows that single water molecules on perfect free standing graphene sheets do not cause any doping. The same result is obtained for the adsorption of single NH_3 molecules on graphene (Wehling 2006, Leenaerts et al. 2008, Ribeiro et al. 2008).

A 3×3 graphene supercell corresponds to an adsorbate concentration of $n = 2$ nm^{-2}, which is well inside the range of concentrations (1–10 nm^{-2}) found experimentally in Ref. (Moser et al. 2008). Therefore, the doping effects found experimentally (Novoselov et al. 2004, Schedin et al. 2007) are very likely due to more complicated mechanisms than the interactions of graphene with single water molecules.

The following possible scenarios are being discussed here: The experiments dealing with the effect of water on graphene were carried out using graphene on top of the SiO_2 substrates, which might be crucial for achieving doping. In addition, for finite concentrations of H_2O on graphene, H_2O clusters might form or

the H_2O might bond to defects in the graphene sheet and cause doping in this way. The latter scenario has been investigated in (Wehling 2006). For H_2O, NO_2, and NH_3 adsorbates at vacancies in graphene, structural relaxations (employing 5×5 graphene supercells) have been performed. Besides the physisorbed states, similar to those occurring on perfect graphene sheets, chemisorbed configurations exist for all of these adsorbates.

H_2O and NH_3 partially dissociate when bonding to the defect, probably because these are closed-shell molecules. In contrast, the NO_2 molecule having one unpaired electron remains almost unchanged on adsorption. The N–O bonds are elongated from 1.20 Å for the free molecule to 1.24 Å in the adsorbed configuration, while the O–N–O bond angle decreases from 134° to 123°. As discussed in Kostov et al. (2005), there exist various metastable adsorption geometries of H_2O on graphene, and a similar situation can be expected for NH_3. To date, no analysis exploring the electronic properties of all these adsorption geometries exists. However, various attempts for full relaxations in Wehling (2006) lead to similar electronic supercell band structures. (Examples are shown in Figure 24.10).

As the sublattice degeneracy is broken, a gap of about 2 eV opens around the Fermi level in all cases. The only states within ± 1 eV of E_F are the impurity bands that have major contributions from either carbon atoms in the vicinity of the defect or the adsorbate molecules. Doping of graphene is determined by the filling of these impurity bands. All cells exhibit two impurity bands that are well separated from all other graphene bands. For the pure vacancy, H_2O and NH_3 adsorbates fully occupy the

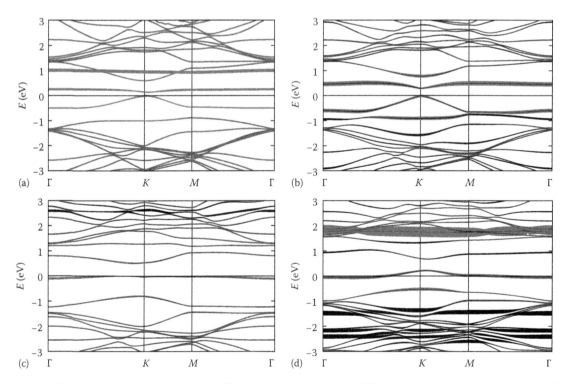

FIGURE 24.10 Band structures of 5×5 graphene supercells containing a vacancy. Jahn–Teller reconstructed pure vacancy (a) and the vacancies with the adsorbed molecules: (b) H_2O, (c) NH_3, and (d) NO_2. Contributions from the carbon atoms that are nearest neighbors of the vacancy are marked as light gray fat bands.

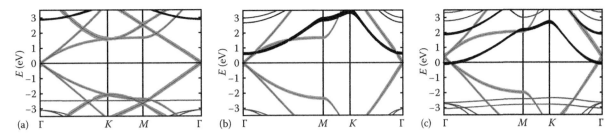

FIGURE 24.11 Bandstructures of supercells with single molecules (a), a bilayer (b), and a tetralayer (c) of water on graphene. The graphene π bands are marked in light gray, the nearly free electron bands in blue. Due to the H_2O dipole moments, graphene's nearly free electron band is shifted with respect to its π bands. (From Wehling, T.O. et al., *Appl. Phys. Lett.*, 93, 202110, 2008a. With permission.)

lower of these bands, while the upper one is completely empty in the ground state. As the cells containing H_2O and NH_3 adsorbates do not exhibit any additional impurity bands in the immediate vicinity of the Fermi level, these configurations do *not* lead to the doping of graphene. This is in contrast to the case of the NO_2 adsorbates (see Figure 24.10d), where one of the impurity bands is only half-filled and the other one is empty. The POMO of the free NO_2 molecule hybridizes with the graphene bands and is fully occupied in the adsorbed configuration (see blue marked states around 1.4 eV below E_F in Figure 24.10e). Thus, adsorption of NO_2 molecules at vacancies in graphene results in *p*-doping, with one hole per adsorbate molecule.

Therefore, the adsorption at the vacancies leads to similar doping effects as the adsorption on the perfect graphene—in particular, no single molecule effects for H_2O and NH_3. Moreover, adsorption at vacancies as the dominant interaction mechanism for graphene with gas molecule adsorbates seems unlikely since doping on the order of 10^{12}e cm^{-2} can be achieved with chemical doping (Schedin et al. 2007); but the amount of vacancies or similar defects has been estimated to be on the order of $10\,\mu m^{-2} = 10^9$ cm^{-2} (Geim 2008).

Cluster formation and substrate effects have been addressed in a recent DFT study (Wehling et al. 2008a). To model doping of graphene by large water clusters, bi-layers and four layers of ice Ih adsorbed on graphene have been considered. These overlayer structures have been proposed as the basis of ice growth on hexagonal metal surfaces (Meng et al. 2007) and they fit with a lattice mismatch of 0.23 Å as $(\sqrt{3} \times \sqrt{3})$ R30° overlayers on the graphene unit cell.

The supercell band structures (Figure 24.11b and c) show that the electric field by proton-ordered ice on top of graphene changes the energy of graphene's nearly free electron bands. In contrast to pristine graphene, where these bands start at 3 eV above the Dirac point, in the case of single H_2O adsorbates on graphene (Figure 24.11a), their bottom is 0.6 eV above and 0.1 eV below the Dirac point for a bi- and a tetralayer of ice Ih on top of graphene, respectively. This shift is due to electrostatic fields induced by the H_2O dipole moments and results in hole doping for the tetralayer of ice on graphene. The water adlayers cause a change in the contact potential that can be compared to the electrostatic force microscopy experiment (Moser et al. 2008). While only the tetralayer causes doping, the corresponding change in contact potential

turns out to exceed the experimental value from Ref. (Moser et al. 2008) by more than a factor of 4. Therefore, doping due to multiple fully oriented ice overlayers as in Figure 24.11 is probably not the most important interaction mechanism of water with graphene.

We now turn to the effect of the SiO_2 substrate in the water–graphene interplay. The SiO_2 surfaces used in the experiments on graphene gas sensing were amorphous (Schedin et al. 2007, Moser et al. 2008). In a simplified approach (Wehling et al. 2008a), the (111) surface of the crystalline SiO_2 in the β-cristobalite form has been chosen as the model system for studying graphene on an SiO_2 substrate. For a fully passivated SiO_2 substrate, no additional states occur in the vicinity of the Fermi level of graphene, as can be expected for two inert systems in contact with each other. However, this situation changes if the substrate is defective. The (111) surface of β-cristobalite allows for creating the so-called Q_3^0 defects (Wilson and Walsh 2000) having one under-coordinated silicon atom that are well known to occur on amorphous SiO_2 surfaces.

These defects lead to additional states in the vicinity (±1 eV) around the Fermi level (see Figure 24.12a and d). In the presence of these defects, the H_2O adsorbates influence both doping of graphene and electron scattering in graphene. The avoided crossing in Figure 24.12a indicates significant hybridization of the defect and the graphene bands, and it is strongly reduced in the presence of H_2O. Furthermore, the H_2O dipole moments cause local electrostatic fields that allow shifting the impurity bands significantly with respect to the graphene bands and lead to doping in this way (see Figure 24.12b, c, e, and f).

Although no exhaustive treatment of all possible defects on SiO_2 interacting with water and graphene is available to date, the examples discussed here indicate that the substrate is crucial for achieving sensitivity of graphene to H_2O. Future investigations need to clarify whether this is also the case for other closed shell adsorbates like NH_3 or CO.

24.4 Outlook and Discussion

We have elaborated on a strong difference between NO_2 and other adsorbates like H_2O, NH_3, or N_2O_4 regarding the doping strength and doping mechanisms. This is a manifestation of a general difference between the doping by open- and closed-shell

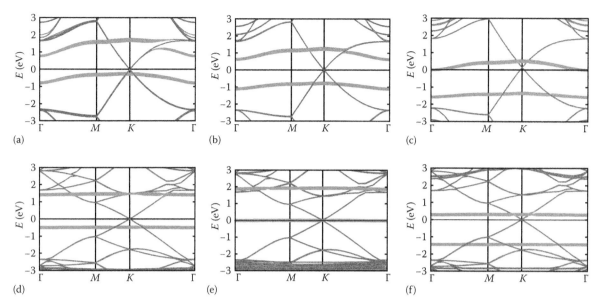

FIGURE 24.12 Band structures for graphene on defective SiO$_2$ substrates. (a–c) 2 × 2 and (d–f) 4 × 4 graphene supercells with every second, (a–c), or eighth, (d–f) surface Si atom forming a Q_3^0 defect. Spin up and down bands are shown at the same time. Contributions at the defect site are marked as light gray fat bands. (a,d) Without water adsorbates. (b) With water on top of graphene. (c,e,f) With water between graphene and the substrate. (c,f) H$_2$O dipole moment pointing tilted toward the substrate. (e) H$_2$O dipole moment pointing toward graphene. (From Wehling, T.O. et al., *Appl. Phys. Lett.*, 93, 202110, 2008. With permission.)

adsorbates. The latter type of impurities act generally as rather weak dopants or do not give rise to any doping at all, whereas the open-shell impurities cause strong doping. The reason for this is that the closed shell molecules are chemically rather inert and exhibit HOMO/LUMO gaps typically of the order $E_{HLc} \approx 5 - 10\,eV$ (Zhan et al. 2003). Thus, for doping by diamagnetic adsorbates, a chemical potential mismatch between graphene and the adsorbate on that order or more complicated mechanisms like cluster formation or substrate effects are required.

Open-shell adsorbates are qualitatively different. As one orbital is only partially populated, occupied and unoccupied states are separated by a Hund exchange of the order of $E_{HLo} \approx 1\,eV$. Thus, any open-shell molecule will give rise to doping as long as the chemical potential mismatch $\Delta\mu$ between the adsorbate and graphene exceeds half the Hund exchange splitting—that is, $\Delta\mu \gtrsim 0.5\,eV$.

This displays a close relation between graphene's DOS and its sensitivity to adsorbates, in general. As there is no gap in the spectrum, a small mismatch in the chemical potential can be sufficient to provide an active donor or acceptor level. The presence of a gap like in conventional semiconductors means that the chemical potential mismatch has to exceed half the value of that gap for obtaining any doping. Therefore, graphene gas sensors will be generally more sensitive than those built from usual semiconductors (Wehling et al. 2008b). Regarding applications, there is another important point about some gas adsorbates at graphene: the occupancy of the NO$_2$ acceptor orbital on graphene is independent of temperature, which is very different from the normal semiconductors where the acceptor states are inside the gap and populated due to thermal excitation. Therefore, graphene electronic devices could be operated at arbitrarily low temperatures in contrast to usual semiconductor devices.

Although the band theory seems to allow for a consistent understanding of the impact of gas molecule adsorbates on graphene, further tests of this model are needed. For establishing the importance of the substrate in sensing of closed shell molecules by graphene, experiments like those on water adsorption on CNTs (Kim et al. 2003) should be repeated for graphene and a broad class of adsorbates. In experiments similar to those in Refs. (Du et al. 2008, Morozov et al. 2008), it needs to be tested whether graphene on top of PMMA or free hanged graphene is mainly insensitive to closed-shell adsorbates and whether open-shell molecules like NO$_2$ cause doping in all cases.

Recently, scanning tunneling spectroscopy experiments on graphene on different substrates have been reported (Rutter et al. 2007, Zhang et al. 2008). Local probe experiments on graphene with gas adsorbates will be able to study adsorbate induced electronic states directly and may be inspiring to understand the puzzle of electron scattering in graphene (Wehling et al. 2007). Predictions like adsorbate induced gap opening in graphene (Ribeiro et al. 2008) or bonding of OH-groups (Boukhvalov and Katsnelson 2008) and hydrogen (Boukhvalov et al. 2008) to graphene can be studied in this way.

Graphene gas sensors have already been demonstrated to have highest sensitivities. An important question for future applications is to which extent selective gas sensors can be realized. Today's graphene gas sensors can distinguish between two classes of adsorbates: those causing hole or electron doping. A more precise classification of adsorbates is highly desirable. Possible routes for achieving this goal include chemical functionalization of graphene or functionalization of the substrate.

References

Boukhvalov, D. W. and Katsnelson, M. I. (2008). Modeling of graphite oxide, *JACS* **130**: 10697–10701.

Boukhvalov, D. W., Katsnelson, M. I., and Lichtenstein, A. I. (2008). Hydrogen on graphene: Electronic structure, total energy, structural distortions and magnetism from first-principles calculations, *Phys. Rev. B* **77**: 035427.

Castro Neto, A. H., Guinea, F., Peres, N. M. R., Novoselov, K. S., and Geim, A. K. (2009). The electronic properties of graphene, *Rev. Mod. Phys.* **81**: 109–162.

Chang, H., Lee, J. D., Lee, S. M., and Lee, Y. H. (2001). Adsorption of NH_3 and NO_2 molecules on carbon nanotubes, *Appl. Phys. Lett.* **79**: 3863–3865.

Collins, P. G., Bradley, K., Ishigami, M., and Zettl, A. (2000). Extreme oxygen sensitivity of electronic properties of carbon nanotubes, *Science* **287**: 1801–1804.

Du, X., Skachko, I., Barker, A., and Andrei, E. Y. (2008). Approaching ballistic transport in suspended graphene, *Nat. Nano.* **3**: 491–495.

Geim, A. K. (2008). Private communication.

Geim, A. K. and Novoselov, K. S. (2007). The rise of graphene. *Nat. Mater.* **6**: 183–191.

Gilje, S., Han, S., Wang, M., Wang, K., and Kaner, R. (2007). A chemical route to graphene for device applications, *Nano Lett.* **7**: 3394–3398.

Harris, P. (2005). New perspectives on the structure of graphitic carbons, *Crit. Rev. Solid State.* **30**(4): 235–253.

Iijima, S. (1991). Helical microtubules of graphitic carbon, *Nature* **354**: 56–58.

Katsnelson, M. I. and Novoselov, K. S. (2007). Graphene: New bridge between condensed matter physics and quantum electrodynamics, *Solid State Commun.* **143**: 3.

Kim, W., Javey, A., Vermesh, O., Wang, Q., Li, Y., and Dai, H. (2003). Hysteresis caused by water molecules in carbon nanotube field-effect transistors, *Nano Lett.* **3**: 193–198.

Kong, J., Franklin, N. R., Zhou, C., Chapline, M. G., Peng, S., Cho, K., and Dai, H. (2000). Nanotube molecular wires as chemical sensors, *Science* **287**: 622–625.

Kostov, M. K., Santiso, E. E., George, A. M., Gubbins, K. E., and Nardelli, M. B. (2005). Dissociation of water on defective carbon substrates, *Phys. Rev. Lett.* **95**: 136105.

Kresse, G. and Hafner, J. (1994). Norm-conserving and ultrasoft pseudopotentials for first-row and transition elements, *J. Phys.: Condes. Matter* **6**: 8245–8257.

Kroto, H. W.; Heath, J. R., O'Brien, S. C., Curl, R. F., and Smalley, R. E. (1985). C60: Buckminsterfullerene, *Nature* **318**: 162–163.

Leenaerts, O., Partoens, B., and Peeters, F. M. (2008). Adsorption of H_2O, NH_3, CO, NO_2, and NO on graphene: A first-principles study, *Phys. Rev. B* **77**: 125416.

Meijer, E. J. and Sprik, M. (1996). A density-functional study of the intermolecular interactions of benzene, *J. Chem. Phys.* **105**: 8684–8689.

Meister, J. and Schwarz, W. H. E. (1994). Principal components of ionicity, *J. Phys. Chem.* **98**: 8245–8252.

Meng, S., Kaxiras, E., and Zhang, Z. (2007). Water wettability of close-packed metal surfaces, *J. Chem. Phys.* **127**: 244710.

Moreh, R., Finkelstein, Y., and Shechter, H. (1996). NO_2 adsorption on Grafoil between 297 and 12 K, *Phys. Rev. B* **53**: 16006–16012.

Morozov, S. V., Novoselov, K. S., Katsnelson, M. I., Schedin, F., Elias, D. C., Jaszczak, J. A., and Geim, A. K. (2008). Giant intrinsic carrier mobilities in graphene and its bilayer, *Phys. Rev. Lett.* **100**: 016602.

Moser, J., Verdaguer, A., Jimenez, D., Barreiro, A., and Bachtold, A. (2008). The environment of graphene probed by electrostatic force microscopy, *Appl. Phys. Lett.* **92**: 123507.

Novoselov, K. S., Geim, A. K., Morozov, S. V. et al. (2004). Electric field effect in atomically thin carbon films, *Science* **306**: 666–669.

Novoselov, K. S., Geim, A. K., Morozov, S. V. et al. (2005). Two-dimensional gas of massless dirac fermions in graphene, *Nature* **438**: 197–200.

Peng, S. and Cho, K. (2000). Chemical control of nanotube electronics, *Nanotechnology* **11**: 57–60.

Ribeiro, R. M., Peres, N. M. R., Coutinho, J., and Briddon, P. R. (2008). Inducing energy gaps in monolayer and bilayer graphene: Local density approximation calculations, *Phys. Rev. B* **78**: 075442.

Robinson, J. T., Perkins, F. K., Snow, E. S., Wei, Z., and Sheehan, P. E. (2008). Reduced graphene oxide molecular sensors, *Nano Lett.* **8**: 3137–3140.

Rutter, G. M., Crain, J. N., Guisinger, N. P., Li, T., First, P. N., and Stroscio, J. A. (2007). Scattering and interference in epitaxial graphene, *Science* **317**: 219–222.

Santucci, S., Picozzi, S., Gregorio, F. D. et al. (2003). NO_2 and CO gas adsorption on carbon nanotubes: Experiment and theory, *J. Chem. Phys.* **119**: 10904–10910.

Schedin, F., Geim, A. K., Morozov, S. V. et al. (2007). Detection of individual gas molecules adsorbed on graphene, *Nat. Mater.* **6**: 652–655.

Segall, M. D., Pickard, C. J., Shah, R., and Payne, M. C. (1996). Population analysis in plane wave electronic structure calculations, *Mol. Phys.* **89**: 571.

Sjovall, P., So, S. K., Kasemo, B., Franchy, R., and Ho, W. (1990). NO_2 adsorption on graphite at 90 K, *Chem. Phys. Lett.* **172**: 125–130.

Wallace, P. R. (1947). The band theory of graphite, *Phys. Rev.* **71**: 622.

Wehling, T. O. (2006). Impurity effects in graphene, Fachbereich Physik, Universität Hamburg, Hamburg, Germany. Diplomarbeit.

Wehling, T. O., Balatsky, A. V., Katsnelson, M. I., Lichtenstein, A. I., Scharnberg, K., and Wiesendanger, R. (2007). Local electronic signatures of impurity states in graphene, *Phys. Rev. B* **75**: 125425.

Wehling, T. O., Katsnelson, M. I., and Lichtenstein, A. I. (2008a). First-principles studies of water adsorption on graphene: The role of the substrate, *Appl. Phys. Lett.* **93**: 202110.

Wehling, T. O., Novoselov, K. S., Morozov, S. V. et al. (2008b). Molecular doping of graphene, *Nano Lett.* **8**: 173–177.

Wilson, M. and Walsh, T. R. (2000). Hydrolysis of the amorphous silica surface. i. structure and dynamics of the dry surface, *J. Chem. Phys.* **113**: 9180–9190.

Zhan, C.-G., Nichols, J., and Dixon, D. (2003). Ionization potential, electron affinity, electronegativity, hardness, and electron excitation energy: molecular properties from density functional theory orbital energies, *J. Phys. Chem. A* **107**: 4184–4195.

Zhang, Y., Brar, V. W., Wang, F. et al. (2008). Giant phonon-induced conductance in scanning tunneling spectroscopy of gate-tunable graphene, *Nat. Phys.* **4**: 627.

Zhang, Y., Tan, Y.-W., Stormer, H., and Kim, P. (2005). Experimental observation of the quantum Hall effect and Berry's phase in graphene, *Nature* **438**: 201–204.

Zhao, J., Buldum, A., Han, J., and Lu, J. P. (2002). Gas molecule adsorption in carbon nanotubes and nanotube bundles, *Nanotechnology* **13**: 195–200.

Zhou, S. Y., Siegel, D. A., Fedorov, A. V., and Lanzara, A. (2008). Metal to insulator transition in epitaxial graphene induced by molecular doping, *Phys. Rev. Lett.* **101**: 086402.

Graphene Cones

Henning Heiberg-Andersen
Institute for Energy Technology

Gavin Stuart Walker
University of Nottingham

Ame Torbjørn Skjeltorp
Institute for Energy Technology

and

University of Oslo

Stine Nalum Naess
Norwegian University of
Science and Technology

25.1 Introduction

Graphene cones may be regarded as the noblest realization of the cone shape in graphite. Macroscopically they differ from a sheet of graphite—graphene—only by their curvature. In graphene, however, the carbon atoms are arranged in a flat hexagonal network, like a honeycomb, and this network cannot be embedded seamlessly on a conical surface without deformation of the hexagonal faces. Obviously, if such an arrangement was possible to force into place, it would decay back to its flat equilibrium state the moment the force was gone. So how are the carbon atoms in the graphene cones arranged? Although direct examination through a microscope is not easily done, there is strong experimental and theoretical evidence that their seamless carbon networks are hexagonal, except at the tips, where there are one or more non-hexagonal faces.

Of the major new forms of graphite, graphene and graphene cones are the newest. Of course, graphene has been around ever since graphite was discovered, but it was in 2004 that the first single sheet was successfully produced (Novoselov et al. 2004). Graphene cones were synthesized for the first time a decade earlier (Ge and Sattler 1994). The new scientific and technological possibilities offered by the availability of graphene are tremendous. However, nanoelectronic applications of graphene involve, at least at the connection points, deviations from the idealized theoretical model of a perfect periodic lattice. The impact of defects on the electronic properties of a hexagonal carbon network is therefore a hot topic right now, and so the graphene cones have attracted renewed attention. Consequently, important new results have appeared since the last comprehensive review on

graphene cones (Heiberg-Andersen 2006) was written. The present chapter is not only a timely update, but it is also an attempt to present the status of this research field to readers without a background in theoretical and computational methods. This attempt is challenging, because the study of graphene cones has now reached an advanced theoretical level. For this reason, a section of this chapter is devoted to the underlying concepts and approximations shared by all models of the atomic and electronic structure of graphene cones. On the other hand, we will not quote tedious derivations of more advanced theory developments but will try to explain the results and physical ideas in clear language.

Although advanced theoretical work has been done on the atomic structure and electronic properties of graphene cones, equally vital aspects are still largely unexplored. The nucleation and growth mechanisms, for example, are poorly understood, thus making it very difficult to reproducibly grow cones in the laboratory. Outside the laboratory, there is no evidence that graphene cones have ever occurred. Except from a few types of seashells, Nature does not seem to produce hollow cones by herself. Another feature of the cones that is not understood yet is their extraordinary ability to store hydrogen gas reversibly at ambient temperatures. To pursue this and other suggested applications, substantial progress must be made on synthesis and separation.

Graphene cones are an unconventional material. They are not crystals, but molecules too large to be modeled accurately atom by atom. They are not periodic like graphene, yet they possess a high degree of regularity. This raises the question of which scientific toolbox is the most appropriate, the solid state physicist's or the chemist's. Maybe unconventional ideas are what the field needs most. Therefore, we hope this chapter will attract the interest of readers from other disciplines and industry as well.

25.2 Background

Since carbon is essential to all known forms of life, it is both abundant and extensively studied. Nevertheless, the flexibility of graphite was not fully recognized before the end of the last century. Graphene cones belong to the series of new carbon structures that would be unthinkable before the discovery of the soccer ball molecule C_{60} that has 12 pentagonal faces in the carbon network. This molecule is popularly called the *Bucky ball*, in honor of Buckminster Fuller, an architect with a passion for geodesic domes. In the same spirit, the whole family of closed carbon networks discovered since is called *fullerenes*. The connection between curvature and non-hexagonal faces follows from a centuries old theorem of Euler, and its constraints on the atomic structure of a graphene cone will be demonstrated in Section 25.2.2. To understand the reality of these constraints, however, the basic carbon chemistry of Section 25.2.1 is a prerequisite. A historical review of the research on graphene cones is given in Section 25.2.3, and the underlying concepts of the atomic and electronic structure models applied to graphene cones are explained in Section 25.2.4.

25.2.1 Basic Carbon Chemistry

The electronic configuration of an isolated carbon atom is $1s^2 2s^2 2p^2$. In the presence of other atoms, the electrons in the $1s$ shell do not take part in any covalent bonds, so, from a chemical point of view, they contribute nothing but charge. Although the $2p$ shell has two vacancies, carbons do not bind to each other simply by sharing p electrons. Since the $2p$ and $2s$ shells of carbon have nearly the same energy, it turns out that a lower bonding energy can be obtained by activating all the four electrons in these two shells. The two periodic forms of carbon, diamond and graphite, originate from two different ways of sharing these four electrons between the carbon atoms. In diamond, each carbon in the lattice can be regarded as an atom with four hybridized orbitals, made up from one s orbital, and a linear combination of the three p orbitals. The name of this bonding type is therefore sp^3 hybridization. In chemical jargon, the diamond structure consists of four-coordinate centers, since each carbon atom is covalently bound to four other carbons atoms. The bonds are identical, except from their orientation in space. For graphite, on the other hand, the carbon is three-coordinate. The covalent bonds cut the graphene plane in three sectors of angle $2\pi/3$. They are made up from one s orbital and two p orbitals confined in the plane, so the name of this hybridization is sp^2. Along the bond-axis, these orbitals look like atomic s orbitals, and are therefore also called σ orbitals. The Greek letter for s is applied to prevent confusing atomic with molecular orbitals. Likewise, electrons originating the p orbitals normal to the plane, which do not participate in the sp^2 hybridization, are said to form π orbitals. The most energetic electrons of a graphene sheet occupy exclusively

π orbitals. Thus, the study of the electronic properties of graphene is a study of the π orbitals.

The concept of π electrons moving around in an inert lattice held together by inert σ electrons has been very successfully applied in organic chemistry. With the discovery of fullerenes, however, came the question about the reality of this separation for the electrons of curved graphitic networks. Curvature requires deviations from the sp^2 bond angle, and the π-orbitals are no longer normal to the three σ-bonds. These problems are solved by postulating a rehybridization of the orbitals, a curvature dependent mixture of sp^2 and sp^3. This is also called pyramidalization, due to the resulting out-of-plane shift of the atomic positions. According to standard perturbation theory, orbitals separated by a large energy gap mix poorly. Thus, for moderate curvatures, the concept of π orbitals still make sense for the most energetic electrons, although the pyramidal distortions of the lattice on which they move must be taken into account if accurate electronic structure predictions are sought.

The length of a bond between two carbons in a flat graphene sheet is 1.41 Å, where a nanometer (nm) equals 10 Å and 10^{-9} m. In order to stay flat, an isolated graphene sheet must be saturated by hydrogens along the edges, such that the outermost carbons are bound to two other carbons and one hydrogen. These carbon–hydrogen bonds have no π component, and their length is 1.09 Å.

25.2.2 The Hard Constraints of Geometry and Topology

The seemingly trivial exercise of mentally constructing a cone from a flat disk of radius R will soon prove useful. This must be done by cutting a sector from the sheet, and then joining the two sector boundaries of the residual sheet. The size of the sector is defined by the sector angle θ. The circumference of the cone base is $2\pi r$, where r is the base radius. On the other hand, this circumference must equal $(2\pi - \theta)R$ assuming that θ is given in radians. Thus the apex angle ϕ of the cone is given by

$$\arcsin(\phi/2) = \frac{r}{R} = \frac{2\pi - \theta}{2\pi}. \quad (25.1)$$

The sector angle θ is called the *disclination* angle, a crystallographic term for the corresponding line defect. Let us now replace the continuous disk surface with the hexagonal lattice of a graphene sheet. It is then observed from Figure 25.1 that a new seamless network of three-coordinated atoms can be achieved after removal of a sector *if and only if the disclination angle θ is an integer multiple of $\pi/3$—half the bond angle of the flat hexagonal network*. This statement has also been proved rigorously (Klein 2002). Since a maximum of five sectors of disclination angle $\pi/3$ can be removed—if there shall be anything left of the sheet—this implies that *the graphene cones have only five possible apex angles*. Let the *disclination number n* denote the number of such sectors removed, and we obtain from (25.1) the values of Table 25.1.

However, application of (25.1) to a real graphene sheet, as is done here, is an oversimplification. Embedding the residual

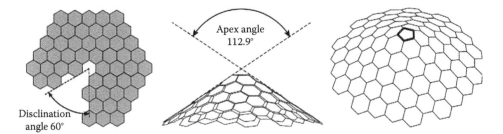

FIGURE 25.1 Formation of a cone from a graphene sheet with disclination angle of 60°. (From Ekşioğlu, B. and Nadarajah, A., *Carbon*, 44, 360, 2006. With permission.)

TABLE 25.1 Disclination Numbers and Apex Angles of Graphene Cones

n	1	2	3	4	5
ϕ	112.9°	83.6°	60.0°	38.9°	19.2°

graphene sheet on a conical surface implies bringing the carbon atoms closer to each other, especially close to the apex and for small apex angles. The sp^2 bonds will also be bent. Thus, the atomic coordinates of these theoretical graphene cones can hardly be the physical equilibrium positions. Still, the differences between the real apex angles and those of Table 25.1 are expected to be smaller than the error bounds of the measurements done in the laboratory.

Anyway, it is clear that a cone with apex angle far from any of those in Table 25.1 cannot be a graphene cone. But it can, for example, be a helically wound graphene sheet. In this case, the connection between apex angle and degree of overlap is given by (25.1). The first reported carbon cones were of this type, and will be discussed in the next section.

When $n = 1$, as in Figure 25.1, the three-coordinate carbon network is preserved by the occurrence of a pentagon at the tip. For higher disclination numbers, however, there is a plethora of topologically possible facial combinations at the tip. To figure these out, we apply Euler's theorem for convex polyhedra. A convex polyhedron can be stretched to a planar graph without crossing edges. The carbon networks of fullerenes and graphene cones are convex polyhedra, while networks on toroidal surfaces or Möbius' strips are not. According to this theorem, the number of faces, F, edges, E, and vertices, V, is given by

$$F - E + V = 2. \tag{25.2}$$

The vertices and edges correspond here to the carbon atoms and the bonds between them, respectively. It simplifies our derivation to regard the cones as parts of closed carbon networks. Then the three edges meet in each vertex, so $E = 3V/2$. Since each vertex also belongs to three faces we have $V = \Sigma_k k \cdot F_k/3$, where F_k is the number of faces with k edges, such that $F = \Sigma_k F_k$. Substituting these relations into (25.2) gives

$$\sum_k (6 - k) F_k = 12 \tag{25.3}$$

for a closed carbon network. The celebrated message of this equation is that a fullerene, no matter how big or small, must have exactly 12 pentagons. Closed cages with faces different from hexagons and pentagons are commonly called quasi-fullerenes. It is further seen from (25.3) that two pentagons are equivalent to a square, three pentagons are equivalent to a triangle, etc. These equivalences apply at the three-coordinated tip of a graphene cone although it is open at the base. Thus, a graphene cone with disclination number $n = 4$, for example, can, from a purely topological point of view, both have four pentagons or two squares at the tip. These and other possibilities can readily be verified by varying the locations of the disclination sectors of disclination angle $\pi/3$. The physical consensus, however, is that configurations of pentagonal faces at the tips are most likely. Smaller faces imply increased deviations from the optimum sp^2 bond angle, resulting in increased bond stress.

25.2.3 History of Graphene Cones and Related Structures

It could be argued that the corannulene molecule, $C_{20}H_{10}$, which has five hexagonal carbon rings around a central pentagon, is the oldest known graphene cone. It was synthesized for the first time in 1966 (Barth and Lawton 1966). The possible existence of more extended graphene cones was suggested in a theoretical paper in 1994 (Balaban and Klein 1994). Right thereafter, the first synthesis of graphene cones and cone-shaped fullerenes were reported (Ge and Sattler 1994). Both cone types had apex angles close to 19°, which correspond to $n = 5$, according to Table 25.1. The popularity of fullerenes had its peak back then, so the fullerene cone got most of the attention. A remarkable spinoff of this discovery happened in the field of HIV research, where the protein shells encapsulating the viral genome of the synthesized virus had unexpected conical shapes (Ganser et al. 1999). The dominant geometry was found to coincide with the reported fullerene cone, and thus the researchers were guided to the correct topology of the protein shell.

In 1997, graphene cones with all the five possible apex angles were synthesized by accident in a distorted version of an industrial process for the production of carbon-black (Krishnan et al. 1997). Carbon-blacks are amorphous particles with higher surface to volume ratio than soot, and their main application is rubber reinforcement in car tires. The anomalous sample

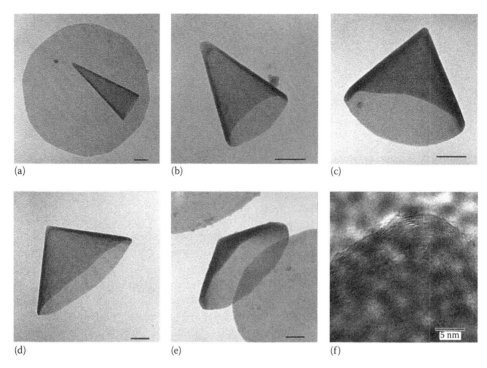

FIGURE 25.2 HRTEM images of the five cone shapes synthesized by Krishnan et al. 1997. (From Krishnan, A. et al., *Nature*, 388, 451, 1997. With permission.)

contained about 20% cones, 80% flat disks and 10% carbon-black and soot. The apex angle distribution peaked sharply at the five values of Table 25.1, and it was concluded that graphene cones, rather than helically wound sheets had been formed. Figure 25.2 shows images of the cones, taken with a high resolution transmission electron microscope (HRTEM), which also reveals the layered structure at the apex. Natural occurrences of carbon cones were reported more recently (Jaszczak et al. 2003), and their apex angle distribution is shown in Figure 25.3. This distribution is much broader, and the cones have ripples on their surface, which, according to Section 25.2.2, supports the conclusion that these cones are helically wound graphene sheets while the former are seamless graphene cones.

Synthetic helically grown cones were reported much earlier (Double and Hellawell 1974). These structures are in fact the first known forms of cone-shaped graphite. More recently, an extremely sharp version of these cones, a nanopipette was synthesized and considered for high-precision drug injection (Mani et al. 2003). Other variants of cone-shaped graphite are tubular and herringbone cones. The tubular cones (Zhang et al.

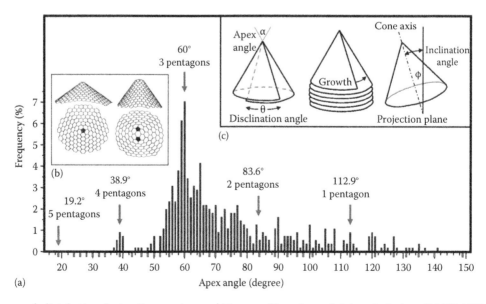

FIGURE 25.3 Apex angle distribution of naturally occurring graphitic cones. (From Jaszczak, J.A. et al., *Carbon*, 41, 2085, 2003. With permission.)

2003) are multilayered nanotubes where the length of the layers decreases gradually from the inner to the outer layer, giving the outer geometry of a cone. The herringbone cones (Dimovski et al. 2002) have apex angles of only a few degrees, and it was suggested that they had grown from a nanotube variant where the graphene layers are stacked in a herringbone pattern. The conclusion of a later work (Ekşioğlu and Nadarajah 2006) was that the herringbone pattern and the "forbidden" apex angles (see previous section) benchmarks helical cone growth.

The endcap of a closed nanotube is, ideally, a complete hemisphere with six pentagons, as can be understood from the facial sum rule (25.3) in the previous section. If one of the six pentagons is placed far from the other five, the end of the tube narrows to a cone with a 19.2° apex angle. These twiner-like structures are called *nanohorns*. They were synthesized for the first time by laser ablation of carbon at room temperature (Iijima et al. 1999), with an output as high as 10 g per h.

Next to carbon in the periodic table we find boron and nitride that possess the same sp^2 hybridization ability. There exist BN analogs to fullerenes, nanotubes, and cones, but pentagonal faces in BN networks are unfavorable. This follows since pentagons imply the presence of BB or NN bonds, which are both weaker than the BN bonds. The tip morphology of the discovered BN cones (Terauchi et al. 2000), is not established yet.

25.2.4 Atomic and Electronic Structure Models

With one exception, all the theoretical results on atomic and electronic structure of graphene cones lack experimental confirmation, simply because the relevant experiments have not yet been done. Consequently, the predictions can only be judged by the quality of the methods, which, therefore, will be explained in some depth here.

25.2.4.1 Fundamentals

Here we point out the simplifying approximations to the full quantum mechanical many-body problem that are common to most practiced models of electronic structure in chemistry and solid state physics. In both disciplines, we are dealing with a system of N atomic nuclei surrounded by a cloud of n electrons. Let $\mathbf{r} \equiv \mathbf{r}_1, \mathbf{r}_2, \dots \mathbf{r}_n$ and $\mathbf{R} \equiv \mathbf{R}_1, \mathbf{R}_2, \dots \mathbf{R}_N$ denote the electronic and nuclear coordinates, respectively. For a given total energy E_T, this many-body system is completely described by the wave function $\Psi(\mathbf{r}, \mathbf{R})$. To obtain it, we must solve the full Schrödinger equation

$$H_T \Psi(\mathbf{r}, \mathbf{R}) = E_T \Psi(\mathbf{r}, \mathbf{R}), \tag{25.4}$$

where the Hamiltonian H_T contains the kinetic energy operators for all the electrons and nuclei in addition to the three Coulomb interaction terms $U_{ee}(\mathbf{r})$, $U_{ne}(\mathbf{r}, \mathbf{R})$, and $U_{nn}(\mathbf{R})$, which are, respectively, electron–electron repulsions, nucleus–electron attractions, and nucleus–nucleus repulsions. The kinetic energy operators are second derivatives acting on the electronic and nuclear coordinates, and the full Schrödinger equation is

intractable for a molecule or solid. The first simplification to be made is known as the Born–Openheimer approximation. In a molecule or solid, the nuclei oscillate around their average lattice positions $\mathbf{R}^0 \equiv \mathbf{R}_1^0, \mathbf{R}_2^0, \dots, \mathbf{R}_N^0$ at a much slower pace than the motion of the electrons. Therefore, derivatives of the total wave function with respect to the nuclear coordinates are relatively small, and the Taylor-expanding both sides of (25.4) justifies the following separation of electronic and nuclear motion:

$$\Psi(\mathbf{r}, \mathbf{R}) \simeq \psi(\mathbf{r}; \mathbf{R}^0)\xi(\mathbf{R} - \mathbf{R}^0), \tag{25.5}$$

where

  ξ is an oscillatory wave function describing the nuclear motions around $\mathbf{R}^0$
  $\psi(\mathbf{r}; \mathbf{R}^0)$ is the wave function of the electron cloud moving in the fixed lattice of nuclei in the $\mathbf{R}^0$ positions

Assuming $\mathbf{R}^0$ is given, the many electron wave function is found by solving the Schrödinger equation

$$H\psi(\mathbf{r}; \mathbf{R}^0) = E\psi(\mathbf{r}; \mathbf{R}^0), \tag{25.6}$$

where E is the total energy of the electron cloud and the simpler Hamiltonian operator is given by

$$H = -\frac{\hbar^2}{2m} \sum_{i=1}^{n} \nabla_i^2 + U_{ee}(\mathbf{r}) + U_{ne}(\mathbf{r}; \mathbf{R}^0), \tag{25.7}$$

where

  $\hbar$ is the Planck constant divided by 2π
  m is the electron mass
  ∇_i is the derivative with respect to the coordinates of electron i

Since U_{ne} is just the sum of electron interactions with the nuclear lattice, this Hamiltonian would be separable in individual electron coordinates if it was not for the second term, U_{ee}. This motivates the so-called *mean field* approximation, which is to simulate the effect of U_{ee} by an average potential $\bar{U}(r_i)$ exerted on any electron i by all the other electrons in the cloud. The mean field Hamiltonian $\bar{H}$ is then separable:

$$\bar{H} = \sum_{i=1}^{n} \bar{H}_i, \tag{25.8}$$

The individual Hamiltonians $\{\bar{H}_i\}$ are identical:

$$\bar{H}_i = -\frac{\hbar^2}{2m} \nabla_i^2 + \bar{U}(r_i) - \frac{e^2}{4\pi\varepsilon_0} \sum_{k=1}^{N} \frac{Z_k}{\left|\mathbf{r}_i - \mathbf{R}_k^0\right|}. \tag{25.9}$$

where

  e is the electron charge
  ε_0 is the electric constant
  Z_k is the atomic number of the kth nucleus

It follows that the mean field approximation to the many-electron wave function $\psi(\mathbf{r}; \mathbf{R}^0)$ is a product of n solutions to the individual Schrödinger equation

$$\bar{H}_i \phi_i(\mathbf{r}_i; \mathbf{R}^0) = E_i \phi_i(\mathbf{r}_i; \mathbf{R}^0). \tag{25.10}$$

If the actual system is a molecule, the solutions $\{\phi_i(\mathbf{r}_i; \mathbf{R}^0)\}$ are called *molecular orbitals*, and E_i is the energy of an electron in the ith molecular orbital. For a solid, the mean field approximation gives a periodic potential in the one-electron Hamiltonian. According to Bloch's theorem, the solutions can then be expressed in the form

$$\phi_i(\mathbf{k}, \mathbf{r}_i; \mathbf{R}^0) = \exp(i\mathbf{k} \cdot \mathbf{r}) u_i(\mathbf{k}, \mathbf{r}_i; \mathbf{R}^0), \tag{25.11}$$

and are called *Bloch waves*. The periodic Bloch function u_i oscillates with the same period as the static nuclear coordinates $\mathbf{R}^0$, and the electron energy is periodic in the wave vector $\mathbf{k}$:

$$E_i(\mathbf{k} + \mathbf{K}) = E_i(\mathbf{k}). \tag{25.12}$$

Here $\mathbf{K}$ is a vector in the *reciprocal lattice* of the static nuclear coordinates $\mathbf{R}^0$, which means that for any of these positions, say $\mathbf{R}_j^0$, we have

$$\exp\left(i\mathbf{K} \cdot \mathbf{R}_j^0\right) = 1. \tag{25.13}$$

Since the nuclear coordinates of a solid are periodic, so is the reciprocal space and the unit cell repeated throughout this space is called the *Brillouin zone*, or, sometimes, the *first* Brillouin zone. It then follows from (25.12) that all possible electron energies of the ith band are found by solving (25.10) for wave vectors within the (first) Brillouin zone. The reason why the discrete molecular energy levels are smeared out in bands in a solid is the absence of boundary conditions in the model of a solid as an *infinite* periodic arrangement of nuclear coordinates. This is entirely analogous to solving the quantum mechanical exercise problem of a particle trapped in a box: It is the boundaries of the box that remove ambiguities and leave a discrete set of energy solutions, and this spectrum turns into a continuum if the box is made infinitely large. The influence of end effects on the electronic properties of the nanosized solids is indeed a central problem in theoretical nanoscience.

25.2.4.2 Molecular Mechanics

This is the fastest method for obtaining the static nuclear coordinates of a molecule or solid, or simulating a dynamic system of interacting atoms. In molecular mechanics (MM), one attempts to incorporate the effect of the electron cloud on the interacting nuclei by introducing effective internuclear potentials, and the motion of the nuclei is described by classical mechanics. Static coordinates are then found by energy minimization, and dynamic processes are simulated by solving the classical equations at successive time steps at constant total energy. For sp^2 hybridized carbon structures, the Brenner–Tersoff potential (Brenner 1990) is most frequently used. For two interacting nuclei, i and j, this potential, V_{ij}, has the form

$$V_{ij} = V_R(R_{ij}) + b_{ij} V_A(R_{ij}), \tag{25.14}$$

where R_{ij} is the internuclear distance and the terms V_R and V_A are repulsive and attractive, respectively. The constant b_{ij} is the *empirical bond order* between the nuclei. The scaling of the attractive interaction with bond order is the way of simulating, making and breaking of bonds in MM. The attractive and repulsive terms of (25.14) have been tuned to reproduce a large database of solid state and molecular properties of carbon. In actual simulations, a standard van der Waals term, which will not be quoted here, is added.

25.2.4.3 Density Functional Theory

Density functional theory (DFT) is a modern quantum chemical method to obtain simultaneously the average nuclear coordinates, $\mathbf{R}^0$, and the electron density ρ of a molecule or solid. Since the computational expense is lower than for other methods of comparable accuracy, this method has been widely used in simulations of the relatively large structures appearing in nanoscience. The basis of DFT is the Hohenberg–Kohn theorems, the first of which states that two different sets of fixed nuclear coordinates cannot give identical ground state electron densities. The coordinate set $\mathbf{R}^0$ is thus a functional of ρ. In mathematical symbols, this is expressed as $\mathbf{R}^0[\rho]$. The value of ρ at a point $\mathbf{r}'$ is given by

$$\rho(\mathbf{r}') = \sum_{i=1}^{n} \int d\mathbf{r}_1 \int d\mathbf{r}_2 \ldots \int d\mathbf{r}_n \delta(\mathbf{r}' - \mathbf{r}_i) \mid \psi(\mathbf{r}_1, \mathbf{r}_2, \ldots \mathbf{r}_n; \mathbf{R}^0[\rho]) \mid^2, \tag{25.15}$$

where

δ is the Dirac delta function
ψ is the solution of (25.5), the many-electron Schrödinger equation in the Born–Openheimer approximation

By the second Hohenberg–Kohn theorem, this implies that the ground state electronic energy is completely determined by ρ. This facilitates an efficient energy minimization procedure by expressing ρ in terms of the so-called Kohn–Sham (KS) orbitals $\left\{\phi_i^{KS}\right\}$:

$$\rho(\mathbf{r}') = \sum_{i=1}^{n} \mid a_i \phi_i^{KS}(\mathbf{r}') \mid^2. \tag{25.16}$$

These orbitals are determined by the effective Schrödinger equation

$$H_{KS} \phi_i^{KS} = \varepsilon_i \phi_i^{KS}, \tag{25.17}$$

where H_{KS} is the KS Hamiltonian and the total electron energy to be minimized is the sum of the KS energies $\{\varepsilon_i\}$ for the occupied orbitals. The KS Hamiltonian

$$H_{KS} = -\frac{\hbar^2}{2m}\nabla_i + U_{eff}(\mathbf{r}_i, \rho) \qquad (25.18)$$

reflects iterative nature of this scheme through the effective potential, U_{eff}:

$$U_{eff}(\mathbf{r}_i, \rho) = -\frac{e^2}{4\pi\varepsilon_0}\sum_{k=1}^{N}\frac{Z_k}{|\mathbf{r}_i - \mathbf{R}_k^0[\rho]|} + \frac{e^2}{4\pi\varepsilon_0}\int d\mathbf{r}'\frac{\rho(\mathbf{r}')}{|\mathbf{r}_i - \mathbf{r}'|} + U_{xc}(\mathbf{r}_i, \rho) \qquad (25.19)$$

where U_{xc} accounts for the exchange correlations of the original many-electron problem. Although the KS orbitals are closely related to the molecular orbitals originating from the mean field approximation, this procedure *involves no formal approximations of the many-electron problem*. In practical calculations however, approximations have to be done, since U_{xc} *is unknown*. Consequently, a long list of papers deals with the development and testing of exchange functionals. We will not discuss the various types here, but assume instead that U_{xc} is known. If so, the iterative solution of the KS equations (25.17) would give the electronic ground state energy, the correct electron density, and, according to the first Hohenberg–Kohn theorem, the belonging set of nuclear coordinates. These coordinates need not correspond to the nuclear configuration of lowest energy; the iteration will stop when a local minimum of the total energy as a functional of ρ is found.

25.2.4.4 Tight-Binding and the Hückel Model

Electronic structure models that implicitly or explicitly involve the mean field approximation are called *independent electron theories*, since the many-electron wave function in this approximation becomes a product of independent solutions $\{\phi_i\}$ of the same effective one-electron Hamiltonian H. The tight-binding (TB) model is such a theory, and the solution of the one-electron Schrödinger equation

$$H\phi = E\phi \qquad (25.20)$$

is sought as a linear combination of atomic orbitals situated on the N nuclei. For illustration, we consider the simplest TB scheme applied to a fullerene. In this case, only the single atomic p orbital of each carbon atom that does not take part in the sp^2-hybridization is included. Since this p orbital, call it φ, has the same functional form for all the carbon atoms, the solution takes the form

$$\phi(\mathbf{r}) = \sum_{k=1}^{N}c_k\varphi\left(\mathbf{r} - \mathbf{R}_k^0\right), \qquad (25.21)$$

where $\mathbf{r}$ is the electron coordinate. It remains to determine the constants $\{c_k\}$ for each one-electron energy E. If the small overlap of p orbitals from different carbons is ignored, this amounts to solving the following set of linear equations:

$$\sum_{l=1}^{N}H_{kl}c_l = Ec_k, \qquad (25.22)$$

where

$$H_{kl} = \int d\mathbf{r}\varphi^*(\mathbf{r} - \mathbf{R}_k)H\varphi(\mathbf{r} - \mathbf{R}_l). \qquad (25.23)$$

The Hückel model (Hückel 1931) takes this simplification a step further for the π-electronic orbitals of planar hydrocarbons. In the Hückel matrix M, the elements H_{kl} are set to zero if k and l are to non-adjacent atoms, and are equal to an empirically determined negative constant β for all adjacent carbons. The diagonal entries H_{kk} of M are all equal to the negative constant α, which is also empirical. The N independent one-electron wave functions and energies are then found by solving the eigenproblem of

$$M = \alpha I + \beta A, \qquad (25.24)$$

where

    I is the identity matrix

    A is the *adjacency* matrix, which has entries of value 0 or 1 and defines the σ bonded network of carbon atoms

Notice that this model is entirely topological—the results are determined entirely by how the carbon atoms are connected by σ bonds, beyond that, it does not matter what their spatial coordinates are. Despite its stunning simplicity, the Hückel model has been successfully applied in organic chemistry since its invention. Negative eigenvalues of M correspond to *bonding orbitals* and positive eigenvalues to *anti-bonding orbitals*. If the eigenvalue is zero, the orbital is said to be *un-bonding*.

Things get far more complicated if the $2s$ electrons are included in the linear combination of atomic orbitals. The model Hamiltonians of such more comprehensive TB schemes involve hopping amplitudes between different orbitals at both equal and different atomic sites.

25.2.4.5 Bloch Waves and Dirac Electrons in Graphene

In the case of graphene, where the atoms form a hexagonal lattice, the Brillouin zone is hexagonal too. It is found that the valence and conduction bands cross the Fermi level only at two opposite corners, $\mathbf{K}_+$ and $\mathbf{K}_-$, called the Fermi points of the Brillouin zone. A linear combination of the degenerate Bloch eigenstates at $\mathbf{K}_\pm$ is thus a solution of the one-electron Hamiltonian. The hexagonal graphene lattice can be divided into two sublattices A and B, such that atoms in A are adjacent only to atoms in B, and *vice versa*. The importance of this is that in the TB model of the π orbitals, the amplitudes of the two degenerate Bloch eigenstates vanish at different sublattices, so they can be denoted as $\phi_A(\mathbf{K}_\pm, \mathbf{r})$, which is nonzero only on sublattice A, and $\phi_B(\mathbf{K}_\pm, \mathbf{r})$, which is nonzero only on sublattice *B*. Via an expansion called the *effective mass theory* (EMT) (Peierls 1933, Luttinger and Kohn 1955); it was then found that, for energies *E* around the Fermi level, the electron motion could be described by the wavefunction (DiVincenzo and Mele 1984)

$$\psi(\mathbf{r}) = f_A(\mathbf{r})\phi_A(\mathbf{K}_{\pm},\mathbf{r}) + f_B(\mathbf{r})\phi_B(\mathbf{K}_{\pm},\mathbf{r}), \qquad (25.25)$$

where the envelope functions f_A and f_B are found by solving the equation

$$\hbar v_F \begin{pmatrix} 0 & \dfrac{\partial}{\partial x} - i\dfrac{\partial}{\partial y} \\[2mm] \dfrac{\partial}{\partial x} + i\dfrac{\partial}{\partial y} & 0 \end{pmatrix} \begin{pmatrix} f_A(\mathbf{r}) \\ f_B(\mathbf{r}) \end{pmatrix} = E\begin{pmatrix} f_A(\mathbf{r}) \\ f_B(\mathbf{r}) \end{pmatrix}, \quad (25.26)$$

where u_F is the Fermi velocity. One needs a background in quantum electrodynamics to get excited over this; the last equation is algebraically identical to the Dirac equation for a massless particle with spin, moving freely with speed u_F. Note, however, that "spin-up" and "spin-down" refer here to sub-lattice *A(B)* and *B(A)*, and not to the electron spin. This remarkable connection to relativistic quantum mechanics has resulted in much attention to the electronic properties of graphene, and several attempts have been made to transfer these ideas to graphene cones. As we shall see in Section 25.3.4.2, the latter is far from straightforward, to say the least. Notice that Equation 25.26 has no reference to the nature of the graphene sheet; the massless electron moves on a homogeneous continuous surface. For this reason, band structure models based on this equation are called *continuum models*.

25.3 State of the Art

25.3.1 Synthesis and Characterization

The first graphene and fullerene cones (Ge and Sattler 1994) was synthesized together with nanotubes by heating a pure foil of carbon to more than 3000°C, and then letting the carbon vapor condense on a graphite substrate. The amount of cones formed this way was small, and, as mentioned, they had exclusively the smallest of the five possible apex angles. Their heights and base diameters were up to 24 and 8 nm, respectively. This method has not been further developed for higher cone yield. It was proposed (Sattler 1995) that the observed cones had grown from seeds, in the form of incomplete hemispheres containing the required five pentagons.

The unintended synthesis of graphene cones with all the five possible apex angles occurred in a quite different environment. The Norwegian industrial conglomerate Kvaerner holds a patent on emission free production of carbon-black from heavy oil (Kvaerner 1998). In *Kvaerner's Carbon-Black and Hydrogen* (KCB) process, a beam of heavy oil is sprayed into a hot reactor vessel, where the carbon-black products are separated from the remaining hydrocarbons and by-product H_2 gas. The latter

are fed back into the reactor until the plasma is completely separated into carbon-black and H_2 gas. As mentioned, an accidental distortion of the setup caused the formation of disks and cones. The disk radii are in the 250–2000 nm range, typically larger than the distance from tip to base along the surface of the cones in the sample. Most of the disks are 10–30 nm thick (Garberg et al. 2008), while the cones are typically 20–50 nm thick with 50–150 layers (Helgesen et al. 2008). Large efforts are devoted to enhance the output of cones and separating disks and cones of various apex angles from the existing samples, but there are still no definitive breakthroughs to report in this respect.

More recently, stacked graphene cones have also been reported to grow on palladium catalyzed carbon nanofibers (Terrones et al. 2008). Many of the observed cones were open at the tip, like lamp shades. These had the "forbidden angles" 30°, 50°, and 70°, and the deviations from the values of Table 25.1 was attempted to be explained by larger flexibility due to the open tip. In accordance with the derivation of Section 25.2.2, this point of view has been disputed, and it has been argued that these cones are helically wound sheets (Ekşioğlu and Nadarajah 2006). However, HRTEM images of the sample showed some cones with closed tips. This is not consistent with helically grown cones, which would be expected to have an open or amorphous tip. Thus it seems more likely that both graphene cones and helically wound sheets are formed. Some of the palladium particles were found to have conical shape, and were interpreted as initiators of cone growth. No attempted improvement of this method has been reported, but it is found that further heating gives coalescent cones (Mũnoz-Navia et al. 2005b). The cones from the KCB process, on the other hand, attain a faceted shape after heating, as shown in the scanning electron microscope (SEM) image of Figure 25.4.

Since no high grade cone samples are available yet, characterization experiments have only been carried out on the raw KCB sample and are partly or completely obscured by signals from

FIGURE 25.4 SEM image of a cone from the heat-treated sample. (Courtesy of J.P. Pinheiro, n-TEC.)

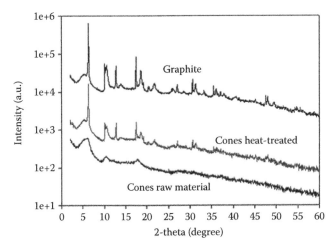

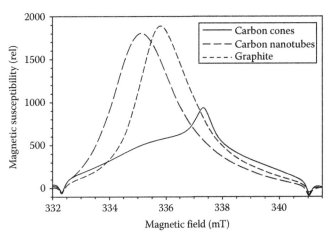

FIGURE 25.5 X-ray diffraction of graphite, and the raw and heat-treated KCB samples. The curves are shifted vertically for comparison. (From Helgesen, G. et al., *Mater. Res. Soc. Symp. Proc.*, 1057, II10, 2008. With permission.)

FIGURE 25.7 Magnetic response of graphite, nanotubes and the raw KCB sample. (From Helgesen, G. et al., *Mater. Res. Soc. Symp. Proc.*, 1057, II10, 2008. With permission.)

the more abundant disks. Thus, with one exception, no decisive tests of the theoretical predictions in the following sections have yet been possible. The exception is graphene cones with one pentagon, where the electronic structure has been studied closely (An et al. 2001) by scanning tunneling microscopy (STM). This technique will be described together with the relevant theoretical results in Section 25.3.4.1. The raw KCB sample still shows some important and interesting characteristics. Figure 25.5 compares high resolution X-ray (wavelength = 0.375 Å) diffraction data on graphite and the raw and the heat-treated KCB sample. The similarities and differences of these three curves makes sense in the light of the faceting of the heated cones; the semi-pyramidal structures of Figure 25.4 have larger portions of planar graphene, which results in sharper Bragg peaks. The peaks in the Raman shifts of graphite, nanotubes, and the KCB sample shown in Figure 25.6 correspond to specific vibrational

frequencies characterizing the type of bonding in the material. It has been suggested that relative smearing of the KCB peaks originates from sp^3 re-hybridization at the cone tips. The most striking deviations from nanotubes and graphite are seen in the magnetic response of the KCB sample, shown in Figure 25.7. It is not clear whether this behavior is caused by magnetic moments of unpaired electrons or by anomalous magnetic currents in the KCB sample.

Since the nucleation mechanisms are poorly understood, synthesis of carbon nanostructures is still something of a black art. For non-catalytic nucleation of tubes and cones, there are two competing theories, the "pentagon-road" and the ring-stacking model. The first name is somewhat misleading, because the major ingredient in this theory is the formation of a graphene sheet, which subsequently curves to a more closed structure with pentagonal faces in order to get rid of dangling bonds. The formal construction of a cone in Section 25.2.2, minus the removal of the disclination sector, is the "pentagon road" to a graphene cone, and was recently put forward as the real nucleation mechanism (Tsai and Fang 2007). The ring-stacking model postulates carbon rings as nucleation seeds. Hence, for the formation of graphene cones, the distribution of apex angles is related to the number of reaction pathways and the formation energy for the various nucleation seeds (Treacy and Kilian 2001). It was found that large ring seeds favored nucleation of cones with multiple pentagons through the rearrangement of bonds. However, a theoretical reproduction of the experimental apex angle distribution has not yet been done via any of the two nucleation models.

25.3.2 Geometry, Topology and Stability

The connection between the curvature and topology of a graphene cone was discussed in Section 25.2.2; the seamless three-coordinated network of carbon atoms is curved by the presence of non-hexagonal faces, and for any combination of

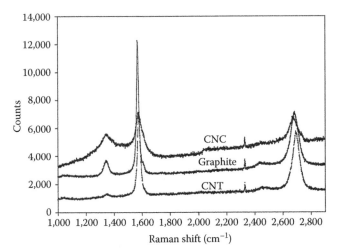

FIGURE 25.6 Raman data for graphite, nanotubes (CNT) and the raw KCB sample (CNC). (From Helgesen, G. et al., *Mater. Res. Soc. Symp. Proc.*, 1057, II10, 2008. With permission.)

faces, the apex angle of the cone must be one of the five given in Table 25.1. Two questions then arise naturally: For a given apex angle, which kind of tip topology gives the most stable cone? Are there any relationships between the infinite numbers of possible tip topologies? The last question is largely answered by Klein and Balaban (2006), regarding the cone tip as a combination of non-hexagonal faces in an infinite graphene sheet. This yields eight distinct classes of cones for the most relevant case, that is, that all non-hexagonal faces are pentagons. It is impossible to transform cones from different classes into each other by a finite number of bond rearrangements. By a complicated topological derivation, there are two classes for each of the cones with—two to four pentagons and one class for a cone with one or five pentagons. The question about stability was addressed by computing and comparing the energies of different isomeric cone tips corresponding to the same apex angle (Han and Jaffe 1998). DFT and other methods not discussed previously in this chapter gave convergent results. The tightest possible non-adjacent configuration of pentagons gave the most stable cones. From the fullerene field, it is known that adjacent pentagons undermine stability. Other non-hexagonal faces than pentagons were not considered in this analysis, and the increased energy of cones with larger distance between the pentagons was attributed to increased bond stress. The geometries of cones with one pentagon, square, or triangle have also been determined by DFT (Compernolle et al. 2004). The most important results of that work, however, concerned electronic properties and will be discussed in Section 25.3.4.3. MM calculations of different cones gave the anticipated result that the energy increases with the decreasing apex angle (Tsai and Fang 2007).

The stability question is also related to how the orbitals are filled. If all the bonding orbitals contain two electrons of opposite spin and all the un-bonding and anti-bonding orbitals are empty, the system is chemically inert, and stable in the sense that Jahn–Teller distortions do not occur. The Jahn–Teller effect is an energetically favored reduction of the symmetry of the atomic lattice of systems with incompletely filled orbitals, leading to bond stress and obvious points of chemical attack. The "isolated pentagon rule" from the fullerene field originates from such considerations, and it is found that the tightest non-adjacent configuration of the pentagons at the cone tips promote chemical stability in the Hückel model (Heiberg-Andersen and Skjeltorp 2005). This is in agreement with the above DFT simulations that implicitly account for the Jahn–Teller effect in the total energy minimization.

25.3.3 Mechanical Properties

In solid mechanics, the stiffness of a material is given by the Young's modulus, which is simply the stress/strain ratio along the axis of a rod of the material. This macroscopic concept anticipates a linear relationship between force and deformation, as in Hooke's law, before further increase of the load gives plastic deformation of the material. It is not obvious that this measure of stiffness generally applies at the nanoscale; if there is no Hook's

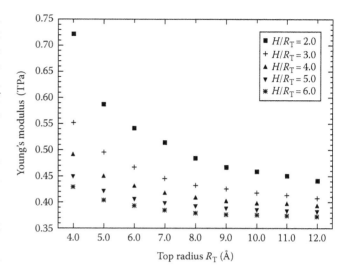

FIGURE 25.8 Simulated Young's modules of open cones with 19.2° apex angle and different ratios of height to top radius. (From Wei, J.X. et al., *Appl. Phys. Lett.*, 91, 261906, 2007. With permission.)

law regime of strain, the Young's modulus cannot be defined. Carbon nanotubes definitely have a Young's modulus. They are recognized as the ultimate future construction material; a lightweight reinforcer that is 10 times stronger than steel. How do the graphene cones compare? This question was addressed in a few recent papers.

By molecular dynamics simulations, using the Brenner–Tersoff bond-order potential described in Section 25.2.4.2, it was shown that nanotubes and graphene cones deform proportionally to the load until carbon bonds start to break (Wei et al. 2007). Thus a Young's modulus could be defined, and was calculated from continuum elastic theory, as for an isotropic solid. Within this model, the dependence of the Young's modulus on the apex angle φ could be determined. The ratio of the Young's modulus of a graphene cone to that of a nanotube of the same length and radius equal to the average cone radius was found to be cos 4φ. The model cones for the elastic calculations were open at the apex, and the obtained Young's moduli depended on the radius of this opening, as well as the height. Figure 25.8 illustrates this dependence for a cone with a 19.2° apex angle.

The same molecular dynamics scheme was used to simulate the buckling of cones under compression (Liew et al. 2007). Figure 25.9 shows the deformations of a 19.2° cone under various amounts of axial and in-plane compressional strain. It was found that under axial compression, the buckling load of a nanotube with the same height/radius ratio was almost twice that of the cone. The in-plane stiffness of a cone was found to be larger than that of a tube with comparable radius. Apart from the latter result, the message of these computational works is that the cones do not challenge the nanotubes as reinforcement materials.

Cones with larger apex angles show unique response to simulated compression by an indenter directed against the tip

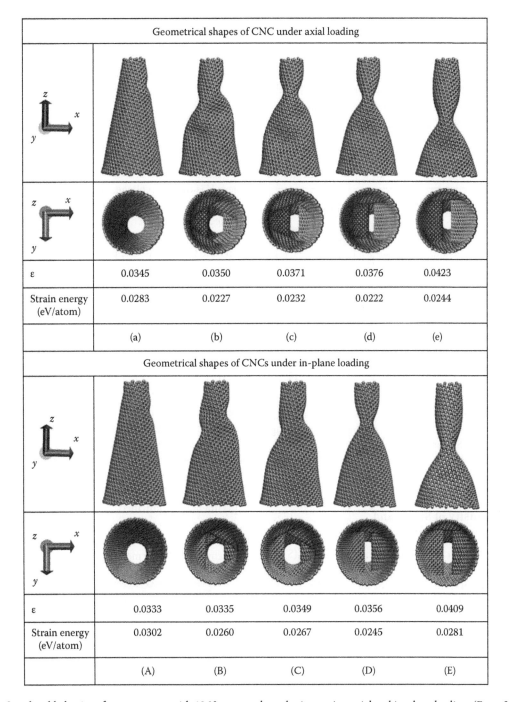

FIGURE 25.9 Simulated behavior of an open cone with 19.2° apex angle under increasing axial and in-plane loading. (From Liew, K.M. et al., *Phys. Rev. B*, 75, 195435, 2007. With permission.)

(Jordan and Crespi 2004), as shown in Figure 25.10. Beyond the Hooke's law regime, the cone starts to invert, and the local structures of both the upward and downward sloping regions corresponds to that of the un-deformed cone. The mechanical stress is focused in the transition region between upward and downward sloping, and this region grows linearly with increasing deformation. Thus, the elastic energy is a linear function of the indenter displacement in this regime, as shown in Figure 25.11. An interesting consequence of this behavior is that a cone

with large apex angle can be transformed to its chiral inverse (i.e., flipped inside-out) without any reorganizing of the chemical bonds. An earlier work (Shenderova et al. 2001) predicted metastable wavy geometries for cones with one central pentagon and radii above 14 Å. During the simulated inversion, no such states were observed. A suggested explanation for this was the different boundary conditions brought on by the indenter. Both works applied the Brenner–Tersoff potential in the MM simulation of the cones.

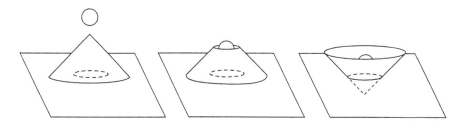

FIGURE 25.10 Schematic of a simulated compression of a large angle cone by a spherical indenter. (From Jordan, S.P. and Crespi, V.H., *Phys. Rev. Lett.*, 93, 255504, 2004. With permission.)

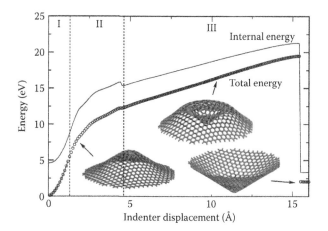

FIGURE 25.11 Internal (cone) and total (cone+indenter+substrate) energy as a function of indenter displacement during inversion through the apex. (From Jordan, S.P. and Crespi, V.H., *Phys. Rev. Lett.*, 93, 255504, 2004. With permission.)

25.3.4 Electronic Properties

25.3.4.1 Local Density of States and Petal Superstructures

With the invention of the scanning tunneling microscope (STM) in 1981, it became possible to investigate the electronic properties of molecules and solids in terms of the local density of states (LDOS) near the Fermi level. For any given point $\mathbf{r}$ of the structure under investigation, the LDOS $\rho(\mathbf{r}, E)$ gives the probability of finding an electron of energy E within an infinitesimal volume element around $\mathbf{r}$. As the name indicates, STM exploits the quantum mechanical tunneling principle, namely that the probability of penetrating a constant potential barrier U that exceeds the kinetic energy K of the particle is not zero, as in classical mechanics, but proportional to $\exp(-2L\sqrt{2m(U-K)}/\hbar)$, where L is the length of the barrier and m is the mass of the particle. The principle applies of course to non-constant barriers too, but the operational expression is then more complicated. The STM image is formed by the variation in the tunneling current I of electrons between the scanning tip of the microscope and the sample. If $\mathbf{r}$ measures the distance from the tip to a point in the sample, I is related to the LDOS and the Fermi level E_F as

$$I \propto \int_{E_F}^{E_F + eV} dE \rho(\mathbf{r}, E), \qquad (25.27)$$

where V is the bias voltage. This expression reflects the fact that there must be vacant states above the Fermi level that the tunneling electrons can enter. According to the foregoing discussion, I will fall sharply with increasing distance between the microscope and the sample. The physical size of the scanning tip limits the resolution, but this technique can yet reveal both the atomic sites and the accuracy of the theoretical electronic wave functions used to calculate the LDOS.

The calculated LDOS of a model graphene cone with one pentagon at the tip and chemical formula $C_{500}H_{50}$ displayed a petal superstructure (Kobayashi 2000), as shown in Figure 25.12. This simulated STM image was later confirmed experimentally (An et al. 2001). Since both DFT and EMT (see Section 25.2.4.5) gave very similar LDOS, an interpretation of the superstructures could be given in terms of EMT. The petals were found to originate from the phases of the EMT amplitude and not from the interference of scattered low energy Bloch waves, a common phenomenon around impurities in graphite (Mizes and Foster 1989). As mentioned, this is the only theoretical prediction about electronic properties of graphene cones, which has been experimentally confirmed.

Also the field emission properties of a graphene cone or a capped nanotube are determined by the tip's LDOS around E_F. Field emission (FEM) means release of electrons from the emitter

FIGURE 25.12 Simulated STM image of a graphene cone with one central pentagon. (From Kobayashi, K., *Phys. Rev. B*, 61, 8496, 2000. With permission.)

under the influence of an external electric field. For both graphene cones and capped nanotubes, the various theoretical models generally show enhanced LDOS around the pentagons at the tip. Graphene cones have been considered as emitters in display technologies (Charlier and Rignanese 2001), also due to the notion that their aspect ratios might give them longer lifetimes than emitting nanotubes. With the current state of art of synthesis and separation of graphene cones, however, such applications are not possible yet. Nanotube displays, on the other hand, already exist.

25.3.4.2 Continuum Models and TB Results

In addition to DFT and EMT, the LDOS at the cone tips have also been calculated by various TB and continuum models. Although enhancement of LDOS and electronic charge at the tips are predicted by all models, there are significant discrepancies between the predicted positions of the peaks in the LDOS. The continuum models vary in the way they account for the non-hexagonal faces as "topological defects" in the perfect graphene sheet. A common feature, though, is that the tip, and thus the non-hexagonal faces, is treated as a singular point on an otherwise continuous surface. With this line of attack, topological features like the relative positions of the non-hexagonal faces, appear difficult to incorporate properly. To outline the competing continuum models for cones would require lengthy declarations of concepts from quantum field theory and topology. The status, however, is that three different gauge transformations of the Dirac Hamiltonian have been put forward to make the transition from a flat graphene sheet, and the results are conflicting. An unique feature of the first continuum model (Lammert and Crespi 2000), is that the relative configuration of the pentagons at the tip are reflected in the gauge fields. This model predicts a nonzero apex density of extended states at the Fermi level for cones with two pentagons in the closest non-adjacent configuration. According to the next continuum model (Osipov and Kochetov 2001), on the other hand, only cones with three pentagons should have nonzero apex density of states at the Fermi level, unless the states are localized. Finally, the newest continuum model (Sitenko and Vlasii 2007) rejects both these assertions as consequences of inappropriate treatment of irregular eigenmodes of the transformed one-particle Hamiltonian, and, in the case of three pentagons, entwinement of the two graphene sublattices. For the time being, these claims have not been confronted by the proponents of the preceding continuum models.

There is a recursion method (Haydock et al. 1972) that enables TB calculations of the local density of states for infinite graphene sheets and cones, provided only π orbitals are taken into account. Due to this restriction, the recursive TB results for cones (Tamura et al. 1997) should be consistent with the continuum models, according to the outline of Section 25.2.4.5. As predicted by the model of Lammert and Crespi, the calculated apex density of a cone with two pentagons in the closest non-adjacent configuration does not vanish at the Fermi level. This result is thus in open disagreement with the two later continuum models. The shape of the TB apex density is however not as predicted by Lammert and Crespi.

The continuum model of Lammert and Crespi, if correct, has some subtle but measurable implications. The fictitious gauge flux through the apex depends on the relative configuration of the pentagons, and affects the response of the cone to a magnetic or electric field. These effects arise from residual phases acquired by the electronic wave functions under rotation around the singular point representing the apices with multiple pentagons. Characteristic interference of waves traveling in opposite direction around the apex should show up in STM images, and a magnetic field should set up characteristic Landau levels (quantized magnetic orbits). These predicted phenomena are analogous to the Aharanov–Bohm effect, where charged particles are affected by a magnetic or electric field even when they are confined to regions where this field vanishes. The Aharanov–Bohm effect has been observed in nanotubes (Shaver et al. 2006), and its theoretical impact on cones with two pentagons was derived in Lammert and Crespi (2000, 2004).

Two different TB calculations including σ electrons have been performed on finite cones by Charlier and Rignanese (2001) and Muñoz-Navia et al. (2005a). It is noteworthy that both these schemes give a peak in the apex density of states at or near the Fermi level for cones with three pentagons in the configuration shown in Figure 25.13b. For two pentagons the results of

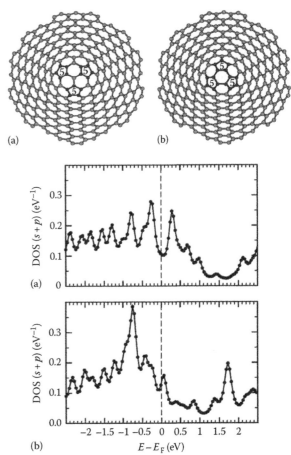

FIGURE 25.13 TB calculation of apex densities of states for two different configurations of the three pentagons of a 60° cone. (From Muñoz-Navia, M. et al., *Phys. Rev. B*, 72, 235403, 2005a. With permission.)

Charlier and Rignanese (2001) are quite different from those obtained for infinite cones with the recursion method. It is not clear if this is due to finite size effects or σ electrons. Figure 25.13, taken from Muñoz-Navia et al. (2005a), shows in addition the equally important point that the peak positions are sensitive to the relative configuration of the pentagons at the tip. This work also showed the importance of symmetry to the electronic properties of cones with open tips.

25.3.4.3 DFT Results

The formal derivation of DFT was outlined in Section 25.2.4.3. In practical calculations, the KS orbitals are expressed as linear combinations of atomic orbitals, and the accuracy of the results depend on the quality of these basis sets and the choice of exchange functional. As a rule, however, DFT predictions of electronic properties are more reliable than those of the various TB models applied to graphene cones.

One could expect the mixing of σ and π orbitals due to curvature to be small for the highest occupied molecular orbital (HOMO) and the lowest unoccupied molecular orbital (LUMO), since these most important orbitals (chemically speaking) have much higher energy than the σ orbitals in a planar hydrocarbon. If so, these orbitals should be ruled primarily by topology, as in the Hückel model. Figure 25.14 shows the comparison between the Hückel HOMO and the highest occupied KS orbital of a DFT simulation of cone with four pentagons at the tip (Heiberg-Andersen et al. 2008). The agreement is quite good, and the differences may be explained largely by the absence of Coulomb repulsion between the two electrons occupying the HOMO in the Hückel model.

In addition to the exact geometry, the electronic properties of cones with one pentagon, square, or triangle at the tip has

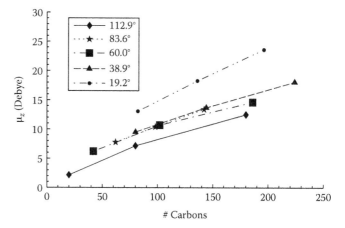

FIGURE 25.15 Simulated dipole moments versus # carbons for the five different cone angles. (From Heiberg-Andersen, H. et al., *J. Non-Cryst. Solids*, 354, 5247, 2008. With permission.)

been studied by DFT (Compernolle et al. 2004). It was found that the mixing of π and σ states in the apex density of states decreased with the size of the non-hexagonal face. This implies a decrease of σ/π mixing with decreasing curvature, as expected. Less obvious was the finding that the HOMO–LUMO energy gap was largely independent of the size of the non-hexagonal face. Further, the charging effects of the square and the triangle were found to be different than those obtained for infinite cones (Tamura et al. 1997) with the TB model which includes only π orbitals. It was also concluded that the local density of states of different cones within one of Klein's eight topology classes have common characteristics.

Except for the smallest apex angle, it is possible to place the pentagons symmetrically at the tip, such that the only vanishing component of the static dipole moment μ, which we call μ_z, is parallel to the longitudinal (z) axis. Since the synthesized graphene cones are quite large, it was found worthwhile to calculate the dipoles over a size range accessible to DFT in order to obtain dipole/mass curves for the cones (Heiberg-Andersen et al. 2008), although direct extrapolation of these curves into the mesoscopic region is speculative. As shown in Figure 25.15, the dipoles have a nearly linear relationship on the number of carbons in the cones, and the curves for cones with two and four pentagons nearly coincide. The computed dipoles were found to be very sensitive to the quality of the basis sets used in the DFT simulations, and therefore dipole predictions for larger cones in terms of simpler electronic models were not attempted.

25.3.5 Hydrogen Storage

Rapid depletion of the world's oil fields has put the search for alternative fuels into high gear. Hydrogen is often touted as the fuel of the future, but there are two obstacles that have not yet been overcome. First, hydrogen does not exist as a free gas, but is found primarily in compounds such as water. Hence, viable amounts of hydrogen gas need to be generated from these

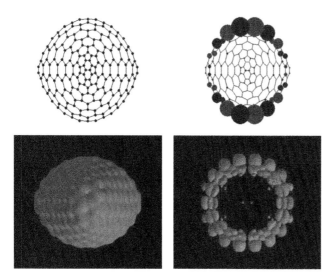

FIGURE 25.14 Upper panel: Molecular graph and Hückel HOMO of a cone with four pentagons at the tip. Lower panel: DFT simulation of the total electron density and HOMO of the same cone. (From Heiberg-Andersen, H. et al., *J. Non-Cryst. Solids*, 354, 5247, 2008. With permission.)

compounds in an energy efficient manner. Second, the hydrogen must be stored, and subsequently released for use, with minimal additional energy inputs. The last stop of the hydrogen route, the fuel cell, is the least problematic. The performance of a modern fuel cell is excellent (having efficiencies of 50%–60%); the main problem is the price tag due to the platinum catalyst. A well-known and very attractive feature of the hydrogen technology is that it is pollution free at the user's end, that is, the only outputs are electricity and water.

To avoid the inherent problems of the high pressure hydrogen tank or the cooled liquid hydrogen tank, the car industry needs a cheap, lightweight adsorptive material for on-board storage. Hydrogen adsorption on carbon surfaces is an old idea, and the first experiments were done with activated carbon in 1967 (Kidnay and Hiza 1967). As the carbon nanostructures were discovered, hydrogen storage was among the first attempted applications. After a period with reports of wildly divergent hydrogen weight percents, the consensus converged against values well below the implementation limit for vehicles, which is currently set to 6% by the U.S. Department of Energy. Then the focus changed to hydrogen storage in metal-doped nanotubes. For graphene cones and nanohorns, however, the research on the pristine structures is far from completed.

In 2001 hydrogen storage experiments were carried out on the raw KCB sample. Figure 25.16 shows the hydrogen desorption curve measured at ambient pressure. The sharp peak to the left is the so-called physisorption peak; the increase in hydrogen pressure is from the release of molecular hydrogen that is weakly attached to the adsorbent by the van der Waals forces. Chemisorption is the jargon for covalent attachment of hydrogen. The physisorption peak is seen at low temperatures for all storage materials. It is the second, broad peak around room temperature that is astonishing with this desorption curve. Since conventional CH bonds start to break first at temperatures above ~300°C, the origin of this peak is not understood. The potential for on-board storage of hydrogen, on the other hand is clear, and this discovery is the subject of a U.S. Patent

(Maeland and Skjeltorp 2001). The broad desorption peak is confirmed by recent experiments on the KCB sample (Yu et al. 2008). Since there is a mixture of cones with all the five possible apex angles, in addition to flat disks and soot, the storage capacity of the cones, in terms of hydrogen weight percent, cannot be derived from these experiments. Neither is it clear which of the cones—or, for that matter, the disks—is responsible for the broad desorption peak.

At 20 K, right above the boiling point of hydrogen, the adsorption isotherms show that hydrogen is packed with higher density within a single-walled carbon nanohorn than in the liquid form (Fernandez-Alonso et al. 2007). In atomistic simulations, the H_2 gas interacts stronger with nanohorns than nanotubes (Tanaka et al. 2004), and the hydrogen concentration is higher in the cone-shaped tip of the horn than in the tubular part. If the performance of the nanohorns originates from the cone-shaped tip, as these results suggest, graphene cones may be even better suited for hydrogen storage.

25.3.6 Miscellaneous Results

This section contains original results that do not fit into one of the previous categories of more extended research activity.

25.3.6.1 Graphene Cones in Electrorheological Fluid

An electrorheological (ER) fluid changes from liquid to solid-like behavior under the influence of an electromagnetic field. The resulting variation of viscosity with field strength (Winslow 1949) is called the *electrorheological effect*. Recently, experiments were performed with an ER fluid consisting of the KCB sample of graphene cones and disks dispersed in silicon oil (Svåsand et al. 2007). Figure 25.17 shows the formation of chain structures

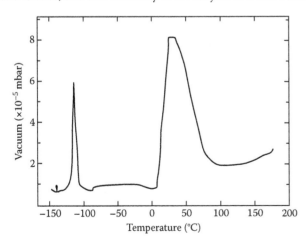

FIGURE 25.16 Hydrogen desorption curve for the raw KCB sample. (Courtesy of A.T. Skjeltorp, Institute for Energy Technology, Kjeller, Norway.)

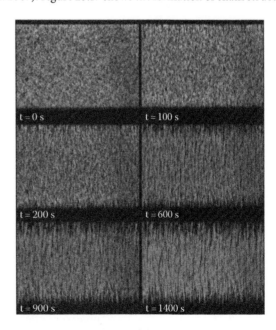

FIGURE 25.17 Time evolution of chain structures in the KCB sample suspended in silicon oil. (From Svåsand, E. et al., *Colloids Surf., A*, 308, 67, 2007. With permission.)

in a fluid with a weight per cent of 0.05 cones/disk in a 50 Hz ac field of 139 V/mm. Despite increased structural order, the current through the field was too small ($\ll 1\,\mu A$) to be measured. When a dc field was applied, the carbon particles did not form chains, but moved immediately to the positive electrode, possibly an indicator of excessive electrons.

25.3.6.2 Substitutional Atoms and Tip Functionalization

The effect on HOMO–LUMO gaps and bonding energies of substitution of boron and nitrogen atoms at the cone tips have been investigated by DFT (Azevedo 2004). The analysis was restricted to cones with one or two pentagons, and it was found that boron substitution reduced the energetic cost of pentagons while nitrogen enhanced it. Both substituents reduced the HOMO–LUMO gap, but nitrogen had the largest effect.

Functionalization of nanotubes by methyl radicals has been done and leads to improved solubility in organic solvents. The

idea behind the recent computational work (Trzaskowski et al. 2007) on functionalization of graphene cones is that a more site specific functionalization can be achieved for cones than for tubes. DFT and MM simulations showed that methyl attachment to carbons at the tips was energetically favored, and the reaction was, as for nanotubes, exothermic. Thus, stable cones with functionalized tips appear to be achievable, and one of the suggested applications was fullerene trapping as shown in Figure 25.18.

25.3.6.3 Conic Radicals

An unique feature of graphene cones with odd numbers of pentagons is that they contain symmetric radicals; an odd number of carbons gives for these cones an odd number of terminating hydrogens along the edges, and thus unpaired electrons. The spectral systematics of these radicals were studied by the Hückel method, and it was found that the radicals with one or five pentagons at the tip have a vacancy in the highest bonding orbital, and those with three pentagons have the most energetic electron in the lowest anti-bonding orbital (Heiberg-Andersen and Skjeltorp 2007).

25.4 Discussion

The SEM images of the KCB cones, the five peaks in the apex angle distribution and the X-ray diffraction data are strong evidence that they consist of stacked sheets of curved graphene, and not helically wound sheets, as in naturally occurring graphite cones. Also the low conductivity of the KCB sample dispersed in silicon oil (Section 25.3.6.1) points in this direction; a helically wound sheet should have better conductivity than individual stacked sheets. However, this difference may be completely overruled by the conductive properties of the more abundant disks.

The nucleation of the graphene cones have been approached theoretically by both the "pentagon road" and the ring-stacking models. The weakness of the "pentagon road" is the assumed pre-formation of a flat sheet with many dangling bonds along the edges. For nanotubes, the problem with this assumption was clarified by an analysis showing that the stress resulting from bending was negligible compared to the energy cost of a dangling bond (Fan et al. 2003). The formation of an unsaturated graphene sheet with the proper geometry for rolling into a cone thus seems quite unlikely.

Due to the ongoing controversy of extended versus localized states near the Fermi level at the cone tips, it is not clear which of the three proposed continuum models for cones, if any, is the correct one. It should be mentioned, however, that the model of Lammert and Crespi has not produced results in direct conflict with compatible TB results. This is also the only proposed continuum model that incorporates the relative configuration of the non-hexagonal faces at the cone tips, which importance is proved by both TB and DFT calculations of the local density of states.

Through its ability to reproduce STM images, DFT is recognized as a reliable tool for obtaining nuclear coordinates and

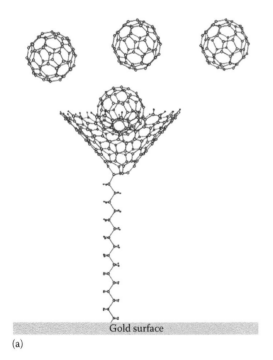

Gold surface

(a)

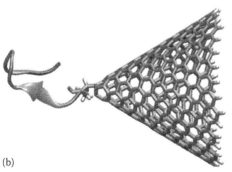

(b)

FIGURE 25.18 Schematic representation of a fullerene trap (a) and a peptide-cone complex (b). (From Trzaskowski, B. et al., *Chem. Phys. Lett.*, 444, 314, 2007. With permission.)

electronic properties of carbon nanostructures. Since experimental results are still scarce in this field, it is therefore reasonable to regard the DFT results as "simulated experiments" for the time being. Unfortunately, the computational expense (high memory requirements is the main bottleneck) prevents DFT computation of the electronic properties of real cones in the micrometer range. TB calculations of the LDOS of cones of various sizes indicate the importance of end effects, so it is difficult to relate the DFT results on small model systems to the properties of the large synthesized cones.

The broad hydrogen desorption peak near room temperature still remains a mystery. It simply does not seem compatible with physisorption nor chemisorption. Some type of physisorption was suggested in (Yu et al. 2008), mainly on the ground that no known types of CH bonds are supposed to break in this temperature region, but it is far from obvious that the weak diffusion forces are able to compose a hydrogen trap so far above the normal physisorption peak. As long as the storage mechanism is not understood, it cannot be argued theoretically that the hydrogen is trapped in the cones and not in the disks. Thus, further experiments on samples with higher cone fractions are needed to clarify the properties of the cones.

25.5 Summary

There is little doubt that graphene cones actually have been synthesized in quantities sufficient to justify extended research efforts and tentative technological applications. The nucleation mechanism is still not understood, but it appears that a quantitative nucleation theory must be based on the ring-stacking model rather than the "pentagon road". The body of theoretical works presented brings evidence for fascinating electronic and mechanical consequences of the unique topology and geometry of graphene cones. One of the predicted effects of a non-hexagonal face in the carbon network, the origin of petal superstructures in the STM image, has been confirmed by experiment. Although the theoretical models differ in their predictions of peak positions in the apex density of states, the general agreement of enhanced density of states at the apex makes the cones interesting for FEM applications. Another promising application is hydrogen storage for fuel cell driven vehicles. Experiments done on the raw KCB sample shows a unique capability of storing hydrogen reversibly at ambient temperatures.

25.6 Future Perspective

The main obstacle to increased understanding and technological applications of graphene cones is the low quality of existing samples, which contain either a mixture of graphene cones of all angles in addition to disks and soot or a mixture of cones with different origin and tip morphology. Currently this problem is attacked on the two fronts of synthesis methods and separation of existing samples (HYCONES 2006). A breakthrough here would give an array of highly needed experimental results. The hydrogen storage capacity of the cones, for example, could then be unambiguously measured, and other applications, like field emission, could be pursued. Finally, the theoretical controversies, especially between the various continuum models, could be firmly resolved. Within the current modeling paradigms, it is hard to imagine substantial and unambiguous theoretical progress on more realistically sized model cones without more experimental results on their atomic and electronic structures. To increase the chances for such data to appear, it could perhaps be a good idea to direct more of the theoretical efforts towards the challenging problems of synthesis and separation of graphene cones.

Acknowledgments

This work is funded by the European Commission (Project No. NMP3-CT-2006-032970/HYCONES). We thank J. P. Pinheiro at n-Tec for providing unpublished SEM images of heat-treated cones.

References

An, B., Fukuyama, S., Yokogawa, K. et al. 2001. Single pentagon in a hexagonal carbon lattice revealed by scanning tunneling microscopy, *Applied Physics Letters* **78**: 3696.

Azevedo, S. 2004. Effects of substitutional atoms in carbon nanocones, *Physics Letters A* **325**: 283–286.

Balaban, A. T. and Klein, D. J. 1994. Graphitic cones, *Carbon* **32**: 357–359.

Barth, E. and Lawton, G. 1966. Dibenzo[ghi,mno]fuoranthene, *Journal of the American Chemical Society* **88**: 380–381.

Brenner, D. W. 1990. Empirical potential for hydrocarbon for use in simulating the chemical vapor deposition of diamond films, *Physical Review B* **42**: 9458–9471.

Charlier, J.-C. and Rignanese, G.-M. 2001. Electronic structure of carbon nanocones, *Physical Review Letters* **86**: 5970–5973.

Compernolle, S., Kiran, B., Chibotaru, L. F., Nguyen, M. T., and Ceulemans, A. 2004. *Ab initio* study of small graphitic cones with triangle, square and pentagon apex, *Journal of Chemical Physics* **121**: 2326–2336.

Dimovski, S., Libera, J. A., and Gogotsi, Y. 2002. A novel class of carbon nanocones, *Materials Research Society Symposium Proceedings* **706**: Z6.27.1–Z6.27.6.

DiVincenzo, D. P. and Mele, E. J. 1984. Self-consistent effective-mass theory for intralayer screening in graphite intercalation compounds, *Physical Review B* **29**: 1685–1694.

Double, D. D. and Hellawell, A. 1974. Cone-helix growth forms of graphite, *Acta Metallurgica* **22**: 481–487.

Ekşioğlu, B. and Nadarajah, A. 2006. Structural analysis of conical carbon nanofibers, *Carbon* **44**: 360–373.

Fan, X., Buczko, R., Puretzky, A. A. et al. 2003. Nucleation of single-walled carbon nanotubes, *Physical Review Letters* **90**: 145501.

Fernandez-Alonso, F., Bermejo, F. J., Cabrillo, C., Loutfy, R. O., Leon, V., and Saboungi, M. L. 2007. Nature of the bound states of molecular hydrogen in carbon nanohorns, *Physical Review Letters* **98**: 215503.

Ganser, B. K., Li, S., Klishko, Y., Finch, J. T., and Sundquist, I. S. 1999. Assembly and analysis of conical models of the hiv-1 core, *Science* **283**: 80–83.

Garberg, T., Naess, S. N., Helgesen, G., Knudsen, K. D., Kopstad, G., and Elgsaeter, A. 2008. A transmission electron microscope and electron diffraction study of carbon nanodisks, *Carbon* **46**: 1535–1543.

Ge, M. and Sattler, K. 1994. Observation of fullerene cones, *Chemical Physics Letters* **220**: 192–196.

Han, J. and Jaffe, R. 1998. Energetics and geometrics of carbon nanoconic tips, *Journal of Chemical Physics* **108**: 2817.

Haydock, R., Heine, V., and Kelly, M. J. 1972. Electronic structure based on the local atomic environment for tight-binding bands, *Journal of Physics C* **5**: 2845–2858.

Heiberg-Andersen, H. 2006. Carbon nanocones, in M. Rieth and W. Schommers (eds.), *Handbook of Theoretical and Computational Nanotechnology*, American Scientific Publishers, Stevenson Ranch, CA, pp. 507–536.

Heiberg-Andersen, H. and Skjeltorp, A. T. 2005. Stability of conjugated nanocones, *Journal of Mathematical Chemistry* **38**: 589–604.

Heiberg-Andersen, H. and Skjeltorp, A. T. 2007. Spectra of conic carbon radicals, *Journal of Mathematical Chemistry* **42**: 707–727.

Heiberg-Andersen, H., Skjeltorp, A. T., and Sattler, K. 2008. Carbon nanocones: A variety of non-crystalline graphite, *Journal of Non-Crystalline Solids* **354**: 5247–5249.

Helgesen, G., Knudsen, K. D., Pinheiro, J. P. et al. 2008. Carbon cones—A structure with unique properties, *Materials Research Society Symposium Proceedings* **1057**: II10–I46.

Hückel, E. 1931. Quantentheoretische beiträge zum benzolproblem, *Zeitschrift für Physik* **70**: 204–286.

HYCONES 2006. Project information. http://www.hycones.eu.

Iijima, S., Yudasaka, M., Bandow, S. et al. 1999. Nano-aggregates of single-walled graphitic carbon nano-horns, *Chemical Physics Letters* **309**: 165–170.

Jaszczak, J. A., Robinson, G. W., Dimovski, S., and Gogotsi, Y. 2003. Naturally occurring graphite cones, *Carbon* **41**: 2085–2092.

Jordan, S. P. and Crespi, V. H. 2004. Theory of carbon nanocones: Mechanical chiral inversion of a micron-scale three-dimensional object, *Physical Review Letters* **93**: 255504.

Kidnay, A. J. and Hiza, M. J. 1967. High pressure adsorption isotherms of neon, hydrogen and helium at 76 k, *Advances in Cryogenic Engineering* **12**: 730–740.

Klein, D. J. 2002. Topo-combinatoric categorization of quasi-local graphitic defects, *Physical Chemistry Chemical Physics* **4**: 2099–2110.

Klein, D. J. and Balaban, A. T. 2006. The eight classes of positive-curvature graphitic nanocones, *Journal of Chemical Information and Modeling* **46**: 307–320.

Kobayashi, K. 2000. Superstructure induced by a topological defect in graphitic cones, *Physical Review B* **61**: 8496–8500.

Krishnan, A., Dujardin, E., Treacy, M. M. J., Hugdahl, J., Lynum, S., and Ebbesen, T. W. 1997. Graphitic cones and the nucleation of curved carbon surfaces, *Nature* **388**: 451–454.

Kvaerner 1998. Method for production of microdomain particles by use of a plasma process. Kvaerner's Patent PCT/NO98/0093.

Lammert, P. E. and Crespi, V. H. 2000. Topological phases in graphitic cones, *Physical Review Letters* **85**: 5190–5193.

Lammert, P. E. and Crespi, V. H. 2004. Graphen cones: Classification by fictitious flux and electronic properties, *Physical Review B* **69**: 035406.

Liew, K. M., Wei, J. X., and He, X. Q. 2007. Carbon nanocones under compression: Buckling and post-buckling behaviors, *Physical Review B* **75**: 195435.

Luttinger, J. M. and Kohn, W. 1955. Motion of electrons and holes in perturbed periodic fields, *Physical Review* **97**: 869–883.

Maeland, A. J. and Skjeltorp, A. T. 2001. Hydrogen storage in carbon material. U.S. Patent 6,290,753 B1.

Mani, R. C., Li, X., Sunkara, M. K., and Rajan, K. 2003. Carbon nanopipettes, *Nano Letters* **3**: 671–673.

Mizes, H. A. and Foster, J. S. 1989. Long-range electronic perturbations caused by defects using scanning tunneling microscopy, *Science* **244**: 559–562.

Muñoz-Navia, M., Dorantes-Dávila, J., Terrones, M., and Terrones, H. 2005a. Ground state electronic structure of nanocones, *Physical Review B* **72**: 235403.

Mūnoz-Navia, M., Dorantes-Dávila, J., Terrones, M. et al. 2005b. Synthesis and electronic properties of coalesced graphitic nanocones, *Chemical Physics Letters* **407**: 327–332.

Novoselov, K. S., Geim, A. K., Morozov, S. V. et al. 2004. Electric field effect in atomically thin carbon films, *Science* **306**: 666–669.

Osipov, V. A. and Kochetov, E. A. 2001. Dirac fermions on graphite cones, *JETP Letters* **73**: 562–565.

Peierls, R. 1933. Zur theorie des diamagnetismus von leitungselektronen, *Zeitschrift für Physik* **80**: 763–791.

Sattler, K. 1995. Scanning tunneling microscopy of carbon nanotubes and nanocones, *Carbon* **33**: 915–920.

Shaver, J., Kono, J., Portugall, O. et al. 2006. Magneto-optical spectroscopy of ecitons in carbon nanotubes, *Physica Status Solidi (B)* **243**: 3192–3196.

Shenderova, O. A., Lawson, B. L., Areshkin, D., and Brenner, D. W. 2001. Predicted structure and electronic properties of individual carbon nanocones and nanostructures assembled from nanocones, *Nanotechnology* **12**: 191–197.

Sitenko, Y. A. and Vlasii, N. D. 2007. Electronic properties of graphene with a topological defect, *Nuclear Physics B* **787**: 241–259.

Svåsand, E., Helgesen, G., and Skjeltorp, A. T. 2007. Chain formation in a complex fluid containing carbon cones and disks in silicon oil, *Colloids and Surfaces A* **308**: 67–70.

Tamura, R., Akagi, K., Tsukada, M., Itoh, S., and Ihara, S. 1997. Electronic properties of polygonal defects in graphitic carbon sheets, *Physical Review B* **56**: 1404–1411.

Tanaka, H., Kanoh, H., El-Merraoui, M. et al. 2004. Quantum effects on hydrogen adsorption in internal nanospaces of single-wall carbon nanohorns, *Journal of Physical Chemistry B* **108**: 17457–17465.

Terauchi, M., Tanaka, M., Suzuki, K., Ogino, A., and Kimura, K. 2000. Production of zigzag-type BN nanotubes and BN cones by thermal annealing, *Chemical Physics Letters* **324**: 359–364.

Terrones, H., Hayashi, T., Munoz Navia, M. M. et al. 2008. A transmission electron microscope and electron diffraction study of carbon nanodisks, *Carbon* **46**: 1535–1543.

Treacy, M. M. J. and Kilian, J. 2001. Designability of graphitic cones, *Materials Research Society Symposium Proceedings* **675**: W2.6.1–W2.6.6.

Trzaskowski, B., Jalbout, A. F., and Adamowicz, L. 2007. Functionalization of carbon nanocones by free radicals: A theoretical study, *Chemical Physics Letters* **444**: 314–318.

Tsai, P.-C. and Fang, T.-H. 2007. A molecular dynamics study of the nucleation, thermal stability and nanomechanisms of carbon nanocones, *Nanotechnology* **18**: 105702–105709.

Wei, J. X., Liew, K. M., and He, X. Q. 2007. Mechanical properties of carbon nanocones, *Applied Physics Letters* **91**: 261906.

Winslow, W. M. 1949. Induced fibration of suspensions, *Journal of Applied Physics* **20**: 1137.

Yu, X., Tverdal, M., Raaen, S., Helgesen, G., and Knudsen, K. D. 2008. Hydrogen adsorption on carbon nanocone material studied by thermal desorption and photoemission, *Applied Surface Science* **255**: 1906–1910.

Zhang, G., Jiang, X., and Wang, E. 2003. Tubular graphitic cones, *Science* **300**: 472–474.

IV

Indentation
and Patterning

26

Theory of Nanoindentation

Zhi-Qiang Feng
Université d'Évry-Val d'Essonne

Qi-Chang He
Université Paris-Est

Qingfeng Zeng
Northwestern Polytechnical University

Pierre Joli
Université d'Évry-Val d'Essonne

26.1 Introduction

The development of nanomaterials used in nanotechnology, for example, assembly of structures, manufacture of nanotubes, etc., requires the knowledge of mechanical properties such as Young's modulus, the yield stress, or the buckling loadings. The evaluation by nanoindentation of these properties represents a real challenge for the researchers because it is difficult to carry out experiments at such length scales. Moreover, to validate an experimental result, it is necessary to be ensured of its reproducibility, which requires many samples and thus involves a high cost.

Recent developments in science and technology have advanced the capability to fabricate and control materials and devices with nanometer grain/feature sizes to achieve better mechanical properties in materials and more reliable performance in microelectromechanical systems (MEMS) and nanoelectromechanical systems (NEMS) (Bhushan, 1999; Miller and Tadmor, 2002; Vashishta et al., 2003; Tambe and Bhushan, 2004; Liu et al., 2005; Jian et al., 2006). Nanoindentation instrument provides a valid approach to investigate the mechanical characterizations of nanomaterials, such as the hardness, dislocation motion, and Young's modulus, which are required to design structural/functional elements in micro- and nanoscale devices. Many powerful capabilities in in situ and *ex situ* imaging, acoustic emission detection, and high-temperature testing are now being used to probe nanoscale phenomena such as defect nucleation and dynamics, mechanical instabilities or strain localization, and phase transformations (Wang et al., 2003; Aouadia, 2006; Schuh, 2006; Gouldstone et al., 2007; Lin et al., 2007; Snyders et al., 2007; Szlufarska et al., 2007). With the same pace of experimental work, analytical theory and computer simulation, especially the latter, are developed to reproduce or even predict the intrinsic

phenomena during nanoindentation (Vashishta et al., 2003; Szlufarska, 2006). Among these simulation methods, first principles calculations (FP) (Aouadia, 2006; Snyders et al., 2007), molecular dynamics (MD) (Noreyan et al., 2005; Liu et al., 2007), finite element method (FEM) (Liu et al., 2005; Feng et al., 2007; Zhong and Zhu, 2008), and their hybrid methods such as first principle molecular dynamics (FPMD) (Schneider et al., 2007) and quasicontinuum method (QC) (Miller and Tadmor, 2002; Dupuy et al., 2005), from the trade-off between efficiency and accuracy point of view, are mostly applied to investigate the evolution of structure and properties in different spatial and temporal scales during nanoindentation.

We try to present in this chapter a review of the theoretical study and computational modeling of nanoindentation. The chapter is organized as follows. Section 26.2 is devoted to the basics of nanoindentation. Some classic solutions of indentation problems are summarized. In Section 26.3, solutions of contact and plasticity problems by the finite element method are presented with numerical examples. In Section 26.4, a survey on atomistic and multiscale approaches for nanoindentation modeling is provided.

The material of this chapter comes from the teaching notes and research works of the authors. We also used other research articles in the literature to supplement the content of various topics discussed in this chapter.

26.2 Principle and Methods of Nanoindentation

Nanoindentation is an important and popular mechanical experimental technique used in nanomaterials science and nanotechnology. Compared with other mechanical experimental characterization methods, nanoindentation entails a quite simple setup and specimen preparation. In addition, nanoindentation leaves only a small imprint and can be thus considered as nondestructive. However, a correct and accurate exploitation of experimental data from nanoindentation tests necessitates a full understanding of the nanoindentation principle and of the assumptions made in carrying out nanoindentation tests. Since there exist a good few reviews on the experimental procedure of nanoindentation and on the interpretation of nanoindentation test data (Bhushan, 2004; Cheng and Cheng, 2004; Fischer-Cripps, 2004, 2006; Sharpe, 2008), we focus on a clear presentation of the nanoindentation principle and a thorough discussion of the assumptions underlying nanoindentation tests.

For us to quickly grasp the fundamental principle of indentation tests, we first recall the result obtained by Love (1939) who studied the problem of a linearly isotropic elastic half-space impressed by a rigid axisymmetric conical indenter (Figure 26.1).

Denoting the force applied to the indenter by P, and the resulting displacement or penetration depth of the indenter by h, Love (1939) derived for the first time the contact pressure $p(r)$ and the relation between P and h as follows:

$$p(r) = \frac{P}{\pi a^2} \cosh^{-1}\left(\frac{a}{r}\right) \quad \text{and} \quad P = \frac{2E^* \tan\alpha}{\pi} h^2 \quad (26.1)$$

where

　　α symbolizes the semiangle of the conical indenter
　　the combined (or reduced) elastic modulus E^* is defined in terms of the Young modulus E and Poisson ratio ν of the material constituting the half-space by

$$E^* = \frac{E}{1 - \nu^2} \quad (26.2)$$

Sneddon (1948) employed the integral transformation method to derive the same formula as (26.1) and gave a solution to the

more general problem of a linearly isotropic elastic half-space indented by a rigid axisymmetric punch of arbitrary profile (Sneddon, 1965). It follows from Equation 26.1 that the tangent stiffness, viz., the slope of the P–h curve, is given by

$$\frac{dP}{dh} = \frac{4E^* \tan\alpha}{\pi} h \quad (26.3)$$

Next, note that the penetration depth h is related to the contact depth h_c, i.e., the height of the part of the indenter which is actually in contact with the elastic half-space (Figure 26.1) by (see, for example, Sneddon, 1948)

$$h = \frac{\pi}{2} h_c \quad (26.4)$$

In addition, the projected contact area A is provided by

$$A = \pi a^2 = \pi h_c^2 \tan^2\alpha = \frac{4}{\pi} h^2 \tan^2\alpha \quad (26.5)$$

In the above equation, a is the radius of the contact circle separating the contacting and noncontacting parts of the indenter (Figure 26.1), and use is made of formula (26.4) in writing the last equality of (26.5). Introducing (26.5) into (26.1) and (26.3) yields

$$\frac{P}{A} = \frac{E^* \cot\alpha}{2} \quad (26.6)$$

$$\frac{dP}{dh} = 2E^* \sqrt{\frac{A}{\pi}} \quad (26.7)$$

The formula (26.7) expressing the tangent stiffness in terms of the projected contact area is derived by considering a conical indenter. It is in fact valid for all axisymmetric indenters such cylindrical circular punches and spherical indenters (Bulychev et al., 1975; Pharr et al., 1992). To specify this important point, we recall below the results concerning the indentation of a half-space by a rigid flat cylindrical punch (Figure 26.2) and by a rigid spherical indenter (Figure 26.3).

In the situation of a circular flat punch of radius a (see Maugis, 1999), the pressure and the relation between the applied load P and the penetration depth h read

$$p(r) = \frac{P}{2\pi a^2}\left(1 - \frac{r^2}{a^2}\right)^{-0.5} \quad \text{and} \quad P = 2aE^* h \quad (26.8)$$

and the projected contact area is the same as the contact area, which is constant and equal to $A = \pi a^2$. It is easy to verify that (26.7) holds in this case. When a rigid spherical indenter of radius R is pressed into a linearly elastic half-space, we have the formulas of Hertz (see for example Johnson, 1985):

$$p(r) = \frac{3P}{2\pi a^2}\left(1 - \frac{r^2}{a^2}\right)^{0.5} \quad \text{and} \quad P = \frac{4}{3}E^* \sqrt{Rh^3} \quad (26.9)$$

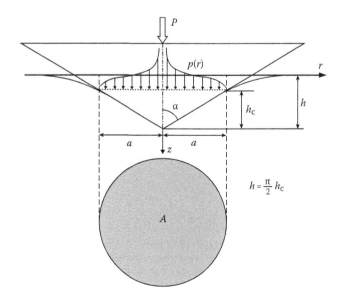

FIGURE 26.1　Indentation of an elastic half-space by a rigid cone.

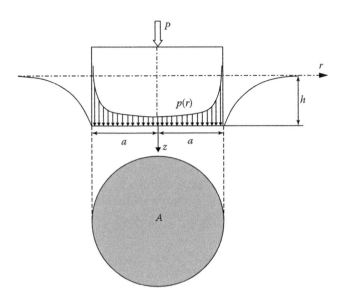

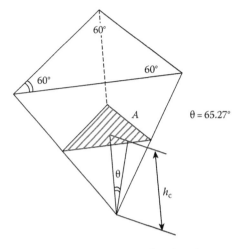

FIGURE 26.4 Geometry of a three-sided Berkovich indenter.

FIGURE 26.2 Indentation of a half-space by a rigid circular flat punch.

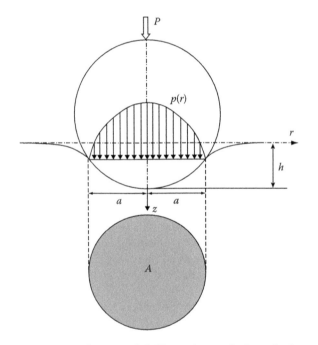

FIGURE 26.3 Indentation of a half-space by a rigid spherical indenter.

$$a^2 = Rh \qquad (26.10)$$

where a is the radius of the circular contact surface. With (26.9) and (26.10), it is straightforward to check that (26.7) is satisfied.

Theoretically speaking, the reduced modulus E^* of a linearly elastic isotropic solid would be easily identified by using (26.1) or (26.3) if the $P–h$ curve could be experimentally determined for a rigid conical impressed into this solid. However, the experimental reality is much more complex, so that (26.1) cannot be directly applied. We now proceed to discuss the main aspects of real nanoindentation tests.

The indenter being most widely used for nanoindentation is not a conical one but the three-sided Berkovich indenter (Figure 26.4). The face semiangle θ of a typical Berkovich indenter is equal to 65.27°. The reason why the three-sided pyramidal Berkovich indenter is employed in nanoindentation rather than the more familiar four-sided Vickers one is that it is much easier to grind the three faces of the former than the four faces of the latter so as to meet at a point when both are made of diamond. Clearly, the geometry of the three-sided Berkovich indenter is quite different from that of a cone. Nevertheless, the forgoing results for a conical indenter are in practice assumed to be applicable to a three-sided pyramidal indenter by requiring that these two indenters have the same ratio of the projected contact area A to the contact depth h_c. More precisely, for the Berkovich indenter, the projected contact area is calculated in terms of the contact depth h_c by

$$A = 3\sqrt{3}h_c^2 \tan^2\theta \qquad (26.11)$$

Setting A obtained by (26.11) to be equal to A given by (26.5) in terms of h_c for $\theta = 65.27°$, it follows that the semi-angle of the conical indenter "equivalent" to the Berkovich indenter is

$$\alpha = 70.30° \qquad (26.12)$$

Introducing this value into (26.5) yields

$$A = 24.5h_c^2 = 9.93h^2 \qquad (26.13)$$

In nanoindentation, the indenter tip is usually made of diamond, which is strongly resistant to deformations but cannot be treated ultimately as rigid. In general, it is known from contact mechanics (Johnson, 1985) that when the indenter is deformable, the reduced (or combined) elastic modulus E^* involved in formula (26.1) should be defined by

$$\frac{1}{E^*} = \frac{1-\nu^2}{E} + \frac{1-\nu_i^2}{E_i} \qquad (26.14)$$

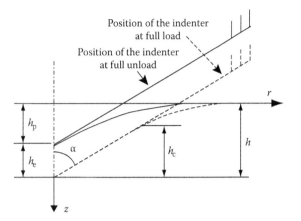

FIGURE 26.5 Schematic of the nanoindentation of an elasto-plastic solid by a conical cone at full load and unload.

where E_i and v_i are the Young's modulus and Poisson ratio of the indenter. When the tip of an indenter consists of diamond, the values $E_i = 1141\,\mathrm{GPa}$ and $v_i = 0.07$ are commonly adopted.

In a nanoindentation test where a sharp indenter such as the Berkovich indenter is used, plastic deformations occur even at the very early stage of loading. The plastic zone in the indented specimen is enlarged with increasing loading. The experimental load–displacement curve, or P–h curve, relative to loading is in general quite different from the parabolic load–displacement curve predicted by formula (26.1). Figure 26.5 is a schematic of the indentation of an elastoplastic specimen by a conical indenter, and Figure 26.6 is a schematic of the corresponding load–displacement curve for loading and unloading. In these two figures, h_{max} is the maximum penetration depth associated with the applied maximum load P_{max}; h_p is the plastic or residual penetration depth at full unloading while $h_e = h_{max} - h_p$ is the elastic penetration depth recovered at full unloading.

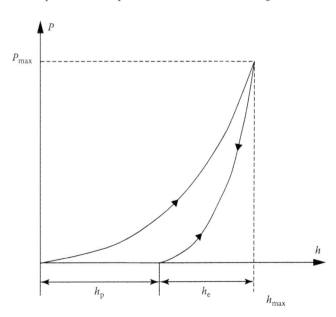

FIGURE 26.6 Schematic of the load–displacement curve corresponding to the nanoindentation depicted by Figure 26.5

The identification of the reduced modulus E^* by nanoindentation is based on sensing the load–displacement relationship relative to unloading rather than the one associated to loading. Indeed, it is known from the theory of elastoplasticity that the plastic effects are well confined during unloading. So, it is reasonable to assume that the load–displacement curve for unloading can be correctly described by formula (26.1) together with formulas (26.4) and (26.5) upon replacing h by h_e. In other words, the unloading curve can be treated as it were an elastic loading curve. However, the assumptions underlying formulas (26.1), (26.4) and (26.5) do not take into account a few good factors identified as relevant to nanoindentation tests. Among these factors, the following two appear to be particularly important:

- The tip of an indenter, such as the Berkovich indenter, is not an idealized point but has a finite radius.
- At full unload, there is in general a residual imprint, which forms an angle with the indenter due to radial plastic deformations (Figure 26.5).

Clearly, even though the reloading path can be reasonably considered as being elastic or following the previous unloading path, it is different from the one predicted by formula (26.1). In analysis and interpretation of nanoindentation experiments, formula (26.7) is modified as follows:

$$\left.\frac{dP}{dh}\right|_{h_{max}} = 2\beta E^* \sqrt{\frac{A}{\pi}} \tag{26.15}$$

where β is a dimensionless correction or fitting parameter accounting for the effects of the factors that are relevant to nanoindentation tests but not considered in the theoretical derivation of (26.7). The numerical value of β, which should be adopted is a debatable issue. This is not surprising, because the factors whose effects are supposed to be captured by β are numerous. For the three-sided Berkovich indenter, Oliver and Pharr (2004) proposed that $1.0226 \leq \beta \leq 1.085$, and the value $\beta = 1.034$ or $\beta = 1.05$ is often used.

With the help of (26.15), the reduced elastic modulus E^* can be deduced from the initial slope $dP/dh\,|_{h_{max}}$ of an unloading curve, provided the corresponding projected contact area A is available. In principle, A can be calculated by (26.5) in which we pose $h = h_{max}$. The question is how to accurately determine $dP/dh\,|_{h_{max}}$ and A. To this end, Oliver and Pharr (1992) proposed an identification procedure, which is now widely used in nanoindentation. A key relation needed for the Oliver–Pharr procedure is the following:

$$h_c = h_{max} - \beta\xi \frac{P_{max}}{dP/dh\,|_{h_{max}}} \tag{26.16}$$

where

β is the aforementioned correction factor
$\xi = 2(\pi - 2)/\pi \approx 0.72$ for a conical indenter

When setting $\beta = 1$, formula (26.16) can be rigorously deduced by combining (26.1), (26.3) and (26.4). In the Oliver–Pharr procedure, the load–displacement curve is first obtained by using a power-law function

$$P = \eta(h - h_{\mathrm{p}})^m \tag{26.17}$$

to fit experimental data. In (26.17), m is the power-law index, η is a constant parameter, and h_{p} is the residual penetration depth measured experimentally. With (26.17), it is easy to compute

$$\left.\frac{\mathrm{d}P}{\mathrm{d}h}\right|_{h_{\max}} = m\eta(h_{\max} - h_{\mathrm{p}})^{m-1} \tag{26.18}$$

Substituting (26.18) and (26.16) into (26.5) gives the projected contact area A. Finally, the reduced modulus E^* is calculated by (26.15).

Though becoming standard, the Oliver–Pharr procedure appears open to criticism with respect to the following two points. First, when the value of m is quite different from 2, the power law (26.17) used to fit experimental data may be incompatible with formula (26.1), which is the starting point of the method on which the Oliver–Pharr procedure is based; even when $m = 2$, the identification of η by a fitting method implies the determination of E^*. Second, the projected contact area A corresponding to the maximum load $P_{\max}$ can be determined directly by formula (26.5) and via the measurement of $h_{\max}$; the calculation of A by means of (26.16) is much more complicated, because the determination of the derivative $\mathrm{d}P/\mathrm{d}h\,|_{h_{\max}}$ is much more vulnerable to error than the measurement of $h_{\max}$.

26.3 Finite Element Analysis of Contact and Plasticity

The analysis of nanoindentation can be very complex because of two principal strongly nonlinear phenomena:

- Elastic–plastic deformation of the indented materials
- Frictional contact between the indenter and materials

Different analytical models were proposed in the case of thin metal layers (with elastoplastic behavior) (Oliver and Pharr, 1992) or polymers (with hyperelastic behavior and pile-up) (Bucaille et al., 2003). In the case where analytical derivation of the mechanical properties is not feasible, numerical modeling may therefore help to clarify and put in evidence the different contributions of thin films and substrates. Knapp et al. (1999) used the finite element approach to determine hardness properties of thin films in two dimensions. They suggested a fit procedure to evaluate mechanical properties of thin layers and substrates. This kind of procedure reveals some success but with several limitations: two dimensional geometry, hundred of nanometers length scales, geometrically perfect indenter, and the neglecting of interface interactions (film/substrate). For smaller length scale, it seems to be necessary to develop more complex interaction film-substrate models, as in the work of Bull et al. (Bull et al., 2004; Berasetgui et al., 2004) and Raabe et al. (Wang et al., 2004; Zaafarani et al., 2006).

The aim of this section is to describe the main numerical approaches in solving a three-dimensional elastoplastic contact problem.

26.3.1 Modeling of Elastoplastic Materials at Finite Strains

Usually, behavior laws of solids are differential equations that connect the rate of stress to the rate of strain. Nagtegaal (1982) and Hughes and Winget (1980) have proposed several integration schemes. The concept of consistent linearization introduced by Nagtegaal (1982) and extended by Simo and Taylor (1985) made it possible to write, in the case of small or large deformations, effective consistent tangent stiffness matrices.

At low temperature, or when the loading speeds are relatively low, the metallic materials present inelastic time-independent deformations. Within the framework of large deformation, the total rate of deformation tensor $\mathbf{d}$ is split into two components: an elastic part $\mathbf{d}^{\mathrm{e}}$, connected linearly to the Lie derivative of the Kirchhoff stress tensor τ, and a plastic part $\mathbf{d}^{\mathrm{p}}$:

$$\mathcal{L}_{\mathbf{v}}\tau = \mathbb{D} : \mathbf{d}^{\mathrm{e}} = \mathbb{D} : (\mathbf{d} - \mathbf{d}^{\mathrm{p}}) \tag{26.19}$$

where $\mathbb{D}$ denotes the fourth-order elasticity tensor.

It is supposed that there exists a convex field (elastic domain) in the space of stresses inside of which there is no plastic flow. The yield function $f = 0$ represents the loading surface delimiting the elastic domain. In the case of isotropic plasticity considered here, we use the von Mises criterion as follows:

$$f(\tau, \boldsymbol{\alpha}, R) = \|\boldsymbol{\eta}\| - \sqrt{\frac{2}{3}} R(p) \quad \text{with} \quad \boldsymbol{\eta} = \mathrm{dev}(\tau) - \boldsymbol{\alpha} \tag{26.20}$$

where $R(p)$ and $\boldsymbol{\alpha}$ represent, respectively, the radius and the center position of the elastic domain. $\mathrm{dev}(\bullet)$ stands for the deviator tensor of $(\bullet)$ and $\|\boldsymbol{\eta}\| = \sqrt{\boldsymbol{\eta} : \boldsymbol{\eta}}$. In the case of associated plasticity, the normality rule is written by

$$\mathbf{d}^{\mathrm{p}} = \dot{\lambda}\mathbf{n} \quad \text{with} \quad \mathbf{n} = \frac{\boldsymbol{\eta}}{\|\boldsymbol{\eta}\|} \tag{26.21}$$

where $\dot{\lambda}$ denotes the plastic multiplier. The plastic loading and unloading condition can be expressed in terms of the Kuhn–Tucker condition

$$\dot{\lambda} \geq 0, \quad f(\tau, \boldsymbol{\alpha}, R) \leq 0, \quad \dot{\lambda}f(\tau, \boldsymbol{\alpha}, R) = 0 \tag{26.22}$$

The isotropic law of strain hardening is defined by the evolution of the radius R with respect to the cumulated plastic strain p by

$$R(p) = R_{\mathrm{s}} - (R_{\mathrm{s}} - R_0)\mathrm{e}^{-\gamma p} \tag{26.23}$$

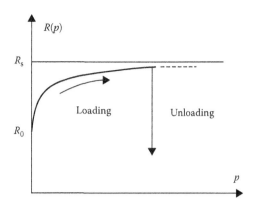

FIGURE 26.7 Loading and unloading.

where R_s, R_0, and γ are material constants. R_s represents the saturated radius of the elastic domain, and R_0 the initial one as shown in Figure 26.7.

The rate of the cumulated plastic strain is given by

$$\dot{p} = \sqrt{\frac{2}{3}} \left\| \mathbf{d}^{\mathrm{P}} \right\| = \sqrt{\frac{2}{3}} \dot{\lambda} \qquad (26.24)$$

The linear Prager kinematic hardening is defined by the evolution of back stress $\boldsymbol{\alpha}$ of the elastic domain with respect to the plastic strain rate:

$$\mathcal{L}_v \boldsymbol{\alpha} = \frac{2}{3} H \mathbf{d}^{\mathrm{P}} \qquad (26.25)$$

where H is the slope of the kinematic work hardening. The isotropic strain hardening and the kinematic hardening can be respectively interpreted as the expansion and the translation of the elastic domain, as shown in Figure 26.8.

The integration of behavior laws plays a very important role in a finite element code. Indeed, it determines the precision of the solution. Errors on the estimates of the variables, once made, are not retrievable any more. Moreover when the estimates depend on the history of the loading, these errors can be propagated from an increment to another. The results deviate more and

more from the solution. In this study, the implicit integration algorithm has been chosen. Let us consider a plastically admissible state (corresponding to the load step n) and the known characteristics of this state are: $\boldsymbol{\tau}_n$, $\boldsymbol{\alpha}_n$, p_n verifying $f(\boldsymbol{\tau}_n, \boldsymbol{\alpha}_n, R(p_n)) = 0$. Integrating the behavior lows consists in finding the characteristics of the state $n + 1$. The most used method to integrate the laws of plastic behavior is undoubtedly the "radial return mapping" (RRM), initially introduced by Wilkins (1964), Krieg and Krieg (1977) for models of perfectly plastic behavior. The extension of this method to the case of the models with nonlinear kinematic work hardening was carried out by Simo and Taylor (1986). The principle of this method is to calculate the final stress $\boldsymbol{\tau}_{n+1}$ as the projection of a test stress $\boldsymbol{\tau}^{\mathrm{E}}$ onto the yield surface according to the normal passing by $\boldsymbol{\tau}^{\mathrm{E}}$ (Figure 26.8). The test stress is calculated by supposing that the increment of strain is entirely elastic. A standard stress update is outlined at following (for details, see Simo and Hughes, 1998, Chapter 8).

Say the position vectors in the "deformed" and "undeformed" state are represented by $\mathbf{x}$ and $\mathbf{X}$, respectively and the displacement vector $\mathbf{u} = \mathbf{x} - \mathbf{X}$, the total deformation gradient is defined as

$$\mathbf{F} = \frac{\partial \mathbf{x}}{\partial \mathbf{X}} = \mathbf{I} + \nabla \mathbf{u} \qquad (26.26)$$

One of the major challenges while integrating the rate constitutive equations at finite strains is to achieve incremental objectivity. To this end, it is common to define an intermediate configuration between load steps, n and $n + 1$:

$$\mathbf{x}_{n+\theta} = (1-\theta)\mathbf{x}_n + \theta\mathbf{x}_{n+1} \quad \text{and} \quad \mathbf{F}_{n+\theta} = (1-\theta)\mathbf{F}_n + \theta\mathbf{F}_{n+1}$$

$$\text{with} \quad \theta \in [0,1] \qquad (26.27)$$

The relative deformation gradients $\mathbf{f}_{n+\theta}$ and $\tilde{\mathbf{f}}_{n+\theta}$, the relative incremental displacement gradient $\mathbf{h}_{n+\theta}$ and the incremental Eulerian strain tensor $\tilde{\mathbf{e}}_{n+\theta}$ are defined by

$$\mathbf{f}_{n+\theta} = \mathbf{F}_{n+\theta}\mathbf{F}_n^{-1}, \quad \tilde{\mathbf{f}}_{n+\theta} = \mathbf{f}_{n+1}\mathbf{f}_{n+\theta}^{-1}, \quad \mathbf{h}_{n+\theta} = \frac{\partial \mathbf{u}(\mathbf{x}_{n+\theta})}{\partial \mathbf{x}_{n+\theta}},$$

$$\tilde{\mathbf{e}}_{n+\theta} = \frac{1}{2}\tilde{\mathbf{f}}_{n+\theta}^T \left[\mathbf{I} - (\mathbf{f}_{n+1}\mathbf{f}_{n+1}^T)^{-1} \right] \tilde{\mathbf{f}}_{n+\theta} \qquad (26.28)$$

Within the interval $[t_n, t_{n+1}]$, the deformation gradient can be expressed as

$$\mathbf{d}_{n+\theta} = \frac{1}{2\Delta t} \left[\mathbf{h}_{n+\theta} + \mathbf{h}_{n+\theta}^T + (1-2\theta)\mathbf{h}_{n+\theta}^T \mathbf{h}_{n+\theta} \right] \qquad (26.29)$$

The objective approximation for the Lie derivative of the Kirchhoff stress is given by

$$\mathcal{L}_v \boldsymbol{\tau}_{n+\theta} = \frac{1}{\Delta t} \mathbf{f}_{n+\theta} \left[\mathbf{f}_{n+1}^{-1} \boldsymbol{\tau}_{n+1} \mathbf{f}_{n+1}^{-T} - \boldsymbol{\tau}_n \right] \mathbf{f}_{n+\theta}^T \qquad (26.30)$$

By evaluating (26.19) at the intermediate configuration and using the flow rule (26.21), we obtain

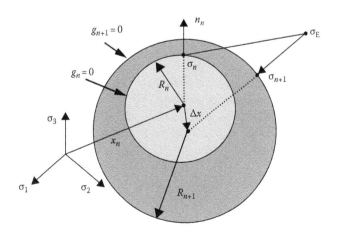

FIGURE 26.8 Radial return mapping.

TABLE 26.1 Radial Return Mapping Algorithm

1. Elastic predictor

$$\boldsymbol{\tau}_{n+\theta}^{E} = \mathbf{f}_{n+\theta}\boldsymbol{\tau}_{n}\mathbf{f}_{n+\theta}^{T} + \mathbb{D}:\tilde{\mathbf{e}}_{n+\theta}; \quad \boldsymbol{\alpha}_{n+\theta}^{E} = \mathbf{f}_{n+\theta}\boldsymbol{\alpha}_{n}\mathbf{f}_{n+\theta}^{T}$$

$$p_{n+\theta}^{E} = p_{n}; \quad \boldsymbol{\eta}_{n+\theta}^{E} = \mathrm{dev}\left(\boldsymbol{\tau}_{n+\theta}^{E}\right) - \boldsymbol{\alpha}_{n+\theta}^{E}; \quad \mathbf{n}_{n+\theta} = \frac{\boldsymbol{\eta}_{n+\theta}^{E}}{\left\|\boldsymbol{\eta}_{n+\theta}^{E}\right\|}$$

2. Plastic corrector

$$f_{n+\theta}^{E} = \left\|\boldsymbol{\eta}_{n+\theta}^{E}\right\| - \sqrt{\frac{2}{3}}R\left(p_{n+\theta}^{E}\right)$$

if $f_{n+\theta}^{E} < 0$ then

$$\mathrm{set}(\bullet) = (\bullet)_{n+\theta}^{E}$$

else

$$\Delta\lambda = \frac{f_{n+\theta}^{E}/2\mu}{1 + \dfrac{H+R'}{3\mu}} \quad \text{with} \quad R' = \frac{dR}{dp}$$

$$\boldsymbol{\tau}_{n+1} = \boldsymbol{\tau}_{n+\theta}^{E} - 2\mu\Delta\lambda\mathbf{n}_{n+\theta}$$

$$\boldsymbol{\alpha}_{n+1} = \boldsymbol{\alpha}_{n+\theta}^{E} + \frac{2}{3}\Delta\lambda H\mathbf{n}_{n+\theta}$$

$$p_{n+1} = p_{n+\theta}^{E} + \sqrt{\frac{2}{3}}\Delta\lambda$$

$$\Delta t \mathcal{L}_{v}\boldsymbol{\tau}_{n+\theta} = \mathbb{D}:[\Delta t\mathbf{d}_{n+\theta} - \Delta\lambda\mathbf{n}_{n+\theta}] \quad \text{with} \quad \Delta\lambda = \Delta t\dot{\lambda}_{n+\theta} \quad (26.31)$$

Similarly, we have

$$\boldsymbol{\alpha}_{n+\theta} = \mathbf{f}_{n+\theta}\boldsymbol{\alpha}_{n}\mathbf{f}_{n+\theta}^{T} + \frac{2}{3}\Delta\lambda H\mathbf{n}_{n+\theta} \quad (26.32)$$

Table 26.1 summarizes the predictor–corrector step in the RRM integration algorithm developed above.

26.3.2 Modeling of Contact Problems with Friction

The analysis of contact problems with friction is of great importance in many engineering applications (Johnson, 1985). The numerical treatment of the unilateral contact with dry friction is certainly one of the nonsmooth mechanics topics for which many efforts have been made in the past. In the literature, many approaches have been developed to deal with such problems using the finite element method. A large literature base is available for a variety of numerical algorithms (Zhong, 1993; Wriggers, 2002). The bipotential method proposed by DeSaxcé and Feng (1998) has been successfully applied to solve contact problems between elastic or hyperelastic bodies (Feng et al., 2005, 2006). In the present work, this method will be applied to solve the nanoindentation problem involving the contact between the rigid indenter and the elastoplastic film.

26.3.2.1 Governing Equations

The finite element method is often used in computational mechanics. Without going into details, quasistatic nonlinear behaviors of solid media, discretized by N_{e} finite elements, are governed by the following equilibrium equations:

$$\mathbf{F}_{\mathrm{in}} - \mathbf{F}_{\mathrm{ex}} - \mathbf{R} = \mathbf{0} \quad (26.33)$$

where
$\mathbf{F}_{\mathrm{ex}}$ denotes the vector of external loads
$\mathbf{R}$ is the vector of contact reaction forces

The vector of internal forces $\mathbf{F}_{\mathrm{in}}$ is calculated by

$$\mathbf{F}_{\mathrm{in}} = \mathop{\wedge}_{e=1}^{N_{e}} \mathbf{F}_{\mathrm{in}}^{e} \quad \text{with} \quad \mathbf{F}_{\mathrm{in}}^{e} = \int_{V^{e}} \mathbf{B}^{T}\boldsymbol{\sigma}dV \quad (26.34)$$

The symbol $\wedge$ denotes a standard finite element assembly operator. The Cauchy stress tensor $\boldsymbol{\sigma}$ is related to the Kirchhoff stress tensor $\boldsymbol{\tau}$ by $\boldsymbol{\sigma} = \boldsymbol{\tau}/\det(\mathbf{F})$ and the latter is obtained from the integration of the constitutive laws as shown in Section 26.3.1. It is noted that Equation 26.33 is strongly nonlinear with respect to the nodal displacements $\mathbf{U}$, because of finite strains and large displacements of solid. Moreover, the constitutive laws of contact with friction are usually represented by inequalities and the contact potential is even nondifferentiable. A typical solution procedure for this type of nonlinear analysis is obtained by using the Newton–Raphson iterative procedure (Joli and Feng, 2008):

$$\mathbf{K}_{T}^{i}\Delta\mathbf{U}^{i} = \mathbf{F}_{\mathrm{ex}} + \mathbf{R}^{i} - \mathbf{F}_{\mathrm{in}}^{i}$$
$$\mathbf{U}^{i+1} = \mathbf{U}^{i} + \Delta\mathbf{U}^{i} \quad (26.35)$$

where i and $i + 1$ are the iteration numbers at which the equations are computed. $\mathbf{K}_{T} = \partial\mathbf{F}_{\mathrm{in}}/\partial\mathbf{U}$ stands for the tangent stiffness matrix and $\Delta\mathbf{U}$ the vector of nodal displacements correction.

It is noted that Equation 26.35 cannot be solved directly because $\Delta\mathbf{U}$ and $\mathbf{R}$ are both unknown. The key idea is to determine first the reaction vector $\mathbf{R}$ in a reduced system which only concerns the contact nodes. Then, the displacement increments can be computed in the whole structure using contact reactions as external loading. In the following, we focus our attention on describing how to determine the contact forces. Let us begin with the general description of contact kinematics.

26.3.2.2 Contact Kinematics

First of all, basic definitions and notations used are described. For the sake of simplicity, we consider two deformable bodies Ω^{a} (Figure 26.9), $a = 1, 2$, coming into contact. Each body is decomposed by finite elements and the nodal positions in the global coordinate frame are represented by the vector $\mathbf{x}_{a}$. The boundary Γ^{a} of each body is assumed to be sufficiently smooth everywhere such that an outward unit normal vector can be defined at any point P_{a} on Γ^{a}.

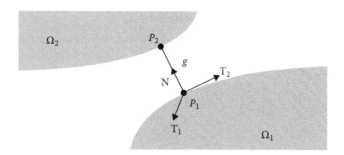

FIGURE 26.9   Contact kinematics.

We consider only the case with N_c contact nodes P_1^α ($\alpha = 1, N_c$) defined on Γ^1 and P_2^α are target points defined by the normal projection of P_1^α onto Γ^2. We can build the relative position between P_1^α and P_2^α by

$$\mathbf{x}^\alpha = \mathbf{x}\left(P_1^\alpha\right) - \mathbf{x}\left(P_2^\alpha\right) \qquad (26.36)$$

We consider a local orthogonal reference frame by means of three vectors $\mathbf{t}_1^\alpha$, $\mathbf{t}_2^\alpha$, and $\mathbf{n}^\alpha$ which are defined with respect to the global reference frame. We set the following notation:

$$\mathbf{x}^\alpha = \mathbf{x}_t^\alpha + x_n^\alpha \mathbf{n}^\alpha = x_{t_1}^\alpha \mathbf{t}_1^\alpha + x_{t_2}^\alpha \mathbf{t}_2^\alpha + x_n^\alpha \mathbf{n}^\alpha \qquad (26.37)$$

We can easily define the transformation matrix $\mathbf{H}_\alpha$ between the local and the global reference frames such as

$$\mathbf{x}^\alpha = \mathbf{H}_\alpha \mathbf{X} \quad \text{with} \quad \mathbf{X}^T = \left\{\mathbf{x}_1^T \mathbf{x}_2^T\right\} \quad \text{and} \quad \mathbf{x}^{\alpha T} = \left\{x_{t_1}^\alpha x_{t_2}^\alpha x_{t_n}^\alpha\right\}$$
$$(26.38)$$

The incremental form of Equation 26.38 gives the gap vector between P_1^α and P_2^α

$$\mathbf{x}_{i+1}^\alpha = \mathbf{H}_\alpha \Delta \mathbf{X}_i + \mathbf{g}_i^\alpha \quad \text{with} \quad \mathbf{g}_i^\alpha = \mathbf{x}_i^\alpha \qquad (26.39)$$

where $\mathbf{g}_0^\alpha = (0\ 0\ g)$ represents the initial gap vector, which is determined by a contact collision detector. If we opted to carry out this operation at the beginning of each load step, then we neglect the variation of the normal at the contact point during one load step. If this assumption is not satisfied, it is possible to reduce the load step or to perform the collision detection at each iteration. In the case of discontinuous curvature of contact surfaces, special smoothing techniques can be used as in Heegaard and Curnier (1993). In the local reference frame the contact force $\mathbf{r}^\alpha$ can be defined by

$$\mathbf{r}^\alpha = \mathbf{r}_t^\alpha + r_n^\alpha \mathbf{n}^\alpha = r_{t_1}^\alpha \mathbf{t}_1^\alpha + r_{t_2}^\alpha \mathbf{t}_2^\alpha + r_n^\alpha \mathbf{n}^\alpha \qquad (26.40)$$

Using the same transformation matrix $\mathbf{H}^\alpha$ previously defined, we can have the contact force in the global reference frame $\mathbf{R}^\alpha = \mathbf{H}_\alpha^T \mathbf{r}^\alpha$. The global vector of contact reaction force is defined by $\mathbf{R} = \overset{N_c}{\underset{\alpha=1}{\wedge}} \mathbf{R}^\alpha$.

Remark 1: We deal with the quasistatic contact problem with friction but we need a numerical approximation of the tangential relative velocity at each local point α to deal with the dry friction Coulomb laws. For a given time (or loading) history $\theta \in [0, T]$, where $[0, T]$ is a time interval which can be partitioned into N subintervals of size $\Delta\theta$, we adopt a backward Euler time discretization of the time derivative $\dot{\mathbf{x}}_t$ as follows:

$$\dot{\mathbf{x}}_{t(i+1)} \approx \frac{\mathbf{x}_{t(i+1)} - \mathbf{x}_{t(0)}}{\Delta\theta} \qquad (26.41)$$

In quasistatic cases, N is the total number of load steps and we can set $\Delta\theta = 1$. As we have discussed above, at each load step, a contact detection is performed. In this way, we have $\mathbf{x}_{t_0} = \mathbf{0}$. Then, Equation 26.41 reduces to $\dot{\mathbf{x}}_t \approx \mathbf{x}_t$ by omitting the underscript iteration number.

Remark 2: In the total Lagrangian formulation we have $\Delta\mathbf{X}_i = \Delta\mathbf{U}_i$.

26.3.2.3 Signorini Conditions and Coulomb Friction Laws

To simplify the notations, the superscript α is omitted also the underscript i in the description of the contact laws. The unilateral contact law is characterized by a geometric condition of nonpenetration, a static condition of no-adhesion and a mechanical complementary condition. These three conditions are known as Signorini conditions expressed, for each contact point, in terms of the signed contact distance x_n and the normal contact force r_n by

$$\text{Signor}(x_n, r_n) \Leftrightarrow x_n \geq 0, \quad r_n \geq 0 \quad \text{and} \quad x_n r_n = 0 \qquad (26.42)$$

The classic Coulomb friction rule is defined by

$$\text{Coul}(\mathbf{x}_t, \mathbf{r}_t) \Leftrightarrow \text{if } \|\mathbf{x}_t\| = 0 \quad \text{then} \quad \|\mathbf{r}_t\| \leq \mu r_n \quad \text{else} \quad \mathbf{r}_t = -\mu r_n \frac{\mathbf{x}_t}{\|\mathbf{x}_t\|}$$
$$(26.43)$$

where μ is the coefficient of friction. The so-called Coulomb K_μ (Figure 26.10) is the convex set of admissible forces which is defined by

$$K_\mu = \{\mathbf{r} \in \mathbb{R}^3 \quad \text{such that} \quad \|\mathbf{r}_t\| - \mu r_n \leq 0\} \qquad (26.44)$$

The complete contact law (Signorini conditions + Coulomb friction laws) is thus a complex non smooth dissipative law including three statuses:

- No contact: $x_n > 0$ and $\mathbf{r} = 0$

- Contact with sticking: $\|\mathbf{x}_t\| = 0$ and $\mathbf{r} \in \text{int}(K_\mu)$

- Contact with sliding: $\|\mathbf{x}_t\| \neq 0$ and $\mathbf{r} \in \text{bd}(K_\mu)$ $\qquad (26.45)$

    with $\quad \mathbf{r}_t = -\mu r_n \dfrac{\mathbf{x}_t}{\|\mathbf{x}_t\|}$

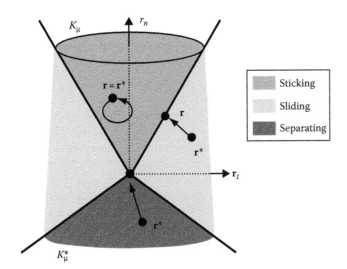

FIGURE 26.10 Coulomb cone and contact projection operators.

where "int(K_μ)" and "bd(K_μ)" denote the interior and the boundary of K_μ, respectively. The multivalued character of the law lies in the first and the second parts of the statement. If r_n is null then **x** is arbitrary but its normal component x_n should be positive. In other words, one single element of $\mathbb{R}^3$ (**r** = 0) is associated with an infinite number of gap vectors $\mathbf{x} \in \mathbb{R}^3$. The same arguments can be developed for the second part of the statement. For this reason, the contact forces cannot be derived from a potential function of gap vectors.

26.3.2.4 Bipotential Method

De Saxcé and Feng (1998) have demonstrated that the complete contact dry friction law is strictly equivalent to the following projection operator:

$$\mathbf{r} = \mathrm{Proj}_{K_\mu}(\mathbf{r}^*) \qquad (26.46)$$

where $\mathbf{r}^*$ is the so-called augmented contact forces vector and is given by

$$\mathbf{r}^* = \mathbf{r} - \rho\mathbf{x}^* \quad \text{with} \quad \mathbf{x}^* = \mathbf{x} + \mu\|\mathbf{x}_t\|\mathbf{n} \qquad (26.47)$$

where ρ is an arbitrary positive parameter. This operator has been established from convex analysis and augmented Lagrangian techniques. It can be proved that **r** and $\mathbf{x}_t$ are differential inclusions of a unique pseudopotential function called bipotential.

The three possible contact statuses as mentioned in Equation 26.45 are illustrated in Figure 26.10. Within the bipotential framework, these statues can be stated as: $\mathbf{r}^* \in K_\mu$ (contact with sticking), $\mathbf{r}^* \in K_\mu^*$ (separating), and $\mathbf{r}^* \in \mathbb{R}^3 - (K_\mu \cup K_\mu^*)$ (contact with sliding). K_μ^* is the polar cone of K_μ. Consequently, the projection operation can be explicitly defined by

$$\mathrm{Proj}_{K_\mu}(\mathbf{r}^*) = \mathbf{r}^* \quad \text{if } \|\mathbf{r}_t^*\| < \mu r_n^*,$$

$$\mathrm{Proj}_{K_\mu}(\mathbf{r}^*) = 0 \quad \text{if } \mu\|\mathbf{r}_t^*\| < -r_n^*, \qquad (26.48)$$

$$\mathrm{Proj}_{K_\mu}(\mathbf{r}^*) = \mathbf{r}^* - \left(\frac{\|\mathbf{r}_t^*\| - \mu r_n^*}{1+\mu^2}\right)\left(\frac{\mathbf{r}_t^*}{\|\mathbf{r}_t^*\|} - \mu\mathbf{n}\right) \text{ otherwise.}$$

26.3.2.5 Equilibrium Equations of Contact Points

We still omit the underscript i, the system of equations to be solved at each Newton–Raphson iteration can be summarized as follows:

$$(\alpha = 1, N_c)\begin{cases} \mathbf{K}_T\Delta\mathbf{U} = \mathbf{F}_{ex} + \overset{N_c}{\underset{\alpha=1}{\wedge}}\mathbf{R}^\alpha - \mathbf{F}_{in} \\ \mathbf{x}^\alpha = \mathbf{H}_\alpha\Delta\mathbf{U} + \mathbf{g}^\alpha \\ \mathbf{r}^\alpha = \mathrm{Proj}_{K_\mu}(\mathbf{r}^{*\alpha}) \end{cases} \qquad (26.49)$$

Eliminating $\Delta\mathbf{U}$ leads to the following reduced system of $6 \times N_c$ equations:

$$(\alpha = 1, N_c)\begin{cases} \mathbf{x}^\alpha = \sum_{\beta=1}^{N_c}\mathbf{W}_{\alpha\beta}\mathbf{r}^\beta + \tilde{\mathbf{x}}^\alpha \\ \mathbf{r}^\alpha = \mathrm{Proj}_{K_\mu}(\mathbf{r}^{*\alpha}) \end{cases} \qquad (26.50)$$

with

$$(\alpha = 1, N_c)\begin{cases} \tilde{\mathbf{x}}^\alpha = \mathbf{H}_\alpha\mathbf{K}_T^{-1}(\mathbf{F}_{ex} - \mathbf{F}_{in}) + \mathbf{g}^\alpha \\ \mathbf{W}_{\alpha\beta} = \mathbf{H}_\alpha\mathbf{K}_T^{-1}\mathbf{H}_\beta^T \end{cases} \qquad (26.51)$$

26.3.2.6 Global Solution: Nonlinear Gauss–Seidel-Like Algorithm

Jourdan et al. (1998) have applied a nonlinear Gauss–Seidel-like algorithm to simulate deep drawing problems. Signorini conditions and Coulomb friction laws were derived from two distinct pseudopotentials. As we will see below, this algorithm can be readily extended to be applied to the case of the bipotential formulation.

The principle of this algorithm is to decompose the global solution of the algebraic system (26.50) into N_c successive local solutions of the six following equations:

$$\begin{cases} \mathbf{x}^\alpha = \mathbf{W}_{\alpha\alpha}\mathbf{r}^\alpha + \mathbf{x}^{\alpha\beta} \\ \mathbf{r}^\alpha = \mathrm{Proj}_{K_\mu}(\mathbf{r}^{*\alpha}) \end{cases} \qquad (26.52)$$

with

$$\mathbf{x}^{\alpha\beta} = \sum_{\beta=1,\beta\neq\alpha}^{N_c} \mathbf{W}_{\alpha\beta}\mathbf{r}^{\beta} + \tilde{\mathbf{x}}^{\alpha} \tag{26.53}$$

where $\mathbf{x}^{\alpha\beta}$ represents the part of the relative position at the contact point α due to the initial gap, the external forces, and contact forces of $N_c - 1$ other contact nodes β. This contribution is "frozen" during each local solution. One series of N_c local solutions corresponds to one iteration k of the algorithm. The iterative process is successively applied for each contact point ($\alpha = 1, N_c$) until the convergence of the solution. The contact convergence criterion is stated as

$$\frac{\left\| \mathbf{r}^{(k+1)} - \mathbf{r}^{(k)} \right\|}{\left\| \mathbf{r}^{(k+1)} \right\|} \leq \varepsilon_g \tag{26.54}$$

where
  $\mathbf{r} = \{\mathbf{r}^1 \mathbf{r}^2 \cdots \mathbf{r}^{N_c}\}$ is the vector of contact reactions of all contact nodes
  ε_g is a user-defined tolerance. The initial condition is given by $\mathbf{r}^{(0)} = \mathbf{0}$

In the bipotential formulation, the usual approach to solve the local implicit equations (26.52) is to use a predictor/corrector Uzawa algorithm. Many examples have been successfully treated by Feng et al. (Feng, 1995; Feng et al., 2003).

26.3.2.7 Local Solution: Uzawa Algorithm

The numerical solution of the implicit equation (26.52) can be carried out by means of the Uzawa algorithm, which leads thus to an iterative process involving one predictor–corrector step:

$$\text{Predictor } \mathbf{r}^{\alpha*(k+1)} = \mathbf{r}^{\alpha(k)} - \rho^{(k)}\left(\mathbf{x}^{\alpha\beta(k)} + \mu\left\|\mathbf{x}_t^{\alpha\beta(k)}\right\|\mathbf{n}\right)$$
$$\text{Corrector } \mathbf{r}^{\alpha(k+1)} = \text{Proj}_{\mathbf{K}_\mu}(\mathbf{r}^{\alpha*(k+1)}) \tag{26.55}$$

where k and $k + 1$ are the iteration numbers at which the contact reactions are computed. The corrector step is explicitly given by Equation 26.48. In view of Equation 26.52, the gap vector is updated by

$$\mathbf{x}^{\alpha(k+1)} = \mathbf{W}_{\alpha\alpha}\mathbf{r}^{\alpha(k+1)} + \mathbf{x}^{\alpha\beta(k)} \tag{26.56}$$

It is noted that the solution is controlled by a global convergence criterion (iteration k) as stated in Equation 26.54. The advantage of this approach is the simplicity of programming and the numerical robustness but it needs more iterations when compared with the implicit Newton algorithm presented recently in (Joli and Feng, 2008). However, the latter is more time-consuming at each iteration.

26.3.3 Numerical Examples

26.3.3.1 Boussinesq–Love Indentation Problem

The Boussinesq–Love indentation problem concerns the frictionless contact between a rigid conical indenter and an elastic half-space as shown in Figure 26.1. In order to validate the numerical approach, we have developed a finite element model of the Boussinesq–Love indentation problem is created by using axisymmetric isoparametric elements. The characteristics of the problem are: $E = 10\,\text{MPa}$, $v = 0.495$, and $\alpha = 60°$. Figure 26.11 shows the initial mesh and the deformed shape of the half-space together with the distribution of the von Mises stress. We observe a stress concentration at the tip zone of the indenter as predicted by Equation 26.1. The maximum value of the von Mises stress is $13.868\,\text{MPa}$. Figure 26.12 plots the evolution of the total contact force versus the displacement of the indenter. The numerical solution is compared with the analytical one given by Equation 26.1. As we can see, when the displacement is relatively small, good concordance is observed between numerical and analytical solutions. However, the difference increases with respect to the displacement. This can be explained by the fact that the analytical solution is determined under the assumption of small deformations. In the numerical algorithm developed above, large deformations are taken into account by means of the calculation of the internal force vector and the tangent stiffness matrix.

26.3.3.2 Nanoindentation

The elastoplastic contact model described above has been implemented in an in-house finite element program (Feng et al., 2007). In this study, the nanoindentation problem is solved with the program. A tetrahedral indenter of Berkovich type comes in contact onto a thin layer surface (thickness of $300\,\text{nm}$) on a substrate. The thin layer is supposed to have an elastoplastic behavior. The indenter is supposed to be rigid with a blunted

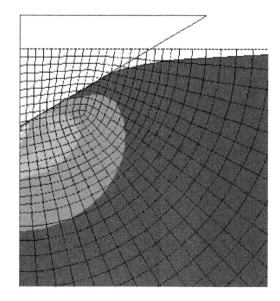

FIGURE 26.11 Deformed shape with stress.

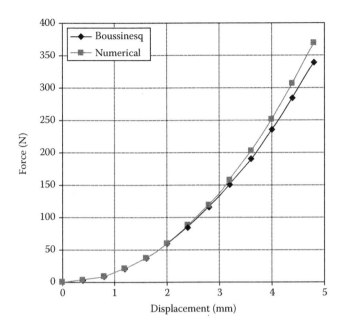

FIGURE 26.12   Load–displacement curve.

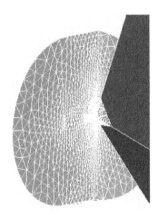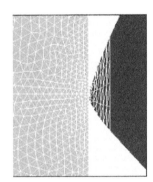

FIGURE 26.13   3D and 2D view of the mesh.

point of radius of curvature of approximately 50 nm. Figure 26.13 shows the model with a grid in 3D as well as a more detailed view in 2D. It should be noted that the symmetry of the model is taken into account in order to reduce the computing time. As discussed in Section 26.2, we needed to have a triangular base pyramid with the tip rounded. The tip of the indenter has been created with a 3D animation program (LightWave 7.5). It has several special features that have allowed us to realize the rounded tip with a radius of 50 nm with a tolerance of 2% as shown in Figure 26.13. Then we have applied a mesh on this 3D object. The mesh of the film and the substrate are generated with an appropriate level of meshing to have the minimum number of contact elements between the tip and our meshed material, as we can see in Figure 26.13.

Figure 26.14 shows the print left on the surface and the distribution of stresses around the zone of indentation. The triangular form of the print accurately reproduces what we observe in experiments on this scale.

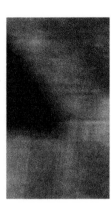

FIGURE 26.14   Numerical and experimental results.

Recently, a micro/nanoscale computational contact model is proposed in (Sauer and Li, 2007) to study the adhesive contact between deformable bodies.

26.4 Atomistic and Multiscale Approaches

The basic concept of modeling of natural phenomena is based on computing the energy of a physical structure, since both the static and dynamic properties of the structure are related with the first or higher order derivative of energy with respect to structure positions and time. People always try to predict properties of materials with the aid of as few experimental data as possible. Generally, the fewer the experimental data are input, the higher the computational resources cost. Some atomistic calculations ask for powerful supercomputers or clusters with thousands of CPUs but still cover a structure size around 100 nm (Vashishta et al., 2003). Consequently, multiscale approaches are required for practical nanoindentation modeling.

26.4.1 First Principles Calculations for Nanoindentation Modeling

First principles calculations use the laws of quantum mechanics based solely on a small number of physical constants, the speed of light, Planck's constant, the masses and charges of electrons and nuclei, to obtain the energy by solving the Schrödinger equation (Leach, 2001):

$$\left\{ \frac{-h^2}{8\pi^2 m} \nabla^2 + \mathbf{V} \right\} \Psi(\mathbf{r},t) = \frac{ih}{2\pi} \frac{\partial \Psi(\mathbf{r},t)}{\partial t} \qquad (26.57)$$

here

Ψ is the wave function
$\mathbf{r}$ is the position of the particle
t is the time
h is the Planck's constant
m is the mass of the particle
∇^2 is the Laplacian operator
$\mathbf{V}$ is the potential field in which the particle is moving
i is the imaginary unit

Exact solutions to the equation are not trivial, so a variety of mathematical transformation and approximation techniques are proposed. Born–Oppenheimer approximation separates nuclear and electronic motions. Nonrelativistic approximation treats the mass of the particles independent of their speeds. The orbital approximation makes the total wave function of the many-particle system constructable from one-electron wave functions. The nonrelativistic, time-independent form of the Schrödinger description is as follows:

$$H\Psi(\mathbf{r}) = E\Psi(\mathbf{r}) \qquad (26.58)$$

where

E is the energy of the system

H is the Hamiltonian operator, equal to

$$H = \sum_i \frac{-h^2}{8\pi^2 m_i} \nabla^2 + \mathbf{V} \qquad (26.59)$$

The resistance of a material to deformations is related to its elastic constants. The linear elastic constants form a 6 × 6 symmetric matrix, three tensile and three shear components, having 21 different components, such that $\sigma_i = D_{ij}\varepsilon_j$ for stresses, σ, and strains, ε. Consequently, elastic constants can be evaluated by calculating the stress tensor for a number of slightly distorted structures. In this case, the internal coordinates are optimized by minimizing the total energy in each run, while keeping the lattice parameters fixed. Strains can be applied in the x, y, or z directions or shear strains can be applied. Other properties such as the bulk modulus (response to an isotropic compression), Poisson coefficient, Lame constants, and so forth can be computed from the values of D_{ij}.

26.4.2 Molecular Dynamics Calculations for Nanoindentation Modeling

Molecular dynamics (MD) simulation studies were initiated in the late 1950s at the Lawrence Radiation Laboratory (LRL) by Alder and Wainwright (Alder and Wain wright, 1959, 1960) in the field of equilibrium and nonequilibrium statistical mechanics. The key idea of the molecular dynamic model is quite simple and it consists in computing trajectories of a finite set of atoms from numerical integration of equations of motion. We present here the basic principle of molecular dynamic by considering a triangular network in which atoms are connected by interaction forces derived from potential functions (harmonic or Lennard-Jones). The scenario of the simulation consists in applying stress σ or velocity $\mathbf{v}$ conditions onto the boundary of the network (Figure 26.15).

For each atom of position $\mathbf{x}_i$, we establish the dynamic equations following the second Newton law. Only the contributions of the external forces due to the potential interaction of the j connected atom and the local damping force are taken into account as follows:

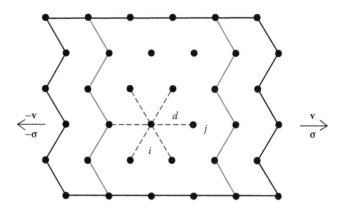

FIGURE 26.15   Network of atoms.

$$m_i\ddot{\mathbf{x}}_i = \sum_j \mathbf{F}_{ij} - \eta\dot{\mathbf{x}}_i \qquad (26.60)$$

where m_i and η are respectively the mass and the damping parameter of the atom i. The interaction force between two atoms is defined by

$$\mathbf{F}_{ij} = g(d) \frac{\mathbf{x}_i - \mathbf{x}_j}{\|\mathbf{x}_i - \mathbf{x}_j\|} \qquad (26.61)$$

This force can be differentiated from potential functions giving several mathematical formulations of $g(d)$, such as

$$\text{Harmonic potential:} \quad g(d) = -k(d-a) \qquad (26.62)$$

$$\text{Lennard Jones potential:} \quad g(d) = \frac{ka}{6}\left[\left(\frac{a}{d}\right)^{13} - \left(\frac{a}{d}\right)^7\right]$$

$$(26.63)$$

where a and d represent respectively the resting length and the distance during the simulation between two atoms in interaction. The parameter k denotes the stiffness between two atoms. Zhang et al. has used the Lennard-Jones potential function for numerical simulation of mechanical behaviors of carbon nanotubes (Zhang et al., 2007).

Classically, the ordinary differential equations (26.60) are integrated by using an explicit numerical scheme such as the Verlet numerical scheme (Verlet, 1967). The time step of the numerical integration is constant and to prevent any problem of numerical instability, it must be very small as compared to the lowest characteristic period of the physical system.

In the simulation of nanoindentation, the load on the indenter P can be calculated by summing the forces acting on the atoms of the indenter in the indentation direction. Indentation depth h can be calculated as the displacement of the tip of the indenter relative to the initial surface of the indented solid (Szlufarska, 2006). The coordinates and velocities at a later time can be determined, once the initial coordinates and velocities of the atoms are known. A load–displacement (P–h) response and the deformation structures in a material can be analyzed by monitoring the trajectories of the atoms.

Since most of the empirical potentials for MD calculations need the comparison with experimental data, this limits the application for new materials or for materials where experimental data do not exist. Another drawback of MD is that there is no universal force field can be applied to an atom in different chemical environment. For example, potentials of carbon in diamond and C_{60} have different values and/or forms (Brenner, 1990; Smith and Beardmore, 1996; Christopher et al., 2001).

26.4.3 Multiscale Approaches for Nanoindentation Modeling

Because discrete atomic effects become important only in the vicinity of defects, interfaces and surfaces, multiscale approaches aim to model an atomistic system without explicitly treating every atom in the problem (Miller and Tadmor, 2002; Zeng et al., 2009). There are mainly two types of multiscale methods (Liu et al., 2004): concurrent and hierarchical. Concurrent methods, such as quasicontinuum (QC) method (Miller and Tadmor, 2002), simultaneously solve a fine-scale model in some local region of interest and a coarser scale model in the remainder of the domain. Hierarchical, or serial coupling methods, such as heterogenous multiscale method (HMM) (Weinan et al., 2003), use results of a fine-scale model simulation to acquire data for a coarser-scale model that is used globally, e.g., to determine parameters for constitutive equations. Currently, both quantum mechanics and empirical interatomic potentials are using to describe the atomic details depends on the nature of practical problems and the computation cost (Abraham et al., 1998; Ogata et al., 2001).

In the case of concurrent methods, lots of efforts have been made to solve the "pad" or "handshake" region between the coarse and fine scales. A bridging domain method was proposed in Xiao and Belytschko (2004) for coupling continua with molecular dynamics. The purpose of the handshake region is to assure smoother coupling between the atomistic and continuum regions. These methods were reviewed and further improved in (Karpov et al., 2006). In the case of hierarchical methods, scale separation is exploited so that coarse-grained variables can be evolved on macroscopic spatial/temporal scales using data that are predicted based on the simulation of the microscopic process on microscale spatial/temporal domains (Weinan et al., 2003). As for a real application, concurrent and hierarchical methods can be combined to yield optimal efficiency.

26.5 Conclusion and Perspective

Nanoindentation becomes more and more important because of the development of nanomaterials and MEMS/NEMS. Modeling and simulation are good approaches to "look at" the details during the deformation of a material or devices under external forces.

We have presented in this chapter a review of the principle and methods of nanoindentation. Two principal nonlinear problems presented in nanoindentation (contact and plasticity) are solved by means of the finite element method with effective algorithms. The state of the art on atomistic and multiscale approaches for nanoindentation modeling is presented and discussed.

It seems that the theoretical models could be extended to take into account the friction and the numerical models could be improved by considering the adhesion in the unilateral contact model. To identify the properties of materials, an optimization procedure could be carried out to minimize the difference between the experimental $P - h$ curve and the numerical one.

With the development of realistic computational models, efficient algorithms and powerful computer resources, multiscale phenomena of nanoindentation can be easily understood by freely adjusting parameters beyond the scope of laboratory experiments. This information is of importance to design high quality and multifunctional MEMS/NEMS, and more generally, in nanoscience and nanotechnology.

References

Abraham, F. F., Broughton, J. Q., Bernstein, N., and Kaxiras, E. (1998). Spanning the length scales in dynamic simulation. *Comput. Phys., 12(6)*, 538–546.

Alder, B. and Wainwright, T. (1959). Studies in molecular dynamics: I. General method. *J. Chem. Phys., 31*, 459.

Alder, B. and Wainwright, T. (1960). Studies in molecular dynamics: II. Behavior of a small number of elastic spheres. *J. Chem. Phys., 33*, 1439.

Aouadia, S. M. (2006). Structural and mechanical properties of TaZrN films: Experimental and ab initio studies. *J. Appl. Phys., 99*, 053507.

Berasetgui, E. G., Bull, S. J., and Page, T. F. (2004). Mechanical modelling of multilayer optical coatings. *Thin Solid Films, 447–448*, 26–32.

Bhushan, B. (1999). Nanoscale tribophysics and tribomechanics. *Wear, 225–229*, 465–492.

Bhushan, B. (2004). (ed.) *Springer Handbook of Nanotechnology*, 2nd edition. Berlin, Germany: Springer.

Brenner, D. (1990). Empirical potential for hydrocarbons for use in simulating the chemical vapor deposition of diamond films. *Phys. Rev. B, 42*, 9458.

Bucaille, J. L., Stauss, S., Felder, E., and Michler, J. (2003). Determination of plastic properties of metals by instrumented indentation using different sharp indenters. *Acta Mater., 51*, 1663–1678.

Bull, S. J., Berasetgui, E. G., and Page, T. F. (2004). Modelling of the indentation response of coatings and surface treatments. *Wear, 256*, 857–866.

Bulychev, S. I., Alekhin, V. P., Shorshorov, M. K., Ternovskii, A. P., and Shnyrev, G. D. (1975). Determining Young's modulus from the indenter penetration diagram. *Zavodskaya Laboratoriya, 41*, 1137–1140.

Cheng, Y. T. and Cheng, C. M. (2004). Scaling dimensional analysis and indentation measurements. *Mater. Sci. Eng. R*, *44*, 91–149.

Christopher, D., Smith, R., and Richter, A. (2001). Nanoindentation of carbon materials. *Nucl. Instrum. Methods Phys. Res. B*, *180*, 117–124.

DeSaxcé, G. and Feng, Z.-Q. (1998). The bi-potential method: A constructive approach to design the complete contact law with friction and improved numerical algorithms. *Math. Comput. Model.*, *28(4–8)*, 225–245.

Dupuy, L. M., Tadmor, E. B., Miller, R. E., and Phillips, R. (2005). Finite-temperature quasicontinuum: Molecular dynamics without all the atoms. *Phys. Rev. Lett.*, *95*, 060202.

Feng, Z.-Q. (1995). 2D or 3D frictional contact algorithms and applications in a large deformation context. *Commun. Numer. Methods Eng.*, *11*, 409–416.

Feng, Z.-Q., Peyraut, F., and Labed, N. (2003). Solution of large deformation contact problems with friction between Blatz-Ko hyperelastic bodies. *Int. J. Eng. Sci.*, *41*, 2213–2225.

Feng, Z.-Q., Joli, P., Cros, J.-M., and Magnain, B. (2005). The bi-potential method applied to the modeling of dynamic problems with friction. *Comput. Mech.*, *36*, 375–383.

Feng, Z.-Q., Peyraut, F., and He, Q.-C. (2006). Finite deformations of Ogden's materials under impact loading. *Int. J. Non-Linear Mech.*, *41*, 575–585.

Feng, Z.-Q., Zei, M., and Joli, P. (2007). An elasto-plastic contact model applied to nanoindentation. *Comput. Mater. Sci.*, *38*, 807–813.

Fischer-Cripps, A. C. (2004). *Nanoindentation*. Berlin, Germany: Springer.

Fischer-Cripps, A. C. (2006). Critical review and interpretation of nanoindentation test data. *Surf. Coat. Technol.*, *200*, 4153–4165.

Gouldstone, A., Chollacoop, N., Dao, M., Li, J., Minor, A. M., and L., S. Y. (2007). Indentation across size scales and disciplines: Recent developments in experimentation and modeling. *Acta Mater.*, *55*, 4015–4039.

Heegaard, J. H. and Curnier, A. (1993). An augmented Lagrangian method for discrete large slip contact problems. *Int. J. Numer. Methods Eng.*, *36*, 569–593.

Hughes, T. J. L. and Winget, J. (1980). Some computational aspects of elastic-plastic large strain analysis. *Int. J. Numer. Methods Eng.*, *15*, 1862–1867.

Jian, S. R., Fang, T. H., Chuu, D. S., and Ji, L. W. (2006). Atomistic modeling of dislocation activity in nanoindented GaAs. *Appl. Surf. Sci.*, *253*, 833–840.

Johnson, K. L. (1985). *Contact Mechanics*. Cambridge, U.K.: Cambridge University Press.

Joli, P. and Feng, Z.-Q. (2008). Uzawa and Newton algorithms to solve frictional contact problems within the bi-potential framework. *Int. J. Numer. Methods Eng.*, *73*, 317–330.

Jourdan, F., Alart, P., and Jean, M. (1998). A Gauss-Seidel like algorithm to solve frictional contact problems. *Comput. Methods Appl. Mech. Eng.*, *155*, 31–47.

Karpov, E. G., Yu, H., Park, H. S., Liu, W. K., Wang, Q. J., and Qian, D. (2006). Multiscale boundary conditions in crystalline solids: Theory and application to nanoindentation. *Int. J. Solids Struct.*, *43*, 6359–6379.

Knapp, J. A., Follstaedt, D. M., Myers, S. M., Barbour, J. C., and Friedmann, T. A. (1999). Finite-element modeling of nanoindentation. *J. Appl. Phys.*, *58*, 1460–1474.

Krieg, R. D. and Krieg, B. D. (1977). Accuracies of numerical solution method for the elastic-perfectly plastic model. *ASME, J. Press. Vessel Pip. Div.*, *99*, 510–515.

Leach, A. R. (2001). *Molecular Modelling: Principles and Applications*. Englewood Cliffs, NJ: Prentice-Hall.

Lin, Y. H., Chen, T. C., Yang, P. F., Jian, S. R., and Lai, Y. S. (2007). Atomic-level simulations of nanoindentation-induced phase transformation in mono-crystalline silicon. *Appl. Surf. Sci.*, *254*, 1415–1422.

Liu, C. L., Fang, T. H., and Lin, J. F. (2007). Atomistic simulations of hard and soft films under nanoindentation. *Mater. Sci. Eng., A 452–453*, 135–141.

Liu, W. K., Karpov, E. G., Zhang, S., and Park, H. S. (2004). An introduction to computational nanomechanics and materials. *Comput. Methods Appl. Mech. Eng.*, *193*, 1529–1578.

Liu, Y., Wang, B., Yoshino, M., Roy, S., Lu, H., and Komanduri, R. (2005). Combined numerical simulation and nanoindentation for determining mechanical properties of single crystal copper at mesoscale. *J. Mech. Phys. Solids*, *53*, 2718–2741.

Love, A. E. H. (1939). Boussinesq's problem for a rigid cone. *Q. J. Math.*, *10*, 161.

Maugis, D. (1999). *Contact, Adhesion and Rupture of Elastic Solids*. Berlin, Germany: Springer.

Miller, R. E. and Tadmor, E. B. (2002). The quasicontinuum method: Overview, applications and current directions. *J. Comput.-Aided Mater. Des.*, *9*, 203–239.

Nagtegaal, J. C. (1982). On the implementation of inelastic constitutive equations with special reference to large deformation problems. *Comput. Methods Appl. Mech. Eng.*, *33*, 469–484.

Noreyan, A., Amar, J. G., and Marinescua, I. (2005). Molecular dynamics simulations of nanoindentation of beta-SiC with diamond indenter. *Mater. Sci. Eng., B 117*, 235–240.

Ogata, S., Lidorikis, E., Shimojo, F., Nakano, A., Vashishta, P., and Kalia, R. (2001). Hybrid Finite-element/Molecular-dynamics/Electronic-density-functional approach to materials simulations on parallel computers. *Comput. Phys. Commun.*, *138*, 143–154.

Oliver, W. C. and Pharr, G. M. (1992). An improved technique for determining hardness and elastic modulus using load and displacement sensing indentation experiments. *J. Mater. Res.*, *7*, 1564–1583.

Oliver, W. C. and Pharr, G. M. (2004). Measurement of hardness and elastic modulus by instrumented indentation: Advances in understanding and refinements to methodology. *J. Mater. Res.*, *19*, 3–20.

Pharr, G. M., Oliver, W. C., and Brotzen, F. R. (1992). On the generality of the relationship among contact stiffness, contact area and the elastic modulus during indentation. *J. Mater. Res.*, *7*, 613–617.

Sauer, R. A. and Li, S. (2007). An atomic interaction-based continuum model for adhesive contact mechanics. *Finite Elem. Anal. Des.*, *43*, 384–396.

Schneider, J. M., Sigumonrong, D. P., Music, D., Walter, C., Emmerlich, J., Iskandar, R., et al. (2007). Elastic properties of Cr$_2$AlC thin films probed by nanoindentation and ab initio molecular dynamics. *Scri. Mater.*, 57, 1137–1140.

Schuh, C. A. (2006). Nanoindentation studies of materials. *Mater. Today*, 9, 32–40.

Sharpe, W. N. (2008). *Springer Handbook of Experimental Solid Mechanics*. Berlin, Germany: Springer.

Simo, J. C. and Hughes, T. J. R. (1998). *Computational Inelasticity*. New York: Springer-Verlag.

Simo, J. C. and Taylor, R. L. (1985). Consistent tangent operator for rate-independent elasto-plasticity. *ASME, J. Press. Vessel Pip. Div.*, 48, 101–118.

Simo, J. C. and Taylor, R. L. (1986). A return mapping algorithm for plane stress elastoplasticity. *Int. J. Numer. Methods Eng.*, 22, 649–670.

Smith, R. and Beardmore, K. (1996). Molecular dynamics simulations of particle impacts with carbon-based materials. *Thin Solid Films*, 272, 255–270.

Sneddon, I. N. (1948). Boussinesq's problem for a rigid cone. *Mathematical Proceedings of the Cambridge Philosophical Society*, 44, 492–507.

Sneddon, I. N. (1965). The relation between load and penetration in the axisymmetric Boussinesq problem for a punch of arbitrary profile. *Int. J. Eng. Sci.*, 3, 47–57.

Snyders, R., Music, D., Sigumonrong, D., Schelnberger, B., Jensen, J., and Schneider, J. M. (2007). Experimental and ab initio study of the mechanical properties of hydroxyapatite. *Appl. Phys. Lett.*, 90, 193902.

Szlufarska, I. (2006). Atomistic simulations of nanoindentation. *Mater. Today*, 9(5), 42–50.

Szlufarska, I., Kalia, R. K., Nakano, A., and Vashishta, P. (2007). A molecular dynamics study of nanoindentation of amorphous silicon carbide. *J. Appl. Phys.*, 102, 023509.

Tambe, N. S. and Bhushan, B. (2004). Scale dependence of micro/nano-friction and adhesion of MEMS/NEMS materials, coatings and lubricants. *Nanotechnology*, 15, 1561–1570.

Vashishta, P., Kalia, R. K., and Nakano, A. (2003). Multimillion atom molecular dynamics simulations of nanostructures on parallel computers. *J. Nanoparticle Res.*, 5, 119–135.

Verlet, L. (1967). Computer experiments on classical fluids, I: Thermodynamic properties of Lennard-Jones molecules. *Phys. Rev.*, 159, 98–103.

Wang, W., Jiang, C. B., and Lu, K. (2003). Deformation behavior of Ni$_3$Al single crystals during nanoindentation. *Acta Mater.*, 51, 6169–6180.

Wang, Y., Raabe, D., Klüber, C., and Roters, F. (2004). Orientation dependence of nanoindentation pile-up patterns and of nanoindentation microtextures in copper single crystals. *Acta Mater.*, 52, 2229–2238.

Weinan, E., Engquist, B., and Huang, Z. (2003). Heterogeneous multi-scale method - a general methodology for multi-scale modeling. *Phys. Rev. B*, 67 (9), 092101.

Wilkins, M. L. (1964). Calculation of elastic-plastic flow. In *Methods of Computational Physics*, B. Alder, S. Fernbach and M. Retenberg (Eds.), New York: Academic Press, 3:211–263.

Wriggers, P. (2002). *Computational Contact Mechanics*. Chichester, U.K.: John Wiley & Sons.

Xiao, S. P. and Belytschko, T. (2004). A bridging domain method for coupling continua with molecular dynamics. *Comput. Methods Appl. Mech. Eng.*, 193, 1645–1669.

Zaafarani, N., Raabe, D., Singh, R. N., Roters, F., and Zaefferer, S. (2006). Three-dimensional investigation of the texture and microstructure below a nanoindent in a Cu single crystal using 3D EBSD and crystal plasticity finite element simulations. *Acta Mater.*, 54, 1863–1876.

Zeng, Q., Zhang, L., Xu, Y., Cheng, L., and Yan, X. (2009). A unified view of materials design: Two-element principle. *Mater. Des.*, 30, 487–493.

Zhang, H. W., Wang, J. B., Ye, H. F., and Wang, L. (2007). Parametric variational principle and quadratic programming method for van der Waals force simulation of parallel and cross nanotubes. *Int. J. Solids Struct.*, 44, 2783–2801.

Zhong, Y. and Zhu, T. (2008). Simulating nanoindentation and predicting dislocation nucleation using interatomic potential finite element method. *Comput. Methods Appl. Mech. Eng.*, 197, 3174–3181.

Zhong, Z. H. (1993). *Finite Element Procedures for Contact-Impact Problems*. Oxford, NY: Oxford University Press.

27

Nanoindentation on Silicon

Tong Hong Wang
National Cheng Kung University

and

*Advanced Semiconductor
Engineering, Inc.*

Te-Hua Fang
National Formosa University

Yu-Cheng Lin
National Cheng Kung University

27.1 Introduction

Silicon is a chemical element that has the symbol Si and atomic number 14. As one of the most common elements in the universe, silicon occasionally occurs as a pure, free element in nature, but is more widely distributed in dust, planetoids, and planets as various forms of silicon dioxide or silicate. On earth, silicon is the second most abundant element in the crust, making up 25.7% of the crust by mass.

Silicon is an important material having a wide range of applications. Elemental silicon is the principal component of most semiconductor devices, most importantly integrated circuits or microchips. Silicon is widely used in semiconductors because it remains a semiconductor at higher temperatures than the semiconductor germanium and because its native oxide is easily grown in a furnace and forms a better semiconductor/dielectric interface than almost all other material combinations. Numerous researchers have studied the material properties of bulk silicon. However, related studies in the micro- to nanometer scale are limited [1].

There is wide interest in the mechanical characterization of thin films and small volume of material using depth-sensing indentation tests [1]. Usually, the primary goal of such testing is to determine elastic modulus and hardness of a test material through experimental readings of indenter load and depth of penetration [2–6]. These load versus displacement readings (Figure 27.1) provide an indirect measure of the area of contact at full load, from which the mean contact pressure, and thus the hardness, can be estimated. In addition, indentation techniques can also be used to calculate other elastic and plastic properties such as stiffness, creep, fracture toughness, adhesion, strain-hardening exponent, and others [1]. For thin films on substrates, measurements of film properties with indentation depths of

less than 10% of the total film thickness is recommended so as to eliminate the substrate effect [7,8]. Besides, a nanoindentation technique on a clamped freestanding thin film is also available [9,10], and it avoids the complicated role of the substrate whose properties strongly influence the test for very thin films.

A typical nanoindentation test is carried out using a specific indenter to indent a bottom-fixed test material, which can be either a bulk substrate or a film on a substrate. Moreover, due to its good resolution in the nanoscale, further delicate investigations such as phase transformation, pop-in, and pop-out, and other nanoscale phenomena are available.

The remainder of this chapter is organized as follows. In Section 27.2, we present phase transformation of silicon after nanoindentation. Section 27.3 features material characteristics through experimental measurement in Section 27.3.1 and through finite element analysis in Section 27.3.2. Conclusions are presented in Section 27.4.

27.2 Phase Transformation

Indentation on silicon often shows a "kink-back" or "pop-out" phenomenon, which reveals itself as a phase transformation accompanied by a sudden volume release in the unloading part of the load–displacement curve. During loading to around 12 GPa, silicon transforms from the cubic diamond phase (cd or Si-I) to a metallic phase with the β-tin structure (Si-II) [11–13]. This phase is unstable, and on pressure release, amorphous silicon and/or a mixture of high-pressure crystalline phases can be formed from Si-II depending on unloading rates [14–16]. On fast unloading, the metallic phase transforms to amorphous silicon whereas on slow unloading it transforms to a mixture of high-pressure polycrystalline phases (Si-III and Si-XII).

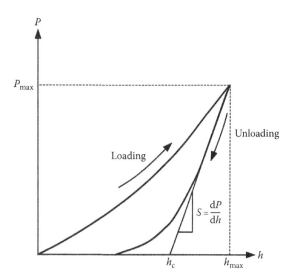

FIGURE 27.1 Load versus displacement for elastic–plastic loading followed by elastic unloading, where h_{max} is the depth from the original sample surface at load of P_{max}, h_c is the contact depth or plastic depth, and S is the stiffness of the test material, which can be determined from the slope of the initial unloading (typically one-third of the curve) by evaluating the maximum load and the maximum depth, dP/dh. (From Wang, T.H. et al., *Nanotechnology*, 18, 135701, 2007. With permission.)

The metastable phases include the body-centered cubic phase (bc8 or Si-III) and its rhombohedral distortion (r8 or Si-XII), the hexagonal diamond phase (hd or Si-IV), and two simple tetragonal phases (Si-VIII and Si-IX) [17–19].

Domnich et al.'s [16] work demonstrated that Raman microspectroscopy could be successfully used for phase analysis of nanoindentation and confirmed phase transformation of Si during nanoindentation tests.

Three-typical load–displacement curves are shown in Figure 27.2. Normally, slower loading/unloading rates (1 mN/s) and higher maximum loads (50 mN) lead to the appearance of pop-outs in the unloading curves (Figure 27.2a) while at faster rates (3 mN/s) and lower loads (30 mN), elbows were more often observed (Figure 27.2c). Also, several indentations showed a mixed response, when the change in slope of the unloading curve was followed by a sudden decrease in indentation depth (Figure 27.2b). This tendency remained essentially the same for different devices used and did not depend on the crystallographic orientation and surface finish of the samples. The latter fact implies that in the nanoindentation experiments, the response of silicon on pressure release is affected mostly by its transformation (metallization) during the loading cycle, and to a lesser extent by the state of the wafer surface prior to indentation. Ruffell et al. [20] further observed that pop-outs were observed in both crystalline silicon and amorphous silicon and at approximately the same position on the unloading curves.

Three characteristic types of spectra (Figure 27.3) were observed. The spectrum in Figure 27.3a shows a typical Raman response of Si-III and Si-XII phases [21–23]. In the literature, bands at 166, 382, and 433 cm^{-1} have been assigned to the Si-III phase, while the bands at 350 and 394 cm^{-1} can serve as a

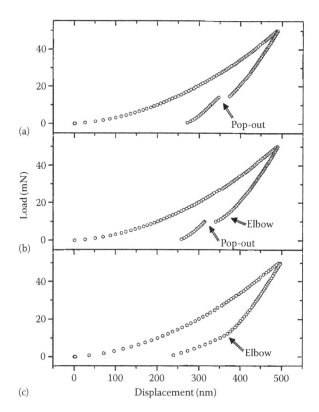

FIGURE 27.2 Three types of load–displacement curves in the nanoindentation of Si, obtained at the loading rate of 3 mN/s and maximum load of 50 mN. (From Domnich, V. et al., *Appl. Phys. Lett.*, 76, 2214, 2000. With permission.)

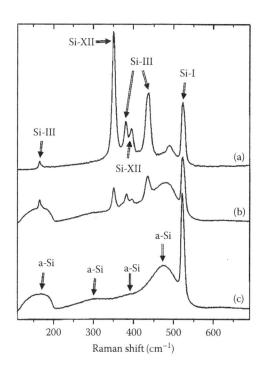

FIGURE 27.3 Three types of Raman spectra from nanoindentation on Si, revealing (a) metastable phases, (b) a mixture of a-Si, Si-III, and Si-XII, and (c) amorphous Si. (From Domnich, V. et al., *Appl. Phys. Lett.*, 76, 2214, 2000. With permission.)

fingerprint of the Si-XII phase [22,24]. The origin of the peak at 490 cm⁻¹ has not yet been unambiguously identified.

Another typical Raman spectrum from nanoindentations is presented in Figure 27.3c. Broad peaks around 170, 300, 390, and 470 cm⁻¹ have been identified as transverse acoustic-like, longitudinal acoustic-like, longitudinal optical-like, and transverse optical-like bands of amorphous silicon, respectively [25], while a peak at 520 cm⁻¹ represents the pristine Si-I as before. Finally, the Raman spectra of several nanoindentations revealed traces of both amorphous silicon and metastable Si-III and Si-XII phases (Figure 27.3b).

The pop-out is a sudden volume increase, and hence the uplift of the material surrounding an indenter. It is a consequence of an abrupt phase transformation from metallic Si-II to either Si-III or Si-XII. The elbow, the gradual change in slope of the unloading curve, is the result of the expansion of the material during slow transformation from metallic to the amorphous semiconducting phase, which contributes to the indenter uplift.

Phase transformation from Si-II to a metastable crystalline state takes place in the pressure range of 5.0–8.5 GPa, with the mean value of 6.62 GPa. In the range of 4.5–5.5 GPa, both partial amorphization and phase transformation of metallic silicon to metastable phases can take place. At pressure less than 4.5 GPa, only amorphous silicon is being formed. Gilman [26] claimed that phase transformations during indentation may be induced by shear (bond bending) rather than compression (bond shortening), thus one would expect lower numbers of transition pressures obtained from indentation experiments as compared to purely hydrostatic conditions in diamond anvil studies.

27.3 Material Characteristics

27.3.1 Experimental Measurement

27.3.1.1 Theoretical Details

Hardness is defined as the resistance to local plastic deformation. It is expressed as the maximum indentation load, P_{max}, divided by the contact area, A [27]:

$$H = \frac{P_{max}}{A} \qquad (27.1)$$

where the contact area is a function of the contact depth, h_c, and can be determined according to the following form:

$$A = 24.5 h_c^2 \qquad (27.2)$$

The constant of 24.5 is used for a perfect Berkovich indenter tip.

Young's modulus of the test material, E, can be obtained by the following equation:

$$E = (1 - v^2)\left(\frac{1}{E^*} - \frac{1 - v_i^2}{E_i}\right)^{-1} \qquad (27.3)$$

where

v is the Poisson's ratio of the test material

E_i and v_i denote Young's modulus and Poisson's ratio of the indenter, respectively

The indenter properties used in this calculations are $E_i = 1140$ GPa and $v_i = 0.07$ and it was assumed that $v_j = 0.3$ for the silicon. E^* is the reduced modulus of the system and can further be defined as

$$E^* = \frac{\sqrt{\pi} S}{2\beta\sqrt{A}} \qquad (27.4)$$

where

S is the stiffness of the test material, and can be determined from the initial unloading slope by evaluating the maximum load and maximum depth, which is $S = dP/dh$

β is a shape constant that depends on the geometry of the indenter, which is 1.034 for a Berkovich tip.

The process of undertaking the test requires considerable experimental skill. Creep and thermal are substantially two types of drift affecting the reading. Creep may manifest itself when the load is held constant. Another is an observed change in depth with constant load that is virtually indistinguishable from specimen creep. It is a change in dimensions of the instrument due to thermal expansion or contraction of the apparatus. To correct for thermal drift, nanoindentation instruments allow for holding to be accumulated at either maximum load or at the end of the unloading from maximum load.

27.3.1.2 Effect of Indentation Load

Four maximum loads, 500, 1000, 2000, and 3000 μN, were chosen for indentation at a loading rate of 500 μN/s. The load–displacement curves and their atomic force microscopic (AFM) images for different maximum loads are shown in Figure 27.4. As expected, increasing the load increased the indentation depth and a more obvious indentation mark was obtained. The elbow was also shown up during the unloading.

By implementing Equations 27.1 and 27.3, information from the load–displacement curves (Figure 27.4a), Young's modulus and hardness can be retrieved, as shown in Figure 27.5. Both the Young's moduli and hardnesses were linearly decreased as the indentation loads increased. For the indentation loads between 500 and 3000 μN, Young's moduli varied between 118 and 140 GPa while hardnesses varied between 12 and 13 GPa. The variations are insignificant, i.e., representing their own intrinsic properties.

27.3.1.3 Effect of Creep

Creep, which describes the tendency of a material to move or to deform permanently to relieve stress. Material deformation occurs as a result of long-term exposure to levels of stress that are below the yield or ultimate strength of the material. There are three ways to measure creep: indentation-load relaxation experiments [27], constant rate of loading tests [28], and constant-load indentation tests [29]. Of these, the constant-load indentation test is the most widely used.

Here creep characteristics under an identical maximum load of 2000 μN and a loading rate of 500 μN/s with different holding times of 0, 15, 30, and 60 s. Figure 27.6 shows the

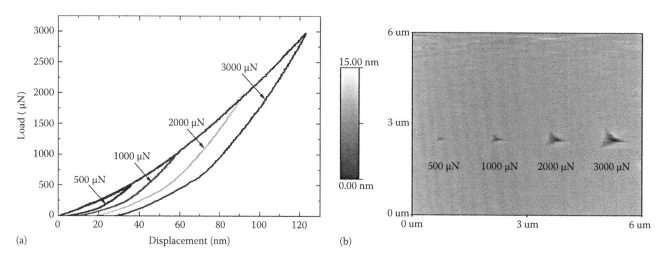

FIGURE 27.4 (a) Load–displacement curves and (b) AFM images for different maximum loads. (From Fang, T.H. et al., *Microelectron. Eng.*, 77, 389, 2005. With permission.)

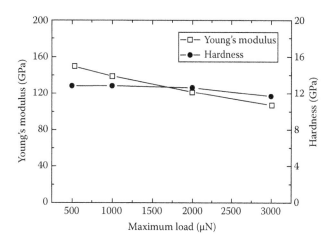

FIGURE 27.5 Young's moduli and hardnesses for different maximum loads. (From Fang, T.H. et al., *Microelectron. Eng.*, 77, 389, 2005. With permission.)

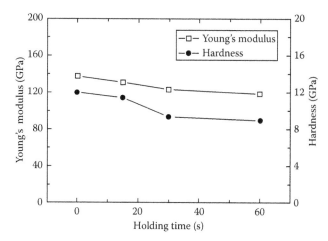

FIGURE 27.7 Young's moduli and hardnesses for different holding times. (From Fang, T.H. et al., *Microelectron. Eng.*, 77, 389, 2005. With permission.)

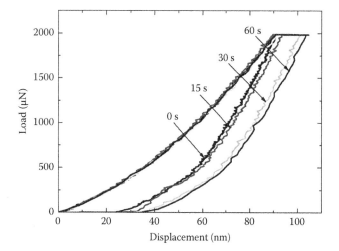

FIGURE 27.6 Load–displacement curves for different holding times. (From Fang, T.H. et al., *Microelectron. Eng.*, 77, 389, 2005. With permission.)

load–displacement curves for different holding times. From the result we can see that, the longer the holding time at maximum load is, the deeper the indenter can displace. Slopes at unloading curves are about the same representing that Young's moduli should be similar. Young's moduli and hardnesses for different holding times are shown in Figure 27.7. Young's moduli and hardnesses were ranged between 118 and 137 GPa and between 9 and 12 GPa. Both Young's moduli and hardnesses were slightly decreased and creep were stabilized at and after holding time of 30 s.

27.3.1.4 Effect of Indentation Cycle

Mechanical reliability of a component is often characterized through a cyclic loading, where the maximum induced stress is less than its yield stress. The progressive and localized structural damage is namely fatigue failure.

To study the fatigue effect, we let the indenter to indent 20 cycles on the silicon at an identical location. Maximum load

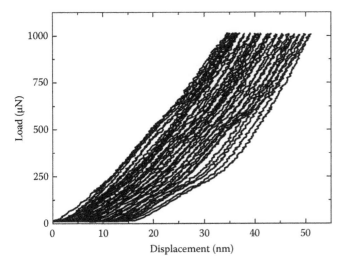

FIGURE 27.8 Load–displacement curve for 20 cyclic indentations. (From Fang, T.H. et al., *Microelectron. Eng.*, 77, 389, 2005. With permission.)

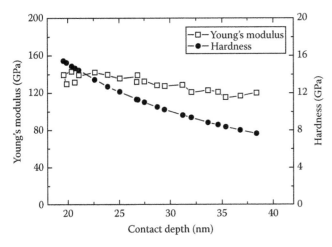

FIGURE 27.9 Young's moduli and hardnesses for different contact depths. (From Fang, T.H. et al., *Microelectron. Eng.*, 77, 389, 2005. With permission.)

and loading rate were set at 1000 μN and 100 μN/s, respectively. Figure 27.8 shows the load–displacement curve for the 20 cyclic indentations. The curve is not overlapped and continues deeper. It remained at a depth of 16 nm after the 20-cycle indentation indicating that the cyclic load induces an unrecoverable fatigue damage. Therefore, the contact depth was increased incrementally for every indent and its related property: hardness as well as Young's modulus were decreased, as shown in Figure 27.9.

27.3.2 Numerical Analysis

27.3.2.1 Finite Element Modeling

There are some experimental difficulties to characterize the effects such as geometry and roundness of the tip, friction, and etc. To estimate the effects of these issues, finite element analysis remains the preferred tool for evaluating the behaviors of

materials because it provides convenient and direct information of the material response and it allows varying the studied factors systematically.

The finite element model was constructed based on a 0.5 μm high diamond Berkovich indenter with a simplified 70.3° effective cone angle on a 5 μm thick silicon substrate. Pyramid shaped indenters are generally treated the same as conical indenters with a cone angle that provides the same area to depth relationship. Due to symmetry, only one half of the indenter and the substrate, Figure 27.10a, were modeled. The one-half two-dimensional finite element model under the assumption of axisymmetry as shown in Figure 27.10b contains 1885 elements and 5540 nodes. A quadratic quadrilateral element in which each element is defined by eight nodes was employed for the substrate, while a rigid line element was used for the indenter. Surface-to-surface contact elements were applied to

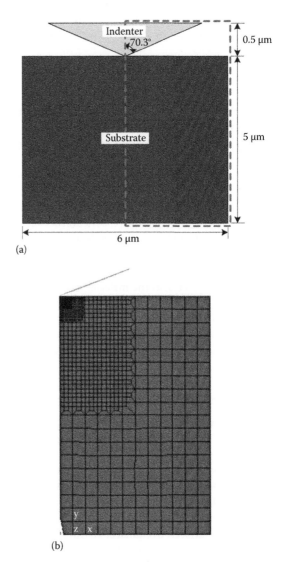

FIGURE 27.10 (a) Modeling region of the Berkovich indenter and the silicon substrate and (b) the one-half finite element model. (From Wang, T.H. et al., *Mater. Sci. Eng. A*, 447, 244, 2007. With permission.)

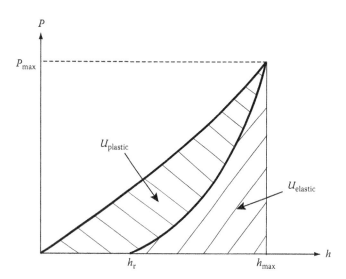

FIGURE 27.11 Plastic energy and elastic energy of indentation. (From Wang, T.H. et al., *Mater. Sci. Eng. A*, 447, 244, 2007. With permission.)

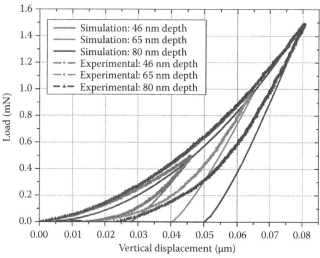

FIGURE 27.12 Load–displacement curves for experiment and simulation. (From Wang, T.H. et al., *Mater. Sci. Eng. A*, 447, 244, 2007. With permission.)

the exposed surfaces for which there were a possibility of touching each other. The horizontal displacement on the symmetric boundary of the indenter and the substrate was constrained while the vertical displacement at the bottom of the substrate was constrained. The static analysis including the large deflection was carried out using the commercial finite element package ANSYS v. 10.0.

The silicon has a Young's modulus of 168 GPa, a Poisson's ratio of 0.278, and a yield stress of 7 GPa without strain hardening at room temperature. This way the silicon behaves as elastic and perfectly plastic. Since the uncertainty of the friction between touching surfaces makes it difficult to measure the contact friction coefficient it was assumed to be 1 as per Coulomb's law of friction.

The application of the load to the indenter and the resulting displacement or vice versa, as the material recovers, represents work done in the system, and is shown schematically in Figure 27.11. The net area enclosed by the load–displacement curve represents the energy lost through plastic deformation, $U_{plastic}$, and from stored elastic energy, $U_{elastic}$, from residual stresses within the test material.

27.3.2.2 Verification with Experiment

Both experimental and numerical indentations were conducted with depths of 46, 65, and 80 nm for the verification purpose. Figure 27.12 shows the comparison of the load–displacement curves. The comparison shows that the simulation curves match well with the experimental ones. Discrepancies were found on the unloading curves. This may be due to the uncertainty of tip bluntness, surface roughness of the silicon and the friction between touching surfaces. Domnich et al. [16] claimed that the "elbow" behavior, i.e., the gradual slope change of the experimental unloading curve was due to the amorphization of the silicon when the pressure was released. However, this is impossible to characterize on a mechanical model. At any rate, the comparisons have a good qualitatively trend.

In this section, we will discuss the effect of the different coefficients of friction, tip roughness, and geometry of the indenter to the silicon substrate.

27.3.2.3 Effect of Coefficient of Friction

The effect of friction between touching surfaces on the induced response of silicon is examined by varying the coefficient of friction, μ from 0, 025, 0.5, 0.75, and to 1. The maximum depth of 100 μm was indented for each coefficient of friction. Figure 27.13 shows the load–displacement curves of different μ and enlarged the regions of interest. Table 27.1 tabulates the $U_{plastic}$, $U_{elastic}$, E, H, and Δe for each cell. Figure 27.14a and b shows the indentation surface profile at maximum load, and the residual surface profile after unloading, for different coefficients of friction,

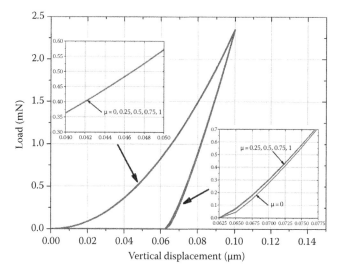

FIGURE 27.13 Load–displacement curves for different coefficients of friction. (From Wang, T.H. et al., *Mater. Sci. Eng. A*, 447, 244, 2007. With permission.)

TABLE 27.1 $U_{plastic}$, $U_{elastic}$, E, H, and Δe for Different Coefficients of Friction

μ	$U_{plastic}$ (pJ)	$U_{elastic}$ (pJ)	E (GPa)	H (GPa)	Δe (%)
0	39.00	38.29	193.11	17.00	32.80
0.25	38.66	38.78	197.64	17.87	34.62
0.5	38.62	38.80	198.12	17.93	34.75
0.75	38.61	38.80	197.77	17.95	34.80
1	38.60	38.80	198.09	17.96	34.83

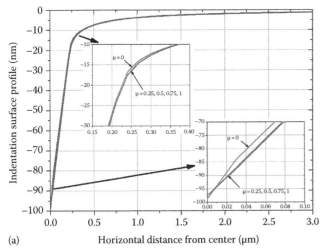

(a)

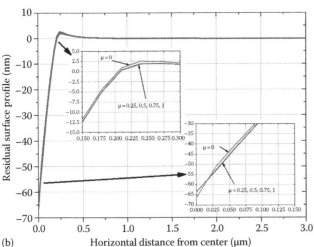

(b)

FIGURE 27.14 (a) Indentation surface profiles at maximum load and (b) residual surface profile after unloading for different coefficients of friction. (From Wang, T.H. et al., *Mater. Sci. Eng. A*, 447, 244, 2007. With permission.)

respectively. From these figures, no significant differences can be found among the investigated μ in terms of load–displacement curve, indentation surface, or residual surface profile. A lesser sideward squeeze on both surface profiles is observed for the frictional indenter compared to the frictionless one. Moreover, a zero friction surface results in greater vertical displacement during unloading and deeper depth of surface profile due to its sliding characteristic. Slight discrepancies in $U_{plastic}$, $U_{elastic}$, E, H, and

Δe are found as well, as shown in Table 27.1. It is worth noting that the plastic energy, the elastic energy, and the elastic recovery reach a steady state when μ equals to 0.25, and there is a slight decrease after that until μ equals 1.

Figure 27.15 shows the von Mises stress distributions of the coefficient of friction at 0 and 0.5. No significant differences are found in the von Mises stress distributions and magnitude at the maximum loading depth among cells, except for after the unloading between frictionless and frictional surfaces. The magnitude of the residual von Mises stress after unloading are 7.066 GPa at maximum loading depth, and 7 GPa for the coefficient of friction at 0, while they are 7.109 GPa at maximum loading depth and 4.741 GPa for the coefficient of friction at 0.5. The magnitude of the frictional surface, after unloading, is about 32.3% less than that of the frictionless one. For the distribution after unloading, the maximum von Mises stress of the silicon on frictionless surfaces is concentrated at the center of the silicon surface while that on frictional surfaces is located at the inner region of the silicon slightly towards the edge where the indenter touched. In addition to the von Mises stress distributions after unloading, μ was also determined to be minor in this particular study. A similar finding of friction was also found by Fujisawa et al. [30] with a rigid spherical indenter.

27.3.2.4 Effect of Tip Roundness

In practice, a sharp Berkovich indenter having a zero tip radius is not realistic. A typical indenter has a tip radius around 100 nm. The longer the tip is used the blunter it becomes, and the tip radius gets larger. The effect of tip roundness on the induced response of silicon is examined by varying the tip radii through 0, 50, 100, 150, and 200 nm, respectively. The maximum depth of 100 μm was indented for each tip radius. Figure 27.15 shows the load–displacement curves of different tip radii, and Table 27.3 tabulates the $U_{plastic}$, $U_{elastic}$, E, H, and Δe of each cell. It is evident that from Figure 27.16 and Table 27.2 that the load, $U_{plastic}$, $U_{elastic}$, E, H, and Δe increase with the increase in tip radius. This is as expected, since a tip with a larger diameter touches a larger area of the silicon at the identical depth than the tip with the smaller diameter, and consequently induces a greater response. This shows why a blunter tip will result in a higher load, higher plastic energy, and higher elastic energy. Considering the fact that the bluntness of the indenter will vary over its service life, a careful correction factor [27] should be settled on when making experimental measurements.

Figure 27.17a and b shows the indentation surface profile during maximum loading depth and the residual surface profile after unloading of different tip radii over the entire horizontal distance. Each figure on the horizontal axis up to the initial 0.5 mm horizontal distance from the center has been enlarged in order to determine the slight differences from one another. It can be seen from Figure 27.16 that a larger tip radius leads to a slightly deeper indentation surface profile and residual surface profile along the straight surface of the tip. A blunter tip not only pushes the silicon downward, but also pushes it sideward.

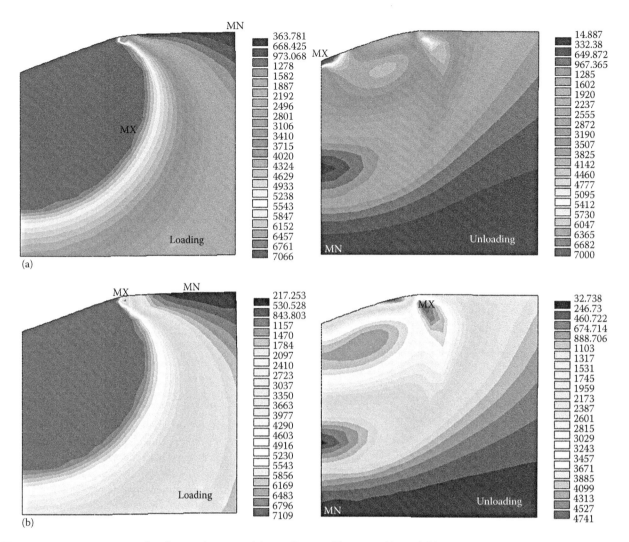

FIGURE 27.15 Von Mises stress distributions (Unit: MPa) for coefficients of friction at (a) 0 and (b) 0.5. (From Wang, T.H. et al., *Mater. Sci. Eng. A*, 447, 244, 2007. With permission.)

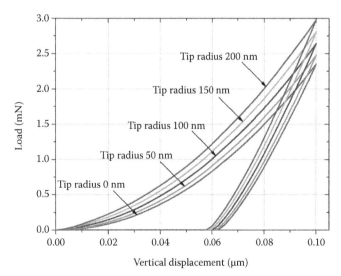

FIGURE 27.16 Load–displacement curves for different tip radii. (From Wang, T.H. et al., *Mater. Sci. Eng. A*, 447, 244, 2007. With permission.)

TABLE 27.2 $U_{plastic}$, $U_{elastic}$, E, H, and Δe for Different Tip Radii

Tip Radius (nm)	$U_{plastic}$ (pJ)	$U_{elastic}$ (pJ)	E (GPa)	H (GPa)	Δe (%)
0	38.60	38.80	198.09	17.96	34.83
50	41.86	41.83	213.85	20.17	35.81
100	45.94	46.04	234.76	23.50	37.19
150	50.38	50.30	259.66	27.48	38.53
200	54.81	54.88	287.70	32.30	39.85

However, within a 25 nm horizontal distance from the center of the residual surface profile, which is the location of the tip, the larger tip radius shows a residual surface profile that is shallower than the tip with a smaller radius. This indicates that the spherical shape of the tip causes a smooth transition zone form elastic to elastic–plastic contact resulting in lesser permanent damage. Figure 27.18 shows the von Mises stress distribution for indenter tip radii of 0 and 150 nm. It was found that the von Mises stress distributions are similar with little difference in magnitude among studied tip radii. The finding of a lesser von Mises stress with a larger tip radius is inversely proportional to the elastic recovery's trend.

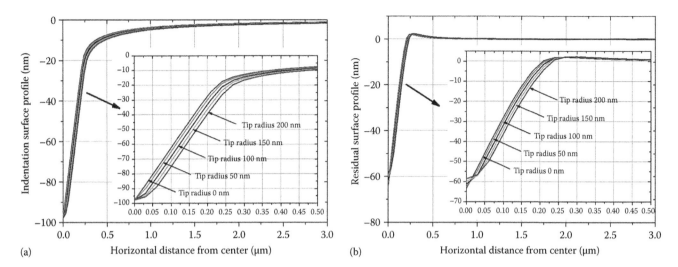

(a)

(b)

FIGURE 27.17 (a) Indentation surface profiles at maximum load and (b) residual surface profile after unloading for different tip radii. (From Wang, T.H. et al., *Mater. Sci. Eng. A*, 447, 244, 2007. With permission.)

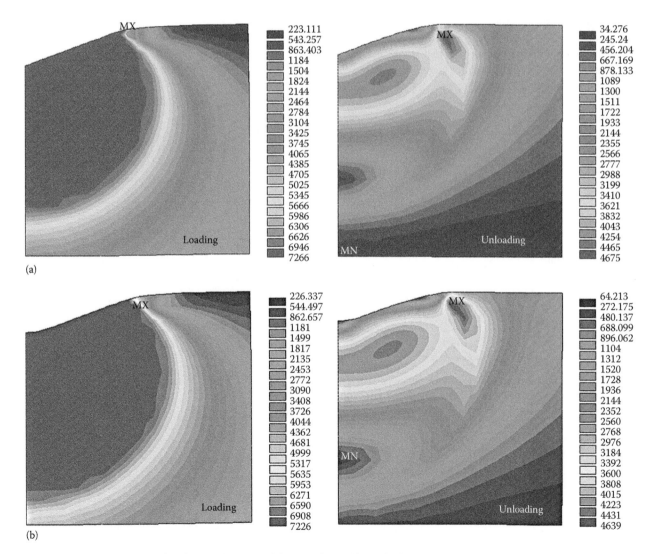

(a)

(b)

FIGURE 27.18 Von Mises stress distributions (Unit: MPa) for tip radius at (a) 0 and (b) 150 nm. (From Wang, T.H. et al., *Mater. Sci. Eng. A*, 447, 244, 2007. With permission.)

27.3.2.5 Effect of Indenter Geometry

The effect of the indenter geometry on the induced response of silicon is examined by comparing the Berkovich, conical, Knoop, and spherical tip, respectively. The conical and the Knoop tip have a 65° and 77.64° effective cone angle, respectively, while the spherical tip has a 0.5 μm radius. The maximum depth of 100 mm was indented for each tip. Figure 27.19 shows the load–displacement curves of the different indenter geometries, and Table 27.3 summarizes the $U_{plastic}$, $U_{elastic}$, E, H, and Δe of each cell. Different projected areas [27] for E and H calculations were implemented for different indenter geometries. It is evident that the conical tip exhibits the lowest load during maximum loading depth while the Knoop tip has the highest load. This indicates that the more surface area is in contact, the greater the load response from the tip. However, the tip with the sharpest shape, such as the conical tip, results in the higher stress concentration and induces the greatest residual von Mises stress after unloading. The lesser Δe listed in Table 27.3 shows an identical trend.

Figure 27.20a and b shows the indentation surface profile during maximum loading depth and the residual surface profile after unloading of different indenter geometries, respectively. Based on the concept that the tip with the larger diameter results in the deeper indentation depth along the straight surface of

the tip (Section 27.3.2.4), it is reasonable that the spherical tip incurred a deeper indentation surface profile and residual surface profile than the Berkovich and conical tip, while the Knoop tip in which the effective angle of 77.64° is the critical angle to get beyond the touching area at a depth identical to the spherical tip. In addition, the pile-up of the silicon induced by the spherical tip on the residual surface profile is the greatest while that induced by the Knoop tip is the lowest due to the fact that the Knoop tip has the greatest effective angle compared to the Berkovich and the conical tip.

Figure 27.21 shows the von Mises stress distributions of silicon of different indenter geometries. The greatest effective angle of the Knoop tip results in the lowest residual von Mises stress and the greatest Δe on silicon, even greater than the spherical tip. However, the maximum von Mises stress affected zone is large beneath the Knoop tip.

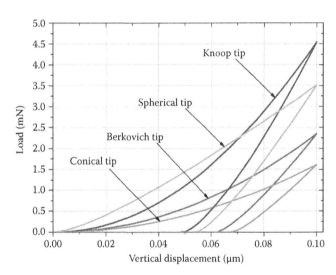

FIGURE 27.19 Load–displacement curves for different indenter geometries. (From Wang, T.H. et al., *Mater. Sci. Eng. A*, 447, 244, 2007. With permission.)

TABLE 27.3 $U_{plastic}$, $U_{elastic}$, E, H, and Δe for Different Indenter Geometries

Indenter Geometry	$U_{plastic}$ (pJ)	$U_{elastic}$ (pJ)	E (GPa)	H (GPa)	Δe (%)
Berkovich	38.60	38.80	198.09	17.96	31.98
Cone	30.23	22.99	218.89	23.70	28.28
Knoop	52.96	96.84	219.60	17.18	48.73
Sphere	82.36	70.66	204.53	309.49	44.35

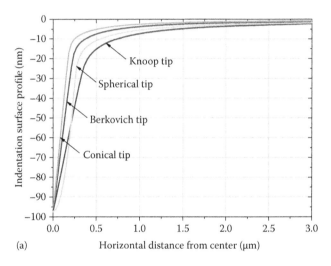

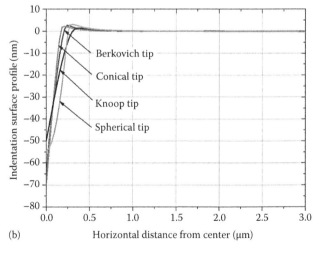

FIGURE 27.20 (a) Indentation surface profiles at maximum load and (b) residual surface profiles after unloading for different indenter geometries. (From Wang, T.H. et al., *Mater. Sci. Eng. A*, 447, 244, 2007. With permission.)

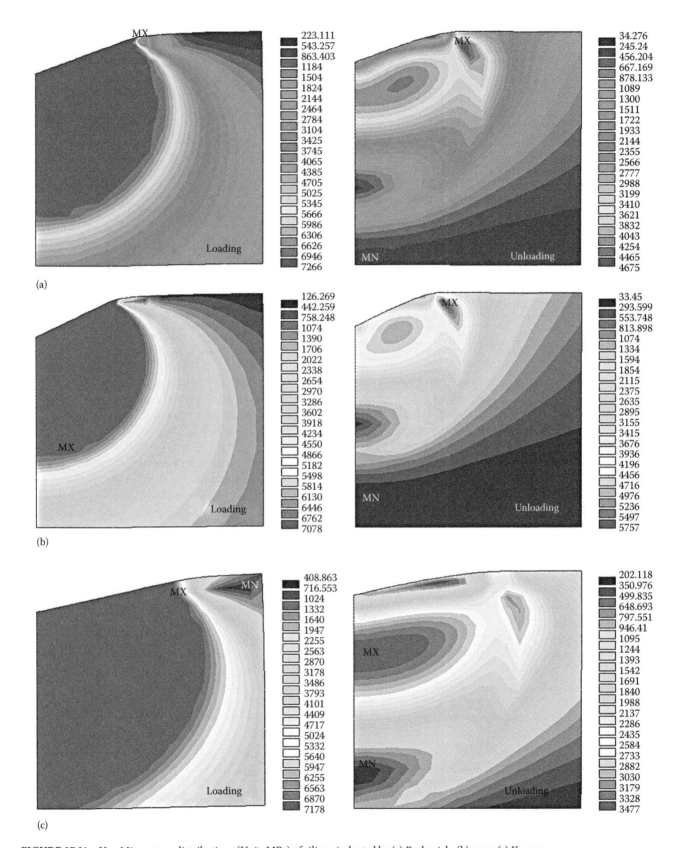

FIGURE 27.21 Von Mises stress distributions (Unit: MPa) of silicon indented by (a) Berkovich, (b) cone, (c) Knoop

(*continued*)

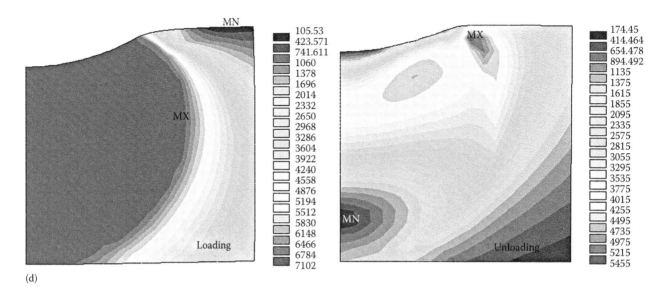

FIGURE 27.21 (continued)   (d) spherical tip. (From Wang, T.H. et al., *Mater. Sci. Eng. A*, 447, 244, 2007. With permission.)

27.4 Conclusion

The nanomechanical properties of Si were measured and analyzed using both the experimental nanoindentation technique and the finite element method. Si in general sinks-in during maximum depth loading and piles-up after unloading. Young's modulus and the hardness for Si decreased as the indentation loads, hold times, and indentation cycles increased.

No significant differences can be found when varying the coefficient of friction, except the von Mises stress distributions after unloading, which are substantially different between frictionless and frictional surfaces. Greater stress affected zones are also found in deeper indentations which incurs a comparatively larger plastic deformation in the touched surface area and squeezes out a higher pile-up. A greater tip radius leads to a slightly deeper indentation surface profile and residual surface profile along the straight surface of the tip, but with a shallower residual surface profile around the region of the tip radius. A sharper shape, such as the conical tip results in a larger stress concentration and induces a greater residual von Mises stress after unloading.

References

1. Fisher-Cripps, A. C. 2002. *Nanoindentation*. New York: Springer.
2. Oliver, W. C. and Pharr, G. M. 1992. An improved technique for determining hardness and elastic-modulus using load and displacement sensing indentation experiments. *J. Mater. Res.* **7**: 1564–1583.
3. Doerner, M. F. and Nix, W. D. 1986. A method for interpreting the data from depth-sensing indentation instruments. *J. Mater. Res.* **1**: 601–609.
4. Cheng, Y. T. and Cheng, C. M. 1998. Relationships between hardness, elastic modulus, and the work of indentation. *Appl. Phys. Lett.* **73**: 614–616.
5. Fang, T. H., Chang, W. J., and Lin, C. M. 2005. Nanoindentation and nanoscratch characteristics of Si and GaAs. *Microelectron. Eng.* **77**: 389–398.
6. Wang, T. H., Fang, T. H., and Lin, Y. C. 2007. A numerical study of factors affecting the characterization of nanoindentation on silicon. *Mat. Sci. Eng. A.* **447**: 244–253.
7. Wang, T. H., Fang, T. H., and Lin, Y. C. 2007. Analysis of the substrate effects of strain-hardening thin films on silicon under nanoindentation. *Appl. Phys. A* **86**: 335–341.
8. Wang, T. H., Fang, T. H., and Lin, Y. C. 2008. Finite-element analysis of the mechanical behavior of Au/Cu and Cu/Au multilayers on silicon substrate under nanoindentation. *Appl. Phys. A* **90**: 457–463.
9. Wang, T. H., Fang, T. H., Kang, S. H., and Lin, Y. C. 2007. Nanoindentation characteristics of clamped freestanding Cu membranes. *Nanotechnology* **18**: 135701.
10. Wang, T. H., Fang, T. H., Kang, S. H., and Lin, Y. C. 2008. Creep characteristics of clamped Cu membranes subjected to indentation. *Jpn. J. Appl. Phys.* **47** (2): 1019–1021.
11. Minomura, S. and Drickamer, H. G. 1962. Pressure-induced phase transformations in Si, Ge, and some III-V compounds. *J. Phys. Chem. Solids* **23**: 451.
12. Gupta, M. C. and Ruoff, A. L. 1980. Static compression of silicon in the [100] and in the [111] directions. *J. Appl. Phys.* **51**: 1072–1075.
13. Hu, J. Z. and Spain, I. L. 1984. Phases of silicon at high-pressure. *Solid State Commun.* **51**: 263–266.
14. Bradby, J. E., Williams, J. S., Wong-Leung, J., Swain, M. V., and Munroe, P. 2000. Transmission electron microscopy observation of deformation microstructure under spherical indentation in silicon. *Appl. Phys. Lett.* **77**: 3749–3751.
15. Bradby, J. E., Williams, J. S., Wong-Leung, J., Swain, M. V., and Munroe, P. 2001. Mechanical deformation in silicon by micro-indentation. *J. Mater. Res.* **16**: 1500–1507.

16. Domnich, V., Gogotsi, Y., and Dub, S. 2000. Effect of phase transformations on the shape of the unloading curve in the nanoindentation of silicon. *Appl. Phys. Lett.* **76**: 2214–2216.

17. Crain, J., Ackland, G. J., Maclean, J. R., Piltz, R. O., Hatton, P. D., and Pawley, G. S. 1994. Reversible pressure-induced structural transitions between metastable phases of silicon. *Phys. Rev. B* **50**: 13043–13046.

18. Weill, G., Mansot, J. L., Sagon, G., Carlone, C., and Besson, J. M. 1989. Characterization of Si-III and Si-IV, metastable forms of silicon at ambient pressure. *Semicond. Sci. Technol.* **4**: 280–282.

19. Zhao, Y. X., Buehler, F., Sites, J. R., and Spain, I. L. 1986. New metastable phases of silicon. *Solid State Commun.* **59**: 679–682.

20. Ruffell, S., Bradby, J. E., and Williams, J. S. 2006. High pressure crystalline phase formation during nanoindentation: Amorphous versus crystalline silicon. *Appl. Phys. Lett.* **89**: 091919.

21. Kobliska, R. J., Solin, S. A., Selders, M., Chang, R. K., Alben, R., Thorpe, M. F., and Weaire, D. 1972. Raman scattering from phonons in polymorphs of Si and Ge. *Phys. Rev. Lett.* **29**: 725.

22. Hanfland, M. and Syassen K. 1990. Raman modes of metastable phases of Si and Ge. *High Press. Res.* **3**: 242–244.

23. Piltz, R. O., Maclean, J. R., Clark, S. J., Ackland, G. J., Hatton, P. D., and Crain, J. 1995. Structure and properties of silicon-XII—A complex tetrahedrally bonded phase. *Phys. Rev. B* **52**: 4072–4085.

24. Kailer, A., Gogotsi, Y. G., and Nickel, K. G. 1997. Phase transformations of silicon caused by contact loading. *J. Appl. Phys.* **81**: 3057–3063.

25. Marinov, M. and Zotov, N. 1997. Model investigation of the Raman spectra of amorphous silicon. *Phys. Rev. B* **55**: 2938–2944.

26. Gilman, J. J. 1992. Insulator-metal transitions at microindentations. *Mater. Res. Soc. Symp. Proc.* **276**: 191.

27. Wu, T. W., Mosherf, M., and Alexopoulos, P. S. 1990. The effect of the interfacial strength on the mechanical-properties of aluminum films. *Thin Solid Films* **187**: 295–307.

28. Mayo, M. J. and Nix, W. D. 1988. A micro-indentation study of superplasticity in Pb, Sn, and Sn-38 Wt-percent-Pb. *Acta Metall.* **36**: 2183–2192.

29. Raman V. and Berriche, R. 1992. An investigation of the creep processes in tin and aluminum using a depth-sensing indentation technique. *J. Mater. Res.* **7**: 627–638.

30. Fujisawa, N., Li, W., and Swain, M. V. 2004. Observation and numerical simulation of an elastic-plastic solid loaded by a spherical indenter. *J. Mater. Res.* **19**: 3474–3483.

Nanohole Arrays on Silicon

Hidetaka Asoh
Kogakuin University

Sachiko Ono
Kogakuin University

28.1 Introduction

Nanostructured materials not only have potential technological applications in various fields, but are also of fundamental interest in that the properties and functions of a material can change with the transition between the bulk and atomic/molecular scales. In particular, silicon with ordered pores of dimensions ranging from submicrometer to nanometer range, so-called porous silicon, has generated considerable interest in both basic research and commercial applications because of the electrochemical properties of silicon. Therefore, many investigations have been carried out on the surface chemistry and semiconductor electrochemistry of silicon and the development of novel techniques enabling the fabrication of unique nano-/microstructures. For a general background to the use of electrochemistry in silicon technology, see the important reference books listed in the bibliography [1,2].

In this chapter, we describe two techniques for the fabrication of nanohole arrays on a silicon surface based on the principles of electrochemistry:

1. The fabrication of silicon nanohole arrays using self-ordered anodic porous alumina as a mask for the localized anodization of a silicon substrate
2. The fabrication of porous silicon by metal-assisted chemical etching

These techniques are characterized by the templating of self-organized materials. Although studies have been performed on the formation of porous silicon by conventional anodization in hydrofluoric acid (HF), very few attempts have been made to realize a novel fabrication process without external bias. The period of research discussed in this chapter is from approximately 2000 to the present day. In particular, we focus on our attempts to fabricate various porous structures and inverse patterns including column arrays, as described below.

28.2 Fabrication of Nanohole Arrays on Silicon Using Anodic Alumina Mask

28.2.1 Nanopatterning of Silicon Surface

One of the most significant issues in the development of nanotechnology both in basic research and for commercial applications is the fabrication and design of semiconductor materials with ordered nano-/microstructures because of their potential applications in various devices, including quantum electronic devices, photoelectronic devices, functional electrodes, photoelectrochemical cells, bioreactors, sensors, and data storage devices. The techniques commonly used in fabricating these functional devices with nanometer dimensions involve conventional lithographic technologies, that is, top-down patterning techniques using optical, electron, or x-ray beams. Although these techniques have many advantages, they are not suitable for large-scale patterning with sub-100 nm feature sizes due to the limit of their resolution and low throughput. Therefore, a novel patterning technique that can be used easily and efficiently is required. In terms of simple and inexpensive methods, nanopatterning techniques based on the templating of self-organized materials have recently attracted considerable attention as a key fabrication method owing to their relative simplicity and low cost. Several studies on nanopatterning the surface of semiconductors using self-ordered anodic porous alumina have been performed [3–21].

28.2.2 Anodic Porous Alumina

Porous anodic oxide film, so-called anodic porous alumina, which is formed by the anodization of Al, is a typical self-ordered nanoporous material. It is often referred to as a nanochannel or nanohole structure depending on the dimensions of the porous anodic oxide film. This porous material has thus far been widely used as a key material in the fabrication of several types of devices [22–43] because of its unique solid geometry, that is, honeycomb structure at the nanometer scale.

Anodic porous oxide film can be formed on Al in various electrolytes on the basis of the following reaction:

$$2Al + 3H_2O \rightarrow Al_2O_3 + 6H^+ + 6e^-$$

From the many morphological studies of anodic porous alumina, it has been known that the anodic oxide film on Al consists of two regions: a thick outer region of porous-type oxide and a thin inner region of barrier-type oxide lying adjacent to the Al. In 1953, Keller, Hunter, and Robinson reported on the structural features of porous alumina film determined by electron microscopy, and were the first to propose a geometrical cell model, as shown in Figure 28.1a [44]. Moreover, they suggested that cell dimensions, such as pore and cell diameters and barrier layer thickness, primarily depend on formation voltage, and they increase linearly with voltage. As for the general background of the growth mechanism of anodic films, see Refs. [44–46].

We will now focus on the applications of anodic porous alumina. Over the past several decades, although studies have been carried out on the protection or design of Al surfaces for commercial applications, little attention has been given to the regularity of the self-organized cell structure in anodic porous

alumina. In 1995, Masuda and Fukuda reported the formation of ordered pore configurations with 100 nm periodicity in anodic porous alumina under optimized anodization conditions [47]. Since then, there has been renewed interest in anodic porous alumina as a template or host structure for the fabrication of several types of nanodevices.

28.2.3 Anodization of Aluminum on Silicon Substrate

In recent years, to apply anodic porous alumina as a template structure for the patterning of metal and semiconductor surfaces, the formation of anodic porous alumina using evaporated or sputtered films of Al on various substrates including semiconductor [3–19] and conductive oxide materials (e.g., indium tin oxide, ITO) [20,21] has been widely studied. In this process, the resulting porous structure can be directly used as a host or a template to fabricate nanoscale devices, and can also be used as an etching mask to transfer a closely packed hexagonal array of pores into various substrates.

To clarify the structural transformation of interface structures during anodization, we studied the anodization behavior of Al films sputtered on a silicon substrate using oxalic acid ($H_2C_2O_4$) solution, a kind of dicarboxylic acid and a commonly used solution for anodization [48]. Figure 28.1b shows a typical scanning electron microscopy (SEM) cross-sectional image of porous alumina. The anodized specimen was mechanically cleaved, producing cracks in the oxide film. The growth of straight parallel channels perpendicular to the substrate can be observed.

Figure 28.2 shows the current density transient for the anodization of an Al film sputtered on a silicon substrate in oxalic acid at 40 V. During the first constant-current stage up to point a, the Al on the silicon substrate was undergoing anodization. Then the current decreased suddenly and a visible change in the color of the $Al_2O_3/Al/Si$ structure was observed (point b). This implies that the Al film sputtered on the silicon substrate was consumed gradually at the silicon surface and changed to a transparent oxide film. When the Al film was consumed completely, the current reached a minimum value (point c). Further anodization

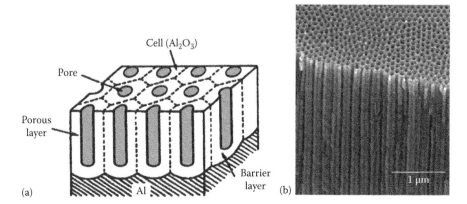

FIGURE 28.1   (a) Geometrical cell model of anodic porous alumina and (b) cross-sectional view of anodic oxide film formed in 0.3 mol dm^{-3} oxalic acid solution at 20°C and at 40 V.

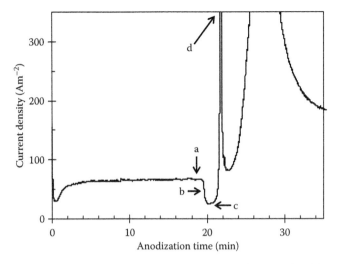

FIGURE 28.2 Current density–time curve for anodization of Al film sputtered on silicon in 0.3 mol dm⁻³ oxalic acid at 20°C and at 40 V. Arrows mark the average anodization times for the images in Figure 28.4a through d. (From Asoh, H. et al., *Appl. Phys. Lett.*, 83, 4408, 2003.)

resulted in a subsequent sharp increase in the current (point d), implying the localized anodization of the underlying silicon substrate. After the first sharp peak of the current, a second broad peak was observed. This indicates that the silicon substrate was anodized over the entire area of the specimen following the destruction of the porous film by extensive gas evolution. When anodization was conducted in sulfuric acid (H_2SO_4) at 25 V, a similar current density transient was confirmed, as shown in Figure 28.3 [49].

The current density transient for the transformation of the nanostructure of anodic porous alumina grown on the silicon substrate is comparable to that observed during the formation of anodic alumina film. After removing the anodic alumina film, the geometric pattern that remained on the silicon substrate was observed by SEM, as shown in Figure 28.4a through d. When the Al film remained on the silicon substrate, a self-ordered cell configuration with scalloped pattern on the Al film was confirmed, as shown in Figure 28.4a. This pattern was in good agreement with the hemispherical barrier layer of the porous film. The interval between the scallops, that is, the size of the cells, was approximately 100 nm, for a formation voltage of 40 V. As the barrier layer of the porous film reached the silicon surface, isolated Al dots were observed at the cell junction as shown in Figure 28.4b. With further anodization, the Al film was consumed completely as indicated by the flat silicon substrate shown in Figure 28.4c. These results imply that the silicon oxide, which can be produced only at the Al_2O_3/Si interface, is expanded radially from the center of each cell followed by an increase in their contacting area and decrease in remaining Al. With overanodization, convex features, which are thought to be silicon oxide, were observed as nanodots at the center of each cell as shown in

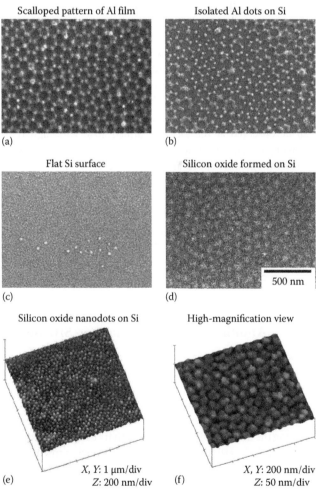

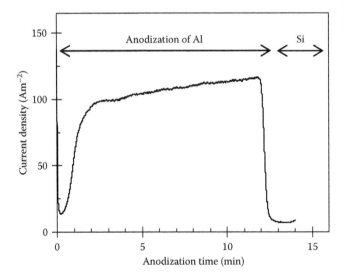

FIGURE 28.3 Current density–time curve for anodization of Al film sputtered on silicon in 0.3 mol dm⁻³ sulfuric acid at 20°C and at 25 V. (From Asoh, H. et al., *Appl. Surf. Sci.*, 252, 1668, 2005.)

FIGURE 28.4 (a–d) SEM images of silicon surface after removal of anodic porous alumina, corresponding to the points marked in Figure 28.2. (From Asoh, H. et al., *Appl. Phys. Lett.*, 83, 4408, 2003.) (e) Low-magnification and (f) high-magnification AFM images of silicon oxide nanodot arrays formed on silicon substrate.

Figure 28.4d. These convex features were arranged hexagonally over the entire area of the specimen, corresponding to the cell configuration of the anodic porous alumina as shown in Figure 28.4e and f.

These results indicate that porous alumina directly formed on a silicon substrate can act as a mask for the localized anodization of the underlying silicon substrate without additional through-hole treatment of the barrier layer, namely, silicon oxide can be produced only in the conductive area between the barrier layer of a porous film and a silicon surface. Chen et al. have reported that nanodot arrays of titanium oxide can be prepared from TiN/Al films on a silicon substrate by the localized anodization of a TiN layer using anodic alumina as a mask [50,51]. Their result indicates that the formation of oxide nanodot arrays, which are formed using alumina as a mask for the localized anodization of the underlying substrate, can be achieved even in the case of using an underlying layer other than a silicon substrate. In brief, the localized anodization of the underlying substrate through a barrier layer of anodic alumina can be regarded as a flexible and reproducible patterning technique.

In addition, the formation mechanism of the silicon oxide pattern on the silicon substrate by localized anodization is similar to that induced by lithography techniques using a conductive probe in a scanning probe microscope (SPM) [52–55]. In the case of SPM lithography using anodization, when a negative bias is applied to the tip, a local anodization process is carried out on the silicon surface when moist air is passed between the tip and the sample. As a result, the growth of silicon oxide can be observed at the locally scanned area. Although there are many advantages of using SPM lithography, improvements in the lithographic speed and the area over which fabrication can be performed are necessary for the realization of industrial applications. On the other hand, our proposed method based on the templating of self-ordered anodic porous alumina has the advantage of possible use for large-area patterning in a single step. In principle, there is no limitation to the area that can be patterned, because it is unnecessary to use a special operating system such as a piezoelectric scanner.

28.2.4 Transfer of Nanoporous Pattern of Anodic Porous Alumina onto Silicon

To examine the pattern transfer of anodic porous alumina onto a silicon substrate, an anodized specimen was immersed in HF solution to remove the silicon oxide at the center of each cell. In the case of chemical etching in HF solution, the selective removal of silicon oxide from the silicon substrate was easily realized by utilizing the difference in chemical reactivity between silicon and silicon oxide for HF treatment [48]. After removing silicon oxide by wet chemical etching in 47 wt% HF for 90 s, an array of shallow nanoholes was observed on the silicon substrate by atomic force microscopy (AFM), as shown in Figure 28.5. The anodization was stopped immediately at the minimum current density (point c in Figure 28.2). The diameter and depth of the concave features were 60–80 and ~10 nm, respectively. The arrangement and shape of this hole array were similar to those of the porous

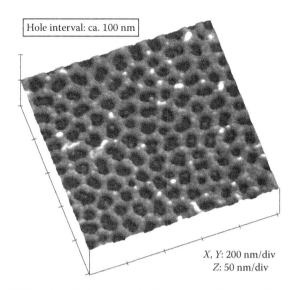

Hole interval: ca. 100 nm

X, Y: 200 nm/div
Z: 50 nm/div

FIGURE 28.5 AFM image of silicon nanohole array fabricated by selective removal of silicon oxide in HF. (From Oide, A. et al., *Electrochem. Solid-State Lett.*, 8, G172, 2005.)

pattern of the upper anodic porous alumina. Namely, the transfer of the nanoporous pattern of anodic alumina onto the silicon substrate was achieved by removing the silicon oxide, which was produced by the anodic oxidation of the local part of the silicon substrate underneath the barrier layer corresponding to the pore base, as schematically shown in Figure 28.6.

The periodicity of holes in an array on a silicon substrate is basically determined by the pore interval of the upper anodic porous alumina, which is known to be strongly dependent on anodization conditions, such as the type of electrolyte and the formation voltage. Namely, the dimensions of the resultant nanohole array can be adjusted easily by controlling the anodization conditions. In fact, nanohole arrays with a 60 nm hole periodicity were fabricated on a silicon substrate by anodizing an Al film in sulfuric acid at 25 V and subsequent chemical etching in HF, as shown in Figure 28.7. More detailed results on the reduction of the dimensions of nanopatterns are reported in one of our previous papers [49].

28.2.5 Fabrication of Silicon Nanocolumn Arrays

On the basis of the above patterning mechanism, the reverse patterning of a silicon substrate can be achieved by chemical etching in potassium hydroxide (KOH) solution using silicon oxide patterns formed by anodization as etching masks. Several studies have been performed on the direct patterning of silicon by SPM lithography [53] and nanoelectrode lithography [56,57]. These techniques also consist of two continuous processes, that is, localized anodization and subsequent chemical etching. Because the dissolution rate of silicon in KOH solution is much faster than that in silicon oxide, the exposed silicon surface can be etched selectively. Similarly to these techniques, our proposed patterning process has the potential to control

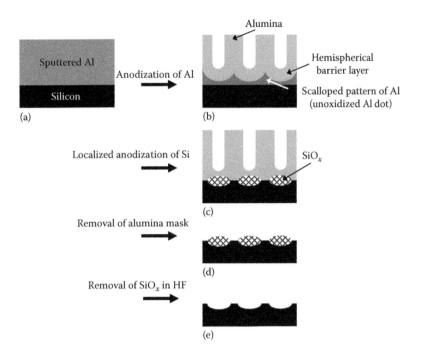

FIGURE 28.6 Schematic of patterning process by localized anodization of silicon: (a) Al film sputtered on silicon substrate; (b,c) transformation of anodic porous alumina grown on silicon substrate during anodization; (d) silicon oxide nanodot arrays formed after removal of alumina film; (e) nanohole arrays formed on silicon after chemical etching in HF. (From Asoh, H. et al., *Appl. Surf. Sci.*, 252, 1668, 2005.)

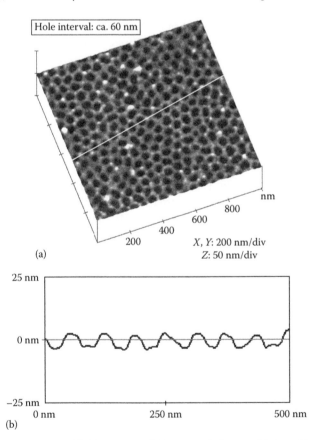

FIGURE 28.7 (a) AFM image of silicon nanohole array fabricated by selective removal of silicon oxide in HF and (b) line-scan image of cross section along white line in (a). (From Asoh, H. et al., *Appl. Surf. Sci.*, 252, 1668, 2005.)

the reverse geometric pattern of semiconductor materials at the nanometer scale using a chemical reaction with a positive or negative resist.

To examine the ability of silicon oxide patterns fabricated by localized anodization to function as wet-etching masks, anodized silicon was immersed in 1 wt% KOH solution for 20 min [58]. The anodization was stopped when the current density decreased (point b in Figure 28.2). Figure 28.8a shows a typical AFM image of the silicon column array. The cross-section analysis of the AFM image revealed that the convex features have a top diameter of 60 nm, a bottom diameter of 100 nm, and a height of ~20 nm (Figure 28.8b). This result indicates that the silicon oxide pattern formed on a silicon substrate can act as a mask for the chemical etching of the underlying silicon substrate. From the AFM image in Figure 28.8a, it was confirmed that the silicon surface was preferentially dissolved at the boundary between adjacent cells. In addition, the arrangement of this convex pattern was in agreement with that of the hole array structure shown in Figure 28.5. The most noteworthy point is the reversibility of the patterning. The silicon column array obtained by chemical etching in KOH solution had an inverse structure to that of the silicon nanohole array formed on the silicon substrate using HF treatment. The key factor for the fabrication of column arrays seems to be the limitation of the dimensions of the silicon oxide produced by the localized anodization in addition to the subsequent etching conditions. This is because when anodization was stopped at the minimum current density (point c in Figure 28.2) similar to the case of the hole array formation, no column array was obtained even after KOH treatment. On the basis of this approach, two nanostructures with positive and

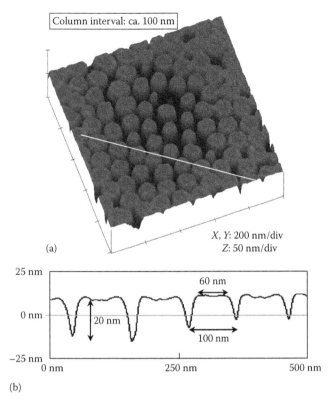

(a)   X, Y: 200 nm/div
       Z: 50 nm/div

(b)

FIGURE 28.8 (a) AFM image of silicon column array obtained by selective etching of silicon in KOH and (b) line-scan image of cross section along white line in (a). (From Oide, A. et al., *Electrochem. Solid-State Lett.*, 8, G172, 2005.)

negative patterns can be obtained arbitrarily by the selection of appropriate local anodization time and etching conditions.

28.2.6 Electrochemical Etching of Silicon through Anodic Porous Alumina

Silicon with a regular porous structure of submicron to nanometer order has received much attention owing to its optical properties, such as photonic band gap (i.e., a frequency range

where photons are not allowed to propagate in any direction) [59,60]. The techniques commonly used in fabricating porous silicon with a regular macropore structure involve a combination of standard lithography and electrochemical etching in HF [61–63]. However, many patterning procedures, such as electron-beam lithography, are costly and are not suitable for the production of an ordered nanohole array structure over a large area (more than 1 cm²).

Using our process described in the previous section, the transfer of the nanoporous pattern of anodic porous alumina onto a silicon substrate can be achieved by the selective removal of silicon oxide from the silicon substrate by wet chemical etching in HF solution. However, the obtained porous structure is an extremely shallow nanohole array with a depth of ~10 nm. Therefore, we attempted to fabricate ordered nanohole arrays with high-aspect ratio structures onto a silicon substrate by electrochemical etching through anodic porous alumina [64].

After electrochemical etching through anodic porous alumina, the geometric pattern on the silicon substrate was observed by AFM. The experimental process was reported in detail in one of our previous papers [64]. Figure 28.9a shows an AFM image of a typical nanohole array in a silicon substrate. The periodicity of the obtained hole array, which was basically determined by the pore interval of the upper anodic porous alumina, was approximately 100 nm, when the formation voltage was 40 V in oxalic acid. The holes in the array were arranged hexagonally over the entire area of the specimen corresponding to the pore configuration of anodic porous alumina. This result indicates that the porous alumina formed on a silicon substrate can act as a mask, even for localized electrochemical etching in HF. That is, etch pits are only generated on the exposed silicon surface through anodic porous alumina. Although some dissolution of the oxide wall of alumina due to chemical etching in HF was observed during the electrochemical etching of silicon, it was confirmed that the as-anodized oxide film could act as a mask if the etching time was shorter than 60 s under these etching conditions. Prolonged immersion in HF caused the destruction of the alumina mask and induced the formation of disordered

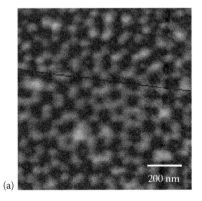
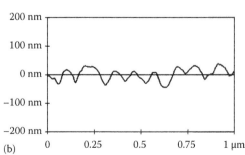

(a)   (b)

FIGURE 28.9 (a) AFM image of surface of silicon nanopore array after removal of anodic porous alumina and (b) typical height profile. The silicon substrate was electrochemically etched in HF:H₂O:isopropanol (3:9:26 wt%) solution at a current density of 7.6 mA cm⁻² for 50 s. (From Asoh, H. et al., *Electrochem. Commun.*, 7, 953, 2005.)

pores. The cross-sectional analysis of the AFM image reveals that the diameter and depth of typical pores were ~100 and ~50 nm, respectively (Figure 28.9b). The depth of the obtained nanohole array was increased to approximately fivefold that of the array obtained using chemical etching shown in Figure 28.5. On the basis of the present method, it is thought that a nanohole array with a high aspect ratio can be fabricated because etch pit generation and pore formation proceed uniformly at regular intervals.

To produce a nanohole array with a high aspect ratio, it is necessary to improve the electrochemical etching conditions such as the type of acid, electrolyte temperature, current density, substrate parameters, resistivity, doping density, and illumination intensity. Further experiments are now in progress to fabricate novel devices based on the specific optical properties of porous silicon/silicon electrodes.

28.3 Fabrication of Nano-/Microhole Arrays on Silicon Using Metal-Assisted Chemical Etching

28.3.1 Formation of Porous Silicon by Metal-Assisted Chemical Etching

Regarding Si microstructures, silicon with a regular porous structure of submicron to nanometer order has received much attention as described above. The electrochemical etching of silicon in HF is a promising technique for micromachining and has been applied to the fabrication of three-dimensional (3D) silicon structures, such as pillars, tubes, and macropores [61,65,66]. However, many patterning procedures using conventional lithography are costly and not suitable for the production of an ordered porous structure over a large area required for industrial applications. In terms of the simplicity and efficiency of the fabrication process, the metal-assisted chemical etching of silicon substrates has been developed since 2000 for applications in silicon-based optoelectronics [67,68]. In these studies, microporous silicon layers with a thickness of ~3 μm, which were formed in HF with hydrogen peroxide (H_2O_2) [67] and HF without H_2O_2 [68], were examined focusing on the light-emitting properties and conversion efficiency of solar cells containing the porous silicon layers. In 2005, Tsujino and Matsumura reported that deep cylindrical nanoholes with a depth greater than 100 μm in silicon can be obtained by immersing a silicon substrate loaded with Ag nanoparticles in a solution containing HF and H_2O_2 for a long time [69]. The common result of applying these methods is that a microporous silicon layer is formed without external bias [70–77]. In the following section, we describe the fabrication of ordered silicon microstructures, including hole arrays, by a combination of colloidal crystal templating and metal-assisted chemical etching using patterned noble metals such as Ag and Pt–Pd as a catalyst.

28.3.2 Formation of Silicon Column Arrays Using Colloidal Crystal Templating

The nano-/micropatterning of solid substrates using self-assembled colloidal particles as a mask, which is often referred to as colloidal lithography or nanosphere lithography, has attracted considerable attention as a key fabrication method owing to its relative simplicity and low cost [78–82]. Therefore, we have focused on the usefulness of electrochemical processes for fabricating an ordered pattern using colloidal crystals as a mask, and we proposed a novel method for fabricating a size- and shape-controlled nano-/micropattern on a silicon substrate based on a combination of colloidal crystal templating and anodization [83,84]. We then continued our preliminary work and proposed a novel micromachining technique based on a combination of colloidal crystal templating and metal-assisted chemical etching.

The principle of pattern transfer for fabricating column and hole arrays is schematically shown in Figure 28.10 [85–88].

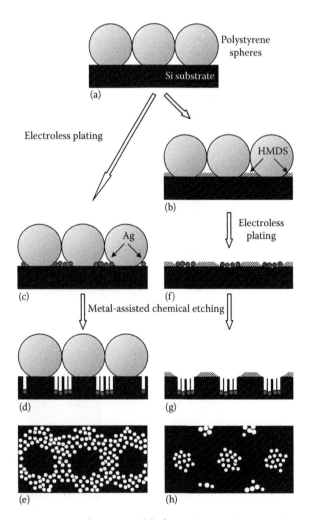

FIGURE 28.10 Schematic model of site-selective chemical etching of silicon: (a) colloidal crystals on silicon substrate; (b) HMDS coating; (c,f) electroless plating; (d,g) chemical etching of silicon using Ag particles as catalyst; (e,h) top view after removal of colloidal crystals. (From Asoh, H. et al., *Electrochem. Commun.*, 9, 535, 2007.)

Figure 28.11a and b shows scanning electron microscope (SEM) images of Ag particles deposited on a silicon substrate. In the case without a mask, the fine Ag particles spread out over the whole silicon surface, as shown in Figure 28.11a. The sizes and distribution of the particles on the silicon surface are in agreement with previous results [69]. On the other hand, when a 2D hexagonal array of polystyrene spheres of 3 μm diameter was used as a direct mask during electroless plating (i.e., metal deposition that occurs without the use of external electric source), a Ag honeycomb pattern was obtained, as shown in Figure 28.11b. Using colloidal crystals as a mask, Ag particles were deposited selectively among the spheres. In other words, selective metal deposition can proceed only on the exposed parts of the silicon surface, which are located in the voids among the spheres on the silicon substrate. This result indicates that colloidal crystals can act as a mask for localized electroless plating in aqueous solution. The details of the electroless plating process were reported in one of our previous papers [89]. In Figure 28.11b, the center-to-center distance between the holes in the Ag honeycomb pattern, which was basically determined by the diameter of the polystyrene spheres, was approximately 3 μm. The framework of the Ag honeycomb pattern was composed of an aggregation of fine Ag particles. The sizes of the Ag particles

were mostly between 50 and 100 nm, but were scattered in the range from 10 to 250 nm.

After the deposition of the Ag particles on the silicon substrate, the specimens were immersed in a mixed solution of HF and H_2O_2 to form porous silicon by metal-assisted chemical etching. Figure 28.11c shows an SEM image of the etched silicon surface using a patterned Ag catalyst. The periodicity of the obtained silicon convex arrays was approximately 3 μm, corresponding to the diameter of the polystyrene spheres used as a mask for electroless plating. The configuration of silicon convex arrays, which were arranged hexagonally over the entire area of the specimen, has an inverse relation to the honeycomb pattern of the Ag particles, as shown in Figure 28.11b. That is, chemical etching proceeds only on the Ag-coated Si surface, in agreement with the proposed mechanism [67,70] and, consequently, the contact area between the polystyrene spheres and the underlying Si substrate has a disklike shape. However, prolonged chemical etching in a HF-containing solution caused the destruction of silicon microstructures owing to the excessive dissolution of the horizontal plane. Nevertheless, deep straight holes that had grown vertically downward from the surface and were deeper than 30 μm were found in a fractured cross section, as shown in Figure 28.12 [87]. The diameter

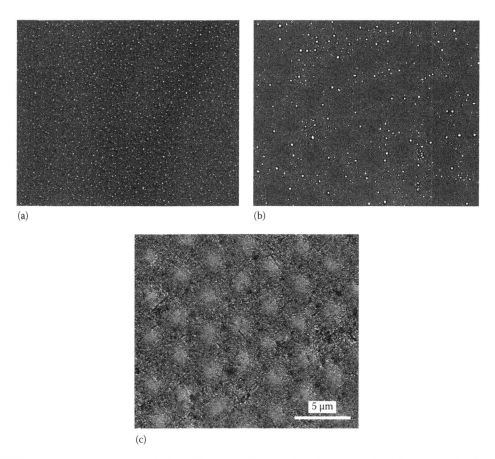

(a)

(b)

(c)

FIGURE 28.11 SEM images of Ag particles deposited on silicon: (a) without mask and (b) with colloidal crystal mask. Electroless plating was conducted in AgClO$_4$/NaOH for 20 min. (c) SEM image of surface of silicon etched in 5 mol dm^{-3} HF/1 mol dm^{-3} H$_2$O$_2$ for 1 min. (From Asoh, H. et al., *Electrochem. Commun.*, 9, 535, 2007.)

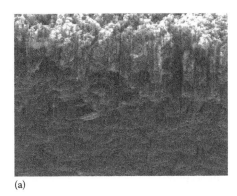
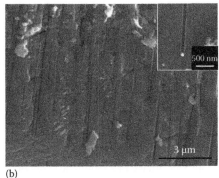

FIGURE 28.12 SEM images of a fractured section of the silicon substrate obtained after etching specimen with the Ag honeycomb pattern shown in Figure 28.11 for 30 min: (a) near-surface region and (b) intermediate area revealing deep straight holes that had grown vertically downward from the silicon surface. The inset shows a deep hole with an Ag particle at the tip. (From Ono, S., et al., *Electrochim. Acta*, 52, 2898, 2007.)

of the long holes appears to be approximately 100 nm, which is in agreement with the size of the Ag particles. Ag particles were found at the tip of straight long holes as shown in the inset in Figure 28.12b, similar to those described in an earlier report [69].

A mechanism involving a localized electrochemical process has been proposed to explain the mechanism of metal-assisted chemical etching as follows [67,70]:

Cathode reaction (at noble metal as a local cathode):

$$H_2O_2 + 2H^+ \rightarrow 2H_2O + 2h^+,$$

$$2H^+ \rightarrow H_2\uparrow + 2h^+.$$

Anode reaction (at silicon surface):

$$Si + 4h^+ + 4HF \rightarrow SiF_4 + 4H^+,$$

$$SiF_4 + 2HF \rightarrow H_2SiF_6.$$

Overall reaction:

$$Si + H_2O_2 + 6HF \rightarrow 2H_2O + H_2SiF_6 + H_2\uparrow.$$

The formation of porous silicon by metal-assisted chemical etching can proceed not only using Ag particles but also using other noble metals, such as Au, Pt, Au–Pd, and Pt–Pd.

28.3.3 Formation of Silicon Hole Arrays by Colloidal Crystal Templating

To expand the range of applications of ordered silicon microstructures, it is necessary to control the silicon surface morphology. By colloidal crystal templating, it is possible to fabricate negative and positive patterns by changing the configuration of the Ag particles used as a catalyst. To fabricate a metal pattern that is the reverse of the Ag honeycomb pattern shown in Figure 28.11b, two-step replication was applied. First, a colloidal crystal mask was formed on the silicon substrate, as described above. Second, specimens were placed in hexamethyldisilazane (HMDS) vapor overnight (Figure 28.10b). HMDS is a popular reagent for forming hydrophobic surfaces based on the immobilization of trimethylsilyl groups on the surface [90,91], namely, areas of HMDS-coated silicon exhibit hydrophobicity and are thought to inhibit Ag deposition. Finally, electroless plating and chemical etching were conducted as described above (Figure 28.10f through h).

Figure 28.13a shows the isolated patterns of Ag particles deposited by selective electroless plating using HMDS-coated

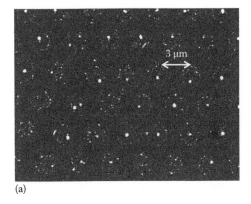
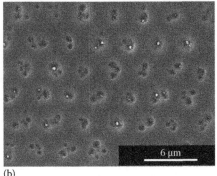

FIGURE 28.13 (a) Isolated patterns of Ag particles formed on silicon and (b) silicon nanohole array. The electroless plating and etching conditions were the same as those for Figure 28.11. (From Asoh, H. et al., *Electrochem. Commun.*, 9, 535, 2007.)

silicon. The deposition is restricted to well-defined bare silicon surfaces, and does not occur on the HMDS-coated silicon. This result indicates that the HMDS-coated areas, which are located in the voids among the spheres on the silicon substrate, possess sufficient hydrophobicity and can act as a mask for localized electroless plating in aqueous solution. In Figure 28.13a, the center-to-center distance between the island microarrays of Ag particles was approximately 3 μm. The isolated Ag patterns were composed of an aggregation of Ag particles with sizes in the range of 50–100 nm.

Figure 28.13b shows an SEM image of a nanohole array on a silicon substrate. The aggregation of nanoholes with sizes in the range of 50–100 nm was arranged hexagonally over the entire area of the specimen, corresponding to the 2D hexagonal array of polystyrene beads used as the original mask for the formation of the HMDS honeycomb pattern. Some Ag particles, which were detected as bright circular spots, were observed at the bottom of the pores due to the short etching time. The sizes of the particles observed in Figure 28.13b coincided with those of the deposited Ag particles shown in Figure 28.13a. These results indicate that chemical etching proceeds only on the Ag-coated silicon surface, and consequently the HMDS-coated silicon parts remain in a honeycomb pattern.

Figure 28.14 shows silicon hole arrays with different periodicity after site-selective chemical etching using isolated Ag

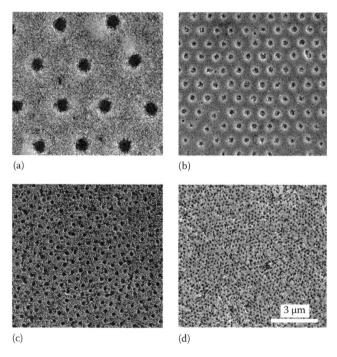

(a)   (b)

(c)   (d)

3 μm

FIGURE 28.14 SEM images of silicon hole arrays after removal of mask: (a) 3 μm periodicity, (b) 1 μm periodicity, (c) 500 nm periodicity, and (d) 200 nm periodicity. Electroless plating was conducted in 10^{-3} mol dm^{-3} AgClO$_4$ and 10^{-3} mol dm^{-3} NaOH for 20 min. The chemical etching times in 5 mol dm^{-3} HF/1 mol dm^{-3} H$_2$O$_2$ were (a) 5 min and (b–d) 30 s. (From Asoh, H. et al., *ECS Trans.*, 6, 431, 2007; Arai, F. et al., *Electrochemistry*, 76, 187, 2008.)

patterns. The magnification of each image was the same. The periodicity of the holes was basically determined by the diameter of the polystyrene spheres used as a mask. In each case, chemical etching proceeds only on the Ag-coated silicon surface, and consequently the areas of HMDS-coated silicon remain in a honeycomb pattern. The shortest hole periodicity, which was attained by the optimization of the etching time, was approximately 200 nm, as shown in Figure 28.14d. This indicates that the formation of silicon hole arrays with a periodicity of less than 1 μm can be achieved by the process described in this section.

28.3.4 Silicon Microwell Arrays

By using circular metal thin films as a catalyst instead of metal nanoparticles, the formation of silicon microwells with micrometer-scale openings was also achieved by metal-assisted chemical etching. The experimental process was reported in detail in one of our previous papers [92]. After the deposition of Pt–Pd thin films on a silicon substrate, the specimens were immersed in a mixed solution of HF and H$_2$O$_2$ to etch the silicon substrate by metal-assisted chemical etching. The etching conditions were the same as those using Ag nanoparticles as a catalyst (Figure 28.11). During chemical etching, the central part of the silicon substrate surrounded by the honeycomb mask gradually sagged downward. Figure 28.15a through c shows plane-view and tilted-view SEM images of the silicon surface etched using the patterned Pt–Pd catalyst.

From the tilted view shown in Figure 28.15c, it was confirmed that the pores were conical. The diameter of the opening of each silicon microwell was approximately 3 μm due to the chemical dissolution of the horizontal plane. The depth of each silicon microwell was estimated to be approximately 2 μm. The crest and side walls of the silicon microwells were extremely smooth. In addition, the most noteworthy point is that the circular Pt–Pd thin films used as the catalyst remained at the bottom of each well. The SEM images shown in Figure 28.15b and c revealed that the residual detected as bright contrast was the Pt–Pd catalyst.

Figure 28.16 shows the relationship between the pore depth and the etching time using patterned noble-metal thin films as a catalyst [93]. In addition to the results for the metal-assisted chemical etching carried out using the honeycomb mask, the results obtained without using the honeycomb mask are also plotted for each metal species such as Pt–Pd, Au, and Pt. Etching rate increases in the following order: Au < Pt ≤ Pt–Pd. If the metal catalyst species are the same, the etching rate without using the mask is faster than that obtained using the mask. Etching rate and the morphology of the resultant porous structure were assumed to be affected by the difference in the shape of metal catalyst and the diffusion behavior of injected positive holes at the silicon–metal interface.

Figure 28.17 shows macroporous silicon with a high aspect ratio formed using circular Pt–Pd thin films as a catalyst.

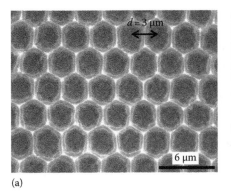
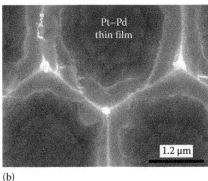
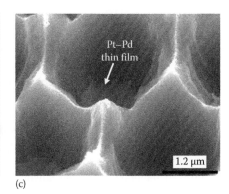

(a)   (b)   (c)

FIGURE 28.15 (a) SEM image of silicon microwell array containing Pt–Pd film with 3 μm periodicity, (b) high-magnification view, and (c) tilted (45°) view. Chemical etching was conducted in 5 mol dm⁻³ HF/1 mol dm⁻³ H₂O₂ for 3 min. (From Asoh, H. et al., *Appl. Phys. Exp.*, 1, 067003, 2008.)

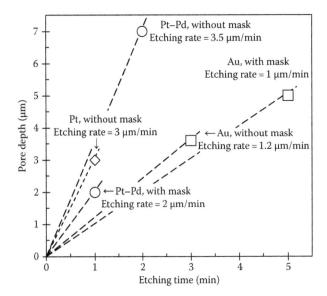

FIGURE 28.16 Relationship between depth of pores formed by metal-assisted chemical etching and corresponding etching time. (Redrawn from Asoh, H. et al., *Electrochim. Acta*, 54, 5142, 2009.)

When metal-assisted chemical etching was conducted in HF with a high concentration of 10 mol dm⁻³, the morphology of the resultant porous structure was significantly different from that of the silicon microwells formed in HF with a relatively low concentration of 5 mol dm⁻³ [94]. In the case of low-concentration HF, injected positive holes are expected to diffuse into silicon bulk and oxidize silicon at locations away from the metal-coated silicon surface. On the other hand, in the case of high-concentration HF, the diffusion of positive holes is thought to be suppressed. One of the notable features of the obtained porous structure is that the diameter of each pore was hardly increased during chemical etching. Namely, the dissolution of silicon oxide is accelerated locally at the silicon–metal interface in the direction of the pore depth, resulting in the formation of macroporous silicon with a high aspect ratio. Further research on metal-assisted chemical etching using metal thin films as a catalyst would clarify the relationship between the mechanism for controlling the morphology of the resultant pattern and the etching conditions, such as the composition and concentration of etchant, substrate parameters, resistivity, and doping density.

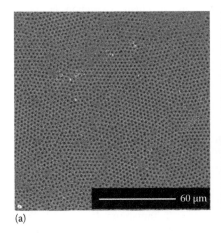

(a)   (b)

FIGURE 28.17 (a) Top-view and (b) cross-sectional SEM images of macroporous silicon formed by metal-assisted chemical etching. Chemical etching was conducted in 10 mol dm⁻³ HF/1 mol dm⁻³ H₂O₂ for 2 min using a Pt–Pd catalyst.

28.4 Summary

In this chapter, we have mainly focused on the fabrication of nanohole arrays on a silicon substrate by a combination of an electrochemistry technique and templating. The regularity and periodicity of hole arrays in the silicon substrate is basically determined by the geometry of the mask, that is, the anodic porous alumina and colloidal crystals used in the initial template. Therefore, improving the regularity of the mask structure may be essential for expanding the range of applications of ordered nanohole arrays on silicon. In addition, some of the recent applications discussed herein have the advantage that they can be used for large-area patterning using a simple electrochemical process with high throughput and low cost, which cannot be achieved using conventional lithographic techniques.

We are also now studying the formation of nano-/microhole arrays on compound semiconductors, such as GaAs [95,96] and InP, with the expectation that these hole arrays can be used for novel electronic devices and chemical-sensing applications. Such studies will be very useful for developing new functional materials. In addition to device applications, there is strong interest in the fundamental study of semiconductor electrochemistry. Finally, we are convinced that the silicon hole arrays and various fabrication techniques described here will be increasingly applied in the near future to many research fields that require an ordered surface morphology or a periodic porous structure with dimensions ranging from the submicrometer to nanometer scale.

Acknowledgments

This work was partly financially supported by a Grant-in-Aid for Scientific Research from the Japan Society for the Promotion of Science and the Light Metal Education Foundation of Japan. Thanks are also due to the "High-Tech Research Center" Project for Private Universities: matching fund subsidy from the Ministry of Education, Culture, Sports, Science, and Technology.

References

1. X. G. Zhang, *Electrochemistry of Silicon and Its Oxides*, Kluwer Academic/Plenum Publishers, New York (2001).
2. V. Lehmann, *Electrochemistry of Silicon*, Wiley-VCH, Weinheim, Germany (2002).
3. S. Gavrilov, S. Lemeshko, V. Shevyakov, and V. Roschin, *Nanotechnology*, **10**, 213 (1999).
4. S. Shingubara, O. Okino, Y. Sayama, H. Sakaue, and T. Takahagi, *Solid-State Electron.*, **43**, 1143 (1999).
5. T. Iwasaki, T. Motoi, and T. Den, *Appl. Phys. Lett.*, **75**, 2044 (1999).
6. D. Crouse, Y.-H. Lo, A. E. Miller, and M. Crouse, *Appl. Phys. Lett.*, **76**, 49 (2000).
7. S. Shingubara, O. Okino, Y. Murakami, H. Sakaue, and T. Takahagi, *J. Vac. Sci. Technol. B*, **19**, 1901 (2001).
8. W. Hu, D. Gong, Z. Chen, L. Yuan, K. Saito, C. A. Grimes, and P. Kichambare, *Appl. Phys. Lett.*, **79**, 3083 (2001).
9. H. Masuda, K. Yasui, Y. Sakamoto, M. Nakao, T. Tamamura, and K. Nishio, *Jpn. J. Appl. Phys.*, **40**, L1267 (2001).
10. S. Shingubara, Y. Murakami, H. Sakaue, and T. Takahagi, *Jpn. J. Appl. Phys.*, **41**, L340 (2002).
11. A. Cai, H. Zhang, H. Hua, and Z. Zhang, *Nanotechnology*, **13**, 627 (2002).
12. M. T. Wu, I. C. Leu, J. H. Yen, and M. H. Hon, *Electrochem. Solid-State Lett.*, **7**, C61 (2004).
13. M. T. Wu, I. C. Leu, and M. H. Hon, *J. Mater. Res.*, **19**, 888 (2004).
14. L. Pu, Y. Shi, J. M. Zhu, X. M. Bao, R. Zhang, and Y. D. Zheng, *Chem. Commun.*, **8**, 942 (2004).
15. N. V. Myung, J. Lim, J.-P. Fleurial, M. Yun, W. West, and D. Choi, *Nanotechnology*, **15**, 833 (2004).
16. H. Shiraki, Y. Kimura, H. Ishii, S. Ono, K. Itaya, and M. Niwano, *Appl. Surf. Sci.*, **237**, 369 (2004).
17. F. Müller, A.-D. Müller, S. Schulze, and M. Hietschold, *J. Mater. Sci.*, **39**, 3199 (2004).
18. M. Kokonou, A. G. Nassiopoulou, and K. P. Giannakopoulos, *Nanotechnology*, **16**, 103 (2005).
19. M. Tian, S. Xu, J. Wang, N. Kumar, E. Wertz, Q. Li, P. M. Campbell, M. H. W. Chan, and T. E. Mallouk, *Nano Lett.*, **5**, 697 (2005).
20. S. Z. Chu, K. Wada, S. Inoue, and S. Todoroki, *J. Electrochem. Soc.*, **149**, B321 (2002).
21. S. Z. Chu, K. Wada, S. Inoue, and S. Todoroki, *Chem. Mater.*, **14**, 266 (2002).
22. S. Kawai and R. Ueda, *J. Electrochem. Soc.*, **122**, 32 (1975).
23. H. Daimon, O. Kitakami, O. Inagoya, and A. Sakemoto, *Jpn. J. Appl. Phys.*, **30**, 282 (1991).
24. D. Al-Mawlawi, N. Coombs, and M. Moskovits, *J. Appl. Phys.*, **70**, 4421 (1991).
25. C. A. Huber, T. E. Huber, M. Sadoqi, J. A. Lubin, S. Manalis, and C. B. Prater, *Science*, **263**, 800 (1994).
26. M. Zheng, L. Menon, H. Zeng, Y. Liu, S. Bandyopadhyay, R. D. Kirby, and D. J. Sellmyer, *Phys. Rev. B*, **62**, 12 282 (2000).
27. K. Nielsch, R. B. Wehrspohn, J. Barthel, J. Kirschner, U. Gösele, S. F. Fischer, and H. Kronmüller, *Appl. Phys. Lett.*, **79**, 1360 (2001).
28. M. Saito, M. Kirihara, T. Taniguchi, and M. Miyagi, *Appl. Phys. Lett.*, **55**, 607 (1989).
29. C. K. Preston and M. Moskovits, *J. Phys. Chem.*, **97**, 8495 (1993).
30. Y. Du, W. L. Cai, C. M. Mo, J. Chen, L. D. Zhang, and X. G. Zhu, *Appl. Phys. Lett.*, **74**, 2951 (1999).
31. H. Masuda, M. Ohya, H. Asoh, M. Nakao, M. Nohtomi, and T. Tamamura, *Jpn. J. Appl. Phys.*, **38**, L1403 (1999).
32. Y. Li, G. W. Meng, L. D. Zhang, and F. Phillip, *Appl. Phys. Lett.*, **76**, 2011 (2000).

33. K. Itaya, S. Sugawara, K. Arai, and S. Saito, *J. Chem. Eng. Jpn.*, **17**, 514 (1984).

34. M. Konno, M. Shindo, S. Sugawara, and S. Saito, *J. Membr. Sci.*, **37**, 193 (1988).

35. S. K. Dalvie and R. E. Baltus, *J. Membr. Sci.*, **71**, 247 (1992).

36. T. Sano, N. Iguchi, K. Iida, T. Sakamoto, M. Baba, and H. Kawaura, *Appl. Phys. Lett.*, **83**, 4438 (2003).

37. F. Matsumoto, K. Nishio, T. Miyasaka, and H. Masuda, *Jpn. J. Appl. Phys.*, **43**, L640 (2004).

38. H. Masuda, K. Nishio, and N. Baba, *Thin Solid Films*, **223**, 1 (1993).

39. D. Al-Mawlawi, C. Z. Liu, and M. Moskovits, *J. Mater. Res.*, **9**, 1014 (1994).

40. D. Routkevitch, T. Bigioni, M. Moskovits, and J. M. Xu, *J. Phys. Chem.*, **100**, 14037 (1996).

41. P. Hoyer, K. Nishio, and H. Masuda, *Thin Solid Films*, **286**, 88 (1996).

42. J. Liang, H. Chik, A. Yin, and J. Xu, *J. Appl. Phys.*, **91**, 2544 (2002).

43. M. S. Sander, A. L. Prieto, R. Gronsky, T. Sands, and A. M. Stacy, *Adv. Mater.*, **14**, 665 (2002).

44. F. Keller, M. S. Hunter, and D. L. Robinson, *J. Electrochem. Soc.*, **100**, 411 (1953).

45. J. W. Diggle, T. C. Downie, and C. W. Goulding, *Chem. Rev.*, **69**, 365 (1969).

46. J. P. O'Sullivan and G. C. Wood, *Proc. R. Soc. Lond. A*, **317**, 511 (1970).

47. H. Masuda and K. Fukuda, *Science*, **268**, 1466 (1995).

48. H. Asoh, M. Matsuo, M. Yoshihama, and S. Ono, *Appl. Phys. Lett.*, **83**, 4408 (2003).

49. H. Asoh, A. Oide, and S. Ono, *Appl. Surf. Sci.*, **252**, 1668 (2005).

50. P.-L. Chen, C.-T. Kuo, T.-G. Tsai, B.-W. Wu, C.-C. Hsu, and F.-M. Pan, *Appl. Phys. Lett.*, **82**, 2796 (2003).

51. P.-L. Chen, C.-T. Kuo, F.-M. Pan, and T.-G. Tsai, *Appl. Phys. Lett.*, **84**, 3888 (2004).

52. E. S. Snow, P. M. Campbell, and P. J. McMarr, *Appl. Phys. Lett.*, **63**, 749 (1993).

53. H. Sugimura, T. Yamamoto, N. Nakagiri, M. Miyashita, and T. Onuki, *Appl. Phys. Lett.*, **65**, 1569 (1994).

54. M. Ara, H. Graaf, and H. Tada, *Appl. Phys. Lett.*, **80**, 2565 (2002).

55. M. Cavallini, P. Mei, F. Biscarini, and R. Garcia, *Appl. Phys. Lett.*, **83**, 5286 (2003).

56. A. Yokoo, *Jpn. J. Appl. Phys.*, **42**, L92 (2003).

57. A. Yokoo and S. Sasaki, *Jpn. J. Appl. Phys.*, **44**, 1119 (2005).

58. A. Oide, H. Asoh, and S. Ono, *Electrochem. Solid-State Lett.*, **8**, G172 (2005).

59. U. Grüning, V. Lehmann, and C. M. Engehardt, *Appl. Phys. Lett.*, **66**, 3254 (1995).

60. U. Grüning, V. Lehmann, S. Ottow, and K. Busch, *Appl. Phys. Lett.*, **68**, 747 (1996).

61. V. Lehmann and H. Föll, *J. Electrochem. Soc.*, **137**, 653 (1990).

62. V. Lehmann, *J. Electrochem. Soc.*, **140**, 2836 (1993).

63. H. W. Lau, G. J. Parker, R. Greef, and M. Hölling, *Appl. Phys. Lett.*, **67**, 1877 (1995).

64. H. Asoh, K. Sasaki, and S. Ono, *Electrochem. Commun.*, **7**, 953 (2005).

65. P. Kleimann, J. Linnros, and R. Juhasz, *Appl. Phys. Lett.*, **79**, 1727 (2001).

66. S. Matthias, F. Müller, C. Jamois, R. B. Wehrspohn, and U. Gösele, *Adv. Mater.*, **16**, 2166 (2004).

67. X. Li and P. W. Bohn, *Appl. Phys. Lett.*, **77**, 2572 (2000).

68. S. Yae, Y. Kawamoto, H. Tanaka, N. Fukumuro, and H. Matsuda, *Electrochem. Commun.*, **5**, 632 (2003).

69. K. Tsujino and M. Matsumura, *Adv. Mater.*, **17**, 1045 (2005).

70. S. Chattopadhyay, X. Li, and P. W. Bohn, *J. Appl. Phys.*, **91**, 6134 (2002).

71. S. Chattopadhyay and P. W. Bohn, *J. Appl. Phys.*, **96**, 6888 (2004).

72. S. Cruz, A. Hönig-d'Orville, and J. Müller, *J. Electrochem. Soc.*, **152**, C418 (2005).

73. K. Tsujino and M. Matsumura, *Electrochem. Solid-State Lett.*, **8**, C193 (2005).

74. K. Tsujino, M. Matsumura, and Y. Nishimoto, *Sol. Energy Mater. Sol. Cells*, **90**, 100 (2006).

75. K. Tsujino and M. Matsumura, *Electrochim. Acta*, **53**, 28 (2007).

76. T. Hadjersi, *Appl. Surf. Sci.*, **253**, 4156 (2007).

77. C. Chartier, S. Bastide, and C. Lévy-Clément, *Electrochim. Acta*, **53**, 5509 (2008).

78. H. W. Deckman and J. H. Dunsmuir, *Appl. Phys. Lett.*, **41**, 377 (1982).

79. J. C. Hulteen and R. P. V. Duyne, *J. Vac. Sci. Technol. A*, **13**, 1553 (1995).

80. C. Haginoya, M. Ishibashi, and K. Koike, *Appl. Phys. Lett.*, **71**, 2934 (1997).

81. B. Gates, S. H. Park, and Y. Xia, *Adv. Mater.*, **12**, 653 (2000).

82. K. H. Park, S. Lee, K. H. Koh, R. Lacerda, K. B. K. Teo, and W. I. Milne, *J. Appl. Phys.*, **97**, 024311 (2005).

83. H. Asoh, A. Uehara, and S. Ono, *Jpn. J. Appl. Phys.*, **43**, 5667 (2004).

84. H. Asoh, A. Oide, and S. Ono, *Electrochem. Commun.*, **8**, 1817 (2006).

85. H. Asoh, F. Arai, and S. Ono, *Electrochem. Commun.*, **9**, 535 (2007).

86. H. Asoh, F. Arai, and S. Ono, *ECS Trans.*, **6**, 431 (2007).

87. S. Ono, A. Oide, and H. Asoh, *Electrochim. Acta*, **52**, 2898 (2007).

88. F. Arai, H. Asoh, and S. Ono, *Electrochemistry*, **76**, 187 (2008).

89. H. Asoh, S. Sakamoto, and S. Ono, *J. Coll. Interface Sci.*, **316**, 547 (2007).

90. A. Ivanisevic and C. A. Mirkin, *J. Am. Chem. Soc.*, **123**, 7887 (2001).

91. M. Maccarini, M. Himmelhaus, S. Stoycheva, and M. Grunze, *Appl. Surf. Sci.*, **252**, 1941 (2005).

92. H. Asoh, F. Arai, K. Uchibori, and S. Ono, *Appl. Phys. Express*, **1**, 067003 (2008).

93. H. Asoh, F. Arai, and S. Ono, *Electrochim. Acta*, **54**, 5142 (2009).

94. S. Ono, F. Arai, and H. Asoh, *ECS Trans.*, **19**, 393 (2009).

95. Y. Yasukawa, H. Asoh, and S. Ono, *Electrochem. Commun.*, **10**, 757 (2008).

96. Y. Yasukawa, H. Asoh, and S. Ono, *ECS Trans.*, **13**, 83 (2008).

29

Nanoindentation of Biomaterials

Jin Tong
Jilin University

Jiyu Sun
Jilin University

Jiang Zhou
Jilin University

29.1 Introduction

Natural biomaterials have special structures and functions through the evolution of exchanging material, energy, and information with natural surroundings over millions of years (Tong et al. 2001). The quantitative measurement of the mechanical properties of natural biomaterials will help one to understand biological structures and biological functions of living things, and to develop bionic materials including bionic smart materials, advanced biomedical materials, and industrial materials. The development of the value-added transformation of agricultural biomass materials also needs the understanding of their structure and mechanical properties. The mechanical properties of a material in micro- and nanoscale are different from those in macroscopic scales.

Nanoindentation is an almost nondestructive testing technique, which has many advantages, such as small size of indent (100–1000 nm), low load (even nN), high spatial resolution (≤1 μm), and no special requirements for specimen size and shape. Recently, nanoindentation technique has been used for investigating the mechanical properties of the varied biomaterials and their structures at micro- and nanoscales (Sarikaya et al. 2002, Haque 2003). With quantitative and controllable nanoindentation on surfaces and the convenience of automatic operation, nanoindenter provided a powerful tool for assessing mechanical properties in microelectronic parts and films. According to its working principles, a nanoindenter can be regarded as one tool in the scanning probe microscope (SPM) family. Nanoindentation techniques were developed rapidly over the past two decades and widely used for examining the

properties of metal and film or coatings as well as the mechanical properties of biomaterials, such as bone (Rho and Pharr 1999, Haque 2003, Oyen and Ko 2008), tooth (Kinney et al. 2003, Marshall et al. 2003, Ho et al. 2004, Angker and Swain 2006), and cartilage (Ebenstein and Pruitt 2006), since the nanoindenter can measure materials' properties in situ and without disruption of the microstructure. Some studies of nanoindentation properties of insect materials were carried out (Arzt et al. 2002, Enders et al. 2004, Tong et al. 2004). Nanoindentation has become an effective technique to measure nanomechanical properties in diverse biomaterials ranging from mineralized to soft tissues. It can be expected that nanoindentation tests will be more and more in the applications of biomaterials in the future.

The quantitative measurements of mechanical properties of natural biomaterials in micro- and nanoscales and the details of the measuring methods were mainly described. An optimal method for determining the holding time and the loading rate as two important testing parameters in tests of biomaterials was presented. The effects of the surfaces roughness of the specimens, the sample hydration, and the selection of nanoindenter tips on measuring results of biomaterials and the measuring method of the cross section of the biomaterials were reviewed. The nano-dynamic analysis of biomaterials was discussed. The state of the art of the nanoindentation of some biomaterials was presented, including beetles *Copris ochus* Motschulsky, *Potosia (Liocola) brevitarsis lewis*, *Cybister*, *Allornyrina dichotoma*, and *Holotrichia trichophora*; dragonfly *Anax parthenope julius* Brauer, cicada Homóptera, and Cicàdidae; *Drosophila melanogaster*, tooth, bone (cartilage), and others; and some critical discussion was given as well. The future development of the nanoindentation techniques and their applications in the development of the bionic materials and biomedical materials were put forward.

29.2 Background

The researchers from different fields study nanoindentation of biomaterials for different objectives. For example, medical scientists and doctors study the nanomechanical properties of biomaterials for developing new biomedical materials; scientists and engineers in bionic engineering study the relationship of the nanoindentation of biomaterials with their micro- and nanoscale structures for developing advanced materials learning from natural materials; biologists study the nanoindentation of biomaterials for understanding the structures and functions of living things. There were not widely accepted definitions of biomaterial so far. Even so, biomaterials generally include two aspects (Ratner et al. 2000, Black and Brozino 2003): one is biological (natural) materials, which are formed through biological processes, such as structural protein (e.g., collagen fiber, silk), structural composites (e.g., bamboo stem), and mineralized materials (e.g., bone, dentin, mollusk shell); the other is biomedical material. The latter is narrow in the sense of biomaterial and can be natural or man-made materials used for replacing the whole or part of a human organ with pathological changes.

The characterization of material properties is an important step before utilizing the material for any purpose. The materials used for a component should be without failure within the lifespan of the component. Mechanical properties are very important among most of the properties for the applications of materials in human body. Biomaterials possess distinguished specific mechanical properties, such as specific strength and specific Young's modulus. Hard tissues such as bone have higher hardness and Young's moduli, while soft tissues such as cartilage usually have high viscoelasticity and their mechanical response is strongly dependent on loading rate and time (Franke et al. 2008). The conventional testing methods for the mechanical properties of biomaterials include tensile testing, bending testing, impacting testing, hardness testing, and microhardness testing (Vickers and Knoop indentation), etc. However, these conventional methods are difficultly used to measure the mechanical properties of such biomaterials as cuticle of dung beetles since it is very thin (with thickness in micro or nanoscale). Nanoindenter provides a tool to resolve this problem (Tong et al. 2004, Ebenstein and Pruitt 2006).

Theoretically, nanohardness is generally dependent on the characteristic of the indenter and the response of the specimen material. Nanohardness is the instantaneous force that can be endured at the unit area of the indent projective surface. Therefore, the nanoindentation testing method is more suitable for biomaterials than the conventional methods, and it can be used for testing extremely tiny regions and very thin specimens without considerably damaging them. There are some limitations in nanoindentation test although it has many advantages. Therefore, the nanoindentation testing method is a useful

technique to complement to the conventional methods rather than to replace them.

Nanoindentation techniques were applied to the studies of the head–neck joint and the wing-locking mechanism in beetles. The foot-substrate adhesion in flies and geckos has been examined using both nanoindentation technique and theoretical contact mechanics for investigating biological attachment systems in insects, which may lead to the design of artificial attachment systems potentially useful in microtechnology (Enders et al. 2004).

The concave–convex mechanism in the elytra couple was investigated in order to develop lightweight bionic composite structures; the structural characteristics of the forewings of two species of beetles were also studied (Chen et al. 2007). The nanoindentation properties of insect wing were examined in order to provide a clue to design bionic aerofoil materials used for micro-air-vehicles (Song et al. 2004, 2007; Tong et al. 2007). The nanoindentation properties of the attachment pads of animals were discussed by Scherge and Gorb (2001). Nanoindentation also was used to determine the compressibility of a cell wall. The stiffness of the bacterial cell wall ($\approx 0.42\,N/m$) was determined by subtracting the cantilever deflection on a bacterium from the deflection on the hard substrate (Franke et al. 2008). A quasi-static nanoindentation technique was used to measure cuticle stiffness of live *Drosophila melanogaster* when it was pupal, larval, and early adult in vivo (Kohane et al. 2003).

29.3 Details of Biomaterial Nanoindentation Methods

The Oliver–Pharr method is based on the assumption that specimen materials are elastic and isotropic and adhesion between indenter and testing specimen is negligible (Oliver and Pharr 1992, 2004). Many biomaterials display viscoelastic or time-dependent behavior, such as the cuticle of dung beetle *Copris ochus* Motschulsky, as shown in Figure 29.1. When the nanoindentation method is used in the examination of biomaterials, some additional steps are needed to acquire the right results. It is suggested that to increase the holding time before unloading and unloading rate are effective method to reduce viscoelastic effects on results during unloading (Tang et al. 2007).

29.3.1 Loading Methods (Holding Time, Loading Rate, and Load Value)

The test results of nanoindentation of materials are directly related to the testing method used. The holding time and the loading rate are two important testing parameters in the tests of biomaterials since the viscoelastic deformation takes place in most of biomaterials (Ngan and Tang 2002, Miyajima et al. 2003, Mittra et al. 2006, Vanleene et al. 2006). It was shown that the values of reduced modulus (E_r) and nanohardness (H) of the cuticle of dung beetle *Copris ochus* Motschulsky were beyond 49% and 130% to the stable values respectively when the effect

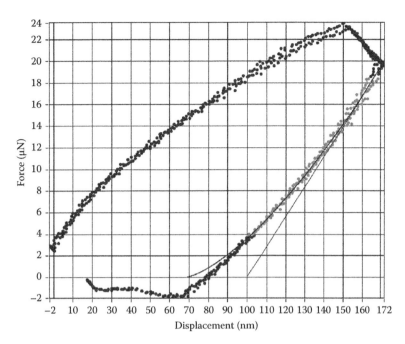

FIGURE 29.1 Force–displacement curve of a single indentation measurement on foreleg femur cuticle of dung beetle *Copris ochus* Motschulsky. (From Tong, J. et al., *J. Bionics Eng.*, 1, 221, 2004. With permission.)

of viscoelastic deformation was ignored (Tong et al. 2004). A similar result was presented by Fan and Rho (2003). They found that there is a strong effect of the indentation rate on the derived elastic moduli of osteonal lamellae that is proportional to the strain rate and there is a statistically significant effect of the strain rate on hardness. These indicated that the interstitial bone shows a viscoplastic behavior at the micrometer scale (Vanleene et al. 2006).

There is nose phenomenon in initial unloading segment when a triangle type of loading–unloading mode is used, i.e., no holding time. In this case, the creep effects can affect the measurement accuracy of the contact stiffness and contact area (Tong et al. 2004, Ngan et al. 2005). A thorough discussion of this nose phenomenon was presented by Briscoe et al. (1998). The curves of force–time, force–displacement, and displacement–time by a single indentation measurement on elytron cuticle of dung beetle *Copris ochus* Motschulsky were given in Figure 29.2. Figure 29.2a shows the triangle type of loading–unloading mode (force–time curve), i.e., the unloading was performed without a hold time at peak load; the displacement increases slightly in the initial portion of the unloading section because the creep rate of the material is initially higher than the unloading rate. This phenomenon results in a negative and changing slope in the initial unloading region, making it impossible to use the compliance method to the reduced elastic modulus. Figure 29.2b shows the force–displacement curve corresponding to Figure 29.2a. In order to eliminate this nose, a trapezoidal-type loading function can be utilized in nanoindentation tests to allow the material to approach equilibrium prior to unloading (see Figure 29.2c). The nanoindentation properties of some biomaterials were examined under various holding times,

shown in Table 29.1, in which the creep phenomena were dissipated prior to unloading. An appropriate hold time can be determined based on the creep and unloading rates used in the experiments (Bembey et al. 2006a).

For investigations of the factors influencing the holding time and loading rate, a series of tests were required to determine the optimal values (Tong et al. 2004). The holding time can be changed for determining the suitable value of loading time through experiments, and then the threshold of holding time can be determined. Figure 29.2d shows the force–displacement curve corresponding to Figure 29.2c.

Experimental optimization design methods, including orthogonal designs and regressive designs, can be used to determine the optimal holding time and loading rate of nanoindentation tests. For example, the optimal holding time and loading rate for the tests of nanoindentation of stigma of the dragonfly *Anax parthenope julius* Brauer were determined by an experimental optimization design method with an arrangement of the test scheme as shown in Table 29.2 (Tong et al. 2007). The orthogonal experimental design scheme $L_8(4^2 \times 2^2)$ was used, which reformed from the scheme $L_8(2^7)$, using the parataxis method (Ren 2001). In that work, both the factors of holding time and loading rate were considered in four-level values, which were 0, 20, 40, 60 s and 3, 23, 53, 73 μN/s respectively. So, the total number of the testing points was just eight although the total testing times were over eight since one-point datum was the average value of several repeated tests. There will be 16 datum points if the experimental optimization design method was not applied. Therefore, experimental optimization design method can effectively reduce the experimental times in measurements of nanoindentation properties of biomaterials. In the

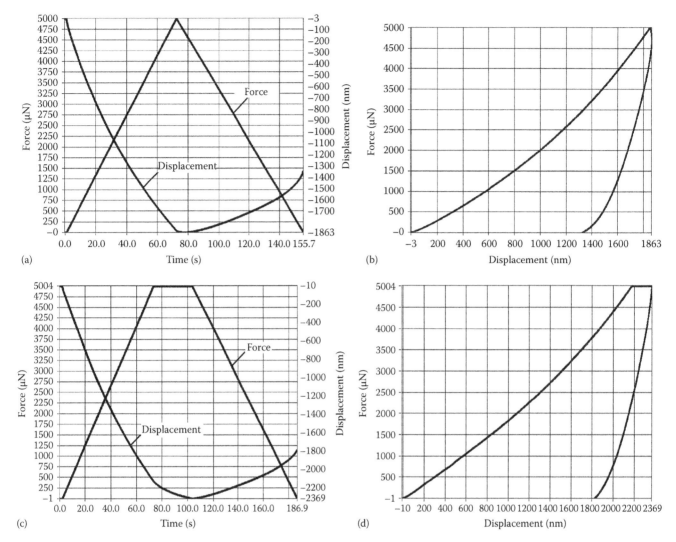

FIGURE 29.2 Force–time, force–displacement, and displacement–time curves of a single indentation measurement on elytra cuticle of dung beetle *Copris ochus* Motschulsky, showing (a) the nose effect resulting from indenting a viscoelastic material with utilizing triangle-type loading function; (b) the force–displacement curve under the triangle-type loading function; (c) the elimination of the nose effect by indenting a viscoelastic material with utilizing a trapezium-type loading function; and (d) the force–displacement curve under the trapezium-type loading function.

work on the nanoindentation of stigma of the dragonfly *Anax parthenope julius* Brauer, 10 repeating indentations were conducted to determine the average values of E_r. The test results were shown in y column in Table 29.2, where, y is the reduced modulus (E_r), y_{j1}, y_{j2}, y_{j3} and y_{j4} are the sum of the y at the same level for the factor A and factor B respectively. For the factor A, the y_{j1} is equal to the sum of y of the test 1 and test 2 and the rest may be deduced by analogy; $\bar{y}_{j1}, \bar{y}_{j2}, \bar{y}_{j3}$, and $\bar{y}_{j4}$ are the average value of y_{j1}, y_{j2}, y_{j3}, and y_{j4} respectively; R_j is the range of reduced modulus that is equal to the maximum value subtracting the minimum value of $\bar{y}_{j1}, \bar{y}_{j2}, \bar{y}_{j3}$, and $\bar{y}_{j4}$. In the experimental optimization methods, the value of R_j is the direct reflection of importance of the factors and a higher value of R_j means a greater influence to the test results (Ren 2001). Therefore, the factor A ($R_j = 0.10$) has more effect than the factor B ($R_j = 0.08$) on

the experimental results of the dragonfly stigma, which means the value of the holding time is more important to the stigma test. The value of y_{j1} is larger than that of the others (y_{j2}, y_{j3}, and y_{j4}), almost 50% difference, while, the value of y_{j2}, y_{j3}, and y_{j4} are close. These suggest that the results are unbelievable when the factor A is 0 s and the results are believable only when the factor A should be over 20 s. Therefore, the accurate values of E_r can be obtained as long as the holding time was more than 20 s. This means that the influence of viscoelastic phenomenon on the test results of the biomaterials' reduced modulus and hardness can be eliminated when the holding time is over 20 s. It was known that the least value of $\bar{y}_{j1}$ is the best. In the work on the nanoindentation of the stigma of the dragonfly *Anax parthenope julius* Brauer, as for the factor A, $\bar{y}_{j2}$ was 0.1145 and $\bar{y}_{j3}$ was 0.1093. Both the values were close, especially when they were chosen in the

TABLE 29.1 Nanoindentation Properties of Some Biomaterials Examined Using Various Holding

Specimen	Holding Time (s)	References
Human dental enamel	3	Habelitz et al. (2001a)
Vascular tissues	5	Ebenstein and Pruitt (2004)
Trabecular bone	5, 60	Mittra et al. (2006), Zysset et al. (1999)
Cortical bone	60	Zysset et al. (1999)
Cartilage	10	Ebenstein et al. (2004)
Insect cuticle	20	Tong et al. (2004), Sun et al. (2006)
Dragonfly wing	20	Tong et al. (2007)
Dentine	30	Angker et al. (2005)
Osteons	100	Rho et al. (1999)
Bovine femur	100	Rho and Pharr (1999)
Bone	10, 120	Hengsberger et al. (2002), Bushby et al. (2004)

TABLE 29.2 Measuring Scheme and Result Analysis of Nanoindentation Parameters of the Stigma of the Dragonfly *Anax parthenope julius* Brauer Using Scheme $L_8(4^2 \times 2^2)$

The Test Number	The Factors		
	Holding Time (s) A	Loading Rate (µN/s) B	E_r (GPa) y
1	1(0)	1(3)	0.24
2	1(0)	4(73)	0.18
3	4(60)	1(3)	0.12
4	4(60)	4(73)	0.12
5	2(20)	2(23)	0.12
6	2(20)	3(53)	0.10
7	3(40)	2(23)	0.11
8	3(40)	3(53)	0.10
y_{j1}	0.42	0.36	
y_{j2}	0.22	0.23	
y_{j3}	0.21	0.20	
y_{j4}	0.24	0.30	
$\bar{y}_{j1}$	0.21	0.18	
$\bar{y}_{j2}$	0.11	0.11	
$\bar{y}_{j3}$	0.10	0.10	
$\bar{y}_{j4}$	0.12	0.15	
R_j	0.11	0.08	
Primary and secondary factors	A, B		
Combined optimization	A_2B_3		

two significant figures (both were equal to 0.11 approximately). Therefore, the most optimal experimental scheme was A_2B_3 (the holding time is 20 s, the loading rate is 53 µN/s, and the time taken for one test is 31.32 s under the experimental conditions). While the holding time is 30 s, the loading rate is 53 µN/s and the time taken for one test is 51.32 s under the experimental conditions for the scheme A_3B_3. Because the investigated specimens were living biomaterial, the testing time was an important factor affecting the experimental results. The shorter the testing time is, the better the result is.

It was found that the viscoelastic effects on the calculated reduced modulus and tip–specimen contact depth in mice bone can be reduced effectively when S (the apparent unloading stiffness) was substituted by S_e (the real elastic contact stiffness) in Oliver–Pharr method and, so, the accuracy of the measured elastic modulus and hardness can be improved (Tang et al. 2007).

Load value can affect the test results. The study of the effects of embedding material, loading rate and load value, and penetration depth in nanoindentation of trabecular bone showed that the load within 250–750 µN did not affect the measured material modulus and hardness (Mittra et al. 2006). On the other hand, a maximum load below 10 µN will result in a significant error due to the effects of the sampling speed and drift of the transducer (Kaufman and Klapperich 2005). The estimations of the contribution of material volume to measured elastic modulus showed that the surface layer influences the measured modulus at low loads (Bushby et al. 2004). Therefore, to avoid the influence of the substrate materials on the measurements of the nanomechanical properties of a specimen, the penetration depth should be less than 10% of the specimen thickness (Habelitz et al. 2001a, Fischer-Cripps 2002). The maximum load used in investigation will be limited to ensure that the indents were not too deep. In addition, the effect of indentation size is believed from the hierarchal structure of the materials, i.e., the tip with different load values will probe different levels of the hierarchal structure. The suitable load value that should reach a more or less steady state of modulus and hardness can be determined through experiments.

29.3.2 Surface Roughness and Tip Selection

The effects of surface roughness of natural biomaterials on their nanoindentation tests should be taken into account. The surface roughness of biomaterials will affect the precision of the results of nanoindentation testing considerably. The effects of surface roughness on the calculated values of elastic modulus and hardness are not significant when a Berkovich indenter with tip radius of 100 nm is used (Fischer-Cripps 2002). The tips with too-small radius make narrow indent and affect the results of elastic modulus and hardness. Usually, the tip radius has to be at least 10 times larger than the surface roughness (Fischer-Cripps 2002). In particular, the tip selection is a critical issue when the nanoindentation tests of soft biomaterials were performed (Ebenstein and Pruitt 2004). For the nanoindentation tests of hard biomaterials, the Berkovich tip, the cube-corner tip and

TABLE 29.3 Tip Selection of Indenter and Applications in Some Biomaterials

Tip Types	Applications
Berkovich	Insect cuticle, teeth, bone, glassy polymer, mineralized tissue, insect wing, nacre, cartilage
Spherical tip	Soft polymer tissue, teeth
Cylindrical flat punch	Viscoelastic material
Conical tip	Scratch tests

three-sided pyramidal tip are commonly used since they have sharp points. Table 29.3 shows the tip selections and applications in some biomaterials.

Because the value of the projected contact area is affected by the contact depth, some rough surfaces of biomaterial specimens need to be pretreated before nanoindentation tests. The ultrasonic cleanout of the specimens for surface tests was conducted before indenting. Then, pre-scanning method can be used to decide the point to indent into the material at that place where the surface is smoother. Figure 29.3a shows a pre-scanning image of an elytron cuticle of dung beetle *Copris ochus*

Motschulsky using a conical tip with 2 μm radius before indenting, and Figure 29.3b shows the image of a surface zone including an in situ indent. Note that the overlapping of two indenting points should be avoided; otherwise, the testing results will be affected. In general, the interval between two adjacent indents is required to be five times more than the indenter tip radius. The scratch on the specimen surface during the operation of nanoindentation is unavoidable. In general, the effects of scratch on the hard biomaterials are lower than the soft biomaterials. It was found that the microscopy-sliced bone and tooth surfaces show lower surface roughness as compared with polished bone and tooth surfaces (Xu et al. 2003, Enders et al. 2004), and cryo-microtoming may be an option for reducing surface roughness of soft tissues (Ebenstein and Pruitt 2006).

An investigation into the nanoindentation properties of biomaterials on the cross section is often required for examining the distribution of the nanoindentation properties through the cross section. The specimens for nanoindentation on cross section can be prepared by the following procedure: a piece of a biomaterial is cut carefully to a flat cross-section surface and then polished to create a smooth surface for nanoindentation tests (Ho et al. 2004, Mohanty et al. 2006). In another way, a piece of a biomaterial can be directly embedded in epoxy resin or other resin and then polished to form a smooth cross-section surface for nanoindentation tests (Rho and Pharr 1999, Zysset et al. 1999, Povolo and Hermida 2000, Ho et al. 2004, Sun et al. 2006). Figure 29.4 shows the lamellar structure of the polished cross-sections of a dentin specimen with a lower surface roughness (Ho et al. 2004). By embedding the elytron of the dung beetle *Copris ochus* Motschulsky in epoxy resin and through polishing procedures, the root-mean-square (RMS) of the profiles of cross-section surface of the elytron can reach to 4.35 nm, which is smooth enough for nanoindentation tests (shown in Figure 29.5).

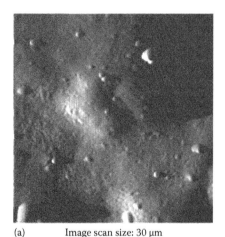

(a)   Image scan size: 30 μm

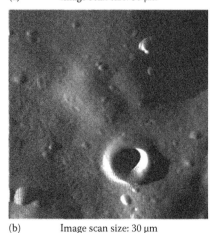

(b)   Image scan size: 30 μm

FIGURE 29.3 Elytra surface of dung beetle *Copris ochus* Motschulsky, showing (a) a pre-scanning image and (b) the image with an in situ indent for the same zone as (a).

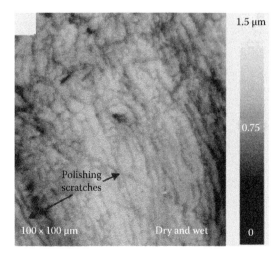

FIGURE 29.4 The lamellar structure in polished specimens had a smeared globular appearance (as compared to the ultrasectioned specimens). (From Ho, S.P. et al., *Biomaterials*, 25, 4847, 2004. With permission.)

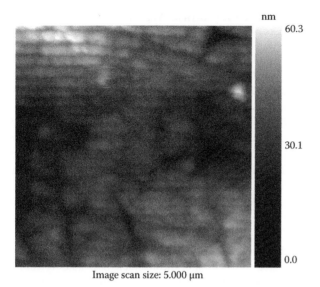

Image scan size: 5.000 μm

FIGURE 29.5 The cross-section morphology of elytra cuticle of dung beetle *Copris ochus* Motschulsky.

29.3.3 Sample Hydration

Most biomaterials are naturally hydrated. Nanoindentation properties of those biomaterials are required to measure in vivo. Despite most biological materials are naturally hydrated, much of the early work in bone or dentin indentation was performed using dehydrated specimens embedded in epoxy resin to facilitate specimen preparation and some studies demonstrated that the elastic modulus and hardness were found to much largely increase after dehydrate (Arzt et al. 2002, Bushby et al. 2004, Sun et al. 2007) and, meanwhile, a decrease in viscous deformation was found (Schöberl and Jäger 2006). Hydration and the hydration degree are less important issues for the hard tissues of biomaterials, but they are important issues for some soft tissues since soft tissues have a water contents up to 80% (Franke et al. 2008). The indentation moduli of bone as a hard tissue is increased from 11% to 28% after dehydration (Rho et al. 1999, Hengsberger et al. 2002, Bushby et al. 2004, Hoffler et al. 2005) and further increased after embedding (Bushby et al. 2004, Hoffler et al. 2005). This effect was found in nanoindentation experiments at varying hydration degrees (Bembey et al. 2006a). The elastic moduli of fully demineralized dentin (more comparable to a soft tissue) is increased with four orders of magnitude as compared with dehydration and the full demineralized dentin is still three times stiffer than the original dentin even after rehydration (Balooch et al. 1998). The application of a dynamic mechanical analyzer (DMA) as a tool for investigating the viscoelastic properties of bone was explored and, furthermore, the effects of various test parameters were examined and a reliable technique for using the DMA for bone research were established by Yamashita et al. (2001). They performed isothermal tests of human cortical bone at the body temperature using the DMA and the loss tangent and storage modulus were measured,

which are the representative measures of viscoelastic properties of bone (Yamashita et al. 2001).

The following methods for the measurement of naturally hydrated biomaterials can be used: the utilization of appropriate hydrating fluid (Habelitz et al. 2002, Ho et al. 2004) for treating the specimens, the utilization of the submerged specimens in a fluid cell (Rho and Pharr 1999, Bushby et al. 2004), the utilization of water-absorbing polymeric foam (Ebenstein and Pruitt 2004), designing the microfluidic platform for nanoindentation of continuous hydration of hydrogel samples (Kaufman and Klapperich 2005). A schematic representation of the test fixture is shown in Figure 29.6. These methods provide solutions for the measurements of elastic modulus and nanohardness under conditions close to in vivo. A simple and valid technique is that the nanoindentation is performed over a short time before dehydration occur (Ebenstein and Pruitt 2006, Schöberl and Jäger 2006, Sun et al. 2007).

Nanoindentation is a valuable technique for examining biomaterials, not only because vacuum conditions are not required, but also because the measuring conditions near in vivo are achievable (e.g., the nanoindentation tests can be performed under fluids) (Guidoni et al. 2006). The nanoindentation properties of dehydrated dentin immersed in Hank's balanced salt solution (HBSS) were nearly their nanoindentation properties under in vivo conditions (Guidoni et al. 2006).

In conclusion, the problems in the measurements of a hydrated biomaterial can be solved by

1. The utilization of an appropriate hydrate fluid
2. Conducting tests through submerging the specimen in a fluid cell
3. Running the tests within a very short time after sampling

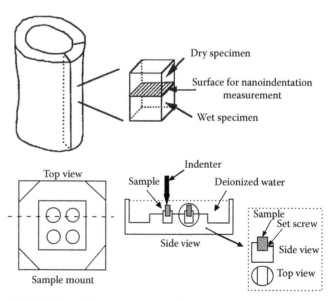

FIGURE 29.6 Schematic representation of specimen location for wet and dry testing and mounting stage used for testing wet specimens. (From Rho, J.Y. and Pharr, G.M., *J. Mater. Sci.: Mater. Med.*, 10, 485, 1999. With permission.)

The approach that the entire specimen submerged in the hydrating liquid has several disadvantages. The completely submerging specimen obscures viewing to the specimen surfaces when using the optic microscope attached to the nanoindenter; the utilization of specialized "extender" tips that are longer than standard tips is required and these tips prevent wicking of the hydrating fluid up to the transducer, which makes data analysis more complicate because a capillary force changing with the position of the indenter is added (Mann and Pethica 1996). To avoid these problems, the hydration system developed for this study can utilize the specimens by placing water-absorbing polymeric foam in contact with the specimens from its edges (Ebenstein and Pruitt 2004).

Another very important issue for testing biological systems is sample aging (Scherge and Gorb 2001). Temporal factors (i.e., storage time before the indentation) have demonstrated significant impact on tissue modulus. So, caution must be made to interpret indentation results if indentations are performed in different time periods even for the same specimen (Mittra et al. 2006).

The material structure of the gula plate and the head part of the head-to-neck articulation system in the beetle *Pachnoda marginata* was examined by scanning electron microscopy (SEM) and transmission electron microscopy (TEM) and the local mechanical properties (hardness and elastic modulus) of the gula material were determined through nanoindentation experiments (Barbakadze et al. 2006). To understand the effects of an outer wax layer and desiccation on the mechanical behavior of the gula material, the specimens were tested in fresh, dry, and chemically treated (lipid extraction in organic solvents) conditions using nanoindenter. It was found that the desiccation of specimens has strong influence on their nanoindentation results. A decrease of water content of about 15%–20% of the cuticle mass resulted in an increase of hardness from 0.1 to 0.49 GPa and an elastic modulus from 1.5 to 7.5 GPa and the lipid extraction caused a slight further hardening (to 0.52 GPa) (Barbakadze et al. 2006).

For investigating the testing duration in natural condition, the mass of the detached parts was taken as a function of time. Pieces of elytron cuticle were cut. The initial mass was taken as 100%. Each datum point is the average values from the three measurements. It was found that the mass of specimens lost ~12% of their initial mass after 100 min (Sun et al. 2007). This value is comparable with data of approximately 18% for the gula material (Barbakadze et al. 2006). The curve of the elytron cuticle tends to change to flat and the water loss approaches about 50% after 18 h. This certainly impacts the microstructure of elytron cuticle. It can be concluded that the whole test duration controlled within 1–1.5 h is better (Sun et al. 2007).

It was demonstrated that it is important to avoid prolonged dehydration during specimen preparation and to select an appropriate hydrating fluid (Ebenstein and Pruitt 2006). A quasi-static nanoindentation technique was used to measure cuticle stiffness of live Drosophila melanogaster during its pupal, larval, and early adult development in vivo (Kohane et al. 2003).

29.3.4 Nano Dynamic Mechanical Analysis

For viscoelastic materials, it is very difficult to obtain meaningful and accurate nanoindentation data using quasi-static testing because of the effects of the selection of the loading function and the type of tip utilized for measurement of nanoindentation properties due to creep- and strain-rate effects.

Some studies suggested that the holding time before unloading and the unloading rate can be increased as effective methods to reduce viscoelastic effects during unloading (Rho and Pharr 1999, Bushby et al. 2004, Oliver and Pharr 2004). When the nanoindentation properties of very soft materials including most biological tissues are performed, the important things are the determination of the holding time before unloading and the unloading rate, which should be effectively suppressing viscoelastic effects since the viscoelasticity closely depends on a complicated convolution of the peak load, the holding duration before unloading, and the unloading rate (Ngan and Tang 2002). A recently developed method for reducing the viscoelastic effects during nanoindentation was applied to mice bone samples (Tang et al. 2007). Static indents can be analyzed to enable the determination of the viscoelastic response, either by simply assuming that different indentation rates will give varying responses according to the viscoelasticity of the polymer (Nowicki et al. 2003) or by modeling the unloading process to account for a time-dependent response as well as time-independent elastic deformation (Lu et al. 2003). Both of these approaches were based on the same principle, which termed as "creep compliance" (Hayes et al. 2004).

However, the viscoelastic response of such materials as polymers is difficultly used to analyze from the load–displacement curve alone. Another approach to determine the viscoelastic properties is dynamic nanoindentation (Loubet et al. 1995, 2000, Odegard et al. 2005). The application of an alternating current force modulation in nanoindentation was used to measure the dynamic properties of viscoelastic materials such as loss modulus and storage modulus (Pethica and Oliver 1987, Asif and Pethica 1998). The storage modulus relates to the stiffness or the phase response of the material to the applied force. This modulus relates to the elastic recovery of the specimen and the amount of energy recovered from the specimen subsequent to a loading cycle. The loss modulus relates to the damping behavior of the material and can be observed from the time lag between the maximum force and the maximum displacement. Basically, dynamic nanoindentation involves the application of an oscillatory load to the indentation tip during a static indent, and the dynamic properties can be determined through monitoring the phase lag between displacement and load measurement (Hayes et al. 2004). The dynamic nanoindentation was used to determine the dynamic properties of polymeric materials (Lee et al. 2004, Park et al. 2004), bone (Yamashita et al. 2001), nacre (Mohanty et al. 2006, Stempfle et al. 2007), and soft tissue (Franke et al. 2008). Some commercialized nanoscale dynamic techniques, such as nano-DMA, were developed to provide a nanoscale analog to DMA testing.

Using nanoindentation to measure viscoelastic properties through DMA method has been proven in principle on the cortical bone specimens from human femora. It was found that DMA can be used as an effective tool to test bone (Yamashita et al. 2001). It was demonstrated from the examination of four forelimb bone samples from one mouse that single cylindrical brass block mounted by unsaturated polyester resin can be used to prepare mounting specimens so as to quickly dissipate the heat generated during the polymerization reaction of the mounting resin to minimize the effect of temperature changes on the specimens (Tang et al. 2007).

29.4 Presentation of the State of the Art

Mineralized tissues have extensively been studied using nanoindentation technique and some work about dentin (Kinney et al. 2003), bone (Haque 2003), and soft tissues (Ebenstein and Pruitt 2006, Franke et al. 2008) were reviewed recently. An increasing interest in bionic design of materials based on structures of natural biomaterials has led to the nanomechanical characterization of acellular biomaterials.

Nanoindentation has been used in conjunction with computational modeling to determine the mechanical properties of the aragonite blocks and interfacial proteins that make up the complex microstructure of nacre and the fracture resistance of the mother-of-pearl lining mollusk shells (Katti et al. 2001, Bruet et al. 2005). In bionic engineering, the applications of nanoindentation include characterization of structure–property relationships in specific regions of insect cuticle (e.g., exoskeleton) (Enders et al. 2004, Barbakadze et al. 2006, Tong et al. 2007), mapping the variation in modulus and hardness across the 200–600 μm diameter sponge spicule (Sarikaya et al. 2002) and measurement of anisotropic mechanical properties of spider silk fibers (Ebenstein and Pruitt 2006), which give an insight into natural biomaterials and provide clues to develop bionic materials. Based on the understanding the microscale and nanoscale structures and chemical and mechanical properties of natural biomaterials, materials scientists and engineers could design new composite materials with properties higher on the macroscale.

29.4.1 Insects

Some studies of the nanoindentation properties of the insect materials have been carried out recently. Nature has created a hierarchal structure to conform over a range of size scales to achieve the amazing adhesion ability, such as head–neck joint and the wing-locking mechanism, and the foot–substrate adhesion in flies, beetle, spider, and gecko, to climb and stick to almost varied surfaces. The nanoindentation adhesion-test technique was extended to measure the biological adhesion properties (Arzt et al. 2002, Enders et al. 2004, Northen and Turner 2006). As mentioned above, to understand the effect of desiccation and an outer wax layer of specimens on the mechanical behavior of the biomaterial, the gula plate of the beetle *Pachnoda marginata* were tested in fresh, dry, and chemically treated conditions by using nanoindenter and it was found that desiccation has strong influence on the nanoindentation results (Barbakadze et al. 2006). The exoskeleton of a ground beetle *Scarites subterraneus* was tested using nanoindentation and it was found that the hardness and reduced elastic modulus of its mandible materials were much higher than those of the abdomen materials (Michelle hysitron web page). This may be due to an increased level of heavy metals and halogens, such as Zn and Mn incorporated into the nanoscale structure of the exoskeleton during maturation, and likely due to the different functions of fighting and protection of the two areas (mandible and abdomen).

Insect cuticle exhibits viscoelastic behavior. The nanoindentation method for the cuticle of the dung beetle *Copris ochus* Motschulsky was investigated to eliminate the viscoelastic effect on the testing result (Tong et al. 2004). Considering the multilayer structure of the cuticle, the nanoindentation properties of the cross section was investigated. The reduced modulus (E_v) and nanohardness (H_v) of the surface cuticle in the vertical direction found by nanoindentation was 3.54 ± 0.12 GPa and 0.20 ± 0.01 GPa respectively. The nanoindentation results showed that the reduced modulus (E_t) and nanohardness (H_t) of each layer in the transverse direction was gradually decreased from the outer layer to the inner layer in the transverse direction. The elastic modulus at the outer layer was largest, reaching 7.06 ± 0.54 GPa and E_v was less than E_t. This phenomenon may have resulted from the composite effects of the multilayer. An experimental model was proposed to describe the nanomechanical properties of elytra cuticle of the dung beetle *Copris ochus* Motschulsky without consideration of the anisotropy of chitin (Sun et al. 2006).

The nanoindentation properties of the stigma of dragonfly *Anax parthenope julius* Brauer were investigated at position-1, -2, and -3, as shown in Figure 29.7 (Tong et al. 2007), and the test results are illustrated in Figure 29.8. It can be found that the E_r and H at the position-1 are the maximum. This is because position-1 is near the leading edge. E_r at the position-2 is the minimum, while the H at the position-3 is the minimum.

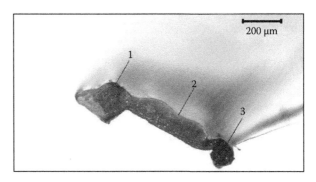

FIGURE 29.7 The stereoscopic photograph of the cross section of the dragonfly (*Anax parthenope julius* Brauer) stigma showing the three measuring positions. (From Tong, J. et al., *J. Mater. Sci.*, 42, 2894, 2007. With permission.)

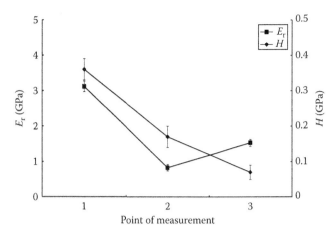

FIGURE 29.8 The nanomechanical results at the three positions of two specimens of the dragonfly (*Anax parthenope julius* Brauer) stigma. (From Tong, J. et al., *J. Mater. Sci.*, 42, 2894, 2007. With permission.)

The realization to the mechanical properties of stigma of dragonfly should help to deeply understand its flying mechanism and would provide an inspiration to design some new materials and structures of mechanical parts. The reduced modulus and nanohardness of the elytron cuticle of four species of beetles (*Potosia (Liocola) brevitarsis lewis, Cybister, Allornyrina dichotoma, Holotrichia trichophora*) measured by the Nano Indenter SA2 (MTS) are 9.08, 8.21, 4.76, 6.00 GPa and 0.44, 0.48, 0.18, 0.18 GPa respectively (Yang et al. 2007).

The mechanical properties of the cicada (Homóptera, Cicàdidae) wing were investigated by nanoindentation and tensile testing separately (Song et al. 2004). The results indicated that the mean Young's modulus, hardness, and yield stress of the membranes of the wings were approximately 3.7 GPa, 0.2 GPa, and 29 MPa, respectively, and the mean Young's modulus and strength of the veins along the direction of the venation of wings were approximately 1.9 GPa and 52 MPa, respectively. These results provide a clue to conduct bionic design of the aerofoil materials of micro-air vehicles.

The in vivo test was seldom performed in insect due to the limitation of the specimen preparation, although the cuticle stiffness of live *Drosophila melanogaster* was measured for its pupal, larval, and early adult development in vivo (Kohane et al. 2003).

29.4.2 Teeth

A primary study of teeth was focused on mapping mechanical properties across the dentin–enamel junction (DEJ) to examine how the less-mineralized dentin to transit to the more-mineralized enamel with higher modulus (Fong et al. 2000, Urabe et al. 2000, Marshall et al. 2001). An example of this examination is shown in Figures 29.8 (Fong et al. 2000). The enamel of the human teeth is a highly mineralized hard tissue that exhibits a complex hierarchical structure and provides an interesting model for bionic design of new materials. The nanoindentation properties are different considerably at different zones due to

the complex hierarchical structural organization and distinct composition of dentin and enamel, so DEJ and other similar calcified tissues were examined (Cui et al. 2007). The hierarchical structure of biomaterials is essential to their function. However, this brings forward a challenge in testing multiple-length scales (Franke et al. 2008). The nanohardness and modulus of the isolated domains within isolated prisms and the surrounding sheaths of teeth were tested by AFM and nanoindentation system (Ge et al. 2005). Biological materials cannot be considered as homogeneous material at classical-length scales due to their hierarchical structure and, therefore, it is important to the application of highly localized techniques to characterize the material nanoindentation properties (Franke et al. 2008). The mechanical properties of single enamel rods at different orientations were investigated (Habelitz et al. 2001a).

The hardness and elastic modulus were quantified using nanoindentation as a function of distance from the DEJ (Fong et al. 2000, Urabe et al. 2000, Marshall et al. 2001, Park et al. 2008). However, large variations in the proposed width of the DEJ is better when the different measuring techniques of nanoindentation and a width of 12–20 mm was revealed using nanoindentation measurements (Fong et al. 2000, Urabe et al. 2000, Marshall et al. 2001). The nanoscratches are used as a useful method to determine functional dimensions of interfaces in hard biological tissues, such as enamel and dentin, and across the DEJ of human teeth. The friction coefficient profiles were recorded in order to determine the functional width of the DEJ (Habelitz et al. 2001b), which might assist in better understanding the mechanical behavior of the dental fillers and thus facilitate the design of robust fillers with excellent mechanical properties (Al-Haik et al. 2008).

It was reported that the mechanical property gradient was smooth across the DEJ, with highest hardness and elastic modulus values in the enamel, intermediate across the DEJ, and lowest in the bulk dentin (Urabe et al. 2000, Marshall et al. 2001) and, in particular, with regards to the potential changes, the mechanical property gradient arises with aging (Park et al. 2007, 2008). It was, in contrast, reported that the hardness and the associated elastic modulus are lowest at the DEJ (Tesch et al. 2001).

Storage in different solutions (such as deionized water, calcium chloride-buffered saline solution and Hank's balanced salts solution (HBSS)) will result in the different nanomechanical properties of dentin and enamel due to the surface mineralization (Habelitz et al. 2002).

A comparison between the structure, chemical composition, and mechanical properties of collagen fibers at three regions within a human periodontium enabled one to define a novel tooth attachment mechanism (Ho et al. 2007).

29.4.3 Bone

Mechanical properties of bone have been extensively studied since the importance in medical engineering. The nanoindentation properties of bone were reviewed by Haque (2003). It was found that the nanomechanical properties of bone change at different locations (trabecular and cortical bones, osteonal bone) (Rho et al. 1999, Roy et al. 1999, Zysset et al. 1999, Hoffler et al. 2005), along different directions (Roy et al. 1999, Van Eijden et al. 2004) and at varied structural levels (Rho et al. 2002, Donnelly et al. 2006, Joo et al. 2007). Bone has anisotropic mechanical properties due to its complex hierarchical structure (Rho and Pharr 1999, Rho et al. 2001, Hengsberger et al. 2003, Fan et al. 2006, Katz et al. 2007). Presumably, the anisotropy may be mainly due to a combination of many factors, e.g., orientations along the arrangement of the collagen fibers and their assembling patterns. The fibrils were found to be assembled as bundles or aligned as arrays in different patterns, which resulted in different mechanical properties in all three orthogonal directions (Cui et al. 2007). It was known that the material behavior is sensitive both to the material microstructure and to the chemical environment (Bembey et al. 2006b). The nanomechanical properties of bone have relationship with age-related changes (Rho et al. 2002, Wang et al. 2007a), the degree of mineralization (Mulder et al. 2008), protein intake (Hengsberger et al. 2005), and whether it was hydrated or not (Rho and Pharr 1999, Yamashita et al. 2001, Hengsberger et al. 2002, Bushby et al. 2004, Hoffler et al. 2005, Schöberl and Jäger 2006); all of these can be investigated by nanoindentation system at ultrastructural levels. A finite element model of an inhomogeneous contact problem was developed and used to interpret experimental nanoindentation data on bone and dentin (Oyen and Ko 2008). Nanoindentation characteristics were served as a means to distinguish clinical types of osteogenesis imperfecta (OI) bone (Fan et al. 2006). The nanomechanical properties of bovine cortical bone and mouse bone were measured using nanoindentation (Kavukcuoglu et al. 2007, Wang et al. 2007b). In conclusion, the above studies will be helpful to understand the role of local tissue properties and hierarchical structure and deformation mechanisms of bone in the macroscale (Ebenstein and Pruitt 2006).

29.4.4 Others

Studies on the nanoindentation of other biomaterials have been carried out, such as the work on dental mouth guards (Low et al. 2002), calcified explanted heart valves, gel, and artery (Ebenstein et al. 2004). The potential applications of nanoindentation are unrestricted in theory. The dimension, shape, or form of specimens limits the utilization of conventional methods and nanoindentation technique is required. The mechanical properties of a specific area of biomaterials can be evaluated by nanoindentation.

Nacre (the pearly internal layer of molluskan shells) is an attractive nanocomposite displaying high mechanical properties, low density, and a good biocompatibility with human bone. The mechanical properties of polished specimens of nacre were measured (Li et al. 2004, Bruet et al. 2005, Sun et al. 2007) and the fracture mechanisms were investigated (Stempfle et al. 2007). The viscoelastic properties of nacre were studied using dynamic nanoindentation (Mohanty et al. 2006). The

viscoelasticity of nacre was found to increase as the testing frequency was increased. The results at lower indentation depths indicated the viscoelastic nature of the inorganic phase. The thin sheets of calcite, termed folia (that make up much of the shell of an oyster), are composed of foliated lath. The fracture mechanisms were discussed through the analysis of the correlation between preferred orientation and structural characteristics during cracking of the folia (Lee et al. 2008). The folia has a hardness of about 3 GPa and elastic modulus of about 73 GPa, while the nacre has a hardness between 9.7 and 11.4 GPa and elastic modulus between 60 and 143 GPa. Some studies were focused on the sensitivity of soft tissue of biomaterials, such as cartilage. The nanoindentation tests of biomaterials with very low moduli require different calibrations from regular quasistatic nanoindentation testing and the sensitivity is mainly dependent upon the total mass of the indenter and the damping of the system (Franke et al. 2008). The adhesion of biomaterial to the indenter is still a sensitive topic (Carrillo et al. 2005). A smaller size of the probe in nanoindentation allows mapping mechanical properties in cartilage from smaller animal models and smaller joints, such as, rabbit knee (Ebenstein et al. 2004) and rabbit finger-joint cartilage (Li et al. 2006). The nanomechanical properties of alginate-recovered chondrocyte matrices were considered as a function of ex vivo incubation time (Tomkoria et al. 2007). Some advanced synthetic materials may be developed based on the morphology and the polymorphism of biomaterials.

29.5 Critical Discussion

Although the applications of the instrumented indentation techniques to biomaterials have been increased in recent years, some limitations in nanoindentation tests restrict its extensive applications. These limitations include the following: nanoindentation is sensitive to specimen preparation (i.e., tests under dry and physiological conditions resulted in remarkable difference of the results), the instruments and their parts (optical and mechanical) and the response time. Therefore, the nanomechanical properties of some complex biomaterials, such as trabecular bone, are still difficult to characterize. Although the mechanical properties are easily measured on certain intermediate hierarchical levels by nano-, micro-, and macro-indentations, it is difficult to relate these bulk mechanical behaviors to the structures perfectly due to the complexity of biomaterials. Many studies were based on the Oliver–Pharr methodology to measure the nanoindentation properties of the hard, mineralized biological tissues. However, the load-displacement analysis is based on the elastic contact theory in the Oliver–Pharr methodology and, so the method used to investigate materials with significant viscoelastic behavior will result in some error. A recently developed method for correcting the viscoelastic effects during nanoindentation is applied to mice bone examination. The real elastic contact stiffness (S_e) was used to substitute apparent elastic contact stiffness (S) in Oliver–Pharr method. This can effectively remove the effects of viscoelasticity of biomaterials on the calculated reduced modulus and tip–specimen contact depth and hence can improve the measuring accuracy of the elastic modulus and hardness (Tang et al. 2007). For nanoindentation performed with a spherical tip, the methods of contact mechanics based on the Johnson–Kendall–Roberts (JKR) adhesion model were shown to reach a more accurate measurement of specimen modulus (Carrillo et al. 2005, Ebenstein and Wahl 2006). A method modified from the Oliver–Pharr Method, called the modified slopes method (MSM), was compared with the Oliver–Pharr Method for the examination of bone specimens. It was found that, for a given material, the values of E or H computed using the two methods are not significantly different. Even so, MSM was considered preferable because it is straightforward. The nanoindentation measurements and the values of constants used are just dependent upon geometry of the nanoindenter used (Lewis et al. 2006).

More attention has been paid to the dynamic indentation (e.g., DMA) and some models were applied in the evaluation of the testing results. The classical approach consisting of a dashpot in parallel alignment to a spring, also known as the Kelvin or Voigt model, can describe most viscoelastic materials, but there is a lack of the ability to accommodate instantaneous elastic recovery using the approach (Franke et al. 2008). The model is too simple to be used for the all biomaterials.

Some further explorations are required, such as a combination of multiple spring and damping elements to form various models. Another important issue is real in vivo nanoindentation testing method, such as the fixation method of living specimens and testing stability.

29.6 Summary

The details of testing methods for nanoindentation of biomaterials were introduced in this chapter, including the determination of critical measuring parameters such as holding time, loading rate and load value; the effects of surface roughness and specimen hydration on testing results; and the selection of the nanoindenter tip. The main research results of such biomaterials as insects' cuticle, teeth, and bone were also reviewed.

The history of researches of nanoindentation of biomaterials is short mainly due to the limitation of nanoindentation instrumentation. However, with the improvements of nanoindentation technique, it has become a frontier technology and the nanoindentation of biomaterials should be a hot subject. Nanoindentation technique provides an effective tool for exploring the mechanical properties of biomaterials and the relationship to their microstructure. The nanoindentation technique will be utilized well in biomedical and bionic materials.

The case studies introduced in this chapter are only parts of the research results of nanoindentation of biomaterials. Even so, the basic knowledge of nanoindentation of biomaterials described in this chapter assuredly provides a clue for scientists and engineers to study the nanoindentation of biomaterials in different research areas.

29.7 Future Perspective

It can be believed that with developments of microelectronic technology, computer technology and information technology the instruments of nanoindentation will further be reformed. The miniaturization of nanoindenter may become a developing trend, which will be beneficial to perform the nanoindentation tests of biomaterials in vivo outdoors. For example, a portable nanoindenter would be developed for investigating the real-time change of biomechanical properties of crop stalk during the growing process.

The present theory and methods of nanoindentation are mainly used for testing elastic property. The development of the theory and methods of nanoindentation is required for easily examining the viscoelastic and plastic properties of materials including biomaterials. The comprehensive understanding of biomaterials should be reached if a more effective theory and, as a result, a more effective method is developed for directly measuring viscoelastic and plastic properties of biomaterials.

The nanoindentation of biomaterials should be extended to more fields. The nanoindentation behaviors of dentin, bone, and cartilage are mainly involved in medical applications. Other biomaterials, such as blood vessel wall materials and skin of some mammalian animals, can be examined by nanoindenter to help medical experts and doctors to diagnose human heavy malady and invent some new therapeutics.

Bionics, learning from nature, is an everlasting frontier of science and technology and an important resource of technical innovation (Lu 2004). Learning from the structures and functions of natural biomaterials needs a deep understanding of properties and microstructure of natural biomaterials. The nanoindentation properties of natural biomaterials, such as cuticle and wing of insects, should be helpful to reveal the properties, biomechanics, bio-tribology and microstructure of natural biomaterials, and their relationships, and provide some important and useful information for developing bionic composite materials for micro-aircraft, bionic tribology, bionic medical apparatus, and bionic organs (tissue engineering).

Acknowledgments

This work was supported by the National Hi-tech Project (863 Project) (Grant No. 2009AA043603–4, 2009AA04364–2), by National Foundations of Agricultural Technology Transformation of China (Grant No. 2009GB23600507), the National Natural Science Foundation of China (Grant No. 30600131, 50675087), the Special Research Fund for the Doctoral Program of High Education of China (Grant No. 20060183067), the National Science Fund for Distinguished Young Scholars of China (Grant No. 50025516), and "Project 985" of Jilin University. The authors would like to thank Elsevier Limited, Springer Publishing Company, and the editorial office of the *Journal of Bionic Engineering* for the copyright permissions for some figures in their publications in this chapter.

References

Al-Haik, M., Trinkle, S., Sumali, H., Garcia, D., Yang, F., Martinez, U., Miltenberger, S. 2008. Investigation of the nanomechanical and tribological properties of tooth-fillings materials. *Proceedings of the ASME International Mechanical Engineering Congress and Exposition, IMECE 2007*, vol. 2, Seattle, WA, pp. 145–151.

Angker, L., Swain, M. V. 2006. Reviews: Nanoindentation: Application to dental hard tissue investigations. *Journal of Materials Research* 21: 1893–1905.

Angker, L., Swain, M., Kilpatrick, N. 2005. Characterising the micro-mechanical behaviour of the carious dentine of primary teeth using nano-indentation. *Journal of Biomechanics* 38: 1535–1542.

Arzt, E., Enders, S., Gorb, S. 2002. Towards a micromechanical understanding of biological surface devices. *Materials Research and Advanced Techniques* 93: 345–351.

Asif, S. A. S., Pethica, J. B. 1998. Nano-scale viscoelastic properties of polymer materials. In *Thin Films—Stresses and Mechanical Properties VII*, R. C. Cammarata, M. Nastasi, E. P. Busso, W. C. Oliver. (eds.), *Material Research Society Symposium Proceedings*, Warrendale, PA, vol. 505, p. 103.

Balooch, M., Wu, M. I. C., Balazs, A., Lundkvist, A. S., Marshall, S. J., Marshall, G. W., Siokaus, W. J., Kinney, J. H. 1998. Viscoelastic properties of demineralized human dentin measured in water with an atomic force microscope (AFM) based indentation. *Journal of Biomedical Materials Research* 40: 539–544.

Barbakadze, N., Enders, S., Gorb, S., Arzt, E. 2006. Local mechanical properties of the head articulation cuticle in the beetle *Pachnoda marginata* (Coleoptera, Scarabaeidae). *Journal of Experimental Biology* 209: 722–730.

Bembey, A. K., Oyen, M. L., Bushby, A. J., Boyde, A. 2006a. Viscoelastic properties of bone as a function of hydration state determined by nanoindentation. *Philosophical Magazine* 86: 5691–5703.

Bembey, A. K., Bushby, A. J., Boyde, A., Ferguson, V. L., Oyen, M. L. 2006b. Hydration effects on the micro-mechanical properties of bone. *Journal of Materials Research* 21: 1962–1968.

Black, J. B., Brozino, J. D. 2003. *Biomaterials: Principles and Applications*. Boca Raton, FL: CRC Press.

Briscoe, B. J., Fiori, L., Pelillo, E. 1998. Nano-indentation of polymeric surfaces. *Journal of Physics-London-D: Applied Physics* 31: 2395–2405.

Bruet, B. J. F., Panas, R., Tai, K., Ortiz, C., Qi, H. J., Boyce, M. C. 2005. Nanoscale morphology and indentation of individual nacre tablets from the gastropod mollusc *Trochus niloticus*. *Journal of Materials Research* 20: 2400–2419.

Bushby, A. J., Ferguson, V. L., Boyde, A. 2004. Nanoindentation of bone: Comparison of specimens tested in liquid and embedded in polymethylmethacrylate. *Journal of Materials Research* 19: 249–259.

Carrillo, F., Gupta, S., Balooch, M., Marshall, S. J., Marshall, G. W., Pruitt, L., Puttlitz, C. M. 2005. Nanoindentation of polydimethylsiloxane elastomers: Effect of crosslinking, work of adhesion, and fluid environment on elastic modulus. *Journal of Materials Research* 20: 2820–2830.

Chen, J. X., Ni, Q. Q., Xu, Y. L., Iwamoto, M. 2007. Lightweight composite structures in the forewings of beetles. *Composite Structures* 79: 331–337.

Cui, F. Z., Cheng, Z. J., Ge, J. 2007. Nanomechanical properties of tooth and bone revealed by nanoindentation and AFM. *Key Engineering Materials*, vol. 353–358, PART 3, *Progresses in Fracture and Strength of Materials and Structures* (Selected peer reviewed papers from the *Asian Pacific Conference Fracture and Strength 2006* (*APCFS'06*)), pp. 2263–2266.

Donnelly, E., Baker, S. P., Boskey, A. L., van der Meulen, M. C. H. 2006. Effects of surface roughness and maximum load on the mechanical properties of cancellous bone measured by nanoindentation. *Journal of Biomedical Materials Research—Part A* 77: 426–435.

Ebenstein, D. M., Pruitt, L. A. 2004. Nanoindentation of soft hydrated materials for application to vascular tissues. *Journal of Biomedical Materials Research—Part A* 69: 222–232.

Ebenstein, D. M., Pruitt, L. A. 2006. Nanoindentation of biological materials. *Nano Today* 1: 26–33.

Ebenstein, D. M., Wahl, K. J. 2006. A comparison of JKR-based methods to analyze quasi-static and dynamic indentation force curves. *Journal of Colloid and Interface Science* 298: 652–662.

Ebenstein, D. M., Kuo, A., Rodrigo, J. J., Reddi, A. H, Ries, M., Pruitt, L. A. 2004. Nanoindentation technique for functional evaluation of cartilage repair tissue. *Journal of Materials Research* 19: 273–281.

Enders, S., Barbakadse, N., Gorb, S. N., Arzt, E. 2004. Exploring biological surfaces by nanoindentation. *Journal of Materials Research* 19: 880–887.

Fan, Z., Rho, J. Y. 2003. Effects of viscoelasticity and time-dependent plasticity on nanoindentation measurements of human cortical bone. *Journal of Biomedical Materials Research—Part A* 67: 208–214.

Fan, Z., Swadener, J. G., Rho, J. Y., Roy, M. E., Pharr, G. M. 2006. Anisotropic properties of human tibial cortical bone as measured by nanoindentation. *Journal of Orthopaedic Research* 20: 806–810.

Fischer-Cripps, A. C. 2002. *Nanoindentation*. New York: Springer-Verlag.

Fong, H., Sarikaya, M., White, S. N., Snead, M. L. 2000. Nano-mechanical properties profiles across dentin–enamel junction of human incisor teeth. *Materials Science & Engineering C* 7: 119–128.

Franke, O., Göken, M., Hodge, A. M. 2008. The nanoindentation of soft tissue: Current and developing approaches. *JOM* 60: 49–53.

Ge, J., Cui, F. Z., Wang, X. M., Feng, H. L. 2005. Property variations in the prism and the organic sheath within enamel by nanoindentation. *Biomaterials* 26: 3333–3339.

Guidoni, G., Denkmayr, J., Schöberl, T., Jäger, I. 2006. Nanoindentation in teeth: Influence of experimental conditions on local mechanical properties. *Philosophical Magazine* 86: 5705–5714.

Habelitz, S., Marshall, S. J., Marshall, G. W. Jr., Balooch, M. 2001a. Mechanical properties of human dental enamel on the nanometre scale. *Archives of Oral Biology* 46: 173–183.

Habelitz, S., Marshall, S. J., Marshall, G. W. Jr., Balooch, M. 2001b. The functional width of the dentino-enamel junction determined by afm-based nanoscratching. *Journal of Structural Biology* 135: 294–301.

Habelitz, S., Marshall, G. W. Jr., Balooch, M., Marshall, S. J. 2002. Nanoindentation and storage of teeth. *Journal of Biomechanics* 35: 995–998.

Haque, F. 2003. Application of nanoindentation to development of biomedical materials. *Surface Engineering* 19: 255–268.

Hayes, S. A., Goruppa, A. A., Jones, F. R. 2004. Dynamic nanoindentation as a tool for the examination of polymeric materials. *Journal of Materials Research* 19: 3298–3306.

Hengsberger, S., Kulik, A., Zysset, P. 2002. Nanoindentation discriminates the elastic properties of individual human bone lamellae under dry and physiological conditions. *Bone* 30: 178–184.

Hengsberger, S., Enstroem, J., Peyrin, F., Zysset, P. 2003. How is the indentation modulus of bone tissue related to its macroscopic elastic response? A validation study. *Journal of Biomechanics* 36: 1503–1509.

Hengsberger, S., Ammann, P., Legros, B., Rizzoli, R., Zysset, P. 2005. Intrinsic bone tissue properties in adult rat vertebrae: modulation by dietary protein. *Bone* 36: 134–141.

Ho, S. P., Goodis, H., Balooch, M., Nonomura, G., Marshall, S. J., Marshall, G. 2004. The effect of sample preparation technique on determination of structure and nanomechanical properties of human cementum hard tissue. *Biomaterials* 25: 4847–4857.

Ho, S. P., Marshall, S. J., Ryder, M. I., Marshall, G. W. 2007. The tooth attachment mechanism defined by structure, chemical composition and mechanical properties of collagen fibers in the periodontium. *Biomaterials* 28: 5238–5245.

Hoffler, C. E., Guo, X. E., Zysset, P. K., Goldstein, S. A. 2005. An application of nanoindentation technique to measure bone tissue lamellae properties. *Journal of Biomechanical Engineering-Transactions of the ASME* 127: 1046–1053.

Joo, W. K., Kim, B. I., Bae, S. I., Kim, C. S., Song, J. I. 2007. Mechanical properties on nanoindentation measurements of osteonic lamellae in a human cortical bone. *Key Engineering Materials*, vol. 353–358, PART 3, *Progresses in Fracture and Strength of Materials and Structures* (Selected peer reviewed papers from the *Asian Pacific Conference Fracture and Strength 2006 (APCFS'06)*), pp. 2248–2252.

Katti, D. R., Katti, K. S., Sopp, J. M., Sarikaya, M. 2001. 3D finite element modeling of mechanical response in nacre-based hybrid nanocomposites. *Computational and Theoretical Polymer Science* 11: 397–404.

Katz, J. L., Misra, A., Spencer, P., Wang, Y., Bumrerraj, S., Nomura, T., Eppell, S. J., Tabib-Azar, M. 2007. Multiscale mechanics of hierarchical structure/property relationships in calcified tissues and tissue/material interfaces. *Materials Science and Engineering C, Next Generation Biomaterials* 27: 450–468.

Kaufman, J. D., Klapperich, C. M. 2005. Nanomechanical testing of hydrated biomaterials: Sample preparation, data validation and analysis. *Materials Research Society Symposium Proceedings* 841: 69–74.

Kavukcuoglu, N. B., Denhardt, D. T., Guzelsu, N., Mann, A. B. 2007. Osteopontin deficiency and aging on nanomechanics of mouse bone. *Journal of Biomedical Materials Research—Part A* 83: 136–144.

Kinney, J. H., Marshall, S. J., Marshall, G. W. 2003. The mechanical properties of human dentin: A critical review and re-evaluation of the dental literature. *Critical Reviews in Oral Biology & Medicine* 14: 13–29.

Kohane, M., Daugela, A., Kutomi, H., Charlson, L., Wyrobek, A., Wyrobek, J. 2003. Nanoscale *in vivo* evaluation of the stiffness of *Drosophila melanogaster* integument during development. *Journal of Biomedical Materials Research Part A: Applied Biomaterials* 66A: 633–642.

Lee, C. S., Jho, J. Y., Choi, K., Hwang, T. W. 2004. Dynamic mechanical behavior of ultra-high molecular weight polyethylene irradiated with gamma rays. *Macromolecular Research* 12: 141–143.

Lee, S. W., Kim, G. H., Choi C. S. 2008. Characteristic crystal orientation of folia in oyster shell, *Crassostrea gigas*. *Materials Science and Engineering C* 28: 258–263.

Lewis, G., Xu, J., Dunne, N., Daly, C., Orr, J. 2006. Critical comparison of two methods for the determination of nanomechanical properties of a material: Application to synthetic and natural biomaterials. *Journal of Biomedical Materials Research Part B: Applied Biomaterials* 78B: 312–317.

Li, X., Chang, W., Chao, Y. J., Wang, R., Chang, M. 2004. Nanoscale structural and mechanical characterization of a natural nanocomposite material: The shell of red abalone. *Nano Letters* 4: 613–617.

Li, C., Pruitt, L., King, K. 2006. Nanoindentation differentiates tissue-scale properties of native articular cartilage. *Journal of Biomedical Materials Research* 78: 729–738.

Loubet, J. L., Lucas, B. N., Oliver, W. C. 1995. Some measurements of viscoelastic properties with the help of nanoindentation. National Institute of Standards Special Publication 896, *Conference Proceedings International Workshop on Instrumented Indentation*, San Diego, CA, pp. 31–34.

Loubet, J. L., Oliver, W. C., Lucas, B. N. 2000. Measurement of the loss tangent of low-density polyethylene with nanoindentation technique. *Journal of Materials Research* 15: 1195–1198.

Low, D., Sumiib, T., Swain, M. V., Ishigamid, K., Takedad, T. 2002. Instrumented indentation characterisation of mouth-guard materials. *Dental Materials* 18: 211–215.

Lu, Y. 2004. Significance and progress of bionics. *Journal of Bionics Engineering* 1: 1–3.

Lu, H., Wang, B., Ma, J., Huang, G., Viswanathan, H. 2003. Measurement of creep compliance of solid polymers by nanoindentation. *Mechanics of Time-Dependent Materials* 7: 189–207.

Mann, A. B., Pethica, J. B. 1996. Nanoindentation studies in a liquid environment. *Langmuir* 12: 4583–4586.

Marshall, G. W. Jr., Balooch, M., Gallagher, R. R., Gansky, S. A., Marshall, S. J. 2001. Mechanical properties of the dentinoenamel junction: AFM studies of nanohardness, elastic modulus, and fracture. *Journal of Biomedical Materials Research Part A* 54: 87–95.

Marshall, S. J., Balooch, M., Habelitz, S., Balooch, G., Gallagher, R., Marshall, G. W. 2003. The dentin–enamel junction—A natural, multilevel interface. *Journal of the European Ceramic Society* 23: 2897–2904.

Michelle, D. Mechanical Properties of an Arthropod Exoskeleton. http://www.hysitron.com/page_attachments/0000/0413/Mechanical_Properties_of_an_Arthropod_Exoskeleton.pdf

Mittra, E., Akella, S., Qin, Y. X. 2006. The effects of embedding material, loading rate and magnitude, and penetration depth in nanoindentation of trabecular bone. *Journal of Biomedical Materials Research Part A* 79: 86–93.

Miyajima, T., Nagata, F., Kanematsu, W., Yokogawa, Y., Sakai, M. 2003. Elastic/ plastic surface deformation of porous composites subjected to spherical nanoindentation. *Key Engineering Materials* 240–242: 927–930.

Mohanty, B., Katti, K. S., Katti, D. R., Verma, D. 2006. Dynamic nanomechanical response of nacre. *Journal of Materials Research* 21: 2045–2051.

Mulder, L., Koolstra, J. H., den Toonder, J. M., van Eijden, T. M. 2008. Relationship between tissue stiffness and degree of mineralization of developing trabecular bone. *Journal of Biomedical Materials Research Part A* 84: 508–515.

Ngan, A. H. W., Tang, B. 2002. Viscoelastic effects during unloading in depth-sensing indentation. *Journal of Materials Research* 17: 2604–2610.

Ngan, A. H. W., Wang, H. T., Tang, B., Sze, K. Y. 2005. Correcting power-law viscoelastic effects in elastic modulus measurement using depth-sensing indentation. *International Journal of Solids and Structures* 42: 1831–1846.

Northen, M. T., Turner, K. L. 2006. Meso-scale adhesion testing of integrated micro- and nano-scale structures. *Sensors and Actuators, A: Physical* 130–131: 583–587.

Nowicki, M., Richter, A., Wolf, B., Kaczmarek, H. 2003. Nanoscale mechanical properties of polymers irradiated by UV. *Polymer* 44: 6599–6606.

Odegard, G. M., Gates, T. S., Herring, H. M., Gates, T. S. 2005. Characterisation of viscoelastic properties of polymeric materials through nanoindentation. *Experimental Mechanics* 45: 130–136.

Oliver, W. C., Pharr, G. M. 1992. Improved technique for determining hardness and elastic modulus using load and displacement sensing indentation experiments. *Journal of Materials Research* 7: 1564–1583.

Oliver, W. C., Pharr, G. M. 2004. Measurement of hardness and elastic modulus by instrumented indentation: Advances in understanding and refinements to methodology. *Journal of Material Research* 19: 3–20.

Oyen, M. L., Ko, C. 2008. Indentation variability of natural nanocomposite materials. *Journal of Materials Research* 23: 760–767.

Park, K., Mishra, S., Lewis, G., Losby, J., Fan, Z. F., Park, J. B. 2004. Quasi-static and dynamic nanoindentation studies on highly crosslinked ultra-high-molecular-weight polyethylene. *Biomaterials* 25: 2427–2436.

Park, S., Wang, D. H., Zhang, D., Arola, D. 2007. Property gradients in human enamel: A nanoscopic evaluation. *Proceedings of the SEM Annual Conference and Exposition on Experimental and Applied Mechanics 2007*, vol. 2, Springfield, MA, pp. 1225–1229.

Park, S., Wang, D. H., Zhang, D. S., Romberg, E., Arola, D. 2008. Mechanical properties of human enamel as a function of age and location in the tooth. *Journal of Materials Science: Materials in Medicine* 19: 2317–2324.

Pethica, J. B., Oliver, W. C. 1987. Tip surface interactions in STM and AFM. *Physica Scripta* T19A: 61–66.

Povolo, F., Hermida, E. B. 2000. Measurement of the elastic modulus of dental pieces. *Journal of Alloys and Compounds* 310: 392–395.

Ratner, B. D., Hoffman, A. S., Schoen, F. J., Lemons, J. E. 2000. *Biomaterials Science: An Introduction to Materials in Medicine.* San Diego, CA: Academic Press.

Ren, L. Q. 2001. *Optimum and Analysis of Test Design.* Changchun, China: Jilin Science & Technology Press. (in Chinese).

Rho, J. Y., Pharr, G. M. 1999. Effects of drying on the mechanical properties of bovine femur measured by nanoindentation. *Journal of Materials Science: Materials in Medicine* 10: 485–488.

Rho, J. Y., Roy, M. E., Tsui, T. Y., Pharr, G. M. 1999. Elastic properties of microstructural components of human bone tissue as measured by nanoindentation. *Journal of Biomedical Materials Research* 45: 48–54.

Rho, J. Y., Currey, J. D., Zioupos, P., Pharr, G. M. 2001. The anisotropic Young's modulus of equine secondary osteones and interstitial bone determined by nanoindentation. *Journal of Experimental Biology* 204: 1775–1781.

Rho, J. Y., Zioupos, P., Currey, J. D., Pharr, G. M. 2002. Microstructural elasticity and regional heterogeneity in human femoral bone of various ages examined by nanoindentation. *Journal of Biomechanics* 35: 189–198.

Roy, M. E., Rho, J. Y., Tsui, T. Y., Evans, N. D., Pharr, G. M. 1999. Mechanical and morphological variation of the human lumbar vertebral cortical and trabecular bone. *Journal of Biomedical Materials Research* 44: 191–197.

Sarikaya, M., Fong, H., Sopp, J. M., Katti, K. S., Mayer, G. 2002. Biomimetics: Nanomechanical design of materials through biology. *15th ASCE Engineering Mechanics Conference*, June 2–5, 2002, Columbia University, New York.

Scherge, M., Gorb, S. S. 2001. *Biological Micro- and Nanotribology: Nature's Solutions*. Berlin, Germany: Springer-Verlag.

Schöberl, T., Jäger, I. L. 2006. Wet or dry-hardness, stiffness and wear resistance of biological materials on the micron scale. *Advanced Engineering Materials* 11: 1164–1169.

Song, F., Lee, K. L., Soh, A. K., Zhu, F., Bai, Y. L. 2004. Experimental studies of the material properties of the forewing of cicada (Homóptera, Cicàdidae). *Journal of Experimental Biology* 207: 3035–3042.

Song, F., Xiao, K. W., Bai, K., Bai, Y. L. 2007. Microstructure and nanomechanical properties of the wing membrane of dragonfly. *Materials Science and Engineering A* 457: 254–260.

Stempfle, P., Pantale, O., Njiwa, R. K., Rousseau, M., Lopez, E., Bourrat, X. 2007. Friction-induced sheet nacre fracture: Effects of nano-shocks on cracks location. *International Journal of Nanotechnology* 4: 712–729.

Sun, J. Y., Tong, J., Zhou, J. 2006. Application of nano-indenter for investigation of the properties of the elytra cuticle of the dung beetle (*Copris ochus* Motschulsky). *IEE Proceedings: Nanobiotechnology* 153: 129–133.

Sun, J. Y., Tong, J., Zhang, Z. J. 2007. The specimen preparation methods for nanoindentation testing of biomaterials: A review. *Proceedings of SPIE* 6831: 68310H.1–68310H.6.

Tang, B., Ngan, A. H. W., Lu, W. W. 2007. An improved method for the measurement of mechanical properties of bone by nanoindentation. *Journal of Materials Science: Materials in Medicine* 18: 1875–1881.

Tesch, W., Eidelman, N., Roschger, P., Goldenberg, F., Klaushofer, K., Fratzl, P. 2001. Graded microstructure and mechanical properties of human crown. *Dentin Calcified Tissue International* 69: 147–149.

Tomkoria, S., Masuda, K., Mao, J. 2007. Nanomechanical properties of alginate-recovered chondrocyte matrices for cartilage regeneration. *Proceedings of the Institution of Mechanical Engineers, Part H: Journal of Engineering in Medicine* 221: 467–473.

Tong, J., Ma, Y., Ren, L. 2001. Naturally biological materials and their tribology: A review. *Tribology* 21: 235–240. (in Chinese).

Tong, J., Sun, J. Y., Chen, D. H., Zhang, S. J. 2004. Factors impacting nanoindentation testing results of the cuticle of dung beetle *Copris ochus* Motschulsky. *Journal of Bionics Engineering* 1: 221–230.

Tong, J., Zhao, Y. R., Sun, J. Y., Chen, D. H. 2007. Nanomechanical properties of the stigma of dragonfly *Anax parthenope julius* Brauer. *Journal of Materials Science* 42: 2894–2898.

Urabe, I., Nakajima, S., Sano, H., Tagami, J. 2000. Physical properties of the dentin-enamel junction region. *American Journal of Dentistry* 13: 129–135.

Van Eijden, T. M. G. J., van Ruijven, L. J., Giesen, E. B. W. 2004. Bone tissue stiffness in the mandibular condyle is dependent on the direction and density of the cancellous structure. *Calcified Tissue International* 75: 502–508.

Vanleene, M., Mazeran, P. E., Thoa, M. C. H. B. 2006. Influence of strain rate on the mechanical behavior of cortical bone interstitial lamellae at the micrometer scale. *Journal of Materials Research* 21: 2093–2097.

Wang, X., Yoon, Y. J., Ji, H. 2007a. A novel scratching approach for measuring age-related changes in the *in situ* toughness of bone. *Journal of Biomechanics* 40: 1401–1404.

Wang, X. J., Li, Y. C., Hodgson, P. D., Wen, C. E. 2007b. Nano- and macro-scale characterisation of the mechanical properties of bovine bone. *Materials Forum* 31: 156–159.

Xu, J., Rho, J. Y., Mishra, S. R., Fan, Z. 2003. Atomic force microscopy and nanoindentation characterization of human lamellar bone prepared by microtome sectioning and mechanical polishing technique. *Journal of Biomedical Materials Research Part A* 67: 719–726.

Yamashita, J., Furman, B. R., Rawls, H. R., Wang, X., Agrawal, C. M. 2001. The use of dynamic mechanical analysis to assess the viscoelastic properties of human cortical bone. *Journal of Biomedical Materials Research Part B* 58: 47–53.

Yang, Z. X., Wang, W. Y., Yu, Q. Q., Dai, Z. D. 2007. Measurements on mechanical parameters and studies on microstructure of elytra in beetles. *Fuhe Cailiao Xuebao/ Acta Materiae Compositae Sinica* 24: 92–98.

Zysset, P. K., Guo, X. E., Hoffler, C. E., Moore, K. E., Goldstein, S. A. 1999. Elastic modulus and hardness of cortical and trabecular bone lamellae measured by nanoindentation in the human femur. *Journal of Biomechanics* 32: 1005–1012.

30

Writing with Nanoparticles

Debdulal Roy
National Physical Laboratory

30.1 Introduction

Novel properties of a material appear when particle size approaches the nanometer scale. This makes nanotechnology a new field of research, and motivates scientists and technologists to utilize new properties of nanoparticles for a wide range of applications. For example, semiconducting materials exhibit optical absorption and emission properties similar to that of atoms/molecules, that can be used in medical diagnosis, nanoelectronics, and optics. As the size of platinum particles reduces to a few nanometer, the particles start acting as a catalyst for hydrogenation. Gold changes its optical absorption properties and colors depending on its particle-size. Technologies are being developed with gold nanoparticles for biodiagnostics, cancer treatment, and memory devices. Gold nanoparticles have been used for long-lasting coloring of stain glasses long before the arrival of nanotechnology. Novel properties of these nanoparticles were not utilized primarily because of insufficient understanding of their properties and absence of mechanisms of directing them to the desired locations. Creating features on a surface needs a specific tool or a pen to direct the ink to specific locations. Before human civilization started to use symbols, engraved features on stones and walls were the main method of communication, and the tools that were used for hunting were used as the pen for writing these features on large chunk of stones. A pen with a sharp metal or bone as stylus and iron salt as ink, similar to the fountain pen used today, existed in ancient Greece. In today's context, we know more about the properties of nanoparticles than we knew 20 years ago; however, writing with nanoparticles is still a challenge, especially for industrial applications. In other words, we need a suitable "pen" to put these particles where we wish to create the features. A few examples of applications that require nanoparticles as ink are nanoelectronics [1], nano-biosensors [2], and nanoscopic light sources [3].

In today's life, it is difficult to imagine spending a day without using any transistor; it has changed our quality of life, methods of communication, and level of comfort in the last century. Transistors are being used in medical diagnosis, mobile phones, computers, and music and video players. Since 1950s, the size of transistors has fallen from ~1 cm to few tens of nanometers; in fact, the size of transistors have halved every 18 months. It was first observed by Gordon Moore in 1960s, and even today it is following the same trend (Figure 30.1). According to the plot shown in Figure 30.1, the expected size of transistors in the next 10 years is going to be few nanometers or the size of a few molecules. In all these years, transistors have been made by surface patterning techniques, e.g., photolithography and electron-beam (e-beam) lithography. In these cases, patterning starts from a silicon wafer of millimeters to centimeters in size. The wafer is etched and doped, or new layers are deposited to make the structures for the devices; this is called the top-down approach. An example of a top-down approach that I always refer to is the palace in the city of Petra in Jordan (Figure 30.2), where the entire palace was cut out from a hill. Another example of such a structure on a micrometer scale, prepared by focused ion beam (FIB), is also shown in Figure 30.2b.

Today, we no longer build houses or palaces by cutting a big chunk of stone; we use small building blocks such as bricks to build tall buildings of hundreds of meters in height. To keep Moore's law valid, the electronic industry will move toward using building blocks such as molecules or nanoparticles to fabricate the whole integrated circuit instead of using the top-down approach. There are significant research activities on the fundamental understanding of single-molecule electronics or quantum electronics; however, there are very few techniques to deterministically place the building blocks, i.e., nanoparticles or molecules for fabricating the devices. In this chapter, we will concentrate on placing nanoparticles at a specific position

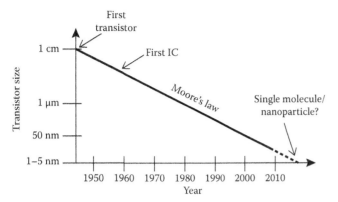

FIGURE 30.1 Schematic plot of change in the size of the transistor over 60 years.

(a)

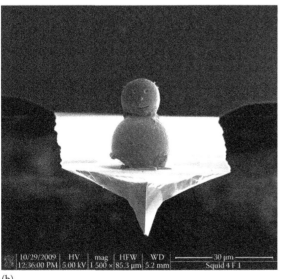

(b)

FIGURE 30.2 (a) Photograph of a palace in the city of Petra: an example of a large-scale structure made by the top-down approach. (Courtesy of Dr. Alex Shard). (b) Scanning electron micrograph of a snowman prepared by FIB.

with reference to the entire device, unlike self-assembled structures where position of particles/molecules with reference to the nearby neighbors is considered [4]. For this reason, only a limited number of techniques can be used to place the nanoparticles at the required positions. The techniques can be divided into two categories: (1) direct writing and (2) guided writing. There are number of variations of the techniques based on the writing steps and the types of instruments used. Each of the methods will be discussed in detail and their pros and cons will be highlighted. Some of the intuitive applications of the devices that can be fabricated using nanoparticles will be detailed in the subsequent sections (Table 30.1).

30.2 Overview of Nanoparticles

Nanoparticles are unique as materials because their properties are in between those of atoms and bulk materials. There are two important characteristics of nanoparticles that have attracted attention in the last two decades: (1) large surface area to volume ratio and (2) electronic properties that are different from bulk properties. If a bulk sphere of radius R is divided into N number of small spheres of radius r, then the total surface area of the material will increase by R/r. Therefore, smaller the size of the particle it is divided into, larger is the ratio. Surface area is important since it defines the boundary between the particle and the surrounding environment. A large surface area provides more access to the particles for various activities such as chemical reactivity and coupling with electrons. For example, due to the large surface area/volume ratio, nanoparticles are used in sensing applications with very high sensitivities. Large surface area/volume also increases the total surface energy of the material. These changes affect other physical parameters such as melting point. The melting point of a nanoparticle lowers as the particle size becomes smaller. Electronic properties of semiconducting and metal nanoparticles are immensely interesting for applications in electronics, optics, and sensing (Figure 30.3).

If we consider nanoparticles as a collection of atoms, the electronic states of the nanoparticles will be the resultant of the interactions between various electrons within the entire crystal. In case of bulk sample, the electronic structure is described as the superposition of atomic orbitals of the atoms positioned periodically within the bulk crystal, and the boundaries are approximated at infinity. This is called Bloch function. For bulk semiconducting materials, the electronic bands are split into valence band and conduction band. In case of nanoparticles, the size of the particles is comparable to de Broglie wavelength of the electrons and the electrons are confined within the boundaries of the surface. The electronic structure is no longer two broad energy bands like that in the bulk, but they are split into discrete energy levels with higher density of states and smaller energy gap. Because of these discrete energy states, semiconducting nanoparticles are also called quantum dots. Metal nanoparticles, on the other hand, go through metal to insulator transition

TABLE 30.1 Different Lithography Techniques

	E-Beam Lithography	Photolithography	Soft Lithography	Dip-Pen Nanolithography
Process	Electron beam is used to create features on a surface by partly exposing it through a mask	UV to visible range light is commonly used to create features by partly exposing it through a mask	An elastomeric stamp or a mould is used to transfer relief features or patterns to a surface	An AFM-like probe is used to deliver the materials onto a surface
Materials that can be patterned	Polymer-based e-beam resists (e.g., PMMA), organometallic compounds for direct writing, nanoparticles for direct sintering	Photosensitive polymer-based resists (photoresists), organometallic compounds, nanoparticles	Self-assembled monolayers (SAM) on suitable surfaces, polymers, beads and nanoparticles, biological molecules, sol-gel materials, conducting polymers	Hydrophilic and amphiphilic molecules, nanoparticles dispersible in water, biological molecules, polymers
Types of surfaces that can be patterned	Planar surfaces	Planar surfaces	Planar and nonplanar surfaces	Planar and nonplanar surfaces (within a limit)
Suitable feature size	10 nm to few micrometers	100 nm to hundreds of micrometers	30 nm to few micrometers	10 nm to few hundreds of nanometers
Application areas	Solid materials, e.g., electronics, MEMS	Solid materials, e.g., electronics, MEMS	Suitable for soft materials and chemically sensitive surfaces, e.g., polymer electronics, biosurfaces	Suitable for soft materials and chemically sensitive surfaces, e.g., biosensors
Cost of production	High investment Expensive	Moderate investment Low	Low investment Low	Moderate investment Low

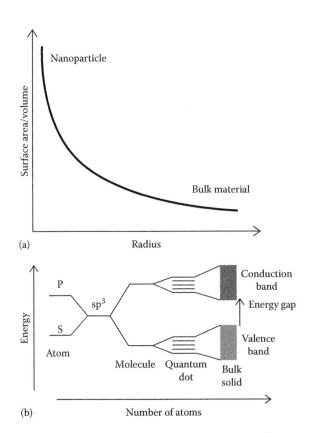

(a)

(b)

FIGURE 30.3 (a) Schematic plot showing surface area/volume ratio with the radius (size) of a particle. (b) Schematic of electronic band structure of an atom, semiconducting nanoparticle (quantum dot), and bulk semiconductor.

when the size is varied; this is termed as size-induced metal to insulator transition (SIMIT) [5,6].

For the technologists, these are interesting materials to exploit the size dependence electronic properties of nanoparticles. To make a device, it is important to understand how much energy is required to add or extract an electron to, or from, a nanoparticle. The concept is similar to chemical affinity of molecules—the amount of energy required to oxidize or reduce a molecule. This is called "charging energy." In a bulk conductor or semiconductor, the number of carriers is infinitely large, and we are not concerned with adding or extracting a single electron. However, in case of nanoparticles, since there are only finite number of electrons, manipulating a single electron is possible and it can be exploited to fabricate devices. The flow of electron in such devices occurs via quantum mechanical tunneling, and the devices are termed as single electron transistors (SET). The charging energy of a nanoparticle depends upon the capacitance of the system and energies of the available energy bands. For example, for a smaller particle, the capacitance (C) can be so small that the available thermal energy at room temperature is enough to inject an electron (charging energy $E_C = e^2/2C$, e is the charge of an electron); therefore, such energy barriers are only observed at low temperatures. This effect is called Coulomb blockade. The first observation of Coulomb blockade was reported by Neugebauer and Webb [7] and then 25 years later by Barner and Ruggiero [8]. Since charging and discharging of the particle depend on the capacitance, now it is possible to design a single electron transistor, which can be controlled by capacitance, similar to field effect transistor that is controlled by a gate. A schematic of such an SET device is given in Figure 30.4 [1] and an example of a nanoparticle-based device will be discussed in Section 30.5.

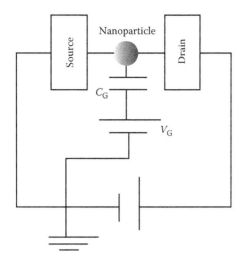

FIGURE 30.4 Schematic circuit diagram for single electron transistor with single nanoparticle.

30.3 Lithography with Nanoparticles

In this section, a brief description of the commonly used techniques for guiding nanoparticles on a surface is given. The techniques have been divided into two groups: (1) direct writing techniques and (2) guided writing techniques. Nanoparticles are directly placed/created on the required position by the technique described in the first group. There are fewer intermediate steps required for these techniques. Since the subject on writing with nanoparticles has just nucleated few years ago, some of the techniques have been elucidated with fewer examples. In the second group, the techniques modify the chemistry of a surface first, and guide the nanoparticles to the specific regions relying on chemical affinity, charge-based interactions, or self-assembling properties. Eventually, both the groups of techniques aim to place nanoparticles in a controlled manner on a surface.

30.3.1 Direct Writing Techniques

30.3.1.1 Dip-Pen Nanolithography

Dip-pen nanolithography (DPN), a direct writing tool, uses a scanning probe microscope tip as the stylus, and uses molecules or nanoparticles in a solvent as ink. Even solid thin films of metals have been used as the source ink and a heated probe to write with. This is different from using AFM/STM tip to manipulate atoms [9,10], molecules or particles [11]. Repositioning of xenon atoms has been demonstrated by Eigler et al. way back in 1990, which was the first demonstration of building structures with atoms (Figure 30.5). Repositioning of nanoparticles has also been demonstrated by Wang et al. [11].

Similar to repositioning nanoparticles using scanning probe microscopy (SPM), optical tweezers have also been used to pattern surfaces by repositioning nanoparticles using an optical trap [12]. These methods are slow and not always suitable for deterministically positioning nanoparticles to make devices. Therefore, there were needs for developing tools to write directly

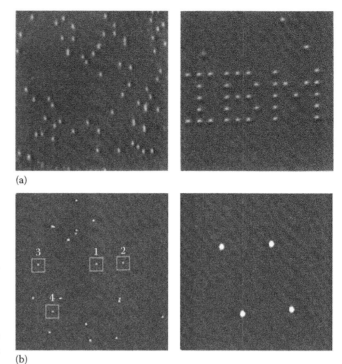

FIGURE 30.5 (a) Patterning with xenon atoms [9] (upper left and right) and (b) repositioning nanoparticles using scanning probe microscopy [11] (lower left and right). (From Wang, Y. et al., *Appl. Phys. Lett.*, 90(13), 133102-1, 2007. With permission.)

on a surface using SPM. When first discovered, an AFM tip of silicon or silicon nitride, coated with molecules or particles, was used as the stylus. The layer of the ink on the AFM tip-shaft acted as the reservoir. Recently, developments have taken place to increase the number of tips as well as improve the ink-delivery system. It is now possible to write larger size wafers at much faster speed, scalable to industrial production line.

Atomic force microscope, a versatile tool in nanotechnology for imaging nanoscale structures, works based on a cantilever force sensor. Deflection of a thin foil of silicon, attached with a sharp tip, is measured by laser deflection. An XYZ-scanning stage is used to control the force between the tip-end and the sample surface. This allows scanning a surface without damaging the tip or the surface. At ambient conditions, there are always layers of water on the tip as well as the surface. These molecules, if dissolvable in water, tend to flow through the tip to the sample surface. The chemical nature of the sample surface plays an important role. The first report of DPN showed writing with thiol-terminated mercaptohexadecanoic acid (MHA) on a gold surface [13]. Thiol-terminated molecules form strong bonds between gold and the sulfur molecules. Such molecules are commonly used in DPN and in forming self-assembled monolayer. A wide range of surfaces are used for writing with various other molecules; in all the cases, chemical affinity between the surface and the ink molecule is very important.

For example, thiol-terminated 16-mercaptohexanoic acid and 1-octadecanthiol are used to write on gold, silazanes to write on

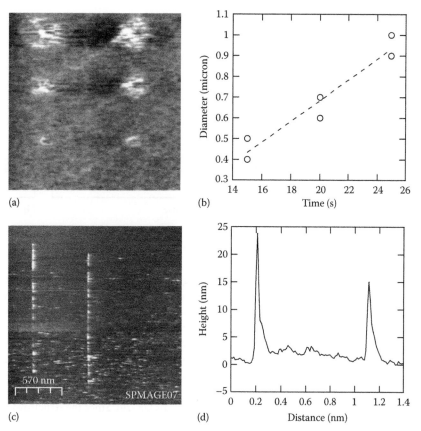

FIGURE 30.6 (a) Dependence of feature size on writing time: (a) features written with CdSe/ZnS nanoparticles for different time, scan size 5 μm × 5 μm. Line leveling was applied to the image (b) variation of diameter of the dots with time. (c) An example of a thin line written with CoFe$_2$O$_4$ nanoparticles, scan size 10 μ × 10 μm. (d) A line cross section of the lines shown in (c). The height and width of the lines in the image are ~25 and ~35 nm, respectively. (From Roy, D. et al., *Appl. Surf. Sci.*, 254(5), 1394, 2007. With permission.)

silicon oxide [15], and alkynes to write on silicon [16]. Molecules of different types such as small-chain organic molecules, long-chain organic molecules, DNA, and protein molecules have been used to write using DPN. There are few attempts to write with nanoparticles on mica [14], silicon, and silicon oxide surfaces [17]. We used CoFe$_2$O$_4$ magnetic nanoparticles to write on freshly cleaved mica surfaces and CdSe/ZnS quantum dots on gold surface (Figure 30.6). Features of different sizes and shapes have been written using DPN primarily to demonstrate the feasibility of deterministically depositing nanoparticles at specific locations. Diffusion of nanoparticles on the surface plays an important role in determining the sizes of the features: higher diffusivity of the particles broadens the features even though they are delivered with a sharp tip for a short period of time (few tens of microseconds).

When molecules are used as the ink, they diffuse from the tip to the surface through a meniscus formed at the tip end, and nanoparticles are believed to behave in a similar way. According to Fick's law of diffusion

$$C(x,t) = C_s \mathrm{erfc}\left[\frac{x}{2\sqrt{Dt}}\right] \qquad (30.1)$$

where

$C(x, t)$ is the concentration at a distance x at time t
C_s is the concentration at the source (which is the tip end)
D is the diffusion coefficient

The size of the dots are expected to vary with $\sqrt{t}$, and in case of writing lines, the width of the lines will depend on the speed of writing. However, for short period of writing time ($t < 25\,s$) in this experiment, the sizes of the dots were observed to be proportional to the time of writing. Clearly the spot size depends on the wettability of the surface; freshly prepared gold being hydrophilic, a larger spot size is plausible. Wu et al. performed similar experimentation with citrate capped gold nanoparticles and amine functionalized silicon surface [18]. They found a nonlinear relationship between the spot size with time: $d = 0.47 \times t^{0.184}$. Considering the deposition mechanism, it is expected that the spot size will depend on the size of the probe, wettability of the surface, mobility of the nanoparticles, dwell time, and evaporation rate. If the nanoparticles are bound to the surface either chemically or electrostatically, the spot-size will be smaller. On the other hand, if the rate of evaporation of water is fast enough (dependant on humidity of the environment), there will be lesser time for the particles to diffuse. Since flow of molecules or nanoparticles are taking place through the water meniscus at the tip-end, wettability of water plays an important role; therefore, prior to diffusion of the ink on the surface, the spot-size depends on both the probe-size and wettability of the surface. The description above assumes that the ink molecules are water soluble/dispersible. For amphiphilic molecules, it is reported that transport of the molecules takes place through the meniscus interface [19]. Nafday et al. showed formation of hollow dots when MHA was used as the ink, whereas filled dots are formed when written with ODT [19]. Although no work has yet been reported where amphiphilic nanoparticles were used, and further investigation is required to understand the transport mechanism.

30.3.1.2 E-Beam Lithography

E-beam lithography has made fabrication of structures of sizes below the diffraction limit possible and structures of below 20 nm have been reported regularly since 1990. This is one of the mask-less techniques, which has helped to continue reducing the size of the structures in accordance with Moor's law for more than a decade. Typically, an electron beam sensitive resist is used to form a thin film on a substrate, and then parts of it are exposed to the e-beam. Then, either the exposed parts or the unexposed parts are dissolved in a suitable solvent to develop the structures. The most important parameter in e-beam lithography is the electron dose, measured in Coulomb/unit area, calculated from the total charge of incident electrons. FIB can also be used to form the structures exactly the same way—by using ions instead of electrons. Since ions are heavier than electrons, they can penetrate more inside a film and, therefore, the dose required to form the structures is lower than that required in e-beam lithography. Direct writing of nanoparticles with e-beam and FIB has been demonstrated by many research groups [20–24]. The basic principle of directly forming nanoparticles is by exposing either a thin layer of nanoparticles or by reducing a layer of organometallic compound.

Unlike using resist for e-beam lithography for developing the patterns, often nanoparticles or organometallic compounds are directly used to write the structures (Figures 30.7 and 30.8).

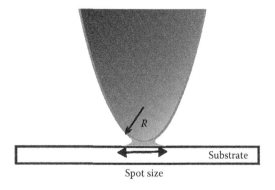

FIGURE 30.7 Wetting of the surface at the tip–substrate interface. The spot size depends on the size of the probe and wettability of water on the surface.

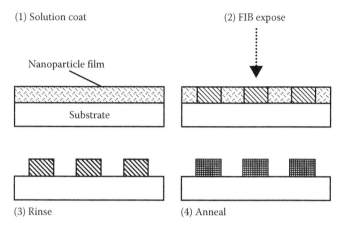

FIGURE 30.8 Schematics of the steps for forming nanoscale islands made with nanoparticles with FIB.

Organometallic compounds or complexes form a bond or a coordination bond between a metallic atom and carbon atoms. These bonds are unstable to e-beam, ion beam, deep-UV radiation, or thermal energy. Organometallic compounds used in this way for e-beam lithography are often described as positive resist. Few examples of nanostructures formed with organometallic compounds and nanoparticles are given in Table 30.2. Nanoparticles capped with various organic ligands are also used as source materials [25–29]. Since the structures are formed by decomposing the compounds (organometallic compounds or the ligands) during writing, there is always some shrinkage of the structures. This, in fact, is favorable in some cases, where making smaller feature size is the primary aim.

Lin et al. used 1-dodecanethiol coated gold nanoparticles to form a monolayer on silicon nitride substrate surface [27]. Upon exposure to e-beam, the ligands were removed from the nanoparticles, leaving the particles adhere to the substrate. When the substrate surface was washed with solvent, the nanoparticles with ligands attached to the core could be removed, leaving the exposed areas as patterns of gold nanostructures.

TABLE 30.2 Examples of Structures Written by E-Beam Lithography and FIBs

Nanostructure	Source of Material	Method of Writing
Gold nanostructure	Dodeca-(triphenylphosphine), hexa(chloro)pentapentacontagold $Au_{55}(PPh_3)_{12}Cl_6$ [30]	FIB
	Gold stabilized with $C_{16}H_{33}S$ [28,31]	E-beam
	Gold stabilized with 1-dodecanethiol [27]	E-beam
	Gold and silver passivated with thiols [24,25]	E-beam
Palladium nanostructure	Palladium acetate [32]	FIB
	Palladium stabilized with $(C_8H_{17})_4N_1Br^-$ [26]	E-beam
Silver nanostructure	Oleic acid stabilized Ag [20]	FIB

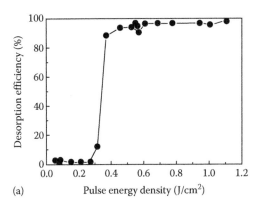
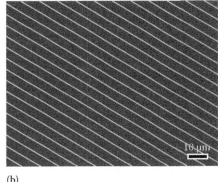

(a) (b)

FIGURE 30.9 (a) Plot showing the critical energy required for desorption of nanoparticles. (b) Scanning electron micrograph of lines written with nanoparticles. (From Kim, H. et al., *J. Appl. Phys.*, 102(8), 083505, 2007. With permission.)

Reetz et al. used $(C_8H_{17})_4NBr^-$ stabilized Pd nanoparticles (2 nm diameter) to form a film of nanoparticles by spin-coating on GaAs substrate [26]. An electron dose of 200,000 μC/cm² resulted in the formation of lines having width of 132 nm, subsequently reduced to 67 nm by thermal annealing.

Harriot et al. used palladium acetate to form a film (thickness ~120 nm) on oxidized silicon by spin coating [32]. FIB with 20 kV Ga ions was used to write features by decomposing the organometallic compound. During decomposition, the features shrink by a factor of 8.

Hoffmann et al. used the gold-cluster compound dodeca-(triphenylphosphine), hexa(chloro)pentapentacontagold, and $Au_{55}(PPh_3)_{12}Cl_{16}$ to prepare a film on thermally oxidized silicon [30]. It was dissolved in dichloromethane and coated by spinning the substrate to make films having thickness of 150 nm. Writing was performed at varying speeds in the range of 50–2000 μm/s using FIB. For electrical applications, resistivity of the structures is an important parameter. There has been much endeavor in improving the resistivity by altering the writing process, increasing the dose, and thermal annealing after writing the structures.

30.3.1.3 Photolithography

Photolithography is one of the most commonly used techniques for microfabrication. First, the substrate is coated with thin film of a photoresist by spinning it. Optical radiation (UV to visible range) is used to transfer the structures from a photomask to the photoresist by exposing the resist to the radiation. A series of chemical treatments then form the desired structures on the surface by removing the unwanted areas of the film. For conventional photolithography, the structure is then used for further etching or deposition of other substances. Photolithography is used in a different way for writing with nanoparticles. Similar to e-beam and ion beam lithography, it is often used to decompose light sensitive organometallic compounds or remove layers of nanoparticles by desorption from the surface.

Kim et al. utilized laser-induced desorption of silver nanoparticles to pattern a surface [33]. They used Nd:YAG pulsed laser (wavelength 1064 nm) with varying pulse time to study the desorption behavior. It was observed that there was a step

function on the desorption efficiency as a function of pulse energy density (Figure 30.9). The efficiency was measured by comparing the area density of nanoparticles before and after irradiation by a single 10 ns pulse.

As discussed before, metallization of organometallic compound is often used as a source of metal for forming nanoparticles [34,35]. Yin et al. used metal complex as a precursor, palladium (II) bis(acetylacetonato), Pd-(acac)₂ to form Pd nanoparticles in PMMA using UV radiation. The organometallic compounds were vaporized in a nitrogen atmosphere at 180°C and absorbed into a polymer film. UV radiation decomposes the compound and forms Pd nanoparticles.

UV laser of wavelength shorter than 300 nm was used to pattern the surface. Examples of such nanoparticles formed in PMMA are shown in Figure 30.10. Similar to e-beam and opti-

FIGURE 30.10 Formation of Pd nanoparticles using reduction of an organometallic compound using UV radiation. TEM image of microscale, patterned polymer surface (a), magnified image of exposed and masked regions showing the nanoparticles (b and c). (From Yin, D.H. et al., *Langmuir*, 21(20), 9352, 2005. With permission.)

cal radiation, thermal decomposition is also exploited to form nanoparticles [36].

30.3.2 Guided Writing Techniques

Unlike direct writing techniques, guided writing involves more than one step: the features are normally written using lithography techniques for defining the structures by chemical modification or modification of surface charge, and then self-assemble prefabricated nanoparticles at the defined sites.

An example of the complex structures created with gold nanoparticles has been demonstrated by Liu et al. in Figure 30.11 [37]. This indirect method of writing involved self-assembly of monolayers of *n*-octadecyltrichlorosilane with CH$_3$-group at the end; this group is then modified using tip-mediated local oxidation to –COOH group and subsequently to –NH$_2$ termination. Nanoparticles were fabricated ex-situ with citrate cap on them; these particles are negatively charged due to the very chemical nature of the citrate group. When self-assembled, they prefer to assemble themselves onto the NH$_2$ functionalized sites.

Garno et al. have demonstrated another method of indirect writing—instead of deposition of the chemical functional group, this method relies on the partial removal of the resist film from the surface, often described as nanoshaving, and then assemble the nanoparticles at the exposed sites [38]. This way, one can deposit nanoparticles at the specific sites and suppress diffusion of the particles on the surface. Nanoshaving has been used by many researchers to create nanoscale structures and as an alternative method to other lithography techniques such as e-beam lithography and photolithography. In this method, a self-assembled monolayer is formed on a surface, and then structures are created by ploughing through the monolayer and exposing the surface underneath (Figure 30.12). Subsequently, the exposed surface is chemically modified and the nanoparticles are self-assembled onto the modified regions. In these steps, surface chemistry of the monolayer, the surface underneath, as well as the chemistry

of the nanoparticles play very important roles. For example, the chemistry of the monolayer and the nanoparticles are designed in such a way that nanoparticles are preferentially deposited onto the exposed surface underneath. This usually involves four steps: (1) preparation of self-assembled monolayer, (2) creation of the structures using nanoshaving, (3) assembly of nanoparticles, and (4) removal of excess nanoparticles. Garno et al., however, used another technique (Figure 30.9b), which is a combination of nanoshaving and DPN to deposit the nanoparticles as the tip ploughs through the monolayer and exposes the surface underneath [38]. This technique has been termed as nano-pen-read-write (NPRW) due to its capability to perform both imaging and depositing the nanoparticles.

30.3.2.1 Charge-Based Writing

Similar to chemical patterning, surface charge can also be patterned on a surface and subsequently direct the suitably charged nanoparticles onto the defined areas. This is often termed as electrostatic funneling. When nanoparticles are dispersed, often they are capped with charged molecules, which deter agglomeration of the nanoparticles in a colloidal solution. For example, gold nanoparticles are often capped with citrate molecules, which are negatively charged. When the substrate is patterned with positive and negative charges, due to electrostatic interaction, negatively charged nanoparticles will be preferentially placed onto the positively charged areas (Figure 30.13).

Huang et al. have shown that single gold nanoparticles could be placed in an individual APTES functionalized spots (Figure 30.14) [39]. Once one nanoparticle occupies its place in a positively charged spot, it repels another negatively charged particle to come to the same spot. Charge of the particle, APTES, and SiO$_2$—all three are important in ensuring that only one particle sits in one spot. Debye length, the distance over which electrostatic interactions become greatly attenuated, plays an important role in placing the particles: if the Debye length is too short,

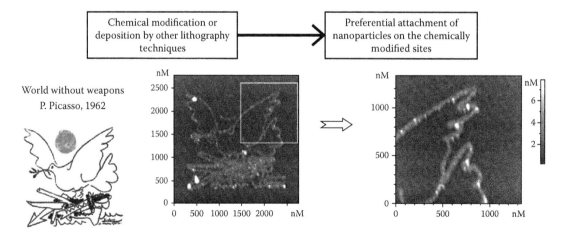

FIGURE 30.11 Demonstration of forming more complex structures by Liu et al. [37]. Intended complex structure (world without weapons by Picasso) to be patterned with nanoparticles (left), topography image of structure formed with nanoparticles (middle and right). For details, see the reference. (From Liu, S.T. et al., *Nano Lett.*, 4(5), 845, 2004. With permission.)

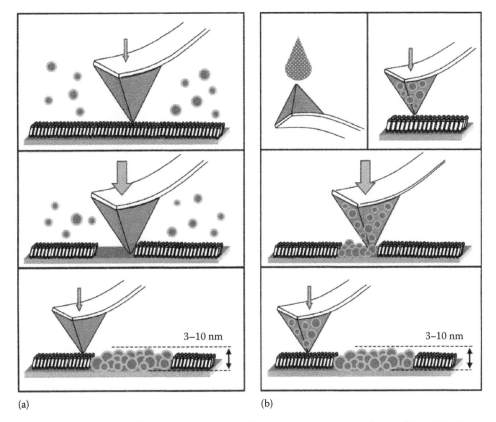

(a)   (b)

FIGURE 30.12 Garno et al. have demonstrated the precise positioning of nanoparticles using nanoshaving followed by deposition of nanoparticles [38]. Method (a) relies on positioning the nanoparticles from colloidal solution onto the shaved regions and method (b) uses nanoparticles as the ink, similar to DPN, and deposits onto the shaved region directly. (From Garno, J.C. et al., *Nano Lett.*, 3(3), 389, 2003. With permission.)

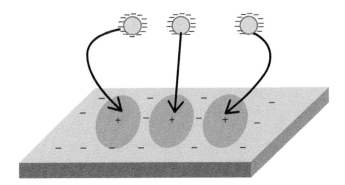

FIGURE 30.13 Schematic diagram showing patterned surfaces with positive and negative charges and preferential deposition of negatively charged nanoparticles.

the particles will be pulled to the surface until they are very close to it, if it is too long, then the particles will be screened by the negatively charged surfaces.

Debye length is obtained as $\lambda = \sqrt{\left(\varepsilon k T / 2e^2 I\right)}$, where ε is permittivity, k is Boltzmann constant, T is temperature in K, e is charge of an electron, and I is the charge of the particle. Generally, the Debye length reaches few tens of nanometers. Once the particle is pulled onto a spot, then it moves to a region to minimize the energy of interaction and it will repel any other

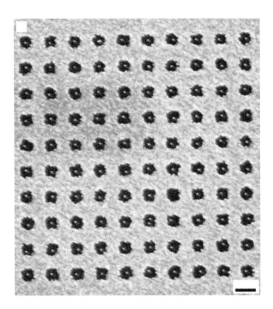

FIGURE 30.14 Assembly of single gold nanoparticles using the electrostatic funneling technique. The bright spots inside the dark circles are individual gold nanoparticles; dark circles are APTES-functionalized silicon oxide and the bright regions are MHA-functionalized gold surface. (Reprinted from Huang, H.W. et al., *Appl. Phys. Lett.*, 93(7), 073110, 2008. With permission.)

particle coming to that spot. This way it is possible to place single nanoparticle to a spot defined a priori.

30.3.2.2 Microcontact Printing

Although photolithography and e-beam lithography are by far very popular techniques for microfabrication, microcontact printing is finding its application in soft lithography, especially with sensitive materials such as biological molecules and organic polymers. An elastomeric block with a patterned relief structure is used to transfer molecules to the surface. Polydimethylsiloxane (PDMS) is commonly used to make the stamp. Silicon rubber, polyeurathenes, and polyimides are also used for making the stamp. PDMS is elastic in nature and makes conformal contact, which helps to form the monolayer. Its elasticity and low interfacial energy ($\sim 22\,mJ/m^2$) help to avoid sample damage [40]. The other advantages of using PDMS are isotropic nature, thermal stability, anhygroscopic nature, and permeability to gases. These make PDMS stamp durable and usable for a long time.

A schematic of the steps of forming structures with nanoparticles using microcontact printing is shown in Figure 30.15. An elastomeric stamp is prepared by casting the polymer in a mold having the relief structure on its surface (often a silicon structure or diffraction grating) and then curing it. Normally there is a small amount of shrinkage ($\sim 1\%$) of the polymer and peeling off is not very difficult. First, organic molecules are attached to the elastomeric stamp and transferred to the surface by stamping. Alkanethiol molecules having –SH groups in them are commonly used to print a monolayer of the molecules on gold [41].

The final stage of printing with nanoparticles is the attachment of the nanoparticles at the printed regions by self-assembling process. It relies on chemical affinity between the capping molecules on the nanoparticles and those printed on the surface, and wettability of the printed region by the solvent containing the nanoparticles. If the solvent wets the printed surface, a 2D monolayer is formed. Wettability of the bare surface (outside the printed region) also plays an important role in confining the nanoparticles onto the printed region. If the chemical affinity of the nanoparticles to the printed region is not strong enough, then the particles would tend to form a 3D superlattice [42], depending on their size distribution. Clearly, chemistry of the printed surface and that of the nanoparticles needs to be matched for stronger interaction. Interactions between the nanoparticles and the surface would be the most crucial factor for successfully placing the nanoparticles in the desired locations.

It is worth mentioning that there are various other soft-lithography techniques such as replica molding, microtransfer molding, micromolding in capillaries, solvent assisted micromolding, phase-shift photolithography, cast molding, embossing, and injection molding that are also used for modifying surface chemistry. A very good review has been done by Xia and Whitesides [40].

30.3.2.3 Chemical Modification of a Self-Assembled Monolayer by E-Beam and UV Radiation and Electrical Discharge

Similar to creating patterned self-assembled monolayer, e-beam assisted chemical modification of self-assembled monolayer has also been used for guiding nanoparticles onto specific locations [43,44].

Deep-UV sources such as ArF (193 nm) and KrF (248 nm) lasers, low and high pressure mercury lamps are often used to irradiate a monolayer and chemically modify the functional group (Figure 30.16). Calvert et al. formed self-assembled monolayer of phenyltrochlorosilane (PTCS) and chloromethylphenyltrimethoxysilane (CMPTS) from nonaqueous solvent on thermally oxidized silicon [45,46]. In this case, upon exposure to UV radiation, cleavage of Si–C-bonds led to local modification

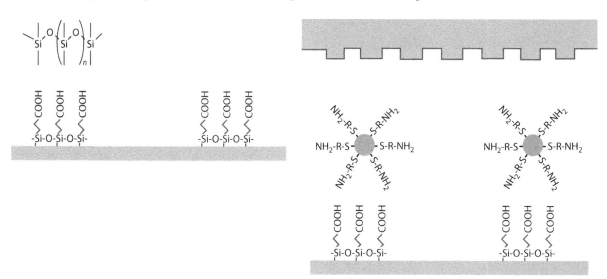

FIGURE 30.15 Molecular structure of PDMS (upper left); schematic diagram of a PDMS stamp (upper right); schematic diagram of a chemically modified surface by printing with a PDMS stamp (lower left); attachment of nanoparticles on the chemically modified surface (lower right).

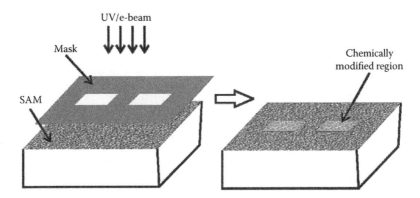

FIGURE 30.16 Schematic diagram of the steps of chemical modification of a SAM surface by optical or e-beam lithography.

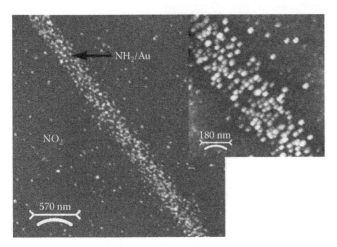

FIGURE 30.17 Topography image of gold nanoparticles assembled onto e-beam-irradiated 3-(4-nitrophenoxy)-propyltrimethoxysilane (NPPTMS) monolayer. (Reprinted from Mendes, P.M. et al., *Langmuir*, 20(9), 3766, 2004. With permission.)

of the exposed region. Mendes et al. used 3-(4-nitrophenoxy)-propyltrimethoxy-silane (NPPTMS) in anhydrous THF under an Ar atmosphere to form a self-assembled monolayer on oxidized silicon [47]. E-beam was used to chemically modify the $-NO_2$ group to NH_2 group. Finally, this modified surface was exposed to an acidic solution of colloidal gold nanoparticles, and the particles were selectively adhered to the NH_2 functionalized regions (Figure 30.17).

Other than chemical attachment, patterns have been created by placing nanoparticles by physical means in trenches created by resist structures and e-beam or optical lithography [48]. It is not difficult to anticipate that the arrangement of such structures with nanoparticles occurs purely by the nature of the interactions between particles and the physical shape of the space.

Field-induced oxidation (FIO), similar to e-beam and optical radiation, is another method of modifying a self-assembled monolayer [49]. Yonezawa et al. have reported fabrication of 1D nanostructure with silver nanoparticles using FIO by an AFM-tip. They used trimethoxysilyl-*n*-propyl-*N,N,N*-trimethylammonium chloride (TTCl) and *n*-octyltrichloride (OTS) to form hydrophilic and hydrophobic SAM, respectively, on silicon. Using the silicon tip as negative electrode, the monolayer is oxidized to

silicon. Then the patterned samples were dipped in aqueous dispersion of silver nanoparticles. Due to non-wetting of the solution on the hydrophobic surface, no nanoparticle was attached onto the hydrophobic part of the surface. Silver nanoparticles self-assembled along the hydrophilic (oxidized) regions.

Another way of forming nanoparticles using e-beam lithography is through using resist and depositing the materials. Haynes et al. used a positive resist (ZEP520) to form patterns with the resist first and then deposited gold or silver to form nanoparticles of the metals for studying plasmon behavior (Figure 30.18) [50].

30.4 Technological Implications

Supra-molecule-like properties, possibilities of tuning the optoelectronic and chemical properties by changing size and shape, and above all, extremely high surface area to volume ratio make nanoparticles attractive for technological applications. There are demands for miniaturization of devices, reduction of energy consumption, and making less negative impact to the environment. On top of that, there are urgent requirements for developing technologies such as detecting traces of disease elements, faster computing systems, and replacing existing technologies that are less sustainable. Technologies with nanoparticles have started developing less than 10 years ago and some of the technologies have been demonstrated where nanoparticles were placed in an unguided manner. These technologies will require novel lithographic techniques to position the particles. In this section, we will discuss applications where deterministic positioning of nanoparticles will be required. These examples are obviously not exhaustive. In fact, it is up to one's imagination to find novel applications of nanoparticles in developing advanced technologies.

30.4.1 Nanoelectronics

Electrical measurements on single nanoparticles have been performed in the past using a scanning tunneling microscope (STM) [51–53]. It is easier to identify a single particle with an STM for electrical measurements than placing a single particle

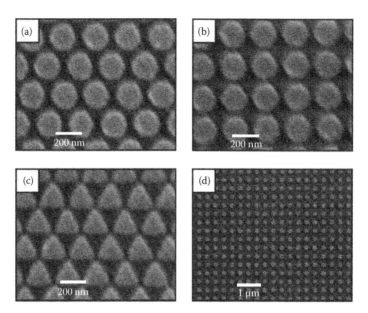

FIGURE 30.18 Scanning electron microscope images of nanostructures written by electron beam lithography by directly irradiating ZEP520, dissolving the unexposed parts, and then depositing silver or gold for studying the plasmon behaviors. (From Haynes, C.L. et al., *J. Phys. Chem. B*, 107(30), 7337, 2003. With permission.)

between two electrodes. However, for development of nanoelectronics using nanoparticles, it will be required to position nanoparticles at the right locations, e.g., between the electrodes. Klein et al. performed Coulomb blockade experiments with quantum dots placed in between two gold electrodes (Figure 30.19). They observed different Coulomb blockade regions by varying the gate voltage. They could measure the energy required to add/extract an electron to/from the nanoparticle. SET has been demonstrated by Sato et al. using thiol linked gold nanoparticles [54]. Junno et al. performed the measurements by preparing the sample using e-beam lithography [55]. A review on single electron transistor published by Likharev is recommended for further detail [56].

This is probably the beginning of SET research, and no doubt, there is a long way toward fabricating millions of devices that exist in a computer chip. Lithography techniques will be essential to place nanoparticles connected to the electrodes and control it by the gate. Carbon nanotubes are another candidate for nanoelectronics. The first CNT transistor was fabricated in 1998 by Tans et al. [58]. In the initial days, CNTs were spread on metal electrodes. The electrical contact between the nanotubes and the electrodes were poor in these cases. In later reports, the electrodes were deposited onto the nanotubes to improve the contact between CNT and the metal electrodes [59]. Deposition of CNT to make transistors is still an important issue for development of CNT-based integrated circuits. There are many efforts to guide CNT to the regions between the electrodes [60,61]. An example of array of nanotubes prepared by guided self-assembly is shown in Figure 30.20. A good review of the recent development of carbon nanotube

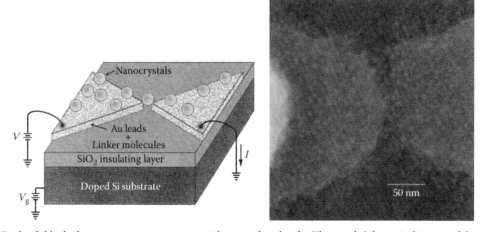

FIGURE 30.19 Coulomb blockade measurements on nanoparticles as undertaken by Klein et al. Schematic diagram of the experimental setup (left) and SEM image of the electrodes and nanoparticles (right). (From Klein, D.L. et al., *Nature*, 389(6652), 699, 1997. With permission.)

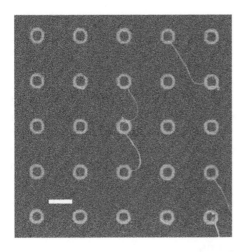

FIGURE 30.20 Topography image of array of structures of single-wall carbon nanotubes patterned by guided self-assembly. This technique can potentially be used to fabricate devices using carbon nanotubes. (From Wang, Y.H. et al., *Proc. Natl. Acad. Sci. USA*, 103(7), 2026, 2006. With permission.)

has been published by Avouris and Chen [62]. There is no report of directly writing carbon nanotubes for developing the devices yet.

30.4.2 Biosensors

Probably the biggest technological applications of nanoparticles beside that in materials science is in biodiagnostics. Nanoparticles hold very high potential for usage in biosensors

due to its very high surface area to volume ratio, which can increase the sensitivity of detection. Velev and Kaler have demonstrated an electronic immunosensor by guiding nanoparticles into the gap between two electrodes using dielectrophoresis [2]. Lee et al. used DPN to functionalize a gold surface with 16-MHA. MHA is negatively charged and it can specifically attach a positively charged IgG. After attachment, it retains its biological activity to its target antigen. The bare surface (other than the modified areas) and unmodified MHA were passivated with BSA to avoid unwanted interaction with the antigen (Figure 30.21). This way, HIV-1 nucleic acid target from as small as a 1 μL solution could be detected.

There are not many published works where nanoparticles were selectively placed for electronic or electrochemical sensing. However, it is clear that there are opportunities for developing highly sensitive biosensors using soft lithography and positioning nanoparticles at high density arrays.

30.4.3 Nanoparticles as Memory Elements

There are strong needs for higher densities in static and dynamic memory devices. Due to their size, nanoparticles can potentially increase the memory density manifold, reduce energy consumption significantly, and make the speed of operation faster. An example of an array of ~10 nm iron oxide nanoparticles patterned by Xiaogang et al. on an MHA functionalized surface is shown in Figure 30.22 [64].

There is also interest in using phase change materials such as gallium, which undergoes structural transition when excited

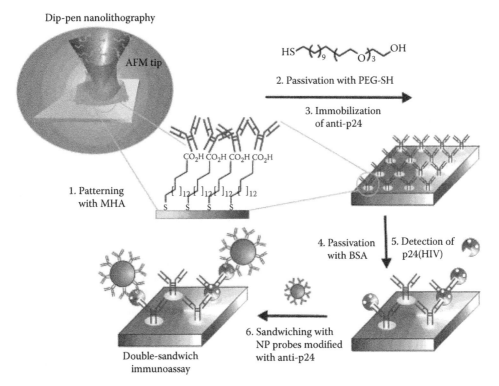

FIGURE 30.21 Schematic of fabricating immunoassays by chemical modification using DPN for the detection of HIV-1 p24 antigen with anti-p24 antibody. (From Lee, K.B. et al., *Nano Lett.*, 4(10), 1869, 2004. With permission.)

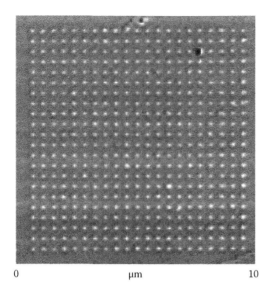

0   μm   10

FIGURE 30.22 Topography image of manganese ferrite nanostructures fabricated using DPN. (From Xiaogang, L. et al., *Adv. Mater.*, 14(3), 231, 2002. With permission.)

optically. The change in reflectivity is detected by another laser beam. It is expected that such techniques will require precise positioning tools to fabricate arrays of particle to manufacture nanoscale memory or optical devices [65–67]. Recently, Novembre et al. have shown charge trapping in gold nanoparticles and pentacine, which can potentially be used as a memory device [68].

30.5 Conclusions and Future Perspectives

It may be apparent now that we will not be able to build nanoscale devices with nanoparticles and take advantage of their novel properties unless we are able to control their positions on a surface. At this stage, there are very few technologies, which use nanoparticles as the building block to fabricate devices. It is anticipated that it will facilitate the development of novel devices that have not been feasible by other methods. There is no doubt that the bottom-up approach will have many advantages as well as challenges. Probably the biggest challenge of using nanoparticles for fabrication of devices is the precise positioning of the nanoparticles on a surface. In this chapter, we have discussed direct writing methods such as DPN, e-beam and optical lithography, and guided methods such as chemical modification of SAM by e-beam, UV radiation, electrochemical oxidation, as well as microcontact printing. Further developments are required in control over deposition of specific number of nanoparticles and precision on the position in the nanometer scale. The choice of technique will very much depend on the type of devices, and physicochemical nature of the surface as well as that of the nanoparticles. It is clear that existing fabrication laboratories can adopt some of the techniques, which involve chemical modification of surfaces, direct formation of nanoparticles by decomposition of organometallic compounds or even conventional lithography steps such as exposing a resist, developing it and then depositing a thin layer of material. Currently, directed self-assembly appears to be most easily adopted by the research laboratories. In future, direct writing techniques such as dip-pen nanolithography, direct formation of nanoparticles by e-beam or UV radiation will have significant advantages. Further research and development are required to improve these methods. The need and development of technologies for writing with nanoparticles will also depend on understanding the physical and chemical properties of nanoparticles and identification of their potential for applications. It needs involvement of physicists, chemists, biologists, material scientists, and engineers to make best use of the novel properties of nanoparticles.

References

1. D.L. Klein, R. Roth, A.K.L. Lim, A.P. Alivisatos, and P.L. McEuen, A single-electron transistor made from a cadmium selenide nanocrystal. *Nature*, 1997, 389(6652): 699–701.
2. O.D. Velev and E.W. Kaler, In situ assembly of colloidal particles into miniaturized biosensors. *Langmuir*, 1999, 15(11): 3693–3698.
3. X.F. Duan, Y. Huang, R. Agarwal, and C.M. Lieber, Single-nanowire electrically driven lasers. *Nature*, 2003, 421(6920): 241–245.
4. M. Brust, D. Bethell, D.J. Schiffrin, and C.J. Kiely, Novel gold-dithiol nano-networks with nonmetallic electronic-properties. *Advanced Materials*, 1995, 7(9): 795–797.
5. G. Nimtz, P. Marquardt, and H. Gleiter, Size-induced metal-insulator transition in metals and semiconductors. *Journal of Crystal Growth*, 1988, 86(1–4): 66–71.
6. A. Snow and H. Wohltjen, Size-induced metal to semiconductor transition in a stabilized gold cluster ensemble. *Chemistry of Materials*, 1995, 10(4): 947–949.
7. C.A. Neugebauer and M.B. Webb, Electrical conduction mechanism in ultrathin, evaporated metal films. *Journal of Applied Physics*, 1962, AIP 33: 74–82.
8. J.B. Barner and S.T. Ruggiero, Observation of the incremental charging of Ag particles by single electrons. *Physical Review Letters*, 1987, 59(7): 807.
9. D.M. Eigler and E.K. Schweizer, Positioning single atoms with a scanning tunneling microscope. *Nature*, 1990, 344(6266): 524–526.
10. J.A. Stroscio and D.M. Eigler, Atomic and molecular manipulation with the scanning tunneling microscope. *Science*, 1991, 254(5036): 1319–1326.
11. Y. Wang, Y. Zhang, B. Li, J.H. Lu, and J. Hu, Capturing and depositing one nanoobject at a time: Single particle dip-pen nanolithography. *Applied Physics Letters*, 2007, 90(13): 133102.
12. J.P. Hoogenboom, D.L.J. Vossen, C. Faivre-Moskalenko, M. Dogterom, and A. van Blaaderen, Patterning surfaces with colloidal particles using optical tweezers. *Applied Physics Letters*, 2002, 80(25): 4828–4830.

13. R.D. Piner, J. Zhu, F. Xu, S. Hong, and C.A. Mirkin, "Dip-pen" nanolithography. *Science*, 1998, 283: 661.

14. D. Roy, M. Munz, P. Colombi, S. Bhattacharyya, J.P. Salvetat, P.J. Cumpson, and M.-L. Saboungi, Directly writing with nanoparticles at the nanoscale using dip-pen nanolithography. *Applied Surface Science*, 2007, 254(5): 1394–1398.

15. A. Ivanisevic and C.A. Mirkin, "Dip-Pen" nanolithography on semiconductor surfaces. *Journal of the American Chemical Society*, 2001, 123(32): 7887–7889.

16. P.T. Hurley, A.E. Ribbe, and J.M. Buriak, Nanopatterning of alkynes on hydrogen-terminated silicon surfaces by scanning probe-induced cathodic electrografting. *Journal of the American Chemical Society*, 2003, 125(37): 11334–11339.

17. G. Gundiah, N.S. John, P.J. Thomas, G.U. Kulkarni, C.N.R. Rao, and S. Heun, Dip-pen nanolithography with magnetic Fe_2O_3 nanocrystals. *Applied Physics Letters*, 2004, 84(26): 5341–5343.

18. B. Wu, A. Ho, N. Moldovan, and H.D. Espinosa, Direct deposition and assembly of gold colloidal particles using a nanofountain probe. *Langmuir*, 2007, 23(17): 9120–9123.

19. O.A. Nafday, M.W. Vaughn, and B.L. Weeks, Evidence of meniscus interface transport in dip-pen nanolithography: An annular diffusion model. *Journal of Chemical Physics*, 2006, 125(14): 144703.

20. D.S. Kong, J.S. Varsanik, S. Griffith, and J.M. Jacobson, Conductive nanostructure fabrication by focused ion beam direct-writing of silver nanoparticles. *Journal of Vacuum Science & Technology B*, 2004, 22(6): 2987–2991.

21. E.M. Hicks, S.L. Zou, G.C. Schatz, K.G. Spears, R.P. Van Duyne, L. Gunnarsson, T. Rindzevicius, B. Kasemo, and M. Kall, Controlling plasmon line shapes through diffractive coupling in linear arrays of cylindrical nanoparticles fabricated by electron beam lithography. *Nano Letters*, 2005, 5(6): 1065–1070.

22. D.H. Yin, S. Horiuchi, M. Morita, and A. Takahara, Tunable metallization by assembly of metal nanoparticles in polymer thin films by photo- or electron beam lithography. *Langmuir*, 2005, 21(20): 9352–9358.

23. M.K. Corbierre, J. Beerens, and R.B. Lennox, Gold nanoparticles generated by electron beam lithography of gold(I)-thiolate thin films. *Chemistry of Materials*, 2005, 17(23): 5774–5779.

24. Y. Chen, R.E. Palmer, and J.P. Wilcoxon, Sintering of passivated gold nanoparticles under the electron beam. *Langmuir*, 2006, 22(6): 2851–2855.

25. S. Griffith, M. Mondol, D.S. Kong, and J.M. Jacobson, Nanostructure fabrication by direct electron-beam writing of nanoparticles. *Journal of Vacuum Science & Technology B*, 2002, 20(6): 2768–2772.

26. M.T. Reetz, M. Winter, G. Dumpich, J. Lohau, and S. Friedrichowski, Fabrication of metallic and bimetallic nanostructures by electron beam induced metallization of surfactant stabilized Pd and Pd/Pt clusters. *Journal of the American Chemical Society*, 1997, 119(19): 4539–4540.

27. X.M. Lin, R. Parthasarathy, and H.M. Jaeger, Direct patterning of self-assembled nanocrystal monolayers by electron beams. *Applied Physics Letters*, 2001, 78(13): 1915–1917.

28. T.R. Bedson, R.E. Palmer, T.E. Jenkins, D.J. Hayton, and J.P. Wilcoxon, Quantitative evaluation of electron beam writing in passivated gold nanoclusters. *Applied Physics Letters*, 2001, 78(13): 1921–1923.

29. M.H.V. Werts, M. Lambert, J.P. Bourgoin, and M. Brust, Nanometer scale patterning of Langmuir-Blodgett films of gold nanoparticles by electron beam lithography. *Nano Letters*, 2002, 2(1): 43–47.

30. P. Hoffmann, G. Benassayag, J. Gierak, J. Flicstein, M. Maarstumm, and H. Vandenbergh, Direct writing of gold nanostructures using a gold-cluster compound and a focused-ion beam. *Journal of Applied Physics*, 1993, 74(12): 7588–7591.

31. J.L. Plaza, Y. Chen, S. Jacke, and R.E. Palmer, Nanoparticle arrays patterned by electron-beam writing: Structure, composition, and electrical properties. *Langmuir*, 2005, 21(4): 1556–1559.

32. L.R. Harriot, K.D. Cummings, M.E. Gross, and W.L. Brown, Decomposition of palladium acetate films with a microfocused ion beam. *Applied Physics Letters*, 1986, 49: 1661.

33. H. Kim, H. Shin, J. Ha, M. Lee, and K.S. Lim, Optical patterning of silver nanoparticle Langmuir-Blodgett films. *Journal of Applied Physics*, 2007, 102(8): 083505.

34. A.Y. Cheng, S.B. Clendenning, G.C. Yang, Z.H. Lu, C.M. Yip, and I. Manners, UV photopatterning of a highly metallized, cluster-containing poly(ferrocenylsilane). *Chemical Communications*, 2004, (7): 780–781.

35. W.Y. Chan, S.B. Clendenning, A. Berenbaum, A.J. Lough, S. Aouba, H.E. Ruda, and I. Manners, Highly metallized polymers: Synthesis, characterization, and lithographic patterning of polyferrocenylsilanes with pendant cobalt, molybdenum, and nickel cluster substituents. *Journal of the American Chemical Society*, 2005, 127(6): 1765–1772.

36. H.F. Hamann, S.I. Woods, and S.H. Sun, Direct thermal patterning of self-assembled nanoparticles. *Nano Letters*, 2003, 3(12): 1643–1645.

37. S.T. Liu, R. Maoz, and J. Sagiv, Planned nanostructures of colloidal gold via self-assembly on hierarchically assembled organic bilayer template patterns with in-situ generated terminal amino functionality. *Nano Letters*, 2004, 4(5): 845–851.

38. J.C. Garno, Y.Y. Yang, N.A. Amro, S. Cruchon-Dupeyrat, S.W. Chen, and G.Y. Liu, Precise positioning of nanoparticles on surfaces using scanning probe lithography. *Nano Letters*, 2003, 3(3): 389–395.

39. H.W. Huang, P. Bhadrachalam, V. Ray, and S.J. Koh, Single-particle placement via self-limiting electrostatic gating. *Applied Physics Letters*, 2008, 93(7): 073110.

40. Y. Xia and G.M. Whitesides, Soft lithography. *Annual Review of Materials Science*, 1998, 28: 153–184.

41. A. Kumar and G.M. Whitesides, Features of gold having micrometer to centimeter dimensions can be formed through a combination of stamping with an elastomeric stamp and an alkanethiol ink followed by chemical etching. *Applied Physics Letters*, 1993, 63(14): 2002–2004.

42. R.L. Whetten, M.N. Shafigullin, J.T. Khoury, T.G. Schaaff, I. Vezmar, M.M. Alvarez, and A. Wilkinson, Crystal structures of molecular gold nanocrystal arrays. *Accounts of Chemical Research*, 1999, 32(5): 397–406.

43. U. Schmelmer, A. Paul, A. Kuller, M. Steenackers, A. Ulman, M. Grunze, A. Golzhauser, and R. Jordan, Nanostructured polymer brushes. *Small*, 2007, 3(3): 459–465.

44. U. Schmelmer, A. Paul, A. Kuller, R. Jordan, A. Golzhauser, M. Grunze, and A. Ulman, Surface-initiated polymerization on self-assembled monolayers: Effect of reaction conditions. *Macromolecular Symposia*, 2004, 217: 223–230.

45. J.M. Calvert, M.S. Chen, C.S. Dulcey, J.H. Georger, M.C. Peckerar, J.M. Schnur, and P.E. Schoen, Deep ultraviolet patterning of monolayer films for high-resolution lithography. *Journal of Vacuum Science & Technology B*, 1991, 9(6): 3447–3450.

46. J.M. Calvert, M.S. Chen, C.S. Dulcey, J.H. Georger, M.C. Peckerar, J.M. Schnur, and P.E. Schoen, Deep ultraviolet lithography of monolayer films with selective electroless metallization. *Journal of the Electrochemical Society*, 1992, 139(6): 1677–1680.

47. P.M. Mendes, S. Jacke, K. Critchley, J. Plaza, Y. Chen, K. Nikitin, R.E. Palmer, J.A. Preece, S.D. Evans, and D. Fitzmaurice, Gold nanoparticle patterning of silicon wafers using chemical e-beam lithography. *Langmuir*, 2004, 20(9): 3766–3768.

48. D.Y. Xia and S.R.J. Brueck, A facile approach to directed assembly of patterns of nanoparticles using interference lithography and spin coating. *Nano Letters*, 2004, 4(7): 1295–1299.

49. T. Yonezawa, T. Itoh, N. Shirahata, Y. Masuda, and K. Kournoto, Positioning of cationic silver nanoparticle by using AFM lithography and electrostatic interaction. *Applied Surface Science*, 2007, 254(2): 621–626.

50. C.L. Haynes, A.D. McFarland, L.L. Zhao, R.P. Van Duyne, G.C. Schatz, L. Gunnarsson, J. Prikulis, B. Kasemo, and M. Kall, Nanoparticle optics: The importance of radiative dipole coupling in two-dimensional nanoparticle arrays. *Journal of Physical Chemistry B*, 2003, 107(30): 7337–7342.

51. H. Vankempen, J.G.A. Dubois, J.W. Gerritsen, and G. Schmid, Small metallic particles studied by scanning-tunneling-microscopy. *Physica B*, 1995, 204(1–4): 51–56.

52. S. Devarajan and S. Sampath, Single electron charging events on Au-Ag alloy nanoclusters. *Chemical Physics Letters*, 2006, 424(1–3): 105–110.

53. K. Schouteden, N. Vandamme, E. Janssens, P. Lievens, and C. Van Haesendonck, Single-electron tunneling phenomena on preformed gold clusters deposited on dithiol self-assembled monolayers. *Surface Science*, 2008, 602(2): 552–558.

54. T. Sato, H. Ahmed, D. Brown, and B.F.G. Johnson, Single electron transistor using a molecularly linked gold colloidal particle chain. *Journal of Applied Physics*, 1997, 82(2): 696–701.

55. T. Junno, M.H. Magnusson, S.B. Carlsson, K. Deppert, J.O. Malm, L. Montelius, and L. Samuelson, Single-electron devices via controlled assembly of designed nanoparticles. *Microelectronic Engineering*, 1999, 47(1–4): 179–183.

56. K.K. Likharev, Single-electron devices and their applications. *Proceedings of the IEEE*, 1999, 87(4): 606–632.

57. Y.H. Wang, D. Maspoch, S.L. Zou, G.C. Schatz, R.E. Smalley, and C.A. Mirkin, Controlling the shape, orientation, and linkage of carbon nanotube features with nano affinity templates. *Proceedings of the National Academy of Sciences of the United States of America*, 2006, 103(7): 2026–2031.

58. S.J. Tans, A.R.M. Verschueren, and C. Dekker, Room-temperature transistor based on a single carbon nanotube. *Nature*, 1998, 393(6680): 49–52.

59. P. Avouris, Carbon nanotube electronics. *Chemical Physics*, 2002, 281(2–3): 429–445.

60. J.B. Hannon, A. Afzali, C. Klinke, and P. Avouris, Selective placement of carbon nanotubes on metal-oxide surfaces. *Langmuir*, 2005, 21(19): 8569–8571.

61. S. Auvray, V. Derycke, M. Goffman, A. Filoramo, O. Jost, and J.P. Bourgoin, Chemical optimization of self-assembled carbon nanotube transistors. *Nano Letters*, 2005, 5(3): 451–455.

62. P. Avouris and J. Chen, Nanotube electronics and optoelectronics. *Materials Today*, 2006, 9(10): 46–54.

63. K.B. Lee, E.Y. Kim, C.A. Mirkin, and S.M. Wolinsky, The use of nanoarrays for highly sensitive and selective detection of human immunodeficiency virus type 1 in plasma. *Nano Letters*, 2004, 4(10): 1869–1872.

64. L. Xiaogang, L. Fu, S. Hong, V.P. Dravid, and C.A. Mirkin, Arrays of magnetic nanoparticles patterned via "Dip-Pen" nanolithography. *Advanced Materials*, 2002, 14(3): 231–234.

65. N.I. Zheludev, Single nanoparticle as photonic switch and optical memory element. *Journal of Optics A: Pure and Applied Optics*, 2006, 8(4): S1–S8.

66. B.F. Soares, F. Jonsson, and N.I. Zheludev, All-optical phase-change memory in a single gallium nanoparticle. *Physical Review Letters*, 2007, 98(15): 153905.

67. B.F. Soares, M.V. Bashevoy, F. Jonsson, K.F. MacDonald, and N.I. Zheludev, Polymorphic nanoparticles as all-optical memory elements. *Optics Express*, 2006, 14(22): 10652–10656.

68. C. Novembre, D. Guerin, K. Lmimouni, C. Gamrat, and D. Vuillaume, Gold nanoparticle-pentacene memory transistors. *Applied Physics Letters*, 2008, 92(10): 103314.

31

Substrate Self-Patterning

Jens Falta
University of Bremen

Thomas Schmidt
University of Bremen

31.1 Introduction

The implementation of Ge nanostructures into Si-based devices has great potential for future high-speed devices, due to advantages like enhanced carrier mobilities and smaller bandgap and hence is still attracting an increasing interest in fundamental and applied research [1–3]. With respect to devices based on quantum dots, a high density of quantum dots and a uniform distribution of their size are required. Such a uniform size distribution can be achieved by a uniform distribution of the inter-dot distances. This correlation between an ordered arrangement and a sharp size distribution has been demonstrated for stacked layers of Ge islands separated by Si spacer layers [4–6].

For stacked nanoisland layers, self-organized ordering, i.e., a sharpening of the islands' distance distribution, can be explained in terms of strain [7]. After an overgrowth of Ge islands with a spacer layer of Si, the lattice mismatch between Ge and Si leads to the formation of strain fields emerging from the islands. This results in a strain modulation at the surface of the spacer layer and heterogeneous nucleation conditions for subsequently deposited Ge. For these Ge atoms, it is energetically more favorable to form 3D islands in regions where the lateral lattice parameter is closer to the Ge bulk value as compared to regions with a lateral lattice constant close to Si. For very thin spacer layers, this effect leads to the reproduction of the island pattern in the subsequently growing layer. For large spacer layers (compared to the average lateral island distance), the strain fields overlap significantly and the surface strain modulation contrast fades. In this case, a homogeneous nucleation is observed, i.e., a rather random arrangement of islands in the next layer of nanoislands. For an intermediate spacer layer thickness, however, the beginning overlap of strain fields leads to a weakening of vertical ordering

and, simultaneously, this overlap induces an increasing lateral ordering with every layer of nanoislands deposited [7]. Apart from Ge/Si, this phenomenon has also been observed for many different material systems like InAs/GaAs(001) [8,9], InP/InGaP(001) [10], PbSe/PbTe(111) [11,12], and CdSe/ZnSe(001) [13,14].

To use the concept of strain-induced spatial arrangement with an increasing number of quantum-dot layers stacked on top of each other, a rather high number of layers are needed to obtain significant ordering, as the bottom layers do hardly show any lateral correlation. For an improvement, regular spacing between quantum dots already in the first layer would be desirable. Various approaches for the realization of such well-ordered Ge nanostructure arrays have been developed, most prominently by the use of lithography. While optical lithography, due to the diffraction limit, cannot provide sufficient resolution for the fabrication of Ge quantum dots with dimensions of only a few 10 nm, electron-beam lithography in conjunction with reactive ion etching has been shown to enable the selective growth of ordered Ge dot arrays [15,16]. Alternatively, a pre-patterning of the Si substrate by scanning probe techniques is a viable concept [17]. Although these artificial patterning techniques offer a very precise control of the island nucleation sites, they require a sequential substrate manipulation and, therefore, these techniques are very time consuming and expensive. In this respect, concepts relying on self-organization are much better suited for technological applications, since they enable parallelized fabrication.

Different self-organized growth techniques have been employed in order to generate regularly spaced nanostructures in a single layer, i.e., without (or prior to) stacking. Heterogeneous nucleation conditions are generated by a strain modulation of the substrate in most of these methods. For instance, Ge growth on $Si_{1-x}Ge_x$ buffer layers with misfit dislocations can be used to obtain spatially ordered Ge islands [18]. In this case, the strain fields emerging from the misfit dislocations impose a pattern of preferred nucleation sites. Alternatively, pseudomorphically strained Ge_xSi_{1-x} buffer layers can be used. For appropriate alloy compositions and miscut orientations, step bunching occurs for GeSi growth on Si(001) [19], leading to a periodically rippled surface morphology, where the Ge concentration in the troughs is smaller than on the ripple peaks [20]. This compositional modulation is accompanied by a modulation of the lateral lattice constant, again imposing strain-induced selective growth of Ge nanoislands [21,22]. In both cases, for dislocated as well as rippled $Si_{1-x}Ge_x$ buffers, the strain modulation period and thus the average distance between the subsequently grown Ge 3D nanoislands depend on the Ge content x of the buffer layer; therefore, this parameter can hardly be used for additional tailoring of the electronic properties.

In the following we will present an alternative concept employing surface adsorbates in order to combine the strengths of self-assembly and self-ordering.

In order to get there, we will guide the reader through the basic concepts of surface adsorption, reconstructions, and growth, as well as the role of surface defects. Thereafter the schemes of surface patterning will be developed.

31.2 Background (History and Definitions)

31.2.1 Surface Adsorption

The deposition and sticking of atoms or molecules to surfaces is referred to as adsorption. The adsorption process has been one of the most intensely investigated issues in surface science over the last 30 years. The adsorption process leads to the buildup of a bond between the surface and the adsorbate, which can occur at different levels of bond strength and is strongly dependent on the actual system of adsorbate, surface, and adsorption temperature. Generally, two regimes are distinguished (Figure 31.1):

a. Physisorption, governed by van der Waals–like interactions, which by nature are much weaker than chemical bonds, and
b. Chemisorption, leading to chemical, i.e., strong, bonds between the surface and the adsorbate.

Physisorption is found if a nonreactive atom or molecule is placed on an inert surface, as, e.g., is the case for Helium adsorption on an Au surface. In most cases, for physisorption low temperatures are mandatory in order to suppress desorption processes which become activated as soon as the thermal energy of the adsorbate approaches the activation energy for the desorption path. For the same reason, physisorption can easily be reversed, just by increasing the sample temperature.

Chemisorption is likely to occur if conditions are provided such that a reactive atom or molecule adheres on a reactive

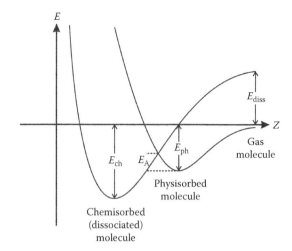

FIGURE 31.1 Schematic energy diagram for the adsorption of a molecule at a surface. A local minimum E_{ph} for the physisorbed state (van der Waals interaction) is present at larger distance z from the surface. Even taking into account the dissociation energy E_{diss}, an energetically more stable configuration is achieved for the chemisorbed state (covalent, metallic, or ionic). For the transition from the physisorbed to the chemisorbed state, an activation barrier E_A might have to be overcome.

surface. Examples for prototype chemisorption can be found for adsorption of aggressive gases like carbon oxide (CO) on transition metal surfaces or the adsorption of reactive atoms with high (or low) electron affinity on a semiconductor surface which usually exhibits unsaturated bonds, i.e., unpaired electrons.

As physisorbed adsorbates can easily be displaced on the surface or even be removed, these cannot be used to change the growth process on a surface and we will not go into further details of physisorption.

Chemisorption, however, can alter the surface properties significantly, and we will return to this issue after introducing the concept of surface reconstructions.

31.2.2 Surface Reconstructions and Surface Morphology

Both metal and semiconductor surfaces show the phenomenon that the actual structure of the surface is different from a crystal upon bulk termination along the corresponding crystal lattice plane. If only the vertical arrangement of atoms in the surface layer is affected, this rearrangement is called surface relaxation, and if the lateral position of atoms also changes, the apparent structure is called a surface reconstruction.

Before we introduce the first example for surface reconstruction, we would like to briefly review the crystal plane notations used in this context. A sketch of a simple cubic crystal is shown in Figure 31.2. In general, crystal planes are labeled by so-called Miller indices. For crystals with a cubic unit cell, like Si and Ge, these Miller indices are given by the Cartesian coordinates of the crystal plane normal vector, which is defined to consist of the smallest possible triple integer numbers. The simplest example is the (001) surface sketched by the shaded surface of the cube

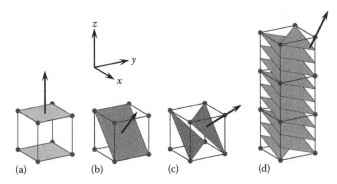

FIGURE 31.2 Crystal planes of a simple cubic crystal, from left to right: (a) (001) plane, (b) (101) plane, (c) (111) plane, and (d) (113) planes.

in Figure 31.2a. If a plane of interest cuts the crystal more tilted, this is reflected by more nonzero components of the corresponding surface vector. Three examples can be seen in Figure 31.2b through d. For instance, Figure 31.2c shows (111) planes which cut the cube at three corners. The (113) planes, as sketched in Figure 31.2d, have an orientation closer to (001). Its density of atoms is smaller than that found for (111); however, the distance between neighboring planes is smaller. Further details and exact definitions on how we determine the surface vectors can be found in the standard literature of solid-state physics and x-ray diffraction (XRD).

One of today's most important tools for surface investigations is the scanning tunneling microscope (STM) for the development of which Binnig and Rohrer received the Nobel Prize of Physics in 1986. The basic part of an STM is a scanning unit which allows the movement of a very sharp tip with sub-Ångström (picometer) accuracy. A schematic is shown in Figure 31.3a. Upon approaching a biased tip to a surface a small current can be measured before the tip contacts the sample. This tunneling current is of quantum mechanical nature and reflects the wave properties of the electrons. In Figure 31.3b, the electronic situation of scanning tunneling microscopy is shown. An electron

can tunnel the potential barrier between the tip and the sample if empty states are available at its energy inside the sample and vice versa. For metallic tips and surfaces this is always the case. For semiconducting samples this requires sufficiently large tip biases in order to overcome the band gap of the material. For more details, the reader is referred to the literature [23–26].

A prominent, maybe the most prominent, example for a surface reconstruction is the Si(111)-7 × 7 structure. The abbreviation 7 × 7 denotes the size of the surface unit cell to be seven times larger than the truncated-bulk unit cell in both surface directions. A scanning tunneling microscopy image and a schematic drawing according to the so-called dimer-adatom-stacking fault (DAS) model [27] of the 7 × 7 surface reconstruction are shown in Figure 31.4.

A comparison of the STM image and the schematic immediately shows that not all of the atoms in the structure are visible in the STM image. However, some distinct features of the 7 × 7 reconstruction are quite apparent. First of all, these are the so-called corner holes which span the parallelogram that borders the 7 × 7 unit cell. Here, the absence of atoms in the top three surface layers leads to recording deep depressions in the STM image. Second, the so-called adatoms can easily be recognized as protrusions inside the triangular network of corner holes. And third, strongly dependent on the actual tunneling conditions, a slight contrast is visible between adjacent triangles. This contrast can be attributed to a difference in the electronic structure of the faulted and unfaulted half of the 7 × 7 unit cell. Not visible are the three dimers which are connecting the corner holes in the schematic drawing. From three of these constituents the name of this structure is derived: DAS is the abbreviation for dimer-adatom-stacking fault model.

As chemisorption leads to the formation of new bonds between the adsorbate and the surface, this process is likely to also affect the reconstruction of the surface. Ga adsorption on Si(111) will serve as an example in the following.

Upon deposition of up to one monolayer Ga onto Si(111), different surface structures can be observed, depending on

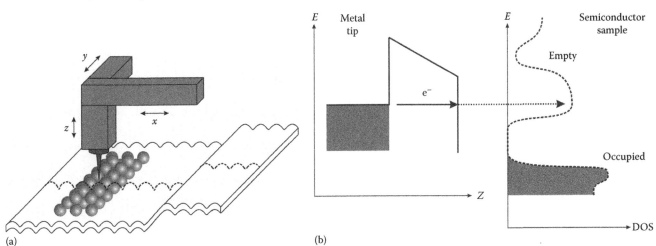

FIGURE 31.3 Work principle of a scanning tunneling microscope. (a) The tunneling current is recorded while scanning a tip across the sample surface by the use of piezo crystal movement (x, y, z). (b) Tunneling of electrons from a metallic tip to a semiconductor surface requires the availability of empty states on the semiconductor side. This is accomplished by applying a sufficiently high bias between tip and surface.

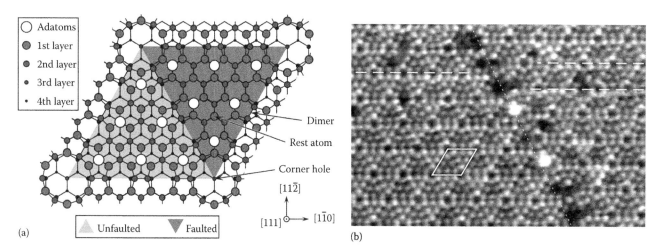

(a)

(b)

FIGURE 31.4 (a) Unit cell of the Si(111)-7 × 7 surface reconstruction according to the DAS model as determined by transmission electron diffraction by Takayanagi et al. [27]. For details refer to the text. (b) Scanning tunneling micrograph of a Si(111)-7 × 7 surface, recorded at +0.7 V tip bias (i.e., probing occupied states) and 0.25 nA tunneling current. A unit cell is marked by solid white lines. Note that at these tunneling conditions, the faulted half of the unit cell appears slightly brighter than the unfaulted half. The dotted line indicates a domain boundary of the reconstruction. Its presence is deduced from the fact that the rows of the corner holes (dashed white lines) do not match on the left and right side of the domain boundary.

the temperature of deposition and the conditions of annealing which may subsequently be applied. For temperatures below 670 K, the 7 × 7 periodicity of the surface remains intact. Within the 7 × 7 unit cells, however, the buildup of small Ga clusters can be observed by STM [28,29]. A sketch of these "magic clusters" and a corresponding STM image can be found in Figure 31.5. The name "magic cluster" refers to the fact that all clusters are identical in size and shape.

For higher temperatures the issue concerning the activation energy for lifting the 7 × 7 surface reconstruction can be overcome and the formation of a 6.3 × 6.3 surface structure has been reported [30,31]. In this structure the bulk-terminated Si(111) surface is mimicked; however, Ga atoms replace the Si in the outermost surface layer, see Figure 31.6a, and form

an sp²-like configuration. This configuration is flatter than the bulk-terminated Si surface and it minimizes the number of dangling bonds on the surface. If the Ga atom in this site would require the same space as Si, it would form a 1 × 1 unit cell, just replacing Ga for Si. This is, however, not the case. As the covalent radius of Ga is larger than that of Si, this structure is under strong compressive stress which leads to a strained surface layer. As a result, the topmost Ga-Si double layer is not in registry with the underlying crystal and an incommensurate surface structure is formed as shown in Figure 31.6a. A surface completely covered with this adsorbate-induced structure will consist of small patches of 6.3 × 6.3 domains. At their merging lines, i.e., at the domain boundaries, defects are likely to be found. Here open surface bonds can be expected. This is the

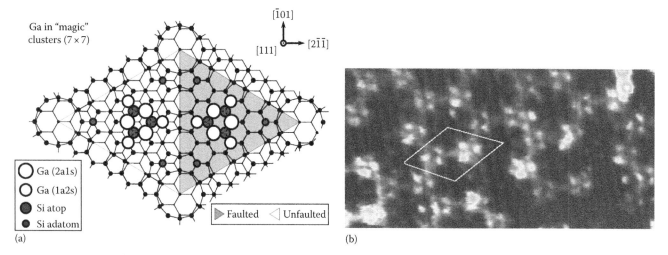

(a)

(b)

FIGURE 31.5 Magic clusters, a low-coverage and low-temperature phase of Ga on Si(111). (a) Structural model according to Lai and Wang [28]. Six Ga atoms reside in each half of the unit cell. Three of them bond to one Si atop atom (i.e., Si atoms in the reconstruction residing on top of the last complete Si bilayer) and two Si atoms of the first layer (1a2s), while the other three bond to two Si atop atoms and one Si atom of the first layer (2a1s). (b) STM image (unoccupied states). The unit cell is marked by solid lines. The Ga 2a1s atoms are the most prominent features.

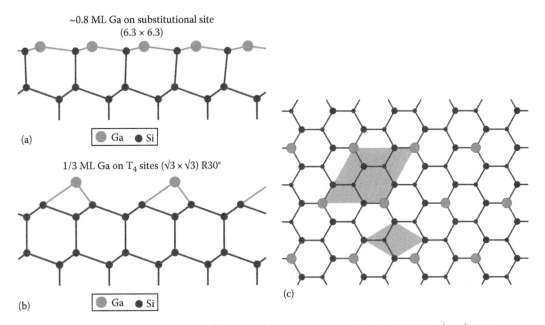

FIGURE 31.6 Adsorption geometry of Ga on Si(111) for (a) the Ga:Si(111)-6.3 × 6.3 and (b) the Ga:Si(111)-√3 × √3-R30° reconstruction. A top view of the latter structure is shown in (c). The unit cell of the √3 × √3-R30° structure as well as the underlying 1 × 1 substrate unit cell are indicated by shaded rhomboids, making the rotation by 30° apparent. This adsorption site is called T_4 as the adsorbate has four nearest neighbors, i.e., the three neighbors in the topmost layer as well as the atom underneath the adsorbate in the second layer.

first type of surface defects, which will be discussed in detail after describing a third surface reconstruction of Ga on Si(111).

If the sample is heated to temperatures between approximately 800 and 900 K during deposition or annealing, most of the Ga desorbs from the surface except for a maximum coverage of 1/3 ML of Ga which, however, is rearranged and forms another surface structure called √3 × √3-R30° (in the following referred to as √3 structure) as depicted in Figure 31.6b and c. The Ga-induced √3-reconstruction has a unit cell, the side length of which is longer than that of the bulk-terminated surface by a factor of √3 in both surface directions. In addition, the unit vectors of the reconstructed unit cell are rotated by 30° with respect to the underlying 1 × 1 unit cell vectors of the bulk-terminated surface. The unit cell and the underlying hexagonal lattice of the (111) surface are also displayed in Figure 31.6c.

In this structure the Ga is found to bond to three underlying Si atoms in the upper half of the top Si bilayer. Here the Ga and its Si neighbors form an sp^3-like configuration. In adsorption-induced bending measurements the stress measured for the Si(111)-Ga-√3 × √3 reconstruction was determined to be compressive in the order of 4.5 eV/1 × 1 surface unit cell [32]. For the similar system of Sb-induced Si(111)-√3 × √3, a tensile stress of 1.5 eV is observed [33] as the Sb favors a tetrahedral bonding configuration with the underlying Si atoms.

No matter how much effort is put into the alignment of a physical surface to a desired crystal lattice plane, all real surfaces will exhibit surface steps. Step edges are the most important type of surface defects and their density can easily be tuned by changing the miscut angle between the nominal and the physical surface orientation. For high step densities and a surface miscut aligned to a specific crystallographic direction,

the surface Miller indices may be determined and the surface is labeled vicinal. In addition to line defects like surface steps and domain boundaries, surfaces will also show point defects like missing atoms, misplaced atoms, or surface impurities, i.e., surface atoms of a different element or molecules adsorbed to the surface.

31.2.3 Epitaxy: Growth on Surfaces

All methods of epitaxy proceed by adding atoms to the surface of a substrate. These atoms may come from a liquid (liquid phase epitaxy) or the gas phase (vapor pressure epitaxy, chemical vapor deposition). Also sources like Knudsen cells or electron-beam evaporators may be used (molecular beam epitaxy), just to name the most prominent examples of growth methods. In most cases, the growth proceeds through surface processes. These have been studied intensively since the 1950s when Bauer [34] developed the first classification of growth. If the incorporation of material B deposited onto a substrate of material A proceeds

a. in a layer-by-layer-like fashion, this growth mode is named Franck–van der Merve growth,
b. by the immediate formation of three-dimensional (3D) islands, we talk about Vollmer–Weber Growth, and
c. by formation of a wetting layer of material B and subsequent growth of 3D islands, Stranski–Krastanov growth is occurring.

A sketch of these growth modes is shown in Figure 31.7. This classification is not complete and the growth mode present may depend on many parameters like growth temperature and deposition rate.

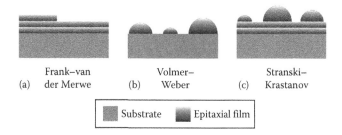

(a) Frank–van der Merwe (b) Volmer–Weber (c) Stranski–Krastanov

☐ Substrate ■ Epitaxial film

FIGURE 31.7 Growth modes close to thermodynamic equilibrium, according to Bauer [34].

For example, Franck-van der Merwe growth may proceed very smoothly in a step-flow fashion. At high growth temperatures the surface atoms are highly mobile and the large surface diffusion length leads to a growth mode in which the surface expands by adding atoms one after the other to step edges, leading to the observation of steps moving across a surface, as can be seen from Figure 31.8a. Since all step edges proceed in the same manner, the surface morphology does not change in this case. At lower growth temperatures, the mobility of the deposited atoms decreases, leading to the formation of stable nuclei for the next layer on the terraces between surface steps, as depicted in Figure 31.8b. At these temperatures the growth proceeds through surface morphologies which are alternatingly smooth and rough. At even lower deposition temperatures, the surface roughness increases during growth and a smooth intermediate surface is no longer observed, as sketched in Figure 31.8c. This is the case when the size of stable nuclei in the new surface layers becomes very small or the activation energy for mobile surface atoms crossing surface steps, the Ehrlich–Schwoebel barrier, becomes large in comparison to the thermal energy of the atoms diffusing on the surface. At even lower deposition temperatures, incorporation at non-lattice sites occurs and no epitaxial growth is observed. Further details on the physics of epitaxial growth can be found in the literature [35].

If a third species C is added to the deposition process, the picture changes again.

For a proper choice of C, the growth mode may also be changed. Germanium, for example, is known to follow a Stranski–Krastanov growth scheme if deposited onto a silicon substrate. Germanium forms large 3D islands where the lattice-mismatch-induced strain is released plastically, i.e., by introduction of defects. However, for many applications it is highly desirable to fabricate Ge films of uniform thickness with a low defect density. This can be achieved by the use of a so-called surfactant (surface active agent) or adsorbate mediation of growth. An example for surfactant-mediated epitaxy is the deposition of flat, defect-free Germanium films on Bi- or Sb-terminated Si(111) [36–43]. In a first step, before Ge deposition, a monolayer of Bi or Sb is adsorbed to the surface. In a second step, Ge is deposited onto this surface and the incorporation of Ge atoms into a uniform Ge film proceeds through site exchanges between surface Ge atoms and surfactant atoms. The driving force for this process is the minimization of the surface free energy. The site exchange does proceed through a complicated reaction path as calculated by Schroeder et al. [44]. The key for the formation of a smooth Ge film is the incorporation of Ge under the surfactant layer. Step by step the Ge film grows pseudomorphically, i.e., the Ge atoms occupy Si crystal lattice sites, except for an elastic relaxation in the vertical direction which reduces the strain. With increasing film thickness, pseudomorphic growth leads to a progressive accumulation of strain energy. Accordingly, for larger Ge film thicknesses, plastic relaxation processes are induced. In Ge films on Si(111) substrates, dislocations that can glide in the (111) plane and almost fully relax the Ge film are generated. At the interface of the silicon substrate a highly regular network of dislocations can be found which even lead to the observation of superstructure spots in diffraction experiments [38,41,42,45] and periodic surface undulations in scanning tunneling microscopy [36,37,40].

31.2.4 Versatile Surface Tools: Low-Energy Electron Microscopy and Diffraction

A very versatile experimental method for monitoring growth and self-assembly is low-energy electron microscopy (LEEM) and related methods like x-ray photoemission electron microscopy (XPEEM). The experiments described in the following were carried out with the spectroscopic photoemission and low-energy electron microscope (SPELEEM) at the nanospectroscopy beamline at ELETTRA [46,47]. A LEEM instrument is similar to a transmission electron microscope (TEM) run at an electron energy of 20 keV; however, it contains an immersion lens that decelerates the electrons immediately ahead of the sample surface to a few electron volts only. Thus, the electrons cannot penetrate the sample but are reflected and subsequently reaccelerated to the operation voltage of the microscope. The primary electron beam and the reflected electrons are separated by use of a magnetic biprism. Finally a magnified image of the surface can be recorded at video rate. The best lateral resolution demonstrated today

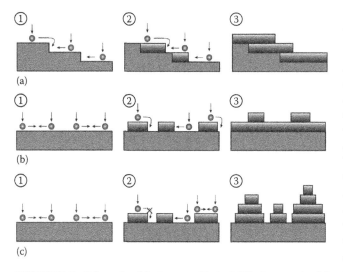

FIGURE 31.8 Schematic of the impact of limited diffusion on epitaxial growth. (a) Step flow growth (large diffusion length), (b) layer-by-layer growth, and (c) multi-layer growth (no diffusion across step edges).

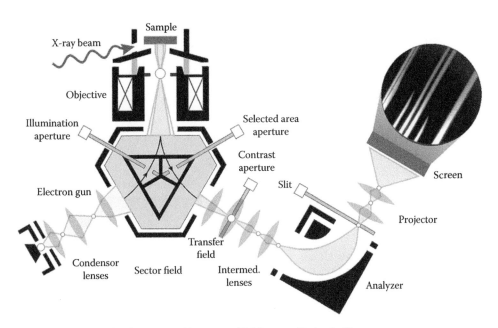

FIGURE 31.9 LEEM and PEEM principle of operation. (Courtesy of F. Meyer zu Heringdorf.)

(using additional electron optics for spherical aberration correction) is around 3 nm [48]. In order to achieve contrast, different mechanisms can be employed by choosing an appropriate energy or part of the diffraction pattern of the surface for imaging. Details about this method can be found in the literature [49,50].

A LEEM system in principle can also be used to perform photoemission electron microscopy (PEEM). For this purpose the electron beam is switched off and the sample is irradiated with light. This can be ultraviolet light from a conventional He-discharge lamp or synchrotron radiation in the UV to soft-x-ray region. The latter has the advantage of a tunable wavelength. The illumination leads to the emission of photoelectrons which can be used for image creation in the imaging column of the LEEM. The principle setup for LEEM and PEEM is sketched in Figure 31.9. For PEEM, the use of an energy analyzer is advantageous for the resolution and it opens further opportunities like imaging electrons of a specific energy loss.

Closely linked and actually developed much earlier than LEEM is the method of low-energy electron diffraction (LEED). Here, the wave nature of electrons is used in order to perform diffraction experiments similar to x-ray diffraction. For kinetic energies in the order of 10–150 eV (electron volts), the de Broglie wavelength of electrons is in the order of interatomic distances, i.e., 1–4 Å as that of x-rays. Distinct from the x-ray case electrons show a much stronger interaction with matter leading to a drastically smaller penetration depth of a few atomic layers only. On the one hand this makes electrons a highly surface-sensitive probe, while on the other hand this small penetration depth affects the structure in reciprocal space and the diffraction patterns themselves. Perpendicular to the surface direction the structures in reciprocal space strongly elongate due to the finite size of the diffracting structure in real space, i.e., the few surface layers in which all electrons are diffracted. This leads to the case of 2D diffraction and the appearance of diffraction spots

in LEED at all electron wavelengths while x-rays are diffracted by many atomic layers within the x-ray penetration depth of microns. Hence for x-rays 3D diffraction is found and x-ray diffraction spots can only be observed at distinct sets of x-ray wavelength and diffraction angles. A schematic comparing the reciprocal space for x-ray diffraction and low-energy diffraction can be found in Figure 31.10. In case of a surface reconstruction, the period length of the real surface may be larger than that of the bulk-terminated (i.e., nonreconstructed) surface. In reciprocal space, this will lead to an increased density of diffraction rods. For the simple case of an ($n \times m$) reconstruction, additional spots appear, the position of which is p/n along the reciprocal (10) surface direction and (q/m) along the reciprocal (01) surface direction, for all integer p and q. This results in $n \times m$

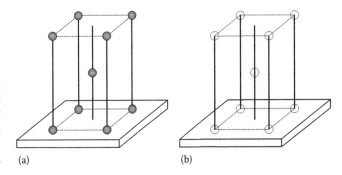

FIGURE 31.10 Comparison of reciprocal space as observed by (a) X-ray diffraction (XRD) and (b) low-energy electron diffraction (LEED) low-energy electron diffraction. In XRD, the diffraction originates from a large volume. Hence, the corresponding structure in reciprocal space is found to consist of Bragg points. In LEED, the diffracting area extends large laterally. However, due to the very small free path length of low-energy electrons the thickness of the diffracting volume is very finite. This leads to the observation of Bragg rods in LEED which are oriented vertical to the surface.

surface rods per surface Brillouin zone instead of just one as for the nonreconstructed case. An example will be given later (cf. Figure 31.17).

31.3 Adsorbate-Induced Surface Self-Patterning

In the following, we will focus on the approach to exploit adsorbate-induced surface self-patterning for selective growth of Ge nanoislands on Ga-terminated Si surfaces. The two examples shown here, Ge/Ga:Si(111) and Ge/Ga:Si(113), illustrate that even for such similar physical systems quite different mechanisms can be observed that lead to self-ordering.

31.3.1 Flatland Patterning: Ge Growth on Ga-Terminated Si(111)

Ge growth on Ga-terminated Si(111) surfaces is of specific interest, because it was already shown that depending on Ga coverage, different morphologies can be observed for subsequent Ge epitaxy. Whereas on a Ga:Si(111)-6.3 × 6.3 surface a smooth Ge film evolves, Ge growth on Si(111)-√3 × √3-R30° leads to the formation of 3D islands, similar to Ge deposition on bare Si(111), however, with a sharper island size distribution [51,52]. The different behavior for the 6.3 × 6.3 structure on the one hand and the √3 × √3 structure on the other hand is attributed to changes occurring in the bonding geometry upon the initial deposition of Ge. As outlined in the preceding section, Ge is incorporated underneath the Ga layer, which floats at the surface. While Ga can occupy substitutional sites in a Ge surface layer, and therefore can form a 6.3 × 6.3 reconstruction on the growing Ge film, the Ga T_4 adsorption site, characteristic of the Si(111)-√3 × √3 structure (for details see Figure 31.6), is not stable on Ge. Instead, when starting with a Si(111)-√3 × √3 surface, a change toward 6.3 × 6.3 occurs upon Ge deposition [52]. As the 6.3 × 6.3 reconstruction requires a higher local Ga coverage (about 0.8 ML as compared to 1/3 ML for the √3 reconstruction), Ga-covered 6.3 × 6.3 domains and domains free of Ga evolve in this case. The interplay of Ga-terminated surface areas (6.3 × 6.3 as well as √3 × √3) and bare surface regions is addressed in more detail for the case of an only partially Ga-covered starting surface in the following.

At higher substrate temperatures, the Ga adsorption process initially proceeds at surface defects, i.e., domain boundaries and surface step edges. LEEM images of the initial adsorption of Ga on Si(111) are shown in Figure 31.11 [53]. The electron energy for these LEEM images was chosen in order to maximize the contrast at the later stages of the experiment. Only a faint contrast is visible from the bare Si(111)-7 × 7 surface at the beginning. This contrast originates from step edges (running top down) and domain boundaries of the 7 × 7 reconstruction (running from left to right).

Upon start of the Ga deposition, dark contrast arises both at step edges as well as at domain boundaries, as can be seen from Figure 31.11b recorded after deposition of 0.08 ML Ga, with 1 ML = $7.83 × 10^{14}$ atoms/cm². This contrast is due to the nucleation of

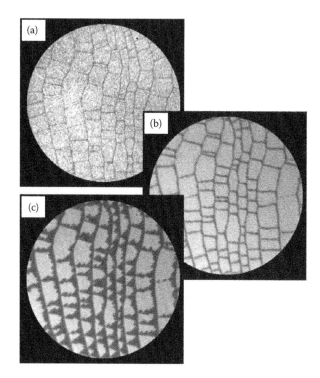

FIGURE 31.11 Bright-field LEEM images during Ga deposition on Si(111) at 650°C, (a) prior to Ga adsorption, (b) at 0.08 ML, and (c) at 0.13 ML coverage. The gray scale contrast in (a) has been increased by a factor of three to make step edges (extending from the top to the bottom) and domain boundaries (from left to right) of the Si(111)-7 × 7 surface better visible. The field of view is 5 μm in each frame. The electron energy was 3.1 eV.

Ga/Si(111)-√3 × √3 domains as confirmed from dark field LEEM using a √3 × √3 superstructure spot for imaging. These have a *local* coverage of 1/3 ML. With increasing Ga coverage, the √3 × √3 patches grow in size. In Figure 31.11c, after deposition of 0.13 ML Ga, a typical shape of the √3 × √3 domains becomes visible reflecting the threefold symmetry of the system. Since no nucleation at the interior of the 7 × 7 domains takes place, a 2D phase separation results, and a nanopattern is achieved when the Ga deposition is stopped before the √3 × √3 reconstruction extends over the entire surface. This nanopattern will serve as a modified substrate for subsequent growth of Ge nanostructures.

The evolution of the surface structure and morphology during Ge deposition onto a Ga/Si(111) nanopattern can be viewed in Figure 31.12. After deposition of 0.3 ML Ge (Figure 31.12a), the surface has already undergone a transformation which can directly be seen from the drastic change in contrast. Figure 31.11c, for comparison, was obtained at the same electron energy. When switching to observation of the LEED pattern at this stage of deposition (Figure 31.12b), only a diffuse 7 × 7-like pattern is found. Conclusively, the √3 × √3 patches have vanished already during initial Ge deposition. This indicates that Ge is incorporated preferentially at the former √3 × √3 patches, which leads to a site exchange of Ge and Ga, and thus to the transition from a Ga/Si(111)-√3 × √3 reconstruction to a Ga/Ge/Si(111)-(6.3 × 6.3) structure [52].

FIGURE 31.12 Bright-field LEEM images (a) and (c) through (h), as well as LEED pattern (b), all recorded during Ge growth at 450°C on a partially covered Ga:Si(111) surface. The Ge deposit is indicated in each frame. (Field of view: 5 μm, electron energy: 3.1 eV.)

This incommensurate structure has a larger local Ga coverage as compared to Ga/Si(111)-√3 × √3. Since there is no further Ga supply as the Ga deposition had been disrupted, this leads to the formation of a patchwork of 6.3 × 6.3 Ga-terminated domains as well as Ga-depleted areas in between [52]. In the LEEM image in Figure 31.12a, the spotty appearance of the former √3 × √3 regions is a further indication for such a patchwork with a length scale near the resolution limit of the microscope.

Upon further Ge deposition, the contrast becomes weaker (see Figure 31.12c and d), which can be attributed to a high density of mobile Ge adsorbate atoms on the surface [54]. In Figure 31.12e, small bright spots appear which are identified as Ge islands as will be shown below. The nucleation of these Ge islands is almost complete in Figure 31.12f, and the contrast in the image increases again. This points to a lattice relaxation of the islands which makes them stable against decay and energetically attractive to Ge adatoms [55], leading to a reduced adatom density and thus to an enhanced image contrast [54]. From Figure 31.12f through h, the number of islands remains almost constant and the lateral size of the islands increases only slightly which is indicative for a 3D island growth. A large-scale view of the surface as shown in Figure 31.13 clearly demonstrates the quality of alignment which can be achieved.

In order to unambiguously determine the nature of the bright spots referred to as Ge islands so far, a LEEM micrograph is compared to x-ray photoemission electron micrographs (XPEEM) in Figure 31.14. The image shown in Figure 31.14b was obtained using Ge 3*d* photoelectrons and therefore directly provides chemical contrast. Taking a slight drift between the two images into account (see black circles), virtually every bright spot in the LEEM image can be identified as a Ge-rich region. This proves that the spots observed in LEEM indeed are Ge islands. The rather weak overall contrast of the XPEEM image in Figure 31.14b is attributed to the existence of a Ge-wetting layer between the islands. When comparing the XPEEM images in Figure 31.14b and c, a clear correlation between the Ge 3D islands and Ga-rich

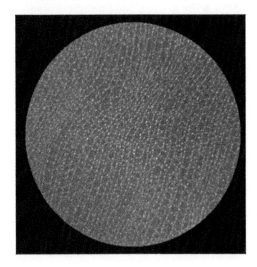

FIGURE 31.13 Bright-field LEEM image of Ge nanoislands grown at 450°C on a partially covered Ga:Si(111) surface. This large-scale view (20 μm diameter) reveals the strong alignment of the Ge islands.

regions is revealed. Because of the surface sensitivity of the photoemission signal, this shows that Ga resides at the surface of the Ge 3D islands, which in turn leads to the conclusion that the nucleation of such islands in Ga-rich regions is energetically favored as compared to nucleation on Ga-depleted areas, and Ga segregates to the surface during Ge 3D island growth.

It should be noted that in the presence of step bunches, a preferential nucleation of 3D islands has also been reported for the growth on bare Si(111) substrates [56,57]. However, such an alignment mechanism can be ruled out for the Ge islands on regularly stepped, bare Si(111) surfaces [57,58] as in the present case.

From a closer inspection of the LEEM images (see Figure 31.15), a kinetic limitation of the selective growth process becomes obvious. The 3D islands do not only nucleate at step edges and domain boundaries, but also start to evolve at the centers of the initial 7 × 7 domains. However, there are pronounced

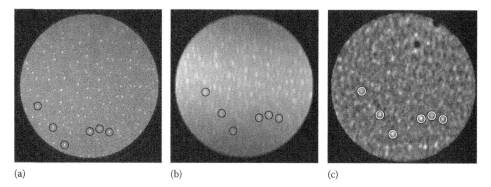

FIGURE 31.14 Images with 5 μm field of view, each depicting the same surface area. (a) Bright-field LEEM. (b) XPEEM using Ge 3*d* photoelectrons. (c) XPEEM with Ga 3*d* photoelectrons. Black circles mark identical objects on the surface.

FIGURE 31.15 Bright-field LEEM image (945 nm × 1260 nm). Ge islands are predominantly found in Ga-rich regions (light grey), whereas no islands are found in the Ga-depleted regions in between, provided these domains are small enough (see region A). In larger Ga-depleted regions (see B + C), some additional Ge islands are present. However, there is a denuded zone B, the width of which can be interpreted as a diffusion length. Ge atoms impinging in this region can attach to Ge islands in the Ga-rich regions.

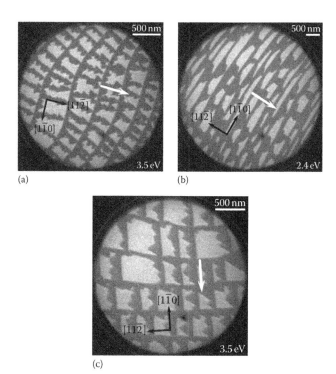

FIGURE 31.16 Bright-field LEEM images of partially Ga-covered Si(111) surfaces, for samples with different miscut orientation. The step-down direction is indicated by a white arrow in each frame.

denuded zones at the borders of these domains. The width of the denuded zones, which can be interpreted as an effective diffusion length [59], is found to be about 90 nm for the growth conditions used here. Within the smallest of the initial 7 × 7 domains, virtually no 3D islands are found.

The domain shape of the nanopattern depends on the miscut orientation of the substrate, as shown in Figure 31.16. For a miscut toward [11$\bar{2}$] (Figure 31.16a), the Ga-terminated $\sqrt{3} \times \sqrt{3}$-R30° domains exhibit a "Christmas-tree" arrangement, whereas for the opposite direction (Figure 31.16b), a strip pattern

is observed. If the step-down direction points to [$\bar{1}$10] (Figure 31.16c), the Ga-rich domains form a "key-bit" pattern. Though the pattern shape appears quite different for the different miscut directions, there are common findings for all the cases shown in Figure 31.16. The $\sqrt{3} \times \sqrt{3}$ domains nucleate at step edges and at the boundaries of the initial 7 × 7 reconstruction of the clean Si(111) surface. At the step edges, they always grow toward the lower terrace side, irrespective of the miscut orientation. From the initial domain boundaries, growth toward both adjacent domains is possible. The Ga domains prefer a triangular shape with straight edges along the $\langle 1\bar{1}0 \rangle$ directions that connect the corner holes of the adjacent 7 × 7 domains of the clean substrate [60]. Whenever such a $\langle 1\bar{1}0 \rangle$ direction is parallel to the step edges (Figure 31.16b) or initial domain boundaries (Figure 31.16c), the growth perpendicular to this direction is very slow, giving rise to strip and key-bit patterns. It is the highly anisotropic line energy of the boundaries between √3 × √3 and 7 × 7 domains that impedes kink formation and drives the pattern formation. As the line energy and its anisotropy is mainly based on the high-tensile surface strain of the 7 × 7 reconstruction [32,61], the growth scheme observed here is not limited to Ga adsorption, but is more generic. It has also been verified for In on Si(111) [60]. Obviously, the absolute miscut angle governs the average step edge density. The density of initial domain boundaries can, to some extent, be controlled during the preparation of the clean Si(111) substrate. Together with the miscut orientation, the influence of which has been discussed above, these degrees of freedom can be utilized to tailor the size and the shape of the adsorbate-induced nanopattern for subsequent selective growth.

31.3.2 Employing Self-Created Ridges: Ge Nanostructures on Facetted Ga/Si(113)

Most of the high-index Si surfaces are energetically instable against faceting. The Si(113) surface, however, has a low surface free energy [62] and therefore is a stable facet of Si. Technologically, Si(113) might become an alternative to Si(001) [63], the latter still being the substrate for most device applications. Distinct from Si(111), the lack of rotational symmetry of the Si(113) surface facilitates the growth of highly anisotropic Ge islands. In a narrow temperature and coverage range, nanowires extending in [33$\bar{2}$] direction can be grown, see refs. [64,65]. It is still under debate if this strong anisotropy is stabilized by a reduction of the strain energy originating from the lattice mismatch between Ge and Si [66,67]. Also, nanowires have been found to be metastable and to be transformed into elongated islands by ripening at growth temperature [68]. Hence, growth kinetics and especially diffusion anisotropy might also play an important role.

Regarding the spatial correlation of such Ge nanowires (as well as that of similar Ge$_x$Si$_{1-x}$ nanowires), vertical ordering in stacked heterostructures has been observed [69–71]. For Si(113) substrates, however, no reports on lateral ordering of Ge nanostructures are available so far.

Similar to the findings on the Si(111) surface, as described above, a nanopattern of Ga-covered regions at the step edges of Si(113) substrates and bare Si(113)-3 × 2 regions in between can be obtained upon submonolayer Ga deposition at high temperatures. Opposed to Si(111), however, this nanopattern cannot be exploited for subsequent growth of aligned Ge islands [72]. Nevertheless, ordered arrays of Ge nanoislands are obtained after saturating the Si(113) substrate with Ga beforehand.

The evolution of the surface reconstruction during Ga adsorption up to saturation coverage is shown in the LEED patterns in Figure 31.17a through c. At a coverage of 0.42 ML, the initial 3 × 2 pattern has changed to a superposition of contributions from 3 × 2 and 2 × 2 domains, corresponding to a coexistence of bare Si(113)-3 × 2 domains and Ga:Si(113)-2 × 2 areas. After exposure to about 0.6 ML, the entire surface is 2 × 2 reconstructed corresponding to a complete Ga termination at this stage. Nevertheless, the evolution of the surface structure is not yet complete. Upon further Ga deposition, the LEED pattern changes again, as shown in Figure 31.17c for a Ga coverage of 1.1 ML. The nature of this rather complex LEED pattern is revealed in Figure 31.18.

Before analyzing the LEED pattern of Figure 31.18 in more detail we would like to introduce the effects of inclined surface areas, so-called surface facets, as sketched in Figure 31.19. If the surface consists of areas which are tilted with respect to the nominal orientation (average orientation) of the surface this leads to distinct features in the structure of the reciprocal space and the corresponding LEED images. In addition to or substituting the Bragg rods of the nominal surface, inclined Bragg rods are present which lead to LEED spots moving with respect to the Bragg peaks of the nominal surface when changing the electron energy which determines the z-component of the scattering vector.

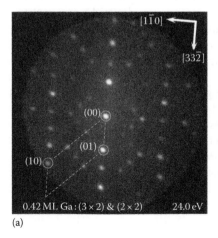 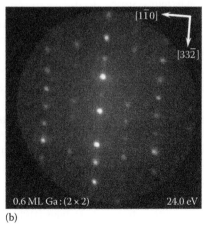 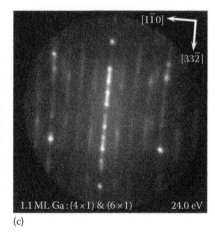

(a)   (b)   (c)

FIGURE 31.17 LEED patterns recorded during Ga adsorption on Si(113). (a) Coexistence of bare Si(113)-3 × 2 and Ga:Si(113)-2 × 2 domains, (b) after deposition of 0.6 ML Ga and formation of the Ga:Si(113)-(2 × 2), i.e., the surface is completely terminated with Ga, and (c) facets with (4 × 1) and (6 × 1) reconstruction have formed after Ga saturation.

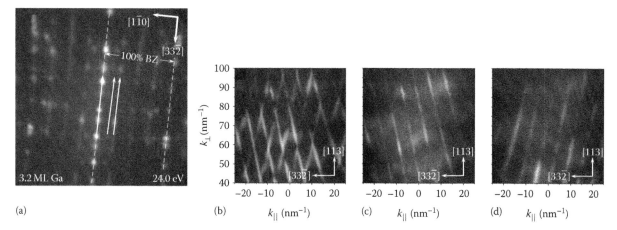

(a)   (b) $k_{||}$ (nm⁻¹)   (c) $k_{||}$ (nm⁻¹)   (d) $k_{||}$ (nm⁻¹)

FIGURE 31.18 LEED pattern (a) of a Ga saturated Si(113) surface, and reciprocal space maps (b) to (d) obtained from cross sections for different electron energies along the three lines in [33$\bar{2}$] direction indicated by arrows in (a). The coordinate of these cross sections in [1$\bar{1}$0] direction were (b) zero, (c) 1/6 Brillouin zone (BZ), and (d) 1/4 BZ.

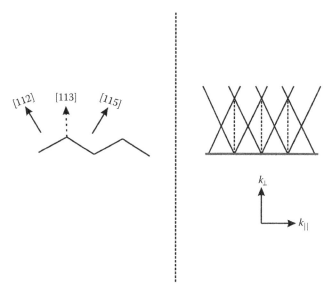

FIGURE 31.19 Surface facets in real space (left) and reciprocal space as observed by LEED.

This is exactly what is found in Figure 31.18. The LEED pattern presented in Figure 31.18a was obtained after Ga saturation. It shows a large variety of diffraction spots. These spots are moving in reciprocal space if the electron energy is changed. Reflections occur at lines parallel to the [33$\bar{2}$] direction, i.e., the (almost) vertical direction in the LEED pattern in Figure 31.18a. The [1$\bar{1}$0]-position of these lines, i.e., the position along the (almost) horizontal direction, is found in two series of diffraction lines, one at 1/6, 2/6, 3/6, 4/6, and 5/6 as compared to the next fundamental diffraction spot in this surface orientation and one at 1/4, 2/4, and 3/4 position. Line scans have been taken as a function of electron energy along the [33$\bar{2}$] direction for (i) the high symmetry line through the (00) reflection and (ii) and (iii) the lines at 1/6 and 1/4 position as indicated by the arrows. Three reciprocal space maps in the $k_{||}$–$k_\perp$ plane which

have been compiled from these data are shown in Figure 31.18b through d.

Figure 31.18b shows a plot of the diffraction peak position of the line scan along (00) as a function of the vertical scattering vector which is proportional to the square root of the electron energy. Figure 31.18c and d shows the same plot for the peaks found at 1/6 and 1/4 of the Brillouin zone in [1$\bar{1}$0] as indicated by the arrows in Figure 31.18a.

Remarkably, there are no reciprocal lattice rods running along the [113] direction, as would have been expected for a flat surface. Instead, all reciprocal lattice rods are inclined. This can only be observed for a completely facetted surface as sketched in Figure 31.19.

In Figure 31.18b, two different types of rods are found with opposite inclination angles of about 10°. From these inclination angles, the crystallographic orientation of the facets can be determined to be (112) facets and (115) facets, respectively. Whereas the experimental data in Figure 31.18b is well explained by the model of the reciprocal space depicted in Figure 31.19, the reciprocal space maps in Figure 31.18c and d show rods that are inclined toward only one direction each. This means that only the (112) facets contribute to the intensity in Figure 31.18c, and only the (115) facets yield intensity in the region of reciprocal space mapped in Figure 31.18d. As the data in Figure 31.18c and d were taken at 1/6 and 1/4 of the surface Brillouin zone, this finding corroborates that different surface reconstructions are present on each facet type, a (6 × 1)-reconstruction on the (112) facets, and a (4 × 1)-reconstruction on the (115) facets.

With this information the interpretation of the LEEM images obtained after Ga saturation, such as shown in Figure 31.20, becomes straightforward. The dark and bright stripes extending along [1$\bar{1}$0] are identified as (112) and (115) facets. From real-space images, it is also possible to directly obtain the average spacing between facets of common orientation, i.e., the periodicity of the surface morphology along the [33$\bar{2}$] direction. For this periodicity, a value of about 40 nm is found. Already at first

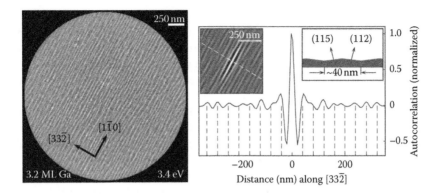

FIGURE 31.20 Left: Bright-field LEEM image of a Ga saturated Si(113) surface. The dark and bright contrast is caused by (115) and (112) facet pairs. Right: Cross section along [33$\bar{2}$] direction through the 2D autocorrelation function (inset) of the LEEM image. The sketch in the top right inset illustrates the surface morphology.

inspection, the facet array in the LEEM image in Figure 31.20 appears quite regularly arranged. This is confirmed by the autocorrelation along [33$\bar{2}$] (also shown in Figure 31.20), in which correlation peaks up to eighth order are clearly resolved.

This facet structure has a strong impact on the surface morphology after Ge deposition. As can be seen from the LEEM image in Figure 31.21, a very high density (about 1.3×10^{10} cm^{-2}) of small Ge 3D islands is formed at 500°C. Again, the Ge islands have an anisotropic shape, but here they are elongated along [1$\bar{1}$0]. This is the opposite shape anisotropy as found for Ge growth on bare or partially Ga-covered Si(113) [72]. A quantitative analysis of the average Ge island diameter yields a value of about 25 nm in [33$\bar{2}$] direction, in contrast to about 40 nm along [1$\bar{1}$0], as can be seen in the histograms in Figure 31.21. The same type of anisotropy is observed for the Ge island spacing. From the autocorrelation along [1$\bar{1}$0], as depicted in Figure 31.21, an average nearest-neighbor distance of 100 nm is derived, whereas in [33$\bar{2}$] direction, a value of about 50 nm is found, the latter value matching the facet periodicity of about 40 nm within the experimental uncertainty. In addition, more pronounced maxima are observed in the autocorrelation along [33$\bar{2}$] as compared to that along [1$\bar{1}$0] (cf. Figure 31.21), pointing to a better ordering

in [33$\bar{2}$] direction. Both, the spacing and the ordering anisotropy, are strong indications for an alignment of the Ge islands with the facets. This is further supported by the measured value of 25 nm for the Ge island width in [33$\bar{2}$] direction, as this value is in good agreement with the average facet width, i.e., half the facet array periodicity. Due to different surface free energies of the different facets, one can expect that the Ge islands preferentially nucleate at one type of facet (which, however, could not be resolved in our experiments). The facet boundary then acts as a growth barrier for the Ge islands; thus, the width of these islands is expected to be limited by the facet widths.

In addition to this impact from surface energetics, surface kinetics also favors the reversal of the shape anisotropy. This can be seen from Figure 31.22, where a surface region with a large contaminant island has been selected for LEEM imaging. In the vicinity of this contamination along the ⟨1$\bar{1}$0⟩ directions, almost no Ge islands are found within a distance of about 300–400 nm. The presence of these denuded zones shows that the contamination has an attractive chemical potential for Ge species. The fact that the width of the denuded zones along ⟨33$\bar{2}$⟩ is much smaller proves that the Ge diffusion along the facet trenches in [1$\bar{1}$0] direction is much faster than in the perpendicular direction.

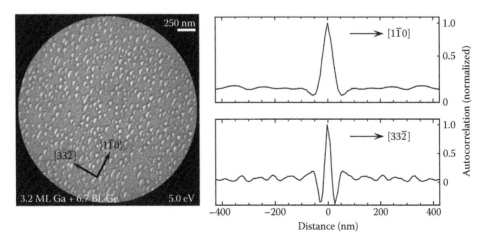

FIGURE 31.21 Left: Bright-field LEEM image after Ge growth at 500°C on a faceted Ga:Si(113) surface. The bright spots are small Ge 3D islands. Right: Cross sections through the autocorrelation function of the LEEM image along [1$\bar{1}$0] (top) and [33$\bar{2}$] direction (bottom).

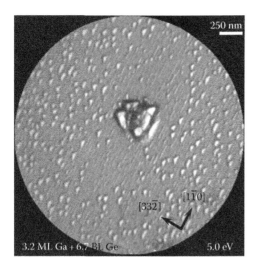

FIGURE 31.22 Bright-field LEEM image after Ge growth at 500°C on a faceted Ga:Si(113) surface. A large contaminant island is visible near the center of the image, surrounded by small Ge islands. In the neighborhood of the contaminant island, denuded zones along the $\langle 1\bar{1}0 \rangle$ directions can be seen, where almost no Ge islands are found.

31.4 Future Perspective

Both vertical and lateral ordering play a key role in leading the way to well-ordered 3D arrays of nanostructures. For stacks of quantum-dot material embedded in spacer layers of the host crystal, state-of-the-art techniques take advantage of the built-in strain fields to enforce a vertical ordering. The size of the nanostructures, the spacer layer thickness, the surface orientation in conjunction with the elastic anisotropy of the spacer material and its chemical composition, as well as corrugated surface morphologies, may influence the vertical ordering. Moreover, as the elastic interaction fades with distance, the resulting ordering is of short-range type. Nevertheless, for proper choice of the parameters, an excellent alignment along the growth direction can be achieved today. For Ge quantum dots in Si, a spread of the lateral quantum-dot position in subsequent layers of about 3 nm has been reported, resulting in a disorder-induced shift of the electronic quantum-dot ground state of only 1 meV [73]. Hence, hardly any application will require an even more precise alignment along the vertical direction.

The situation is more complicate regarding the lateral arrangement. Similar to the vertical dot–dot correlation, lateral ordering can be achieved employing elastic interactions. However, this mechanism is rather weak and requires a huge number of quantum-dot layers [4]. Therefore, a seed layer is required with a well-ordered 2D arrangement of nanostructures.

Optimum control over the nucleation sites of quantum dots in such a seed layer is certainly given by lithographic tools, in particular electron beam [16] or ion beam [74,75] lithography. These techniques are very time consuming as they write the pattern sequentially. In addition, today there is much progress in the attempt to prepare nanoscale patterns with optical lithography.

Using extreme ultraviolet light, patterns with a periodicity of a few tens of nanometers are possible. Common to all lithographic approaches is that they can provide a long-range spatial correlation with positional fluctuations in the order of 1 nm only. The size spread of quantum dots grown on such lithographically prepatterned substrates is typically significantly below 10%. Such an ultimate control over the size and especially the positions of quantum dots is probably necessary for the realization of novel device concepts relying on individual addressing and electronic coupling of quantum dots, e.g., in solid-state-based quantum computing [76,77]. For a large variety of less-ambitious applications, however, self-ordering will be a (cost-) efficient alternative.

Using rather well-established methods like strained or dislocated buffer layers [18,21,22,78] and/or step bunching [21,57], typical Ge_xSi_{1-x} island sizes of about 50–200 nm are obtained, with a size spread of significantly more than 10% (estimated from the literature). In some cases [18,79], average quantum-dot diameters below 50 nm are reported, which might be suitable for charge carrier confinement.

Such small Ge islands can also be achieved by adsorbate-induced substrate prepatterning, as presented in the previous sections, with a similar size spread. For Ge growth on partially Ga-terminated Si(111), a scheme for an even more pronounced island alignment may be deduced: substrates with larger vicinity could be applied in order to completely suppress the 3D island nucleation within the 7 × 7 domains and to tailor the size of the pattern. In case of selective growth on Ga-terminated Si(113), the period length of the facet array depends on the preparation parameters, especially on the substrate temperature during Ga adsorption [80]. This can be used to tune the Ge island size. The approach of an adsorbate prepatterned surface may in principle also be applied to $Si_{1-x}Ge_x$ alloys. Superior to strain-based approaches, the pattern size could then be tuned independently of the chemical composition, leaving this parameter free for tailoring of the electronic properties.

References

1. G. Abstreiter, P. Schittenhelm, C. Engel, E. Silveira, A. Zrenner, D. Meertens, and W. Jäger, *Semicond. Sci. Technol.*, **11** (1996) 1521.
2. E. Kasper, *Appl. Surf. Sci.*, **102** (1996) 189.
3. J. Konle, H. Presting, H. Kibbel, K. Thinke, and R. Sauer, *Sol.-State Electron.*, **45** (2001) 1921.
4. C. Teichert, M.G. Lagally, L.J. Peticolas, J.C. Bean, and J. Tersoff, *Phys. Rev. B*, **53** (1996) 16334.
5. O. Kienzle, F. Ernst, M. Rühle, O.G. Schmidt, and K. Eberl, *Appl. Phys. Lett.*, **74** (1999) 269.
6. V. Holý, J. Stangl, S. Zerlauth, G. Bauer, N. Darowski, D. Lübbert, and U. Pietsch, *J. Phys. D*, **32** (1999) A234.
7. J. Tersoff, C. Teichert, and M.G. Lagally, *Phys. Rev. Lett.*, **76** (1996) 1675.
8. Q. Xie, A. Madhukar, P. Chen, and N.P. Kobayashi, *Phys. Rev. Lett.*, **75** (1995) 2542.

9. F. Heinrichsdorff, A. Krost, D. Bimberg, A.O. Kosogov, and P. Werner, *Appl. Surf. Sci.*, **123–124** (1998) 725.

10. A. Fantini, F. Phillipp, C. Kohler, J. Porsche, and F. Scholz, *J. Cryst. Growth*, **244** (2002) 129.

11. G. Springholz, V. Holý, M. Pinczolits, and G. Bauer, *Science*, **282** (1998) 734.

12. G. Springholz, J. Stangl, M. Pinczolits, V. Holý, P. Mikulik, P. Mayer, K. Wiesauer et al., *Physica E*, **7** (2000) 870.

13. Th. Schmidt, T. Clausen, J. Falta, G. Alexe, T. Passow, D. Hommel, and S. Bernstorff, *Appl. Phys. Lett.*, **84** (2004) 4367.

14. Th. Schmidt, E. Roventa, G. Alexe, T. Clausen, J.I. Flege, S. Bernstorff, A. Rosenauer et al., *Phys. Rev. B*, **72** (2005) 195334.

15. Z. Zhong, A. Halilovic, H. Lichtenberger, F. Scheffler, and G. Bauer, *Physica E*, **23** (2004) 243.

16. O.G. Schmidt, N.Y. Jin-Phillipp, C. Lange, U. Denker, K. Eberl, R. Schreiner, H. Gräbeldinger et al., *Appl. Phys. Lett.*, **77** (2000) 4139.

17. A. Hirai and K.M. Itoh, *Physica E*, **23** (2004) 248.

18. E.V. Pedersen, S.Y. Shiryaev, F. Jensen, J.L. Hansen, and J.W. Petersen, *Surf. Sci. Lett.*, **399** (1998) L351.

19. J. Tersoff, Y.H. Phang, Z. Zhang, and M.G. Lagally, *Phys. Rev. Lett.*, **75** (1995) 2730.

20. T. Walther, C.J. Humpheys, and A.G. Cullis, *Appl. Phys. Lett.*, **71** (1997) 809.

21. A. Ronda, M. Abdallah, J.M. Gay, J. Stettner, and I. Berbezier, *Appl. Surf. Sci.*, **162–163** (2000) 576.

22. A. Ronda and I. Berbezier, *Physica E*, **23** (2004) 370.

23. J.A. Kubby and J.J. Boland, *Surf. Sci. Rep.*, **26** (1996) 61.

24. R. Wiesendanger and H. Güntherodt (Eds.), *Scanning Tunneling Microscopy I*, Springer Verlag, Berlin, Heidelberg, Germany, 1992.

25. R. Wiesendanger and H. Güntherodt (Eds.), *Scanning Tunneling Microscopy II*, 2nd edition, Springer Verlag, Berlin, Heidelberg, Germany, 1995.

26. R. Wiesendanger and H. Güntherodt (Eds.), *Scanning Tunneling Microscopy III*, 2nd edition, Springer Verlag, Berlin, Heidelberg, Germany, 1996.

27. K. Takayanagi, Y. Tanishiro, M. Takahashi, and S. Takahashi, *J. Vac. Sci. Technol. A*, **3** (1985) 1502.

28. Y.M. Lai and Y.L. Wang, *Phys. Rev. B*, **64** (2001) 241404.

29. S. Gangopadhyay, Th. Schmidt, and J. Falta, *Surf. Sci.*, **552** (2004) 63–69.

30. M. Otsuka and T. Ichikawa, *Jpn. J. Appl. Phys.*, **24** (1985) 1103.

31. D.M. Chen, J.A. Golovchenko, P. Bedrossian, and K. Mortensen, *Phys. Rev. Lett.*, **61** (1988) 2867.

32. R.E. Martinez, W.M. Augustyniak, and J.A. Golovchenko, *Phys. Rev. Lett.*, **64** (1990) 1035–1038.

33. P. Kury, P. Zahl, and M. Horn-von Hoegen, *Anal. Bioanal. Chem.*, **379** (2004) 582–587.

34. E. Bauer, *Z. Krist.*, **110** (1958) 372.

35. J.A. Venables, *Introduction to Surface and Thin Film Processes*, Cambridge University Press, Cambridge, U.K., 2000.

36. G. Meyer, B. Voigtländer, and N.M Amer, *Surf. Sci.*, **274** (1992) L541.

37. B. Voigtländer and A. Zinner, *J. Vac. Sci. Technol. A*, **12** (1994) 1932.

38. M. Horn-von Hoegen, A. Al-Falou, H. Pietsch, B.H. Müller, and M. Henzler, *Surf. Sci.*, **298** (1993) 29.

39. M. Horn-von Hoegen and M. Henzler, *Phys. Stat. Sol. A*, **146** (1994) 337.

40. Th. Schmidt, J. Falta, G. Materlik, J. Zeysing, G. Falkenberg, and R.L. Johnson, *Appl. Phys. Lett.*, **74** (1999) 1391.

41. A. Janzen, I. Dumkow, and M. Horn-von Hoegen, *Appl. Phys. Lett.*, **79** (2001) 2387.

42. Th. Schmidt, R. Kröger, T. Clausen, J. Falta, A. Janzen, M. Kammler, P. Kury et al., *Appl. Phys. Lett.*, **86** (2005) 111910.

43. J. Falta, Th. Schmidt, G. Materlik, J. Zeysing, G. Falkenberg, and R.L. Johnson, *Appl. Surf. Sci.*, **162–163** (2000) 256–262.

44. K. Schroeder, A. Antons, R. Berger, and S. Blügel, *Phys. Rev. Lett.*, **88** (2002) 046101-1.

45. M. Horn-von Hoegen, M. Pook, A.Al Falou, B.H. Müller, and M. Henzler, *Surf. Sci.*, **284** (1993) 53.

46. T. Schmidt, S. Heun, J. Slezak, J. Diaz, and K. Prince, *Surf. Rev. Lett.*, **5** (1998) 1287.

47. A. Locatelli, L. Aballe, T.O. Mentes, M. Kiskinova, and E. Bauer, *Surf. Interface Anal.*, **38** (2006) 1554–1557.

48. Elmitec, Private communication.

49. E. Bauer, *Rep. Prog. Phys.*, **57** (2004) 895.

50. E. Bauer, *Surf. Rev. Lett.*, **5** (1998) 1275.

51. J. Falta, M. Copel, F.K. LeGoues, and R.M Tromp, *Appl. Phys. Lett.*, **62** (1993) 2962.

52. J. Falta, Th. Schmidt, A. Hille, and G. Materlik, *Phys. Rev. B*, **54** (1996) R17288.

53. Th. Schmidt, J.I. Flege, S. Gangopadhyay, T. Clausen, A. Locatelli, S. Heun, and J. Falta, *Phys. Rev. Lett.*, **98** (2007) 066104; *Virt. J. Nanoscale Sci. Technol.*, **15** (2007).

54. R.M. Tromp and M.C. Reuter, *Phys. Rev. B*, **47** (1993) 7598.

55. F.M. Ross, J. Tersoff, and R.M. Tromp, *Phys. Rev. Lett.*, **80** (1998) 984.

56. H. Omi and T. Ogino, *Thin Solid Films*, **369** (2000) 88.

57. A. Sgarlata, P.D. Szkutnik, A. Balzarotti, N. Motta, and F. Rosei, *Appl. Phys. Lett.*, **83** (2003) 4002.

58. B. Voigtländer and A. Zinner, *Appl. Phys. Lett.*, **63** (1993) 3055.

59. B. Voigtländer, A. Zinner, T. Weber, and H.P. 0Bonzel, *Phys. Rev. B*, **51** (1995) 7583.

60. Th. Schmidt, S. Gangopadhyay, J.I. Flege, T. Clausen, A. Locatelli, S. Heun, and J. Falta, *New J. Phys.*, **7** (2005) 193-(1–11).

61. J.B. Hannon and R.M. Tromp, *JVSTA*, **19** (2001) 2596.

62. Y.P. Feng, T.H. Wee, C.K. Ong, and H.C. Poon, *Phys. Rev. B*, **54** (1996) 4766.

63. H.J. Müssig, J. Dabrowski, K.E. Ehwald, P. Gaworzewski, A. Huber, and U. Lambert, *Microelectron. Eng.*, **56** (2001) 195.

64. H. Omi and T. Ogino, *Appl. Phys. Lett.*, **71** (1997) 2163.

65. H. Omi and T. Ogino, *Phys. Rev. B*, **59** (1999) 7521.

66. D.J. Bottomley, H. Omi, and T. Ogino, *J. Cryst. Growth*, **225** (2001) 16–22.

67. K. Sumitomo, H. Omi, Z. Zhang, and T. Ogino, *Phys. Rev. B*, **67** (2003) 035319.

68. Z. Zhang, K. Sumitomo, H. Omi, and T. Ogino, *Surf. Sci.*, **497** (2002) 93.

69. A.A. Darhuber, J. Zhu, V. Holý, J. Stangl, P. Mikulik, K. Brunner, G. Abstreiter et al., *Appl. Phys. Lett.*, **73** (1998) 1535.

70. H. Omi and T. Ogino, *Appl. Surf. Sci.*, **130–132** (1998) 781.

71. J. Zhu, K. Brunner, G. Abstreiter, O. Kienzle, F. Ernst, and M. Rühle, *Phys. Rev. B*, **60** (1999) 10935.

72. Th. Schmidt, T. Clausen, J.I. Flege, S. Gangopadhyay, A. Locatelli, T.O. Mentes, F.Z. Guo et al., *New J. Phys.*, **9** (2007) 392-1–392-13.

73. V. Holy, J. Stangl, Th. Fromherz, R. Lechner, E. Wintersberger, G. Bauer, Ch. Dais, E. Müller, and D. Grützmacher, *Phys. Rev. B*, **79** (2009) 035324.

74. J.L. Gray, S. Atha, R. Hull, and J.A. Floro, *Nano Lett.*, **4** (2004) 2447.

75. A. Karmous, A. Cuenat, A. Ronda, I. Berbezier, S. Atha, and R. Hull, *Appl. Phys. Lett.*, **85** (2004) 6401.

76. B.E. Kane, *Nature*, **393** (1998) 133.

77. K.L. Wang, *J. Nanosci. Nanotechnol.*, **2** (2002) 235.

78. C. Teichert, C. Hofer, K. Lyutovich, M. Bauer, and E. Kasper, *Thin Solid Films*, **380** (2000) 25.

79. Berbezier, A. Ronda, and A. Portavoce, *Appl. Phys. Lett.*, **83** (2003) 4833.

80. M. Speckmann, Th. Schmidt, J.I. Flege . S. Gangopadhyay, T. Clausen, P. Sutler and J. Falta, *Phys. Rev. B*, 2010, submitted.

V

Nanosensors

Nanoscale Characterization with Fluorescent Nanoparticles

Lionel Aigouy
*Laboratoire de Physique
et d'Etude des Matériaux*

Michel Mortier
*Laboratoire de Chimie de la
Matière Condensée de Paris*

32.1 Introduction

Luminescence or fluorescence equally refers to the emission of light by a material after absorption of energy. This effect, which has been known for a very long time, has now become a powerful and useful tool of characterization in physics, chemistry, and biology. With the massive and rapid development of nanotechnologies, fluorescent materials are now synthesized under different forms, from micropowders to nanoparticles, crystalline or amorphous, to perfectly spherical or with a complicated shape. This miniaturization, associated with the development of very sensitive detection techniques, can enable us to develop new instruments and, in particular, to use a single luminescent particle as a tool to determine some specific physical properties of its local environment. In this chapter, we describe some particular applications of luminescent materials that concern the development of nanoscale sensors. It is organized in two sections. In Section 32.2, we describe the optical properties and the synthesis of some luminescent materials. We will limit ourselves to inorganic rare-earth-doped fluoride particles that exhibit intense light emission. In Section 32.3, we describe some applications of these particles for the characterization of structures. We first show that a single submicron luminescent particle can be manipulated and fixed at the end of an atomic force microscope tip. We then show that this particle can act as a nanodetector of light, being able to image evanescent fields localized near metallic or dielectric nanostructures. Afterward, since luminescence is a strongly temperature-dependent effect, we show that this particle can act as a nanoscale thermometer, being able to measure the temperature of an electronic device with a submicron lateral resolution.

32.2 Luminescent Nanoparticles: Synthesis and Optical Properties

Many inorganic materials can emit some luminescence. This property is due to the existence, in the material, of electronic levels that are separated by an energy gap equal to the energy of the emitted photon, e.g., a violet photon with a wavelength around 400 nm corresponds to an energy gap of 25,000 cm^{-1}. The origin of these levels can be either intrinsic or extrinsic. For instance, the reduction of size of a semiconducting material that can induce the creation of energy levels by physical confinement is an intrinsic effect. Such nanoparticles are called quantum dots (QD). In fact, most of the inorganic luminescent materials, minerals, are insulating materials. It means that they are transparent to visible light because of a very wide energy gap between the two main states of their electrons, either in the valence band or in the conducting band. Without intrinsic optical properties, they need the addition (doping) of optically active ions that will add localized energy levels inside this wide gap to become luminescent. Many different ions can confer such properties, and the two main families are the so-called 3d transition metal ions, e.g., Cr^{3+}, Mn^{2+}, Fe^{2+}, Co^{2+}, and Ni^{2+}, and the so-called 4f rare earth ions, e.g., Nd^{3+}, Er^{3+}, and Eu^{3+}. Those inorganic materials can be either ordered, and are said to be crystalline, or fully disordered or amorphous such as glasses. Crystalline materials or glasses can be of various compositions, such as oxides based on oxygen, and halides based on fluorine, chlorine, or bromine. Some inorganic materials can also emit luminescence, e.g., dye molecules such as rhodamine or coumarine.

We have chosen to focus on rare-earth-doped fluoride materials for their efficient and versatile fluorescence and their convenient optical excitability. The chemical synthesis of various fluoride materials is described after the introduction of specific electronic level schemes and the optical properties of rare earths ions.

32.2.1 Rare Earth Ions

We describe here the electronic configuration of the family of lanthanides, or rare earth ions, that defines their optical properties.

32.2.1.1 Energy-Level Scheme

The trivalent rare earth ions (RE^{3+}) offer a unique opportunity of diversity in their optical emission spectra extending from the vacuum ultraviolet (VUV) to the infrared spectral range. They generally exhibit a discrete and sharp line-shaped emission spectrum. These properties are due to their N electrons distributed on the open $4f$ shell, which completes their 54 electrons, filling the xenon-like shell $1s^2 2s^2 2p^6 3s^2 3p^6 4s^2 3d^{10} 4p^6 4d^{10} 5s^2 5p^6$. The rare earth family extends from La^{3+} with $N = 0$ to Lu^{3+} with $N = 14$. However, only the 13 rare earths having a partly filled $4f$ shell correspond to the light-emitting ions that we consider here, from Ce^{3+} to Yb^{3+}. Their emission spectra consist of lines originating from electronic transitions within the set of electronic levels (manifold) of the $4f$ configuration. The electronic transitions are governed by so-called selection rules that depend on the type of transition, electric dipole, magnetic dipole,... The most intense ones should correspond to electric dipole transitions but they are forbidden by the parity selection rule because the various levels of a same configuration have the same parity. Their narrow spectra reflect the weak interaction of the rare earth ions with the crystal field imposed by the surrounding material thanks to an efficient shielding by outer $5s^2 5p^6$ electronic shells. So, the energy levels involved in the absorption and emission spectra could be described satisfactorily with a quasi one-ion model. These energy levels of the free atom, and the case of the ion in a crystalline environment, can be determined through the use of a Hamiltonian accounting for mutual Coulomb repulsion and spin–orbit interaction that are of similar order of magnitude. A typical spacing of few thousands wavenumbers (cm^{-1}) is observed between main line groups corresponding to this intermediate coupling case. These levels, called spectroscopic terms and labeled $^{2S+1}L_J$, have a degeneracy of $2J + 1$, with J the total kinetic momentum, L the total orbital momentum, and S the total spin momentum (Figure 32.1).

A remarkable feature of the optical spectra of rare earth ions in a crystal or a glass reside in the existence of groups of sharp lines separated by few hundreds wavenumbers (cm^{-1}) and reflecting the splitting induced by the so-called crystal field of surrounding charges of the ions of the material. Acting as a perturbation of the spectroscopic states, the crystal field removes partially the degeneracy as a function of

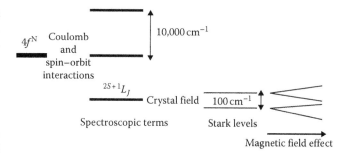

FIGURE 32.1 Schematic energy-level scheme of the $4f$ shell of rare earth ions.

its local symmetry and the spectroscopic terms are splitted into Stark levels. For a half integer value of J, these levels keep degeneracy greater or equal to 2 and they are called Kramers doublets.

A complete description of the energy level calculation has been described in several textbooks for free ions or ions embedded in crystals (Dieke, 1968; Hüfner, 1978). The crystal field effects have been specifically studied by several authors and reflect a very well defined local order described by group theory (Holsa and Porcher, 1981; Malta, 1982).

When the environment of the rare earth ions is assumed to come up only through the crystal field splitting that weakly affect the general optical properties, a clear difference can be observed between the spectra of rare earths in a crystal and in an amorphous material such as a glass. In ordered materials, one single or few distinct sites are offered to the rare earth ion and the individual linewidths are then mainly homogeneously broadened by phonons. At room temperature, such widths are not bigger than wavenumbers. In disordered compounds, a huge number of different sites can be occupied giving rise to a large inhomogeneous broadening, typically exceeding $100 \, cm^{-1}$. In glasses, the crystal field effect has also been studied, but most of the theories fail to describe the right number of observed transitions or use a very low local symmetry order, which is not corresponding to the general idea of isotropy in such amorphous material. Another description accounting for descending symmetry principles in group theory has, quite recently, allowed a satisfactory explanation of a complete degeneracy removal together with a high global symmetry but accompanied by low symmetry perturbations (Mortier et al., 2000). In any case, the crystal field splitting of few hundreds wavenumbers allows the use of those splitted energy levels to be highly temperature sensitive and then to make local temperature sensing as described in Section 32.3.4. Under an external magnetic field effect, each Kramer doublet of the rare earth ions splits into two Zeeman levels following a proportional law with field intensity (Figure 32.1). Such an effect can be used for magnetic field sensing through optical spectra modification or fluorescence intensity changes (Tikhomirov et al., 2009). This effect is also used for electron spin resonance measurements.

32.2.1.2 Intensity of Transitions: Energy Transfers

The interaction with crystal field and lattice vibrations can mix states of opposite parity and then allows transitions which have been previously described as forbidden by parity selection rule. Also, some lines originate from magnetic dipole transitions, which are allowed within a same electronic configuration. In every case, the intensities remain quite low because of the outer electronic shielding limiting the wavefunction mixing in case of electric transitions or because of the low intensity of magnetic transitions.

A peculiar and very useful property of rare earths lies in their ability to give rise to energy transfers both in crystals and glasses. Indeed, the optical properties of the rare-earth-doped materials mainly result from individual properties of ions, and not from collective states, according to the previous description of their electronic levels. In such a case, in a material (crystal or glass) containing different rare earth ion species, one type of ion can be first excited in a given energy domain before the observation of light emission by a second type. The first one is generally described as a donor or sensitizer and the second one as an acceptor. When no energy mismatch appears between the levels of the two species, the transfer is said to be a resonant one. In other cases, the nonresonant transfer is assisted by an energy compensation originating from the lattice vibration energy, either given or received. Many energy transfers can occur and some of them can even involve simultaneously several ions but their probability is lower. These transfers can be radiative or nonradiative, i.e., without any intermediary photon but through interaction, typically dipole–dipole. This latter case is more efficient than the radiative case that supposes emission and absorption. Radiative transfer depends only from the distance between ions but does require that an emitted photon meets the cross section of a second ion (Auzel, 2004).

The first advantage that can be obtained from energy transfer is the use of an efficient absorbing ion, due to higher cross section or broader band, to sensitize an accepting ion that has an interesting emission spectrum but a weak direct absorption capability. That is, for instance, the respective case of Yb^{3+} and Er^{3+} ions. The broad absorption band of Yb^{3+} enhances greatly the Er^{3+} excitation capability thanks to a maximum cross section, which is five times greater than that of Er^{3+} at around 980 nm (Dantelle et al., 2005).

The second advantage of energy transfers resides in the high ability to allow anti-Stokes fluorescence. Because the emission probability of rare earth ions is quite limited by their mainly electric forbidden character, their excited states offer quite a long lifetime and they are then susceptible to allow excited state absorption (Figure 32.2). The lifetimes can also be lengthened thanks to low energy vibration modes materials. When the excited state absorption cross section is also high enough, the process can occur and the ions can emit anti-Stokes emissions. Many other up-conversion processes can give rise to anti-Stokes emission through energy transfers, such as up-conversion,

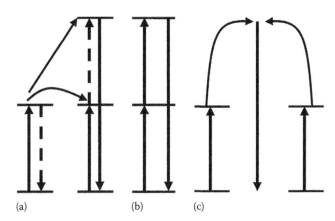

FIGURE 32.2 Three types of energy transfers: (a) APTE, (b) excited state absorption, and (c) cooperative fluorescence.

cooperative sensitization, and cooperative luminescence. See Auzel (2004) for a review of the up-conversion and anti-Stokes processes. One of the most efficient is the so-called APTE or ETU up-conversion process that allows an easy and strong visible emission by Er^{3+} ions for instance after excitation around 980 nm (Figure 32.3). The up-conversion processes involve several photons, typically two photons for a green emission of erbium; so they offer a nonlinear response to the excitation intensity and can be used as a quadratic detector. Also, the infrared excitation of the fluorescent material prevents any spurious emission that could occur with classical Stokes emission arising from blue or UV excitation. The up-conversion scheme using Yb^{3+} and Er^{3+} ions is used in Section 32.3.

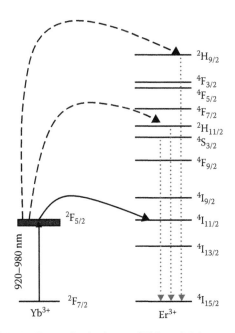

FIGURE 32.3 Energy-level scheme of Yb^{3+} and Er^{3+} ions. The solid arrow corresponds to a resonant energy transfer (sensitization) between Yb^{3+} and Er^{3+} when the dashed arrows correspond to up-conversion processes, giving rise to three main visible emission bands from Er^{3+}.

A very appreciable property of the fluorescence of rare earth ions, when using excitation and fluorescence in their ground $4f^N$ configuration, is the total absence of bleaching under any optical excitation and also the absence of any blinking whatever the intensities. It means that, because of an absolute electronic stability, the luminescence intensity of the material remains constant without any interruption and will not vanish definitively after a certain time of excitation.

32.2.2 Fluoride Materials

The fluoride compounds offer simultaneously a low frequency of phonon spectrum (atomic vibrations), typically <600 cm^{-1}, and a large solubility for the rare earth ions used as optical active centers in a chemically stable compound with regard to the other halide compounds. The low phonon cut-off frequency is due to the ionic character of the fluoride materials and is a crucial advantage for the fluorescence arising from only few hundreds of active ions in a nanoparticle. A low phonon frequency reduces the probability of nonradiative transition for the rare earths ions and lengthens their excited state lifetime. Nonradiative transitions transform electronic excitation into heat instead of light emission. When the phonon energy can be lower in the other halide compounds such as chlorides and bromides, their reactivity, enhanced by the specific surface of nanoparticles, makes them highly sensitive to moisture and species adsorption that are poisonous for fluorescence. Fluoride compounds then offer a fairly good compromise between chemical stability and phonon frequency. They also exhibit a wide transparency window from the UV to the IR domain, offering the best conditions for doping by optically active ions.

32.2.2.1 Synthesis of Glasses

In the first approach, fluorescent particles have been obtained by grinding of bulk materials, either fully amorphous, fluoride glass, or partially crystallized, glass–ceramic. A glass is a solid material that has kept the high disorder of a liquid and it is consequently a quite perfectly isotropic medium. The luminescent properties keep the same character. Such isotropy allows preparing isotropic nanoprobes, which can be useful to avoid selectivity effects in the detection of local fields, as shown in Section 32.3. A glass–ceramic possesses the same global isotropy than a glass because it is mainly a glassy material but with a nanostructure that insures locally a crystalline environment to the rare earth ions and then a more intense fluorescence.

The synthesis of fluoride glasses is obtained by melting raw powders of high purity in a graphite crucible. The furnace is installed inside an argon glove box that prevents any reaction with moisture or oxygen during melting and quenching. The melt is obtained by heating at a temperature higher than the theoretical melting temperature of the blend as determined by way of differential thermal analysis (example of DTA curve

reported on Figure 32.5a). After a homogenization stage of several minutes at this high temperature, the liquid is poured between two cold copper pieces to ensure a strong quenching of the melt. A solid but liquid like state is then obtained. To release the strains induced by this strong quenching, a thermal treatment is done just below the glass transition temperature of the glass. This glass transition temperature indicates the beginning of the softening of the glass and is obtained by thermal analysis. After the thermal release of the strains, the glass is ready for cutting and polishing bulk samples or grinding to prepare particles.

The main job to prepare glasses is to determine the composition that will confer the best optical properties to the rare earth ions used in any cases studied here. First, the glass stability has to be insured by the choice of glass forming compounds (AlF$_3$, ZrF$_4$, GaF$_3$). A former glass compound is able to give rise alone to a glass network just by melting followed by a fast cooling. The addition of glass modifiers compounds (KF, LiF, NaF) will allow opening the rigid glass network of the forming compound and then the insertion of other species such as rare earth ions. Also, several intermediate compounds (LaF$_3$, YF$_3$, "Ln"F$_3$, BaF$_2$) behave between the network forming and modifier; they can lower the melting temperature. These different compounds will determine the vibration spectrum and the transparency window of the glass. They will also determine the stability of the glass with regard to crystallization tendency, for instance, during grinding or heating. Indeed, the ionic character of the fluoride glass is mainly fixed by the F$^-$ anions, but the weight of the cations will also influence the cutoff frequency of vibration. The weakly covalent character of fluorine compounds (smaller than 10%) insures a weaker chemical bonding, and then a lower vibration frequency, than in oxide materials for instance, The ionic character of the material and the polarizability of the ions will also determine the refraction index of the glass. For these reasons, the glass formulation is really the deciding factor (Mortier et al., 2001, 2003).

We have then used a specific composition, allowing glass stability, high capability of rare earth doping without any clustering risk and limited vibration cutoff frequency. For the first time, a glass only doped with erbium (ZBEAN) has been prepared (Fragola et al., 2003). Also, Yb^{3+} and Er^{3+} codoping has been used to insure a more efficient absorption and up conversion emission as explained in Section 32.2.1. The ZBYbAN molar glass composition is 45.5 ZrF$_4$, 23 BaF$_2$, 11YbF$_3$, 3 ErF$_3$, 3 AlF$_3$, 0.5 InF$_3$, and 14 NaF (Aigouy et al., 2003). The 11/3 ratio between Yb and Er ensures a high probability of sensitization of erbium ions by ytterbium. 5 g of powder are melted during 45 min at 950°C before quenching at 240°C between preheated copper pieces. After heating at 260°C for 1 h, the glass is slowly cooled down to room temperature. The obtained sample is transparent and has a pink color.

After strong grinding of bulk pieces of the as obtained glass, particles with submicron size are obtained (Aigouy et al., 2003). This method is quite rough but ensures the synthesis of isotropic materials in low size. Also, a constant quality of the particles, with regular fluorescent centers content is guaranteed by the unique origin of a bulk sample and very homogeneous materials. The grinding does not affect the amorphous nature of the particles.

32.2.2.2 Synthesis of Glass–Ceramics

Similar to glass, the formulation is the main job to prepare glass–ceramic materials (Mortier and Patriarche, 2000, 2). Indeed, glass–ceramics are obtained by the voluntary and controlled devitrification of a part of a glass. It supposes that the composition permits to crystallize a particular phase inside a matrix that stays amorphous. For optical applications, the most popular systems are oxyfluoride materials in which a fluoride minor phase is crystallized inside an oxide glassy matrix. The ideal case is to get a system in which the rare earth ions are segregated inside the fluoride minor crystal phase. The material is mainly an oxide glass but the optical properties of the rare earths are those obtained in a fluoride crystal with low phonon energy and low nonradiative rate, high solubility and then high doping rate. The inhomogeneous linewidths observed in glasses are strongly reduced because of a crystalline environment in the glass–ceramic. This linewidth reduction induces a strong enhancement of the maximum absorption or emission cross-section (Figure 32.4a) which enhances greatly the optical efficiency of the glass–ceramic when compared to the initial glass. The excited states lifetime are dramatically lengthened by the segregation of the entire rare earths content inside the nanocrystalline phase (Figure 32.4b). Also, such glasses are easily prepared in the air and an adapted thermal cycle induces the crystallization of the minor phase as nanocrystallites. The system remains macroscopically homogeneous and isotropic, and the transparency is kept thanks to nanometer particle sizes.

In this frame, we have developed a family of glasses based on germanium and lead oxides, lead and rare earths fluorides. The molar composition is tuned around 50 GeO_2, 40 PbO, 10 PbF_2, yYbF$_3$,

xErF$_3$ with $0 < y < 5$, $0 < x < 3$. The glass is obtained by melting of the high commercial powders in a platinum crucible for 20 min at 1050°C in the air. The melt is poured between two copper plates heated at 150°C to reduce the thermal shocks. Transparent glasses of 2–3 mm thickness are obtained and heated below the glass transition temperature to release the strains. The devitrification occurs after a 10 h thermal treatment at temperature just above the glass transition (labeled T_g on Figure 32.5a). The low phonon cutoff frequency of lead fluoride (336 cm^{-1}) ensures a very low nonradiative probability. Also, the segregation of the rare earths inside the minor phases reinforces the energy transfers (Dantelle et al., 2005) that are researched between ytterbium and erbium for the purpose of our applications.

In the same way with bulk glasses, submicron size particles are obtained by grinding of the bulk glass–ceramic sample. Free standing and efficiently fluorescent nanoparticles have also been obtained by the selective chemical attack of an oxyfluoride glass–ceramic (Figure 32.6). With this method, the oxide glassy matrix is used similarly to a template, but an inorganic one, to drive the synthesis of the fluoride nanoparticles that are finally separated by hydrofluoric acid etching of the oxide glass (Mortier and Patriarche, 2006; Tikhomirov et al., 2008).

32.2.2.3 Direct Synthesis of Nanoparticles

Beside previous techniques used to prepare fluorescent particles from bulk samples, several different routes have been successfully employed to prepare crystalline nanoparticles. The smallest size particles (several nanometers to tens of nanometers) are quite easy to prepare when the intermediary size (hundred nanometers to half a micron) require a more complex synthesis.

The classical solid state chemistry route, using thermal treatment, is not used to prepare submicron size particles because of a thermally induced growth of the particles during the reaction. However, a solid state reaction can be used successfully in the frame of the mechanical allowing, also called mechanosynthesis. In such case, the simultaneous grinding and mechanically induced

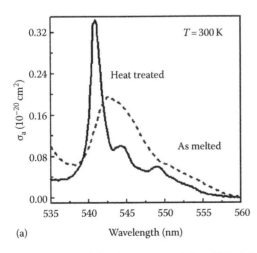
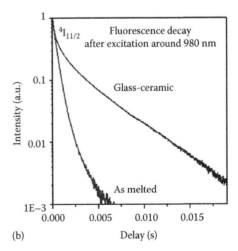

FIGURE 32.4 Optical properties of glass (50 GeO_2, 40 PbO, 10 PbF_2, 2 ErF$_3$) (dotted line) and corresponding glass–ceramic (solid line) at 300 K. (a) Absorption cross-section of the $^4I_{15/2} \rightarrow {}^4S_{3/2}$ transition of erbium ions. (b) Fluorescence decay of the $^4I_{11/2}$ level of erbium ions.

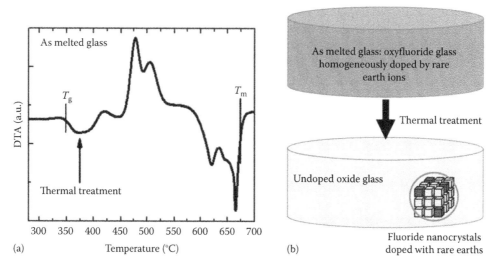

(a)

(b)

FIGURE 32.5 (a) Thermal analysis curve (DTA) of as-melted glass: T_g corresponds to the glass transition temperature and T_m to the melting temperature. (b) Schematic principle of the transformation of homogeneous as-melted glass in a glass–ceramic composed of rare-earth-doped fluoride nanocrystals embedded inside the oxide glass.

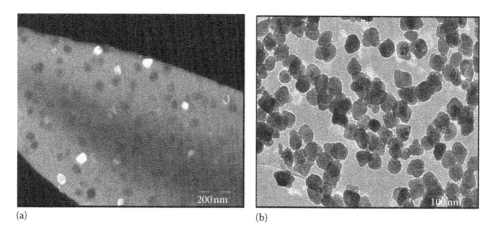

(a)

(b)

FIGURE 32.6 Transmission electron micrographs: (a) glass–ceramic (From Dantelle, G. et al., *J. Solid State Chem.*, 179, 1995, 2006. With permission.) corresponding to Figure 32.4b and (b) free standing particles obtained by chemical etching of this bulk glass–ceramic.

solid state reaction allow the preparation of spherically shaped particles with sizes varying from several nanometers to few hundreds.

Soft chemistry methods are widely used to prepare oxide nanoparticles. Some of them have been transposed to fluoride synthesis. The fluorine chemistry is generally considered to be a peculiar one because of the high reactivity of fluorine and the difficulty to use safely hydrofluoric acid. However, fluoride compounds can also be synthesized in aqueous media without oxide or hydroxyl contamination.

The first one consists in a classical reverse microemulsion method, often called reverse micellar system (Figure 32.7). Microemulsions are made by dissolving a surfactant in cyclohexane inside a Teflon beaker. This mixture is stirred until it is homogeneous and then aqueous solution of soluble salts (rare earth nitrates for instance) is poured in slowly. In a similar way, a hydrofluoric acid microemulsion is made by substituting the aqueous nitrate solution with aqueous HF solution. These two microemulsions are then mixed and stirred at room temperature in order to allow the inter-micellar exchange leading to the formation of fluoride nanoparticles (Bensalah et al., 2006; Dantelle et al., 2006; Labéguerie et al., 2008). A nonaqueous method has also been developed to obtain fluoride nanoparticles using a micellar like principle but excluding any water use (Labéguerie et al., 2006).

Another possible way consists in the simple co-precipitation of the fluoride nanoparticles by a slow adding of a solution containing a soluble salt to another solution containing dissolved hydrogen fluoride. The solution containing soluble salts is added to the dissolved hydrogen fluoride solution using an automatic pipette in order to control the growth of the particles from the size drop and the number of drops per unit time (Bensalah et al., 2006; Labéguerie et al., 2008).

These two methods give small size particles (few nanometers) in the case of several different simple fluoride compounds, such as CaF_2, PbF_2, LaF_3, and YF_3 either doped by rare

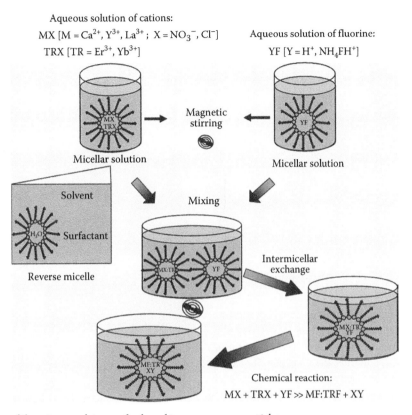

Aqueous solution of cations:
$MX [M = Ca^{2+}, Y^{3+}, La^{3+}; X = NO_3^-, Cl^-]$
$TRX [TR = Er^{3+}, Yb^{3+}]$

Aqueous solution of fluorine:
$YF [Y = H^+, NH_4FH^+]$

Magnetic stirring

Micellar solution

Micellar solution

Solvent

Surfactant

Mixing

Reverse micelle

Intermicellar exchange

Chemical reaction:
$MX + TRX + YF \gg MF{:}TRF + XY$

FIGURE 32.7 Description of the microemulsion method used to prepare nanoparticles.

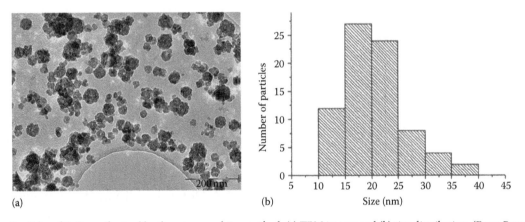

(a)

(b)

FIGURE 32.8 Particles of CaF_2 synthesized by the microemulsion method. (a) TEM images and (b) size distribution. (From Bensalah, A. et al., *J. Solid State Chem.*, 179, 2636, 2006. With permission.)

earth ions or not (Figure 32.8). Higher temperature synthesis can also been realized in order to get bigger sizes (few tens of nanometers). However, it is possible to attain really big size particles (several hundreds of nanometers) when preparing more complex compounds such as $xKF (1 - x)YF_3$ compounds (Figure 32.9).

Either glassy or crystalline, the fabrication method, the composition of the fluoride material and the rare earth ions can be chosen according to the exact need of size or fluorescence expressed by the nanoscale characterization described in Section 32.3.

32.3 Nanoscale Luminescent Sensors

32.3.1 General Idea

The fluorescent particles described in Section 32.2 constitute a fascinating nanosource of light, very sensitive to the local environment. In chemistry, or more precisely when they are in aqueous solution, their luminescence depends on many properties of the fluid, like its pH or its contamination by specific ions (Valeur, 2002). Similarly, other nanostructures, like for instance, carbon nanotubes, recently proved to be very sensitive pH indicators

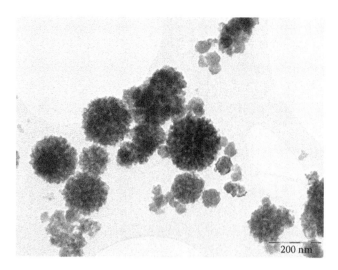

FIGURE 32.9 TEM image of KYF_4 particles synthesized by the coprecipitation method.

(Cognet et al., 2007), their fluorescence being strongly dependent on the concentration of H^+ ions. In this section, we will describe some applications of fluorescent particles, but in the domain of physics or material science. We will show that a single submicron-sized fluorescent particle can be used to characterize the local optical properties of nanostructured surfaces and the local thermal properties of microelectronic devices. Recent progresses in nanomanipulations now makes it possible to settle a single particle at the end of an atomic force microscope tip. We can develop a sensor whose position is controlled with nanometer precision that has the ability to form images and to reveal the desired effect.

32.3.2 Sensor Development: Tip Fabrication

In order to use a single fluorescent particle as a movable sensor, we have to glue it at the end of a support. This support is a sharp atomic force microscope tip, obtained by the electrochemical erosion of a tungsten wire. The tip extremity is first immersed in a polymer glue to a depth of 1 or 2 µm. The polymer, which forms a thin layer around the tip, remains sticky until photoreticulation under ultraviolet radiation. The tip is afterward moved near to a surface on which the fluorescent particles were placed. The approach is made by means of piezoelectric actuators, under a classic optical microscope and under strong magnification. A series of images showing the bonding of a particle of 250 nm in diameter at the end of a tip is shown in Figure 32.10 (Aigouy et al., 2004). In this particular case, the particle is gold and has a diameter of about 250 nm. The same procedure is used to set the fluorescent particles to the tip extremity. The minimal size of the objects that can be manipulated with this system is about 100 nm. To glue smaller objects, one must use other techniques like positive dielectrophoresis or use manipulators placed in a scanning electron microscope (SEM) (de Jonge and van Druten, 2003; Gan, 2007; Peng et al., 2008).

We show in Figure 32.11 two SEM images of a nanoparticle glued at the end of a tungsten tip. The particles are either a fragment of amorphous glass (diameter ~200 nm) or a PbF_2 nanocrystal (diameter ~100–150 nm). This particle will be excited by a

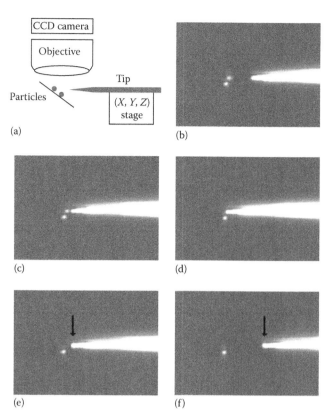

FIGURE 32.10 (a) Description of the experimental nanomanipulation set-up; (b) to (f) optical images showing the gluing of a single particle at the end of a sharp tip. The particle has a diameter of 250 nm. The image size is 16.5×11.7 µm. (From Aigouy, L. et al., *Appl. Opt.*, 43, 3829, 2004. With permission.)

laser beam and will act as a nanoscale light source or detector. In contact with a nanostructured surface, the particle fluorescence will be affected by the local properties of the sample, such as local electromagnetic fields or other physical effects like temperature, static electric or magnetic fields. In the next section, we describe an application where the particle is a local optical sensor.

32.3.3 Detection of Local (Evanescent) Fields

32.3.3.1 What Is an Evanescent Field?

In far-field optical microscopy, we are used to observe or characterize large objects, whose sizes are typically larger than the wavelength of light. The contrast mechanisms are well known and are linked to the reflection or the transmission coefficients of the material, or to its ability to absorb, scatter, or polarize light. In fact, when a surface or an object is illuminated by an incident radiation, some light is scattered in free space and some light remains localized near it (see Figure 32.12). This local electromagnetic field is commonly called an evanescent field. It contains information about the local optical and physical properties of the materials. The structure of evanescent fields is usually very complex since it depends on the dielectric constant of the sample, on its topography, but also on the way light impinges on the

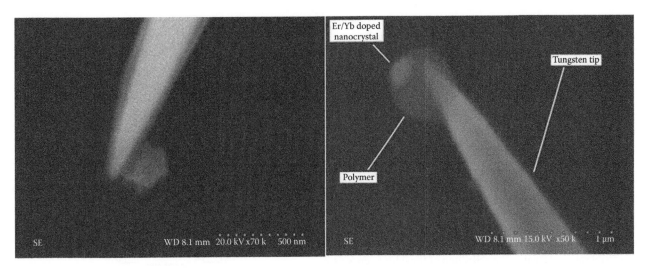

FIGURE 32.11 SEM images of a fluorescent particle glued at the end of a tungsten tip: amorphous particle (left), PbF_2 nanocrystal (right). Both particles are codoped with erbium and ytterbium ions.

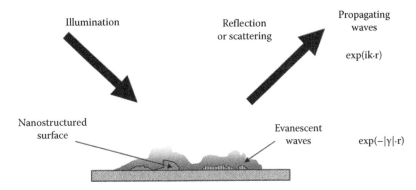

FIGURE 32.12 Illumination of a nanostructured surface: the surface reflects or scatters the incident light. An evanescent field is localized at a subwavelength distance from the surface.

structure, its polarization direction, its angle of incidence, and of course its wavelength.

In contrast to propagating fields, evanescent fields rapidly vanish when getting far from the structure. Their intensity decays exponentially and, in the visible spectral range, they almost do not exist anymore at a distance larger than a wavelength. The detection of these evanescent components is essential to characterize the local optical properties of nanostructures. To detect them, new microscopes, namely scanning near-field optical microscopes (SNOM), have been developed (Pohl et al., 1984; Greffet and Carminati, 1997; Hecht et al., 2000). Like other scanning probe microscopes, SNOMs are made of a sharp probe (a thin optical fiber, or an apertureless tip) whose aim is to locally detect these photons. Fluorescent particles can also efficiently detect local fields. When placed near an illuminated nanostructure, the particles may absorb the photons on the surface, and their fluorescence is a direct measurement of the local field intensity.

In Sections 32.3.3.2 and 32.3.3.3, we describe two examples of visualization of local fields with a fluorescent particle. The first example is related to the field localized near a metallic nanoparticle. The second application concerns the visualization of surface plasmon polaritons, which are waves that propagate at a metal/dielectric interface, but are evanescent in a direction perpendicular to the surface.

32.3.3.2 Imaging Localized Fields

Metallic nanoparticles are well known for their ability to efficiently scatter light in the far-field. For gold and silver particles, a maximum scattering occurs for a certain wavelength, which depends on the particle size. For some wavelengths, the incident photons may interact with the free electrons of the particle and induce collective oscillations, known as surface plasmons (Xia and Halas, 2005). This resonance explains why some particles have a specific color. For instance, spherical gold particles whose diameter is 50 nm appear green when illuminated with white light. Their color shifts toward the red wavelength when their size increases.

In addition to efficient scattering, metallic nanoparticles also have a strong electromagnetic field in their immediate vicinity. This intense field may be used to enhance the fluorescence of molecules, to increase Raman signals and second-harmonic generation (Kneipp et al., 1999; Parfenov et al., 2003). This local field is also of interest for several applications in nano-lithography and can be used to induce surface modifications (Koenderink et al., 2007). Visualizing the field distribution near metallic nanostructures may help to understand quantitatively the mechanisms of all these effects and optimize their possible applications. We present in Figure 32.13 the experimental setup used to observe the near-field optical distribution with a fluorescent particle attached to the end of a tungsten tip. In that case, the sample is illuminated in a transmission mode with a near-infrared laser beam ($\lambda = 975$ nm). Upon illumination, a complex field distribution is created, which depends on the local dielectric properties and on the topography. This local field is absorbed by the scanning particle that fluoresces in free space. As explained in Section 32.2, the particle is excited by an anti-Stokes process at $\lambda = 975$ nm (by up-conversion) and we collect the light emitted in the visible at $\lambda = 550$ nm. The collection of the fluorescence as a function of the position of the tip on the surface gives a representation of the near-field optical distribution.

In order to show the possibility of the technique, we have chosen to describe an experiment performed on an elementary sample structure composed of gold and latex nanospheres deposited on a SiO_2 surface (see Figure 32.14). The diameter of the particles is 250 and 220 nm for gold and latex, respectively. The topography and the fluorescence images of the structure are represented in Figure 32.14. The topography shows three spheres with similar shape and height. The optical image is completely different. A fluorescence enhancement is well visible on one of the beads but remains unaffected by the two other beads. The explanation

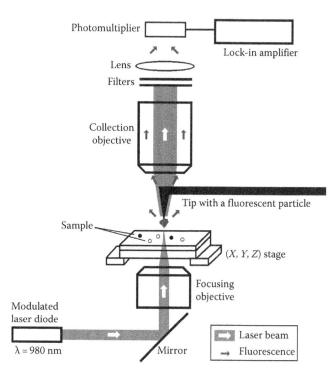

FIGURE 32.13 Experimental setup for the observation of near-field optical distributions with a fluorescent particle. (From Aigouy, L. et al., *J. Appl. Phys.*, 97, 104322, 2005. With permission.)

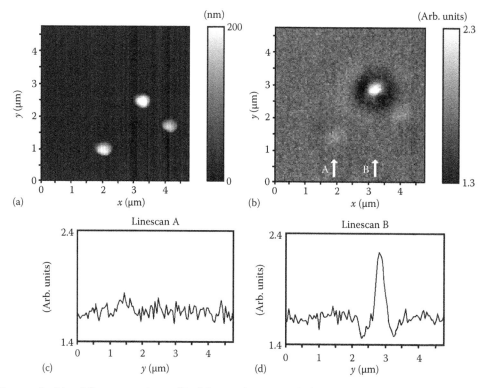

FIGURE 32.14 Topography (a) and fluorescence image (b) of the sample composed of metallic and dielectric nanospheres. Two vertical cross-sections along the arrows A and B are given in (c) and (d). (From Aigouy, L. et al., *J. Appl. Phys.*, 97, 104322, 2005. With permission.)

is that one bead is made of gold, whereas the others are made of latex. The fluorescence enhancement, visible on the cross sections of Figure 32.14, is induced by the strong field located near the metallic nanoparticle (Aigouy et al., 2005). Near the dielectric spheres, the local field does not vary enough to induce a fluorescence enhancement.

From this example, we can see that fluorescent particles are very efficient near-field optical detectors, sensitive to the nature of the materials on a surface, and are able to quantitatively measure the intensity of the local field near nanostructures. In the following section, we describe another application of the technique, which concerns the characterization of surface waves created by nanoapertures in a thin metallic film.

32.3.3.3 Imaging Surface Plasmon Polaritons

32.3.3.3.1 What Is a Surface Plasmon Polariton?

In the preceding section, we have seen that an intense electromagnetic field can be localized near metallic nanostructures. This field enhancement is provoked by the resonant oscillation of charges on the sphere, known as localized surface plasmons. Similarly, on a flat metallic surface or for instance at the vacuum/gold interface, an electromagnetic wave may exist and propagate on several tens of micrometers. This wave, which is also linked to the oscillation of surface charges, is known as a surface plasmon polariton (SPP) (Zayats et al., 2005). A schematic description of an SPP wave is given in Figure 32.15. Its amplitude decreases exponentially with the distance to the surface. It has a transverse magnetic field, and two electric field components. Its wave vector is given by the simple relation:

$$k_{sp} = \frac{\omega}{c} \sqrt{\frac{\varepsilon_d \varepsilon_m(\omega)}{\varepsilon_d + \varepsilon_m(\omega)}} \qquad (32.1)$$

where
ε_d and ε_m are the optical constants of the dielectric (the air for instance) and the metal respectively
ω is the frequency
c is the speed of light in vacuum

In the visible or near-infrared range of the electromagnetic spectrum, ε_m is negative and the modulus of k_{sp} is larger than k_o.

This implies that an SPP at the metal/air interface cannot be excited by simply illuminating the metallic surface from the air side.

A well known technique to create an SPP consists in illuminating the thin metallic layer through the dielectric surface that supports it (for instance, a glass prism as represented in Figure 32.15). For a given angle of incidence, the in-plane component of the incident wave vector may match the value of k_{sp} and create the SPP. The resonance condition is given by the simple relation:

$$k_x = \frac{2\pi}{\lambda_0} n_{prism} \sin\theta_{res} = k_{sp} \qquad (32.2)$$

where
n_{prism} is the refractive index of the prism
θ_{res} is the angle for which k_x equals k_{sp}

There are also other ways to create SPPs, like the use of a grating on the metallic surface, the local illumination of the metallic surface by the optical fiber of a scanning near-field optical microscope. The controlled nanostructuration of the surface by nanoapertures, and ridges can also be used to efficiently create SPP waves.

32.3.3.3.2 Observation of Surface Plasmon Polaritons Created by Nanoslits in a Gold Film

Since the SPP intensity exponentially decays with the distance to the surface, it is not possible to observe them with a classical far-field optical microscope, except if some defects on the surface transform them into propagating waves. A very simple technique has recently been implemented to detect SPPs. Fluorescent molecules deposited on the metallic surface may absorb the surface photons and emit light by fluorescence. The detection of this emission allows to indirectly visualize the surface electromagnetic field (Ditlbacher et al., 2002). Another way to detect SPPs is to use a near-field probe, as described in Section 32.3.3.1. In that case, the tip (an optical fiber for instance) acts as a nanodetector of the surface electromagnetic field. The fluorescent particle glued at the end of a sharp tip can also be used to efficiently detect SPPs created by nanostructures, and in particular by nanoapertures in a thin metallic film. Apertures in a thin metallic film have recently been widely investigated since the observation of unexpected spectacular effects like

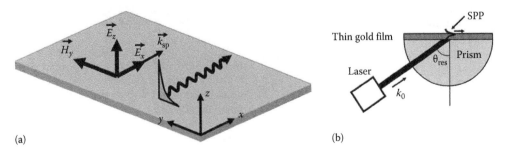

FIGURE 32.15 (a) Structure of a SPP wave at the gold/air interface; (b) example of illumination used to generate SPP waves on a metallic thin film.

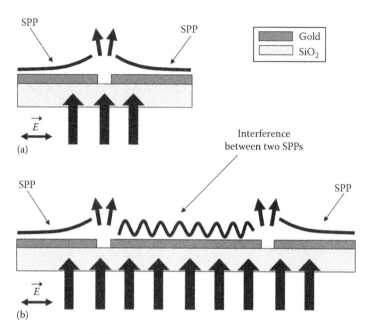

FIGURE 32.16 (a) Single-slit aperture in a thin gold film illuminated in transmission by a plane wave. The incident polarization is perpendicular to the slit axis. The aperture diffracts the light in the far field and generates SPPs that propagate on the metal surface; (b) same situation but with two parallel slits: the two SPPs interfere and create a fringe pattern.

extraordinary optical transmission (EOT) (Ebbesen et al., 1998; Liu and Lalanne, 2008). SPP waves may play an important role in such effect, and their direct observation is of great importance to understand and optimize EOT.

We consider a very simple nanostructure, composed of two adjacent slit apertures perforated in a thin gold film deposited on a SiO_2 substrate. By illuminating a slit in transmission, some light is transmitted and diffracted in the far-field. When the incident beam is linearly polarized in a direction perpendicular to the slit long axis, additional waves are created such as SPPs. The situation is described in Figure 32.16. If two parallel slits are illuminated, each aperture may create the surface waves, and in the space between the two slits, an interference pattern may appear due to the interaction of two surface waves. If one looks at the structure with a classical optical microscope, one will see the light directly transmitted through the apertures, but the SPP and the interference pattern will not be visible due to their exponential decay in the z-direction.

We show in Figure 32.17 the images obtained when the fluorescent particle is scanning the exit of two 300 nm-wide nanoslit apertures separated by 6.24 μm. The images are obtained when the tip scans the sample in the (y, z) and (x, y) planes, respectively. The interference pattern between two counterpropagating SPPs is clearly visible between the apertures as well as the light divergence in free space and the vertical extension of the pattern. The nanoscale optical characterization of such structures is important to understand the mechanisms of SPP generation, to evaluate the amount of SPP created, and to analyze the interaction of these waves with other adjacent nanostructures (Aigouy et al., 2007).

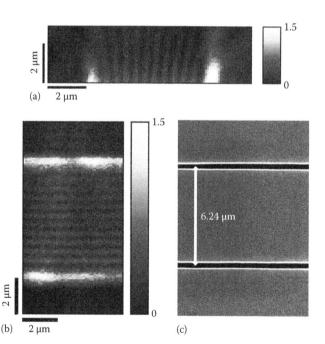

FIGURE 32.17 Fluorescence images measured at the exit of two slits made in a thin gold film. (a) Scan in a plane perpendicular to the surface [(y, z) plane]; (b) scan in a plane parallel to the surface [(x, y) plane]. The interference pattern is due to the interference between two counterpropagating SPPs. A SEM image of the structure is given in (c). (Reprinted from Aigouy, L. et al., *Phys. Rev. Lett.*, 98, 153902, 2007. With permission.)

32.3.4 Nanoscale Thermometer

In the following sections, we describe another application of fluorescent particles, in a domain of thermal imaging. Since fluorescence is a strongly temperature-dependent effect, a fluorescent material can be a nanoscale thermal sensor, which unlike other thermal imaging techniques (thermocouples or thermoresistive probes (Mills et al., 1998; Shi et al., 2000; Pollock and Hammiche, 2001)), does not need to be powered by an electrical current. There are several methods to extract the temperature from fluorescence variations (Allison and Gillies, 1997). They include the intensity and lifetime variations (Zhang et al., 1993; Van Keuren et al., 2005; Li et al., 2007), the peak wavelength shift (Löw et al., 2008), and the comparison of the relative intensity of optical transitions from two adjacent thermalized energy levels (Berthou and Jörgensen, 1990; Maurice et al., 1995). In particular, erbium/ytterbium-codoped fluorescent particles possess several energy levels whose relative populations are in thermal equilibrium. We now explain how we can benefit from this specific property to determine the temperature of a device.

32.3.4.1 Measurement Principle

As detailed in Section 32.2, erbium/ytterbium codoped materials possess, in the visible range, several fluorescence lines. Two transitions, $^4S_{3/2} \rightarrow {}^4I_{15/2}$ (emission at $\cong 550$ nm) and $^2H_{11/2} \rightarrow {}^4I_{15/2}$ (emission at $\cong 520$ nm), are very interesting for thermics because their relative intensity is directly linked to temperature by a simple relation (Berthou and Jörgensen, 1990; Maurice et al., 1995):

$$\frac{I_{520}}{I_{550}} = A \exp\left(-\frac{\Delta E}{k_b T}\right) \tag{32.3}$$

where

ΔE is the energy separation between the two lines
T is the temperature
k_b is the Boltzmann constant
A is a parameter that depends on the host material

Therefore, the knowledge of the intensity ratio directly allows the determination of the absolute temperature of the particle and of its local environment. We have represented in Figure 32.18a the photoluminescence spectra of an amorphous fluoride glass particle as a function of temperature. The spectra, normalized by the value at 520 nm, clearly show that the 550 nm peak decreases more rapidly than the 520 nm one. The evolution of the fluorescence intensity ratio (FIR) I_{520}/I_{550} as a function of the inverse temperature is represented in Figure 32.18b. The logarithm of the experimental data is well fitted by a linear curve, in agreement with expression (32.3). An example of thermal characterization of a metallic stripe heated by an electrical current is described in the next section.

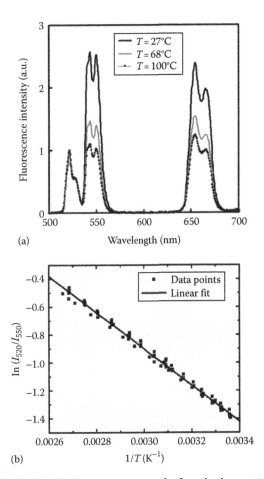

(a)

(b)

FIGURE 32.18 (a) Fluorescence spectra for fluoride glass particles as a function of temperature. The spectra are normalized by the value at 523 nm. (b) Evolution of the logarithm of the FIR as a function of the inverse temperature.

32.3.4.2 Example of Heating of Microelectronic Devices

To determine the absolute temperature of the particle, we have to measure the relative intensity of two emission lines. This can be done with the experimental configuration described in Figure 32.19. Since devices are often fabricated on an opaque substrate, the illumination of the particle has to be performed laterally. The incident laser beam is intensity-modulated at low frequency (around 620 Hz). The particle fluorescence is collected by a high numerical aperture objective, split into two directions, and sent to two photomultiplier tubes. Two bandpass filters, having a 10 nm bandwidth and centered on 520 and 550 nm, allow sending the two emission lines on the detectors. Finally, the fluorescence signals are detected by two lock-in amplifiers, synchronized to the modulation frequency on the incident beam.

The sample is powered by a direct (dc) electrical current and moved with a piezoelectric translator to form an image. Three images are constructed: the stripe topography and the fluorescence images of the particle in the two spectral regions at 520

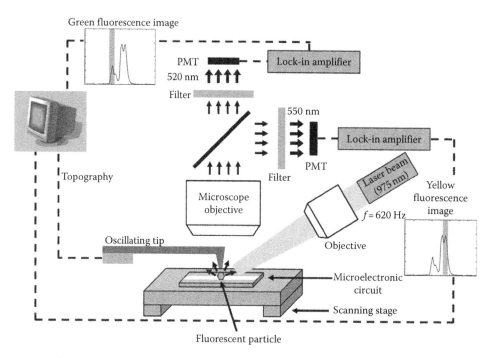

FIGURE 32.19 Experimental setup for thermal imaging in the direct current mode. (From Samson, B. et al., *J. Appl. Phys.*, 102, 024305, 2007. With permission.)

and 550 nm. The ratio of the two fluorescence images provides the particle temperature when it scans the sample surface.

We show in Figures 32.20 and 32.21 the topography and the evolution of the FIR of the particle when scanning a 9 μm long and 1.25 μm wide aluminum stripe grown on an oxidized silicon surface (Samson et al., 2007). In Figure 32.21, the current in the stripes varies from 0 to 9 mA. When no current is flowing through the stripe, the FIR remains constant on the whole surface. When the current increases, the FIR increases in the central part of the image, on the thin part of the device where

the current density is the highest. The FIR enhancement indicates that the intensity of the 550 nm line has diminished compared to the 520 nm one. This observation is a direct evidence of a temperature rise in the stripe. From these images, the lateral spatial resolution is estimated to be around 200 nm, in the range of the fluorescent particle size. For that special experiment, the temperature sensitivity was around 5°C. This sensitivity and the signal/noise ratio could be largely improved by increasing the time constant and reducing the scanning speed.

We have seen that a small fluorescent particle can be used as an efficient thermal sensor that can measure the temperature of an operating device, showing its eventual defects and hot spots. In the next section, we show that this characterization technique can also be adapted to the observation of heating under an alternating electrical current.

32.3.4.3 Heating under Alternating Excitation

The thermal characterization of microelectronic devices and circuits powered by an alternating current (ac) is of great interest. The use of a modulation technique strongly enhances the signal/noise ratio, and the study of the heat diffusion as a function of frequency may help to determine the thermal conductivity of materials (Lefèvre and Volz, 2005). A sketch of the experimental configuration is represented in Figure 32.22. An offseted square alternating current (for instance varying between 0 and 3 mA) is applied to the device. The resulting Joule effect modulates the stripes temperature that alternatively perturbs the fluorescence. Using a lock-in amplifier synchronized to the electrical current modulation, we can observe the heating of the device.

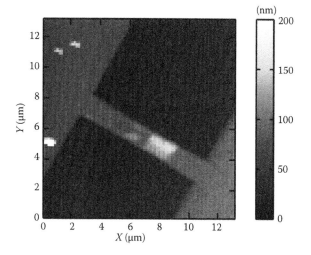

FIGURE 32.20 Topography of the aluminum stripe. The stripe is connected to two large pads for electrical connections. (From Samson, B. et al., *J. Appl. Phys.*, 102, 024305, 2007.)

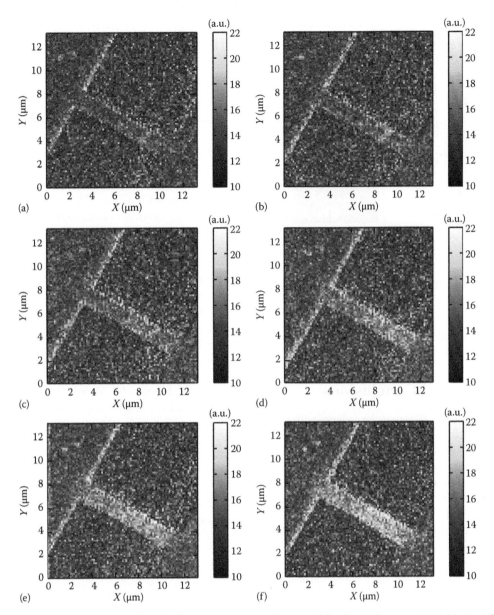

FIGURE 32.21 Evolution of the FIR measured on the aluminum stripe as a function of the electrical current: 0 mA (a), 3 mA (b), 5 mA (c), 7 mA (d), 8 mA (e), and 9 mA (f). The Joule heating is clearly appearing in the stripe with increasing currents. The maximum temperature measured is ~45°C in the middle of the stripe. (From Samson, B. et al., *J. Appl. Phys.*, 102, 024305, 2007. With permission.)

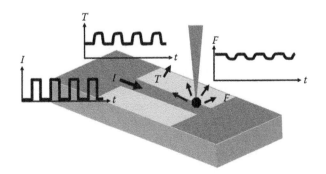

FIGURE 32.22 Schematic description of the thermal characterization of a device powered by an alternating current. The current modulates the temperature that periodically modifies the fluorescence.

In the case of an alternating current excitation, we only detect the most intense emission line and measure its variations as a function of the tip position on the surface. An example of measurement is illustrated in Figure 32.23.

A 200 nm-wide nickel stripe is heated by a 1.5 mA current modulated at 326 Hz. When the particle is on the stripe, the fluorescence is modulated and a contrast is visible. The lateral heat diffusion in the substrate and in the insulating SiO_2 layer is clearly apparent. However, although the images obtained in the ac-mode are well contrasted, they remain purely qualitative. In contrast to the images obtained in the dc-mode, the image represents the modulation of the fluorescence induced by the temperature, and not the temperature itself. This ac technique is therefore

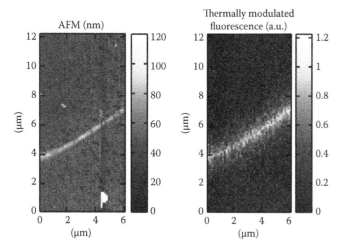

FIGURE 32.23 Topography (left) and thermally modulated fluorescence images (right) of a 200 nm wide nickel stripe powered by an ac electrical current.

rather intended to indicate where the temperature gradients are the most important in the devices (Samson et al., 2008).

32.4 Summary and Conclusions

Fluoride glasses or nanocrystalline particles codoped with erbium and ytterbium ions are inorganic compounds that exhibit intense room temperature fluorescence. These materials are robust, and can work in liquids and at high temperatures. We have shown that a single particle glued at the end of an atomic force microscope tip can be used to characterize several local properties of nanostructures and devices. The particle is a nanoscale optical sensor that can detect the electromagnetic fields localized near metallic nanoparticles or propagating at a metal/dielectric interface. The particle is also a nanoscale thermal sensor that can measure the temperature of a material or a device with a submicron lateral resolution. We have restrained ourselves to the optical and thermal characterization, but one can imagine using some particles sensitive to other external effects like electric fields by Stark effect or magnetic fields by Zeeman effect. The nanoscale characterization with fluorescent particles is an emerging domain with numerous exciting applications.

References

Aigouy, L., De Wilde, Y., and Mortier, M. 2003. Local optical imaging of nano-holes using a single fluorescent erbium-doped glass particle as a probe. *Applied Physics Letters* 83: 147–149.

Aigouy, L., De Wilde, Y., Mortier, M., Gierak, J., and Bourhis, E. 2004. Fabrication and characterization of fluorescent rare-earth-doped glass-particle-based tips for near-field optical imaging applications. *Applied Optics* 43: 3829–3837.

Aigouy, L., Mortier, M., Gierak, J. et al. 2005. Field distribution on metallic and dielectric nanoparticles observed with a fluorescent near-field optical probe. *Journal of Applied Physics* 97: 104322.

Aigouy, L., Lalanne, P., Hugonin, J. P., Julié, G., Mathet, V., and Mortier, M. 2007. Near-field analysis of surface waves launched at nano-slit apertures. *Physical Review Letters* 98: 153902.

Allison, S. W. and Gillies, G. T. 1997. Remote thermometry with thermographic phosphors: Instrumentation and applications. *Review of Scientific Instruments* 68: 2615–2650.

Auzel, F. 2004. Upconversion and anti-Stokes processes with f and d ions in solids. *Chemical Reviews* 104: 139–173.

Bensalah, A., Mortier, M., Patriarche, G., Gredin, P., and Vivien, D. 2006. Synthesis and optical characterizations of undoped and rare earth doped CaF_2 nano-particles. *Journal of Solid State Chemistry* 179: 2636–2644.

Berthou, H. and Jörgensen, C. K. 1990. Optical-fiber temperature sensor based on upconversion-excited fluorescence. *Optics Letters* 15: 1100–1102.

Cognet, L., Tsyboulski, D. A., Rocha, J.-D. R., Doyle, C. D., Tour, J. M., and Weisman, R. B. 2007. Stepwise quenching of exciton fluorescence in carbon nanotubes by single-molecule reactions. *Science* 316: 1465–1468.

Dantelle, G., Mortier, M., Vivien, D., and Patriarche, G. 2005. Nucleation efficiency of erbium and ytterbium fluorides in transparent oxyfluoride glass-ceramics. *Journal of Materials Research* 20(2): 472–481.

Dantelle, G., Mortier, M., Patriarche, G., and Vivien, D. 2006. Er^{3+}-doped PbF_2: Comparison between nanocrystals in glass-ceramics and bulk single-crystals. *Journal of Solid State Chemistry* 179: 1995–2003.

de Jonge, N. and van Druten, N. J. 2003. Field emission from individual multiwalled carbon nanotubes prepared in an electron microscope. *Ultramicroscopy* 95: 85–91.

Dieke, G. H. 1968. *Spectra and Energy Levels of Rare Earth Ions in Crystals*. New York: Interscience Publishers.

Ditlbacher, H., Krenn, J. R., Felidj, N. et al. 2002. Fluorescence imaging of surface plasmon fields. *Applied Physics Letters* 80: 404–406.

Ebbesen, T. W., Lezec, H. J., Ghaemi, H. F., Thio, T., and Wolff, P. A. 1998. Extraordinary optical transmission through subwavelength hole arrays. *Nature* 391: 667–669.

Fragola, A., Aigouy, L., De Wilde, Y., and Mortier, M. 2003. Upconversion fluorescence imaging of erbium-doped fluoride glass particles by apertureless SNOM. *Journal of Microscopy* 210: 198–202.

Gan, Y. 2007. A review of techniques for attaching micro- and nanoparticles to a probe's tip for surface force and near-field optical measurements. *Review of Scientific Instruments* 78: 081101.

Greffet, J.-J. and Carminati, R. 1997. Image formation in near-field optics. *Progress in Surface Science* 56: 133–237.

Hecht, B., Sick, B., Wild, U. P. et al. 2000. Scanning near-field optical microscopy with aperture probes: Fundamentals and applications. *Journal of Chemical Physics* 112: 7761–7774.

Holsa, J. and Porcher, P. 1981. Free ion and crystal field parameter for REOCl-Eu^{3+}. *Journal of Chemical Physics* 75: 2108–2117.

Hüfner, S. 1978. *Optical Spectra of Transparent Rare Earth Compounds*. New York: Academic Press.

Kneipp, K., Kneipp, H., Itzkan, I., Dasara, R. R., and Feld, M. S. 1999. Ultrasensitive chemical analysis by Raman spectroscopy. *Chemical Reviews* 99: 2957–2976.

Koenderink A. F., Hernandez J. V., Robicheaux F., Noordam, L. D., and Polman A. 2007. Programmable nanolithography with plasmon nanoparticle arrays. *Nano Letters* 7: 745–749.

Labéguerie, J., Gredin, P., Mortier, M., Patriarche, G., and De Kozak, A. 2006. Synthesis of fluoride nanoparticles in non-aqueous nanoreactors. Luminescence study of Eu^{3+}:CaF$_2$. *Zeitschrift fur Anorganische und Allgemeine Chemie* 632: 1538–1543.

Labéguerie, J., Dantelle, G., Gredin, P., and Mortier M. 2008. Luminescence properties of PbF$_2$:Yb-Er nanoparticles synthesized by two different original routes. *Journal of Alloys and Compounds* 451: 563–566.

Lefèvre, S. and Volz, S. 2005. 3ω-Scanning thermal microscope. *Review of Scientific Instruments* 76: 033701.

Li, S., Zhang, K., Yang, J.-M., Lin, L., and Yang, H. 2007. Single quantum dots as local temperature markers. *Nano Letters* 7: 3102–3105.

Liu, H. and Lalanne, P. 2008. Microscopic theory of the extraordinary optical transmission. *Nature* 452: 728–731.

Löw, P., Kim, B., Takama, N., and Bergaud, C. 2008. High-spatial-resolution surface-temperature mapping using fluorescent thermometry. *Small* 4: 908–914.

Malta, O. L. 1982. Theoretical crystal field parameters for the YOCl-Eu^{3+} system—A simple overlap model. *Chemical Physics Letters* 88: 353–356.

Maurice, E., Monnom, G., Dussardier, B., Saïssy, A., Ostrowsky, D. B., and Baxter, G. W. 1995. Erbium-doped silica fibers for intrinsic fiber-optic temperature sensors. *Applied Optics* 34: 8019–8025.

Mills, G., Zhou, H., Midha, A., Donaldson, L., and Weaver, J. M. R. 1998. Scanning thermal microscopy using batch-fabricated probes. *Applied Physics Letters* 72: 2900–2902.

Mortier, M. and Patriarche G. 2000. Structural characterisation of transparent oxyfluoride glass-ceramics. *Journal of Materials Science* 35: 4849–4856.

Mortier, M. and Patriarche, G. 2006. Oxide glass used as inorganic template for fluorescent fluoride nanoparticles synthesis. *Optical Materials* 28: 1401–1404.

Mortier, M., Huang, Y. D., and Auzel, F. 2000. Crystal field analysis of Er^{3+}-doped oxide glasses: Germanate, silicate and ZBLAN. *Journal of Alloys and Compounds* 300–301: 407–413.

Mortier, M., Monteville, A., and Patriarche, G. 2001. Devitrification of fluorozirconate glasses: From nucleation to spinodal decomposition. *Journal of Non-Crystalline Solids* 284: 85–90.

Mortier, M., Goldner, P., Féron, P., Stéphan, G. M., Xu, H., and Cai, Z. 2003. New fluoride glasses for laser applications. *Journal of Non-Crystalline Solids* 326–327: 505–509.

Parfenov, A., Gryczynski, I., Milicka, J., Geddes, C. D., and Lakowicz, J. R. 2003. Enhanced fluorescence from fluorophores on fractal silver surfaces. *Journal of Physical Chemistry B* 107: 8829–8833.

Peng, Y., Luxmoore, I., Forster, D., Cullis, A. G., and Inkson, B. J. 2008. Nanomanipulation and electrical behaviour of a single gold nanowire using in-situ SEM-FIB-nanomanipulators. *Journal of Physics: Conference Series* 126: 012031.

Pohl, D. W., Denk, W., and Lanz, M. 1984. Optical stethoscopy: Image recording with resolution λ/20. *Applied Physics Letters* 44: 651–653.

Pollock, H. M. and Hammiche, A. 2001. Micro-thermal analysis: Techniques and applications. *Journal of Physics D: Applied Physics* 34: 23–53.

Samson, B., Aigouy, L., Latempa, R. et al. 2007. Scanning thermal imaging of an electrically excited aluminum microstripe. *Journal of Applied Physics* 102: 024305.

Samson, B., Aigouy, L., Löw, P., Bergaud, C., Kim B., and Mortier, M. 2008. AC thermal imaging of nanoheaters using a scanning fluorescent probe. *Applied Physics Letters* 92: 023101.

Shi, L., Plyasunov, S., Bachtold, A., McEuen, P. L., and Majundar, A. 2000. Scanning thermal microscopy of carbon nanotubes using batch-fabricated probes. *Applied Physics Letters* 77: 4295–4297.

Tikhomirov, V. K., Mortier, M., Gredin, P., Patriarche, G., Görller-Walrand, C., and Moshchalkov, V. V. 2008. Preparation and up-conversion luminescence of 8 nm rare-earth doped fluoride nanoparticles. *Optics Express* 16: 14544–14549.

Tikhomirov, V. K., Chibotaru, L. F., Saurel, D., Gredin, P., Mortier, M., and Moshchalkov, V.V. 2009. Er^{3+}-doped nanoparticles for optical detection of magnetic field. *Nano Letters* DOI: 10.1021/nl803244v.

Valeur, B. 2002. *Molecular Fluorescence*. Weinheim, Germany: Wiley-VCH.

Van Keuren, E., Cheng, M., Albertini, O., Luo, C., Currie, J., and Paranjape, M. 2005. Temperature profiles of microheaters using fluorescence microthermal imaging. *Sensors and Materials* 17: 1–6.

Xia, Y. and Halas, N. J. 2005. Shape-controlled synthesis and surface plasmonic properties of metallic nanostructures. *MRS Bulletin* 30: 338–343.

Zayats, A. V., Smolyaninov, I. I., and Maradudin, A. A. 2005. Nano-optics of surface plasmon polaritons. *Physics Reports* 408: 131–314.

Zhang, Z., Grattan, K. T. V., and Palmer, A. W. 1993. Temperature dependence of fluorescence lifetimes in Cr^{3+}-doped insulating crystals. *Physical Review B* 48: 7772–7778.

Optochemical Nanosensors

Yong-Eun Koo Lee
University of Michigan

Raoul Kopelman
University of Michigan

33.1 Introduction

An optochemical sensor is a device that detects and/or quantifies the presence of specific target chemicals by optical methods. A good sensor should be sensitive to the target analyte but not to the other analytes, and it should not influence the quantity of the target analyte. The performance of a sensor is determined by various factors listed as in Table 33.1. Compared to other traditional sensors such as electrochemical sensors, optical sensors have significant advantages such as high sensitivity, wide dynamic range, and freedom from electromagnetic interference. Sensors made of optical fibers have been a typical form of optochemical sensors, with advantages such as multiplexing ability and applicability for remote sensing, which were not possible with traditional electrode sensors. These optical fiber sensors, however, are not suitable for intracellular applications because of their relatively large size, typically hundreds of micrometers in diameter, in comparison to cells that have a typical dimension of tens of micrometers. They have been miniaturized to the nanoscale, down to 20 nm (Cullum and Vo-Dinh 2000), by limiting the sensor area to the tip of a finely pulled fiber, especially for application to intracellular measurements in single mammalian cells (Barker and Kopelman 1998; Barker et al. 1998; Tan et al. 1992, 1999). Still, their use has been limited by their not insignificant physical invasiveness. While the fiber tip is very small compared to traditional sensors, the pulled fiber shaft is not small enough to prevent physical perturbation to samples of micrometer dimensions, such as live cells or their subcompartments. For example, the inserted volume of a fiber sensor with a 200 nm tip was found to be nearly 4% of the volume of a mouse oocyte cell,

which is larger than most mammalian cells (Buck et al. 2004). On the other hand, like larger chemical sensors, these sensors contain a chemically inert matrix (like silica or hydrogel) that minimizes the chemical perturbations of the sample.

Avoiding the above-described physical invasiveness, fluorescent molecular probes have so far played a major role for intracellular sensing and imaging (Haugland 2005). While minimizing physical perturbation, these probes are not chemically inert. Overall, these molecular probes have several drawbacks against a reliable intracellular measurement. For instance, the probe molecules have to be in a cell-permeable form that often requires the proper derivatization of the indicator molecules, which in itself might interfere with their function. The intracellular measurement is also often skewed by sequestration to specific organelles inside the cell, or by nonspecific binding to proteins and other cell components. The indicator dye is usually not "ratiometric" by itself, but just loading into the cell a separate reference dye is not a guarantor of ratiometric measurements, because of the aforementioned sequestration and nonspecific binding phenomena. Furthermore, the cytotoxicity or chemical perturbation effects of the available indicator dyes is often a problem, as the mere presence of these dye molecules may chemically interfere with the cell's processes. As noted above, chemical interference or ratiometric measurement has not been a problem for traditional fiber optical sensors. The impacts of these drawbacks of "naked" molecular probes are even more severe concerning their *in vivo* application (Koo et al. 2006). Only a small portion of the molecular probes would reach a specific location of interest within the body, which has also been an issue for drug delivery. The same goes for the case of crossing Biological

TABLE 33.1 Key Terms for Sensor Performance

Term	Definition
Sensitivity	Minimum analyte quantity needed for a detectable output change or output change per unit of analyte quantity.
Selectivity	Ability to determine a particular analyte in a complex mixture without interference from other components in the mixture.
Dynamic range	Working range of a sensor where the signal varies in a monotonic manner with the analyte quantity. Typically expressed as the maximum and minimum values of analyte quantity that can be measured or as the ratio between them.
Resolution	Smallest detectable change in analyte quantity.
Response time	Time required for a sensor's output to change from its previous state to a correct new state.
Reversibility	Ability to reproduce the same output when a sensor is reused to measure the same quantity of a particular analyte after being exposed to different quantities of the analyte.
Stability	Ability to produce stable outputs against external factors such as light, temperature, or time.
Photobleaching	Photochemical destruction of a fluorophore. It bleaches the color and the fluorescence, influencing the optical sensor response. High photobleaching means low photo-stability.

barriers, for example, the blood–brain barrier (BBB) as well as the issue of multiple drug resistance (MDR), caused by the ability of certain tumor cells to pump back out small drug, or probe, molecules. In view of the above, this exposition is focused on the new generation of nanosensors that is based on nanoparticles.

Recent advances in nanotechnology have made many types of nanoparticles available as platforms for constructing new types of bioanalytical nanosensors. Indeed, a nanoparticle has many advantages as a building block for bioanalytical and biomedical sensors as follows:

First, the nanoparticles are physically noninvasive. The nanoparticles are in the dimension range of 1–1000 nm, that is, from a few atoms to mitochondria size, thus resulting in minimal physical interference to cells. For example, a single spherical nanoparticle of 20–600 nm in diameter takes up only 1 ppm to 1 ppb of a mammalian cell volume (Buck et al. 2004). It should be noted that due to the small size of the nanoparticle, the nanoparticle sensors require much less sample volume than traditional sensors and usually also have very fast response times.

Second, most of the nanoparticles are inert and nontoxic, and therefore are chemically noninvasive. The nanoparticle matrix not only protects cellular contents from the incorporated sensing components but also protects the sensing components from membrane/organelle sequestration or protein binding (Cao et al. 2004; Park et al. 2003; Sumner et al. 2002).

Third, the nanoparticles have a high surface-to-volume ratio, resulting in high accessibility of analytes to the sensing components within the nanoparticle or on the surface.

Fourth, the nanoparticle can be engineered for high loading of single or multiple sensing components within the matrix or on its surface. With high amounts of dye per nanoparticle, high signal to background ratio and high sensitivity can be achieved. Moreover, signal amplification can be obtained for nanoparticles loaded with a high amount of indicator dyes, either within the nanoparticle matrix or on its surface, due to interaction of analytes with multiple sensing dyes that are located in close proximity to each other (Montalti et al. 2005). With loading of multiple components per nanoparticle, ratiometric measurements, multiplex sensing, or sophisticated synergistic designs can be made.

Fifth, the nanoparticle can be engineered to have biological molecules or hydrophilic coating on the surface for targeting, for crossing the BBB, or for prolonged plasma lifetime that enables *in vivo* sensing at specific cells or areas of interest. The targeting efficiency of nanoparticles with multiple surface-conjugated targeting moieties is reported to be higher than that of an individual free targeting moiety, indicating another advantage of nanoparticle sensors over molecular probes (Hong et al. 2007; Montet et al. 2006). With surface-conjugated synthetic peptides or a surface-coat layer of polysorbate, nanoparticles can be transported across the BBB (Costantino et al. 2005; Kreuter et al. 2003). With PEG coating on the surface, the nonspecific binding to biological molecules and the RES uptake of nanoparticles are reduced, resulting in a longer residing of the nanoparticles in the blood circulation and, therefore, enhanced delivery of the nanoparticles to the *in vivo* location of interest.

Sixth, the nanoparticles are especially advantageous for *in vivo* sensing in tumors. The nanoparticles can avoid the MDR effect, as they are not pumped back out of cancer cells. Furthermore, the nanoparticles are accumulated preferentially at tumor sites through an effect called enhanced permeability and retention (EPR) (Maeda 2001), which further enhances the targeting efficiency of nanoparticles conjugated with tumor-specific targeting moieties.

Seventh, some types of nanoparticles possess unique but controllable optical properties that are superior to those of molecular probes and can be utilized as sensing components. For example, semiconductor nanoparticles, commonly called quantum dots (QDs), have large fluorescence quantum yields, resistance to photobleaching, and good chemical stability. The optical properties of QDs are tunable by controlling the size, composition, and preparation procedures. Metallic nanoparticles (metal nanoparticle or metal nano-shell/film coated on nonmetallic nanoparticle) have localized surface plasmon resonances (LSPR) and induce surface-enhanced Raman scattering (SERS), which are free from photobleaching (Kneipp et al. 2002). The LSPR wavelength of the metallic nanoparticles can be tuned by changing the shape, size, and composition of the metal nanoparticle or metal shell thickness (Jensen et al. 2000; Link and El-Sayed 1999; Oldenburg et al. 1998).

The first nanoparticle sensor for intracellular measurements, called PEBBLE (photonic explorer for biomedical use with biologically localized embedding), was introduced over a decade ago (Clark et al. 1998; Sasaki et al. 1996). Since then, a variety of nanoparticle-based sensors has been developed for intracellular and *in vivo* measurements of important chemical analytes such as ions, small molecules, as well as tumor biomarker proteins. The nanoparticles have also been applied to develop sensors for laboratory assays for proteins, DNAs, and pathogenic species like bacteria and viruses.

Below, the basic structure of optical nanoparticle sensors is described, and then the designs and applications of nanoparticle-based optochemical sensors are presented.

33.2 Optical Nanoparticle Sensor Structure

The basic structure of a nanoparticle sensor is made of a nanoparticle platform loaded with sensing components. Loading of the sensing components into the nanoparticles is made during synthesis, or after the formation of the nanoparticles, by encapsulation, covalent linkage, bio-affinity interaction such as streptavidin—biotin, or physical adsorption through a charge—charge or hydrophobic interaction. The covalent linkage can be made by simple coupling reactions, between the nanoparticle, functionalized typically with amine-, carboxyl-, or thiol- groups, and the molecules containing similar functional groups.

33.2.1 Nanoparticle Platform

The nanoparticle platform can exhibit excellent chemical stability and biocompatibility. A wide variety of nanoparticle matrices have been used as sensor platforms, which includes polymer nanoparticles made of organic, inorganic, or organic–inorganic hybrid materials, polymer-capped liposomes or micelles, QDs, as well as metallic nanoparticles. The matrix properties of the nanoparticle platform, such as pore size, hydrophobicity, and charge play important roles in the loading of sensing components and accessibility of analytes, affecting the sensor properties (Koo et al. 2007). For example, polyacrylamide (PAA) nanoparticles have served as a good matrix for a variety of ion sensors (Clark et al. 1999a,b; Park et al. 2003; Sumner and Kopelman 2005; Sumner et al. 2002, 2005) due to their neutral and hydrophilic nature, while a hydrophobic nanoparticle matrix is preferred for oxygen sensors (Cao et al. 2004; Koo et al. 2004).

33.2.2 Sensing Component

The sensing components include an analyte "recognizer" that binds or interacts with the target analyte and a transducer that signals binding. The most common and simplest form of a sensing component is an analyte-sensitive molecular probe that serves as an analyte recognizer as well as an optical signal transducer. However, sensing components made of a distinct analyte recognizer and an optical transducer have been also utilized to construct sophisticated synergistic nanoparticle sensors (see Section 33.3). The optical signals that have been used for nanoparticle sensors include the fluorescence intensity, fluorescence anisotropy, and fluorescence lifetime, SERS, and absorption/scattering of LSPR.

33.2.2.1 Fluorescence

Fluorescence is the most popular optical modality. It is highly sensitive and allows simultaneous analysis of multiple fluorescent reporters. The intensity, anisotropy, or lifetime of fluorescence has been used as the signal for sensors. The intensity measurement of an indicator in the presence of an analyte is the most common sensing method, which requires a relatively simple instrumental set-up. However, the measurements can be affected quite a lot by excitation intensity, absolute concentration, and sources of optical loss. A ratiometric mode of operation is essential to get rid of such problems for intensity-based measurements. Fluorescence intensity measurements are often limited by photobleaching and interference from autofluorescence from cellular components. Fluorescence lifetime or anisotropy measurements are independent of probe concentration and less affected by photobleaching, as well as by scattering or decay characteristics of the background, but require more complex instrumentation than intensity measurements.

The fluorescence anisotropy, r, is the change in the polarization of the fluorescent emission of a fluorophore relative to the excitation polarization, which is given by

$$r = \frac{I_{VV} - G \cdot I_{VH}}{I_{VV} + 2 \cdot G \cdot I_{VH}}$$

where
  I_{VV} and I_{VH} are the parallel and perpendicular polarized fluorescence intensities measured with the vertically polarized excitation light
  I_{HV} and I_{HH} are the same fluorescence intensities measured with the excitation light horizontally polarized
  G, the "G-factor" is I_{HV}/I_{HH}, a measure of the system sensitivity to the emission polarization in the horizontal and vertical orientations

Fluorescence anisotropy depends on the intrinsic properties of a molecule as well as the environment in which it resides (Lippitsch 1993). Rotational diffusion of the fluorophore in the solvent, and changes to the lifetime of the excited state of the fluorophore are two of the major mechanisms to induce the changes in fluorescence anisotropy and have been utilized as the sensing mechanisms for two kinds of nanoparticle sensors based on fluorescence anisotropy measurements. The first kind is based on binding of an analyte to the fluorescent indicator. When the fluorescent molecule forms a complex with another substance, its rotational rate decreases and the fluorescence-anisotropy value increases. The degree of variation depends on the strength of the binding interaction and the size of the complex. The second kind sensor is based on dynamic fluorescence quenching, that is, reduction in lifetime, by analytes.

Fluorescence lifetime has been also utilized as a signal for sensors based on dynamic fluorescence quenching (Lackowicz 2006). There are two options for excitation and detection on fluorescence lifetimes: time-domain pulse methods and frequency domain or phase-resolved methods. In time-domain methods, employing pulsed excitation, lifetimes are measured directly from the fluorescence signal or by photon-counting detection. In the frequency-domain or phase-resolved method, sinusoidally modulated light is used as an excitation source, and the lifetimes are determined from the phase shift or modulation depth of the fluorescence emission signal.

33.2.2.2 SERS

SERS is a surface-sensitive technique that results in the enhancement of Raman scattering by Raman-active molecules adsorbed on SERS-active metal substrates. Increases in the intensity of Raman signal have been regularly observed on the order of 10^4–10^6, and can be as high as 10^8 and 10^{14} for some systems (Kneipp et al. 1999; Moskovits 1985). The examples of SERS-active metal substrates include nanometer-sized silver or gold structures.

33.2.2.3 LSPR

The free electrons of the metal behave like a gas of free charge carriers (plasma) and can be excited to sustain propagating plasma waves. Plasma waves are longitudinal electromagnetic charge density waves and their quanta are referred to as plasmons. The plasmon exists in two forms: bulk plasmons in the volume of a plasma and surface plasmons that are bound to the interface of plasma and a dielectric. When light interacts with metal nanostructure (nanoparticle or nano shell/film), much smaller than the incident wavelength, the excitation of surface plasmons, so called surface plasmon resonance (SPR), is localized on the metal nanostructure and denoted as LSPR. A metal nanostructure can be seen as a resonator for surface plasmons and, like any (moderately damped) resonator, if excited resonantly, there is strong enhancement of LSPR compared to the exciting electromagnetic field, by orders of magnitude (Hohenau et al. 2007). As a result of these LSPR modes, the nanoparticles absorb and scatter light so intensely that single nanoparticles are easily observed by eye using dark-field (optical scattering) microscopy (Anker et al. 2008; Willets and Van Duyne 2007).

33.3 Examples of Optochemical Nanosensor Designs

33.3.1 Sensors Using Nanoparticle as a Platform Only

Here, the nanoparticle serves only as a platform to be loaded with a variety of types of single or multiple sensing components, as shown in Figure 33.1A.

33.3.1.1 Nanoparticle Platform with Incorporated Molecular Probes

In this design, the analyte-sensitive molecular probes and/or reference dyes are loaded into the nanoparticle core or its

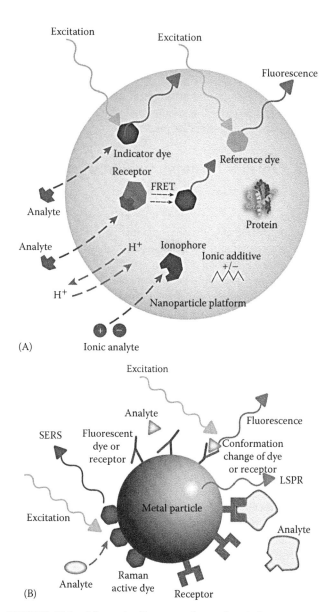

FIGURE 33.1 Schematic diagrams of optochemical nanosensors: (A) nanosensors using nanoparticle as a platform only; (B) nanosensors using metallic nanoparticle as a platform and a part of sensing components.

surface coated layer. The sensors use fluorescence as an optical signal by loading organic fluorescent dyes or fluorescent proteins specific to certain analytes such as DsRed (Sumner et al. 2005) or FLIPPi (Sun et al. 2008). The nanoparticle matrixes so far used include polymeric nanoparticles such as PAA (Clark et al. 1998, 1999a,b; King and Kopelman 2003; Park et al. 2003; Sumner and Kopelman 2005; Sumner et al. 2002, 2005; Xu et al. 2002), silica (Peng et al. 2007a; Xu et al. 2001), organically modified silica (ormosil) (Cao et al. 2005; Koo et al. 2004), polydecylmethacrylate (Cao et al. 2004) and polystyrene (Schmälzlin et al. 2005, 2006), fluorescent europium nanoparticles coated with polyelectrolyte (Brown and McShane 2005), and polymerized liposome (Cheng and

Aspinwall 2006). The nanoparticle sensors based on this design have been developed to sense ions (H^+, Ca^{2+}, Mg^{2+}, Zn^{2+}, Fe^{3+}, K^+, $Cu^{+/2+}$, phosphate, and H^+) (Almdal et al. 2006; Brown and McShane 2005; Clark et al. 1999a,b; Horvath et al. 2002; Park et al. 2003, Peng 2007a; Sumner and Kopelman 2005; Sumner et al. 2002, 2005; Sun et al. 2008), hydroxyl radical (King and Kopelman 2003), and small molecules (O_2, singlet oxygen, hydrogen peroxide) (Cao et al. 2004, 2005; Guice et al. 2005; Horvath et al. 2002; Kim 2010; Koo et al. 2004; Schmälzlin et al. 2005, 2006; Xu et al. 2001). Most of these sensors rely on fluorescence intensity measurements, but some use fluorescence anisotropy (Horvath et al. 2002) and lifetime measurements (Schmälzlin et al. 2005, 2006).

33.3.1.2 Nanoparticle Platform with Incorporated Enzymes and Molecular Probes

The design utilizes enzymatic reaction of analytes catalyzed by the encapsulated enzyme within nanoparticle core, causing fluorescence change either by fluorescence product or by coembedded fluorescent indicator dye for depleted reactant. Examples include a glucose sensor and two hydrogen peroxide sensors. The glucose sensor was developed by incorporating glucose oxidase, an oxygen sensitive ruthenium-based dye, and a reference dye within PAA nanoparticles (Xu et al. 2002). The enzymatic oxidation of glucose to gluconic acid results in the local depletion of oxygen that increases the fluorescence of the encapsulated oxygen sensitive dye. The hydrogen peroxide nanoparticle sensors were developed by encapsulating horseradish peroxidase within either PAA (Poulsen et al. 2007) or PEG (Kim et al. 2005) nanoparticles. The PAA-based sensor has coembedded fluorescein dye, the fluorescence intensity of which decreases when HRP catalyzes the oxidation of dye in the presence of ROS (Poulsen et al. 2007). The PEG-based sensor uses externally introduced Amplex Red (10-acetyl-3,7-dihydroxyphenoxazine) for H_2O_2 measurement (Kim et al. 2005). The encapsulated HRP catalyzes the reaction between Amplex Red and H_2O_2, forming red fluorescent product, resorufin, in proportion to the amount of H_2O_2.

33.3.1.3 Nanoparticle Platform with Ion-Correlation Sensing Components

In this design of ion sensing nanoparticle sensors, the hydrophobic polymeric nanoparticle is loaded with three lipophilic components: a highly selective but optically silent ionophore as an analyte recognizer, a pH sensitive fluorescent dye as an indirect optical signal transducer for analyte ions, and an ionic additive. When a selective ionophore binds the ions of interest, thermodynamic equilibrium-based ion exchange (for sensing cations) or ion co-extraction (for sensing anions) occurs simultaneously. This changes the local pH in proportion to the concentration of the ion of interest within the nanoparticle matrix, which is optically detected by the pH-sensitive dye. The ionic additive is used to maintain ionic strength. The ion-correlation nanoparticle sensors have been developed for K^+, Na^+, and

Cl^- ions (Brasuel et al. 2001, 2002, 2003; Dubach et al. 2007; Ruedas-Rama and Hall 2007).

33.3.1.4 Nanoparticle Platform with Optically Silent Analyte-Sensitive Ligands/Receptors and Analyte-Insensitive Fluorescent Dyes

The sensors were prepared by loading the nanoparticles with fluorescent dyes and analyte-specific nonfluorescent ligands/receptors. There are two different types of sensors of this group. In the first type, the fluorescent dyes and analyte-specific ligands/receptors are located close enough or coupled together to cause FRET (Fluorescence Resonance Energy Transfer). The sensitivity of the sensors was tunable by the ratio of ligand to dye (Arduini et al. 2005; Brasola et al. 2003; Rampazzo et al. 2005). The sensors have been developed for the detection of Cu^{2+} (Arduini et al. 2005; Brasola et al. 2003; Meallet-Renault et al. 2004, 2006; Montalti et al. 2005; Rampazzo et al. 2005), Co^{2+} (Montalti et al. 2005), Ni^{2+} (Montalti et al. 2005) ions utilizing silica (Arduini et al. 2005; Brasola et al. 2003; Montalti et al. 2005; Rampazzo et al. 2005), and latex nanoparticles (Meallet-Renault et al. 2004, 2006). In the second type, the nanoparticles loaded with fluorescent dyes were surface-modified with biomolecules such as antibodies, peptides, aptamers, or DNAs in order to achieve selective detection of analytes. There is no interaction or energy transfer between surface-attached molecules and fluorescent dyes. There is no change in intensity or fluorescence peak positions after the sensors bind with the analytes either. However, the fluorescence anisotropy of the nanoparticle sensors can be changed due to the complex formation between fluorescent nanosensors and the bioanalytes. The measurements have been performed based on intensity (Wang et al. 2005, 2007; Zhou and Zhou 2004; Zhao et al. 2004) as well as fluorescence anisotropy (Deng et al. 2006). It should be noted that a separation step is required for intensity-based measurements, which usually involves washing out unbound nanoparticle sensors for samples fixed on a support, or centrifugation, or magnetic separation for floating samples. The sensors have been developed for the detection of bacteria (Wang et al. 2005, 2007; Zhao et al. 2004), DNAs (Zhou and Zhou 2004), and proteins (Deng et al. 2006). Some of these nanoparticle sensors were applied for multiplex assay (Wang et al. 2005, 2007; Zhou and Zhou 2004). Silica nanoparticles (Deng et al. 2006; Wang et al. 2005, 2007; Zhao et al. 2004) and gold core-silica shell nanoparticles (Zhou and Zhou 2004) have been used.

33.3.2 Nanoparticle as a Platform as well as a Part of Sensing Components

In this design, the nanoparticle or nano shell/film serves not only as a platform but also as a part of sensing components. The nanoparticles that have been used for this group of sensors include QDs, commercial FRET nanoparticles, metallic nanoparticles or nanofilms, peroxalate ester nanoparticles, and titanium dioxide nanoparticles.

33.3.2.1 QD Sensors

QDs have strong and stable fluorescence but are not sensitive to specific analytes. QDs have been utilized to prepare the sensors just as silica-based nanosensors described in Section 33.3.1.4 above. In one case, the fluorescence of the QDs becomes sensitive (quenches or restores) to specific analytes when QDs are surface-conjugated with the analyte-selective molecules or receptors. Here, the QDs make a FRET pair with either surface-conjugated molecules or their complexes with analytes. The QD sensors have been used for the detection of Cu^{2+} (Chen and Rosenzweig 2002; Gattas-Asfura and Leblane 2003), Ag^+ (Gattas-Asfura and Leblane 2003), and maltose (Medintz et al. 2003; Sandros et al. 2005, 2006). In other cases, the QDs are conjugated with biomolecules specific to DNAs or proteins and there is no change in the fluorescence of QD by the binding event. The QD sensors have been developed for the detection of proteins (Cai et al. 2006; Ghazani et al. 2006; Gao et al. 2004; Goldman et al. 2004; Klostranec et al. 2007; Mulder et al. 2006; Wang et al. 2004) and viruses (Agrawal et al. 2005; Bentzen et al. 2005). We note that commercial FRET nanoparticles were also made into sensors for the detection of multiple viruses (Agrawal et al. 2005).

33.3.2.2 Metallic Nanosensors

The metallic nanoparticles have three unique properties: (1) fluorescence quenching, (2) SERS of surfaced-bound Raman-active molecules, and (3) LSPR. Nanosensors based on these properties have been developed by functionalizing the metallic surface with organic molecules and/or analyte-specific organic or biological molecules (see Figure 33.1B).

The fluorescence quenching property has been utilized for constructing FRET sensors, as in the case of QDs. The metallic nanoparticles are surface-conjugated with analyte-selective molecules or ligands that contain organic fluorophores or QDs. The distance between metallic nanoparticle surface and the fluorescent molecules changes upon binding with the specific analytes, leading to analyte sensitive quenching or restoring of the fluorescence. FRET-based metallic nanoparticle sensors have been developed to detect Hg^{2+} (Huang and Chang 2006), Cu^{2+} (He et al. 2005) ions, as well as DNAs (Dubertret et al. 2001; Maxwell et al. 2002; Wu et al. 2006) and single proteins (Huang et al. 2007; Peng et al. 2007b; You et al. 2007).

The extinction and scattering of LSPR show distinctive spectral shift, in response to small changes in the local refractive index. Most organic molecules have a higher refractive index than buffer solution; thus, when they bind to metallic nanoparticles or nanostructures, the local refractive index increases, causing the extinction and scattering spectrum to be red shifted (Anker et al. 2008). The LSPR sensors have been developed for detection of DNAs (Endo et al. 2005) and proteins (Endo et al. 2006; Haes et al. 2005) or multiple proteins in an array. These LSPR sensors have been so far mostly made of metal film over polymeric nanoparticles or metal nanoparticles on a chip or glass slide; that is, containing large ensembles of nanoparticle sensors, rather than a single nanoparticle sensor. The LSPR sensors made of a single nanoparticle have been applied for *in vitro* monitoring of actin rearrangement in live fibroblasts (Kumar et al. 2007) and *in vitro/in vivo* detection of receptor proteins on the surface of tumor cells for cancer diagnosis and/or photothermal therapy (Aaron et al. 2007; El-Sayed et al. 2005, 2006; Loo et al. 2005). It should be noted that the LSPR sensors aggregate on the cell surfaces due to closely located receptors and show the red shift in absorbance. This allows the use of the wavelength significantly longer than that of isolated LSPR sensors, resulting in more selective and efficient imaging and therapy (Larson et al. 2007).

SERS has also been utilized for designing nanoparticle sensors by labeling the metallic nanoparticles with Raman active dyes and/or analyte-specific ligands (Bishnoi et al. 2006; Cao et al. 2002; Talley et al. 2004) or with self-assembled monolayer that could create a pocket for specific molecules (Lyandres et al. 2005; Shah et al. 2007; Stuart et al. 2006). The metallic SERS sensors have been developed for the detection of hydrogen ion (Bishnoi et al. 2006; Talley et al. 2004), DNAs and RNAs (Cao et al. 2002), glucose (Lyandres et al. 2005; Stuart et al. 2006), and lactate (Shah et al. 2007) *in vitro* and *in vivo*.

LSPR and SERS are complementary sensing modalities available for the same metallic nanoplatforms. The SERS enhancement results from Raman excitation and emission coupled with the nanoparticle LSPR modes, and it is greatest when the LSPR λ_{max} falls between the excitation wavelength and the wavelength of the scattered photon (McFarland et al. 2005). LSPR is very sensitive to changes in the characteristics of nanoparticles and therefore can be tuned by changing the size and shape of the nanoparticles. The SERS glucose sensor is an example that utilizes the complementary nature of LSPR and SERS to improve the signals (Lyandres et al. 2005; Stuart et al. 2006).

33.3.2.3 Peroxalate Nanoparticle Sensor

A nanoparticle sensor for hydrogen peroxide was developed utilizing the reaction between peroxalate ester and hydrogen peroxide. The sensor was made of polymeric nanoparticles containing peroxalate ester in the backbone and encapsulated fluorescent dyes (Lee et al. 2007). The detection of the hydrogen peroxide was made from the chemiluminescence produced by the following mechanism: the peroxalate ester groups react with hydrogen peroxide, generating a high-energy dioxetanedione that then chemically excites the encapsulated fluorescent dyes, leading to chemiluminescence.

33.3.2.4 Titania Nanoparticle Sensor

Phosphate-modified titanium dioxide ($P-TiO_2$) nanoparticles were utilized for a laboratory assay for the selective detection of catechol and its derivatives with nM detection limits (Wu et al. 2007). The assay uses a mixture of $P-TiO_2$ nanoparticles and fluorescein dyes. The sensing is based on fluorescence quenching of fluorescein due to the strong absorbance band of the complex between catechol derivatives and $P-TiO_2$.

33.4 Examples of Applications

33.4.1 Sensing/Imaging in Live Cells

The sensors for ions and small molecules have been developed mostly for intracellular measurements. Successful intracellular measurements have been demonstrated by the sensors for pH (Clark et al. 1999a, 1998; Peng et al. 2007a; Talley et al. 2004), Mg^{2+} (Martin-Orozco et al. 2006; Park et al. 2003), Ca^{2+} (Clark et al. 1999a,b), oxygen (Guice et al. 2005; Koo et al. 2004; Schmälzlin et al. 2005, 2006; Xu et al. 2001), singlet oxygen (Kim 2010), K^+ (Brasuel et al. 2001), Na^+ (Brasuel et al. 2002), and Cl^- (Brasuel et al. 2003) ions. We note that some of the sensors have been developed for intracellular measurements but have not been applied as the sensitivity is not high enough to detect the analytes inside cells. For example, there have been many nanoparticle sensors for Cu^{2+} (Arduini et al. 2005; Brasola et al. 2003; Chen and Rosenzweig 2002; Gattas-Asfura and Leblane 2003; Meallet-Renault et al. 2004, 2006; Montalti et al. 2005; Rampazzo et al. 2005; Sumner et al. 2005), but none of them have been applied for intracellular studies. The detectable range of Cu^{2+} by these sensors was in the range of 1 nM to 1000 μM while the normal unbound copper ion level inside cells is only femtomolar.

The delivery of nanoparticle sensors into cells has been done by standard delivery methods. The intracellular delivery methods of nanoparticles, through the plasma membrane barrier, include microinjection (Schmälzlin et al. 2005, 2006), gene gun delivery (Brasuel et al. 2001, 2002, 2003; Clark et al. 1999a; Koo et al. 2004; Park et al. 2003; Xu et al. 2001), liposome incorporation (Clark et al. 1999a), nonspecific or receptor-mediated endocytosis with surface conjugated translocating proteins/peptides (Akerman et al. 2002; Chan and Nie 1998; Guice et al. 2005, Reddy et al. 2006; Peng et al. 2007a; Talley et al. 2004), and membrane penetrating TAT peptides (Kim 2010; Kumar et al. 2007). We note that there is negligible physical and chemical perturbation to the cell by these delivery techniques. For example, cell viability after gene gun delivery was found to be about 99% compared to the control cells (Clark et al. 1999a). Once delivered into cells, these sensors can be used with conventional microscopy techniques, enabling high spatial and temporal resolution. Intravenous injection is a typical method of *in vivo* delivery of the nanosensors.

The ratiometric Mg^{2+} ion PEBBLE sensor made of PAA nanoparticle with encapsulated coumarine 343 and Texas Red dyes (Martin-Orozco et al. 2006; Park et al. 2003) is a representative example of the nanoparticle sensors applied for intracellular measurements. Coumarine 343 is a very sensitive Mg^{2+} ion probe with a very high selectivity for magnesium ions over calcium ions that severely interfere with the response of commercially available probes. It is a small hydrophilic dye that does not penetrate the cell membrane by itself but can be delivered into cells after being encapsulated within the nanoparticles. The average hydrodynamic diameter of the nanoparticle sensor was 40 nm and the sensor showed very little leaching of both sensing and reference dyes. The nanoparticle sensor response was reversible and barely affected by nonspecific binding of proteins while that of the free

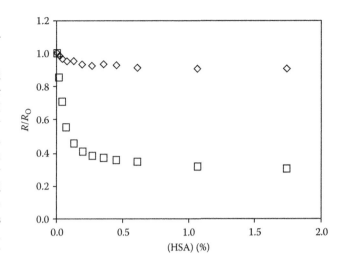

FIGURE 33.2 Response of Mg^{2+} PEBBLE sensor and free molecular probes in the presence of human serum albumin (HSA). Peak ratio of C343/TRD PEBBLE sensor (diamonds) is compared to the peak ratio of a free C343 and TRD dye solution (squares). (Adapted from Park, E.J. et al., *Anal. Chem.*, 75(15), 3784, 2003. With permission.)

dyes was affected a lot with ~70% reduction in fluorescence intensity, as shown in Figure 33.2. This demonstrates that the nanoparticle matrix efficiently blocks the proteins or large biological molecules from interfering in the incorporated molecular probes. The dynamic range of these sensors was 1–30 mM, with a linear range from 1 to 10 mM, with a response time of <4 s.

The sensors were applied for two different studies. In one case (Park et al. 2003), the sensors were delivered to C6 glioma cells via gene gun bombardment. The spectra of the sensors were monitored before and after the addition of an aliquot of KCl to the cells (Figure 33.3). The addition of KCl causes ion channels to open, thus allowing magnesium to enter the intracellular space, resulting in

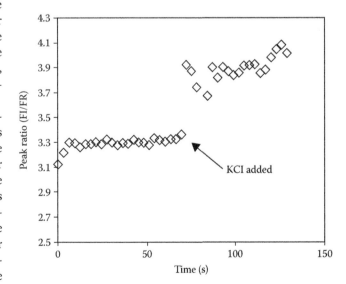

FIGURE 33.3 Response of Mg^{2+} PEBBLE sensors injected into C6 Glioma cells. An aliquot of KCl was added at $t = 69$ s. (Adapted from Park, E.J. et al., *Anal. Chem.*, 75(15), 3784, 2003. With permission.)

an increase of free magnesium. The peak ratio of the PEBBLE sensors increases as a result of the increase in free magnesium. A control experiment was conducted using PEBBLEs in solution in order to verify that the increase in peak ratio was not actually a response to the increase in potassium. In the other case (Martin-Orozco et al. 2006), the sensors were applied to find out the role of the Mg^{2+} for intracellular survival of *Salmonella* species inside the phagosomal vacuoles, more specifically, to find out if the Mg^{2+} ion is one of the potential regulation factors of PhoPQ protein that prevents fusion of the *Salmonella*-containing vacuoles (SCV) with host cell lysosomes in the first hours of invasion (Garvis et al. 2001; Vescovi et al. 1996). The Mg sensors were phagocytosed into RAW264.7 macrophages, together with a *Salmonella* strain that expresses GFP constitutively in order to monitor the induction of PhoPQ. The concentration of Mg^{2+} within the SCV was monitored by the Mg PEBBLEs within live macrophages while the concentration of Mg^{2+} in the medium was varied from 0.5 to 10 mM. As shown in Figure 33.4, the concentration of Mg^{2+} within the SCV

is rapidly regulated and stabilizes around 1 mM, demonstrating that the concentration of Mg^{2+} within the SCV has no effect on the early induction of PhoPQ.

33.4.2 *In Vivo* Sensing/Imaging/Monitoring

One of the important issues for *in vivo* sensing by optical measurements is limited tissue penetration-depth by photons. To avoid such problem, it is necessary to use the near-infrared (NIR) light in the spectral range of 650–900 nm, which is separated from the major absorption peaks of blood and water (Cheong et al. 1990; Delpy and Cope 1997; Ntziachristos et al. 2003). The use of NIR excitation also helps reduce the cellular autofluorescence. The nanoparticle sensors with NIR absorption, scattering, or emission have been developed using metallic nanoparticles, QDs, and polymeric nanoparticles loaded with NIR dyes. It should be noted that the LSPR of metallic nanoparticles or fluorescence of QDs can be tailored for the NIR spectral range by controlling the size or shape.

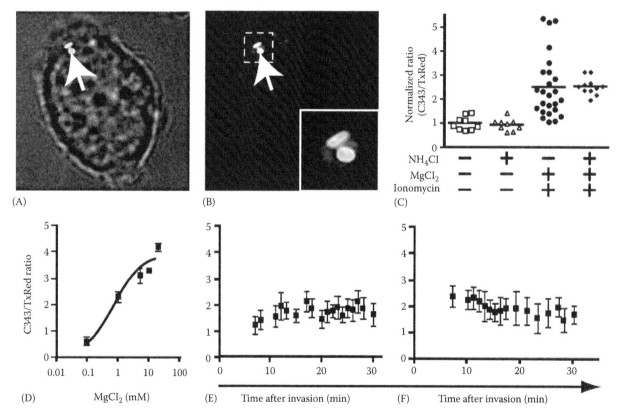

FIGURE 33.4 Measurements of $[Mg^{2+}]$ in SCV using Mg^{2+} PEBBLE sensor. (A) An overlay of the DIC and fluorescence images of RAW264.7 macrophages loaded with PEBBLEs and a *Salmonella* strain, illustrating the location of bacteria (arrow) within an infected cell. (B) An overlay image of the bacterial fluorescence (from GFP, bright white color in the image) and PEBBLE's fluorescence (from Texas Red (TxRed), gray color in the image). The area denoted by the dotted line is enlarged in the inset. (C) The response of the PEBBLE sensor inside RAW264.7 macrophages under various conditions. Cells were initially bathed in a NaCl buffer deprived of Mg^{2+} (leftmost column). The cells were then treated with either 10 mM NH_4Cl, a combination of 10 mM $MgCl_2$ and 1 μM ionomycin, or all three agents, as indicated. Data from three experiments are illustrated. The fluorescence ratio of C343 to TxRed was normalized to facilitate comparison between experiments. (D) *In vitro* calibration of PEBBLEs. The fluorescence ratio of the C343 and TxRed signals (ordinate) is presented as a function of $[Mg^{2+}]$ (abscissa). (E and F) Determination of free $[Mg^{2+}]$ in the SCV. SCV were loaded with PEBBLEs during infection and their fluorescence ratio was monitored over time by digital imaging, while identifying the location of the bacteria by their DAPI fluorescence. The $[Mg^{2+}]$ of the medium in E and F was 0.5 and 2 mM, respectively. Data in E and F are means ± SE of 10 determinations for each time point, obtained from four independent experiments. (Adapted from Martin-Orozco, N. et al., *Mol. Biol. Cell*, 17(1), 498, 2006. With permission.)

Most of the *in vivo* applications by nanoparticle sensors have been focused on the cancer imaging where nanoparticle sensors behave as a targeted contrast enhancement agent. Current imaging techniques are not sensitive enough to determine the malignancy of a tumor whereas biopsies, the current standard method for cancer diagnosis, are highly invasive. The cancer imaging using nanoparticle sensors, however, may provide enough sensitivity to become a diagnostic tool due to their preferential accumulation in tumor due to the EPR effect as well as surface-attached receptor molecules specific to overly expressed biomarker proteins found in malignant tumor cells or vasculature. It should be pointed out that the optical nanoparticles without any surface conjugated with bioreceptors specific to cancer biomarker proteins have been also applied with success for *in vivo*

cancer imaging due to the EPR effect. However, these nanoparticles are not considered as sensors as they do not have any specificity, and therefore are not covered here. Fluorescence and LSPR have been used as optical modalities for nanoparticle-based cancer imaging *in vivo*. For examples, real-time *in vivo* fluorescent imaging of human prostate cancer growth (Gao et al. 2004), tumor vasculatures of subcutaneous implanted U87MG human glioblastoma tumors (Cai et al. 2006), and KPL-4 human breast tumors (Tada et al. 2007), in nude mice, have been obtained with QDs conjugated with antibodies or RGD peptides. LSPR reflectance imaging by topically delivered bioconjugated gold nanoparticles showed a more than tenfold stronger contrast on precancerous epithelium than normal one in an *in vivo* hamster model as in Figure 33.5 (Aaron et al. 2007).

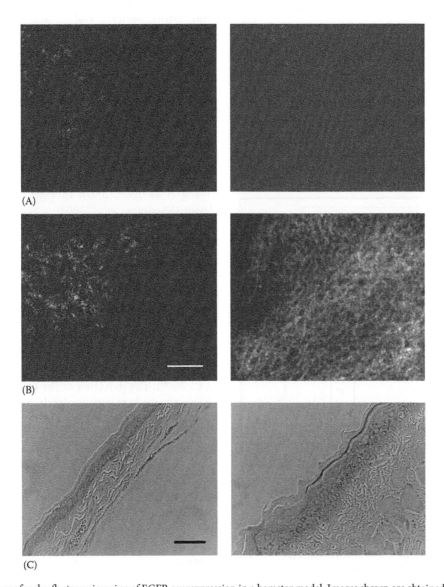

(A)

(B)

(C)

FIGURE 33.5 *In vivo* confocal reflectance imaging of EGFR overexpression in a hamster model. Images shown are obtained from the same hamster before the beginning of the treatment with carcinogen DMBA (A) and after 3 weeks of DMBA treatment (B). The images were taken immediately before [(A), left, and (B), left] and after [(A), right, and (B), right] topical application of 25 nm gold nanoparticles conjugated with anti-EGFR monoclonal antibodies. EGFR immuno-histochemical staining (C) reveals elevated EGFR levels in the DMBA-treated animal [(C), right] and very low levels of EGFR expression before DMBA treatment [(C), left]. The scale bars are ~50 μm. (From Aaron, J. et al., *J. Biomed. Opt.*, 12, 034007, 2007. With permission.)

Multifunctional nanoparticle sensors have been also developed for targeted bimodal imaging such as fluorescence imaging and MRI (Mulder et al. 2006; Veiseh et al. 2005), fluorescence/LSPR scattering imaging, (Aaron et al. 2007), as well as integrated targeted optical imaging and therapy (Loo et al. 2005), using iron oxide nanoparticles, QDs, gold nanoparticles, and gold nanoshell over silica nanoparticles, respectively.

The nanoparticle sensors have been also applied for *in vivo* sensing of metabolites such as hydrogen peroxide and glucose. The peroxalate nanoparticle sensors for hydrogen peroxide were used for the *in vivo* detection of hydrogen peroxide, externally injected as well as endogenously produced in the peritoneal cavity of mice during lipopolysaccharide-induced inflammatory response (Loo et al. 2005). The *in vivo* glucose sensing was performed using SERS implantable glucose sensor (Stuart et al. 2006). The sensor was made of silver film over latex polystyrene nanospheres whose surface is functionalized with a mixed SAM consisting of decanethiol (DT) and mercaptohexanol (MH). DT/MH has dual hydrophobic and hydrophilic properties; it selectively partitions glucose from interfering analytes and brings glucose closer to the nanostructured surface. The sensors show good reversibility and fast response time of less than 30 s in bovine plasma (Lyandres et al. 2005) as well as *in vivo* (Stuart et al. 2006). The sensor was subcutaneously implanted in a Sprague–Dawley rat and was able to monitor *in vivo* glucose concentration, as shown in Figure 33.6.

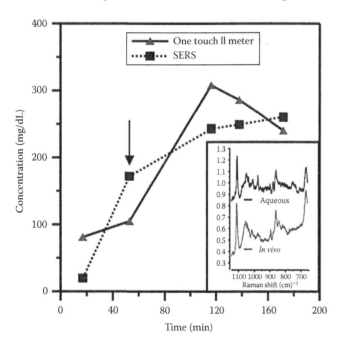

FIGURE 33.6 Time course of the *in vivo* glucose measurement. Glucose infusion was started at $t = 60$ min. Triangles (▲) are electrochemical measurements made using One Touch II blood glucose meter, and squares (■) are measurements made using the SERS sensor. A sharp rise in glucose concentration is detected by both SERS and electrochemical techniques after the start of the glucose infusion ($t = 60$ min). The inset shows a typical *in vivo* spectrum compared to a typical *ex vivo* spectrum of the same surface ($\lambda_{ex} = 785$ nm, $P = 50$ mW, $t = 2$ min). (Adapted from Stuart, D.A. et al., *Anal. Chem.*, 78, 7211, 2006. With permission.)

33.4.3 Laboratory Assay for Disease Biomarkers or Pathogens

The nanoparticle sensors have been applied for laboratory assay of body fluids or food/water samples to diagnose and predict disease or to test the food/water safety. Rapid detection of analytes at ultra low concentrations, without time-consuming sample preparation procedures, as well as the simultaneous multiplex detection, of single or multiple analytes, for high throughput assays, are major goals of laboratory assaying. With the advantages such as high signal per nanoparticle, high photostability, and low sample volume requirement, the nanoparticle sensor achieved such goals.

The nanoparticle sensor based assays have been so far successful in detecting single or multiple disease biomarker proteins, DNAs, bacteria, and viruses, as summarized in Table 33.2. Simultaneous detection of multiple analytes were possible by using the relevant oligonucleotides or antibodies conjugated to nanoparticles of distinct fluorescence wavelengths (Ghazani et al. 2006), Raman fingerprints (Cao et al. 2002), fluorescence response pattern (You et al. 2007), or optically bar-coded nano- or microsphere with unique fluorescence signatures (Klostranec et al. 2007, Wang et al. 2005, 2007). Optical barcodes have been prepared from a combination of QDs (Klostranec et al. 2007) or that of multiple fluorophores that can be excited at a single wavelength but produce fluorescence at different wavelengths (Wang et al. 2005, 2007). In some application, microfluidic device (Klostranec et al. 2007) or microcapillary flow system (Agrawal et al. 2005) was utilized for sequential, high-throughput readout of single barcodes.

These nanoparticle-based assays showed high sensitivity, without sample enrichment and amplification steps, thus presenting great promise for the diagnosis of disease at the earliest stage of its development or fast detection of food/water safety. For example, simultaneous detection of three pathogenic bacterial species was achieved within 30 min, with high sensitivity and specificity, using silica NPs with three different color-codes, conjugated with three specific antibodies (Wang et al. 2007). The detection of the respiratory syncytial virus (RSV) was made at very low levels, and possibly down to single virus particles, in solution, by monitoring co-localization of two nanoparticles with distinctive emission wavelengths that bind to the target virus at the same time, using a confocal microscope with a microcapillary flow system (Agrawal et al. 2005) or by imaging on a confocal microscope (Bentzen et al. 2005). The detection limit of proteins is typically in a few ng/mL (Deng et al. 2006; Huang et al. 2007; You et al. 2007). An array of sensors to detect seven different proteins (BSA, β-galactosidase, acid phosphatase and alkaline phosphatase, cytochrome c, Lipase, and subtilisin A) is an interesting example (You et al. 2007). The sensors in this array were made of six noncovalent complexes between a highly fluorescent anionic polymer and six different surface-modified gold nanoparticles. The gold nanoparticles of this array are surface-modified with different organic moiety instead

TABLE 33.2 Examples of Optochemical Nanosensors Applied for Diagnostic Assay

NP Type	Analytes	Optical Signal	Analytes: Single/Multiple	References
Silica	Bacteria	Fluorescence intensity	Single	Zhao et al. (2004)
	Bacteria	Fluorescence intensity	Multiple	Wang et al. (2005, 2007)
	Proteins	Fluorescence anisotropy	Single	Deng et al. (2006)
	DNAs	Fluorescence intensity	Multiple	Zhou and Zhou (2004)[a]
Commercial FRET NP	Viruses	Fluorescence intensity	Multiple	Agrawal et al. (2005)
QD	Viruses	Fluorescence intensity	Single	Agrawal et al. (2005) and Bentzen et al. (2005)
	Proteins	Fluorescence intensity	Multiple	Klostranec et al. (2007), Goldman et al. (2004), and Ghazani et al. (2006)[b]
	Proteins	Fluorescence intensity	Single	Wang et al. (2004)[b]
Titanium dioxide NP	Proteins	Fluorescence intensity	Single	Wu et al. (2007)
Gold nanoparticles	Proteins	Fluorescence intensity	Single	Peng et al. (2007b) and Huang et al. (2007)
	Proteins	Fluorescence intensity	Multiple	You et al. (2007)
	DNA	Fluorescence intensity	Single	Dubertret et al. (2001), Maxwell et al. (2002), and Wu et al. (2006)
	DNA, RNA	SERS	Multiple	Cao et al. (2002)
Ensemble of gold nanoparticles on mica substrate	Proteins	LSPR	Single	Haes et al. (2005)
Gold nanolayer over silica nanoparticles	DNA	LSPR	Single	Endo et al. (2005)
	Proteins	LSPR	Multiple	Endo et al. (2006)

[a] Au core-silica shell particles.

[b] This assay was applied for tissue samples. All the others are for liquid samples.

of biological receptors unlike most of the protein sensors. The presence of proteins disrupts the nanoparticle–polymer interaction, producing highly repeatable and characteristic fluorescence response patterns for individual proteins at nanomolar concentrations.

33.5 Critical Issues for Optochemical Nanosensors

The applications of nanoparticle-based optochemical sensing to *in vitro/in vivo* sensing and laboratory assays have been successfully demonstrated as above. However, there are several things to be considered or improved.

First, the intracellular measurements of small molecules or ions have been limited to a few analytes while nanoparticle sensors designed for analytes that are more diverse are available. In order to achieve a wider application of nanoparticle sensors, the performance of many nanoparticle sensors with respect to signal intensity, sensitivity, and selectivity should be improved. The development of more sensitive NIR sensing dyes to eliminate auto-fluorescence from cellular components or receptors that are more selective toward analytes/targets would be one way to improve the sensitivity and selectivity of the sensors. The adoption of a MOON (MOdulated Optical Nanoprobe)-type sensor design for S/N enhancement would be another way to improve the sensitivity. MOONs are metallically half-capped fluorescent nanoparticles, which can be either magnetically rotated (Anker and Kopelman 2003; Anker et al. 2003), that is,

MagMOONs, or using Brownian rotation (Behrend et al. 2004). The magnetically induced periodic motion of the MagMOON (or random thermal motion in the case of Brownian MOONs) modulates the fluorescence signal, enabling the separation of the signal from background, leading to the enhancement of the signal-to-noise (S/N) or, strictly, signal-to-background ratios by several orders of magnitude, up to 4000 times (Anker and Kopelman 2003). The MOON design can be applied for any fluorescent nanoparticle sensor by adding a metal coating on one hemisphere of the nanoparticle.

Second, for *in vivo* applications, there is a safety issue with respect to toxicity and/or bioelimination of the nanoparticle sensors. This is especially important for nanoparticle sensors used as contrast enhancement agents in clinical applications. Toxicity is not a problem for most of the nanoparticle matrices. However, it is a challenging issue for QDs (Azzazy et al. 2007). The bioelimination profile of nanoparticle sensors will have a different pattern from that of small molecules, which constitute the typical form of currently used drugs or imaging agents. Such profiles should be studied well for long-term health effects.

33.6 Summary and Future Perspective

The nanoparticle has many advantages as a building block for bioanalytical and biomedical sensors, due to its small size, nontoxicity, excellent engineerable properties, and, for a certain group of nanoparticles, like QDs or metallic nanoparticles, their distinct optical properties as well. Optochemical nanosensors of a variety of designs have been developed using various types of nanoparticles, for detecting ions (H^+, Ca^{2+}, Mg^{2+}, K^+, Na^+, Cl^-, phosphate ion, Fe^{3+}, Zn^{2+}, Co^{2+}, Ni^{2+}, Cu^+/Cu^{2+}, Ag^+, and Hg^{2+}), radicals ($OH^\bullet$), small molecules (oxygen, singlet oxygen, glucose, hydrogen peroxide, maltose, and lactate), DNAs, proteins, as well as pathogens like viruses and bacteria. The nanoparticle sensors have been applied, in a single untethered form or as a large ensemble on a rigid substrate, in three important bioanalytical and biomedical areas: intracellular sensing/imaging, *in vivo* sensing/imaging/monitoring, and as laboratory assays for disease diagnosis or food safety. The intracellular sensing, using free nanosensors, has been so far successful in noninvasive measurements of ions (H^+, Ca^{2+}, Mg^{2+}, K^+, Na^+, and Cl^-) and small molecules (O_2, singlet oxygen, and H_2O_2). *In vivo* measurements of small metabolite molecules (glucose and H_2O_2) were also achieved either using free nanosensors or an implanted device containing a large number of fixed nanoparticle sensors. The *in vivo* imaging of tumor biomarker proteins, using free nanosensor-based targeted contrast agents, shows great promise as a tumor diagnostic tool. Nanosensor-based assays of body fluids or food/water samples utilized both free nanosensors and nanosensor-ensembles that were fixed on a substrate. The detection of single or multiple disease biomarker proteins, DNAs, and pathogens (bacteria and viruses) was achieved, working rapidly and with high sensitivity, but without the time-consuming sample preparation and signal amplification steps.

With all the above-described successful applications of the nanosensors so far developed, continued research efforts on more sensitive and selective optochemical nanosensors are expected. Such intracellular measurements will be pursued for single species detection of more diverse analytes, as well as for simultaneous sensing of multiple analytes that are related to each other in an important metabolic process. Single cell chemical analysis, that is, measurements of chemical analytes inside a single cell or a specific location within a single cell, would be one of the important future applications of optochemical nanosensors. It is expected that the single cell analysis would differentiate between individual cells, leading to diagnosing disease at an early stage, when changes on a tissue level are not yet evident but chemical changes within cells are observable. The ultimate goal for nanosensors, used as targeted contrast agents and diagnostic laboratory assays, is their application in a clinical setting. The diagnostic accuracy of such applications would not only be enhanced by improving the selectivity and sensitivity of the nanosensors, but also by finding better disease biomarkers. It is expected that some of the nanosensor-based laboratory assays could enter the mainstream of clinical diagnostic laboratories in the near future. The potential for near-term clinical applications of *in vivo* cancer diagnostic imaging with nanoparticle-based targeted contrast agents is also high, provided that the safety issue will be satisfactorily cleared.

Acknowledgments

This work was supported by NIH Grants 1R01-EB-007977-01, R21/R33 CA125297 and 1R41 CA 130518-01A1 as well as NSF Grant DMR 0455330.

References

Aaron, J., Nitin, N., Travis, K. et al. 2007. Plasmon resonance coupling of metal nanoparticles for molecular imaging of carcinogenesis in vivo. *J. Biomed. Opt.* 12: 034007-1–034007-11.

Agrawal, A., Tripp, R. A., Anderson, L. J., Nie, S. 2005. Real-time detection of virus particles and viral protein expression with two-color nanoparticle probes. *J. Virol.* 79: 8625–8628.

Akerman, M. E., Chan W. C. W., Laakkonen, P., Bhatia, S. N., Ruoslahti, R. 2002. Nanocrystal targeting in vivo. *Proc. Natl. Acad. Sci. U. S. A.* 99(20): 12617–12621.

Almdal, K., Sun, H. H., Poulsen A. K. et al. 2006. Fluorescent gel particles in the nanometer range for detection of metabolites in living cells. *Polym. Adv. Technol.* 17: 790–793.

Anker, J. N., Kopelman, R. 2003. Magnetically modulated optical nanoprobes. *Appl. Phys. Lett.* 82(7): 1102–1104.

Anker, J. N., Behrend, C., Kopelman, R. 2003. Aspherical magnetically modulated optical nanoprobes (MagMOONs). *J. Appl. Phys.* 93(10): 6698–6700.

Anker, J. N., Hall, W. P., Lyandres, O., Shah, N. C., Zhao J, Van Duyne, R. P. 2008. Biosensing with plasmonic nanosensors. *Nat. Mater.* 7(6): 442–453.

Arduini, M., Marcuz, S., Montolli, M. et al. 2005. Turning fluorescent dyes into Cu(II) nanosensors. *Langmuir* 21(20): 9314–9321.

Azzazy, H., M. E., Mansour, M. M., Kazmierczak, S. C. 2007. From diagnostics to therapy: Prospects of quantum dots. *Clin. Biochem.* 40: 917-927.

Barker, S. L. R., Kopelman, R. 1998. Development and cellular applications of fiber optic nitric oxide sensors based on a gold-adsorbed fluorophore. *Anal. Chem.* 70: 4902–4906.

Barker, S. L. R., Thorsrud, B., Kopelman, R. 1998. Nitrite and chloride selective fluorescent nano-optodes and in vitro application to rat conceptuses. *Anal. Chem.* 70: 100–104.

Behrend, C. J., Anker, J. N., Kopelman, R. 2004. Brownian modulated optical nanoprobes. *Appl. Phys. Lett.* 84(1): 154–156.

Bentzen, E. L., House, F., Utley, T. J., Crowe, Jr., J. E., Wright, D. W. 2005. Progression of respiratory syncytial virus infection monitored by fluorescent quantum dot probes. *Nano Lett.* 5: 591–595.

Bishnoi, S. W., Rozell, C. J., Levin, C. S. et al. 2006. All-optical nanoscale pH meter. *Nano Lett.* 6(8): 1687–1692.

Brasola, E., Mancin, F., Rampazzo, E., Tecilla, P., Tonellato, U. 2003. A fluorescence nanosensor for Cu2+ on silica particles. *Chem. Commun.* 24: 3026–3027.

Brasuel, M., Kopelman, R., Miller, T. J., Tjalkens, R., Philbert, M. A. 2001. Fluorescent nanosensors for intracellular chemical analysis: Decyl methacrylate liquid polymer matrix and ion-exchange-based potassium PEBBLE sensors with real-time application to viable rat C6 glioma cells. *Anal. Chem.* 73(10): 2221–2228.

Brasuel, M., Kopelman, R., Kasman, I., Miller, T. J., Philbert, M. A. 2002. Ion concentrations in live cells from highly selective ion correlations fluorescent nano-sensors for sodium. *Proc. IEEE Sensors* 1: 288–292.

Brasuel, M. G., Miller, T. J., Kopelman, R., Philbert, M. A. 2003. Liquid polymer nano-PEBBLES for Cl- analysis and biological applications. *Analyst* 128(10): 1262–1267.

Brown, J. Q., McShane, M. J. 2005. Core-referenced ratiometric fluorescent potassium ion sensors using self-assembled ultrathin films on europium nanoparticles. *IEEE Sensors J.* 5(6): 1197–1205.

Buck, S. M., Xu, H., Brasuel, M., Philbert, M. A., Kopelman, R. 2004. Nanoscale probes encapsulated by biologically localized embedding (PEBBLEs) for ion sensing and imaging in live cells. *Talanta* 63: 41–59.

Cai, W., Shin, D., Chen, K. et al. 2006. In vivo targeting and imaging of tumor vasculature using arginine-glycine-aspartic acid (RGD) peptide-labeled quantum dots. *Nano Lett.* 6: 669–676.

Cao, Y. C., Jin, R., Mirkin, C. A. 2002. Nanoparticles with Raman spectroscopic fingerprints for DNA and RNA detection. *Science* 297: 1536–1540.

Cao, Y., Koo, Y. L., Kopelman, R. 2004. Poly(decyl methacrylate)-based fluorescent PEBBLE swarm nanosensors for measuring dissolved oxygen in biosamples. *Analyst* 129: 745–750.

Cao, Y., Koo, Y. L., Koo, S., Kopelman, R. 2005. Ratiometric singlet oxygen nano-optodes and their use for monitoring photodynamic therapy nanoplatforms. *Photochem. Photobiol.* 81(6): 1489–1498.

Chan, W. C. W., Nie, S. M. 1998. Quantum dot bioconjugates for ultrasensitive nonisotopic detection. *Science* 281: 2016–2018.

Chen, Y. F., Rosenzweig, Z. 2002. Luminescent CdS quantum dots as selective ion probes. *Anal. Chem.* 74(19): 5132–5138.

Cheng, Z. L., Aspinwall, C. A. 2006. Nanometre-sized molecular oxygen sensors prepared from polymer stabilized phospholipid vesicles. *Analyst* 131(2): 236–243.

Cheong, W. F., Prahl, S. A., Welch, A. J. 1990. A review of the optical properties of biological tissues. *IEEE J. Quantum Electron.* 26: 2166–2185.

Clark, H. A., Barker, S. L. R., Brasuel, M. et al. 1998. Subcellular optochemical nanobiosensors: Probes encapsulated by biologically localised embedding (PEBBLEs). *Sens. Actuators B* 51: 12–16.

Clark, H. A., Hoyer, M., Parus, S., Philbert, M. A., Kopelman, R. 1999a. Optochemical nanosensors and subcellular applications in living cells. *Mikrochim. Acta* 131: 121–128.

Clark, H. A., Hoyer, M., Philbert, M. A., Kopelman, R. 1999b. Optical nanosensors for chemical analysis inside single living cells. 1. Fabrication, characterization, and methods for intracellular delivery of PEBBLE sensors. *Anal. Chem.* 71(21): 4831–4836.

Costantino, L. F., Gandolfi, F., Tosi, G., Rivasi, F., Vandelli, M. A., Forni, F. 2005. Peptide-derivatized biodegradable nanoparticles able to cross the blood-brain barrier. *J. Control. Release* 108: 84–96.

Cullum, B. M., Vo-Dinh, T. 2000. The development of optical nanosensors for biological measurements. *Trends Biotechnol.* 18(9): 388–393.

Delpy, D. T., Cope, M. 1997. Quantification in tissue near-infrared spectroscopy. *Philos. Trans. R. Soc. Lond. B* 352: 649–659.

Deng, T., Li, J., Jiang, J., Shen, G., Yu, R. 2006. Preparation of near-IR fluorescent nanoparticles for fluorescence-anisotropy-based immunoagglutination assay in whole blood. *Adv. Funct. Mater.* 16:2147–2155.

Dubertret, B., Calame, M., Libchaber, A. J. 2001. Single-mismatch detection using gold-quenched fluorescent oligonucleotides. *Nat. Biotechnol.* 19: 365–370.

Dubach, J. M., Harjes, D. I., Clark, H. A. 2007. Fluorescent ion-selective nanosensors for intracellular analysis with improved lifetime and size. *Nano Lett.* 7(6): 1827–1831; Ion-selective nano-optodes incorporating quantum dots. *J. Am. Chem. Soc.* 129: 8418–8419.

El-Sayed, I. H., Huang, X., El-Sayed, M. A. 2005. Surface plasmon resonance scattering and absorption of anti-EGFR antibody conjugated gold nanoparticles in cancer diagnostics: Applications in oral cancer. *Nano Lett.* 5: 829–834.

El-Sayed, I. H., Huang, X., El-Sayed, M. A. 2006. Selective laser photo-thermal therapy of epithelial carcinoma using anti-EGFR antibody conjugated gold nanoparticles. *Cancer Lett.* 239: 129–135.

Endo, T., Kerman, K., Nagatani, N. et al. 2006. Multiple label-free detection of antigen-antibody reaction using localized surface plasmon resonance-based core-shell structured nanoparticle layer nanochip. *Anal. Chem.* 78: 6465–6475.

Endo, T., Kerman, K., Nagatani, N., Takamura, Y., Tamiya, E. 2005. Label-free detection of peptide nucleic acid-DNA hybridization using localized surface plasmon resonance based optical biosensor. *Anal. Chem.* 77(21): 6976–6984.

Gao, X., Cui, Y., Levenson, R. M., Chung, L. W. K., Nie, S. 2004. In vivo cancer targeting and imaging with semiconductor quantum dots. *Nat. Biotechnol.* 22: 969–976.

Garvis, S. G., Beuzon, C. R., Holden, D. W. 2001. A role for the PhoP/Q regulon in inhibition of fusion between lysosomes and *Salmonella*-containing vacuoles in macrophages. *Cell Microbiol.* 3: 731–744.

Gattas-Asfura, K. M., Leblane, R. M. 2003. Peptide-coated CdS quantum dots for the optical detection of copper(II) and silver(I). *Chem. Commun.* 21: 2684–2685.

Ghazani, A. A., Lee, J. A., Klostranec, J. et al. 2006. High throughput quantification of protein expression of cancer antigens in tissue microarray using quantum dot nanocrystals. *Nano Lett.* 6(12): 2881–2886.

Goldman, E. R., Clapp, A. R., Anderson, G. P. et al. 2004. Multiplexed toxin analysis using four colors of quantum dot fluororeagents. *Anal. Chem.* 76: 684–688.

Guice, K. B., Caldorera, M. E., McShane, M. J. 2005. Nanoscale internally referenced oxygen sensors produced from self-assembled nanofilms on fluorescent nanoparticles. *J. Biomed. Opt.* 10: 064031-1–064031-10.

Haes, A. J., Chang, L., Klein, W. L., Van Duyne, R. P. 2005. Detection of a biomarker for Alzheimer's disease from synthetic and clinical samples using a nanoscale optical biosensor. *J. Am. Chem. Soc.* 127: 2264–2271.

Haugland, R. P. 2005. *The Handbook: A Guide to Fluorescent Probes and Labeling Technologies.* 10th edn. Eugene, OR: Molecular Probes Inc.

He, X., Liu, H., Li, Y. et al. 2005. Nanoparticle-based fluorometric and colorimetric sensing of copper(II) ions. *Adv. Mater.* 17(23): 2811–2815.

Hohenau, A., Leitner, A., Aussenegg, F. R. 2007. Near-field and far-field properties of nanoparticle arrays. In *Surface Plasmon Nanophotonics.* M. L. Brongersma and P. G. Kik, eds., pp. 11–25. Berlin, Germany: Springer.

Hong, S., Leroueil, P. R., Majoros, I. J., Orr, B. G., Baker, Jr. J. R., Banaszak Holl, M. M. 2007. The binding avidity of a nanoparticle-based multivalent targeted drug delivery platform. *Chem. Biol.* 14: 107–115.

Horvath, T., Monson, E., Sumner, J., Xu, H., Kopelman, R. 2002. Use of steady-state fluorescence anisotropy with PEBBLE nanosensors for chemical analysis. *Proc. SPIE (Int. Soc. Opt. Eng.)* 4626: 482–492.

Huang, C., Chang, H. 2006. Selective gold-nanoparticle-based "turn-on" fluorescent sensors for detection of mercury(II) in aqueous solution. *Anal. Chem.* 78: 8332–8338.

Huang, C., Chiu, S., Huang, Y., Chang, H. 2007. Aptamer-functionalized gold nanoparticles for turn-on light switch detection of platelet-derived growth factor. *Anal. Chem.* 79(13): 4798–4804.

Jensen, T. R., Malinsky, M. D., Haynes, C. L., Van Duyne, R. P. 2000. Nanosphere lithography: Tunable localized surface plasmon resonance spectra of silver nanoparticles. *J. Phys. Chem. B* 104: 10549–10556.

Kim, S. H., Kim, B., Yadavalli, V. K., Pishko, M. V. 2005. Encapsulation of enzyme within polymer spheres to create optical nanosensors for oxidative stress. *Anal. Chem.* 77(21): 6828–6833.

Kim, G., Koo, Lee, Y., Xu, H., Philbert, M. A., Kopelman, R. 2010. Nanoencapsulation method for high selectivity sensing of hydrogen peroxide inside live cells. *Anal. Chem.* 82: 2165–2169.

King, M., Kopelman, R. 2003. Development of a hydroxyl radical ratiometric nanoprobe. *Sens. Actuators B* 90: 76–81.

Klostranec, J. M., Xiang, Q., Farcas, G. A. et al. 2007. Convergence of quantum dot barcodes with microfluidics and signal processing for multiplexed high-throughput infectious disease diagnostics. *Nano Lett.* 7 (9): 2812–2818.

Kneipp, K., Kneipp, H., Itzkan, I., Dasar, R. R., Feld, M. S. 1999. Ultrasensitive chemical analysis by Raman spectroscopy. *Chem. Rev.* 99: 2957–2975.

Kneipp, K., Kneipp, H., Itzkan, I., Dasari, R. R., Feld, M. S. 2002. Surface-enhanced Raman scattering and biophysics. *J. Phys.: Condens. Matter* 14: R597–R624.

Koo, Y. L., Agayan, R., Philbert, M. A., Rehemtulla, A., Ross, B. D., Kopelman, R. 2007. Photonic explorers based on targeted multifunctional nanoplatforms: In vitro and in vivo biomedical applications. In *New Approaches in Biomedical Spectroscopy*, K. Kneipp, R. Aroca, H. Kneipp, and E. Wentrup-Byrne, eds., pp. 200–218. New York: American Chemical Society.

Koo, Y. L., Cao, Y., Kopelman, R., Koo, S., Brasuel, M., Philbert, M. A. 2004. Real-time measurements of dissolved oxygen inside live cells by Ormosil (organically modified silicate) fluorescent PEBBLE nanosensors. *Anal. Chem.* 76: 2498–2505.

Koo, Y. L., Reddy, G. R., Bhojani, M. et al. 2006. Brain cancer diagnosis and therapy with nano-platforms. *Adv. Drug Deliv. Rev.* 58: 1556–1577.

Kreuter, J., Ramge, P., Petrov, V. 2003. Direct evidence that polysorbate-80-coated poly(butyl cyanoacrylate) nanoparticles deliver drugs to the CNS via specific mechanisms requiring prior binding of the drug to the nanoparticles. *Pharm. Res.* 20: 409–416.

Kumar, S., Harrison, N., Richards-Kortum, R., Sokolov, K. 2007. Plasmonic nanosensors for imaging intracellular biomarkers in live cells. *Nano Lett.* 7(5): 1338–1343.

Lackowicz, J. R. 2006. *Principles of Fluorescence Spectroscopy*, 3rd edn., New York: Springer.

Larson, T. A., Bankson, J., Aaron, J., Sokolov, K. 2007. Hybrid plasmonic magnetic nanoparticles as molecular specific agents for MRI/optical imaging and photothermal therapy of cancer cells. *Nanotechnology* 18(32): 325107-1–325107-8.

Lee, D., Khaja, S., Velasquez-Castano, J. C. et al. 2007. In vivo imaging of hydrogen peroxide with chemiluminescent nanoparticles. *Nat. Mater.* 6(10): 765–769.

Link, S., El-Sayed, M. A. 1999. Size and temperature dependence of the plasmon absorption of colloidal gold nanoparticles. *J. Phys. Chem. B* 103(21): 4212–4217; Spectral properties and relaxation dynamics of surface plasmon electronic oscillations in gold and silver nanodots and nanorods. *J. Phys. Chem. B* 103(40): 8410–8426.

Lippitsch, M. E. 1993. Optical sensors based on fluorescence anisotropy. *Sens. Actuators B* 11: 499–502.

Loo, C., Lowery, A., Halas, N., West, J., Drezek, R. 2005. Immunotargeted nanoshells for integrated cancer imaging and therapy. *Nano Lett.* 5: 709–711.

Lyandres, O., Shah, N. C., Yonzon, C. R., Walsh, J. T., Glucksberg, M. R., Van Duyne, R. P. 2005. Real-time glucose sensing by surface-enhanced Raman spectroscopy in bovine plasma facilitated by a mixed decanethiol/mercaptohexanol partition layer. *Anal. Chem.* 77: 6134–6139.

Maeda, H. 2001. The enhanced permeability and retention (EPR) effect in tumor vasculature: The key role of tumor-selective macromolecular drug targeting. *Adv. Enzyme Regul.* 41: 189–207.

Martin-Orozco, N., Touret, N., Zaharik, M. L. et al. 2006. Visualization of vacuolar acidification-induced transcription of genes of pathogens inside macrophages. *Mol. Biol. Cell* 17(1): 498–510.

Maxwell, D. J., Taylor, J. R., Nie, S. 2002. Self-assembled nanoparticles probes for recognition and detection of biomolecules. *J. Am. Chem. Soc.* 124: 9606–9612.

McFarland, A. D., Young, M. A., Dieringer, J. A., Van Duyne, R. P. 2005. Wavelength-scanned surface-enhanced Raman excitation spectroscopy. *J. Phys. Chem.* B 109: 11279–11285.

Meallet-Renault, R., Pansu, R., Amigoni-Gerbier, S., Larpent, C. 2004. Metal-chelating nanoparticles as selective fluorescent sensor for Cu^{2+}. *Chem Commun.* 20: 2344–2345.

Meallet-Renault, R., Herault, A., Vachon, J. J., Pansu, R. B., Amigoni-Gerbier, S., Larpent, C. 2006. Fluorescent nanoparticles as selective Cu(II) sensors. *Photochem. Photobiol. Sci.* 5(3): 300–310.

Medintz, I. L., Clapp, A. R., Mattoussi, H., Goldman, E. R., Fisher, B., Mauro, J. M. 2003. Self-assembled nanoscale biosensors based on quantum dot FRET donors. *Nat. Mater.* 2(9): 630–638.

Montalti, M., Prodi, L., Zaccheroni, N. 2005. Fluorescence quenching amplification in silica nanosensors for metal ions. *J. Mater. Chem.* 15: 2810–2814.

Montet, X., Funovics, M., Montet-Abou, K., Weissleder, R., Josephson, L. 2006. Multivalent effects of RGD peptides obtained by nanoparticle display. *J. Med. Chem.* 49(20): 6087–6093.

Moskovits, M. 1985. Surface-enhanced spectroscopy. *Rev. Mod. Phys.* 57: 783–826.

Mulder, W. J. M., Koole, R., Brandwijk, R. J. et al. 2006. Quantum dots with a paramagnetic coating as a bimodal molecular imaging probe. *Nano Lett.* 6: 1–6.

Ntziachristos, V., Bremer, C., Weissleder, R. 2003. Fluorescence imaging with nearinfrared light: New technological advances that enable in vivo molecular imaging. *Eur. Radiol.* 13: 195–208.

Oldenburg, S. J., Averitt, R. D., Westcott, S. L., Halas, N. J. 1998. Nanoengineering of optical resonances. *Chem. Phys. Lett.* 288: 243–247.

Park, E. J., Brasuel, M., Behrend, C., Philbert, M. A., Kopelman, R. 2003. Ratiometric optical PEBBLE nanosensors for real-time magnesium ion concentrations inside viable cells. *Anal. Chem.* 75(15): 3784–3791.

Peng, J., He, X., Wang, K., Tan, W., Wang, Y., Liu, Y. 2007a. Noninvasive monitoring of intracellular pH change induced by drug stimulation using silica nanoparticle sensors. *Anal. Bioanal. Chem.* 388: 645–654.

Peng, Z., Chen, Z., Jiang, J., Zhang, X., Shen, G., Yu, R. 2007b. A novel immunoassay based on the dissociation of immunocomplex and fluorescence quenching by gold nanoparticles. *Anal. Chim. Acta* 583: 40–44.

Poulsen, A. K., Scharff-Poulsen, A. M., Olsen, L. F. 2007. Horseradish peroxidase embedded in polyacrylamide nanoparticles enables optical detection of reactive oxygen species. *Anal. Biochem.* 366(1): 29–36.

Rampazzo, E., Brasola, E., Marcuz, S., Mancin, F., Tecilla, P., Tonellato, U. 2005. Surface modification of silica nanoparticles: A new strategy for the realization of self-organized fluorescent chemosensors. *J. Mater. Chem.* 15: 2687–2796.

Reddy, G. R., Bhojani, M. S., McConville, P. et al. 2006. Vascular targeted nanoparticles for imaging and treatment of brain tumors. *Clin. Cancer Res.* 12(22): 6677–6686.

Ruedas-Rama, M. J., Hall, E. A. H. 2007. K^+-selective nanospheres: Maximizing response range and minimizing response time. *Analyst* 131(12): 1282–1291.

Sandros, M. G., Gao, D., Benson, D. E. 2005. A modular nanoparticle-based system for reagentless small molecule biosensing. *J. Am. Chem. Soc.* 127(35): 12198–12199.

Sandros, M. G., Shete, V., Benson, D. E. 2006. Selective, reversible, reagentless maltose biosensing with core-shell semiconducting nanoparticles. *Analyst* 131(2): 229–235.

Sasaki, K., Shi, Z., Kopelman, R., Masuhara, H. 1996. Three-dimensional pH microprobing with an optically-manipulated fluorescent particle. *Chem. Lett.* 25: 141–142.

Schmälzlin, E., Van Dongen, J. T., Klimant, I. et al. 2005. An optical multifrequency phase-modulation method using microbeads for measuring intracellular oxygen concentrations in plants. *Biophys. J.* 89(2): 1339–1345.

Schmälzlin, E., Walz, B., Klimant, I., Schewe, B., Löhmannsröben, H. G. 2006. Monitoring hormone-induced oxygen consumption in the salivary glands of the blowfly, *Calliphora vicina*, by use of luminescent microbeads. *Sens. Actuators B* 119(1): 251–254.

Shah, N. C., Lyandres, O., Walsh, J. T., Glucksberg, M. R., Van Duyne, R. P. 2007. Lactate and sequential lactate-glucose sensing using surface-enhanced Raman spectroscopy. *Anal. Chem.* 79: 6927–6932.

Stuart, D. A., Yuen, J. M., Shah, N. et al. 2006. In vivo glucose measurement by surface-enhanced Raman spectroscopy. *Anal. Chem.* 78: 7211–7215.

Sumner, J. P., Kopelman, R. 2005. Alexa Fluor 488 as an iron sensing molecule and its application in PEBBLE nanosensors. *Analyst* 130(4): 528–533.

Sumner, J. P., Aylott, J. W., Monson, E., Kopelman, R. 2002. A Fluorescent PEBBLE nanosensor for intracellular free zinc. *Analyst* 127: 11–16.

Sumner, J. P., Westerberg, N., Stoddard, A. K., Fierke, C. A., Kopelman, R. 2005. Cu^+ and Cu^{2+} sensitive PEBBLE fluorescent nanosensors using Ds Red as the recognition element. *Sens. Actuators B* 113(2): 760–767.

Sun, H. H., Scharff-Poulsen, A. M., Gu, H. et al. 2008. Phosphate sensing by fluorescent reporter proteins embedded in polyacrylamide nanoparticles. *ACS Nano* 2(1): 19–24.

Tada, H., Higuchi, H., Wanatabe, T. M., Ohuchi, N. 2007. In vivo real-time tracking of single quantum dots conjugated with monoclonal anti-HER2 antibody in tumors of mice. *Cancer Res.* 67: 1138–1144.

Talley, C. E., Jusinski, L., Hollars, C. W., Lane, S. M., Huser, T. 2004. Intracellular pH sensors based on surface-enhanced Raman scattering. *Anal. Chem.* 76: 7064–7068.

Tan, W., Shi, Z., Smith, S., Birnbaum, D., Kopelman, R. 1992. Submicrometer intracellular chemical optical fiber sensors. *Science* 258: 778–781.

Tan, W., Kopelman, R., Barker, S. L. R., Miller, M. T. 1999. Ultrasmall optical sensors for cellular measurements. *Anal. Chem.* 71: 606A–612A.

Vescovi, E. G., Soncini, F. C., Groisman, E. A. 1996. Mg^{2+} as an extracellular signal: Environmental regulation of *Salmonella* virulence. *Cell* 84(1): 165–174.

Veiseh, O., Sun, C., Gunn, J. et al. 2005. Optical and MRI multifunctional nanoprobe for targeting gliomas. *Nano Lett.* 5: 1003–1038.

Wang, H. Z., Wang, H. Y., Liang, R. Q., Ruan, K. C. 2004. Detection of tumor marker CA125 in ovarian carcinoma using quantum dots. *Acta Biochim. Biophys. Sin.* 36(10): 681–686.

Wang, L., Yang, C., Tan, W. 2005. Dual-luminophore-doped silica nanoparticles for multiplexed signaling. *Nano Lett.* 5: 37–43.

Wang, L., Zhao, W. J., O'Donoghue, M. B., Tan, W. 2007. Fluorescent nanoparticles for multiplexed bacteria monitoring. *Bioconjug. Chem.* 18(2): 297–301.

Willets, K. A., Van Duyne, R. P. 2007. Localized surface plasmon resonance spectroscopy and sensing. *Annu. Rev. Phys. Chem.* 58: 267–297.

Wu, Z., Jiang, J., Fu, L., Shen, G., Yu, R. 2006. Optical detection of DNA hybridization based on fluorescence quenching of tagged oligonucleotide probes by gold nanoparticles. *Anal. Biochem.* 353: 22–29.

Wu, H., Cheng, T., Tseng, W. 2007. Phosphate-modified TiO_2 nanoparticles for selective detection of dopamine, levodopa, adrenaline, and catechol based on fluorescence quenching. *Langmuir* 23: 7880–7885.

Xu, H., Aylott, J. W., Kopelman, R. 2002. Fluorescent nano-PEBBLE sensors designed for intracellular glucose imaging. *Analyst* 127(11): 1471–1477.

Xu, H., Aylott, J. W., Kopelman, R., Miller, T. J., Philbert, M. A. 2001. A real-time ratiometric method for the determination of molecular oxygen inside living cells using sol-gel-based spherical optical nanosensors with applications to rat C6 glioma. *Anal. Chem.* 73: 4124–4133.

You, C., Miranda, O. R., Gider, B. et al. 2007. Detection and identification of proteins using nanoparticle-fluorescent polymer 'chemical nose' sensors. *Nat. Nanotechnol.* 2(5): 318–323.

Zhao, X., Hilliard, L. R., Mechery, S. J. et al. 2004. A rapid bioassay for single bacterial cell quantitation using bioconjugated nanoparticles. *Proc. Natl. Acad. Sci. U. S. A.* 101: 15027–15032.

Zhou, X., Zhou, J. 2004. Improving the signal sensitivity and photostability of DNA hybridizations on microarrays by using dye-doped core-shell silica nanoparticles. *Anal. Chem.* 76: 5302–5312.

Quantum Dot Infrared Photodetectors and Focal Plane Arrays

Xuejun Lu
University of Massachusetts Lowell

34.1 Introduction

Quantum dots (QDs) are semiconductor nanocrystallites that have dimensions smaller than the de Broglie wavelength of electrons in semiconductors [1,2]. There are generally two size groups of quantum dots obtained from different methods. The first is colloidal QDs such as CdSe and PbS [3–5]. The colloidal QDs can be synthesized in various sizes and forms and can also be combined with conductive polymers. The colloidal QDs have a dimension of 3–5 nm diameter. Their working wavelengths (emission or absorption) are in visible or near infrared (IR) regions. The second type is epitaxial QDs such as InAs. Epitaxial QDs are self-assembled nanocrystallites grown by molecular beam epitaxy (MBE) or metal organic chemical vapor deposition (MOCVD) through the Stranski–Krastanow (S-K) growth mode [6,7]. The epitaxial QDs have dimensions of usually ~20–40 nm at the base and are 5–8 nm high [2]. They work at near infrared (NIR) through far infrared (FIR) regions. In this chapter, we will be focusing on the inter-subband transition in InAs QDs for LWIR optoelectronic devices.

Epitaxially grown semiconductor QDs have shown great promise in various optoelectronic devices such as QD lasers [8–11] and QD infrared photodetectors (QDIP) [9–17]. Due to the nanoscale quantum confinement, QDs exhibit atomic-like properties, including discrete energy levels within the conduction (*c*) and valence (*v*) bands, delta-function-like density of states (DOS) [18], reduced electron–phonon scattering [8], and enhanced overlap of wavefunctions [19–21]. These quantum-confined properties open a new area of possibilities for unipolar optoelectronic devices covering a broad wavelength range from middle infrared (MIR) through terahertz (THz) [14,22,23] with significantly improved performance features, including low threshold current, high photoconductive gain, and large EO coefficients [24,25]. These advantages make semiconductor QD nanocrystallites one of the most promising artificial materials for a great variety of optoelectronic devices in IR through THz wavelength regions. Devices and technologies in these wavelength ranges are of great importance to many civilian and homeland security applications such as target detection and tracking, remote sensing, chemical analysis, and medical diagnostics [26–28]. This chapter focuses on reviews of the properties of 3-D-confined QDs and the transitions between the subbands (inter-subband transitions) and their applications in quantum dot infrared photodetector and focal plane arrays (FPA). Device physics, fabrication, and characterization of QDIP will be presented.

34.2 Properties of QDs

Due to the three-dimension confinement of carriers, QDs show complete discrete energy levels within the conduction band and valence bands. For an initial understanding, we start with a simple quantum box with dimensions L_x, L_y, and L_z. Here, we focus on the conduction band. The wavefunction and energy state of a box-shaped QD are given in Equations 34.1 and 34.2, respectively:

$$\Psi_{c,nlm} = \left[\frac{2}{L}\right]^{3/2} \sin\left[\frac{n\pi}{L_x}x\right]\sin\left[\frac{l\pi}{L_y}y\right]\sin\left[\frac{m\pi}{L_z}z\right]U_c(\vec{r}), \qquad (34.1)$$

$$E_{c,nlm} = E_C + \frac{\hbar^2\pi^2}{2m_c^*L_x^2}n^2 + \frac{\hbar^2\pi^2}{2m_c^*L_y^2}l^2 + \frac{\hbar^2\pi^2}{2m_c^*L_z^2}m^2, \qquad (34.2)$$

where

   $n, l, m = 1, 2, 3 \ldots$
   $U_c(\vec{r})$ is the wavefunction of a unit cell
   m_c^* is the effective mass of the electrons in conduction band
   $\hbar = h/2\pi$, where h is the Planck's constant

The 3-D quantum confinement leads to a split of the conduction band into atomic-like discrete energy levels. The separation between these energy levels depends on the dimensions of the quantum box.

The atomic-like discrete energy levels show a delta-function-like density of state (DOS), $\rho_{QD}(E)$, which can be written as

$$\rho_{QD}(E) = g(E_n)\delta(E - E_n), \qquad (34.3)$$

where $g(E_n)$ is the degeneracy of the energy level E_n. Figure 34.1a through c shows the schematic drawing of the DOS of bulk materials, quantum well (QW) and QDs, respectively.

The atomic-like discrete energy level not only shows a completely different DOS than bulk semiconductors but also has a longer excited-state lifetime. This is primarily due to the nonradiative relaxation caused by the thermally activated electron–LO phonon-scattering process [29], as shown Figure 34.2a. In a QW structure, the energy states are quantized only along the growth direction (*z*-direction) while the in-plane energy states are quasi-continuous. These quasi-continuous energy states make it easy to achieve resonant electron–LO phonon scattering. The LO scattering nonradiatively depopulates electrons from the upper states to the lower states at a much faster (2300 times) rate than the radiative emission processes [30,31]. Such a fast nonradiative relaxing rate leads to short excited-state lifetime in quantum dots, since discrete energy levels are off-resonance with the LO phonons. Thus, the LO phonon nonradiative relaxation can be substantially reduced, which leads to long excited-state lifetime. This is typically referred to as "phonon bottleneck" effect [19,32]. The long excited-state lifetime was predicted and experimentally observed using time-resolved photoluminescence (PL) technique [19,33].

The transition rates between the inter-subbands can be described using Fermi's Golden rule:

$$w_{jm} = \frac{2\pi}{\hbar}\left|\langle\psi_j|H'|\psi_m\rangle\right|^2\delta(E_{jm} - \hbar\omega), \qquad (34.4)$$

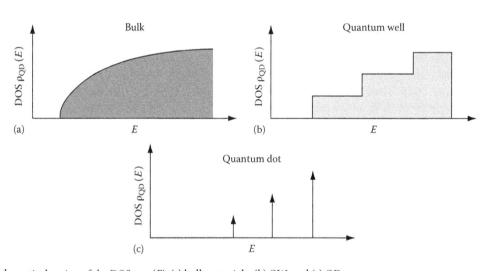

FIGURE 34.1 Schematic drawing of the DOS, $\rho_{QD}(E)$: (a) bulk materials, (b) QW, and (c) QDs.

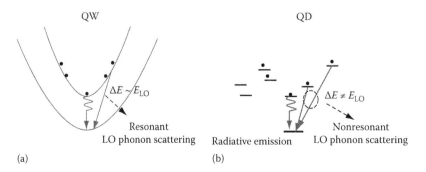

FIGURE 34.2 LO phonon-scattering process: (a) resonant LO phonon scattering in QWs and (b) nonresonant LO phonon scattering in QDs.

where

   ψ_m and ψ_j, are the wavefunctions of the subbands
   H' is the interaction Hamiltonian of the incident light with QDs

Under electric dipole approximation [34], H' can be written as [34,35]

$$H' = -e\vec{r} \cdot \vec{E}, \tag{34.5}$$

where

   $-e\vec{r}$ is the electric dipole moment
   $\vec{E}$ is the electric field

Equations 34.4 and 34.5 also include the quantum selection rule for the inter-subband transitions, i.e., the non-zero matrix element $H'_{jm} = \langle \psi_j | H' | \psi_m \rangle \neq 0$. For normal incidence, the electric field is along the x- or y-direction. Assuming $\vec{E}$ is along the x-direction, for the simple quantum box example, the matrix element H'_{jm} can be written as

$$H'_{jm} = \langle \psi_j | -exE | \psi_m \rangle = C \left\langle \sin\left[\frac{j\pi}{L_x}x\right] \middle| -exE \middle| \sin\left[\frac{m\pi}{L_x}x\right] \right\rangle, \tag{34.6}$$

where C is a constant containing the integration of unit cell wavefunction $U_c(\vec{r})$. Due to the quantization in the x-direction, the normal incidence along the x-direction has a non-zero matrix element H'_{jm}, indicating normal incidence absorption and detection capability. A real QD usually has a lens or pyramid shape rather than a cube or box shape. This would cause the mixture of wave-functions. Nevertheless, due to the quantization in the x-direction, it has the normal incidence light absorption and detection capability. Such normal incidence absorption eliminates requirement for surface gratings in quantum well (QW)-based photodetectors and thus greatly simplifies the fabrication complexity of a large format ($1 \times 1\,\text{K}$) photodetector array and focal plane array (FPA).

34.3 Structure of a QD Infrared Photodetector and Its Advantages

As discussed in the previous section (Section 34.2), due to the 3-D quantum confinement, QDs show complete discrete energy levels within the conduction band and valence bands. The transitions between the inter-subbands can be used for infrared detection. The schematic structure and simplified band diagram of a QD infrared photodetector (QDIP) are shown in Figure 34.3. It consists of vertically-stacked InAs quantum dots layers with GaAs capping layers. The electrons are excited by the normal incident light and subsequently collected through the top electrode and generate photocurrent. This is a uni-polar photodetector. Only conduction band is involved in the photocurrent generation process.

Due to the unique properties of QDs, QDIP offers several advantages as listed in the following:

1. Normal incidence light detection capability

As discussed in the previous section (Section 34.2), due to the 3-D quantization, the normal incidence along the x-direction has a non-zero matrix element H'_{jm}. This enables normal incidence absorption and detection. Such normal incident light detection capability is especially suitable for two-dimensional (2-D) focal plane array (FPA).

2. Low dark current

Delta-function like density of state (DOS), the total states of QDs for a given energy interval ΔE will be much less than those of bulk materials, or quantum wells. The major source of photodetector noise currents comes from the thermal excitation process (thermal-electrons). The thermal excitation processes for unbiased and biased scenarios are shown in Figure 34.4a and b, respectively. Under zero bias (Figure 34.4a, the thermal excitation process is balanced by the relaxation, and the thermal excitation rate equals the relation rate, i.e.,

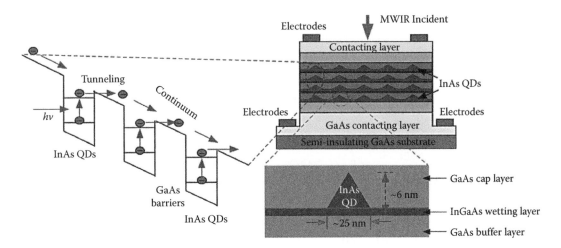

FIGURE 34.3 Schematic structure and simplified band diagram of a QDIP.

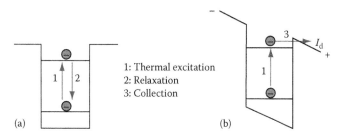

FIGURE 34.4 Thermal excitation, relaxation, and collection processes in a photodetector: (a) unbiased and (b) biased.

$$R_{\text{th}} = R_{\text{relax}} = \frac{N}{\tau}, \qquad (34.7)$$

where

R_{th} is the thermal excitation rate
R_{relax} is the relaxation rate, which is related to the excited-state lifetime τ
N is the number of electrons on the excited state

Under a high external bias, the thermally excited electrons can be effectively collected before they relax to the ground states. Thus the collection rate equals the thermal excitation rate. The electrons collected by the external bias form the dark current. The dark current can therefore be written as

$$I_d = qR_{\text{collection}} = qR_{\text{th}} = q\frac{N}{\tau}, \qquad (34.8)$$

For QWIP, the number of electrons on the excited states N_{QW} can be written as

$$N_{\text{QW}} = \sum_m \int_{E_m}^{\infty} \frac{N(E)}{1 + \exp\left[\dfrac{q(E - E_F)}{kT}\right]} dE, \qquad (34.9)$$

where

m is the index of the different energy bands in QWs
E_F is the Fermi-level of the quantum well, which depends primarily on the doping concentration (N_0)
$N(E)$ is the density of state of the QW

The dark current density in quantum wells can thus be written as

$$I_{d,\text{QW}} = \frac{qdN_{\text{QW}}}{\tau_{\text{QW}}} = \frac{qd}{\tau_{\text{QW}}} \sum_m \int_{E_m}^{\infty} \frac{N(E)}{1 + \exp\left[\dfrac{q(E - E_F)}{kT}\right]} dE, \qquad (34.10)$$

where d is the thickness of the active layers. For QDs, the number of electrons on the excited states N_{QW} can be written as

$$N_{\text{QD}} = \sum_m \frac{g_m}{1 + \exp\left[\dfrac{q(E_m - E_F)}{kT}\right]}, \qquad (34.11)$$

where

E_F is the Fermi-level of the quantum dots, which depends primarily on the doping concentration (N_0) of the dots
E_m is the energy of the energy level m of the dots
N_{QD} is the volume density of the dots
q is the charge of an electron
g_m is the degeneracy of the QD energy level

The thermally generated dark current in quantum dots $I_{d,\text{QD}}$ can thus be expressed as

$$I_{d,\text{QD}} = \frac{qdN_{\text{QD}}}{\tau_{\text{QD}}} = \frac{qd}{\tau_{\text{QD}}} \sum_m \frac{g_m}{1 + \exp\left[\dfrac{q(E_m - E_F)}{kT}\right]}, \qquad (34.12)$$

From Equations 34.10 and 34.12, one can see that QDIP shows lower noise current than QWIPs [36] due to primarily two reasons:

1. The excited-state lifetime of QDs is much longer than that of QWs, which makes the thermal excitation rate much smaller in QDs.
2. The density of states of QDs are much lower than that of QWs, which allows QDs to hold less thermally-generated electrons for dark current

Figure 34.5 shows the dark current density of the QDIPs made from different MBE growths. Also shown in Figure 34.5 is the dark current density of a QWIP reported in [36].

As indicated in Figure 34.5, the QDIPs from different MBE growths show very similar level of the dark current density.

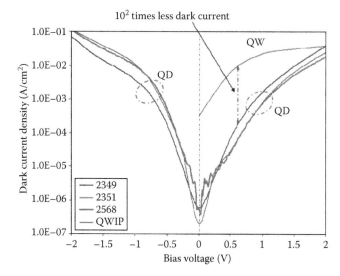

FIGURE 34.5 Dark current density of the QDIPs made from different MBE growths.

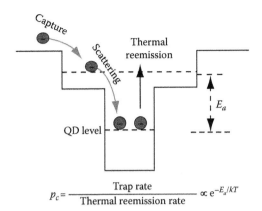

$$p_c = \frac{\text{Trap rate}}{\text{Thermal reemission rate}} \propto e^{-E_a/kT}$$

FIGURE 34.6 Temperature-dependent electron trap and thermal reemission process.

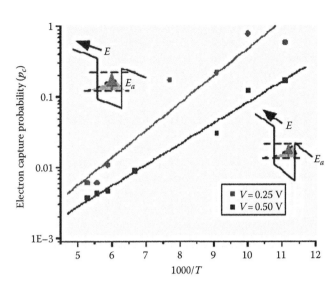

FIGURE 34.7 Temperature-dependent electron capture probability. (From Lu, X. et al., *Appl. Phys. Lett.*, 91, 051115, 2007.)

Over two orders of magnitude of lower dark current can be obtained in QDIPs.

3. High photoconductive gain

The electron capture and reemission processes are shown in Figure 34.6. Electrons are captured into QDs mainly through the wetting layer or QW due to the continuous energy levels and the large density of states of the wetting layer and QW [32]. Because the electron de Broglie wavelength is comparable to the length scale of QD heterostructures, the electron capture into the QW depends strongly on the well thickness. An oscillation capture time as a function of the well thickness has been observed [33,37].

Once captured into the wetting layer or quantum well, the electrons are relaxed into the QDs through phonon scattering [19]. The thermal reemission out of the QDs follows the temperature-dependent Arrhenius equation $e^{(-E_b/kT)}$ with activation energy (or barrier energy) E_b [19,29]. Electron can also be thermally reemitted out of the QD. The net electron capture probability is proportional to $\exp(-E_a/kT)$, where E_a is the activation energy (or barrier energy), T is the absolution temperature, and k is Boltzmann's constant.

The electron capture and reemission processes also temperature-dependent properties [30]. The electron capture probability p_c is related to the electron relaxing into a QD and reemission out of the QD by [38]

$$p_c = \frac{\tau_{em}}{\tau_{relax}}, \tag{34.13}$$

where $1/\tau_{relax}$ and $1/\tau_{em}$ represent electron relaxing and reemission rates, respectively. The Arrhenius-type p_c and temperature T relation has been experimentally verified [30].

Figure 34.7 shows such p_c and temperature T relations at different biases. The slightly lower activation energy (E_a) at higher bias is attributable to the quantum-fined Stark effect [39,40].

When p_c is small, photoconductive gain G_{ph} of a quantum dot infrared photodetector is related to the carrier capture probability p_c by the follow equation [19].

$$G_{ph} = \frac{I_{collect}}{I_{ph}} \approx \frac{1}{FNp_c}, \tag{34.14}$$

where

N is the number of QD layers

F is the QD filling factor, which is usually 0.35

The bias voltage dependent photoconductive (PC) gain at different temperatures is shown in Figure 34.8. A high PC gain of over 100 can be obtained. The high PC gain leads to large

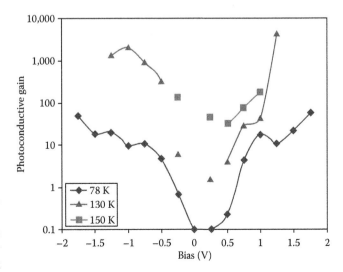

FIGURE 34.8 Photoconductive gain at different temperatures.

photoresponsivity [38]. However, one should note that the PC gain would also increase the dark current level as well. Under background noise limited situation, the PC gain would increase the signal to noise ratio (SNR) [36–41].

34.4 QD Growth and Characterization

Stranski–Krastanow (S-K) mode epitaxial growth of QDs by using MBE or MOCVD [6,7] are generally used to obtain high-quality, defect-free QDs with very good size uniformity. This method utilizes highly lattice-mismatched growth of In(Ga)As on GaAs. Due to the strain built up in the lattice-mismatched epitaxial growth, self-assembled QDs will form after a few mono layers (i.e., the critical thickness) of the layer-by-layer growth. The typical growth temperature for GaAs is around 620°C. The temperature can be measured by an optical pyrometer. QDs are around 470°C by depositing a few monolayers (ML) of InAs on GaAs or InGaAs. Different QD size and density can be obtained by changing the QD growth temperature and the amount of InAs deposited.

The QD size and density can be measured using atomic force microscopy and cross-section transmission electron microscopy (XTEM). The AFM images of the QDs are shown in Figure 34.9. The lateral size and the density of typical QDs grown by MBE are ~25 nm and ~2.9 × 10^10 cm^−2, respectively. A cross-sectional transmission electron microscopy (XTEM) image of typical QDs grown by MBE is shown in Figure 34.10. The layered QDs can be clearly seen. The height of the QDs is typically ~6 nm. Note that the QDs were aligned across all these layers, which clearly indicates that the QD growth was strain driven.

The energy states of QDs can be measured by measuring the photoluminescence (PL) emission from the QDs. Figure 34.11 shows the simplified band diagram and the principle of the PL. The PL process starts with the excitation of electro-hole pairs (EHP) generated by a pump laser in the conduction and valence bands. The electrons and holes subsequently diffuse and are captured by QDs. The recombination of EHP on the ground and excited states of QDs give reemissions (photoluminescence) at longer wavelength than the pump laser.

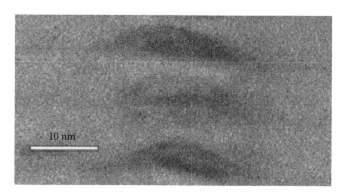

FIGURE 34.10 XTEM image of QDs grown by MBE.

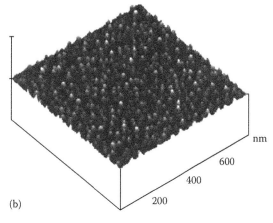

FIGURE 34.9 AFM images of typical QDs grown by MBE: (a) top view and (b) side view. (From Lu, X. et al., *Semicond. Sci. Technol.*, 22, 993, 2008. With permission.)

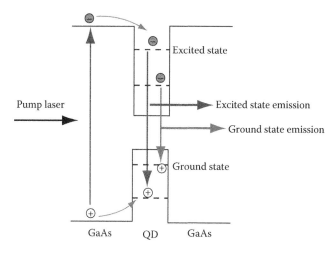

FIGURE 34.11 Simplified band diagram in the PL process.

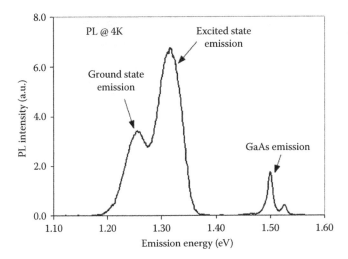

FIGURE 34.12 Example PL from InAs QDs in GaAs.

Figure 34.12 shows an example PL of InAs/GaAs QDs. Since the PL comes from EHP recombination at QD energy levels, the PL contains important information energy states of QDs. The peak wavelength and the width (full width half magnitude (FWHM)) reveal the energy levels and the uniformity of the QDs, respectively.

34.5 Fabrication of QDIP

After the growth, the wafer was processed into 100 μm diameter circular mesas using standard photo-lithography and wet-etching procedures. Figure 34.13 shows the flowchart of the fabrication process. It starts spin-coating of a thin-layer (~0.3 μm thick) of photoresist on top of the MBE grown QD wafer (Figure 34.13b). The photoresist is then patterned in individual 100 μm-diameter circular pixels using standard photolithography technology (Figure 34.13c). The patterned wafer is then wet-etched using the H_2SO_4:H_2O_2:H_2O (1:80:80) solution for ~120 s to reach the bottom contact layer (Figure 34.13d). The depth of the etching is measured by using a profilometer or an optical profiler. The top and bottom electrodes can be formed simultaneously on top of and surrounding the mesas by standard E-beam metal-evaporation deposition, lift-off, and rapid thermal annealing of n-type (Ni/Ge/Au) alloys (Figure 34.13e through j).

The pictures of 12 × 12 QDIP array and the individual QDIP are shown in Figure 34.14a and b, respectively. The QDIP is then wire-bonded and mounted in an infrared (IR) dewar with a ZnSe IR window that has more than 60% transmission over the 3–14 μm broad-band IR region. The packaged QDIP in a liquid nitrogen (LN_2) dewar is shown in Figure 34.15.

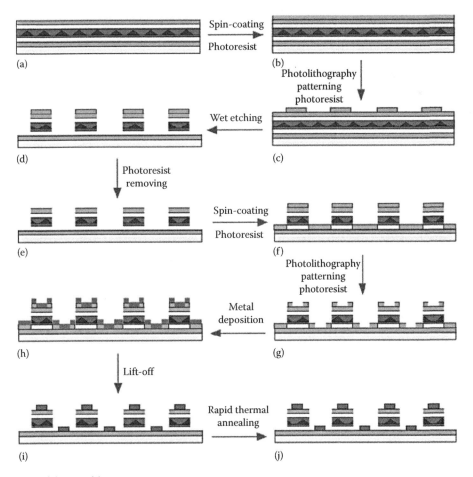

FIGURE 34.13 Flowchart of the FPA fabrication process.

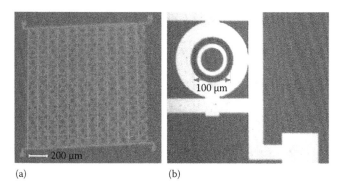

(a)   (b)

FIGURE 34.14 Fabricated QDIPs: (a) 12 × 12 QDIP array and (b) an individual QDIP with an 100 × 100 μm bonding pad.

FIGURE 34.15 Photography of a packaged QDIP in an LN₂ dewar.

34.6 QDIP Characterization

QDIP characterization includes the following: (1) detection spectrum; (2) dark current and noise current; (3) noise and photoconductive gain calculation; (4) photocurrent measurement; (5) photoresponsivity and photodetectivity calculation, (6) temperature dependent performance.

    1. Detection spectrum

The spectral response of the QDIP is typically measured by using a Fourier transform infrared (FTIR) spectrometer. Figure 34.16 shows the diagram of an FTIR spectrometer. A typical FTIR spectrometer consists of a blackbody infrared (IR) source, a 50% beam splitter, a fixed mirror and a mirror and an IR detector.

    The working principle of the FTIR spectrometer is briefly described as the following: the 50% beam splitter divides the incoming IR beam into two optical beams that reflect back from the fixed mirror and the moveable mirror respectively. The two reflected beams are combined by the beam splitter and pass through a sample. Because of the constantly changing position of the moveable mirror, the interference of the two reflected

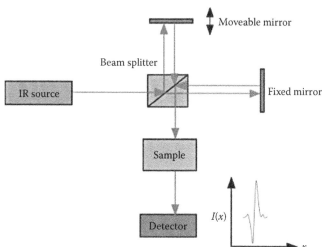

FIGURE 34.16 Principle of the FTIR spectrometer.

beams generates an interferogram. Every data point (a function of the moving mirror position) of the interferogram contains the information about every infrared frequency. The frequency information can thus be obtained by doing an inverse Fourier transform as shown schematically in Figure 34.17. A transmission FTIR spectrum of a GaAs wafer is shown in Figure 34.17 as an example.

The interference of the two reflected beams can be written as

$$I(x) = \left| E_1 + E_2 \right|^2 = E_1^2 + E_2^2 + 2E_1E_2\cos(\theta)$$

$$= \frac{I_0}{2}\left[1 + \cos\left(\frac{2\pi}{\lambda}x\right)\right], \tag{34.15}$$

where

    I_0 is the intensity of the incident IR beam
    λ is the wavelength
    x is the position

For non-monochromatic IR source:

$$I(x) = \int_0^\infty \frac{I_0}{2}\left[1 + \cos\left(\frac{2\pi}{\lambda}x\right)\right]G(\lambda)\mathrm{d}\lambda, \tag{34.16}$$

where $G(\lambda)$ is the transmission profile of the sample.

$$I(x) = \int_0^\infty \frac{I_0}{2}G(\lambda)\mathrm{d}\lambda + \int_0^\infty \frac{I_0}{2}\cos\left(\frac{2\pi}{\lambda}x\right)G(\lambda)\mathrm{d}\lambda$$

$$= \frac{I_0}{2} + \int_0^\infty \frac{I_0}{2}\cos\left(\frac{2\pi}{\lambda}x\right)G(\lambda)\mathrm{d}\lambda. \tag{34.17}$$

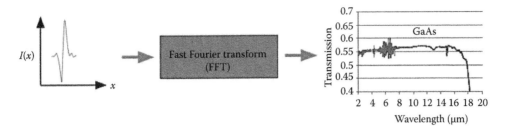

FIGURE 34.17 Principle of the FTIR spectrometer.

The AC part of $I(x)$, is actually Fourier transform of the transmission profile of the sample, $G(\lambda)$, which be obtained by performing a reverse Fourier transform.

$$G(\lambda) = \int_{-\infty}^{\infty} I(x) \exp\left(-j\left(\frac{2\pi}{\lambda}x\right)\right) dx. \qquad (34.18)$$

In practical situation, the position x can not move from $-\infty$ to ∞. The resolution of the FTIR system is determined by the maximum movement x_{max}:

$$\text{Resolution} \propto \frac{1}{2\pi x_{max}}. \qquad (34.19)$$

Figure 34.18 shows the QDIP spectral response measurement setup using an FTIR spectrometer. The spectral measurement uses the QDIP to replace the original IR detector of the FTIR, since the IR source is broad band. The resulting spectrum is the spectrum response of the QDIP.

Figure 34.19 shows an FTIR photocurrent spectrum of the QDIP with the $In_{0.20}Ga_{0.80}As$ capping layers at both positive and negative biases. The inset of Figure 34.19 shows the QDIP heterostructure. The spectrum is peaked at 9.7 μm. The spectral width (FWHM) is ~1.2 μm. The $\Delta\lambda/\lambda$ was measured to be ~12%, which indicates the photocurrent was generated by the electron transitions between bounded states of the QDs.

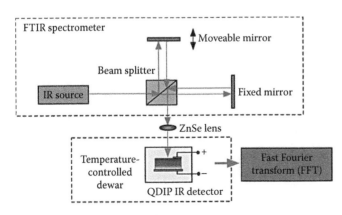

FIGURE 34.18 QDIP spectral response measurement setup.

2. Dark current and noise current measurement

Dark current (I_d) of the QDIP can be measured using a semiconductor parameter analyzer or a source meter. Figure 34.20 shows the dark current as a function of bias voltages at different temperatures.

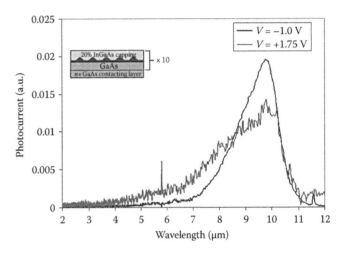

FIGURE 34.19 Photocurrent spectrum of a QDIP with $In_{0.20}Ga_{0.80}As$ capping layers at different biases. (From Meisner, M. et al., *Semicond. Sci. Technol.*, 23, 095016, 2008. With permission.)

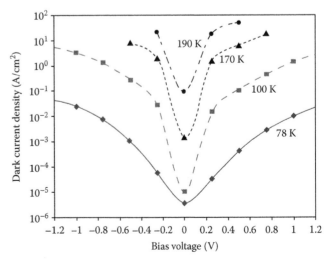

FIGURE 34.20 Dark current of the QDIP as a function of bias voltages at different temperatures. (From Lu, X. et al., *Appl. Phys. Lett.*, 91, 051115, 2007. With permission.)

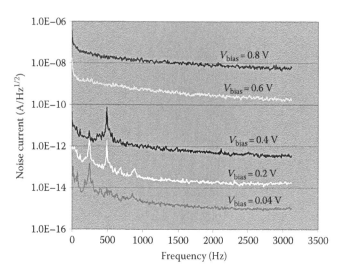

FIGURE 34.21 Noise current spectrum at different bias voltages.

The spectral density of dark current (i_{noise}, in A/Hz$^{1/2}$), i.e., noise current can be measured by obtaining the spectrum of the dark current at certain bias voltages. The spectrum of the dark current can be obtained by a fast Fourier transform (FFT) spectrum analyzer. Figure 34.21 shows the noise current for different bias voltages. Note that the $1/f$ noise dominates at low frequency range from DC to around 700 Hz. To avoid $1/f$ noise, the noise currents (i_{noise}) at high frequency (1 kHz) were used for generation-recombination (GR) noise analysis. The dark current induced noise is a white noise. The noise floor can be measured starting $f = 1000$ Hz.

3. Photoconductive gain and photoresponsivity measurement

The noise current i_{noise} contains GR noise current and thermal noise (Johnson noise) current I_{th}:

$$i_{\text{noise}}^2 = 4eG_n I_d + I_{\text{th}}^2, \tag{34.20}$$

where
  G_n is the noise gain
  e is the charge of an electron (1.6×10^{-19} C)
  I_d is the dark current of the QDIP

The thermal noise current can be calculated using

$$I_{\text{th}} = \sqrt{\frac{4kT}{R}}, \tag{34.21}$$

where
  k is Boltzmann's constant
  T is the absolute temperature
  R is the differential resistance of the QDIP, which can be extracted from the slope of the dark current

The noise gain G_n can thus be calculated from Equation 34.20:

$$G_n = \frac{i_{\text{noise}}^2 - 4kT/R}{4eI_d}. \tag{34.22}$$

As a good approximation, when the electron-capture probability into a QD is small [41], the photoconductive gain and noise gain are equal in a conventional photoconductor. The photoconductive gain can thus be calculated. The calculated photoconductive (PC) gain G for different bias voltages is shown in Figure 34.8. Note that the PC gain is larger at higher temperature, indicating a lower capture probability at high temperature.

The photoresponsivity of the QDIP array was measured using the calibrated cavity blackbody source at 1000 K. The total number of incident photons on the QDIPs can be determined by calculating the blackbody emission at the detector's peak wavelength and the spectral width. The photoresponsivity can be written as

$$\Re = \frac{4\pi I_{\text{ph}}}{P_d \Delta\Omega \Delta\nu}, \tag{34.23}$$

where
  I_{ph} is the photocurrent of the detector
  P_d is the power spectral density of the cavity blackbody at 1000 K
  $\Delta\Omega$ is the solid angle of the photodetector
  $\Delta\nu$ is the spectral width of the detector

The solid angle of the photodetector can be written as

$$\Delta\Omega = \frac{\pi D^2}{R_0^2}, \tag{34.24}$$

where
  D is the diameter of the photodetector
  R_0 is the distance from the blackbody to the photodetector

Figure 34.22 shows an example of the photoresponsivity of an LWIR QDIP at different bias voltages. For QDIP, a high photoresponsivity of >7.9 A/W photoresponsivity can be obtained due to the high PC gain of QDIP. Figure 34.22 also clearly shows a temperature dependent photoresponsivity [38].

4. Photodetectivity (D^*) calculation

The photodetectivity D^* can be calculated using [36,41]

$$D^* = \frac{\Re\sqrt{A}}{\sqrt{i_{\text{noise}}^2 + i_{B,\text{noise}}^2}} = \frac{(eG_{\text{ph}}\eta/h\nu)\sqrt{A}}{\sqrt{4G_{\text{noise}}I_d + 4G_{\text{ph}}^2\Phi_B}}, \tag{34.25}$$

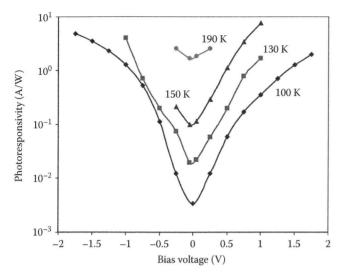

FIGURE 34.22 Photoresponsivity of the QDIP at different bias voltages. (From Lu, X. et al., *Appl. Phys. Lett.*, 91, 051115, 2007. With permission.)

where

A is the detector area

$i_{B,noise}$ is the noise current due to the background radiation

η is the absorption quantum efficiency

Φ_B is the background photon absorbed by the QDIP

h is the Planck's constant

ν is the frequency of the incident optical beam

Figure 34.23 shows the bias-dependent photodetectivity D^* of a QDIP at different temperatures. At 77 K, A high photodetectivity D^* of 4.8×10^9 cm $Hz^{1/2}$/W can be obtained at the bias voltage of -1.5 V. At 190 K, a photodetectivity D^* of is $\sim 1.3 \times 10^8$ cm $Hz^{1/2}$/W at the bias voltage of -0.25 V.

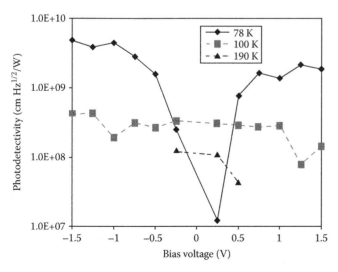

FIGURE 34.23 Photodetectivity (D^*) of a QDIP as a function of bias voltages for different temperatures.

34.7 QD Focal Plane Array Development and Characterization

In infrared image acquisition, focal plane arrays (FPA) offer great advantages over point-to-point or line-by-line scanning based imaging systems. It allows a faster frame rate and enables staring mode image acquisition. A complete FPA consists of IR photodetector array and integrated silicon read out circuits (ROIC) via the flip-chip bonding hybridization technique. This section will present fabrication and characterization of a QD FPA. Note that due to its surface normal IR incidence detection capability, no surface grating is required in QD FPA. This would simplify the FPA fabrication.

1. 320 × 256 focal plane array fabrication and hybridization

320 × 256 focal plane array (FPA) is fabricated using the same photolithography, wet etch and lift-off metallization procedures as shown in Figure 34.24. Each pixel of the 320 × 256 FPA has the dimension of 28 × 28 μm on a 30 μm pitch. A picture of the fabricated 320 × 256 FPA and a zoom in view are shown in Figure 34.24a and b, respectively.

The FPA is subsequently hybridized with an Indigo 9705 ROIC using the standard indium evaporation, flip-chip bonding and substrate removal techniques. Figure 34.25 shows a picture of the ROIC chips.

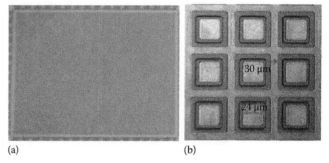

(a)   (b)

FIGURE 34.24 Fabricated 320 × 256 focal plane array: (a) the 320 × 256 FPA chip and (b) zoom-in view of the pixels.

FIGURE 34.25 Picture of ROIC chips.

The flip-chip bonding hybridization process is schematically shown in Figure 34.26.

Figure 34.27a and b shows the fully-packaged FPA in a ceramic chip carrier and a thermal image of a researcher taken using the FPA without using any filter. The FPA temperature was 67 K and the bias of the FPA was −0.7 V. The integration time was set to 16.7 ms.

Two-point non-uniformity correction at extended area blackbody temperatures of 20°C and 30°C was used to obtain the image.

Figure 34.28a and b shows images at MIR (3–5 μm) and LWIR (8–12 μm) bands obtained at different biases, indicating imaging with on-demand voltage-tunable multi-spectral detection band selections.

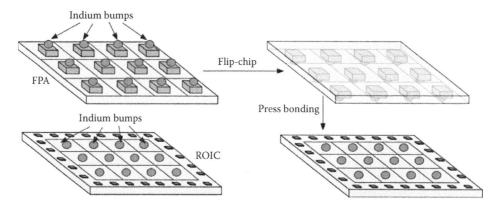

FIGURE 34.26 Schematic illustration of the flip-chip bonding hybridization process.

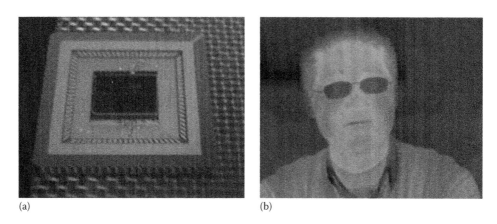

(a)   (b)

FIGURE 34.27 (a) Fully packaged FPA with ROIC and (b) a thermal image of a researcher taken using the FPA at an estimated FPA temperature of 67 K. (From Vaillancourt, J. et al., *Semicond. Sci. Technol.*, 24, 045008, 2009. With permission.)

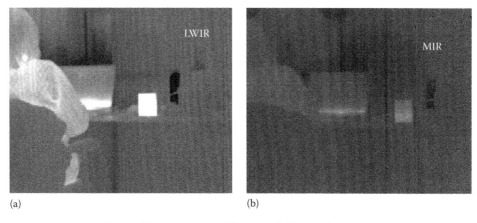

(a)   (b)

FIGURE 34.28 FPA images at different biases: (a) LWIR (8–12 μm) band, and (b) MIR (3–5 μm) band, indicating imaging with on-demand voltage-tunable multi-spectral detection band selections.

The FPA performance is measured at blackbody temperatures of 10°C 20°C, 30°C, and 50°C, respectively. The sensitivity of each pixel of the FPA can be calculated by

$$S\ (\mathrm{mV/^{\circ}C}) = \frac{mV(T_2) - mV(T_1)}{T_2 - T_1}, \qquad (34.26)$$

where

T_2 and T_1 are initial and final temperatures of the blackbody, respectively

$mV(T_2)$ and $mV(T_1)$ are the voltages of the pixel at temperature of T_2 and T_1, respectively

Figure 34.29 shows the measured sensitivity map of the QD FPA at 30°C. The average sensitivity is 8.3 mV/°C with a standard deviation σ of 8.3 mV/°C. The sensitivity map shows quite uniform areas. The histogram of the FPA sensitivity is illustrated in Figure 34.30. The narrow distribution indicates the uniformity of the FPA sensitivity.

The noise equivalent temperature difference (NEΔT) can be estimated as

$$\mathrm{NE}\Delta T = \frac{i_n}{I_p}\left(\frac{kT_B^2\lambda}{hc}\right), \qquad (34.27)$$

where

T_B is the blackbody temperature
c is the speed of light
λ is the detection wavelength
k is the Boltzmann's constant
i_n is the noise current
I_p is the photocurrent

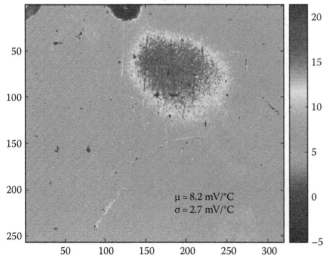

FIGURE 34.29 **(See color insert following page 22-8.)** Measured sensitivity map of the 320 × 256 FPA at 30°C. The average sensitivity is 8.3 mV/°C with a standard deviation σ of 8.3 mV/°C.

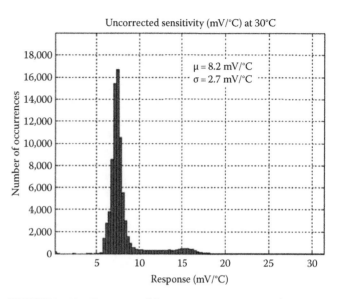

FIGURE 34.30 Histogram of the 320 × 256 FPA at 30°C. The narrow distribution indicates the uniformity of the FPA sensitivity.

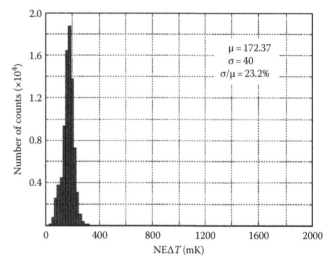

FIGURE 34.31 Histogram of the NEΔT of the FPA. An average NEΔT value of 172 mK was obtained at the blackbody temperature of 30°C with a standard deviation σ of 40 mK.

In the FPA measurement, the noise current i_n is the average of noise currents at blackbody temperatures of 20°C, 30°C, and 50°C. Figure 34.31 shows the histogram of the NEΔT of the FPA. An average NEΔT value of 172 mK can be obtained at the blackbody temperature of 30°C and the FPA bias voltage of −0.7 V. The standard deviation σ of the NEΔT is calculated to be 40 mK. Since NEΔT scales with the square of F#, a lower NEΔT can be expected for a smaller F# lens system.

The cross-talk of the FPA can be characterized by using a fixed area blackbody source. The fixed area blackbody source will be imaged on part of the FPA. By changing the temperature of the blackbody and observing the responses of the pixels near the part of the FPA that correspond to the blackbody source, one can determine the cross-talk between the pixels. Figure 34.32 shows

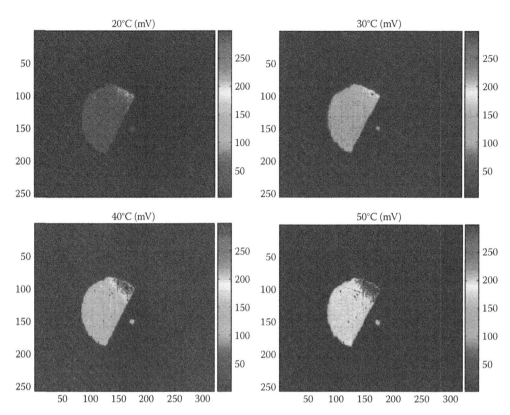

FIGURE 34.32 (**See color insert following page 22-8.**) Principle of the cross-talk measurement by changing the fixed area blackbody temperature from 20°C to 50°C.

the principle of the cross-talk measurement. As the blackbody temperature is increased from 20°C to 50°C, the responses of the pixels near the edge show no obvious change. This indicates low cross-talk between pixels.

34.8 Summary and Conclusion

Due to the three-dimensional (3D) quantum confinement of carriers, quantum dot infrared photodetectors (QDIPs) offer great advantages for infrared detection and sensing, including intrinsic sensitivity to normal incident radiation and long excited-state lifetime, which allows efficient collection of photo-excited carriers and ultimately leads to high photoconductive (PC) gain, high photoresponsivity. The normal incidence detection capability greatly simplifies the fabrication complexity for a large format (1 × 1 K) FPA. The high photoconductive (PC) gain and high photoresponsivity provide a promising way for low-level IR detection. Compared with quantum well infrared photodetectors (QWIP), QDIP also shows lower dark current due to the 3D quantum confinement. The long excited-state lifetime together with low dark current makes QDIPs promising for high-temperature operation. QDIPs with high operating temperatures of $T = 190\,\text{K}$ and $T = 300\,\text{K}$ have been reported working at both middle MWIR and LWIR. Voltage-tunable multi-spectral QDIPs have also been demonstrated using stacked QD layers with

different capping layers, i.e., InGaAs, GaAs and AlGaAs for short-wave (SWIR), middle-wave (MWIR) and long-wave (LWIR) infrared detections, respectively. These works clearly show the potentials of QDIP technology for highly sensitive, high operation temperature and multi-spectral IR detection and imaging.

References

1. A. Miyamoto and Y. Suematsu, Gain and the threshold of three-dimensional quantum-box lasers, *IEEE J. Quantum Electron.*, 22, 1915–1921 (1986).
2. P. Bhattacharya, S. Ghosh, and A. D. Stiff-Roberts, Quantum dot opto-electronic device, *Annu. Rev. Mater. Res.*, 34, 1–40 (2004).
3. W. U. Huynh, J. J. Dittmer, and A. P. Alivisatos, Hybrid nanorod-polymer solar cells, *Science*, 295, 2425–2427 (2002).
4. V. I. Klimov, A. A. Mikhailovsky, S. Xu, A. Malko, J. A. Hollingsworth, C. A. Leatherdale, H. J. Eisler, and M. G. Bawendi, Optical gain and stimulated emission in nanocrystal quantum dots, *Science*, 290, 314–317 (2000).
5. P. T. Guerreiro, S. Ten, N. F. Borrelli, J. Butty, G. E. Jabbour, and N. Peyghambarian, PbS quantum-dot doped glasses as saturable absorbers for mode locking of a Cr:forsterite laser, *Appl. Phys. Lett.*, 71, 1595–1597 (1997).

6. I. N. Stranski and V. L. Krastanow, Abhandlungen der mathematisch-naturwissenschafilichen klasse, *Akad. Wiss. Lit. Mainz Math.-Natur. KI. IIb*, 146, 797 (1939).

7. J. A. Venables, G. D. T. Spiller, and M. Hanbrücken, Nucleation and growth of thin films, *Rep. Prog. Phys.*, 47, 399–459 (1984).

8. D. Bimberg, N. Kirstaedter, N. N. Ledentsov, Zh. I. Alferov, P. S. Kop'ev, and V. M. Ustinov, InGaAs-GaAs quantum-dot lasers, *IEEE J. Sel. Top. Quantum Electron.*, 3, 196–205 (1997).

9. D. L. Huffaker, G. Park, Z. Zou, O. B. Shchekin, and D. G. Deppe, 1.3 mm room-temperature GaAs-based quantum-dot laser, *Appl. Phys. Lett.*, 73, 2564–2566 (1998).

10. D. Klotzkin, K. Kamath, and P. Bhattacharya, Quantum capture times at room temperature in high-speed InGaAs–GaAs self-organized quantum-dot lasers, *IEEE Photon. Technol. Lett.*, 9, 131–1303 (1997).

11. T. C. Newell, D. J. Bossert, A. Stintz, B. Fuchs, K. J. Malloy, and L. F. Lester, Gain and linewidth enhancement factor in InAs quantum-dot laser diodes, *IEEE Photon. Technol. Lett.*, 11, 1527–1529 (1999).

12. H. C. Liu, Quantum dot infrared photodetector, *Opto-Electron. Rev.*, 11, 1–5 (2003).

13. E.-T. Kim, A. Madhukar, Z. Ye, and J. C. Campbell, High detectivity InAs quantum-dot infrared photodetectors, *Appl. Phys. Lett.*, 84, 3277–3279 (2004).

14. P. Bhattacharya, X. H. Su, S. Chakrabarti, G. Ariyawansa, and A. G. U. Perera, Characteristics of a tunneling quantum-dot infrared photodetector operating at room temperature, *Appl. Phys. Lett.*, 86, 191106-1–191106-3 (2005).

15. D. Pan, E. Towe, and S. Kennerly, Normal-incidence inter-subband (In, Ga)As/GaAs quantum dot infrared photodetectors, *Appl. Phys. Lett.*, 73, 1937–1939 (2003).

16. S. Krishna, Quantum dots-in-a-well infrared photodetectors, *J. Phys. D: Appl. Phys.*, 38, 2142–2150 (2005).

17. L. Jiang, S. S. Li, N.-T. Yeh, J.-I. Chyi, C. E. Ross, and K. S. Jones, $In_{0.6}Ga_{0.4}As$/GaAs quantum-dot infrared photodetector with operating temperature up to 260 K, *Appl. Phys. Lett.*, 82 1986–1988 (2003).

18. J. Jiang, S. Tsao, T. O'Sullivan, W. Zhang, H. Lim, T. Sills, K. Mi, M. Razeghia, G. J. Nrown, and M. Z. Tidrow, High detectivity InGaAs/InGaP quantum-dot infrared photodetectors grown by low pressure metalorganic chemical vapor deposition, *Appl. Phys. Lett.*, 84, 2166–2168 (2004).

19. U. Bockelmann and G. Bastard, Phonon scattering and energy relaxation in two-, one-, and zero-dimensional electron gases, *Phys. Rev. B*, 42, 8947–8951 (1990).

20. S. Schmitt-Rink, D. A. B. Miller, and D. S. Chemla, Theory of the linear and nonlinear optical properties of semiconductor microcrystallites, *Phys. Rev. B*, 35, 8113–8115 (1987).

21. E. Hanamura, Very large optical nonlinearity of semiconductor microcrystallites, *Phys. Rev. B*, 37, 1273–1279 (1988).

22. G. Ariyawansa, A. G. U. Perera, X. H. Su, S. Chakrabarti, and P. Bhattacharya, Multi-color tunneling quantum dot infrared photodetectors operating at room temperature, *Infrared Sci. Technol.*, 50, 156–161 (2007).

23. C. Kammerer, S. Sauvage, G. Fishman, P. Boucaud, G. Patriarche, and A. Lemaître, Mid-infrared intersub-level absorption of vertically electronically coupled InAs quantum dots, *Appl. Phys. Lett.*, 87, 173113–173115 (2005).

24. G. Gohosh, A. S. Lenihan, M. V. G. Dut, D. G. Steel, and P. Bhattacharya, Nonlinear optical and electro-optic properties of InAs/GaAs self-orgainzed quantum dots, *J. Vac. Sci. Technol. B*, 19, 1455–1458 (2001).

25. O. Qasaimeh, K. Kamath, P. Bhattacharya, and J. Phllips, Linear and quadratic electro-optic coefficients of self-orgainzed $In_{0.4}Ga_{0.6}As$/GaAs quantum dots, *Appl. Phys. Lett.*, 72, 1275–1277 (1998).

26. General Accounting Office (GAO), Missile Defense: Review of Results and Limitations of an Early National Missile Defense Flight Test, GAO-02-124, Washington, DC, February (2002).

27. D. A. Reago, S. Horn, J. Campbell, and R. Vollmerhausen, Third generation imaging sensor system concept, *Proceeding of SPIE*, 3701, 108 (1999).

28. J. M. Mooney, V. E. Vickers, M. An, and A. K. Brodzik, High-throughput hyperspectral infrared camera, *J. Opt. Soc. Am. A*, 14, 2951 (1997).

29. B. S. Williams, H. Callebaut, S. Kumar, Q. Hu, and J. L. Reno, 3.4-THz quantum cascade laser based on longitudinal-optical-phonon scattering for depopulation, *Appl. Phys. Lett.*, 82, 1015 (2003).

30. J. Faist, F. Capasso, D. L. Sivco, C. Sirtori, A. L. Hutchinson, and A. Y. Cho, Quantum cascade laser, *Science*, 264, 553–556 (1994).

31. C.-F. Hsu, O. Jeong-Seok, P. Zory, and D. Botez, Intersubband quantum-box semiconductor lasers, *IEEE J. Sel. Top. Quantum Electron.*, 6, 491 (2000).

32. H. Benisty, C. M. Sotomayor Torres, and C. Weisbuch, Intrinsic mechanism for the poor luminescence properties of quantum-box systems, *Phys. Rev. B*, 44, 10945 (1991).

33. S. Krishna and P. Bhattacharya, Intersubband gain and stimulated emission in long-wavelength ($\lambda = 13 \mu m$) intersubband In(Ga)As-GaAs quantum-dot electroluminescent devices, *IEEE J. Sel. Top. Quantum Electron.*, 37, 1066 (2001).

34. D. A. B. Miller, *Quantum Mechanics for Scientists and Engineers*, Cambridge University Press, Cambridge, NY (2008).

35. S. L. Chuang, *Physics of Optoelectronic Devices*, John Wiley & Sons, Inc., New York (1995).

36. B. F. Levine, Quantum-well infrared photodetectors, *J. Appl. Phys.*, 74, R1–R81 (1993).

37. P. W. M. Blom, C. Smit, J. E. M. Haverkort, and J. H. Wolter, Carrier capture into a semiconductor well, *Phys. Rev. B*, 47, 2072–2081 (1993).

38. X. Lu, J. Vaillancourt, and M. J. Meisner, Temperature-dependent photoresponsivity and high-temperature (190 K) operation of a quantum-dot infrared photodetector, *Appl. Phys. Lett.*, 91, 051115 (2007).

39. D. A. B. Miller, D. S. Chemla, T. C. Damen, A. C. Gossard, W. Wiegmann, T. H. Wood, and C. A. Burrus, Band-edge electroabsorption in quantum well structures: The quantum-confined Stark effect, *Phys. Rev. Lett.*, 53, 2173–2176 (1984).

40. D. A. B. Miller, D. S. Chemla, T. C. Damen, A. C. Gossard, W. Wiegman, T. H. Wood, and C. A. Burrus, Electric field dependence of optical absorption near the band gap of quantum-well structures, *Phys. Rev. B*, 32, 1043–1060 (1985).

41. H. C. Liu, Noise gain and operating temperature of a quantum well infrared photodetectors, *Appl. Phys. Lett.*, 61(22), 2703–2705 (1992).

42. X. Lu, J. Vaillancourt, and M. J. Meisner, A modulation-doped longwave infrared quantum dot photodetector with high photoresponsivity, *Semicond. Sci. Technol.*, 22, 993–996 (2008).

43. M.J. Meisner, J. Villancourt, and X. Lu, Voltage-tunable dula-band InAs quantum-dot infrared photodetectors based on InAs quantum dots with different capping layers, *Semicond. Sci. Technol.*, 23(9), 095016 (2008).

44. J. Vaillancourt, P. Vasinajindakaw, X. Lu, A. Stintz, J. Bundas, R. Cook, D. Burrows, K. Patnaude, R. Dennis, A. Reisinger, and M. Sundaram, A voltage-tunable multispectral 320′ 256 InAs/GaAs quantum-dot infrared focal plane array, *Semicond. Sci. Technol.*, 24(4), 045008 (2009).

VI

Nano-Oscillators

35

Nanomechanical Resonators

Josef-Stefan Wenzler
Boston University

Matthias Imboden
Boston University

Tyler Dunn
Boston University

Diego Guerra
Boston University

Pritiraj Mohanty
Boston University

35.1 Introduction to Nanomechanical Systems

This chapter is intended to be an introduction to micro/nanoelectromechanical systems (MEMS/NEMS). Since both micro scale and nano scale structures have the same classical description, the formalisms and the analyses in this chapter can be applied to a range of devices. Therefore, without any loss of generality, we refer to the entire class of systems as nanomechanical systems. Here, the focus is on the design and the operation of nanomechanical resonators in the context of specific applications.

Nanoelectromechanical devices integrate mechanical elements, such as resonators, actuators, and sensors, with electronic circuits. These elements consist of structures that come in numerous shapes and forms, ranging from single doubly clamped beams, disks, rings, gears, and squares to more elaborate geometries. Nanomechanical structures are typically characterized by their dimensions, resonance frequencies, and quality factors Q, a measure of energy loss in the system. Typical dimensions of these devices range from several hundreds of microns down to a few nanometers, spanning typical MEMS (microelectromechanical systems) to more recent NEMS (nanoelectromechanical systems). Resonance frequencies range from hundreds of kilohertz to gigahertz. Quality factors of up to hundreds of thousands have been achieved at megahertz frequencies (Nguyen, 2005).

Nanomechanical systems can be used for many applications in single or multiple function configurations. Micromechanical accelerometers and gyroscopes are currently being employed in consumer applications to detect acceleration and angular rotation, respectively. Integrated micromechanical microphones are also finding use in consumer electronics because of the increasing requirement of small size. In communication systems, micro and nanomechanical systems can be used as front-end receiver components such as switches, filters, mixers, and timing oscillators. Micromechanical systems such as IC-compatible capacitors, inductors, switches, filters, reference oscillators, and mixers have been demonstrated by Nguyen et al. at MHz–GHz frequencies (Nguyen, 2005).

Today, micro and nanomechanical systems find application in aerospace, automotive, biological, instrumentation, robotics, manufacturing, and other industries. Micromechanical accelerometers and gyroscopes can be found in airplanes for stability control and in cars as airbag release triggers. Their use in other consumer products includes Wii controls, ink-jet printers, cameras, and pressure transducers. The micromechanical and nanomechanical industry is estimated to be at $8 billion by 2009. Analog Devices, Phillips, Akustica, Discera, SiTime, Sand 9, Epcos, Fujitsu, Agilent, and Infineon are some of the companies that manufacture these products today.

In addition to their useful applications, nanomechanical systems also present great opportunities for research in the basic sciences. Recently, Shim, Imboden, and Mohanty demonstrated synchronization of two coupled nanomechanical oscillators with implications for neuro-computing and microwave communication (see Figure 35.1) (Shim et al., 2007). In 2005, Badzey and Mohanty reported the observation of stochastic resonance in bistable nanomechanical silicon oscillators,

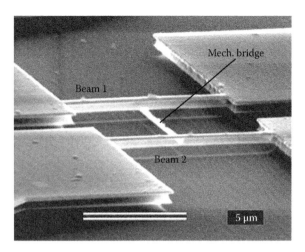

FIGURE 35.1 Two doubly clamped beams coupled by a mechanical bridge, forming an H-structure, used to study synchronization at ultralow temperatures. (Adapted from Shim, S. et al., *Science*, 316(5821), 95, 2007.)

FIGURE 35.2 Nanomechanical torsional resonator used for the detection of the torque produced by electron spin flip. (Adapted from Zolfagharkhani, G. et al., *Nat. Nanotechnol.*, 3, 720, 2008.)

devices that could play a role in developing a controllable high-speed nanomechanical memory cell (Badzey et al., 2004; Badzey and Mohanty, 2005; Guerra et al., 2008, 2009) and non-linear sensors of small forces. Several other groups (Knobel and Cleland, 2003; LaHaye et al., 2004; Gaidarzhy et al., 2005; Etaki et al., 2008) have been using high-frequency nanomechanical resonators at ultralow temperatures in search for macroscopic quantum mechanical behavior. For example, LaHaye et al. (2004) have integrated a nanomechanical resonator and a radio-frequency superconducting single electron transistor (RF-SSET) for the best displacement sensitivity of a nanomechanical resonator to date, achieving a position resolution factor of 4.3 above the quantum limit. In another experiment, Etaki et al. (2008) successfully integrated a nanomechanical resonator and a superconducting quantum interference device (SQUID), achieving a position resolution that is 36 times the quantum limit. Furthermore, current state-of-the-art atomic force microscopes utilize nanomechanical resonators as their detection tips, leading the way in microscopy at the atomic level (Giessibl, 2003). Zolfagharkhani et al. recently measured spin currents in a nanomechanical torsional resonator (see Figure 35.2) with possible applications in spintronics and precision measurements of charge and parity-violating forces. The ultrahigh torque sensitivity may also enable experiments measuring the untwisting of DNA and torque-generating molecules (Mohanty et al., 2004; Zolfagharkhani et al., 2008). In addition to the detection of fundamental symmetries, nanomechanical systems are being considered as sensors for the detection of quantities of fundamental importance such as gravity waves (Braginsky and Khalili, 1996).

Other research groups have used nanomechanical systems to develop ultrasensitive mass sensors as potential detectors of hazardous chemicals and biological agents (Villarroya et al., 2006; Yang et al., 2006). Moreover, Madou has discussed the impact of nanomechanical systems on biomedical engineering with

specific applications in diagnostics, responsive drug delivery, biocompatibility, and self-assembly (Madou, 2003).

An introduction of the theory of micro/nanomechanical resonators including subsections on the damped driven harmonic oscillator, elasticity theory, quality factor Q, noise sources in a typical measurement setup, and a subsection on nanomechanical resonator design considerations. Followed by Section 35.2 discussing and comparing three different measurement techniques, namely, the magnetomotive, electrostatic, and piezoelectric techniques. Finally, Section 35.3 discusses several useful applications of micro and nanomechanical devices today.

35.1.1 Damped Driven Harmonic Oscillator Model

The equation of motion of a nanomechanical system, such as a beam, reflects the dynamics of all points along the extended object. This is described by the time-dependent elasticity theory, a complete theory which is discussed in all detail elsewhere (Cleland, 2003). Here, however, we introduce a simplified yet very powerful and explanatory model.

The basic assumption of this model is that each point along a micro/nanomechanical resonator can be described as a damped, driven harmonic oscillator with an effective mass m_{eff} and effective spring constant $k_{eff} = \omega_0^2 m_{eff}$, where ω_0 is the resonant angular frequency of the resonator (see Figure 35.3). Its displacement $y(t)$ at a certain point x along the resonator at a certain time t, can be described by the differential equation,

$$m_{eff}\frac{d^2 y(t)}{dt^2} + \gamma m_{eff}\frac{dy(t)}{dt} + k_{eff}y(t) = F(t) = F_o e^{i\omega t}, \quad (35.1)$$

where the damping coefficient γ represents the loss of energy due to the coupling to internal and external degrees of freedom. The effective mass m_{eff} will depend on the resonator geometry

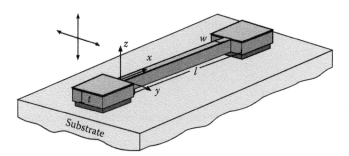

FIGURE 35.3 Schematic diagram of the classical doubly clamped suspended beam. When a periodic force F is applied at the resonance frequency of the beam, it causes the resonator to oscillate.

and resonance mode shape. As for all inhomogeneous differential equations, the solution is given by the general solution to the homogeneous equation added to a particular solution of the inhomogeneous equation. Consider the ansatz:

$$y(t) = y_o e^{i(\omega t + \phi)} \tag{35.2}$$

$$\frac{dy(t)}{dt} = i\omega y_o e^{i(\omega t + \phi)} \tag{35.3}$$

$$\frac{d^2 y(t)}{dt^2} = -\omega^2 y_o e^{i(\omega t + \phi)}. \tag{35.4}$$

Solving for the real and imaginary parts, one obtains two equations for the two unknowns y_o and ϕ:

$$\frac{F_o}{m_{\text{eff}}} \cos(\phi) = y_o \left(\omega_o^2 - \omega^2 \right) \tag{35.5}$$

$$\frac{F_o}{m_{\text{eff}}} \sin(\phi) = y_o \omega \gamma. \tag{35.6}$$

One can thus solve for ϕ,

$$\tan(\phi) = \frac{\sin(\phi)}{\cos(\phi)} = \frac{\omega \gamma}{\omega_o^2 - \omega^2} \rightarrow \phi = \arctan\left(\frac{\omega \gamma}{\omega_o^2 - \omega^2} \right). \tag{35.7}$$

The amplitude y_o can be determined using $\sin^2(\phi) + \cos^2(\phi) = 1$, so that

$$\frac{F_o^2}{m_{\text{eff}}^2} \left(\sin^2(\phi) + \cos^2(\phi) \right) = y_o^2 \left[\left(\omega_o^2 - \omega^2 \right)^2 + \omega^2 \gamma^2 \right]. \tag{35.8}$$

Thus, y_o is described by the typical Lorentzian line shape

$$y_o = \frac{F_o}{m_{\text{eff}}} \frac{1}{\sqrt{\left(\omega_o^2 - \omega^2 \right)^2 + \omega^2 \gamma^2}}. \tag{35.9}$$

The maximum amplitude occurs on resonance when $\omega \simeq \omega_o$:

$$y_o \simeq \frac{F_o}{m_{\text{eff}} \gamma \omega_o}. \tag{35.10}$$

Finally, this expression can be rewritten by introducing a parameter that quantifies damping, a measure of energy loss in the system. This parameter is called the quality factor Q and is defined as the ratio of total energy of the resonator W to the amount of energy lost to its environment during one oscillation ΔW. (We will discuss dissipation and quality factor in detail in Section 35.1.3.)

$$Q \equiv 2\pi \frac{W}{\Delta W} \simeq \frac{\omega_o}{\gamma}. \tag{35.11}$$

Thus, the final expression for the maximum value of y_o is given by

$$y_o \simeq \frac{F_o Q}{k_{\text{eff}}}. \tag{35.12}$$

This displacement is then detected employing either the magnetomotive, electrostatic, or piezoelectric technique discussed in Sections 35.2.1 through 35.2.3, respectively. Though the above-mentioned detection techniques are the most commonly used, there are other ways of detecting the motion of nanomechanical resonators, in particular piezoresistive and optical detection techniques that are discussed elsewhere (Ekinci, 2005).

35.1.2 Elasticity Theory of Continuum Mechanics

In Section 35.1.1, we discuss a model that essentially reduced the nanomechanical resonator to a point particle behaving like a damped driven harmonic oscillator. This description, however useful, is incomplete, as it does not explain the deformation of the oscillator as a whole.

This section will introduce a more complete theory of nanomechanical resonators that treats the structure as a single extended elastic object. This is commonly referred to as the standard elasticity theory of continuum mechanics for classical nanomechanical systems. This description is valid as long as the basic assumptions of the continuum approximation hold. In particular, the fact that the structure is made of atoms is ignored. This assumption holds when all the dimensions are much larger than atomic length scales. Here, we consider a beam resonator for simplicity, even though this theory can be applied to many different mechanical shapes. A more detailed analysis of this theory can be found in many classic textbooks (Thomson, 1953; Landau and Lifshitz, 1959; Timoshenko and Goodier, 1971; Cleland, 2003).

Consider a beam structure with length l, thickness t, and width w (see Figure 35.3). If the displacement along the length of the beam in the y direction is designated by $y(x, t)$, and the displacement in the z direction as $z(x, t)$, then the equations of motion are given by the famous Euler Bernoulli equations

$$\frac{\partial^4 y}{\partial x^4} = -\frac{\rho A}{EI_y}\frac{\partial^2 z}{\partial t^2} \tag{35.13}$$

$$\frac{\partial^4 z}{\partial x^4} = -\frac{\rho A}{EI_z}\frac{\partial^2 z}{\partial t^2}, \tag{35.14}$$

where

$A = wt$ is the cross section

ρ is the material mass density

E is the Young's modulus

I_y and I_z are the moments of inertia for dynamics in the y- and z-direction of the resonator

For example, in the z-direction

$$I_z = \int_{-t/2}^{t/2} z^2 w(z)\mathrm{d}z = \frac{1}{12}At^2. \tag{35.15}$$

Similarly, one finds that $I_y = (1/12)Aw^2$. To solve these equations one has to consider several boundary conditions. The boundary conditions for the out-of-plane motion of multiple beam types are listed below:

- *Doubly clamped beam*: For a doubly clamped beam, there is no motion at both ends of the resonator. Hence, the displacement u and its first derivative $\mathrm{d}u/\mathrm{d}x$ must be zero:

$$z = 0 \quad \text{for } x = 0, l \tag{35.16}$$

$$\frac{\partial z}{\partial x} = 0 \quad \text{for } x = 0, l. \tag{35.17}$$

Equivalent conditions are true for in-plane motion.

- *Singly clamped beam*: For a beam free to move on one end but clamped on the other, the boundary conditions are slightly different. At the fixed end, both the displacement and its first derivative must again be zero; however, on the free end, we have no moment and no strain. Thus, the boundary conditions become

$$z = 0, \quad \frac{\partial z}{\partial x} = 0 \quad \text{for } x = 0 \tag{35.18}$$

$$\frac{\partial^2 z}{\partial x^2} = 0, \quad \frac{\partial^3 z}{\partial x^3} = 0 \quad \text{for } x = l. \tag{35.19}$$

Again, for in-plane motion replace z and y.

- *Free beam*: Finally, for the beam free to move on both ends, the boundary conditions for the out-of-plane motion are given by no moment or strain on either end:

$$\frac{\partial^2 z}{\partial x^2} = 0 \quad \text{for } x = 0, l, \tag{35.20}$$

$$\frac{\partial^3 z}{\partial x^3} = 0 \quad \text{for } x = 0, l, \tag{35.21}$$

with equivalent boundary conditions for in-plane motion.

The equation of motion can be solved by using separation of variables, $z(x, t) = z'(t)z''(x)$, obtaining solutions of the following form:

$$z(x, t) = \underbrace{e^{-i\omega t}}_{z'(t)}\underbrace{\left[a_1\cos(\beta x) + a_2\sin(\beta x) + a_3\cosh(\beta x) + a_4\sinh(\beta x)\right]}_{z''(x)}. \tag{35.22}$$

The boundary conditions discretize the solutions for the allowed frequencies $\omega \to \omega_n$ and the allowed values for $\beta \to \beta_n$. Moreover, the coefficients a_i are also determined from the boundary conditions. To incorporate dissipation that is discussed in detail in Section 35.1.3, one replaces the time dependent part of the solution according to $e^{-i\omega_n t} \to e^{-i\omega_n t - \gamma_n t}$, where $\gamma_n = \omega_n/2Q$ with Q being the quality factor of the resonator. The general solution can be written as

$$z(x, t) = e^{-i\omega_n t - \gamma_n t}\big[a_1\cos(\beta_n x) + a_2\sin(\beta_n x) + a_3\cosh(\beta_n x)$$
$$+ a_4\sinh(\beta_n x)\big]. \tag{35.23}$$

Note, that the spatial solution $z''(x)$ describes the shape of the resonator along the x-direction as a whole, rather than treating it as a point particle as discussed in Section 35.1.1. The specific shapes of the first four modes for the doubly and singly clamped beam are discussed and depicted in Chapter 37 by A. Cleland of this book.

Expressions for shear modes and other geometries are determined in similar fashion, although they are more mathematically involved. These modes are calculated elsewhere (Thomson, 1953; Landau and Lifshitz, 1959; Timoshenko and Goodier, 1971; Cleland, 2003; Merono, 2007). A collection of expressions for the resonant mode frequencies for a variety of resonators are given in Table 35.1. The first column of Table 35.1 shows a sketch of the resonator geometry. Expressions for up to three fundamental resonance modes of each resonator are listed in the second column. Finally, constants for expressions in column two or geometry constraints are given in column three. More details on each geometry and its analytical expression for resonance modes can be found by consulting the corresponding reference given in the first column.

35.1.3 Damping and Dissipation

Dissipation, the loss of energy produced by nonconservative forces, is one of the most important parameters of a resonator. It partially determines its mechanical resistance R_M (more on that later), its usefulness for RF applications and the signal size.

TABLE 35.1 Resonance Frequencies of Nanomechanical Resonators with Different Shapes

Geometry	Res. Frequency	Constants
Clamped–free beam (Merono, 2007)	Flexural $f_o = \dfrac{1}{2\pi} \dfrac{C_n^2}{2\sqrt{3}} \dfrac{t}{l^2} \sqrt{\dfrac{E}{\rho}}$	$C_1 = 1.875\ C_2 = 4.694\ C_3 = 7.854$
Clamped–clamped beam (Lin et al., 2004)	Flexural $f_o = C_n \dfrac{t}{l^2} \sqrt{\dfrac{E}{\rho}}$	$C_1 = 1.03\ C_2 = 2.83\ C_3 = 5.55$
Free–free beam (Wang et al., 2000)	Flexural $f_o = 1.03 \dfrac{t}{l^2} \sqrt{\dfrac{E}{\rho}}$	$l_s = \dfrac{1}{4 f_o} \sqrt{\dfrac{G\gamma}{\rho J_s}}$ $J_s = h w_s \dfrac{h^2 + w_s^2}{12}$ $\gamma = 0.229 h w_s^3$
Paddle resonator (Merono, 2007)	Flex./Tors./Bulk $f_{\text{Flex.}} = \dfrac{1}{2\pi} \sqrt{\dfrac{2 E t^3 w^a}{m_{\text{eff}} l^3}}$ $f_{\text{Tors.}} = \dfrac{1}{2\pi} \sqrt{\dfrac{24 \beta^6 w^3 G}{\rho w_1^3 w_2 l}}$ $f_{\text{Bulk}} = \dfrac{1}{4l} \sqrt{\dfrac{E}{\rho(1 - v^2)}}$	
Disk radial contour mode (Wang et al., 2004)	Bulk $f_o = \dfrac{C_n}{r} \sqrt{\dfrac{E}{\rho}}$	$C_1 = 0.342\ C_2 = 0.903\ C_3 = 1.440$ for silicon
Wine-glass resonator (Lin et al., 2004)	Bulk $f_o = \dfrac{\xi}{2\pi r} \sqrt{\dfrac{E}{2\rho(1 + v)}}$	ξ Fitting parameter

(continued)

TABLE 35.1 (continued) Resonance Frequencies of Nanomechanical Resonators with Different Shapes

Geometry	Res. Frequency	Constants
Square stem resonator (Demirci and Nguyen, 2006) 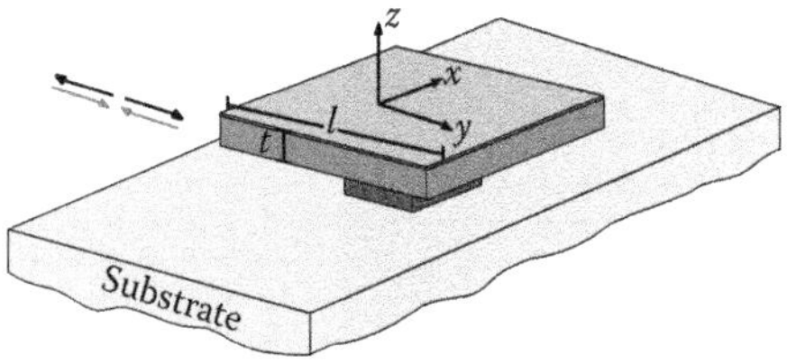	Bulk $f_o = 0.9697 \dfrac{t}{l^2} \sqrt{\dfrac{E}{\rho}}$	
Ext. square corner resonator (Kaajakari et al., 2004) 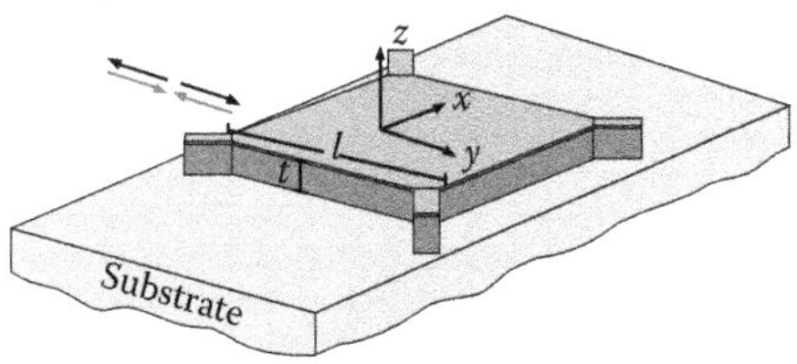	Bulk $f_o = \dfrac{l}{2l} \sqrt{\dfrac{E_{2D}}{\rho}}$	$E_{2D} = \left(C_{11} + C_{12} - \dfrac{2C_{12}^1}{C_{11}} \right)$ C_{ij} are mechanical stiffness coefficients

Note: The arrows to the left of each mode designate oscillation directions.
ᵃ w is the width of the central beam.
ᵇ β is a function of w/t.

Here, the most common causes of dissipation in nanomechanical resonators are discussed and quantified by the quality factor Q, a figure of merit of oscillators. Q is a measure to quantify dissipation. It is given by

$$Q = \frac{\omega_o}{\Delta\omega_o} \simeq \frac{\omega_o}{\gamma}, \tag{35.24}$$

where

ω_o is the resonance frequency
γ is the damping coefficient of Equation 35.1
$\Delta\omega_o$ is the full width at half maximum of the Lorentzian fit of the resonance response

Thus, Q is inversely proportional to dissipation. There are many sources of dissipation. Assuming that all sources are independent one can add all the contributions from various sources as follows. The total quality factor Q_{tot} is then given by

$$\frac{1}{Q_{tot}} = \frac{1}{Q_{Air}} + \frac{1}{Q_{Anchor}} + \frac{1}{Q_{Surface}} + \frac{1}{Q_{TED}} + \frac{1}{Q_{Others}}. \tag{35.25}$$

Each term in Equation 35.25 is introduced in the paragraphs below, though a more detailed description of these dissipation mechanisms is reported elsewhere (Mohanty et al., 2002).

35.1.3.1 Q_{Air} Contribution

This term accounts for damping of the resonator due to the ambient gas pressure or liquid in which the resonator is immersed. This contribution can only be neglected if the device is studied in vacuum (<mTorr). An expression for the loss of energy due to air or fluid damping was derived and given by Hosaka et al. (1995)

$$\frac{1}{Q_{Air}} = \frac{\rho h^2 b \omega_n}{\beta}, \tag{35.26}$$

where

ρ is the mass density
h is the thickness
b is the width
ω_n is the resonance frequency of mode n of the resonator

β is a parameter that depends on the fluid characteristics and for a beam it is given by

$$\beta = 3\pi\mu h + \frac{3}{4}\pi h^2 \sqrt{(2\rho\mu\omega_n)}, \tag{35.27}$$

where μ is the fluid viscosity.

35.1.3.2 Q_{Anchor} Contribution

This term accounts for clamping losses near the anchors of the resonator. It is typically non-negligible and can be improved by placing anchors at the nodal positions of a given resonant mode. In general, the smaller the resonator becomes, the more the clamping losses will contribute to the total dissipation of the system. Approximate expressions for the dissipation associated with clamping losses in clamped-free beams were derived by Jimbo et al. (1968) and Hosaka et al. (1995), and later expressions for the doubly clamped beam were derived by (Cross and Lifshitz, 2001)

$$Q_{CL\ \text{in plane}}^{-1} \approx \alpha \frac{w^3}{L^3} \tag{35.28}$$

$$Q^{-1}_{CL \text{ out of plane}} \approx \beta \frac{w}{L}. \tag{35.29}$$

Note, that dissipation for in-plane motion is much smaller than dissipation for out-of-plane motion as long as $w \ll l$.

35.1.3.3 Q_{Surface} Contribution

The third term in Equation 35.25 accounts for losses due to surface imperfections. The high surface to volume ratio in nanomechanical devices is responsible for the significant contribution to the total dissipation due to surface defects such as impurities or defect crystal terminations. Surface defects are unavoidable, but they can also be enhanced during fabrication steps, such as reactive ion etching, or over time due to absorption of water, oil, hydrocarbons, and even metallic particles. It has been shown that surface losses can be reduced by thermally annealing the resonator (Yang et al., 2002).

35.1.3.4 Q_{TED} Contribution

This term explains dissipation due to thermal expansion and contraction of different areas across the resonator. This effect is generally only important for bigger structures. However, as structure dimensions decrease this contribution to the total dissipation declines as well and can typically be neglected for nanomechanical devices.

35.1.3.5 Q_{Others} Contribution

The last term in Equation 35.25 incorporates all other mechanisms that in general can be neglected, such as intrinsic dissipation due to quantum friction that is only important at sub-Kelvin temperatures (Mohanty et al., 2002), or losses due to air squeezing that is only relevant when the resonator is not in vacuum and close to a rigid wall (which is usually only the case for the electrostatic technique.) There are other contributions that are less important to most nanomechanical devices, which are not mentioned here but can be found elsewhere (Mohanty et al., 2002; Cleland, 2003).

35.1.4 Noise Analysis

Noise analysis is essential for any sensitive measurement. In particular, devices consisting of single or multiple nanomechanical resonators are prone to both extrinsic and intrinsic noise sources. The objective in a typical noise analysis is to compare all relevant noise sources to a given signal size and determine if the noise is bigger or smaller than the signal. Naturally, the signal cannot be measured for the former case. Therefore, either the noise must be reduced or the signal has to be enhanced in order to conduct such a measurement.

This section introduces common noise sources in nanomechanical devices. There are several sources of noise in almost any nano or micromechanical system, some of which are mechanical and others electrical in nature. For a given setup, care needs to be taken that the signal size exceeds the biggest noise source by preferably an order of magnitude. The most common sources for noise arise from the interaction of the resonator with the thermal bath (Johnson noise), from the discrete nature of the charge carriers moving through the device (shot noise), from pre-amplifiers commonly used to enhance small signals originating from the nanomechanical system (preamp noise), and from the collisions of the resonator with ambient gas molecules (gas noise). Each of these noise sources are introduced in the paragraphs below.

35.1.4.1 Mechanical Johnson Noise

Here, we denote the thermal Johnson noise by either N^{th}_x or N^{th}_V, the root mean square values of the displacement x and voltage V, respectively. It arises from the thermal bath actuating each resonance of the resonator with an energy $E = k_B T$. The expression for this noise source is given by (Mohanty et al., 2002)

$$N^{\text{th}}_x = \sqrt{\frac{4k_B T Q}{m_{\text{eff}} \omega_o^3}} \Delta f, \tag{35.30}$$

where

T is the temperature
k_B is the Boltzmann constant
Q is the quality factor
m_{eff} is the effective mass
ω_o is the resonance frequency of the resonator
Δf is the bandwidth of the measurement setup

For example, consider a $100 \times 100 \times 10\,\mu\text{m}^3$ silicon resonator, with an effective mass 3×10^{-10} kg. If this resonator has a resonance frequency $\omega_o = 2\pi \cdot 10^6$ with a quality factor $Q = 10^4$, then one can expect thermal motion on the order of $N^{\text{th}}_x \sim 1 \times 10^{-15}$ m at $T = 0.3$ K with a 1 Hz bandwidth. If the measurement is made at room temperature (300 K), the thermal motion is on the order of $N^{\text{th}}_x \approx 30$ fm. Since in this field voltage is typically measured, the amplitude noise must then be converted into a voltage signal which varies depending on the detection technique. Mechanical Johnson noise becomes stronger for light or low-frequency resonators and it is less relevant for high-frequency or heavy resonators.

35.1.4.2 Electrical Johnson Noise

This noise source arises from voltage fluctuations across any electrical resistor R in the measurement circuit due to the coupling to the thermal bath. It mainly depends on ambient temperature and the overall resistance of the circuit. The exact expression for electrical Johnson N^{th}_V noise is given by

$$N^{\text{th}}_V = \sqrt{4k_B T R \Delta f}, \tag{35.31}$$

where

T is the temperature
k_B is the Boltzmann constant
Δf is the bandwidth of the measurement setup

Typical values for electrical Johnson noise in the measurement circuit ($R = 50\,\Omega$) at $T = 300$ K and $T = 0.3$ K are $N^{\text{th}}_V \sim 10^{-11}$ V and $N_V^{th} \sim 10^{-12}$ V, respectively, for a bandwidth of 1 Hz. As many high-frequency setups operate at $R = 50\ \Omega$, this noise source is often neglectable, especially at low temperatures; however, for high impedance measurements at room temperature, electrical Johnson noise can become a dominant noise source.

35.1.4.3 Shot Noise

Any electrical circuit with current I and resistance R exhibits shot noise N_V^{sh} due to the discrete nature of the electric charge according to

$$N_V^{sh} = \sqrt{2eIR\Delta f}, \tag{35.32}$$

where e is the electric charge. Note that shot noise increases with increasing currents in the system. Since nanomechanical devices are typically operated with small currents, shot noise is not usually a dominant noise source. For a typical electrical circuit with $R = 50\ \Omega$ and a current of $I = 1\ \mu A$, this noise source is on the order of $N_V^{sh} = 10^{-12}$ V. Comparing the electrical Johnson noise to the shot noise, one finds that as long as $2k_BT > eI$ the electrical Johnson noise dominates. Of course, if the opposite is the case shot noise will dominate.

35.1.4.4 Gas Molecule Collision Noise

The noise introduced by collisions between the resonator with attack area A, quality factor Q, and effective spring constant k_{eff}, and the ambient gas molecules of mass m_G and velocity $v = \sqrt{3k_BT/m_G}$ at pressure P is given by the following expression:

$$N_V^G = \sqrt{\frac{m_G vPAQ^2}{k_{eff}^2}\Delta f}. \tag{35.33}$$

This expression depends on the ambient gas molecules, the ambient pressure, and the size and quality factor of the resonator. However, in most cases this contribution can be neglected, especially if the resonator is operated in a vacuum. Typical values for a $100 \times 100 \times 10\,\mu m$ resonator in nitrogen gas range from 10^{-11} V $> N_V^G > 10^{-16}$ V depending on the ambient pressure (Mohanty, 2002).

35.1.4.5 Pre-Amplifier and Resonance Noise

Many detection setups involve a preamplifier (preamp) to enhance signal size. Often, the pre-amplifier introduces the biggest noise source. Therefore, it is crucial that the signal size is larger than the preamplifier input noise. A relatively quiet pre-amplifier still introduces noise on the order of $1\,nV/\sqrt{Hz}$. Table 35.2 summarizes the noise sources and their expressions, as mentioned above.

35.1.5 Resonator Design

In this section, we discuss resonator design concepts with specific examples. However, we will not discuss fabrication and processing techniques, since there are excellent textbooks and handbooks on the topic (Senturia, 2001).

TABLE 35.2 Summary of Expressions for Typical Noise Sources in MEMS and NEMS Devices

Noise	Expression
Electric. therm. noise	$N_V^{th} = \sqrt{4k_BTR\Delta f}$
Shot noise	$N_V^{sh} = \sqrt{2eIR\Delta f}$
Mech. therm. noise	$N_x^{th} = \sqrt{\dfrac{4k_BTQ}{m_{eff}\omega_o^3}\Delta f}$
Gas noise	$N_V^G = \sqrt{\dfrac{m_G vPAQ^2}{k_{eff}^2}\Delta f}$
Preamp noise	$\sim 1\dfrac{nV}{\sqrt{Hz}}$

For any given application, the design of the nanomechanical device is critical. The parameters such as resonant frequency, Q factor, resonance amplitude, and actuation and detection methods determine the way a nanomechanical device needs to be designed. For example, among the most important parameters for any resonator are its mode frequencies that can be determined by solving the Euler–Bernoulli equations of the structure. Hence, the desired mode frequencies often influence the dimensions of the resonator. Sometimes a particular application dictates the design of the nanomechanical device, as is the case, for instance, for a nanomechanical gyroscope (see Section 35.2.2). In other instances, the desired actuation and detection methods restrict the possible designs of the resonator. No matter which parameters determine the design, it is always helpful to consult an analytical model for the given structure.

Another useful tool commonly used to design a nanomechanical device is called the finite element method (FEM). FEM software can prove to be very useful, particularly, when an analytical model is not available. Frequently used FEM simulation programs are ANSYS, Coventor, and Comsol. Finite element simulation analysis numerically computes the coupled equations of motion of a mesh of points strategically distributed across the entire structure of the resonator. Using the given boundary conditions, the program calculates the fundamental modes and the corresponding resonance frequency, mode shape, effective mass, and effective spring constant.

The choice of materials is also critical to the design of a resonator. Since resonance frequency is typically proportional to the sound velocity $\sqrt{E/\rho}$, a material-dependent constant (see Table 35.3), the choice of material will depend greatly on the desired resonance frequency. For example, diamond has the highest velocity of sound of all the materials listed in Table 35.3. It results in a resonance frequency of a given structure that is four times higher than for the same structure made out of GaAs. Moreover, the material is important with respect to the actuation and detection method. For example, AlN is a piezoelectric material and therefore can be actuated and detected piezoelectrically, while diamond cannot be used for piezoelectric actuation. Finally, it is also important to decide which metals one

TABLE 35.3 Collection of Material Constants for Commonly Used MEMS/NEMS Materials

Material	E (GPa)	ρ (kg/m³)	$\sqrt{\dfrac{E}{\rho}}$ (m/s)	ϵ	ν	G (GPa)
Diamond	900	3,500	16,040	5.5–10	0.2	478
AlN	320	3,300	10,000	8.9	0.24	131
SiN	250	3,440	8,520	8.4	0.27	120
Si	120	2,330	7,180	11.7	0.22	79
GaAs	85	5,320	4,000	13.1	0.31	—
SiO₂	75	2,200	5,840	3.7	0.17	36
Al₂O₃	390	4,000	9,870	9	0.21	—
Cr	140	7,190	4,410	—	0.21	115
Au	78	19,280	2,011	—	0.44	27
Ti	110	4,500	4,944	—	0.32	41.4
Al	70	2,700	5,092	—	0.35	—
Ag	83	10,490	2,813	—	0.37	—

Note: E is the Young's modulus, ρ is the mass density, ϵ is the dielectric constant, ν is the Poisson ratio, and G is the shear modulus of the material. Values given in this table may vary strongly depending on how the materials are prepared (e.g., crystalline vs. amorphous materials) and can differ quite substantially for thin films as compared to bulk shapes.

wants to utilize for the electrodes as some metals can seriously alter resonance frequencies due to mass loading. As an example, usage of gold will reduce the resonance frequency more than a metal like titanium as a result of the different sound velocities and densities of these two metals. On the other hand, gold is an inert metal and thus can be stored in air, while titanium corrodes over time resulting in unstable resonator performance. On this account, storage considerations also play a role when designing a nanomechanical resonator. A good example of successful sample design is given below.

Example. Antenna Structure

In 2007, Gaidarzhy et al. designed a nanomechanical doubly clamped beam type resonator for the detection of quantum signatures in a nanomechanical system. To access the quantum regime, the resonant frequency needed to be in the gigahertz range (Gaidarzhy et al., 2007). As derived in Section 35.1.2, resonant frequency is $1/l^2$, where l is the length of the beam. Thus, the logical approach would be to decrease the length. However, decreasing the length also reduces the displacement of the resonating beam, making it undetectable at gigahertz frequencies.

However, by simulating and designing a doubly clamped antenna-like structure as shown in Figure 35.4 and by switching resonator material from silicon to diamond, they obtained a resonator with collective resonant modes in the gigahertz range. This collective mode is described by a number of smaller cantilevers oscillating in phase at their resonant frequency. Since the individual resonant frequency of the cantilevers is much higher in the collective mode, the set of cantilevers force the entire structure to vibrate at the resonance frequency determined by the cantilevers.

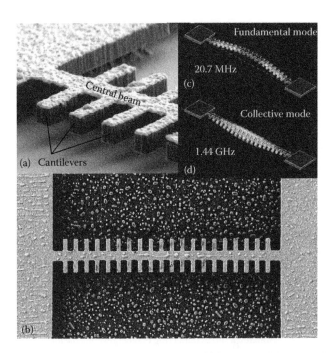

FIGURE 35.4 (a) Close-up 3D image of the suspended diamond antenna structure depicting the central beam and the cantilevers. The lighter top layer is a layer of evaporated gold, while the darker, thicker material is diamond. (b) Top view of the antenna structure consisting of 10.7 μm long central beam with a set of 20 cantilevers on each side. (c) and (d) depict the fundamental (20.67 MHz) and collective (1.44 GHz) modes of the antenna structure, respectively. (Adapted from Gaidarzhy, A. et al., *Appl. Phys. Lett.*, 91(20), 203503, 2007.)

35.2 Actuation and Detection of Nanomechanical Resonators

Many different techniques are used to actuate and detect nanomechanical resonators. The most prominent are the optical, electrostatic, piezoelectric, piezoresistive, and magnetomotive techniques. Though one of the most common detection methods for MEMS devices, optical detection techniques become problematic for scaled-down nanomechanical resonators. Further, the requirement of lasers makes these approaches unsuitable for commercialization. For these reasons, we do not discuss them here. Another technique that is not discussed in this section is the piezoresistive technique, which is difficult to use at high frequencies as a result of high impedances (Ekinci, 2005). The other three techniques (magnetomotive, electrostatic, and piezoelectric) are introduced in detail in Sections 35.2.1 through 35.2.4. In addition, each section demonstrates the equivalence of the nanomechanical resonator to an electrical RLC circuit.

35.2.1 Magnetomotive Technique

The Magnetomotive technique exploits the Lorentz force for actuation and measures an emf induced by the Faraday effect for the detection of motion in the resonating device. A simple analysis for a doubly clamped beam is performed below. However, this

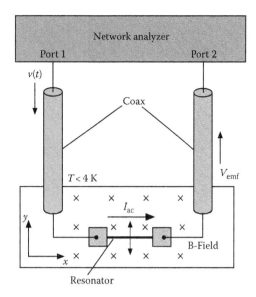

FIGURE 35.5 Schematic of the measurement setup used for the magnetomotive technique, consisting of a network analyzer to actuate and sense the resonator, a superconducting magnet responsible for the magnetic field and the resonator to be measured.

approach can be adopted for almost any other resonator. Also, it is important to note that for symmetry reasons, this technique only allows for the detection of odd resonance modes.

Consider a doubly clamped beam of length l, width w, and thickness t, as shown in Figure 35.3. While a magnetic field B is applied perpendicular to both the beam and the current, the beam is excited by an ac current $I_{ac}(t)$ due to the Lorentz force $\vec{F} = l\vec{I}_{ac}(t) \times \vec{B}$. The ensuing motion of the beam in the magnetic field induces an emf V_{emf} that is monitored by a vector network analyzer in the frequency domain (see Figure 35.5). Assuming sinusoidal oscillation, the induced emf of the resonating beam can be written as follows (Mohanty et al., 2002):

$$V_{emf}(x,t) = \frac{dy(x,t)}{dt} lB \sin[\theta(x)], \qquad (35.34)$$

where
$\theta(x)$ is the angle between the magnetic field and the electrode on the beam
$y(x, t)$ is the displacement of the beam given by Equation 35.9

An average V_{emf} can be derived by integrating Equation 35.34 over the entire length of the beam:

$$V_{avg}(t) = \xi lB \frac{dy(t)}{dt}. \qquad (35.35)$$

The constant ξ contains information about the resonator shape and resonance mode, and they must be calculated separately. For the first harmonic mode of a double clamped beam, $\xi \simeq 0.53$. In practice, the average induced emf is measured by a frequency domain vector network analyzer. Thus, $V_{avg}(t)$ needs to be

Fourier transformed into frequency space yielding the well-known Lorentzian line shape with resonance frequency $\omega_o = \sqrt{k_{eff}/m_{eff}}$:

$$V_{avg}(\omega) = \frac{i\omega\omega_o^2\xi l^2 B^2}{\omega_o^2 - \omega^2 + i\gamma\omega} \frac{I_{ac}(\omega)}{k_{eff}}. \qquad (35.36)$$

Using Equation 35.11, we can rewrite Equation 35.36 near resonance $\omega \simeq \omega_o$ in terms of the quality factor Q (Braginskii, 1985):

$$V_{avg}(\omega) = \frac{\xi l^2 B^2 \omega_o}{k_{eff}} Q I_{ac}. \qquad (35.37)$$

Thus, the detection signal using the magnetomotive technique for a doubly clamped beam is proportional to B^2 and l^2 and increases linearly with Q.

Equivalent Electrical RLC Circuit

It can be shown that any mechanical resonator, no matter how complex, can be modeled by some form of an electrical RLC circuit (though the mathematical rigor increases with the complexity of the resonator geometry). For a simple doubly clamped beam measured using the magnetomotive technique described above, the components R_M, L_M, and C_M of an equivalent electrical parallel RLC circuit can be found as follows. One can define a mechanical resistance of the beam as $V_{avg} = R_M I_{ac}$, resulting in

$$R_M = \frac{\xi l^2 B^2 \omega_o}{k_{eff}} Q. \qquad (35.38)$$

Further, it can be shown (Cleland and Roukes, 1999) that the capacitance and the inductance of the equivalent parallel RLC circuit take on the following expressions.

$$C_M = \frac{k_{eff}}{\xi l^2 B^2 \omega_o^2}, \qquad (35.39)$$

$$L_M = \frac{\xi l^2 B^2}{k_{eff}}. \qquad (35.40)$$

In general, expressions for R_M, L_M and C_M and their arrangement in an equivalent RLC circuit depend on the transduction technique. For example, for the magnetomotive technique R_M, L_M and C_M are arranged in a parallel RLC circuit, while for the electrostatic and piezoelectric method the arrangement is a series RLC circuit as shown in Figure 35.6. This analysis is quite general in the sense that mechanical motion can be described in terms of equivalent electrical circuit parameters, and the physical meaning of the parameters also extends to the description of the mechanical system. The resistance R_M, for example, is the dissipative part, naturally connected to the quality factor Q. The capacitance C_M is associated with compliance, $1/k_{eff}$ and L_M with effective mass m_{eff}, and as such C_M and L_M represent potential and kinetic energy storage in the resonator, respectively.

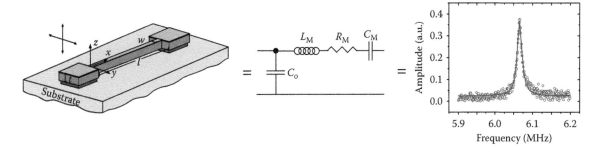

FIGURE 35.6 Schematic view of the equivalence between the mechanical resonator (left) and an equivalent electrical RLC circuit, assuming employment of the electrostatic or piezoelectric method (middle). The data on the right depicts a typical resonance response of a $20\,\mu m \times 500\,nm \times 80\,nm$ beam (circles) with the Lorentzian fit (curve).

In addition, the capacitance C_M can be important in describing the short-circuiting of the high-frequency field in relation to the true parasitic capacitance of the system.

35.2.2 Electrostatic Technique

The electrostatic technique exploits the electrostatic force between two capacitor plates and a nanomechanical resonator. With the beam held at constant voltage V relative to the excitation and detection electrodes the resonator is actuated by a radiofrequency voltage signal $\upsilon(t)$ applied to the nearby excitation electrode (see Figure 35.7), which capacitively forces the beam (force F). Subsequent motion of the beam induces a current $i(t)$ on the detection electrode (Guerra et al., 2008). This current is then amplified by a transimpedance amplifier and detected.

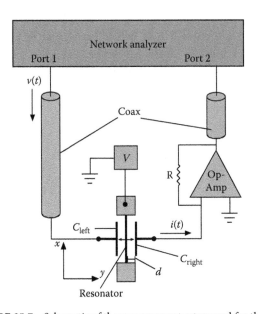

FIGURE 35.7 Schematic of the measurement setup used for the electrostatic technique consisting of a network analyzer to actuate and sense the resonator, a DC-voltage supply to DC-bias the resonator, and the resonator itself sandwiched between the left and right gate, used for actuation and detection.

Here, we derive expressions for the displacement of the beam as a function of frequency $x(\omega)$ and the output current $i(t)$ in the parallel plate approximation. The schematic of the measurement setup of the electrostatic method is depicted in Figure 35.7. In addition, we will investigate the dependence of both resonance amplitude and frequency on bias voltage V.

Consider the capacitor formed by one of the gates and the beam. The standard expression for the capacitance is $C = \epsilon_o (A/(d - x))$, where A is the area of the plates and d is their separation. The equation of motion can be written as

$$m\ddot{x} + \alpha\dot{x} + kx = F_{\text{left}} + F_{\text{right}} \approx \frac{1}{2}V_{\text{left}}^2\frac{dC_{\text{left}}}{dx} + \frac{1}{2}V_{\text{right}}^2\frac{dC_{\text{right}}}{dx}. \tag{35.41}$$

To the first order in x, Equation 35.41 can be approximated

$$m\ddot{x} + \alpha\dot{x} + kx \approx \frac{1}{2}V_{\text{left}}^2\left[C'_{\text{left}} + C''_{\text{left}} \cdot x\right] + \frac{1}{2}V_{\text{right}}^2\left[C'_{\text{right}} + C''_{\text{right}} \cdot x\right]. \tag{35.42}$$

Using the symmetry relations $C'_{\text{left}} = -C'_{\text{right}}(= \pm C')$, $C''_{\text{left}} = C''_{\text{right}} = C''$, and assuming small excitation signals $V_{\text{left}} = V$ and $V_{\text{right}} = V + \upsilon(t)$, we can rewrite Equation 35.42 to first order in υ

$$\ddot{x} + \gamma\dot{x} + \left[\frac{k}{m} - \frac{V^2C''}{m}\right]x = \frac{F}{m} = -\frac{\upsilon(t)VC'}{m}, \tag{35.43}$$

where
$\gamma = \alpha/m$
$F = \upsilon(t)VC'$

As for all inhomogeneous differential equations, the solution is found by adding a particular solution to the solution of the homogeneous differential equation. If we apply the following ansatz,

$$\upsilon(t) = \upsilon_o\cos(\omega t) \quad x(t) = x(\omega)\cos(\omega t + \phi(\omega)), \tag{35.44}$$

the solutions can be found

$$x(\omega) = \frac{C'\upsilon_o V}{m\sqrt{(\omega_o^2 - \omega^2)^2 + \omega^2\gamma^2}}; \quad \omega_o = \sqrt{\frac{k}{m} - \frac{C''V^2}{m}};$$

$$\phi(\omega) = \arctan\left(\frac{\omega\gamma}{\omega_o^2 - \omega^2}\right). \tag{35.45}$$

Note, the displacement grows linearly with V while resonance frequency decreases as $\omega_o(V) = \sqrt{\gamma - \delta V^2}$ as seen in inset (b) of Figure 35.8. However, for most experiments $x(\omega)$ is not measured directly. Instead, we usually measure a current $i(t)$ associated with the charge transfer between the electrodes and the resonating beam. This output current depends on the resonance frequency of the resonator as well as on the change in capacitance and the bias voltage between the gates and the beam. It is given by

$$i(t) = \frac{dQ}{dt} = V\frac{dC}{dt} + C\frac{dV}{dt} = V\frac{dC}{dt} = V\frac{dC}{dx}\frac{dx}{dt} = VC'\dot{x}. \tag{35.46}$$

Incorporating the gain of the transimpedance amplifier, the output voltage $\upsilon_{out}(t)$ becomes

$$\upsilon_{out}(t) = Ri(t), \tag{35.47}$$

where R is the resistor in the feedback loop of the preamp, as shown in Figure 35.7. Therefore, the detection signal is described by

$$\upsilon_{out}(\omega) = \frac{\omega C'^2\upsilon_o V^2 R}{m\sqrt{\left(\omega_o^2 - \omega^2\right)^2 + \omega^2\gamma^2}}, \tag{35.48}$$

which on resonance $\omega \approx \omega_o$ reduces to

$$\upsilon_{out} = \frac{C'^2 V^2 R\upsilon_o}{m\gamma}. \tag{35.49}$$

Hence, the measured output voltage depends quadratically on the beam bias voltage V as shown in inset (b) of Figure 35.8. It is worth mentioning that Equation 35.49 appears to have unlimited gain as $R \to \infty$. However, Equation 35.49 assumes ideal operation of the amplifier. This assumption fails for very high values of R, therefore infinite gain as $R \to \infty$ is not possible.

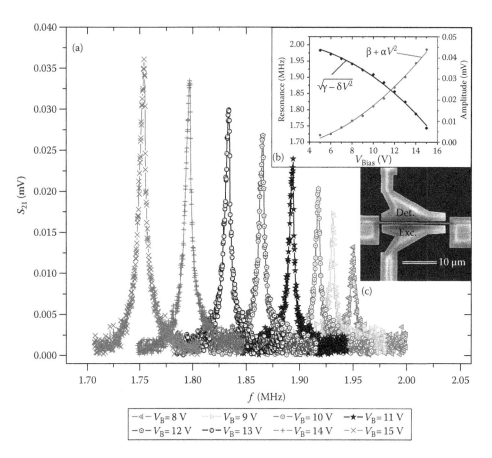

FIGURE 35.8 (a) Typical response of a 20 μm × 300 nm × 150 nm doubly clamped beam resonator using the electrostatic method for both actuation and sensing. The bias voltage dependence of both the resonance amplitude and frequency are depicted in inset (b). A SEM image of the actual device is shown in (c).

Equivalent Electrical RLC Circuit

Again, it is possible to determine expressions for an equivalent electrical RLC circuit for the electrostatic method. As always the mechanical resistance is defined as input voltage $\upsilon(t)$ divided by output current $i(t)$, $R_M = v(t)/i(t)$. On resonance the output current is given by

$$i(t) = V \frac{dC}{dx} \frac{dx}{dt} = VC' \omega_o x_o. \tag{35.50}$$

Displacement on resonance $\left(\omega \approx \omega_0 = \sqrt{k_{eff}/m_{eff}} \right)$ becomes $x_o = QF/k$ with $F = \upsilon(t)V(dC/dx)$. Hence, the mechanical resistance can be expressed as

$$R_M = \frac{\upsilon(t)}{i(t)} = \frac{\sqrt{km_{eff}}}{Q\eta^2} \quad \text{where } \eta = V \frac{dC}{dx}. \tag{35.51}$$

Similarly, one can find expressions for the equivalent mechanical capacitance and inductance (Lin et al., 2004):

$$C_M = \frac{\eta^2}{k}, \tag{35.52}$$

$$L_M = \frac{m_{eff}}{\eta^2}. \tag{35.53}$$

For parallel capacitor plates, these equations can further be expanded by utilizing $dC/dx \approx C(1/d) = \epsilon(A/d^2)$, where A is the area of the parallel capacitors, d the separation between the two plates, and ϵ the dielectric constant of the gap (ϵ_o in vacuum). The equivalent electrical elements for the resonator thus become (Lin et al., 2004)

$$R_M = \frac{\sqrt{km_{eff}}}{Q} \frac{d^4}{V^2 \epsilon^2 A^2}, \tag{35.54}$$

$$C_M = \frac{V^2 \epsilon^2 A^2}{d^4 k}, \tag{35.55}$$

$$L_M = \frac{m_{eff} d^4}{V^2 \epsilon^2 A^2}. \tag{35.56}$$

For typical nanomechanical resonators, it is desired to match R_M with the 50 Ω impedance lines of the measurement setup, while keeping V as low as possible. Though a perfect impedance matching is a challenging task, careful choice of d, A, V, and even ϵ can minimize the mechanical resistance R_M and therefore the losses in the system. Note that the analysis here can easily be applied to other shapes. For instance, to derive expressions for a disk, one has to simply change $\eta = V(dC/dx) = V\pi rt$, where r is the radius and t is the thickness of the disk, and plug η back into Equations 35.51 through 35.53 to obtain the new expressions for R_M, C_M, and L_M.

35.2.3 Piezoelectric Technique

Piezoelectric materials have the ability to generate an electric potential when a mechanical stress is applied to the material. Conversely, they deform when an electric potential is applied. The latter effect is exploited to actuate mechanical motion in a resonator and the former to sense this motion.

To demonstrate the basic concept of the piezoelectric actuation and detection technique, consider a simple block of piezoelectric aluminum nitride of length l, width w, and thickness t. The block has input (actuation), output (sensor), and ground electrodes as shown in Figure 35.9. To actuate the bulk resonator, an electric potential V_{IN} is applied at the input electrode. Given the mass density ρ and the elasticity modulus of AlN (depends on orientation) E_p, the applied strain S in both the transverse and longitudinal directions can be determined:

$$S_x = \frac{\Delta l}{l} = d_{31} E_z = d_{31} \frac{V_{IN}}{t}, \tag{35.57}$$

$$S_y = \frac{\Delta w}{w} = d_{31} E_z = d_{31} \frac{V_{IN}}{t}, \tag{35.58}$$

$$S_z = \frac{\Delta t}{t} = d_{33} E_z = d_{33} \frac{V_{IN}}{t}. \tag{35.59}$$

Here, d_{ij} are elements of the piezoelectric matrix. Often the transverse components are of interest when it comes to sensing and actuating, hence for the rest of this section we will focus on x and y directions using d_{31} only. In general, if V_{IN} is applied at resonance of any of the modes of the resonator, it forces the piezoelectric material to deform creating pockets of strain throughout the piezoelectric block. This strain induces a charge proportional to the strain at the output electrode, where it is measured. The transmission of electrical to mechanical energy between the electrical circuit and the piezoelectric resonator is ideal. In fact, this assumption is not valid and losses associated with the conversion are accounted for by the inclusion of an electromechanical coupling coefficient κ^2. Typical values for total electromechanical coupling coefficient for a two-port setup

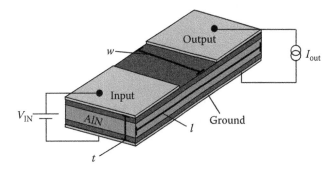

FIGURE 35.9 Schematic of a two-port piezoelectric bulk resonator, consisting of two ports (input and output) sandwiching the piezoelectric material. A voltage supplied to the input port induces strain in the resonator, which can be measured in the form of a current at the output electrode.

such as the one seen in Figure 35.9 depend strongly on the material and for AlN range from 1% to 5%. The effective displacement of an actuated AlN block can thus be estimated using

$$\delta l = \kappa^2 d_{31} l \frac{V_{\text{IN}}}{t}. \tag{35.60}$$

Finally, the following example gives an order of magnitude estimate for piezoelectric displacement and force values. Consider a $100 \times 100 \times 10\,\mu\text{m}$ AlN block with $d_{31} \approx -2(pC/N)$ and elasticity constant $E_p = 300\,\text{GPa}$, being actuated by an ac input signal $V_{\text{IN}} = 10\,\text{V}$. The electromechanical coupling coefficient is estimated to be $\kappa^2 \approx 10^{-2}$. Hence, the estimated length scale associated with the resonators' deformation due to the input signal is of the order

$$\Delta l = \kappa^2 d_{31} l \frac{V_{\text{IN}}}{t} = 10^{-2} \cdot 2 \times 10^{-12} \cdot 10^{-4} \frac{10}{10^{-6}} \approx 20 \text{ pm}. \tag{35.61}$$

Note, a voltage applied to the input electrode causes the beam to oscillate with typical amplitudes of a couple of picometers. This oscillation induces a charge on the detection electrode due to strain caused by the oscillation of the piezoelectric material. The charge on the detection electrode can be found by integrating the electric displacement D across the surface of the detection electrode, where D consists of two terms, one related to the applied electric field across the piezoelectric material and the other due to the strain in the piezoelectric resonator. It is the latter that describes the mechanical motion of the device. On that account, we find

$$Q = \iint_A \vec{D} \cdot d\vec{A} = \iint_A \left(\epsilon \vec{E} + d_{31} E_p S_x \right) \cdot d\vec{A}. \tag{35.62}$$

This charge can be converted into an output current by taking its time derivative:

$$I_{\text{out}} = \frac{\partial}{\partial t} \left(\iint_A \vec{D} \cdot d\vec{A} \right). \tag{35.63}$$

The overall admittance Y can then be calculated using

$$Y = \frac{I_{\text{out}}}{V_{\text{IN}}}. \tag{35.64}$$

This analysis can be applied quite generally to many different types of structures. In the two examples below, we go through specific calculations for a free piezoelectric block and a doubly clamped beam (Weaver et al., 1990).

Example 1. Free Piezoelectric Block

In this example, we follow Piazza (Piazza, 2005) and discuss how a piezoelectric material can be used to transduce nanomechanical motion. The following is a 1D analysis for a piezoelectric

block resonating along its length. Consider a block of initial length l, width w, and thickness t. Let us furthermore assume that, as a voltage V_{in} is applied to the block at the input electrode, the length of the block changes by u. The material of the block has density ρ and elastic modulus E_p. The fundamental equation of motion is given by Weaver et al. (1990):

$$\rho \frac{\partial^2 u}{\partial t^2} - E_p \frac{\partial^2 u}{\partial x^2} = 0. \tag{35.65}$$

We assume we can separate time and space variables with a sinusoidally time varying solution and concentrate on the space dependent part of the solution, which is given by

$$u(x) = C_1 \sin(kx), \tag{35.66}$$

where
$k = \omega l/2c$, with ω being the frequency of the excitation
$c = \sqrt{E_p/\rho}$ the velocity of the sound of the material

At the end surfaces of the block, the stress is zero, and the strain is given by the piezoelectric term

$$S_x = S_y = \frac{\partial u}{\partial x} = d_{31} \frac{V_{\text{in}}}{t}, \tag{35.67}$$

where
V_{in} is the voltage applied to the block
d_{31} is the contour piezo coefficient

One can then solve for C_1:

$$C_1 = \frac{d_{31} V_{\text{in}}}{kT \cos\left(\dfrac{kl}{2}\right)}. \tag{35.68}$$

We can then find the current, I_{out}, using Equation 35.63:

$$I_{\text{out}} = \frac{\partial}{\partial t'} \left(\iint_A D \cdot dA \right) = i\omega W \int_{-l/2}^{l/2} \left(e_{31} \frac{\partial u}{\partial x} + \epsilon E \right) dx \tag{35.69}$$

$$= i\omega \left[2w \frac{e_{31} d_{31}}{kt} \tan\left(\frac{kl}{2}\right) + \epsilon \frac{wl}{t} \right] V_{\text{in}}. \tag{35.70}$$

The overall admittance Y can thus be determined as

$$Y = \frac{I_{\text{out}}}{V_{\text{in}}} = i\omega \left[2\frac{w}{t} \frac{d_{31} e_{31}}{k} \tan\left(k\frac{l}{2}\right) + \epsilon \frac{wl}{t} \right]. \tag{35.71}$$

The second term in the expression above is the capacitance between the two electrodes of the device, C_o (in this case, a parallel plate capacitor). The first term represents the motional part of the resonator, which includes damping if we replace k with

$$k = \frac{\omega l}{2c}\left(1 + i\frac{1}{2Q}\right), \qquad (35.72)$$

where Q is the quality factor of the resonator.

Equivalent Electrical RLC Circuit

It can be shown in this simplified example ($e_{31} = E_p d_{31}$) that the first term in Equation 35.71 corresponds to an electrical series RLC circuit with the following expressions for resistor R_M, capacitor C_M, and inductor L_M (Rosenbaum, 1988)

$$R_M = \frac{\pi^2 t}{8w}\frac{\sqrt{\rho}}{E_p^{3/2}Q d_{31}^2}, \qquad (35.73)$$

$$C_M = \frac{8}{\pi^2}\frac{wl}{t}d_{31}^2 E_p, \qquad (35.74)$$

$$L_M = \frac{\rho}{8}\frac{lt}{w}\frac{1}{d_{31}^2 E_p^2}. \qquad (35.75)$$

For piezoelectric resonators, the equivalent mechanical circuit components depend strongly on their shapes.

Example 2. Doubly Clamped Piezoelectric Beam

This analysis is adapted from De Voe (De Voe, 2001). Consider a doubly clamped beam of length *l*, width *w*, and thickness *t* with small strips of piezoelectric material on both ends of the beam as seen in Figure 35.10. If we apply an input voltage across the piezoelectric strips at one of the two ends of the beam, its deformation will induce strain along the whole beam. This strain can then be measured piezoelectrically via the strip at the other end of the beam. More precisely, for a beam oscillating with amplitude y'_o, the output current I_{out} can be calculated as follows:

$$I_{out} = 2\pi i f_o \int_0^l\int_0^w D\,dydx = 2\pi i f_o \int_0^l\int_0^w (\epsilon E + d_{31}E_p S_1)\,dydx, \qquad (35.76)$$

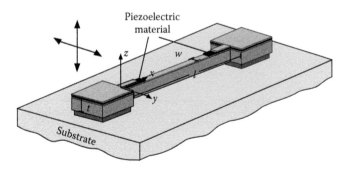

FIGURE 35.10 Schematic of a doubly clamped beam equipped with piezoelectric actuation and detection electrodes. Piezoelectric material on both ends of the beam is used to actuate and detect the resonating beam.

where

> D is the electric displacement
> ϵ is the dielectric constant
> E_p is the Young's modulus
> d_{31} is the piezoelectric constant
> S_1 is the strain in the system

On the detection side, the electric field $E = 0$. Therefore,

$$I_{out} = 2\pi i f_o \int_0^l\int_0^w d_{31}E_p S_1\,dydx. \qquad (35.77)$$

The strain has extensional and bending strain components given by

$$S_1^{bend} = -\frac{t}{2}\frac{d^2z'(x)}{dx^2} \quad S_1^{ext} = \frac{1}{2l}\int_0^l\left(\frac{dz'(x)}{dx}\right)^2 dx. \qquad (35.78)$$

For small deformations $S_1^{bend} \gg S_1^{ext}$, and thus the extensional strain can be neglected. Furthermore, it can be shown that

$$z'(x) = z'_o\phi(x), \qquad (35.79)$$

where

$$\phi(x) = \sinh(\beta x) - \sin(\beta x) + \alpha\cosh(\beta x) - \alpha\cos(\beta x), \quad (35.80)$$

and

$$\frac{d\phi(x)}{dx} = \beta\big(\cosh(\beta x) - \cos(\beta x) + \alpha\sinh(\beta x) + \alpha\sin(\beta x)\big). \qquad (35.81)$$

Using previously calculated expressions (Section 35.1.2) for β and α,

$$\beta l = 4.73 \quad \text{and} \quad \alpha = -1.018 \qquad (35.82)$$

we find

$$I_{out} = -2\pi i f_o d_{31}E_p w\frac{t}{2}z'_o\int_0^{0.224l}\frac{d^2\phi(x)}{dx^2}dx, \qquad (35.83)$$

which can further be simplified to

$$I_{out} = 2\pi i f_o d_{31}E_p w\frac{t}{2}z'_o\frac{d\phi(0.224l)}{dx}, \qquad (35.84)$$

where the electrode is cut-off at $x = 0.224l$, where $d\phi(x)/dx$ is maximized. Given an actuation method R_M, C_M and L_M can be established as described previously. Results of similar calculations for two other commonly used resonators are summarized in Table 35.4. The table includes sketches of resonator geometries, equivalent RLC circuits, and expressions for R_M, L_M,

TABLE 35.4 Collection of Bulk Mode Resonators for Piezoelectric Actuation and Detection and Corresponding RLC Circuit Parameters

Design	Equiv. Circuit	Results
Plate contour mode[a]		$\eta \approx 2E_p d_{31} \dfrac{w}{2}$ $M_{eq} \approx \dfrac{\rho w l t}{2}$ $R_M = \dfrac{\pi t}{8w} \dfrac{\sqrt{\rho}}{E_p^{3/2} Q d_{31}^2}$ $C_M = \dfrac{8}{\pi^2} \dfrac{wl}{t} d_{31}^2 E_p$ $L_M = \dfrac{\rho}{8} \dfrac{lt}{w} \dfrac{1}{d_{31}^2 E_p^2}$ $C_o = \epsilon \dfrac{wl}{t}$ $f_o = \dfrac{1}{2l} \sqrt{\dfrac{E_p}{\rho(1-v^2)}}$ $N = \dfrac{\eta_1}{\eta_2}$
Disk contour mode[b,c,d]		$\eta \approx \beta E_p d_{31} \dfrac{\pi r}{2}$ $M_{eq} \approx \alpha \rho \pi r^2 t$ $R_M = \dfrac{4\alpha \xi t}{\pi \beta^2 r} \dfrac{1}{\rho^{1/2} E_p^{3/2} (2+2v)^{1/2}}$ $L_M = \dfrac{4\alpha t}{\pi \beta^2 d_{31}^2 E_p^2}$ $C_M = \dfrac{\beta^2}{\alpha \xi^2} \dfrac{\pi}{2} \dfrac{r}{t} d_{37}^2 E_p \rho(1+v)$ $C_{in} = C_{out} = \epsilon \dfrac{\pi r^2}{4t}$ $f_o = \dfrac{\xi}{2\pi r} \sqrt{\dfrac{E_p}{2\rho(1+v)}}$ $N = \dfrac{\eta_1}{\eta_2}$
Ring contour mode[e]		$R_M = \dfrac{\pi t}{8\lambda} \dfrac{\sqrt{\rho(1-v^2)}}{E_p^{3/2} Q d_{31}^2}$ $C_M = \dfrac{8}{\pi^2} \dfrac{w\lambda}{t} \dfrac{d_{31}^2 E_p}{(1-v^2)}$ $L_M = \dfrac{\rho}{8} \dfrac{wt}{\lambda} \dfrac{(1-v^2)^2}{d_{31}^2 E_p^2}$ $C_o = \epsilon \dfrac{w\lambda}{t}$ $f_o \approx \dfrac{1}{2w} \sqrt{\dfrac{E_p}{\rho(1-v^2)}}$

Sources: Piazza, G. et al., *IEEE/ASME J. Microelectromech. Syst.*, 15(6), 1406, 2006. Piazza, G., Piezoelectric aluminium nitride vibrating RF MEMS for radio frequency front end technology. PhD thesis, University of California, Berkeley, CA, 2005.

[a] Adjust tether length to $\lambda/4$-wavelength.
[b] $\beta \approx 0.29$ for AlN.
[c] $\alpha \approx 0.71$ for AlN.
[d] ξ is a fit parameter.
[e] λ is the average circumference of the ring.

C_M, and the idealized electromechanical coupling coefficients η_1 at the input and η_2 at the output electrode not to be confused with κ^2. A more detailed derivation of these expressions can be found elsewhere (Piazza, 2005).

35.2.4 Summary of Measurement Techniques

In Sections 35.2.1 through 35.2.3, three measurement techniques are discussed. The magnetomotive technique is advantageous for detecting odd flexural modes at high frequencies, as it typically has the smallest impedance of all the three measurement techniques. However, it is vastly limited by the requirement of relatively strong magnetic fields (several Tesla), which are usually achieved with superconducting coils at liquid helium temperature. Though this approach is useful for low temperature experiments in fundamental research, this technique cannot be used for commercial applications due to high costs and large measurement setups.

The electrostatic method does not depend on a magnetic field, hence it can be used to measure resonances of nanomechanical devices at room temperature without the requirement of a large measurement setup. In fact, the entire setup can be integrated on a single chip, making it a low cost detection technique and therefore a prominent detection technique of commercially available products today. Another advantage of the electrostatic method is that it can be used for both bulk and flexural modes. However, high impedances and parasitic losses make it difficult to use this technique for high frequency devices, i.e. for smaller resonators. This limitation on resonator size of the electrostatic method is not shared by the magnetomotive method.

Finally, the piezoelectric method can be used to detect both flexural and bulk modes. It can be implemented without a large measurement setup. In addition, out of the three detection techniques, piezoelectrically actuated and measured resonators tend to have the highest signal size and frequency, making them especially attractive for the telecommunication industry. A drawback of this technique is its limitation to a small number of materials

that exhibit piezoelectric properties. A summary of advantages and disadvantages of the three detection methods is presented in Table 35.5 (Mohanty et al., 2002).

35.3 Applications of Micromechanical and Nanomechanical Resonators

Micromechanical and nanomechanical resonators can be used in a wide range of device applications. Here, we discuss a handful of applications that are being commercially produced and used in consumer products. Most of the resonators in these devices are still tens or hundreds of micrometer large, although their scaling down to sub-100 nm scale for high frequencies and better sensitivity is inevitable.

Starting with accelerometers, the most commercially successful application to date, the following sections also introduce gyroscopes, timing or reference oscillators, and mechanical circuit elements such as filters, mixers, switches and tunable capacitors, and inductors. These devices are used in a wide range of industries such as aerospace, automotive, biotechnology, instrumentation, robotics, manufacturing, and mobile communication. Sections 35.3.1 through 35.3.5 introduce each of these applications in detail.

35.3.1 Accelerometers

Accelerometers are the most successful MEMS sensor in terms of commercial applications. Millions of MEMS accelerometers are sold each year to the automotive industry. Today MEMS accelerometers come fully IC-integrated and they cost far less than accelerometers did decades ago. These MEMS devices are actuated and detected either magnetomotively, optically, piezoelectrically, electromagnetically, or capacitively, the latter being the most common due to its simplicity. Other reasons for capacitive or electrostatic actuation include low power consumption, high temperature stability, and the lack of requirement of exotic metals. One disadvantage is the presence of high parasitic capacitances, since the parasitics make the operation of high-frequency accelerometers difficult.

Accelerometers are typically characterized by their bandwidth, noise floor, sensitivity, drift, dynamic range, shock resistance, and power consumption. The resonant frequency that depends on the shape of the resonator (see Table 35.1) sets an upper limit to useful frequency range and sensitivity usually measured in displacement d per g of acceleration.

Major noise sources in a typical accelerometer are the resonator itself, the readout circuit, mechanical damping, and electrical resistors in the circuit. The Johnson noise of the mechanical resistance of the resonator can be expressed in terms of Brownian force F_B (Gabrielson, 1993):

$$F_B = \sqrt{4k_B T D}, \tag{35.85}$$

TABLE 35.5 Summary of Features of the Three Detection Techniques

Feature	Detection Technique		
	Magnetomotive	Electrostatic	Piezoelectric
High-frequency operation	>1 GHz	Limited by parasitic cap.	>1 GHz
Operation environment	Vacuum < 4 K	Vacuum/atm. press. 300 K	Atm. pressure 300 K
Detection modes	Mostly flexural	Flexural/bulk	Flexural/bulk
Operating cost	Expensive	Cheap	Cheap
Typical R_M values	10^1–$10^2\ \Omega$	>$10^4\ \Omega$	10^2–$10^4\ \Omega$
Other features	Needs high mag. fields	Resonator size limited	Need piezoelectric material

which translates into Brownian motion x_B, and hence Brownian acceleration:

$$g_B = \frac{1}{g}\sqrt{\frac{4k_B T \omega_o}{m_{eff} Q}}, \qquad (35.86)$$

where

- T is temperature
- k_B is the Boltzmann constant
- D is the damping coefficient of the effective mass m_{eff} supported by the effective spring constant k_{eff}
- ω_o is the resonant frequency of the resonator
- g is gravitational acceleration
- $Q = (\omega_o m_{eff})/D$ is the quality factor of the resonator

One can see that higher Q and higher m_{eff}, lower the Johnson noise of the mechanical resistor.

35.3.2 Gyroscopes

Another successful commercial application of MEMS/NEMS resonators are gyroscopes. MEMS gyroscopes can be found in airplanes, cars, and even Wii controllers (Song, 1997). MEMS gyroscopes are micromechanical sensors, designed to detect the Coriolis force $\vec{F}$ acting on an object with effective mass m_{eff}, moving at constant velocity $\vec{v}$ in a rotating frame of reference with angular velocity $\vec{\omega}$. The standard expression for the Coriolis force is

$$\vec{F} = 2m_{eff}\vec{v} \times \vec{\omega}. \qquad (35.87)$$

The most common gyroscopes are tuning fork gyroscopes that typically have two modes of oscillation perpendicular to one another (Figure 35.11). The Coriolis force in these devices is exploited when one mode is excited and placed in the plane of a rotating reference frame and the other is actuated by the resulting Coriolis force. On that account, the resonant frequencies of the two modes need to be identical, isolated from other modes and ideally decoupled from one another in order to extract a strong and clean output signal.

Other possible gyroscope devices called vibrating wheel gyroscopes consist of a wheel driven to oscillate about its center of symmetry. When the device rotates about either of the two in-plane axes, tilting results that can be detected capacitively by electrodes underneath the wheel. Putty et al. have demonstrated a third kind of gyroscope, the wine-glass resonator gyroscope which is described in more detail elsewhere (Putty, 1995).

35.3.3 Timing Oscillator—Mechanical Clock

Mechanical timing oscillators or clocks play a crucial role in modern electronics. While time measurement is their primary function, they are also employed as references in integrated circuits to analyze rf signals (radio), to synchronize logic operations (computer) and to transmit rf signals (cellphone), to name just

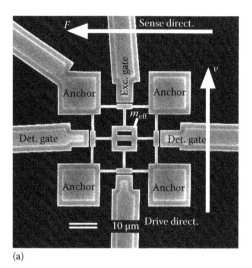

(a)

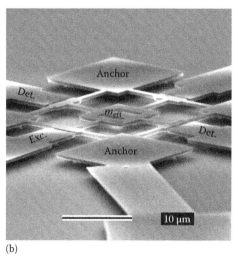

(b)

FIGURE 35.11 (a) SEM photograph of a simple tuning fork gyroscope. This gyroscope is actuated and detected using the electrostatic method. The design focuses on decoupling the drive and the sense modes. (b) 3-D image of the same gyroscope clearly demonstrating the suspended and anchored parts.

a few applications. In general, the more accurate and stable the clock, the better the performance of the underlying devices that reference it. A classic example is the performance of a GPS system that primarily depends on the quality of the synchronization between the clock in the satellite and the GPS system (Nguyen, 2007). While the best clocks (atomic clocks) are not economical, MEMS reference oscillators offer an attractive alternative, due to their precision, stability, CMOS integratability, size, and affordability.

MEMS resonators with Q's in the tens of thousands and operating in the gigahertz frequency range have already been demonstrated (Nguyen, 2007). They have the potential to become the new standard for consumer electronics. In fact, a MEMS clock meeting GSM requirements has been demonstrated by Lin et al. (2005), based on 9 wine-glass disk array resonators with a resonance frequency of 64 MHz and quality factor of $Q = 118,900$. A successful demonstration of a 428 MHz

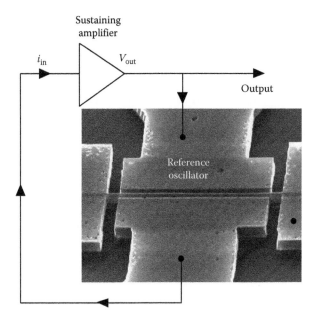

FIGURE 35.12 Schematic diagram of a self-sustaining reference oscillator. An ideally high-frequency and high Q resonator is placed in a feedback control loop to create a stable self-sustaining reference oscillator.

self-sustaining ultrahigh-frequency nanomechanical oscillator, operating at low temperature and high fields, is reported by Feng et al. (2008). An oversimplified sketch of such a self-sustaining nanomechanical oscillator is shown in Figure 35.12.

35.3.4 Filters, Mixers, and Switches

Many micromechanical and nanomechanical resonators that could potentially replace and improve existing electronic circuit elements have been proposed, fabricated, and measured. Wang and Nguyen (1999) have demonstrated a 340 kHz micromechanical filter with 403 Hz bandwidth and insertion loss of less than 0.6 dB. In a more elaborate micromechanical array of coupled disk resonators, Nguyen et al. have demonstrated a filter at 146 MHz, with a 201 kHz bandwidth, and less than 2 dB insertion loss (Li et al., 2006). They have also fabricated a micromechanical mixer–filter that mixes a 240 MHz signal down to 37 MHz and then filters. This device based on two coupled doubly clamped resonators has a combined insertion loss of 9.5 dB and a bandwidth of roughly 0.63 MHz (Wong and Nguyen, 2004). A low-power and high-speed mechanical switch integrated on-chip with silicon circuitry has been demonstrated by Guerra et al. (2008), which exploits the nonlinearity of a simple doubly clamped resonator with two distinct states in the hysteretic nonlinear regime. They utilize electrostatic actuation and sensing of the switching device with 100% fidelity by phase modulating the drive signal. A different micromechanical switch has been demonstrated by Yao et al. in which a micromechanical beam is physically forced between the open and closed positions by an electrostatic force. This kind of switch has lower insertion

loss than its FET-based counterparts, although it operates at lower frequencies (Yao et al., 1999). Tunable capacitors and micromechanical engineered inductors for high-frequency tunable biasing and matching applications have also been demonstrated by Young and Yoon, with capacitances ranging from 1 to 4 pF and inductances ranging from 1 to 4 nH (Young, 1996; Yoon, 1999).

35.3.5 Biomedical Applications

Micromechanical and nanomechanical resonators are being used in a number of biomedical applications for research and development. In general, mechanical resonators designed as biosensors either detect stresses caused by the bending of the resonator or they detect frequency shifts in the resonant frequency of the resonator due to mass loading or surface strain. The rest of this section summarizes some key results in this field.

In 2006, Shekhawat et al. used singly clamped beams, also referred to as micromechanical cantilever, to detect stress caused by the bending of the cantilevers (Shekhawat et al., 2006). This bending occurs as molecules (e.g., biotin and antibodies) bind to the top layer of the cantilever covered with a layer of receptors. A MOSFET transistor integrated at the base of the cantilever detects a change in the drain current caused by a change in electron mobility as the cantilever gives in (typically tens of nm) to the weight of the molecules that bind to the top layer of receptors. They achieve a displacement resolution of 5 nm for a $250 \times 50 \times 1.5\,\mu m$ cantilever (Shekhawat et al., 2006). In a similar experiment conducted by Wu et al., optical detection of MEMS cantilevers allowed for the detection of prostate-specific antigen (PSA) in a background solution of human serum albumin (HSA) to simulate detection of prostate cancer under clinical conditions (Wu et al., 2001). Su et al. used a different approach to detect DNA strands using micromachined cantilevers (Su et al., 2003). They measured the shift in resonance frequency of their cantilever as gold nano-particle modified DNA attached to the surface of the resonator. They were able to detect concentrations of target DNA down to 0.05 nM. Along the same lines, Lee et al. have detected the frequency shift of a cantilever resonator as PSA coupled to their resonator; however, they used a piezoelectric detection technique, and they achieved a resolution of 10 pg/mL (Lee et al., 2005). An overview of micro and nanomechanical resonators as fundamental elements of bioMEMS, their advantages and disadvantages over other medical sensors, and their prospects for the future has been discussed by Bashir in 2004 (Bashir, 2004).

References

Badzey, R. and P. Mohanty. Coherent signal amplification in bistable nanomechanical oscillators by stochastic resonance. *Nature*, 437:995–998, 2005.

Badzey R., G. Zolfagharkhani, A. Gaidarzhy, and P. Mohanty. A controllable nanomechanical memory element. *Appl. Phys. Lett.*, 85(16):3587–3589, 2004. doi: 10.1063/1.1808507.

Bannon F., C. Nguyen, and J. Clark. High-q hf microelectromechanical filters. *IEEE J. Solid State Circuits*, 35(4):512–526, 2004. doi: 10.1109/4.839911.

Bashir R. Biomems: State-of-the-art in detection, opportunities and prospects. *Adv. Drug Deliv. Rev.*, 56(11):1565–1586, 2004. ISSN 0169-409X. doi: 10.1016/j.addr. 2004.03.002. Intelligent Therapeutics: Biomimetic Systems and Nanotechnology in Drug Delivery.

Braginskii V. *Systems with Small Dissipation*. University of Chicago Press, Chicago, IL, 1985.

Braginsky V. B. and F. Ya. Khalili. Quantum nondemolition measurements: The route from toys to tools. *Rev. Mod. Phys.*, 68(1):1–11, January 1996. doi: 10.1103/RevModPhys.68.1.

Cleland A. *Foundations of Nanomechanics: From Solid-State Theory to Device Applications*. Springer, Heidelberg, Germany, 2003. ISBN 978-3-540-43661-4.

Cleland A. N. and M. L. Roukes. External control of dissipation in a nanometer-scale radiofrequency mechanical resonator. *Sens. Actuators A: Phys.*, 72(3):256–261, 1999. ISSN 0924-4247. doi: 10.1016/S0924-4247(98)00222-2.

Cross M. C. and R. Lifshitz. Elastic wave transmission at an abrupt junction in a thin plate with application to heat transport and vibrations in mesoscopic systems. *Phys. Rev. B*, 64(8):085324, August 2001. doi: 10.1103/PhysRevB.64.085324.

Demirci M. and C. Nguyen. Mechanically corner-coupled square microresonator array for reduced series motional resistance. *IEEE J. Solid State Circuits*, 15(6):1419–1436, 2006. doi: 10.1109/JMEMS.2006.883573.

De Voe D. L., Piezoelectric thin film micromechanical beam resonators. *Sens. Actuators A: Phys.*, 88:263–272, 2001.

Ekinci K.. Electromechanical transducers at the nanoscale: Actuation and sensing of motion in nanoelectromechanical systems (NEMS). *Small*, 1:786–797, 2005.

Etaki S., M. Poot, I. Mahboob, K. Onomitsu, H. Yamaguchi, and H. S. J. van der Zant. Motion detection of a micromechanical resonator embedded in a d.c. squid. *Nat. Phys.*, 4(10):785–788, 2008.

Feng X., C. White, A. Hajimri et al. A self-sustaining ultra-high-frequency nanoelectromechanical oscillator. *Nat. Nanotechnol.*, 3:342–346, 2008. doi: 10.1038/nnano.2008.125.

Gabrielson T. Mechanical-thermal noise in micromachined acoustic and vibration sensors. *IEEE Trans. Eletron. Devices*, 40(5):903–909, 1993. doi: 10.1109/16.210197.

Gaidarzhy A., G. Zolfagharkhani, R. Badzey, and P. Mohanty. Evidence for quantized displacement in macroscopic nanomechanical oscillators. *Phys. Rev. Lett.*, 94(3):030402, January 2005. doi: 10.1103/Phys-RevLett.94.030402.

Gaidarzhy A., M. Imboden, P. Mohanty, J. Rankin, and B. Sheldon. High quality factor gigahertz frequencies in nanomechanical diamond resonators. *Appl. Phys. Lett.*, 91(20):203503, 2007. doi: 10.1063/1.2804573.

Giessibl F. Advances in atomic force microscopy. *Rev. Mod. Phys.*, 75(3):949–983, July 2003. doi: 10.1103/RevModPhys.75.949.

Guerra D., M. Imboden, and P. Mohanty. Electrostatically actuated silicon-based nanomechanical switch at room temperature. *Appl. Phys. Lett.*, 93(3):033515, 2008. doi: 10.1063/1.2964196.

Guerra D., T. Dunn, and P. Mohanty. Signal amplification by 1/f noise in silicon-based nanomechanical resonators. *Nano Lett.*, 9(9):3096–3099, 2009.

Hosaka H., K. Itao, and S. Kuroda. Damping characteristics of beam-shaped micro-oscillators. *Sens. Actuators A: Phys.*, 49(1–2):87–95, 1995. ISSN 0924-4247. doi: 10.1016/0924-4247(95)01003-J.

Jimbo Y., K. Itao, and J. Horolog. Energy loss of a cantilever vibrator. (Japanese translation) *Inst. Jpn.*, 46:1–15, 1968.

Kaajakari V., T. Mattila, A. Oja et al. Square-extensional mode single-crystal silicon micromechanical resonator for low-phase-noise oscillator applications. *IEEE Electron. Device Lett.*, 25(4):173–175, 2004. doi: 10.1109/LED.2004.824840.

Knobel R. and A. Cleland. Nanometre-scale displacement sensing using a single electron transistor. *Nature*, 424:291–293, 2003. doi: 10.1038/nature01773.

LaHaye M. D., O. Buu, B. Camarota, and K. C. Schwab. Approaching the quantum limit of a nanomechanical resonator. *Science*, 304(5667):74–77, 2004. doi: 10.1126/science.1094419.

Landau L. and E. Lifshitz. *Theory of Elasticity*. Pergamon, London, U.K., 1959.

Lee J., Kyo S. Hwang, J. Park, K. Yoon, D. Yoon, and T. Kim. Immunoassay of prostate-specific antigen (PSA) using resonant frequency shift of piezoelectric nanomechanical microcantilever. *Biosens. Bioelectron.*, 20(10):2157–2162, 2005. ISSN 0956-5663. doi: 10.1016/j.bios.2004.09.024. Selected Papers from the Eighth World Congress on Biosensors, Part II.

Li S., Y. Lin, Z. Ren et al. Disk-array design for suppression of unwanted modes in micromechanical composite-array filters. *19th IEEE International Conference on Micro Electro Mechanical Systems, 2006*, Istanbul, Turkey, pp. 866–869, January 2006.

Lin Y., S. Lee, S. Li et al. Series-resonant VHF micromechanical resonator reference oscillator. *IEEE J. Solid. State Circuits*, 39(12):2477–2491, 2004.

Lin Y., S. Li, Z. Ren et al. Low phase noise array-composite micromechanical wine-glass disk oscillators. *IEEE International Electron Devices Meeting, 2005. IEDM Technical Digest*, Washington, DC, pp. 287–290, December 2005.

Madou M. The impact of MEMS and NEMS on biotechnology in the 21st century (invited). IEEE Computer Society, *Proceedings of the Symposium on Design, Test, Integration and Packaging of MEMS/MOEMS*, Washington, D.C., p. 1, May 2003.

Merono J. Integration of CMOS-MEMS resonators for radio-frequency applications in the VHF and UHF bands. PhD thesis, Universitat Autònoma de Barcelona, Barcelona, Spain, 2007.

Mohanty P., D. A. Harrington, K. L. Ekinci, Y. T. Yang, M. J. Murphy, and M. L. Roukes. Intrinsic dissipation in high-frequency micromechanical resonators. *Phys. Rev. B*, 66(8):085416, August 2002. doi: 10.1103/PhysRevB.66.085416.

Mohanty P., G. Zolfagharkhani, S. Kettemann, and P. Fulde. Spin-mechanical device for detection and control of spin current by nanomechanical torque. *Phys. Rev. B*, 70(19):195301, November 2004. doi: 10.1103/Phys-RevB.70.195301.

Nguyen C. Rf MEMS in wireless architectures (invited). Annual ACM IEEE Design Automation Conference, *Proceedings of the 42nd Annual Design Automation Conference*, Anaheim, California, 233:416–420, June 2005.

Nguyen C. MEMS technology for timing and frequency control. *IEEE Trans. Ultrason., Ferroelectr. Freq. Control*, 54(2):251–270, 2007. doi: 10.1109/TUFFC.2007.240.

Piazza G., Piezoelectric aluminium nitride vibrating RF MEMS for radio frequency front end technology. PhD thesis, University of California, Berkeley, CA, 2005.

Piazza G., P. J. Stephano, and A. P. Pisano. Piezoelectric aluminium nitride vibrating contour-mode MEMS resonators. *IEEE/ASME J. Microelectromech. Syst.*, 15(6):1406–1408, 2006.

Putty M. *A Micromachined Vibrating Ring Gyroscope*. University of Michigan, Ann Arbor, MI, 1995.

Rosenbaum J. *Bulk Acoustic Wave Theory*. Virginia Polytechnic Institute and State University, Blacksburg, VA, 1988.

Senturia S. *Microsystem Design*. Kluwer Academic Publishers, Boston, MA/Dordrecht, the Netherlands/London, U.K. 2001.

Shekhawat G., T. Soo-Hyun, and P. D. Vinayak. MOSFET-Embedded microcantilevers for measuring deflection in biomolecular sensors. *Science*, 311(5767):1592–1595, 2006. doi: 10.1126/science.1122588.

Shim S., M. Imboden, and P. Mohanty. Synchronized oscillation in coupled nanomechanical oscillators. *Science*, 316(5821):95–99, 2007. doi: 10.1126/science.1137307.

Song C. Commercial vision of silicon based inertial sensors. *International Conference on Solid State Sensors and Actuators, 1997*, Chicago, IL, Vol. 2, pp. 839–842, 1997.

Su M., S. Li, and V. P. Dravid. Microcantilever resonance-based DNA detection with nanoparticle probes. *Appl. Phys. Lett.*, 82(20):3562–3564, 2003. doi: 10.1063/1.1576915.

Thomson W. T. *Mechanical Vibrations*. Prentice Hall, Englewood Cliffs, NJ, 1953.

Timoshenko S. P. and J. N. Goodier. *Theory of Elasticity*. McGraw-Hill, New York, 1971.

Villarroya M., J. Verd, J. Teva et al. System on chip mass sensor based on polysilicon cantilevers arrays for multiple detection. *Sens. Actuators A: Phys.*, 132(1):154–164, 2006. ISSN 0924-4247. doi: 10.1016/j.sna.2006.04.002. The 19th European Conference on Solid-State Transducers.

Wang K. and Nguyen C. High-order medium frequency micromechanical electronic filters. *IEEE/ASME J. Microelectromech. Syst.*, 8(4):534–557, 1999. doi: 10.1109/84.809070.

Wang J., J. Butler, T. Feygelson et al. 1.51-GHz polydiamond micromechanical disk resonator with impedance-mismatched isolating support. *Proceedings, MEMS'04*, Maastricht, the Netherlands, pp. 641–644, January 2004.

Wang K., A.-C. Wong C.T.-C. Nguyen, VHF free-free beam high Q micromechanical resonators, *IEEE/ASME J. Microelectromech. Syst.*, 9(3):347, 2000.

Weaver Jr. W., S. Timoshenko, and D. Young. *Vibration Problems In Engineering*. John Wiley & Sons, New York, 1990.

Wong A. and C. Nguyen. Micromechanical mixer-filters (mixlers). *IEEE/ASME J. Microelectromech. Syst.*, 13(1):100–112, 2004. doi: 10.1109/JMEMS.2003.823218.

Wu G., R. Datar, and K. Hansen. Bioassay of prostate-specific antigen (PSA) using microcantilevers. *Nat. Biotechnol.*, 19:856–860, 2001. doi: 10.1109/84.767108.

Yang J., T. Ono, and M. Esashi. Energy dissipation in submicrometer thick single-crystal silicon cantilevers. *J. Microelectromech. Syst.*, 11(6):775–783, 2002. doi: 10.1109/JMEMS.2002.805208.

Yang Y. T., C. Callegari, X. L. Feng, K. L. Ekinci, and M. L. Roukes. Zeptogram-scale nanomechanical mass sensing. *Nano Lett.*, 6(4):583–586, 2006. doi: 10.1021/nl052134m.

Yao Z. J., S. Chen, S. Eshelman, D. Denniston, and C. Goldsmith. Micromachined low-loss microwave switches. *IEEE/ASME J. Microelectromech. Syst.*, 8(2):129–134, 1999. doi: 10.1109/84.767108.

Yoon J. B., C. H. Han, and C. K. Kim. Monolithic overhang inductors fabricated on silicon and glass substrates. *IEEE IEDM Electron Device Meeting*, Washington DC, pp. 753–756, 1999.

Young D. J. and B. E. Boser. A micromachine–based RF low-noise voltage controlled oscillator. *Technical Digest, IEEE Solid State Sensor and Actuator Workshop*, Hilton Head Island, SC; June 1996.

Zolfagharkhani G., A. Gaidarzhy, P. Degiovanni, S. Kettemann, P. Fulde, and P. Mohanty. Nanomechanical detection of itinerant electron spin flip. *Nat. Nanotechnol.*, 3:720–723, 2008. doi: 10.1038/nnano.2008.311.

36

Mechanics of Nanoscaled Oscillators

Duangkamon Baowan
Mahidol University

Ngamta Thamwattana
University of Wollongong

Barry J. Cox
University of Wollongong

James M. Hill
University of Wollongong

36.1 Introduction

The discovery of carbon nanotubes by Iijima (1991) has given rise to the possible creation of many new nanodevices. Due to their unique mechanical properties such as high strength, low weight, and flexibility, both multi-walled and single-walled carbon nanotubes promise many new applications for nanomechanical systems. One device that has attracted much attention is the nanoscale oscillator, or the so-called gigahertz oscillator. While there are difficulties for micromechanical oscillators, or resonators, to reach frequencies in the gigahertz range, it is possible for nanomechanical systems to achieve this. Cumings and Zettl (2000) experimented on multi-walled carbon nanotubes and removed the cap from one end of the outer shell to attach a moveable nanomanipulator to the core in a high-resolution transmission electron microscope. By pulling the core out and allowing it to retract back into the outer shell, they found an ultralow frictional force against the intershell sliding. They also observed that the extruded core, after release, quickly and fully retracts inside the outer shell due to the restoring force resulting from the van der Waals interaction acting on the extruded core. These results led Zheng and Jiang (2002) and Zheng et al. (2002) to study molecular gigahertz oscillators, for which the sliding of the inner-shell inside the outer-shell of a multi-walled carbon nanotube can generate oscillatory frequencies up to several gigahertz.

Based on the results of Zheng et al. (2002), the shorter the inner core nanotube, the higher is the frequency. As a result, instead of using multi-walled carbon nanotubes, Liu et al. (2005) investigated the high frequencies generated by using a C_{60} fullerene that is oscillating inside a single-walled carbon nanotube. Furthermore, in contrast to the multi-walled carbon nanotube oscillator, the C_{60}–nanotube oscillator tends not to suffer from a rocking motion, as a consequence of the reduced frictional effect. While Qian et al. (2001) and Liu et al. (2005) use molecular dynamics simulations to study this problem, this chapter employs elementary mechanical principles to provide models for the C_{60}–single-walled carbon nanotube, double-walled carbon nanotubes, and single-walled carbon nanotube–nano-bundle oscillators. The oscillatory behavior of these systems is investigated by utilizing the continuum approximation arising from the assumption that the discrete atoms can be smeared across each surface. The chapter provides a synopsis of the work of the authors appearing in Baowan and Hill (2007), Baowan et al. (2008), Cox et al. (2007a–c, 2008), Hilder and Hill (2007), and Thamwattana and Hill (2008).

36.2 Interaction Energy

For two separate molecular structures (i.e., non-bonded), the interaction energy E can be evaluated using either a discrete atom–atom formulation or a continuous approach. Thus, the nonbonded interaction energy may be obtained either as a summation of the interaction energy between each atom pair, namely,

$$E = \sum_i \sum_j \Phi(\rho_{ij}), \qquad (36.1)$$

where Φ is the potential function for atoms i and j located at a distance ρ_{ij} apart on two distinct molecular structures and it is assumed that each atom on the two molecules has a well-defined coordinate position. Alternatively, the continuum approximation assumes that the atoms are uniformly distributed over the entire surface of the molecule, and the double summation in (36.1) is replaced by a double integral over the surface of each molecule, thus

$$E = \eta_1 \eta_2 \iint \Phi(\rho) dS_1 dS_2, \qquad (36.2)$$

where

  η_1 and η_2 are the mean surface density of atoms of the two interacting molecules

  ρ represents the distance between the two typical surface elements dS_1 and dS_2 located on the two interacting molecules

Note that the mean atomic surface density is determined by dividing the number of atoms that make up the molecule by the surface area of that molecule. In this chapter, the mean surface densities for graphene sheet, carbon nanotubes, and C_{60} fullerene are taken to be 0.3812, 0.3812, and 0.3787 Å^{-2}, respectively.

The continuum approach is an important approximation, and Girifalco et al. (2000) state that

> From a physical point of view the discrete atom-atom model is not necessarily preferable to the continuum model. The discrete model assumes that each atom is the centre of a spherically symmetric electron distribution while the continuum model assumes that the electron distribution is uniform over the surface. Both of these assumptions are incorrect and a case can even be made that the continuum model is closer to reality than a set of discrete Lennard-Jones centres.

One such example is a C_{60} fullerene, in which the molecule rotates freely at high temperatures so that the continuum distribution averages out the effect. Qian et al. (2003) suggest that the continuum approach is more accurate for the case when "the C nuclei do not lie exactly in the centre of the electron distribution, as is the case for carbon nanotubes." However, one of the constraints of the continuum approach is that the shape of the molecule must be reasonably well defined to be able to evaluate the integral analytically, and therefore the continuum approach is mostly applicable to highly symmetrical structures, such as spheres, cylinders, and cones. Hodak and Girifalco (2001) point out that the continuum approach ignores the

effect of chirality and that nanotubes are only characterized by their diameters. The continuum or continuous approximation has been successfully applied to a number of systems, including C_{60}–nanotube, C_{60}–C_{60}, and nanotube–nanotube (see for example, Baowan and Hill, 2007; Baowan et al., 2008; Cox et al., 2007a–c, 2008; Thamwattana and Hill, 2008). For the graphite-based and C_{60}-based potentials, Girifalco et al. (2000) state that calculations using the continuum and discrete approximations give similar results, such that the difference between equilibrium distances for the atom–atom interactions is less than 2% (see also Hilder and Hill, 2007).

36.2.1 Interaction Force

The van der Waals force refers to the attractive or repulsive force between molecules, or the intermolecular force. It is named after the Dutch scientist Johannes Diderik van der Waals (1837–1923), since he was one of the first to propose an intermolecular force. For his work on the equation of state for gases and liquids, he won the 1910 Nobel Prize in Physics. The van der Waals force is visible in nature; for example, the gecko climbs walls making use of the van der Waals force between the wall and the gecko's setae, or hair-like structures, on their feet. The van der Waals interaction force between two typical nonbonded atoms of two molecules is given by

$$F_{vdW} = -\nabla E, \qquad (36.3)$$

where

  the energy E is given by either (36.1) or (36.2)

  the symbol ∇ refers to the vector gradient

The gradient in Cartesian coordinates is given by

$$\nabla E(x, y, z) = \frac{\partial E}{\partial x}\hat{\mathbf{i}} + \frac{\partial E}{\partial y}\hat{\mathbf{j}} + \frac{\partial E}{\partial z}\hat{\mathbf{k}}, \qquad (36.4)$$

where $(\hat{\mathbf{i}}, \hat{\mathbf{j}}, \hat{\mathbf{k}})$ denote unit vectors in the (x, y, z) directions, respectively. So, for example, the resultant axial force (z-direction) is obtained by differentiating the integrated interaction energy with respect to z, and therefore (36.3) simplifies to become

$$F_z = -\frac{\partial E}{\partial z}. \qquad (36.5)$$

36.2.2 Lennard-Jones Potential

Sir John Lennard-Jones (1894–1954), a mathematician knighted in 1946, is regarded as the father of modern computational chemistry due to his contributions to theoretical chemistry. In particular, Sir Lennard-Jones had interests in quantum mechanics and intermolecular forces. These interests led to his most notable contribution, the interatomic interaction potential proposed in 1924, and which now bears his name.

The Lennard-Jones potential is a simple mathematical model that describes the interaction between two nonbonded atoms, and is given by

$$\Phi(\rho) = -A\rho^{-m} + B\rho^{-n}, \qquad (36.6)$$

where

A and B are referred to as the attractive and repulsive constants, respectively

ρ is the distance between the atoms

In many cases, the values $m = 6$ and $n = 12$ are adopted, and this is commonly referred to as the 6–12 potential. For hydrogen-bonding interactions, a 10–12 potential is used and it should be noted that there are a number of other empirically motivated potentials in the literature, such as the Morse potential. Alternatively, the Lennard-Jones potential can be written in the form

$$\Phi(\rho) = 4\varepsilon \left[-\left(\frac{\sigma}{\rho}\right)^6 + \left(\frac{\sigma}{\rho}\right)^{12} \right], \qquad (36.7)$$

where

σ is the van der Waals diameter

ε denotes the energy well depth, $\varepsilon = A^2/(4B)$

From (36.7), the equilibrium distance ρ_0 for two atoms is given by $\rho_0 = 2^{1/6}\sigma = (2B/A)^{1/6}$, which is shown in Figure 36.1.

The van der Waals force is a short-range force, and therefore on using the Lennard-Jones potential to represent the interaction between molecular structures it is only necessary to include the nearest neighbor interactions. For example, investigating the behavior of a molecule near the open end of a carbon nanotube, we can consider the tube to be semi-infinite in length.

The Lennard-Jones potential is believed to apply between nonbonded (i.e., atoms on separate molecules) and nonpolar

TABLE 36.1 Lennard-Jones Constants in the Graphitic Systems

	A (eV × Å⁶)	B (eV × Å¹²)	ρ_0 (Å)	$|\varepsilon|$ (meV)
Graphene–graphene	15.2	24.1 × 10³	3.83	2.39
C_{60}–C_{60}	20.0	34.8 × 10³	3.89	2.86
C_{60}–graphene	17.4	29.0 × 10³	3.86	2.62

Source: Girifalco, L.A. et al., *Phys. Rev. B*, 62, 13104, 2000.

(i.e., nonelectrostatic) atomic interactions and has been successfully applied to a number of molecular configurations of carbon nanostructures. For example, two identical parallel carbon nanotubes, between two C_{60} fullerenes, and between a carbon nanotube and a fullerene C_{60} (both inside and outside the tube). Numerical values of the Lennard-Jones constants for carbon–carbon atoms and atoms in graphene–graphene, C_{60}–C_{60}, and C_{60}–graphene are shown in Table 36.1, and these parameter values are used throughout this chapter.

36.2.3 Acceptance Condition and Suction Energy

The suction energy (W) is defined as the total work performed by van der Waals interactions on a molecule entering a carbon nanotube. In certain cases, the van der Waals force becomes repulsive as the entering particle crosses the tube opening. In these cases, the acceptance energy (W_a) is defined as the total work performed by the van der Waals interactions on the particle entering the nanotube, up until the point that the van der Waals force once again becomes attractive.

36.2.3.1 Acceptance Condition

For a particular molecule, with a center of mass located at a distance Z from the nanotube end (negative Z in Figure 36.2), to be accepted into the interior of a nanotube (positive Z) by the van der Waals force alone, the sum of its kinetic energy and the work done from $-\infty$ to Z_0, the positive root of the interaction force as shown in Figure 36.2, must be positive. The acceptance

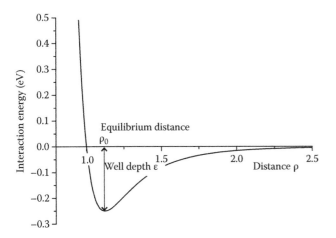

FIGURE 36.1 Graph of Lennard-Jones potential.

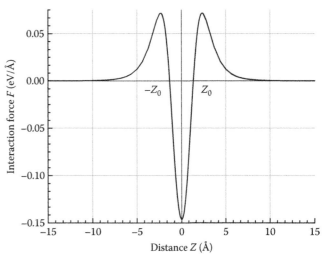

FIGURE 36.2 Typical interaction force.

condition becomes $E_K + W_a > 0$, where E_K is the kinetic energy of the entering molecule and W_a is the work done from $-\infty$ to Z_0, and is termed the acceptance energy. Therefore, for a molecule assumed initially at rest and located on the tube axis, to be accepted into the interior of a nanotube, the condition becomes $W_a > 0$ and the following inequality must hold true:

$$\int_{-\infty}^{Z_0} F(Z)dZ > 0, \qquad (36.8)$$

where again Z_0 is the positive root of the axial interaction force $F(Z)$, as shown in Figure 36.2, and this is the formal condition for a molecule to enter the nanotube.

36.2.3.2 Suction Energy

The suction energy W is defined as the total energy or work done generated by van der Waals interactions acquired by a particular molecule as a consequence of being sucked into the nanotube, or more formally with reference to Figure 36.2

$$W = \int_{-\infty}^{\infty} F(Z)dZ, \qquad (36.9)$$

which is the work done that is transformed into kinetic energy. Alternatively, the suction energy can be written as

$$W = -\int_{-\infty}^{\infty} \frac{dE}{dz} dz = E(-\infty) - E(\infty). \qquad (36.10)$$

36.3 Oscillatory Behavior

Imagine an oscillating system such as the pendulum of a clock. If we displace the pendulum in one direction from its equilibrium position, the gravitational force will push it back toward equilibrium. If we displace it in the other direction, the gravitational force still acts toward the equilibrium position. No matter what the direction of the displacement, the force always acts in a direction to restore the system to its equilibrium position. Similarly, in a nano-oscillator, the inner structure wants to relocate to the equilibrium location from which it is perturbed.

The frequency f is a measure of the number of occurrences of a repeating event per unit time. The period T is the duration of one cycle in a repeating event and the period is the reciprocal of the frequency $f = 1/T$. The period is measured in time units, seconds (s), while frequency is measured in the SI unit of hertz (Hz). For a constant velocity V, we have $V = X/T$, where X is the total displacement, so that we may derive $f = V/X$.

In the following section, the analysis for the oscillatory behavior of a C_{60} fullerene inside a carbon nanotube is given. Subsequently, similar techniques are employed to study the oscillatory behavior of double-walled carbon nanotubes and a carbon nanotube in a nano-bundle.

36.4 Oscillation of a Fullerene C_{60} inside a Single-Walled Carbon Nanotube

36.4.1 Interaction Energy

The interaction between an approximately spherical fullerene and a cylindrical carbon nanotube is modeled using the continuum approximation obtained by averaging atoms over the surface of each entity. First, the calculation for the interaction energy between a carbon atom and a C_{60} fullerene is reviewed utilizing the Lennard-Jones potential function. The potential energy for an atom on the tube interacting with all atoms of the sphere with radius a is given by

$$E(\rho) = -Q_6(\rho) + Q_{12}(\rho),$$

where Q_n is defined by

$$Q_n = C_n \eta_f \int_S \frac{1}{\rho^n} dS,$$

and ρ denotes the distance between a typical tube surface element and the center of the fullerene, and is given by $\rho^2 = b^2 + (Z - z)^2$. The constants C_6 and C_{12} are the Lennard-Jones potential constants A and B, respectively, and η_f represents the atomic surface density of a C_{60} fullerene. Upon integrating, we obtain

$$Q_n(\rho) = \frac{2C_n \eta_f \pi a}{\rho(2-n)} \left(\frac{1}{(\rho+a)^{n-2}} - \frac{1}{(\rho-a)^{n-2}} \right). \qquad (36.11)$$

By substituting Equation 36.11 into $E(\rho)$ and simplifying, the interaction energy between an atom and a C_{60} fullerene can be obtained and is given by

$$E(\rho) = \frac{\eta_f \pi a}{\rho} \left[\frac{A}{2} \left(\frac{1}{(\rho+a)^4} - \frac{1}{(\rho-a)^4} \right) - \frac{B}{5} \left(\frac{1}{(\rho+a)^{10}} - \frac{1}{(\rho-a)^{10}} \right) \right].$$

$$(36.12)$$

The total potential energy between a C_{60} fullerene and a carbon nanotube is obtained by performing another surface integral on (36.12) over a cylindrical nanotube.

From Figure 36.3, the van der Waals interaction force between the fullerene molecule and an atom on the tube is of the form $F_{vdW} = -\nabla E$, and therefore, the axial force is obtained by differentiating the energy with respect to the axial direction z and is given by

$$F_z = -\frac{(Z-z)}{\rho} \frac{dE}{d\rho}.$$

The total axial force between the entire carbon nanotube and the fullerene is obtained by integrating the van der Waals force

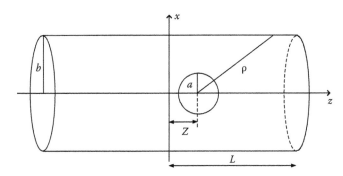

FIGURE 36.3 Geometry for the fullerene C_{60} oscillation.

over the cylindrical nanotube surface. In this case, such a force is independent of the cylindrical angle, and therefore we may deduce

$$F_z^{tot}(Z) = -2\pi b \eta_g \int_0^\infty \frac{dE}{d\rho} \frac{(Z-z)}{\rho} dz, \qquad (36.13)$$

where η_g is the mean atomic density of carbon atoms in a graphene structure such as a carbon nanotube, and since $\rho^2 = b^2 + (Z-z)^2$, we have $d\rho = -[(Z-z)/\rho]dz$. Thus, (36.13) can be simplified to give

$$F_z^{tot}(Z) = 2\pi b \eta_g \int_{\sqrt{b^2+z^2}}^\infty \frac{dE}{d\rho} d\rho,$$

$$= -2\pi^2 \eta_f \eta_g a^2 b \left[\frac{A}{2\rho a} \left(\frac{1}{(\rho+a)^4} - \frac{1}{(\rho-a)^4} \right) \right.$$
$$\left. - \frac{B}{5\rho a} \left(\frac{1}{(\rho+a)^{10}} - \frac{1}{(\rho-a)^{10}} \right) \right]_{\rho=\sqrt{b^2+z^2}}. \qquad (36.14)$$

Now by placing the fractions over common denominators, expanding and reducing to fractions in terms of powers of $(\rho^2 - a^2)$, it can be shown that

$$\frac{A}{2\rho a} \left(\frac{1}{(\rho+a)^4} - \frac{1}{(\rho-a)^4} \right) = -4A \left(\frac{1}{(\rho^2-a^2)^3} + \frac{2a^2}{(\rho^2-a^2)^4} \right), \qquad (36.15)$$

$$\frac{B}{5\rho a} \left(\frac{1}{(\rho+a)^{10}} - \frac{1}{(\rho-a)^{10}} \right) = -\frac{4B}{5} \left(\frac{5}{(\rho^2-a^2)^6} + \frac{80a^2}{(\rho^2-a^2)^7} + \frac{336a^4}{(\rho^2-a^2)^8} \right.$$
$$\left. + \frac{512a^6}{(\rho^2-a^2)^9} + \frac{256a^8}{(\rho^2-a^2)^{10}} \right). \qquad (36.16)$$

Substituting these identities into (36.14) gives a precise expression for the z component of the van der Waals force experienced by a fullerene located at a position Z on the z-axis as

$$F_z^{tot}(Z) = \frac{8\pi^2 \eta_f \eta_g b}{a^4 \lambda^3} \left[A \left(1 + \frac{2}{\lambda} \right) - \frac{B}{5a^6 \lambda^3} \left(5 + \frac{80}{\lambda} + \frac{336}{\lambda^2} \right. \right.$$
$$\left. \left. + \frac{512}{\lambda^3} + \frac{256}{\lambda^4} \right) \right], \qquad (36.17)$$

where $\lambda = (b^2 - a^2 + Z^2)/a^2$. Analytically determining the roots of $F_z^{tot}(Z)$ is not a simple task due to the complexity of the expression and the order of the polynomial involved, but the roots can be determined numerically. The function for the spherical C_{60} fullerene inside a carbon nanotube is demonstrated in Figure 36.4, there will be at most two real roots of the form $Z = \pm Z_0$, and these roots will only exist when the value of b is less than some critical value b_0 for some particular value of the parameter a. In the case of a C_{60} fullerene, ($a = 3.55$ Å) and $b_0 \approx 6.509$ Å.

The integral of $F_z^{tot}(Z)$, as defined by (36.17), represents the work imparted onto the fullerene and equates directly to the kinetic energy. Therefore, the integral of (36.17) from $-\infty$ to Z_0 represents the acceptance energy (W_a) for the system and would need to be positive for a nanotube to accept a fullerene by suction forces alone. If the acceptance energy is negative, then this represents the magnitude of initial kinetic energy needed by the fullerene in the form of the inbound initial velocity for it to be accepted into the nanotube.

To calculate this acceptance energy, we make a change of variable $Z = \sqrt{b^2 - a^2} \tan\psi$ so that $\lambda = (b^2 - a^2)\sec^2\psi/a^2$ and $dZ = \sqrt{b^2 - a^2} \sec^2\psi d\psi$ and the limits of the integration change to $-\pi/2$ and $\psi_0 = \tan^{-1}\left(Z_0/\sqrt{b^2 - a^2}\right)$, which yields

$$W_a = \frac{8\pi^2 \eta_f \eta_g b}{a^2 \sqrt{b^2 - a^2}}$$

$$\times \left[A(J_2 + 2J_3) - \frac{B}{5a^6} (5J_5 + 80J_6 + 336J_7 + 512J_8 + 256J_9) \right] \qquad (36.18)$$

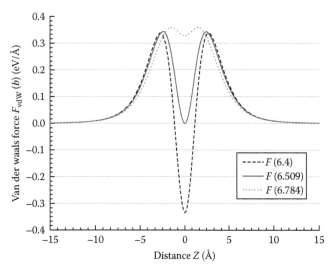

FIGURE 36.4 Force experienced by a C_{60} fullerene due to van der Waals interaction with a semi-infinite carbon nanotube.

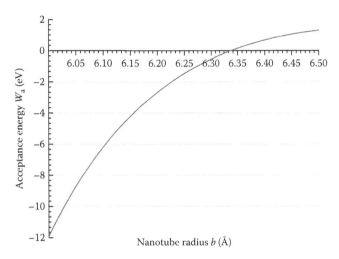

FIGURE 36.5 Acceptance energy threshold for a C_{60} fullerene to be sucked into a carbon nanotube.

where $J_n = a^{2n}(b^2 - a^2)^{-n} \int_{-\pi/2}^{\psi_0} \cos^{2n}\psi \, d\psi$. In this case, a value of Z_0 cannot be specified explicitly and must be determined numerically. Once determined, it can be substituted in the expression for W_a for any value of parameters where $b < b_0$. In Figure 36.5, the acceptance energy is graphed for a C_{60} fullerene and a nanotube of radii in the range $6.1 < b < 6.5$ Å, using the values of Z_0, as graphed in Figure 36.6. The graphs show that $W_a = 0$ when $b \approx 6.338$ Å, and nanotubes that are smaller than this will not accept C_{60} fullerenes by suction forces alone. Therefore, this model predicts that a $(10, 10)$ nanotube ($b = 6.784$ Å) will accept a C_{60} fullerene from rest, but a $(9, 9)$ nanotube ($b = 6.106$ Å) will not. As an example, for the $(8, 8)$ nanotube with a radius $b = 5.428$ Å, the acceptance energy predicted by this model is $W_a = -252$ eV. This equates to firing the C_{60} fullerene at an unlikely speed greater than 8200 m s^{-1} for the molecule to be accepted into the nanotube interior. Also, when $b > b_0$, the force

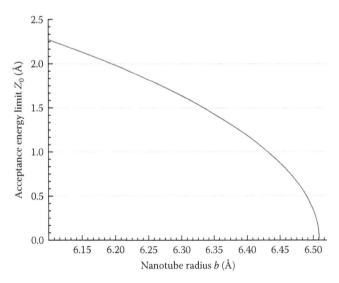

FIGURE 36.6 Upper limit of integration Z_0 used to determine the acceptance energy for a C_{60} fullerene and carbon nanotube.

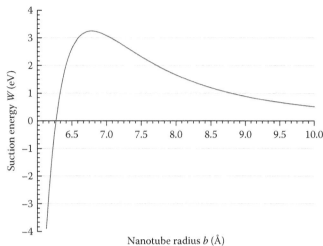

FIGURE 36.7 Suction energy for a C_{60} fullerene entering a carbon nanotube.

graph does not cross the axis and therefore Z_0 is not real, and in this case the fullerene will always be accepted by the nanotube.

The suction energy (W) for a fullerene can be determined from the same integral that is used to obtain the acceptance energy (W_a) (36.18) except that the upper limit of integration is changed from ψ_0 to $\pi/2$. In this case, the value for J_n becomes

$$J_n = \frac{a^{2n}}{(b^2 - a^2)^n} \frac{(2n-1)!!}{(2n)!!} \pi,$$

where $!!$ represents the double factorial notation such that $(2n - 1)!! = (2n - 1)(2n - 3)\ldots 3 \cdot 1$ and $(2n)!! = (2n)(2n - 2)\ldots 4 \cdot 2$. Substitution and simplification gives

$$W = \frac{\pi^3 \eta_f \eta_g a^2 b}{(b^2 - a^2)^{5/2}} \left[A(3 + 5\mu) \right.$$
$$\left. - \frac{B(315 + 4620\mu + 18018\mu^2 + 25740\mu^3 + 12155\mu^4)}{160(b^2 - a^2)^3} \right],$$
$$(36.19)$$

where $\mu = a^2/(b^2 - a^2)$. In Figure 36.7, the suction energy W is plotted for a C_{60} fullerene entering a nanotube with radii in the range $6 < b < 10$ Å. The graph shows that W is positive whenever $b > 6.27$ Å and has a maximum value of $W = 3.243$ eV when $b = b_{max} = 6.783$ Å. It is worth noting that a $(10, 10)$ carbon nanotube with $b \approx 6.784$ Å is almost exactly the optimal size to maximize W, and therefore it is possible to have a C_{60} fullerene accelerate to a maximum velocity upon entering the nanotube.

36.4.2 Oscillatory Behavior

In an axially symmetric cylindrical polar coordinate system (r, θ, z), it is assumed that a fullerene C_{60} is located inside a carbon nanotube of length $2L$, centered around the z-axis and of radius b.

TABLE 36.2 Suction Energy and Velocity for Various Oscillator Configurations

Oscillator Configuration	Tube Radius b (Å)	Energy W (eV)	Velocity v (m/s)
C_{60}–(10, 10)	6.784	3.243	932
C_{60}–(11, 11)	7.463	2.379	798
C_{60}–(12, 12)	8.141	1.512	636
C_{60}–(13, 13)	8.820	0.982	513

As shown in Figure 36.3, it is also assumed that the center of the C_{60} molecule is on the z-axis. This is justified for the carbon nanotube (10, 10) with the radius $b = 6.784$ Å (Table 36.2).

From the symmetry of the problem, only the force in the axial direction needs to be considered. As a result, Newton's second law, neglecting the frictional force, gives

$$M\frac{d^2Z}{dt^2} = F_z^{tot}(Z), \qquad (36.20)$$

where

M is the total mass of a C_{60} molecule which is 1.196×10^{-24} kg
$F_z^{tot}(Z)$ is the total axial van der Waals interaction force between the C_{60} molecule and the carbon nanotube, given by

$$F_z^{tot}(Z) = 2\pi b \eta_g [E(\rho_2) - E(\rho_1)], \qquad (36.21)$$

where $E(\rho)$ is the potential function given by (36.12), which is restated here

$$E(\rho) = \frac{\eta_f \pi a}{\rho}\left[\frac{A}{2}\left(\frac{1}{(\rho+a)^4} - \frac{1}{(\rho-a)^4}\right) - \frac{B}{5}\left(\frac{1}{(\rho+a)^{10}} - \frac{1}{(\rho-a)^{10}}\right)\right],$$

where

$$\rho_1 = \sqrt{b^2 + (Z+L)^2}$$
$$\rho_2 = \sqrt{b^2 + (Z-L)^2}$$

Here, it is assumed that the effect of the frictional force may be neglected, which is reasonable for certain chiralities and diameters of the tube. For example, the preferred position of the C_{60} molecule inside the carbon nanotube (10, 10) is when the center is located on the z-axis. In this case, the molecule tends to move along in the axial direction and tends not to suffer a rocking motion.

In Figure 36.8, $F_z^{tot}(Z)$, as given by (36.21), is plotted for the case of the carbon nanotube (10, 10) ($b = 6.784$ Å) with a length $2L = 100$ Å. It can be observed that the force is very close to zero everywhere except at both ends of the tube, where there is a pulse-like force that attracts the buckyball back toward the center of the tube. For $a < b \ll L$, it is found that $F_z^{tot}(Z)$ can be approximated by using the Dirac delta function $\delta(x)$ and thus (36.20) reduces to give

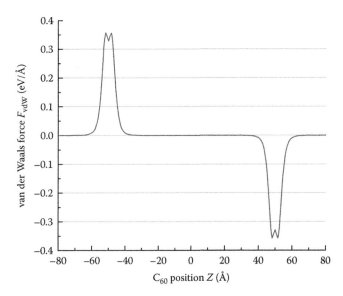

FIGURE 36.8 Plot of $F_z^{tot}(Z)$ as given by (36.21) for the buckyball oscillating inside the carbon nanotube (10, 10).

$$M\frac{d^2Z}{dt^2} = W[\delta(Z+L) - \delta(Z-L)], \qquad (36.22)$$

where W is the suction energy for the C_{60} molecule that is given by (36.19). Now, multiplying (36.22) by dZ/dt on both sides produces

$$M\frac{d^2Z}{dt^2}\frac{dZ}{dt} = W[\delta(Z+L) - \delta(Z-L)]\frac{dZ}{dt}. \qquad (36.23)$$

From $dH(x)/dx = \delta(x)$, where $H(x)$ is the usual Heaviside unit step function, (36.23) can be written as

$$\frac{M}{2}\frac{d}{dt}\left(\frac{dZ}{dt}\right)^2 = W\frac{d}{dZ}[H(Z+L) - H(Z-L)]\frac{dZ}{dt}. \qquad (36.24)$$

By integrating both sides of (36.24) with respect to t, from $t = 0$, (assuming the buckyball is at infinity) to any time t, where the ball is located at Z, and subsequently using that $H(Z + L) - H(Z - L) = 1$ for $-L \le Z \le L$ and zero elsewhere, yields

$$\frac{M}{2}\left(\frac{dZ}{dt}\right)^2 = W[H(Z+L) - H(Z-L)] + \frac{M}{2}v_0^2, \qquad (36.25)$$

where v_0 is the initial velocity that the ball is fired on the z-axis toward the open end of the carbon nanotube in the positive z-direction. From (36.25) for $-L \le Z \le L$, we have

$$\frac{M}{2}\left(\frac{dZ}{dt}\right)^2 = W + \frac{M}{2}v_0^2, \qquad (36.26)$$

which implies that the buckyball travels inside the carbon nanotube at a constant speed $dZ/dt = v = (2W/M + v_0^2)^{1/2}$. An initial velocity v_0 may be necessary for the case where the C_{60} molecule is not sucked into the carbon nanotube due to the strong repulsion force.

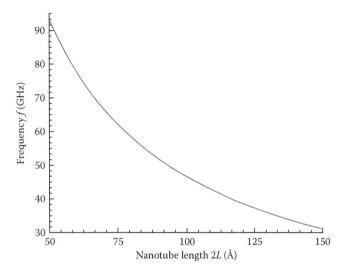

FIGURE 36.9 The variation of the oscillatory frequency of the buckyball with respect to the length of the carbon nanotube (10, 10).

On using (36.26), we obtain a velocity $v = 932\,\text{m s}^{-1}$ for the case when the C_{60} molecule is initially at rest outside the carbon nanotube (10, 10) and the molecule gets sucked into the tube due to the attractive force. This gives rise to a frequency $f = v/(4L) = 46.6\,\text{GHz}$, where the length of the nanotube is assumed to be $2L = 100\,\text{Å}$. In Figure 36.9, the oscillatory frequency is plotted with respect to the tube length. Also considered is the case where the C_{60} molecule is fired on the tube axis toward the open end of the carbon nanotube of radius $a < 6.338\,\text{Å}$, which does not accept a C_{60} molecule by suction forces alone due to the strong repulsive force of the carbon nanotube. For the carbon nanotube (9, 9) ($b = 6.106\,\text{Å}$), an initial velocity v_0 needs to be approximately $1152\,\text{m s}^{-1}$ for the C_{60} molecule to penetrate into the tube. The C_{60} molecule cannot penetrate into either (8, 8), (7, 7), (6, 6), or (5, 5) nanotubes, even though it is fired into the tube with an initial velocity as high as $1600\,\text{m s}^{-1}$. In addition, it is found from the present model that for an (8, 8) nanotube, with $a = 5.428\,\text{Å}$, the minimum initial velocity must be approximately $8210\,\text{m s}^{-1}$ for the C_{60} molecule to penetrate into the tube.

36.5 Oscillation of Double-Walled Carbon Nanotubes

36.5.1 Interaction Energy

With reference to a rectangular Cartesian coordinate system (x_1, y_1, z_1) with its origin located at the center of the outer tube, a typical point on the surface of the inner tube has the coordinates $(a\cos\theta_1, a\sin\theta_1, z_1)$, where a is the assumed radius of the inner tube. Similarly, with reference to a rectangular Cartesian coordinate system (x_2, y_2, z_2) with the origin located at the center of the outer tube, a typical point on the surface of the outer tube has the coordinates $(b\cos\theta_2, b\sin\theta_2, z_2)$, where b is the assumed radius of the outer tube, as shown in Figure 36.10. Now assuming that the two tubes are concentric and that the distance between their centers is Z, the distance ρ between two typical points is given by

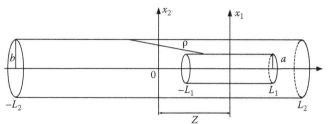

FIGURE 36.10 Double-walled carbon nanotubes of lengths $2L_1$ and $2L_2$. (From Baowan, D. and Hill, J.M., *Z. Angew. Math. Phys.*, 58, 857, 2007. With permission.)

$$\rho^2 = a^2 + b^2 - 2ab\cos(\theta_1 - \theta_2) + (z_2 - z_1)^2.$$

The total potential energy of all atoms of the inner tube interacting with all atoms of the outer tube is given by

$$E^{\text{tot}} = \eta_g^2 ab \int_0^{2\pi}\int_0^{2\pi} (-AI_6 + BI_{12})d\theta_1 d\theta_2, \qquad (36.27)$$

where

η_g represents the mean surface density of the carbon nanotubes

a and b are the radii of the inner and outer tubes, respectively

We define the integrals I_n ($n = 6, 12$) as follows:

$$I_n = \int_{-L_2}^{L_2}\int_{Z-L_1}^{Z+L_1} \frac{dz_1 dz_2}{\rho^n} = \int_{-L_2}^{L_2}\int_{Z-L_1}^{Z+L_1} \frac{dz_1 dz_2}{\left[\lambda^2 + (z_2 - z_1)^2\right]^{n/2}}, \qquad (36.28)$$

where λ^2 denotes $a^2 + b^2 - 2ab\cos(\theta_1 - \theta_2)$. On letting $x = z_2 - z_1$, the integral I_6 becomes

$$I_6 = \int_{Z-L_1}^{Z+L_1}\int_{-L_2-z_1}^{L_2-z_1} \frac{dx\,dz_1}{(\lambda^2 + x^2)^3} = \frac{1}{\lambda^5}\int_{Z-L_1}^{Z+L_1}\int_{-\tan^{-1}\left[(L_2+z_1)/\lambda\right]}^{\tan^{-1}\left[(L_2-z_1)/\lambda\right]} \cos^4\phi\,d\phi\,dz_1,$$

where the final line follows by making the substitution $x = \lambda\tan\phi$ to obtain

$$I_6 = \frac{1}{\lambda^5}\int_{Z-L_1}^{Z+L_1}\left\{\frac{3}{8}\tan^{-1}\left(\frac{L_2-z_1}{\lambda}\right) + \frac{3}{8}\frac{\lambda(L_2-z_1)}{\left[\lambda^2 + (L_2-z_1)^2\right]}\right.$$
$$+ \frac{1}{4}\frac{\lambda^3(L_2-z_1)}{\left[\lambda^2 + (L_2-z_1)^2\right]^2} + \frac{3}{8}\tan^{-1}\left(\frac{L_2+z_1}{\lambda}\right)$$
$$\left.+ \frac{3}{8}\frac{\lambda(L_2+z_1)}{\left[\lambda^2 + (L_2+z_1)^2\right]} + \frac{1}{4}\frac{\lambda^3(L_2+z_1)}{\left[\lambda^2 + (L_2+z_1)^2\right]^2}\right\}dz_1.$$

Finally, using the two substitutions $x = (L_2 - z_1)/\lambda$ and $y = (L_2 + z_1)/\lambda$ gives

$$I_6 = \sum_{i=1}^{4}(-1)^{i+1}\left\{\frac{3}{8}\frac{(Z-\ell_i)}{\lambda^5}\tan^{-1}\left(\frac{Z-\ell_i}{\lambda}\right) - \frac{1}{8\lambda^2\left[\lambda^2+(Z-\ell_i)^2\right]}\right\},$$

(36.29)

and by precisely the same method, I_{12} becomes

$$I_{12} = \sum_{i=1}^{4}(-1)^{i+1}\left\{\frac{63}{256}\frac{(Z-\ell_i)}{\lambda^{11}}\tan^{-1}\left(\frac{Z-\ell_i}{\lambda}\right) - \frac{21}{256\lambda^8\left[\lambda^2+(Z-\ell_i)^2\right]}\right.$$
$$-\frac{21}{640\lambda^6\left[\lambda^2+(Z-\ell)^2\right]^2} - \frac{3}{160\lambda^4\left[\lambda^2+(Z-\ell_i)^2\right]^3}$$
$$\left. -\frac{1}{80\lambda^2\left[\lambda^2+(Z-\ell_i)^2\right]^4}\right\},$$

(36.30)

where the four lengths ℓ_i ($i = 1, 2, 3, 4$) are defined by $\ell_1 = -(L_1 + L_2)$, $\ell_2 = -(L_2 - L_1)$, $\ell_3 = L_1 + L_2$, and $\ell_4 = L_2 - L_1$, and these are the locations for the four critical positions for the oscillation, as shown in Figure 36.11.

Thus, from (36.29) and (36.30) there are two types of integrals that need to be determined, and they are given by

$$K_n^* = \int_0^{2\pi}\int_0^{2\pi}\frac{d\theta_1 d\theta_2}{\lambda^m(\lambda^2+P_i^2)^n}, \quad L_n^* = \int_0^{2\pi}\int_0^{2\pi}\frac{1}{\lambda^n}\tan^{-1}\left(\frac{P_i}{\lambda}\right)d\theta_1 d\theta_2,$$

(36.31)

where

    m and n are certain positive integers

    P_i ($i = 1, 2$) is the abbreviation used for $P_1 = Z + L_1$ and $P_2 = Z - L_1$

Their evaluation is detailed in Appendix 36.A, and (36.31) can be integrated to yield

$$K_n^* = \frac{4\pi^2}{(a+b)^m\left[(a+b)^2+P_i^2\right]^n}\sum_{j=0}^{\infty}\frac{(1/2)_j(m/2)_j}{(j!)^2}\left(\frac{4ab}{(a+b)^2}\right)^j$$
$$\times F\left(\frac{1}{2}+j, n; 1+j; \frac{4ab}{(a+b)^2+P_i^2}\right),$$

$$L_n^* = 4\pi^2\sum_{k=0}^{\infty}\sum_{j=0}^{\infty}\frac{P_i^{2k+1}(2k)!}{2^{2k}(k!)^2(2k+1)(a+b)^n\left[(a+b)^2+P_i^2\right]^{k+1/2}}\frac{(1/2)_j(n/2)_j}{(j!)^2}$$
$$\times\left(\frac{4ab}{(a+b)^2}\right)^j F\left(\frac{1}{2}+j, k+\frac{1}{2}; 1+j; \frac{4ab}{(a+b)^2+P_i^2}\right),$$

where $F(a, b; c; z)$ denotes the usual hypergeometric function. Although complicated, numerical values for these integrals

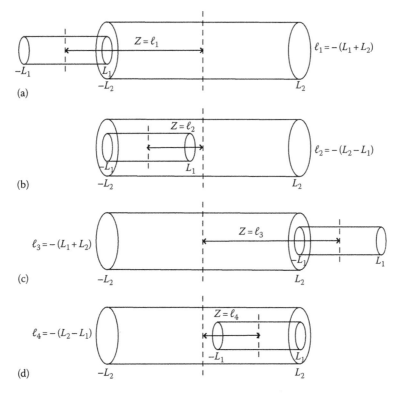

FIGURE 36.11 Four critical positions for two concentric nanocylinders. (From Baowan, D. and Hill, J.M., *Z. Angew. Math. Phys.*, 58, 857, 2007. With permission.)

may be readily evaluated using the algebraic computer package MAPLE or a similar mathematical software package.

Using the algebraic computer package MAPLE, the potential function and van der Waals force versus the difference between the centers of the tubes Z and varying inner tube lengths are shown in Figures 36.12 and 36.13, respectively. As a result of the four critical positions for the distance between the centers of the tubes, there are three regions for the inner tube behavior, (ℓ_1, ℓ_2), (ℓ_2, ℓ_4), and (ℓ_4, ℓ_3). In terms of the potential function, the inner tube will travel with decreasing potential energy in the first region to reach a constant minimum energy in the second region, after which the energy will increase until it becomes zero at the position that the inner tube leaves the outer tube. For the force distribution, the force is almost zero in the second region, and there are two strong attractive forces in the first and the third regions that tend to keep the inner tube inside. This

means that once inside the outer cylinder, the inner tube will tend to oscillate rather than escape from the outer tube, because the forces at the tube extremities tend to reverse the direction of the motion. However, not every inner tube will necessarily be sucked in by the inter-atomic van der Waals force alone, and it may be necessary to initiate the oscillatory motion either by initially extruding the inner cylinder or by giving the inner tube an initial velocity. We also observe that when $L_1 \ll L_2$, we obtain the peak-like forces that are similar to those obtained for a C_{60} fullerene oscillating inside a single-walled nanotube. The force distribution for double-walled carbon nanotubes, as shown in Figure 36.13, may be approximated by the Heaviside function $H(Z)$ to obtain

$$F_z^{\text{tot}} = W\Big[H(Z + L_2 + L_1) - H(Z + L_2 - L_1) - H(Z - L_2 + L_1)$$
$$+ H(Z - L_2 - L_1) \Big],$$

where $W = -E^{\text{tot}}(Z)$ is the suction energy where $E^{\text{tot}}(Z)$ denotes the interaction potential for two parallel carbon nanotubes given by (36.27). We refer the reader to Baowan and Hill (2007) for further details of this section.

36.5.2 Oscillatory Behavior

Newton's second law is also adopted here to describe the oscillation behavior of double-walled carbon nanotubes with the inner tube oscillating. We determine the frequency of the oscillation for the case where the inner tube is pulled out at a distance d, as shown in Figure 36.14, and released. The frictional force is assumed negligible throughout the movement of the inner tube.

We approximate the van der Waals force experienced by the oscillating nanotube as a Heaviside function and we assume that the inner nanotube travels through three regions, as illustrated in Figure 36.15. Newton's second law for region one can be written as

$$M \frac{d^2 Z}{dt^2} = -W,$$

where
  M denotes the mass of the inner tube
  W is the suction energy for double-walled carbon nanotubes

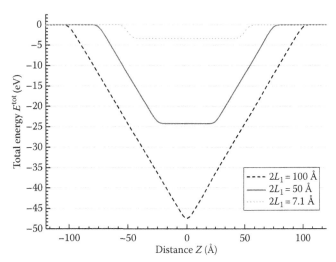

FIGURE 36.12 Total potential energy of a $(5, 5)$ nanotube with various $2L_1$ entering into a $(10, 10)$ nanotube with length $2L_2 = 100$ Å.

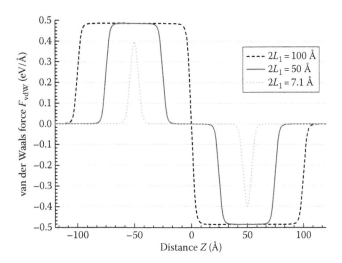

FIGURE 36.13 Force distribution of a $(5, 5)$ nanotube with various $2L_1$ entering into a $(10, 10)$ nanotube with length $2L_2 = 100$ Å.

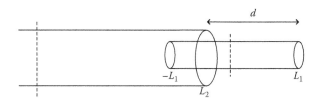

FIGURE 36.14 The extrusion distance d for the inner tube oscillating inside the outer tube. (From Baowan, D. and Hill, J.M., *Z. Angew. Math. Phys.*, 58, 857, 2007. With permission.)

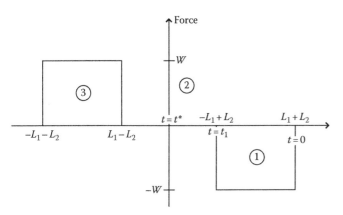

FIGURE 36.15 Three regions for the idealized van der Waals force for double-walled carbon nanotubes oscillator.

We assume that the inner tube is initiated at rest at $Z = Z_0 = L_2 - L_1 + d$, where d is the extrusion distance as illustrated in Figure 36.14. We may derive the velocity equation in this region as given by

$$\frac{dZ}{dt} = -\frac{W}{M}t, \tag{36.32}$$

so that the displacement equation can be obtained as

$$Z(t) = \frac{-W}{2M}t^2 + (L_2 - L_1 + d), \quad -L_1 + L_2 < Z < Z_0,$$

and the time for the inner tube to travel the length of this region is

$$t_1 = \sqrt{\frac{2Md}{W}}.$$

The traveling velocity for the inner tube in the second region is a constant that can be determined from the velocity of the first region (36.32) at $t = t_1$. Similarly, the displacement equation for the second region becomes

$$Z(t) = -\sqrt{\frac{2Wd}{M}}t + (L_2 - L_1 + d), \quad L_1 - L_2 < Z < -L_1 + L_2.$$

The period T for the oscillating inner tube can be obtained as $T = 4t^*$, thus $f = 1/(4t^*)$ and t^* can be determined by the above equation at $Z = 0$. Finally, the oscillation frequency in a particular case for $v_0 = 0$ becomes

$$f = \frac{1}{4}\sqrt{\frac{2W}{M}}\left(\frac{\sqrt{d}}{(L_2 - L_1 + 2d)}\right), \tag{36.33}$$

which has a maximum value at $d = (L_2 - L_1)/2$. In addition, the extrusion distance d must be less than the length of the inner tube $2L_1$. Physically, the frequency only applies when the length

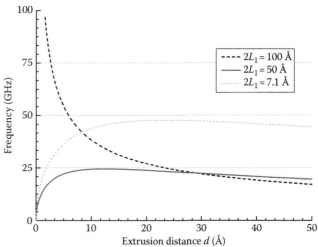

FIGURE 36.16 The frequency profile for $(5,5)$ nanotube of length $2L_1$ oscillating inside $(10,10)$ nanotube versus extrusion distance d when the initial velocity is assumed to be zero.

lies within the limits $L_1 + 2d_{min} < L_2 < 5L_1$, where d_{min} denotes the practical limitation on the minimum extrusion distance.

We observe that the frequency always increases when the initial velocity is increased. Furthermore, the shorter the inner tube, the higher is the frequency. This is because the force is a constant in each case, so that the lighter the weight for the shorter tube, the higher the velocity and, therefore, the higher the frequency. Moreover, the longer extrusion distance tends to increase the oscillatory frequency because it leads to a higher potential energy level. This in turn gives a higher van der Waals force, and a higher velocity, and as a result gives rise to a higher frequency. In the case of equal lengths, when the extrusion distance is increased, this increases the distance for the inner tube to move from one end to the other, which leads to a lower oscillation frequency. The longer length also leads to a larger mass, which tends to slow down the movement as shown in Figure 36.16. Noting that when $L_1 = L_2$, small oscillations occur near a stable equilibrium point. Moreover, when the extrusion and the initial velocity are equal to zero, the system becomes static.

By using this model, we calculate an example of the oscillating $(5,5)$ nanotube $(a = 3.392 \text{ Å})$ in the $(10,10)$ nanotube $(b = 6.784 \text{ Å})$, where both tubes have the same half-length $L_1 = L_2 = 50 \text{ Å}$ and the inner tube is initiated at rest. We obtain $W = 0.4851 \text{ eV Å}^{-1}$, which for an extrusion distance of $d = 50 \text{ Å}$ produces a frequency of $f = 17.12 \text{ GHz}$.

36.6 Oscillation of Nanotubes in Bundles

36.6.1 Interaction Energy

Generally, a nanotube bundle is referred to as a closely packed array of N aligned carbon tubes and the central region can be used to oscillate a single nanotube, which is located and aligned along the central axis (see Figure 36.17).

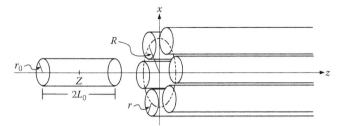

FIGURE 36.17 Geometry of a single nanotube entering a bundle. (From Thamwattana, N. et al., *J. Phys. Condens. Matter.*, 21, 144214, 2009. With permission.)

In this section, we consider a single carbon nanotube oscillating in the middle of a bundle of finite length carbon nanotubes of N-fold symmetry and each of radius r. We assume that the center of the oscillating tube remains on the z-axis during its motion. In a cylindrical polar coordinate system, a typical point on the oscillating tube has the coordinates $(r_0 \cos \theta_0, r_0 \sin \theta_0, z_0 + Z)$, where r_0 is the tube radius, $-L_0 \leq z_0 \leq L_0$, and Z is the distance between the center of the oscillating tube and the origin. The coordinates of a typical surface point of the nanotubes in the bundle are given by

$$\left(R\cos\left(\frac{2\pi(i-1)}{N}\right) + r\cos\theta_i, R\sin\left(\frac{2\pi(i-1)}{N}\right) + r\sin\theta_i, z_i \right),$$
(36.34)

where
$$-L \leq z_i \leq L$$
$$1 \leq i \leq N$$

In this analysis, we again assume that the friction is negligible throughout the motion. Energy dissipation through radial breathing modes and other secondary modes of vibration will have a dampening effect but these are all ignored in the present model.

Due to the assumed symmetry of the problem, we need only consider the interaction between the oscillating tube that is located at the center of the bundle and one of the carbon nanotubes in the bundle surrounding the oscillating tube. As shown in

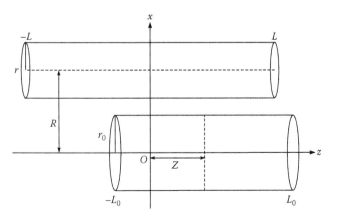

FIGURE 36.18 Geometry of two interacting tubes.

Figure 36.18, we examine the tube with coordinates $(r_0 \cos \theta_0, r_0 \sin \theta_0, z_0 + Z)$, which is in the middle of the bundle and the tube $i = 1$ with coordinates $(r \cos \theta_1 + R, r \sin \theta_1, z_1)$. The total interaction energy of the oscillating nanotube inside the bundle is given by $E^{\text{tot}} = N E$, where E is the energy of the interaction between the two nanotubes and N is the number of tubes in the bundle that are located symmetrically around the inner oscillating nanotube.

Using the Lennard-Jones potential and the continuum approach, we obtain the interaction energy E as

$$E = rr_0\eta_g^2 \int_0^{2\pi}\int_0^{2\pi}\left[\int_{-L}^{L}\int_{-L_0}^{L_0}\left(-\frac{A}{\rho^6} + \frac{B}{\rho^{12}}\right)dz_0 dz_1\right]d\theta_0 d\theta_1,$$
(36.35)

where
η_g is the mean atomic density of a nanotube
ρ denotes the distance between two typical surface elements on each nanotube

which is given by

$$\rho^2 = (r\cos\theta_1 + R - r_0\cos\theta_0)^2 + (r\sin\theta_1 - r_0\sin\theta_0)^2 + (z_1 - z_0 - Z)^2.$$

We can rewrite (36.35) as

$$E = rr_0\eta_g^2 \int_0^{2\pi}\int_0^{2\pi}(-AI_6 + BI_{12})d\theta_0 d\theta_1,$$
(36.36)

where the integrals I_n ($n = 6, 12$) are defined by

$$I_n = \int_{-L}^{L}\int_{Z-L_0}^{Z+L_0}\frac{dz_0^\star dz_1}{\left[\lambda^2 + (z_1 - z_0^\star)^2\right]^{n/2}},$$
(36.37)

where
$$z_0^\star = z_0 + Z$$
$$\lambda^2 = (r\cos\theta_1 + R - r_0\cos\theta_0)^2 + (r\sin\theta_1 - r_0\sin\theta_0)^2$$

We note that Equation 36.37 involves the same integral as Equation 36.28.

We comment that formally for the case $R = 0$, we have the scenario of an oscillating nanotube inside another nanotube, and the integrals I_n can be evaluated analytically in terms of hypergeometric functions, as shown in Section 36.5.1, for the case of double-walled carbon nanotubes. In other words, this system describes the oscillation of double-walled carbon nanotubes. Since $R > 0$, we need to evaluate these integrals numerically.

In Figures 36.19 and 36.20, we plot the total energy E^{tot} and the van der Waals force F_{vdW} for a (5,5) carbon nanotube ($r = 3.392$ Å) oscillating in a 6-fold symmetry (5,5) carbon nanotube bundle. Noting that we use $R = 9.9283$ Å, which is the

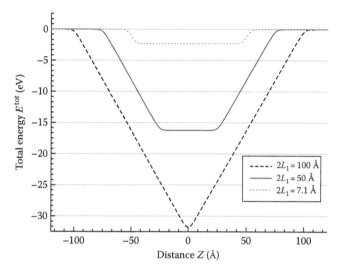

FIGURE 36.19 Potential energy of $(5,5)$ nanotube of length $2L_0$ oscillating inside 6-fold symmetry bundle comprising $(5,5)$ nanotubes of length $2L = 100$ Å.

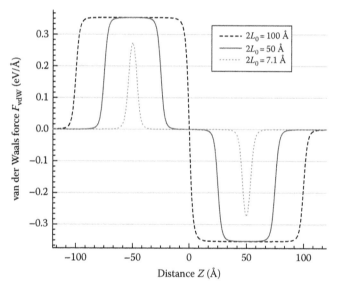

FIGURE 36.20 Force distribution of $(5,5)$ nanotube of length $2L_0$ oscillating inside 6-fold symmetry bundle comprising $(5,5)$ nanotubes of length $2L = 100$ Å.

optimal bundle radius for the interaction of a $(5,5)$ nanotube and a bundle of 6-fold symmetry. A similar behavior to that of the double-walled carbon nanotube oscillators as studied in Section 36.5 is obtained here. A single carbon nanotube has a minimum energy at $Z = 0$ inside the bundle. By pulling the tube away from the minimum energy configuration in either direction, the van der Waals force tends to propel the nanotube back toward the center of the bundle. Accordingly, we obtain an oscillatory motion of the nanotube inside the bundle. We also observe that when $L_0 \ll L$, we obtain peak-like forces that are similar to those obtained for a C_{60} molecule oscillating inside a single-walled nanotube.

36.6.2 Oscillatory Behavior

By adopting the simplified model of motion for oscillating nanotubes presented in Section 36.5, we can show that when $L \geq L_0$ that the van der Waals force F_{vdW} experienced by the oscillating nanotube can be approximated by

$$F_{vdW} = W\big[H(Z + L + L_0) - H(Z + L - L_0) - H(Z - L + L_0)$$
$$+ H(Z - L - L_0)\big],$$

where

$H(z)$ is the Heaviside unit-step function
W is the suction energy per unit length

which is given by

$$W = -NE_{tt}(R), \qquad (36.38)$$

where E_{tt} is the interaction energy for a tube–tube interaction, which is given by (36.47). From this model, we may show that, assuming the nanotube is initially at rest and extruded by a distance d out of the nanotube bundle, then in a similar manner to (36.33) the resulting oscillatory frequency f is given by

$$f = \frac{1}{4}\sqrt{\frac{2W}{M}}\left(\frac{\sqrt{d}}{2d + (L - L_0)}\right), \qquad (36.39)$$

where M is the mass of the oscillating nanotube, which is given by $M = 4\pi r_0 L_0 \eta m_0$, and m_0 is the mass of a single carbon atom, which is 1.993×10^{-26} kg.

As found in Section 36.5, the maximum frequency occurs when the extrusion distance satisfies the relationship $d = (L - L_0)/2$. In this case, the maximum frequency f_{max} is given by

$$f_{max} = \frac{1}{8}\sqrt{\frac{W}{M(L - L_0)}}. \qquad (36.40)$$

However, there are certain limitations on those oscillators that can attain this frequency. Firstly, the extrusion distance d must be less than the length of the oscillating nanotube $2L_0$. This leads to an upper limit on the ratio of the bundle length L to the oscillator length L_0, that is, $L < 5L_0$, and bundle oscillators longer than this will not be able to attain the theoretical maximum frequency f_{max} given in Equation 36.40. Secondly, a sensible lower limit is required on the extrusion distance. The reason for this is that when $L = L_0$, then extremely high frequencies are theoretically achievable by choosing very small extrusion distances d. However, for an oscillator to be practical, the total displacement of the oscillating nanotube needs to be measurable, and this represents a practical limit on the design of such oscillators. We denote this practical limitation on the minimum extrusion distance as d_{min}, and therefore we conclude that Equation 36.40 only applies when the bundle length lies within the limits

$$L_0 + 2d_{min} < L < 5L_0. \tag{36.41}$$

We also comment that the constraints (36.41) have not previously appeared in the literature and they apply equally well to double-walled oscillators.

A comparison of the results of the present model with the molecular dynamics study by Kang et al. (2006) shows reasonable overall agreement considering the assumptions of the model presented here. By using this model, we investigate the oscillation of a (5, 5) nanotube initially at rest in a 6-fold nanotube bundle also comprising of (5, 5) nanotubes. Both the bundle and the oscillating nanotube have the same half-length $L = L_0 = 50$ Å, and R is assumed to be 9.9283 Å. We obtain $W = 0.3998$ eV Å^{-1} that for an extrusion distance of 50 Å produces a frequency of $f = 15.72$ GHz.

Appendix 36.A: Analytical Evaluation of Integrals (36.31)

Analytically two types of integrals for the Lennard-Jones potential function are evaluated. These two integrals are involved in the solution for the total potential energy of the nanostructure system, namely, for the double-walled carbon nanotubes and the nanotubes in bundles, and they are determined in following sections.

36.A.1 Evaluation of the Integral K_n^*

Here, the integral K_n^* is determined. Since λ^2 is an even function of $\theta_1 - \theta_2$, either intermediate integral is independent of the other variables, and therefore this second variable is assigned to have zero value. In this event one integration may be trivially performed and may be deduced

$$K_n^* = 8\pi \int_0^{\pi/2} \frac{dx}{\lambda^m (\lambda^2 + P_i^2)^n},$$

where $\lambda^2 = (a - b)^2 + 4ab \sin^2 x$. Instead of considering the above equation, only the integral K_n is considered, which is given by

$$K_n = \int_0^{\pi/2} \frac{dx}{\lambda^m (\lambda^2 + P_i^2)^n}. \tag{36.42}$$

On letting $\mu + \nu = (b + a)^2$, $\sigma + \nu = (b + a)^2 + P_i^2$, $\beta = (b - a)^2/(b + a)^2$ and $\nu = 4ab$, (36.42) becomes

$$K_n = \int_0^{\pi/2} \frac{dx}{(\mu + \nu \sin^2 x)^{m/2} (\sigma + \nu \sin^2 x)^n}.$$

Making the substitution $t = \cot x$ to obtain

$$K_n = \int_0^\infty \frac{(1 + t^2)^{n + \frac{m}{2} - 1}}{(\nu + \mu + \mu t^2)^{m/2} (\nu + \sigma + \sigma t^2)^n} dt$$

$$= \frac{1}{(\mu + \nu)^{m/2} (\sigma + \nu)^n} \int_0^\infty \frac{(1 + t^2)^{n + \frac{m}{2} - 1}}{(1 + \beta t^2)^{m/2} (1 + \gamma t^2)^n} dt,$$

where
$$\beta = \mu/(\mu + \nu)$$
$$\gamma = \sigma/(\sigma + \nu)$$

Now on writing this integral in the form

$$K_n = \frac{1}{(\mu + \nu)^{m/2} (\sigma + \nu)^n}$$

$$\times \int_0^\infty \frac{1}{\left[1 - (1 - \beta)t^2/(1 + t^2)\right]^{m/2} \left[1 - (1 - \gamma)t^2/(1 + t^2)\right]^n} \frac{dt}{(1 + t^2)},$$

and making the substitution

$$z = \frac{t}{(1 + t^2)^{1/2}}, \quad t = \frac{z}{(1 - z^2)^{1/2}}, \quad dt = \frac{dz}{(1 - z^2)^{3/2}},$$

and the substitution $u = z^2$, obtains

$$K_n = \frac{1}{2(\mu + \nu)^{m/2} (\sigma + \nu)^n} \int_0^1 \frac{u^{-1/2} (1 - u)^{-1/2}}{[1 - (1 - \beta)u]^{m/2} [1 - (1 - \gamma)u]^n} du.$$

Note that $\mu + \nu = (b + a)^2$, $\sigma + \nu = (b + a)^2 + P_i^2$, $\beta = (b - a)^2/(b + a)^2$, and $\gamma = [(b - a)^2 + P_i^2]/[(b + a)^2 + P_i^2]$. According to Bailey (1972, p. 73), the definition of an Appell hypergeometric function of two variables and of the first kind is defined by

$$F_1(\alpha; \beta, \beta'; \gamma; x, y) = \sum_{n=0}^\infty \sum_{m=0}^\infty \frac{(\alpha)_{m+n}(\beta)_m (\beta')n}{m! n! (\gamma)_{m+n}} x^m y^n.$$

Also from Bailey (1972), the expression for the function F_1 in terms of a definite integral and a series involving the ordinary hypergeometric function (pp. 77 and 79) are

$$\frac{\Gamma(\alpha)\Gamma(\gamma - \alpha)}{\Gamma(\gamma)} F_1(\alpha; \beta, \beta'; \gamma; x, y) = \int_0^1 \frac{u^{\alpha-1}(1 - u)^{\gamma-\alpha-1}}{(1 - ux)^\beta (1 - uy)^{\beta'}} du,$$

$$F_1(\alpha; \beta, \beta'; \gamma; x, y) = \sum_{i=0}^\infty \frac{(\alpha)_i (\beta)_i}{i! (\gamma)_i} F(\alpha + i, \beta'; \gamma + i; y) x^i,$$

$$\tag{36.43}$$

so that K_n becomes

$$K_n = \frac{\pi}{2(a+b)^m[(a+b)^2+P_i^2]^n} \sum_{j=0}^{\infty} \frac{(1/2)_j (m/2)_j}{(j!)^2}$$

$$\times F\left(\frac{1}{2}+j, n; 1+j: \frac{4ab}{(a+b)^2+P_i^2}\right)\left[\frac{4ab}{(a+b)^2}\right]^j,$$

where $F(a, b; c; z)$ denotes the usual hypergeometric function.

36.A.2 Evaluation of the Integral L_n^*

Similar to the Section 36.A.1, the integral L_n^* can be written as

$$L_n^* = 8\pi \int_0^{\pi/2} \frac{1}{\lambda^n} \tan^{-1}\left(\frac{P_i}{\lambda}\right) dx,$$

where $\lambda^2 = (a-b)^2 + 4ab\sin^2 x$. For convenience, L_n is defined by

$$L_n = \int_0^{\pi/2} \frac{1}{\lambda^n} \tan^{-1}\left(\frac{P_i}{\lambda}\right) dx.$$

Since $(P_i/\lambda)^2 < \infty$, from Gradshteyn and Ryzhik (2000, p. 59, Equation 1.644.1), it is obtained

$$\tan^{-1}\left(\frac{P_i}{\lambda}\right) = \frac{P_i}{\sqrt{\lambda^2+P_i^2}} \sum_{k=0}^{\infty} \frac{(2k)!}{2^{2k}(k!)^2(2k+1)}\left(\frac{P_i^2}{\lambda^2+P_i^2}\right)^k,$$

and thus

$$L_n = \sum_{k=0}^{\infty} \frac{P_i^{2k+1}(2k)!}{2^{2k}(k!)^2(2k+1)} \int_0^{\pi/2} \frac{1}{\lambda^n(\lambda^2+P_i^2)^{k+1/2}} dx.$$

From the result for K_n in Appendix 36.A.1, it may be deduced

$$L_n = \frac{\pi}{2} \sum_{k=0}^{\infty} \frac{P_i^{2k+1}(2k)!}{2^{2k}(k!)^2(2k+1)} \cdot \frac{1}{(\mu+\nu)^{n/2}(\sigma+\nu)^{k+1/2}}$$

$$\times F_1\left(\frac{1}{2}; \frac{n}{2}, k+\frac{1}{2}; 1; 1-\beta, 1-\gamma\right),$$

and using the reduction of Appell's hypergeometric functions (36.43), the formula for L_n is given by

$$L_n = \frac{\pi}{2} \sum_{k=0}^{\infty} \sum_{j=0}^{\infty} \frac{P_i^{2k+1}(2k)!}{2^{2k}(k!)^2(2k+1)(a+b)^n[(a+b)^2+P_i^2]^{k+1/2}} \cdot \frac{(1/2)_j(n/2)_j}{(j!)^2}$$

$$\times F\left(\frac{1}{2}+j, k+\frac{1}{2}; 1+j; \frac{4ab}{(a+b)^2+P_i^2}\right)\left[\frac{4ab}{(a+b)^2}\right]^j.$$

Commenting that Colavecchia et al. (2001) examine in some detail the numerical evaluation of the usual hypergeometric and the Appell hypergeometric functions.

Appendix 36.B: Derivation of Equation 36.47 for E_{tt}

The interaction energy for a nanotube and a single atom is given by the integral

$$E_{ta} = \eta\left(-AI_3 + BI_6\right),$$

$$I_n = r \int_{-\infty}^{\infty} \int_0^{2\pi} \left[(\rho-r)^2 + 4r\rho\sin^2\frac{\phi}{2} + z^2\right]^{-n} d\phi dz. \tag{36.44}$$

On first evaluating the z integration using the substitution $z = \lambda \tan\psi$, where $\lambda^2 = (\rho-r)^2 + 4r\rho\sin^2(\phi/2)$, we may show that

$$I_n = \frac{r\pi(2n-3)!!}{2^{n-1}(n-1)!} \int_0^{2\pi} \left[(\rho-r)^2 + 4r\rho\sin^2\frac{\phi}{2}\right]^{1/2-n} d\phi.$$

Then by substituting $t = \sin^2(\phi/2)$ we obtain

$$I_n = \frac{r\pi(2n-3)!!}{2^{n-2}(\rho-r)^{2n-1}(n-1)!} \int_0^1 t^{-1/2}(1-t)^{-1/2}\left\{1-\left[1-\left(\frac{\rho+r}{\rho-r}\right)^2\right]t\right\}^{1/2-n} dt,$$

which is the fundamental integral form for the hypergeometric function and therefore

$$I_n = \frac{r\pi^2(2n-3)!!}{2^{n-2}(\rho-r)^{2n-1}(n-1)!} F\left(n-\frac{1}{2}, \frac{1}{2}; 1; 1-\left(\frac{\rho+r}{\rho-r}\right)^2\right),$$

where $F(a, b; c; z)$ is the usual hypergeometric function. Using the transformation from Erdélyi et al. (1953, Section 36.2.9, Equation 36.2), we may show that

$$I_n = \frac{r\pi^2(2n-3)!!}{2^{n-2}(\rho-r)(\rho+r)^{2n-2}(n-1)!} F\left(\frac{3}{2}-n, \frac{1}{2}; 1; 1-\left(\frac{\rho+r}{\rho-r}\right)^2\right).$$

Finally, by using Erdélyi et al. (1953, Section 36.3.2, Equation 36.28), we may transform this expression from a hypergeometric function to a Legendre function and show that

$$I_n = \frac{r\pi^2(2n-3)!!}{2^{n-2}(\rho^2-r^2)^{n-1/2}(n-1)!} P_{n-3/2}\left(\frac{\rho^2+r^2}{\rho^2-r^2}\right), \tag{36.45}$$

where $P_\nu(z)$ is the Legendre function of the first kind of degree ν and by substituting (36.45) into (36.44) gives

$$E_{ta} = \frac{3r\pi^2\eta_g}{4}\left[-\frac{AP_{3/2}(\gamma)}{(\rho^2-r^2)^{5/2}} + \frac{21BP_{9/2}(\gamma)}{32(\rho^2-r^2)^{11/2}}\right], \quad (36.46)$$

where $\gamma = (\rho^2 + r^2)/(\rho^2 - r^2)$.

Using (36.46), we derive an analytical expression for the interaction potential per unit length for two parallel carbon nanotubes E_{tt}. The distance between the axes is denoted by δ and the radii of the tubes are r_1 and r_2. This can be calculated by substituting $\rho^2 = (\delta - r_2)^2 + 4\delta r_2 \sin^2(\theta/2)$ into (36.46) and then integrating over the circumference of the second cylinder. Two analytical expressions are derived for this potential, namely, in the form of a series of associated Legendre functions or alternatively using Appell's hypergeometric functions of two variables that can be expressed as

$$E_{tt} = \frac{3}{2}\eta_g^2 r_1 r_2 \pi^3 \alpha^{-5}\left[-AF_2\left(\frac{5}{2}, -\frac{3}{2}, \frac{1}{2}, 1, 1; -\frac{r_1^2}{\alpha^2}, -\frac{4r_2\delta}{\alpha^2}\right)\right.$$
$$\left. +\frac{21}{32}B\alpha^{-6}F_2\left(\frac{11}{2}, -\frac{9}{2}, \frac{1}{2}, 1, 1; -\frac{r_1^2}{\alpha^2}, -\frac{4r_2\delta}{\alpha^2}\right)\right], \quad (36.47)$$

where

$F_2(\alpha, \beta, \beta', \gamma, \gamma'; x, y)$ is an Appell's hypergeometric function of two variables

α is defined by $\alpha^2 = (\delta - r_2)^2 - r_1^2$

We refer the reader to Cox et al. (2008) for further details of the derivation leading to (36.47).

Acknowledgments

The authors are grateful for the Australian Research Council for support through the Discovery Project Scheme and the provision of Australian Postdoctoral Fellowships for NT and BJC, and an Australian Professorial Fellowship for JMH.

References

Bailey, W.N. (1972). *Generalized Hypergeometric Series*. Hafner Publishing Co., New York.

Baowan, D. and Hill, J.M. (2007). Force distribution for double-walled carbon nanotubes and gigahertz oscillators. *Z. Angew. Math. Phys.*, 58:857–875.

Baowan, D., Thamwattana, N., and Hill, J.M. (2008). Suction energy and offset configuration for double-walled carbon nanotubes. *Commun. Nonlinear Sci. Numer. Simul.*, 13:1431–1447.

Colavecchia, F.D., Gasaneo, G., and Miraglia, J.E. (2001). Numerical evaluation of Appell's F_1 hypergeometric function. *Comput. Phys. Commun.*, 138:29–43.

Cox, B.J., Thamwattana, N., and Hill, J.M. (2007a). Mechanics of atoms and fullerenes in single-walled carbon nanotubes. I. Acceptance and suction energies. *Proc. R. Soc. Lond. A*, 463:461–476.

Cox, B.J., Thamwattana, N., and Hill, J.M. (2007b). Mechanics of atoms and fullerenes in single-walled carbon nanotubes. II. Oscillatory behaviour. *Proc. R. Soc. Lond. A*, 463:477–494.

Cox, B.J., Thamwattana, N., and Hill, J.M. (2007c). Mechanics of fullerenes oscillating in carbon nanotube bundles. *J. Phys. A: Math. Theory*, 40:13197–13208.

Cox, B.J., Thamwattana, N., and Hill, J.M. (2008). Mechanics of nanotubes oscillating in carbon nanotube bundles. *Proc. R. Soc. Lond. A*, 464:691–710.

Cumings, J. and Zettl, A. (2000). Low-friction nanoscale linear bearing realized from multi-walled carbon nanotubes. *Science*, 289:602–604.

Erdélyi, A., Magnus, W., Oberhettinger, F., and Tricomi, F.G. (1953). *Higher Transcendental Functions*, Volume 1. McGraw-Hill, New York.

Girifalco, L.A. (1992). Molecular properties of C_{60} in the gas and solid phases. *J. Phys. Chem.*, 96:858–861.

Girifalco, L.A., Hodak, M., and Lee, R.S. (2000). Carbon nanotubes, buckyballs, ropes, and a universal graphitic potential. *Phys. Rev. B*, 62:13104–13110.

Gradshtcyn, I.S. and Ryzhik, I.M. (2000). *Table of Integrals, Series and Products*, 6th edn. Academic Press, San Diego, CA.

Hilder, T.A. and Hill, J.M. (2007). Continuous versus discrete for interacting carbon nanostructures. *J. Phys. A: Math. Theory*, 40:3851–3868.

Hodak, M. and Girifalco, L.A. (2001). Fullerenes inside carbon nanotubes and multi-walled carbon nanotubes: Optimum and maximum sizes. *Chem. Phys. Lett.*, 350:405–411.

Iijima, S. (1991). Helical microtubules of graphitic carbon. *Nature*, 354:56–58.

Kang, J.W., Song, K.O., Hwang, H.J., and Jiang, Q. (2006). Nanotube oscillator based on a short single-walled carbon nanotube bundle. *Nanotechnology*, 17:2250–2258.

Liu, P., Zhang, Y.W., and Lu, C. (2005). Oscillatory behaviour of C60-nanotube oscillators: A molecular-dynamics study. *J. Appl. Phys.*, 97:094313.

Qian, D., Liu, W.K., and Ruoff, R.S. (2001). Mechanics of C_{60} in nanotubes. *J. Phys. Chem. B*, 105:10753–10758.

Qian, D., Liu, W.K., Subramoney, S., and Ruoff, R.S. (2003). Effect of the interlayer potential on mechanical deformation of multiwalled carbon nanotubes. *J. Nanosci. Nanotechnol.*, 3:185–191.

Thamwattana, N. and Hill, J.M. (2008). Oscillation of nested fullerenes (carbon onions) in carbon nanotubes. *J. Nanopart. Res.*, 10:665–677.

Thamwattana, N., Cox, B., and Hill, J.M. (2009). Oscillation of carbon molecules inside carbon nanotube bundles. *J. Phys. Condens. Matter*, 21:144214–144219.

Zheng, Q. and Jiang, Q. (2002). Multiwalled carbon nanotubes as gigahertz oscillators. *Phys. Rev. Lett.*, 88:045503.

Zheng, Q., Liu, J.Z., and Jiang, Q. (2002). Excess van der Waals interaction energy of a multiwalled carbon nanotube with an extruded core and the induced core oscillation. *Phys. Rev. B*, 65:245409.

Nanoelectromechanical Resonators

Andrew N. Cleland
University of California

37.1 Introduction

Nanometer-scale mechanical devices are undergoing extensive development, both for advancing pure and applied research as well as for potential industrial and commercial applications. Typically such devices are actuated and sensed using electronic methods, and as such are termed nanoelectromechanical system (NEMS) resonators. Potential applications include high-Q electromechanical filters and clocks, surface force probes for atomic force microscopy, mass sensing, local charge, and magnetic sensing, as well as high-speed switches, among other possibilities. There is also an exciting effort among physicists, who are developing nanomechanical resonators for studies and applications in quantum mechanics.

What constitutes a NEMS resonator is a matter of definition. Mechanical resonators with at least one dimension (typically the thickness) that is less than $1\,\mu m$ are often classified as NEMS resonators, although a more accepted definition is that at least two dimensions need to meet this criterion, with at least one dimension approaching an order of $0.1\,\mu m$ (100 nm).

Nanoscale resonators are mostly fabricated using semiconductor processing technology, which includes lithographic patterning, etching, and thin-film deposition in a complex sequence of aligned steps, typically performed on planar silicon wafers or other optically flat substrates compatible with the processing equipment used in the fabrication. Two primary methods that have been used to fabricate NEMS resonators include so-called top-down or subtractive processing, where the mechanical elements are essentially cut out of larger pieces of material, in contrast to the alternative bottom-up approach where the resonator element is formed by self-assembly, and then is integrated with patterned support structures and with the electronic detection system. Note that "top-down" processing is in general much more advanced and allows much more sophisticated construction and design principles, but is limited to size scales determined by the lithographic processing technology (with minimum sizes of a few tens of nanometers possible but challenging). By contrast, the "bottom-up" approach typically affords a direct means of achieving size scales as small as 1 nm for the minimum resonator dimensions, but presents significant challenges in integration and yield. The top-down approach is presently favored for applications of NEMS resonators to measurement or detection, while the bottom-up approach has to date focused on measurements of the nanoscale resonators themselves (with some notable exceptions).

NEMS resonators have been fabricated in the top-down method from a wide range of different materials, including from single-crystal silicon [1], GaAs (gallium arsenide) [2], silicon-on-insulator heterostructures [3], silicon carbide [4], silicon nitride, aluminum nitride [5], polycrystalline silicon [6,7], and polycrystalline metals such as nickel and aluminum [8]. The alternative bottom-up approach has been successful in using intrinsically nanoscale materials such as carbon nanotubes and eutectic-growth based nanowires, silicon and GaAs nanowires being popular choices [9,10]. An image of a set of resonators fabricated using a top-down approach, taken using a scanning electron microscope, is shown in Figure 37.1.

Three key figures of merit for the mechanical response of NEMS resonators are their resonance frequency, stiffness, and mechanical quality factor. The stiffness quantifies the amount of mechanical deflection that is generated by an external force. The quality factor Q represents how much mechanical loss is present at the mechanical resonance frequency, and corresponds roughly to the number of oscillations a resonator will execute once excited by a sudden mechanical actuation. These figures of merit depend strongly on the particular geometry chosen for the resonator, as well as the choice of materials used to fabricate it.

FIGURE 37.1 A set of nanomechanical beams made from aluminum nitride using top-down processing. (From Cleland, A.N. et al., *Appl. Phys. Lett.*, 79, 2070, 2001.)

A very popular geometry for NEMS resonators is the *flexural beam* configuration, which includes *cantilevered* and *doubly clamped* beams. Cantilevered beams are often chosen for their intrinsically low stiffness: Cantilevered beams undergo large displacements of the cantilevered end when subjected to relatively weak forces, making them ideal for applications such as atomic force microscopy and other force-based imaging applications. The doubly clamped beam geometry is often chosen for its ease of fabrication and intrinsically high resonance frequency: For the same physical dimensions and materials, a doubly clamped beam has a fundamental resonance frequency roughly seven times that of a cantilevered beam. Resonance frequencies up to and above 1 GHz can be achieved using the doubly clamped beam geometry, when combined with submicron lithography; however, these high frequencies necessitate relatively stiff beams, achieved by making the aspect ratio of length to thickness L/t smaller, so the structures are not particularly responsive to external forces. To achieve frequencies much above 1 GHz, other geometries, such as film bulk acoustic resonators (FBARs), employing "breathing mode" thickness resonances rather than flexural motion, are more easily employed [11,12].

The resonance frequency of a mechanical structure in general scales as S^{-1}, where S is the length scale of the structure. Millimeter-scale structures thus have resonance frequencies in the kHz to MHz range, while micrometer-scale structures are in the MHz to GHz range. To reach frequencies above 1 GHz, size scales below 1 μm are needed; ~10 GHz film resonators have been made with thicknesses of order ~0.3 μm. By contrast, the mass of a resonator scales with S^3, and the stiffness scales as S.

We will look in more detail at the simplest solid mechanics description of the vibrational properties of flexural resonators, and we will present the fundamental formulas that yield the resonance frequencies and mode shapes of both cantilevered and doubly clamped beam resonators. We will then briefly discuss the means typically employed to actuate and sense the motion of NEMS resonators.

37.2 Doubly Clamped Flexural Resonators

In Figure 37.2 we show the generic structure for a doubly clamped flexural resonator, with length L, width w and thickness t, oriented along the x-axis, driven into flexural resonance with displacement U along the y-axis.

The dynamic motion of an flexural beam, either cantilevered or doubly clamped, is most easily described using the Euler–Bernoulli theory, which applies to beams with aspect ratios $L/t \gg 1$ [13,14], that is, the "thin beam" limit. For an isotropic material, the transverse displacement $U(x, t)$ of the beam centerline, with displacement in the y direction, obeys the differential equation

$$\rho A \frac{\partial^2 U}{\partial t^2}(x, t) = -\frac{\partial^2}{\partial x^2} EI \frac{\partial^2 U}{\partial x^2}(x, t), \qquad (37.1)$$

where

ρ is the material density
$A = wt$ the cross-sectional area
E is the Young's modulus
$I = wt^3/12$ the bending moment of inertia

This equation does not include an external driving force, nor does it include dissipation (energy loss); its solutions therefore correspond to the self-resonance modes (frequencies ω and mode shapes $U(x, t)$) of a lossless resonator.

Note that this equation does not have the form one usually encounters when studying waves in stretched wires (e.g., violin strings), or in bulk systems, for example, electromagnetic waves in vacuum or sound waves in air. In these systems, the equations that describe the wave dynamics involve second-order derivatives in time t and second-order derivatives in position x. The corresponding solutions include traveling sinusoidal waves, for example, $\sin(kx - \omega t)$ for a wave with wavelength $\lambda = 2\pi/k$ and frequency $f = \omega/2\pi$, traveling in the $+x$ direction. The frequency and wavelength satisfy the standard dispersion relation $f\lambda = c$, where c is the speed of the wave.

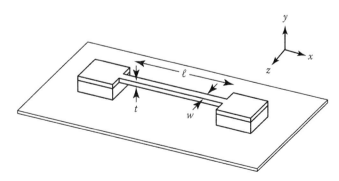

FIGURE 37.2 Doubly clamped flexural resonator, with length ℓ, oriented along the x axis with displacement along the y axis.

TABLE 37.1 Material Properties (Young's Modulus and Density) for a Few Common Materials Used for NEMS Resonators

Material	ρ (kg/m³)	E (GPa)
Silicon	2330	179
GaAs	5320	85.5
Aluminum nitride	3280	320
Silicon nitride	3100	~300
Plastic (generic)	1200	2.1

By contrast, for the Euler "thin beam", Equation 37.1, the dynamic equation has a second-order derivative in time, but a fourth-order derivative in position, leading to markedly different behavior.

The material parameters in Equation 37.1 include the Young's modulus E and the volume density ρ. Values of these parameters for some common materials are given in Table 37.1.

To proceed, we assume that every point on the beam is moving harmonically, that is, that we can describe the displacement as $U(x, t) = U(x)e^{i\omega t}$, where $\omega = 2\pi f$ is the radial frequency of the motion. Our solution will ultimately specify at what frequency the self-resonant motion occurs, but for now the frequency is unknown. Furthermore, for a prismatic beam, the Young's modulus E and the bending moment I do not vary with position x, so the derivative does not act on them. The dynamic equation (37.1) then becomes

$$\rho A\omega^2 U(x) = EI\frac{\partial^4 U(x)}{\partial x^4}. \tag{37.2}$$

We define the parameter β by

$$\beta = \left(\frac{\rho A\omega^2}{EI}\right)^{1/4}, \tag{37.3}$$

which has units of inverse length, and we assume a spatial dependence of the form $U(x) = \exp(\kappa x)$. Inserting this form into (37.2), and canceling out the common exponential terms, we find the *dispersion relations* $\kappa = \pm\beta, \pm i\beta$ (here $i = \sqrt{-1}$). Hence, a general solution will have the form

$$U(x) = Ae^{i\beta x} + Be^{-i\beta x} + Ce^{\beta x} + De^{-\beta x}, \tag{37.4}$$

or, in terms of real functions only,

$$U(x) = a\cos(\beta x) + b\sin(\beta x) + c\cosh(\beta x) + d\sinh(\beta x). \tag{37.5}$$

There are five unknown parameters, a, b, c, d, and β (the frequency ω, which as yet is unknown, determines β); thus, we need four boundary conditions, and an initial condition, to obtain a unique solution.

Our first example is for a doubly clamped beam. The displacements $U(x = 0)$ and $U(x = L)$ are then zero, as are the slopes $dU/dx(x = 0)$ and $dU/dx(x = L)$ (clamping rigidly

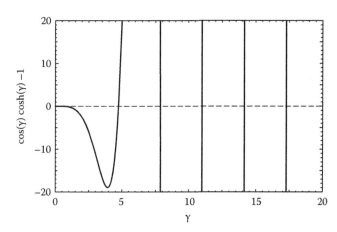

FIGURE 37.3 The function $\cos(\gamma)\cosh(\gamma) - 1$, whose zero crossings give the mode frequencies for the doubly clamped beam.

prevents the beam from bending at its ends). Using the general solution (37.5), these boundary conditions imply that $a = -c$ and $b = -d$, and the allowed values of β form a discrete set. Defining $\gamma = \beta L$, we have the determining equation for the mode values γ_n given by

$$\cos(\gamma_n)\cosh(\gamma_n) - 1 = 0. \tag{37.6}$$

The left-hand side of this equation is plotted in Figure 37.3, showing the zero-crossing points that determine the γ_n.

The zeroes of this function can be found numerically, with $\gamma_n = 4.73004, 7.8532, 10.9956, 14.1372\ldots$ for the lowest four modes. These values then determine the self-resonance frequencies of the beam, through the inverse of (37.3), that is,

$$\frac{\omega_n}{2\pi} = \sqrt{\frac{EI}{\rho A}}\left(\frac{\gamma_n}{L}\right)^2 = \frac{\gamma_n^2}{\sqrt{12}}\sqrt{\frac{E}{\rho}}\frac{t}{L^2}, \tag{37.7}$$

Note that the mode frequencies scale with the beam thickness t and inversely with square of the beam length L. The frequency for the fundamental mode is given by

$$\frac{\omega_1}{2\pi} = 1.027\sqrt{\frac{E}{\rho}}\frac{t}{L^2}, \tag{37.8}$$

and the higher modes are $\omega_n/\omega_1 = 2.757, 5.404,$ and 8.933 for $n = 2, 3,$ and 4.

The functional form for the displacement associated with the nth mode, $U_n(x)$, is given by

$$U_n(x) = a_n\left[\cos\left(\frac{\gamma_n x}{L}\right) - \cosh\left(\frac{\gamma_n x}{L}\right)\right] + b_n\left[\sin\left(\frac{\gamma_n x}{L}\right) - \sinh\left(\frac{\gamma_n x}{L}\right)\right]. \tag{37.9}$$

The relative amplitudes a_n and b_n for the first few modes are found numerically and are given by $a_n/b_n = 1.01781, 0.99923,$

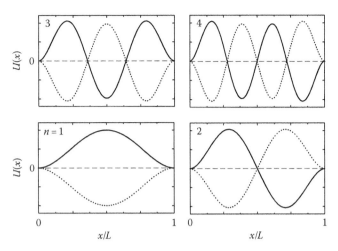

FIGURE 37.4 The four lowest frequency modes for the doubly clamped beam, solved using the Euler thin-beam approximation. The frequencies for these modes are given in the text.

1.0000,… The spatial dependence for the lowest four solutions ($n = 1–4$) for the doubly clamped beam are shown in Figure 37.4.

The solution does not determine the overall amplitude of the beam, set by the magnitude of one of the undetermined parameters, say a_n; this would be determined by the initial conditions, for example, the displacement of the beam midpoint at $t = 0$. The general nature of the solutions is however independent of the initial conditions.

We note that the normal modes $U_n(x)$ that correspond to each mode n can be made to satisfy a set of orthonormality relations,

$$\int_0^L U_n(x)U_m(x)dx = L^3\delta_{nm}, \tag{37.10}$$

equal to L^3 if $n = m$ and zero otherwise. The null integral for different modes are a fundamental property of these types of eigensolutions; making the integral for the same mode equal to L^3 amounts to a correct choice of the absolute amplitudes a_n and b_n (only the relative magnitudes are determined by the mode equations). The same relations can be made to hold for the cantilever modes that we described in Table 37.2.

For a beam clamped at one end, and free at the other, we have the cantilever geometry. The ends of the cantilevered beam are taken to be at $x = 0$ (fixed end) and $x = L$ (free end). The boundary conditions are given by

TABLE 37.2 Resonance Frequencies for a Doubly Clamped Beam Made of Silicon, with Varying Geometric Dimensions

Thickness	Length	$\omega_1/2\pi$	$\omega_2/2\pi$
1 µm	100 µm	0.90 MHz	2.48 MHz
100 nm	10 µm	9.00 MHz	24.8 MHz
10 nm	1 µm	90.0 MHz	248 MHz
10 nm	100 nm	9.0 GHz	24.8 GHz

$$\left.\begin{array}{l} U(0) = 0, \\[6pt] \dfrac{dU}{dx}(0) = 0, \\[6pt] \dfrac{d^2U}{dx^2}(L) = 0, \\[6pt] \dfrac{d^3U}{dx^3}(L) = 0, \end{array}\right\} \tag{37.11}$$

where the third and fourth conditions come from the requirement for zero transverse force and zero torque at the free end. These boundary conditions are applied to the general solution for the displacement given by Equation 37.5.

Proceeding as before, we find that the frequencies are determined by the equation

$$\cos(\gamma_n)\cosh(\gamma_n) + 1 = 0, \tag{37.12}$$

where again $\gamma = \beta L$. This equation has solutions $\gamma_n = 1.875, 4.694, 7.855, 10.996…$. The cantilever self-resonance frequencies are given by the same equation as for the doubly clamped beam, Equation 37.7, but using the values of γ_n given here; the functional dependence of the frequencies on beam length and thickness is the same. The frequency for the fundamental mode of a cantilever is given by

$$\frac{\omega_1}{2\pi} = 0.1615\sqrt{\frac{E}{\rho}}\frac{t}{L^2}, \tag{37.13}$$

and the higher modes are $\omega_n/\omega_1 = 6.267, 17.551,$ and 34.393 for $n = 2, 3,$ and 4.

The relative amplitudes for the mode shape $U(x)$ are given by $a_n = -c_n, b_n = -d_n$, so the general functional form is

$$U_n(x) = a_n\left[\cos\left(\frac{\gamma_n x}{L}\right) - \cosh\left(\frac{\gamma_n x}{L}\right)\right]$$
$$+ b_n\left[\sin\left(\frac{\gamma_n x}{L}\right) - \sinh\left(\frac{\gamma_n x}{L}\right)\right]. \tag{37.14}$$

with $a_n/b_n = -1.3622, -0.9819, -1.008, -1.000…$. In Figure 37.5, we display the first four modes for a cantilevered beam.

In Table 37.3, we give the resonance frequencies for a number of cantilevered beams made of silicon, with varying geometric dimensions.

37.2.1 Dissipation in Mechanical Resonators

We now turn to a discussion of how to include mechanical dissipation in the dynamic equations of motion; this will then allow us to consider the motion of beams driven by an external force. Energy dissipation in mechanical resonators is frequently characterized by the quality factor, or Q, of the resonator. The Q of

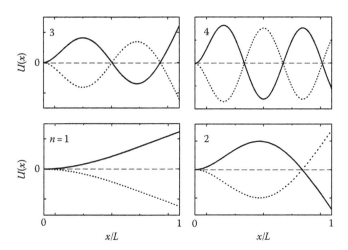

FIGURE 37.5 The four lowest frequency modes for the cantilevered beam, solved using the Euler thin-beam approximation. The frequencies for these modes are given in the text.

TABLE 37.3 Resonance Frequencies for a Cantilevered Beam Made of Silicon, with Varying Geometric Dimensions

Thickness	Length	$\omega_1/2\pi$	$\omega_2/2\pi$
1 μm	100 μm	141 kHz	5.64 MHz
100 nm	10 μm	1.41 MHz	56.4 MHz
10 nm	1 μm	14.1 MHz	564 MHz
10 nm	100 nm	1.41 GHz	56.4 GHz

a resonator is approximately the number of oscillations a resonator can execute after an initial excitation, before the energy of oscillation is reduced by a factor of $e^{2\pi}$ from its initial value. More precisely, if a resonator is excited by a harmonic excitation, the Q for a lightly damped resonator is the ratio of the resonance frequency to the resonance width, defined as the frequency difference between the points where the amplitude response is $1/\sqrt{2}$ its peak value.

Resonators display the highest quality factors (lowest dissipation) when measured at low temperatures in vacuum. When operated at room temperature in vacuum, the Q tends to be lower than at low temperature; when operated in air or a liquid, the Q is substantially lower.

The quality factor of resonators in vacuum has a complex temperature dependence. Typically, the quality factor will fall as the temperature is reduced below room temperature, reaching a minimum at a temperature in the range of 100–150 K. As the temperature is reduced further, the quality factor will increase again, reaching a maximum plateau near a temperature of a few or a few tens of Kelvin. The lowest temperature quality factor is typically higher than the room temperature value, sometimes by a factor of 10 or more. The range of quality factors seen when changing a resonator's temperature over the full range can cover a factor of 5 to a few tens. Discussions of the loss mechanisms understood to contribute to this dependence can be found in the literature [15–17].

When operated in air or in a liquid rather than in vacuum, a resonator experiences additional loss due to viscous effects from the fluid through which it is moving. Precise predictions of the mechanical loss in this situation are difficult, due to the long-time correlation effects in the fluid. In principle, the full Navier–Stokes equations [18] must be used to describe the fluid motion to arrive at an accurate estimate for the fluid damping. The fluid response depends strongly on the detailed geometry and motion of the resonator, making general predictions of limited use.

It is however possible to make estimates for energy loss when a resonator is operated in a moderate vacuum, in the regime known as the "molecular flow regime." In this limit, gas molecules have mean free paths large compared to the minimum dimensions of the resonator, and energy is transferred between the resonator and the gas molecules through elastic collisions, rather than viscous drag. The mean free path of a molecule in a gas at temperature T and pressure p is

$$\ell = \frac{k_B T}{\sqrt{2}\pi d^2 p}, \qquad (37.15)$$

where d is the average diameter of a gas molecule. The mean free path in air at ambient pressure (101 kPa) is about 60 nm; at a pressure of 1 Torr (130 Pa), the mean free path is 50 μm. The quality factor in this regime, for a resonator with mass M and resonance frequency f_0, is approximately

$$Q \sim \frac{M f_0}{4 p A} \sqrt{\frac{k_B T}{m}}, \qquad (37.16)$$

where

m is the average mass of a gas molecule

A is the surface area of the resonator

The Q falls inversely with pressure, and for larger resonators can fall by two to four orders of magnitude in passing from vacuum to atmospheric pressure. A resonator will follow this scaling to a pressure of a few Torr (of order 100 Pa), depending on the size of the resonator; small resonators exhibit molecular flow response at higher pressures, with the smallest resonators exhibiting this response at pressures as high as atmospheric pressure.

The most significant mechanism for *internal* energy loss in a nanomechanical resonator, that is, excluding that from motion in a fluid, is through intrinsic losses in the beam material, both surface and bulk, both of which can be treated phenomenologically using Zener's model for an elastic solids [19]. Other important loss terms include thermoelastic processes [16], mechanical radiation of energy from the beam supports [20], and possibly by coupling through the displacement transducer [3].

In Zener's model, the Hooke's stress–strain relation $\sigma = E\varepsilon$, relating the stress σ to the strain ε, is generalized to allow for mechanical relaxation in the solid:

$$\sigma + \tau_\varepsilon \frac{d\sigma}{dt} = E_R\left(\varepsilon + \tau_\sigma \frac{d\varepsilon}{dt}\right), \qquad (37.17)$$

where E_R is the relaxed value of Young's modulus. Loads applied slowly generate responses with the relaxed modulus, while rapidly varying loads involve a different value for the modulus.

We consider harmonic stress and strain variations, $\sigma(t) = \sigma e^{i\omega t}$ and $\varepsilon(t) = \varepsilon e^{i\omega t}$. At low frequencies $\omega\tau \ll 1$, this becomes the standard Hooke's law relation with $E = E_R$. At high frequencies $\omega\tau \gg 1$, the modulus becomes $E = E_U = (\tau_\sigma/\tau_\varepsilon)E_R$, the *unrelaxed* Young's modulus. For intermediate frequencies, the Young's modulus is complex, of the form

$$E = E_{\text{eff}}(\omega)\left(1 + \frac{i\omega T}{1 + \omega^2\tau^2}\Delta\right), \quad (37.18)$$

with mean relaxation time $\tau = (\tau_\sigma\tau_\varepsilon)^{1/2}$, fractional modulus difference $\Delta = (E_U - E_R)/E_R$, and effective Young's modulus

$$E_{\text{eff}} = \frac{1 + \omega^2\tau^2}{1 + \omega^2\tau_\varepsilon^2}E_R. \quad (37.19)$$

The imaginary part in the Young's modulus leads to energy dissipation.

Equation 37.18 implies that the stress σ will include a component that is 90° out of phase with the strain ε, which causes energy loss at a rate proportional to Δ. For small Δ, we define the inverse quality factor Q^{-1} as the ratio of the imaginary to the real part of E:

$$Q^{-1} = \frac{\omega\tau}{1 + \omega^2\tau^2}\Delta. \quad (37.20)$$

We then use the effective Young's modulus E_{eff} in the Euler–Bernoulli formula, Equation 37.1 at frequency ω,

$$\omega^2\rho A U(x) = E_{\text{eff}}(\omega)I\left(1 + \frac{i}{Q}\right)\frac{\partial^4 U}{\partial x^4}(x). \quad (37.21)$$

The spatial solutions $U(x)$ are the same as for Equation 37.1, but the damped eigenfrequencies ω_n' are given in terms of the undamped frequencies ω_n by

$$\omega_n' = \left(1 + \frac{i}{2Q}\right)\omega_n, \quad (37.22)$$

for small dissipation Q^{-1}. The imaginary part of ω_n' implies that if the beam is excited into its nth eigenmode, its displacement amplitude will decay in time as $\exp(-\omega_n t/2Q)$ (its energy decays as $\exp(-\omega_n t/Q)$).

Characteristic values for the quality factors of doubly clamped and cantilevered resonators tend to fall in the range of 10^3–10^5; the measured quality factors tend to be less than what is expected from bulk material effects, so for nanoscale resonators it is relatively broadly accepted that the quality factors are limited by surface effects, which dominate the resonator behavior more and more as the surface-to-volume ratio is made smaller,

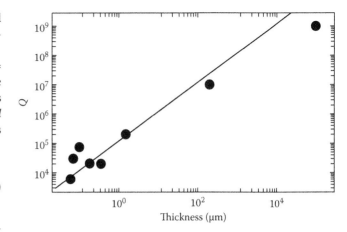

FIGURE 37.6 Quality factor as a function of resonator thickness, for a wide range of both doubly clamped and cantilevered resonators. The line shows a qualitative dependence where Q is found to be roughly proportional to the resonator thickness. (Adapted from Yasumura, K.Y. et al., *J. Microelectromech. Syst.*, 9, 117, 2000.)

as occurs when the overall scale of a resonator is reduced. A number of review articles discussing Q limiting effects in resonators have been published; see, for example, [16,17].

In Figure 37.6, we show a compendium of quality factors measured for resonators with an enormous range of size scales, ranging from meter-scale resonators intended for detecting gravity waves, to submicron-thickness resonators developed for radiofrequency measurements and surface-scanning applications. The general trend is that the quality factor is seen to scale with the thickness of the resonator (the horizontal axis), as exemplified by the solid line that gives the general dependence. This supports the argument that surface effects are dominant, as does the observation by a number of researchers that cleaning the resonator surface with ultra-high vacuum techniques, and maintaining the resonator in a vacuum, can improve the quality factor of the resonator by one to two orders of magnitude.

37.2.2 Driven Damped Beams

We now consider what happens to a damped beam driven by an external force. We consider a harmonic driving force $F(x, t) = f(x)\exp(i\omega_d t)$, where $f(x)$ is the position-dependent force per unit length and ω_d is the drive frequency. We assume that the force is uniform across the beam cross-section and directed along the y direction (the direction of beam motion). For this calculation, we also assume that the drive frequency ω_d is close to the fundamental resonance frequency ω_1.

The equation of motion then becomes [14]

$$\rho A\frac{\partial^2 U}{\partial t^2} + E A\frac{\partial^4 U}{\partial x^4} = f(x)e^{i\omega_d t}. \quad (37.23)$$

Here, the Young's modulus E is assumed to be given by the damped Zener model, Equation 37.19.

We solve this equation for long time t, $\omega_1 t/Q \gg 1$, so any transient motion due to the initial conditions has been damped out. Solutions to (37.23) then have the form $U(x, t) = U(x)e^{i\omega_d t}$, with motion only at the drive frequency ω_d (note this is distinct from the undamped self-resonant motion, which only occurs at the self-resonance frequencies ω_n). The amplitudes $U(x)$ that solve (37.23) may be complex, reflecting the fact that the motion is not necessarily in phase with the driving force $F(x, t)$.

The self-resonance eigenmodes are guaranteed to form a complete set of functions over the beam length, so we can always write the driven mode shape $U(x)$ as a linear superposition of the eigenmodes $U_n(x)$,

$$U(x) = \sum_{n=1}^{\infty} a_n U_n(x), \qquad (37.24)$$

with coefficients a_n determining the relative contribution of each mode. Inserting this in the equation of motion (37.23), we find

$$-\omega_d^2 \rho A \sum_{n=1}^{\infty} a_n U_n(x) + EA \sum_{n=1}^{\infty} a_n \frac{\partial^4 U_n}{\partial x^4} = f(x). \qquad (37.25)$$

Using the defining relation for the eigenfunctions, Equation 37.1, the dispersion relation, Equation 37.22, and the orthogonality relations, Equation 37.10, this can be written as

$$\left(\omega_n'^2 - \omega_d^2 \right) a_m = \frac{1}{\rho A L^3} \int_0^L U_n(x) f(x) dx, \qquad (37.26)$$

for each term n in the expansion. For ω_d close to ω_1, only the $n = 1$ term in Equation 37.26 has a significant amplitude, given by

$$a_1 = \frac{1}{\rho A L^3} \frac{1}{\omega_1^2 - \omega_d^2 + i\omega_1^2/Q} \int_0^L U_1(x) f(x) dx, \qquad (37.27)$$

for small dissipation Q^{-1}.

We now assume the force is uniform along the beam length, so that $f(x) = f$. The integral in Equation 37.27 is then

$$\eta_1 = \frac{1}{L^2} \int_0^L U_1(x) dx = 0.8309. \qquad (37.28)$$

The amplitude is then

$$a_1 = \frac{\eta_1}{\omega_1^2 - \omega_d^2 - i\omega_1^2/Q} \frac{f}{M}, \qquad (37.29)$$

where $M = \rho A L$ is the mass of the beam, and the corresponding displacement of the beam is $U(x, t) = a_1 U_1(x) \exp(i\omega_d t)$.

The result (37.29) is very similar to the response of a damped simple harmonic oscillator driven by an external harmonic force with a frequency near that of the oscillator. If we imagine

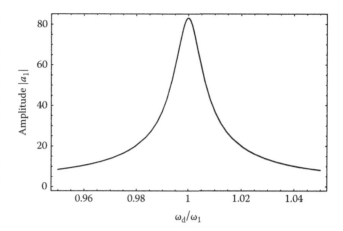

FIGURE 37.7 Displacement amplitude as a function of driving force frequency, for $Q = 100$. The driving force f is scaled to give an amplitude of $\eta_1 Q$ at the resonance frequency.

keeping the magnitude of the force f constant and sweeping the frequency of the force through the resonance, we can plot the magnitude of the response as a function of frequency, as is shown in Figure 37.7. The peak response is at the self-resonant frequency ω_1, and the amplitude at the resonance peak is, for $Q \gg 1$, approximately

$$a_{1,\text{max}} \approx \frac{\eta_1 f}{M\omega_1^2} Q = \frac{f}{k_{\text{eff}}} Q, \qquad (37.30)$$

where we define the effective spring constant $k_{\text{eff}} \equiv M\omega_1^2 / \eta_1$. The displacement of the center point of the beam due to a *static* force per unit length f is approximately f/k_{eff}. When the beam is driven at its fundamental resonance frequency, the displacement a_1 is the static displacement times the quality factor of the resonator.

If the force distribution $f(x)$ is instead chosen to be proportional to the eigenfunction $U_1(x)$, the integral Equation 37.27 is unity, so that η_1 in Equation 37.29 is replaced by the number 1.

We point out that the response function, Equation 37.29, while similar to that of a damped, one-dimensional harmonic oscillator, differs slightly in the Q-dependent denominator, but the difference is only apparent for small values of Q: For values of Q greater than about 15, the fractional difference is less than 1%.

37.3 Noise in Driven Damped Beams

Systems that dissipate energy are necessarily sources of noise; the converse is also often true. This is the basic statement of the fluctuation–dissipation theorem, and is best known in relation to electrical circuits, where it is termed the Nyquist–Johnson theorem. An electrical circuit element with an electrical impedance $Z(\omega)$ that has a non-zero real part, $R(\omega) = \text{Re } Z(\omega)$, will be a source of noise, that is, of fluctuations in the voltage $V(t)$ across the impedance Z, or equivalently in the current $I(t)$ through Z. A voltmeter placed across the circuit element will measure an instantaneous voltage that fluctuates with a Gaussian distribution in amplitude, with zero average value, and a width that is determined only by $R(\omega)$ and the temperature T. A useful way to quantify the noise is to use the average spectral density of the noise in angular frequency space, defined for a noise voltage $V(t)$ by

$$S_V(\omega) = \left\langle \int_{-\infty}^{\infty} V^2(t)\cos(\omega t)dt \right\rangle. \tag{37.31}$$

Here the angle brackets $\langle \ldots \rangle$ indicate that a statistical ensemble average, over many equivalent systems, is to be taken. The spectral density is proportional to the electrical noise power in a unit bandwidth. The Nyquist–Johnson theorem states that this quantity is given by $S_V(\omega) = (2/\pi)R(\omega)\hbar\omega\coth(\hbar\omega/k_BT)$. At high temperatures or low frequencies, such that $k_BT \gg \hbar\omega$, this approaches the classical limit $S_V(\omega) \to 2k_BTR(\omega)/\pi$. The spectral noise density $S_V(f)$, as a function of frequency $f = \omega/2\pi$, is given in the high temperature limit by $S_V(f) = 2\pi S_V(\omega) = 4k_BTR(f)$. The SI units of $S_V(f)$ are V^2/Hz. The corresponding current spectral noise density is $S_I(f) = S_V(f)/R^2(f) \to 4k_BT/R(f)$, in the high temperature limit, with units of A^2/Hz.

The fluctuation–dissipation theorem applies to mechanical resonators with non-zero dissipation, that is, with finite Q, and ensures that the mechanical resonator will also be a source of noise, but due to the resonant nature of the resonator's response, the noise spectral density takes on a somewhat different form. We will only treat the equivalent of the high-temperature limit, $k_BT \gg \hbar\omega$, for the resonator noise.

37.3.1 Dissipation-Induced Amplitude Noise

The displacement of a forced, damped beam driven near its fundamental frequency is given by Equation 37.23. In the absence of noise, this solution represents pure harmonic motion at the drive frequency ω_d. As discussed above, the non-zero value of Q^{-1} and temperature T necessitates the presence of noise, from the fluctuation-dissipation theorem. Regardless of the origin of the dissipation mechanism, it acts to thermalize the motion of the resonator, so that in the presence of dissipation only (no driving force), the mean energy $\langle E_n \rangle$ for each mode n of the resonator will be given by $\langle E_n \rangle = k_BT$, where T is the physical temperature of the resonator. This noise term has been considered by a number of authors [21,22].

The thermalization occurs due to the presence of a noise force $f_N(x,t)$ per unit length of the beam. Each point on the beam experiences a noise force with the same spectral density, but fluctuating independently from other points; the noise at any two points on the beam is uncorrelated. The noise be written as an expansion in terms of the eigenfunctions $U_n(x)$,

$$f_N(x,t) = \frac{1}{L}\sum_{n=1}^{\infty} f_{N_n}(t)U_n(x), \tag{37.32}$$

where the force f_{N_n} associated with the mode n is uncorrelated that for other modes n'; the factor $1/L$ appears because of the normalization of the Y_n.

The noise force $f_{N_n}(t)$ has a white spectral density $S_{f_n}(\omega)$, and a Gaussian distribution with zero mean. The magnitude of the spectral density S_{f_n} may be evaluated by requiring that it achieve thermal equilibrium for each mode n. The spectral density of the noise-driven amplitude a_n of the nth mode is given by

$$S_{a_n}(\omega) = \frac{1}{(\omega_n^2 - \omega^2)^2 + (\omega_n^2/Q)^2}\frac{S_{f_n}(\omega)}{M^2}. \tag{37.33}$$

The SI units for S_{f_n} are $(N/m)^2/(rad/s) = kg^2/(s^3\text{-}rad)$. Those for S_{a_n} are $1/(rad/s)$, because a_n is dimensionless.

The kinetic energy associated with the spectral density S_{a_n} is given by

$$\langle E_n \rangle = \frac{1}{2}\int_0^\infty\int_0^L \rho A\omega^2 S_{a_n}(\omega)U_n^2(x)dx\,d\omega$$

$$= \frac{1}{2}\int_0^\infty \rho AL^3\omega^3 S_{a_n}(\omega)d\omega$$

$$\approx \frac{\pi}{4}\frac{QL^2}{\omega_n}\frac{S_{f_n}(\omega)}{M}, \tag{37.34}$$

where the last equality becomes exact in the limit $Q^{-1} \to 0$. The error in Equation 37.34 for finite Q is less than 1% for $Q > 10$.

In order that this yield thermal equilibrium, the kinetic energy is $\langle E_n \rangle = \frac{1}{2}k_BT$, so the spectral density S_{f_n} must be given by

$$S_{f_n}(\omega) = \frac{2k_BT M\omega_n}{\pi QL^2}. \tag{37.35}$$

The term L^2 appears in Equation 37.35 because f_{N_n} is the force per unit length of beam. An equivalent derivation for a one-dimensional simple harmonic oscillator with resonance frequency ω_0 yields the force density $S_F(\omega) = 2k_BT M \omega_0/\pi Q$. We can write the spectral density of the thermally driven amplitude as

$$S_{a_n}(\omega) = \frac{\omega_n}{(\omega_n^2 - \omega^2)^2 + (\omega_n^2/Q)^2}\frac{2k_BT}{\pi ML^2Q}. \tag{37.36}$$

A plot of this dependence is shown in Figure 37.8. When superposed with a driving force with a carrier frequency $\omega_d = \omega_1$, the power consists of a δ-function peak at the carrier superposed with the Lorentzian given by Equation 37.36.

37.3.2 Allan Variance

The amplitude fluctuations can be viewed equivalently as fluctuations in the resonator frequency. A useful way to quantify the amplitude fluctuations commonly used to compare frequency standards, is the Allan variance $\sigma_A(\tau_A)$ [23,24]. The Allan

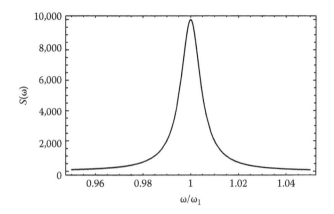

FIGURE 37.8 Spectral density of noise $S_{f_n}(\omega)$, for $Q = 100$. Spectral density is in units of $2k_{\rm B}T\omega_n^3/\pi M\,L^2$.

variance is defined in the time domain, as the variance over time in the measured frequency of a resonator, each measurement averaged over a time interval $\tau_{\rm A}$, with zero dead time between measurement intervals. The defining expression for the square of the Allan variance is

$$\sigma_A^2(\tau_A) = \frac{1}{2f_d^2}\frac{1}{N-1}\sum_{m=2}^{N}\left(\bar{f}_m - \bar{f}_{m-1}\right)^2, \qquad (37.37)$$

where $\bar{f}_m$ is the average frequency measured over the mth time interval, of length $\Delta t = \tau_{\rm A}$, and $f_{\rm d} = \omega_{\rm d}/2\pi$ is the nominal drive frequency.

For a white fractional frequency noise density $S_y(\omega) = C$, the Allan variance is

$$\sigma_A(\tau_A) = \sqrt{\frac{\pi C}{\tau_A}}. \qquad (37.38)$$

For thermally limited motion, the Allan variance in this case is

$$\sigma_A(\tau_A) = \sqrt{\frac{k_B T}{8P_d Q^2 \tau_A}}, \qquad (37.39)$$

where $P_{\rm d}$ is the power in the drive applied to the resonator. Defining the dimensionless drive energy $\varepsilon_{\rm d}$ as the ratio of drive energy per cycle to the thermal energy, $\varepsilon_{\rm d} = 2\pi\,P_{\rm d}/\omega_{\rm d}k_{\rm B}T$, we have

$$\sigma_A(\tau_A) = \frac{1}{Q}\sqrt{\frac{\pi}{4\varepsilon_d \omega_d \tau_A}}. \qquad (37.40)$$

The Allan variance falls inversely with the square root of the product $\omega_{\rm d}\tau_{\rm A}$, and it is also proportional to the dissipation Q^{-1}. Other things being equal, increasing the resonator frequency $\omega_{\rm d}$ lowers the Allan variance, as does increasing the drive power actuating the resonator.

37.4 Actuation and Detection Methods

An important aspect of using NEMS resonators is choosing how to actuate and detect the motion of the mechanical element. There are a wide range of techniques, each applicable to a different set of circumstances and applications. Here, we outline some of the more popular methods and refer the interested reader to the extensive literature on the different techniques.

37.4.1 Displacement Sensing

37.4.1.1 Optical Lever, Optical Beam

A common method used in atomic force microscopes for detecting the displacement of an atomic force cantilever is the *optical lever* or *optical beam*, where a beam of light from a laser is focused on the end of the cantilever, and the reflected beam detected by a photodetector, typically a quadrant or split photodetector that reports changes in the relative illumination of the different sectors of the detector. The beam is aligned to the center of the detector, and changes in the deflection of the cantilever then change the relative amount of light hitting the different sectors, yielding a detection signal. The motion of the cantilever is magnified by the moment arm traveled by the laser beam, so that very small motions (of order 0.1 nm) can be detected in this manner [25]. The implementation is relatively simple, although the laser beam must be aligned to the cantilever and the photodetector, and the alignment maintained over the use of the instrument. The speed of this technique is limited by the response time of the photodetector; the cantilever must, however, be large enough to reflect a good fraction of the laser illumination, which means minimum lateral dimensions of a few microns.

37.4.1.2 Optical Interferometer

By placing two mostly reflecting mirrors with their reflecting surfaces facing one another, and passing a light beam through the mirror pair, the resulting interference of the light reflecting between the two mirrors gives rise to a net transmission through the pair that depends sensitively on the distance between the mirrors; this effect gives rise to for instance *Newton's rings* when two pieces of glass are pressed together and is the basis of an optical instrument known as the *Fabry–Perot interferometer* [26]. This arrangement of mirror faces is the basis of a NEMS displacement sensor, where the mechanically active resonator element forms one face of the mirror pair, and a second mirror, or the polished end of an optical fiber, forms the other face. Light passed through this system (or, in another implementation, reflected back through the first fixed mirror) then generates interference fringes that permit sensitive monitoring of the resonator displacement, with displacement sensitivities of less than 0.01 nm possible [27]. Alignment must be maintained, and is sensitive to unwanted vibrations, but the sensitivities achieved can outweigh the challenges associated with this approach. Again, the speed of the technique is limited by the response time of the photodetector used to sense the transmitted or reflected illumination, and the cantilever must be sufficiently large, and sufficiently reflective, to interfere a significant amount of the illumination.

37.4.1.3 Electrostatic Sensing

The electrical capacitance of two pieces of metal is a function of their separation, forming the basis for the electrostatic displacement sensing method. By placing a metal electrode near a NEMS resonator, and a second electrode on the surface of the NEMS resonator, the capacitance between the electrodes will vary as the resonator moves, a variation that can be detected with the appropriate electronics. This method integrates easily with conventional electronics, allowing a simple interface to the mechanical degree of freedom; however, the sensitivity of this technique tends to be low, and as the capacitance becomes smaller as the size scale of the resonator is reduced, it does not scale well to the smallest resonator sizes. We note however that one extremely sensitive method of detecting motion does use this type of sensing, but the sensitivity relies on closely coupling the mechanical element to an ultra-sensitive charge transducer known as a single electron transistor [28]. Detection noise then approaches the quantum limit in this implementation, limited by "back-action" noise from the transducer.

37.4.1.4 Magnetomotive Sensing

The motion of a metal conductor through a magnetic field will induce an electromotive force (EMF), if the motion is perpendicular to magnetic field lines. In other words, a voltage will develop across the ends of the conductor, as stated by Faraday's law [29]. By placing a NEMS resonator in a transverse magnetic field, the motion of the resonator through the magnetic field can be monitored by measuring the voltage developed across an electrode placed on the surface of the NEMS resonator. This technique has been used by a number of researchers, especially as it combines easily with the magnetomotive actuation method (see below) [1]. It does however work best with strong magnetic fields, which can be obtained either by placing the NEMS resonator on the surface of a strong permanent magnet (SmCo magnets have sufficient coercive fields, of order 1 T, for this application), or by using superconducting magnets in a cryogenic environment (fields in the range of 8–15 T are relatively commonplace using such magnets).

37.4.2 Actuation Techniques

37.4.2.1 Piezoelectric Actuation

Piezoelectric materials are a special class of insulators that generate voltages (internal electric fields) when the material is strained, and, conversely, when voltages are applied to them, strain is generated in the material. These are often used as methods to generate ultrasonic vibrations in a number of applications, and can be used to both actuate and sense the motion of NEMS resonators. Actuation schemes have been implemented where a piezoelectric transducer is placed in mechanical contact with a NEMS resonator, or with a substrate on which a NEMS resonator has been fabricated, and the vibration frequency of the piezoelectric transducer tuned (using an external voltage synthesizer) to the NEMS resonance frequency. Alternatively, the NEMS resonator itself can be fabricated from, or include, a piezoelectric element, so that actuation and sensing can be achieved directly from the piezoelectric properties of the material. This approach provides, for example, the basis for FBAR resonators developed for the mobile phone industry [12,11]; their applicability to very high frequency resonators (>1 GHz) is part of the reason for their appeal and success in this particular application.

37.4.2.2 Electrostatic Actuation

Electrostatic actuation is in a sense the counterpart of electrostatic displacement sensing. If a voltage is applied between two metal electrodes, a force will be generated due to the combination of charge developed on the electrode surfaces and the accompanying electric field; this force is always attractive. This can be used to actuate a NEMS resonator, if one electrode is placed on the NEMS resonator and the other on a nearby rigid support. This method can be implemented relatively simply, without the need for external equipment other than the source of the (oscillating) voltage, and very large forces can be generated if the electrodes can tolerate large electric fields. This method has been used to actuate a wide range of different NEMS geometric designs.

37.4.2.3 Magnetomotive Actuation

This method is the counterpart of the magnetomotive sensing mentioned above. If a NEMS resonator with a metal electrode on its top surface is placed in a transverse magnetic field, passing an electrical current through the electrode will generate a transverse force on the resonator, as described by the Lorentz force law [29]. Sweeping the frequency of the current through the resonance frequency of the resonator will excite the resonator into that mode. When coupled with the magnetomotive sensing, this provides a simple and powerful method to actuate and detect the motion of a NEMS resonator. It does however require a large magnetic field (of order of 1 T or larger), which can cause problems in implementation and reduces portability.

37.4.2.4 Thermal Actuation

This is not an actuation technique *per se*, but instead uses the fact that the thermal noise present in any dissipative system will necessarily actuate motion of the resonator, as described above. Measuring the noise-induced displacement of a resonator with a sufficiently sensitive displacement sensor allows the effect of this thermal noise to be observed, and allows in essence passive measurements of the resonance properties, and effective noise temperature, of the resonator. Very high displacement sensitivity, or extremely flexible resonators (with small effective spring constants, leading to larger thermally driven displacements), are needed to detect this motion, although this has been achieved with a number of different sensing methods.

References

1. A. N. Cleland and M. L. Roukes. Fabrication of high frequency nanometer scale mechanical resonators from bulk Si crystals. *Appl. Phys. Lett.*, 69: 2653–2656, 1996.

2. A. N. Cleland, J. S. Aldridge, D. C. Driscol, and A. C. Gossard. Nanomechanical displacement sensing using a quantum point contact. *Appl. Phys. Lett.*, 81: 1699, 2002.

3. A. N. Cleland and M. L. Roukes. A nanometre-scale mechanical electrometer. *Nature*, 392: 160, 1998.

4. Y. T. Yang, K. L. Ekinci, X. M. H. Huang, L. M. Schiavone, M. L. Roukes, C. A. Zorman, and M. Mehregany. Monocrystalline silicon carbide nanoelectromechanical systems. *Appl. Phys. Lett.*, 78: 162–164, 2001.

5. A. N. Cleland, M. Pophristic, and I. Ferguson. Single-crystal aluminum nitride nanomechanical resonators. *Appl. Phys. Lett.*, 79: 2070–2072, 2001.

6. C. T. C. Nguyen and R. T. Howe. Design and performance of CMOS micromechanical resonator oscillators. *Proceedings of the 1994 IEEE International Frequency Control Symposium*, Boston, MA, pp. 127–134, 1994.

7. K. Wang, A. C. Wong, and C. T. Nguyen. VHF free-free beam high-Q micromechanical resonators. *J. Microelectromech. Syst.*, 9: 347–360, 2000.

8. T. F. Li, Yu. A. Pashkin, O. Astafiev, Y. Nakamura, J. S. Tsai, and H. Im. High-frequency metallic nanomechanical resonators. *Appl. Phys. Lett.*, 92: 043112, 2008.

9. H. B. Peng, C. W. Chang, S. Aloni, T. D. Yuzvinsky, and A. Zettl. Ultrahigh frequency nanotube resonators. *Phys. Rev. Lett.*, 97: 087203, 2006.

10. V. Sazonova, Y. Yaish, H. Ustunel, D. Roundy, T. A. Arias, and P. L. McEuen. A tunable carbon nanotube electromechanical oscillator. *Nature*, 431: 284–287, 2004.

11. K. Nam, Y. Park, B. Ha, D. Shim, I. Song, J. Park, and G. Park. Piezoelectric properties of aluminum nitride for thin film bulk acoustic wave resonator. *J. Korean Phys. Soc.*, 47: S309–S312, 2005.

12. R. C. Ruby, P. Bradley, Y. Oshmyansky, A. Chien, and J. D. Larson III. *IEEE Ultrasonics Symposium*, Atlanta, GA, p. 813, 2001.

13. A. N. Cleland. *Foundations of Nanomechanics*. Springer, New York, 2002.

14. S. Timoshenko, D. H. Young, and Jr. W. Weaver. *Vibration Problems in Engineering*. John Wiley & Sons, New York, 1974.

15. S. Evoy, A. Olkhovets, D. W. Carr, L. Sekaric, J. M. Parpia, and H. G. Craighead. Temperature dependent internal friction in silicon nanoelectromechanical systems. *Appl. Phys. Lett.*, 77: 2397–2399, 2000.

16. R. Lifshitz and M. L. Roukes. Thermoelastic damping in micro- and nanomechanical systems. *Phys. Rev. B*, 61: 5600–5609, 2000.

17. K. Y. Yasumura, T. D. Stowe, E. M. Chow, T. Pfafman, T. W. Kenny, B. C. Stipe, and D. Rugar. Quality factors in micron- and submicron-thick cantilevers. *J. Microelectromech. Syst.*, 9: 117–125, 2000.

18. P. K. Kundu and I. M. Cohen. *Fluid Mechanics*, 3rd edn. Elsevier, New York, 2004.

19. A. S. Nowick and B. S. Berry. *Anelastic Relaxation in Crystalline Solids*. Academic Press, New York, 1972.

20. D. M. Photiadis and J. A. Judge. Attachment losses of high q oscillators. *Appl. Phys. Lett.*, 85: 482–484, 2004.

21. T. R. Albrecht, P. Grütter, D. Horne, and D. Rugar. Frequency modulation detection using high-Q cantilevers for enhanced force noise microscope sensitivity. *J. Appl. Phys.*, 69: 668–673, 1991.

22. T. B. Gabrielson. Mechanical and thermal noise in micromechanical acoustic and vibration sensors. *IEEE Trans. Electron Devices*, 40: 903–909, 1993.

23. D. W. Allan. Statistics of atomic frequency standards. *Proc. IEEE*, 54: 221–230, 1966.

24. W. F. Egan. *Frequency Synthesis by Phase Lock*. John Wiley & Sons, New York, 1981.

25. G. Meyer and N. M. Amer. Novel optical approach to atomic force microscopy. *Appl. Phys. Lett.*, 53: 1045–1047, 1988.

26. E. Hecht and A. Zajac. *Optics*, 3rd edn. Addison-Wesley, Reading, MA, 1997.

27. H. J. Mamin and D. Rugar. Sub-attonewton force detection at millikelvin temperatures. *Appl. Phys. Lett.*, 79(20): 3358–3360, 2001.

28. R. G. Knobel and A. N. Cleland. Nanometre-scale displacement sensing using a single electron transistor. *Nature*, 424: 291–293, July 2003.

29. D. J. Griffiths. *Introduction to Electrodynamics*, 3rd edn. Benjamin Cummings, San Francisco, CA, 1999.

38

Spin-Transfer Nano-Oscillators

Stephen E. Russek
National Institute of Standards and Technology

William H. Rippard
National Institute of Standards and Technology

Thomas Cecil
National Institute of Standards and Technology

Ranko Heindl
National Institute of Standards and Technology

38.1 Introduction

John Slonczewski (1996) and Luc Berger (1996) predicted that electron currents in magnetic multilayer devices can transport angular momentum from one magnetic layer to another, thereby exerting a torque on the local magnetization (see Figure 38.1a). Under the correct conditions, they predicted that the magnetization can undergo sustained oscillations at microwave frequencies. The necessary conditions include: (1) the spacer layers between the magnetic layers are thin, <50 nm, so that spins do not depolarize as they go from one layer to another; (2) the device is sufficiently small (<100 nm) so that the amount of spin momentum transported by the electron current is a significant fraction of the angular momentum of the magnetic element; (3) there are sufficient nonlinearities in the configuration to stabilize the precessional orbits. Microwave oscillations in magnetic multilayers were first detected by Tsoi et al. (1998, 2000) in a point-contact geometry. Subsequently, microwave emission was directly observed in nanoscale device structures by Kiselev et al. (2003) and Rippard et al. (2004). These structures are commonly referred to as spin-torque oscillators, spin-transfer oscillators, or spin-transfer nano-oscillators (STNOs). The most common device configuration, shown in Figure 38.1a, uses two conducting magnetic layers: a polarizer whose magnetization is fixed and a free layer whose magnetization is free to rotate in response

to the current generated spin torque. The layers are separated by a metallic or an insulating spacer layer. An example of microwave emission from an STNO for various bias currents is shown in Figure 38.1b.

The advantages of spin-transfer oscillators are that they are highly tunable by current and magnetic field, they are among the smallest microwave oscillators yet developed, they are relatively easy to fabricate in large quantities, they are compatible with standard silicon processing, and they operate over a broad temperature range. STNOs are closely related to giant magnetoresistance (GMR) and tunneling magnetoresistance (TMR) devices that have been developed for magnetic recording read heads and magnetic random access memory. STNOs utilize both spin-transfer torque, to set the magnetization into oscillation, and the GMR or TMR effect, to produce an output voltage. Challenges still remain before widespread applications of STNOs are possible. These challenges include increasing the output power above the present value of ~0.5 μW (Deac et al. 2008, Houssameddine et al. 2008), removing the need for large applied magnetic fields, understanding and controlling the oscillator linewidth, and reducing device-to-device variations.

This article presents a tutorial on spin-transfer oscillators; no prior knowledge of spin transfer or magnetic devices is assumed. For more in-depth discussion of spin-transfer effects the reader is referred to reviews by Sun (2006), Silva and Rippard (2008), and Ralph and Stiles (2008), and Bertotti et al. (2008).

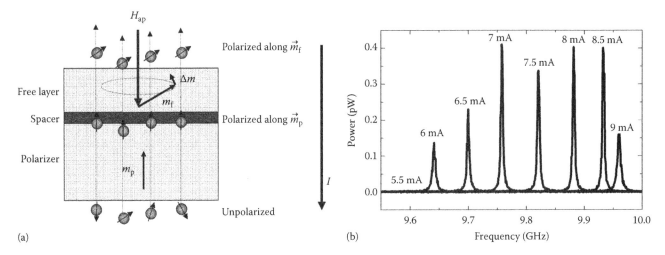

FIGURE 38.1   (a) Schematic showing electrons being spin polarized along $\vec{m}_\mathrm{p}$ by the bottom magnetic layer and, in the act of rotating toward the new polarization direction $\vec{m}_\mathrm{f}$ in the top magnetic layer, transferring spin momentum to the top layer magnetic moment. The transferred spin moment balances out damping, which tries to align $\vec{m}_\mathrm{f}$ with the applied field $\vec{H}_\mathrm{ap}$ and stabilizes a precessional orbit. (b) Microwave emission from an STNO for several different bias currents.

38.2 Physics of Giant Magnetoresistance, Tunneling Magnetoresistance, and Spin Transfer

The physics of spin-transfer torques is similar to that giving rise to the GMR and TMR effects. These effects arise from the fact that, in addition to charge, electrons have an additional degree of freedom, the electron spin. In ferromagnets (and sometimes in normal metals and semiconductors), the valence electrons can be polarized: there are more spins pointing in one direction rather than another. The spin density polarization along a particular axis $\hat{p}$ is given by $P = (n_\uparrow - n_\downarrow)/(n_\uparrow + n_\downarrow)$, where $n_\uparrow$, $n_\downarrow$ are the number of valence electrons with their spins aligned and anti-aligned along $\hat{p}$. The properties of the conduction electrons, including their momentum state and scattering rates, can depend on the spin direction. Since an electron has a charge e, the electron spin $\vec{S}$ is associated with a magnetic moment $\vec{\mu} = \gamma\vec{S}$, where $\gamma = g(e/2m_e)$ is the gyromagnetic ratio, m_e is the electron mass, and g is the gyromagnetic constant, which for the materials considered here, is close to the free electron value of $g = 2$. In general, the magnetic moment has both an orbital as well as a spin component. For the materials currently used in STNOs, the orbital moment is not important and we consider just the moment due to spin. The charge on an electron is negative so γ is negative and $\vec{S}$ and $\vec{\mu}$ are antiparallel. Since the electron has a spin of $s = \hbar/2$ the electron moment is nearly equal to the Bohr magneton $\mu_e = (ge/2m_e)(\hbar/2) \approx \mu_B = (\hbar e/2m_e) = 9.274 \times 10^{-24}\,\mathrm{A\,m^2}$.* For an electron in a metal the gyromagnetic ratio is approximately (it will vary depending on the material)

$\gamma = 1.76 \times 10^{11}\,\mathrm{A\,m^2/J\,s}$. Often it is more convenient to remember the value $\gamma/2\pi = 28\,\mathrm{GHz/T}$ since this will be the constant relating the electron spin precession and STNO oscillator frequency to the effective magnetic field. To characterize magnetic devices it is useful to evaluate the local magnetization $\vec{M}$ which is the moment per unit volume: $\vec{M} = 1/V \sum_i \vec{\mu}_i$, where V is the volume of the element being considered and i indexes over all of the moments in V. Often, as in Figure 38.1a, we refer to the normalized moment of a device layer using lower case symbols $\vec{m} = \vec{M}/|\vec{M}|$, where $\vec{m}$ is a unit vector in the direction of the magnetization.

Polarization of electrons is most prominent in magnetic materials, which have a spontaneous magnetic moment: there is a net alignment of the electron spins on each atom and a well-defined ordering of moments from atom to atom. The alignment of the electron spins is due to a quantum mechanical effect that requires the electron wave function to be anti-symmetric. The anti-symmetry of the wave function leads to an effective interaction, the exchange interaction, that lowers the system energy if the electron spins on a given atom, if given a choice, are aligned. This energy reduction is due to the requirement that the wave function for the electrons must go to zero if two electrons with the same spin occupy the same location. The vanishing of the probability that two electrons with the same spin state occupy the same location reduces the Coulomb repulsion energy. For electrons on the same atom, this effect leads to the alignment of spins in partially filled d and f orbitals, as described by Hund's rules. For electrons on different atoms, there are similar effects; however, the exchange interaction can be either positive (favoring spin alignment) or negative (favoring spin anti-alignment), depending on the relative distance between the atoms and their relative orientation. If the inter-atomic exchange interaction favors alignment then the

---

* In this article all equations and quantities are expressed in Système International (SI) units.

material, below a Curie temperature T_C, is a ferromagnet and if the inter-atomic exchange interaction favors anti-alignment then the material, below a Néel temperature T_N, is an antiferromagnet.

For STNO, GMR, and TMR devices, we are most interested in conducting ferromagnets that have both a spontaneous polarization as well as mobile electrons. The simplest model of a conducting ferromagnetic is the *sd* model proposed by Mott (1936). In this model, we consider transition metals to consist of partially filled *d* shells, which contain most of the polarization, and mobile *s* electrons that carry most of the current as illustrated in Figure 38.2. The *d* electrons are relatively localized; they have large effective masses and low velocities. The *d* electrons are polarized since the shell is partially filled and the exchange interaction promotes alignment of the electron spins. For Fe, Co, Ni metals (which are the most common materials used in the magnetic layers shown in Figure 38.1a), the moment per atom at 0 K is of 2.22, 1.72, 0.61 μ_B, respectively, corresponding to roughly 0.5–2 more electron spins pointing parallel to the magnetization axis than antiparallel. The saturation magnetization M_s and Curie temperatures T_C for Fe, Co, Ni metals at 0 K are 1.71×10^6, 1.42×10^6, 0.48×10^6 A/m, and 770°C, 1131°C, and 358°C, respectively.

Calculated spin-dependent band density of states and conductivities for face centered cubic (fcc) Co are shown in Figure 38.2b. Most of the polarization is due to *d*-like electrons well below the Fermi surface. Electrons with magnetic moments parallel or antiparallel to the magnetization direction are referred to as majority or minority electrons, respectively. Note that for Co at the Fermi surface there are more minority spins than majority spins since the minority *d* bands are at the Fermi surface. However, the conductivity at the Fermi surface is dominated by the majority *s*-like electrons (designated as *sp* electronics in Figure 38.2c,d). This can be intuitively understood by realizing that most of the electron-scattering events in clean metals preserve the spin direction, i.e., scattering events that cause spin flips are relatively rare. Hence, majority electrons must scatter into other majority states and similarly minority electrons must scatter into other minority states. Since there is a larger density of states at the minority Fermi surface there are more states to scatter into for this channel. This excess of scattering, referred to as *sd* scattering, causes the conductance in the minority state *s*-band to be considerably smaller than for electrons in the majority state.

In a rigid band picture, the states for the minority electrons are similar to those for the majority electrons except that they are shifted by additional exchange energy. At the Fermi surface, the minority electrons will have smaller wave vectors and velocities. For metallic contacts, this potential barrier, as seen in Figure 38.3, causes scattering at the interface between regions with different magnetizations. In particular, since there is a large potential step for spins that go from being majority to minority, and vice versa, these electrons will be scattered strongly. The strong scattering leads to a higher resistance when the magnetizations of the free and polarizer layers are not aligned (Camley and Barnaś 1989, Grünberg 2008). The device resistance is then a function of the relative angle θ_m between the magnetizations in the two layers and, to a good approximation, is given by $R = R_{av} - (\Delta R/2)\cos(\theta_m)$, as shown in Figure 38.3d, where

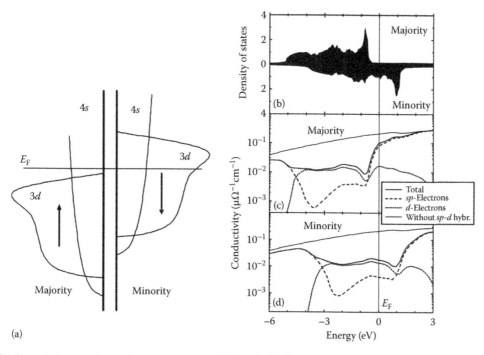

FIGURE 38.2 (a) Schematic density of states for a transition metal. (b) Calculated density of states for majority and minority electrons in fcc Co. (c) Conductivity of majority electrons. (d) Conductivity of minority electrons. At the Fermi surface, most of the current is carried by majority *sp* electrons. (From Tsymbal, E.Yu. and Pettifor, D.G., *Phys. Rev. B*, 54, 15314, 1996. With permission.)

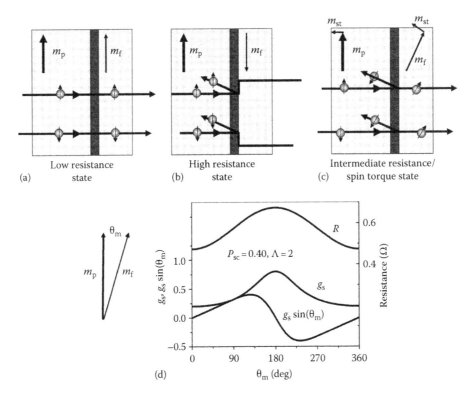

FIGURE 38.3 (a) Magnetic layers in the parallel low resistance state where electrons pass through the interface since they do not need to change the form of their wavefunctions. (b) Magnetic layers in the antiparallel high resistance state where there is a high level of reflection from the interface because minority electrons must become majority electrons and vice versa. (c) Magnetic layers in a non-collinear state where there is a spin torque. Here, conduction electrons must rotate their polarization and in the process apply a torque to the local magnetization. (d) Plot of the spin-torque parameter g_s, the magnitude of the spin torque $g_s \sin(\theta_m)$, and device resistance R as a function of the angle θ_m between the magnetizations in the two magnetic layers.

$R_{av} = (R_{ap} + R_p)/2$, $\Delta R = R_{ap} - R_p$ and R_{ap}, R_p are the device resistances in the antiparallel and parallel states, respectively. The geometry shown in Figure 38.3 is referred to as current perpendicular to the plane (CPP) configuration. GMR originally was discovered in structures that had the current flowing in the plane (CIP) of the layers. Spin-transfer devices require large spin-polarized currents flowing from one layer to another and, hence, must be in the CPP configuration. The resistance for a 100 nm metallic device with a single ferromagnetic interface is typically a few ohms and the resistance change is a few tenths of an ohm. The fractional resistance change, the magnetoresistance (MR) ratio, is $(\Delta R/R_p) = (R_{ap} - R_p)/R_p \approx 2\%$–20%. In practice, it is impossible to obtain a sharp change in the magnetization if there is a strong exchange coupling at the interface between the two magnetic layers so a thin spacer layer is inserted. The states in the spacer layer must match that of the conducting electrodes or there will be unwanted scattering due to the spacer layer. The most common choice for the spacer layer in metallic devices is Cu. The spacer layer thickness must be less than the spin-flip relaxation length, which for copper at room temperature is approximately $\lambda_{sf} \cong 100$–400 nm (Albert et al. 2002).

An alternative to inserting a thin metallic spacer layer is to insert an insulator that is thin enough to allow tunneling through the barrier. The resulting device is referred to as a magnetic tunnel junction (MTJ). The device resistance is now determined by both the tunneling process (how the wavefunctions decay in the barrier) and by the density of available states in the electrodes. If the wave function decay is independent of whether the electron spin is majority or minority the MR ratio is given by Julliere's formula (Julliere 1975), which accounts only for density of states effects:

$$\frac{\Delta R}{R_p} = \frac{2P_1 P_2}{1 - P_1 P_2}$$

where P_1, P_2 are the polarizations of the electrons at the Fermi surface of the two layers that contribute to the tunneling current. For typical ferromagnetic metals with polarizations of 20%–40%, this will give MR ratio values on the order of 10%–40%.

If the polarizations could be made close to 100%, as in a half metal which has only one spin polarization at the Fermi surface, then the MR ratio could get arbitrarily large. However, the MR ratio, as it is conventionally defined, is not a good figure of merit for an STNO. As discussed in Section 38.6, the maximum output power of an STNO is $P_{out} \propto I^2 R_L \Delta R^2 / R_{av}^2$, where R_L is the load resistance. Hence, $\Delta R/R_{av}$, which varies between 0 and 2, is a better figure of merit.

The TMR can be greatly enhanced if the majority and minority wavefunctions decay differently in the tunnel barrier (Butler et al. 2001, Yuasa and Djayaprawira 2007). This is not due to any explicit dependence of tunneling on the electron spin but rather due to the fact that majority and minority electrons have, at the Fermi surface, different spatial wavefunctions with different symmetries and different decay lengths. To exploit this effect, there has to be a high degree of crystalline texture in the electrodes and tunnel barrier and the barrier has to be thick enough to allow for the wave function of one channel, usually the minority spin channel, to decay in magnitude substantially below that of the other channel. The most successful MTJs, to date, have used (001) MgO barriers with recrystallized body centered cubic (bcc) CoFeB electrodes, which have MR ratios of 200%–500% at room temperature. Typical magnetoresistive response, MR ratios, and barrier resistivities are shown in Figure 38.4. For STNO devices, thin barriers are required with resistance area products RA < 10 $\Omega\,\mu m^2$. As seen in Figure 38.4b, the MR ratio falls off when barriers are this thin and the tunnel barriers are very susceptible to breakdown, making the fabrication of MTJs for STNO applications a challenge.

The discussion above describes why the device resistance will depend on the relative orientation of the magnetizations in the two magnetic electrodes. If the local magnetization acts on the flow of the conduction electrons then, conversely, the flow of conduction electrons must act on the local magnetization. This is shown schematically in Figure 38.3c where the polarizer and free layer magnetizations are at an angle θ_m relative to each other. Conduction electrons are polarized by flowing through the polarization layer then, when they are incident on the free layer, their polarization rotates to align with the direction of the free-layer magnetization. This rotation of the conduction electron moment is due to a torque applied by the local magnetization and, hence, the local magnetization feels an equal and opposite torque. The transverse component of the momentum of the conduction electron is transferred to the local magnetization. The spin-transfer torque on the free layer, which is the change in spin momentum of the localized moment per second, is given by two terms, a transverse torque (which is shown in Figure 38.3c) and a field-like torque. The transverse torque is the most important and is given by

$$\vec{\tau}_{st} = \frac{g_s(\theta_m)\mu_e I}{\gamma e}\,\vec{m}_f \times (\vec{m}_f \times \vec{m}_p)$$

where

I/e is the number of electrons incident per second on the free layer

$\mu_e/\gamma = \hbar/2$ is the magnitude of the spin of each electron

$g_s(\theta_m)$ is the spin-torque parameter which is proportional to the polarization of the spin currents

$\vec{m}_f \times (\vec{m}_f \times \vec{m}_p)$ is a vector with magnitude $\sin(\theta_m)$ and direction perpendicular to $\vec{m}_f$ in the $\vec{m}_f$, $\vec{m}_p$ plane.

It will be important to keep track of the sign of the applied current. Here, a positive current will mean that electrons are

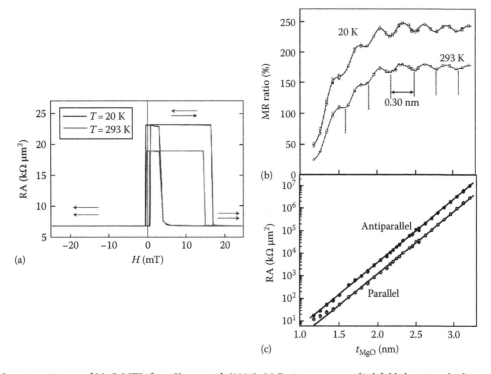

FIGURE 38.4 Magnetoresistance of MgO MTJs from Yuasa et al. (2004). (a) Resistance vs. applied field showing a high resistance antiparallel state and a low resistance parallel state. (b) The MR as a function of MgO barrier thickness. (c) The RA product vs. MgO barrier thickness. (From Yuasa, S. et al., *Nat. Mater.*, 3, 868, 2004. With permission.)

flowing from the free layer into the polarizer and a negative current will mean that electrons are flowing from the polarizer into the free layer. For this convention, *a negative current means that the free-layer magnetic moment is pushed toward that of the polarizer*, as is the case in Figure 38.3c; *a positive current means that the magnetic moment is pushed away from that of the polarizer.*

While we have been focusing on the torque applied to the free layer by electrons flowing from the polarizer, there is also a torque on the polarizer caused by the flow of electrons. This can be seen in Figure 38.3c, which shows that the electrons that are reflected from the free layer form a polarized current incident on the polarizer, which will apply a torque to that layer. The current induced torques will drive the magnetizations in each layer in the same direction: if the current is negative, as shown in Figure 38.3c, the free layer magnetization will be pushed toward the polarizer and the polarizer will be pushed away from the free layer magnetization. In fact, the only thing that distinguishes the polarizer from the free layer is that, in device construction, we try to make the polarizer magnetization fixed and unresponsive to the current induced torques. Some STNO designs utilize the fact that all the layers will have a torque to set multiple layers into motion (Tsoi et al. 2004).

A complete theory of spin-dependent transport must encompass both the spin torque and GMR/TMR. A useful theory by Slonczewski (2002) predicts the magnitude of the spin torque for symmetric metallic devices in terms of two dimensionless parameters P_{sc}, Λ:

$$g_s(\theta_m) = \frac{P_{sc}\Lambda^2}{(\Lambda^2+1)+(\Lambda^2-1)\cos(\theta_m)}$$

$$\Lambda^2 = GR_{ap}; \quad P_{sc} = \frac{\delta R}{R_{ap}}$$

where $G = Ae^2k_F^2/4\pi^2\hbar$ is a conductance with a magnitude of ~5 S for a 75 nm device with a copper spacer layer, A is the area of the device and k_F is the Fermi wave vector of the spacer layer. P_{sc} is the spin current polarization (the difference between the currents in the two spin channels divided by the total current), which is proportional to $\delta R = R_{ap}\sqrt{1-R_p/R_{ap}}$. Given R_{ap} and R_p one can calculate Λ and P_{sc} and, hence, the spin torque. This model explicitly demonstrates that spin torque, device resistance, and magnetoresistance are intimately related. Typical values for Λ and P_{sc} for the materials currently used in STNOs are 1.5–4 and 0.2–0.6, respectively. Plots of g_s, $g_s \sin(\theta_m)$, and R are shown in Figure 38.3d for a standard all-metallic device. Xiao et al. (2004) have shown that the above expression for $g_s(\theta_m)$ is in reasonable agreement with more quantitative calculations using realistic band structures and Boltzmann transport theory and have extended the theory to include asymmetric structures. A few things to note: there is no spin torque when the layers are aligned since $\sin(\theta_m)$ goes to zero; g_s is typically larger near 180° when the polarizer and free layer are close to being anti-aligned. The asymmetry in the spin torque can be understood

by realizing that the polarized electrons need not be transmitted to create a torque. When the layers are close to anti-aligned, the torque can be very large since most of the electrons are reflected and there are many more spin-transfer events per transmitted electron.

38.3 Single-Domain Equation of Motion and Phase Diagrams

38.3.1 Equation of Motion without Damping

A localized spin i obeys the torque equation, which says that the rate of change of angular momentum $\vec{S}_i$ is equal to the torque applied $\vec{\tau}_i$

$$\frac{d\vec{S}_i}{dt} = \vec{\tau}_i$$

In the case of a conservative system, a system that does not exchange energy with its environment, we can write the torque in terms of an effective field $H_{\text{eff},i}$ acting on the ith spin as:

$$\frac{d\vec{S}_i}{dt} = \mu_0\vec{\mu}_i \times \vec{H}_{\text{eff},i}$$

Since $\vec{\mu} = \gamma\vec{S}$, the equation can be rewritten in terms of the magnetic moment

$$\frac{d\vec{\mu}_i}{dt} = -\mu_0|\gamma|\vec{\mu}_i \times \vec{H}_{\text{eff},i} \quad \text{or} \quad \frac{d\vec{M}(\vec{r})}{dt} = -\mu_0|\gamma|\vec{M}(\vec{r}) \times \vec{H}_{\text{eff}}(\vec{r})$$

where $\vec{M}(\vec{r})$ is the magnetization or the moment per unit volume at a point $\vec{r}$. The effective field is the negative gradient of the free energy density with respect to the magnetization*:

$$\mu_0\vec{H}_{\text{eff}} = -\vec{\nabla}_M U = -\vec{\nabla}_M(U_{ap} + U_{ms} + U_{an} + U_{ex} + \cdots)$$

$$= \mu_0(\vec{H}_{ap} + \vec{H}_{ms} + \vec{H}_{an} + \vec{H}_{ex} + \cdots)$$

For STNO devices the most important energy terms are the interaction with the applied field (Zeeman energy) $U_{ap} = -\mu_0 \vec{M}\cdot\vec{H}_{ap}$, the magnetostatic energy due to dipolar interactions between the spins U_{ms}, the anisotropy energy due to crystalline or interfacial energies U_{an}, and the exchange energy due to spin-dependent quantum mechanical interactions U_{ex}. The effective field is then the sum of the applied magnetic field $\vec{H}_{ap}$, magnetostatic

---

* Two fields that are of interest in magnetism: the magnetic field $\vec{H}$ in units of A/m and the magnetic flux density $\vec{B}$ in units of tesla. $\vec{H}$ and $\vec{B}$ are related through $\vec{B} = \mu_0\vec{H} + \mu_0\vec{M}$. It is convenient to think of $\vec{H}$ as the "driving" field and $\vec{B}$ as the actual local field that a small test spin (such as a proton) would feel. Extending this concept, we view H_{eff} as an effective driving field, not the local flux density. We often quote the magnetic field as $\mu_0\vec{H}$ in tesla since this is often a more convenient unit. It is useful to remember the correspondence: 1 T ⇔ 796 kA/m. A good introduction to magnetostatic energy, demagnetizing factors, anisotropy energies, and exchange energies can be found in O'Handley (2000).

field $\vec{H}_{ms}$, anisotropy field $\vec{H}_{an}$, and exchange field $\vec{H}_{ex}$. In this section, we will assume that the exchange interaction is sufficiently strong that all of the moments in a device layer are aligned and move together. This approximation is often referred to as the single domain or macrospin model. In this model, *the magnitude of the moment is constant in time*. We will discuss the case of nonuniform moments in Section 38.8. We will also assume, for the time being, that the polarization layer moment is truly fixed and only consider the dynamics in the free layer. We write the equation of motion in terms of the normalized moment of the free layer $\vec{m}_f = \vec{M}_f / M_s$:

$$\frac{d\vec{m}_f}{dt} = -\mu_0 |\gamma| \vec{m}_f \times \vec{H}_{eff}$$

The equation says simply that the free layer moment will precess, with constant magnitude, around the effective field always in a direction that is perpendicular to the energy gradient and hence on a constant energy trajectory.

The magnetostatic energy density and field for a uniformly magnetized element can be written as $U_{ms} = \frac{1}{2}\mu_0 M_s^2 \left(N_{xx} m_{fx}^2 + N_{yy} m_{fy}^2 + N_{zz} m_{fz}^2 \right)$ where N_{xx}, N_{yy}, N_{zz} are the shape-dependent demagnetizing factors along the x, y, z directions, respectively. The demagnetizing factors sum to 1 and are smallest for directions along the longest dimensions of the magnetic element. This energy term simply states that the moment likes to lie along the longest dimension of the element. For a sphere,

the demagnetizing factors are $N_{xx} = N_{yy} = N_{zz} = 1/3$ and there is no preferential direction that minimizes the magnetostatic energy, For an infinitely long cylinder, the demagnetizing factors are 1/2 perpendicular to the axis of the cylinder and 0 along the axis so that the moment likes to lie along the axis of the cylinder. For a typical $75 \times 50 \times 3$ nm STNO free layer, shown in Figure 38.5, the demagnetizing factors are $N_{xx} = 0.047$, $N_{yy} = 0.072$, $N_{zz} = 0.881$ and the moment likes to lie in the plane of the film along the long axis of the device. The magnetostatic field is obtained by differentiating the energy with respect to the magnetization: $\vec{H}_{ms} = -M_s(N_{xx} m_{fx} \hat{x} + N_{yy} m_{fy} \hat{y} + N_{zz} m_{fz} \hat{z})$. The magnetostatic field along a particular axis is proportional to the magnetization along that axis and points opposite to the direction of the magnetization.

The magnetostatic energy difference between the x and y directions at $T = 300$ K for the device shown in Figure 38.5, normalized to $k_B T$, is $\Delta U_{ms} V / k_B T = 27$. It is a general feature of these nanoscale devices that the characteristic energies are not much greater than the thermal energies and hence thermal fluctuations are an important part of device operation.

The anisotropy energy may be due either to the preference for the moment to lie along certain crystalline directions or for the moment to lie perpendicular to interfaces. An example of the former is in L10 FePt films in which the magnetization likes to lie along the c-axis of the face-centered tetragonal cell. An example of the latter are Co/Ni multilayers in which the magnetization likes to be perpendicular to the thin film interfaces.

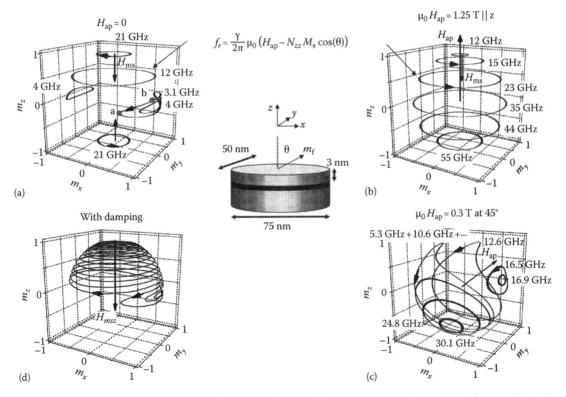

FIGURE 38.5 Free precessional orbits of the free layer of a magnetic device with a $3 \times 50 \times 75$ nm, $M_s = 800$ kA/m free layer (a) with no damping and no applied field, (b) with no damping and 1.25 T applied field in the z-direction, (c) with no damping and 0.3 T applied field 45° off the z-axis, and (d) with damping ($\alpha = 0.02$) and no applied field.

Here, we consider only uniaxial anisotropy energies of the form $U_{an} = -\frac{1}{2}\mu_0 M_s H_{an}(\vec{m} \cdot \hat{z})^2$, where $\pm\hat{z}$ are the easy axis directions (the low energy directions assuming H_{an} is positive). The anisotropy field is given by $\vec{H}_{an} = M_s m_{fz}\hat{z}$. Typical anisotropy energies and anisotropy fields are $U_{an} = 0.5–1 \times 10^5\,\text{J/m}^3$ and $\mu_0 H_{an} = 1.25–2.5\,\text{T}$. The form of the uniaxial anisotropy energy term is similar to that for the z-component of the magnetostatic energy.

If the magnetization is oriented along a particular direction and allowed to evolve according to the torque equation, the magnetization will precess about the effective fields and trace out a constant energy orbit. Some typical "free precession" orbits are shown in Figure 38.5 for a magnetic device without perpendicular anisotropy and in Figure 38.6 for a magnetic device with perpendicular anisotropy. Also shown are the precession frequencies. These "free precession" orbits will be approximately those sampled by the spin-transfer driven oscillations discussed later. In Figure 38.5a, there are no applied fields and the moment precesses around the internal magnetostatic fields, while in Figure 38.5b and c the moment precesses around a combination of internal fields and applied fields. For the case of weak fields and no perpendicular anisotropy (Figure 38.5a), there are two different types of orbits: in-plane orbits and out-of-plane orbits. In addition, for high symmetry geometries there can be degenerate orbits, i.e., two orbits with the same shape and frequency. For a circular orbit, the frequency is given by the precession speed divided by the orbit circumference $f_r \cong (\gamma\mu_0/2\pi m_t)|\vec{m}_f \times H_{eff}|$,

where m_t is the orbit radius. For an elliptical orbit, the precession frequency is proportional to the geometric mean of the precession speed at the major and minor axes of the orbit divided by the orbit circumference

$$f_r \cong \frac{\gamma\mu_0}{2\pi\sqrt{m_{ta}m_{tb}}}\sqrt{\left|\vec{m}_f \times \vec{H}_{eff}\right|_a \left|\vec{m}_f \times \vec{H}_{eff}\right|_b},$$

where m_{ta} and m_{tb} are the major and minor axis lengths, respectively. For the perpendicular orbits shown in Figure 38.5a and b, which are nearly circular, the resonant frequencies are given by

$$f_r \cong \frac{\gamma}{2\pi}\mu_0(H_{ap} - N_{zz}M_{fz}) = \frac{\gamma}{2\pi}\mu_0(H_{ap} - N_{zz}M_s\cos(\theta)),$$

where θ is the angle between the moment and the z axis and we have assumed $H_{ap}\|z$ and N_{xx}, N_{yy} are close to zero. For zero applied field, the frequency is negative denoting clockwise rotation (as viewed from the positive z direction) while, for large applied fields the frequency is positive denoting counterclockwise rotation. For zero applied field, the frequency decreases as θ increases and the magnetostatic field decreases. Near $\theta = 90°$ the orbit frequencies get small and the orbits go in-plane. For large z-axis applied fields, $H_{ap} > M_s$, the frequency will increase as θ increases going to

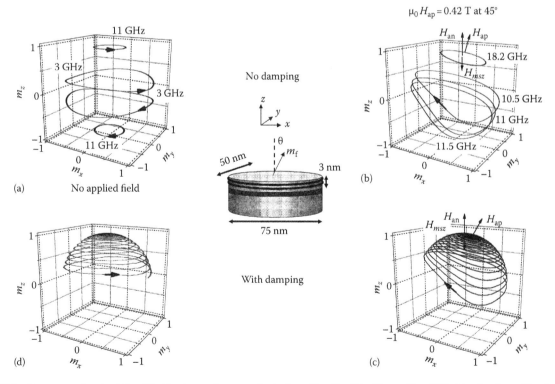

FIGURE 38.6 Free precessional orbits of the free layer of a magnetic device with a $3 \times 50 \times 75\,\text{nm}$, $M_s = 800\,\text{kA/m}$ free layer. The free layer has a perpendicular anisotropy with an anisotropy field of $\mu_0 H_{an} = 1.25\,\text{T}$. (a) With no damping and no applied field, (b) with no damping and 0.3 T applied field in the x-direction and y-direction, (c) with damping ($\alpha = 0.02$) and 0.3 T applied field in the x-direction and y-direction, and (d) with damping ($\alpha = 0.02$) and no applied field.

$$f_r \cong \frac{\gamma}{2\pi} \mu_0 H_{ap}$$

at $\theta = 90°$ and

$$f_r \cong \frac{\gamma}{2\pi} \mu_0 (H_{ap} + N_{zz} M_s)$$

at $\theta = 180°$.

For in-plane orbits with an applied field along the x-direction the resonance frequencies are

$$f_r \cong \frac{\gamma}{2\pi} \mu_0 \sqrt{(H_{apx} - N_{xx} M_{fx} + N_{zz} M_{fx})(H_{apx} - N_{xx} M_{fx} + N_{yy} M_{fx})}$$

The first term under the radical corresponds to the torques applied at point b in Figure 38.5a due to the applied field and the x and z components of the demagnetizing fields. The second term under the radical corresponds to the torques applied at point a in Figure 38.5a due to the applied field and the x and y components of the demagnetizing fields. These equations, referred to as Kittel equations (Kittel 1948), illustrate a general property of magnetic devices, since the effective field depends on the magnetization vector, the frequency of the precessional orbit varies for different orbits.[*] For the case shown in Figure 38.5a where there is no applied field, the frequency formula reduces to

$$f_r \cong \frac{\gamma}{2\pi} \mu_0 M_{fx} \sqrt{(N_{zz} - N_{xx})(N_{yy} - N_{xx})}$$

indicating that the frequency will decrease as the orbit opens up and M_{fx} decreases. The variation of the resonant frequency is critical to the operation of the STNO since the device will try to vary its orbit and orbit frequency to balance out the damping and spin torques that will be discussed in the next sections. Note that f_r is the fundamental frequency of the orbit and, depending on the configuration, there can be higher harmonics. For in-plane orbits, the component along the axis of the orbit m_{fx} will, as seen by inspecting the in-plane orbits in Figure 38.5a, oscillate at twice this frequency.

When the field is not applied along a high symmetry direction, as shown in Figure 38.5c, the orbits can have a distorted shape and there will be many higher harmonics present. This illustrates the non-linearity of the torque equation due to the dependence of the effective field on the moment direction. The magnetization components will in general not be simple functions of the form $\cos(\omega t + \phi)$.

An alternative geometry is to induce a perpendicular anisotropy using the correct magnetic materials. Orbits for a device with perpendicular anisotropy $\mu_0 H_{an} = 1.25\,T$ and no damping are shown in Figure 38.6a and b. When the anisotropy field is larger than the magnetization the equilibrium state of the moment points perpendicular to the plane of the device. In Figure 38.6a there is no external field applied and the moment precesses around the sum of the magnetostatic and anisotropy fields. The effective field now has the opposite sign compared to Figure 38.5a and the rotational direction is reversed. Since the anisotropy and z-component of the magnetostatic field have the same form but opposite signs, we can use the previous formulas for resonant frequencies with the substitution $(-N_{zz} M_{fz}) \Rightarrow (H_{an} m_{fz} - N_{zz} M_{fz})$. The frequency is given by

$$f_r \cong \frac{\gamma}{2\pi} \mu_0 \left[H_{ap} + (H_{an} - N_{zz} M_s) \cos(\theta) \right]$$

and again, as $m_z = \cos(\theta)$ decreases and the orbit opens up, the orbit frequency will decrease until m_z changes sign and then it will increase, rotating in the opposite sense. In Figure 38.6b an external field at an angle $\theta = 45°$ is applied with components $\mu_0 H_{apx} = 0.3\,T$ and $\mu_0 H_{apy} = 0.3\,T$. The larger orbits are again non-elliptical with substantial higher harmonic components.

38.3.2 Equation of Motion with Damping

The equation of motion discussed above assumes that the local moment does not interact with its environment except through a reversible applied field. In actuality, the moment interacts strongly with its environment, particularly the conduction electrons and phonons. This interaction leads to dissipation and the relaxation of the moment into a low energy state. To account for these interactions the equation of motion has to be modified:

$$\frac{d\vec{m}_f}{dt} = -\mu_0 \frac{|\gamma|}{1 + \alpha^2} \vec{m}_f \times \vec{H}_{eff} - \alpha \frac{\mu_0 |\gamma|}{1 + \alpha^2} \vec{m}_f \times (\vec{m}_f \times \vec{H}_{eff})$$

This equation, referred to as the Landau Lifshitz Gilbert (LLG) equation,[†] has three modifications. First, we have to add a damping term (second term on the right) that causes the moment to relax in a direction toward the effective field, which moves the system toward a low energy state. The triple cross product in the damping term pulls out the component of $\vec{H}_{eff}$ that is normal to the free layer moment so that the magnitude of the moment is still conserved. The strength of the relaxation is determined by the dimensionless damping parameter α. Second, we have to add the factor $1/(1 + \alpha^2)$ to both the first and second terms. This causes viscous damping: in the limit of large α the magnetization responds slowly. The third change is buried within H_{eff}. We need to add a random thermal fluctuation field H_{th} to H_{eff}.

---

[*] The dependence of precession frequency of a ferromagnetic element on the orbit amplitude and orientation is quite different than what is typically found in nuclear magnetic resonance (NMR) where there are no internal effective fields and the frequency of the orbit is independent of the amplitude of the orbit. This is because the torque is proportional to the transverse component of the moment times the applied field. As the orbit grows in size the torque increases with the size of the orbit thus maintaining the frequency.

[†] For a recent comprehensive review of the LLG equation see Bertotti et al. (2008).

H_{th} is typically chosen to be a random Gaussian-distributed field with a root-mean-square (rms) average value of

$$\mu_0 H_{th,rms} = \sqrt{\frac{2kT\alpha}{VM_s\gamma\Delta t}},$$

where Δt is the period over which the thermal field is applied. It is this random field that gives rise to paramagnetic response when the magnetic element becomes sufficiently small and gives the classic Langevin magnetic behavior. The damping term describes phenomenologically the average interaction of the magnetic device with its environment while the thermal field describes the stochastic part. The magnitude of the damping constant for materials used in magnetic devices is typically between 0.005 and 0.05 and is always positive.

When we incorporate damping into the equations of motion, we see (Figures 38.5d and 38.6c,d) that the orbits relax slowly to the nearest energy minimum. The relaxation time is on the order of 1–5 ns. In addition, there are thermal fluctuations that cause some random motion of the orbits. Since the damping constant is typically small for materials of interest, the damping term can be considered a small perturbation to the precessional term. As the moment relaxes, it will sample a family of nearly constant energy orbits. To a first approximation, these are the orbits that are stabilized by the transfer of spin from the applied current.

38.3.3 Equation of Motion with an Applied Current

John Slonczewski (1996) realized that the LLG equation would have to be modified in magnetic devices since the electron currents transport angular momentum from one layer to another

$$\frac{d\vec{m}_f}{dt} = -\mu_0 \frac{|\gamma|}{1+\alpha^2} \vec{m}_f \times \vec{H}_{eff} - \alpha \frac{\mu_0|\gamma|}{1+\alpha^2} \vec{m}_f \times (\vec{m}_f \times \vec{H}_{eff})$$
$$+ \frac{g(\theta_m)\mu_e I}{eM_s V} \vec{m}_f \times (\vec{m}_f \times \vec{m}_p)$$

The new term accounts for the angular momentum transferred by a current I from a layer with polarization direction $\vec{m}_p$, as discussed in Section 38.2. Here, we only consider the transverse torque and ignore the smaller field-like torque. The magnitude of the spin-transfer term normalized to the precessional term is approximately $g(\theta_m)\mu_e I/\mu_0\gamma eM_s^2 V \approx 0.03$, where we have assumed that $H_{eff} \sim M_s$, an applied current of 1 mA, and the same device dimensions as used in Figures 38.5 and 38.6. For a device of this size the maximum current before catastrophic failures is approximately 10 mA. Since the spin-transfer term is small, on the order of the damping term, it can also be viewed as a small perturbation to the precessional term. The equation of motion with the spin-torque term is referred to as the Landau–Lifshitz–Gilbert–Slonczewski (LLGS) equation and it has been extensively studied both numerically (Sun 2000) and analytically (Bertotti et al. 2005).

Depending on the relative orientation of the polarization direction to the effective field, the spin-transfer term can either add energy or remove energy from the magnetic system. To understand how these terms affect the magnetization dynamics it is useful to look at a simple high symmetry configuration. In the case of perpendicular applied field and polarization, as shown in Figure 38.8, the LLGS equations can be rewritten as

$$\frac{d\vec{m}_f}{dt} = -\mu_0 \frac{|\gamma|}{1+\alpha^2} \vec{m}_f \times \vec{H}_{eff} - \left(\alpha \frac{|\gamma|\mu_0 H_{eff}}{1+\alpha^2} - \frac{g(\theta_m)\mu_e I}{eM_s V} \right) \vec{m}_f \times (\vec{m}_f \times \hat{z})$$

We can see that the spin-torque term opposes the damping term if I is positive and adds to the damping if I is negative. When the current becomes greater than a critical current, the effective damping becomes negative and the system will move toward orbits of higher energy. The critical current IC is given by setting the effective damping rate

$$\alpha_{eff} = \alpha \frac{|\gamma|\mu_0 H_{eff}}{1+\alpha^2} - \frac{g_s(\theta_m)\mu_e I}{eM_s V}$$

to 0 giving

$$I_c = \alpha \frac{|\gamma|\mu_0 H_{eff}}{1+\alpha^2} \frac{eM_s V}{g_s(\theta_m)\mu_e}$$

To obtain a small critical current to set the free layer in motion, devices need free layers with small damping constants, low saturation magnetizations, small volumes, and high polarizations. Reducing H_{eff} is, in general, not feasible since a large H_{eff} is required to maintain a high operation frequency. Conversely, when designing a fixed polarizer layer, we require a large magnetization, large damping, and large effective field so that the current is not sufficient to get the polarizer precessing. The above discussion applies strictly to the case when everything is symmetric about the z-axis. In general, the effective damping will not be zero everywhere on the orbit. Stable orbits are obtained by having the integrated damping and spin-torque terms balance on the orbit.

Having the effective damping go to zero is not sufficient to obtain a stable orbit. For stability, we require positive damping if the orbit is perturbed to a higher energy level and negative damping if the orbit is perturbed to a lower energy level. For the case of a circular device with the applied field and polarization directions along the z axis, as shown in Figure 38.8, the effective field is given by $H_{eff} = H_{ap} - N_{zz}M_s \cos(\theta)$. This term gives rise to nonlinear damping and causes the damping to increase when the orbits increase in size. When the current is increased to make the effective damping go negative the orbit begins to grow, θ increases, and the effective damping increases until it becomes zero. If the orbit grows beyond this point, the damping goes positive pushing the orbit back to the 0-damping orbit. In this geometry, the LLGS equation becomes similar to the van der Pol equation that has non-linear damping. A STNO, similar to a van der Pol oscillator, is an auto-oscillator that has a resonator driven by a nonlinear damping term that adjusts the amount of energy being put in and taken out of the system to stabilize a particular oscillation mode. It should be pointed out that, unlike the van der Pol equation, the LLGS equation has a nonlinear precessional term so that the oscillation frequency, as seen in Figures 38.5 and 38.6, varies depending on oscillation amplitude.

It is instructive to write down the z-component of the LLGS equation in this geometry

$$\frac{dm_{fz}}{dt} = \left(\alpha |\gamma| \frac{H_{ap} - N_{zz}M_s \cos(\theta)}{1+\alpha^2} - \frac{g_s(\theta_m)\mu_e I}{eM_s V} \right) \sin^2(\theta).$$

Since the terms in this equation are small the z-component of the magnetization, unlike the oscillating transverse components, will evolve slowly. The right side of the equation is plotted as a function of the polar angle in Figure 38.7 for a z-axis applied field of 1.25 T and typical STNO parameter values. If $dm_{fz}/dt > 0$, then the orbit is pushed toward smaller angles, if $dm_{fz}/dt < 0$, the orbit is pushed toward larger angles and if $dm_{fz}/dt = 0$ then stationary solutions are found. If the current is zero, the moment is pushed to a static state at $\theta = 0$ as expected. As the current is increased, this point becomes unstable and the system will evolve to a stable precessional state. At large currents the system will be pushed to a static state at $\theta = 180°$, i.e., the device switches and no stable dynamics are observed. The energy of the STNO is also plotted in Figure 38.7 as a function of the free-layer polar angle. Since the energy can wander by several $k_B T$ due to thermal fluctuations, the orbit angle will also wander

giving rise to variations of the STNO frequency and a finite linewidth. If the precessional state has a large slope at the zero crossing then the restoring force for the orbit will be large and a sharp emission peak will be observed. If the slope at the zero crossing is small, then the orbit will have a small restoring force and the frequency will tend to wander.

The spin-torque parameter, g_s, is also a function of angle if $\Lambda \neq 1$, increasing monotonically as the angle θ_m between the free layer and polarization layer moments varies from 0° to 180°. This dependence will make the system asymmetric: inverting both the sign of the current and orientation of the polarizer will not give the same orbits. Starting with the free layer moment parallel to the polarizer will cause $g_s(\theta_m)$ to increase as the orbit opens up and may prohibit stable orbits, while starting with the free layer magnetization antiparallel to the polarizer will cause a decreasing $g_s(\theta_m)$ which will stabilize the orbits. Other device parameters, such as the damping coefficient α, may also be a function of the magnetization angle. Further, the spin transfer and damping torques will, in general, be along different directions with varying amplitudes making a general analysis of orbit stability difficult.

Examples of current stabilized orbits are shown in Figure 38.8a for a device with a perpendicular applied field, perpendicular polarizer, and $\Lambda = 2$. The orbits shown are numerically calculated from the LLGS equation with an applied field of -1.5 T so that the magnetization starts antiparallel to the polarizer and the current is positive, which pushes the free layer moment toward the polarizer. This orientation has the largest region of stable precession. Figure 38.8c,d shows "phase diagrams" of the steady state response of the system to different applied fields and currents. Figure 38.8c plots the average rms value of m_{fx} using an intensity scale: black is zero and indicates a quiescent state, white is 0.7 and represents a state of maximum oscillation. Figure 38.8d plots the average value of m_{fz}, which shows the orientation of the magnetization as a function of applied field and current. For a given field value, as the current is ramped, the magnetization goes from being aligned with the polarizer's direction at negative currents to being anti-aligned with the polarizer at large positive currents. This magnetization rotation is reflected in a change in device resistance from its minimum value to its maximum value. A region of oscillation is often identified by a region of intermediate device resistance.

The high-symmetry configurations, while being useful to understand how STNOs function, often do not make practical STNOs. In the case shown in Figure 38.8, the angle θ_m between the free layer and polarizer remains constant on the precessional orbit so there is no time-varying voltage output. A third magnetic reference layer is required to cause a changing resistance and an output voltage.

It can be seen in Figure 38.8 that the orbits are not perfectly sharp and, in fact, there are dynamical excitations present when the current is zero. This is due to the thermal field which continually agitates the system and gives rise to a finite linewidth, as discussed in Section 38.7.

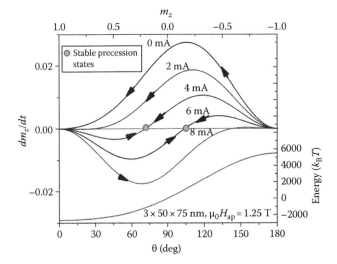

FIGURE 38.7 Plot of rate of change of m_{fz} (orbit axis) as a function of the polar angle for various injection currents. The applied field is 1.25 T along the z axis, $M_s = 800$ kA/m, and $\Lambda = 1$. The stable precessional states are shown by the gray circles. The bottom plot shows the device energy as a function of the polar angle.

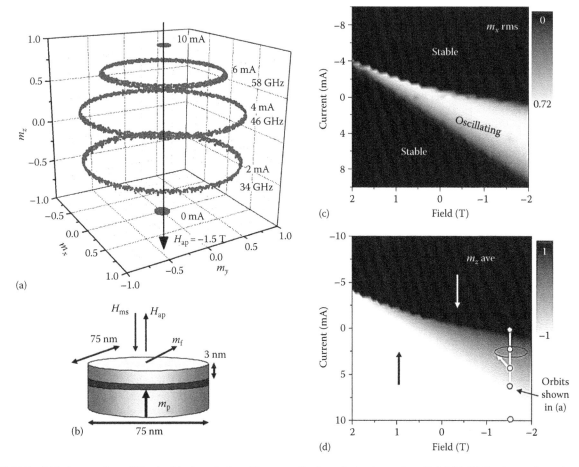

FIGURE 38.8 (a) Spin-transfer stabilized orbits for a device with perpendicular polarization and applied field. (b) Device configuration showing perpendicular polarizer and perpendicular applied field. (c) Phase diagram showing the rms average of m_x as a function of applied current and field. (d) Phase diagram showing average of m_z as a function of applied current and field. The simulations were done using $\alpha = 0.02$, $M_s = 800\,kA/m$, $T = 300\,K$, $\Lambda = 2$, $P_{sc} = 0.4$ and the direction of the polarizer was fixed and not allowed to vary with applied magnetic field.

38.4 Device Configurations and Fabrication

Many different STNO configurations have been explored. The configurations, as seen in Figure 38.9, can be classified by the type of spacer layer, the patterning geometry, and the magnetic geometry. The main requirements are that there is efficient generation of spin torque to induce oscillations without damaging the device with large electrical currents and there is a large magnetoresistance in the correct geometry to convert the time dependent magnetization into a useful output voltage.

38.4.1 Barrier Type and Stack Configuration

Devices that have metallic spacer layers, as shown in Figure 38.9a, are referred to as spin valves and those with thin insulating layers, as shown in Figure 38.9b, are called MTJs. The term "spin valve" was coined by Dieny et al. (1991) and refers to a system with two magnetic layers separated by a metallic spacer that lets spins through when the moments are aligned and blocks electrons when the layer magnetizations are anti-aligned.

A typical stack is shown in Figure 38.9a and consists of a 2.5 nm Ta adhesion layer, a 50 nm Cu layer that serves as a high conductivity base electrode and a seed layer that nucleates (111) texture, a 20 nm $Co_{90}Fe_{10}$ layer that serves as the spin polarizer, a 5 nm Cu spacer layer, a low moment 5 nm $Ni_{80}Fe_{20}$ free layer, and a top Cu capping layer. The free layer is made from a magnetically soft Permalloy alloy that can easily respond to both applied fields and spin torques. The polarizing layer, in this design, is thicker and has a higher moment than the free layer to prevent it from undergoing spin-torque induced oscillations. CoFe alloys are used to obtain fcc texture which is consistent with the desired (111) texture in the stack. Pure Co is likely to grow in an hexagonal close packed (hcp) structure which is undesirable for these devices.

All metallic devices, such as spin valves, have low resistances on the order of 2–5 Ω. Accurate measurement of the intrinsic impedance of an all metallic device is difficult since the resistance of the electrodes and contacts contribute a considerable fraction of the device resistance. A typical device may have a $\Delta R = 0.2\,\Omega$, intrinsic resistance of $R_{int} \approx 1\,\Omega$, and a total resistance of $R_{av} = 5$–$10\,\Omega$. The devices have a low resistance area product of RA $\cong 10^{-2}\,\Omega\,\mu m^2$. Given the low resistance of all

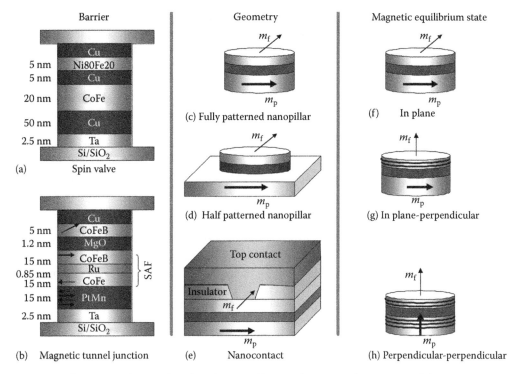

FIGURE 38.9 Different configurations of STNOs based on the type of barrier, the patterned geometry, and the magnetic state.

metallic devices, it is easy to apply large currents to initiate spin-transfer oscillations. However, since the resistance change is relatively small, the output power is also relatively small. The maximum observed output powers range from 1 to 10 nW.

Tunneling devices have thin insulating barriers. The most successful barrier material to date is MgO (Yuasa et al. 2004). For STNO applications, the junctions must have a high MR and a high transparency (low RA product). The high transparency is required because the TMR and spin torque fall off at voltages above a few hundred mV and the junction breakdown voltages are typically 0.5–1 V. To safely get 1 mA, a typical critical current, through a 100 nm junction requires RA < $(1\,V \cdot 10^{-2}\,\mu m^2/1\,mA)$ = $10\,\Omega\,\mu m^2$. This RA product corresponds to MgO barriers that are less than 0.7 nm thick. To obtain a high polarization and ΔR, the MgO barrier requires a (001) crystalline texture which is achieved by annealing the MTJ stack at high temperatures (300°C–400°C). MTJs have the advantage that ΔR and R_{av} can be much larger, typically 100 and 200 Ω, respectively.

In addition to metallic and MTJ devices, a third class of devices has been developed using porous nano-oxide films as the spacer layer. These devices can be viewed as an intermediate structure between a MTJ and a metallic device and has the advantage of a higher device resistance than an all metallic device and it can accommodate more current than an MTJ.

The stack structure for either a spin valve or a MTJ can be a simple bilayer, as shown in Figure 38.9a or a pinned bilayer as shown in Figure 38.9b. The simple bilayer usually has a fixed layer that is thicker or has a higher magnetization material to stabilize the layer to prevent it from oscillating. A pinned stack uses an antiferromagnetic layer (usually $Ir_{20}Mn_{80}$ or $Pt_{50}Mn_{50}$) and a synthetic antiferromagnetic (SAF) to fix the direction of the polarizer. The antiferromagnetic layer couples to the adjacent ferromagnetic layer (pinned layer) through the exchange interaction. Since the antiferromagnet has no moment it does not respond to an applied field and it is difficult to rotate the magnetic order. The direction of the pinned ferromagnetic layer is set by depositing or annealing in a magnetic field. This orients the pinned layer in the desired direction and then allows the antiferromagnet to relax into a low energy state in which the antiferromagnetic moments line up with the ferromagnetic moments. When the ferromagnetic moment attempts to rotate away from the preferred direction, it experiences an exchange bias field that tries to keep it in its pinned direction. The exchange bias is an interfacial energy proportional to the area of contact whereas the Zeeman energy is proportional to the volume or total moment of the magnetic element. This causes the exchange bias field to vary inversely with the moment of the layer, so thin layers are required to get large pinning fields.

A SAF is often used for a pinned layer instead of a single ferromagnetic layer. A SAF consists of two ferromagnetic layers separated by a thin layer of Ru, which promotes strong antiparallel orientation. The effective moment of the SAF is the difference between the two layer moments. For the device shown in Figure 38.9b, the effective moment of the SAF is close to zero and the pinning strength can be very high 0.1–0.5 T.

The device stacks are usually deposited in one step using a computerized sputter system with 6–10 cathodes. The sputter system should have base pressures of $<10^{-6}\,Pa$ (0.75×10^{-8} torr) since small amounts of residual gases can affect the microstructure of the various layers. The deposition conditions, such as deposition pressure and power, need to be carefully controlled to insure that the interfaces are sharp and flat, the crystalline texture is correct, and the magnetic properties are optimized. Often, small partial pressures of reactive gases (N_2 or O_2) may be used during one or more deposition steps to control the film microstructure.

38.4.2 Patterning Geometry and Magnetic Configuration

There are three basic patterning geometries: nanopillars in which the magnetic structure is patterned down to nm dimensions (Figure 38.9c), half patterned nanopillars in which the polarizer is left unpatterned (Figure 38.9d), and nanocontacts in which a small (<100 nm) contact is made to a larger mesa (Figure 38.9e). The magnetic structure of a nanopillar is very sensitive to the shape of the element and there is a strong interaction between the free layer and polarizer. A half patterned nanopillars reduces the interaction between the two layers and makes the polarizer more likely to be fixed in a magnetic field since it has a higher volume. The nanocontacts have the advantage that there are no device edges to induce large magnetostatic fields that may inhibit the ability of the device magnetization to precess uniformly. However, as will be seen in the next section, the exchange coupling of the oscillating region to the larger mesa gives rise to spinwaves.

Each of these geometries can use a variety of magnetic orientations with the free and fixed layer being designed to be in-plane or perpendicular to the plane. Figure 38.9f shows a simple in-plane device where the magnetizations like to lie in plane due to the magnetostatic fields. Figure 38.9g shows a device with a perpendicular free layer, which is shown schematically as a multilayer. The fixed layer still lies in plane. This is an efficient geometry since it is easy to get the free layer to precess and there is a large angular variation between the free layer and the polarizer leading to a large MR and output voltage. Figure 38.9h shows a geometry in which both the free layer and polarizer are perpendicular. More advanced devices can have three or more magnetic layers and can use complex bit shaping to obtain different modes of operation.

Etching of STNO devices still remains challenging due to the large number of different materials in the device stacks and the lack of selective etches. Etching is usually done with an ion beam etch or a reactive ion etch process. Care must be taken to prevent redeposition on the devices' edges during etching or oxidation of the device edges as this can alter both the electron conduction though the device and the magnetic properties of the free layer. The nanocontact geometry avoids many of the complexities involved with the etch process required for nanopillar geometry.

38.5 Dependence of Frequency on Current and Field

STNO frequencies vary with both applied field and current. As the magnetic field is increased, the precession frequency will generally increase as the square root of the applied field at low fields and linearly at high fields as predicted by the Kittel equation. An example of the frequency and field variation of a spin valve nanocontact oscillator with in-plane applied fields is shown in Figure 38.10, along with the predicted single domain

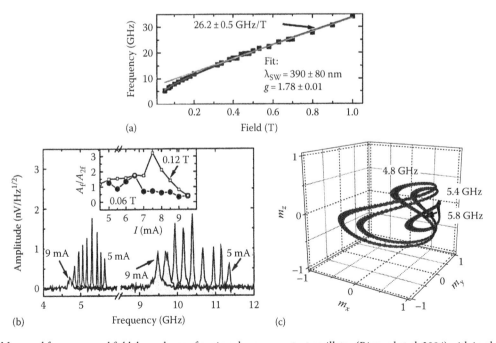

FIGURE 38.10 Measured frequency and field dependence of a spin valve nanocontact oscillator (Rippard et al. 2004) with in plane applied fields (a) field dependence, (b) output spectra for various applied currents with 60 mT applied field showing first and second harmonics, (c) predicted orbits from the single domain model with 60 mT applied field.

orbits. A red shift (frequency decrease) is seen as the current is increased as expected from the discussion in Section 38.3. A strong second harmonic signal is observed since the output voltage is predominantly proportional to m_{fx}.

As the applied field is moved out of plane, we expect to see spectra that red shift then blue shift (increase frequency) with current, as seen in Figure 38.5c, and then at large, nearly out-of-plane fields we expect to see solely a blue shift, as seen in Figure 38.5b. This trend can be seen in the data shown in Figure 38.11, which plots the oscillator frequency as a function of applied current and field for a spin valve nanocontact oscillator when the magnetic field is applied 10° off normal. At low fields, one sees a slight red shift then blue shift as the current is increased. At high fields, there is a strong blue shift with an average magnitude of $\cong 1.5\,\mathrm{GHz/mA}$. There are, however, sharp jumps in the frequency suggesting that the device abruptly changes oscillation modes at certain fields and currents. These modal jumps have not yet been fully explained. Figure 38.11 shows data for both increasing and decreasing currents highlighting the fact that, for magnetic devices, there can be hysteresis and the behavior of a device

may depend on its current and field history. Also shown in the color maps are the STNO linewidth (Figure 38.11a) and power output (Figure 38.11b). For spin-valve nanocontact STNOs the linewidths vary from a few MHz to several hundred MHz and the maximum output power is on the order of 1 nW. Output power and linewidths will be discussed in more detail in the next sections.

The output frequency from a spin valve nanocontact STNO with a perpendicular free layer and an in-plane polarizer is shown in Figure 38.12a. Here $\mu_0(H_{\mathrm{an}}-M_{\mathrm{s}}) \cong 0.20\,\mathrm{T}$. As previously discussed, for this geometry, we expect a red shift as the current increases (as seen in Figure 38.6a). The moment initially points vertically at low currents with no precession. As the current is increased, the orbit should open up and the effective field $H_{\mathrm{ap}} + (H_{\mathrm{an}} - M_{\mathrm{s}})\cos(\theta)$ will decrease causing the frequency to decrease. The precessional angle, shown in Figure 38.12b, can be calculated from the data knowing the saturation magnetization and the perpendicular anisotropy field. In this configuration with an in-plane polarizer, the orbit at larger currents will saturate near 90° and will begin to deviate from the "free-precession" orbits. The maximum output power of 1 nW is, as expected, observed near 90° when the orbit obtains its maximum amplitude. One advantage of this type of perpendicular oscillator is that it can operate with little or no applied field.

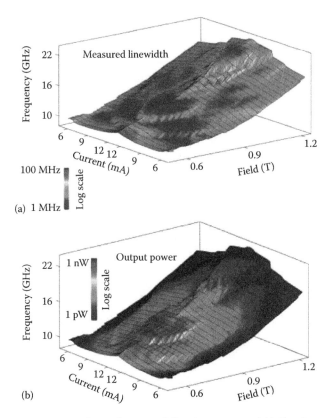

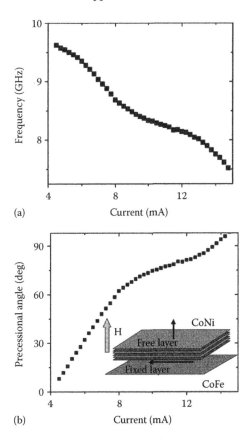

FIGURE 38.11 **(See color insert following page 22-8.)** (a) Plot showing the measured frequency of a spin valve nanocontact STNO as a function of applied current and field. For these plots the device current is increased from 6 to 12 mA and then decreased from 12 to 6 mA to investigate any regions that may be hysteretic with current. The oscillator linewidth is shown by the color map with blue being small linewidths (~1 MHz) and red being large linewidths (~100 MHz). (b) is the same as (a) except the total output power is shown on the color map. (Adapted from Rippard, W.H. et al., *Phys. Rev. B*, 74, 224409, 2006.)

FIGURE 38.12 (a) Measured oscillator frequency vs. current for a spin valve nanocontact perpendicular-free layer STNO with 300 mT vertical field applied and $\mu_0(H_{\mathrm{an}}-M_{\mathrm{s}}) \cong 0.20\,\mathrm{T}$. (b) Precession angle vs. current calculated from the measured frequency.

38.6 Output, Circuit Model, and Device Measurement

For an oscillator to be useful there needs to be an output. There are two possible outputs generated by an STNO: (1) a microwave voltage and (2) strong oscillations in the magnetic field close to the oscillator. Using the strong local microwave magnetic fields for data recording applications has been proposed by Zhu et al. (2008). The microwave fields within a few tens of nanometers from the device can be on the order of a 1–2 T. There are very few devices that are capable of producing such large microwave magnetic fields. These fields can resonantly drive magnetic data storage bits to assist in their switching.

More conventionally, we are interested in voltage and current outputs. A STNO may be viewed as a time dependent resistor as shown in Figure 38.13. The STNO resistance is given by

$$R = R_{av} + \frac{\Delta R}{2}(\vec{m}_f \cdot \vec{m}_p) \cong R_{av} + \frac{\Delta R}{2}\cos(\omega t).$$

In Figure 38.13, we have the STNO connected through a bias tee to a load R_L. We assume the inductor is sufficiently large and close to the device so that the drive current remains constant. The changing STNO resistance alternately directs the current through the STNO when its resistance is low or through the load when its resistance is high. The optimal device would be one that would change its resistance from zero to an infinitely large value so that no power would be dissipated in the STNO and all of the power would be transmitted to the load. The time-averaged power delivered is

$$P_{out} = \frac{V_{out}^2}{2R_L} = \frac{I^2}{8}\frac{\Delta R^2 R_L}{(R_{av}+R_L)^2},$$

which for a metallic STNO with $I = 10\,\text{mA}$, $\Delta R = 0.1\,\Omega$, $R_{av} = 5\,\Omega$ and $R_L = 50\,\Omega$ is 2 nW and for an MTJ STNO with $I = 1\,\text{mA}$, $\Delta R = 150\,\Omega$, $R_{av} = 450\,\Omega$, and $R_L = 50\,\Omega$ is 0.5 μW. If we

could put 10 metallic STNOs in series and have them phase lock (as discussed in Section 38.9) we could get out 62 pW. If we could put 10 MTJs in parallel and have them phase lock we could get 16 μW. Of course, in both cases we have increased the drive power by a factor of 10 and have increased the output power by slightly more than this because we have impedance matched to the load. The efficiency

$$\frac{P_{out}}{P_{in}} = \frac{V_{out}^2}{2I^2 R_{av} R_L} = \frac{1}{8}\frac{\Delta R^2 R_L}{R_{av}(R_{av}+R_L)^2}$$

in both cases is low, ranging from 0.5×10^{-3} to 5×10^{-3}. This can be compared to the efficiency of Gunn oscillators which is typically 0.03–0.05. Strategies to increase the efficiency will be discussed in Section 38.9, which discusses phase locking.

All devices will have a parallel shunt capacitance, as seen in Figure 38.13, that could potentially limit high-frequency operation. Here again, nanoscale dimensions come to the rescue and intrinsic parallel capacitances are on the order of 1 aF, which will allow operation up to 300 GHz. Care still must be taken when designing the circuit layout to avoid parasitic capacitances from the electrodes.

38.7 Linewidth

Two of the most important parameters characterizing an STNO are the oscillator's full width at half maximum (FWHM) linewidth Δf of the power spectra and the quality factor $Q = f_r/\Delta f$. Figure 38.14 shows an STNO emission spectra with $f_r = 13.15\,\text{GHz}$, $\Delta f = 7.1\,\text{MHz}$, and $Q = 1800$. The linewidth can vary widely depending on the device and operation conditions. As shown in Figure 38.11a using the color scale, the linewidth for a single device can vary from a few MHz to 100 MHz as the current and applied field are varied and the linewidth variation can be very complex. One trend that can be observed in Figure

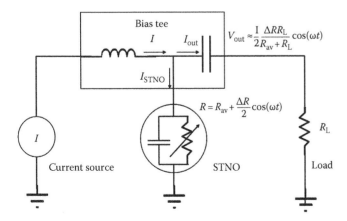

FIGURE 38.13 Circuit diagram of an STNO being driven by a current source through a bias tee. The microwave signal is output to a load R_L.

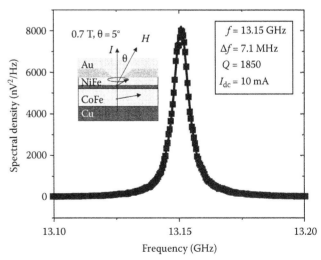

FIGURE 38.14 An example spin-valve nanocontact STNO spectra showing a narrow linewidth at room temperature. The STNO geometry is shown in the inset, the contact area is approximately 75 nm.

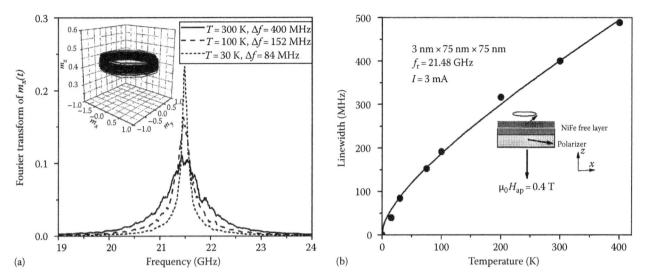

FIGURE 38.15 (a) LLGS simulation of STNO emission at 30, 100, and 300 K to show how thermal fluctuations give rise to the STNO linewidth. The Fourier transforms of m_{fx} is plotted and the linewidths listed are the half width at half maximum which correspond to the power spectra FWHM linewidths. (b) Calculated temperature dependence of the STNO linewidth along with a fit which shows a linear behavior at high temperatures. The device configuration is shown in the inset.

38.11a is that the linewidth tends to be larger near regions of rapid frequency variation and smaller in plateau regions where the variation of frequency is small. In general, linewidths for nanocontacts are smaller than that for nanopillars, where linewidths can often approach 1 GHz.

One contribution to the observed linewidth is due to thermal fluctuations that perturb the amplitude and phase of the orbit. LLGS simulations of STNO emission at $T = 30$, 100, and 300 K are shown in Figure 38.15 along with an inset showing the thermally perturbed trajectories. The Fourier transform of m_{fx} is plotted, which is proportional to the output voltage since, for this simulation, the polarization is predominantly along the x direction. The peak width for the voltage spectra is twice that for the power spectra and therefore half width at half maximum are listed to correspond to the power spectra widths. The calculated linewidth is approximately linear in temperature with the linewidth going to zero at $T = 0$ K.

Experimentally, STNO devices show a decrease in linewidth with temperature, however, the linewidth tends to saturate at temperatures near 100 K. The linewidth may be limited by micromagnetic effects caused by device edges, defects, or inhomogeneous current injection that cause different parts of the device to oscillate at slightly different frequencies. Experimentally, many parameters vary with temperature, including M_s and α, causing the orbit to change as a function of temperature. As indicated in Section 38.3, the linewidth is sensitive to the stability of the particular orbit and different orbits can have widely varying linewidths. This makes the analysis of the temperature dependence of the linewidth difficult. As indicated, point contact devices typically show smaller linewidths than nanopillar devices. This may be due to a larger effective area (the coherent oscillating region is larger than that of the contact) and reduced inhomogeneities near the region of oscillation that reduce the inhomogeneous linewidth broadening.

Most of the oscillator linewidth is due to phase noise, the random deviation of the oscillator phase from that of an ideal reference oscillator. The amplitude noise is relatively small, however, since the amplitude of oscillation changes the frequency, the amplitude and phase noise are coupled. In addition to thermal magnetic fluctuations, which are associated with the damping constant α, there are also Johnson current fluctuations associated with the device resistance R. The Johnson current noise is typically small in metallic devices while, in MTJs both Johnson noise and shot noise may contribute to the STNO linewidth. These noise sources explain the correlation between the observed linewidth and the change in frequency with current and field: if the frequency changes rapidly with field and current then fluctuations in these quantities will cause considerable phase noise and an increased linewidth.

38.8 Micromagnetics, Vortex Oscillators, and Spinwaves

In the previous sections, we considered the magnetic system to consist of a uniformly magnetized free layer and a fixed polarization layer. These assumptions are not always justified. The magnetization in both (or several) layers may be nonuniform and both layers can undergo dynamic response. The nonuniformities can be driven by the magnetostatic energy which causes the magnetization to rotate at the device edges and the current induced magnetic fields (sometimes called the Oersted fields) that promote a vortex structure. The LLGS equations of motion can be generalized to accommodate spatially nonuniform magnetization and dynamics in multiple layers by discretizing the system, as shown in Figure 38.16, and solving a large set of coupled integro/differential equations

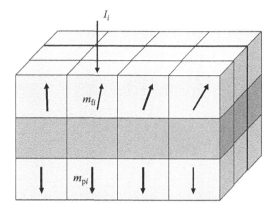

FIGURE 38.16 Discretized device structure used for micromagnetic simulations. Typical cell sizes are on the order of 3 nm.

$$\frac{d\vec{m}_{fi}}{dt} = -\mu_0 \frac{|\gamma|}{1+\alpha_i^2} \vec{m}_{fi} \times \vec{H}_{effi} - \alpha_i \frac{\mu_0 |\gamma|}{1+\alpha_i^2} \vec{m}_{fi} \times (\vec{m}_{fi} \times \vec{H}_{effi})$$
$$+ \frac{g(\theta_m)\mu_e I_i}{eM_{si}Vi} \vec{m}_{fi} \times (\vec{m}_{fi} \times \vec{m}_{pi})$$

Here, $\vec{m}_{fi}$, $\vec{H}_{effi}$, I_i refer to the moment, effective field, and current in the ith element, respectively. Each element now feels a strong exchange field due to the neighboring elements. The magnetostatic fields must be calculated by integrating over the entire sample. The properties of each element such as saturation magnetization, damping, or anisotropy can also vary. This complex set of equations can be numerically integrated by software packages such as the Object Oriented MicroMagnetic Framework (OOMMF http://math.nist.gov/oommf/).

FIGURE 38.17 Micromagnetic simulation of two nano-contact STNOs that are 500 nm apart. The oscillating magnetization under the contacts can generate spinwaves that can couple the two oscillators. The presence of spinwaves can be verified by making a cut using a focused ion beam and seeing the interactions between the oscillators disappear. The inset shows a secondary electron micrograph of the actual device.

An example of micromagnetic effects is shown in Figure 38.17 where two nanocontact STNOs are oscillating on the same magnetic mesa. When the region under the contacts oscillates, spin waves are generated that cause the oscillators to interact. The spin waves can be detected by using one STNO as an emitter and one as a detector (Pufall et al. 2006). When both STNOs are oscillating and the frequencies become close, the two oscillators can phase lock. Figure 38.17 shows the mesa structure after a line was cut by a focused ion beam which then causes the spinwave coupling to disappear. The simulation shown in Figure 38.17 does not account for the Oersted fields generated by the currents. These fields can cause the radiation pattern to be very asymmetric. The detailed spin wave radiation pattern generated from these nanocontact devices has not yet been experimentally determined.

Another example of micromagnetic effects is the vortex oscillator. A vortex oscillator in the nanocontact geometry, after Pufall et al. (2007), is shown in Figure 38.18. By adjusting the thickness of the magnetic layers and going to large device currents, which create large circumferential magnetic fields, vortices can be formed in either the free layer or the polarizer layer (Pribiag et al. 2007). A vortex has a core in which the moment points out of the plain of the device and the magnetization surrounding the core is in a circumferential in-plane direction. The spin torque causes the vortex to move around the region of current injection. Vortex oscillators usually have low frequencies

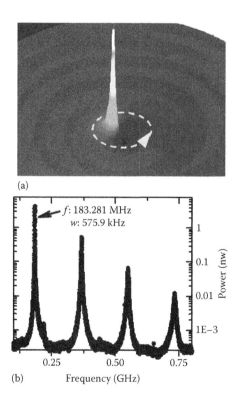

FIGURE 38.18 (a) Schematic illustrating a vortex rotating around a current injection site. (b) Measured output from a vortex oscillator with an applied field of 2.5 mT showing several nW output power. The fundamental frequency is 183.28 MHz and the FWHM is 575.9 kHz.

near 1 GHz and show little dependence of the frequency on drive current. The oscillator shown in Figure 38.18 is highly nonlinear with large harmonic amplitudes indicating that the vortex is not undergoing simple harmonic motion. One advantage of vortex oscillators is that they require small or no applied magnetic field for operation.

There are many further examples of using nonuniform dynamical states in STNOs. The reader is referred to Silva and Rippard (2008) for more complete discussion of micromagnetic effects.

38.9 Phase Locking

An important aspect of an oscillator is to control the phase of the oscillator with respect to a reference oscillator or other oscillators in an array. This is required if one wants to adjust the phase of an oscillator in a phased array antenna or to lock a set of oscillators together to enhance the total power output. It is a general property of non-linear auto-oscillators that the oscillator will try to adjust its frequency to match that of an injected signal. In the case of STNOs, the injected signal can be either a microwave current or magnetic field. Figure 38.19 shows an example of phase locking to an injected microwave current. The device geometry is shown in the inset in Figure 38.19a. With no applied current the frequency increase with applied current as shown in Figure 38.19a. When a microwave current is injected, the oscillator will lock over a given range of applied current. This locking range is inversely dependent on the linewidth of the oscillator and is typically 0.5–2 mA. When the signal is locked, there will be a low frequency voltage that

appears, seen in Figure 38.19f, due to the mixing of the applied microwave signal with the oscillator output. When the oscillators are locked, the frequencies will be the same but there will be a phase difference between the two oscillators that varies over the locking range. This property is useful for adjusting the phase of the STNO relative to a reference oscillator by adjusting the bias current.

Another important example of phase locking is the coupling of two or more STNOs to have them oscillate at the same frequency with an adjustable relative phase. An example of mutual phase locking of two point contact STNOs is shown in Figure 38.20 (Kaka et al. 2005, Mancoff 2005). The geometry is similar to that shown in Figure 38.15. When the oscillators are far apart, the frequency of the oscillators can be swept and there is little interaction between them. In the experiment shown, the current through oscillator B was swept and oscillator A was kept at a fixed current of 11.5 mA. As the oscillators are brought close in space, 500 nm as shown in Figure 38.20b, and frequency they interact and phase lock. In this case the phase locking is due to a spin-wave coupling between the oscillators. Figure 38.21 shows the spectral outputs before, during, and after phase locking. When the oscillators are locked, the linewidth narrows and the power increases. In this case, the signals are taken out independently and the output power is given by $P_{out} = P_1 + P_2 + 2\sqrt{P_1 P_2} \cos(\theta)$. When the two oscillators are in phase a factor of 4 increase in power is observed. The possibility of obtaining an N^2 increase in power output from N phase locked oscillators is being pursued as a method to dramatically increase the output power. To get an N^2 power increase, the signal cannot be connected simply in parallel or series. The outputs need to be extracted independently and then combined in phase.

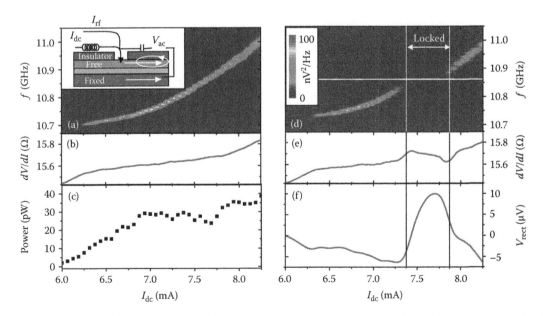

FIGURE 38.19 (**See color insert following page 22-8.**) Measurement of an STNO phase locking to an injected microwave signal. The device geometry is shown in the inset (a); (a), (b) and (c) show the measured frequency, device resistance, and output power as a function of bias current with no injected microwave signal. (d), (e), and (f) show the measured frequency, device resistance, and dc voltage as a function of bias current with an injected microwave signal. (Adapted from Rippard, W.H. et al., *Phys. Rev. Lett.*, 95, 067203, 2005.)

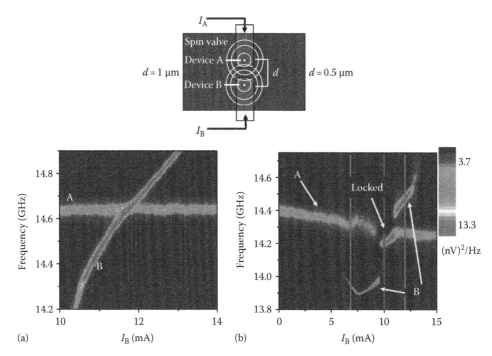

FIGURE 38.20 **(See color insert following page 22-8.)** Demonstration of mutual phase locking of two STNOs (Kaka et al. 2005). (a) The emission spectra of two oscillators as a function of the current through oscillator B when the oscillators are far apart showing no interaction. (b) The emission spectra of two oscillators as a function of the current through oscillator B when the oscillators are close showing strong interactions and phase locking.

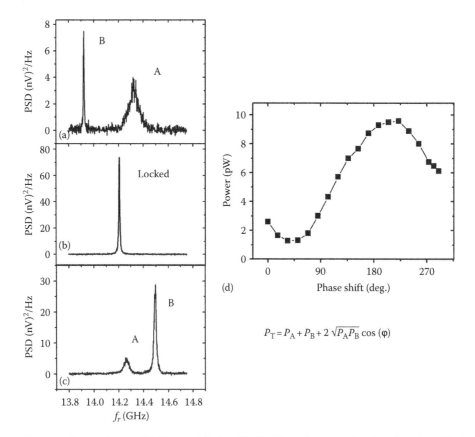

$$P_{T} = P_{A} + P_{B} + 2\sqrt{P_{A}P_{B}}\cos(\varphi)$$

FIGURE 38.21 Spectral output of a two element STNO array (a) when STNO B has a frequency less than STNO A, (b) when the oscillators are locked showing an increase in output power and decrease in linewidth and (c) when STNO B has a frequency greater than STNO A, (d) the output power as a function of a relative phase shift before the signals are combined. (From Kaka, S. et al., *Nature*, 437, 389, 2005.)

38.10 Applications

Several applications have been suggested for STNOs. Since power output, in general, scales with the size of the oscillator and the linewidth inversely proportional to the size, a single nano-oscillator will never be a high-power ultra-narrow-linewidth oscillator. Making phase locked arrays may solve this problem if high power is required. A better set of applications are those that make use of the nanoscale aspects and large-range high-speed tunability of STNOs. These applications include local on chip clocks for VLSI applications, high-density massively-parallel microwave signal processors, small phased array transmitters, chip-to-chip micro-wireless communications, and local excitation sources for nanosensors. Figure 38.22 illustrates an array of STNOs being used for rapid demodulation of an incoming signal. This application relies on a large array of variable frequency STNO demodulators working simultaneously to analyze an incoming signal. Figure 38.23 illustrates a nanoscale STNO phased array antenna that could have possible applications in chip-to-chip communications. The relative phase of each

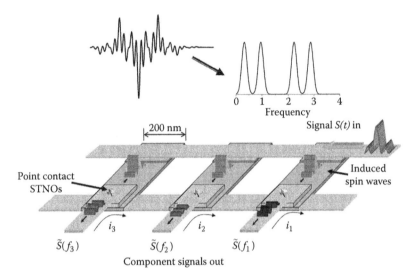

FIGURE 38.22 Schematic showing a possible application using an array of STNOs for rapid signal processing of an incoming signal. The application is enabled by the ability to have a large number of STNOs on the same chip each operating at a different frequency.

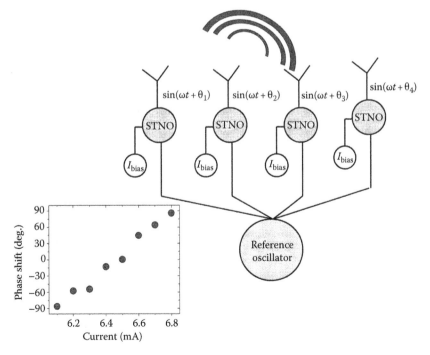

FIGURE 38.23 Schematic of a phased array antenna using STNOs. The ability to have a large number of oscillators with precisely controlled phase may allow beam steering in micro-wireless applications. The inset on the lower left shows the phase of an STNO as the current is varied through its locking range.

oscillator can be adjusted by varying the bias current as seen in the plot in the lower left. The use of a small phased array fabricated on a standard CMOS wafer would allow efficient beaming of transmit and receive signals for a local microwave communication bus.

38.11 Summary and Outlook

STNOs are exciting new nanoscale spin-based devices that offer many unique capabilities: nanoscale size, wide frequency range, high tunability, operation over a wide temperature range, nonlinear operation, and compatibility with standard CMOS processing. Many challenges remain before these devices can be incorporated in nanotechnologies. Experimentally the demonstrated frequency range is 0.5–250 GHz;* however, to obtain high frequencies, large magnetic fields have been required, which is not practical for applications. The highest frequency that has been demonstrated without an applied field is ~5 GHz. Several strategies exist to boost the frequency in the absence of an applied field. These include introducing stronger magnetic anisotropies, exchange coupling to a fixed layer, and using antiferromagnets as the oscillation layer. At present, the maximum power output from STNOs is on the order of a microwatt. The power needs to be boosted into the mW range for many applications, although, for local clocks and nanoscale signal processing the existing power levels may be sufficient. There are still problems with device-to-device reproducibility, controlling the oscillator linewidths, and understanding the detailed mode structure of the oscillator. It is anticipated that most of these problems will be solved in the next few years and that STNOs will become a useful nanoscale device for a variety of applications.

References

Albert, F. J. et al. 2002. Quantitative study of magnetization reversal by spin-polarized current in magnetic multilayer nanopillars, *Phys. Rev. Lett.* 89, 226802.

Berger, L. 1996. Emission of spin waves by a magnetic multilayer traversed by a current, *Phys. Rev. B* 54, 9353.

Bertotti, G. et al. 2005. Magnetization switching and microwave oscillations in nanomagnets driven by spin-polarized currents, *Phys. Rev. Lett.* 94, 127206.

Bertotti, G., Mayergoyz, D. I., Serpico, C. 2008. *Nonlinear Magnetization Dynamics in Nanosystems*, Elsevier Science & Technology, Amsterdam, the Netherlands.

Butler, W. H. et al. 2001. Spin-dependent tunneling conductance of Fe|MgO|Fe sandwiches, *Phys. Rev. B* 63, 054416.

Camley, R. E., Barnaś, J. 1989. Theory of giant magnetoresistance effects in magnetic layered structures with antiferromagnetic coupling, *Phys. Rev. Lett.* 63, 664.

Deac, A. M. et al. 2008. Bias-driven high-power microwave emission from MgO-based tunnel magnetoresistance devices, *Nat. Phys.* 4, 803–809.

Dieny, B. et al. 1991. Giant magnetoresistance in soft ferromagnetic multilayers, *Phys. Rev. B* 43, 1297; Dieny, B. et al. U.S. Patent 5,159,513.

Grünberg, P. A. 2008. Nobel Lecture: From spin waves to giant magnetoresistance and beyond, *Rev. Mod. Phys.* 80, 1531–1540.

Houssameddine, D. et al. 2008. Spin transfer induced coherent microwave emission with large power from nanoscale MgO tunnel junctions, *Appl. Phys. Lett.* 93, 022505.

Julliere, M. 1975. Tunneling between ferromagnetic films, *Phys. Lett.* 54A, 225.

Kaka, S. et al. 2005. Mutual phase-locking of microwave spin torque nano-oscillators, *Nature* 437, 389.

Kiselev, S. I. et al. 2003. Microwave oscillations of a nanomagnet driven by a spin-polarized current, *Nature* 425, 380.

Kittel, C. 1948. On the theory of ferromagnetic resonance absorption, frequency of uniform precession, *Phys. Rev.* 73, 155.

Mancoff, F. B. et al. 2005. Phase-locking in double-point-contact spin-transfer devices, *Nature* 437, 393.

Mott, N. F. 1936. The electrical conductivity of transition metals, *Proc. R. Soc.* 153&156, 699.

O'Handley, R. O. 2000. *Modern Magnetic Materials: Principles and Applications*, John Wiley & Sons, New York.

OOMMF, 2006. Object Oriented MicroMagnetic Framework http://math.nist.gov/oommf/.

Pribiag, V. S. et al. 2007. Magnetic vortex oscillator driven by d.c. spin-polarized current, *Nat. Phys.* 3, 498–503.

Pufall, M. R. et al. 2006. Electrical measurement of spin-wave interactions of proximate spin transfer nanooscillators, *Phys. Rev. Lett.* 97, 087206.

Pufall, M. R. et al. 2007. Low-field current-hysteretic oscillations in spin-transfer nanocontacts, *Phys. Rev. B* 75, 140404.

Ralph, D. C., Stiles, M. D. 2008. Spin transfer torques, *J. Magn. Magn. Mater.* 320, 1190–1216.

Rippard, W. H. et al. 2004. Direct-current induced dynamics in $Co_{90}Fe_{10}/Ni_{80}Fe_{20}$ point contacts, *Phys. Rev. Lett.* 92, 027201.

Rippard, W. H. et al. 2005. Injection locking and phase control of spin transfer nano-oscillators, *Phys. Rev. Lett.* 95, 067203.

Rippard, W. H. et al., 2006. Comparison of frequency, linewidth, and output power in measurement of spin-transfer nanocontact oscillators, *Phys. Rev. B* 74, 224409.

Silva, T. J., Rippard, W. H. 2008. Developments in nano-oscillators based upon spin-transfer point-contact devices, *J. Magn. Magn. Mater.* 320, 1260–1271.

Slonczewski, J. C. 1996. Current-driven excitation of magnetic multilayers, *J. Magn. Magn. Mater.* 159, L1–L7.

Slonczewski, J. C. 2002. Currents and torques in metallic magnetic multilayers, *J. Magn. Magn. Mater.* 247, 324–338.

Sun, J. Z. 2000. Spin-current interaction with a monodomain magnetic body: A model study, *Phys. Rev. B* 62, 570.

---

* High-frequency operation (100–250 GHz) of STNOs can be inferred from features in the high-field differential resistance data (Tsoi et al. 1998) although direct detection of microwave emission above 60 GHz has not yet been demonstrated.

Sun, J. Z. 2006. Spin angular momentum transfer in current perpendicular nanomagnetic junctions, *IBM J. Res. & Dev.* 50, 81–100.

Tsoi, M. et al. 1998. Excitation of a magnetic multilayer by an electric current, *Phys. Rev. Lett.* 80, 4281.

Tsoi, M. et al. 2000. Generation and detection of phase-coherent current-driven magnons in magnetic multilayers, *Nature* 406, 46–48.

Tsoi, M., Sun, J. Z., Parkin, S. S. P. 2004. Current-driven excitations in symmetric magnetic nanopillars, *Phys. Rev. Lett.* 93, 036602.

Tsymbal, E. Yu., Pettifor, D. G. 1996. Effects of band structure and spin-independent disorder on conductivity and giant magnetoresistance in Co/Cu and Fe/Cr multilayers, *Phys. Rev. B* 54, 15314.

Xiao, J., Zangwill, A., Stiles, M. D. 2004. Boltzmann test of Slonczewski's theory of spin-transfer torque, *Phys. Rev. B* 70, 172405.

Yuasa, S., Djayaprawira, D. D. 2007. Giant tunnel magnetoresistance in magnetic tunnel junctions with a crystalline MgO(0 0 1) barrier, *J. Phys. D: Appl. Phys.* 40, R337–R354.

Yuasa, S. et al. 2004. Giant room-temperature magnetoresistance in single-crystal Fe/MgO/Fe magnetic tunnel junctions, *Nat. Mater.* 3, 868.

Zhu, J., Zhu, X., Tang, Y. 2008. Microwave assisted magnetic recording, *IEEE Trans. Magn.* 44, 125.

VII

Hydrogen Storage

39

Endohedrally
Hydrogen-Doped Fullerenes

Lemi Türker
Middle East Technical University

Çağlar Çelik Bayar
Middle East Technical University

39.1 Introduction

39.1.1 Fullerenes

The favored targets for organic chemists in 1960s and 1970s were to synthesize nonbenzenoid aromatics. In their minds there was a prevailing dogma that due to the possible cyclic delocalization of π-electrons, aromaticity was best realized in planar molecules of cyclic nature. In those days, the aromaticity tacitly remained as a two-dimensional concept. Remaining benzene as the archetypal superstar of aromatic molecules, [18]-annulene with its D_{6h} symmetry attracted attention due to its aromaticity.

The synthesis of corannulene, which has a bowl-shaped aromatic structure, led scientists to ponder about some other nonplanar aromatic compounds having three-dimensional delocalization of the π-electrons. The I_h-symmetric football-shaped C_{60} and its bigger homologue soccer ball-shaped C_{70} were subjected to theoretical interest. They are members of a family called fullerenes. The pentagons are essential for the curved structure of fullerenes. The building principle of the fullerenes is a consequence of the Euler theorem that says that for the closure of each spherical network of n hexagons, 12 pentagons are required, with the exception of $n = 1$. Thus, carbon allotropes C_{76}, C_{78}, C_{82}, C_{84}, C_{90}, C_{94}, and C_{96} are also all fullerenes. Of the fullerenes, C_{60} has been the most extensively studied compound.

39.1.2 Hydrogen and Fullerenes

It is clear that in the twenty-first century much more effort will have to be directed at alternative energy sources, of which hydrogen is one of the most promising clean and renewable energy source. However, the storage, transport, and delivery of hydrogen are important elements in a hydrogen energy system. In mobile applications of hydrogen systems, the storage becomes essential to a sustainable energy economy. Light weight, safe, and high energy density storage will enable the use of hydrogen as a transportation fuel. Efficient and cost-effective stationary hydrogen storage will permit photovoltaic and wind sources to serve as base load power systems.

The storage of hydrogen in fullerenes or fullerene-like systems is at the moment a crawling field of research. Most of the work in this area still has been mainly theoretical. However, it could be an alternative to the metal hydride storage systems.

39.2 Experimental Studies on the Endohedrally Hydrogen-Doped Fullerenes

The preparation of endohedral fullerenes, the spherical carbon molecules incorporating atom(s) or a molecule inside the framework (Bethune et al. 1993; Nagase et al. 1996; Kadish and Ruoff 2000; Liu and Sun 2000; Shinohara 2000; Akasaka and Nagase 2002), has so far relied on hardly controllable physical processes, such as covaporization of carbon and metal atoms (Kadish and Ruoff 2000; Shinohara 2000) or high-pressure/high-temperature treatment (650°C/3000 atm) with noble gases (Saunders et al. 1994a,b, 1996), which yield only limited quantities (e.g., only a few milligrams) of a pure product after laborious isolation procedures. This situation has been a severe obstacle to the development of fundamental as well as application-oriented studies on these molecules of great importance.

To bring about a breakthrough to this situation and to make their science developed, an entirely new approach for their production is highly desired. In this regard, a molecular surgery approach to endohedral fullerenes proposed by Rubin (Rubin 1997, 1999; Nierengarten 2001) is quite appealing because of its great potential and versatility. This method is comprised of steps for opening an orifice on the surface of fullerene, insertion of a small atom or a molecule through the orifice, and closure of the orifice, making use of rational techniques of organic synthesis. In this way, an efficient production of various endohedral fullerenes in a much larger amount can be expected.

As the first step of the molecular surgery approach, Hummelen et al. pioneered an efficient route to open an 11-membered ring orifice on the surface of C_{60} (**1**) (see Figure 39.1) (Hummelen et al. 1995a,b). However, even a small atom, such as helium, was found to be difficult to pass through this orifice (Rubin 1999). Rubin et al., for instance, reported the synthesis of cobalt(III) complex **2** (see Figure 39.1), whose cobalt atom was ideally located above a 15-membered ring orifice, but the insertion of this metal atom into the C_{60} cage through the orifice was not possible even by the application of such high pressures as 40,000 atm in a solid

state (Arce et al. 1996; Edwards et al. 1998). A great progress in this research field was brought about again by Rubin's group, who found an elegant strategy to synthesize open-cage fullerene derivative **3** (see Figure 39.1) with a 14-membered ring orifice (Schick et al. 1999). Although the shape of the orifice is rather elliptic, the second step of the molecular surgery was first achieved using **3**, that is, insertion of a helium atom (1.5% yield) or a hydrogen molecule (5% yield) in the hollow cavity of **3** through the orifice under the conditions of 288–305°C/ca. 475 atm and 400°C/100 atm, respectively (Rubin et al. 2001). Recently, Iwamatsu et al. reported a fullerene derivative **4** (see Figure 39.1) with a surprisingly huge orifice, with its molecular shape almost looking like a bowl, and showed that a water molecule can get inside the cage even at room temperature under a normal pressure (Iwamatsu et al. 2004).

In recent years, many developments have been made in this subject. Murata et al. (2006) reported the details of their study to synthesize a new endohedral fullerene, $H_2@C_{60}$, in more than 100 mg quantities by the closure of the 13-membered ring orifice of an open-cage fullerene using the four-step organic reactions shown in Figure 39.2. The 13-membered ring orifice,

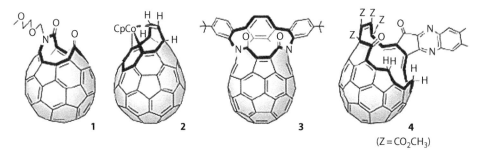

FIGURE 39.1 Synthesized orifices **1** to **4**. (From Murata, M. et al., *J. Am. Chem. Soc.*, 128, 8024, 2006. With permission.)

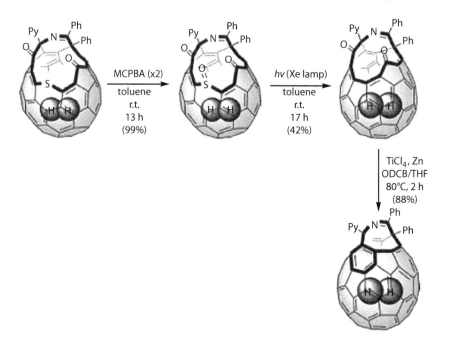

FIGURE 39.2 Synthetic pathway of a new endohedral fullerene, $H_2@C_{60}$. (From Murata, M. et al., *J. Am. Chem. Soc.*, 128, 8024, 2006. With permission.)

in a previously synthesized open-cage fullerene incorporating hydrogen in 100% yield, was reduced to a 12-membered ring by extrusion of a sulfur atom at the rim of the orifice, and the ring was further reduced into an 8-membered ring by reductive coupling of two carbonyl groups also at the orifice. The final closure of the orifice was completed by a thermal reaction.

Purification of $H_2@C_{60}$ was accomplished by recycle HPLC. A gradual downfield shift of the NMR signal for the encapsulated hydrogen observed on reduction of the orifice size was interpreted based on the gauge-independent atomic orbital and the nucleus-independent chemical shift calculations. The spectral as well as electrochemical examination of the properties of $H_2@C_{60}$ has shown that the electronic interaction between the encapsulated hydrogen and the outer C_{60} π-system is quite small but becomes appreciable when the outer π-system acquires more than three extra electrons.

This methodology has also been applicable to C_{70}. Komatsu et al. (2005) have performed hydrogen insertion into an open-cage C_{70}, which has given a mixture of open-cage C_{70} containing one molecule of hydrogen and that containing two molecules of hydrogens. The behavior of two hydrogen molecules in the open-cage C_{70} has been elucidated by a low-temperature DNMR study.

Three additional fullerene derivatives, **I**, **II**, and **III**, Figure 39.3, were synthesized by the Bingel reaction (Camps and Hirsch 1997), benzyne addition (Hoke et al. 1992), and Prato reaction (Maggini et al. 1993) in order to further investigate the tendency in the shift of the NMR signal of the inner hydrogen. It was found to be quite similar to that observed for the ³He NMR signal of the corresponding derivatives of ³He@C_{60}.

The encapsulation of molecular hydrogen into an open-cage fullerene having a 16-membered ring orifice has been investigated by Iwamatsu et al. (2005) having a synthetic pathway shown in Figure 39.4. The pressurization of H_2 is achieved at 0.6–13.5 MPa to afford endohedral hydrogen complexes of open-cage fullerenes in up to 83% yield. The efficiency of encapsulation is dominantly dependent on both the H_2 pressure and the temperature. Hydrogen molecules inside the C_{60} cage are observed in the range of –7.3 to –7.5 ppm in ¹H NMR spectra and the formations of hydrogen complexes are further confirmed by mass spectrometry.

The trapped hydrogen is released by heating. The activation energy barriers for this process are determined to be 22–24 kcal/mol. The DSC measurement of the endohedral H_2 complex reveals that the escape of H_2 from the C_{60} cage corresponds to an exothermic process, indicating that encapsulated H_2 destabilizes the fullerene.

Matsuo et al. have synthesized the organic and organometallic derivatives of dihydrogen-encapsulated [60]fullerene **2–4** (Figure 39.5) in good yield and showed that the uniquely upfield-shifted singlet signal of the encapsulated dihydrogen can act as a sensitive probe for inside and outside the environment of the fullerene cage (Matsuo et al. 2005). It is interesting to note that the ¹H NMR chemical shift of the dihydrogen in the potassium cyclopentadienide **2** (Figure 39.5) (δ = –9.79) is the most deshielded among all related compounds. While it is difficult to rationalize the origin of this deshielding, they ascribe this to the location of the

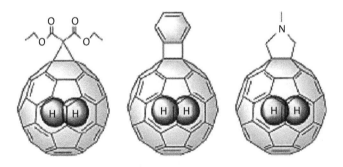

FIGURE 39.3 Three fullerene derivatives **I**, **II**, and **III**. (From Murata, M. et al., *J. Am. Chem. Soc.*, 128, 8024, 2006. With permission.)

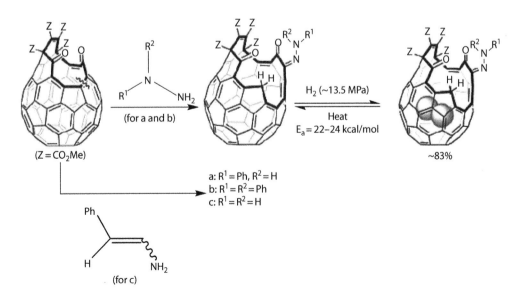

FIGURE 39.4 Synthetic pathway of a 16-membered ring orifice. (From Iwamatsu, S. et al., *J. Org. Chem.*, 70, 4820, 2005. With permission.)

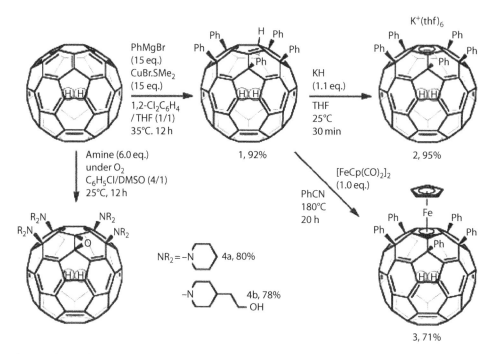

FIGURE 39.5 Synthetic pathway of organic and organometallic derivatives of dihydrogen-encapsulated [60]fullerene. (From Matsuo, Y. et al., *J. Am. Chem. Soc.*, 127, 17148, 2005. With permission.)

dihydrogen relative to the anionic 6π-cyclopentadienide moiety in **2**, that is, one of the hydrogen atoms may be located in a deshielding region of the cage.

Solid-state NMR studies of endohedral H_2–fullerene complexes, including ^1H and ^{13}C NMR spectra, ^1H and ^{13}C spin relaxation studies, and the results of ^1H dipole–dipole recoupling experiments were presented by Carravetta et al. (2007). The available data involves three different endohedral H_2-fullerene complexes, **I**: H_2@ATOCF (aza-thia-open-cage-fullerene), **II**: H_2@C_{60}, and **III**: H_2@di(ethoxycarbonyl) methanofullerene studied over a wide range of temperatures and applied magnetic fields. The symmetry of the cage influences strongly the motionally averaged nuclear spin interactions of the endohedral H_2 species, as well as its spin relaxation behavior. In addition, the nonbonding interactions between fullerene cages are influenced by the presence of endohedral hydrogen molecules. The review also presents several pieces of experimental data that are not yet understood, one example being the structured ^1H NMR line shapes of endohedral H_2 molecules trapped in highly symmetric cages at cryogenic temperatures. This review demonstrates the richness of the NMR phenomena displayed by H_2-fullerene complexes, especially in the cryogenic regime.

Oksengorn (2003) has performed the preparation of endohedral complexes H_2@C_{60}, together with hydrogenated C_{60}. Analyses by ^1H NMR and by infrared and vacuum ultraviolet spectra have been carried out, which proved the existence of these two kinds of compounds in the samples. A quantitative measurement has been done, giving values for the endohedral complexes of about 18%, and for the hydrogenated C_{60} of about 35%.

39.3 Theoretical Studies on the Endohedrally Hydrogen-Doped Fullerenes and Fullerene-Like Species

39.3.1 Fullerene Nanocage Capacity for Hydrogen Storage

39.3.1.1 nH_2@C_{60} Geometries and Energies

Mauser et al. have presented first systematic comparison of the chemical behavior of the convex outer and the concave inner surfaces of C_{60} by analyzing the results of semiempirical and density functional theory (DFT) calculations on exohedral and endohedral complexes with H– and F– atoms as well as with the methyl radical (Mauser et al. 1997). All semiempirical calculations in this study used the VAMP 6.1 program with the PM3 Hamiltonian within the unrestricted Hartree–Fock (UHF) formalism and the DFT calculations were performed with the UB3LYP/D95* Method in Gaussian 94 using the PM3 geometry.

In case of H@C_{60}, they observed that hydrogen was quite free inside C_{60} and barriers between the bound and the endohedral minimum structure of H@C_{60} were very low (2.32 kcal/mol). The exact location of the endohedral minimum of H@C_{60} depends on the method of calculation, but it is consistently slightly off-center (1.1 Å from the center with the PM3-Hamiltonian). Also it was indicated that HC$_{60}$ (exo) structure was more stable than the endo one.

Neek-Amal et al. (2007) have numerically calculated the ground-state energies and electron distributions of a simple atom confined by C_{60} fullerene within the diffusion Monte Carlo method. They modeled the C_{60} fullerene with attractive confining well and obtained the physical quantities of one or two electrons of the atom (H or He) endohedrally. Their results for a hydrogen atom located at the origin are in excellent agreement with exact calculations. Within the diffusion Monte Carlo method, they obtained the ground-state energy for off-centered H and He atoms for different confining well potentials. They described the electron distributions for endohedral case as a function of nucleus position and the confining well potential also. Moreover, electron distributions were more influenced by changing the nucleus position. They believed that for an accurate quantitative calculations of off-centered effects, the real physical model for the C_{60} molecule should be deformed in the presence of endohedral atoms, and deviation from spherical symmetry should be considered.

Possible transition structures for the formation of endohedral $X@C_{60}$ complexes have been proposed by Sanville and BelBruno (2003) using Gaussian 98 suite of programs. The proposed structures were those expected for the direct insertion mechanism (ring penetration or direct insertion of the species by passage through a pentagonal or hexagonal ring) and windowing mechanism (breaking of sufficient carbon–carbon bonds to create a large opening in the cage that subsequently recloses after endohedral insertion).

Transition state structures were obtained using partial geometry optimizations with the HF formalism and 3–21 G* basis set. This yielded theoretical values for the energy barriers for formation of $X@C_{60}$, where X = H, He, N, NO, P, and As. Direct insertion was possible for H, He, and (possibly) N with calculated barriers of 3.27, 10.98, and 16.12 eV, respectively. The possible transition structure for the windowing mechanism was examined for all endohedral species using a semiempirical (PM3) approach with complete optimization. On the basis of the computed energy barriers, the direct insertion mechanism was found to be plausible for H and N with calculated barriers of 3.01 and 5.72 eV, respectively. (The remaining systems would not converge to structures with endohedral atoms, but instead formed heterofullerenes.)

Pupysheva et al. studied the properties of $H_n@C_{60}$ structures using density functional theory (Pupysheva et al. 2008). The calculations were carried out by SIESTA code program. Their formation energy, average relative elongation of the fullerene C–C bonds, and hydrogen pressure inside the fullerene nanocage were calculated as functions of the number of hydrogen atoms n. Although structures with a large amount of encapsulated hydrogen were highly endothermic, they still corresponded to the local minima of the potential energy surface. It was found that for large n, some hydrogen atoms can be chemisorbed on the inner surface of the carbon cage, that is, they can form covalent C–H bonds. The maximum number of hydrogen atoms inside C_{60}, which can form a metastable structure, that is, corresponds to an energy minimum, was determined to be $n = 58$. The mechanism of $H_{58}@C_{60}$ breaking was studied by ab initio molecular dynamics simulations at room temperature. It was shown that the hydrogen chemisorption, which weakens the fullerene C–C bonds, plays the key role in the opening of the nanocage.

They also derived a general relation between the internal pressure and average relative C–C bond elongation in fullerene cages of various radii. Using the correspondence between the density and the pressure of encapsulated hydrogen at either zero or finite temperature, they provided an estimate for the amount of hydrogen necessary to be confined at given strain in any fullerene nanocage. Their data demonstrated the excellent mechanical properties of the fullerene cages, which make them efficient nanocontainers with high theoretical capacity for hydrogen storage.

Endohedrally hydrogen-doped C_{60} systems, $nH_2@C_{60}$ ($n = 9$, 12, 15, 19, 21, and 24), have been theoretically investigated by Türker and Erkoç (2003) at the level of AM1 (restricted Hartree-Fock (RHF)) type quantum chemical treatment. All the computations were performed by using the Hyperchem (release 5.1) and ChemPlus (2.0) package programs.

Table 39.1 shows certain energies of $nH_2@C_{60}$ type ($n = 9$, 12, 15, 19, 21, and 24) systems. As seen there, both of these systems are stable but highly endothermic. As expected, core–core interaction energy becomes more repulsive parallel to increase in the number of hydrogen molecules inserted into C_{60} cage. Also, the binding energy gets less negative but the heat of formation value more positive as n approaches to 24. Parallel to increase of n, the highest occupied molecular orbital (HOMO) energy rises up,

TABLE 39.1 Some Energies of the Systems Considered

Energy (kcal/mol)	$9\,H_2@C_{60}$	$12\,H_2@C_{60}$	$15\,H_2@C_{60}$	$19\,H_2@C_{60}$	$21\,H_2@C_{60}$	$24\,H_2@C_{60}$
Total	−181,785	−183,313	−184,766	−186,525	−187,356	−188,564
Binding	−9,886.10	−9,837.02	−9,713.67	−9,370.56	−9,150.41	−8,781.66
Isolated atomic	−171,899	−173,476	−175,052	−177,155	−178,206	−179,783
Electronic	−2,725,710	−2,848,150	−2,969,335	−3,127,169	−3,205,119	−3,316,797
Core–core interaction	2,543,925	2,664,838	2,784,569	2,940,643	3,017,763	3,128,233
Heat of formation	1,305.14	1,666.83	2,102.79	2,862.71	3,291.27	3,972.63
LUMO[a]	−2.9313	−2.9082	−2.9003	−2.9705	−2.8894	−2.9183
HOMO[a]	−9.5554	−9.4758	−9.4098	−9.0423	−8.8470	−8.9039

Source: Türker, L. and Erkoç, Ş., *J. Mol. Struct. (Theochem)*, 638, 37, 2003.
[a] Energies in eV.

but for the lowest unoccupied molecular orbital (LUMO) energy level, such generalization is not possible.

39.3.1.2 Debate on the Hydrogen Storage Capacity of C_{60}

Fullerene C_{60} has a relatively large and robust cage structure that promises a wide range of applications by filling it with atoms or molecules. The storage of hydrogen molecules inside C_{60} raises some questions such as how the C_{60} and H_2 molecules interact and what the energy spectrum looks like. More important than those is may be the number of H_2 molecules endohedrally doped in a C_{60} cage, that is, whether it is energetically efficient to enclose more hydrogens in the carbon cage, whether there is a maximum capacity.

There are a few theoretical papers considering endohedral C_{60} fullerene containing more than one hydrogen molecule (Barajas-Barraza and Guirado-López 2002; Türker and Erkoç 2003, 2006; Koi and Oku 2004a,b; Dolgonos 2005; Dodziuk 2005, 2006, 2007; Ren et al. 2006). The results of semiempirical calculations (Barajas-Barraza and Guirado-López 2002; Türker and Erkoç 2003; Koi and Oku 2004a,b; Ren et al. 2006) followed by density functional (Barajas-Barraza and Guirado-López 2002; Ren et al. 2006) or HF (Koi and Oku 2004a,b) energy calculations and force field methods (Dodziuk 2005; Dolgonos 2005) yielded contradictory results. The maximum number of hydrogens to form stable $H_n@C_{60}$ composite was calculated to be 23 (Barajas-Barraza et al. 2002; Ren et al. 2006), 24 (Türker and Erkoç 2003), or 25 (Koi and Oku 2004a,b) molecules. Dolgonos and Dodziuk (Dodziuk 2005, 2006, 2007) insisted (based on simple geometrical reasonings and neglecting suitable orientation of H_2 molecules, the effect of compression on H_2 molecules, and expansion of C_{60} cage) that there was not enough space for more than one hydrogen molecule inside C_{60}. Türker and Erkoç (2006) objected to that there is also a disagreement regarding the possibility of hydrogen chemisorption on the inner walls of the fullerene, which was reported in references Koi and Oku (2004a,b), while in the others, all the encapsulated hydrogens were found in the molecular form (Barajas-Barraza and Guirado-López 2002; Türker and Erkoç 2003; Ren et al. 2006).

Barajas-Barraza and Guirado-López (2002) considered various fullerenes and single-walled nanotubes for endohedral H_2 doping. They fully optimized the considered structures using the semiempirical modified neglect of differential overlap (MNDO) method. Then, based on these geometries, single-point DFT calculations were performed in order to analyze the form of stored hydrogens and to find out the maximum storage capacities of the different configurations. The number of internal hydrogen molecules studied was in the range of 1–23 for C_{60} and 1–35 for C_{82}, the later numbers defining the maximum storage capacity for each one of the carbon fullerenes. They found that besides the repulsive interactions present in the systems, there are also some confinement effects that are of crucial importance in determining the ground-state structure and stability of these compounds. When six H_2 molecules are present in C_{60} structure, they arranged themselves in well-defined configurations (octahedral and icosahedral clusters). However, for 19 H_2 molecules, a considerably distorted

"icosahedra-like" structure was found. They checked the reliability of their structural optimizations by comparing their equilibrium geometry obtained for $(H_2)_6@C_{60}$ with those obtained by means of ab initio HF and DFT (LDA) calculations. The results were within the HF and DFT approaches. The structures obtained do not differ significantly from that optimized at the MNDO level.

They also calculated the intercalation energy per atom. The results indicate that incorporating a few H_2 molecules inside both C_{60} and C_{82} cavities seems to be difficult because of strong repulsive interactions. However, their conjecture was obtained by performing low-energy H-ion bombardment on a C_{60} sample; H-atoms may get trapped in the fullerene cage.

Narita and Oku (2002) used molecular dynamics to investigate the storage and discharge of H_2 in C_{60}. The C_{60} included a H_2 molecule kept stable at $T = 298$ K and $P = 0.1$ MPa. They found that although the H_2 molecule vibrated in C_{60} cage, it was not discharged from the cage. Some H_2 molecules were stored in the C_{60} cage when the pressure was 5 MPa. The H_2 molecules passed through the hexagonal rings of the C_{60} cage at 0.5 ps. They were stable in C_{60}.

Ren et al. (2006) studied the state of hydrogen molecules confined in C_{60} and carbon nanocapsules. Their calculations were PM3 type geometry optimization followed by B3LYP/6–31G(d) type single-point energy evaluation. They have found that in the case of C_{60}, with the increase in the number of encapsulated H_2 molecules, the average H_2–H_2 distance first rises until $n(H_2) = 5$, at which the distance reached is 1.90 Å. Following this, the average distance drops to the minimum of 1.69 Å, when $n(H_2) = 15$. When $n(H_2) = 16$, a special configuration is observed. Even though the average repulsive energy for $n(H_2) = 16$ is smaller than that of $n(H_2) = 15$, the average space is larger. The $n(H_2) = 16$ configuration possesses highly symmetric structure than that of $n(H_2) = 15$ configuration. They stored H_2 in C_{60} up to 23 molecules. It was generally found that for the systems possessing very high symmetries, the energy of the system converged to a much lower energy state than those with highly disordered arrangements.

Yang, having studied hydrogen-filled C_{60} by using DFT approach, found that the encapsulation reaction was endothermic for all cases. As more H_2 molecules incorporated, the formation energy gradually approached a saturation value just below 4 eV (Yang 2007). The C_{60} was able to accommodate at most 29 H_2 molecules, one more enclosed H_2 caused damage of the carbon cage. Also it was observed that as more H_2 molecules were packed inside the C_{60} cage, interaction between the hydrogens and the cage took place.

Moreover, the further calculations showed that the HOMO–LUMO gap varies within the range of 0.12 and 0.29 eV for the number of encapsulated H_2 between 8 and 15, with $H_{20}@C_{60}$ having the smallest gap and $H_{30}@C_{60}$ the largest. Further increase of encapsulated H_2 widened the HOMO–LUMO gap.

Pupysheva et al. (2008) modeled fullerene nanocages filled with hydrogen and studied the capacity of such endohedral fullerenes to store hydrogen. It was shown by using DFT that for large number of encapsulated hydrogen atoms, some of them became chemisorbed on the inner surface of the cage. In contrast

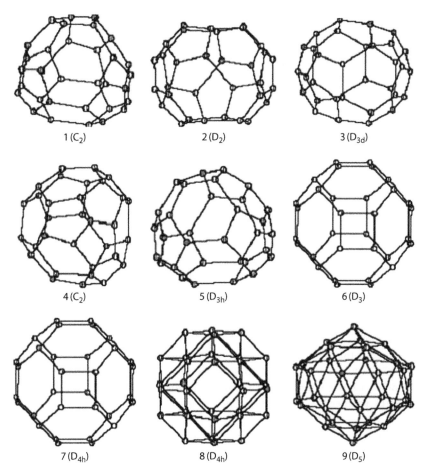

FIGURE 39.6 Nine structural isomers calculated for the C_{32} cluster. (From Sun, Q. et al., *J. Phys.: Condens. Matter*, 13, 1931, 2001. With permission.)

to claims of Dolgonos (2005) and Dodziuk (2005, 2006), a maximum of 58 hydrogen atoms inside a C_{60} cage was found to still remain a metastable structure and the mechanism of its breaking was studied by ab initio molecular dynamics simulations. Hydrogen pressure inside the fullerene nanocage was estimated. The calculations revealed that with increase of n (the number of hydrogens), the nanocage expands slowly. It was shown that the hydrogen chemisorption, which weakens the fullerene C–C bonds, plays the key role in the opening of the nanocage. The conclusion of the authors was that the fullerene cages should have excellent mechanical properties that make them efficient nanocontainers with high theoretical capacity for hydrogen storage.

39.3.2 Other Fullerenes

Sun et al. (2001) studied pure and doped C_{32} clusters using a powerful ab initio ultrasoft pseudopotential scheme with a plane wave basis (the Vienna Ab Initio Simulation Program) (Kresse and Hafner 1993, 1994; Kresse and Furthmüller 1996). Among the nine structural isomers (Figure 39.6), the fullerene structure with D_3 symmetry was found to be the most stable. Table 39.2 lists the total binding energies and gaps between the HOMO and the LUMO of the nine isomers. Isomer 6 in the fullerene

TABLE 39.2 Total Binding Energies E (eV) and the HOMO-LUMO Gaps Δ (eV) for Nine Isomers

Isomer	E (eV)	Δ (eV)
1	−266.244	0.426
2	−265.542	0.272
3	−265.673	0.572
4	−267.413	0.894
5	−265.077	1.280
6	−268.497	1.291
7	−262.280	0.680
8	−188.453	0.700
9	−188.452	0.100

Source: Sun, Q. et al., *J. Phys.: Condens. Matter*, 13, 1931, 2001. With permission.

structure was the most stable with the largest HOMO–LUMO gap (1.29 eV), which was quite close to the experimental value (1.30 eV) (Kietzmann et al. 1998), and the average binding energy per atom was 8.39 eV, smaller than the corresponding value of 8.72 eV for C_{60} (Baştuğ et al. 1997).

For the most stable D_3 structure, they considered two configurations: the H_2 structure was orientated perpendicular and

TABLE 39.3   Some Calculated Energies of the Species Considered

Species	Energy (kJ/mol)					
	Total	Binding	Isolated Atomic	Electronic	Core–Core Interactions	Heat of Formation
Se@C_{60}	−702,037	−39,500	−662,537	−10,190,174	9,488,137	3,627
(Se + 3H_2)@C_{60}	−710,894	−40,789	−670,105	−10,744,307	10,033,414	3,646
(Se + 4H_2)@C_{60}	−713,693	−41,065	−672,628	−10,918,077	10,204,382	3,806
(Se + 5H_2)@C_{60}	−716,458	−41,308	−675,151	−11,098,102	10,381,645	3,999
(Se + 7H_2)@C_{60}	−721,766	−41,570	−680,196	−11,449,411	10,727,647	4,609
(Se + 8H_2)@C_{60}	−724,336	−41,617	−682,719	−11,624,156	10,899,822	3,998
(Se + 9H_2)@C_{60}	−726,882	−41,640	−685,242	−11,799,357	11,072,474	5,411

Source: Türker, L., *J. Mol. Struct. (Theochem)*, 717, 5, 2005.

parallel to the axis; their total binding energies were −274.232 and −274.248 eV, respectively. Parallel configuration was more stable. For the equilibrium bond length of the free H_2 molecule, they got the value of 0.765 Å, in agreement with the experimental result of 0.75 Å (Huber and Herzberg 1979). However, after encapsulation, the bond length was reduced to 0.74 Å, similar to what has been found for H_2@C_{60} (Cioslowski 1991). The formation energy was defined as the difference between the total binding energies: $\Delta E = E(H_2$@$C_{32}) - E(C_{32}) - E(H_2)$; for H_2@C_{60}, $\Delta E = 0.05$ eV (Cioslowski 1991); but due to the smaller size of the C_{32} cage, more energy was needed, and for H_2@C_{32}, $\Delta E = 1.00$ eV.

In order to calculate the vibration frequency of H_2 in a C_{32} cage, the host atoms were kept fixed at their equilibrium positions; the light mass of the hydrogen atom justified this approach. Anharmonic effects were important in H_2: for the free molecule, the harmonic frequency was 4400 cm^{-1} (Huber and Herzberg 1979), while the anharmonic effects lowered this value to 4161 cm^{-1} (Stoicheff 1957). In calculating the vibration frequency of H_2, they considered the anharmonic effect up to the fourth order of H–H bond displacement. Around the equilibrium bond length r_0, the potential energy was expressed as $E(r) = E(r_0) + (M\omega_h^2/2)(r - r_0)^2 + \alpha(r - r_0)^3 + \beta(r - r_0)^4$ (Landau and Lifshitz 1977).

From the harmonic frequency ω_h, the vibration frequency ω was obtained as follows:

$$\omega = \omega_h + \frac{3h}{(2\pi)^2 cM}\left[-\frac{5}{2}\left(\frac{\alpha}{M\omega_h^2}\right)^2 + \frac{\beta}{M\omega_h^2}\right]$$

where c and M denoted the speed of light and the reduced mass, respectively. By fitting $E(r)$ for several points, they obtained $\omega_h =$ 4411 and 4676 cm^{-1}, $\omega = 4201$ and 4452 cm^{-1} for free and encapsulated H_2, respectively. Compared with the free H_2 molecule, on encapsulation in C_{32}, the bond length was reduced, while the vibrational frequency was increased, similar to the situation for H_2@C_{60} (Cioslowski 1991).

Certain endohedrally Se and hydrogen-doped C_{60} systems (Se + nH$_2$)@C_{60} have been quantum chemically investigated by Türker (2005), using PM3 self-consistent fields molecular orbital

(SCF MO) method at the RHF level. All these computations were performed by using the Hyperchem (release 5.1).

The results of the calculations (see Table 39.3) have revealed that all the structures are stable (the total and binding energies) but endothermic (the heats of formation values) in nature. As seen in Table 39.3, an increase in the number of hydrogens adds up the stabilities but also increases the core–core repulsion energies and the endothermic heats of formation. The presently performed calculations indicate that Se@C_{60} structure cannot accommodate more than nine hydrogen molecules. Any attempt to insert more hydrogens result in expulsion of hydrogen(s) from the cage during the geometry optimization process. Table 39.4 shows some energies of the group of endohedral hydrogens in (Se + nH$_2$)@C_{60} systems calculated by using MM+ (molecular mechanics) method. As seen there, as the number of hydrogens increases, all the energies except the electrostatic energies regularly increase. The electrostatic energy has the most negative value for $n = 4$, then becomes less and less negative. For $n = 9$ (which stands for the maximum number of hydrogens could be doped in Se@C_{60}), it has the minimum absolute value. All these can be considered as the indication of capability of Se@C_{60} system as hydrogen storage molecular device at moderate conditions.

In the study of Türker (2002), endohedrally beryllium-doped C_{60} structure, Be@C_{60}, yet nonexistent, in the form of (Be + nH$_2$)@C_{60} has been considered for hydrogen storage, where $n = 1$–5. They are considered for semiempirical molecular orbital treatment at the level of AM1 (RHF) method. All these computations were

TABLE 39.4   Some Calculated Energies of Endohedral Hydrogens in (Se + nH$_2$) Systems

n	Energy (kcal/mol)			
	Total	Bond	van der Waals	Electrostatic
3	294.18	7.64	297.32	−10.77
4	391.08	9.68	393.93	−12.52
5	532.31	11.02	530.59	−9.30
7	906.23	14.49	900.09	−8.36
8	1138.86	16.60	1124.79	−2.53
9	1402.20	18.34	1385.49	−1.63

Source: Türker, L., *J. Mol. Struct. (Theochem)*, 717, 5, 2005.

TABLE 39.5 Some Calculated Properties of $(Be + nH_2)@C_{60}$ Systems

Property	n					
	0	1	2	3	4	5
Area	542.38	541.67	545.00	545.42	547.00	546.53
Volume	1204	1205	1206	1208	1212	1213
Polarizability	113.77	114.54	115.32	116.00	116.86	117.64
Dipole moment	0.0000	0.2334	7.2948	6.7278	9.2728	0.2968
Charge on Be	−0.07	−0.23	0.28	0.11	0.38	0.11
Symmetry	—	C_s	C_s	C_1	C_1	C_1

Source: Türker, L., *J. Mol. Struct. (Theochem)*, 577, 205, 2002.

Note: Area, volume, polarizability, and dipole moment values are in the order of 10^{-20} m^2, 10^{-30} m^3, 10^{-30} m^3, and 10^{-30} Cm, respectively.

performed by using the Hyperchem release (5.1) and Chemplus (2.0) package programs.

Table 39.5 shows some calculated properties of the presently considered structures. As seen there, the area, volume, and polarizability values slightly change depending on the number of hydrogen molecules inserted into the cage. However, dipole moments are subjected to great variation that arises from the interaction of atomic orbitals of Be atom ($1s^2 2s^2$) with π-molecular orbitals of the C_{60} cage and orientation of the Be atom in the cage. Although, in Be@C_{60} case, Be atom resides at the intersection of axes of inertia, as hydrogen atoms inserted into the cage, it moves from its original position toward the inner surface of the C_{60} cage (see Figure 39.7).

Present calculations indicate that for $n = 0$ and 1, Be atom acquires negative charge that is unusual for a metal atom like Be. However, $n = 2$ onward, Be atom possesses positive charges. All these imply that in the first group of structures, C_{60} cage acts as an electron donor, whereas in the later group, it gets the role of an electron acceptor. This variation in the character of C_{60} cage is due to the presence of hydrogen molecules and (as a result of their presence) off-centered position of Be atom in the cage.

In the case of $n = 5$, the situation is rather different. The hydrogen molecule nearby the Be atom is characterized with very elongated bond (see Figure 39.7 for $n = 5$). The bond length in hydrogen molecule is 0.742×10^{-10} m (Huhey 1978), whereas the bond length in the above-mentioned H_2 in the C_{60} cage is calculated to be 2.46×10^{-10} m (the bond lengths in the others are in between 0.68 and 0.71×10^{-10} m). Also, the charge calculations indicate that the charge on hydrogen atoms is about −0.28 unit that implies that certain degree of hydride formation occurs between the Be atom and the hydrogen molecule of consideration. In other words, in $n = 5$ case, BeH_2 assembly occurs, in which H–Be–H angle is 160°. Note that the corresponding angle in BeH_2 is 180° (Durant and Durant 1970; Schriver et al. 1990).

As for the energetics of these systems, calculations of AM1 type reveal (see Figure 39.8) that all the $(Be + nH_2)@C_{60}$ composite structures should be stable (negative total and binding energies) but highly endothermic (positive heat of formation data) in nature. Figure 39.8 shows the variation of some energies of these structures per molecule of hydrogen, based on Be@C_{60} reference

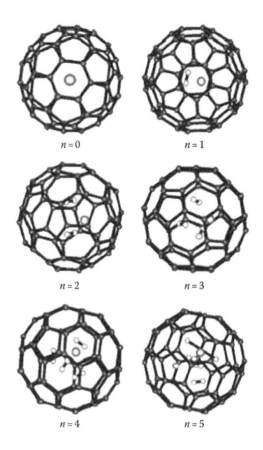

$n = 0$ $n = 1$

$n = 2$ $n = 3$

$n = 4$ $n = 5$

FIGURE 39.7 Structures of the systems considered. (From Türker, L., *J. Mol. Struct. (Theochem)*, 577, 205, 2002.)

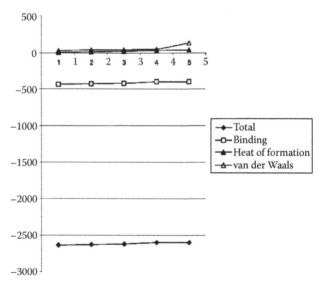

FIGURE 39.8 Variations of some energies of $(Be + nH_2)@C_{60}$ system per molecule of hydrogen, based on Be@C_{60} structure. (From Türker, L., *J. Mol. Struct. (Theochem)*, 577, 205, 2002.)

system, that is $(E_n − E_0)/n$, where E_n and E_0 are the energy under consideration for the system having n molecules of hydrogen and Be@C_{60} system, respectively. The graph indicates that the increased number of hydrogen molecules destabilizes the system on relative basis.

TABLE 39.6   Stabilities of Some (Li + nH$_2$)@C$_{60}$ Structures

Energy (kJ/mol)	n					
	0	1	2	3	4	5
Total	−738,647	−741,331	−743,949	−746,520	−749,024	−751,477
Binding	−38,721	−38,205	−39,624	−39,996	−40,300	−40,554
Isolated atomic	−699,927	−702,126	−704,325	−706,524	−708,723	−710,922
Electronic	−9,854,591	−10,034,332	−10,211,299	−1,037,794	−10,561,960	−10,735,529
Core–core repulsion	9,115,944	9,293,000	9,467,348	9,641,349	9,812,935	9,984,052
Heats of formation	4,340	4,292	4,309	4,373	4,504	4,686

Source: Türker, L., *Int. J. Hydrog. Energy*, 28, 223, 2003.

AM1 (UHF) type semiempirical calculations have been carried out by Türker (2003) on (Li + nH$_2$)@C$_{60}$ systems having $n = 0$–5. All the computations were performed by using the Hyperchem (release 5.1) and ChemPlus (2.0) package programs.

Table 39.6 shows the various energies of these systems. The data reveal that they are all stable but highly endothermic structures. It is noted that (Li + H$_2$)@C$_{60}$ has the lowest heat of formation (ΔH_f) value among the structures considered.

Table 39.7 shows the coordinates and the charge of the Li atom in these systems. The data reveal that as the number of hydrogen molecules in the cage increases, the charge on the lithium decreases up to $n = 4$ then again increases to 5. Even in the case of Li@C$_{60}$ in which there is no hydrogen, the lithium atom is not at the center of the C$_{60}$ cage but appreciably moved toward one of the hemispheres. This shows that there exist some interaction between the atomic orbitals of Li and π-skeleton of the C$_{60}$ cage.

TABLE 39.7   Charge and Coordinates of Li Atom in (Li + nH$_2$)@C$_{60}$ Systems

n	Charge	X	Y	Z
0	0.6757	−5.8283	6.4084	−2.5875
1	0.5393	−5.8687	6.4014	−2.6489
2	0.4608	−4.7381	6.2534	−2.7519
3	0.4138	−4.5139	5.9037	−2.4445
4	0.3596	−4.2859	5.8205	−2.3204
5	0.9695	−4.5508	7.0596	−2.7638

Source: Türker, L., *Int. J. Hydrogen Energy*, 28, 223, 2003.
Note: Charges and coordinates are in unit of charge and atomic unit (a.u.), respectively.

Insertion of hydrogen molecules into the cage does not affect the surface area and volume of the cage much; however, positions of the carbon atoms in space should have changed. Some slightly positively charged hydrogen atoms in the hydrogen molecules are also an indication for the presence of certain interaction between the atomic orbitals of lithium and σ-molecular orbital of the hydrogen molecules, whereas some of the hydrogens are just polarized resulting in some dipoles. However, all these are weak interactions contrary to some (M + nH$_2$)@C$_{60}$ cases (e.g., (Be + 5 H$_2$)@C$_{60}$ in which strong hydridic type interaction should occur between the metal and hydrogens (Türker 2002)). Most probably, Li atom, which is an open shell system, should interact more strongly with the π-skeleton of the C$_{60}$ cage rather than σ-skeleton(s) of the hydrogen molecules around it.

Türker has studied AM1 type semiempirical quantum chemical calculations at the level of RHF on endohedrally Be and various numbers of hydrogen (0, 5–10)-doped C$_{70}$ system (Be + nH$_2$)@C$_{70}$ (Türker 2004). All these computations were performed by using the Hyperchem (release 5.1) and ChemPlus (2.0) package programs.

C$_{70}$ is a rugby ball-shaped molecule (Dresselhause et al. 1996) and it is the next higher stable and the isolated pentagon rule satisfying fullerene (Hirsch 1994), whose systematic name is [5,6]-fullerene-70-D$_{5h}$. Endohedrally Be-doped C$_{70}$ structure, Be@C$_{70}$ has not been synthesized yet.

Table 39.8 shows some energies of these composite systems. As seen in the table, all these structures are stable (the total and binding energies are negative) but endothermic (the heat of formation values) in nature. As the number of hydrogen molecules inserted increases, all the energies except the core–core interaction and the heat of formation values become more and

TABLE 39.8   Some Calculated Energies of the Systems Presently Considered

Energy (kJ/mol)	n						
	0	5	6	7	8	9	10
Total	−863,730	−876,885	−879,427	−881,902	−884,365	−886,737	−889,069
Binding	−45,390	−47550	−47,893	−48,169	−48,432	−48,605	−48,738
Isolated atomic	−818,340	−829335	−831,534	−833,733	−835,933	−838,132	−840,331
Electronic	−12,664,650	−13,652,024	−13,851,145	−14,045,115	−14,235,252	−14,425,491	−14,614,474
Core–core interaction	11,800,922	12,775,139	12,971,717	13,163,236	13,350,887	13,538,753	13,725,405
Heats of formation	4,982.17	5,002.19	5,095.03	5,255.45	5,428.03	5,691.27	5,994.10

Source: Türker, L., *J. Mol. Struct. (Theochem)*, 668, 225, 2004.

more negative, whereas the lastly mentioned energies tend to be more positive.

It was reported that in the case of Be@C_{60} system, the results of AM1 (RHF) type calculations infer that hydrogens in the system undergo some interaction with the endohedrally doped Be atom resulting in quasi-hydride formation (Türker 2002). In contrast to that, Be@C_{70} system does not exhibit such kind of interaction up to $n \leq 10$.

Türker and Gümüş (2005) have investigated the hydrogen storage capacity of single-walled and endohedrally Mg-doped C_{120} composite system theoretically. In the present study, the initial structure of the tube C_{120} was constructed starting from C_{60} [(5,6-fullerene-60-I_h), which was excerpted from the Hyperchem library. The halves of C_{60} structure were suitably used as the caps of an armchair type tube having 10 benzenoid rings as the peripheral belt.

The geometry optimizations of all the structures leading to energy minima were achieved by using first MM+ and then PM3 SCF MO method at the RHF level. All the computations were performed by using the Hyperchem (release 5.1) and ChemPlus (2.0) package programs.

After the construction of the C_{120} tube as mentioned above, it is first endohedrally doped by Mg atom and then geometry optimized (Figure 39.9). Figure 39.9 shows the side and the end views of the composite structure. As can be seen in the figure, the Mg atom almost occupies the center of the capped tube.

The geometry-optimized Mg@C_{120} nanotube is then doped with $n \leq 13$ number of hydrogen atoms to obtain $(nH_2 + Mg)@$ C_{120} ($n = 6$–13) type structures. Figure 39.10 shows the geometry-optimized structures of the presently considered composite structures for $n = 8$–10. In each case, the dipole moment of the structure is from somewhere on the surface to the center of the cage structure, which means that the Mg atom polarizes the composite structure. In all the structures, Mg atom is negatively charged, whereas the hydrogens possess positive, negative, or zero charge development. A negatively charged Mg is a rare case but occurs when empty valance shell orbitals of the metal are occupied. In the present case, π-orbitals of the tube structure should interfere with empty valance shell of Mg atom resulting in some negative charge development on Mg atom.

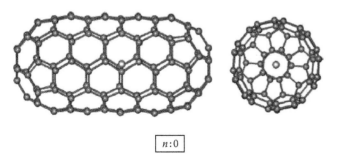

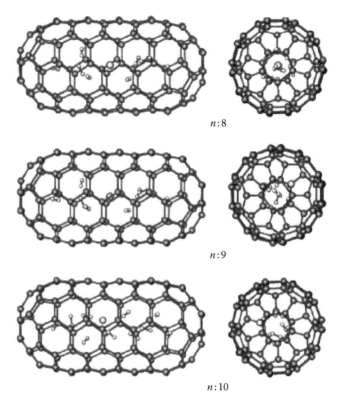

FIGURE 39.10 The geometry-optimized structures of $(nH_2 + Mg)@$ C_{120} systems ($n = 8$–10). (From Türker, L. and Gümüş, S., *J. Mol. Struct. (Theochem)*, 719, 103, 2005.)

Table 39.9 shows some energies of the systems of present interest. The results of PM3 (RHF) type calculations indicate that all the structures considered are stable (the total and binding energies) but endothermic (the heat of formation). As can be seen from Table 39.9, among the homologous series, $(7H_2 + Mg)@$ C_{120} is the least endothermic structure. But no relation between the number of hydrogen molecules and the heat of formation is observed. And also it was indicated that the van der Waals energy increases smoothly as the number of hydrogens in the structure increases.

A single-walled, capped, and selenium-doped carbon nanotube, Se@C_{120}, which is yet nonexistent, was doped endohedrally with hydrogen molecules to obtain a series of $(nH_2 + Se)@C_{120}$ ($n = 0$–11) systems in Türker's publication (Türker 2007).

The geometry optimizations of all the structures leading to energy minima were achieved by using MM+ molecular mechanics method then by PM3 SCF MO method at the RHF level. All the computations were performed by using the Hyperchem (release 7.5) package program.

In the present study, the initial structure of the tube C_{120} was constructed starting from C_{60} [(5,6)-fullerene-60-I_h], which was excerpted from the Hyperchem Library. The halves of C_{60} structure were suitably used as the caps of an armchair type tube having 10 benzenoid rings as the peripheral belt. The resultant tube possesses 120 carbon atoms and both ends are closed. The system contains 10 pentagons originating from the fullerene

$n : 0$

FIGURE 39.9 The geometry-optimized structure of the Mg-doped C_{120} tube. (From Türker, L. and Gümüş, S., *J. Mol. Struct. (Theochem)*, 719, 103, 2005.)

TABLE 39.9   Some Energies of the Present Systems

n	Energy (kJ/mol)					
	Total	Binding	Isolated Atomic	Electronic	Core–core Interaction	Heat of Formation
0	−1,370,040	−79,991	−1,290,049	−28,776,541	27,406,501	5,955
6	−1,388,099	−82,913	−1,305,185	−30,369,881	28,981,782	5,649
7	−1,393,003	−85,295	−1,307,708	−30,660,674	29,267,671	3,703
8	−1,393,969	−83,738	−1,310,231	−30,903,744	29,509,774	5,696
9	−1,397,261	−84,507	−1,312,754	−31,199,628	29,802,367	5,363
10	−1,400,181	−84,904	−1,315,276	−31,469,596	30,069,416	5,402
11	−1,402,742	−84,942	−1,317,799	−31,700,737	30,297,996	5,800
12	−1,406,013	−85,690	−1,320,322	−32,004,272	30,598,259	5,488
13	−1,408,519	−85,674	−1,322,845	−32,260,963	30,852,444	5,941

Source: Türker, L. and Gümüş, S., *J. Mol. Struct. (Theochem)*, 719, 103, 2005.

structure (C_{60}) used to generate the caps. Therefore, the system can be considered in between a fullerene and a small nanotube. Then, a Se atom was inserted into the cage to get $Se@C_{120}$ system and then geometry optimized. In Figure 39.11, the side and end views of $Se@C_{120}$ system are shown. As seen there, Se atom is not

at the center of the tube but close to the interior surface. This is due to some interactions existing between the dopant atom (Se) and the tube cage that result some flattening of the cage wall nearby the Se atom (Figure 39.11). Then, hydrogen molecules were inserted into the cage. The calculated volume steadily increases from 1059×10^{-30} m³ for $n = 1$ case to 1129.6×10^{-30} m³ for $n = 11$ case.

As hydrogen molecules are doped into $Se@C_{120}$ structure, the Se atom is pushed toward the end of tube, whereas hydrogen molecules prefer to be located more or less along the principle axis of the tube having different orientations of molecular axis of hydrogens (Figure 39.11). However, an interesting thing is that when the number of hydrogen molecules reaches to 10, the composite electrostatic potential field leads to the elongation of H–H bond in the molecule nearest to the Se atom (also the same thing happens in the case of $(11 H_2 + Se)@C_{120}$). Although, the length of σ-bond in the other hydrogen molecules are about 0.70–0.72×10^{-10} m, the H–H distance in the hydrogen molecule nearest to the Se atom is 2.09×10^{-10} m (2.13×10^{-10} m in the case of $(11 H_2 + Se)@C_{120}$). Thus, the bond is broken. The distance between the Se atom and the hydrogen atoms of the above-mentioned H_2 molecule (which tends to dissociate) is 1.49×10^{-10} m, whereas H–Se bond in H_2Se is 1.6×10^{-10} m (Durant and Durant 1970).

The Se atom is always negatively charged, although the magnitude of negative charge decreases as the number of hydrogen molecules increases (Table 39.10). Probably, atomic orbitals of Se atom and π-orbitals of the tube overlap at certain interior parts of the tube, thus some electron population is transferred to Se. When hydrogens are inserted, interactions between the cell wall and the

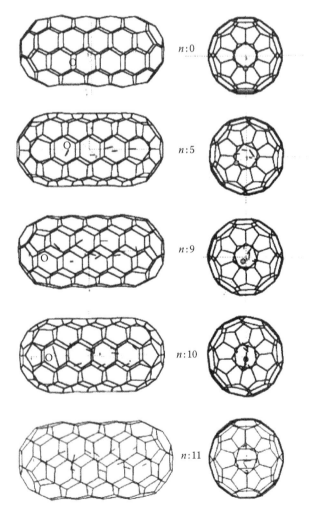

FIGURE 39.11   The geometry-optimized structures of $(nH_2 + Se)@C_{120}$ systems. (From Türker, L., *Int. J. Hydrogen Energy*, 32, 1933, 2007.)

TABLE 39.10   Total Unit of Charge of Dopants in Some $(nH_2 + Se)@C_{120}$ Systems

	n			
	0	5	9	10
Charge on Se	−0.39	−0.44	−0.43	−0.14
Total charge on hydrogens	—	0.18	0.22	0.34

Source: Türker, L., *Int. J. Hydrogen Energy*, 32, 1933, 2007.

TABLE 39.11 Some Calculated Energies of the Structures Considered

Energy (kJ/mol)	n					
	0	1	2	3	4	5
Total	−1,387,249	−1,390,329	−1,393,315	−1,396,325	−1,399,327	−1,402,318
Binding	−80,776	−81,333	−81,797	−82,284	−82,762	−83,231
Isolated atomic	−1,306,473	−1,308,996	−1,311,519	−1,314,041	−1,316,564	−1,319,087
Electronic	−29,344,810	−29,636,294	−29,919,500	−30,190,140	−30,455,568	−30,705,848
Core–core repulsion	27,957,562	28,245,965	28,526,185	28,793,815	29,056,241	29,303,532
Heat of formation	5,252.1	5,130.9	5,102.9	5,052.1	5,009.2	4,976.1
	n					
	6	7	8	9	10	11
Total	−1,405,271	−1,408,215.9	−1,411,155	−1,414,141	−1,417,128	−1,420,028
Binding	−83,660.8	−84,083.4218	−84,500	−84,963	−85,428	−85,804
Isolated atomic	−1,321,610	−1,324,132.48	−1,326,655	−1,329,176	−1,331,701	−1,334,224
Electronic	−30,990,326	−31,255,200	−31,531,038	−31,765,262	−32,042,883	−32,340,016
Core–core repulsion	29,585,055	29,846,984	30,119,882	30,351,122	30,625,753	30,919,989
Heat of formation	4,982.8	4,996.1	5,015.4	4,988.2	4,959.9	5,019.4

Source: Türker, L., *Int. J. Hydrogen Energy*, 32, 1933, 2007.

hydrogens partially localize the π-electrons of the tube causing less fraction of electron population to be transferred to Se atom.

Table 39.11 shows various energies of the composite systems presently considered. As seen in the table, all the systems are stable (the total and binding energies), but they are all endothermic (the heat of formations data). The heat of formations data indicate that depending on the number of hydrogen molecules, the endothermic nature of the composite structures varies. Although, within the set of compounds, there exist local minimum and maximum for the heat of formation values, $n = 10$ case possesses the lowest heat of formation value among the set (Table 39.11). It seems that elongation of H–H bond and orientation of all the other dopants collectively lower the heat of formation values.

Figure 39.12 shows the calculated van der Waals energies of the composite systems studied. As seen there, for certain number of hydrogen molecules, the energy of the system is less than the energy of the system(s) having less number of hydrogen molecules. This shows the presence of some stabilizing effect in those cases.

In the treatise of Türker et al. (2004), PM3 (RHF) type semi-empirical quantum chemical calculations have been carried out on $(nH_2 + Be)@C_{120}$ systems, where C_{120} is a capped tube and $n \leq 15$. The initial structure of the tube C_{120} was constructed starting from C_{60} (5,6-fullerene-60-I_h), which was excerpted from the Hyperchem library. The halves of C_{60} structure were suitably used as the caps of an armchair type tube having 10 benzenoid rings as the peripheral belt.

The optimizations were obtained by the application of the steepest-descent method followed by conjugate gradient methods, Fletcher-Reeves and Polak-Ribiere, consecutively (convergence limit of 4.18×10^{-4} kJ/mol (0.0001 kcal/mol) and RMS gradient of 4.18×10^7 (kJ/M mol) (0.001 kcal/A mol)). All these computations were performed by using the Hyperchem (release 5.1) and ChemPlus (2.0) package programs.

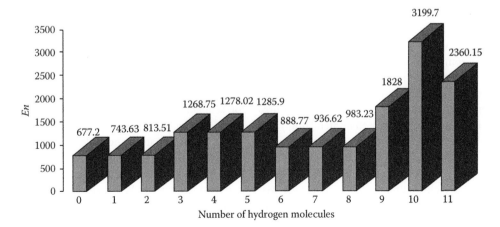

FIGURE 39.12 The change of van der Waals energy as the number of hydrogen molecules increases. (From Türker, L., *Int. J. Hydrogen Energy*, 32, 1933, 2007.)

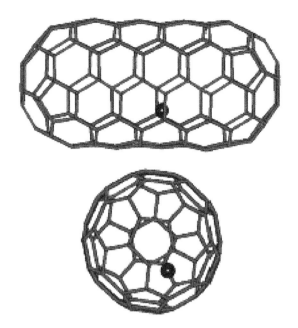

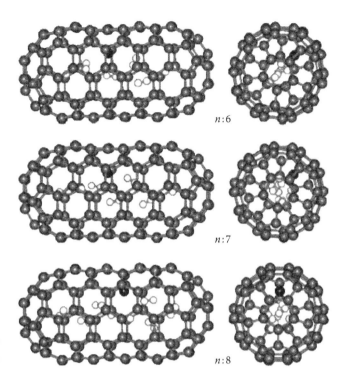

FIGURE 39.13 The side and end view of geometry-optimized Be@C$_{120}$ structure. (From Türker, L. et al., *Int. J. Hydrogen Energy*, 29, 1643, 2004.)

FIGURE 39.14 The geometry-optimized structures for $(n\text{H}_2 + \text{Be})$@C$_{120}$ with $n = 6$–8. (From Türker, L. et al., *Int. J. Hydrogen Energy*, 29, 1643, 2004.)

In the study, the above-mentioned capped tube C$_{120}$ was first endohedrally Be doped to obtain Be@C$_{120}$ structure and then geometry optimized (Figure 39.13). Figure 39.13 shows the side and end views of Be@C$_{120}$ structure. As it is seen there, Be atom does not occupy the center of axes of inertia of the structure that implies some interaction between the wall of the tube and the endohedral substituent Be atom.

Secondly, certain number of hydrogen molecules were added into the hallow of Be@C$_{120}$ to obtain $(n\text{H}_2 + \text{Be})$@C$_{120}$ type endohedrally substituted composite system, where $n = 15$. All these structures were also geometry optimized. Figure 39.14 shows the composite structures for $n = 6$–8.

The direction of the dipole moment in every case of $n \leq 15$ (including $n = 0$) is from somewhere on the carbon cage to the intersection of axes of inertia that means Be atom polarizes the carbon cage. The three-dimensional charge density map of hydrogen and Be atom in $(n\text{H}_2 + \text{Be})$@C$_{120}$ systems for $n = 6$–8 shows that the Be atom is positively charged, whereas hydrogen atoms possess zero, positive, or negative charges.

Table 39.12 shows some energies of the systems of present interest. The results of PM3 (RHF) type calculations indicate that all the structures considered are stable (the total and binding energies) but endothermic (the heat of formation) in nature. The results shown in Table 39.12 indicate that interestingly the structures become less and less endothermic as n increases up to $n = 7$, but for $n = 8$ onward up to 15, the corresponding composite systems become again more and more endothermic with the increasing value of n. Note that for $n = 7$, the composite system possesses the smallest volume among the others. All these results are indicative of the fact that orientation of hydrogen molecules in $(7\text{H}_2 + \text{Be})$@C$_{120}$ structure is optimal, leading to least endothermic structure among the others.

Molecular mechanics calculations done by Dodziuk (2005) indicate that the closure of 2H_2@C$_{70}$ with the preservation of two guests inside the host cage can prove very difficult. C$_{76}$ of D_2 symmetry seems to be the smallest fullerene, for which the endohedral complex with two guest molecules can be obtained. The results of the calculations also suggest that to get an endohedral fullerene complex with three guests, the cage must be larger than that of C$_{80}$. The calculations of the complexes were carried out using Hyperchem program with MM+ force field.

39.4 Hydrogen Storage Capacity for Fullerene Derivatives Having an Orifice

39.4.1 C$_{58}$ and Its Derivatives

Türker have considered endohedrally various numbers (1–5) of hydrogen molecules-doped C$_{58}$H$_4$ nanovesicles, $(n\text{H}_2$@C$_{58}$H$_4)$, for semiempirical molecular orbital treatment at the level of AM1 (RHF) method as well as molecular mechanics calculations (MM+) to investigate the possible usage of these, yet nonexistent, structures as hydrogen storage media (Türker 2001). Molecular mechanics calculations at the level of MM+ force field were used to obtain the van der Waals energies for the structures. All the computations were performed by using the Hyperchem (release 5.1) and ChemPlus (2.0) package programs. The presently considered nanovesicle, C$_{58}$H$_4$, possesses the geometry-optimized structure shown in Figure 39.15.

TABLE 39.12 Some Calculated Energies of Certain $(nH_2 + Be)@C_{120}$ Type Structures

	Energy (kJ/mol)					
n	Total	Binding	Isolated Atomic	Electronic	Core–Core Interaction	Heat of Formation
0	−1,371,066	−80,731	−1,290,335	−28,818,162	27,447,097	5,391.3
1	−1,374,138	−81,280	−1,292,858	−29,084,867	27,710,728	5,278.1
5	−1,386,186	−83,238	−1,302,949	−30,154,900	28,768,714	5,064.9
6	−1,389,164	−83,693	−1,305,471	−30,424,537	29,035,374	5,045.6
7	−1,392,148	−84,154	−1,307,994	−30,702,845	29,310,698	5,020.4
8	−1,395,106	−84,589	−1,310,517	−30,970,604	29,575,499	5,021.6
10	−1,401,007	−85,445	−1,315,562	−31,517,218	30,116,212	5,037.8
15	−1,415,565	−87,388	−1,328,176	−32,871,245	31,455,679	5,273.8

Source: Türker, L. et al., *Int. J. Hydrogen Energy*, 29, 1643, 2004.

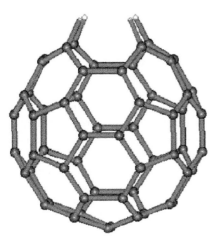

FIGURE 39.15 The geometry-optimized structure of $C_{58}H_4$. (From Türker, L., *Int. J. Hydrogen Energy*, 26, 843, 2001.)

As seen in Table 39.13, although all these structures are characterized with negative total and binding energies (thus, they should be stable structures), they all have endothermic heats of formation. Note that as n increases up to 2, the structures become less endothermic (even less than $C_{58}H_4$) but $n = 3$ onward endogenecity increases with increasing number of endohedral hydrogen molecules. Note that the calculations indicate that the sixth hydrogen

molecule cannot be inserted into the cage (it is expelled out during the geometry optimization process). As seen in Table 39.14, the geometry optimizations on $nH_2@C_{58}H_4$ composite systems cause some small changes in the area and the volume of the systems as n varies. As a result, the thermodynamic stability of the hydrogen assembly inside the cage has been investigated for $n = 1$–5.

Figure 39.16 is the plot of three-dimensional electrostatic potential isosurfaces for some of the systems considered. The presence of hydrogen molecules inside the cage does not change the main appearance of the plot for $C_{58}H_4$. On the contrary, Figure 39.17 shows the three-dimensional electrostatic potential isosurfaces engendered only by the endohedrally doped hydrogens for $H_2@C_{58}H_4$ and $5H_2@C_{58}H_4$ systems.

Analysis of the data so far gathered and presented above shows that the hydrogen molecules inside the $C_{58}H_4$ cage should be mainly under the influence of the van der Waals type interactions with the atoms of the cage. Figure 39.18 shows the plot of van der Waals energy versus the number of hydrogen molecules. This fact is also reflected to the HOMO and the LUMO energies, so that the number of hydrogen molecules inside the cage has almost no effect on the above-mentioned energies as compared with the corresponding values for the empty $C_{58}H_4$ vesicle (see Table 39.13). Thus, $C_{58}H_4$ nanovesicles can play the role of a physical adsorbent for certain number of hydrogen

TABLE 39.13 Some Energies of Endohedrally Hydrogen-Doped $C_{58}H_4$ Vesicles

Energy (kJ/mol)	Number of Hydrogen Molecules					
	0	1	2	3	4	5
Total	−719,107	−721,749	−724,386	−726,999	−729,581	−732,119
Binding	−38,591	−39,034	−39,472	−39,886	−40,268	−40,608
Isolated atomic	−680,516	−682,715	−684,914	−687,113	−689,312	−691,512
Electronic	−9,436,290	−9,612,246	−9,789,710	−9,965,779	−10,141,654	−10,310,426
Core–core repulsion	8,117,183	8,890,496	9,065,324	9,238,780	9,412,074	9,578,307
Heat of formation	3,751	3,744	3,743	3,765	3,818	3,914
LUMO[a]	−4.4492	−4.4478	−4.4461	−4.4405	−4.4269	−4.4277
	(A_1)	(A_1)	(A)	(A)	(A)	(A)
HOMO[a]	−14.6890	−14.6891	−14.6863	−14.6691	−14.6784	−14.6748
	(B_1)	(B_1)	(B)	(A)	(A)	(A)

Source: Türker, L., *Int. J. Hydrogen Energy*, 26, 843, 2001.

[a] Energies in the order of 10^{-19} J. Symmetries of the orbitals in parenthesis.

TABLE 39.14 Some Properties of $n\mathrm{H}_2@\mathrm{C}_{58}\mathrm{H}_4$ Systems

	n					
	0	1	2	3	4	5
Area	554	549	548	549	550	549
Volume	1212	1212	1213	1213	1216	1219
Refractivity	227	229	231	232	234	236
Polarizability	109	110	111	112	112	113

Source: Türker, L., *Int. J. Hydrogen Energy*, 26, 843, 2001.

Note: Area, volume, refractivity, and polarizability values are in the order of $10^{-20}\ \mathrm{m}^2$, $10^{-30}\ \mathrm{m}^3$, $10^{-30}\ \mathrm{m}^3$, and $10^{-30}\ \mathrm{m}^3$, respectively.

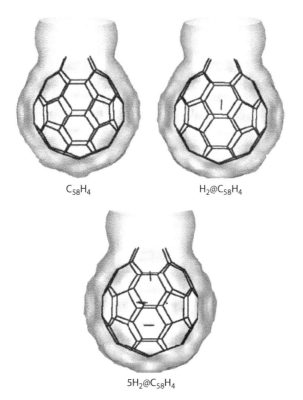

$\mathrm{C}_{58}\mathrm{H}_4$ $\mathrm{H}_2@\mathrm{C}_{58}\mathrm{H}_4$

$5\mathrm{H}_2@\mathrm{C}_{58}\mathrm{H}_4$

FIGURE 39.16 Three-dimensional electrostatic potential isosurface plots for some of the structures considered. (From Türker, L., *Int. J. Hydrogen Energy*, 26, 843, 2001.)

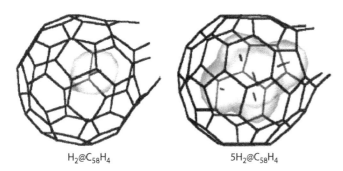

$\mathrm{H}_2@\mathrm{C}_{58}\mathrm{H}_4$ $5\mathrm{H}_2@\mathrm{C}_{58}\mathrm{H}_4$

FIGURE 39.17 Three-dimensional electrostatic potential isosurfaces for hydrogen(s) in the $\mathrm{C}_{58}\mathrm{H}_4$ cage. (From Türker, L., *Int. J. Hydrogen Energy*, 26, 843, 2001.)

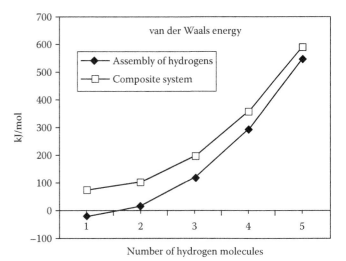

FIGURE 39.18 Plot of van der Waals energy versus the number of hydrogen molecules. (From Türker, L., *Int. J. Hydrogen Energy*, 26, 843, 2001.)

molecules depending on the temperature and the exerted pressure. However, as the applied pressure is increased to fill the vesicles with more of the hydrogen, then some concomitant chemical adsorption could be initiated as a result of some, more effective orbital–orbital interactions between $\mathrm{C}_{58}\mathrm{H}_4$ and H_2 molecules.

C_{60} can destabilize the nonpolar (or weakly polar) molecules, such as H_2 and CO, trapped inside its cage (Cioslowski 1991). The destabilization energies of CO and H_2 were 11.2 and 1.2 kcal/mol, respectively (Cioslowski 1991). These results indicated that the formation of the endohedral complexes, consisting of H_2 (or CO) and C_{60}, is an endothermic process. Hu et al. investigated the endohedral complexes of C_{58} with H_2 (or CO) using the ab initio HF method of Gaussian 94 with 3–21G and 6–31G(*d*) basis sets for geometry optimizations and energy calculations (Hu and Ruckenstein 2004). They chose C_{58} upon C_{60} because it also contains a seven-member ring in addition to the five- and six-member rings. This study revealed how a seven-member ring affected the structures and energies of endohedral complexes.

It was demonstrated that the formation of these complexes was endothermic with destabilization energies of 3.3 kcal/mol for H_2 (and 18.6 kcal/mol for CO). Furthermore, the H_2 (and CO) molecules had different orientations in the C_{58} cage, namely, the orientation of the molecular axis of the former was normal to the face of the seven-member ring, while that of the latter was parallel to that face. It was concluded that the orientation of the nonpolar or weakly polar molecule was mainly dependent on the guest molecular size. In addition, the H–H bond of the H_2 molecule was shortened inside the cage, whereas the length of the C–O bond remained unchanged.

39.4.2 C_{56} Systems

Endohedrally hydrogen-doped C_{56} systems ($n\mathrm{H}_2@\mathrm{C}_{56}$, obtained from C_{60} structure) have been theoretically investigated by Türker and Gümüş (2004). The geometry optimizations of the structures

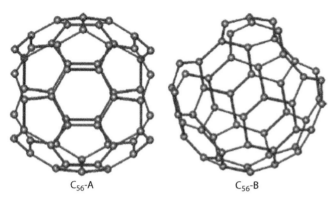

C$_{56}$-A   C$_{56}$-B

FIGURE 39.19 The geometry-optimized structures of the present C$_{56}$ systems. (From Türker, L. and Gümüş, S., *J. Mol. Struct. (Theochem)*, 681, 21, 2004.)

under consideration leading to energy minima were achieved by using first MM+ method and then AM1 SCF MO method at the UHF level (singlet states). All the computations were performed by using Hyperchem (release 5.1) package program.

The present study considers initially two isomers of C$_{56}$ systems, namely, C$_{56}$-A and C$_{56}$-B. All these molecules were obtained by the removal of two C–C bonds that connect the two corresponding five-membered rings in C$_{60}$, [(5,6)-fullerene-60-I$_h$], which was excerpted from the Hyperchem library. In the case of C$_{56}$-A, the apertures locate at the opposite sides of the system, whereas C$_{56}$-B structure was obtained by deleting two C–C bonds that create two apertures perpendicularly located in the system. These systems, especially C$_{56}$-A, can be considered a special type of open-ended carbon nanotube. Hence, it has been thought as having some potency to hold hydrogen gas as some other already reported carbon nanotubes. After obtaining the desired structures, hydrogen doping was performed and at most six hydrogen molecules could be doped in C$_{56}$-A system, whereas the maximum number of hydrogens dopable in C$_{56}$-B was found to be only 3. The geometry-optimized structures of the two C$_{56}$ systems can be seen in Figure 39.19.

The geometry-optimized structures for the hydrogen-doped C$_{56}$ isomers can be seen in Figure 39.20. As seen there, the alignment of hydrogen molecules (in C$_{56}$-A series) nearby the apertures are parallel or nearly so to the molecular axis passing through the apertures. C$_{56}$-A systems do not possess any dipole moment except for $n = 3$ and 5, at which also accommodation of hydrogens is not symmetrical, whereas all C$_{56}$-B type endohedrally hydrogen-doped systems have dipole moments that are almost equal to each other. The dipole moments for the C$_{56}$-B

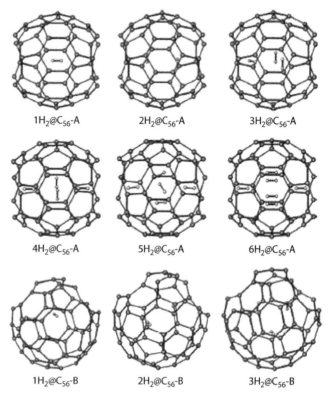

1H$_2$@C$_{56}$-A   2H$_2$@C$_{56}$-A   3H$_2$@C$_{56}$-A

4H$_2$@C$_{56}$-A   5H$_2$@C$_{56}$-A   6H$_2$@C$_{56}$-A

1H$_2$@C$_{56}$-B   2H$_2$@C$_{56}$-B   3H$_2$@C$_{56}$-B

FIGURE 39.20 The geometry-optimized structures of nH$_2$@C$_{56}$-A ($n = 1$–6) and nH$_2$@C$_{56}$-B ($n = 1$–3) systems. (From Türker, L. and Gümüş, S., *J. Mol. Struct. (Theochem)*, 681, 21, 2004.)

systems are greater than the corresponding C$_{56}$-A systems. The direction of the dipole moments for all structures is from the surface of the molecule to the center of the composite structure.

Table 39.15 (C$_{56}$-A) and Table 39.16 (C$_{56}$-B) tabulate some calculated energies of the systems considered. As seen from the tables, all the structures are stable (the total and binding energies) but endothermic (heat of formation) in nature.

When the two series are compared, C$_{56}$-B type structures are more endothermic and less stable than the C$_{56}$-A series having the same number of dopants.

39.4.3 Hydrogenated C$_{60}$: C$_{60}$H$_{60}$

Cioslowski (1991) found that C$_{60}$ can destabilize nonpolar (or weakly polar) molecules trapped inside its cage, such as H$_2$ and CO, and stabilize polar molecules, such as LiF. In contrast to

TABLE 39.15 Some Calculated Energies of C$_{56}$-A ($n = 0$–6) Systems

Energy (kJ/mol)	n						
	0	1	2	3	4	5	6
Total	−687,917	−690,560	−693,204	−695,819	−698,430	−701,015	−703,528
Binding	−35,114	−35,558	−36,002	−36,418	−36,831	−37,216	−37,530
Heat of formation	4,925	4,918	4,909	4,929	4,953	5,003	5,125

Source: Türker, L. and Gümüş, S., *J. Mol. Struct. (Theochem)*, 681, 21, 2004.

TABLE 39.16 Some Calculated Energies of C_{56}-B ($n = 0$–3) Systems

Energy (kJ/mol)	n			
	0	1	2	3
Total	−687,805	−690,445	−693,083	−695,714
Binding	−35,002	−35,442	−35,881	−36,313
Heat of formation	5,038	5,033	5,030	5,034

Source: Türker, L. and Gümüş, S., *J. Mol. Struct. (Theochem)*, 681, 21, 2004.

the unsaturated C_{60} cage, the saturated $C_{60}H_{60}$ cage does not possess π electrons (Scueria 1991). Furthermore, the cavity of $C_{60}H_{60}$ is larger than that of C_{60}. Therefore, $C_{60}H_{60}$ might exhibit different behavior than C_{60} in the formation of endohedral complexes with small guest molecules. In the study of Hu and Ruckenstein (2005a), the interaction between the fullerene cage and an internal guest molecule H_2 is calculated using the B3LYP hybrid DFT method with the 6–31$G(d)$ basis set for the geometry optimization and the energy. Furthermore, the Boys-Bernardi counterpoise procedure is employed to correct the basis set superposition error (BSSE). In addition, the molecular volume, which is defined as the volume inside a contour with a 0.001 e/bohr³ density, was also calculated. All calculations have been carried out using the GAUSSIAN 03 program.

In the system H_2–C_{60}, H_2 is located near the center of the C_{60} cage. Furthermore, the energy calculations showed that H_2 was destabilized inside the C_{60} cage, with destabilization energy (after the BSSE correction) of 4.1 kcal/mol. The result is consistent with the HF calculations of Cioslowski (1991). In H_2–$C_{60}H_{60}$ system, H_2 is also located near the center of the $C_{60}H_{60}$ cage. In contrast to the C_{60} cage, the $C_{60}H_{60}$ cage has almost a zero interaction energy (after the BSSE correction) with H_2.

For nonpolar guest molecules, the dipole-induced dipole attractive interaction is negligible, and the main interaction is due to the electron cloud overlap between the guest molecule and cage, which is repulsive. Consequently, the formation of endohedral complexes between the fullerene cages and nonpolar or weakly polar molecules is an endothermic process. However, when the empty cavity of the cage increases in size, the electron cloud overlap should decrease, leading to a decrease of the repulsive interaction between the cage and the guest molecule. Indeed, because $C_{60}H_{60}$ has a larger empty cavity (3.82 Å diameter) than C_{60} (2.85 Å diameter) (Hu and Ruckenstein 2005b), the endohedral complexes of $C_{60}H_{60}$ have smaller repulsive interactions with the nonpolar (or weakly polar) guest molecules inside the cage than the corresponding complexes of C_{60}. This explains why the endohedral complexes of $C_{60}H_{60}$ with H_2 are more stable than those of C_{60}. Furthermore, because the electron cloud diameter of H_2 (3.36 Å) is larger than the cavity diameter (2.85 Å) of C_{60}, but smaller than the cavity diameter of $C_{60}H_{60}$ (3.82 Å), H_2 is destabilized by C_{60}, but there is a negligible interaction between the $C_{60}H_{60}$ cage and H_2 inside it. The bond lengths of H_2 inside the C_{60} and $C_{60}H_{60}$ cages are not much different from those of the free molecule.

Finally, the calculations showed that H_2 molecule inside either of the cages orients almost along the axis connecting two carbon atoms on opposite sites of the cage, deviating from this axis by about 5°. Cioslowski (1991) demonstrated that guest molecules with different orientations inside the C_{60} cage have almost the same interaction energy with the cage. Because the $C_{60}H_{60}$ cage has a larger cavity than C_{60}, the orientation of the molecule inside should have an even smaller effect on the interaction energy between the guest molecule and the cage.

39.5 Conclusion

Endohedrally hydrogen-doped fullerenes have been the subject of ongoing research both theoretically and experimentally. At the moment, the theoretical studies on the subject (as usually be the case in most fields) are far ahead of the experimental studies. However, by means of all these studies the physics and chemistry of fullerenes undeniably have gotten much insight about their nature and potential behavior in applications.

References

Akasaka, T. and Nagase, S. Eds. 2002. *Endofullerenes: A New Family of Carbon Clusters*. Dordrecht, the Netherlands: Kluwer Academic Publisher.

Arce, M.-J., Viado, A. L., An, Y.-Z., Khan, S. I., and Rubin, Y. 1996. Triple scission of a six-membered ring on the surface of C_{60} via consecutive pericyclic reactions and oxidative cobalt insertion. *J. Am. Chem. Soc.* 118: 3775–3776.

Barajas-Barraza, R. E. and Guirado-López, R. A. 2002. Clustering of H_2 molecules encapsulated in fullerene structures. *Phys. Rev. B* 66: 155426–155437.

Baştuǧ, T., Kürpick, P., Meyer, J., Sepp, W.-D., Fricke, B., and Rosén, A. 1997. Dirac-Fock-Slater calculations on the geometric and electronic structure of neutral and multiply charged C_{60} fullerenes. *Phys. Rev. B* 55: 5015–5020.

Bethune, D. S., Johnson, R. D., Salem, J. R., de Vries, M. S., and Yannoni, C. S. 1993. Atoms in carbon cages: The structure and properties of endohedral fullerenes. *Nature* 366: 123–128.

Camps, X. and Hirsch, A. 1997. Efficient cyclopropanation of C_{60} starting from malonates. *J. Chem. Soc., Perkin Trans.* 1: 1595–1596.

Carravetta, M., Danquigny, A., Mamone, S. et al. 2007. Solid-state NMR of endohedral hydrogen–fullerene complexes. *Phys. Chem. Chem. Phys.* 9: 4879–4894.

Cioslowski, J. 1991. Endohedral chemistry: Electronic structures of molecules trapped inside the C_{60} cage. *J. Am. Chem. Soc.* 113: 4139–4141.

Dodziuk, H. 2005. Modeling complexes of H_2 molecules in fullerenes. *Chem. Phys. Lett.* 410: 39–41.

Dodziuk, H. 2006. Reply to the "Comment on 'Modelling complexes of H_2 molecules in fullerenes' by H. Dodziuk [*Chem. Phys. Lett.* 410 (2005) 39]" by L. Turker and S. Erkoc. *Chem. Phys. Lett.* 426: 224–225.

Dodziuk, H. 2007. Modeling the structure of fullerenes and their endohedral complexes involving small molecules with nontrivial topological properties. *J. Nanosci. Nanotechnol.* 7: 1102–1110.

Dolgonos, G. 2005. How many hydrogen molecules can be inserted into C_{60}? Comments on the paper 'AM1 treatment of endohedrally hydrogen doped fullerene, $nH_2@C_{60}$' by L. Türker and Ş. Erkoç [*J. Mol. Struct. (Theochem)* 638 (2003) 37–40]. *J. Mol. Struct. (Theochem)* 723: 239–241.

Dresselhause, M. S., Dresselhaus, G., and Eklund, P. C. 1996. *Science of Fullerenes and Carbon Nanotubes*, pp. 132, 228. New York: Academic Press.

Durant, P. J. and Durant, B. 1970. *Introduction to Advanced Inorganic Chemistry*. London, U.K.: Longman.

Edwards, C. M., Butler, I. S., Qian, W., and Rubin, Y. 1998. Pressure-tuning vibrational spectroscopic study of (η^5-C_5H_5) Co($C_{64}H_4$). Can endohedral fullerenes be formed under pressure? *J. Mol. Struct.* 442: 169–174.

Hirsch, A. 1994. *The Chemistry of Fullerenes*. Stuttgart, Germany: Georg Thieme.

Hoke, S. H. II, Molstad, J., Dilettato, D. et al. 1992. Reaction of fullerenes and benzyne. *J. Org. Chem.* 57: 5069–5071.

Hu, Y. H. and Ruckenstein, E. 2004. Endohedral complexes of C_{58} cage with H_2 and CO. *Chem. Phys. Lett.* 390: 472–474.

Hu, Y. H. and Ruckenstein, E. 2005a. Density functional theory calculations for endohedral complexes of non-π $C_{60}H_{60}$ cage with small guest molecules. *J. Chem. Phys.* 123: 144303–144305.

Hu, Y. H. and Ruckenstein, E. 2005b. Endohedral chemistry of C_{60}-based fullerene cages. *J. Am. Chem. Soc.* 127: 11277–11282.

Huber, K. P. and Herzberg, G. 1979. *Constants of Diatomic Molecules (Molecular Spectra and Molecular Structure IV)*. New York: Van Nostrand Reinhold.

Huhey, E. J. 1978. *Inorganic Chemistry*, p. 842. New York: Harper and Row.

Hummelen, J. C., Prato, M., and Wudl, F. 1995a. There is a hole in my bucky. *J. Am. Chem. Soc.* 117: 7003–7004.

Hummelen, J. C., Knight, B., Pavlovich, J., González, R., and Wudl, F. 1995b. Isolation of the heterofullerene $C_{59}N$ as its dimer ($C_{59}N)_2$. *Science* 269: 1554–1556.

Iwamatsu, S.-i., Uozaki, T., Kobayashi, K., Re, S., Nagase, S., and Murata, S. 2004. A bowl-shaped fullerene encapsulates a water into the cage. *J. Am. Chem. Soc.* 126: 2668–2669.

Iwamatsu, S., Murata, S., Andoh, Y. et al. 2005. Open-cage fullerene derivatives suitable for the encapsulation of a hydrogen molecule. *J. Org. Chem.* 70: 4820–4825.

Kadish, K. M. and Ruoff, R. S. (Eds.) 2000. Fullerenes: Chemistry, Physics and Technology. In *Endohedral Metallofullerenes: Production, Separation, and Structural Properties*, pp. 357–393. New York: John Wiley & Sons.

Kietzmann, H., Rochow, R., Ganteför, G. et al. 1998. Electronic structure of small fullerenes: Evidence for the high stability of C_{32}. *Phys. Rev. Lett.* 81: 5378–5381.

Koi, N. and Oku, T. 2004a. Molecular orbital calculations of hydrogen storage in carbon and boron nitride clusters. *Sci. Technol. Adv. Mater.* 5: 625–628.

Koi, N. and Oku, T. 2004b. Hydrogen storage in boron nitride and carbon clusters studied by molecular orbital calculations. *Solid State Commun.* 131: 121–124.

Komatsu, K., Murata, M., Maeda, S., and Murata, Y. Organic encapsulation of H_2 in C_{60} and in open-cage C_{70}. Paper presented at *230th ACS National Meeting*, Washington, DC, August 28–September 1, 2005.

Kresse, G. and Furthmüller, J. 1996. Efficient iterative schemes for ab initio total-energy calculations using a plane-wave basis set. *Phys. Rev. B* 54: 11169–11186.

Kresse, G. and Hafner, J. 1993. Ab initio molecular dynamics for liquid metals. *Phys. Rev. B* 47: 558–561.

Kresse, G. and Hafner, J. 1994. Ab initio molecular-dynamics simulation of the liquid-metal–amorphous-semiconductor transition in germanium. *Phys. Rev. B* 49: 14251–14269.

Landau, L. D. and Lifshitz, E. M. 1977. *Quantum Mechanics*, 3rd edn. Oxford, NY: Pergamon.

Liu, S. and Sun, S. 2000. Recent progress in the studies of endohedral metallofullerenes. *J. Organomet. Chem.* 599: 74–86.

Maggini, M., Scorrano, G., and Prato, M. 1993. Addition of azomethine ylides to C_{60}: Synthesis, characterization, and functionalization of fullerene pyrrolidines. *J. Am. Chem. Soc.* 115: 9798–9799.

Matsuo, Y., Isobe, H., Tanaka, T. et al. 2005. Organic and organometallic derivatives of dihydrogen-encapsulated [60]fullerene. *J. Am. Chem. Soc.* 127: 17148–17149.

Mauser, H., Hirsch, A., van Eikema Hommes, N. J. R., and Clark, T. 1997. Chemistry of convex versus concave carbon: The reactive exterior and the inert interior of C_{60}. *J. Mol. Model.* 3: 415–422.

Murata, M., Murata, Y., and Komatsu, K. 2006. Synthesis and properties of endohedral C_{60} encapsulating molecular hydrogen. *J. Am. Chem. Soc.* 128: 8024–8033.

Nagase, S., Kobayashi K., and Akasaka, T. 1996. Endohedral metallofullerenes: New spherical cage molecules with interesting properties. *Bull. Chem. Soc. Jpn.* 69: 2131–2142.

Narita, I. and Oku, T. 2002. Molecular dynamics calculation of H_2 gas storage in C_{60} and $B_{36}N_{36}$ clusters. *Diam. Relat. Matters* 11: 945–948.

Neek-Amal, M., Tayebirad, G., and Asgari, R. 2007. Ground-state properties of a confined simple atom by C_{60} fullerene. *J. Phys. B: At. Mol. Opt. Phys.* 40: 1509–1521.

Nierengarten, J.-F. 2001. Ring-opened fullerenes: An unprecedented class of ligands for supramolecular chemistry. *Angew. Chem. Int. Ed.* 40: 2973–2974.

Oksengorn, B. 2003. Preparation of endohedral complex of molecular hydrogen-fullerene C_{60}, associated with hydrogenated C_{60}. *C. R. Chim.* 6: 467–472.

Pupysheva, O. V., Farajian, A. A., and Yakobson B. I. 2008. Fullerene nanocage capacity for hydrogen storage. *Nano Lett.* 8: 767–774.

Ren, Y. X., Ng, T. Y., and Liew, K. M. 2006. State of hydrogen molecules confined in C_{60} fullerene and carbon nanocapsule structures. *Carbon* 44: 397–406.

Rubin, Y. 1997. Organic approaches to endohedral metallofullerenes: Cracking open or zipping up carbon shells? *Chem. Eur. J.* 3: 1009–1016.

Rubin, Y. 1999. Ring opening reactions of fullerenes: Designed approaches to endohedral metal complexes. *Top. Curr. Chem.* 199: 67–91.

Rubin, Y., Jarrosson, T., Wang, G.-W. et al. 2001. Insertion of helium and molecular hydrogen through the orifice of an open fullerene. *Angew. Chem. Int. Ed.* 40: 1543–1546.

Sanville, E. and BelBruno, J. J. 2003. Computational studies of possible transition structures in the insertion and windowing mechanisms for the formation of endohedral fullerenes. *J. Phys. Chem. B* 107: 8884–8889.

Saunders, M., Jiménez-Vázquez, H. A., Cross, R. J., Mroczkowski, S., Freedberg, D. I., and Anet, F. A. L. 1994a. Probing the interior of fullerenes by ^{3}He NMR spectroscopy of endohedral ^{3}He@C_{60} and ^{3}He@C_{70}. *Nature* 367: 256–258.

Saunders, M., Jiménez-Vázquez, H. A., Cross, R. J. et al. 1994b. Incorporation of helium, neon, argon, krypton, and xenon into fullerenes using high pressure. *J. Am. Chem. Soc.* 116: 2193–2194.

Saunders, M., Cross, R. J., Jiménez-Vázquez, H. A., Shimshi, R., and Khong, A. 1996. Noble gas atoms inside fullerenes. *Science* 271: 1693–1697.

Schick, G., Jarrosson, T., and Rubin, Y. 1999. Formation of an effective opening within the fullerene core of C_{60} by an unusual reaction sequence. *Angew. Chem. Int. Ed.* 38: 2360–2363.

Schriver, D. F., Atkins, P. W., and Langford, C. H. 1990. *Inorganic Chemistry*. Oxford, U.K.: ELBS.

Scueria, G. E. 1991. Ab initio theoretical predictions of the equilibrium geometries of C_{60}, $C_{60}H_{60}$ and $C_{60}F_{60}$. *Chem. Phys. Lett.* 176: 423–427.

Shinohara, H. 2000. Endohedral metallofullerenes. *Rep. Prog. Phys.* 63: 843–892.

Stoicheff, B. P. 1957. High resolution Raman spectroscopy of gases: IX. Spectra of H_2, HD, and D_2. *Can. J. Phys.* 35: 730–741.

Sun, Q., Wang, Q., Yu, J. Z., Ohno, K., and Kawazoe, Y. 2001. First-principles studies on pure and doped C_{32} clusters. *J. Phys.: Condens. Matter* 13: 1931–1938.

Türker, L. 2001. Endohedrally hydrogen doped $C_{58}H_4$ vesicles–a theoretical study. *Int. J. Hydrogen Energy* 26: 843–847.

Türker, L. 2002. Certain endohedrally hydrogen doped Be@C_{60} systems–a theoretical study. *J. Mol. Struct. (Theochem)* 577: 205–211.

Türker, L. 2003. AM1 treatment of $(Li + nH_2)_{n=0-5}$@C_{60} systems. *Int. J. Hydrogen Energy* 28: 223–228.

Türker, L. 2004. AM1 treatment of $(Be + nH_2)$@C_{70} systems. *J. Mol. Struct. (Theochem)* 668: 225–228.

Türker, L. 2005. PM3 treatment of some endohedrally Se and H_2 doped C_{60} systems. *J. Mol. Struct. (Theochem)* 717: 5–8.

Türker, L. 2007. Hydrogen storage capability of Se@C_{120} system. *Int. J. Hydrogen Energy* 32: 1933–1938.

Türker, L. and Erkoç, Ş. 2003. AM1 treatment of endohedrally hydrogen doped fullerene, nH_2@C_{60}. *J. Mol. Struct. (Theochem)* 638: 37–40.

Türker, L. and Gümüş, S. 2004. AM1 treatment of endohedrally hydrogen doped C_{56} systems, nH_2@C_{56}. *J. Mol. Struct. (Theochem)* 681: 21–25.

Türker, L. and Gümüş, S. 2005. Hydrogen storage capacity of Mg@C_{120} system. *J. Mol. Struct. (Theochem)* 719: 103–107.

Türker, L. and Erkoç, Ş. 2006. Comment on 'Modelling complexes of H_2 molecules in fullerenes' by H. Dodziuk [*Chem. Phys. Lett.* 410 (2005) 39]. *Chem. Phys. Lett.* 426: 222–223.

Türker, L., Eroğlu, İ., Yücel, M., and Gündüz, U. 2004. Hydrogen storage capability of carbon nanotube Be@$_{C120}$. *Int. J. Hydrogen Energy* 29: 1643–1647.

Yang, C. K. 2007. Density functional calculation of hydrogen-filled C_{60} molecules. *Carbon* 45: 2445–2458.

40

Molecular Hydrogen in Carbon Nanostructures

Felix Fernandez-Alonso
Science and Technology
Facilities Council

and

University College London

Francisco Javier Bermejo
Consejo Nacional de
Investigaciones Científicas

and

University of the Basque Country

Marie-Louise Saboungi
Centre National de la Recherche
Scientifique—Université d'Orléans

40.1 Introduction: The Hydrogen Storage Challenge

The use of hydrogen, the most abundant element in nature, is being actively pursued as a potential energy vector that could replace fossil fuels. Its combustion produces only heat and water, and it can be efficiently combined with oxygen in a fuel cell to produce electricity, a clean and convenient carrier of energy that has many possible applications including consumer devices, lighting, refrigeration and transportation. Its use as a power source for commercial transport vehicles has been considered for many decades [1], and prototype vehicles with satisfactory performance have already been built. However, realizing the promise of hydrogen as a widely used fuel requires substantial development of the means for its production, storage, and subsequent use. This chapter addresses the problem of hydrogen storage with particular application to transportation and portable power applications, which require a medium combining high hydrogen density with fast kinetics for charging and discharging at moderate temperatures and pressures, i.e., close to ambient.

The obvious method of containment—storing it in pressure vessels at either ambient or cryogenic temperatures—is clearly unsatisfactory for widespread transportation and portable

applications. Similarly, the physics and chemistry of the storage of hydrogen in metallic hosts are well known. Molecular hydrogen under certain conditions of pressure and temperature binds to the metallic lattice by means of a chemical reaction, usually highly exothermic, yielding a stable metallic hydride. However, the heat released upon hydride formation represents a serious problem. In practical terms, a hydrogen-powered car would need to have an on-board heat-exchange system, which is capable of dissipating substantial amounts of energy during refueling, and the rate of hydrogen uptake is necessarily limited. Alternative methods involve physical adsorption of molecular H_2 on porous or laminar substrates. This process produces much less heat than hydride formation and the charge/discharge cycles are much faster. However, it is still necessary to achieve the required storage capacity in terms of mass and volume ratios. The present goals as currently set by the U.S. Department of Energy (DOE) for 2010 are 6 wt % of hydrogen and a mass density of 45 g/L, per the requirements of a typical vehicle with a range of around 500 km [2].

Nanostructured materials have opened up new possibilities for solving the storage problem. Their high porosities make them promising candidates for reaching the required storage capacities, and the small dimensions involved reduce the time taken for the hydrogen molecules to travel from their resident site to the surface. Owing to their light weight and high porosities, carbon-based nanostructured materials have been intensively investigated as potential hydrogen storage media over the past two decades. Several update reports and reviews already exist in the literature [3–18], where the emphasis has been placed on a detailed account of the ultimate hydrogen-storage performance of a vast range of substrates and synthesis protocols, to the detriment of building a consistent and robust microscopic picture of how molecular hydrogen interacts with these materials. This tutorial review is intended to fill the above gap by shifting the focus to the hydrogen molecule. As such, the intent is to be illustrative rather than exhaustive in our exposition of the state of the art in the field. To this end, Section 40.2 discusses in detail both the molecular and bulk properties of hydrogen and the various ways it interacts with solid-state media. Both experimental and theoretical tools are also introduced in some detail, particularly those that can offer novel insight into the hydrogen-storage problem. These topics are illustrated in detail by working out specific examples of direct relevance to the use of carbon-based nanomaterials. Section 40.3 makes extensive use of the fundamental principles introduced earlier on. First, it explores the topological consequences of rolling and bending graphite the basis of constructing novel forms of carbon. The effects of topology and chemical doping are then treated in detail by presenting two case studies taken from our recent work. Section 40.4 concludes this tutorial by providing an assessment of where we stand at the present time, the lessons learnt, and the challenges ahead.

A note on units. The literature is far from being consistent in its use of scientific units, particularly to denote energy. To simplify our exposition, the following choices have been made:

- Energy: electron volt (eV = 1.602×10^{-19} J) or millielectron volt (1 meV = 10^{-3} eV).
- Distance: Angstrom (1 Å = 10^{-10} m = 0.1 nm).
- Time: picosecond (1 ps = 10^{-12} s).
- Pressure: bar (1 bar = 10^5 Pa = 0.1 MPa).

Useful conversion factors for commonly used energy units are 1 eV = 8066 cm^{-1} = 23.1 kcal/mol = 96.6 kJ/mol = 11605 K = 0.0367 au [19,20].

40.2 Fundamentals

40.2.1 Hydrogen Molecule

40.2.1.1 Molecular Properties

Atomic hydrogen (H) is unstable under normal conditions on earth but exists as a stable homonuclear diatom (H_2). Within a molecular-orbital description, the lowest-energy (1s) H atomic orbitals give rise to symmetric (bonding) and antisymmetric (antibonding) linear combinations in H_2 [21]. It is the full occupancy of the $1\sigma_g$ bonding molecular orbital ($E = -11.7$ eV) by two paired electrons and the large energy gap to the lowest-lying antibonding molecular orbital $1\sigma_u$ ($E \sim +2$ eV) that gives

rise to a strong covalent bond characterized by a binding energy of 4.751 eV [22].

To first order, the potential energy of ground-state H_2 in vibrational level v and rotational level J is given by [23]

$$E(v,J) = \omega_{vib}(v + 1/2) + B_{rot}J(J+1) \qquad (40.1)$$

where $\omega_{vib} = 546$ meV and $B_{rot} = 7.54$ meV, the largest known ground-state vibrational and rotational constants of any stable molecular species [22]. ω_{vib} is related to the force constant k holding the nuclei together via $\omega_{vib} = \sqrt{k/\mu}$. μ is the reduced mass for relative motion of two protons of mass m_H given by $\mu = m_H m_H/(m_H + m_H) = m_H/2$ ($m_H = 1.0078$ amu; 1 amu = 1.6605×10^{-27} kg). The energy scale for rotational motions is set by $B_{rot} = \hbar^2/2I$ where $I = \mu r_{HH}^2$ is the moment of inertia and $r_{HH} = 0.741$ Å is the H–H bond distance. Since $B_{rot}/\omega_{vib} \ll 1$, these two motions are effectively uncoupled from each other justifying the use of Equation 40.1. Access to spectral features due to molecular vibrations and rotations thus offers a sensitive means of not only measuring k and r_{HH}, respectively, but also how these evolve as H_2 is forced to interact with other atoms and molecules.

As a homonuclear diatom possessing a center of inversion symmetry, H_2 lacks a permanent dipole moment but has a permanent quadrupole moment $Q_{H_2} = 0.13$ eÅ^2, where e is the electron charge ($e = 1$ au = 1.602×10^{-19} Coulomb) [24]. This electric multipole moment originates from the additional electron density between the two protons due to the presence of a stable chemical bond. An excess of positive charge thus surrounds each H nucleus while negative charge tends to concentrate midway along the H–H axis. The static dipole polarizabilities parallel and perpendicular to the intermolecular axis are $\alpha_{par} = 0.93$ Å^3 and $\alpha_{per} = 0.67$ Å^3, defining an isotropically averaged polarizability $\alpha_{iso} = \alpha_{par} + 2\alpha_{per} = 0.76$ Å^3 [24]. This value is intermediate between the low polarizability characteristic of small atoms such as He (0.20 Å^3) and those typical of larger diatomic molecules such as N_2 (1.74 Å^3) and O_2 (1.57 Å^3) [25]. $\alpha_{par} \sim \alpha_{per}$ also implies that H_2 is characterized by a largely symmetric electron density distribution at typical intermolecular distances in condensed matter (>2 Å). It is for this reason that, to a very good degree of approximation, H_2 may be regarded as being spherical rather than dumbbell shaped. Owing to the simplicity of this molecule, it should come as no surprise that the single-molecule properties of H_2 have been extensively investigated since the early days of quantum mechanics [26] as a reference molecular system. Nowadays, sophisticated first-principles calculations are able to reach "spectroscopic accuracy" of meV or better, corresponding to relative errors in the total energy for ground and excited states of 1 part in 10^{4-5} [27–30].

There are two distinct nuclear-spin isomers of H_2, differing in the relative orientation of the nuclear spin of each proton ($I_H = 1/2$), giving rise to a total nuclear spin of the molecule of $I_{tot} = 0$ or 1. In the so-called *ortho*-hydrogen form (*o*-H_2), the spins of the two protons are parallel and form a triplet state

(I_{tot} = 1), whereas in the *para*-hydrogen form (*p*-H_2) the spins are antiparallel and form a singlet state (I_{tot} = 0). The distinction between *ortho* and *para* states has profound consequences on the energy-level structure of H_2 as the Pauli exclusion principle dictates that the exchange of two protons (fermions) must result in a change of the sign of the total wavefunction [31]. For freely rotating H_2, this means that even(odd) rotational levels must be *para(ortho)* states exclusively. Conversion between these two species requires a nuclear-spin flip and is forbidden under normal circumstances, leading to anomalously slow conversion between them. As a result, samples of enriched *para* or *ortho* species can be prepared and maintained for prolonged periods of time. Very similar considerations apply to other isotopic cousins such as D_2 where the *para/ortho* progression is reversed owing to the bosonic character of the D nucleus (I_D = 1). Beyond H_2, the molecular properties of N_2 ($I_{14\,N}$ = 1) and O_2 ($I_{16\,O}$ = 0), the major constituents of the air we breathe, are also profoundly affected by these subtle quantum-mechanical effects, particularly in $^{16}O_2$, where half of the expected rotational energy levels cannot exist.

The above considerations are not restricted to isolated H_2 (e.g., gas phase) but also apply to the solid and liquid state where rotational motions of individual H_2 molecules are almost undistorted owing to its exceptionally small moment of inertia [22]. In this situation, symmetry considerations (the Wigner–Eckart theorem) impose strict limits on the number and type of electric and magnetic multipole moments allowed for a stationary state of well-defined rotational quantum number J and wavefunction [32])

$$|J, M\rangle = Y_{JM}(\Theta, \phi) \qquad (40.2)$$

where

Y_{JM} are spherical harmonics
(Θ, ϕ) are spherical-polar coordinates

Thus, *p*-H_2 in $J = 0$ cannot have any nonzero multipole moments as it must correspond to a spherically symmetric charge distribution $\left(|0,0\rangle = Y_{00}(\Theta, \phi) = 1/\sqrt{4\pi}\right)$. Similarly *o*-$H_2$ in $J = 1$ can only possess permanent electric quadrupole and magnetic dipole moments because $|1, M\rangle = Y_{1M}(\Theta, \phi)$ transforms as a vector under rotations [32]. Spectroscopists have made extensive use of the noninteracting nature of *p*-H_2 ($J = 0$) to study the energy-level structure of matrix-isolated molecules [33,34]. Likewise, H_2 diffusion in the liquid is very sensitive to increasing amounts of *p*-H_2 owing to a dramatic switch-off of intermolecular forces [35].

Nuclear-spin isomerism in H_2 also transcends a purely academic interest in this molecule and the ultimate consequences of quantum mechanics. At ambient temperature, *normal* hydrogen (*n*-H_2) is an equilibrium mixture of ca. 25% *para* and 75% *ortho* states. On cooling, the *ortho* isomer slowly converts to the *para* counterpart. This conversion has important implications for the storage of this molecule as the energy released by nuclear-spin conversion ($2B_{rot}$ = 14.7 meV per molecule, cf. Equation 40.1) greatly exceeds the heat of evaporation of hydrogen (9.2 meV per molecule). The

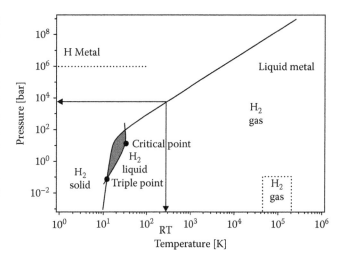

FIGURE 40.1 Hydrogen phase diagram. (Reproduced from Zhou, L., *Renew. Sust. Energy Rev.*, 9, 395, 2005. With permission.)

consequence is loss of liquid by boil-off and the need for greater energy to be expended in the liquefaction process [1].

40.2.1.2 Bulk Properties and Phase Diagram

In spite of its apparent simplicity, understanding the thermodynamic properties of condensed hydrogen has been long, difficult, and full of pitfalls. Comprehensive reviews on the bulk thermodynamic properties of H_2 exist in the literature [36,37]. In what follows, we will restrict ourselves to those features of direct relevance to hydrogen storage at moderate temperatures and pressures, as shown in a simplified phase diagram of Figure 40.1. A solid phase is present at low temperatures, characterized by a low mass density of ~0.07 g/cm³ at 10 K, more than an order of magnitude less dense than liquid water at ambient conditions ~1 g/cm³. It is noteworthy that this drop in density is comparable to the ratio of molecular weights between water and H_2, suggesting that their effective sizes in condensed matter are very similar to each other in spite of their very different chemical compositions.

As the temperature is raised, the phase diagram is clearly dominated by a gaseous phase of much lower density, approaching densities of ~0.09 kg/m³ at 300 K—i.e., a volume of ~10 m³ would be necessary to store 1 kg of H_2 at ambient temperature and pressure. The narrow region corresponding to liquid H_2 is bounded by the triple point T = 13.96 K, P = 0.07 bar, and the critical point at T = 33.18 K, P = 13 bar, the latter characterized by a much lower number density of ~0.03 g/cm³. This liquid phase displays a solid-like density of ~0.07 g/cm³ at $T \sim$ 20 K and its range is severely limited by the low critical temperature of H_2. The weak pairwise intermolecular potential $V(R)$ between H_2 molecules is at the heart of this behavior, as shown in Figure 40.2. To a good approximation, $V(R)$ can be written as a Lennard–Jones potential of the form

$$V(R) = -4\epsilon\left[\left(\frac{\sigma}{R}\right)^6 - \left(\frac{\sigma}{R}\right)^{12}\right] \qquad (40.3)$$

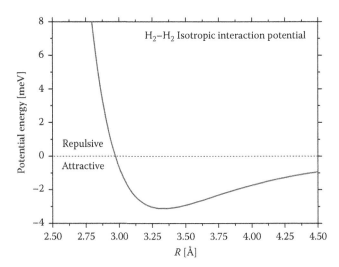

FIGURE 40.2 Isotropically averaged interaction potential between two H_2 molecules using the parameters of Ref. [38].

where

  ϵ and σ are the energy-well depth and diameter of the interaction, respectively

  R is the distance between molecular centers of mass

H_2–H_2 interactions amount to a tiny well depth of ~3 meV at an intermolecular separation of 3.4 Å. Also, the potential becomes repulsive just under 3 Å, further reinforcing our earlier conclusions on the relatively large effective size of the H_2 molecule. Quantum-mechanical effects are also important to describe bulk physical properties such as the thermal conductivity [39] or the diffusivity [40]. To quantify these, two dimensionless quantities can be defined in terms of the values of σ and ϵ of the intermolecular H_2–H_2 potential introduced above, namely,

$$\lambda_t = \hbar(m\epsilon\sigma^2)^{-1/2}, \quad \lambda_r = (2B_{\text{rot}}/V_r)^{1/2} \qquad (40.4)$$

where λ_t stands for the so-called *de Boer* parameter [41], which quantifies the importance of quantum effects for translational degrees of freedom. The quantum rotational parameter λ_r, defined in terms of the ratio of the rotational constant B_{rot} to the energy barrier V_r, serves as an indicator of the quantum character of angular-momentum variables. Typical values for these parameters serve to set a comparison between H_2 and isotopomers and well-known quantum systems such as ^{4}He. For the latter, $\lambda_t = 0.42$, to be compared with 0.27 for H_2 and 0.19 for D_2. In contrast, $\lambda_t = 0.03$ for Ar, an archetypal "classical" rare gas. λ_r for both hydrogen isotopes at temperatures around melting at ambient pressure amount to 4.8 and 3.1, respectively, indicating that molecular rotations are quasi-free and, therefore, quantized, i.e., H_2 is a rotationally disordered quantum solid. While exchange interactions in quantum fluids such as ^{4}He are dominant and a description in terms of Bose statistics at zero temperature is adequate [42], the condensed hydrogens constitute challenging problems since exchange forces play a minor role [43] and Boltzmann statistics apply. However, quantum effects in these "Boltzmann fluids," as inferred from the above values for λ_t and λ_r, still contribute significantly to their behavior at finite temperatures and their description requires recourse to sophisticated (and costly) computational methodologies. These will be explained in more detail in Section 40.2.3.2.2 dealing with nuclear quantum effects.

H_2's rather unfavorable thermodynamic properties have immediate consequences for its storage in sufficient quantities. Efficient storage above the critical temperature necessarily implies a reduction of enormous volumes of H_2 gas—m³/kg at room temperature. Because of its low density, hydrogen's volumetric energy density is poor compared to its mass energy density. The challenge is therefore one of bulk rather than mass. Further, any increase in storage densities requires that either work must be applied to compress H_2 gas or the temperature must be decreased below the critical temperature in order to reach bulk-like densities. Alternatively, suitable materials can be produced such that their interaction with H_2 can overcome the above physical obstacles.

40.2.2 Physisorption, Chemisorption, and No-Man's Land

The interaction of molecules with porous materials may be classified in terms of the strength of the intermolecular forces at play. In this section, we present simple physical arguments to build up a consistent picture of the hierarchy of intermolecular forces between H_2 and carbon-based substrates.

Physical adsorption (physisorption) corresponds to weak adsorbate–substrate binding forces. In this case, the physicochemical properties of the molecular adsorbate such as bond distance, polarizability, or electron-charge density remain largely unperturbed compared to the isolated-molecule case and estimates of interaction strengths can be readily made by recourse to the single-molecule data presented in Section 40.2.1.1 and the theory of intermolecular forces [44,45]. As a starting point, let us consider the binding of H_2 to an uncharged atom B. The leading term in the long-range interaction potential is caused by van der Waals (dispersive) forces. This attractive energy term, V_{vdW}, can be written as

$$V_{\text{vdW}} = -\frac{3}{2}\left[\frac{E_{H_2}E_B}{E_{H_2}+E_B}\right]\frac{\alpha_{H_2}\alpha_B}{R^6} \qquad (40.5)$$

where

  R is the distance between site B and the H_2 center of mass

  α_{H_2} and α_B are the isotropic dipole polarizabilities of H_2 (cf. Section 40.2.1.1) and B, respectively

E_{H_2} and E_B correspond to mean electronic excitation energies and, as such, they set the overall energy scale of the interaction. A commonly used approximation is to set them to the corresponding ionization potential ($IP_{H_2} = 15.4$ eV) [22]. As a first order-of-magnitude estimate, we can consider

$\alpha_B = \alpha_{H_2} = 0.76\,\text{Å}$ and $R = 3.4\,\text{Å}$, the location of the energy minimum for H_2–H_2 interactions as shown in Figure 40.2. Using Equation 40.5, these figures lead to $V_{vdW} \sim -4\,\text{meV}$, a characteristic energy scale which corresponds to cryogenic temperatures around $T \sim 40\,\text{K}$. This is a small binding energy as it involves the induction of instantaneous (dynamic) dipole moments between two well-separated charge distributions. Equation 40.5 also suggests that to increase the value of V_{vdW}, it would be necessary to increase either the polarizability α_B or the mean electronic excitation energy E_B. In practice, however, these two requirements oppose each other because larger values of α imply lower mean atomic excitation energies E. Thus, these effects will tend to cancel each other out, particularly in the typical limit of $E_{H_2} \gg E_B$ where V_{vdW} reads

$$V_{vdW} = -\frac{3}{2}E_B\frac{\alpha_{H_2}\alpha_B}{R^6} \tag{40.6}$$

For graphitic carbon, $\alpha = 1.6\,\text{Å}^3$ [46] and IP = 4.39 eV [47], leading to a value of $V_{vdW} \sim -4\,\text{meV}$ at $R = 3.4\,\text{Å}$, virtually identical to characteristic H_2–H_2 energies (cf. Figure 40.2). A further enhancement of interaction energies may be achieved either via an increase in the number of atoms interacting with H_2 at any one time or, alternatively, by activating a different physisorption mechanism. These two different approaches (as well as their multiple variations) form the basis and rationale for most research work carried to date on H_2 adsorption in carbon-based materials.

The first case can be readily illustrated by considering the potential energy resulting from the cumulative interaction of H_2 with a bulk graphitic surface given by

$$V_{vdW} = -\frac{3}{2}\left[\frac{E_{H_2}E_C}{E_{H_2} + E_C}\right]\sum_i\left[\frac{\alpha_{H_2}\alpha_C}{R_i^6}\right] \tag{40.7}$$

If this surface is sufficiently far from the H_2 molecule, the sum in Equation 40.7 may be approximated by an integral yielding [48]

$$V_{vdW} = -\frac{\pi}{4}\left[\frac{E_{H_2}E_C}{E_{H_2} + E_C}\right]\tilde{\alpha}_C\left[\frac{\alpha_{H_2}}{R_s^3}\right] \tag{40.8}$$

where
  R_s is the perpendicular distance from the molecule to the surface
  $\tilde{\alpha}_C$ is the polarizability of graphite per unit volume $\tilde{\alpha}_C = \rho\alpha_C = 0.17$

For our reference value of $R_s = 3.4\,\text{Å}$, V_{vdW} amounts to $-9\,\text{meV}$, a modest increase of about a factor of 2 over the single-site case. More importantly, the R dependence of the interaction is longer-ranged than previously, now decaying as R^{-3} as opposed to the much-steeper R^{-6} characteristic of the van der Waals interaction. Based on this mechanism, further (and quite significant) enhancements of interaction energies become possible,

particularly if the geometries involved are commensurate with characteristic intermolecular distances. To illustrate this latter effect, we follow the same procedure as that leading to Equation 40.8 for the case of an infinite hollow cylinder of radius R_s embedded in a graphitic matrix. Integration over all possible interactions leads to

$$V_{vdW} = -\frac{3\pi^2}{4}\left[\frac{E_{H_2}E_C}{E_{H_2} + E_C}\right]\tilde{\alpha}_C\left[\frac{\alpha_{H_2}}{R_s^3}\right] \tag{40.9}$$

yielding $V_{vdW} \sim -85\,\text{meV}$ for $R_s = 3.4\,\text{Å}$, that is, an order of magnitude enhancement with respect to the case of a flat surface. These limits of extreme molecular confinement will be discussed in greater detail in Section 40.3 after we take a closer look at the various topological features present in nanostructured carbons. For the time being, it suffices to mention that carbon nanostructures satisfy the above criteria by providing structural motifs in the nanometer range, thereby offering the exciting prospects of modifying the interaction with adsorbates in a fundamental and unique way.

In addition to purely geometric effects, it is also possible to alter the mechanism whereby H_2 binds to the material. To this end, let us consider the overall effect of introducing a net positive charge q at a given binding site. Recalling the single-molecule properties of H_2 discussed in Section 40.2.1.1, the two leading terms in the long-range attractive potential are due to charge-quadrupole V_{qQ} and charge-induced-dipole interactions $V_{q\alpha}$. The first of these is of the form

$$V_{qQ} = C\frac{qQ_{H_2}}{R^3}P_2(\cos\theta) \tag{40.10}$$

where
  θ is the angle between R and the H–H bond direction
  $P_2(x) = \frac{1}{2}(3x^2 - 1)$ is the second Legendre polynomial
  $C = 14.4\,\text{ÅeV}$

For $q > 0$ and $Q_{H_2} > 0$, the minimum energy is attained at $\theta = 90°$ leading to a T-shaped configuration of the H_2-charge complex. Similarly,

$$V_{q\alpha} = -C\frac{q\alpha_{H_2}}{2R^4} \tag{40.11}$$

For $q = 1$ and $R = 3.4\,\text{Å}$, $V_{qQ} \sim -24\,\text{meV}$ and $V_{q\alpha} \sim -41\,\text{meV}$. Thus, the combined effects of charge-quadrupole and charge-induced-dipole interactions can also lead to an order-of-magnitude increase of interaction energies. Moreover, as in the case of topological effects, their characteristic R-dependences are far more favorable than in the van der Waals case, and their strength is dependent on neither the polarizability α nor mean electronic excitation energy E of the substrate. Significant gains can thus be attained via an increase in charge q or a decrease in ionic radius. Hereafter, we refer to this case as "enhanced physisorption."

In addition to the energetic considerations discussed, physisorptive capacities are also dependent on available substrate area and number density of sorption sites. Whereas carbon-based materials can be tailored to provide exceptionally high porosities exceeding $1000 \, \text{m}^2/\text{g}$, energetic requirements represent a far more stringent challenge for room-temperature applications given H_2's low critical temperature ($T \sim 33 \, \text{K}$), above which capillary condensation cannot occur and uptake is limited to coverages of the order of one monolayer. It is also important to emphasize that, for the purposes of gas storage, molecular physisorption ensures that H_2 remains a distinct entity between adsorption and desorption cycles, thus guaranteeing good reversibility of the process as well as prolonged operating lifetimes. The major challenge in this case lies in the enhancement of H_2–substrate interactions to attain operation around room temperature. We also note that quantum-mechanical effects such as H_2 zero-point-energy motions have been neglected so far. Full incoporation of these subtle yet potentially important effects require state-of-the-art computational tools, described in Section 40.2.3.2.

Whereas in the physisorptive limit, H_2 remains a distinct molecular entity, disruption of the H_2 molecular-orbital structure is also possible via a variety of "chemisorption" mechanisms. For undoped graphitic materials, work to date indicates that the process is fully dissociative, that is, it leads to complete breaking of the H–H bond, a process that requires energies of around 4.5 eV and leads to the formation of chemical hydrides with binding energies well above 1 eV. For example, it is known that hydrogen molecules can react with bucky-balls at high pressures and temperatures ($P > 500 \, \text{bar}$, $T > 500 \, \text{K}$) [49,50]. Similarly, high-pressure uptake studies on nanotubes show hydrogen desorption above 400 K, indicative of a chemisorbed atomic-H layer on the outer tube walls [64]. This case presents the opposite problem to physisorption, as the extreme temperature–pressure conditions needed in H_2 uptake–release cycles severely limits its use in practical applications. In this context, we note that other target materials for hydrogen storage such as the metal hydrides share a similar problem, with binding energies in the range ~1.0 eV, leading to high desorption temperatures and poor uptake–release kinetics.

From the above discussion, it becomes clear that physisorption and chemisorption approach the H_2 containment problem from two opposing limits, starting from either too low or too high H_2 binding energies. Costly protocols to store hydrogen at low temperatures (physisorption) or at high temperatures and pressures (chemisorption) therefore become mandatory. What is then the optimal binding energy for room temperature H_2 storage? Insightful estimates may be obtained by recourse to simple thermodynamic arguments. Consider a mixture of adsorbed and gaseous H_2 held at a temperature T and pressure P. The condition of thermodynamic equilibrium between these two phases requires

$$\Delta G = \Delta H - T\Delta S = 0 \qquad (40.12)$$

where ΔG is the difference in Gibbs free energy between adsorbed and gaseous H_2. From this condition, the enthalpy change is simply given by

$$\Delta H = T\Delta S \qquad (40.13)$$

Since the most significant contribution to the change in entropy ΔS arises from the translational entropy of the gas, it might be approximated by the Sackur–Tetrode equation to read [31]

$$\Delta S \sim S_{\text{gas}}^{\text{trans}} = k_{\text{B}} \left[\frac{5}{2} + \ln \frac{k_{\text{B}}T}{P\Lambda^3} \right] \qquad (40.14)$$

where

$k_{\text{B}} = 1.3806 \times 10^{-23}$ J/K is Boltzmann's constant
Λ is the thermal de Broglie wavelength given by

$$\Lambda = \left[\frac{2\pi\hbar^2}{m_{H_2} k_{\text{B}} T} \right]^{1/2} \qquad (40.15)$$

At $T = 300 \, \text{K}$ and $P = 1 \, \text{bar}$, $\Delta S \sim 14 k_{\text{B}}$ resulting in $\Delta H = T\Delta S \sim 370 \, \text{meV}$. Whereas ΔH contains an explicit (linear) dependence on temperature, its pressure dependence is far weaker, as shown in Equation 40.14. Consequently, a significant increase in gas pressure to, say, $P = 100 \, \text{bar}$ results in a relatively modest decrease in the required energies ($\Delta H \sim 250 \, \text{meV}$). As a consistency check, Equations 40.13 through 40.15 predict $\Delta H \sim 40$–70 meV at cryogenic temperatures $T \sim 50$–80 K and $P = 1$–100 bar, in good agreement with best current estimates for H_2 interaction energies with polycylic aromatic hydrocarbons, such as benzene and coronene, and flat graphite surfaces [51]. These thermodynamic considerations are not only insightful for the uninitiated but are also in line with the currently accepted range of 200–500 meV defining the "no-man's land" region of binding energies needed for room-temperature hydrogen storage applications [52].

Both geometric confinement and enhanced physisorption represent a means of approaching these target binding energies by offering order-of-magnitude enhancements. Another possibility closer to the chemisorption limit is afforded via Kubas-type binding schemes, first observed in d-metal complexes in the early 1980s [53]. Chemical bonding in these species does not fall into any of the standard schemes (e.g., ionic, covalent, etc.). Rather, it involves charge donation from the highest-occupied H_2 molecular (bonding) orbital to empty d-metal orbitals followed by subsequent back-donation from filled metal d-orbitals into the lowest unoccupied ligand (antibonding) orbital [54,55]. The net result from this redistribution of electron density between bonding and antibonding H_2 orbitals results in a significant weakening of the H–H bond. In the limit of strong back-donation from an electron-rich metal, bond cleavage to form a dihydride species results. The H–H distance r_{HH} is a convenient molecular parameter to quantify these chemical changes, and it is experimentally accessible either via neutron scattering [56] or nuclear magnetic resonance (NMR) techniques [57]. Different transition metals and peripheral ligands can be used to tune the chemical interaction between H_2 and the metal center. Of the several hundred examples reported to date, the majority are characterized by $r_{HH} < 1$ Å, in contrast to conventional dihydride and polyhydride complexes with $r_{HH} > 1.5$ Å. A quantitative description of these "dihydrogen complexes" still represents a challenge to theory because of the need for an accurate description of the electronic structure of transition-metal centers. The use of these novel chemical bonding concepts to engineer new carbon-based materials is still at a relatively early stage of development and it is to be noted that the advantages of enhanced binding energies must always be balanced against an increase in substrate density caused by the introduction of heavy transition metals. In this context, theoretical calculations have suggested that Kubas-like concepts can be also extended to nontransition metals such as Li, Be, and Mg in highly ionic environments [58].

40.2.3 Tools

40.2.3.1 Experimental: Bulk vs Microscopic Probes

The most straightforward way to ascertain the storage capacity of a target material is to measure the uptake of H_2 gas at a given temperature and pressure. The two most common methods are based on either weight-based (gravimetric) or volume-based (volumetric) techniques [59]. In addition, electrochemical procedures have also been developed for the storage and subsequent measurement of hydrogen capacities, but the results tend to yield systematically larger capacities owing to a different storage mechanism and will not be discussed any further here (see, e.g., Ref. [15]). Gravimetric methods are the most direct measurement of uptake. As a consequence, they tend to be less susceptible to hard-to-control systematic errors inherent to volumetric techniques, where a measured volume of H_2 gas (normally at standard temperature/pressure conditions) needs to be subsequently converted to mass, a task which might not be easy to achieve in practice if absolute values are required. Common sources of error include: pressure changes arising from temperature fluctuations around ambient, temperature changes following gas compression, and leaks in the gas manifold. Tibbetts et al. [60] have analyzed in detail these common sources of experimental error, casting serious doubts on many of the claims for gravimetric hydrogen uptake in carbon-based materials greater than 1 wt % around room temperature. Recent comparative studies on nanotubes over the temperature range 77–300 K [61] infer that pores up to 5–7 Å can absorb H_2 at ambient conditions and ultimate sorption capacities depend on both available surface area and adsorption energy of available sites.

The pressure dependence of H_2 uptake in nanoporous carbons at a fixed temperature typically follows the predictions of the Langmuir model [65,66], also denoted "Type-I" isotherm in the Brunauer–Deming–Deming–Teller (BDDT) classification scheme [69,70]. This behavior is characteristic of adsorbents with a predominantly microporous structure where only monolayer coverages can be attained above the critical temperature of H_2 ($T = 33$ K). Qualitative deviations from the generic Type-I behavior described below should always be the reason of caution as they imply either a very different sorption mechanism or experimental errors unaccounted for in the determination of uptake.

Within the remit of the Langmuir model, molecular adsorption is treated as an equilibrium process between free molecules in the gas phase, a fixed number of adsorption sites on the surface (as provided by a given amount of substrate), and an adsorbate phase [71]. At thermodynamic equilibrium, the fractional number of occupied adsorption sites θ at fixed temperature T is given by

$$\theta = \frac{bP}{1+bP} \qquad (40.16)$$

where

P is the gas pressure

The equilibrium constant b is temperature dependent

$\theta = n/n_{\max}$ where n is the amount adsorbed (e.g., mol/g) and $n_{\max}$ is the adsorption capacity of the material. The validity of the Langmuir model is ascertained experimentally by recasting Equation 40.16 in linear form, namely,

$$\frac{P}{n} = \frac{1}{bn_{\max}} + \frac{P}{n_{\max}} \qquad (40.17)$$

A linear dependence of P/n vs P in experimental isotherm data is typically taken as sufficient justification for the validity of the Langmuir model. Deviations from this behavior at high pressures arise from the increasing importance of adsorbate-adsorbate interactions. In thermodynamic terms, b is related to the Gibbs free energy and hence to the enthalpy change for the process via $b \propto \exp[-\Delta H_{ads}/RT]$, where $\Delta H_{ads} = (E_{des} - E_{ads})$, and $E_{des}(E_{ads})$ corresponds to the activation energy for the desorption(adsorption) step; we also note that b is only a constant if the enthalpy of adsorption is independent of coverage, a major assumption of the Langmuir model. Qualitatively speaking, Langmuir isotherms follow a linear behavior of the form $\theta = bP$ at low pressures where the number of adsorption sites greatly exceeds the available adsorbant, but saturate beyond a given pressure due to an increasing occupancy of surface sites, as shown in Figure 40.3a. Generalization of the Langmuir isotherm to incorporate the effects of multilayer adsorption follows the so-called BET prescription of Brunauer et al. [67].

Both temperature and pressure strongly influence H_2 uptake. At a given pressure, the extent of adsorption is determined by the value of b which, in turn, it is strongly dependent on both temperature and adsorption enthalpy. Thus, higher values of b can be achieved by either decreasing the system temperature or increasing the strength of the interaction via ΔH_{ads}. Given the above, it is also possible to determine ΔH_{ads} from a study of the pressure–temperature dependence of gas uptake by

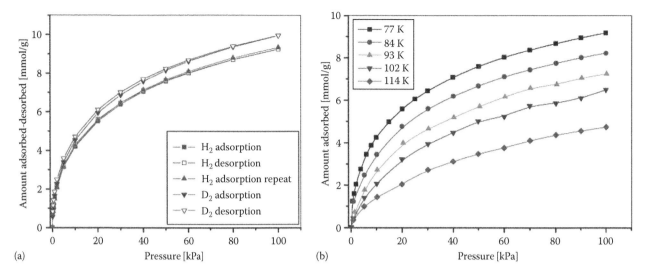

FIGURE 40.3 Hydrogen adsorption isotherms on a porous activated carbon sample: (a) H_2 and D_2 uptake at $T = 77\,K$; (b) H_2 temperature series over the range $T = 77$–$114\,K$ (1 bar = 100 kPa). (Reproduced from Zhao, X.B. et al., *J. Phys. Chem. B*, 109, 8880, 2005. With permission.)

recourse to a Clausius-Clapeyron-type relation, which at constant coverage reads

$$\frac{\partial \ln P}{\partial T^{-1}} = \frac{\Delta H_{ads}}{R} \qquad (40.18)$$

where a negative slope is indicative of an energetically favored adsorption process ($\Delta H_{ads} < 0$). Figure 40.3b shows a representative example of the temperature dependence of H_2 adsorption on a typical porous carbon, leading to an isosteric heat of adsorption of ~41 meV. It is important to note that the value obtained for the adsorption enthalpy corresponds to a given gas uptake, but the procedure can be repeated for different surface coverages enabling a determination of adsorption enthalpies over a whole range of coverages. This feature has been, for example, used with success in order to distinguish adsorption sites in carbon nanohorns [62]. H_2 adsorption in nanoporous carbons shows a linear increase up to pressures of several bar [63]. Physisorption on carbon nanotubes at cryogenic temperatures and pressures of ~100 bar is capable of attaining up to ~7 wt% H_2 uptake but this value dramatically falls by an order of magnitude at room temperature [64].

While uptake experiments provide a relatively straightforward means of characterizing new target materials, they offer limited insight into the nature, relative abundance, or strength of H_2 binding sites. This deficiency of bulk uptake measurements therefore calls for microscopic physical probes capable of unraveling the details of the H_2 potential energy landscape. However, the spectroscopic detection of ground-state H_2 is a difficult task. To appreciate the extent of this problem, one may consider the possibility of using photon-based probes. The lack of a permanent dipole moment in H_2 does not permit the use of conventional infrared spectroscopy via electric-dipole-allowed transitions. Instead, very weak infrared quadrupole [23,72]) and Raman [73] transitions become mandatory. A slightly more favorable case is

that of the of the HD isotopomer but its dipole moment is still so small (~10^{-4} Debye) that only absorption experiments using very long path lengths (tens of meters) and multipass spectrometers have been possible [74]. The situation is marginally better for the case of electronic spectroscopy using visible and ultraviolet light sources because all excited-state transitions to bound electronic states lie in the hard-to-reach vacuum ultraviolet ($\lambda < 200\,nm$) [75]. To these intrinsic limitations, one needs to add the presence of a substrate in any H_2 uptake experiment; its contribution to the total signal at these short wavelengths will overwhelm the spectral response, making the isolation of the H_2 signal a very cumbersome task. In spite of these difficulties, Raman scattering of H_2, D_2, and HD on nanotubes at pressures of a few bar have been performed recently [76,77]. The small spectral shifts observed in the rotational spectrum in the range 12–75 meV were interpreted as conclusive evidence for the physisorptive character of the adsorption process. Similarly, ^{1}H NMR has also been utilized by some authors to study H_2 adsorption in nanotubes [78–81]. NMR spectral shifts were found by Yu et al. [79] to be inversely proportional to temperature, a signature of the presence of metal catalysts in their specimens. More recently, Pietraß and Shen have found evidence for trapping of H_2 in the interior of carbon nanotubes at high pressures.

The above difficulties are circumvented if one considers the use of thermal neutrons as a structural and spectroscopic probe. The interaction of neutrons with matter is mediated by the strong nuclear force, it is therefore element-specific, and insensitive to the surrounding electronic environment. Further, there exists an inherent sensitivity to hydrogen over all other elements, a feature which can be readily exploited in our particular case. As shown in Figure 40.4, two types of scattering processes are to be distinguished. First, coherent neutron scattering carries information on interparticle correlations (e.g., Bragg peaks), just as in conventional x-ray diffraction experiments [84]. Most elements, including hydrogen, have similar coherent neutron

Nuclide	σ_{coh}	σ_{inc}	Nuclide	σ_{coh}	σ_{inc}
^1H	1.8	80.2	V	0.02	5.0
^2H	5.6	2.0	Fe	11.5	0.4
C	5.6	0.0	Co	1.0	5.2
O	4.2	0.0	Cu	7.5	0.5
Al	1.5	0.0	^{36}Ar	24.9	0.0

FIGURE 40.4 Selected list of coherent and incoherent neutron scattering cross sections.

scattering cross sections lying in the range 1–10 barn (1 barn = 1×10^{-28} m²). This behavior is to be contrasted with x-rays whose scattering probability is proportional to the number of electrons in a given atom, thus becoming increasingly insensitive to the lighter elements. Second, incoherent neutron scattering gives access to single-molecule dynamical properties including translational diffusion (mobility) as well as quantized molecular motions (vibrations and rotations) and their geometries [82–84]. This information is accessible in spectroscopic experiments that measure both the scattering angle as well as the energy transfer between neutron and sample. It is a remarkable (and very convenient) fact that the incoherent cross section for hydrogen is the largest of all elements and that of carbon approaches zero. This means that in a material containing both hydrogen and carbon, neutrons will mostly see the protons, with sensitivities reaching micromolar levels with state-of-the-art instrumentation.

Neutron spectroscopy is conceptually identical to inelastic light scattering (e.g., dynamic light scattering or Raman spectroscopy) [85]. A schematic diagram of the high-resolution neutron spectrometer IRIS [86] is shown in Figure 40.5. To attain high spectral resolution, energy transfers are analyzed with graphite and mica neutron mirrors after scattering has taken place, the so-called inverted-geometry configuration. The energy resolution varies from few µeV for elastic scattering to ~100–150 µeV at energy transfers above 10 meV. In the context of H_2 dynamics, the range of interest goes from so-called quasielastic events with characteristic energy transfers below 1 meV, up to tens of meV. In the former case, quasielastic scattering is sensitive to adsorbate mobility and diffusion. At higher energy transfers, it becomes possible to access the position and width of rotational transitions starting at $2B_{rot} = 14.7$ meV in the free-rotor limit. This transition is optically forbidden because it involves a nuclear-spin flip between ground-state p-H_2 ($J = 0$) and rotationally excited o-H_2 ($J = 1$) (cf. Section 40.2.1). Figure 40.6 shows the low-energy inelastic neutron scattering (INS) spectrum of solid p-H_2. The most intense feature corresponds to the aforementioned *para → ortho* rotational transition and its width is resolution-limited. Additional features at 7 and 20 meV are fundamental and combination bands corresponding to the most intense lattice vibrational frequencies in the solid and the simultaneous excitation of rotational and lattice vibrational modes, respectively [87]. Most noteworthy is the enhancement of the scattered intensity by over an order of magnitude above 15 meV.

In addition to energy-resolved spectra, the finite mass of the neutron also gives access to the geometry of motions via the momentum-transfer dependence of the scattering process. To illustrate this unique feature of neutron scattering, let us consider a neutron with an energy of ~25 meV ($T = 300$ K). Its associated de Broglie wavelength is $\lambda \sim 1.8$ Å, corresponding to a linear momentum $k = 2\pi/\lambda = 3.5$ Å^{-1}. Elastic scattering at this incident wavelength results in a momentum-transfer range $Q = k\sqrt{2(1-\cos\theta)}$ where θ is the scattering angle 0° for forward scattering (no particle deflection), and 180° for backward scattering (maximal particle deflection). The available

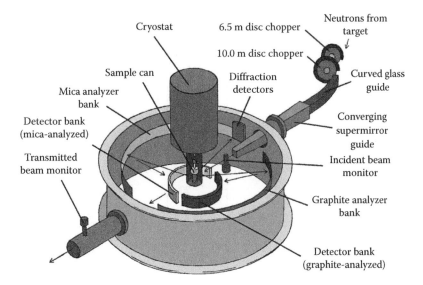

FIGURE 40.5 Schematic diagram of the high-resolution neutron spectrometer IRIS, located at the ISIS Facility, Rutherford Appleton Laboratory, United Kingdom.

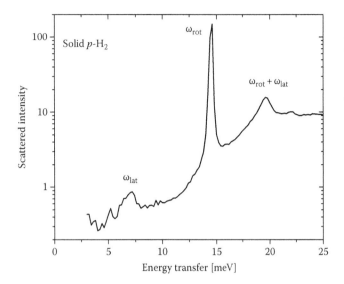

FIGURE 40.6 INS spectrum of solid p-H_2 at $T = 12\,K$ [93,94], measured on the TOSCA [95] inelastic neutron spectrometer (ISIS Facility, Rutherford Appleton Laboratory, United Kingdom). Note the logarithmic intensity scale.

Q range is therefore $Q = 0–7\,Å^{-1}$, corresponding to characteristic distances around $d = 2\pi/Q_{max} \sim 1\,Å$. Figure 40.7 illustrates the unique merits of quasielastic neutron scattering (QENS) spectroscopy for the case of liquid H_2. The Q-dependence of the spectral linewidth follows the expectations of a jump-diffusion model, which considers translational mass diffusion as resulting from a series of stochastic jumps [88]. Within the remit of such a model, QENS energy widths Γ follow a momentum-transfer dependence of the form

$$\Gamma(Q) = \frac{D_T Q^2}{1 + (D_T Q^2 / E_0)}, \qquad (40.19)$$

allowing a direct measurement of the translational diffusion coefficient D_T and an associated residence time $\tau = \hbar/E_0$. From these data, $D_T = 0.43 \pm 0.01\,Å^2/ps$ and $\tau = 0.20 \pm 0.02\,ps$, in good agreement with literature values [89,90]. Additional insight concerning the *geometry* of particle motions is also given by the analysis of the Q-dependence of the QENS intensities. For the bulk liquid, the incoherent intensity of the QENS signal arises entirely from o-H_2 scattering and it is given by an expression that describes a freely rotating homonuclear diatom [91]. Mathematically,

$$I_{rot}(Q) = j_0^2\left(\frac{Qr_{HH}}{2}\right) + 2j_2^2\left(\frac{Qr_{HH}}{2}\right), \qquad (40.20)$$

which is given in terms of the internuclear distance of H_2 ($r_{HH} = 0.741\,Å$); j_n stands for a spherical Bessel function of order n (see inset in Figure 40.7b).

The exploitation of both inelastic and quasielastic neutron scattering in the context of H_2 adsorption in carbon nanostructures will be illustrated in more detail in the two case studies presented in Sections 40.3.2 and 40.3.3.

40.2.3.2 Computer Modeling

Computer simulations have become an integral part of our attempts to unravel the details of H_2 adsorption in storage materials. The enormous calculational power afforded by modern computers can be used to make definite predictions about the adsorption behavior of a given material, compare and benchmark different models against experimental data, as well as interrogate the details of a particular model beyond what is possible in the laboratory. This last feature is a particularly useful one as it affords us with sophisticated tools as opposed to heuristic and phenomenological models of limited general validity. The essence of any "computational experiment" is

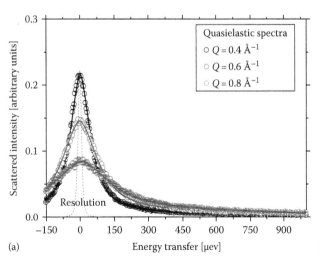

(a)

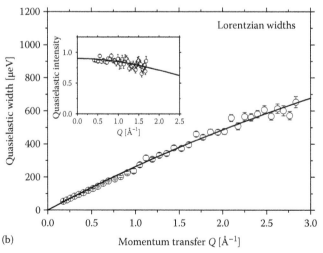

(b)

FIGURE 40.7 (a) QENS spectra of liquid n-H_2 at $T = 15\,K$ and data fits using a Lorentzian line shape. (b) Energy widths as a function of Q^2, from which a translational diffusion coefficient can be directly obtained by recourse to Equation 40.19. The inset shows the associated form factor compared to theoretical expectations (cf. Equation 40.20). (Adapted from Fernandez-Alonso, F. et al., *Phys. Rev. Lett.*, 98, 215503, 2007.)

relatively simple to understand and consists of the following generic steps:

- Create a virtual description of your system in a computer. This could be an ensemble of H_2 molecules at chemical equilibrium with the substrate at a temperature T, where interactions are described by intermolecular potential functions of the type described in Section 40.2.
- Let the system evolve or converge by solving the relevant many-body equations. In this case, we might choose to track the temporal evolution of our H_2 ensemble by solving Newton's equations of motion, if time-dependent properties are required (molecular dynamics [MD] method). Alternatively, we might want to extract thermodynamic information such as coverage or enthalpy of adsorption, a task which requires Monte-Carlo (MC) methods. For a detailed account of these two methodologies, see Ref. [96].
- Compute observables and, preferably, interrogate their validity and predictive power via a comparison with other simulations or (ideally) experimental data.

The above programme is quite generic and it applies to the two major categories of computer modeling work of relevance in our context, namely: (a) the statistical mechanics of H_2 adsorption in heterogeneous media, mostly based on so-called semi-empirical approaches; and (b) first-principles computations using wave-function-based or density-functional-theory electronic structure methods. We discuss these two in turn, noting that the term "multiscale modeling" has been coined to describe the sequential use of these two approaches.

40.2.3.2.1 Semiempirical Simulations

Molecular adsorption on solid substrates is an inherently heterogeneous process characterized by the coexistence of several phases in the same system. In such a case, the grand canonical (GC) ensemble (constant chemical potential μ, volume V, and temperature T) is the most natural choice for the calculation of thermodynamic quantities, including adsorbate fraction, density profiles, configurational energies, and isosteric heats of adsorption. Since at chemical equilibrium μ and T are the same for bulk and adsorbed phases, knowledge of the former enables the gas pressure to be calculated from an equation of state. Likewise, the ensemble average of the number of molecules in the system (equal to the amount adsorbed) can be obtained directly from a GCMC simulation where random particle creation, deletion, and displacement ensure the proper generation of thermodynamic microstates. Moreover, a series of GCMC calculations at different values of μ gives simulated adsorption isotherms. Isosteric heats of adsorption are given by the following fluctuation formula:

$$q_{st} = k_B T - \frac{\langle NU \rangle - \langle N \rangle \langle U \rangle}{\langle N^2 \rangle - \langle N \rangle \langle N \rangle} \quad (40.21)$$

with N and U being the number of particles and total energy of a given configuration, and $\langle \rangle$ denotes a configurational average. The method is discussed in detail by Nicholson and Parsonage [97].

If time-dependent properties such as mean-square displacements and diffusion coefficients are of interest, MD is the method of choice. In MD, the forces and torques acting on each adsorbed molecule are propagated in time using Newton's equations of motion. In this case, the microcanonical ensemble (N,V,E) is the most convenient choice as it conserves both the total number of molecules and energy of the system. Alternative methods based on the use of thermostats are available for running in the canonical (N,V,T) ensemble. Calculation of μ, however, requires running simulations at several N and, therefore, the MD method is less convenient than GCMC for the calculation of adsorption isotherms.

As a first approximation, the system is regarded as a classical one. H_2–H_2 pairwise interactions are modeled using intermolecular potentials such as that of Lennard-Jones as shown by Equation 40.3 in Section 40.2.1.2. This potential form can be used to model the H_2 molecule as a single Lennard-Jones site (spherical model) or as a two-site model where each individual H atom in the molecule interacts independently with the surroundings. For the spherical model, popular parameters are $\epsilon_H = 2.95$ meV and $\sigma_H = 2.96$ Å [98]. The interaction between H_2 and the substrate also requires a choice of potential function. One such choice is the "10–4–3" Steele potential function [99] where the model pore is assumed to be bounded by infinite stacked planes of graphitic sheets separated by a distance Δ. This potential only depends on the distance Z to a given graphitic surface and is designed to model dispersion forces between adsorbate and adsorbent. It is of the form

$$V(Z) = 2\pi \rho_s \epsilon_s \sigma_s^2 \Delta \times \left[\frac{2}{5} \left(\frac{\sigma_s}{Z} \right)^6 - \left(\frac{\sigma_s}{Z} \right)^4 - \frac{\sigma_s^4}{3\Delta(0.61\Delta + Z)^3} \right]$$

$$(40.22)$$

where

- $\rho_s = 0.114$ Å^{-3} is the number of carbon atoms per unit volume in graphite
- ϵ_s and σ_s are analogous energy and hard-sphere parameters as those already defined for H_2–H_2 interactions.

Their values are obtained by recourse to the Lorentz–Berthelot rules [100]. $\epsilon_s = \sqrt{\epsilon_H \epsilon_C}$ and $\sigma_s = \frac{1}{2}(\sigma_H + \sigma_C)$ where $\epsilon_C = 2.41$ meV and $\sigma_C = 3.40$ Å are Lennard-Jones parameters for graphite. Little work, however, has been reported on the applicability of these combination rules for the case of hydrogen on carbon [101]. Moving beyond idealized models for the carbon substrate necessarily involves the inclusion of the inherent atomic-scale structure and corrugation of real materials, but only at the expense of great computational cost. Sophisticated procedures based on reverse MC methods to generate realistic 3D structural models from x-ray and neutron-diffraction data have been recently implemented to this end [114].

The recipe shown above forms the basis of much of the simulation work performed to date. Maximal H_2 capacities in model graphitic pores are predicted to be ca. 1.5 wt % at room temperature irrespective of whether H_2 is treated as a sphere or a dumb-bell [102]. This figure is to be contrasted with earlier experimental observations of 40 wt % in herringbone graphitic nanofibers at 110 bar and 298 K [103,104]. Adsorption in carbon-nanotube geometries has also been investigated using curvature-dependent potential functions [105–107]. H_2 adsorption is found to take place preferentially outside the tubes (exohedral sites). The average adsorption energy increases with decreasing tube diameter over the range 4–12 Å, reaching maximal values as high as 150–250 meV, promising values for vehicular onboard applications. Gas loading is also accompanied by a substantial lattice expansion of the substrate lattice [108] and H–H bond elongation. While encouraging, these predictions still await experimental verification.

40.2.3.2.2 Quantum Effects

At sufficiently low temperatures, H_2 and its isotopomers can be no longer treated classically owing to their low mass, e.g., their associated de Broglie wavelength for nuclear motions become comparable to intermolecular distances (cf. Section 40.2.1.2). An adequate description of these quantum-mechanical effects requires recourse to the so-called path-integral formalism [109] where a quantum particle of mass m is characterized by a spread in position proportional to $\sqrt{1/mk_BT}$ due to the Heisenberg uncertainty principle. The interested reader may consult Refs. [110,160]) for a succinct exposition of this formalism in the context of hydrogen storage.

For H_2 adsorbed in carbon-based materials, Wang and Johnson [110] have shown that nuclear quantum effects for H_2 in nanotubes and slit pores are significant at cryogenic temperatures ($T < 100$ K) and can persist up to room temperature, particularly for adsorption at interstitial sites where nuclear delocalization effects account for a 40% drop in uptake. H_2 zero-point-energy effects are also responsible for the phenomenon of *quantum sieving* in carbon nanotubes and the interstices of tube bundles, where heavy isotopes are preferentially adsorbed over light isotopes owing to the larger effective size of the latter [111]. For a harmonically bound molecule in the limit of zero applied pressure, the isotope selectivity $S_{A/B}$ between species A and B is given by

$$S_{A/B} = \left(\frac{m_A}{m_B}\right)\exp\left[-\frac{E_A}{k_BT}\left(1-\sqrt{\frac{m_A}{m_B}}\right)\right] \qquad (40.23)$$

where E_A is the ground-state energy for isotope A. At low temperatures $T \sim 20$ K, it is found that nanotube radii below 10 Å can achieve $S_{T2/H2} > 1000$. Trasca et al. [112] have applied a similar approach to investigate *ortho-para* H_2 selectivity in the interstices and grooves of nanotube bundles and finds selectivities for the *ortho* species exceeding unity up to 100 K. More recently, Garberoglio et al. [113] have analyzed in detail how different

interaction potentials affect sieving phenomena and find these to be very sensitive to the Lennard-Jones diameter σ. These authors also explore the contribution of rotational degrees of freedom to isotope selectivity and find enhancement factors of up to 10^{3-4}.

A more severe limitation to either the use of idealized substrate geometries as presented earlier, or the inclusion of nuclear quantum-mechanical effects, relates to the general validity and transferability of the underlying interatomic potentials. An accurate description of atomic and molecular interactions using parametrizable potential functions and combination rules are typically the result of long and laborious benchmarking exercises using available (and usually sparse) experimental data and, therefore, agreement with experimental observation in a particular problem renders them of limited applicability. A good example of the limitations inherent to the use of empirical potentials is given by the theoretical work of Hamel and Côté [116] on the H_2–benzene complex. Atom–atom model potentials like the popular WS77 parametrization [115] were found to yield H_2–benzene intermolecular distances close to the value of 3.41 Å obtained from first-principles calculations (see below). However, WS77 predicts a ground-state geometry where H_2 sits parallel to the benzene plane, in stark disagreement with more accurate theoretical work.

40.2.3.2.3 First-Principles (Ab Initio) Modeling

The properties of materials are ultimately determined by the interaction of electrons and nuclei. As such, an adequate description of these requires a full quantum-mechanical treatment of the problem via the solution of the many-body Schrödinger equation. For a system of N_n nuclei at positions $\{R_{N_i}\}$ and N_e electrons at $\{r_{ej}\}$ where $i = 1...N_N$ and $j = 1...N_e$ it reads [117,118]

$$\hat{H}\Psi\left(\{R_{N_i}\},\{r_{ej}\}\right) = E\,\Psi\left(\{R_{N_i}\},\{r_{ej}\}\right) \qquad (40.24)$$

where

$\hat{H}$ is the Hamiltonian operator

E is the total energy

Ψ is the total wavefunction of the system, a complicated multi-dimensional function depending on the positions and momenta of all particles

$\hat{H}$ is given by

$$\hat{H} = \hat{T}_N + \hat{T}_e + \hat{V}_{NN} + \hat{V}_{Ne} + \hat{V}_{ee} \qquad (40.25)$$

and contains both nuclear and electronic kinetic energy terms (T), as well as potential energy contributions (V) arising from Coulomb interactions: nuclear–nuclear (NN, repulsive), nuclear–electron (Ne, attractive), and electron–electron (ee, repulsive). If a solution to Equation 40.24 can be found, the calculation of any observable follows. The only input required in the so-called ab initio (in Latin "from the beginning") approach is the composition of the material as well as a recipe to solve the Schrödinger equation. Numerical approaches to obtain approximate solutions of the so-called electronic-structure problem fall into two broad categories: wavefunction-based (WFB) [119] vs. density-functional-theory (DFT) [120] methods. In either case, solutions are most often than not within the Born–Oppenheimer approximation, that is, the total electronic wavefunction is obtained for a static configuration of the nuclei ("frozen-nuclei" approximation). Given this separation of electronic and nuclear motions, the latter may either be treated classically or quantum-mechanically in an analogous manner to the semiempirical methods discussed previously. WFB and DFT methods are discussed in some detail below, along with a few illustrations of their use in the context of H_2 storage.

The WFB approach [119] attempts a direct solution of Equation 40.24 by computing an approximate electronic wavefunction Ψ using a basis set of atom-centered electron orbitals. The starting point is to replace the exact Hamiltonian $\hat{H}$ by an effective one where each electron moves in an average mean field originating for all other charges. This simplification leads to a set of atom-centered, one-electron orbitals obeying a system of coupled differential equations, whose solution must be obtained iteratively (self-consistent-field Hartree–Fock method). Most errors associated with the Hartree–Fock method are caused by the neglect of electron correlation, for example, it is unable to describe covalent bonding in seemingly simple molecules such as F_2, it fails at describing free electrons as those found in the metallic state, and long-range dispersion forces are missing. Better approximations falling under the label of "correlated-electron methods" are necessary to reach chemical accuracy typically defined as 0.1 eV or better. Inclusion of electron correlation effects requires the implementation of Moller-Pleset perturbation theory (MP), configuration-interaction (CI), or coupled-cluster (CC) methods. However, their use comes at a very high computational cost, particularly as the number of atoms is increased – $\sim N^5$ where N is the number of atoms. Accurate studies are therefore restricted to small molecules or clusters. In spite of these limitations, the value of these calculations cannot be underestimated as they can be quite insightful in the rational design of hydrogen storage materials. Lochan and Head-Gordon [123] have examined H_2 binding affinities for a comprehensive set of molecular ligands, metals, and metal–ligand complexes using uncorrelated Hartree–Fock as well as correlated methods (second-order MP, MP2, and CC singles and doubles, or CCSD). Use of the latter allows a reasonable description of not only electrostatic and orbital interactions but also of the much weaker and subtle dispersion forces. This work emphasizes the combined role of electrostatic, inductive, and covalent charge-transfer effects in order to achieve the ideal H_2 binding energy range for room-temperature storage. Both positively and negatively charged ions provide suitable H_2 binding sites, including lightweight species such as F^- (binding energy 0.35 eV) and Li^+ (0.25 eV). Highly charged ions like Mg^{2+} bind H_2 far too strongly (0.95 eV) but interaction strengths can also be tuned down with reasonable control upon complexation with molecular ligands. Correlated wavefunction-based methods have also been employed to determine accurate H_2–graphite interaction potentials (124). These first-principles

results have also been used to create H_2–graphite Lennard-Jones-type potentials (cf. Equation 40.3), for subsequent use in quantum-mechanical simulations of H_2 adsorption between two graphene sheets kept at a distance d. For $d = 6$–7 Å, very favorable adsorption free energies result in a dramatic increase of the internal H_2 pressure. Under these conditions, it is possible to start approaching current volumetric DOE targets for volumetric storage, with H_2 densities of 62 kg/m³ at pressures of ~100 bar. These interlayer distances between graphene layers are not present in pure graphite ($d = 3.35$ Å) but it is possible to engineer them via the use of intercalation compounds.

As opposed to WFB electronic-structure calculations, DFT [120] gives up any attempt to calculate the electronic wavefunction Ψ and focuses on the direct determination of the electron density n. Its theoretical foundations are based on the premise that the total ground-state energy is uniquely defined by n [121]. The mathematical machinery to compute ground-state energies, the Kohn–Sham formalism [122], is formally identical to the Hartree–Fock method presented earlier in the context of WFB approaches. It is for this reason that the development of DFT software packages into mature modeling tools has been remarkably fast following its original inception in the mid 1960s. As described above, it is important to emphasize that DFT only applies to the electronic ground state of the system. This limitation is not present in WFB methods and can also be overcome within DFT using time-dependent methodologies to calculate electronically excited states. To solve for n, the Kohn–Sham formalism introduces a fictitious system of N_e noninteracting electrons where the total energy may be written as

$$E[n] = T[n] + E_H[n] + E_{ext}[n] + E_{xc}[n] \qquad (40.26)$$

The precise functional forms of all terms on the right-hand side except the exchange-correlation functional $E_{xc}[n]$ are known. This unknown term is at the heart of the DFT method as it embodies all errors made if one assumes that electron–electron interactions are classical (no exchange and correlation allowed) and kinetic energies correspond to the noninteracting case. The two most celebrated approximations to E_{xc} are the local-density-approximation (LDA) and the generalized-gradient approximations (GGA), both of widespread use in the literature. LDA is the simplest approximation possible whereby at every point in space the density is that of a uniform electron gas and $E_{xc}[n]$ becomes a function of n alone. To account for spatial variations of n, GGA and meta-GGA introduce first and higher-order derivatives of the density in the description of $E_{xc}[n]$ in much the same way Taylor-series expansions are used in differential calculus to approximate functions. Extensive computational experimentation over the past 30 years has shown that the LDA works surprisingly well, and provides an excellent description of covalent, metallic, and ionic bonds. Hydrogen-bonded systems or van der Waals interactions are poorly described by the LDA, and binding energies tend to be overestimated by as much as 50%. This last deficiency is reduced significantly by recourse to the GGA or so-called hybrid functionals such as B3LYP. DFT methods offer excellent scaling with system size, particularly in the case of periodic systems (crystalline solids) where plane-waves are the natural basis-set to use, allowing electronic structure calculations for much larger systems than WFB techniques. Their use to model realistic materials has exploded in the past decade and, therefore, there already exists an extensive data base to compare and benchmark new calculations.

While DFT is amply recognized as an efficient and relatively unbiased tool to model materials properties, there are also serious deficiencies in the treatment of graphitic materials because van der Waals interactions are not explicitly included in time-independent DFT. Recent calculations on graphite by Ooi et al. [125] show that both the LDA and GGA give a good description of planar bonding and electronic band structure, while interplanar binding between graphene planes is completely missing within the GGA approximation. The LDA, however, does reproduce interplanar bonding of the right magnitude, leading to a calculated *c*-axis spacing of 6.58 Å, to be compared with an experimental value of 6.71 Å, although it has to be remarked that in this particular case the LDA might be getting the right answer for the wrong reasons.

To date, extensive DFT calculations have been performed to calculate H_2 binding energies to carbon nanotubes, where the behavior of the exchange functional plays an important role (for detailed reviews, see Refs. [126,127]). Mpourmpakis et al. [128] have investigated the effects of nanotube curvature and chirality on H_2 adsorption using DFT and GCMC simulations. Binding energies using the B3LYP exchange functional are found to be similar to those using CCSD WFB methods and derived Langmuir isotherms agreed with existing experimental data. The difficulties inherent in dealing with purely graphitic materials are greatly relaxed in doped graphite-based structures where charging of the graphene planes activates enhanced physisorption mechanisms leading to stronger interplanar binding forces than those afforded by dispersive interactions alone (cf. Section 40.2.2). This behavior has been found in carbon lithium-doped nanotubes [129–131], alkali- [132,133] and alkaline-earth-doped [134] nanoporous carbons as well as charged fullerenes [135]. Our second case study in Section 40.3.3 will deal with this more favorable case in more detail.

40.3 Hydrogen Adsorption in Carbon Nanostructures

40.3.1 Exploring the Consequences of Rolling and Bending Graphite Sheets

Since the pioneering molecular-beam experiments by Kroto et al. [136] leading to the discovery of Buckminsterfullerene or "buckyball" (C_{60}), it is known that natural carbon can exist in other allotropic varieties than those exhibited by diamond or graphite. In fullerenes such as the nearly spherical C_{60}, carbon atoms are arranged into 12 pentagonal faces and two or more

hexagonal or heptagonal faces where all carbon–carbon bonds are sp^2 hybridized, as in bulk graphite. In contrast to graphite, however, the presence of nonhexagonal topological motifs allows for the formation of nonplanar structures. This added flexibility leads to a rich variety of fullerene topologies including spheres, ellipsoids, and tubes. The discovery of such highly unexpected class of carbonous materials stimulated vigorous research efforts into the preparation of unusual forms of carbon, mostly carried out under strong nonequilibrium conditions such as arc discharges, plasmas, and high-power laser ablation. A key result was the production of tubular structures via arc-discharge synthetic routes [137–139] almost 20 years ago, although the discovery may have happened well before [140]. The resulting material was characterized as carbonous tubules formed by rolling graphene sheets. As typical tube diameters were in the nanometer range, they became known as "carbon nanotubes."

The structure of nanotubes and derived nanostructures is best understood by exploring the consequences of rolling and bending isolated graphene, single planar sheets of sp^2-bonded carbon and also the main constituent of graphite. Depending upon the way such a graphene lattice is rolled up, a whole range of nanotube topologies can be achieved [141]. To visualize this process, we consider how to roll a rectangular sheet of sides *A* and *B* inscribed onto a hexagonal graphene sheet, as shown in Figure 40.8. We inscribe this rectangle on the graphene lattice by making one of its vertices coincide with the origin of a reference system *O*, whose origin is placed on top of a given sp^2 carbon. The reference frame is thus uniquely defined by two unit vectors joining the apexes of two adjacent rings ($\mathbf{a}_1, \mathbf{a}_2$) defining a planar Bravais lattice. Depending upon the way we cut the honeycomb plane, an angle $0 < \theta < \pi/6$ can be defined by one side of our *AB* rectangle and the shorter basis vector. From topological arguments [142], a chirality vector $\mathbf{C}_n = m\mathbf{a}_1 + n\mathbf{a}_2$ can be defined in terms of the (*m*, *n*) coordinates of the rectangle with respect to the basis vectors. The usefulness of such a description stems

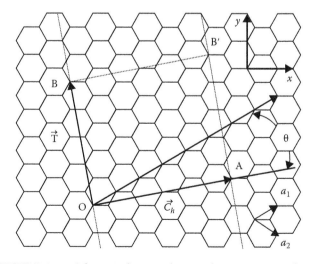

FIGURE 40.8 Schematic diagram showing the construction of carbon nanotubes from graphene. For details, see the text.

from the ease of expressing several important geometric parameters such as the diameter of the carbon nanotube:

$$d = \frac{\tau \left[m^2 + mn + n^2 \right]^{1/2}}{\pi} \qquad (40.27)$$

where $\tau = \sqrt{3} \times 1.421$ Å is the lattice constant of the graphene sheet given in terms of the carbon–carbon bond length. The chirality angle is also given by a simple trigonometric relation, namely,

$$\theta = \tan^{-1} \left[-\frac{\sqrt{3}n}{2m + n} \right]. \qquad (40.28)$$

Theory predicts [143] that the electronic conduction properties of such nanostructures, that is, whether they behave as metallic, narrow-gap semiconducting, or wide-gap semiconducting materials are related to the conditions $m = n$, $n - m = 3I$, and $n - m \neq 3I$, respectively, with I being an integer.

The rolling-up of graphene is not only limited to one sheet, leading to an important distinction between single-wall (SWNT) vs. multiwall (MWNT) nanotubes. SWNT's have diameters of ca. 10 Å whereas MWNTs can contain as many as 40–50 coaxial tubes leading to inner diameters of 10–100 Å and outer diameters up to 300 Å. The interlayer distances in MWNTs (3.4–3.6 Å) are similar to those found in bulk graphite.

The broad range of topological variations afforded by carbon-based nanostructured materials is immediately suggestive of their potential as H_2 storage media. Carbon nanotubes differ from traditional porous graphites in the presence of a well-defined curvature of the graphene sheets. Moreover, the inner cavities have dimensions approaching just a few molecular diameters, opening up the possibility of a significant overlap of potential fields from opposite walls and a subsequent enhancement of physisorption energies (cf. Section 40.2.2). Historically, a worldwide race to find the most suitable carbon-based nanostructured material started in 1997 with the work of Dillon et al. [144]. In this work, the H_2 storage capacity was estimated to be 5–10 wt % in a sample containing 0.1–0.2 mass % SWNTs. A number of conflicting reports did however follow [146] concerning both total H_2 uptake as well as its adsorption and release characteristics. One of the main focus of contention has been the difficulty in reproducing previous results, mostly due to the rather different properties and history of the nanotube samples. Reported storage values have shown a wide spread in values, from an insignificant ~0.01 wt % up to values of 20% and above in lithium-doped nanotubes [145], even though such high values have not been verified experimentally thereafter. In fact, the most common routes of production, which rely on the dielectric breakdown of a graphite electrode, laser ablation under pulsed or continuum regimes, or chemical reactions involving high-temperature carbon ions, yield samples of relatively low purity (of the order of 60%) together with a good number of residual

pollutants such as metal catalysts or amorphous carbon, the effects of which have been very difficult to ascertain. Recent experiments [147] have provided some clarification to the above confusion by showing that H_2 desorption in typical samples can originate from titanium-alloy particles introduced during ultrasonic treatment, and it is not intrinsic to the carbon nanotubes. Likewise, no general consensus exists about the nature and relative distribution of adsorption sites in carbon nanotubes. Numerous molecular simulations (cf. Section 40.2.3.2) have identified several adsorption sites including inner-, outer- and intra-tube adsorption, as well as intra-sheet uptake in MWNTs. However, these have proven quite elusive to differentiate in real materials, most likely owing to very similar adsorption characteristics or the intrinsic inability of bulk uptake studies to probe the microscopic details of hydrogen binding to these materials. Several reviews detailing a vast number of hydrogen uptake studies in carbon nanostructures exist in the literature [4,7,12]. The reader is referred to these as further illustration of a long and intricate history spanning more than a decade of research work, where perhaps too much emphasis has been placed in searching for new target materials to the detriment of a systematic approach rooted on sound physicochemical principles.

To fill the above gap, two recent case studies aimed at providing a consistent microscopic picture of the effects of geometric confinement and chemical doping in carbon nanostructures will be presented in detail below. The objective is to introduce the reader to the state of the art in the field as well as the implications of these results to the hydrogen-storage problem. To this end, we shall make extensive use of the material presented in Section 40.2.

40.3.2 Case Study I: Geometric Confinement in Carbon Nanohorns

Single-walled carbon nanohorns (SWNH) are graphitic structures formed by aggregation of individual nanotubes [151]. They display a dahlia-like shape and a size of the order of 80–100 nm (see electron micrograph in Figure 40.9). SWNHs are synthesized by laser ablation methods using moderate powers (500–600 W) on a graphite substrate via a self-assembly process characterized by reaction rates comparable to catalyst-assisted growth [152]. Moreover, no metal catalyst is required for their synthesis thus enabling low-cost, large-scale production of high-purity samples. Typical yields are 20 g/h and purity exceeds 90% [153]. These nanostructures exhibit very large surface areas exceeding 1500 m²/g [154] and are therefore attractive candidates for gas and liquid storage.

Structurally, the individual tubular structures in SWNHs are like SWNTs with the important exception of their ends. The latter show conical rather than the spherical caps characteristic of SWNTs and have rather narrow opening angles $\theta_c \approx 20°$. From geometric arguments [155], such cones can be formed by cutting a wedge from a graphene sheet and then connecting the exposed edges in a seamless manner. The angle formed by the wedge, known

FIGURE 40.9   Transmission electron micrograph of SWNHs.

as the disclination angle in solid-state physics, is given by $n\pi/3$, $0 \le n \le 6$ and this condition sets the opening angle of the cone to

$$\theta_c = 2\sin^{-1}\left(1 - \frac{n}{6}\right). \qquad (40.29)$$

The observed value thus corresponds to a disclination of $5\pi/3$. Using Euler's rule one expects that the end cap of a SWNH should have $n = 5$ pentagons replacing the graphene hexagons. Such pentagons can also be thought of as defects embedded within an all-hexagon structure and, as such, they may carry a net electrical charge. These pentagonal sites are also responsible for the electronic structure of the cone and it is precisely the presence of such details at the cone ends what explains the striking electrical response of SWNH to gas adsorption [156]. Moreover, detailed

comparisons between experimental data for gas adsorption and computer simulations incorporating nuclear quantum effects show that there is a strong preference for H_2 adsorption at the SWNH conical ends. Even more remarkable are the strong isotopic effects for H_2 vs. D_2 [62], understood on the basis of the differences in thermal de Broglie wavelength of these two isotopes (cf. Equation 40.15 in Section 40.2.2).

The merits of SWNHs compared to other carbon nanostructures were already pointed out in 2002 by Murata and coworkers [157], demonstrating a storage capacity of about 70 g/L at 77 K and 50 bar of applied pressure. SWNHs offer very large surface areas approaching 2000 m²/g and both internal and external tubes surfaces can be accessed by means of oxidation and pressing processes. These processes as well as others including metal decoration make these materials capable of consistently storing up to 3.5 wt % at 77 K and up to 0.8 wt % at 300 K [158]. While still sensibly below the stipulated DOE targets, these values represent a significant step forward for the use of these materials in practical applications.

Previous work [148,150] carried out on SWNTs with a diameter of 14 Å suggests that most of the hydrogen initially adsorbed into SWNT's at pressures reaching 100 bar and cooled down to 10 K escapes from the adsorption sites once the temperature is brought back to ambient, and only a small fraction remains adsorbed if the temperature is brought down again to 10 K. Adsorption of H_2 in SWNHs of similar dimensions is also known to be reversible after temperature cycling, whereby H_2 is released at ambient temperature following low-temperature adsorption [150]. This behavior is particularly evident in neutron scattering spectra, as these data provide direct access to the quantized rotational-energy level structure of the adsorbed H_2 phase via excitation of the lowest-lying *para* → *ortho* transition. In the absence of intermolecular interactions or center-of-mass motions, this transition has a vanishingly narrow linewidth. The observation of line broadenings or spectral shifts can therefore be utilized as a sensitive probe of the potential-energy landscape of the adsorbed H_2.

Figure 40.10a displays a comparison of the spectral line shapes of H_2 adsorbed in SWNT and SWNH, respectively. H_2 spectra

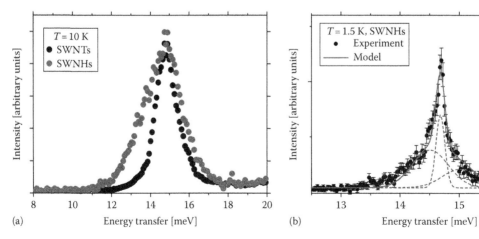

FIGURE 40.10   (a) Neutron rotational spectra of the *para* → *ortho* transition of hydrogen in SWNT and SWNHs at $T = 10$ K. (b) High-resolution neutron spectrum of H_2 in SWNHs at $T = 1.5$ K. For further details see the text.

for the SWNT samples display a single and narrow (resolution-limited) contribution centered at 14.7 meV, indicative of molecular free rotation. What this means is that the interaction between H_2 and the carbon matrix is too weak to be felt by our spectral probe. In contrast, the spectrum for the SWNH sample displays a distorted line shape suggestive of inhomogeneous broadening, that is, the spectrum comprises more than one single band. Moreover, this broad peak is markedly asymmetric with a bias towards lower frequencies. Such differences in the spectral shape are also in agreement with the higher binding energies found by Tanaka et al. [62] who, on the basis of data derived from adsorption experiments, have reported isosteric heats of adsorption for H_2 on SWNHs nearly three times as large as those on SWNTs, corresponding to H_2 binding energies in the range 100–120 meV. This increase in H_2 binding energy has been attributed to strong solid–fluid interactions at the conical tips. Weaker H_2 adsorption has also observed away from the SWNH tips, with an interaction very similar to what has been found for SWNTs.

Neutron spectra of H_2 in SWNHs at a higher energy resolution are shown in Figure 40.10b. These data allow a quantitative assessment of the potential energy landscape felt by the H_2 molecule in SWNHs. To analyze these results, we consider a Stark splitting of the *para → ortho* rotational transition. Physically, it is ascribed to the presence of electrostatic interactions between H_2 and the carbon matrix, mostly due to the defect structure of pentagons forming the conical tips. To lowest order, the interaction potential takes the form

$$V(\Theta,\phi) = V_\Theta \cos^2 \Theta \qquad (40.30)$$

where

V_Θ stands for the height of the potential barrier

Θ is the azimuthal angle between the molecular axis and the field direction

In the limit $V_\Theta \to \infty$, this angular potential leads to two-dimensional rotation of H_2 on a plane perpendicular to the quantization axis, and a *para → ortho* rotational transition located at B_{rot} instead of $2B_{rot}$ characteristic of a three-dimensional diatomic rotor. Using the free-rotor wavefunctions introduced in Equation 40.2 (Section 40.2.1.1), the total energy in the presence of the additional angular term $V(\Theta,\phi)$ is given by

$$\langle J',M' | \hat{H}_{tot} | J,M \rangle = B_{rot} J(J+1)\delta_{J'J}\delta_{M'M}$$
$$+ \langle J',M' | V(\Theta,\phi) | J,M \rangle \qquad (40.31)$$

Since any arbitrary angular potential $V(\Theta,\phi)$ can always be expanded in terms of spherical harmonics according to

$$V(\Theta,\phi) = \sum_{J_v M_v} V_{J_v M_v} Y_{J_v M_v}(\Theta,\phi) \qquad (40.32)$$

where V_{J_v,M_v} are a set of complex constants, Equation 40.31 can be rewritten as

$$\langle J',M' | \hat{H}_{tot} | J,M \rangle = B_{rot} J(J+1)\delta_{J'J}\delta_{M'M}$$
$$+ \sum_{J_v M_v} V_{J_v M_v} \langle J',M' | Y_{J_v M_v}(\Theta,\phi) | J,M \rangle \qquad (40.33)$$

The last term in brackets is nothing more than an integral over three spherical harmonics, easily calculated by recourse to angular-momentum algebra [32] as shown below:

$$\langle J',M' | Y_{J_v M_v}(\Theta,\phi) | J,M \rangle = (-1)^{M'} \sqrt{\frac{(2J'+1)(2J_v+1)(2J+1)}{4\pi}}$$
$$\times \begin{pmatrix} J' & J_v & J \\ 0 & 0 & 0 \end{pmatrix} \begin{pmatrix} J' & J_v & J \\ -M' & M_v & M \end{pmatrix} \qquad (40.34)$$

where the last two terms in parenthesis are Wigner 3-j symbols [159]. Diagonalization of the Hamiltonian matrix described by Equation 40.34 leads to definite predictions of the expected energy-level structure and *para → ortho* transition energies. For the functional form given by Equation 40.30, this procedure is performed numerically as a function of the rotational barrier V_Θ. This calculation assumes that, owing to the electrostatic interaction with the carbon substrate, the three Zeeman sublevels $M = 0$, ± 1 of the *ortho* state split into two components corresponding to the $M = \pm 1$ (maximal) and 0 (minimal) projections of the total angular momentum along the quantization axis. In the limit of a small perturbation, that is, when $V_\Theta/B_{rot} \to 0$, both intensities I_M and energy positions relative to the free-rotor energy ΔE_M approach a constant ratio given by $I_{M=\pm1}/I_{M0} = \Delta E_{M=0}/\Delta E_{M=\pm1} = 2$. By constraining our data-fitting procedures to obey these restrictions, it is possible to account for the observed INS spectrum shown in Figure 40.10b. The unperturbed rotational transition gives rise to the narrow component centered at 14.68 ± 0.01 meV with a resolution-limited half-width of 0.07 ± 0.01 meV. In addition, two broad satellites appear at 14.51 ± 0.03 meV and 15.02 ± 0.04 meV with linewidths of about 0.31 meV. The intensity ratio of the Stark-split component relative to the unperturbed line is roughly 3.3, that is, about 75% of the molecules are interacting with the carbon lattice strongly enough to perturb their high-frequency internal rotation. This ratio agrees with the amount of immobile H_2 obtained from QENS data, as shown below. These results indicate that the most energetically favorable SWNH adsorption sites leads to a solid-like H_2 phase characterized by a significant angular anisotropy in the interaction potential. The observed spectral splitting is about 0.5 meV, corresponding to a temperature of about 6 K and a hindering barrier $V_\Theta = 1.3 \pm 0.3$ meV.

Complementary information on the way in which H_2 adsorbs to the carbon matrix is provided by looking at QENS data

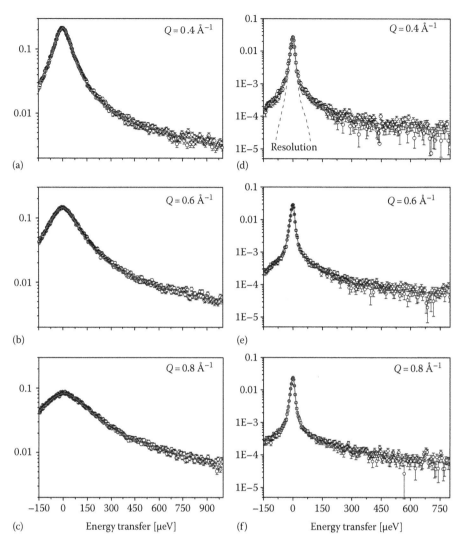

FIGURE 40.11 QENS spectra at $T = 15$ K for bulk liquid hydrogen (left column) and for hydrogen adsorbed on carbon nanohorns (right column).

(cf. Section 40.2.3.1). Figure 40.11 shows a comparison of the QENS spectra of bulk liquid H_2 at $T = 15$ K and hydrogen within SWNH at the same temperature. There we see that there is a rather substantial difference in spectral width between data for the bulk liquid and SWNHs. An obvious difference between the two sets of data is the presence of a strictly elastic component in the H_2-in-SWNH spectra, signaling the presence of species with a highly reduced mobility. This behavior occurs at temperatures and pressures where bulk hydrogen is a liquid and thus shows that a significant fraction of the H_2 molecules have very restricted mobility due to interactions with the SWNH structure. To quantify these findings, QENS spectra were further analyzed with a Bayesian algorithm that infers from the data the minimum number of spectral components needed to account for the experimental observations [161]. The results show that the spectra for the bulk liquid are described adequately with a single Lorentzian line, whereas those for H_2 in SWNH require an additional elastic line as shown in Figure 40.11. Furthermore, the momentum-transfer dependence and the intensity of the QENS components for H_2 in SWHNs is presented in Figure 40.12.

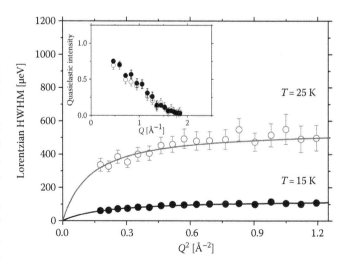

FIGURE 40.12 Q dependence of the QENS linewidths and amplitudes (insets) for H_2 in SWNHs. Solid lines are fits to these data as detailed in the text.

To extract microscopic information from the Q-dependence of the spectral linewidths, the QENS data were analyzed further by recourse to the jump-diffusion model introduced in Section 40.2.3.1. Analysis of the H_2-SWNH data required an additional elastic (immobile) contribution with an intensity about four times larger than the Lorentzian (mobile) component. Such an elastic component arises from a large fraction of molecules which are immobile within our time window (1–75 ps), while the quasi-elastic contribution to the signal can be assigned to translational motions of the mobile fraction. The wave vector dependence of the latter at $T = 25\,K$ shown in Figure 40.12 shows that at a temperature that is about 5 K above the boiling point of the liquid at a pressure of 1 bar, H_2 is still tightly bound to the SWNH matrix. Further analysis of the QENS linewidths yields values for D_T of $0.96 \pm 0.1\,Å^2/ps$ and $6.5 \pm 1.7\,Å^2/ps$ for $T = 15$ and $25\,K$, respectively. This compares to a value of $0.43 \pm 0.01\,Å^2/ps$ for the bulk liquid. The values for the inverse of the residence time are $E_0 = 0.13 \pm 0.01\,meV$ and $0.55 \pm 0.045\,meV$ for 15 and 25 K, respectively, and are to be compared with $E_0 = 3.38 \pm 0.27\,meV$ for the bulk liquid at $T = 15\,K$. These results show that the mobile fraction of H_2 within the SWNH undergoes jump motions that are some 2.3 times faster than those found in the bulk liquid at the same temperature, whereas the residence time increases by a factor of about 25 due to interaction with the carbon matrix. From the diffusion coefficient and the residence time, one can derive estimates of the length of a diffusive step between two different adsorption sites from $l = \sqrt{6 D_T \tau}$ to yield values, which are $l = 0.7 \pm 0.07\,Å$ and $5.44 \pm 0.41\,Å$ for the bulk liquid and hydrogen in SWNHs at $T = 15\,K$, respectively, as well as $6.81 \pm 0.56\,Å$ for H_2 in SWNH at $T = 25\,K$.

Further insight about the bound H_2 fraction is provided by the elastic form factors as shown in Figure 40.13. These data may be parameterized via extension of Equation 40.20 to read

$$I_{elast}(Q) = I_{rot}(Q) I_{com}(Q) \qquad (40.35)$$

Here the term $I_{com} = e^{-\langle u^2 \rangle Q^2/3}$ accounts for an effective mean-square amplitude of vibrations $\langle u^2 \rangle$ describing motions of the center of mass of the H_2 molecule, and it is the only free parameter in our description of the data. Values derived from least-squares fits of the elastic intensity data to the above expression reveals an increase of $\langle u^2 \rangle$ from $0.77 \pm 0.06\,Å^2$ at $T = 5\,K$ up to $1.47 \pm 0.08\,Å^2$ for $T = 25\,K$. We can also consider the observed temperature dependence of $\langle u^2 \rangle$ in terms of a harmonic oscillator with energy levels $E_n = \hbar \omega_c \left(n + \frac{1}{2} \right)$ and Bose population factors $P_n(T) = e^{-\beta E_n} / \Sigma_n e^{-\beta E_n}$, and thus

$$I_{elast}(Q) = I_{rot}(Q) \Sigma_n P_n(T) e^{-\langle u^2 \rangle Q^2/3} \qquad (40.36)$$

with $\langle u^2 \rangle = \left(n + \frac{1}{2} \right) \hbar / M \omega_c$. Here, $\beta = 1/k_B T$, M is the mass of a H_2 molecule and ω_c stands from a characteristic frequency involving the center-of-mass motions of the molecular center of mass. From the lowest-temperature experiment one obtains $\hbar \omega_c = 1.36\,meV$ and also, from the inset of Figure 40.13, we see that such approximation provides an adequate and unified description of all data.

Because no detailed information on the phonon spectrum of SWNH is available, an assignment of the vibrational energy ω_c must remain tentative. This energy is below the phonon spectrum of hydrogen [37] so it must be related to low-energy modes of the SWNH substrate. In spite of its relevance to explain our experimental findings quantitatively, the role of carbon–substrate motions has been largely neglected in simulation work to date [62,160]. Also, since $\hbar \omega_c \sim V_\Theta$, the coupling between internal and center-of-mass degrees of freedom is likely to play an important role in dictating the precise nature of H_2 motions on the SWNH substrate.

The neutron experiments presented in this section provide unambiguous experimental signatures of an enhanced interaction between H_2 and SWNHs. More importantly, the character of this interaction is quantitatively different from what is known for H_2 in SWNTs. The use of high-resolution neutron spectroscopy proves essential at quantifying the effects of molecular confinement in SWNHs, characterized by barriers to free rotation at least four times larger than those characteristic of H_2 adsorption in SWNTs [92]. Both the stochastic and vibrational dynamics of free and bound species are strongly altered due to their adsorption on the carbon substrate. The mobile fraction exhibits a remarkably higher mobility than in the weakly interacting bulk phase. The vibrations of the SWNH–H_2 complex also display large mean-square displacements, indicative of strong quantum effects that persist at temperatures well above the boiling point of the bulk liquid. These neutron results also explain findings from previous studies reporting the presence of two preferential adsorption sites with significantly different mobilities [62].

More recently, multiscale simulations have predicted an appreciable enhancement of H_2–substrate interactions solely via geometric confinement in carbon nanoscrolls, novel

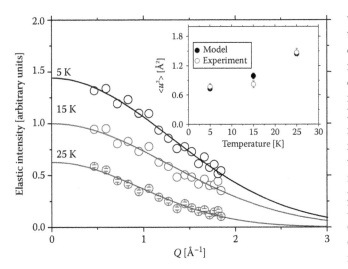

FIGURE 40.13 Q dependence of the elastic intensity for H_2 in SWNH. Solid lines are fits using Equation 40.36. The inset shows the temperature dependence of the expected and observed values for the effective $\langle u^2 \rangle$ mean-square amplitude of center-of-mass vibration.

nanostructures made up of rolled graphene sheets [162]. These theoretical predictions still await experimental confirmation.

40.3.3 Case Study II: Alkali-Doped Graphite Intercalates

Even though nanoscale confinement effects can be effective at increasing the rather feeble interaction between H_2 and graphitic carbon, experimental and theoretical work performed to date suggest that such effects may prove insufficient for room-temperature applications (cf. Sections 40.2.3.2 and 40.3.2). In view of the above, the introduction of suitable dopants (particularly metals) into carbon nanostructures represents a viable and relatively unexplored way forward to enhance sorption energies. Whereas abundant theoretical work exists to support these considerations [132,133,162], experiments probing H_2 binding in metal-doped carbon nanostructures are scarce. Careful gas dosing experiments by Yang [146] have shown a 2% weight uptake by Li-doped nanotubes. More recent studies on alkali-doped carbon nanostructures [164] also report an enhancement of H_2 adsorption compared to the undoped species but the microscopic nature of H_2 binding sites remains largely unexplored.

Graphite intercalation compounds (GICs) are ideal materials to explore in detail the physical chemistry of H_2 uptake by metal-doped carbon nanostructures [165,166]. A wide range of atoms and molecules can be intercalated in graphite to form lamellar structures of regular lattice spacing, with a *c*-axis superlattice in which *n* graphite layers separate each guest layer (so-called stage-*n* GICs). These compounds can adsorb H_2 in a variety of ways, in some cases leading to the breaking of the H–H bond, as observed in KC_8 [167]. They are, therefore, excellent systems to explore, within the same family of materials, the transition between reversible molecular physisorption and dissociative chemisorption. In particular, the series of stage-2

compounds of stoichiometry MC_{24} (M = K, Rb, Cs) are known to adsorb H_2 up to $\sim 2H_2/M$ [168]. At temperatures below 100 K, MC_{24} adopts a commensurate $\sqrt{7} \times \sqrt{7}$ M superlattice bounded by a higher metal density in the interdomain regions. H_2 uptake by RbC_{24} [169,170] and CsC_{24} [171] has been investigated by INS. The observed spectroscopic features were attributed to the presence of two distinct adsorption sites. More recent work on $(H_2)_xKC_{24}$ shows no preferential site occupancy for $x = 1.0$ and 1.5 [172,173], where *x* represents the amount of adsorbed H_2. This finding strongly suggests the existence of a single adsorption site in KC_{24}, making it a superb candidate for more detailed work. In this case study, we explore the H_2–KC_{24} potential energy landscape via a detailed analysis of high-resolution neutron data and plane-wave density-functional-theory (PW-DFT) calculations [174].

Figure 40.14a shows the dependence of the $(H_2)_xKC_{24}$ (003) neutron Bragg reflection with *x*, following the relaxation of *n*-H_2 to the *para* ground state over an interval of several hours. The intensity of the diffraction peak at 2.90 Å (corresponding to pristine KC_{24}) diminishes with *x* while a second feature appears above 2.97 Å, with its position moving steadily toward higher lattice spacings up to $x = 2.0$. It is to be noted that both peaks remain discrete and coexist at all explored *x*, signaling the existence of well-separated domains of pure and hydrogenated KC_{24}. Moreover, the hydrogenated phase is characterized by a $\sim 3\%$ expansion of the GIC galleries at saturation coverages, going from 5.35 ± 0.01 Å to 5.64 ± 0.01 Å at $x = 2$. There is no significant shift of the higher-lying diffraction peak for $x > 2$, indicative of no further adsorption at well-defined sites. According to the Langmuir model introduced in Section 40.2.3.1, single-site adsorption with a coverage-independent binding energy should obey an exponential dependence of the form e^{-x} and $(1 - e^{-x})$ for vacant and filled sites, respectively. Such dependences can be used to describe satisfactorily the neutron diffraction data of Figure 40.14a.

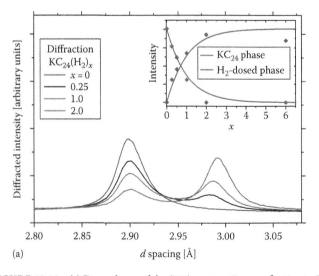

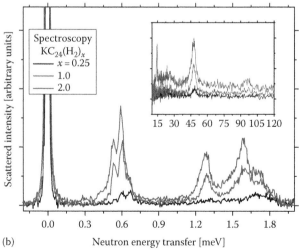

(a)   *d* spacing [Å]   (b)   Neutron energy transfer [meV]

FIGURE 40.14 (a) Dependence of the (003) neutron Bragg reflection in KC_{24} as a function of H_2 coverage *x*. Integrated intensities (symbols) and accompanying fits (lines) are shown in the inset (see text for details); (b) INS data from at low and intermediate (inset) energy transfers. (Adapted from Lovell, A. et al., *Phys. Rev. Lett.*, 101, 126101, 2008.)

Figure 40.14 also shows INS spectra as a function of x. All spectral features shown display the same x dependence, in line with the presence of a single adsorption site and the diffraction data presented above. Two sets of excitations centered at 0.6 and 1.5 meV are present at low energies. As discussed in Section 40.2.3.1, neutron scattering from *para*-H_2 can only be incoherent provided that the *total* nuclear spin changes upon scattering, leading to a dominant response at and above the lowest *para* $\rightarrow$ *ortho* transition [91]. In bulk *para*-H_2 [cf. Figure 40.6] and in carbon-only nanostructures [cf. Figure 40.10], this transition is observed around the free-rotor value of 14.7 meV, whereas in KC_{24} this spectral bandhead has shifted all the way down to 0.6 meV, signaling a strong hindering of H_2 rotations. More quantitative estimates of the hindering potential can be obtained by considering the M-level splitting of H_2 rotational eigenstates in the presence of a potential $V(\Theta, \phi)$, as already described for the case of H_2 in SWHNs. In this case, the observation of well-defined spectral features below the two-dimensional limit B_{rot} is indicative of a strong pinning of the H_2 molecule along a single axis characterized by an orientational potential $V(\Theta, \phi)$ of the form

$$V(\Theta) = V_\Theta \sin^2\Theta \qquad (40.37)$$

Using the formalism leading to Equation 40.34, the observed bandhead at 0.6 meV corresponds to $V_\Theta = 137$ meV. The upper state of this spectral transition therefore correlates with the singly degenerate $|10\rangle$ level in the free-rotor basis, thereby constraining H_2 to lie preferentially along the quantization axis. It is also noteworthy that the V_Θ value for KC_{24} is ~100 times higher than in nanotubes [148,149] and nanohorns [92]. These calculations also predict a doubly degenerate transition at 51 meV, in excellent agreement with the dominant feature observed at 48.5 meV (see inset of Figure 40.14). Moreover, the sudden appearance of the H_2 free-rotor peak at $x = 2$ marks the saturation of available GIC adsorption sites, in line with the diffraction data. In the absence of other librational features below 100 meV,

both the fine structure at 0.6 meV and the manifold centered at 1.5 meV must arise from the simultaneous excitation of H_2 rotational and translational modes, as also observed in the spectrum of the bulk solid shown in Figure 40.6.

First-principles calculations can be used at this point to scrutinize the details of H_2–GIC complex in more depth. To this end, the PW-DFT code CASTEP [175] was utilized within the GGA approximation and the Perdew-Burke-Erzerhof (PBE) functional. Our methodology was first benchmarked against the triatomic K^+–H_2 complex. The rationale behind this choice originates from the extreme deviations from free-rotor behavior observed in the INS data, indicative of a strong interaction with the material. Figure 40.15 evidences a clear preference for a T-shaped geometry ($\Theta = 90°$) of the K^+–H_2 complex displaying a minimum at $R_{eq} = 2.92$ Å. This finding is in line with the dominance of ion–quadrupole interactions presented in Section 40.2.2. The first few radial eigenenergies were computed using the Numerov algorithm [178]) and yield a zero-point energy $E_{ZPE} = 13.5$ meV, as well as a dissociation energy of $E_D = 78$ meV. These values are in good agreement with the ab initio electronic structure calculations of Vitillo et al. [177] yielding $R_{eq} = 2.92$–2.94 Å and $E_D = 68$–78 meV. The linear triatom lies ~85 meV above the ground state, an energy approaching the value of V_Θ inferred from the neutron spectra. However, the predicted rotational spectrum is characterized by a doubly degenerate bandhead starting at 9.6 meV, reflecting a reversal of M-level splittings in favor of H_2 alignment normal to the quantization axis and preferential confinement of the H_2 molecule to lie on the plane perpendicular to the K^+–H_2 axis.

Following our discussion in Section 40.2.2, the two dominant contributions to the attractive intermolecular potential are those expected for the interaction of H_2 with a charged site, namely: (a) ion-quadrupole forces, of the form $(Q_{H_2}/R_{eq}^3)P_2(\cos\Theta)$; and (b) ion-induced-dipole, of the form $-(1/R_{eq}^4)[A + BP_2(\cos\Theta)]$, where A and B are related to the H_2 polarizability tensor via $A = \alpha_{iso}/2$ and $B = (\alpha_{par} - \alpha_{per})/3$. This last expression is a generalization of Equation 40.11 in Section 40.2.2 to account quantitatively

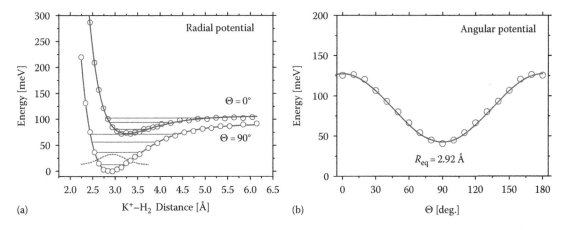

(a) K$^+$–H$_2$ Distance [Å]   (b) Θ [deg.]

FIGURE 40.15 (a) K^+–H_2 radial potential energy curves and ground state wavefunction amplitude (dashed line). (b) Shows the angular potential at R_{eq} for rotation of H_2 by an angle Θ on the plane defined by the K^+–H_2 bond and the H_2 molecular axis. Symbols correspond to PW-DFT calculations whereas the solid line in (b) corresponds to the long-range expansions of the angular potential detailed in the text.

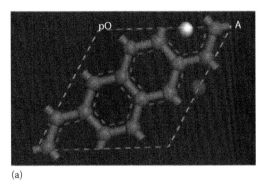
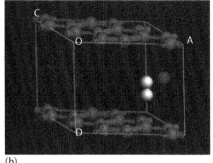

(a)   (b)

FIGURE 40.16 Geometry of the minimum-energy configuration of the H_2–GIC complex from PW-DFT calculations: (a) view along the GIC c-axis; (b) side view along the GIC gallery (gray: K atoms; white: H).

for the angular dependence of this interaction term. Quadrupole and polarizability values have been taken from Ref. [24] (cf. Section 40.2.1.1). As shown in Figure 40.15, this long-range expansion provides an excellent description of the PW-DFT calculations. Further, it highlights the dominant role of ion–quadrupole interactions in dictating the geometry of the K^+–H_2 complex, characteristic of much-stronger binding forces than those found in undoped graphite [124].

To model the GIC, a hexagonal K superlattice was chosen as representative of the material. As shown in Figure 40.16, this model structure has a KC_{14} unit cell because it omits the empty graphite galleries where H_2 cannot enter. For KC_{14}, PW-DFT rightly predicts the transfer of one electron from the alkali metal to the graphene layers [165]. Following these calculations, H_2 was inserted in the unit cell and allowed to relax to the energy minimum (see Figure 40.16). Neither the H_2 bond distance ($\Delta r_{HH}/r_{HH} = +0.003$) nor its internal frequency of vibration ($\Delta\omega_{vib}/\omega_{vib} = -0.02$) changed appreciably during adsorption, both clear indicators of physisorptive uptake. Inside the GIC, H_2 sits 2.87 Å away from the closest K atom, and approximately below the center of the carbon rings. Further, the H_2 axis is perpendicular to the graphite layers, adopting a T-shaped configuration with respect to the alkali, as in the case of K^+–H_2. The electron-density-difference map in Figure 40.17

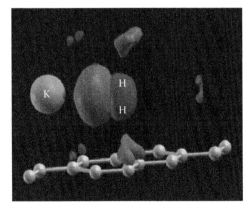

FIGURE 40.17 Difference map obtained by subtracting the individual H_2 and KC_{14} electron densities from that of H_2KC_{14}. Light (dark) gray indicates electron density gain (loss). For clarity, the position of K and H atoms has been superimposed on this map (see also Figure 40.16).

shows the distinct appearance of an induced dipole moment on H_2 as a result of charge migration towards K. Mirroring these changes, there is also charge redistribution in the graphene planes, with negative charge now appearing above and below H_2. These findings neatly account at a quantitative level for the expansion of the GIC galleries upon the addition of H_2, as shown in Figure 40.18 where a direct comparison between the experimental lattice parameters and PW-DFT calculations is presented.

Encouraged by the exquisite agreement between experiment and first-principles calculations, we can further interrogate the latter in order to explore in greater detail the potential-energy surface of the adsorbate complex. H_2 radial and angular scans inside the GIC are shown in Figure 40.19. For $R < 3.5$ Å, the shape of the H_2–GIC curve ($E_{ZPE} = 13.9$ meV) is very similar to that of K^+–H_2, further reinforcing the notion that the energy landscape around the minimum is indeed dominated by strong interactions with a single alkali atom. As shown in Figure 40.19, the PW-DFT orientational potential $V(\Theta, \phi)$ can be written as

$$V(\Theta,\phi) = V_\Theta\left[1 - \left(1 - \frac{V_\phi}{V_\Theta}\sin^2\phi\right)\sin^2\Theta\right] \qquad (40.38)$$

with $V_\Theta = 126$ meV and $V_\phi = 26$ meV. This potential has been constructed via extension of Equation 40.37 to include XY-plane anisotropy, with X aligned with the GIC c-axis and Y running along the GIC gallery plane. As before, the Z-axis is still defined by the line intercepting the K atom and originating at the H_2 center of mass.

Comparison of Figures 40.19 and 40.15 shows that the resulting H_2–GIC orientational potential $V(\Theta, \phi)$ is qualitatively similar to that of K^+–H_2 except for a small preference for H_2 alignment along the GIC c-axis. This feature is the result of a much weaker interaction with the graphite layers. As already discussed in Section 40.2.3.2, the GGA approximation in DFT may underestimate dispersive forces. Recent work, however, shows that H_2–graphite interactions are still described with reasonable accuracy [116] and, in particular, rotational barriers agree within several meV with correlated CCSD(T) methods. In comparison with the experimental INS data, the above

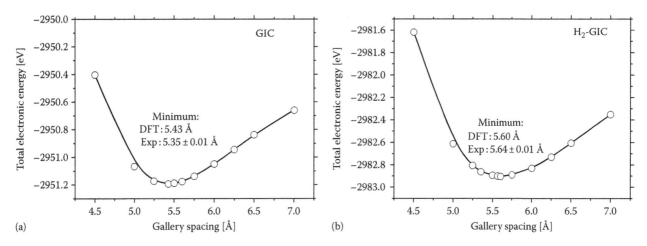

FIGURE 40.18 Total electronic energy as a function of gallery spacing for the GIC (a) and H$_2$–GIC (b) systems. The distance corresponding to the minimum energy in each case is compared with that obtained from the neutron-diffraction data of Figure 40.14a.

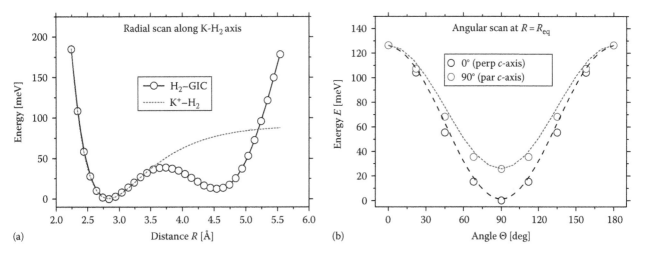

FIGURE 40.19 (a) Radial scan for the H$_2$–GIC complex (solid line) and comparison with the triatom K$^+$–H$_2$ (dashed line). (b) Angular scans for the H$_2$–GIC complex; dashed lines in (b) correspond to fits using Equation 40.38 in the text.

$V(\Theta, \phi)$ gives rise to a spectral bandhead at 4.71 meV, followed by three well-resolved librational transitions at 16, 36, and 38 meV. These three lines are to be contrasted with the single spectral feature we observe at 48.5 meV. Such a rich spectral progression is the result of a significant departure from cylindrical symmetry in the model H$_2$–GIC complex, reflecting sizable energy differences for H$_2$ rotation about the quantization axis and the plane perpendicular to it. None of these transitions, however, are observed in the experimental data. Thus, whilst PW-DFT computations provide insightful estimates for H$_2$–GIC interaction energies, it cannot account for the underlying symmetry of the adsorption site. The above discrepancies therefore call for a revision of currently accepted theoretical models.

An underlying assumption in our analysis has been the localization of H$_2$ at a single adsorption site. As shown in Figure 40.20, closer inspection of the H$_2$–GIC structure reveals the presence of three identical adsorption sites defining an equilateral triangle and located ~1 Å away from the centre of the triangular

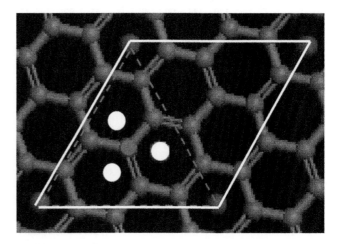

FIGURE 40.20 View along the *c*-axis (white lines define the unit cell). The trigonal subunit cell and its center are shown in dark gray. White circles denote the three adjacent H$_2$ sites.

subunit cell. If the H_2 center of mass is fully delocalized in the quantum-mechanical sense across these three sites, the effective orientational potential including quantum delocalization (QD) effects can be approximated by an average over available sites

$$V_{QD} = \frac{1}{3}\sum_{i=1,3} V_i(\Theta_i, \phi_i) \qquad (40.39)$$

To compute V_{QD}, the potential $V(\Theta_i, \phi_i)$ at each site (cf. Equation 40.38), is first expressed in terms of spherical harmonics Y_{JM} to read $V(\Theta, \phi) = A + BY_{20} + C(Y_{2+2} + Y_{2-2})$. To effect the averaging implied by Equation 40.39, we make recourse to the well-known transformation properties of Y_{JM}'s under rotation [32]

$$R(\phi, \Theta, \chi) Y_{JM} = \sum_{M'} D^J_{M'M}(\phi, \Theta, \chi) Y_{JM'} \qquad (40.40)$$

where

   (ϕ, Θ, χ) are Euler angles
   $D^J_{M'M}$ denotes a Wigner rotation matrix element

Our original reference frame at each site as implied by Equation 40.38 needs to be first rotated by 90° via the use of Equation 40.40 so as to make Z coincide with the GIC c-axis at each site. $V_{Ri}(\Theta, \phi)$ then denotes the potential in this new frame of reference at site i, and it is still of the form $V_R = A_R + B_R Y_{20} + C_R(Y_{2+2} + Y_{2-2})$. At this point, all three sites share the same Z-axis. Following this rotation, we can write

$$V_{QD} = \frac{1}{3}\sum_{i=1,3} V_{R1}\left(\Theta_1, (i-1)\frac{2\pi}{3}\phi_1\right) \qquad (40.41)$$

As $V_{Ri}(\Theta, \phi)$ only contains Y_{20} and $Y_{2\pm2}$ terms, the averaging process implied by Equation 40.39 reduces to

$$V_{QD} = \langle V \rangle \sin^2\Theta \qquad (40.42)$$

with $\langle V \rangle = \frac{1}{2}(V_\Theta + V_\phi)$. Thus, quantum-mechanical delocalization of H_2 across three sites obeying C_3 symmetry recovers the same functional form as deduced from the INS data. To fully appreciate the implications of this result, Figure 40.21 shows contour maps of the orientational potentials deduced from PW-DFT calculations and INS experiments. Assuming a

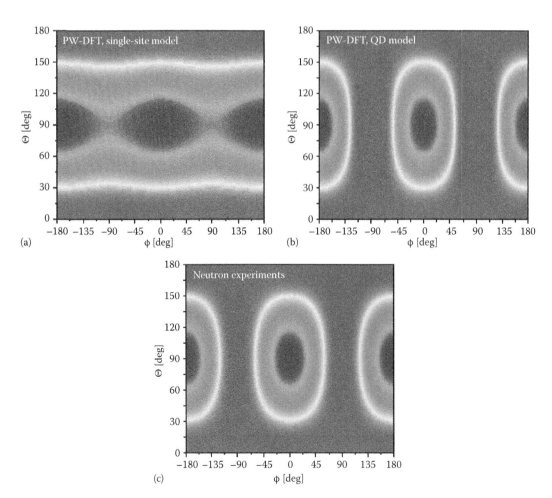

(a)

(b)

(c)

FIGURE 40.21 **(See color insert following page 22-8.)** H_2 orientational potentials deduced from PW-DFT calculations (a,b) and neutron experiments (c). Red/blue denote regions of high/low potential energy.

single-site model is clearly unsatisfactory as quasi-two-dimensional H$_2$ rotations about the ϕ-axis are largely unhindered. Inclusion of QD, on the other hand, forces the H$_2$ to librate about a single axis, in agreement with experimental observation. At a more quantitative level, calculations yield $\langle V \rangle = 76$ meV. This figure is sensibly below the experimental value of 137 meV, possibly reflecting the tendency of our PW-DFT approach to underestimate the strength of H$_2$–graphite interactions [116]. Alternatively, significant coupling between H$_2$ rotational and translational degrees of freedom might also affect the effective height of orientational barriers. Full and very costly quantum-dynamics simulations would be, however, required to further quantify these effects.

This novel binding scenario involving QD can be invoked to explain why diffraction techniques have not been successful at locating the adsorbant structure. It also provides a microscopic mechanism for the maximum uptake of 2H$_2$/K in this material. Given the hexagonal symmetry of available adsorption sites surrounding a given alkali, a coverage approaching 6H$_2$/K would be expected. QD in the GIC reduces this value to the observed ~2H$_2$/K as three adsorption sites are effectively blocked by a single H$_2$ molecule. Similar mechanisms are likely to be at play in the Cs and Rb GIC analogues but the presence of multiple adsorption sites makes the interpretation of experimental and theoretical data in these systems more involved than in the present case.

In summary, the QD scenario proposed above is able to reconcile INS data with first-principles calculations in a remarkably holistic manner. The results strongly suggest that quantum-mechanical effects may not be neglected in modeling H$_2$ in carbon-based nanostructures, possibly extending to all nondissociating hydrogen storage interactions. In particular, QD can reduce the adsorbed H$_2$ density in doped graphites, placing a further limit on the maximal capacity of these storage materials. Not only H$_2$'s rather unfavorable thermodynamic properties but also quantum mechanics appear to conspire against our repeated attempts to tame and exploit the properties of this seemingly simple molecule as the fuel of future generations.

40.4 Outlook

The safe and efficient storage of hydrogen will remain an important scientific endeavor for some years to come. Progress in the last decade has been significant yet not sufficient to meet the technological needs of the hydrogen economy. To date, no single material has been found to meet the stringent requirements for practical applications as identified by government agencies such as the U.S. DOE or the automotive industry. In this work, we have emphasized the pressing need to combine sophisticated experimental and computational tools such as neutron scattering and first-principles materials modeling in order to provide much-needed physical insight into a very challenging scientific problem.

In the case of carbon-based nanostructured materials, the physisorptive nature of the adsorption process will always ensure excellent reversibility. A major limitation at present, however, is the weak interaction of these substrates with H$_2$, limiting storage temperatures below 100 K or so. Operation around room temperature will require an increase of adsorption energies by at least a factor of three (cf. Equation 40.13 in Section 40.2.2). Recent work has identified adsorption sites in carbon-only substrates such as nanohorns approaching these targets (cf. Section 40.3.2). A major difficulty for materials science, however, lies in the engineering and subsequent synthesis of nanoporous substrates with a sufficiently high site density so as to bring H$_2$ uptakes to the desired levels. In this context, nanoporous polymers appear to be promising storage media owing to the presence of metal-free, hypercrosslinked networks with a high fraction of pores small enough for selective H$_2$ uptake. Sorption enthalpies as high as 180–190 meV have been reported recently [179]. Similar considerations apply to metallorganic frameworks where metal ions or clusters are connected through organic bridging ligands to form extended one-, two-, and three-dimensional macromolecular structures. Adsorption enthalpies for these novel architectures now exceed 100 meV [180].

As extensively discussed in Section 40.3.3, chemical doping appears to be the most straightforward way of increasing adsorption energies. For the case we have explored in detail in

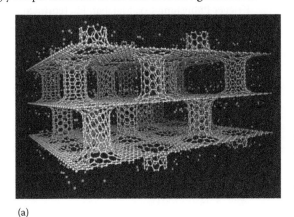

(a)

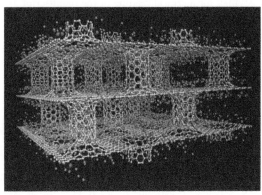

(b)

FIGURE 40.22 GCMC snapshots of H$_2$ adsorbed in pure (a) and lithium-doped (b) pillared graphene nanostructures at $T = 77$ K and $P = 3$ bar (lithium: dark gray, H$_2$: light gray). (From Dimitrakakis, G.K. et al., *Nano Lett.*, 8, 3166, 2008; Dimitrakakis, G.K. et al., *New Sci.*, 2683, 27, 2008. With permission.)

this work, namely KC_{24}, up to two H_2 molecules per formula unit (ca. 1.2 wt %) can be stored up to temperatures of 110–130 K at ambient pressure, and this desorption temperature could be increased further to ca. 200 K by using moderate pressures below 100 bar. A further increase in storage density will require a higher dopant concentration. Higher operating temperatures, on the other hand, may be achieved by an increase of dopant charge via the use, for example, of alkaline-earth and transition-metal intercalants [181]. More basic research into these possibilities is not only timely but also a prerequisite to the rational and efficient design of next-generation storage materials.

An equally important consequence of doping is the expansion of the interlayer spacing in graphitic materials. This process leads to a significant enhancement of H_2 uptake. We have also seen in Section 40.3.3 that H_2 adsorption may be accompanied by a further expansion of the host material. This process necessarily involves an energy penalty for H_2 uptake, with detrimental consequences for the ultimate performance of the material as a storage medium. From these remarks, it should therefore come as no surprise that the fine-tuning of this parameter is now recognized as being key to improve storage capacities via so-called pillaring mechanisms. Figure 40.22 provides a glimpse at what is now possible to model on a computer and could be achievable in the nanomaterials synthesis laboratory in the foreseeable future. The pillars in this case are made up of nanotube structures of variable height. Multiscale computer simulations show that lithium doping results in an uptake of 41 g H_2 L^{-1} under ambient conditions [182], a value comparable to present DOE targets. Without a doubt, only through synergistic efforts involving the development of new routes to nanomaterials synthesis, advanced physical characterization techniques, and state-of-the-art materials modeling, it will be possible to tackle with success a formidable and still unresolved technological problem.

Acknowledgments

The authors wish to thank A. Lovell and D. Colognesi for permission to use Figures 40.5 and 40.6, and N.T. Skipper, M.A. Adams and M. Krzystyniak for a critical reading of the manuscript and insightful comments. F.F.-A. gratefully acknowledges financial support from the U.K. Science and Technology Facilities Council. This work was partially funded by grant MAT-2007-65711C04-01 from the Spanish Ministry of Science and Innovation.

References

1. D.A.J. Rand and R.M. Dell, *Hydrogen Energy: Challenges and Prospects*. RSC Publishing, Cambridge, U.K. (2008), Ch. 5.
2. S. Satyapal, J. Petrovic, and G. Thomas, *Sci. Am.* **296**, 80 (2007); S. Satyapal, J. Petrovic, C. Read, G. Thomas, and G. Ordaz, *Catal. Today* **296**, 246 (2007); URL: www1.eere.energy.gov/hydrogenandfuelcells/
3. H.W. Langmi and G.S. McGrady, *Coord. Chem. Rev.* **251**, 925 (2007).
4. M. Felderhoff, C. Weidenthaler, R. von Helmolt, and U. Eberle, *Phys. Chem. Chem. Phys.* **9**, 2643 (2007).
5. R. von Helmolt and U. Eberle, *J. Power Sources* **165**, 833 (2007).
6. K.M. Thomas, *Catal. Today* **120**, 389 (2007).
7. R. Ströbel, J. Garche, P.T. Moseley, J. Jörissen, and G. Wolf, *J. Power Sources* **159**, 781 (2006).
8. D.K. Ross, *Vacuum* **80**, 1084 (2006).
9. M. Fichtner, *Adv. Eng. Mater.* **7**, 443 (2005).
10. C. Liu and H.-M. Cheng, *J. Phys. D: Appl. Phys.* **38**, R231 (2005).
11. L. Zhou, *Renew. Sust. Energy Rev.* **9**, 395 (2005).
12. A. Züttel, *Naturwissenschaften* **91**, 157 (2004).
13. R.G. Ding, J.J. Finnerty, Z.H. Zhu, Z.F. Yan, and G.Q. Lu, in H.S. Nalwa (Ed.) *Encyclopedia of Nanoscience and Nanotechnology*, Vol. 4, American Scientific Publishers, Los Angeles, CA (2004), pp. 13–33.
14. H.G. Schimmel, G.J. Kearley, M.G. Nijkamp, C.T. Visser, K.P. de Jong, and F.M. Mulder, *Chem. Eur. J.* **9**, 4764 (2003).
15. A. Züttel, P. Sudan, Ph. Mauron, T. Kiyobayashi, Ch. Emmenegger, and L. Schlapbach, *Int. J. Hydrogen Energy* **27**, 203 (2002).
16. V.V. Simonyan and J.K. Johnson, *J. Alloys Compd.* **330–332**, 659 (2002).
17. M.G. Nijkamp, J.E.M.J. Raaymakers, A.J. van Dillen, and K.P. de Jong, *Appl. Phys. A* **72**, 619 (2001).
18. H.-M. Cheng, Q.-H. Yang, and C. Liu, *Carbon* **39**, 1447 (2001).
19. H. Shull and G. G. Hall, *Nature* **184**, 1559 (1959).
20. E.R. Cohen, *The Physics Quick Reference Guide*. AIP Press, New York (1996), Chs. 2 and 4.
21. J.D. Graybeal, *Molecular Spectroscopy*. McGraw-Hill, New York (1988), Ch. 6.
22. K.P. Huber and G. Herzberg, *Molecular Spectra and Molecular Structure IV. Constants of Diatomic Molecules*. Van Nostrand, New York (1979).
23. G. Herzberg, *Molecular Spectra and Molecular Structure*. Krieger Publishing Co., Malabar, FL, reprinted in 1989.
24. A.D. McLean and M. Yoshimine, *J. Chem. Phys.* **45**, 3676 (1966); W. Kolos and L. Wolniewicz, *J. Chem. Phys.* **46**, 1426 (1967).
25. P.W. Langhoff and M. Karplus, in *The Padé Approximant in Theoretical Physics*. Academic Press, New York (1970).
26. L. Pauling and E.B. Wilson, *Introduction to Quantum Mechanics with Applications to Chemistry*. Dover Publications Inc., New York (1985), Ch. XII.
27. W. Cencek and K. Szalewicz, *Int. J. Quantum Chem.* **108**, 2191 (2008).
28. M. Cafiero, S. Bubin, and L. Adamowicz, *Phys. Chem. Chem. Phys.* **5**, 1491 (2003).
29. B. Chen, *J. Chem. Phys.* **102**, 2802 (1995).
30. L. Wolniewicz, *J. Chem. Phys.* **103**, 1792 (1995).
31. D.A. McQuarrie, *Statistical Mechanics*. Harper & Row Publishers, New York (1976), Ch. 6.

32. R.N. Zare, *Angular Momentum: Understanding Spatial Aspects in Chemistry and Physics*. Wiley, New York (1988).

33. Y.-P. Lee, Y.-J. Wu, R.M. Lees, L.-H. Xu, and J.T. Hougen, *Science* **311**, 365 (2006).

34. J.T. Hougen and T. Oka, *Science* **310**, 1913 (2005).

35. F.J. Bermejo, F. Fernandez-Alonso, and C. Cabrillo, *Condens. Matter Phys.* **11**, 95 (2008); W.P.A. Haas, N.J. Poulis, and J.J.W. Borleffs, *Physica* **27**, 1037 (1961).

36. I.F. Silvera, *Rev. Mod. Phys.* **52**, 393 (1980).

37. V.G. Manzhelii and Y.-A. Freiman, *Physics of Cryocrystals*. American Institute of Physics, New York (1997).

38. A. Michels, W. de Graaff, and C.A. Ten Seldam, *Physica* **26**, 393 (1960).

39. D.E. Diller, *J. Chem. Phys.* **42**, 2089 (1965); H.M. Roder and D.E. Diller, *J. Chem. Phys.* **52**, 5928 (1970).

40. T.F. Miller III and D.E. Manolopoulos, *J. Chem. Phys.* **122**, 184503 (2005).

41. J. de Boer, *Physica* **14**, 139 (1948).

42. L.D. Landau and E.M. Lifshitz, *Statistical Mechanics*. Pergamon Press, Oxford, U.K. (1959).

43. K.A. Gernoth, T. Lindenau, and M.L. Ristig, *Phys. Rev. B* **75**, 174204 (2007); J.W. Kim, F.J. Bermejo, M.L. Ristig, and J.W. Clark, Momentum distribution in liquid parahydrogen, in M. Belkacem and P.M. Dinh (Eds.) *Condensed Matter Theories*, Vol. 19, Nova Science Publishers, New York (2005), pp. 3–16.

44. D. Tabor, *Gases, Liquids, and Solids*, 3rd Ed. Cambridge University Press, Cambridge, U.K. (2000).

45. H. Margenau and N.R. Kestner, *Theory of Intermolecular Forces*, 2nd Ed. Pergamon Press, Oxford, U.K. (1971).

46. D. Nicholson, *Surf. Sci.* **181**, L189 (1987).

47. H. Kuroda, *Nature* **201**, 1214 (1954).

48. J.N. Israelachvili, *Intermolecular and Surface Forces with Applications to Colloidal and Biological Systems*. Academic Press, London, U.K. (1985), p. 125.

49. C. Jin, R. Hettich, D. Compton, D. Joyce, J. Blencoe, and T. Burch, *J. Phys. Chem.* **98**, 4215 (1994).

50. A.I. Kolesnikov, V.E. Antonov, I.O. Bashkin, J.-C. Li, A.P. Moravsky, E.G. Ponyatovsky, and J. Tomkinson, *Physica B* **263**, 436 (1999).

51. F. Tran, J. Weber, T.A. Wesolowski, T.A. Cheikh, Y. Ellinger, and F. Pauzat, *J. Phys. Chem. B* **106**, 8689 (2002); O. Hübner, A. Glöss, M. Fichtner, and W.J. Klopper, *J. Phys. Chem. A* **108**, 3019 (2004); T. Heine, L. Zhechkov, and G. Seifert, *Phys. Chem. Chem. Phys.* **6**, 980 (2004).

52. J. Li, T. Furata, H. Goto, Y. Ohashi, Y. Fujiwara, and S. Yip, *J. Chem. Phys.* **119**, 2376 (2003).

53. G.J. Kubas, R.R. Ryan, B.I. Swanson, P.J. Bergamini, and H.J. Wasserman, *J. Am. Chem. Soc.* **106**, 451 (1984).

54. G.J. Kubas, *Metal Dihydrogen and σ-Bond Complexes: Structure, Theory, and Reactivity*. Kluwer Academic/Plenum Publishers, New York (2001).

55. V.I. Bakhmutov, *Dihydrogen Bonds: Principles, Experiments, and Applications*. John Wiley & Sons, Hoboken, NJ (2008).

56. G.J. Kubas, *Chem. Rev.* **107**, 4152 (2007).

57. D.M. Heinekey, A. Lledós, and J.M. Lluch, *Chem. Soc. Rev.* **33**, 175 (2004).

58. C.A. Nikolaides and E.D. Simandiras, *Comments Inorg. Chem.* **18**, 65 (1996); E.D. Simandiras and C.A. Nikolaides, *Chem. Phys. Lett.* **223**, 233 (1994); C.A. Nikolaides and E.D. Simandiras, *Chem. Phys. Lett.* **196**, 213 (1992); E.D. Simandiras and C.A. Nikolaides, *Chem. Phys. Lett.* **185**, 529 (1991).

59. S. Lowell, J.E. Shields, M.A. Thomas, and M. Thommes, *Characterization of Porous Solids and Powders: Surface Area, Pore Size and Density*. Springer, Dordrecht, the Netherlands (2004), Chs. 13–14.

60. G.G. Tibbetts, G.P. Meisner, and C.H. Olk, *Carbon* **39**, 2291 (2001).

61. A. Ansón, M. Benham, J. Jagiello, M.A. Callejas, A.M. Benito, W.K. Maser, A. Züttel, P. Sudan, and M.T. Martínez, *Nanotechnology* **15**, 1503 (2004); A. Ansón, J. Jagiello, J.B. Parra, M.L. San Juan, A.M. Benito, W.K. Maser, and M.T. Martinez, *J. Phys. Chem. B* **108**, 15820 (2004); A. Ansón, A.M. Benito, W.K. Maser, M.T. Izquierdo, B. Rubio, J. Jagiello, M. Thommes, J.B. Parra, and M.T. Martinez, *Carbon* **42**, 1243 (2004).

62. H. Tanaka, H. Kanoh, M. Yudasaka, S. Iijima, and K. Kaneko, *J. Am. Chem. Soc.* **127**, 7511 (2005).

63. N. Texier-Mandoki, J. Dentzer, T. Piquero, S. Saadallah, P. David, and C. Vix-Guterl, *Carbon* **42**, 2744 (2001).

64. C. Liu, Y.Y. Fan, M. Liu, H.T. Cong, H.M. Cheng, and M.S. Dressehaus, *Science* **286**, 1127 (1999).

65. I. Langmuir, *J. Am. Chem. Soc.* **38**, 2221 (1916).

66. R.T. Yang, *Adsorbents: Fundamentals and Applications*. Wiley-Interscience, New York (2003), Ch. 3.

67. S. Brunauer, L.S. Deming, W.S. Deming, and E. Teller, *J. Am. Chem. Soc.* **62**, 1723 (1940).

68. X.B. Zhao, B. Xiao, A.J. Fletcher, and K.M. Thomas, *J. Phys. Chem. B* **109**, 8880 (2005).

69. S. Brunauer, P.H. Emmet, and E. Teller, *J. Phys. Chem.* **60**, 309 (1938).

70. S.G. Gregg and K.S.W. Sing, *Adsorption, Surface Area, and Porosity*. Academic Press, New York (1982).

71. S.R. Morrison, *The Chemical Physics of Surfaces*, 2nd Ed. Plenum Press, New York (1990), Ch. 7.

72. S.L. Bragg, J.W. Brault, and W.H. Smith, *Astrophys. J.* **263**, 999 (1982).

73. D.K. Veirs and G.M. Rosenblatt, *J. Mol. Spec.* **121**, 401 (1987).

74. P. Essenwanger and H.P. Gush, *Can. J. Phys.* **62**, 1680 (1984).

75. F. Fernandez-Alonso, B.D. Bean, J.D. Ayers, A.E. Pomerantz, and R.N. Zare, *Z. Phys. Chem. (Munich)* **214**, 1167 (2000); F. Fernandez-Alonso, PhD thesis, Stanford University, Stanford, CA (2000).

76. K.A. Williams, B.K. Pradhan, P.C. Eklund, M.K. Kostov, and M.W. Cole, *Phys. Rev. Lett.* **88**, 165502 (2002).

77. B.K. Pradhan, G.U. Sumanasekera, K.W. Adu, H.E. Romero, K.A. Williams, and P.C. Eklund, *Physica B* **323**, 115 (2002).

78. T. Pietraß and K. Shen, *Solid State Nucl. Mag.* **29**, 125 (2006).

79. I. Yu, J. Lee, and S.G. Lee, *Physica B* **329–333**, 421 (2003).

80. M. Schmid, S. Kramer, G. Goze, M. Mehring, S. Roth, and P. Bernier, *Synth. Met.* **135–136**, 727 (2003).

81. M. Shiraishi and M. Ata, *J. Nanosci. Nanotechnol.* **2**, 463 (2002).

82. D.L. Price, M.-L. Saboungi, and F.J. Bermejo, *Rep. Prog. Phys.* **66**, 407 (2003).

83. F.R. Trouw and D.L. Price, *Annu. Rev. Phys. Chem.* **50**, 571 (1999).

84. S.W. Lovesey, *Theory of Neutron Scattering from Condensed Matter*, Vol. I. Oxford University Press, Oxford, NY (1984).

85. P.C.H. Mitchell, S.F. Parker, A.J. Ramirez-Cuesta, and J. Tomkinson, *Vibrational Spectroscopy with Neutrons, with Applications in Chemistry, Biology, Materials Science, and Catalysis*. World Scientific, London, U.K. (2005).

86. C.J. Carlile and M.A. Adams, *Physica B* **182**, 431 (1992). URL: www.isis.rl.ac.uk/molecularSpectroscopy/iris/

87. D. Colognesi, M. Celli, and M. Zoppi, *J. Chem. Phys.* **120**, 5657 (2004).

88. P. A. Egelstaff, *An Introduction to the Liquid State*. Oxford University Press, Oxford, NY (1992), p. 229.

89. D.E. O'Reilly and E.M. Peterson, *J. Chem. Phys.* **66**, 934 (1977).

90. B.N. Esel'son, Yu.P. Blagoi, V.N. Grigorjev, V.G. Manzhelii, S.A. Mikhailenko, and N.P. Noklyudov, *Properties of Liquid and Solid Hydrogen*. Israel Program for Scientific Translations, Jerusalem, Israel (1971).

91. J. Dawidowski et al., *Phys. Rev. B* **73**, 144203 (2006).

92. F. Fernandez-Alonso, F.J. Bermejo, C. Cabrillo, R.O. Loutfy, V. Leon, and M.L. Saboungi, *Phys. Rev. Lett.* **98**, 215503 (2007).

93. *ISIS Database of Inelastic Neutron Scattering Spectra*. URL: www.isis.rl.ac.uk/molecularSpectroscopy/

94. M. Celli, D. Colognesi, and M. Zoppi, *Phys. Rev. E* **66**, 021202 (2002).

95. D. Colognesi et al., *Appl. Phys. A* **74**, S64 (2002).

96. M.P. Allen and D.J. Tildesley, *Computer Simulation of Liquids*. Pergamon Press, Oxford, NY (1987).

97. D. Nicholson and N.G. Parsonage, *Computer Simulation and the Statistical Mechanics of Adsorption*. Academic Press, New York (1982).

98. V. Buch, *J. Chem. Phys.* **100**, 7610 (1994).

99. W.A. Steele, *The Interaction of Gases with Solid Surfaces*. Pergamon Press, Oxford, NY (1974).

100. A.J. Stone, *The Theory of Intermolecular Forces*. Clarendon Press, Oxford, NY (1996), p. 157.

101. S.K. Bhatia and A.L. Myers, *Langmuir* **22**, 1688 (2006).

102. R.F. Cracknell, *Phys. Chem. Chem. Phys.* **3**, 2091 (2001).

103. A. Chambers, C. Park, R.T.K. Baker, and N.M. Rodriguez, *J. Phys. Chem. B* **102**, 4253 (1998).

104. C. Park, P.E. Anderson, A. Chambers, C.D. Tan, R. Hidalgo, and N.M. Rodriguez, *J. Phys. Chem. B* **103**, 10572 (1999).

105. M.T. Knippenberg, S.J. Stuart, and H. Cheng, *J. Mol. Model.* **14**, 343 (2008).

106. F. Huarte-Larrañaga and M. Albertí, *Chem. Phys. Lett.* **445**, 227 (2007).

107. H. Cheng, A.C. Cooper, G.P. Pez, M.K. Kostov, P. Piotrowski, and S.J. Stuart, *J. Phys. Chem. B* **109**, 3780 (2005).

108. M.K. Kostov, H. Cheng, A.C. Cooper, and G.P. Pez, *Phys. Rev. Lett.* **89**, 146105 (2002).

109. R.P. Feynmann and A.R. Hibbs, *Quantum Mechanics and Path Integrals*. McGraw-Hill, New York (1965).

110. Q. Wang and J.K. Johnson, *J. Chem. Phys.* **110**, 577 (1999); Q. Wang and J.K. Johnson, *J. Phys. Chem. B* **103**, 277 (1999).

111. Q. Wang, S.R. Challa, D.S. Sholl, and J.K. Johnson, *Phys. Rev. Lett.* **82**, 956 (1999).

112. R.A. Trasca, M.K. Kostov, and M.W. Cole, *Phys. Rev. B* **67**, 035410 (2003).

113. G. Garberoglio, M.M. DeKlavon, and J.K. Johnson, *J. Phys. Chem. B* **110**, 1733 (2006).

114. S.K. Jain, K.E. Gubbins, R.J.-M. Pellenq, and J.P. Pikunic, *Carbon* **44**, 2445 (2006).

115. A.J. Pertsin and A.I. Kitaigorodsky (Eds.), *The Atom-Atom Potential Method*. Springer, Berlin, Germany (1986), p. 89.

116. S. Hamel and M. Côté, *J. Chem. Phys.* **121**, 12618 (2004).

117. A.P. Sutton, *Electronic Structure of Materials*. Oxford University Press, Oxford, NY, reprinted in 2004.

118. R.M. Martin, *Electronic Structure: Basic Theory and Practical Methods*. Cambridge University Press, Cambridge, U.K. (2004).

119. A. Szabo and N.S. Ostlund, *Modern Quantum Chemistry: Introduction to Advanced Electronic Structure Theory*. McGraw-Hill, New York (1989).

120. W. Koch and M.C. Holthausen, *A Chemist's Guide to Density Functional Theory*. Wiley-VCH, Weinheim, Germany (2002).

121. P. Hohenberg and W. Kohn, *Phys. Rev.* **136**, B864 (1964).

122. W. Kohn and L.J. Sham, *Phys. Rev.* **140**, A1133 (1965).

123. R.C. Lochan and M. Head-Gordon, *Phys. Chem. Chem. Phys.* **8**, 1357 (2006).

124. S. Patchkovskii, J.S. Tse, S.N. Yurchenko, L. Zhechkov, T. Heine, and G. Seifert, *Proc. Natl. Acad. Sci. U.S.A.* **102**, 10439 (2005).

125. N. Ooi, A. Rairkar, and J.B. Adams, *Carbon* **44**, 231 (2006).

126. G. Mpourmpakis and G.E. Froudakis, *J. Nanosci. Nanotechnol.* **8**, 3091 (2008).

127. V. Meregalli and M. Parrinello, *Appl. Phys. A - Mater.* **72**, 143 (2001).

128. G. Mpourmpakis, G.E. Froudakis, G.P. Lithoxoos, and J. Samios, *J. Chem. Phys.* **126**, 144704 (2007).

129. I. Cabria, M.J. López, and J.A. Alonso, *J. Chem. Phys.* **128**, 144704 (2008).

130. G. Mpourmpakis, E. Tylianakis, D. Papanikolaou, and G.E. Froudakis, *J. Nanosci. Nanotechnol.* **6**, 3731 (2006).

131. I. Cabria, M.J. López, and J.A. Alonso, *J. Chem. Phys.* **123**, 204721 (2005).

132. R.J.-M. Pellenq, F. Marinelli, J.D. Fuhr, F. Fernandez-Alonso, and K. Refson, *J. Chem. Phys.* **129**, 22470 (2008).

133. O. Maresca, R.J.-M. Pellenq, F. Marinelli, and J. Conard, *J. Chem. Phys.* **121**, 12548 (2004).

134. M. Cobian and J. Iñiguez, *J. Phys.: Condens. Matter* **20**, 285212 (2008).

135. M. Yoon et al., *Nano Lett.* **7** 2578 (2007).

136. H.W. Kroto, J.R. Heath, S.C. O'Brien, R.F. Curl, and R.E. Smalley, *Nature* **318**, 162 (1985).

137. S. Iijima, *Nature* **354**, 56 (1991).

138. S. Iijima and T. Ichihashi, *Nature* **363**, 603 (1993).

139. D.S. Bethune, C.H. Kiang, M.S. de Vries, G. Gorman, R. Savoy, J. Vazquez, and R. Beyers, *Nature* **363**, 605 (1993).

140. L.V. Radushkevich and V.M. Lukyanovich, *Zh. Fizicheskoi Khimii* **26**, 88 (1952).

141. R. Saito, G. Dresselhaus, and M.S. Dresselhaus, *Physical Properties of Carbon Nanotubes*. Imperial College Press, London, U.K. (1998).

142. N. Hamada, S.I. Sawada, and A. Oshiyama, *Phys. Rev. Lett.* **68**, 1579 (1992).

143. R. Saito, M. Fujita, G. Dresselhaus, and M.S. Dresselhaus, *Phys. Rev. B* **46**, 1804 (1992).

144. A.C. Dillon, K.M. Jones, T.A. Bekkedahl, C.H. Kiang, D.S. Bethune, and M.J. Heben, *Nature* **386**, 377 (1997).

145. P. Chen, X. Wu, and K.L. Tan, *Science* **285**, 91 (1999).

146. R.T. Yang, *Carbon* **38**, 623 (2000); H.M. Cheng, Q.-H. Yang, and C. Liu, *Carbon* **39**, 1447 (2001); M. Hirscher, M. Becher, M. Haluska, F. von Zeppelin, X. Chen, U. Dettleff-Weglikowska, and S. Roth, *J. Alloys Compd.* **356–357**, 433 (2003).

147. M. Hirscher et al., *J. Alloys Compd.* **330–332**, 654 (2002).

148. Y. Reng and D.L. Price, *Appl. Phys. Lett.* **79**, 3684 (2001).

149. P.A. Georgiev et al., *J. Phys.: Condens. Matter* **16**, L73 (2004).

150. V. Leon, PhD thesis, University of Orleans, Orleans, France 2006.

151. M. Ge and K. Sattler, *Chem. Phys. Lett.* **220**, 192 (1994); A. Krishnan, E. Dujardin, M.M.J. Treacy, J. Hugdahl, S. Lynum, and T.W. Ebbesen, *Nature* **388**, 451 (1997).

152. D.B. Geohegan et al., *Phys. Status Solidi B* **244**, 3944 (2007).

153. A.A. Puretzky, D.J. Styers-Barnett, C.M. Rouleau, H. Hu, B. Zhao, I.N. Ivanov, and D.B. Geohegan, *Appl. Phys. A - Mater.* **93**, 849 (2008).

154. H. Wang et al., *Nanotechnology* **15**, 546 (2004); D. Kasuya et al., *J. Phys. Chem. B* **106**, 4947 (2002).

155. S. Berber, Y.-K. Kwon, and D. Tomanek, *Phys. Rev. B* **62**, R2291 (2000).

156. K. Urita et al., *Nano Lett.* **6**, 1325 (2006).

157. K. Murata, K. Kaneko, H. Kanoh, M. Yudasaka, D. Kasuya, K. Takahashi, F. Kokai, M. Yudasaka, and S. Iijima, *J. Phys. Chem. B* **106**, 1132 (2002).

158. D.B. Geohegan et al., *U.S. Department of Energy Project ID STP-6 Report*, Oak Ridge National Laboratory, Oak Ridge, TN (2007).

159. W.J. Thompson, *Angular Momentum: An Illustrated Guide to Rotational Symmetries for Physical Systems*. John Wiley & Sons, New York (1994), Ch. 7.

160. A.V.A. Kumar et al., *J. Phys. Chem. B* **110**, 16666 (2006); *Phys. Rev. Lett.* **95**, 245901 (2005).

161. D.S. Sivia et al., *Physica B* **182**, 341 (1992); D. Sivia and J. Skilling, *Data Analysis: A Bayesian Tutorial*. Oxford University Press, Oxford, NY (2006), Ch. 4.

162. G. Mpourmpakis et al., *Nano Lett.* **7**, 1893 (2007).

163. G.E. Froudakis, *Nano Lett.* **1**, 531 (2001).

164. L. Duclaux et al., *J. Phys. Chem. Solids* **67**, 1122 (2006).

165. M. S. Dresselhaus et al., *Adv. Phys.* **51**, 1 (2002).

166. S. A. Solin and H. Zabel, *Adv. Phys.* **37**, 87 (1988).

167. A. Lovell, N.T. Skipper, S.M. Bennington, and R.I. Smith, *J. Alloys Compd.* **446–447,** 397–401 (2007).

168. K. Watanabe et al., *Nat. Phys. Sci.* **233**, 160 (1971).

169. J.P. Beaufils et al., *Mol. Phys.* **44**, 1257 (1981).

170. A.P. Smith et al., *Phys. Rev. B* **53**, 10187 (1996).

171. W.J. Stead et al., *J. Chem. Soc., Faraday Trans. 2* **84**, 1655 (1988).

172. A. Lovell et al., *Physica B* **385–386**, 163 (2006).

173. A. Lovell, PhD Thesis, University College London, London, U.K. (2007).

174. A. Lovell, F. Fernandez-Alonso, N.T. Skipper, K. Refson, S.M. Bennington, and S.F. Parker, *Phys. Rev. Lett.* **101**, 126101 (2008).

175. S.J. Clark et al., *Z. Kristallogr.* **220**, 567 (2005).

176. D. Vanderbilt, *Phys. Rev. B* **41**, 7892 (1990).

177. J.G. Vitillo et al., *J. Chem. Phys.* **122**, 114311 (2005).

178. B.R. Johnson, *J. Chem. Phys.* **67**, 4086 (1977).

179. J. Germain, F. Svec, and J.M. Fréchet, *Chem. Mater.* **20**, 7069 (2008).

180. M. Dinca and J.R. Long, *J. Am. Chem. Soc.* **129**, 11172 (2007).

181. T.E. Weller, M. Ellerby, S.S. Saxena, R.P. Smith, and N.T. Skipper, *Nat. Phys.* **1**, 39 (2005).

182. G.K. Dimitrakakis, E. Tylianakis, and G.E. Froudakis, *Nano Lett.* **8**, 3166 (2008); *New Sci.* **2683**, 27 (2008).

41

Hydrogen Storage in Nanoporous Carbon

Iván Cabria
Universidad de Valladolid

María J. López
Universidad de Valladolid

Julio A. Alonso
Universidad de Valladolid

and

Universidad del País Vasco

and

Donostia International Physics Center

41.1 Introduction

The world energy crisis is not a problem of the amount of energy that we need or we use, but a problem of the speed at which we use energy. Each person in this planet consumes, on an average, 50–70 kWh/day, which means a power of 2–3 kW. The solar radiation reaching the earth's surface is, on an average, 100 and 250 W/m² in winter and summer, respectively. Taking into account a conversion efficiency of the present-day solar cells at 12%, a person needs a power of 166–250 m²/day on an average from solar cells in winter and 66–100 m² in summer. Our society uses small amounts of energy compared with the energy we receive from the sun. However, we consume too much energy in a very short period of time. Hence, it is a problem of power: energy used per unit of time. The reasons for the wide use and success of fossil fuels in the present economy are that they fit this need of "quick energy" and are cheap. These fuels are not clean energy vectors, are responsible of pollution of the environment and global warming, and the economy of many countries depends on foreign fossil fuels, causing from time to time world economic crisis due to insecurity of supply. Besides, they will be exhausted after some decades in the future. Many efforts have been devoted to find clean and "quick energies."

Our society consumes an important part of the total energy for transporting goods and persons by means of machines based on fossil fuels: cars, trucks, ships, planes, etc. An alternative to the present gasoline car is the hydrogen car. This car uses an electric motor based on a hydrogen fuel cell instead of electric batteries. There are many advantages of hydrogen cars over electric battery and gasoline cars: hydrogen is very abundant and can be produced from water by electrolysis. The only emission from a hydrogen fuel cell is pure water. The main advantage is that the full recharge times of a hydrogen and an electric car with the same autonomy range, 200 km, are 10 min and about 8 h, respectively, using 110 V (3 h using 220 V). However, the efficiency of the engine based on electric batteries is slightly larger (72%) than that of an engine based on hydrogen fuel cells (64%). The efficiency of a gasoline engine is smaller (20%). Hydrogen has the highest energy content per mass (120 MJ/kg) compared with any other fuel. Gasoline has a smaller energy content per mass (44 MJ/kg). But it has a larger energy content per volume (35 MJ/L) than gaseous hydrogen at room temperature and 1 atm pressure (0.01 MJ/L) and liquid hydrogen. Electric batteries also have a smaller energy content per mass, but a larger energy content per volume than hydrogen. For instance, nickel hydride batteries have energy contents of 0.11–0.29 MJ/kg and 0.5–1.1 MJ/L.

The most common method to produce hydrogen is by steam reforming of natural gas. This method releases CO_2, but it is possible to separate this gas and inject it in a gas reservoir, avoiding environmental damages. Other methods are electrolysis and biological production from water by means of algae and sunlight. Gas reforming and electrolysis require energy. If the energy to produce hydrogen comes from nonclean sources, then the reduction of global warming and pollution will not be possible. Hence, the use of electricity from renewable sources, such as the sun, wind, hydroelectric dams, and geothermal heat, has been proposed for this purpose.

Nowadays, an efficient reversible onboard hydrogen storage system is the main bottleneck of the hydrogen car technology. There are many research efforts to find appropriate materials for onboard storage. Nanoporous carbon is a family of new materials that, according to experiments and theory, could be suitable for onboard hydrogen storage and hence could solve the technological bottleneck. These materials also have other applications (Schuth et al., 2002): (a) air filters for gas purification, removing contaminants, odors, gases, volatile organic compounds from painting, dry cleaning, and other processes; (b) textiles coated with nanoporous carbon, with antibacteria and deodorization properties; (c) water filters for groundwater remediation, sewage treatment, and drinking water filtration; (d) filters in medical devices; (e) membranes to separate gases and for catalysis processes; (f) storage of gases; and (g) supercapacitors.

The hydrogen storage properties of nanoporous carbons are better understood if the place of this family of materials in the field of hydrogen storage is explained previously. With that purpose on mind, we present in the following sections some concepts common to different subfields of hydrogen storage and we point out the role of nanoporous carbons. In Section 41.2, we explain how storage methods and/or materials are suitable for onboard hydrogen storage if they have certain specific properties. The efforts of scientists are directed to the goal of finding methods and materials having those specific properties. There are three types or methods of hydrogen storage: in gas and liquid forms, and stored in solid materials. Storage in nanoporous carbons falls into the last category. We dedicate Section 41.3 to present the types of hydrogen storage, with special emphasis on the storage in solid materials. There are three mechanisms of hydrogen storage in solid materials: physisorption, chemisorption, and chemical reactions. The relevant mechanism on nanoporous carbons is physisorption, and we dedicate Section 41.4 to physisorption, especially to the physisorption of hydrogen at the nanoscale. In Sections 41.5 through 41.7, we deal with the two main types of nanoporous carbons: activated carbons and carbide-derived carbons, discussing hydrogen storage in those materials. In Section 41.8 there is a comparison of the storage capacities of nanoporous carbons and other materials. To conclude, in Section 41.9, we summarize the present state of knowledge on hydrogen storage on nanoporous carbons and provide future directions in this field.

41.2 Onboard Hydrogen Storage Goals

In January 2002, the Department of Energy (DOE) of the United States and the U.S. Council for Automotive Research (USCAR) consortium signed the FreedomCAR partnership to get a hydrogen storage system for hydrogen cars. The USCAR consortium is composed by Daimler-Chrysler, Ford, and General Motors.

In February 2002, the FreedomCAR partnership established and issued different targets to get a hydrogen car with capacities similar to those of gasoline cars. They identified goals to be reached in future years (DOE, 2002a,b).

Some of these goals are related to the onboard hydrogen storage system. There are two main targets that the storage system should satisfy by 2010: a volumetric capacity of at least 0.045 kg of hydrogen/L of the storage system and a gravimetric capacity of at least 6 wt % of hydrogen at room temperature and moderate pressures (1–40 bar), both with reversible capacities. These are referred to the storage *system*, that is, container plus stored hydrogen. In the case of storage in solid materials, the storage capacities of the adsorbent *material* alone should be higher, by about 1.2–2 times, because the adsorbent material is itself in a container. The interest in nanoporous carbons arises because they might be one of the supporting materials that could help hydrogen cars reach those targets.

The FreedomCAR partnership used a gasoline weighted average corporate vehicle (WACV) to establish those targets, such that the future hydrogen cars should have the same capacities as that car, especially the same autonomy range. This average car includes minivans, light trucks, economy cars, and SUV/crossover vehicles in proportion to their sales in the United States. The targets and their explanation can be found on the web page of the National Project "On-board Hydrogen Storage" of the DOE (2002a,b). This average car weights 1740 kg, has an autonomy range of 600 km, and uses 75 L of gasoline, which contains 2625 MJ of energy, to cover those 600 km. Its storage tank has a volume of 107 L and weights 74 kg, including gasoline.

The energy density of the average car is, then, 2625/107 = 24.53 MJ/L of the storage system. The FreedomCAR partnership took into account that an electric motor based on a hydrogen fuel cell was about 3.5 times more efficient than a gasoline motor in 2002, and assumed that the relative efficiency would be higher by 2010 to establish the target for the energy density or volumetric capacity: 5.4 MJ/L of the storage system. One kilogram of hydrogen contains 120 MJ, therefore, 5.4 MJ/L is equivalent to 0.045 kg of stored hydrogen/L of the storage system. This means also that the car will need 4.8 kg of hydrogen to cover 600 km.

The volumetric capacity v_c of the supporting or adsorbent material is the mass of stored hydrogen, M_H, divided by the volume of the supporting material:

$$v_c = \frac{M_H}{V_{\text{supporting}}} = \rho_H \rho_{\text{supporting}} v_H, \qquad (41.1)$$

where

v_H is the measured volume of stored hydrogen divided by the mass of the supporting material

ρ_H and $\rho_{\text{supporting}}$ are the densities of hydrogen and the supporting material

v_c is measured, for instance, in kg of H/L of the supporting material.

Note that the volumetric capacity of a material is related to the volume of the supporting material, not to the volume of hydrogen.

The second target refers to the gravimetric capacity or uptake w. This is defined

$$w = \frac{M_H}{M_H + M_{supporting}}, \qquad (41.2)$$

where

M_H is the mass of stored hydrogen
$M_{supporting}$ is the mass of the supporting material

w is measured in wt % units, also called sometimes hydrogen wt % units, after multiplying by 100 in Equation 41.2. The mass of the storage system of the average car, including the gasoline, is 74 kg, which gives a specific energy of 35.5 MJ/kg for the storage system. The higher efficiency of the fuel cell motor lowers the specific energy to 7.2 MJ/kg, which is equivalent to 6 hydrogen wt % taking into account that 1 kg of hydrogen contains 120 MJ. That is, the gravimetric capacity w should be at least 0.06 or 6%.

The targets for 2015 are more strict: 0.081 kg of hydrogen/L and 9 hydrogen wt %. The European Union and Japan have also established targets, but the DOE targets act as a reference for all the hydrogen storage research community.

41.3 Types and Mechanisms of Hydrogen Storage

In order to reach the above DOE targets, different types of hydrogen storage have been studied: gas, liquid, and solid storage. Gaseous hydrogen is usually stored in steel tanks at high pressures of 350–700 bar. The main problems of these tanks are the large volume and weight of the tanks, and the energy required to store hydrogen at high pressures. Other methods are storage in lightweight composite tanks at high pressures and in cryogenic tanks at low temperatures (20–77 K). A novel method consists in storing hydrogen gas at high pressures and 300–600 K in hollow glass microspheres.

Liquid hydrogen occupies much less volume than gaseous hydrogen and is usually stored by cooling it down to cryogenic temperatures, say 20 K. This method provides storage capacities of 20 wt % H_2 and 0.030 kg/L. The main disadvantages are the boil-off losses and the need of superinsulated cryogenic tanks. About one-third of the energy content of hydrogen is lost in their liquefaction.

Another possibility is to store hydrogen indirectly as part of other liquids, such as $NaBH_4$ solutions, rechargeable organic liquids, and ammonia. In this case, the usable hydrogen would be produced by a chemical reaction. Chemical hydrides react with water (hydrolysis) or alcohol (alcoholysis) releasing hydrogen. The most studied hydrolysis reaction is that of sodium borohydride with water: $NaBH_4 + 2 H_2O \rightarrow NaBO_2 + 4 H_2$. Other reactions do not involve water or alcohol, for instance, the ammonia–borane reactions, $NH_3BH_3 \rightarrow NH_2BH_2 + H_2 \rightarrow NHBH + H_2$, and the decalin-to-naphthalene reaction, $C_{10}H_{18} \rightarrow C_{10}H_8 + 5 H_2$. But these reactions are not easily reversible onboard the vehicle. Therefore, the reaction products should be removed and recycled off-board.

The third type is hydrogen storage in solid materials. It is safer than gas and liquid storage and reaches higher volumetric capacities. The main types of solid-storing materials are: carbon-based materials, noncarbon materials, and metallic hydrides. There are three mechanisms of hydrogen storage in solid materials: physisorption, chemisorption, and chemical reactions. Carbon-based materials, such as nanoporous carbons and carbon nanotubes (CNTs), and noncarbon materials, such as metal organic frameworks (MOFs), zeolites, and clathrate hydrates and organic polymers of intrinsic microporosity (PIMs), store hydrogen by molecular physisorption on the walls of their pores. On the other hand, metal hydrides store atomic hydrogen by chemisorption in the bulk. Chemical hydrides, a type of metal hydrides, release hydrogen by means of nonreversible chemical reactions. There is no agreement between different workers whether the stored hydrogen is in the liquid or in the solid state. Depending on the author, the semiliquid chemical hydrides are considered to store hydrogen in the liquid or in the solid state.

In the case of storage in the pores of a solid material, the total stored hydrogen is composed of the hydrogen adsorbed on the surface of the pores of the material and the hydrogen compressed in the empty volume of those pores. The DOE targets for the storage capacities of the adsorbent materials refer to the total stored hydrogen, but in the experimental and theoretical reports on storage in solid materials, and in this chapter from now on, these refer to the adsorbed hydrogen, not to the sum of compressed and adsorbed hydrogen. These adsorbed storage capacities are called sometimes excess storage capacities. The total storage capacities, compressed plus adsorbed, are rarely published (Jordá-Beneyto et al., 2007, 2008).

None of the current hydrogen storage methods and none of the known solid materials meets the DOE 2010 targets. In 1997, Dillon et al. (1997) reported a high excess storage capacity of carbon-based materials, in particular CNTs, of 5–10 wt % at 273 K and 0.04 MPa. Nowadays, after some controversy and many experimental and theoretical investigations (Schlapbach and Züttel, 2001; Züttel, 2003; Ansón et al., 2004a,b), the general consensus is that those high capacities were due to measurement errors (very small samples, less than 1 mg, among other error sources) and impurities, and that the storage capacities of carbon nanotubes are very low and far from the DOE requirements. The experiments by Ansón et al. (2004a), which are between the most careful experiments of hydrogen storage on carbon nanotubes, measured about 1 wt % at 77 K and 1 atm, and 0.01 wt % at 298 K and 1 atm.

The two possible mechanisms of storage of hydrogen on the surface of porous materials are physisorption and chemisorption. Physisorption is due to the weak interaction between the closed-shell H_2 molecule and the surface of the adsorbent material. In this case, hydrogen remains in the molecular state after sorption. This mechanism is reversible, that is, by small changes of temperature and pressure, it is possible to release the stored hydrogen. The relation between physisorption and hydrogen storage will be discussed in detail in the next section. Chemisorption occurs when the H_2 molecule interacts strongly with the substrate. This strong chemical interaction is usually accompanied by the dissociation of the molecule. The two H atoms then form independent chemical bonds with surface substrate atoms. The chemisorption binding energies of hydrogen to different materials are between 1 and 3 eV/H atom.

Dissociative chemisorption occurs on the surface of many metals, and the atoms can subsequently diffuse into the bulk of the metal to form metal hydrides. The chemically bonded hydrogen of metal hydrides can be released by heating the material. High temperatures are necessary to release the H atoms, and not all the hydrogen is released. Nevertheless, reversible storage capacities of 5.0–7.5 wt % can be obtained with many metal hydrides. The main problems are the release temperatures, higher than 200°C, and the slow desorption kinetics (Bogdanovic and Sandrock, 2002; Züttel, 2003). $NaAlH_4$, a sodium alanate, is a metal hydride with a release temperature of 315 K and a storage capacity of 5 wt %.

41.4 Physisorption

Hydrogen physisorption is based on the weakly attractive interaction between the closed-shell hydrogen molecules and the surface of the adsorbent material. The potential energy due to these weakly attractive forces has the minimum at H_2-surface distances of 2.5–3.0 Å, and the depth of this minimum is about 0.1 eV/molecule. Small nanopores, of a size approximately two times that distance, are needed to maximize the interaction and hence the storage (Rzepka et al., 1998; Schlapbach and Züttel, 2001; Züttel, 2003; Cabria et al., 2007). Therefore, hydrogen storage based on physisorption requires porous materials, such as nanoporous carbons and metal–organic frameworks. In particular, it will be optimized using materials with a narrow pore size distribution around the range of 5–7 Å. The adsorption of the first layer of molecules is due to the interaction with the surface. The second and successive adsorbed layers, are stabilized by the interaction with the first adsorbed layer, and their binding energies are equal to the latent heat of sublimation of hydrogen.

Physisorbed molecules can be easily adsorbed and desorbed from the adsorbent surface, and hence this mechanism allows for reversible hydrogen storage and fast adsorption/desorption kinetics. However, physisorption reaches lower storage capacities than chemisorption or chemical bonding mechanisms, and allows to reach high capacities only at low temperatures, say 77 K, and substantial pressures, 0.1–10 MPa.

41.4.1 Specific Surface Area

The amount of physisorbed molecules is proportional to the available surface area of a material sample, but the mass of the sample is also important to obtain a large gravimetric storage capacity, which is a relative amount measured in weight %. Hence, the important quantity for adsorption of molecules is not the surface, but the specific surface area (SSA), a concept of surface chemistry. The specific surface area of a material is defined as its total surface area divided by its mass (usually in units of m²/g), or also as its surface area divided by its volume (usually in units of m²/m³ = m⁻¹).

The gravimetric capacity of a material of surface S is

$$w = \frac{M_H}{M_H + M_{adsorbent}} = \frac{S_{SSA}/S_{SSAH_2}}{S_{SSA}/S_{SSAH_2} + 1} \approx S_{SSA}/S_{SSAH_2}, \qquad (41.3)$$

where $S_{SSA} = S/M_{adsorbent}$ and $S_{SSAH_2} = s/m_{H_2}$ are the specific surface areas of the adsorbent material and of a single hydrogen molecule, respectively. m_{H_2} and s are the mass and the adsorption cross section of a single hydrogen molecule, respectively. S_{SSA}/S_{SSAH_2} is of the order of 0.01, much smaller than unity, and so we can make the approximation $w \approx S_{SSA}/S_{SSAH_2}$.

Some simple examples based on a cube are very useful to understand the relevance for hydrogen storage of the specific surface area compared to the geometric surface area. A solid cube of lateral side l and volumetric density ρ has a surface area of $S = 6l^2$ and a specific surface area of $6/(l\rho)$. The cube can adsorb $6l^2/s$ molecules. Another cube of the same side l but with a density k times smaller has the same surface area, but a specific surface area k times larger and a gravimetric capacity approximately k times larger, as can be seen from Equation 41.3. A larger cube of side $2l$ and the same volumetric density as the initial cube adsorbs four times more molecules, but the specific surface area is two times smaller and also its gravimetric capacity. The same cube sliced in n plates of thickness t, $nt = l$, has a surface area of $l^2(2n + 4)$ and a specific surface area of $(2n + 4)/(l\rho)$. The specific surface area of the sliced cube is larger than the specific surface area of the cube for any value of $n > 1$, and hence it has larger gravimetric capacities than a cube of the same mass and volume. This example visualizes the need of highly porous materials (large values of n) to obtain large specific surface areas and large storage capacities.

The method most often used to measure the specific surface area of pores of less than 50 nm is the Brunauer–Emmett–Teller (BET) method. It consists in measuring the volume V of the adsorbed substance at different pressures P and constant temperature (the isotherms) and then plotting $y = x/(V(1 - x))$ vs. $x = P/P_0$, where P_0 is the vapor pressure at the temperature of the experiments. The BET equation (Brunauer et al., 1938) for adsorption of fluids in a solid surface is $y = mx + n$, with

$$m = \frac{(c-1)}{(V_{ml}c)}, \quad n = \frac{1}{(V_{ml}c)}, \quad c = \frac{\exp(E_1 - E_{lq})}{RT}, \qquad (41.4)$$

where

V_{ml} is the volume of a monolayer of adsorbed fluid
E_1 is the heat of adsorption of the first layer
E_{lq} is the heat of liquefaction

The fitting of the experimental isotherms to the BET equation yields the value of V_{ml}, and then the total surface area of the material is

$$S = \frac{V_{ml}N_A s}{v_m}, \qquad (41.5)$$

where

N_A is the Avogadro number

s is the adsorption cross section of a single molecule of the adsorbed substance (in units of area/molecule)

v_m is the molar volume of the adsorbed substance (the units of molar volume are volume/mol)

The BET-specific surface area is then S/M, where M is the molar weight of the adsorbed substance. The results depend on the type of molecule. Usually N_2 and CO_2 are used to measure the specific surface area. This method takes into account the fine structure and texture of the surface.

41.4.2 Specific Pore Volume

A related concept, but in three dimensions, is the specific pore volume: the sum of the volumes of all the pores of a material divided by the mass of the material. The terms surface area and pore volume are sometimes used in the literature instead of specific surface area and specific pore volume, but their area/mass or volume/mass units indicate that these are specific magnitudes.

The gravimetric capacity of a porous material whose pores occupy a volume V_{pore} is

$$w = \frac{n_{H_2} m_{H_2}}{n_{H_2} m_{H_2} + M_{adsorbent}} = \frac{c v_e / v_{eH_2}}{c v_e / v_{eH_2} + 1} \approx v_e / v_{eH_2}, \quad (41.6)$$

where

$v_e = V_{pore} / M_{adsorbent}$ is the specific pore volume of the adsorbent material

$v_{eH_2} = V_{H_2} / m_{H_2}$ is the specific volume of a single molecule (V_{H_2} and m_{H_2} are the volume and mass of a hydrogen molecule)

n_{H_2} is the number of molecules adsorbed

c is a constant defined by the relation $n_{H_2} = c V_{pore} / V_{H_2}$

Equation 41.6 displays well the correlation existing between the specific pore volume and the gravimetric storage capacity.

The most popular method to measure the specific pore volume is the Dubinin–Radushkevich (DR) method (Dubinin and Radushkevich, 1947; Dubinin and Plavnik, 1968). The experimental isotherms are fitted to the DR equation,

$$\ln V = \ln V_{DR} + B\left(\frac{T^2}{\beta^2}\right) x, \quad (41.7)$$

where

$x = (\ln(P_0/P))^2$

V is the volume of the adsorbed substance measured at pressure P

P_0 is the vapor pressure

V_{DR} is the volume of all the pores of the adsorbent material

β is the affinity constant of the adsorbed substance

B is a parameter that depends on the porous structure

The dimensionless affinity constant β is the ratio between the adsorption energy of the adsorbed substance and the adsorption energy of a reference substance, usually benzene. V_{DR} is obtained

from the slope of the fitted function, and the specific pore volume v_e of the adsorbent material is $V_{DR}/M_{adsorbent}$. This equation is based on the continuous filling of pores and is valid only for nanopores.

41.4.3 Hydrogen Physisorption at the Nanoscale

A simple model for the interaction between two weakly interacting identical atoms or molecules consists in a Lennard-Jones 12-6 potential:

$$V(r) = \frac{a}{r^{12}} - \frac{b}{r^6} = 4\varepsilon\left[\left(\frac{\sigma}{r}\right)^{12} - \left(\frac{\sigma}{r}\right)^6\right]. \quad (41.8)$$

The first term is repulsive and takes into account the hard-core volume of the interacting atoms or molecules. The second term is the attractive part of the potential. σ gives the effective hard-core size of the atom or molecule and ε controls the depth of the potential well. In the case of the interaction between a carbon atom and an inert gas atom X, the Berthelot combining rules can be applied: $\sigma_{CX} = (\sigma_{CC} + \sigma_{XX})/2$ and $\varepsilon_{CX} = \sqrt{\varepsilon_{CC}\varepsilon_{XX}}$. Stan and Cole have applied this simple model to study the adsorption of rare gases, He, Ne, Ar, Kr, and Xe, on graphene layers and carbon nanotubes, in the regime of low coverage (Stan and Cole, 1998). In the case of the interaction between a carbon atom and a hydrogen atom of a hydrogen molecule, $\sigma_{CH} = (\sigma_{CC} + \sigma_{HH})/2$ and $\varepsilon_{CH} = \sqrt{\varepsilon_{CC}\varepsilon_{HH}}$, where ε_{HH} is now the parameter representing the weak interaction between two hydrogen atoms in different hydrogen molecules. Typical values of σ_{CH} and ε_{CH}/k_B are about 3.2 Å and 28–31 K, respectively. The total interaction potential between a carbon surface and a hydrogen molecule can be approximated by adding the individual interactions of the carbon atoms of the surface with the hydrogen molecule, taking into account only the C–H distances below a large cutoff value.

The interaction potential between H_2 and a graphene surface obtained in this model, using $\sigma_{CH} = 3.19$ Å and $\varepsilon_{CH}/k_B = 30.5$ K, is plotted in Figure 41.1. The binding energy is 67 meV/molecule and the equilibrium distance is 3.41 Å. This equilibrium distance is smaller than the intermolecular distance in liquid hydrogen (4.1 Å) and larger than the diameter of the H_2 molecule (2.7 Å). A system formed by two parallel flat graphene layers is a good model for a slitpore. Then, in order to maximize the interaction of a hydrogen molecule with the two parallel walls of a slitpore, the distance between the graphene layers should be about two times the equilibrium distance of the graphene-molecule potential of Figure 41.1. Hence, according to this model, nanopores of about 6.8 Å are required to maximize the interaction with the walls of the nanopores. The important conclusion is that pores of nanometric size are required to optimize the hydrogen storage capacity.

Li et al. considered the equilibrium between the two phases of hydrogen in a material, physisorbed hydrogen and compressed hydrogen, using the Van't Hoff equation (Li et al., 2003). The compressed H_2 phase is the hydrogen stored by compression in the empty space of the material, not by physisorption on the

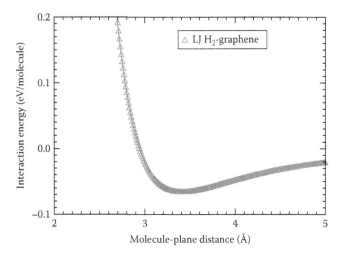

FIGURE 41.1 Interaction energy between a hydrogen molecule and a planar graphene layer (in eV/molecule) as a function of their separation, obtained by summation of Lennard-Jones interactions.

external and internal surfaces of the material. By estimating that the difference between the entropies of the compressed and physisorbed phases was about 10–$15k_B$ per molecule at room temperature and 1 atm, Li et al. showed that an enthalpy difference between those two phases of approximately 0.3–0.4 eV/molecule is necessary to achieve reversible hydrogen storage at that temperature and pressure (Li et al., 2003). We have estimated above that the physisorption energy of hydrogen on a single flat graphitic surface is about 0.1 eV/molecule. Consequently, in order to achieve an interaction energy equal to the above enthalpy change (0.3–0.4 eV/molecule), a hydrogen molecule should interact with a more complex surface, such as the double surface of a slitpore, or with a curved surface, such as in cylindrical or spherical-like pores.

More sophisticated calculations also point to the need of nanoporous carbon materials in order to obtain high storage capacities. Rzepka et al. used Lennard-Jones 12-6 potentials for the interaction between hydrogen molecules and between hydrogen molecules and carbon surfaces in grand canonical Monte Carlo (GCMC) simulations to calculate the storage capacities of slitpores and carbon nanotubes (Rzepka et al., 1998). They found that the optimal pore size was about 7 Å at 300 K and 10 MPa in both cases. Wang and Johnson performed GCMC path integral simulations for molecular hydrogen in slitpores (Wang and Johnson, 1999a) and in arrays of single-wall carbon nanotubes: inside the nanotubes, outside, and in the interstices (Wang and Johnson, 1999a,b). The path integral formalism includes quantum effects in the motion of the confined molecules (Wang and Johnson, 1998). The H_2–H_2 interactions were modeled using the Silvera–Goldman potential (Silvera and Goldman, 1978) and the hydrogen–carbon interactions with the Crowell–Brown potential (Crowell and Brown, 1982). The calculations delivered optimal pore sizes of 6 and 9 Å at 298 and 77 K, respectively. Patchkovskii et al. performed first-principles calculations for the interactions and used a quantum-thermodynamic model, predicting an optimal slitpore width of 6.0–7.5 Å at 300 K (Patchkovskii et al., 2005). An improved

model, using an empirical equation of state for hydrogen and the exact relationship between the equilibrium constant and the equation of state, yielded an optimal slitpore width of 6 Å at 300 K (Cabria et al., 2007). This optimal nanometric size of the pores is confirmed by the experiments: the nanoporous carbons with the highest storage capacities have narrow pore size distributions centered on sizes of 5.0–6.6 Å (de la Casa-Lillo et al., 2002; Gogotsi et al., 2003, 2005; Dash, 2006; Yushin et al., 2006). This can be appreciated in Figure 41.2.

A last remark about hydrogen physisorption can be useful. In order to reach the DOE gravimetric target of 6 wt %, the density of the adsorbed hydrogen, not the density of the compressed hydrogen, should be similar to that of liquid hydrogen. A simple qualitative model will help to understand this comment. Adsorption of one H_2 molecule per 2.61 C atoms is necessary to obtain a gravimetric density of 6 wt %; notice that $2/(2.61 \times 12 + 2) = 0.06$. If the adsorbing material is a graphene sheet and this material adsorbs a hydrogen monolayer of homogeneous density on each side of the sheet, then the H_2–H_2 distance in those two physisorbed monolayers is 3.70 Å, smaller than the intermolecular distance in liquid hydrogen, which is 4.1 Å. The above value of the H_2–H_2 distance is obtained from the following equations:

$$d_{H_2-H_2} = \sqrt{1/\sigma_{H_2}} = \sqrt{S_{2.61}/0.5}, \tag{41.9}$$

$$S_{2.61} = \frac{2.61}{\sigma_C} = \frac{7.83\sqrt{3}d_{C-C}^2}{4}, \tag{41.10}$$

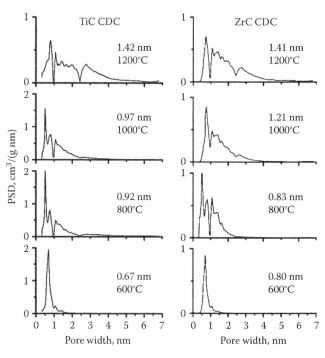

FIGURE 41.2 Pore size distribution of TiC-CDC and ZrC-CDC for different synthesis temperatures of the carbide-derived carbon, CDC. The average pore size is indicated. (Reproduced from Yushin, G. et al., *Adv. Funct. Mater.*, 16, 2288, 2006.With permission.)

where

> d_{C-C} is the nearest neighbor C–C distance
>
> σ_{H_2} is the number of H_2 molecules per unit surface on one side of a graphene sheet
>
> $S_{2.61}$ is the surface occupied by 2.61 C atoms on a graphene sheet
>
> σ_C is the number of carbon atoms per unit surface of a graphene sheet

In a nanoporous carbon, the adsorbed hydrogen would probably form thicker layers of smaller density on the carbon surface, but still near the liquid hydrogen density.

41.5 Properties of Nanoporous Carbons

The International Union of Pure and Applied Chemistry (IUPAC) proposed a definition of micropores, mesopores, and macropores (Sing et al., 1985): these are pores with effective widths $w < 2\,nm$, $2\,nm \leq w \leq 50\,nm$, and $w > 50\,nm$, respectively. However, most authors use the term "nanopores" applied to pores in the nanometer range, and we have used that term throughout this chapter. Nanoporous carbons are then carbon materials with a well-developed porosity and whose porosity is mainly due to nanopores. These materials are not crystalline; they are amorphous with a "worm-like" network of pores. They have large specific surface areas, between 300 and 3800 m²/g.

There are two types of nanoporous carbons, according to the synthesis methods: activated carbons (ACs), synthesized by physical or chemical activation of carbon-organic precursors or coals, and carbide-derived carbons (CDCs), synthesized by halogenation of metal carbides. Activated carbon fibers, carbon nanofibers, and graphitic nanofibers are activated carbons derived from organic polymers or from graphite. Carbon nanotubes and fullerenes are not usually considered nanoporous carbons. Carbon molecular sieves are carbon materials with narrow pores of a precise and uniform size, i.e., with a very narrow pore size distribution (PSD) used as adsorbents for gases and liquids. This term refers to an application and not to a material different from activated carbons and carbide-derived carbons. Figure 41.3 shows a transmission electron microscope image of a nanoporous carbon, in which the porosity of the material can be clearly appreciated.

The adsorption properties and storage capacities of nanoporous carbons depend on four main structural properties: specific surface area, specific pore volume, porosity volume, and pore size distribution. The pore shape and texture are, somehow, included in the specific pore volume. The specific surface area is one of the main structural factors used to explain the hydrogen storage capacities of nonporous and porous adsorbent materials. Nonporous materials show a linear relationship between specific surface area and hydrogen storage capacities. A linear relationship also exists on nanoporous carbons, but not for all samples of activated carbons (Noh et al., 1987; Johansson et al., 2002) and carbide-derived carbons (Dash, 2006; Yushin et al., 2006). The porosity properties should be also taken into account to explain the storage capacities of nanoporous carbons. Many nanoporous carbons have large pores and wide pore size distributions.

FIGURE 41.3 High-resolution transmission electron microscopy (HRTEM) image of a sample of a SiC-CDC. (Reproduced from Johansson, E. et al., *J. Alloys Compd.*, 330, 670, 2002. With permission.)

Among the structural properties, the geometry of the pore is very important in order to achieve large storage capacities. The interaction between two neighbor hydrogen molecules becomes highly repulsive if the distance is smaller than 2.5 Å. Nevertheless, the interaction of those molecules with a curved, concave surface can compensate the repulsive interactions between the molecules accumulated in the curved surface. Large cavities have enough room to store large quantities of hydrogen. However, large pores have a lot of empty space where the interaction energy between H_2 and the pore surface is negligible. The shape of the surface of the adsorbent material is included in the specific pore volume, and this physical magnitude is useful to explain the nonlinear relationship between specific surface area and storage capacities of some nanoporous carbons. In fact, many experimental papers report measurements of both specific surface area and specific pore volume and also a linear relationship between specific pore volume and storage capacities.

The IUPAC also classified the different types of isotherms. The experimental isotherms, storage capacity vs. pressure, of nanoporous solids at low temperatures, say 77 K, are of type I, also called Langmuir type. At room temperature, the isotherms are straight lines, that is, the storage capacities depend linearly on the pressure. The type I isotherms are sometimes fitted to the Langmuir isotherm or Langmuir adsorption equation:

$$c(P) = c_{max} \frac{aP}{1 + aP}. \tag{41.11}$$

The constant a decreases with increasing temperature, and hence this function turns into the observed linear dependence at room temperature.

One of the main goals of the experimental and theoretical studies is to find the relation between the porosity properties and the storage capacities, and since the porosity depends on the synthesis methods used to produce those materials, a second goal is to find experimental methods to tune the desired porosity properties.

Intensive efforts have been dedicated to find methods allowing control of the pore size of the porous carbons. Carbide-derived carbons are a new type of material obtained by chlorination of metal carbides. This method, described below, allows control of the pore size with an accuracy of Angstroms. Since nanoporous carbons are low-weight materials, have a large specific surface area, and have pores in the appropriate size range, they are promising candidates to obtain high hydrogen storage capacities. Another important advantage is that nanoporous carbons are materials of low cost, compared to carbon nanotubes. As of 2008, nanoporous carbons are cheap, between \$5 and \$18/kg, depending on the quality, and activated carbons and carbide-derived carbons have similar prices. Carbon nanotubes are much more expensive. Single-wall carbon nanotubes cost \$25,000–100,000/kg, and multiwall carbon nanotubes \$150–10,000/kg, with a mid-term price target of \$45/kg.

41.6 Activated Carbons

41.6.1 Synthesis

Activated carbon (AC), also called active carbon, activated coal and activated charcoal, is a form of carbon made porous by some processes starting with carbon-organic materials or coal. Activated carbons can be produced from any carbonaceous material such as nut, coconut and palm shells, olive, oilpalm, eucalyptus and peach stones, agricultural waste, wood, pitch, graphite, resin, polymeric compounds and coal, and even from tires (Ariyadejwanich et al., 2003) and chicken waste (Zhang et al., 2007). Activated carbons have specific surface areas between 300 and 3800 m^2/g, specific pore volumes in the range 0.1–0.8 cm^3/g and pore sizes of 4–15 Å. The two main synthesis methods are physical and chemical activation. The physical activation uses gases to process the initial material. There are two main physical activation methods. Carbonization consists in pyrolyzing the precursor at temperatures between 600°C and 900°C in an inert atmosphere, with gases like argon or nitrogen. The oxidation method consists in oxidizing the precursor or the previously carbonized material in atmospheres with carbon dioxide or oxygen at temperatures in the range 600°C–1200°C. Chemical activation consists in an impregnation of the precursor with acids like phosphoric acid, bases like KOH and NaOH, or salts like zinc chloride, followed by carbonization at temperatures in the range 450°C–900°C. Carbonization can be also applied simultaneously to chemical impregnation.

Carbon and graphite nanofibers are synthesized from the decomposition of certain hydrocarbon gases catalyzed on a metal surface at temperatures in the range 400°C–1000°C. This method is called catalytic chemical vapor deposition. Carbon nanofibers consist of graphene layers arranged as stacked cones, cups, or plates. Carbon nanotubes are carbon nanofibers made of graphene layers perfectly wrapped into cylinders. The graphite nanofibers consist of graphite platelets perfectly arranged with respect to the fiber axis. The distance between platelets is 3.4 Å and can be increased by intercalating some molecules between the layers. They have large surface areas (300–700 m^2/g) and diameters between 50 and 1000 Å.

Activated carbons synthesized by carbonizing carbonaceous materials or biomass do not reach easily specific surface areas larger than 500 m^2/g. Polymeric compounds, including cellulose, have a flexible structure that collapses in the pyrolysis process and the structure of the porous carbon obtained is very different from the precursor polymer structure. There is no control of the pore structure in the pyrolysis. Coal is the most used precursor due to its low price and because activated carbon made from coal has a much larger specific surface area. A lot of research is devoted to the production and study of activated carbon made from coal.

41.6.2 Storage Capacities

Kidnay and Hiza reported in 1967 the results of the first experiments on the adsorption of hydrogen on carbon-based materials (Kidnay and Hiza, 1967). They obtained 2 wt % at 25 atm and 76 K on a porous carbon obtained from coconut shell. Carpetis and Peschka were the first to propose the use of carbon-based materials to store hydrogen in 1976 (Carpetis and Peschka, 1976, 1978). They found a storage capacity of 5.2 wt % at 65 K and 41.5 atm for activated carbons. Agarwal et al. studied nine samples of activated carbons, whose specific surface areas were in the range 700–1500 m^2/g and found that the gravimetric capacity was proportional to the specific surface area (Agarwal et al., 1987). They measured gravimetric capacities between 1.1 and 3.1 wt % at 78 K and 40 atm for those samples. The gravimetric capacity at 78 K as a function of the pressure reached very quickly a constant value at about 12 atm, i.e., it saturated. These authors also observed that the storage capacity increased with the surface acidity of the activated carbons. They also reported (Noh et al., 1987) two activated carbon samples with the same BET-specific surface area, 1500 m^2/g, but with different storage capacities, 2.3 and 3.1 wt %. Therefore, a linear relationship between the gravimetric storage capacity and the specific surface area valid for all samples does not exist, and other factors have to be taken into account.

Ströbel et al. also measured the storage capacities of activated carbon samples and obtained a highest capacity of 1.6 wt % at 296 K and 123 atm (Ströbel et al., 1999). Nijkamp et al. studied samples of activated carbon nanofibers with different specific surface areas and pore volumes at 77 K and pressures up to 1 atm and also found that the hydrogen storage capacity was proportional to the specific surface area and to the specific pore volume, with a maximum capacity of 2.13 wt % (Nijkamp et al., 2001; Nijkamp, 2002). The amount of adsorbed hydrogen on activated carbons, activated carbon fibers, carbon nanotubes, and carbon nanofibers is proportional to the specific surface area: about 1.5 wt % per 1000 m^2/g of specific surface area at 77 K and 2.13 wt % per 2100 m^2/g on activated carbons and activated carbon fibers (Nijkamp et al., 2001; Schlapbach and Züttel, 2001). Kadono et al. investigated activated carbon fibers with a narrow pore size distribution around 5 Å and a specific pore volume less than 0.5 cm^3/g and found a maximum capacity of 2.0 wt % at 20 atm and 77 K (Kadono et al., 2003). Lueking et al. reported 1.23 and 0.03 wt % at 20 bar and temperatures of 77 and 300 K, respectively, on a sample of activated graphitic nanofiber with a specific surface area of 555 m^2/g (Lueking et al., 2005).

The porosity of activated carbons is commonly characterized by measuring the N_2 adsorption isotherms at 77 K, because this gas adsorbs in the whole range of pore sizes. However, the slow diffusion of N_2 inside narrow pores (sizes below 10 Å) at 77 K causes measurement errors. CO_2 adsorption at 273 K is also used to characterize activated carbons. The higher saturation pressure P_0 of CO_2 at 273 K induces a more efficient diffusion inside narrow pores (sizes up to 20–30 Å). Texier-Mandoki et al. found a linear relationship between the storage capacities of activated carbons and the specific pore volume measured using CO_2 adsorption isotherms (Texier-Mandoki et al., 2004). They also used N_2 adsorption isotherms. These authors found that a higher porosity results in higher capacities, that narrow pores of less than 7 Å are necessary to obtain high storage capacities, and that larger pores do not contribute significantly. These results are very similar to the ones obtained on carbide-derived carbons, as we discuss below. Finally, they remarked that N_2 adsorption measurements should be complemented with CO_2 measurements to obtain a better characterization of the porosity in the whole range of pore sizes. Lozano et al. found that CO_2 adsorption was more appropriate to characterize activated carbons with different degrees of activation and encouraged the use of this type of adsorption as a complement of N_2 adsorption at 77 K (Lozano-Castelló et al., 2002).

Panella et al. investigated activated carbons and single-walled carbon nanotubes at 77 and 298 K, and found that the storage capacities depend linearly on the pressure at 298 K and also linearly on the specific surface area for all the types of carbon materials studied (Panella et al., 2005). The best material, with a specific surface area of 2560 m^2/g, has a storage capacity of 4.5 wt % at 77 K and 40 bar. Terrés et al. (2005) measured a gravimetric capacity of 2.7 wt % at 77 K and 50 atm on carbons with spherical nanopores and specific surface areas between 946 and 1646 m^2/g.

Linares-Solano et al. measured the storage capacities and porosity properties of several activated carbon nanofibers and activated carbons at 293 K and up to 70 MPa (Panella et al., 2005). The highest capacity was 1 wt % at 10 MPa. They also reported the packing

densities of the samples. In other set of experiments (Jordá-Beneyto et al., 2007) with a series of activated carbons, the best sample had gravimetric capacities of 1.2 and 2.7 wt % at 298 K and pressures of 20 and 50 MPa, respectively, and 5.6 wt % at 77 K and 4 MPa. The volumetric capacities, calculated using the packing density of these materials as $\rho_{supporting}$ in Equation 41.1, are 0.0167 and 0.0372 kg H_2/L at 298 K and pressures of 20 and 50 MPa, respectively, and 0.0388 kg H_2/L at 77 K and 4 MPa. Taking into account the results of different groups, the conclusion is that the activated carbons have gravimetric capacities between 2.0 and 5.6 wt % at 77 K and moderate pressures (1–4 MPa). The same group prepared (Jordá-Beneyto et al., 2008) advanced activated carbon monoliths (ACM), a type of activated carbon with high specific pore volumes up to 1.04 cm^3/g, and measured their storage capacities: 0.0297 kg H_2/L at 77 K and 4 MPa for the best ACM. Because the whole system is under pressure, the packing density of the pressed porous materials should be measured and used as $\rho_{supporting}$ in Equation 41.1 to calculate the volumetric capacities, instead of the crystal density of the unpressed material. The densities of the unpressed and pressed materials are usually called tap and packing densities, respectively. Activated carbons can be pressed up to 5.5 MPa without changing the porous texture. These authors (Jordá-Beneyto et al., 2008) reported the volumetric capacities of activated carbons using measured tap, 0.3–0.66 cm^3/g, and packing densities, 0.4–0.75 cm^3/g. For instance, two of their samples have tap volumetric capacities of 0.0242 and 0.0287 kgH_2/L, respectively, at 77 K and 4 MPa, and the corresponding packing volumetric capacities are 0.0214 and 0.0178 kgH_2/L.

In experiments for activated carbons, activated carbon fibers and nanotubes, the same group found a linear relationship between gravimetric storage capacities and specific pore volume, shown in Figure 41.4. However, in their most recent experiments (Jordá-Beneyto et al., 2008), they have observed that the volumetric capacities are not simply linear with respect to the pore volume; there is a maximum, as shown in Figure 41.5. Hence, adsorbent materials with a good balance between packing density and porosity are necessary to obtain high volumetric

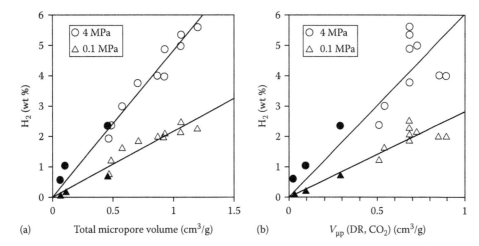

FIGURE 41.4 Gravimetric capacities at 77 K and 0.1 and 4 MPa vs. the specific total pore volume (a) and the specific narrow pore volume (b) of activated carbons (open symbols), carbon nanotubes, and carbon nanofibers (filled symbols). (Reproduced from Jordá-Beneyto, M. et al., *Carbon*, 45, 293, 2007. With permission.)

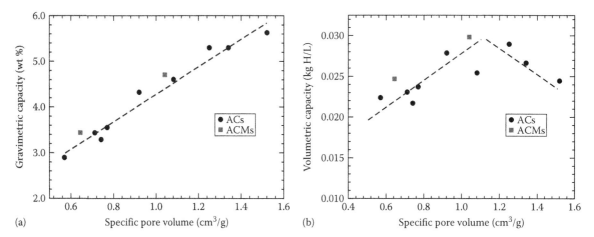

FIGURE 41.5 Gravimetric (a) and volumetric (b) hydrogen storage capacities at 77 K and 4 MPa for activated carbon and ACM powder samples as a function of the specific pore volume of pores below 2 nm. (Data taken from Jordá-Beneyto, M. et al., *Microporous Mesoporous Mater.*, 112, 235, 2008.)

capacities. This balance can be seen from Equation 41.1, which after some algebra becomes

$$v_c = \frac{n_{H_2} m_{H_2} \rho_{supporting}}{M_{supporting}} = \frac{c \rho_{supporting} v_e}{v_{eH_2}}, \quad (41.12)$$

where the symbols have the same meaning as in Equation 41.6. The density $\rho_{supporting}$ decreases as v_e increases, and this causes the maximum on v_c.

41.7 Carbide-Derived Carbons

41.7.1 Synthesis

Carbide-derived carbons (CDCs) form a new class of porous carbons produced by the thermochemical etching of metal or metalloid elements from solid metal carbides like TiC, SiC, ZrC, and B_4C. Cl_2 is the most often used chemical etching agent. Other halogens and halogen compounds can also be used. The synthesis reaction of porous carbon from a metal carbide is the following: $M_aC_b(\text{solid}) + Cl_2(\text{gas}) \rightarrow MCl_x(\text{gas}) + C(\text{solid})$, where M is the metallic element. The metal halide MCl_x is very volatile usually. The metal is, then, just removed as gaseous metal halide and the carbon produced does not collapse. The volume and the external shape of the precursor carbide are preserved, and hence the carbide-derived carbons are highly porous materials, containing open interconnected pores. Up to 80% of the volume is pore volume available for hydrogen storage. The metal is extracted layer by layer from the metal carbide and this allows control of the porosity at the atomic level. HRTEM images indicate that the internal structure of many carbide-derived carbons consists of a three-dimensional network of finite graphene walls. These materials have specific surface areas in the range 500–2300 m²/g, specific pore volumes of 0.2–1.0 cm³/g, porosity volumes of 55%–84%, and pore sizes of 5–20 Å.

It is important to use synthesis methods allowing to tune the porosity (specific surface area, specific pore volume, pore size, and pore size distribution). Most synthesis methods of activated carbons use thermal decomposition of not well-ordered organic and coal precursors, and hence do not allow to tune the porosity properties. The groups of Gogotsi and Fischer have developed a synthesis method to produce nanoporous carbons of tunable pore size, specific surface area, and specific pore volume from some metallic carbides (Gogotsi et al., 2003, 2005; Nikitin and Gogotsi, 2004; Dash, 2006; Yushin et al., 2006).

Other alternative names used for the carbide-derived carbons are amorphous nanoporous carbons (ANPC) and multiwalled nanobarrels (MWNB), depending on the structure of the produced carbon. For instance, SiC-CDC is an ANPC and Al_4C_3-CDC is a MWNB. The structure of the last material looks like an amorphous network of nanobarrels. A nanobarrel usually has 5–10 walls, spaced 3.4 Å, with inner diameters between 5 and 27 nm, outer diameters between 6 and 30 nm, and lengths of 6–50 nm.

41.7.2 Storage Capacities

The first report on the chlorination of a metal carbide (SiC) was published in 1918 and the carbon obtained was considered a waste product, since the researchers were interested in the production of $SiCl_4$. Boehm and Warnecke reported that the N_2 and CO_2 adsorption isotherms for TaC-CDC were of type I (Boehm and Warnecke, 1975). Fedorov et al. studied the porosity of a large number of carbide-derived carbons and found that the benzene adsorption isotherms were also of type I, except for MoC-CDC (Fedorov et al., 1981, 1982a,b). Johansson et al. reported that SiC-CDC and Al_4C_3-CDC had specific surface areas of 1000 and 600 m²/g, respectively, and excess hydrogen gravimetric capacities of 1.5 and 1.8 wt %, respectively, at 77 K and 10 MPa, and 0.25 and 0.3 wt %, respectively, at 300 K and 10 MPa (Johansson et al., 2002). They also noticed that this was another example of a nonlinear relationship between storage capacities and specific surface areas.

The groups of Gogotsi and Fischer studied the relation between the porosity of the nanoporous carbon and the structure of the precursor carbide, and also the relation between porosity and storage capacity (Gogotsi et al., 2003, 2005; Nikitin and Gogotsi, 2004; Dash, 2006; Yushin et al., 2006). These workers produced different carbide-derived carbons by chlorination of B_4C, TiC, SiC, and ZrC, and measured their storage capacities. Those carbons have a high surface area (500–2300 m^2/g), average pore sizes of 5–14 Å, and specific pore volumes of 0.2–1.0 cm^3/g. The pore size can be tuned with an accuracy of Angstroms by changing the precursor and the synthesis temperature. This is important because the pore size distribution can be optimized to obtain the largest storage capacities.

The B_4C carbide does not have a constant interatomic C–C distance in the lattice, and the B_4C-CDC has a broad pore size distribution and contains pores larger than 2 nm, even at the lowest synthesis temperature of 600°C. The pore size distribution practically does not change for synthesis temperatures in the range of 600°C–1200°C. On the other hand, the C–C interatomic distance is almost constant in TiC, and also in SiC and ZrC. The SiC-CDC has a narrow pore size distribution even at the highest synthesis temperature, 1200°C. ZrC-CDC and TiC-CDC have narrow pore size distributions at low synthesis temperatures and the pores are smaller than 2 nm, but the pore size distributions change with the temperature: the pores are larger than 2 nm and the pore size distributions become wider at higher temperatures.

The gravimetric capacity of carbide-derived carbons depends linearly on the specific surface area and also on the pore volume for pores of size below 20 Å, as can be seen in Figures 41.6 and

41.7, respectively; but there is not a clear dependence on the specific surface area and pore volume for pores above 20 Å (Yushin et al., 2006). This explains some reports for samples of activated carbons and carbide-derived carbons whose capacities were not linear with their specific surface areas; those samples probably contained too many pores above 20 Å. Figure 41.7 shows that the gravimetric capacities of CDCs depend also on the synthesis temperature. These experimental results show the importance of having a uniform C–C distance in the structure of the precursor

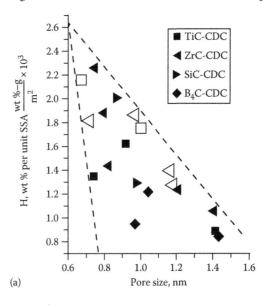

(a)

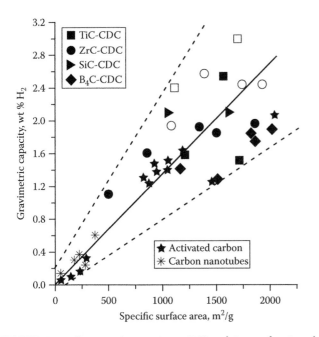

FIGURE 41.6 Gravimetric capacity at 77 K and 1 atm of activated carbons, carbon nanotubes, and carbide-derived carbons vs. the BET-specific surface area. Solid and empty symbols stand for as-produced and hydrogen annealed carbide-derived carbons. The solid line is a linear fit with slope of ca. 1.1 wt % per 1000 m^2/g. (Reproduced from Yushin, G. et al., *Adv. Funct. Mater.*, 16, 2288, 2006. With permission.)

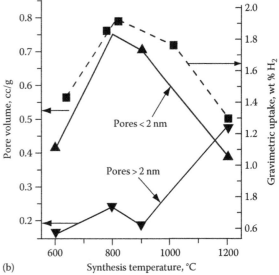

(b)

FIGURE 41.7 Effect of pore size on hydrogen storage. (a) Hydrogen storage normalized to specific surface area vs. pore size for several carbide-derived carbons. The general trend defined by the dashed lines indicates that small pores are more efficient than large pores for a given specific surface area. Solid and empty symbols stand for as-produced and hydrogen-annealed carbide-derived carbons, respectively. (b) Specific pore volume for pores of different size in comparison with the gravimetric capacity vs. the chlorination temperature for B_4C-CDC. (Reproduced from Gogotsi, Y. et al., *J. Am. Chem. Soc.*, 127, 16006, 2005. With permission.)

carbide in order to obtain a narrow pore size distribution, and also an appropriate synthesis temperature. On the other hand, the B_4C lattice parameter is larger than those of other carbides and the pores in B_4C-CDC are also larger.

If carbide nanometric particles are used, then the synthesis temperature is lower and the carbide-derived carbon can be more tunable. The pore size distributions of SiC-CDC produced from nano- and micro-sized particles are different. After synthesizing the carbide-derived carbons, thermal annealing in a hydrogen atmosphere removed Cl_2 molecules trapped in the nanopores, making these pores accessible to hydrogen, and increased the specific surface area, pore volume, and storage capacity.

The experiments (Dash, 2006; Yushin et al., 2006) show that pores in the range 6–7 Å are the most efficient for hydrogen storage (see Figure 41.7) and that pores larger than 20 Å lower the volumetric capacity and contribute little to the gravimetric capacity. This optimal pore size agrees with the theoretical optimal size discussed in Section 41.4. The experiments also showed that these carbide-derived carbons have gravimetric and volumetric capacities up to 3 wt % and 0.0297 kg hydrogen/L at 77 K and 1 atm. There are few experimental results at moderate pressures (Johansson et al., 2002).

41.8 Comparison with the Storage Capacities of Other Materials

The storage capacities of carbon nanotubes, carbon nanofibers, and graphitic nanofibers are several times lower than those of activated carbons and carbide-derived carbons. Zeolites have channels of uniform size and shape, and specific surface areas of about 1000 m²/g, smaller than nanoporous carbons and metal–organic frameworks (Weitkamp et al., 1995; Langmi et al., 2003). Research on hydrogen storage on zeolites has been less intense than on carbon-based materials and metal–organic frameworks, because of their higher weight. The size and shape of the pores in zeolites are controlled by the lattice structure, and the pore volume is less than the pore volume of nanoporous carbons. Organic PIMs form another family of promising adsorbent materials (McKeown et al., 2006, 2007), with specific surface areas of 800 m²/g and pore sizes in the range 6–7 Å. The storage capacities of zeolites and PIMs are similar to those of nanoporous carbons with the same specific surface area. However, in general, their storage capacities are smaller than those of nanoporous carbons, because their specific surface areas are also smaller. Clathrate hydrates, ice lattices encapsulating hydrogen molecules, have very high storage capacities, but these materials are stable only in nonpractical ranges of pressures: above 2300 MPa at 300 K and above 600 MPa at 190 K for $H_2(H_2O)$ (Mao et al., 2002; Mao and Mao, 2004).

As of 2008, the main competitor of nanoporous carbons is a family of materials called metal–organic frameworks (MOFs). These are highly porous materials stable above room temperature, up to 300°C, with a large range of pore sizes, between 4 and 29 Å, specific surface areas between 500 and 4750 m²/g, specific pore volumes between 0.6 and 1.6 cm³/g, porosity volumes between 56% and 91%, and densities between 0.21 and 1.00 g/cm³ (Hailian et al., 1999; Eddaoudi et al., 2002; Chae et al., 2004). Their pore sizes can be tuned and the pores can be functionalized. The largest measured hydrogen capacities at 77 K and 1 atm were 2.4 and 2.59 wt %, obtained for a member of the family known as MOF-505 (Chen et al., 2005) and for a metal–organic framework based on tetracarboxylate complexes of Cu(II) (Lin et al., 2006), respectively. MOF-5 adsorbs 0.17 wt % at 298 K and 68 atm (Panella and Hirscher, 2005). Panella et al. have observed a linear relationship between the storage capacities of metal–organic frameworks and their specific surface areas (Panella et al., 2006). The key to achieve high storage capacities appears to be a good balance between the size of the hydrogen molecule and the size of the pores more than a high specific pore volume (Chun et al., 2005; Lin et al., 2006). The highest capacities of the metal–organic framework family have been reported by the group of Yaghi (Wong-Foy et al., 2006): 7.0 and 6.3 wt %, and 0.032 and 0.034 kgH_2/L on MOF-177 and IRMOF-20, respectively, at 77 K and saturation pressures (70–80 bar). MOF-177 also has the largest BET-specific surface area of porous materials, 4750 m²/g. The metal–organic framework family has gravimetric capacities between 0.9 and 2.6 wt % at 77 K and 1 atm, and volumetric capacities between 0.020 and 0.043 kgH_2/L at 77 K and saturation pressures (70–80 bar).

Some authors (Jordá-Beneyto et al., 2008) have noticed that the volumetric capacities of metal–organic frameworks reported in the literature are not realistic from a practical point of view, because the experimentalists used the density of the metal–organic framework crystals to calculate the volumetric capacity from Equation 41.1, instead of the density of the pressed material or packing density. Using the packing density of MOF-5 (0.3 g/cm³) (Müeller et al., 2006), the volumetric capacity would be 0.0156 kg H_2/L at 77 K and 4 MPa, instead of 0.0306 kgH_2/L obtained using the crystal density of 0.59 g/cm³. That corrected capacity of 0.0156 kgH_2/L is smaller than the volumetric capacities of some activated carbon samples, called AC-5 and AC-6 samples in the original work (Jordá-Beneyto et al., 2008), which have specific surface areas similar to MOF-5, 3500 m²/g, calculated with their packing densities, 0.0214 and 0.0178 kgH_2/L, respectively, at 77 K and 4 MPa.

The saturation of the gravimetric capacity with increasing pressure is due, in a first approximation, to the pore size. The pores are filled with molecules, so above a certain pressure, the saturation pressure, there are no more pores available to H_2, and then, the mass of adsorbed hydrogen reaches a saturation value. Smaller pores are filled at lower saturation pressures. Metal–organic frameworks have, in general, larger pores than nanoporous carbons, and therefore their gravimetric capacities saturate at higher pressures, 70–80 vs. 10–20 bar in case of activated carbons, and are also larger. Nevertheless, the larger specific surface area and specific pore volume of metal–organic frameworks, and not only the larger pore size, have also influence on the larger gravimetric capacities and saturation pressures.

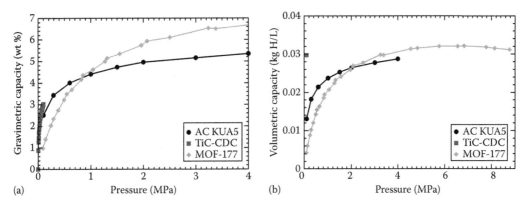

FIGURE 41.8 Gravimetric (a) and volumetric (b) hydrogen storage capacities at 77 K as a function of pressure plotted for the best carbide-derived carbon, activated carbon, and metal–organic framework samples found in the literature: TiC-CDC data taken from Gogotsi et al. (2005), AC-KUA5 data taken from Jordá-Beneyto et al. (2007), and MOF-177 data taken from Wong-Foy et al. (2006).

If we consider the storage capacities of all the samples of activated carbons, carbide-derived carbons and metal–organic frameworks, then these three families of materials have similar gravimetric capacities, between 0.9 and 3.0 wt %, at 77 K and 1 atm. At 77 K and higher pressures, the activated carbon and metal–organic framework families have large gravimetric capacities, 2–5.6 and 2–7 wt %, respectively. Another comparison can be made for the best samples of these three families. The highest activated carbon, carbide-derived carbon, and metal–organic framework storage capacities found in the literature at 77 K are shown in Figure 41.8. TiC-CDC has the largest gravimetric and volumetric capacities at low pressures. At higher pressures, MOF-177 reaches larger gravimetric and volumetric capacities than the best activated carbon sample, KUA5. An explanation of the better performance of some members of the metal–organic framework family can be found in Figure 41.9, showing a plot of the gravimetric capacity vs. the BET-specific surface area for different activated carbon and metal–organic framework samples. It can be observed that the capacity of some activated carbon samples is even larger than the capacity of some metal–organic frameworks with the same specific surface area. However, some metal–organic frameworks have larger specific surface areas (up to 4750 m²/g) than any member of the activated carbon family (these go up to 3800 m²/g), and hence larger gravimetric capacities.

The absolute value of the enthalpy difference between the physisorbed and the compressed H_2 phases is equal to the isosteric heat of adsorption of hydrogen, and hence it measures qualitatively the intensity of the interaction of a hydrogen molecule with the surface of the material. This quantity is not frequently reported in the literature. The isosteric heats of hydrogen adsorption of carbide-derived carbons, metal–organic frameworks, activated carbons, and carbon nanotubes are in the ranges 6–11, 4.4–6.5, 5–7, and 5–7 kJ/mol, respectively (Yushin et al., 2006). The isosteric heat of MOF-177 is 4.4 kJ/mol (Furukawa et al., 2007).

The advantages of metal–organic frameworks with respect to activated carbons and carbide-derived carbons are their higher specific surface areas, higher gravimetric and volumetric capacities at moderate pressures (12–80 bar), well-defined pore sizes, and very narrow pore size distributions. The advantages of activated carbons and carbide-derived carbons are their larger capacities below 1 atm, more mechanical and thermal stability, and larger isosteric heat of hydrogen adsorption. Carbide-derived carbons have tunable and narrow pore size distributions centered on pore sizes of 5–7 Å, and the largest isosteric heats. Carbide-derived carbons have also larger volumetric and gravimetric capacities than activated carbons and metal–organic frameworks at low pressures.

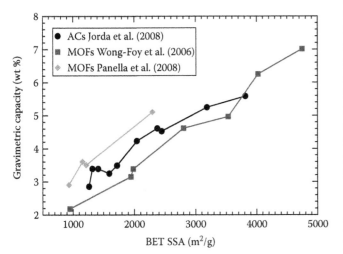

FIGURE 41.9 Gravimetric capacities at 77 K vs. the BET-specific surface area of activated carbons at 4 MPa (Jordá et al., 2008) and of metal–organic frameworks at saturation pressures (Wong-Foy et al., 2006; Panella et al., 2008).

41.9 Summary and Future Directions

Nanoporous carbons are promising materials for onboard hydrogen storage in hydrogen cars. These cars use hydrogen fuel cells instead of electric batteries or gasoline motors and they only release water. They could lower the energy and global warming problems and reduce the fossil fuel dependence and the resultant pollution. The main bottleneck of the hydrogen car technology is the onboard storage system. An efficient storage system requires high hydrogen storage capacities. In 2002, the FreedomCAR partnership established the technological targets for 2007, 2010, and 2015 in order to get a hydrogen car equivalent to the present gasoline cars, especially with the same autonomy range of 600 km. Some of those targets refer to the volumetric and gravimetric storage capacities: for instance, 0.045 kg of hydrogen/L of the storage system and 6 hydrogen wt % by 2010.

There are three types of hydrogen storage: in gas and liquid forms, and absorbed in solids. Storage in solid materials is safer and reaches higher volumetric capacities. There are three mechanisms of storage in solid materials: molecular physisorption on the surface, including the surface of pores, atomic chemisorption in the bulk and also in the surface of pores, and chemical reactions in some metal hydrides. Physisorption is the storage mechanism of nanoporous carbons. Optimal storage by physisorption requires adsorbent materials with high specific surface area and high porosity. According to the most recent experimental and theoretical results, the optimal pore size is between 5 and 7 Å. Simple thermodynamic calculations indicate that to obtain reversible hydrogen storage at room temperature and moderate pressures, the optimal binding energy of a single H_2 molecule to the surface should be between 0.3 and 0.4 eV, a rather narrow range.

Nanoporous carbons exhibit high specific surface area and porosity: specific surface areas between 300 and 3800 m^2/g, specific pore volume between 0.1 and 1.0 cm^3/g, porosity volume in the range 55%–80%, and pores of width between 4 and 20 Å. These properties are the origin of the high storage capacities of these materials at 77 K and moderate pressures (12–80 bar), and can be tuned, especially the pore size distribution, to optimize the storage. Pores below 20 Å are the most efficient to store hydrogen and larger pores contribute very little. Gravimetric and volumetric storage capacities of nanoporous carbons increase with specific surface area and also with the specific pore volume. However, the volumetric capacity as a function of the specific pore volume reaches a maximum, while the density of the adsorbent decreases as the specific pore volume increases. Hence, a compromise between the specific pore volume and the density of the adsorbent material is necessary to optimize the volumetric capacity.

Future research is expected to focus on the design and synthesis of nanoporous materials with three basic properties: pore sizes of 5–7 Å, binding energies of H_2 to the surface of 0.3–0.4 eV/molecule and high values of $v_e \rho_{supporting}$, the product of the specific pore volume and the density of the supporting or adsorbent material.

Acknowledgments

This work was supported by MEC of Spain (Grants MAT2004-23280-E, MAT2005-06544-C03-01 and MAT2008-06483-C03-01), Junta de Castilla y León (Grants VA039A05, VA017A08 and GR23), and the University of Valladolid. I. Cabria acknowledges support from MEC-FSE through the Ramón y Cajal Program. J. A. Alonso acknowledges an Ikerbasque fellowship from the Basque Foundation for Science. We thank Y. Gogotsi and J. E. Fischer for their comments on this manuscript.

References

Agarwal, R. K., Noh, J. S., Schwarz, J. A. et al., 1987. Effect of surface acidity of activated carbon on hydrogen storage. *Carbon* 25, 219–226.

Ansón, A., Benham, M., Jagiello, J. et al., 2004a. Hydrogen adsorption on a single-walled carbon nanotube material: A comparative study of three different adsorption techniques. *Nanotechnology* 15, 1503–1508.

Ansón, A., Callejas, M. A., Benito, A. M. et al., 2004b. Hydrogen adsorption studies on single wall carbon nanotubes. *Carbon* 42, 1243–1248.

Ariyadejwanich, P., Tanthapanichakoona, W., Nakagawab, K. et al., 2003. Preparation and characterization of mesoporous activated carbon from waste tires. *Carbon* 41, 157–164.

Boehm, H. P. and Warnecke, H. H., 1975. Structural parameters and molecular sieve properties of carbons prepared from metal carbides. In *Proceedings of the 12th Biennial Conference on Carbon*, Pittsburgh, PA. Pergamon Press, Oxford, U.K., pp. 149–150.

Bogdanovic, B. and Sandrock, G., 2002. Catalyzed complex metal hydrides. *Mat. Res. Bull.* 27, 712–716.

Brunauer, S., Emmett, P. H., and Teller, E., 1938. Adsorption of gases in multi-molecular layers. *J. Am. Chem. Soc.* 60, 309–319.

Cabria, I., López, M. J., and Alonso, J. A., 2007. The optimum average nanopore size for hydrogen storage in carbon nanoporous materials. *Carbon* 45, 2649–2658.

Carpetis, C. and Peschka, W., 1976. On the storage of hydrogen by use of cryoadsorbents. In *First World Hydrogen Energy Conference*, University of Miami, Coral Gables, FL. Pergamon Press, New York, pp. 9C-45–9C-54.

Carpetis, C. and Peschka, W., 1978. On the storage of hydrogen by use of cryoadsorbents. In *Second World Hydrogen Energy Conference*, Zurich, Switzerland. Pergamon Press, Oxford and New York, pp. 1433–1456.

Chae, H. K., Siberio-Pérez, D. Y., Kim, J. et al., 2004. A route to high surface area, porosity and inclusion of large molecules in crystals. *Nature (London)* 427, 523–527.

Chen, B., Ockwig, N. W., Millward, A. R. et al., 2005. High H_2 adsorption in a microporous metal–organic framework with open metal sites. *Angew. Chem. Int. Ed.* 44, 4745–4749.

Chun, H., Dybtsev, D. N., Kim, H. et al., 2005. High H_2 adsorption by coordination-framework materials. *Chem. Eur. J.* 11, 3521.

Crowell, A. D. and Brown, J. S., 1982. Laterally averaged interaction potentials for $^1H_2$ and $^2H_2$ on the (0001) graphite surface. *Surf. Sci.* 123, 296–304.

Dash, R. K., 2006. Nanoporous carbons derived from binary carbides and their optimization for hydrogen storage, PhD thesis, Drexel University, Philadelphia, PA.

de la Casa-Lillo, M. A., Lamari-Darkrim, F., Cazorla-Amorós, D. et al., 2002. Hydrogen storage in activated carbons and activated carbon fibers. *J. Phys. Chem. B* 106, 10930–10934.

Dillon, A. C., Jones, K. M., Bekkedahl, T. A. et al., 1997. Storage of hydrogen in single-walled carbon nanotubes. *Nature (London)* 386, 377–379.

DOE, 2002a. Multi-year research, development and demonstration plan: Planned program activities for 2003–2010: Technical plan; U. S. Department of Energy. http://www.eere.energy.gov/hydrogenandfuelcells/mypp/pdfs/storage.pdf

DOE, 2002b. Targets for on-board hydrogen storage systems: Current R&D focus is on 2010 targets. http://www1.eere.energy.gov/hydrogenandfuelcells/pdfs/freedomcar_targets_explanations.pdf

Dubinin, M. M. and Plavnik, G. M., 1968. Microporous structures of carbonaceous adsorbents. *Carbon* 6, 183–192.

Dubinin, M. M. and Radushkevich, V. L., 1947. Equation of the characteristic curve of activated charcoal. *Dokl. Akad. Nauk SSSR [Comm. USSR Acad. Sci.]* 55, 331. [In Russian.]

Eddaoudi, M., Kim, J., Rosi, N. et al., 2002. Systematic design of pore size and functionality in isoreticular MOFs and their application in methane storage. *Science* 295, 469–472.

Fedorov, N. F., Ivakhnyuk, G. K., Tetenov, V. V. et al., 1981. Carbon adsorbents based on silicon carbide. *J. Appl. Chem. USSR* 54, 1239–1242.

Fedorov, N. F., Ivakhnyuk, G. K., and Gavrilov, D. N., 1982a. Carbon adsorbents from carbides of the IV–VI groups transition metals. *Zhurnal Prikladnoi Khimii* 55, 272–275.

Fedorov, N. F., Ivakhnyuk, G. K., and Gavrilov, D. N., 1982b. Porous structure of carbon adsorbents from titanium carbide. *Zhurnal Prikladnoi Khimii* 55, 46–50.

Furukawa, H., Miller, M. A., and Yaghi, O. M., 2007. Independent verification of the saturation hydrogen uptake in MOF-177 and establishment of a benchmark for hydrogen adsorption in metal–organic frameworks. *J. Mater. Chem.* 17, 3197–3204.

Gogotsi, Y., Nikitin, A., Ye, H. et al., 2003. Nanoporous carbide-derived carbon with tunable pore size. *Nat. Mater.* 2, 591.

Gogotsi, Y., Dash, R. K., Yushin, G. et al., 2005. Tailoring of nanoscale porosity in carbide-derived carbons for hydrogen storage. *J. Am. Chem. Soc.* 127, 16006–16007.

Hailian, L., Eddaoudi, M., O'Keeffe, M. et al., 1999. Design and synthesis of an exceptionally stable and highly porous metal–organic framework. *Nature (London)* 402, 276–279.

Johansson, E., Hjorvarsson, B., Ekström, T. et al., 2002. Hydrogen in carbon nanostructures. *J. Alloys Compd.* 330, 670–675.

Jordá-Beneyto, M., Suárez-García, F., Lozano-Castelló, D. et al., 2007. Hydrogen storage on chemically activated carbons and carbon nanomaterials at high pressures. *Carbon* 45, 293–303.

Jordá-Beneyto, M., Lozano-Castelló, D., Suárez-García, F. et al., 2008. Advanced activated carbon monoliths and activated carbons for hydrogen storage. *Microporous Mesoporous Mater.* 112, 235–242.

Kadono, K., Kajiura, H., and Shiraishi, M., 2003. Dense hydrogen adsorption on carbon subnanopores at 77 K. *Appl. Phys. Lett.* 83, 3392–3394.

Kidnay, A. J. and Hiza, M. J., 1967. High pressure adsorption isotherm of neon, hydrogen and helium at 76 K. *Adv. Cryog. Eng.* 12, 730.

Langmi, H. W., Walton, A., Al-Mamouri, M. M. et al., 2003. Hydrogen adsorption in zeolites A, X, Y and RHO. *J. Alloys Compd.* 356, 710–715.

Li, J., Furuta, T., Goto, H. et al., 2003. Theoretical evaluation of hydrogen storage capacity in pure carbon nanostructures. *J. Chem. Phys.* 119, 2376–2385.

Lin, X., Jia, J., Zhao, X. et al., 2006. High H_2 adsorption by coordination-framework materials. *Angew. Chem. Int. Ed.* 45, 7358–7364.

Lozano-Castelló, D., Cazorla-Amorós, D., and Linares-Solano, A., 2002. Usefulness of CO_2 adsorption at 273 K for the characterization of porous carbons. *J. Phys. Chem. B* 106, 10930–10934.

Lueking, A. D., Pan, L., Narayanan, D. L. et al., 2005. Effect of expanded graphite lattice in exfoliated graphite nanofibers on hydrogen storage. *J. Phys. Chem. B* 109, 12710–12717.

Mao, W. L. and Mao, H.-K., 2004. Hydrogen storage in molecular compounds. *Proc. Nat. Acad. Sci. USA* 101, 708–710.

Mao, W. L., Mao, H.-K., Goncharov, A. F. et al., 2002. Hydrogen clusters in clathrate hydrate. *Science* 297, 2247–2249.

McKeown, N. B., Ghanem, B., Msayib, K. J. et al., 2006. Towards polymer-based hydrogen storage materials: Engineering ultramicroporous cavities within polymers of intrinsic microporosity. *Angew. Chem. Int. Ed.* 45, 1804–1807.

McKeown, N. B., Budd, P. M., and Book, D., 2007. Microporous polymers as potential hydrogen storage materials. *Macromol. Rapid Commun.* 28, 995–1002.

Müeller, U., Schubert, M., Teich, F. et al., 2006. Metal–organic frameworks—Prospective industrial applications. *J. Mater. Chem.* 16, 626–636.

Nijkamp, M. G., 2002. Hydrogen storage using physisorption: Modified carbon nanofibers and related materials, PhD thesis, Universiteit Utrecht, Utrecht, the Netherlands.

Nijkamp, M. G., Raaymakers, J. E. M. J., van Dillen, A. J. et al., 2001. Hydrogen storage using physisorption—Materials demands. *Appl. Phys. A* 72, 619–623.

Nikitin, A. and Gogotsi, Y., 2004. Nanostructured carbide-derived carbon. In: Nalwa, H. S. (Ed.), *Encyclopedia of Nanoscience and Nanotechnology*, Vol. X. American Scientific Publishers, New York, pp. 1–22.

Noh, J. S., Agarwal, R. K., and Schwarz, J. A., 1987. Hydrogen storage systems using activated carbon. *Int. J. Hydrogen Energy* 12, 693.

Panella, B. and Hirscher, M., 2005. Hydrogen physisorption in metal–organic porous crystals. *Adv. Mater.* 17, 538–541.

Panella, B., Hirscher, M., and Roth, S., 2005. Hydrogen adsorption in different carbon nanostructures. *Carbon* 43, 2209–2214.

Panella, B., Hirscher, M., Pütter, H. et al., 2006. Hydrogen adsorption in metal–organic frameworks: Cu-MOFs and Zn-MOFs compared. *Adv. Funct. Mater.* 16, 520–524.

Patchkovskii, S., Tse, J. S., Yurchenko, S. N. et al., 2005. Graphene nanostructures as tunable storage media for molecular hydrogen. *Proc. Nat. Acad. Sci. USA* 102, 10439–10444.

Rzepka, M., Lamp, P., and de la Casa-Lillo, M. A., 1998. Physisorption of hydrogen on microporous carbon nanotubes. *J. Phys. Chem. B* 102, 10894–10898.

Schlapbach, L. and Züttel, A., 2001. Hydrogen storage materials for mobile applications. *Nature (London)* 414, 353–358.

Schuth, F., Sing, S. W. K., and Weitkamp, J., 2002. *Handbook of Porous Solids*. Wiley-VCH, New York.

Silvera, I. F. and Goldman, V. V., 1978. The isotropic intermolecular potential for H_2 and D_2 in the solid and gas phases. *J. Chem. Phys.* 69, 4209–4213.

Sing, K. S. W., Everett, D. H., Haul, R. A. W. et al., 1985. Reporting physisorption data for gas/solid systems. *Pure Appl. Chem.* 57, 603–619.

Stan, G. and Cole, M. W., 1998. Low coverage adsorption in cylindrical pores. *Surf. Sci.* 395, 280.

Ströbel, R., Jorissen, L., Schliermann, T. et al., 1999. Hydrogen adsorption on carbon materials. *J. Power Sources* 84, 221.

Terrés, E., Panella, B., Hayashi, T. et al., 2005. Hydrogen storage in spherical nanoporous carbons. *Chem. Phys. Lett.* 403, 363–366.

Texier-Mandoki, N., Dentzer, J., Piquero, T. et al., 2004. Hydrogen storage in activated carbon materials: Role of the nanoporous texture. *Carbon* 42, 2744–2747.

Wang, Q. and Johnson, J. K., 1998. Hydrogen adsorption on graphite and in carbon slit pores from path integral simulations. *Mol. Phys.* 95, 299–309.

Wang, Q. and Johnson, J. K., 1999a. Molecular simulation of hydrogen adsorption in single-walled carbon nanotubes and idealized carbon slit pores. *J. Chem. Phys.* 110, 577–586.

Wang, Q. and Johnson, J. K., 1999b. Optimization of carbon nanotube arrays for hydrogen adsorption. *J. Phys. Chem. B* 103, 4809.

Weitkamp, J., Fritz, M., and Ernst, S., 1995. Zeolites as media for hydrogen storage. *Int. J. Hydrogen Energy* 20, 967–970.

Wong-Foy, A. G., Matzger, A. J., and Yaghi, O. M., 2006. Exceptional H_2 saturation uptake in microporous metal–organic frameworks. *J. Am. Chem. Soc.* 128, 3494–3495.

Yushin, G., Dash, R., Jagiello, J. et al., 2006. Carbide-derived carbons: Effect of pore size on hydrogen uptake and heat of adsorption. *Adv. Funct. Mater.* 16, 2288–2293.

Zhang, Y., Cui, H., Ozao, R. et al., 2007. Characterization of activated carbon prepared from chicken waste and coal. *Energy Fuels* 21, 3735–3739.

Züttel, A., 2003. Materials for hydrogen storage. *Mater. Today* 6, 24–33.

42

Hydrogen Adsorption in Nanoporous Materials

Pierre Bénard
*Université du Québec
à Trois-Rivières*

Richard Chahine
*Université du Québec
à Trois-Rivières*

Marc-André Richard
*Université du Québec
à Trois-Rivières*

42.1 Introduction

Hydrogen is a flexible and environmentally friendly energy carrier that can be produced from a variety of feedstock. It offers the possibility to substantially reduce greenhouse gas emissions, improve urban air quality, and enhance security of energy supply. It can be used to power public and dedicated transportation vehicles such as buses and delivery vehicles in the relatively short term, and automobiles on a longer horizon. Through micro-fuel cells, hydrogen can power portable applications such as laptops and cell phones, offering the possibility of better energy density, faster recharging, and longer and more reliable operations compared to current batteries. Finally, hydrogen has promising applications in stationary power production (such as emergency power units) through fuel cells, turbines, and internal combustion engines.

Hydrogen energy technologies face important technical challenges. Hydrogen is a synthetic fuel that has to be produced in an environmentally benign way at high efficiency and low cost and, depending on the application, with a high degree of purity. Mostly, however, the storage density of hydrogen must be improved to meet consumer expectations. While using hydrogen as an automotive fuel competitively with respect to gasoline still requires breakthroughs in storage technologies, it is already being introduced in public transportation as well as in portable and back-up power applications. An efficient mode of hydrogen storage for automotive transportation applications has to satisfy a set of criteria based on net energy density, weight, safety, cost, charging–discharging kinetics, cycling, etc. The most widely quoted criteria have been set by the U.S. Department of Energy (DOE) (Freedom Car and Fuel Partnership 2005). They fixed system targets of 2 kWh/kg (6 wt%), 1.5 kWh/L, and $4 kWh/L for 2010 and of 3 kWh/kg (9 wt%), 2.7 kWh/L, and $2 kWh/L for 2015. These criteria aim at providing on-board hydrogen storage for a driving range greater than 500 km across different vehicle platforms at moderate temperature and pressure without compromising passenger or cargo space, performance, and cost. It is important to stress that these targets include the total weight of the filled storage system. Even though significant progress has been made in the last few years, there is still no single storage material or method that satisfies all these design criteria. The volumetric storage density is also an issue with portable applications, as no conventional hydrogen storage system can meet the volumetric storage density target of 6 g/cc expected by industry partners for micro-fuel cell systems to compete effectively with current battery systems.

42.2 Current Hydrogen Storage Technologies

The combustion of 1 kg of hydrogen can release up to 120 MJ of energy (an amount of 20 MJ/kg notwithstanding, contained in the residual water vapor). A similar among of energy released by the combustion of oil and natural gas would require 2.5–2.75 kg of fuel, making hydrogen one of the most efficient fuels on a gravimetric basis. However, hydrogen, as the lightest of gases, exhibits one of the poorest combustion energy density when expressed on a volumetric basis. In gaseous form under ambient conditions, hydrogen is 8 times lighter than methane, and as a cryogenic liquid, it is 10 times lighter than gasoline. Thus, despite having the largest heat of combustion per unit of mass, hydrogen energy systems suffer from performance issues when volumetric considerations are factored in.

Room-temperature storage of hydrogen in high pressure tanks currently is the most common on-board storage strategy. According to the DOE, the maximum storage density that has been achieved by compression storage is 0.8 kWh/L and 1.6 kWh/kg. The 700 bar tanks are not cost-effective (Lasher 2008) at the moment: at $27/kWh, they are more expensive than most other strategies, and there is no clear pathway to reducing this cost significantly to achieve the DOE goal of $2/kWh for year 2015. In addition, delivering hydrogen in current compressed H_2 tube trailers is expensive: recent case studies (Weinert 2005, Weinert et al. 2006) suggest that delivering compressed hydrogen by road over a distance of 25–50 km costs about 1–2 $/kg.

Hydrogen is also stored in liquid form at 20 K. Liquid hydrogen (LH_2) has a density of 70 kg/m³, compared to 39.6 kg/m³ in gaseous form at room temperature and 70 MPa. To achieve a similar density, hydrogen would need to be compressed at ambient temperature to 182 MPa. The energy necessary to liquefy hydrogen is 22 MJ/kg, about 2–10 times larger than the energy required compressing it to 70 MPa. The efficient liquefaction of hydrogen requires large liquefaction plants, which makes it difficult to implement a decentralized distribution hydrogen system based on liquid hydrogen. In particular, LH_2 is not suitable for the recovery and distribution of the generally decentralized byproduct hydrogen.

Hybrid cryogenic high-pressure tanks for hydrogen storage have recently been proposed. These cryo-compressed tanks are expected to be lighter than hydrides and more compact than ambient-temperature, high pressure vessels, while limiting the energy penalty for liquefaction and lowering losses compared with liquid hydrogen tanks. Designing a storage system that could operate at ambient temperatures and meet the 2015 targets remains beyond classic storage technologies.

42.3 Solid-State Storage of Hydrogen

Designing storage systems for automotive applications that could meet performance expectations requires more than what current conventional storage technologies can offer. Going beyond compression or liquid storage will require taking advantage of the physical and chemical interactions of molecular and atomic hydrogen with other atoms. Consider for instance that hydrogen can be stored efficiently as water, through inter-atomic and intermolecular forces. A given volume of water contains more hydrogen atoms than the same volume of liquid hydrogen. The issue then becomes to restore efficiently hydrogen to its initial form, a formidable task considering the stability of the water molecule. Hydrogen can be stored either irreversibly by chemical binding to other elements in chemical hydrides (as an organic liquid, for instance), or reversibly by forming a reversible metal hydride or by physisorption on materials with large surface areas per unit volume, such as nanocarbons, zeolites, and metal-organic frameworks (MOFs).

Materials-based storage can be characterized in terms of the characteristic binding energy of atomic or molecular hydrogen with a sorbent or other atoms and molecules. The binding energy of atomic hydrogen is about 300–400 kJ/mol in hydrocarbons, 100 kJ/mol and more in chemical hydrides, ranges from 50 to 100 kJ/mol in metal hydrides and is typically below 10 kJ/mol for physisorbed molecular hydrogen on high surface area solids (Zuttel et al. 2002).

The binding energy determines the heat required to release hydrogen from storage as well as the losses through boil off for cryogenic systems. Materials with hydrogen binding energies larger than 50 kJ/mol require more power to release 5 kg of hydrogen than boiling the equivalent mass of water. On the other hand, materials with binding energies less than 10 kJ/mol requires little heat to release the hydrogen, but must operate at cryogenic temperatures to achieve acceptable storage densities. The thermodynamic cycles required to release hydrogen stored in materials from atomic or molecular binding must require acceptable operating conditions. Ideally, this implies storage pressures of 1–100 bars and operating temperatures of 20°C–120°C, based on the assumption that waste heat from fuel cells can be used to release hydrogen from binding. A binding energy lying in the range 10–50 kJ/mol would be best suited in this situation. No materials are currently available in that range. Greater availability of waste heat (from SOFC fuel cells or nuclear power stations) may change these operating conditions. Thus, different material-based storage strategies may have specific niche applications.

Chemical hydrogen storage uses compounds containing hydrogen to react with other hydrogen containing compounds to liberate hydrogen. This approach is referred to as irreversible, since the spent reactant must be reprocessed off-board, even if the reactions are reversible given sufficient energy. An example is the hydrolysis of sodium borohydride.

Reversible metal hydrides have the form $MH_{(1, 2, or 3)}$, where M can be an alkali, alkaline earth, or transition metal. Magnesium-based hydrides have a gravimetric storage density above 7 wt%. However, their decomposition temperature is generally too high to be of practical value for automotive applications. A significant level of R&D efforts has been conducted to lower the decomposition temperature and to accelerate the re/dehydrogenation reactions. This can be accomplished by inducing micro-structural changes through mechanical alloying with elements that can reduce the stability of the hydrides and by using catalysts to improve the absorption/desorption kinetics. A second class of metal hydrides relies on intermetallic compounds. The simplest case is the ternary system AB_xH_n, where A is usually a rare earth or an alkaline earth metal that tends to form a stable hydride, and B a transition metal that tends to form unstable hydrides. Although the decomposition temperature is lower than MH_n compounds, the gravimetric storage is usually generally in the order of a few (wt)%. Even if these hydrides are not appropriate for automotive applications, they can be of interest to low power portable applications where volumetric considerations are more important. Sodium, lithium, and beryllium are the only elements lighter than magnesium that can form solid-state compounds with hydrogen. The hydrogen content of these hydrides can reach a theoretical value of 18 wt% for $LiBH_4$.

A well-known example of a reversible complex hydride is sodium alanate ($NaAlH_4$). It reversibly releases about 3.5 wt% H_2 when doped with titanium catalysts, and can operate at a temperature of 150°C when prepared by ball-milling (Luo and Gross 2004, Srinivasan et al. 2004). These materials could be improved by destabilizing them either chemically, by introducing a new chemical specie in a high capacity hydride, forming an intermediate state that will reduce the heat of formation of the hydride, or by controlling the size distribution of the particles and their joint boundaries.

42.4 Adsorption Storage of Hydrogen on Nanostructures

Molecular hydrogen binding or physisorption in nanoporous materials is envisaged as an alternative to compressed gas storage and is likely the next maturing technology due to many similarities with the latter. Compared to metal- and chemical-hydride storage materials involving chemical binding of atomic hydrogen, physisorption materials have many advantages. They are fully reversible, easily cycled, and have the additional advantages of fast kinetics and low storage pressure. They can also generally be handled safely. Moreover, the heat management penalty is much smaller than for hydrides because the average heat of adsorption in these materials is an order of magnitude lower ($\Delta H \sim 4\,kJ/mol$). However, because of the low ΔH, significant hydrogen storage densities has only been obtained at temperatures much below the targeted temperature range for vehicular applications (−40°C to 85°C), as opposed to most metal and chemical hydrides that are generally regenerated far above the upper limits of the acceptable temperature range.

42.4.1 Physisorption

Adsorption is a phenomenon arising from intermolecular or interatomic interactions between an adsorbate, either a gas or a solute, and a solid surface (the adsorbent) that results in a local increase of the density of the adsorbate close to the surface. Adsorption specifically refers to the process through which the adsorbate concentration is progressively increased close to the surface by changing the thermodynamic conditions of the system (pressure or temperature), whereas desorption refers to the reverse process. Physisorption and chemisorption differ mainly through the intensity of the adsorbate/adsorbent binding. In the case of *physisorption*, weak van der Waals interactions between adsorbate molecules and the substrate are involved, whereas covalent binding typically occurs in the case of *chemisorption*, although the distinction is somewhat arbitrary. Chemisorption can exhibit some measure of irreversible behavior, through a hysteresis between the adsorption and the desorption processes. Adsorption differs fundamentally from absorption, a process in which a molecule or an atom diffuses into a solid or a liquid to form a solution.

For gases, adsorption can occur in one of two regimes: supercritical or subcritical. Supercritical adsorption occurs above the critical temperature of the adsorbent. As such, no liquid phase is possible as a function of pressure. Subcritical adsorption occurs at temperatures below the critical temperature of the adsorbent. The density of the adsorbate close to the surface of the adsorbent at a given subcritical temperature is characterized by a steep increase at pressures close to the liquefaction pressure, associated with the formation of a liquid layer. Due to the low critical temperature of hydrogen (33.18 K), only the supercritical adsorption regime is considered relevant to storage applications.

The adsorption process is usually described in terms of isotherms, which express the functional relation between the adsorbed gas density n_a at a given temperature and the pressure of the bulk gas at thermodynamic equilibrium. The number of adsorbed molecules N_a is the number of molecules within a volume V_a close to the surface of the adsorbent, where the density of the adsorbate is larger than its bulk value, and where the interaction potential between adsorbate molecules and the surface of the adsorbent is appreciable (Figure 42.1). The adsorbed gas density n_a is usually measured relative to the mass of the adsorbent M_{ads}. The adsorbent, on the other hand, is characterized by its surface area S and its porous structure (pore size distribution). The total volume of the pores of the adsorbent is the pore volume V_p. The pore volume can be divided into two contributions: the adsorption volume V_a and the dead volume of the adsorbent V_{dead}, in which the molecular force field of the adsorbent is negligible. The dead volume consists of large pores and interstices. In an ideal microporous adsorbent, $V_p = V_a$.

The adsorbed gas density n_a and the adsorption volume V_a are not readily accessible experimentally. The excess adsorption density n_{ex} is a quantity readily available experimentally through the volumetric method. It is defined as follows:

$$n_{ex} = \frac{N_{tot} - \rho_g V_{Void}}{M_{ads}}$$

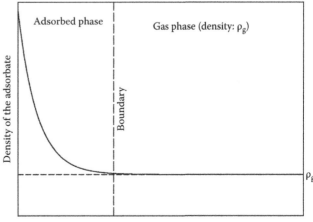

FIGURE 42.1 Definition of the adsorbed volume.

where

N_{tot} is the total number of molecules found within the volume V_{Void} of a measurement cell containing the adsorbent at a given temperature T and pressure P

V_{Void} is the sum of V_p and V_{Empty}, which we define as the empty volume of the cell external to the adsorbent. The quantity ρ_g is the density of the gas at the same pressure P and temperature T. The contributions to $\rho_g V_{Void}$ from the volume of the system that is not subjected to the molecular field of the surface (V_{Empty} and V_{dead}) cancel out in the expression above for n_{ex}. In terms of the net adsorbed density n_a (per unit mass of the adsorbent),

$$ n_{ex} = \frac{n_a - \rho_g V_a}{M_{ads}} $$

The two quantities are comparable (Salem et al. 1998, Myers and Monson 2002) at high temperature and low pressure when

$$ \frac{\rho_g V_a}{M_{ads}} \ll n_{ex} $$

The excess density exhibits a maximum as a function of pressure (Figure 42.2) because n_a, which tends to saturate at high pressure, increases at a slower pace than ρ_g as a function of pressure. The relationship between n_{ex}, n_a, and the amount of adsorbate stored in the volume V_p of the porous structure of an adsorbent is shown in Figure 42.2.

The thermal properties of adsorption are usually described through the differential enthalpy of adsorption Δh, also referred to as the isosteric heat of adsorption. Δh reflects the magnitude of gas–solid interactions and depends on the properties of the adsorbent and the adsorbate (Jagiello et al. 1996):

$$ \Delta h = -R \left[\frac{\partial \ln(P)}{\partial (1/T)} \right]_{n_a} $$

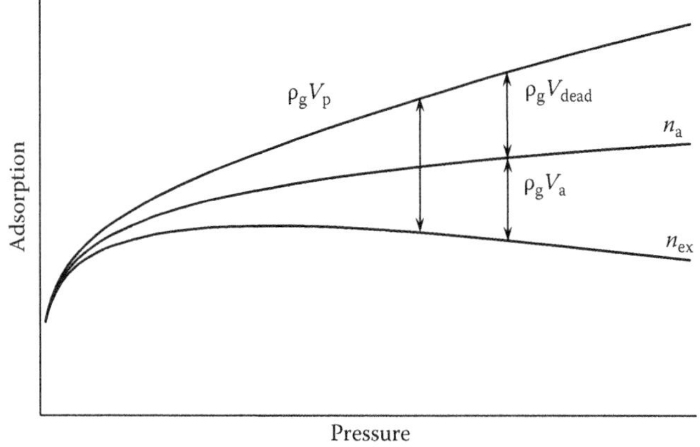

FIGURE 42.2 Excess adsorbed density (lower curve), total adsorbed density in the volume V_a (middle curve), and total stored density in the porous structure of the adsorbent (top curve).

The differential enthalpy of adsorption can be obtained from the absolute adsorption isotherms by calculating the slope of the natural logarithm of the adsorption isotherms as a function of the inverse temperature. The isosteres are curves of the pressure as a function of inverse temperature at constant adsorbed density. It can also be measured by calorimetry (Myers and Monson 2002, Myers 2004). Monson and Myers have derived a formulation of the thermodynamics of adsorption in terms of excess variables. The excess isosteric heat of adsorption evaluated at constant excess density exhibits a singularity at the pressure associated with the maximum of the excess isotherms (Myers and Monson 2002).

42.4.2 Adsorbents for Hydrogen Storage

Finding the best adsorbents for hydrogen storage involves optimizing three parameters: the binding energy of hydrogen with the surface of the material, the surface area accessible to the adsorbate molecules per unit mass of the adsorbent, and the bulk density of the latter. Adsorbents can be optimized volumetrically or gravimetrically. Current emphasis is in maximizing the gravimetric storage density. However, transportation and portable applications have volume restrictions. Optimization of the surface area per unit volume of the adsorbent accessible to the gas molecules should also be taken into account in developing materials for such applications. The binding energy determines the operating temperature and the latency of a hydrogen adsorption-based storage system.

Carbon-based nanoporous materials and MOFs have been the object of intense scrutiny as adsorbents for physisorption based hydrogen storage. The interest behind carbon lies in the large number of highly porous carbon allotropes and their relatively important van der Waals interactions with molecular hydrogen. Carbon-based adsorbents considered for hydrogen storage include activated carbons, single wall nanotubes (SWNTs), carbon nanofibers, carbon foams, carbide-derived carbons, and aerogels. Activated carbons can be described as random arrangements of graphitic planes of various sizes arranged in a highly porous three-dimensional disordered structure, characterized by a pore size distribution. The use of activated carbons as a storage medium for gaseous fuels has been successfully demonstrated for natural gas. At room temperature, methane stored on optimized activated carbon pellets at 35 bars represents 85% of the energy density of methane compressed at more than four times the pressure. For hydrogen, acceptable excess adsorbed densities for storage applications require low temperature operation and highly nanoporous activated carbons. Excess adsorption isotherms of hydrogen on the activated carbon AX-21 (now produced under Maxsorb MSC-30™), exhibit a maximum excess density adsorbed of 54 g/kg (5.4 wt%) at 35 bars and 77 K, corresponding to a storage density of about 8.7 wt%. The storage density is highly temperature dependent, falling below 1% under ambient conditions. SWNTs are tubular carbon nanostructures of 8–20 nm in diameter typically organized in bundles of hundreds of units (Thess et al. 1996). Undoped and untreated SWNTs adsorb less than 1 wt% hydrogen under ambient conditions (Becher et al. 2003) and up to 2.5 wt% at 77 K and 1 atm, depending on sample preparation (Nishimiya et al. 2002, Tarasov et al. 2003, Anson et al. 2004, Poirier et al. 2006). Excess adsorbed densities of 6 and 8 wt% have been obtained at 2 and 40 bars (Ye et al. 1999, Pradhan et al. 2002), respectively, have been obtained at about 77 K. SWNTs are sensitive to heat and acid treatments: a maximum excess adsorbed density of 4.6 wt% has been obtained at 77 K and 1 atm after being treated with hydrofluoric acid and subjected to heat at 873 K (Lafi et al. 2005). Recently, carbon foams based on SWNTs have been proposed that could offer high specific surfaces and better binding than SWNT bundles (Yakobson and Hauge 2007). Carbon nanofibers are nanostructures formed of stacked graphene cones or cups of 1–100 nm in diameter (Melchko et al. 2005). The latest measurements of excess adsorbed density of hydrogen on these structures fail to show significant adsorption under ambient conditions. From 0.1 to 105 atm and 77–295 K, the maximum adsorbed density obtained was less than 1 wt% at 105 bars (Pinkerton et al. 2000, Yang 2000, Poirier et al. 2001, Jordá-Beneyto et al. 2007). Beyond activated carbons, other disordered carbon nanostructures such as carbide-derived carbons and carbon aerogels are currently being investigated. High surface area (~2000 m²/g) carbide-derived carbons are amorphous carbon structures with a pore size distribution that could be tailored to specific applications (Dash 2006). They are produced from carbides through chemical etching. Titanium-carbide derived carbons have excess hydrogen adsorption densities of up to 3% (wt) and 30 kg/m³ (volumetrically). A surprisingly high isosteric heat of hydrogen adsorption of 11 kJ/mol has been reported in these structures, compared to other physisorbents (Dash et al. 2006, Laudisio et al. 2006).

Carbon aerogels are tridimensional networks of interconnected nanometer-sized carbon particles, which form a porous structure that can be synthesized inexpensively. They exhibit a large specific surface area, and their bulk properties such as density, pore size, pore volume, and specific surface can be controlled in principle (Kabbour et al. 2006). Nanoparticles (such as metal dopants) can be incorporated into the aerogels during their synthesis. They can be used as scaffolds for metal hydrides. Because of their higher specific surface area (3200 m²/g), carbon aerogels can store comparatively more hydrogen than activated carbons, yielding an excess adsorbed density of hydrogen of up to 5.3% (wt) at 30 bars and 77 K. The excess adsorbed density under such conditions vary linearly with the specific surface area of the aerogels from 3.5% (wt) to 5.05% (wt) when the specific surface changes from 1460 to 2500 m²/g. The differential enthalpy of adsorption decreases as a function of the specific surface area (from 6.7 to 6.2 kJ/mol), which could be indicative of weaker interactions. This is reminiscent of the behavior of the excess adsorbed density of hydrogen on SWNT bundles as a function of the bundle lattice spacing calculated using grand canonical Monte Carlo simulations. These simulations showed that past a maximum, as the specific surface continues to increase as a function of lattice spacing, the excess adsorbed density decreased due

to weaker average surface–adsorbate interactions (Lachance and Benard 2007).

Although storage densities have been obtained that can approach the DOE targets under cryogenic conditions and mild pressures, the volumetric densities must be improved to achieve the DOE objectives for automotive applications. The storage capacity of carbon-based nanoporous adsorbents seems to scale with the specific surface area (Chahine and Bose 1996), typically storing 1% weight per 500 m²/g. Experimental measurements of hydrogen adsorption on carbon nanostructures have not shown, so far, a clear pathway to increasing the storage density significantly beyond activated carbon. For transportation applications, the U.S. DOE has elected to stop R&D activities on hydrogen storage in pure, undoped SWNTs. Further improvements in volumetric densities and operating temperatures will likely require going beyond pure carbon nanostructures. Avenues being currently pursued include enhancing adsorbate–adsorbent interactions by metal insertion into the carbon matrix, or by dissociating the hydrogen molecule, allowing for hydrogen atoms to bind directly with the carbon nanostructure. Enhanced binding of hydrogen molecules (without molecular dissociation) could be achieved by doping carbon nanostructures with metal, through the formation of so-called Kubas complexes (Kim et al. 2006) with transition metals. Ab initio calculations by Yildirim et al. show that titanium-metalized single-walled nanotubes or C_{60} fullerenes could in principle adsorb up 7% to 8% (wt) hydrogen under ambient conditions (Yildirim and Ciraci 2005, Yildirim et al. 2005). However, Sun et al. showed that clustering of the titanium atoms could lower the density to 2.85% for titanium-doped C_{60} (Sun et al. 2005). Enhancement effects have been predicted for Li and Pt on certain carbon nanostructures. Boron doping has been shown by quantum chemistry calculations and experiments to enhance the binding energy of hydrogen (to 11 kJ/mol) in various carbon nanostructures (Kim et al. 2006, Kleinhammes et al. 2007). A spillover strategy has also been proposed as a mechanism to enhance the storage density of carbon and MOF nanostructures. This approach relies on a supported metal catalyst to dissociate molecular hydrogen. Atomic hydrogen, which binds more strongly with carbon nanostructures than molecular hydrogen, then diffuses through a bridge into the storage substrate. Enhancement by a factor of 2.9 for activated carbon (AX-21) and 1.6 for single-walled nanotubes at 298 K and 1 bar has been obtained using through this strategy, using a palladium catalyst (Lachawiec et al. 2005).

New classes of adsorbents have also been investigated. Metal-organic frameworks (Roswell and Yaghi 2005, Wong-Foy et al. 2006) are highly porous, organized networks of transition metal atoms bridged by organic linkers that have recently been proposed as substrate for physisorption storage of hydrogen. They are being extensively developed for on-board hydrogen storage (Russell and Wheatley 2008), and their use in gas separation and purification is still at the embryonic stage (Férey 2008). Their advantage lies in their large specific surface areas (1000–6000 m²/g) (Roswell and Yaghi 2005) and in the fact that they may be tailored to specific storage applications by changing ligands, transition metals, or by doping. Maximum excess adsorption densities of hydrogen on MOF structures ranging from 2.0 to 7.3 wt% at 77 K have been obtained. The largest uptakes, 7.3 wt% for IRMOF-77 (with a specific surface area in excess of 5500 m^2/g) and 6.7% for IRMOF-20, were obtained at pressures of 70–80 bars. Other related materials, Al-BDC and Cr-BDC (MILS-53), can adsorb 2.8–3.9 wt% at 77 K at pressures of up to 50 bars (Férey et al. 2003). At room temperature (295 K), reversible hydrogen uptakes of less than 1 wt% have been obtained (Poirier et al. 2006). MOFs seem to exhibit useful trends that can guide the development of new sorption substrate: a linear correlation between the excess density adsorbed and the specific surface of MOF structures has been observed at high pressure. At low pressures, however, the low specific surface area materials tend to have the largest excess adsorbed density. This trend is observed in the grand canonical simulations of hydrogen adsorption on carbon nanostructures. It suggests that gravimetric adsorption is determined by the binding energy at low pressure and by the specific surface at high pressure. No clear correlation between specific surface and volumetric storage density can be established for these materials. Unlike activated carbons, where denser materials have lower excess densities but higher volumetric densities, IRMOF-177 exhibits a large excess adsorption density and a large volumetric adsorbed density (about 32 g/L). The adsorption enthalpy of MOF structures is typically lower than pure carbon adsorbents (below 5 kJ/mol).

At 77 K, the maximum excess density of hydrogen stored on carbon and MOF nanostructures obtained experimentally so far lie in the range 6–7.5 wt% at 30–40 bars. In all cases, performances drop significantly as temperature increases. As in the case of carbon nanostructures, spillover strategies are also being applied to MOF. An enhancement factor of about 3 has been reported using spillover for IRMOF-8 MOFs, resulting in a maximum density of 1.8% at 298 K and 10 MPa for IRMOF-8 (Li and Yang 2006).

42.4.3 Adsorption Isotherms

The hydrogen adsorption isotherms are usually determined either volumetrically or gravimetrically. The volumetric method relies on two isothermal cells separated by a valve: an empty reference cell and a measurement cell containing the adsorbent. The reference cell is initially filled with a given amount of hydrogen (determined by its pressure, temperature, and volume). The valve is opened and the gas expands into the measurement cell. After equilibrium is achieved, the amount of hydrogen adsorbed can be obtained through the difference between the initial and final pressures in the system and the known net volume of the cells. The volumetric method yields the excess adsorbed density. The full isotherm is obtained by changing the temperature of the closed system. The gravimetric approach is based on the measurements of mass variation of the adsorbent–adsorbate system during the sorption process. Microbalances are well suited to the small sample masses typically available. The initial

measurements of the adsorbed density on ordered carbon nanostructures such as single-walled nanotubes or carbon nanofibers yielded inconsistent results at room temperature (0–20 wt%), due to the small size of the samples, gas purity, and sample preparation. These issues have since been addressed as high sensitivity gravimetric, and volumetric hydrogen adsorption measurement systems have been developed: gravimetric and volumetric systems can now yield consistent results within about 0.05 wt% on samples weighing from 3 to 25 mg (Poirier et al. 2006).

Several modeling strategies have been used to determine adsorption isotherms from statistical physics or thermodynamic considerations. Excess adsorption isotherms can be expressed as a series expansion in terms of pressure, yielding the so-called virial series expansion (Clark 1970):

$$n_{ex} = B_{AS}\left(\frac{p}{kT}\right) + C_{AAS}\left(\frac{p}{kT}\right)^2 + D_{AAS}\left(\frac{p}{kT}\right)^3 + \cdots$$

The excess adsorbed isotherm depends linearly on pressure in the low temperature limit, which is also known as Henry's law. The second virial coefficient B_{AS} can be obtained experimentally from the low temperature behavior. It expresses the interaction between a single molecule and the surface of the adsorbate. Detailed studies of the low pressure limits of carbon nanostructures are presented in references (Stan and Cole 1998a,b, Stan et al. 2000, Mélançon and Benard 2004).

Statistical physics modeling strategies include gas/surface models using a local or delocalized approach (lattice gases or confined two-dimensional gases). Thermodynamic pore-filling approaches are also extensively used for microporous adsorbents. The Langmuir model is the most basic approach to isotherm modeling in the supercritical régime. It describes the monolayer filling of the surface of an adsorbent by noninteracting adsorbate molecules. The resulting isotherm of the net adsorbed density n_a is a monotonically increasing function of the pressure P, which saturates asymptotically at a value n_m:

$$n_a = n_m \frac{C(P/P^0)}{\left(1 + C(P/P^0)\right)}$$

where

$C = \exp(A/R)\exp(-(B/RT))$

A and B are constants

The Langmuir isotherm does not directly yield the excess adsorption n_{ex}. Modified for excess adsorption, it has been successfully applied to study hydrogen isotherms (Bhatia and Myers 2006). Aranovitch et al. (Aranovitch and Donchue 1995, Bénard and Chahine 2001) proposed self-consistent equations for the excess adsorption isotherms for the adsorption of supercritical gases on activated carbons established through a mean-field approximation of an Ising lattice-based model of the adsorption process. The adsorbent is described as a slit pore of two parallel graphene layers.

Adsorption isotherms have also been studied numerically using the grand canonical Monte Carlo method classically (Lachance and Benard 2007), semiclassically (Lévesque et al. 2002), and through a quantum path integral approach (Wang and Johnson 1999) and classical density functional approaches.

The Dubinin model in its original form is a model of the adsorption isotherm based on a concept of pore filling of microporous adsorbents by subcritical gases. The isotherm is given by the following expression (Dubinin 1975):

$$n_a = n_m \exp\left(-\left(\frac{A}{\varepsilon}\right)^m\right) \quad \text{where } A = RT \ln \frac{P}{P_0}$$

The ratio n_a/n_m represents the degree of filling of the micropores, ε is a characteristic energy of adsorption, and P_0 is the saturation pressure of the subcritical adsorbate. The exponent m is usually 2 for activated carbons. It is typically 2 for most activated carbons (smaller values are associated with narrower pore size distributions). In the original Dubinin model, n_m is assumed to be related to the bulk liquid density of the adsorbate at its normal boiling point corrected for thermal expansion (but this assumption is not always verified). The model has been generalized to the supercritical regime, where the saturation pressure is undefined, by defining an empirical pseudo-saturation fugacity sometimes calculated from the critical conditions (Amankwah and Schwarz 1995). Although it describes very well the overall shape of an adsorption isotherm, the model does suffer from some flaws. Notably it does not obey Henry's law at low pressure, its parameters are not easy to relate to microscopic properties of the system and the pseudo-saturation pressure obtained from best fits to experimental data can be very large. However, at pressures and temperatures of interest to storage applications, the Dubinin model can serve as a basis to provide a relatively simple analytic expression for the experimental adsorption isotherms of hydrogen on a number of microporous adsorbents.

Adsorption storage of hydrogen on microporous adsorption would currently operate at cryogenic temperatures (77 K). In the absence of isothermal control, the adsorption process (endothermic when it desorbs and exothermic when it adsorbs) involves significant temperature variations of the adsorbate–adsorbent system and thermal exchanges with the environment. Predicting the behavior of adsorption-based storage systems therefore requires knowledge of the behavior of the adsorption process over a wide range in temperatures, covering in fact the full supercritical range up to ambient temperatures and a wide range of pressures. The Dubinin approach can be adapted to represent the adsorption of gases over large pressure and temperature ranges in the supercritical region:

$$n_a = n_m \exp\left(-\left(\frac{RT}{\alpha + \beta T}\right)^2 \ln \frac{P}{P_0}\right)$$

where

  R is the universal gas constant
  n_m is the limiting adsorption in mol/kg
  P_0 is interpreted as the pressure associated with the limiting adsorption density of the pores

The characteristic free energy of adsorption of the original model (ε) is replaced by a temperature-dependent expression $\alpha + \beta T$. This substitution allowed a significant reduction of the standard error of fitting for the H_2/AX-21™ system over a wide temperature and pressure range in the supercritical region, and yields excellent interpolations of the adsorption isotherm over the full temperature range by determining the parameters from only two isotherms (77 and 298 K). The model was successfully fitted to describe the adsorption of hydrogen on several microporous systems: the activated carbon AX-21™ (with a BET surface area of 2800 m²/g) over the range 0–6 MPa and 30–298 K as well as the activated carbon CNS-201™ (with a BET surface area of 1440 m²/g) and the MOF $Cu_3(BTC)_2$ (with a BET surface area of 1570 m²/g), both over the range (0–6 MPa and 77–296 K) (Richard et al. 2009a,b). It was also used to model the adsorption isotherms of nitrogen and methane over wide temperature and pressure ranges (Richard et al. 2009a). Figure 42.3 illustrates the adsorption isotherms of hydrogen on AX-21™ as a function of temperature and pressure and the resulting fit to the modified Dubinin isotherm.

The excess density was obtained by considering the adsorption volume V_a as a fitting parameter:

$$n_{ex} = n_a - V_a \rho_g$$

where ρ_g is the bulk gas density of hydrogen.

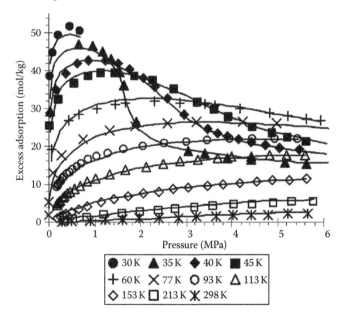

FIGURE 42.3 Modified D-A isotherm fit (solid lines) to the experimental excess adsorption isotherms of hydrogen (shown as symbols) expressed as mol/kg of adsorbent. (From Richard, M.-A. et al., *Adsorption*, 15, 43, 2009a. With permission.)

The parameters of the excess isotherm were found to be $n_{max} = 71.6\,mol/kg$, $\alpha = 3.08\,kJ/mol$, $\beta = 18.9\,J/mol//K$, $P_0 = 1470\,MPa$, and $V_a = 1.43\,L/kg$. The characteristic energy calculated from $\alpha + \beta T$ is in good agreement with other data. The limiting density associated with complete filling of the micropores can be estimated by dividing n_{max} by V_a. A value of 50.2 mol/L is obtained for hydrogen, higher than the liquid density 35.1 mol/L and closer to the density of the solid phase (43.7 mol/L). A similar behavior of the limiting density is observed for nitrogen and methane. The large values of P_0 may be seen as coherent with the high density limit of the adsorbed gas (Richard et al. 2009a).

42.5 Adsorption-Based Storage Systems for Hydrogen

Systems analysis of hydrogen storage units can determine optimal operating conditions for a given adsorbent and provide guidelines on adsorbent material properties to achieve specific performance objectives, i.e., storage density, delivery, fueling, and cost. If the adsorbent completely fills the tank of an adsorption-based storage system, the void volume of the system corresponds to the dead volume v_{dead} of the adsorbent, which is usually expressed per unit mass of the adsorbent. The mass of hydrogen stored in such a system is then given by

$$M_{stored} = M_{Ad}\left(n_a + \rho_g v_{dead}\right) = \rho_{Ad} V_{Tank}\left(n_a + \rho_g v_{dead}\right)$$

where M_{Ad} is the total mass of adsorbent in the system. The stored density of hydrogen in the system is

$$\rho_{stored} = \frac{M_{stored}}{V_{Tank}} = \rho_{Ad}\left(n_a + \rho_g v_{dead}\right)$$

which can be expressed in terms of moles by dividing by the molar mass of hydrogen M_{H_2}. The gravimetric density of hydrogen stored in the system x_{stored} is the ratio of the mass of hydrogen stored to the total mass of the system, obtained by taken into account the mass of the storage system M_{sys}:

$$x_{stored} = \frac{n_a + \rho_g v_{dead}}{1 + M_{sys}/M_{Ad} + \left(n_a + \rho_g v_{dead}\right)}$$

M_{sys} is the mass of the system without adsorbent or hydrogen. This dimensionless number is usually expressed as a percentage. The gravimetric performance targets on hydrogen storage are typically based on this quantity.

The *net storage density* or the *delivery* is the amount of usable hydrogen stored according to the operating conditions of the system. Assuming a minimum discharge pressure of 0.25 MPa below which the system is no longer effective, the storage tank will exhibit a significant residual quantity of hydrogen, especially at

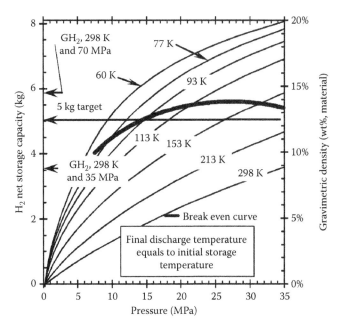

FIGURE 42.4 Calculated net isothermal storage density of an adsorption-based 150 L system using the activated carbon Maxsorb MSC-30™, assuming a residual density at 0.25 MPa and at the stated temperature. (From Richard, M.-A. et al., *AIChE J.*, 55(11), 2985, 2009c. With permission.)

low temperatures (1.36 kg at 77 K in a 150 L reservoir assuming a Maxsorb MSC-30 activated carbon whose isotherms are similar to Figure 42.4). The delivery is calculated by deducting the residual amount to the total amount stored. Figure 42.4 shows the net storage capacity isotherms of a 150 L cryosorption storage unit in kg (left vertical-axis) and in wt% (right vertical-axis) as a function of pressure. The residual storage density becomes increasingly important as the temperature is lowered. The break-even curve on this figure shows where on a given isotherm, a 150 L compressed gas (GH₂) storage tank yields the same amount stored as the adsorption storage unit under the same thermodynamic conditions (above this point cryo-compression storage wins over adsorption). Because of the important residual density, this figure shows that storing 5 kg of hydrogen by cryosorption is more difficult by cryo-compression for a passive storage unit below 93 K.

The net isothermal storage density of the cryosorption system as a function of temperature is shown in Figure 42.5, assuming a pressure of 3 MPa and a residual pressure of 0.15 MPa. This figure shows results from reference Richard et al. (2009c) (full line) and Bhatia and Myers (2006) (dashed line). The optimal operating temperature for an isothermal cryosorption storage unit would be around 100 K using a delivery estimate based on the behavior of the absolute and excess adsorption. The total delivery includes the contribution of the interstices and the large pores of the adsorbent.

The large residual adsorbed density can be retrieved by providing heat to the adsorbent. For example, the residual mass at 0.25 MPa drops from 1.36 kg at 77 K to 0.03 kg at 298 K, a temperature of about 200 K being sufficient to discharge most of

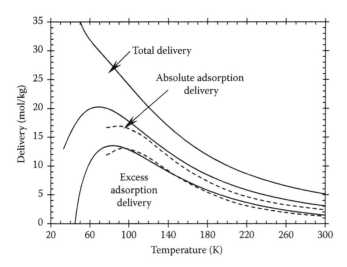

FIGURE 42.5 Calculated isothermal delivery for the cryosorption system as a function of temperature for a storage pressure of 3 MPa and a residual pressure of 0.15 MPa (Richard et al. 2009c). Full lines refer to data from Richard et al. (2009c), dashed lines to data from Bhatia and Myers (2006). (From Richard, M.-A. et al., *AIChE J.*, 55(11), 2985, 2009c. With permission.)

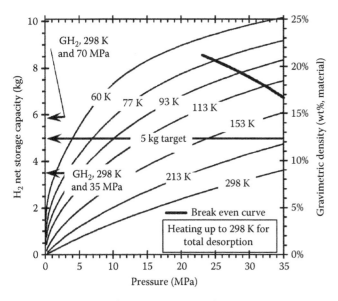

FIGURE 42.6 Calculated net isothermal storage density of an adsorption-based 150 L system using the activated carbon Maxsorb MSC-30™, assuming a residual density at 0.25 MPa and 298 K. (From Richard, M.-A. et al., *AIChE J.*, 55(11), 2985, 2009c. With permission.)

the hydrogen. Figure 42.6 shows the net storage capacity when the system is allowed to heat up to 298 K at the residual pressure of 0.25 MPa. When the residual density can be recovered, a targeted delivery of 5 kg of hydrogen can be reached under a significantly broader range of system pressures and temperature efficiently compared with cryo-compression at the same temperature (below 93 K): the lower the operating temperature of the storage system, the lower pressure required to store a minimum amount of 5 kg.

42.6 Conclusions

Storage of hydrogen by physisorption offers the possibility of fast kinetics, reversibility, and increased density at lower storage pressure. However, in order to exhibit the high capacities required for hydrogen energy applications, a storage system based on adsorption would currently require cooling to temperatures of the order of the temperature of liquid nitrogen. Recent thermodynamic studies of adsorption-based storage system even suggest that pre-cooling of hydrogen to 80 K is insufficient to reach a net stored density of 5 kg for high-surface-area-activated carbons. Significant performance gains can be achieved in such systems by using an active storage system in which the residual density becomes usable by heating the storage tank (for a discharge rate of 1.8 g/s, most of the hydrogen can be withdrawn from the tank if a heat input of approximately 500 W is supplied). Microporous activated carbon, which is easily available, remains the best overall choice in terms of reproducibility, cost, and availability, although new carbon-based materials such as carbon aerogels or carbide-derived carbons and other materials such as MOFs may offer new pathways with somewhat increased storage density and the possibility of doping or scaffolding. Niche storage applications such as distribution of hydrogen may represent short-term technological realization of adsorption storage of hydrogen. Although cryosorption storage offers interesting storage densities on a gravimetric basis, transportation applications may require increasing the volumetric densities currently achievable to meet consumer expectations due to the low density of activated carbons and other adsorbents currently under consideration.

References

Amankwah K.A.G. and J.A. Schwarz. 1995. A modified approach for estimating pseudo-vapor pressures in the application of the Dubinin-Astakhov equation. *Carbon* 33(9):1313–1319.

Anson A., M.A. Callejas, A.M. Benito et al. 2004. Hydrogen adsorption studies on single wall carbon nanotubes. *Carbon* 42:1243–1248.

Aranovitch G.L. and M.D. Donohue. 1995. Adsorption isotherms for microporous adsorbents. *Carbon* 33:1369–1375.

Becher M., M. Haluska, M. Hirscher et al. 2003. Hydrogen storage in carbon nanotubes. *Comptes Rendus Physique* 4:1055–1062.

Bénard P. and R. Chahine. 2001. Determination of the adsorption isotherms of hydrogen on activated carbons above the critical temperature of the adsorbate over wide temperature and pressure ranges. *Langmuir* 17(6):1950–1955.

Bhatia S.K. and A.L. Myers. 2006. Optimum conditions for adsorptive storage. *Langmuir* 22:1688–1700.

Chahine R. and T.K. Bose. 1996. Characterization and optimization of adsorbents for hydrogen storage. *Hydrogen Energy Progress* XI:1259–1263.

Clark A. 1970. *The Theory of Adsorption and Catalysis*. New York: Academic Press.

Dash R.K. 2006. Nanoporous carbons derived from binary carbides and their optimization for hydrogen storage. PhD thesis, Drexel University, Philadelphia, PA.

Dash R.K., J. Chmiola, G. Yushin et al. 2006. Titanium carbide derived nanoporous carbon for energy-related applications. *Carbon* 44:2489–2497.

Dubinin M.M. 1975. Physical adsorption of gases and vapors in micropores. In *Progress in Membrane and Surface Science*, Vol. 9, D.A. Cadenhead, J.F. Danielli, and M.D. Rosenberg (Eds.), pp. 1–70. New York: Academic Press.

Férey G. 2008. Hybrid porous solids: Past, present, future. *Chemical Society Reviews* 37:191–214.

Férey G., M. Latroche, C. Serre, F. Millange, T. Loiseau, and A. Percheron-Guégan. 2003. Hydrogen adsorption in the nanoporous metal-benzenedicarboxylate M(OH)(O$_2$C–C$_6$H$_4$–CO$_2$) (M = Al^{3+}, Cr^{3+}), MIL-53. *Chemical Communications* 53: 2976–2977.

Freedom Car and Fuel Partnership. 2005. *Hydrogen Storage Technologies Roadmap*. U.S. Department of Energy. Available at http://www1.eere.energy.gov/vehiclesandfuels/pdfs/program/hydrogen_storage_roadmap.pdf

Jagiello J., T.J. Bandosz, and J.A. Schwarz. 1996. Characterization of microporous carbons using adsorption at near ambient temperatures. *Langmuir* 12:2837–2842.

Jordá-Beneyto M., F.S. Suárez-García, D. Lozano-Castelló, D. Cazorla-Amorós, and A. Linares-Solano. 2007. Hydrogen storage on chemically activated carbons and carbon nanomaterials at high pressures. *Carbon* 45:293–303.

Kabbour H., T.F. Baumann, J.H. Satcher Jr., A. Saulnier, and C.C. Ahn. 2006. Toward new candidates for hydrogen storage: High-surface-area carbon aerogels. *Chemistry of Materials* 18:6085–6087.

Kim Y.-H., Y. Zhao, A. Williamson, M.J. Heben, and S.B. Zhang. 2006. Nondissociative adsorption of H$_2$ molecules in light-element-doped fullerenes. *Physical Review Letters* 96:016102 (4 pages).

Kleinhammes A., B.J. Anderson, Q. Chen, and Y. Wu. 2007. Characterization of hydrogen adsorption by NMR. In *Poster Presented at DOE Annual Merit Reviews Meeting*, May 15–18, Arlington, VA.

Lachance P. and P. Bénard. 2007. Specific surface effects on the storage of hydrogen on carbon nanostructures. *International Journal of Green Energy* 4:377–384.

Lachawiec A.J. Jr., G. Qi, and R.T. Yang. 2005. Hydrogen storage in nanostructured carbons by spillover: Bridge-building enhancement. *Langmuir* 21(24):11418–11424.

Lafi L., D. Cossement, and R. Chahine. 2005. Raman spectroscopy and nitrogen vapour adsorption for the study of structural changes during purification of single-wall carbon nanotubes. *Carbon* 43:1347–1357.

Lasher S. 2008. Analyses of hydrogen storage materials and onboard systems, project ID #ST1. In *The Proceedings of the DOE Hydrogen Program's Annual Merit Review*, June 9–13, 2008, U.S. Department of Energy, Arlington, VA.

Laudisio G., R.K. Dash, J.P. Singer, G. Yushin, Y. Gogotsi, and J.E. Fischer. 2006. Carbide-derived carbons: A comparative study of porosity based on small-angle scattering and adsorption isotherms. *Langmuir* 22:8945–8950.

Lévesque D., A. Gicquel, F. Lamari Darkrim, and S. Beyaz Kayiran. 2002. Monte Carlo simulations of hydrogen storage in carbon nanotubes. *Journal of Physics: Condensed Matter* 14:9285–9293.

Li Y. and R.T. Yang. 2006. Significantly enhanced hydrogen storage in metal–organic frameworks via spillover. *JACS Communications* 128(3):726–727.

Luo W. and K. Gross. 2004. A kinetics model of hydrogen absorption and desorption in Ti-doped NaAlH$_4$. *Journal of Alloys and Compounds* 385:224–231.

Mélançon E. and P. Bénard. 2004. Theoretical study of the contribution of physisorption to the low-pressure adsorption of hydrogen on carbon nanotubes. *Langmuir* 20:7852–7859.

Melchko A.V., V.I. Merkulov, T.E. McKnight et al. 2005. Vertically aligned carbon nanofibers and related structures: Controlled synthesis and directed assembly. *Journal of Applied Physics* 97:041301-1–041301-39.

Myers A.L. 2004. Characterization of nanopores by standard enthalpy and entropy of adsorption of probe molecules, *Colloids and Surfaces A* 241:9–14.

Myers A.L. and P.A. Monson. 2002. Adsorption in porous materials at high pressure: Theory and experiment. *Langmuir* 18:10261–10273.

Nishimiya N., K. Ishigaki, H. Takikawa et al. 2002. Hydrogen sorption by single-walled carbon nanotubes prepared by a torch arc method. *Alloys and Compounds* 339:275–282.

Pinkerton F.E., B.G. Wicke, C.H. Olk, G.G. Tibbetts, G.P. Meisner, M.S. Meyer, and J.F. Herbst. 2000. Thermogravimetric measurement of hydrogen absorption in alkali-modified carbon materials. *Journal of Physical Chemistry B* 104:9460–9467.

Poirier E., R. Chahine, and T.K. Bose. 2001. Hydrogen adsorption in carbon nanostructures. *International Journal of Hydrogen Energy* 26:831–835.

Poirier E., R. Chahine, P. Bénard, L. Lafi, G. Dorval-Douville, and P.A. Chandonia. 2006. Hydrogen adsorption measurements and modeling on metal-organic frameworks and single-walled carbon nanotubes. *Langmuir* 22(21):8784–8789.

Pradhan B.K., A. Harutyunyan, D. Stojkovic et al. 2002. Large cryogenic storage of hydrogen in carbon nanotubes at low pressures. *Materials Research Society Symposium Proceedings* 706:Z10.3.1–Z10.3.6.

Richard M.-A., P. Bénard, and R. Chahine. 2009a. Gas adsorption process in activated carbon over a wide temperature range above the critical point. Part 1: Modified Dubinin-Astakhov model. *Adsorption* 15:43–51.

Richard M.-A., P. Bénard, and R. Chahine. 2009b. Gas adsorption process in activated carbon over a wide temperature range above the critical point. Part 2: Conservation of mass and energy. *Adsorption* 15:53–63.

Richard M.-A., D. Cossement, P. Chandonia, R. Chahine, D. Mori, and K. Hirose. 2009c. Evaluation of the performance of an adsorption-based hydrogen storage system. *AIChE Journal* 55(11):2985–2996.

Roswell J.L.C. and O.M. Yaghi. 2005. Strategies for hydrogen storage in metal-organic frameworks. *Angewandte Chemie International Edition* 44(30):4670–4679.

Russell E.M. and P.S. Wheatley. 2008. Review of gas storage in nanoporous materials. *Angewandte Chemie International Edition* 47(27):4966–4981.

Salem M.M.K., P. Braeuer, M. Szombathely, M. Heuchel, P. Harting, K. Quitzch, and M. Jaroniec. 1998. Thermodynamics of high-pressure adsorption of argon, nitrogen, and methane on microporous adsorbents. *Langmuir* 14:3376–3389.

Srinivasan S., H. Brinks, B. Hauback, D. Sun, and C. Jensen. 2004. Long term cycling behavior of titanium doped $NaAlH_4$ prepared through solvent mediated milling of NaH and Al with titanium dopant precursors. *Journal of Alloys and Compounds* 377:283–289.

Stan G. and M.W. Cole. 1998a. Low coverage adsorption in cylindrical pores. *Surface Science* 395:280–291.

Stan G. and M.W. Cole. 1998b. Hydrogen adsorption in nanotubes. *Journal of Low Temperature Physics* 110:539–544.

Stan G., M.J. Bojan, S. Curtarolo, S.M. Gatica, and M.W. Cole. 2000. Uptake of gases in bundles of carbon nanotubes. *Physical Review B* 62:2173–2180.

Sun Q., Q. Wang, P. Jena, and Y. Kawazoe. 2005. Clustering of Ti on a C_{60} surface and its effect on hydrogen storage. *Journal of the American Chemical Society* 127:14582–14583.

Tarasov B.P., J.P. Maehlen, M.V. Lototsky, V.E. Muradyan, and V.A. Yartys. 2003. Hydrogen sorption properties of arc generated single-wall carbon nanotubes. *Journal of Alloys and Compounds* 356–357:510–514.

Thess A., R. Lee, and P. Nikolaev et al. 1996. Crystalline ropes of metallic carbon nanotubes. *Science* 273:483–487.

Wang Q. and J.L. Johnson. 1999. Optimization of carbon nanotube arrays for hydrogen adsorption. *Journal of Physical Chemistry B* 103:4809–4813.

Weinert J.X. 2005. A near-term economic analysis of hydrogen fueling stations, UCD-ITS-RR-05-06. University of California, Davis, CA: Institute of Transportation Studies.

Weinert J.X., L. Shaojun, J.M. Ogden, and M. Jianxin. 2006. Hydrogen refueling station costs in Shanghai, paper UCD-ITS-RR-06-04. University of California, Davis, CA: Institute of Transportation Studies.

Wong-Foy A.G., A.J. Matzger, and O.M. Yaghi. 2006. Exceptional H_2 saturation uptake in microporous metal-organic frameworks. *Journal of the American Chemical Society* 128:3494–3495.

Yakobson I. and R. Hauge. 2007. Theoretical models of H_2-SWNT systems for hydrogen storage and optimization of SWNT. In *The Proceedings of the DOE Annual Merit Reviews Meeting*, May 15–18, Arlington, VA.

Yang R.T. 2000. Hydrogen storage by alkali-doped carbon nanotubes—Revisited. *Carbon* 38:623–641.

Ye Y., C.C. Ahn, C. Witham et al. 1999. Hydrogen adsorption and cohersive energy of single-walled carbon nanotubes. *Applied Physics Letters* 74(16):2307–2309.

Yildirim T. and S. Ciraci. 2005. Titanium-decorated carbon nanotubes as a potential high-capacity hydrogen storage medium. *Physical Review Letters* 94:175501 (4 pages).

Yildirim T., J. Iniguez, and S. Ciraci. 2005. Molecular and dissociative adsorption of multiple hydrogen molecules on transition metal decorated C_{60}. *Physical Review B* 72:153403 (4 pages).

Zuttel A., Ch. Nutzenadel, P. Sudan, Ph. Mauron, Ch. Emmenegger, S. Rentsch, L. Schlapbach, A. Weidenkaff, and T. Kiyobayashi. 2002. Hydrogen sorption by carbon nanotubes and other carbon nanostructures. *Journal of Alloys and Compounds* 330–332:676–682.

Index

Printed and bound by CPI Group (UK) Ltd, Croydon, CR0 4YY

18/10/2024

01776253-0019